PPARs in Cellular and Whole Body Energy Metabolism

PPARs in Cellular and Whole Body Energy Metabolism

Special Issue Editors

Walter Wahli
Rachel Tee

MDPI • Basel • Beijing • Wuhan • Barcelona • Belgrade

Special Issue Editors
Walter Wahli
Lee Kong Chian School of Medicine
Nanyang Technological University
Singapore

Center for Integrative Genomics
University of Lausanne
Switzerland

Rachel Tee
Lee Kong Chian School of Medicine
Nanyang Technological University
Singapore

Editorial Office
MDPI
St. Alban-Anlage 66
4052 Basel, Switzerland

This is a reprint of articles from the Special Issue published online in the open access journal *International Journal of Molecular Sciences* (ISSN 1422-0067) from 2017 to 2018 (available at: https://www.mdpi.com/journal/ijms/special_issues/PPARs#)

For citation purposes, cite each article independently as indicated on the article page online and as indicated below:

LastName, A.A.; LastName, B.B.; LastName, C.C. Article Title. *Journal Name* **Year**, *Article Number*, Page Range.

ISBN 978-3-03897-461-1 (Pbk)
ISBN 978-3-03897-462-8 (PDF)

Copyright courtesy of Professor Eric Yap, Lee Kong Chian School of Medicine, Nanyang Technological University Singapore.
"Living Architecture":A single colony of Pseudomonas stutzeri on agar, originally isolated from oral microbiome, demonstrates multicellular organization. From lab of Eric Yap.

Contents

About the Special Issue Editors

Walter Wahli is Professor of Metabolic Disease at the Lee Kong Chian of Medicine, Nanyang Technological University, Singapore. He is also the President of the Council of the Nestle Foundation for the Study of Problems of Nutrition in the World. He is internationalized recognized as a leader in the field of molecular endocrinology and metabolism and as the discoverer of the nuclear hormone receptors Peroxisome Proliferator-Activated Receptors β and γ (PPARβ and PPARγ). He has received several awards for the elucidation of key functions of these receptors, which are activated by fatty acids and fatty acid derivatives. Synthetic PPAR agonists are drugs used mainly for lowering triglycerides and blood sugar. Prior to his present appointments, Walter Wahli was a visiting associate at the National Cancer Institute, National Institutes of Health, Bethesda, USA. He then became Professor and Director of the Institute of Animal Biology at the University of Lausanne in Switzerland. He was Vice-Rector for Research and Continuing Education of the university and founded the Center of Integrative Genomics, which he directed for several years. He was also a member of the Swiss National Science Foundation Research Council and presided over the Biology and Medicine Division.

Rachel Tee is a senior member of Walter Wahli's team and Manager of the NAFLD Investigation Centre (NICE) in Lee Kong Chian School of Medicine, Nanyang Technological University, Singapore. She coordinates research projects and is directly involved in works using various mouse models to investigate the impact of nutrition on the development of fatty liver diseases. She holds a Bachelor of Commerce in Human Resource Management and Management (Double Major), awarded by Murdoch University.

Preface to "PPARs in Cellular and Whole Body Energy Metabolism"

Peroxisome Proliferator-Activated Receptors (PPARs) are transcription factors that belong to the nuclear hormone receptor superfamily. Nuclear receptors are one of the best described classes of regulatory factors that directly control gene expression. Humans and mice have 48 and 49 nuclear receptors, respectively, which modulate many functions such as reproduction, development, metabolism, and whole body homeostasis. They function as ligand-activated factors, and thus convert signalling by small molecules (hormones, vitamins, fatty acids), which control these processes, into appropriate gene responses.

PPARs were first identified in the early 1990s as nuclear receptors for compounds that induce peroxisome proliferation in rodents, which explains their name. Soon thereafter it became clear that fatty acids and many fatty acid derivatives can also directly regulate gene expression through PPARs. The molecular mode of action of PPARs is similar to that of other nuclear hormone receptors. They bind to a short specific response element, called the Peroxisome Proliferator Response Element (PPRE) in the regulatory region of target genes, as heterodimers with the receptor for 9-cis retinoic acid, RXR (retinoid X receptor). As mentioned above, their activity depends on binding ligands, agonists, or antagonists. Three PPAR isotypes have been identified: PPARα, PPARβ (also called δ or now more commonly β/δ), and PPARγ. Some ligands are shared by the three isotypes, while others are more isotype-specific. Drugs bind to and activate PPARα (fibrates) and PPARγ (thiazolidinediones), resulting in different but complementary effects, with the fibrates acting mainly as hypolipidemic agents and the thiazolididiones as insulin sensitizers. These initial findings have instigated an uncommon research activity to unveil the multifaceted roles of PPARs, which has resulted in thousands of publications. In fact, PubMed from the US National Library of Medicine, National Institutes of Health, features more the 30,000 publications in response to a search with the term "peroxisome proliferator-activated receptor". PPARs appear to be highly interesting and sophisticated modulators of a myriad of cellular, organ, and systemic processes, which renders their study especially challenging. For example, their key roles in lipid and carbohydrate metabolism are central for maintaining body homeostasis, particularly under the complex influences of nutrition and physical activity.

Therefore, this book comes at the right time to cover the singular and intricate regulatory roles of all three PPAR isotypes in the ensemble of processes at work in the vertebrate organism. It highlights recent advances of a very active research field by authors at the forefront of their specialties, ranging from pure or fundamental research aiming to improve our understanding of basic PPAR actions to clinical research that determines the safety and efficacy of PPAR drugs and their use in treatment regimens intended for diseased people. Going through the articles in the book reveals, once more, the extraordinary broad scope of the current PPAR research. Tools are described for the identification of novel PPAR ligands, and applications and problems of ligands in drug discovery and development are described. Interestingly, PPAR activity is modulated by post-translational modifications with a wide spectrum of consequences influencing protein stability and co-factor interaction. Two articles in this book address the human response to aerobic training. There are PPAR and PPAR-coactivator gene polymorphisms that can negatively influence glucose metabolism and some genetically-predisposed individuals may gain lesser benefit from exercise-based lifestyle interventions. Furthermore, much attention is given to the exploration of the roles of the three PPAR isotypes in tissue regeneration, metabolic regulation of lipids and fatty acids, functional fine-tuning of different organs (kidneys, heart, adipose tissue, muscle, liver, brain, intestines, vasculature), and whole body energy homeostasis. As underlined in the present book, PPAR functions are often revealed by the study, using different models, of health conditions such as obesity, metabolic syndrome, atherosclerosis, non-alcoholic fatty liver disease, multiple sclerosis and other neurodegenerative diseases, and diabetic nephropathy. Last but not least, inflammation and cancer are linked, and research in recent years has unveiled mechanisms of inflammation-associated carcinogenesis. The roles of PPARs in this process and the control of tumor growth are presented by several authors. Much work remains to be done to clarify the roles of PPARs in tumour progression and metastasis as effects of their promotive and protective nature have been described in this book and elsewhere.

This book would not have been possible without the many authors who have generously shared their knowledge and research to further the knowledge of the PPAR field, thereby promoting its advances for the benefit of all. May they find here the expression of our deepest gratitude for their efforts in supporting this Special Issue through their high-quality contributions. My thanks go also to the reviewers for their constructive and valuable suggestions. We also deeply appreciate the collaboration from the IJMS Editorial Team, with a very special mention going to Ms Reyna Li for assisting us so efficiently in liaising with authors and keeping us updated about the progression of this Special Issue on "PPARs in Cellular and Whole Body Energy Metabolism".

Walter Wahli, Rachel Tee
Special Issue Editors

International Journal of
Molecular Sciences

Article

Molecular Modeling Study for the Design of Novel Peroxisome Proliferator-Activated Receptor Gamma Agonists Using 3D-QSAR and Molecular Docking

Yaning Jian [1,2], Yuyu He [1,2], Jingjing Yang [1,2], Wei Han [1,2], Xifeng Zhai [3], Ye Zhao [1,2] and Yang Li [1,2,*]

1 Biomedicine Key Laboratory of Shaanxi Province, The College of Life Sciences, Northwest University, Xi'an 710069, China; yaningjian001@163.com (Y.J.); heyuyu_66123@163.com (Y.H.); yangjing952017@163.com (J.Y.); 18629670486@163.com (W.H.); zhaoye@nwu.edu.cn (Y.Z.)

2 Key Laboratory of Resource Biology and Biotechnology in Western China, Ministry of Education, Northwest University, Xi'an 710069, China

3 School of Pharmaceutical Sciences, Xi'an Medical University, Xi'an 710021, China; zhaixf@xiyi.edu.cn

* Correspondence: ly2011@nwu.edu.cn; Tel.: +86-29-8830-4569

Received: 30 January 2018; Accepted: 18 February 2018; Published: 23 February 2018

Abstract: Type 2 diabetes is becoming a global pandemic disease. As an important target for the generation and development of diabetes mellitus, peroxisome proliferator-activated receptor γ (PPARγ) has been widely studied. PPARγ agonists have been designed as potential anti-diabetic agents. The advanced development of PPARγ agonists represents a valuable research tool for diabetes therapy. To explore the structural requirements of PPARγ agonists, three-dimensional quantitative structure–activity relationship (3D-QSAR) and molecular docking studies were performed on a series of *N*-benzylbenzamide derivatives employing comparative molecular field analysis (CoMFA), comparative molecular similarity indices analysis (CoMSIA), and surflex-dock techniques. The generated models of CoMFA and CoMSIA exhibited a high cross-validation coefficient (q^2) of 0.75 and 0.551, and a non-cross-validation coefficient (r^2) of 0.958 and 0.912, respectively. The predictive ability of the models was validated using external validation with predictive factor (r^2_{pred}) of 0.722 and 0.682, respectively. These results indicate that the model has high statistical reliability and good predictive power. The probable binding modes of the best active compounds with PPARγ active site were analyzed, and the residues His323, Tyr473, Ser289 and Ser342 were found to have hydrogen bond interactions. Based on the analysis of molecular docking results, and the 3D contour maps generated from CoMFA and CoMSIA models, the key structural features of PPARγ agonists responsible for biological activity could be determined, and several new molecules, with potentially higher predicted activity, were designed thereafter. This work may provide valuable information in further optimization of *N*-benzylbenzamide derivatives as PPARγ agonists.

Keywords: PPARγ; *N*-benzylbenzamide derivatives; 3D-QSAR; CoMFA; CoMSIA; molecular docking

---

1. Introduction

Type 2 diabetes (T2D) is a disease that is generally characterized by relative insulin deficiency caused by insulin resistance in target organs, and pancreatic β-cell dysfunction [1]. In 2014, there were 422-million people with diabetes, with more than 90% estimated to have T2D, worldwide. Unfortunately, this number will increase to approximately 552-million by the year 2030 [2]. Accordingly, T2D is generating a significant socioeconomic burden, as a pandemic disease with a high and increasing fatality [3,4].

The peroxisome proliferator-activated receptor γ (PPARγ) is generally regarded as a molecular target for the thiazolidinedione class of anti-diabetic drugs [5,6], as it plays a key role in the generation and development of diabetes mellitus [7–9]. Recent studies have shown that PPARγ agonists, including rosiglitazone and pioglitazone [10], may be used as insulin sensitizers in target tissues to lower glucose, as well as fatty acid levels in T2D patients.

However, both rosiglitazone and pioglitazone have been withdrawn from the market because of significant hepatotoxicity and cancer development concerns [11]. Hence, there is an urgent need for the development of safer PPARγ modulating drugs. One severe side-effect of known PPARγ agonists, involves sodium and water retention, which may be dangerous for patients suffering from congestive heart conditions [12]. Recently, various new *N*-benzylbenzamide compounds have been shown to act as PPARγ agonists that, not only lowered blood pressure and reduced systemic glucose, triglycerides, and free fatty acid levels, but have also been shown to maintain water and electrolyte homeostasis [13]. Therefore, a variety of *N*-benzylbenzamide compounds have since been identified as safer PPARγ modulators for the treatment of T2D.

Based on CoMFA [14], along with CoMSIA [15], methods involving 3D-QSAR determinations allow for the structure–activity relationship of *N*-benzylbenzamide compounds to be studied. Molecular docking was also applied to reveal the most likely binding modes between the compounds and PPARγ. On the basis of 3D-QSAR and molecular docking results, valuable information can be retrieved for further structured-based drug design, with higher activity. Finally, a series of new potent molecules with a higher predicted activity than the template compound, the latter exhibiting the best activity reported in the literature, have been designed. Our study will potentially provide guidance for the future design of selective and potent PPARγ agonists.

2. Results and Discussion

2.1. CoMFA and CoMSIA Results

The 3D-QSAR models were obtained using a training set of 27 compounds, and a test set of six compounds. The statistical parameters associated with CoMFA and CoMSIA can be found in Table 1. In general, various alignment strategies can lead to different statistical values in the constructed QSAR models. The best CoMFA and CoMSIA models were generated employing a partial least square (PLS) analysis, which produced cross-validated coefficients (q^2). When a cross-validation coefficient, $q^2 > 0.5$, was used, the QSAR model demonstrated statistical significance.

As shown in Table 1, two descriptor fields in CoMFA form all three possible combination models, including steric (S), electrostatic (E) and SE models. The CoMSIA models, with a combination of five descriptor fields, including S, E, hydrophobic (H), hydrogen bond donor (D) and acceptor (A), were developed to generate the optimal 3D-QSAR model. However, some models with a low q^2 value did not meet the criterion ($q^2 > 0.5$), indicating an unacceptable 3D-QSAR model. Still, overfitting seemed to occur for some models (those with a large number of components). From Table 1, we can see that the best established models (CoMFA and CoMSIA) exhibited high q^2 (0.75 and 0.551), r^2 (0.958 and 0.912), and F-values (76.113 and 43.388), along with a low standard error of estimate (SEE) value (0.097 and 0.138), and a suitable number of components (6 and 5), which indicated good statistical correlation of the models. Moreover, the predictive capabilities of the generated models were assessed by calculating their predictive correlation coefficient (r^2_{pred}) involving their corresponding test set molecules. The generated CoMFA and CoMSIA models with maximum external predictive ability (r^2_{pred} 0.722 and 0.682), were considered the best models. The distribution of actual predicted pEC_{50} values of the training and test sets for CoMFA and CoMSIA are shown in Figure 1. The CoMFA and CoMSIA models show a good fit along the diagonal line. Both models also exhibited satisfactory predictive ability throughout the training and test sets.

Table 1. Statistical parameters of the CoMFA and CoMSIA models.

PLS Statistics	ONC	q^2	r^2	SEE	F	Cotribution (%)				
						S	E	H	D	A
CoMFA										
S	8	0.576	0.930	0.132	29.872	100	–	–	–	–
E	7	0.497	0.937	0.122	40.470	–	100	–	–	–
SE	6	0.750	0.958	0.097	76.113	51.6	48.4	–	–	–
CoMSIA										
S	5	0.472	0.804	0.205	17.183	100	–	–	–	–
E	7	0.428	0.911	0.145	27.952	–	100	–	–	–
H	10	0.506	0.958	1.108	36.771	–	–	100	–	–
D	2	−0.051	0.168	0.395	2.418	–	–	–	100	–
A	1	−0.083	0.030	0.418	0.777	–	–	–	–	100
SE	10	0.61	0.949	0.120	29.708	34.0	66.0	–	–	–
SH	3	0.412	0.823	0.186	35.545	36.1	–	63.9	–	–
SD	10	0.505	0.935	0.135	23.071	48.3	–	–	51.7	–
SA	4	0.493	0.803	0.201	22.415	84.9	–	–	–	15.1
EH	5	0.479	0.891	0.153	34.255	–	59.1	40.9	–	–
ED	9	0.352	0.932	0.134	25.959	–	88.1	–	11.9	–
EA	6	0.433	0.876	0.167	23.640	–	89.1	–	–	10.9
HD	10	0.537	0.965	0.100	43.744	–	–	75.1	24.9	–
HA	10	0.525	0.958	0.109	36.225	–	–	90.6	–	9.4
DA	2	−0.036	0.186	0.391	2.740	–	–	–	81.0	19.0
SEH	5	0.541	0.916	0.134	45.732	17.9	51.9	30.2	–	–
SED	10	0.6	0.954	0.113	33.508	32.0	55.9	–	12.1	–
SEA	9	0.607	0.944	0.121	32.005	34.9	60.4	–	–	4.7
SHD	5	0.448	0.913	0.137	43.915	28.9	–	51.0	20.1	–
SHA	5	0.426	0.909	0.140	41.769	33.8	–	59.6	–	6.6
SDA	4	0.501	0.800	0.202	22.018	57.8	–	–	32.2	9.9
EHD	5	0.478	0.885	0.157	32.229	53.1	–	35.9	11.0	–
EHA	5	0.492	0.889	0.154	33.757	–	56.8	38.9	–	4.3
EDA	9	0.368	0.941	0.125	30.157	–	79.6	–	11.8	8.6
HDA	10	0.532	0.964	0.100	43.071	–	–	70.1	23.5	6.4
SEHD	5	0.545	0.912	0.137	43.732	16.1	47.7	27.1	9.1	–
SEHA	5	0.55	0.915	0.135	45.170	17.3	50.5	28.9	–	3.3
SEDA	10	0.593	0.954	0.113	33.520	30.3	52.9	–	11.7	5.1
SHDA	5	0.449	0.905	0.143	39.940	28.1	–	49.5	17.5	4.9
EHDA	5	0.486	0.884	0.159	32.044	–	51.6	35.0	9.7	3.7
SEHDA	5	0.551	0.912	0.138	43.388	15.7	46.7	26.5	8.2	2.8

Optimum number of components (ONC), leave-one-out cross-validated correlation coefficient (q^2), noncross-validated correlation coefficient (r^2), standard error of estimate (SEE), Fischer test values (F), steric field (S), electrostatic field (E), hydrophobic field (H), Hydrogen bond donor field (D), Hydrogen bond acceptor field (A).

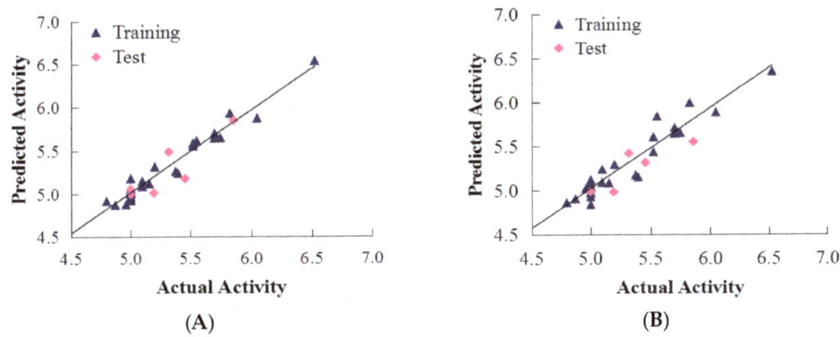

Figure 1. Plots of Actual versus predicted pEC_{50} values, for the training set and test set compounds, for CoMFA (**A**) and CoMSIA (**B**) models.

2.2. CoMFA Contour Map Analysis

The steric and electrostatic fields of the CoMFA model are presented as contour maps in Figure 2. Finally, compound **24c** was selected as the template molecule. The green contours represent regions indicating favorable steric fields, while the yellow contours represent the regions indicating unfavorable steric fields. Moreover, the blue and red contours highlight the positions where electropositive groups and electronegative groups would be favorable, respectively.

2.2.1. Steric Contour Map

The steric contour map in CoMFA (Figure 2A) has a yellow contour near the ortho position of the benzene ring, which indicates that the presence of steric substituents in this region is unfavorable. Furthermore, the yellow contour explains why a $-CF_3$ substituent in ortho position of the benzene ring in compound **2b** is more potent than in compound **10b**, which bears a $-OCF_3$ substituent. Likewise, a small yellow contour map appeared in the para position of the benzene ring, which indicates that the large size of the substituent was not preferred in this area. Moreover, a $-OCH_3$ group in this position in compound **18b**, could be found within the steric field, which led to decreased biological activity. Finally, the large yellow region on the R_2 substitutes may explain why compound **32b**, bearing a phenyl group, was less active than compound **30b** bearing a propyl group.

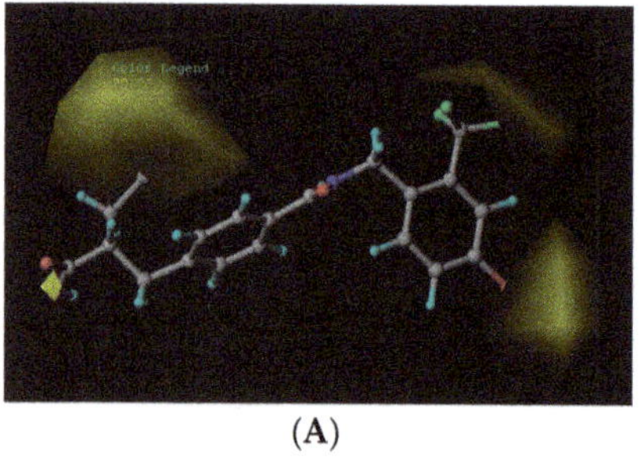

(A)

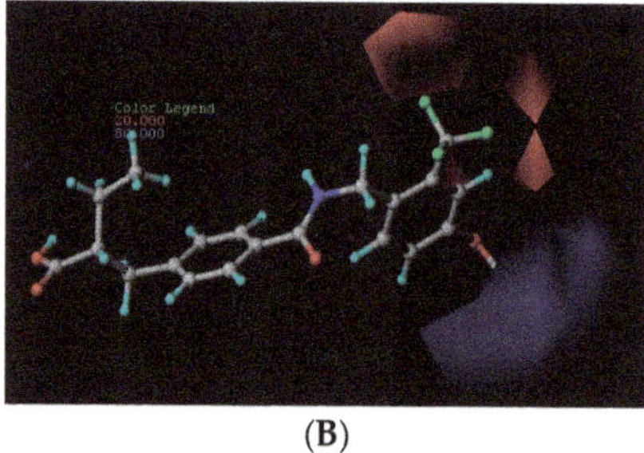

(B)

Figure 2. CoMFA contour maps displayed with most potent compound, **24c**. (**A**) CoMFA steric contour map (green, favored; yellow, disfavored); (**B**) CoMFA electrostatic contour map (blue, electropositive favored; red, electronegative favored).

2.2.2. Electrostatic Contour Map

A large blue contour area near the para position of the benzene ring indicates that the presence of an electropositive group may increase activity (Figure 2B). This assumption becomes even more significant in the case of compounds **20b** and **12b**, as these compounds contain a –Cl and –F substituent, respectively. However, due to the presence of different electron-donating groups, compound **12b** was found to be less biologically active than compound **20b**. The red contours present on the ortho position of the benzene ring suggest that an electron negative group would be favorable in this area, an assumption that proves to be true for compounds **2b** and **6b**, which contain a $-CF_3$ group and a $-CH_3$ group, respectively. However, since a $-CF_3$ group proves to be more electron-withdrawing than a $-CH_3$ group, compound **2b** was determined to be more biologically active than compound **6b**.

2.3. CoMSIA Contour Map Analysis

The CoMSIA steric and electrostatic contour maps were both similar to the CoMFA contour maps discussed above (Figure 3A,B). Thus, only hydrophobic, hydrogen bond donor, as well as hydrogen bond acceptor fields of CoMSIA, were analyzed in this section. The CoMSIA steric, electrostatic, hydrophobic, hydrogen bond donor along with hydrogen bond acceptor contour maps are shown in Figure 3, respectively. Compound **24c** was selected as the corresponding reference molecule.

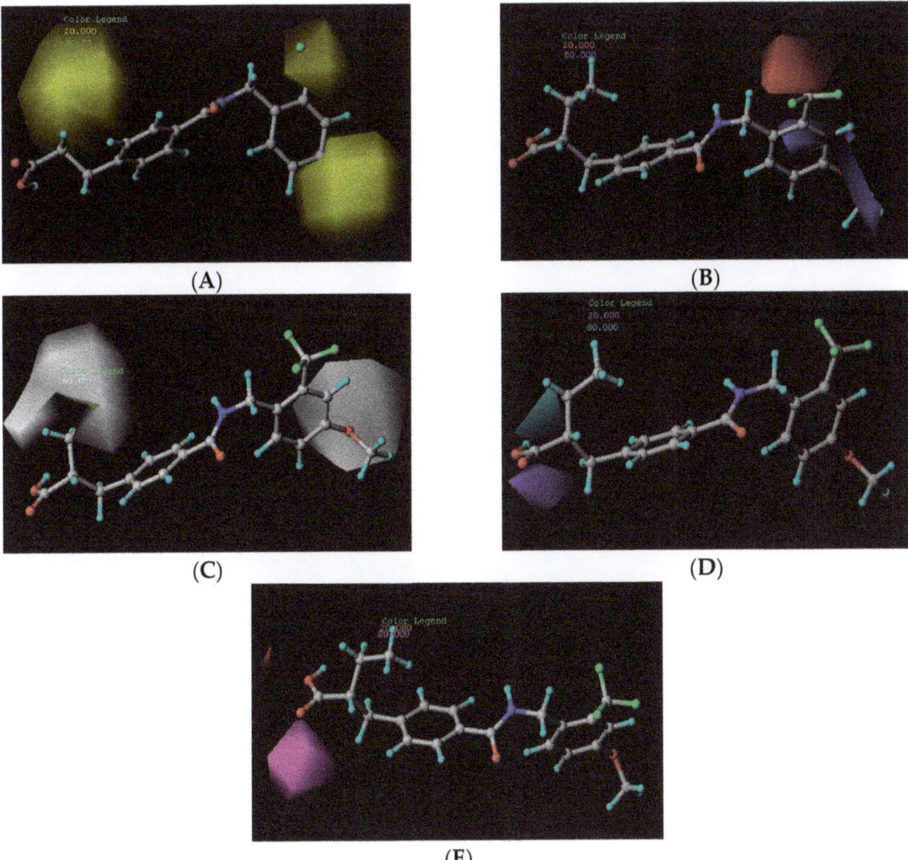

Figure 3. CoMSIA contour maps displayed with most potent compound, **24c**. (**A**) CoMSIA steric contour map (green, favored; yellow, disfavored); (**B**) CoMSIA electrostatic contour map (blue, electropositive favored; red, electronegative favored); (**C**) CoMSIA hydrophobic contour map (yellow, favored; white, disfavored); (**D**) CoMSIA hydrogen donor contour map (cyan, favored; purple, disfavored); (**E**) CoMSIA hydrogen acceptor contour map (magenta, favored; red, disfavored).

2.3.1. Hydrophobic Contour Map

In the hydrophobic contour map (Figure 3C), the yellow contours show favorable hydrophobic regions, while white contours represent unfavorable hydrophobic regions. For the hydrophobic map, one white unfavorable region could be found around the R_2 substitutes, indicating that the addition of hydrophobic substituents in this region would lead to decrease in activity. Further evidence for this notion can be obtained from compound **30b** bearing a propyl group, which is considerably less hydrophobic than a phenyl group. Therefore, compound **30b** proves to be more active than the biologically less active compound **32b**, bearing a phenyl group. The other white contour area observed in the para position of the benzene ring, indicates that hydrophobic substituents were not preferred in this region. This finding can be further explained by the fact that compound **21c** contains a –Cl substituent that generally leads to a higher potency compared to a methoxy group present in compound **19c**.

2.3.2. Hydrogen Bond Donor Map

The contour map for the hydrogen bond donor field is shown in Figure 3D. Cyan and purple contours represent a hydrogen bond donor field favorable region and hydrogen bond donor unfavorable region, respectively. For the hydrogen bond donor map, a cyan contour appeared around the hydroxyl group. This suggests that a hydrogen bond interaction, with the hydrogen atom of the hydroxyl group acting as a hydrogen bond donor, is favorable for increased activity.

2.3.3. Hydrogen Bond Acceptor Map

In the hydrogen bond acceptor contour map (Figure 3E), the magenta contours represent a favorable hydrogen bond acceptor field, while the red contours represent an unfavorable hydrogen bond acceptor field. For the hydrogen bond acceptor map, one favorable polyhedral surface (magenta) is found around the carboxyl group, which suggests that hydrogen bond interactions between the oxygen atom of the carbonyl group, acting as a hydrogen bond acceptor, and a hydrogen atom of the group, lead to an increase in activity.

2.4. Design of More Potent Compounds

Based on CoMFA and CoMSIA models obtained in the present study, the structure–activity relationships of PPARγ agonists could be determined, and several new potent molecules could be designed. The chemical structures of the newly designed compounds, as well as their activity characteristics on PPARγ, were predicted by the CoMFA and CoMSIA models, as seen in Table 2. The predicted activities of the newly designed compounds on PPARγ were all significant. A set of the molecules demonstrated an even better activity than the most active agonist previously reported, further validating the superiority of the models, and indicates that the structure–activity relationships in the work reported herein, may potentially be used in structural modification and optimization.

Table 2. Newly designed compounds and predictive activity.

NO.	R_1	R_2	R_3	CoMFA Predicted	CoMSIA Predicted
N1	CN	C(Me)3	Me	6.942	7.170
N2	CN	C(Me)3	OMe	6.765	7.155
N3	CN	C(Me)3	ET	6.900	7.155
N4	CN	CH(Me)2	OMe	6.982	7.105
N5	CN	C(Me)3	–	6.822	7.083
N6	CN	CH(Me)2	Me	7.064	7.080
N7	CN	CH(Me)2	ET	7.103	7.070
N8	COOH	C(Me)3	Me	6.820	7.063
N9	CN	CH(Me)2	–	7.036	6.999
N10	Cl	C(Me)3	Me	6.742	6.986
N11	CHO	C(Me)3	Me	6.878	6.916
N12	COOH	CH(Me)2	Me	7.073	6.902
N13	CHO	C(Me)3	c-Pr	6.815	6.886
N14	Cl	CH(Me)2	Me	6.781	6.843
N15	–	C(Me)3	–	6.648	6.798

2.5. Docking Analysis

In order to obtain the probable binding conformations between the molecules and the protein, Surflex-dock was carried out to dock the compounds to the binding site of PPARγ. In this study, compound **24c** (template) and a newly designed compound, **N1**, **N9**, and **N12**, were placed in the corresponding binding sites, respectively. The docking score of compound **24c** was 8.913. Meanwhile, the docking scores of compounds **N1**, **N19**, and **N12** were 11.0573, 11.010, and 11.690, respectively. The docking scores of compounds **N1**, **N9**, and **N12** are higher than compound **24c**, which has the highest activity in the training set. This result is in good agreement with corresponding predicted activities of CoMFA and CoMSIA models.

Figure 4 shows the surface of the binding site of PPARγ, the binding modes between compounds **24c**, **N1**, **N9**, and **N12**, and the binding site of the protein. Figure 4A–D, illustrate the surface of the binding site and the conformations of compounds **24c**, **N1**, **N9**, and **N12** (yellow), and the original ligand (purple), as well as the key residues (white) at the binding site. High resemblance between these molecules is observed and they occupied nearly the same binding pocket as PPARγ. It is representative of the active conformation to dock selected compounds. Here, compounds were positioned in the pocket, surrounded by His323, Tyr473, Ser289, Leu453, Ile341, Cys285, etc. As seen in Figure 4E–H, the carbonyl group of the ligand that acts as a hydrogen bond acceptor, formed a hydrogen bond with the backbone O–H of Ser289. The backbone N–H of His323, and O–H of Tyr473, which act as hydrogen bond acceptors and formed hydrogen bonds with the hydroxyl group of the ligand, respectively. Thus, these three hydrogen bond interactions played a major role in the combination of these drugs and the receptor.

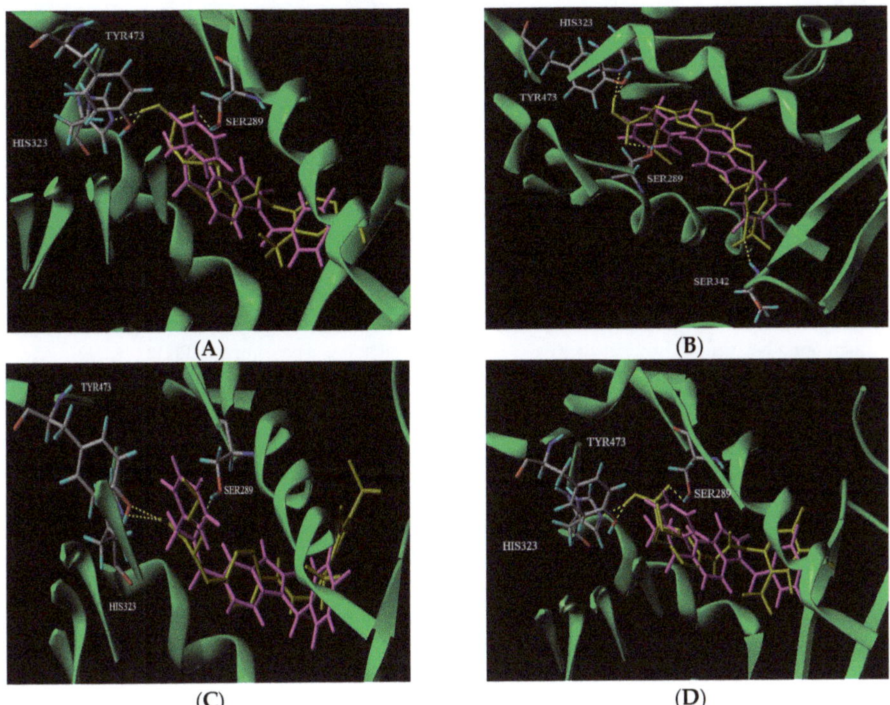

Figure 4. *Cont.*

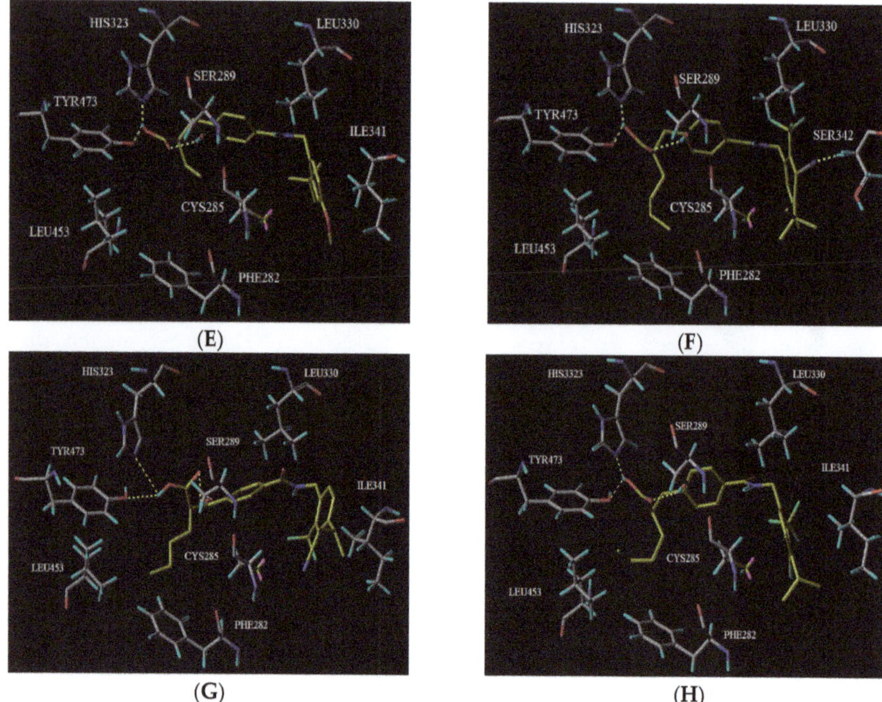

Figure 4. Docking results. (**A**) The surface of the binding site, and the conformation comparison of compound **24c** (yellow), the original ligand (purple), and the key residues (white) at the binding site; (**B**) The surface of the binding site and the comparison of the conformation of, compound **N1** (yellow), the original ligand (purple), and the key residues (white), at the binding site; (**C**) The surface of the binding site, and the comparison of the conformation of compound **N9** (yellow), the original ligand (purple), and the key residues (white), at the binding site; (**D**) The surface of the binding site, and the comparison of the conformation of compound **N12** (yellow), the original ligand (purple), and the key residues (white), at the binding site; (**E**) Interaction between compound **24c** (yellow) and residues (white); (**F**) Interaction between compound **N1** (yellow) and residues (white); (**G**) Interaction between compound **N9** (yellow) and residues (white); (**H**) Interaction between compound **N12** (yellow) and residues (white).

3. Materials and Methods

3.1. Data Set

A set of 33 *N*-benzylbenzamide derivatives and the corresponding activity data were collected from the work of René Blöcher et al. [13] (Table 3). The data set was randomly divided into the training set of 27 compounds (82%) to generate the 3D-QSAR model, and the test set of 6 compounds (18%) to verify the predictive ability of the model. The bioactivities of the compounds were expressed as pEC_{50} ($-logEC_{50}$), which was used as a dependent variable in further investigations. To avoid possible issues during the external validation, the selection of the training and the test set was carried out such that both sets included structurally diverse compounds and all types of activity [16–18].

Table 3. Actual and predicted pEC_{50} values of PPARγ agonists.

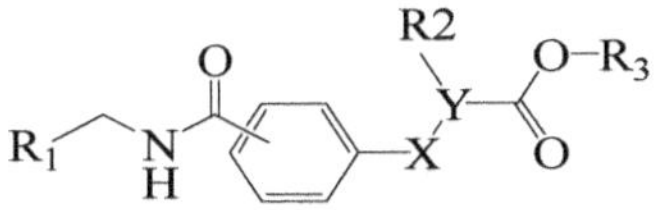

(a) X-Y: CH=C; R_3:CH_2CH_3
(b) X-Y: CH_2-CH; R_3:CH_2CH_3
(c) X-Y: CH_2-CH; R_3:H

NO.	R_1	R_2	Substitution	Actual pEC_{50}	Pred-pEC_{50}	
					CoMFA	CoMSIA
01a *	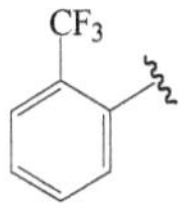 CF_3	Et	para	5.000	4.996	4.992
02b	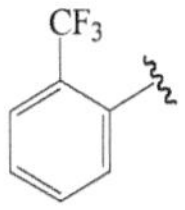 CF_3	Et	para	5.745	5.650	5.655
03c *	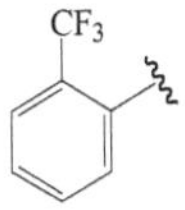 CF_3	Et	para	5.319	5.483	5.420
04b	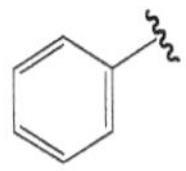	Et	para	4.796	4.915	4.867
05c	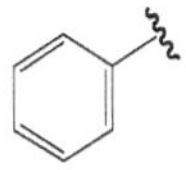	Et	para	4.870	4.869	4.911
06b	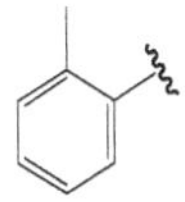	Et	para	5.000	4.941	4.845
07b	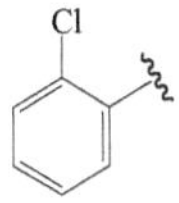 Cl	Et	para	5.000	5.015	5.029
08c	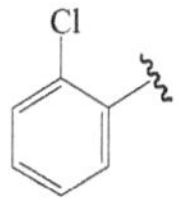 Cl	Et	para	5.000	5.039	5.093
09c	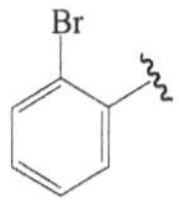 Br	Et	para	5.000	5.019	5.053

Table 3. *Cont.*

NO.	R_1	R_2	Substitution	Actual pEC_{50}	Pred-pEC_{50}	
					CoMFA	CoMSIA
10b *	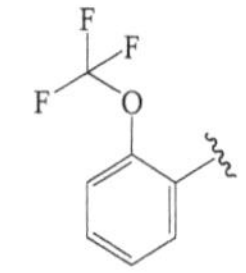	Et	para	5.456	5.174	5.317
11c	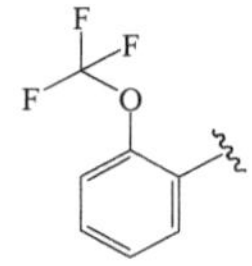	Et	para	5.097	5.090	5.097
12b	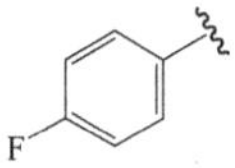	Et	para	4.959	4.874	5.028
13c	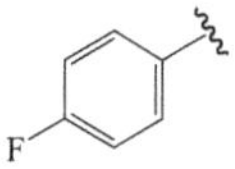	Et	para	5.000	4.924	5.125
14b	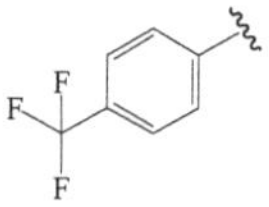	Et	para	5.377	5.261	5.178
15c	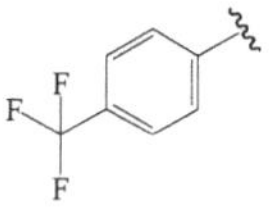	Et	para	5.201	5.315	5.297
16b	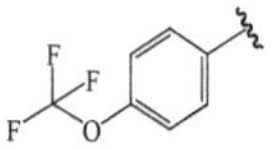	Et	para	5.523	5.588	5.607
17c	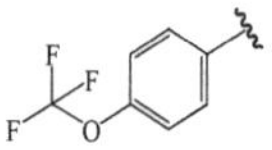	Et	para	5.699	5.645	5.712
18b	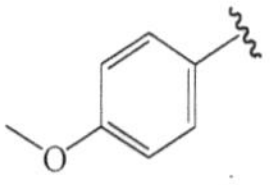	Et	para	5.000	5.060	4.984
19c	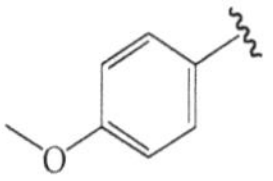	Et	para	5.155	5.121	5.089
20b	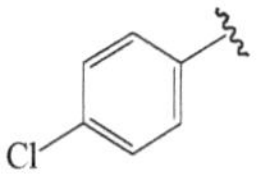	Et	para	5.000	5.182	5.047
21c	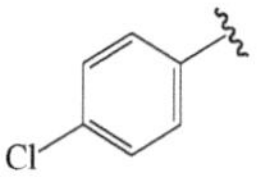	Et	para	5.398	5.239	5.156
22b *	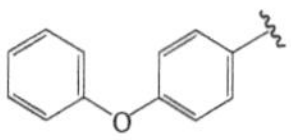	Et	para	5.854	5.846	5.551

Table 3. *Cont.*

NO.	R_1	R_2	Substitution	Actual pEC_{50}	Pred-pEC_{50}	
					CoMFA	CoMSIA
23b	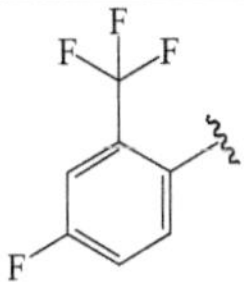 F, F, F, F	Et	para	5.553	5.615	5.841
24c	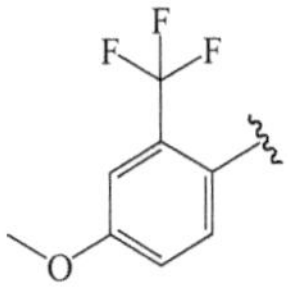 F, F, F, O	Et	para	6.523	6.542	6.349
25b *	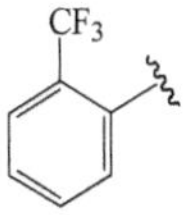 CF_3	Et	meta	5.000	5.052	4.991
26c *	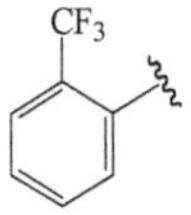 CF_3	Et	meta	5.194	5.008	4.983
27b	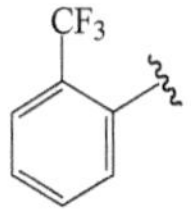 CF_3	H	para	5.000	4.966	4.934
28c	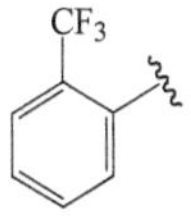 CF_3	H	para	5.000	5.002	4.974
29b	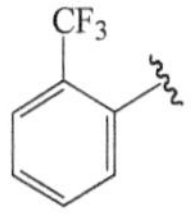 CF_3	Me	para	5.097	5.151	5.245
30b	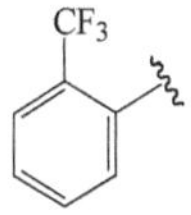 CF_3	Pr	para	6.046	5.875	5.886
31c	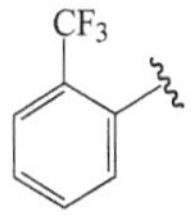 CF_3	Pr	para	5.824	5.933	5.993
32b	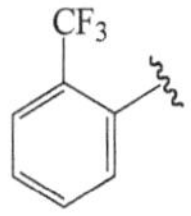 CF_3	Phenyl	para	5.699	5.699	5.648
33c	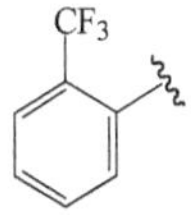 CF_3	Phenyl	para	5.523	5.556	5.439

* Test set molecules.

3.2. Molecular Modeling and Alignment

To obtain the best conformers for each molecule, the Sybyl X-2.1.1 software package was used for all compound modeling and optimization parameters. All structures of the compound series were subjected to preliminary geometry optimization using the Tripos force field with 1000 iterations [19]. Partial atomic charges were calculated by the Gasteiger-Hückel scheme, with an energy gradient convergence criterion of 0.05 kcal/mol Å [20]. Based on the analysis method described above, the lowest energy conformation of each molecule was determined for the definitive QSAR studies. Molecular alignment is one of the most essential steps for the generation of the best CoMFA and CoMSIA models [21]. Thus, molecular alignment was performed using the Distill alignment technique, a user-defined common core of the Sybyl tools [22]. Compound **24c**, exhibiting the highest activity in the complete data set, was selected as the template molecule. The remaining compounds in the Mol2 database were aligned by their corresponding maximum common substructures, as shown in Figure 5A. The rigid body alignment of the molecules is shown in Figure 5B.

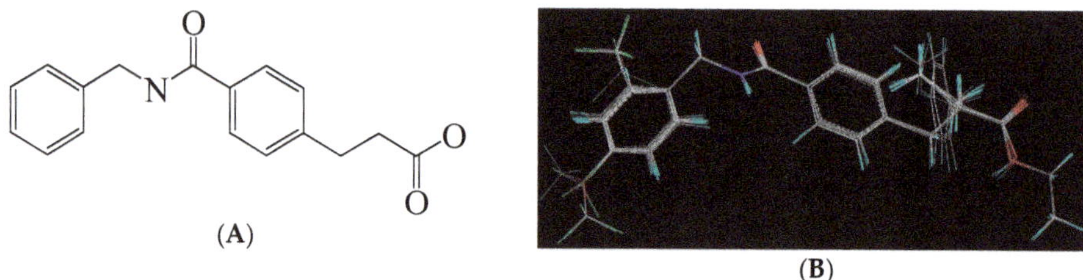

Figure 5. Molecular alignment. (**A**) Common structure retrieved from compound **24**; (**B**) Alignment of the compounds in the training set.

3.3. CoMFA Method

The CoMFA method is often used to describe steric and electrostatic fields. Lennard-Jones and Coulomb potentials were employed to calculate two fields. A 3D cubic lattice, with grid spacing of 2.0 Å, was generated to surround the aligned molecules in all directions. These grid points were generated using the Tripos force field, a sp^3 carbon atom probe with a Van der Waals radius of 1.52 Å, and a charge of +1.00 (default probe atom in Sybyl). Based on the CoMFA method, steric and electrostatic fields were scaled with a default energy cut off of 30 kcal/mol, the latter being the optimal parameter for this model [23].

3.4. CoMSIA Method

The CoMSIA analysis is similar to CoMFA, in regard to the descriptors around the aligned molecules. Three other fields, (i.e., hydrophobic, hydrogen bond donor, and hydrogen bond acceptor fields), were calculated together with the same standard settings used in the CoMFA calculations. More importantly, the distance dependence between the probe atom and each molecule atom was measured by a Gaussian function [24].

3.5. Internal Validation and Partial Least Squares (PLS) Analysis

Partial least square (PLS) regression analysis was performed on the training set to construct the correlation between the QSAR model and activity values [25]. To evaluate the reliability of the models generated from PLS analysis, cross-validation analysis was performed through the leave-one-out (LOO) method, which determines the square of the cross-validation coefficient (q^2) and the optimal number of components (ONC). To obtain the non-cross-validation coefficient (r^2), a final non-cross-validation analysis was performed using the ONC derived from cross validation analysis and the corresponding standard error of estimate (SEE). The value for q^2, a measure of the internal quality of the models, was evaluated as follows:

$$q^2 = \frac{\Sigma(y_{obs} - y_{pre})^2}{\Sigma(y_{obs} - y_{mean})^2}$$

where y_{obs}, y_{pre}, and y_{mean} are observed, predicted, and mean activity in the training set, respectively.

3.6. External Validation of the QSAR Model

To evaluate the predictive ability of CoMFA and CoMSIA models on the test set, the predictive power of the models generated by the CoMFA or CoMSIA analyses with the training set was assessed by calculating the predictive factor r^2 (r^2_{pred}) [26], and measuring the predictive performance of the PLS model. The factor r^2 was calculated as follows:

$$r^2_{pread} = \frac{SD - PRESS}{SD}$$

the sum of squared deviation, (*SD*) between the biological activity of molecules in the test set and the mean biological activity of the training set molecules; the sum of squared deviations between actual and predicted activity values (*PRESS*), for every molecule in the test set. Coefficients and QSAR results in the contour maps were produced with the field type "STDEV*COEFF".

3.7. Molecular Docking

In an effort to explore the interaction mechanism and investigate suitable binding modes, a molecular docking study was performed using the Sybyl package [27]. The crystal structure of PPARγ was retrieved from the RCSB (Research Collaboratory for Structural Bioinformatics) Protein Data Bank (PDB ID: 5TWO) [28]. In the protein preparation phase, the A-chain was used for the docking study. Crystallized ligands and water molecules of the B-chain were deleted and the hydrogen atoms along with the united atom Gasteiger charges were assigned for the receptor [29]. Based on a protomol generation with a threshold parameter of 0.5 and a bloat parameter of 1 Å, the intended active sites where putative ligands could align to and generate potential interactions, were created using the Sybyl package. Binding affinities were presented by Surflex-Dock total scores. In general, conformations of each ligand were ranked by total scores of docking, with the best conformation of the ligand taken into consideration for the corresponding binding interactions. In this study, compound **24c** (template) and the newly designed compound **N1**, were selected and docked to the binding pocket, using the parameters optimized previously.

4. Conclusions

In this paper, 3D-QSAR and molecular docking studies were utilized to investigate the structural requirements for improving the potency of *N*-benzylbenzamide derivatives as PPARγ agonists. The established CoMFA and CoMSIA models were both statistically significant, with high external prediction characteristics, indicating that the models could be used to successfully predict compound activity. Surflex-Dock analysis also demonstrated the binding interactions of the template compound with amino acids. Using the model parameter analysis and contour maps, the corresponding structure-activity relationships were determined (Figure 6). Based on the information derived from the different contour maps, several new compounds with improved activities, were designed, further validating the ability of the generated model. We surmise that this will be helpful for the future development of new PPARγ agonists, in the design and screening of new high-activity compounds.

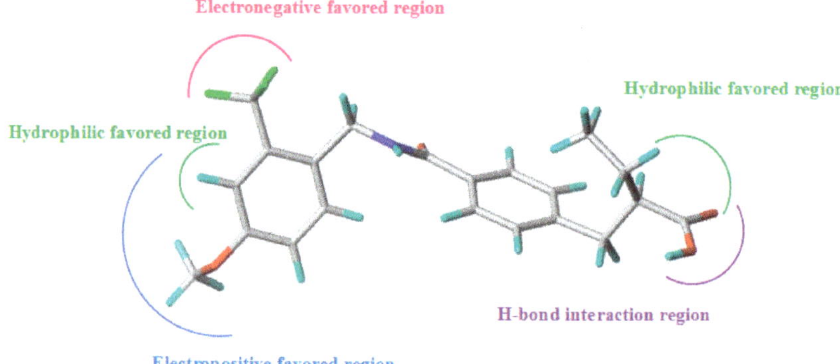

Figure 6. Diagram of structure-activity relationship based on core structure of template compound **24c**.

Acknowledgments: This work was supported by the funding from the National Natural Science Foundation of China (No.31200253), Natural Science Basic Research Plan in Shaanxi Province of China (No.2017JM8049), and Key Science Technology Program of Shaanxi Province (No.2016SF-379).

Author Contributions: Yaning Jian and Yang Li conceived and designed the experiments; Yaning Jian and Yang Li performed the experiments; Yaning Jian, Yuyu He, Jingjing Yang and Wei Han analyzed the data; Ye Zhao and Xifeng Zhai contributed analysis tools; Yaning Jian and Yang Li wrote the paper.

Conflicts of Interest: The authors declare no conflict of interest.

References

1. Chatterjee, S.; Khunti, K.; Davies, M.J. Type 2 diabetes. *Lancet* **2017**, *389*, 2239–2251. [CrossRef]
2. Spurr, S.; Bally, J.; Bullin, C.; Trinder, K. Type 2 Diabetes in Canadian Aboriginal Adolescents: Risk Factors and Prevalence. *J. Pediatr. Nurs.* **2017**, *36*, 111–117. [CrossRef] [PubMed]
3. American diabetes association. Economic Costs of Diabetes in the U.S. in 2012. *Diabetes Care* **2013**, *36*, 1033–1046.
4. Seuring, T.; Archangelidi, O.; Suhrcke, M. The Economic Costs of Type 2 Diabetes: A Global Systematic Review. *PharmacoEconomics* **2015**, *33*, 811–831. [CrossRef] [PubMed]
5. Lehmann, J.M.; Moore, L.B.; Smith-Oliver, T.A.; Wilkison, W.O.; Willson, T.M.; Kliewer, S.A. An antidiabetic thiazolidinedione is a high affinity ligand for peroxisome proliferator-activated receptor gamma (PPARγ). *J. Biol. Chem.* **1995**, *270*, 12953–12956. [CrossRef] [PubMed]
6. Willson, T.M.; Cobb, J.E.; Cowan, D.J.; Wiethe, R.W.; Correa, I.D.; Prakash, S.R.; Beck, K.D.; Moore, L.B.; Kliewer, S.A.; Lehmann, J.M. The Structure-Activity Relationship between Peroxisome Proliferator-Activated Receptor γ Agonism and the Antihyperglycemic Activity of Thiazolidinediones. *J. Med. Chem.* **1996**, *39*, 665–668. [CrossRef] [PubMed]
7. Barroso, I.; Gurnell, M.; Crowley, V.E.; Agostini, M.; Schwabe, J.W.; Soos, M.A.; Maslen, G.L.; Williams, T.D.; Lewis, H.; Schafer, A.J.; et al. Dominant negative mutations in human ppar gamma associated with severe insulin resistance, diabetes mellitus and hypertension. *Nature* **1999**, *402*, 880–883. [CrossRef] [PubMed]
8. Flavell, D.M.; Ireland, H.; Stephens, J.W.; Hawe, E.; Acharya, J.; Mather, H.; Hurel, S.J.; Humphries, S.E. Peroxisome proliferator-activated receptor α gene variation influences age of onset and progression of type 2 diabetes. *Diabetes* **2005**, *54*, 582–586. [CrossRef] [PubMed]
9. Jay, M.A.; Ren, J. Peroxisome proliferator-activated receptor (PPAR) in metabolic syndrome and type 2 diabetes mellitus. *Curr. Diabetes Rev.* **2007**, *3*, 33–39. [CrossRef] [PubMed]
10. Campbell, I.W. The Clinical Significance of PPAR Gamma Agonism. *Curr. Mol. Med.* **2005**, *5*, 349–363. [CrossRef] [PubMed]
11. Ahmadian, M.; Suh, J.M.; Hah, N.; Liddle, C.; Atkins, A.R.; Downes, M.; Evans, R.M. PPARγ signaling and metabolism: The good, the bad and the future. *Nat. Med.* **2013**, *19*, 557–566. [CrossRef] [PubMed]

12. Yang, T.; Soodvilai, S. Renal and vascular mechanisms of thiazolidinedione-induced fluid retention. *PPAR Res.* **2008**, *2008*. [CrossRef] [PubMed]
13. Blöcher, R.; Lamers, C.; Wittmann, S.K.; Merk, D.P.; Hartmann, M.; Weizel, L.; Diehl, O.; Brüggerhoff, A.; Boß, M.; Kaiser, A.; et al. *N*-benzylbenzamides: A novel merged scaffold for orally available dual soluble epoxide hydrolase/peroxisome proliferator-activated receptor γ modulators. *J. Med. Chem.* **2015**, *59*, 61–81. [CrossRef] [PubMed]
14. Sahu, N.K.; Sharma, M.C.; Mourya, V.; Kohli, D.V. QSAR studies of some side chain modified 7-chloro-4-aminoquinolines as antimalarial agents. *Arab. J. Chem.* **2014**, *7*, 701–707. [CrossRef]
15. Klebe, G.; Abraham, U.; Mietzner, T. Molecular Similarity Indices in a Comparative Analysis (CoMSIA) of Drug Molecules to Correlate and Predict Their Biological Activity. *J. Med. Chem.* **1994**, *37*, 4130–4146. [CrossRef] [PubMed]
16. Leonard, J.; Roy, K. On Selection of Training and Test Sets for the Development of Predictive QSAR models. *Qsar. Comb. Sci.* **2010**, *25*, 235–251. [CrossRef]
17. Roy, K.; Paul, S. Exploring 2D and 3D QSARs of 2,4-Diphenyl-1,3-oxazolines for Ovicidal Activity Against Tetranychus urticae. *Qsar. Comb. Sci.* **2009**, *28*, 406–425. [CrossRef]
18. Patel, P.; Chintha, C.; Ghate, M.; Bhatt, H.; Vyas, V.K. 3D QSAR study of 4H-chromen-1,2,3,4-tetrahydropyrimidine-5-carboxylate derivatives as potential anti-mycobacterial agents. *Med. Chem. Res.* **2014**, *23*, 2955–2963. [CrossRef]
19. Wang, F.F.; Zhou, B. Toward the identification of a reliable 3D-QSAR model for the protein tyrosine phosphatase 1B inhibitors. *J. Mol. Struct.* **2018**, *1158*, 75–87. [CrossRef]
20. Liu, J.; Zhang, H.; Xiao, Z.; Wang, F.; Wang, X.; Wang, Y. Combined 3D-QSAR, molecular docking and molecular dynamics study on derivatives of peptide epoxyketone and tyropeptin-boronic acid as inhibitors against the β5 subunit of human 20s proteasome. *Int. J. Mol. Sci.* **2011**, *12*, 1807–1835. [CrossRef] [PubMed]
21. Damre, M.V.; Gangwal, R.P.; Dhoke, G.V.; Lalit, M.; Sharma, D.; Khandelwal, K.; Sangamwar, A.T. 3D-QSAR and molecular docking studies of amino-pyrimidine derivatives as PknB inhibitors. *J. Taiwan Inst. Chem. E* **2014**, *45*, 354–364. [CrossRef]
22. Xie, X.Q.; Chen, J.Z. Data Mining a Small Molecule Drug Screening Representative Subset from NIH Pub Chem. *J. Chem. Inf. Model.* **2008**, *48*, 465–475. [CrossRef] [PubMed]
23. Ståhle, L.; Wold, S. Partial least squares analysis with cross-validation for the two-class problem: A Monte Carlo study. *J. Chemometr.* **1987**, *1*, 185–196. [CrossRef]
24. Xiao, A.; Zhang, Z.; An, L.; Xiang, Y. 3D-QSAR and docking studies of 3-arylquinazolinethione derivatives as selective estrogen receptor modulators. *J. Mol. Model.* **2008**, *14*, 149–159. [CrossRef] [PubMed]
25. Clark, M.; Cramer, R.D.; Opdenbosch, N.V. Validation of the general purpose tripos 5.2 force field. *J. Comput. Chem.* **1989**, *10*, 982–1012. [CrossRef]
26. Golbraikh, A.; Tropsha, A. Beware of *q2*! *J. Mol. Graph. Model.* **2002**, *20*, 269–276. [CrossRef]
27. Ambure, P.S.; Gangwal, R.P.; Sangamwar, A.T. 3D-QSAR and molecular docking analysis of biphenyl amide derivatives as p38α mitogen-activated protein kinase inhibitors. *Mol. Divers.* **2012**, *16*, 377–388. [CrossRef] [PubMed]
28. Yi, W.; Shi, J.; Zhao, G.; Zhou, X.E.; Suinopowell, K.; Melcher, K.; Xu, H.E. Identification of a novel selective pparγ ligand with a unique binding mode and improved therapeutic profile in vitro. *Sci. Rep.* **2017**, *7*, 41487. [CrossRef] [PubMed]
29. Van Westen, G.J.; Overington, J.P.A. ligand's-eye view of protein similarity. *Nat. Methods* **2013**, *10*, 116–117. [CrossRef] [PubMed]

International Journal of
Molecular Sciences

MDPI

Article

Identification of a Novel PPAR-γ Agonist through a Scaffold Tuning Approach

Hyo Jin Gim †, Yong-Sung Choi †, Hua Li, Yoon-Jung Kim, Jae-Ha Ryu and Raok Jeon *

Drug Information Research Institute, Research Institute of Pharmaceutical Sciences, Sookmyung Women's University, Cheongpa-ro 47-gil 100, Yongsan-gu, Seoul 04310, Korea; hjgim83@gmail.com (H.J.G.); uriys2@gmail.com (Y.-S.C.); cooldog227@hotmail.com (H.L.); yoonjungkim73@gmail.com (Y.-J.K.); ryuha@sookmyung.ac.kr (J.-H.R)

* Correspondence: rjeon@sookmyung.ac.kr; Tel.: +82-2-710-9571

† These authors contributed equally to this work.

Received: 31 August 2018; Accepted: 2 October 2018; Published: 4 October 2018

Abstract: Peroxisome proliferator-activated receptors (PPARs) are important targets in metabolic diseases including obesity, metabolic syndrome, diabetes, and non-alcoholic fatty liver disease. Recently, they have been highlighted as attractive targets for the treatment of cardiovascular diseases and chronic myeloid leukemia. The PPAR agonist structure is consists of a polar head, a hydrophobic tail, and a linker. Each part interacts with PPARs through hydrogen bonds or hydrophobic interactions to stabilize target protein conformation, thus increasing its activity. Acidic head is essential for PPAR agonist activity. The aromatic linker plays an important role in making hydrophobic interactions with PPAR as well as adjusting the head-to-tail distance and conformation of the whole molecule. By tuning the scaffold of compound, the whole molecule could fit into the ligand-binding domain to achieve proper binding mode. We modified indol-3-ylacetic acid scaffold to (indol-1-ylmethyl)benzoic acid, whereas 2,4-dichloroanilide was fixed as the hydrophobic tail. We designed, synthesized, and assayed the in vitro activity of novel indole compounds with (indol-1-ylmethyl)benzoic acid scaffold. Compound **12** was a more potent PPAR-γ agonist than pioglitazone and our previous hit compound. Molecular docking studies may suggest the binding between compound **12** and PPAR-γ, rationalizing its high activity.

Keywords: PPAR; agonist; indole; scaffold; tuning

1. Introduction

Peroxisome proliferator-activated receptors (PPARs) are transcription factors that belong to the nuclear receptor superfamily. There are three subtypes of PPARs, designated as PPAR-α, -γ, and -$\delta(\beta)$, which exhibit different tissue expression profiles and modulate specific physiological functions. PPARs play a critical role in the regulation of multiple genes that regulate glucose and lipid metabolism and energy homeostasis [1–5]. Because they are involved in multiple metabolic pathways, PPARs are important molecular targets for the development of new drugs for metabolic diseases, such as obesity, metabolic syndrome (MetS), diabetes, and non-alcoholic fatty liver disease (NAFLD) [6–11]. Recent studies have shown that activation of PPARs not only regulates metabolic pathways but also mediates various biological effects related to inflammation, apoptosis, oxidative stress, and vascular function [12–16]. These effects seem to be beneficial in other disease conditions. It has been reported that the anti–diabetic PPAR agonists are also effective in cardiovascular disease (CVD) [17–19], thyroid [20,21], colorectal [21], and lung cancer [22] and chronic myeloid leukemia (CML) [23–25].

Currently, the PPAR-α agonists, fibrates (e.g., gemfibrozil, Figure 1) are used to treat dyslipidemia, whereas the PPAR-γ agonists, thiazolidinediones (TZDs; e.g., rosiglitazone, Figure 1) are used to treat type 2 diabetes mellitus (T2DM). However, the use of TZDs is associated with various adverse effects, particularly weight gain, bone fractures, cardiovascular complications, and edema [26–28]. No drugs in the market have been identified to target PPAR-δ(β). Continuous efforts are being made by many research groups and pharmaceutical companies worldwide to develop potent and safe therapeutic agents for the treatment of PPAR-associated diseases. In particular, they aim to develop pan, dual, or selective agonists of the three PPAR subtypes [29–33]. Representative PPAR agonists are depicted in Figure 1.

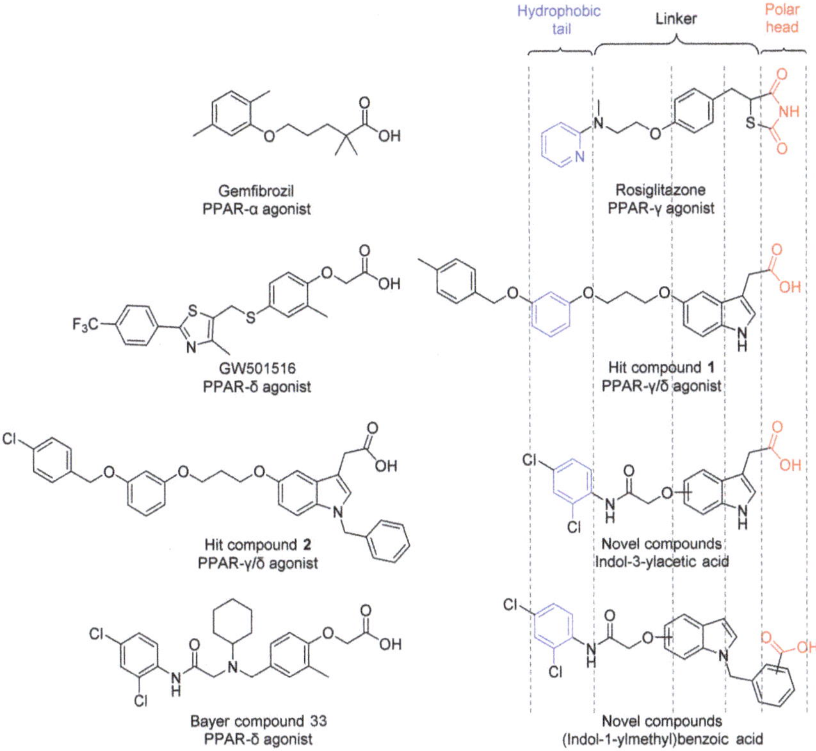

Figure 1. Structures of the representative PPAR agonists and newly designed compounds.

All subtypes of PPAR share structural and functional feature similar to those of other nuclear receptors. Crystal structures of human PPAR-α, PPAR-γ, and PPAR-δ(β) have revealed a common three-dimensional structure of the ligand binding domains (LBDs). An antiparallel sandwich of 12 α-helices (helix 1 to helix 12) and a three-stranded antiparallel β-sheet are forming a large ligand binding cavity in the core of the LBD. The central cavity spans between the AF-2 domain within C-terminal α-helix 12 and the three-stranded antiparallel β-sheet [34]. Interestingly, despite a common general structure of the LBD, ligand binding to PPARs shows both species and isotype specificities. The ligands for PPAR-α or PPAR-γ should be able to adopt a U-shaped conformation and an L-shaped conformation for PPAR-δ(β) [35].

A typical PPAR agonist consists of three parts: a hydrophobic tail moiety, a polar head group (usually bearing a carboxylic acid functionality) and a linker which consists of flexible methylene units and an aromatic ring (Figure 2A). The acidic head group is crucial for PPAR activation. It forms an H-bonding network with a part of the PPAR that mainly contains the critical polar residues, such as Gln286, Ser289, His323, Tyr327, Lys367, His449, and Tyr473 in PPAR-γ (see Figure 2B). The TZD moiety (acidic head group of rosiglitazone) makes several specific interactions with amino acids. The carbonyl groups of the TZD make hydrogen bonds with His323 and His449. His323 forms a secondary hydrogen bond with Tyr473 in the AF-2 domain. The nitrogen of the TZD head group is within hydrogen-bonding distance of the hydroxyl group of Tyr473. Lys367 forms another secondary hydrogen bond with the ligand, at residue His449. The conformation of the TZD head group and the participating amino acids are fixed by these primary and secondary hydrogen bonds [34]. This H-bonding network stabilizes the conformation of the AF-2 domain allowing it to bind with the coactivator proteins. Therefore, the presence of the acidic head in the right position is important in the development of PPAR agonists. Another part of the PPAR agonists is the hydrophobic moiety that mainly interacts with the hydrophobic residues in the LBD. The hydrophobic tail occupies the large cavity of the LBD by interacting with the hydrophobic residues, such as Ile281, Gly284, Ile341, and Leu353 in PPAR-γ (see Figure 2B). The linker wraps the central helix 3 (H3) and interacts with the surrounding hydrophobic residues. In addition, it acts as a linker between the acidic head and hydrophobic tail groups to allow the fitting of the whole compound into the LBD, where proper interactions could be achieved. Therefore, fine tuning of the linker to adjust the atomic distance or three-dimensional arrangement is a powerful approach for structural optimization to increase the biological activity of compounds.

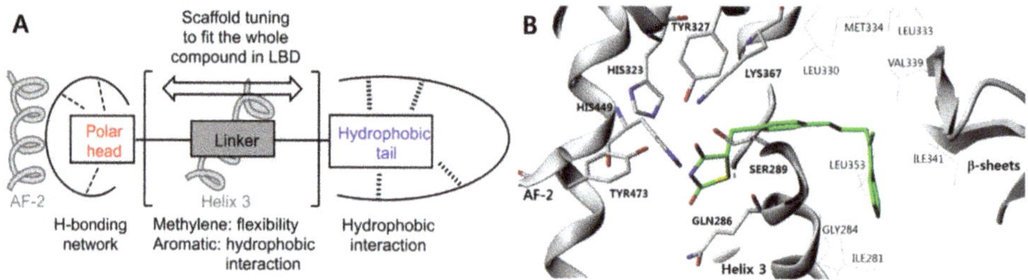

Figure 2. (**A**) Schematic features of PPAR agonists and the concept of the novel target compounds; (**B**) Crystal structure of PPAR-γ LDB and rosiglitazone (PDB code: 2PRG). Only the critical amino acids are displayed and labeled. Polar residues for the key interaction (capped stick) and hydrophobic residues (gray line) are depicted. Rosiglitazone is displayed in green and oxygen, nitrogen and sulfur atom is marked in red, blue and yellow respectively.

Previously, we identified a series of (alkoxyindol-3-yl)acetic acids as PPAR agonists [36–38]. Through stepwise structural modification and optimization, benzyloxy-containing indol-3-ylacetic acid analogs (**1** and **2**) were identified as PPAR-γ/δ dual agonists. Compound **2** lowered blood glucose, insulin, and glycated hemoglobin (HbA1c) levels without causing weight gain; additionally, it reduced the accumulation of lipids and the size of the adipocytes in db/db mice. These findings indicated the potential of the indol-3-ylacetic acid as a core scaffold for anti-diabetic and anti-dyslipidemic drugs. In our previous series of PPAR agonists, the indole group held the critical polar functional group. In addition, the *N*-benzyl group of the indole ring significantly affected the PPAR activity. Hence, we performed further modifications of the indole-based PPAR agonists.

Based on the indol-3-ylacetic acid scaffold, we introduced the dichloroanilide group into the hydrophobic tail to increase the potency and selectivity. Some reported compounds with the 2,4-dichloroanilide moiety showed a significant PPAR agonist potency, and even small changes to the 2,4-dichlorophenyl substitution resulted in a marked change in potency. For example, Bayer compound **33** (Figure 1) showed a potent and selective activity as PPAR-δ agonist with EC_{50} value of 3 nM and >1000-fold selectivity. The 2,4-disubstituted anilines were superior to other substitution patterns [39]. Luckhurst et al. drew inspiration from Bayer compound **33** and combined the anilide portion with isoindoline, tetrahydroisoquinoline, and benzazepine scaffolds. The structure-activity relationship study confirmed that the 2,4-dichloroanilide compounds were the most potent of any other derivatives [40,41]. Although interactions between 2,4-dichloroanilide and LBD of PPAR is not specified in those studies, 2,4-dichloroanilide moiety is multifunctional which can make many kinds of interactions including aromatic pi-interaction, halogen-hydrogen bonding, and hydrogen bonding through aromatic phenyl, dichloro and amide group, respectively. Furthermore, newly designed 2,4-dichloroanilide substituted analogues show proper U-shaped binding mode similar to rosiglitazone in our preliminary molecular modeling study.

For further structural optimization, including re-positioning of the critical acidic head group, we set the change in the hydrophobic tail group as a starting point for our journey (Figure 2).

2. Result and Discussion

The synthesis of dichloroanilide-linked indole compounds is illustrated in Schemes 1–3.

For the introduction of the dichloroanilide moiety as a hydrophobic tail, 2,4-dichloroaniline and 2-chloroacetyl chloride underwent a direct substitution reaction to yield 2-chloro-*N*-(2,4-dichlorophenyl)acetamide (**3**; Scheme 1).

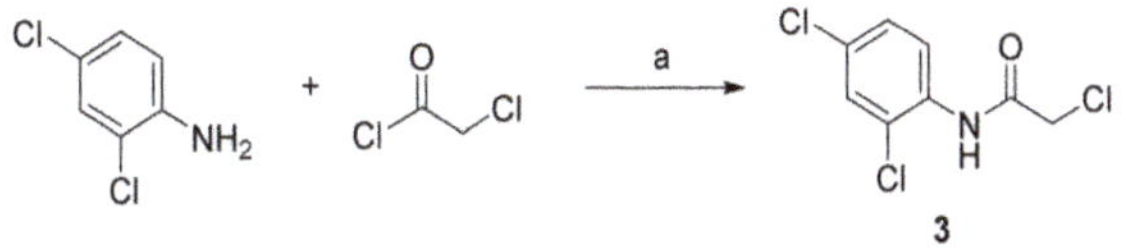

Scheme 1. Reaction conditions: (a) triethylamine, CH_2Cl_2, 0 °C.

Hydroxyindol-3-ylacetic acid methyl ester compounds (**4a**,**4b**) were prepared as previously described [32]. The amino and hydroxyl groups of the indole were benzylated under basic conditions. The *O*-benzyl group was selectively deprotected in the following hydrogenation step. Then, the developed free hydroxyl group was conjugated with compound **3**. The methyl esters were hydrolyzed using LiOH to yield the corresponding carboxylic acid compounds (**5**–**8**; Scheme 2).

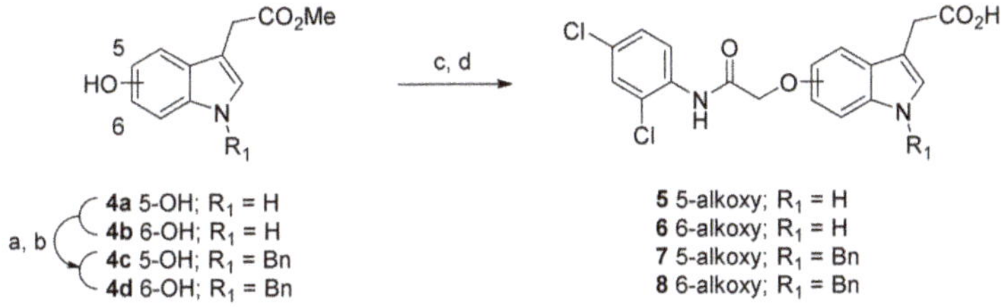

Scheme 2. Reaction conditions: (a) benzyl bromide, Cs_2CO_3, DMF, 80 °C; (b) H_2, Pd/C, EtOAc/EtOH, RT; (c) compound 3, Cs_2CO_3, MeCN, RT; (d) $LiOH{\cdot}H_2O$, THF/MeOH/H_2O, RT.

The starting *O*-benzyl and *tert*-butyldiphenylsilyl (TBDPS)-protected hydroxyindoles were synthesized under typical conditions. Compounds with the 3-benzyl group were prepared by reaction of compound **9a**,**b** with benzyl alcohol, whereas the prenyl group was introduced at the 3-position of the *O*-TBDPS-protected compounds (**9e**–**f**) using prenyl bromide. Compounds with methyl carboxylate groups at various positions of the *N*-linked benzyl group were directly synthesized by substitution reaction between the amino group of the indole and the 2-, 3-, or 4-(methoxycarbonyl)methyl-substituted benzyl chloride. The *O*-benzyl and TBDPS protecting groups were removed by conventional hydrogenation and tetrabutylammonium fluoride (TBAF), respectively. *O*-Alkylation and hydrolysis of the esters yielded the target compounds (**12**–**21**;Scheme 3).

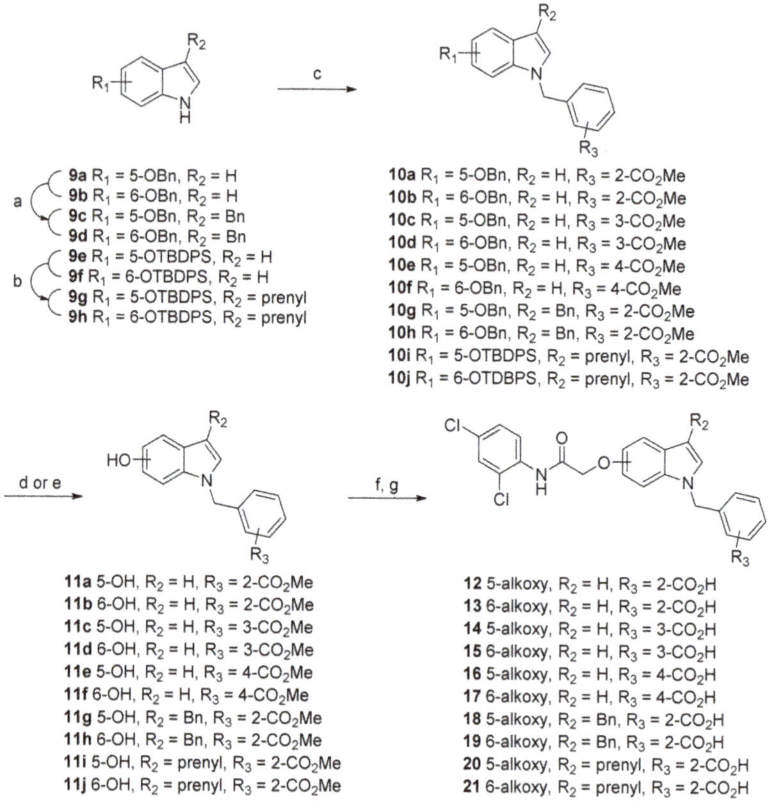

a: **9a** R_1 = 5-OBn, R_2 = H; **9b** R_1 = 6-OBn, R_2 = H → **9c** R_1 = 5-OBn, R_2 = Bn; **9d** R_1 = 6-OBn, R_2 = Bn

b: **9e** R_1 = 5-OTBDPS, R_2 = H; **9f** R_1 = 6-OTBDPS, R_2 = H → **9g** R_1 = 5-OTBDPS, R_2 = prenyl; **9h** R_1 = 6-OTBDPS, R_2 = prenyl

c →

10a R_1 = 5-OBn, R_2 = H, R_3 = 2-CO_2Me
10b R_1 = 6-OBn, R_2 = H, R_3 = 2-CO_2Me
10c R_1 = 5-OBn, R_2 = H, R_3 = 3-CO_2Me
10d R_1 = 6-OBn, R_2 = H, R_3 = 3-CO_2Me
10e R_1 = 5-OBn, R_2 = H, R_3 = 4-CO_2Me
10f R_1 = 6-OBn, R_2 = H, R_3 = 4-CO_2Me
10g R_1 = 5-OBn, R_2 = Bn, R_3 = 2-CO_2Me
10h R_1 = 6-OBn, R_2 = Bn, R_3 = 2-CO_2Me
10i R_1 = 5-OTBDPS, R_2 = prenyl, R_3 = 2-CO_2Me
10j R_1 = 6-OTDBPS, R_2 = prenyl, R_3 = 2-CO_2Me

d or e →

11a 5-OH, R_2 = H, R_3 = 2-CO_2Me
11b 6-OH, R_2 = H, R_3 = 2-CO_2Me
11c 5-OH, R_2 = H, R_3 = 3-CO_2Me
11d 6-OH, R_2 = H, R_3 = 3-CO_2Me
11e 5-OH, R_2 = H, R_3 = 4-CO_2Me
11f 6-OH, R_2 = H, R_3 = 4-CO_2Me
11g 5-OH, R_2 = Bn, R_3 = 2-CO_2Me
11h 6-OH, R_2 = Bn, R_3 = 2-CO_2Me
11i 5-OH, R_2 = prenyl, R_3 = 2-CO_2Me
11j 6-OH, R_2 = prenyl, R_3 = 2-CO_2Me

f, g →

12 5-alkoxy, R_2 = H, R_3 = 2-CO_2H
13 6-alkoxy, R_2 = H, R_3 = 2-CO_2H
14 5-alkoxy, R_2 = H, R_3 = 3-CO_2H
15 6-alkoxy, R_2 = H, R_3 = 3-CO_2H
16 5-alkoxy, R_2 = H, R_3 = 4-CO_2H
17 6-alkoxy, R_2 = H, R_3 = 4-CO_2H
18 5-alkoxy, R_2 = Bn, R_3 = 2-CO_2H
19 6-alkoxy, R_2 = Bn, R_3 = 2-CO_2H
20 5-alkoxy, R_2 = prenyl, R_3 = 2-CO_2H
21 6-alkoxy, R_2 = prenyl, R_3 = 2-CO_2H

Scheme 3. Reaction conditions: (a) benzyl alcohol, KOH, xylene, reflux; (b) prenyl bromide, Na_2CO_3, $MeCN/H_2O$ (9:1), RT; (c) methyl 2-, 3- or 4-(chloromethyl)benzoate, NaH, THF, reflux; (d) H_2, Pd/C, EtOAc/EtOH, RT; (e) TBAF, THF, RT; (f) compound **3**, Cs_2CO_3, MeCN, RT; (g) $LiOH{\cdot}H_2O$, $THF/MeOH/H_2O$, RT.

The structures of all the target compounds (**5–8, 12–21**) were confirmed by their proton nuclear magnetic resonance (^{1}H NMR) spectra and high-resolution mass spectrometry (HRMS), and they were cross-checked against the structures of their methyl ester precursors (**P5–8, P12–21**) which were characterized by the ^{1}H and ^{13}C NMR spectra. Analytical data are provided in the Materials and Methods section.

The PPAR activities of the synthesized dichloroanilide-linked indole analogs were assessed using a standard cell-based transactivation assay in CV-1 cells. GW7647 [42], pioglitazone, and GW0742 [43] were used as positive controls for PPAR-α, -γ, and -δ, respectively. These three positive controls presented the maximum activation for each PPAR subtype. The activity of the compounds was presented as the activation percentage (%) relative to the maximum values of the positive controls at the indicated concentrations. PPAR-α, -γ, and -δ transactivation activities of the tested compounds are summarized in Tables 1 and 2.

Table 1. In vitro cell-based PPAR transactivation activity [a].

Cpd. No.	Alkoxy position	R1	% max [b]		
			PPAR-α	PPAR-γ	PPAR-δ
5	5	H	−38.13 ± 0.06	5.60 ± 0.14	11.84 ± 0.35
6	6	H	1.12 ± 0.24	8.66 ± 0.23	8.27 ± 0.33
7	5	benzyl	21.41 ± 0.21	31.20 ± 0.38	44.35 ± 0.77
8	6	benzyl	31.68 ± 0.22	71.87 ± 0.91	65.41 ± 0.70

[a] The agonistic activity of the compounds (10 μM) was assayed on PPAR-Luc in CV-1 cells; [b] Relative activation with respect to the maximum activation obtained with GW7647 (10 μM), pioglitazone (10 μM), and GW0742 (10 μM) for PPAR-α, -γ, and -δ, respectively.

Table 2. In vitro cell-based PPAR transactivation activity [a].

Cpd. No.	Alkoxy position	R2	R3	% max [b]		
				PPAR-α	PPAR-γ	PPAR-δ
12	5	H	2-CO_2H	105.08 ± 0.41	189.21 ± 1.67	94.86 ± 0.20
13	6	H	2-CO_2H	51.84 ± 0.25	74.54 ± 0.94	86.70 ± 0.53
14	5	H	3-CO_2H	10.33 ± 0.07	42.01 ± 0.52	23.85 ± 0.51
15	6	H	3-CO_2H	78.15 ± 0.04	126.55 ± 0.92	86.09 ± 0.31
16	5	H	4-CO_2H	-19.98 ± 0.08	0.29 ± 0.06	-1.07 ± 0.14
17	6	H	4-CO_2H	-14.44 ± 0.16	-0.84 ± 0.16	4.87 ± 0.24
18	5	benzyl	2-CO_2H	23.79 ± 0.11	93.20 ± 0.59	75.88 ± 0.35
19	6	benzyl	2-CO_2H	7.25 ± 0.12	19.03 ± 0.13	19.40 ± 0.54
20	5	prenyl	2-CO_2H	-5.93 ± 0.14	5.10 ± 0.20	8.92 ± 0.48
21	6	prenyl	2-CO_2H	-2.30 ± 0.19	9.55 ± 0.32	22.16 ± 0.59

[a] The agonistic activity of the compounds (10 μM) was assayed on PPAR-Luc in CV-1 cells; [b] Relative activation with respect to the maximum activation obtained with GW7647 (10 μM), pioglitazone (10 μM), and GW0742 (10 μM) for PPAR-α, -γ, and -δ, respectively.

Compounds **5** and **6** were the simplest indoly-3-ylacetic acids containing the dichloroanilide tail at the 5- and 6- positions, respectively. However, these compounds showed minimum agonistic activities to PPARs. Their *N*-benzylated analogs (**7**,**8**) showed an increase in the activity, which is in line with the trends observed in a previous study. However, their activities were much lower than those of the previous hit compounds **1** and **2** (134.4 and 100.5% activity on PPAR-γ, respectively) [33]. This decline in the activity could be explained by the length and rigidity of the linker between the hydrophobic tail and indole ring. As shown in Figure 1, the linker in compounds **1** and **2** consists of four rotatable bonds between the two planar aromatic ring systems. Rosiglitazone also has three bonds, which are freely rotatable. In contrast, compounds **5**–**8** have three rotatable bonds; however, two of them (same as three atoms) are linked to a planar carbonyl system. This non-flexibility could result in a difference in the docking poses of the compounds.

To better understand the structural features of the synthesized compounds in the active site of the PPAR LBD, a docking study of compounds **1** and **5** was carried out using Surflex Dock interfaced with Sybyl-X (Figure 3). The crystal structure of PPAR-γ LBD complexed with the PPAR-γ selective agonist, rosiglitazone (PDB code: 2PRG) was selected to perform the docking of compounds [34]. The polar head groups of rosiglitazone (green) and compound **1** (magenta) revealed the H-bonding interaction with the polar residues in the hydrophilic pocket. The indole ring acted as an aromatic linker, similar to the benzene ring of rosiglitazone. Thus, the aromatic linker of each compound was well-overlapped in the docking position and made the compound U-shaped. The pyridine ring of rosiglitazone was located in the hydrophobic pocket of PPAR-γ and interacted with the hydrophobic amino acid residues. The benzene ring of compound **1** was well-aligned with the pyridine ring of rosiglitazone. The docking pose of compound **5** (red) might explain its low activity. The acidic head group could form hydrogen bonds, similar to those of compound **1**. The newly introduced 2,4-dichloroanilide moiety was located near the hydrophobic pocket; however, it was misaligned with the pyridine ring of rosiglitazone. Moreover, the benzene ring, a part of the indole moiety, did not superimpose with that of other compounds. This might due to the non-flexibility of the linker. The rigid carbonyl plane was located at the corner; thus, the linker could not turn smoothly. Structural modification was needed to compensate this rigidity. The *N*-benzyl group of compounds **7** and **8** might make additional interaction with the side pocket. Nevertheless, we tried to use this benzyl group as an aromatic linker to hold the polar acidic head.

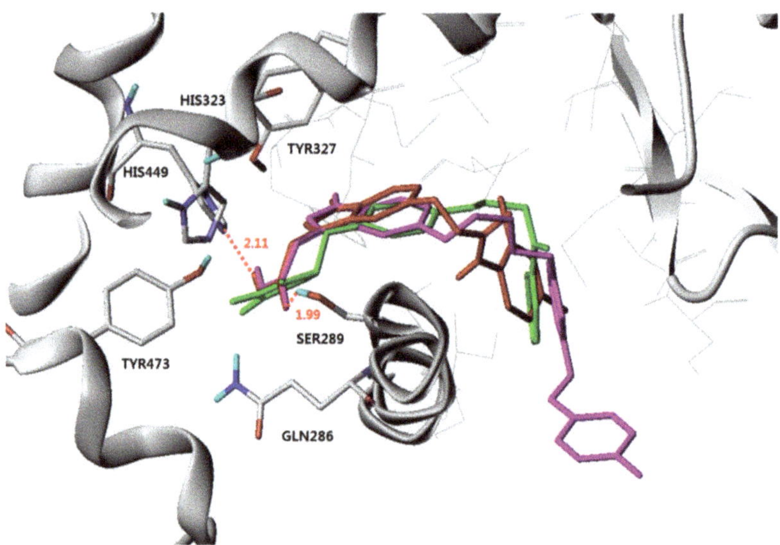

Figure 3. Binding modes of hit compound **1** (magenta) and compound **5** (red) in the active site of PPAR-γ LBD (PDB code: 2PRG; co-crystallized with rosiglitazone (green)). Only the critical amino acids that interacted with the docked ligands are displayed and labeled. Several residues for the key interaction (capped stick), hydrophobic residues (gray line), and hydrogen bonds (red-dotted lines) are depicted. The numbers indicate the atomic distance (Å).

N-Benzylated indole derivatives, which had a carboxylic acid moiety at the *ortho-*, *meta-*, or *para*-position of the benzyl group, were synthesized. Among the 5-alkoxy compounds, the *ortho*-carboxybenzyl-substituted compound **12** was the most active, whereas the *meta*-carboxybenzyl-substituted compound **15** showed the highest activity among the 6-alkoxy series. We supposed that the activity was also affected by the distance between the alkoxy and carboxyl groups. 5-Alkoxy was a little far from the benzyl-substituted amine in the indole structure (*pseudo-para*), and the *ortho*-positioned carboxyl group was the best to fit into the hydrophilic pocket. However, 6-alkoxy was at the *pseudo-meta* position of the indole structure, thus the carboxyl group at the *meta* position showed better activity. Molecular modeling guided us to introduce additional substituents at the 3-position of the indole moiety, which could be placed in the extra back pocket near the indole structure. Thus, 3-benzyl- or prenyl-substituted *N*-(2-carboxybenzyl)indole compounds were also prepared. Unexpectedly, all analogs (**18–21**) with a substitution at the 3-position of the indole moiety exhibited weak PPAR agonistic activity.

A docking study of compound **12** was also carried out using the same system as described above. Compound **12** showed a U-shape binding mode wrapping around H3, similar to that of rosiglitazone (Figure 4). The carboxyl group of compound **12** was located in the hydrophilic pocket and interacted with the key polar residues, Ser289, His323, and Tyr473 by H-bonding. Next to the acidic head group, the benzene ring was positioned in the hydrophobic region formed by Phe282, Phe363, and Leu453. The indole ring acted as an aromatic linker, similar to the benzene ring of rosiglitazone. In addition, the dichloroanilide group was well-fitted in the hydrophobic pocket along with rosiglitazone. The indole group was located close to H3, and dichloroanilide was positioned between H3 and β2. Overall, the dichloroanilide moiety of compound **12** was well-aligned with the hydrophobic tail of rosiglitazone (Figure 4), in contrast to that of compounds **5** (Figure 3). The dichlorophenyl group of compound **12** could interact with the Ile281, Gly284, Ile341, and Leu353 residues, which form the hydrophobic pocket. In particular, the *ortho*-chloro atom was located 3.43 Å away from the carbonyl oxygen of Gly284 to make a halogen bonding interaction. Therefore, the high activity of compound **12** might be attributed to the proper binding and suitable interactions between compound **12** and PPAR-γ.

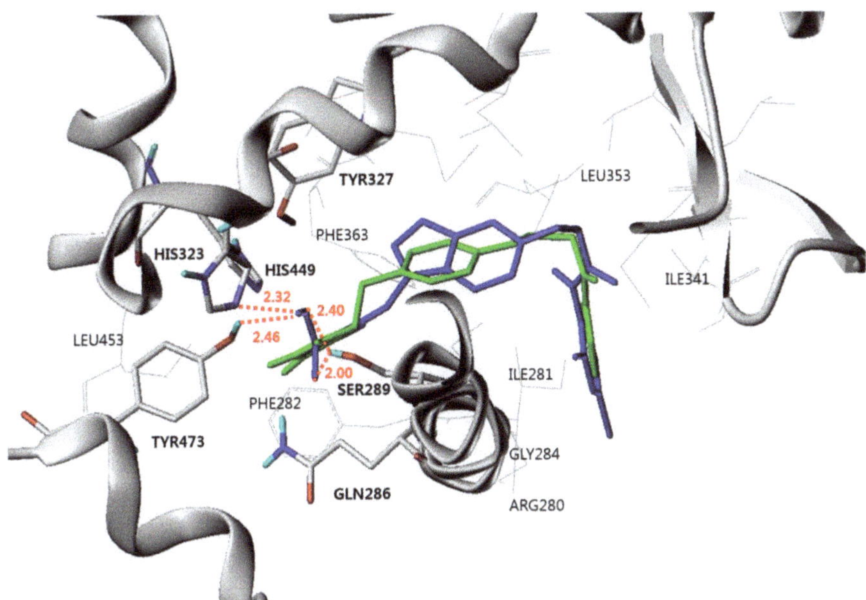

Figure 4. Binding modes of compound **12** (blue) in the active site of PPAR-γ LBD (PDB code: 2PRG; co-crystallized with rosiglitazone (green)). Only the critical amino acids that interacted with the docked ligands are displayed and labeled. Several residues for the key interaction (capped stick), hydrophobic residues (gray line), and hydrogen bonds (red-dotted lines) are depicted. The numbers indicate the atomic distance (Å).

Another potent compound, **15** belonged to the 6-alkoxyindole series. The carboxylic acid at the *meta* position of the *N*-benzyl group made 4 pairs of hydrogen bonds with the Ser289 and Tyr473 residues in the hydrophilic pocket (Figure 5). The carboxylic acid group of compound **15** was aligned similarly to that of compound **12**; however, the other part of compound **15** (from the benzyl group to the 2,4-dichloroanilide tail) was slightly moved back to locate the acid in the center of the hydrophilic region. Because of this movement, compound **15** was possibly less coordinated with the surrounding hydrophobic residues than compound **12**. The 6-alkoxyindole analog with *ortho*-carboxybenzyl group (13) and 5-alkoxyindole analog with *meta*-carboxybenzyl group (**14**) made less H-bonds with the polar residues because their position in 3D was influenced by the conformation of the whole molecule (data not shown). The findings of the docking study could explain the results of the in vitro transactivation assay. The substitution position of carboxylic acid on the benzyl group highly affected the PPAR activity, implying that the optimal positioning and matched combinations of the acidic head and lipophilic tail groups were essential. Taken together, we identified the *N*-(carboxylbenzyl)-substituted indole as a novel and promising scaffold for development of therapeutic agents targeting PPARs.

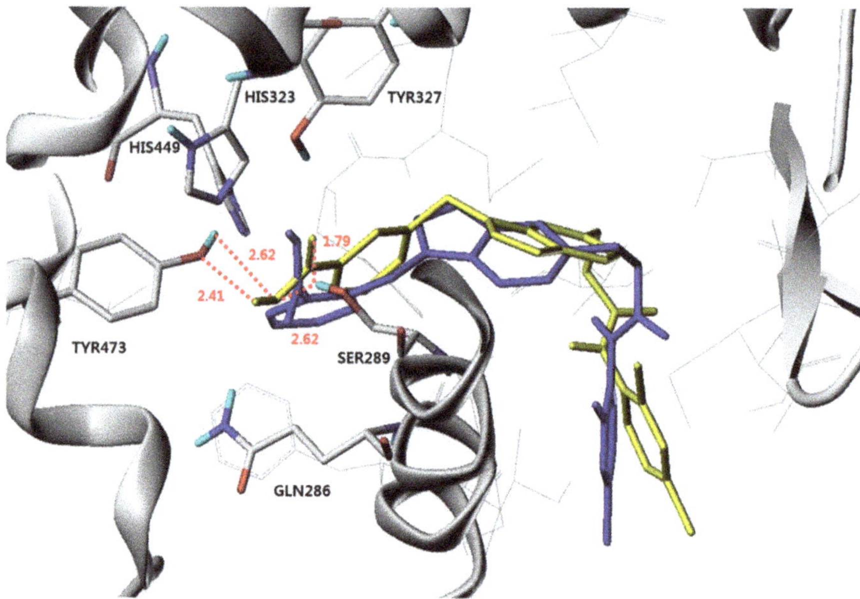

Figure 5. Binding modes of compound **12** (blue) and compound **15** (yellow) in the active site of PPAR-γ LBD (PDB code: 2PRG; co-crystallized with rosiglitazone). Only the critical amino acids that interacted with the docked ligands are displayed and labeled. Several residues for the key interaction (capped stick), hydrophobic residues (gray line), and hydrogen bonds (red-dotted lines) are depicted. The numbers indicate the atomic distance (Å).

PPAR coactivator recruit assay was performed to determine the potency of the 4 most active compounds. Binding of agonist to the PPAR LBD causes a conformational change around helix 12 in the LBD, resulting in higher affinity for the coactivator peptide. When the terbium label on the anti-GST antibody complexed with PPAR is excited at 340 nm, energy is transferred to the fluorescein label on the coactivator peptide and detected as emission at 520 nm. The time-resolved fluorescence resonance energy transfer (TR-FRET) ratio of 520:495 is calculated and used to determine the EC_{50} from a dose response curve of the compound. Based on the biology of the PPAR-coactivator peptide interaction, this ligand EC_{50} is a composite value representing the amount of ligand required to bind to receptor, effect a conformational change, and recruit coactivator peptide. Potency of compounds show similar order to their activity (Table 3). Notably, compound **12** showed EC_{50} of 1.96 nM to PPAR-γ

which was 1,970 fold selective to PPAR-α and 16,600 fold to PPAR-δ. The second potent compound **15** revealed EC_{50} of 32.3 nM to PPAR-γ whereas >100 μM to other PPAR subtypes. These results were huge improvement from our previous hit compounds (**1** and **2**), which exhibited PPAR-γ/δ dual agonistic activity at micromolar concentration. The PPAR-δ ligand binding pocket is significantly smaller than those of PPAR-α and PPAR-γ because of the narrowing of the pocket adjacent to the AF2 domain. Ligands such as TZDs show little or no binding to PPAR-δ. Compound **11** and **15** have *ortho* or *meta* substituted benzoic acid as an acidic head group. Their acidic head groups might seem to be too large to fit within the narrow PPAR-δ pocket. In contrast, the compound **1** and **2** contains an acetic acid head group that complements the narrow PPAR-δ ligand binding. And docking study of compound **11** with crystal structure of PPAR-α (PDB code: 3G8I; co-crystallized with aleglitazar) and PPAR-δ (PDB code: 3TKM; co-crystallized with GW0742) revealed misaligned conformation compared to their co-crystallized ligands, aleglitazar (PPAR-α/γ agonist) and GW0742 (PPAR-δ agonist) respectively (data not shown). The potency and selectivity of compound **11** could be explained by harmonization of steric effect of acidic head group, length and rigidity of hydrophobic tail, and properly tuned aromatic linker system.

Table 3. In vitro PPAR coactivator recruit assay [a].

Cpd. No.	Alkoxy position	R2	R3	EC_{50} (μM)		
				PPAR-α	PPAR-γ	PPAR-δ
12	5	H	2-CO_2H	3.87	0.00196	32.6
13	6	H	2-CO_2H	16.9	0.571	>100
15	6	H	3-CO_2H	>100	0.0326	>100
18	5	benzyl	2-CO_2H	>100	>11.1	28.1

[a] The agonistic activity of the compounds was determined using TR-FRET coactivator recruit assay.

3. Materials and Methods

3.1. General Information

Most reagents and solvents were purchased from commercial sources and used without purification, with the following exceptions. Tetrahydrofuran (THF) was distilled from sodium benzophenone ketyl. Acetonitrile (MeCN), methylene chloride (CH_2Cl_2), dimethylformamide (DMF), and triethylamine were distilled from calcium hydride under nitrogen atmosphere. Column chromatography was performed using silica gel 60 (230–400 mesh, Merck) with the indicated solvents. Thin-layer chromatography (TLC) was performed using Kieselgel 60 F254 plates (Merck). ^1H and ^{13}C NMR spectra were recorded using a Varian Inova 400 or Bruker Avance III HD500 spectrometer using $CDCl_3$, CD_3OD, or $(CD_3)_2SO$ as solvents. Chemical shifts (δ) were expressed in parts per million (ppm) downfield from an internal standard, tetramethylsilane. HRMS data were recorded using an Agilent 6530 QTOF mass spectrometer with an electrospray (ESI) interface with Agilent jet stream technology in the negative ion mode.

3.2. Experimental Procedures and Analytical Data of Compounds

3.2.1. 2-Chloro-*N*-(2,4-Dichlorophenyl)Acetamide (3)

Triethylamine (1.1 equiv.) was added to a solution of 2,4-chloroaniline in CH_2Cl_2 (0.5 M) and stirred at room temperature for 30 min. Then, the reaction mixture was cooled to 0 °C. Chloroacetyl chloride (1.2 equiv.) was added dropwise while the temperature was maintained. After the completion of the reaction, water was added to the reaction mixture to quench the reaction. The organic layer was separated and washed with water thrice. Then, it was dried with brine and $MgSO_4$, filtered, and concentrated under vacuum. The residue was purified by silica gel column chromatography using *n*-hexane and EtOAc (5:1) to yield a quantitative amount of the title compound **3**: ^{1}H NMR (400 MHz, $CDCl_3$) δ 8.89 (s, 1H), 8.34 (d, *J* = 8.8 Hz, 1H), 7.42 (d, *J* = 2.4 Hz, 1H), 7.27 (dd, *J* = 8.8, 2.4 Hz, 1H), 4.23 (s, 2H); ^{13}C NMR (100 MHz, $CDCl_3$) δ 164.0, 132.5, 130.2, 129.1, 128.1, 124.1, 122.0, 43.2.

3.2.2. Synthetic Procedure for *O*-Alkylation of 5- or 6-Hydroxyindole Compounds

The prepared 5- or 6-hydoxyindole compounds were dissolved in anhydrous MeCN (0.5 M), and Cs_2CO_3 (1.5 equiv.) was added and stirred at room temperature for 30 min. Compound **3** (1.2 equiv.) was added to the reaction mixture then stirred until the reaction was complete. Then, the reaction mixture was diluted with water and extracted with EtOAc. The combined organic extracts were washed with water and brine, dried over anhydrous $MgSO_4$, filtered, and concentrated under reduced pressure. The residue was purified by silica gel column chromatography using an appropriate mixture of *n*-hexane and EtOAc as an eluent to yield the corresponding alkoxyindole compounds (the methyl ester form of the final target compounds; yield = 65–86%).

Methyl 2-(5-(2-((2,4-dichlorophenyl)amino)-2-oxoethoxy)-1*H*-indol-3-yl)acetate (P5): ^{1}H NMR (400 MHz, $CDCl_3$) δ 9.12 (s, 1H), 8.46 (d, *J* = 8.9 Hz, 1H), 8.10 (s, 1H), 7.41 (d, *J* = 2.4 Hz, 1H), 7.31 (d, *J* = 8.8 Hz, 1H), 7.28 (dd, *J* = 8.8, 2.3 Hz, 1H), 7.20 (d, *J* = 2.2 Hz, 1H), 7.14 (d, *J* = 2.4 Hz, 1H), 6.96 (dd, *J* = 8.8, 2.4 Hz, 1H), 4.70 (s, 2H), 3.75 (d, *J* = 0.5 Hz, 2H), 3.70 (s, 3H); ^{13}C NMR (100 MHz, $CDCl_3$) δ 172.4, 167.2, 151.7, 132.9, 132.3, 129.6, 129.0, 128.1, 127.9, 124.6, 123.9, 122.1, 112.7, 112.5, 108.7, 103.1, 68.9, 52.2, 31.3.

Methyl 2-(6-(2-((2,4-dichlorophenyl)amino)-2-oxoethoxy)-1*H*-indol-3-yl)acetate (P6): ^{1}H NMR (400 MHz, $CDCl_3$) δ 9.06 (s, 1H), 8.42 (d, *J* = 8.9 Hz, 1H), 8.24 (s, 1H), 7.52 (d, *J* = 8.6 Hz, 1H), 7.38 (d, *J* = 2.3 Hz, 1H), 7.25 (dd, *J* = 8.7, 2.5 Hz, 1H), 7.07 (d, *J* = 2.2 Hz, 1H), 6.88 (dd, *J* = 8.6, 2.3 Hz, 1H), 6.84 (d, *J* = 2.0 Hz, 1H), 4.63 (s, 2H), 3.76 (s, 2H), 3.71 (s, 3H); ^{13}C NMR (100 MHz, $CDCl_3$) δ 172.7, 167.0, 153.8, 136.6, 132.7, 129.6, 129.0, 128.0, 123.8, 123.02, 122.97, 122.0, 120.1, 110.0, 108.5, 96.3, 68.4, 52.2, 31.2.

Methyl 2-(1-benzyl-5-(2-((2,4-dichlorophenyl)amino)-2-oxoethoxy)-1*H*-indol-3-yl)acetate (P7): ^{1}H NMR (400 MHz, $CDCl_3$) δ 9.11 (s, 1H), 8.46 (d, *J* = 8.9 Hz, 1H), 7.40 (d, *J* = 2.4 Hz, 1H), 7.32–7.24 (m, 4H), 7.18 (d, *J* = 8.9 Hz, 1H), 7.15 (d, *J* = 2.4 Hz, 1H), 7.14 (s, 1H), 7.12–7.09 (m, 2H), 6.92 (dd, *J* = 8.9, 2.5 Hz, 1H), 5.26 (s, 2H), 4.69 (s, 2H), 3.74 (d, *J* = 0.6 Hz, 2H), 3.70 (s, 3H); ^{13}C NMR (100 MHz, $CDCl_3$) δ 172.4, 167.2, 151.6, 137.3, 132.9, 132.8, 129.5, 128.98, 128.95, 128.6, 128.5, 128.1, 127.9, 126.9, 123.8, 122.1, 112.4, 111.2, 107.4, 103.4, 68.9, 52.2, 50.5, 31.2.

Methyl 2-(1-benzyl-6-(2-((2,4-dichlorophenyl)amino)-2-oxoethoxy)-1*H*-indol-3-yl)acetate (P8): ^{1}H NMR (400 MHz, $CDCl_3$) δ 9.00 (s, 1H), 8.41 (d, *J* = 8.9 Hz, 1H), 7.55 (d, *J* = 8.6 Hz, 1H), 7.38 (d, *J* = 2.4 Hz, 1H), 7.31–7.22 (m, 4H), 7.13–7.08 (m, 2H), 7.06 (s, 1H), 6.87 (dd, *J* = 8.6, 2.3 Hz, 1H), 6.76 (d, *J* = 2.2 Hz, 1H), 5.22 (s, 2H), 4.61 (s, 2H), 3.75 (s, 2H), 3.70 (s, 3H); ^{13}C NMR (100 MHz, $CDCl_3$) δ 172.5, 166.9, 153.8, 137.0, 132.8, 129.6, 129.0, 128.0, 127.9, 127.2, 126.9, 123.9, 123.8, 122.1, 120.5, 109.6, 107.8, 95.3, 68.5, 52.2, 50.3, 31.2.

Methyl 2-((5-(2-((2,4-dichlorophenyl)amino)-2-oxoethoxy)-1*H*-indol-1-yl)methyl)benzoate (P12): ^{1}H NMR (400 MHz, $CDCl_3$) δ 9.11 (s, 1H), 8.46 (d, *J* = 8.9 Hz, 1H), 8.12–7.98 (m, 1H), 7.39 (d, *J* = 2.3 Hz,

1H), 7.33–7.29 (m, 2H), 7.28–7.25 (m, 1H), 7.21 (d, J = 2.3 Hz, 1H), 7.17–7.12 (m, 2H), 6.90 (dd, J = 8.9, 2.4 Hz, 1H), 6.54 (d, J = 3.0 Hz, 1H), 6.48–6.44 (m, 1H), 5.76 (s, 2H), 4.68 (s, 2H), 3.94 (s, 3H); ^{13}C NMR (100 MHz, $CDCl_3$) δ 167.2, 151.6, 140.1, 133.1, 132.9, 132.7, 131.2, 130.1, 129.1, 129.0, 128.1, 127.8, 127.4, 127.1, 123.8, 122.1, 121.0, 112.2, 111.0, 104.9, 101.8, 95.9, 68.9, 52.4, 49.0.

Methyl 2-((6-(2-((2,4-dichlorophenyl)amino)-2-oxoethoxy)-1*H*-indol-1-yl)methyl) benzoate (P13): ^{1}H NMR (400 MHz, $CDCl_3$) δ 8.99 (s, 1H), 8.39 (d, J = 8.9 Hz, 1H), 8.07–8.02 (m, 1H), 7.59 (d, J = 8.6 Hz, 1H), 7.36 (d, J = 2.4 Hz, 1H), 7.30–7.26 (m, 2H), 7.25–7.22 (m, 1H), 7.08 (d, J = 3.2 Hz, 1H), 6.88 (dd, J = 8.6, 2.3 Hz, 1H), 6.73 (d, J = 2.2 Hz, 1H), 6.55 (dd, J = 3.2, 0.7 Hz, 1H), 6.49–6.44 (m, 1H), 5.73 (s, 2H), 4.59 (s, 2H), 3.95 (s, 3H); ^{13}C NMR (100 MHz, $CDCl_3$) δ 166.9, 153.6, 139.8, 136.9, 133.0, 132.8, 131.2, 129.5, 128.9, 128.7, 128.0, 127.8, 127.4, 127.0, 124.4, 123.8, 122.2, 122.0, 109.9, 102.0, 95.2, 68.5, 52.4, 48.8.

Methyl 3-((5-(2-((2,4-dichlorophenyl)amino)-2-oxoethoxy)-1*H*-indol-1-yl)methyl) benzoate (P14): ^{1}H NMR (400 MHz, $CDCl_3$) δ 9.10 (s, 1H), 8.45 (d, J = 8.9 Hz, 1H), 7.93 (d, J = 7.8 Hz, 1H), 7.89 (s, 1H), 7.38 (d, J = 2.3 Hz, 1H), 7.34 (t, J = 7.7 Hz, 1H), 7.25 (dd, J = 8.9, 2.3 Hz, 1H), 7.21–7.13 (m, 4H), 6.91 (dd, J = 8.8, 2.4 Hz, 1H), 6.51 (d, J = 2.6 Hz, 1H), 5.32 (s, 2H), 4.66 (s, 2H), 3.89 (s, 3H); ^{13}C NMR (100 MHz, $CDCl_3$) δ 167.0, 151.5, 137.8, 132.7, 132.3, 131.1, 130.7, 129.4, 129.3, 129.2, 128.99 128.95, 128.8, 127.92, 127.88, 123.6, 121.9, 112.0, 110.7, 104.9, 101.8, 68.7, 52.2, 50.1.

Methyl 3-((6-(2-((2,4-dichlorophenyl)amino)-2-oxoethoxy)-1*H*-indol-1-yl)methyl) benzoate (P15): ^{1}H NMR (400 MHz, $CDCl_3$) δ 8.98 (s, 1H), 8.39 (d, J = 8.9 Hz, 1H), 7.91 (d, J = 7.7 Hz, 1H), 7.88 (s, 1H), 7.57 (d, J = 8.6 Hz, 1H), 7.36 (d, J = 2.4 Hz, 1H), 7.32 (t, J = 7.7 Hz, 1H), 7.24 (dd, J = 8.9, 2.4 Hz, 1H), 7.18 (d, J = 7.7 Hz, 1H), 7.08 (d, J = 3.2 Hz, 1H), 6.86 (dd, J = 8.6, 2.3 Hz, 1H), 6.75 (d, J = 2.1 Hz, 1H), 6.52 (dd, J = 3.2, 0.8 Hz, 1H), 5.29 (s, 2H), 4.60 (s, 2H), 3.89 (s, 3H); ^{13}C NMR (100 MHz, $CDCl_3$) δ 166.8, 153.6, 137.6, 136.6, 132.7, 131.2, 130.9, 129.5, 129.10, 129.09, 128.9, 128.2, 128.03, 127.99, 124.5, 123.7, 122.3, 122.0, 109.8, 102.2, 95.1, 68.5, 52.3, 50.0.

Methyl 4-((5-(2-((2,4-dichlorophenyl)amino)-2-oxoethoxy)-1*H*-indol-1-yl)methyl) benzoate (P16): ^{1}H NMR (400 MHz, $CDCl_3$) δ 9.10 (s, 1H), 8.46 (d, J = 8.9 Hz, 1H), 7.96 (d, J = 8.2 Hz, 2H), 7.39 (d, J = 2.4 Hz, 1H), 7.27 (dd, J = 8.7, 2.1 Hz, 1H), 7.19 (d, J = 2.4 Hz, 1H), 7.16 (d, J = 3.1 Hz, 1H), 7.15–7.10 (m, 3H), 6.91 (dd, J = 8.9, 2.4 Hz, 1H), 6.53 (d, J = 3.1 Hz, 1H), 5.36 (s, 2H), 4.68 (s, 2H), 3.89 (s, 3H); ^{13}C NMR (100 MHz, $CDCl_3$) δ 167.2, 151.7, 142.6, 132.9, 132.5, 130.3, 129.8, 129.7, 129.5, 129.3, 129.0, 128.1, 126.6, 123.8, 122.1, 120.0, 112.3, 110.9, 105.1, 102.0, 68.8, 52.3, 50.3.

Methyl 4-((6-(2-((2,4-dichlorophenyl)amino)-2-oxoethoxy)-1*H*-indol-1-yl)methyl) benzoate (P17): ^{1}H NMR (400 MHz, $CDCl_3$) δ 8.97 (s, 1H), 8.38 (d, J = 8.9 Hz, 1H), 7.94 (d, J = 8.2 Hz, 2H), 7.58 (d, J = 8.6 Hz, 1H), 7.36 (d, J = 2.3 Hz, 1H), 7.24 (dd, J = 9.1, 2.1 Hz, 1H), 7.11 (d, J = 8.2 Hz, 2H), 7.09 (d, J = 3.2 Hz, 1H), 6.87 (dd, J = 8.6, 2.2 Hz, 1H), 6.72 (d, J = 1.9 Hz, 1H), 6.54 (d, J = 3.1 Hz, 1H), 5.32 (s, 2H), 4.60 (s, 2H), 3.89 (s, 3H); ^{13}C NMR (100 MHz, $CDCl_3$) δ 166.8, 153.7, 142.3, 136.7, 132.7, 130.3, 129.8, 129.6, 128.9, 128.3, 128.0, 126.6, 124.5, 123.7, 122.3, 122.0, 109.9, 102.3, 95.0, 68.5, 52.3, 50.1.

Methyl 2-((3-benzyl-5-(2-((2,4-dichlorophenyl)amino)-2-oxoethoxy)-1*H*-indol-1-yl)methyl) benzoate (P18): ^{1}H NMR (400 MHz, $CDCl_3$) δ 9.06 (s, 1H), 8.43 (d, J = 8.9 Hz, 1H), 8.07–7.99 (m, 1H), 7.38 (d, J = 2.4 Hz, 1H), 7.36–7.22 (m, 7H), 7.20–7.14 (m, 1H), 7.09 (d, J = 8.8 Hz, 1H), 7.02 (d, J = 2.3 Hz, 1H), 6.91–6.84 (m, 2H), 6.52–6.46 (m, 1H), 5.70 (s, 2H), 4.62 (s, 2H), 4.10 (s, 2H), 3.92 (s, 3H); ^{13}C NMR (100 MHz, $CDCl_3$) δ 167.5, 167.2, 151.3, 141.0, 140.3, 133.3, 133.0, 132.9, 131.1, 129.5, 129.0, 128.7, 128.52, 128.48, 128.42, 128.0, 127.8, 127.3, 127.0, 126.1, 123.8, 122.1, 114.9, 112.1, 111.0, 103.6, 68.9, 52.3, 48.8, 31.7.

Methyl 2-((3-benzyl-6-(2-((2,4-dichlorophenyl)amino)-2-oxoethoxy)-1*H*-indol-1-yl)methyl) benzoate (P19): ^{1}H NMR (400 MHz, $CDCl_3$) δ 8.98 (s, 1H), 8.38 (d, J = 8.9 Hz, 1H), 8.06–8.01 (m, 1H), 7.44 (d, J = 8.6 Hz, 1H), 7.37 (d, J = 2.3 Hz, 1H), 7.32–7.22 (m, 7H), 7.22–7.16 (m, 1H), 6.84–6.79 (m, 2H), 6.70 (d, J = 2.0 Hz, 1H), 6.54–6.49 (m, 1H), 5.67 (s, 2H), 4.58 (s, 2H), 4.11 (s, 2H), 3.94 (s, 3H); ^{13}C NMR (100 MHz, $CDCl_3$) δ 167.5, 166.9, 153.8, 141.2, 140.0, 137.5, 133.0, 132.8, 131.2, 129.6, 129.0,

128.8, 128.5, 128.0, 127.9, 127.4, 127.04, 126.95, 126.1, 124.0, 123.8, 122.1, 120.7, 115.4, 109.3, 95.3, 68.6, 52.4, 48.7, 31.7.

Methyl 2-((5-(2-((2,4-dichlorophenyl)amino)-2-oxoethoxy)-3-(3-methylbut-2-en-1-yl)-1*H*-indol-1-yl)methyl) benzoate (P20): ^{1}H NMR (500 MHz, $CDCl_3$) δ 9.01 (s, 1H), 8.40 (d, *J* = 8.9 Hz, 1H), 8.06–8.02 (m, 1H), 7.54 (d, *J* = 8.6 Hz, 1H), 7.38 (d, *J* = 2.3 Hz, 1H), 7.31–7.28 (m, 2H), 7.24 (d, *J* = 2.3 Hz, 1H), 6.86 (dd, *J* = 8.6, 2.2 Hz, 1H), 6.83 (s, 1H), 6.69 (d, *J* = 2.2 Hz, 1H), 6.52–6.48 (m, 1H), 5.67 (s, 2H), 5.45–5.40 (m, 1H), 4.60 (s, 2H), 3.96 (s, 3H), 3.45 (d, *J* = 6.8 Hz, 2H), 1.77 (s, 3H), 1.75 (s, 3H); ^{13}C NMR (125 MHz, $CDCl_3$) δ 167.4, 166.9, 153.7, 140.1, 137.4, 132.9, 132.7, 132.1, 131.1, 129.4, 128.9, 127.9, 127.7, 127.2, 126.9, 125.7, 123.9, 123.7, 122.9, 122.0, 120.4, 115.7, 108.9, 95.2, 68.5, 52.2, 48.5, 25.7, 24.1, 17.8.

Methyl 2-((6-(2-((2,4-dichlorophenyl)amino)-2-oxoethoxy)-3-(3-methylbut-2-en-1-yl)-1*H*-indol-1-yl)methyl) benzoate (P21): ^{1}H NMR (500 MHz, $CDCl_3$) δ 9.13 (s, 1H), 8.47 (d, *J* = 8.9 Hz, 1H), 8.07–8.01 (m, 1H), 7.41 (d, *J* = 2.3 Hz, 1H), 7.31 (dt, *J* = 6.1, 4.1 Hz, 2H), 7.28 (dd, *J* = 8.9, 2.3 Hz, 1H), 7.14 (d, *J* = 2.4 Hz, 1H), 7.09 (d, *J* = 8.8 Hz, 1H), 6.91 (s, 1H), 6.89 (dd, *J* = 8.8, 2.4 Hz, 1H), 6.50–6.46 (m, 1H), 5.70 (s, 2H), 5.43–5.38 (m, 1H), 4.70 (s, 2H), 3.95 (s, 3H), 3.44 (d, *J* = 7.0 Hz, 2H), 1.78 (s, 3H), 1.75 (s, 3H); ^{13}C NMR (125 MHz, $CDCl_3$) δ 167.5, 167.2, 151.1, 140.4, 133.2, 132.9, 132.8, 132.2, 131.0, 129.4, 128.9, 128.4, 127.9, 127.7, 127.3, 127.2, 127.0, 123.7, 122.8, 122.0, 115.2, 111.9, 110.8, 103.4, 68.9, 52.2, 48.6, 25.7, 24.1, 17.9.

3.2.3. Synthetic Procedure for Hydrolysis of the Methyl Esters to Yield the Target Acid Compounds

$LiOH{\cdot}H_2O$ (1.5 equiv.) was added to a solution of the alkoxyindole compounds in THF/MeOH/H_2O (1 M, 2:1:1 *v*/*v*/*v*) and stirred at room temperature. After completing the reaction, the reaction mixture was concentrated under reduced pressure, acidified with 1N HCl solution, and extracted with EtOAc. The combined organic extracts were washed with water and brine, dried over anhydrous $MgSO_4$, filtered, and concentrated under vacuum. The residue was purified by silica gel column chromatography using an appropriate mixture of $CHCl_3$ and MeOH as an eluent to obtain the target compounds **6–9** and **13–22** with 59–92% yield.

2-(5-(2-((2,4-Dichlorophenyl)amino)-2-oxoethoxy)-1*H*-indol-3-yl)acetic acid (5): ^{1}H NMR (400 MHz, $CDCl_3/CD_3OD$) δ 8.14 (d, *J* = 8.8 Hz, 1H), 7.54 (d, *J* = 2.2 Hz, 1H), 7.35 (dd, *J* = 8.8, 2.3 Hz, 1H), 7.31 (d, *J* = 8.8 Hz, 1H), 7.19 (s, 2H), 6.94 (dd, *J* = 8.6, 2.3 Hz, 1H), 4.73 (s, 2H), 3.71 (s, 2H); HRMS(ESI): m/z 391.0263 $[M\text{-}H]^-$ (calcd for $C_{18}H_{13}Cl_2N_2O_4$ = 391.0258).

2-(6-(2-((2,4-Dichlorophenyl)amino)-2-oxoethoxy)-1*H*-indol-3-yl)acetic acid (6):^{1}H NMR (400 MHz, $CDCl_3/CD_3OD$) δ 8.20 (d, *J* = 8.8 Hz, 1H), 7.75 (s, 1H), 7.50 (d, *J* = 8.7 Hz, 1H), 7.47 (d, *J* = 2.3 Hz, 1H), 7.31 (dd, *J* = 8.8, 2.4 Hz, 1H), 7.08 (s, 1H), 6.99 (d, *J* = 2.1 Hz, 1H), 6.83 (dd, *J* = 8.7, 2.3 Hz, 1H), 4.70 (s, 2H), 3.65 (s, 2H); HRMS(ESI): m/z 391.0256 $[M\text{-}H]^-$ (calcd for $C_{18}H_{13}Cl_2N_2O_4$ = 391.0258).

2-(1-Benzyl-5-(2-((2,4-dichlorophenyl)amino)-2-oxoethoxy)-1*H*-indol-3-yl)acetic acid (7): ^{1}H NMR (400 MHz, $CDCl_3/CD_3OD$) δ 8.34 (d, *J* = 8.9 Hz, 1H), 7.50 (s, 1H), 7.45 (d, *J* = 2.3 Hz, 1H), 7.32 (d, *J* = 2.3 Hz, 1H), 7.31–7.26 (m, 2H), 7.24–7.18 (m, 3H), 7.13 (d, *J* = 6.6 Hz, 2H), 6.94 (dd, *J* = 9.2, 2.2 Hz, 1H), 5.30 (s, 2H), 4.72 (s, 2H), 3.74 (s, 2H); HRMS(ESI): m/z 481.0724 $[M\text{-}H]^-$ (calcd for $C_{25}H_{19}Cl_2N_2O_4$ = 481.0727).

2-(1-Benzyl-6-(2-((2,4-dichlorophenyl)amino)-2-oxoethoxy)-1*H*-indol-3-yl)acetic acid (8): ^{1}H NMR (400 MHz, $CDCl_3/CD_3OD$) δ 8.25 (d, *J* = 8.8 Hz, 1H), 7.57 (d, *J* = 8.6 Hz, 1H), 7.44 (d, *J* = 2.2 Hz, 1H), 7.30 (dd, *J* = 8.8, 2.4 Hz, 1H), 7.28–7.21 (m, 3H), 7.16–7.11 (m, 3H), 6.89 (dd, *J* = 8.7, 2.0 Hz, 1H), 6.84 (d, *J* = 2.0 Hz, 1H), 5.29 (s, 2H), 4.66 (s, 2H), 3.74 (s, 2H); HRMS(ESI): m/z 481.0726 $[M\text{-}H]^-$ (calcd for $C_{25}H_{19}Cl_2N_2O_4$ = 481.0727).

2-((5-(2-((2,4-Dichlorophenyl)amino)-2-oxoethoxy)-1*H*-indol-1-yl)methyl)benzoic acid (12): ^{1}H NMR (400 MHz, $CDCl_3/CD_3OD$) δ 8.38 (d, *J* = 8.9 Hz, 1H), 8.03–7.99 (m, 1H), 7.42–7.39 (m, 1H), 7.28 (dd, *J* = 8.9, 2.4 Hz, 1H), 7.24–7.20 (m, 2H), 7.19 (d, *J* = 2.4 Hz, 1H), 7.12 (d, *J* = 3.1 Hz,

1H), 7.10 (d, *J* = 8.9 Hz, 1H), 6.82 (dd, *J* = 8.8, 2.4 Hz, 1H), 6.49 (d, *J* = 2.6 Hz, 1H), 6.42–6.38 (m, 1H), 5.75 (s, 2H), 4.66 (s, 2H); HRMS(ESI): m/z 467.0549 $[M-H]^-$ (calcd for $C_{24}H_{17}Cl_2N_2O_4$ = 467.0571).

2-((6-(2-((2,4-Dichlorophenyl)amino)-2-oxoethoxy)-1*H*-indol-1-yl)methyl)benzoic acid (13): ^{1}H NMR (400 MHz, $CDCl_3/CD_3OD$) δ 8.32 (d, *J* = 8.9 Hz, 1H), 8.09–8.05 (m, 1H), 7.60 (d, *J* = 8.6 Hz, 1H), 7.41 (d, *J* = 2.4 Hz, 1H), 7.39 (d, *J* = 0.7 Hz, 1H), 7.30–7.25 (m, 3H), 7.13 (d, *J* = 3.1 Hz, 1H), 6.89 (dd, *J* = 8.6, 2.2 Hz, 1H), 6.81 (d, *J* = 1.9 Hz, 1H), 6.55 (d, *J* = 3.2 Hz, 1H), 6.52–6.48 (m, 1H), 5.77 (s, 2H), 4.63 (s, 2H); HRMS(ESI): m/z 467.0550 $[M-H]^-$ (calcd for $C_{24}H_{17}Cl_2N_2O_4$ = 467.0571).

3-((5-(2-((2,4-Dichlorophenyl)amino)-2-oxoethoxy)-1*H*-indol-1-yl)methyl)benzoic acid (14): ^{1}H NMR (400 MHz, $CDCl_3/CD_3OD$) δ 8.37 (d, *J* = 8.8 Hz, 1H), 7.99–7.90 (m, 3H), 7.44 (d, *J* = 2.3 Hz, 1H), 7.36 (t, *J* = 7.6 Hz, 1H), 7.30 (dd, *J* = 8.8, 2.4 Hz, 1H), 7.24–7.20 (m, 4H), 6.93 (dd, *J* = 8.9, 2.6 Hz, 1H), 6.51 (d, *J* = 3.2 Hz, 1H), 5.38 (s, 2H), 4.70 (s, 2H); HRMS(ESI): m/z 467.0558 $[M-H]^-$ (calcd for $C_{24}H_{17}Cl_2N_2O_4$ = 467.0571).

3-((6-(2-((2,4-Dichlorophenyl)amino)-2-oxoethoxy)-1*H*-indol-1-yl)methyl)benzoic acid (15): ^{1}H NMR (400 MHz, $CDCl_3/CD_3OD$) δ 8.25 (d, *J* = 8.8 Hz, 1H), 7.94–7.87 (m, 2H), 7.55 (d, *J* = 8.6 Hz, 1H), 7.42–7.38 (m, 1H), 7.28 (s, 1H), 7.26 (dd, *J* = 8.8, 2.4 Hz, 1H), 7.20–7.15 (m, 1H), 7.10 (d, *J* = 2.7 Hz, 1H), 6.86 (dd, *J* = 8.6, 2.2 Hz, 1H), 6.82 (s, 1H), 6.48 (d, *J* = 3.1 Hz, 1H), 5.28 (s, 2H), 4.63 (s, 2H); HRMS(ESI): m/z 467.0557 $[M-H]^-$ (calcd for $C_{24}H_{17}Cl_2N_2O_4$ = 467.0571).

4-((5-(2-((2,4-Dichlorophenyl)amino)-2-oxoethoxy)-1*H*-indol-1-yl)methyl)benzoic acid (16): ^{1}H NMR (400 MHz, $CDCl_3/CD_3OD$) δ 8.17 (d, *J* = 8.9 Hz, 1H), 7.89 (d, *J* = 8.4 Hz, 2H), 7.44 (d, *J* = 2.4 Hz, 1H), 7.28 (dd, *J* = 8.9, 2.3 Hz, 1H), 7.23 (d, *J* = 3.0 Hz, 1H), 7.20–7.16 (m, 2H), 7.11 (d, *J* = 8.1 Hz, 2H), 6.89 (dd, *J* = 8.9, 2.2 Hz, 1H), 6.45 (d, *J* = 3.2 Hz, 1H), 5.39 (s, 2H), 4.68 (s, 2H); HRMS(ESI): m/z 467.0545 $[M-H]^-$ (calcd for $C_{24}H_{17}Cl_2N_2O_4$ = 467.0571).

4-((6-(2-((2,4-Dichlorophenyl)amino)-2-oxoethoxy)-1*H*-indol-1-yl)methyl)benzoic acid (17): ^{1}H NMR (400 MHz, $CDCl_3/CD_3OD$) δ 9.06 (s, 1H), 8.33 (dd, *J* = 8.9, 3.1 Hz, 1H), 7.96 (d, *J* = 8 Hz, 2H), 7.59 (d, *J* = 8.6 Hz, 1H), 7.41 (d, *J* = 2.3 Hz, 1H), 7.28 (dd, *J* = 8.8, 2.3 Hz, 1H), 7.16–7.11 (m, 3H), 6.89 (dd, *J* = 8.6, 2.2 Hz, 1H), 6.79 (d, *J* = 2.0 Hz, 1H), 6.54 (d, *J* = 3.2 Hz, 1H), 5.36 (s, 2H), 4.64 (s, 2H); HRMS(ESI): m/z 467.0549 $[M-H]^-$ (calcd for $C_{24}H_{17}Cl_2N_2O_4$ = 467.0571).

2-((3-Benzyl-5-(2-((2,4-dichlorophenyl)amino)-2-oxoethoxy)-1*H*-indol-1-yl)methyl)benzoic acid (18): ^{1}H NMR (400 MHz, $(CD_3)_2SO$) δ 9.59 (s, 1H), 7.92 (d, *J* = 7.1 Hz, 1H), 7.88 (d, *J* = 9.1 Hz, 1H), 7.70 (s, 1H), 7.45 (d, *J* = 9.0 Hz, 1H), 7.35 (t, *J* = 7.6 Hz, 2H), 7.31–7.19 (m, 6H), 7.14 (d, *J* = 6.7 Hz, 1H), 7.10 (s, 1H), 6.85 (d, *J* = 9.2 Hz, 1H), 6.44 (d, *J* = 7.2 Hz, 1H), 5.72 (s, 2H), 4.71 (s, 2H), 4.03 (s, 2H); HRMS(ESI): m/z 557.1010 $[M-H]^-$ (calcd for $C_{31}H_{23}Cl_2N_2O_4$ = 557.1040).

2-((3-Benzyl-6-(2-((2,4-dichlorophenyl)amino)-2-oxoethoxy)-1*H*-indol-1-yl)methyl)benzoic acid (19): ^{1}H NMR (400 MHz, $(CD_3)_2SO$) δ 10.43 (s, 1H), 7.80 (d, *J* = 8.7 Hz, 1H), 7.63 (d, *J* = 2.1 Hz, 1H), 7.56 (d, *J* = 6.7 Hz, 1H), 7.38 (dd, *J* = 8.7, 2.2 Hz, 1H), 7.30–7.20 (m, 6H), 7.17–7.10 (m, 2H), 7.09–7.00 (m, 2H), 6.70 (d, *J* = 6.5 Hz, 1H), 6.66 (dd, *J* = 8.5, 1.8 Hz, 1H), 5.66 (s, 2H), 4.74 (s, 2H), 3.99 (s, 2H); HRMS(ESI): m/z 557.1010 $[M-H]^-$ (calcd for $C_{31}H_{23}Cl_2N_2O_4$ = 557.1040).

2-((5-(2-((2,4-Dichlorophenyl)amino)-2-oxoethoxy)-3-(3-methylbut-2-en-1-yl)-1*H*-indol-1-yl)methyl)benzoic acid (20): ^{1}H NMR (400 MHz, $CDCl_3/CD_3OD$) δ 8.33 (1H, d, *J* = 8.8 Hz), 8.09 (1H, dd, *J* =5.6, 3.6 Hz), 7.54 (1H, d, *J* = 8.8 Hz), 7.41 (1H, d, *J* = 2.0 Hz), 7.31–7.25 (3H, m), 6.88–6.85 (2H, m), 6.75 (1H, d, *J* = 2.0 Hz), 6.50 (1H, dd, *J* = 5.6, 3.2 Hz), 5.70 (2H, s), 5.43 (1H, t, *J* = 7.2 Hz), 4.62 (2H, s), 3.46 (2H, d, *J* = 6.8 Hz), 1.77 (3H, s), 1.76 (3H, s); HRMS(ESI): m/z 535.1178 $[M-H]^-$ (calcd for $C_{29}H_{25}Cl_2N_2O_4$ = 535.1197).

2-((6-(2-((2,4-Dichlorophenyl)amino)-2-oxoethoxy)-3-(3-methylbut-2-en-1-yl)-1*H*-indol-1-yl)methyl)benzoic acid (21): ^{1}H NMR (400 MHz, $CDCl_3/CD_3OD$) δ 8.31 (1H, d, *J* = 8.8 Hz), 8.08–8.05 (1H, m), 7.55 (1H, d, *J* = 8.4 Hz), 7.42 (1H, d, *J* = 2.4 Hz), 7.31–7.26 (3H, m), 6.89–6.86

(2H, m), 6.78 (1H, d, J = 2.0 Hz), 6.52-6.49 (1H, m), 5.71 (2H, s), 5.44 (1H, t, J = 7.2 Hz), 4.64 (2H, s), 3.47 (2H, d, J = 7.2 Hz), 1.78 (3H, s), 1.77 (3H, s); HRMS(ESI): m/z 535.1177 $[M\text{-}H]^-$ (calcd for $C_{29}H_{25}Cl_2N_2O_4$ = 535.1197).

3.3. In Vitro PPAR Transactivation Assay

Transient transfection and luciferase assay was carried out. CV-1 cells were seeded in 48-well plates at a density of 1.5×10^5 cells/well in Dulbecco's modified Eagle's medium (DMEM) with containing 10% fetal bovine serum (FBS). The cells were transiently transfected with plasmid mixtures containing PPAR-α/γ expression vector and tk-PPRE-luciferase (Luc) vector for 6 h and then treated with the samples for 24 h. HEK293t cells were seeded in a 60-mm dish at a density of 1.5×10^6 cells/dish. After transfection with the plasmid mixtures containing PPAR-δ expression vector and Luc vector for 6 h, the cells were transferred to a 96-well plate and treated with the samples. To normalize the transfection efficiency, a β-galactosidase plasmid was cotransfected. The luciferase activity in the cell lysates was measured using a luciferase assay system (Promega Corp., Madison, WI, USA), and the β-galactosidase activity was determined by measuring the absorbance at 410 nm using an ELISA plate reader. The data were expressed as the relative luciferase activity divided by the β-galactosidase activity. All constructs were kindly gifted by Dr. Ronald M. Evans at The Salk Institute (La Jolla, CA, USA).

3.4. Molecular Modeling

All molecular modeling calculations were carried out using Surflex Dock interfaced with SYBYL-X software, version 2.1.1 running on a system with Window 7 Home Premium K 64-bit OS. In this automated docking program, the flexibility of the ligands, proteins, and biomolecules was considered. The ligand was built in an incremental fashion, where each new fragment was added in all possible positions and conformations to a pre-placed base fragment inside the active site. All the molecules used for docking were sketched in SYBYL, and the energy minimizations were performed using Tripos Force Field and Gasteiger–Huckel charge with 100,000 iterations of the conjugate gradient method with a convergence criterion of 0.05 kcal/mol. To prepare the proteins, all hydrogens and MMFF94 charges were added, and the side-chain amides were fixed. A staged minimization was performed using Tripos Force Field and MMFF94 charge with 10,000 iterations of the Powell method with a convergence criterion of 0.5 kcal/mol without the initial optimization. The 3D coordinates of the active sites were determined from the X-ray crystal structures of PPAR-γ (PDB code: 2PRG), reported as a complex with rosiglitazone.

3.5. In vitro PPAR Coactivator Recruit Assay

EC_{50} of compounds were determined by recruitment of transcriptional coactivators using the TR-FRET based Lanthascreen coactivator assays (Invitrogen, Carlsbad, CA, USA). Which was serviced by Thermo Fisher Scientific (Medicon, WI, USA) according to the LanthaScreen TR-FRET coregulator protocol and assay conditions. Coregulator peptide PGC1a, TRAP220/DRIP2 and C33 was used for the coactivator recruit assay of PPAR-α, -γ, and -δ, respectively.

4. Conclusions

In conclusion, we designed and synthesized several dichloroanilide-linked indole derivatives and examined their potent and selective PPAR activity. The PPAR transcriptional activities of these compounds were tested using a cell-based reporter assay. Compound **12** exhibited significant activity as a PPAR-γ agonist, with approximately 2-fold higher PPAR-γ activity than that of pioglitazone. PPAR coactivator recruit assay revealed EC_{50} of compound **12** with 1.96 nM. And it showed 1970-fold selectivity to PPAR-α and 16,600-fold to PPAR-δ. The docking study provided structural insights of compound **12** in association with PPAR-γ and rationalized its good PPAR-γ agonistic activity. Therefore, the *N*-(carboxybenzyl)indole was suggested as an original framework for the next generation of PPAR ligands. In addition, our scaffold tuning approach could be an expeditious way to optimize the activity of compounds. This novel PPAR-γ agonist and scaffold tuning approach is expected to broaden the bottleneck of new drug discovery for CVD, CML and various types of solid tumors as well as metabolic diseases.

Author Contributions: R.J. conceived and designed the experiments; H.J.G., Y.-J.K. and H.L. performed the experiments; H.J.G., Y.-J.K. and Y.-S.C. analyzed the data; J.-H.R. contributed reagents/analysis tools; Y.-S.C. wrote the paper. Authorship must be limited to those who have contributed substantially to the work reported.

Funding: This research was supported by the National Research Foundation of Korea (NRF) Grant funded by the Korean government (MSIP) (No. 2011-0030074 and 2014R1A2A1A11052761), the Ministry of Education (NRF-2016R1A6A3A11932164 and NRF-2017R1A6A3A11034603) and Sookmyung Women's University Specialization Program Funding.

Conflicts of Interest: The authors declare no conflict of interest.

References

1. Stales, B.; Dallongeville, J.; Auwerx, J.; Schoonjans, K.; Leitersdorf, E.; Fruchart, J.-C. Mechanism of action of fibrates on lipid and lipoprotein metabolism. *Circulation* **1998**, *98*, 2088–2093. [CrossRef]
2. Berger, J.; Moller, D.E. The mechanisms of action of PPARs. *Annu. Rev. Med.* **2002**, *53*, 409–435. [CrossRef] [PubMed]
3. Semple, R.K.; Chatterjee, V.K.K.; O'Rahilly, S. PPAR-γ and human metabolic disease. *J. Clin. Investig.* **2006**, *116*, 581–589. [CrossRef] [PubMed]
4. Wang, Y.-X. PPARs: Diverse regulators in energy metabolism and metabolic diseases. *Cell Res.* **2010**, *20*, 124–137. [CrossRef] [PubMed]
5. Ahmadian, M.; Suh, J.M.; Hah, N.; Liddle, C.; Atkins, A.R.; Downes, M.; Evans, R.M. PPAR-γ signaling and metabolism: The good, the bad and the future. *Nat. Med.* **2013**, *19*, 557–566. [CrossRef] [PubMed]
6. Vamecq, J.; Latruffe, N. Medical significance of peroxisome proliferator-activated receptors. *Lancet* **1999**, *354*, 141–148. [CrossRef]
7. Kersten, S.; Desvergne, B.; Wahli, W. Roles of PPARs in health and disease. *Nature* **2000**, *405*, 421–424. [CrossRef] [PubMed]
8. Berger, J.P.; Akiyama, T.E.; Meinke, P.T. PPARs: Therapeutic targets for the metabolic disease. *Trends Pharmacol. Sci.* **2005**, *26*, 244–251. [CrossRef] [PubMed]
9. Jay, M.A.; Ren, J. Peroxisome proliferator-activated receptor (PPAR) in metabolic syndrome and Type 2 diabetes. *Curr. Diabetes Rev.* **2007**, *3*, 33–39. [CrossRef] [PubMed]
10. Tain, Y.-L.; Hsu, C.-N.; Chan, J.Y.H. PPARs link early life nutritional insults to later programmed hypertension and metabolic syndrome. *Int. J. Mol. Sci.* **2016**, *17*, 20–30. [CrossRef] [PubMed]
11. Gross, B.; Pawlak, M.; Lefebvre, P.; Staels, B. PPARs in obesity-induced T2DM, dyslipidemia and NAFLD. *Nat. Rev. Endocrinol.* **2017**, *13*, 36–49. [CrossRef] [PubMed]
12. Delerive, P.; Fruchart, J.-C.; Staels, B. Peroxisome proliferator-activated receptors in inflammation control. *J. Endocrinol.* **2001**, *169*, 453–459. [CrossRef] [PubMed]
13. Lefebvre, P.; Chinetti, G.; Fruchart, J.-C.; Staels, B. Sorting out the roles of PPAR-α in energy metabolism and vascular homeostasis. *J. Clin. Investig.* **2006**, *116*, 571–580. [CrossRef] [PubMed]
14. Moraes, L.A.; Piqueras, L.; Bishop-Bailey, D. Peroxisome proliferator-activated receptors and inflammation. *Pharmacol. Ther.* **2006**, *110*, 371–385. [CrossRef] [PubMed]
15. Biscetti, F.; Straface, G.; Pitocco, D.; Zaccardi, F.; Ghirlanda, G.; Flex, A. Peroxisome proliferator-activated receptors and angiogenesis. *Nutr. Metab. Cardiovasc. Dis.* **2009**, *19*, 751–759. [CrossRef] [PubMed]
16. Lagana, A.S.; Vitale, S.G.; Nigro, A.; Sofo, V.; Salmeri, F.M.; Rossetti, P.; Rapisarda, A.M.C.; La Vignera, S.; Condorelli, R.A.; Rizzo, G.; et al. Pleiotropic actions of peroxisome proliferator-activated receptors (PPARs) in dysregulated metabolic homeostasis, inflammation and cancer: Current evidence and future perspectives. *Int. J. Mol. Sci.* **2016**, *17*, 999–1008. [CrossRef] [PubMed]
17. Ivanova, E.A.; Parolari, A.; Myasoedova, V.; Melnichenko, A.A.; Bobryshev, Y.V.; Orekhov, A.N. Peroxisome proliferator-activated receptor (PPAR) gamma in cardiovascular disorders and cardiovascular surgery. *J. Cardiol.* **2015**, *66*, 271–278. [CrossRef] [PubMed]

18. Han, L.; Shen, W.-J.; Bittner, S.; Kraemer, F.B.; Azhar, S. PPARs: Regulators of metabolism and as therapeutic targets in cardiovascular disease. Part I: PPAR-α. *Future Cardiol.* **2017**, *13*, 259–278. [CrossRef] [PubMed]
19. Han, L.; Shen, W.-J.; Bittner, S.; Kraemer, F.B.; Azhar, S. PPARs: Regulators of metabolism and as therapeutic targets in cardiovascular disease. Part II: PPAR-β/δ and PPAR-γ. *Future Cardiol.* **2017**, *13*, 279–296. [CrossRef] [PubMed]
20. Copland, J.A.; Marlow, L.A.; Kurakata, S.; Fujiwara, K.; Wong, A.K.C.; Kreinest, P.A.; Williams, S.F.; Haugen, B.R.; Klopper, J.P.; Smallridge, R.C. Novel high-affinity PPARγ agonist alone and in combination with paclitaxel inhibits human anaplastic thyroid carcinoma tumor growth via p21$^{WAF1/CIP1}$. *Oncogene* **2006**, *25*, 2304–2317. [CrossRef] [PubMed]
21. Shimazaki, N.; Togashi, N.; Hanai, M.; Isoyama, T.; Wada, K.; Fujita, T.; Fujiwara, K.; Kurakata, S. Anti-tumour activity of CS-7017, a selective peroxisome proliferator-activated receptor gamma agonist of thiazolidinedione class, in human tumour xenografts and a syngeneic tumour implant model. *Eur. J. Cancer* **2008**, *44*, 1734–1743. [CrossRef] [PubMed]
22. Serizawa, M.; Murakami, H.; Watanabe, M.; Takahashi, T.; Yamamoto, N.; Koh, Y. Peroxisome proliferator-activated receptor γ agonist efatutazone impairs transforming growth factor β2-induced motility of epidermal growth factor receptor tyrosine kinase inhibitor-resistant lung cancer cells. *Cancer Sci.* **2014**, *105*, 683–689. [CrossRef] [PubMed]
23. Prost, S.; Relouzat, F.; Spentchian, M.; Ouzegdouh, Y.; Saliba, J.; Massonnet, G.; Beressi, J.-P.; Verhoeyen, E.; Raggueneau, V.; Maneglier, B.; et al. Erosion of the chronic myeloid leukaemia stem cell pool by PPARγ agonists. *Nature* **2015**, *525*, 380–383. [CrossRef] [PubMed]
24. Winger, B.A.; Neil, P.S. PPARγ: Welcoming the new kid on the CML stem cell block. *Cancer Cell* **2015**, *28*, 409–411.
25. Yousefi, B.; Samadi, N.; Baradaran, B.; Shafiei-Irannejad, V.; Zarghami, N. Peroxisome proliferator-activated receptor ligands and their role in chronic myeloid leukemia: Therapeutic strategies. *Chem. Biol. Drug Des.* **2016**, *88*, 17–25. [CrossRef] [PubMed]
26. Werner, A.L.; Travaglini, M.T. A review of rosiglitazone in type 2 diabetes mellitus. *Pharmacotherapy* **2001**, *21*, 1082–1099. [CrossRef] [PubMed]
27. Nesto, R.W.; Bell, D.; Bonow, R.O.; Fonseca, V.; Grundy, S.M.; Horton, E.S.; Le Winter, M.; Porte, D.; Semenkovich, C.F.; Smith, S.; et al. Thiazolidinedione use, fluid retention, and congestive heart failure. *Circulation* **2003**, *108*, 2941–2949. [CrossRef] [PubMed]
28. Marceille, J.R.; Goins, J.A.; Soni, R.; Biery, J.C.; Lee, T.A. Chronic heart failure-related interventions after starting rosiglitazone in patients receiving insulin. *Pharmacotherapy* **2004**, *24*, 1317–1322. [CrossRef] [PubMed]
29. Henke, B.R.; Adkison, K.K.; Blanchard, S.G.; Leesnitzer, L.M.; Mook, R.A. Jr.; Plunket, K.D.; Ray, J.A.; Roberson, C.; Unwalla, R.; Willson, T.M. Synthesis and biological activity of a novel series of indole-derived PPARγ agonists. *Bioorg. Med. Chem. Lett.* **1999**, *9*, 3329–3334. [CrossRef]
30. Hopkins, C.R.; O'Neil, S.V.; Laufersweiler, M.C.; Wang, Y.; Pokross, M.; Mekel, M.; Evdokimov, A.; Walter, R.; Kontoyianni, M.; Petrey, M.E.; et al. Design and synthesis of novel *N*-sulfonyl-2-indole carboxamides as potent PPAR-γ binding agents with potential application to the treatment of osteoporosis. *Bioorg. Med. Chem. Lett.* **2006**, *16*, 5659–5663. [CrossRef] [PubMed]
31. Le Naour, M.; Leclerc, V.; Farce, A.; Caignard, D.H.; Hennuyer, N.; Staels, B.; Audinot-Bouchez, V.; Boutin, J.A.; Lonchampt, M.; Dacquet, C.; et al. Effect of oxime ether incorporation in acyl indole derivatives on PPAR subtype selectivity. *ChemMedChem* **2012**, *7*, 2179–2193. [CrossRef] [PubMed]
32. Tsakovska, I.; Al Sharif, M.; Alov, P.; Diukendjieva, A.; Fioravanzo, E.; Cronin, M.T.D.; Pajeva, I. Molecular modelling study of the PPARγ receptor in relation to the mode of action/adverse outcome pathway framework for liver steatosis. *Int. J. Mol. Sci.* **2014**, *15*, 7651–7666. [CrossRef] [PubMed]
33. Jian, Y.; He, Y.; Yang, J.; Han, W.; Zhai, X.; Zhao, Y.; Li, Y. Molecular modeling study for the design of novel peroxisome proliferator-activated receptor gamma agonists using 3D-QSAR and molecular docking. *Int. J. Mol. Sci.* **2018**, *19*, 630–644. [CrossRef] [PubMed]
34. Nolte, R.T.; Wisely, G.B.; Westin, S.; Cobb, J.E.; Lambert, M.H.; Kurokawa, R.; Rosenfeld, M.G.; Willson, T.M.; Glass, C.K.; Milburn, M.V. Ligand binding and co-activator assembly of the peroxisome proliferator-activated receptor-γ. *Nature* **1998**, *395*, 137–143. [CrossRef] [PubMed]

35. Xu, H.E.; Lambert, M.H.; Montana, V.G.; Plunket, K.D.; Moore, L.B.; Collins, J.L.; Oplinger, J.A.; Kliewer, S.A.; Gampe, R.T., Jr.; McKee, D.D.; et al. Structural determinants of ligand binding selectivity between the peroxisome proliferator-activated receptors. *Proc. Natl. Acad. Sci. USA* **2001**, *98*, 13919–13924. [CrossRef] [PubMed]
36. Gim, H.J.; Cheon, Y.-J.; Ryu, J.-H.; Jeon, R. Design and synthesis of benzoxazole containing indole analogs as peroxisome proliferator-activated receptor-γ/δ dual agonists. *Bioorg. Med. Chem. Lett.* **2011**, *21*, 3057–3061. [CrossRef] [PubMed]
37. Gim, H.J.; Li, H.; Lee, E.; Ryu, J.-H.; Jeon, R. Design and synthesis of alkoxyindolyl-3-acetic acid analogs as peroxisome proliferator-activated receptor-γ/δ agonists. *Bioorg. Med. Chem. Lett.* **2013**, *23*, 513–517. [CrossRef] [PubMed]
38. Gim, H.J.; Li, H.; Jeong, J.H.; Lee, S.J.; Sung, M.-K.; Song, M.-Y.; Park, B.-H.; Oh, S.J.; Ryu, J.-H.; Jeon, R. Design, synthesis, and biological evaluation of a series of alkoxy-3-indolylacetic acids as peroxisome proliferator-activated receptor γ/δ agonists. *Bioorg. Med. Chem.* **2015**, *23*, 3322–3336. [CrossRef] [PubMed]
39. Weigand, S.; Bischoff, H.; Dittrich-Wengenroth, E.; Heckroth, H.; Lang, D.; Vaupel, A.; Woltering, M. Minor structural modifications convert a selective PPARα agonist into a potent, highly selective PPARα agonist. *Bioorg. Med. Chem. Lett.* **2005**, *15*, 4619–4623. [CrossRef] [PubMed]
40. Luckhurst, C.A.; Stein, L.A.; Furber, M.; Webb, N.; Ratcliffe, M.J.; Allenby, G.; Botterell, S.; Tomlinson, W.; Martin, B.; Walding, A. Discovery of isoindoline and tetrahydroisoquinoline derivatives as potent, selective PPARδ agonists. *Bioorg. Med. Chem. Lett.* **2011**, *21*, 492–496. [CrossRef] [PubMed]
41. Luckhurst, C.A.; Ratcliffe, M.; Stein, L.; Furber, M.; Botterell, S.; Laughton, D.; Tomlinson, W.; Weaver, R.; Chohan, K.; Walding, A. Synthesis and biological evaluation of *N*-alkylated 8-oxybenz[c]azepine derivatives as selective PPARδ agonists. *Bioorg. Med. Chem. Lett.* **2011**, *21*, 531–536. [CrossRef] [PubMed]
42. Brown, P.J.; Stuart, L.W.; Hurley, K.P.; Lewis, M.C.; Winegar, D.A.; Wilson, J.G.; Wilkison, W.O.; Ittoop, O.R.; Willson, T.M. Identification of a subtype selective human PPARα agonist through parallel-array synthesis. *Bioorg. Med. Chem. Lett.* **2001**, *11*, 1225–1227. [CrossRef]
43. Sznaidman, M.L.; Haffner, C.D.; Maloney, P.R.; Fivush, A.; Chao, E.; Goreham, D.; Sierra, M.L.; LeGrumelec, C.; Xu, H.E.; Montana, V.G.; et al. Novel selective small molecule agonists for peroxisome proliferator-activated receptor δ (PPARδ)-synthesis and biological activity. *Bioorg. Med. Chem. Lett.* **2003**, *13*, 1517–1521. [CrossRef]

International Journal of *Molecular Sciences*

Review

The Opportunities and Challenges of Peroxisome Proliferator-Activated Receptors Ligands in Clinical Drug Discovery and Development

Fan Hong [1,2], Pengfei Xu [1,*,†] and Yonggong Zhai [1,2,*]

1 Beijing Key Laboratory of Gene Resource and Molecular Development, College of Life Sciences, Beijing Normal University, Beijing 100875, China; hongfanky@126.com

2 Key Laboratory for Cell Proliferation and Regulation Biology of State Education Ministry, College of Life Sciences, Beijing Normal University, Beijing 100875, China

* Correspondence: pex9@pitt.edu (P.X.); ygzhai@bnu.edu.cn (Y.Z.); Tel.: +86-156-005-60991 (P.X.); +86-10-5880-6656 (Y.Z.)

† Current address: Center for Pharmacogenetics and Department of Pharmaceutical Sciences, University of Pittsburgh, Pittsburgh, PA 15213, USA.

Received: 22 June 2018; Accepted: 24 July 2018; Published: 27 July 2018

Abstract: Peroxisome proliferator-activated receptors (PPARs) are a well-known pharmacological target for the treatment of multiple diseases, including diabetes mellitus, dyslipidemia, cardiovascular diseases and even primary biliary cholangitis, gout, cancer, Alzheimer's disease and ulcerative colitis. The three PPAR isoforms (α, β/δ and γ) have emerged as integrators of glucose and lipid metabolic signaling networks. Typically, PPARα is activated by fibrates, which are commonly used therapeutic agents in the treatment of dyslipidemia. The pharmacological activators of PPARγ include thiazolidinediones (TZDs), which are insulin sensitizers used in the treatment of type 2 diabetes mellitus (T2DM), despite some drawbacks. In this review, we summarize 84 types of PPAR synthetic ligands introduced to date for the treatment of metabolic and other diseases and provide a comprehensive analysis of the current applications and problems of these ligands in clinical drug discovery and development.

Keywords: PPAR; ligand; T2DM; dyslipidemia; TZDs

---

1. Introduction

Peroxisome proliferator-activated receptors (PPARs) are a group of nuclear receptors (NRs) that play essential roles in the regulation of several physiological processes, including cellular differentiation and development, whole-body energy homeostasis (carbohydrate, lipid, protein) and tumorigenesis [1]. PPARs are ligand-activated transcription factors and consist of a DNA binding domain in the N-terminus and a ligand binding domain (LBD) in the C-terminus [2,3]. The family of PPARs comprises three isoforms: PPARα (NR1C1), PPARβ/δ (NR1C2) and PPARγ (NR1C3) [2] and their 3D structures are shown in Figure 1. PPARα is highly expressed in metabolically active tissues and PPARγ which has three forms: PPARγ1, PPARγ2 and PPARγ3 is mainly expressed in white and brown adipose tissue [4]. The least known isoform is PPARβ/δ, which is expressed ubiquitously in virtually all tissues. After interaction with agonists, PPARs are translocated to the nucleus, where they heterodimerize with the retinoid X receptor (RXR) [5]. Then, PPAR-PXR binds to peroxisome proliferator hormone response elements (PPREs) [2] and regulates target genes. All three PPARs have natural agonists, namely, a variety of polyunsaturated long-chain fatty acids and arachidonic acid derivatives.

PPARs regulate genes that are important in cell differentiation and various metabolic processes, especially lipid and glucose metabolism. In both rodents and humans, PPARs are genetic sensors for lipids and modulate genes through the promotion of reverse cholesterol transport, reduction of total triglycerides (TGs) and regulation of apolipoproteins, thermogenesis and glucose metabolism. PPARα regulates the catabolism of fatty acids in the liver by inducing the expression of fatty acid transport protein (FATP) [6], FAT [7], long-chain fatty acid acetyl-CoA synthase (ACS) [8], enoyl-CoA hydratase/dehydrogenase multifunctional enzyme [9] and keto-acyl-CoA thiolase [10] enzymes. PPARγ influences the storage of fatty acids in adipose tissue by regulating the expression of numerous genes, including AP2 [11], PEPCK [12], acyl-CoA synthase [13] and LPL [14]. Furthermore, PPARβ/δ activation also improves lipid homeostasis, prevents weight gain and increases insulin sensitivity [15]. Accordingly, PPARs are considered important targets for the treatment of metabolic syndrome and choreographers of metabolic gene transcription.

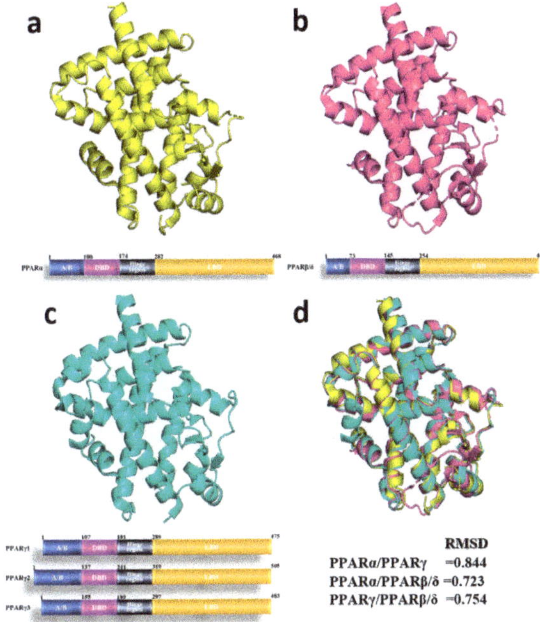

Figure 1. 3D structure and schematic structure of human Peroxisome proliferator-activated receptors (PPARs). 3D structure and schematic structure of PPARα (1I7G [16]) (**a**) PPARβ/δ (1GWX [17]) (**b**) and PPARγ (1FM6 [18]) (**c**,**d**) 3D structure superposition of PPARα (yellow), PPARβ/δ (magenta) and PPARγ (cyan) and RMSD value of three PPARs within pairwise comparison.

PPARs are also called lipid and insulin sensors [2]. Hence, many synthetic agonists of PPARs have different properties and specificities, having been developed for the treatment of different clinical outcomes over the past several decades [19–21]. For example, PPARα activators such as fibrates (fenofibrate, clofibrate) are useful drugs for the treatment of dyslipidemia. They increase HDL, decrease TG and have no effects on low-density lipoprotein (LDL). PPARγ is a target of synthetic insulin sensitizers thiazolidinediones (TZDs), including pioglitazone and rosiglitazone, which were used in the treatment of type 2 diabetes mellitus (T2DM). Dual agonists of PPARα/γ, such as glitazar, have been developed and have recently become available for the combined treatment of T2DM and dyslipidemia. Of course, there are many drugs targeting PPARs for the clinical treatment of various diseases. However, many drugs have been limited or terminated in the clinical stage by their side effect profiles. TZDs are well known to have prompted an alert by the US Food and Drug Administration (FDA) due to adverse effects, such as fluid retention, congestive heart failure (CHF) and adipogenic weight gain [22]. In this review, we summarize the use of some PPAR agonists in therapeutic treatment, with a focus on both the pros and the cons of PPARs as key regulators of glucose and lipid metabolism. Thus far, current clinical data exists for the use of 84 PPAR ligands for the treatment of diabetes mellitus, lipid metabolism disorder and other diseases (Table 1).

Table 1. Medications of PPAR synthetic ligands in currently clinical applications or studies.

Indication	Development Status				Total
	In Market	Withdrawn	Clinical Research	Discontinued in Clinical Research	
Type 2 diabetes	3	1	5	23	32
Diabetic diseases	1	0	5	10	16
Dyslipidemia	7	0	6	8	21
CVDs	0	0	1	1	2
Other diseases	0	0	12	1	13

2. PPAR Ligand Therapeutics in Diabetes Mellitus

Diabetes mellitus is a chronic, "whole-life" condition that increases the body's blood glucose levels. There are three main types of diabetes: type 1 diabetes (insulin dependent), T2DM (insulin resistance or insulin insensitivity) and gestational diabetes [23]. Diabetes mellitus and diabetic complications constitute the most important economic cost of the disease and represent a significant economic burden for the healthcare systems of developed countries [24]. As important modulators of lipid and glucose metabolism, PPAR ligands were used to treat T2DM and diabetes-associated complications.

2.1. *Type 2 Diabetes*

The majority of patients with diabetes are diagnosed with T2DM, which affects at least 250 million people worldwide [25]. Insulin resistance is a major determinant of T2DM, which involves some defects of response to pancreatic insulin in muscle and liver cell [26]. The main treatment for diabetes mellitus is to lower the blood glucose levels to reach as close to normal as possible. Many pharmacological agents are utilized in patients with type 2 diabetes, such as TZDs, biguanide, GLP-1 agonists, DPP-4 inhibitors and SGLT2s. Here, we summarize the market value of the ligands of PPAR-treated type 2 diabetes (Table 2).

TZDs, as PPARγ agonists, are increasingly being used to counteract the effects of diabetes by regulating the transcription of insulin-responsive genes, thereby enhancing insulin sensitivity in adipose tissue, skeletal muscle and liver to help reduce plasma glucose and insulin [26]. TZDs were developed in the late 1990s and have been used to treat up to 26% of people with diabetes mellitus [27]. In the market, the main approved TZD drugs for the treatment of type 2 diabetes are rosiglitazone, pioglitazone, lobeglitazone sulfate and these drugs often used combination with metformin or other antidiabetic drugs. Rosiglitazone (Rosiglitazone Maleate) is a pure ligand of PPARγ without PPARα-binding action [28]. The pharmaceutical company GlaxoSmithKline (Brentford, UK) marketed it as Avandia®, a standalone preparation and combined it with metformin as a compound (Avandamet). Another combination drug approved by the FDA is Avandaryl® (with glimepiride) [29]. Studies on animal models of insulin resistance and diabetes have shown that rosiglitazone prevents the onset of hyperglycemia, proteinuria and pancreatic islet cell degeneration [23]. In patients with T2DM, rosiglitazone reduces fasting plasma glucose (FPG), HbA1c, insulin, C-peptide and postprandial serum glucose [30]. However, in rosiglitazone monotherapy, clinically significant side effects such as edema, anemia and weight gain are frequently reported with a conventional dosage of drug [31]. Moreover, patients with unstable heart failure (HF) and patients with a history of myocardial infarction (MI) should avoid the use of rosiglitazone due to the increased risk of cardiovascular disease (CVD) [32]. Pioglitazone hydrochloride is the hydrochloride salt of thiazolidinedione with antidiabetic properties and potential antineoplastic activity [33]. Pioglitazone monotherapy significantly improves

HbA1c and FPG while producing beneficial effects on serum lipids in patients with type 2 diabetes with no evidence of drug-induced hepatotoxicity [34]. However, raising the dose and time of pioglitazone use increased the risk of bladder cancer and reached statistical significance after 24 months of exposure. Thus, the FDA issued an alert about a potential relation between the incidence of bladder cancer and the prescription of pioglitazone [35]. However, a recent meta-analysis based on 193,099 persons in the bladder cancer cohort conservatively suggested that pioglitazone use was not associated with a statistically significant increased risk of bladder cancer [36]. Given the many adverse effects of troglitazone, combination therapy can better treat type 2 diabetes. Alogliptin benzoate, a dipeptidyl peptidase-4 inhibitor, has a blood glucose-dependent insulinotropic effect via elevated concentrations of glucagon-like peptide-1 [37,38]. Pioglitazone/alogliptin combination therapy was effective and generally well tolerated in Japanese subjects with T2DM and is considered to be useful in clinical settings [39].

Table 2. Approved drugs of PPAR ligands for type 2 diabetes treatment.

Generic Name	Type of PPAR Agonist	Molecular Weight	Company
Rosiglitazone Maleate	PPARγ agonist	473.5	GlaxoSmithKline
Pioglitazone Hydrochloride	PPARγ agonist	392.898	Takeda(Originator) Lilly
Lobeglitazone Sulfate	Dual PPARα/γ agonist	578.61	Chong Kun Dang

Lobeglitazone sulfate, a novel PPARγ agonist, was conceptually designed by modification of the rosiglitazone structure with a substituted pyrimidine [40]. Lobeglitazone has a *p*-methoxyphenoxy group at the 4-position of the pyrimidine moiety [41] and is structurally similar to two well-known TZD drugs, rosiglitazone and pioglitazone. These substituted pyrimidines were selected based on their empirical effects on triglyceride accumulation in adipocytes in vitro and their glucose-lowering and lipid-modulating activities in diabetic mice in vivo [42,43]. In contrast to other TZDs, lobeglitazone is mainly excreted in the feces, reducing the concerns about the risk of bladder cancer in the mice [44] and rats [45]. In the study of lobeglitazone in patients with T2DM, lobeglitazone showed a favorable balance of efficacy and safety during the extension study [46]. In pharmacokinetic studies in healthy adults, lobeglitazone was well tolerated and did not significantly affect the pharmacokinetics of metformin or vice versa [47]. In addition, the glucose-lowering effect of lobeglitazone is more promising in obese patients with inadequate glycemic control, long-term diabetes and severe insulin resistance.

The full activation of PPARγ is related to the phosphorylation of PPARγ Ser273, which results in a series of side effects [48]. Therefore, many new insulin sensitizers based on the pharmacology of the TZDs for clinical use have focused on the selective activation of PPARs in the clinical stage. Here, we summarize the PPAR ligands used to treat type 2 diabetes in the clinical stage (Table 3).

Table 3. Drugs of PPAR ligands for type 2 diabetes treatment in clinical stage.

Generic Name	Type of PPAR Agonist	Molecular Weight	Company	Development Status
Chiglitazar	PPARs agonist	594.61	ChipScreen	Phase III active
KDT-501	PPARα agonists	404.588	KinDex Pharmaceuticals	Phase II active
Naveglitazar	PPAR modulator	422.477	Lilly(Originator)Ligand (Originator)	Phase II Pending
AVE-0897	Dual PPARα/γ agonist	469	Genfit(Originator)Sanofi	Phase I active
ZY-H2	Dual PPARα/γ agonist	unknown	Zydus cadila	Phase I Pending

Chiglitazar is a configuration-restricted non-TZD PPAR pan agonist with AC50 values of 1.2, 0.08 and 1.7 μM in CV-1 cells for PPARα, PPARγ and PPARδ, respectively and is currently in phase III clinical development in China [49]. In animal studies, chiglitazar demonstrated comparable antidiabetic effects to those of rosiglitazone but had fewer adverse effects involving body weight and fat pad weight increases in KKAy and db/db diabetic mouse models. Clinical studies (phase IIa and IIb) also show that the complete dose range of chiglitazar has a well-tolerated safety profile in patients with T2DM [49]. Its overall encouraging profile in terms of efficacy versus toxicity might be related to the balanced activity of chiglitazar towards different PPAR subtypes [49]. KDT-501 is a compound chemically derived from hops that has antidiabetic effects in rodents [50]. Multiplex analysis of gene expression revealed that KDT-501 enhanced the expression of PGC1α and PPARα but showed no evidence of activating PPARγ [51]. The oral administration of KDT-501 in DIO mouse and ZDF rat models of diabetes reduced plasma HbA1c and improved glucose metabolism. A recent study showed that KDT-501 treatment reduced plasma triglyceride levels in an open-label, phase II clinical trial including nine obese, insulin-resistant subjects [52]. Plasma total and high-molecular-weight (HMW) adiponectin were higher and plasma tumor necrosis factor alpha (TNFα) also reduced after KDT-501 treatment [52].

Many other drugs are currently in clinical studies, including naveglitazar (phase II, Lilly (Indianapolis, IN, USA)), AVE-0897 (phase I, Genfit (Originator) Sanofi) and ZY-H2 (phase I, Zydus Cadila (Ahmedabad, Gujarat, India)).

Due to safety and tolerability issues such as weight gain, edema, CHF and bone fracture, many drugs have been terminated during the clinical research stage. For example, a class of pharmaceutical molecules exhibiting PPARα/γ dual effects is known as the "glitazars," including aleglitazar, ragaglitazar, tesaglitazar, sipoglitazar, muraglitazar, cevoglitazar and naveglitazar [53]. They have been investigated for potential use in treating T2DM and dyslipidemia simultaneously. Here, we summarize the "glitazar" drugs for the treatment of T2DM that were terminated in the clinical research stage (Table 4).

Table 4. Dual PPARα/γ agonist "glitazar" for type 2 diabetes treatment.

Generic Name	Type of PPAR Agonist	Molecular Weight	Company	Development Status
Aleglitazar	Dual PPARα/γ agonist	437.51	Roche	Phase III discontinued
Ragaglitazar	Dual PPARα/γ agonist	419.477	Novo Nordisk Pharmaceutical	Phase III discontinued
Imiglitazar	Dual PPARα/γ agonist	470.525	Takeda	Phase III discontinued
Tesaglitazar	Dual PPARα/γ agonist	408.465	AstraZeneca	Phase III discontinued
Peliglitazar	Dual PPARα/γ agonist	530.577	Bristol-Myers Squibb	Phase II discontinued
Farglitazar	Dual PPARα/γ agonist	546.623	GlaxoSmithKline	Phase II discontinued
Sipoglitazar	Dual PPARα/γ agonist; Insulin sensitizer	463.552	Takeda	Phase II discontinued
Reglitazar	Dual PPARα/γ agonist	392.411	Japan Tbacco(Originator) Pfizer	Phase II discontinued
Indeglitazar	Dual PPARα/γ agonist	389.422	Pfizer	Phase II discontinued
Muraglitazar	Dual PPARα/γ agonist	516.55	Bristol-Myers Squibb	NDA Filing US

The dual PPARα/γ agonist alegitazar exerts antihyperglycemic and lipid profile-modifying effects [54], leading to insulin-sensitizing and glucose-lowering activities and favorable effects on lipid profiles and biomarkers of cardiovascular risk [55]. However, the development of aleglitazar was halted because of a lack of cardiovascular efficacy and PPAR-related side effects in patients with T2DM post-acute coronary syndrome [56]. Ragaglitazar was mentioned as carcinogenic to the urinary bladder in Sprague-Dawley male rats exposed to 50 mg/kg/day (approximately 10 times the human exposure) in a 2-year carcinogenicity study [57]. Ragaglitazar was in phase III trials by Novo Nordisk (Copenhagen, Denmark) but was terminated in July 2002 because it caused urinary bladder tumors in mice [58]. Similarly, the development of tesaglitazar was discontinued because it severely increased serum creatinine in diabetic patients. Sipoglitazar, an azolealkanoic acid derivative, exhibits selective

PPAR agonist activities towards PPARs. For example, sipoglitazar was used to treat patients with metabolic syndrome and T2DM through improving peripheral insulin sensitivity, lowering the lipid content of bodies and reducing body weight [59]. Sipoglitazar reached phase II clinical trials by Takeda for the treatment of diabetes; however, this research has been discontinued. The development of reglitazar, a PPARγ agonist that is structurally similar to TZDs and exhibits some degree of PPARα activity, was discontinued due to its lower than expected efficacy after phase II clinical trials [60].

In brief, no "glitazar" drugs, which also include muraglitazar (NDA Filing US, Bristol-Myers Squibb (Ney York, NY, USA), imiglitazar (phase III, Takeda (Tokyo, Japan)), indeglitazar (phase II, Pfizer (Ney York, NY, USA)), farglitazar (phase II, GlaxoSmithKline) and peliglitazar (phase II, Bristol-Myers Squibb (Ney York, NY, USA)), has ever been approved for clinical use due primarily to the concern of cancer risk in animals, despite their promising effects on related metabolism.

In addition to "glitazar," other PPAR agonists for the treatment of T2DM have also halted development in the clinical research stage are lost development, as shown in Table 5.

Balaglitazone is a novel partial agonist of PPARγ that was developed by Dr. Reddy's laboratories in India. As a selective partial PPARγ agonist, balaglitazone presents a better safety profile than full agonists and cuts down HbA1c levels significantly. Balaglitazone provides robust glycemic control as an add-on to insulin therapy and a trend towards less severe side effects was observed in phase III trials [61]. However, the investment was halted in 2011. FK-614, a novel non-TZD PPARγ agonist, was as an antidiabetic agent and displays beneficial effect on improving insulin resistance [62]. FK-614 induces adipocyte differentiation by stimulating PPARγ in Zucker obese rats and altering WAT characteristics and improving systemic insulin sensitivity [63,64]. However, Astellas (Tokyo, Japan) (pharmaceutical company developing FK-614) has discontinued the development of FK-614 for the treatment of type 2 diabetes because its efficacy and safety parameters showed insufficient advantages over competitors [65]. Ciglitazone improves glycemic control by increasing insulin sensitivity [66]. Long-term use of ciglitazone treatment can significantly reduce blood glucose in diabetic db/db mice, accompanied by recovery of glomerular immunopathology and renal tubular disorders [67]. Ciglitazone had been in phase II clinical trials by Takeda for the treatment of diabetes mellitus. However, this research has been discontinued [58]. In addition, many drugs are lost from development in the clinical stage, including rivoglitazone hydrochloride (phase III, Daiichi Sankyo (Tokyo, Japan)), ONO 5129 (phase II, Ono), EML-4156 (phase II, Merck Serono), netoglitazone; isoglitazone (phase II, Mitsubishi Tanabe Pharma (Originator) Perlegen Sciences), PN-2034 (phase II, Wellstat (Originator) Sanofi), Edaglitazone (phase II, Roche (Basel, Switzerland)), darglitazone sodium (phase I, Pfizer), AVE-5376 (phase I, Sanofi), DS-6930 (phase I, Daiichi Sankyo) and E-3030 (phase I, Eisai).

As mentioned above, in many clinical studies of TZDs targeting PPARγ have encountered problems with the adverse effects of TZDs and the use of these drugs has been limited, or they have been withdrawn from the markets in the United States, Europe and other countries [68]. However, the debate on the safety of TZDs continues and some scientists are also attempting to develop new classes of insulin sensitizers. Thus, there is still a need for novel TZDs. The selective modulation of PPARγ provides the opportunity to improve the safety profile while retaining the desirable therapeutic effects.

Table 5. Drugs of PPAR ligands for treatment of type 2 diabetes discontinued in clinical stage.

Generic Name	Type of PPAR Agonist	Molecular Weight	Company	Development Status
Troglitazone	PPARγ agonists	441.542	Daiichi Sankyo (Originator) Pfizer	Withdrawn
Rivoglitazone Hydrochloride	PPARγ agonists	433.907	Daiichi Sankyo (Originator) Santen	Phase III discontinued
Balaglitazone	Partial agonist of PPARγ	395.433	Dr Reddy's Laboratories (Originator) Rheoscience	Phase II discontinued
FK-614	PPARγ agonists; Insulin sensitizer	468.393	Astellas (Originator) Aestus Therapeutics	Phase II discontinued
Ciglitazone	PPAR agonists	333.446	Takeda	Phase II discontinued
ONO 5129	Dual PPARα/γ agonist	unknown	Ono	Phase II discontinued
EML-4156	Dual PPARα/γ agonist	314.381	Merck Serono	Phase II discontinued
Netoglitazone; Isaglitazone	Dual PPARα/γ agonist	381.421	Mitsubishi Tanabe Pharma (Originator) Perlegen Sciences	Phase II discontinued
PN-2034	PPARγ agonist	unknown	Wellstat (Originator) Sanofi	Phase II discontinued
Edaglitazone	PPARγ agonists	464.554	Roche	Phase II discontinued
Darglitazone Sodium	Dual PPARα/γ agonist	442.465	Pfizer	Phase I discontinued
AVE-5376	Dual PPARα/γ agonist	unknown	Sanofi (Originator)	Phase I discontinued
DS-6930	PPARγ agonists	136.129	Daiichi Sankyo	Phase I discontinued
E-3030	Dual PPARα/γ agonist	481.93	Eisai	Phase I discontinued

2.2. Diabetes-Associated Complications

Diabetes increases the risk of cardiovascular disease [25], retinopathy [69], renal failure [70] and peripheral vascular disease. Moreover, diabetes-associated complications and comorbidities also add to the lethality of T2DM [71]. Similarly, PPAR agonists have a good therapeutic effect on diabetes-associated complications, such as diabetic dyslipidemia, hypertension and Alzheimer's disease. Here, we summarize the PPAR ligands used to treat diabetes-associated, as shown in Table 6.

A very common metabolic abnormality associated with diabetes is dyslipidemia, which occurs in over 50% of T2DM patients and is often unresponsive to statin treatment [72,73]. Saroglitazar, a novel glitazar compound, is indicated mainly for T2DM patients for the treatment of diabetic dyslipidemia and hypertriglyceridemia not controlled by statin therapy [74]. Saroglitazar has shown dual PPARα/γ agonism with a predominant PPARα and moderate PPARγ activity has shown encouraging results at all stages of clinical trials. So far, Saroglitazar has been unrelated to any serious adverse events and it has not any adverse effects of weight gain and edema associated with TZDs [74]. Another drug used to improve dyslipidemia is HPP593, an effective selective PPAR δ agonist with no off-target activity. HPP593 exhibits an anti-diabetic effect in animal models of T2DM and also has demonstrated a reduction in LDL cholesterol and TGs and improved HDL cholesterol content. HPP593 is now in phase I clinical trials by High Point Pharmaceuticals (a subsidiary of vTv Therapeutics) for the treatment of diabetes and dyslipidemia [58]. K-111 is a new insulin-sensitizer with PPARα activity but without PPARγ activity [75]. K-111 is structurally unrelated to thiazolidinediones; however, it has been shown to exert antihyperinsulinemic and lipid-lowering activity in rodents [75]. Furthermore, K-111 exhibits various pharmacological therapies for insulin sensitivity [76], dyslipidemia [77] and hypertension [78] in a nonhuman primate model. CLX-0921 is a weak activator of PPAR but retains effective glucose uptake activity in vitro and has equivalent glucose lowering activity in vivo to rosiglitazone. In addition, compared to rosiglitazone, CLX-0921 showed a 10-fold reduction in vitro adipogenic potential and increased glycogen synthesis, which is usually independent of rosiglitazone or pioglitazone [79]. In addition to treatment with diabetes, CLX-0921 has shown an inhibitory effect on lipopolysaccharides-induced TNFα production in human monocytes. Mechanistic studies showed that some of the effects of CLX-0921 are attributable to the inhibition of IκB phosphorylation and subsequent inhibition of NFκB activation, an effect not seen for other thiazolidinediones [80].

Table 6. Drugs of PPAR ligands for treatment of diabetic associated complications in market or clinical stage.

Generic Name	Type of PPAR Agonist	Indication	Molecular Weight	Company	Development Status
Saroglitazar	Dual PPARα/γ agonist	Diabetic dyslipidemia	439.57	Zydus cadila	Approved
AMG-131	PPARγ agonist	Type 2 diabetes; Multiple sclerosis (MS)	672.38	Amgen (Originator) InteKrin Therapeutics	Phase II active
K-111	PPARα agonists	Type 2 diabetes; Hyperlipidemia	379.75	Roche	Phase II Pending
CLX-0921	PPARγ agonist	Type 2 diabetes; Rheumatoid arthritis (RA)	519.568	Theracos	Phase II Pending
HPP 593	PPARδ	Diabetes Dyslipidemia	unknown	vTv Therapeutics LLC	Phase II active
SAR-351034	PPAR agonists	Type 2 diabetes; Dyslipidemia	unknown	Sanofi	Phase I active

Among the patients with T2DM, approximately 10% developed diabetic nephropathy (DN) [81]. DN remains the leading cause of end-stage renal disease (ESRD) in the United States [82]. In the process of diabetic glomerular damage, podocytopathy is extremely important [83]. PPARγ is located in all three types of glomerular cells with prominent expression in podocytes [84]. The endogenous lipid electrophile 10-nitrooctadec-9-enoic acid (nitro-oleic acid, NO2-OA) can target and activate PPARγ. In animal models, NO2-OA has demonstrated benefits in a variety of metabolic and circulatory diseases, including hypertension [85] vascular neointimal proliferation [86], obesity with metabolic syndrome [87] and hyperglycemia in diabetes [88]. NO2-OA improved renal ischemia-reperfusion injury by inhibiting Bax translocation and activation and the subsequent mitochondria-dependent apoptotic cascade by regulating PPAR [89]. AMG-131, a novel, non-TZD, selective PPARγ modulator, is under development by InteKrin Therapeutics, Inc. for the treatment of T2DM and multiple sclerosis (MS). AMG-131 displays robust glucose-lowering activity in rodent models of diabetes while exhibiting a reduced side effect profile compared to marketed TZDs [90]. In phase I and II clinical trials, AMG-131 was well tolerated, without any serious adverse events or reports of fluid retention [91]. In addition, SAR-351034 is also a PPAR agonist intended for the treatment of diabetes and dyslipidemia.

Numerous dual PPAR agonists have been developed; however, because of collateral side effects, none of these agents apart from saroglitazar has been marketed. Here, we summarized the PPAR drugs for the treatment of diabetes-associated complications that were terminated in the clinical research stage (Table 7).

Table 7. Drugs of PPAR ligands for treatment of diabetic associated complications terminated in clinical stage.

Generic Name	Type of PPAR Agonist	Indication	Molecular Weight	Company	Development Status
MK-0767	Dual PPARα/γ agonist	Type 2 diabetes; Dyslipidemia	422.36	Kyorin (Originator) Merck Sharp & Dohme	Phase III discontinued
Cevoglitazar	Dual PPARα/γ agonist	Type 2 diabetes; Lipodystrophy	558.528	Novartis	Phase II discontinued
Sodelglitazar	Pan–PPAR agonists; Insulin sensitizer	Type 2 diabetes; Hyperlipidemia	499.539	GlaxoSmithKline	Phase II discontinued
AVE-0847	Dual PPARα/γ agonist	Type 2 diabetes; Lipodystrophy	unknown	Sanofi	Phase II discontinued
KRP-101	PPARα agonists	Diabetes; Dyslipidemia	451.49	Kyorin	Phase II discontinued
DSP-8658	Dual PPARα/γ agonist	Type 2 diabetes; Alzheimer's disease	unknown	Dainippon Sumitomo	Phase I discontinued
ARH-049020	PPAR agonists	Type 2 diabetes; Insulin resistance	429.51	AstraZeneca	Phase I discontinued
LY-510929	Dual PPARα/γ agonist	Type 2 diabetes; Hyperlipidemia	463.55	Lilly	Phase I discontinued
GSK-376501	PPARγ agonist	Type 2 diabetes; Hypercholesterolemia	531.649	GlaxoSmithKline	Phase I discontinued
Tetradecylthioacetic acid	Pan–PPAR agonists; Lipid Peroxidation inhibitors	Type 2 diabetes; Dyslipidemia	288.49	Badische Anilin-und-Soda-Fabrik	Phase I discontinued

The sulfur-substituted fatty acid analog tetradecylthioacetic acid (TTA) is a pan–PPAR activator that reduces plasma lipids and enhances hepatic fatty acid oxidation in rodents [92]. In rats, TTA causes a significant reduction in plasma triacylglycerol accompanied by increased mitochondrial and peroxisomal β-oxidation in the liver [93,94]. TTA might exert beneficial effects by increasing complete fatty acid oxidation and TAG formation, thereby improving overall energy metabolism and fatty acid handling in T2DM skeletal muscle [95]. However, the development of TTA has been discontinued due to deleterious effects on the heart, including reduced cardiac efficiency, impaired mitochondrial respiratory capacity and reduced functional recovery following ischemia-reperfusion [96]. Cevoglitazar, a dual agonist of PPARα/γ, is currently being developed for the treatment of dyslipidemia and obesity associated with T2DM [97]. Cevoglitazar has demonstrated both antiobesity and antidiabetic properties in mice and monkey models of obesity, providing a potential novel approach for the treatment of human obesity, diabetes and related metabolic disorders by using a single small molecule [98]. In phase I trials, the compound was reportedly more efficacious than fenofibrate in lowering lipids and at last report, it was also in phase IIa trials for the treatment of dyslipidemia [99]. However, Novartis (Basel, Swiss) announced that they had terminated the development of cevoglitazar without providing a reason [99]. The dual PPARα/γ ligand MK-0767, also known as KRP-297, was found to have potent insulin-sensitizing and antihyperglycemic activities in a preclinical model of obese T2DM, ob/ob mice [100,101]. The effects of the compound on triglyceride and cholesterol levels were assessed in hamster and dog, two species that have previously provided predictive data on the beneficial actions of other drugs, such as fibric acid derivatives and statins, currently used to treat human dyslipidemia [102]. However, MK-0767 has been noted to produce urothelial cancer and hemangiosarcoma in rodents and thus, its development has been discontinued [103]. Sodelglitazar is a panagonist active towards all three PPARs. Sodelglitazar reached phase II clinical development for the treatment of T2DM and metabolic syndrome [104]. However, this research has been discontinued because of serious safety concerns [105]. DSP-8658 is a nonthiazolidinedione compound that markedly improves glucose metabolism and increases β-cell volume, reduces adipocyte size and ameliorates plasma TG levels in diabetic mice [106]. DSP-8658 reached phase I clinical trials by Dainippon Sumitomo for the treatment of Alzheimer's disease and type 2 diabetes. However, this research has been discontinued [58].

In addition, many drugs intended for the treatment of diabetes—associated complications have been terminated at the clinical research stage, including AVE-0847 (phase II, Sanofi), KRP-101 (phase II, Kyorin), ARH-049020 (phase I, AstraZeneca), LY-510929 (phase I, Lilly) and GSK-376501 (phase I, GlaxoSmithKline).

3. PPAR Ligand Therapeutics in Lipid Metabolism Disorder

The PPAR family of NRs is implicated in the regulation of lipid homeostasis and represents a valuable therapeutic target for obesity. Obesity, defined as a body mass index (BMI) $\geq 30\ kg/m^2$, is an international public health issue that affects the quality of life, increases the risk of illness and raises healthcare costs in countries in all parts of the world [107–109]. Obesity is strongly associated with insulin resistance [110], nonalcoholic fatty liver disease (NAFLD)/nonalcoholic steatohepatitis [111], dyslipidemia [112] and atherosclerosis [113]. In this metabolic derangement, PPARα agonists, mainly fibrates and omega-3 fatty acids, act as powerful TG-lowering agents. They are used mainly to treat metabolic dyslipidemia [21], which is an abnormal amount of lipids including triglycerides, cholesterol and fat phospholipids in the blood.

3.1. *Dyslipidemia*

Hyperlipidemia, the most common type of dyslipidemia, is a condition of elevated lipid levels and is known to accelerate the process of atherosclerosis, which may prove fatal in the development of various cardiovascular diseases. Increases in lipids, such as LDL, cholesterol and triglycerides, are mainly responsible for hyperlipidemia. The current pharmacotherapy for hyperlipidemia includes statins, niacin, fibric acid derivatives and cholesterol absorption inhibitors [114]. Fibrates, such as PPARα activators, have been used for decades in the management of combined dyslipidemia [115]. Fibrates can lower triglyceride levels by an average of 36% and raise levels of small HDL particles [116]. Fibrates increase the production of apolipoprotein AI (apoAI) and AII in the liver, which in turn stimulates HDL production. Triglyceride synthesis is also decreased and lipoprotein lipase activated in response to treatment with fibrates, reducing VLDL synthesis and enhancing its clearance [117]. In addition to fibrates, these approved drugs improve lipid metabolism, as shown in patients with dyslipidemia treated with bezafibrate [118], fenofibrate [119] and ciprofibrate [120] and to a lesser extent in patients treated with gemfibrozil [121]. The approved PPAR ligand drugs for the treatment of dyslipidemia are shown in Table 8.

Clofibrate, the fibric acid derivative, was first approved for use in the United States in 1967 and was the most universally used lipid-lowering drug for many years [122]. However, after the World Health Organization trial found no reduction in overall cardiovascular events and an increase in overall mortality, the use of clofibrate was declined sharply, in part because of cholecystectomy secondary to death [123]. Many fibric acid analogs have been developed since then. Currently, gemfibrozil and fenofibrate are approved for use in the United States; besides bezafibrate and ciprofibrate are available in Europe [124]. Fenofibrate is an oral prodrug that is converted by esterases into its active metabolite, fenofibric acid [125], which is one of the most widely lipid-lowering agent and usually combines with a statin [126]. Fenofibrate has been used commercially under the brand name Tricor® [127,128] but its use is considerably limited because it has very low bioavailability, chiefly under fasting conditions, due to its poor water solubility and lipophilic nature [129]. Trilipix® (choline fenofibrate, ABT-335) is the newest formulation of a fibric acid derivative approved by the FDA. Trilipix® does not require enzymatic cleavage to become active. Instead, it rapidly dissociates to the active form of free fenofibric acid within the gastrointestinal tract and does not undergo first-pass hepatic metabolism [130]. Fenofibric acid has proven to be safe both as a monotherapy and in combination with statins. In addition, long-term trials have shown that treatment with fenofibric acid combined with statins for up to 2 years in patients with mixed dyslipidemia is safe, in that that no deaths, rhabdomyolysis, or other serious adverse events were reported [126]. The old and well-known lipid-lowering fibric acid derivative bezafibrate is the first clinically tested pan–PPAR activator with a good safety profile [131]. A clinical study, the Bezafibrate Atherosclerosis Coronary Intervention Trial (BECAIT), has shown that the long-term administration of bezafibrate can slow the rate of progression of atherosclerotic lesions in young male post infarction patients and thus reduce the incidence of coronary events [132]. However, from a biochemical point of view, bezafibrate is a PPAR ligand with a relatively low potency. Gemfibrozil, similar to other fibric acid derivatives, has a wide range of potentially favorable effects on lipoprotein metabolism [133]. The VA High-Density Lipoprotein Intervention Trial (VA-HIT), which was conducted with gemfibrozil, is the first lipid intervention trial to show that raising HDL-C concentrations in persons with established coronary heart disease (CHD) and both a low HDL-C and a low LDL-C level will significantly reduce the incidence of major coronary events [116]. Gemfibrozil increases plasma HDL-C by decreasing cholesteryl ester transfer protein-mediated cholesterol exchange from HDL and by directly stimulating hepatic HDL synthesis and secretion [134]. Ciprofibrate is known to decrease TG and TC levels and increase HDL cholesterol levels in hyperlipidemic patients [135]. However, ciprofibrate raises serum creatinine and lowers the activity of hepatic enzymes in the serum [136]. Pemafibrate (K-877) is a novel member of the selective PPARα modulator (SPPARMα) family [137] that was designed to have a higher PPARα agonistic activity and selectivity than existing PPARα agonists (such as fibrates) [138]. Pemafibrate exhibits protective antiatherogenic properties in mice by its TG and remnant lipoprotein-lowering effects, its beneficial effects on HDL metabolism and RCT and its anti-inflammatory activity in macrophages and the arterial wall, resulting in reduced atherosclerosis burden [139]. In phase III clinical trials, compared to fenofibrate, pemafibrate has greater PPARα activation in vitro and lower effects on TGs than fenofibrate. It may become a better choice for patients with metabolic syndrome and T2DM who with residual CV risk [137]. Statins, the favorable agents for lower lipid parameters, combining with fibrates is a better treatment strategy because the two drugs work differently and can complement each other [140,141].The combination of fenofibrate with 20 mg or 40 mg simvastatin was more potent in reducing TG and increasing HDL-C levels than monotherapy with simvastatin or fenofibrate separately [142]. In addition, another drug, pravastatin sodium/fenofibrate, is also on the market for dyslipidemia treatment. However, statin–fibrate combination should be attention due to increasing risk of myopathy and rhabdomyolysis [143].

Table 8. Drugs of PPAR ligands for treatment of dyslipidemia in market.

Generic Name	Type of PPAR Agonist	Indication	Molecular Weight	Company
Clofibrate	PPAR agonists	Hyperlipidemia Hypertriglyceridemia Hypercholesterolemia	242.699	Pfizer
Fenofibrate; Fenomax	PPARα agonists	Hypercholesterolemia Hypertriglyceridemia	360.834	Abbvie
Choline Fenofibrate	PPARα agonists	Hyperlipidemia	421.918	Abbvie
Bezafibrate	Pan–PPAR agonists	Hypertriglyceridemia Hypercholesterolemia Mixed hyperlipidemia	361.822	Unknown
Gemfibrozil	PPAR agonists	Hyperlipidemia; Ischemic heart disorder	250.338	Pfizer
Ciprofibrate	PPAR agonists	Hyperlipidemia	289.152	Unknown
Pemafibrate	PPARα agonists	Dyslipidemia	490.556	Kowa

Forty years after the introduction of the first fibrate in clinical practice, the exact role of these pharmacologic compounds remains ill-defined [144]. Hence, there are still novel PPAR agonists intended for dyslipidemia treatment in the clinical research stage, as shown in Table 9.

Table 9. Drugs of PPAR ligands for treatment of dyslipidemia in clinical stage.

Generic Name	Type of PPAR Agonist	Indication	Molecular Weight	Company	Development Status
Elafibranor	Dual PPARα/δ agonist	Non-alcoholic fatty liver disease (NAFLD); Dyslipidemia; Type 2 diabetes	384.49	Genfit	Phase III active
Icosabutate	PPAR agonists; Cholesterol ester transfer protein inhibitors	Hypertriglyceridemia	374.565	BASF	Phase II active
ZYH-7	PPARα agonists	Dyslipidemia	unknown	Zydus cadila	Phase II active
CER-002	PPARδ agonists	Dyslipidemia	unknown	Nippon Chemiphar	Phase I active
GSK-625019	PPAR agonists	Metabolic Syndrome X; Type 2 diabetes	unknown	GlaxoSmithKline	Phase I Pending
KD-3010	PPARα agonists	Obesity; Diabetes; Dyslipidemia	670.72	Kalypsys	Phase I Pending

Nonalcoholic steatohepatitis (NASH) defines a subgroup of nonalcoholic fatty liver disease where liver steatosis coexists with hepatic cell injury (apoptosis and hepatocyte ballooning) and inflammation [145]. It occurs in close association with obesity, T2DM and cardiometabolic conditions that define the metabolic syndrome [146]. Elafibranor is a selective dual agonist against PPARα/δ that has demonstrated efficacy in disease models of NAFLD/NASH and liver fibrosis [147]. Elafibranor exerts its major effects through the transcriptional regulation of key genes involved in hepatic lipid and glucose metabolism but also modulates hepatic inflammation and collagen turnover [147]. In phase III trials, elafibranor consistently improved plasma lipids and glucose homeostasis, peripheral and hepatic insulin resistance and liver inflammatory markers in dyslipidemic, prediabetic and T2DM patients [148,149]. Three prescription OM3-FAs (eicosapentaenoic acid (EPA) and docosahexaenoic acid (DHA)) have been approved for the management of severe hyperlipidemia [150]. Icosabutate, a first-in-class synthetic, structurally enhanced omega-3 fatty acid derivative, has PPARα activity but with potentially important differences from the fibrates and OM3-FAs. Preclinical observations proved to be consistent with results from an exploratory phase Ib study in hypercholesterolemic subjects, in which icosabutate significantly reduced TGs, ApoC3 and low-density lipoprotein cholesterol (LDL-C) [151]. KD-3010, a dual PPARβ/δ agonist, is under development by Kalypsys. Kalypsys has demonstrated activity in animal models of nonalcoholic steatohepatitis, high fat diet-induced obesity and the ob/ob mouse. Phase Ia safety/tolerability studies have been completed and a phase Ib dose-range study was begun in 2007 [152]. In addition, there are many drugs at the clinical research stage, including ZYH-7 (phase II, Zydus Cadila), CER-002 (phase I, Nippon Chemiphar) and GSK-625019 (phase I, GlaxoSmithKline).

There are also PPAR ligand drugs intended for the treatment of dyslipidemia whose development was terminated in the clinical research stage. We summarize these drugs as follows (Table 10).

The treatment of mixed dyslipidemia is fraught with difficulty because of the need to reduce LDL-C and TG levels while trying to elevate HDL-C levels. For this purpose, combination drug therapy is often the only effective option. Unfortunately, the drug combinations utilized for mixed dyslipidemia potentially increase the risk for adverse events. Rosuvastatin, the newest in its class, is the most potent statin currently available and provides significant reductions in LDL-C and TG and elevations in HDL-C. When used in combination to treat mixed dyslipidemia, rosuvastatin and fenofibrate or rosuvastatin and fenofibric acid demonstrate beneficial effects in this patient population and are well tolerated with no increased risk of adverse events [153]. In addition, many drugs have been terminated at the clinical research stage, including GW-501516 (phase II, GlaxoSmithKline), GFT 14 (phase II, Genfit), GW-544 (phase I, GlaxoSmithKline), DFR-11605 (phase I, Dr Reddys Laboratories), MP-136 (phase I, Mitsubishi Tanabe Pharma), DRF-10945 (phase I, Dr Reddys Laboratories), NS-220 (phase I, Nippon Shinyaku Pharma) and F-16482 (phase I, Pierre Fabre).

Table 10. Drugs of PPAR ligands for treatment of dyslipidemia discontinued in clinical stage.

Generic Name	Type of PPAR Agonist	Indication	Molecular Weight	Company	Development Status
GW-501516	PPARδ agonists	Hyperlipidemia	453.494	GlaxoSmithKline	Phase II discontinued
GFT 14	PPARα agonists	Dyslipidemia	unknown	Genfit	Phase II discontinued
GW-544	Dual PPARα/γ agonist	Hyperlipidemia	510.58	GlaxoSmithKline (Originator)Ligand	Phase I discontinued
DFR-11605	PPAR agonists	Obesity	unknown	Dr Reddys Laboratories (Originator)Perlecan	Phase I discontinued
MP-136	PPARα agonists	Dyslipidemia	unknown	Mitsubishi Tanabe Pharma	Phase I discontinued
DRF-10945	PPARα agonists	Lipid metabolism disorders	unknown	Dr Reddys Laboratories (Originator)Perlecan	Phase I discontinued
NS-220	PPARα agonists	Lipid metabolism disorders	373.449	Nippon Shinyaku Pharma	Phase I discontinued
F-16482	PPAR modulator	Metabolic Syndrome X	unknown	PIERRE FABRE	Phase I discontinued

3.2. Cardiovascular Diseases (CVDs)

Dyslipidemia is one of the major risk factors for CVD and plasma TG levels are a strong predictor of CVD [154]. CVDs are the leading cause of mortality and morbidity, accounting for 31% of all deaths worldwide. Of all deaths due to CVD, approximately 80% are due to CHD or stroke. Numerous studies have shown that blood cholesterol-lowering therapy reduces the occurrence of atherosclerotic cardiovascular disease (ASCVD) [155]. 3-Hydroxy-3-methyl-glutaryl-coenzyme A reductase (HMG-CoA) inhibitors or statins have demonstrated a significant reduction in CVD risk in a large number of landmark trials [156]. However, 70% of risk remains even after the treatment of high LDL-C by statins [157]. To further reduce this risk, fibrates are recommended to manage elevated TG and low HDL-C levels.

Hence, dual therapy of statins with fibrates can improve triglyceride and HDL-C levels more than monotherapy with equivalent dose statins, as shown in Table 11.

Table 11. Drugs of PPAR ligands for treatment of cardiovascular disease (CVD).

Generic Name	Type of PPAR Agonist	Indication	Molecular Weight	Company	Development Status
Gemcabene Calcium	PPAR agonists	Hypercholesterolemia	340.473	Gemphire Therapeutics	Phase II active
KRP-105	PPARα agonists	Hypercholesterolemia	unknown	Kyorin	Phase I discontinued

Pitavastatin is a competitive inhibitor of HMG-CoA reductase, the enzyme that stimulates the production of mevalonate, which is the rate-determining step in cholesterol biosynthesis [158]. The use of drugs that inhibit this enzyme has been associated with reductions in TC and LDL-C in a dose-dependent manner [159]. The co-administration of fenofibrate with pitavastatin for 7 days was found to be safe, well tolerated and without clinically significant PK interactions [160]. Furthermore, low doses of pitavastatin and fenofibrate were both effective in decreasing sd-LDL-C concentration and reduction [161]. In addition to the co-administration of fibrate with stain, there are other drugs in the clinical research stage. Gemcabene calcium is a small molecule, the monocalcium salt of a dialkyl ether dicarboxylic acid with the chemical name 6,6′-oxybis(2,2-dimethylhexanoic acid) monocalcium salt and is currently in late-stage clinical development. In rodents, gemcabene showed varying targets, including apoC-III, apoA-I and peroxisomal enzymes, which are considered to be regulated via PPAR gene activation, suggesting a PPAR-mediated mechanism of action for the observed hypolipidemic effects observed in rodents and humans [162]. By inhibiting interleukin-1 beta (IL-1β) -induced inflammation and CRP production and resulting in improvements in CVD events through inhibiting IL-1β, canakinumab, in the Canakinumab Anti-inflammatory Thrombosis Outcomes Study (CANTOS) study [163] and gemcabene have shown hypolipidemic and anti-inflammatory properties, in addition to LDL lowering activity, which offers an added benefit to CVD patients [164]. KRP-105, developed by Kyorin, is a highly selective PPARα agonist. In addition to improving the lipid metabolism, KRP-105 increased adiponectin, reduced leptin and suppressed weight gain in animal models, suggesting its potential as a unique antidyslipidemia agent. However, KRP-105 was discontinued from development as part of the company's R & D strategy [165].

4. PPAR Ligand Therapeutics in Other Diseases

PPARs are not only drug targets of glucose and lipid metabolism but also can be used to treat other diseases, such as primary biliary cholangitis, gout, cancer, AD and ulcerative colitis. Here, we summarize the PPAR ligand drugs for the treatment of other diseases in the clinical research stage (Table 12).

Functional studies of PPARδ are still in its infancy and there are increasing evidences that ubiquitously expressed PPARδ has multiple effects and can control a variety of physiological processes, mainly including lipid and lipoprotein metabolism regulation [166,167], insulin sensitivity [168], cardiac function [169], epidermal biology [170], neuroprotection [171] and gastrointestinal tract function and disease [172] Primary biliary cholangitis is a progressive cholangitic liver disease that, if untreated, progresses to cirrhosis and death or liver transplantation [173]. Two types of drugs are currently approved for the medical treatment of primary biliary cholangitis (PBC), ursodeoxycholic acid and obeticholic acid [174] but both have certain adverse effects [174,175]. Seladelpar, a selective PPARδ agonist, is a new therapy for PBC through regulating the cholesterol transporter ABCG5/ABCG8 [176]. Seladelpar appeared safe and well tolerated with no specific adverse reaction definitively associated with the drug [176]. Seladelpar reduces the number of macrophages, fibrosis and other markers of stellate cell activity in a mouse model [177]. In patients with mixed dyslipidemia [176] or homozygous familial hypercholesterolemia, seladelpar reduced LDL-C and induced sustained decreases in biochemical markers of cholestasis such as alkaline phosphatase, γ-glutamyl transpeptidase (GGT) and total bilirubin [178]. In phase III trials, seladelpar treatment normalized alkaline phosphatase levels but this treatment was associated with grade 3 increases in aminotransferases and the study was stopped early. Accordingly, the effects of seladelpar at lower doses should be explored.

Gout is the most common cause of inflammatory arthritis and has a major impact on quality of life [179,180]. Chronic hyperuricemia, the biochemical signature of the disease, leads to the deposition of urate crystals in articular structures and the disruption of these crystals is believed to trigger flares [181]. Arhalofenate, a selective partial PPARγ modulator, is a single enantiomer of halofenate and developed as a lipid-lowering agent [182,183]. Recently, arhalofenate was proven to be a uricosuric drug that lowers serum UA by blocking its reabsorption by the inhibition of URAT1 [184] in the proximal tubules of the kidney. Additionally, arhalofenate has been suggested to exert a potent anti-inflammatory effect [184]. In the phase IIb study, arhalofenate at a dosage of 800 mg decreased gout flares significantly compared to allopurinol at a dosage of 300 mg [184]. Another dual PPARα/γ agonist, oxeglitazar, whose development was halted in phase I clinical trials, is also used for gout treatment.

Table 12. Drugs of PPAR ligands for treatment of other diseases in clinical stage.

Generic Name	Type of PPAR Agonist	Indication	Molecular Weight	Company	Development Status
Seladelpar lysine dihydrate	PPARδ agonists	Primary biliary cirrhosis	626.685	Janssen (Originator) CymaBay Therapeutics	Phase III active
Arhalofenate	Partial PPARγ modulators	Chronic gout	415.793	CymaBay Therapeutics	Phase II active
T3D-959	Dual agonist of PPARδ/γ	Alzheimer's disease	443.47	DARA BioSciences	Phase II active
Efatutazone hydrochloride	Selectively activates PPARγ	Thyroid cancer; Non-small cell lung cancer; Colorectal cancer	593.52	Daiichi Sankyo	Phase II Pending
IVA-337	PPAR agonists	Systemic sclerosis	434.92	Abbvie(Originator)Inventiva	Phase II active
Fonadelpar	PPAR agonists	Corneal disorders	504.524	Senju Pharmaceuticals	Phase II active
OMS-403	PPARγ agonists	Opioid abuse Smoking cessation	unknown	Omeros	Phase II active
10-Nitrooctadec-9-enoic acid	PPARγ ligands; Transcription factor modulators; Inflammation mediator modulators	Acute kidney injury Renal failure	327.465	Complexa	Phase I active
GED-0507-34	PPAR modulator	Inflammatory bowel disease	unknown	Giuliani	Phase I active
Macuneos	PPARα agonists	Age-related macular degeneration	unknown	Biophytis	Phase I active
MA-0211	PPARδ modulators	Duchenne muscular dystrophy	unknown	Astellas	Phase I active
Oxeglitazar	Dual PPARα/γ agonist	Gout	314.381	Merck Serono	Phase I Pending
Etalocib sodium	PPARγ agonists; 5-Lipoxygenase inhibitor; Leukotriene B4 receptor antagonist	Pancreatic cancer; Non-small cell lung cancer	566.601	Lilly(Originator)Vernalis	Phase II discontinued

Over the course of several decades of research, evidence has emerged that Alzheimer's disease (AD) is quite complex and is associated with a multitude of cellular, biochemical and molecular abnormalities [185]. In fact, AD could be regarded as a brain form of diabetes, since insulin resistance and deficiency develop early and progress with the severity of neurodegeneration [186]. T3D-959 is a small-molecule dual agonist of PPARδ/γ [185] and has clear effects that preserve spatial learning and memory in an established experimental model of sporadic AD [186]. In a phase IIa trial, T3D-959 significantly improved motor performance and preserved both cortical and normalized white matter structure via the agonism of PPARδ and PPARγ in AD model rats [186].

Lung cancer is one of the highest cancer deaths worldwide and more than 60% of lung cancer patients are already in an incurable stage of diagnosis [187,188]. For many years, platinum-based doublet chemotherapy has become the most common treatment for patients with advanced non-small cell lung cancer (NSCLC) [189]. However, excessively toxic chemotherapy is also a concern for the public. PPARγ has been shown to possess antitumor properties in preclinical models of human cancers, including NSCLC [190,191]. Efatutazone is a novel third-generation thiazolidinedione that selectively activates PPARγ-mediated transcription with little effect on other PPAR subtypes [192]. Efatutazone is at least 50 times more potent than rosiglitazone and 500 times more potent than troglitazone for PPAR response element activation and the inhibition of cancer cell growth [193]. In a phase I study, efatutazone demonstrated acceptable tolerability with evidence of disease control in patients with advanced malignancies [192]. In addition, efatutazone inhibits the proliferation of human pancreatic and anaplastic thyroid tumor-cell cultures [194]. Daiichi Sankyo (originator of efatutazone hydrochloride) reinitiated enrolment in a phase II trial of efatutazone for the treatment of thyroid cancer. Another agonist of PPARγ, etalocib sodium (LY293111), which is a biphenyl-substituted diaryl ether carboxylic acid, is also a potential agent for the medical treatment of NSCLC [195]. In a phase I study, oral LY293111 was generally well tolerated, with a recommended phase II dose of 600 mg orally twice daily [196]. LY has also been found to inhibit pancreatic cancer cell lines as well as human pancreatic xenografts [197]. The development of LY-293111 for NSCLC treatment has subsequently been discontinued; however, clinical research on its effect on pancreatic and other cancers are ongoing.

Recent epidemiological data show that the incidence and prevalence of ulcerative colitis (UC) are increasing in many parts of the world [198]. PPARγ has been shown to be expressed in macrophages [199], dendritic cells (DCs) [200] and T and B lymphocytes [200]. More importantly, rosiglitazone was shown to be effective in the treatment of mild to moderately active UC [201]. (*R*)-(−)-GED-0507-34 has demonstrated 100- to 150-fold higher PPARγ activation than 5-ASA in vitro using Caco-2 cells transfected with PPRE-Luc reporter system [202]. None of these deleterious events has been observed with the new PPARγ modulator GED-0507-34, even when used at high concentrations during toxicological studies performed in rats, dogs and rabbits and no side effects were observed in the phase I study performed in 24 healthy subjects [202]. This new molecule is currently in phase II of clinical trials [203]. IVA337, the pan-PPAR agonist, is a therapeutic agent for systemic sclerosis through improving inflammatory and fibrosis [204]. There are many drugs used in the treatment of other diseases, including OMS-403 (phase II, Opioid abuse, Smoking cessation), fonadelpar (phase II, Corneal disorders), IVA-337 (phase II, Systemic sclerosis), macuneos (phase I, Age-related macular degeneration), MA-0211 (phase I, Duchenne's muscular dystrophy).

5. Discussion

Metabolic abnormalities, including T2DM, dyslipidemia, NAFLD and CVD, are a worldwide epidemic that seriously endangers global health. Considering the wide range of roles involved in energy homeostasis and cell proliferation/apoptosis, PPAR agonists are suggested for the treatment of metabolic disorders. In this study, we comprehensively summarized the roles of PPAR synthetic ligands in current clinical applications or studies for the treatment of T2DM, DN, obesity, CVDs, MS, AD, gout, cancer, PBC, UC et al., as shown in Figure 2.

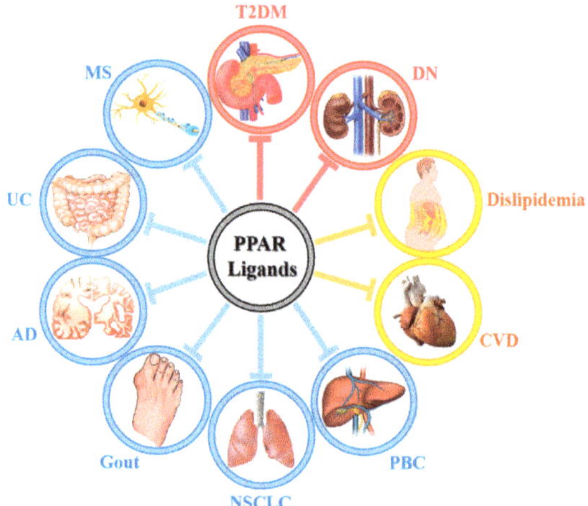

Figure 2. Concept map of the PPAR ligands in various kinds of diseases. T-bar: inhibition.

Diabetes treatment drugs represented by TZDs, which mainly activate PPARγ, have received widespread attention and are focuses for drug development. Over the past decades, in addition to the eight existing TZD drugs that have been approved and used in clinical treatment, many drugs are still in clinical studies or have even been discontinued. The use of TZDs for diabetes treatment in humans has been limited by side effects, including edema, weight gain and worsening of CHF. Thus, an increasing number of partial PPARγ agonists or SSPARMs, such as INT131 and MK0533, have been developed to reduce the side effects while improving insulin sensitivity. In a recent study, we reported that DBZ (danshensu bingpian zhi), a putative PPARγ agonist, simultaneously prevented HFD-induced obesity-related metabolic syndrome and gut dysbiosis. It also has antiatherosclerotic effects that involve inflammation suppression and the promotion of reverse cholesterol transport through concurrent partial activation of both PPARγ and LXRs [4,205,206]. Drugs for treating dyslipidemia via activating PPARα, especially represented by fibrates, are also widely used. Fibrate decreases the level of triglyceride-rich lipoproteins in serum by increasing the gene expression involved in fatty acid-β-oxidation and a decrease in apolipoprotein C-III gene expression [207]. Furthermore, PPARα agonists can increase the stability of atherosclerotic plaques and reduce the accumulation of hepatic fat accumulation, leading the party to NASH/NAFLD and reducing the risk of CVD. PPARα agonists have few adverse effects but do generally increase the plasma levels of homocysteine and creatinine, which must also be emphasized [208]. PPARδ is ubiquitously expressed and a target for management by the different components of metabolic syndrome. Clinical trials on selected PPARδ agonists have assessed both metabolic and vascular outcomes and no severe side effects have been reported to date, except for GW1516, which induced cancer in several organs in rodents [209]. Any differential mechanism of PPARδ action in different tissues should be explored in order to develop new PPARδ agonists with improved efficacy and safety. In addition to modulating lipid and glucose metabolism, PPAR agonists play significant roles in several diseases, including primary biliary cholangitis, gout, AD, non-small cell lung cancer and UC.

Currently used agonists are still at a relatively preliminary stage, the potency is weak (as is the case for PPARα), or there are many side effects (such as in PPARγ). In the past decade, increasing numbers of compounds have been developed, including dual PPAR agonists (PPARα/γ, PPARα/δ and PPARδ/γ) and pan-PPAR agonists or selective modulators. For example, clofibric acid and fenofibric acid are dual activators of PPARα and PPARγ, with a selectivity to PPARγ of about 10-fold. In addition, bezafibrate, another fibric acid that activates all three PPAR subtypes (α, γ and δ), has a broader role [131].

Unfortunately, the development of diverse dual PPAR agonists has not met with the anticipated success. Their development has thus far been halted in late-phase clinical trials because of reported side effects, such as increased cardiovascular risk (muraglitazar), carcinogenicity (ragaglitazar and MK-767), liver toxicity (imiglitazar) and renal injury (tesaglitazar) [210]. In this article, we summarize the current PPAR ligands in clinical drug discovery and development. We hope that more powerful dual PPAR agonists or pan-PPAR agonists will be highly effective in a clinical setting of patients with coexisting relevant lipid and glucose metabolism disorders.

Acknowledgments: This study was supported by grants from National Natural Science Foundation of China (Nos. 31571164 and 31271207 to Y.G. Zhai). We thank Jingwei Xu for guiding drawing of PPARs' 3D structure and Nature Research Editing Service for language editing.

Conflicts of Interest: The authors declare no conflict of interest.

References

1. Derosa, G.; Sahebkar, A.; Maffioli, P. The role of various peroxisome proliferator-activated receptors and their ligands in clinical practice. *J. Cell. Physiol.* **2018**, *233*, 153–161. [CrossRef] [PubMed]
2. Grygiel-Gorniak, B. Peroxisome proliferator-activated receptors and their ligands: Nutritional and clinical implications—A review. *Nutr. J.* **2014**, *13*, 17. [CrossRef] [PubMed]
3. Lagana, A.S.; Vitale, S.G.; Nigro, A.; Sofo, V.; Salmeri, F.M.; Rossetti, P.; Rapisarda, A.M.; La Vignera, S.; Condorelli, R.A.; Rizzo, G.; et al. Pleiotropic actions of peroxisome proliferator-activated receptors (PPARs) in dysregulated metabolic homeostasis, inflammation and cancer: Current evidence and future perspectives. *Int. J. Mol. Sci.* **2016**, *17*, 999. [CrossRef] [PubMed]
4. Xu, P.; Zhai, Y.; Wang, J. The role of PPAR and its cross-talk with car and lxr in obesity and atherosclerosis. *Int. J. Mol. Sci.* **2018**, *19*, 1260. [CrossRef] [PubMed]
5. Amber-Vitos, O.; Chaturvedi, N.; Nachliel, E.; Gutman, M.; Tsfadia, Y. The effect of regulating molecules on the structure of the PPAR-RXR complex. *Biochim. Biophys. Acta* **2016**, *1861*, 1852–1863. [CrossRef] [PubMed]
6. Echeverria, F.; Ortiz, M.; Valenzuela, R.; Videla, L.A. Long-chain polyunsaturated fatty acids regulation of PPARs, signaling: Relationship to tissue development and aging. *Prostaglandins Leukotrienes Essent. Fatty Acids* **2016**, *114*, 28–34. [CrossRef] [PubMed]
7. Glatz, J.F.; Luiken, J.J. From fat to fat (cd36/sr-b2): Understanding the regulation of cellular fatty acid uptake. *Biochimie* **2017**, *136*, 21–26. [CrossRef] [PubMed]
8. Nakamura, M.T.; Yudell, B.E.; Loor, J.J. Regulation of energy metabolism by long-chain fatty acids. *Prog. Lipid Res.* **2014**, *53*, 124–144. [CrossRef] [PubMed]
9. Marcus, S.L.; Miyata, K.S.; Zhang, B.; Subramani, S.; Rachubinski, R.A.; Capone, J.P. Diverse peroxisome proliferator-activated receptors bind to the peroxisome proliferator-responsive elements of the rat hydratase/dehydrogenase and fatty acyl-coa oxidase genes but differentially induce expression. *Proc. Natl. Acad. Sci. USA* **1993**, *90*, 5723–5727. [CrossRef] [PubMed]
10. Zhang, B.; Marcus, S.L.; Miyata, K.S.; Subramani, S.; Capone, J.P.; Rachubinski, R.A. Characterization of protein-DNA interactions within the peroxisome proliferator-responsive element of the rat hydratase-dehydrogenase gene. *J. Biol. Chem.* **1993**, *268*, 12939–12945. [PubMed]
11. Tontonoz, P.; Hu, E.; Graves, R.A.; Budavari, A.I.; Spiegelman, B.M. MPPAR gamma 2: Tissue-specific regulator of an adipocyte enhancer. *Genes Dev.* **1994**, *8*, 1224–1234. [CrossRef] [PubMed]
12. Tontonoz, P.; Hu, E.; Devine, J.; Beale, E.G.; Spiegelman, B.M. PPAR gamma 2 regulates adipose expression of the phosphoenolpyruvate carboxykinase gene. *Mol. Cell. Biol.* **1995**, *15*, 351–357. [CrossRef] [PubMed]
13. Yan, S.; Yang, X.F.; Liu, H.L.; Fu, N.; Ouyang, Y.; Qing, K. Long-chain acyl-coa synthetase in fatty acid metabolism involved in liver and other diseases: An update. *World J. Gastroenterol.* **2015**, *21*, 3492–3498. [CrossRef] [PubMed]
14. Dubois, V.; Eeckhoute, J.; Lefebvre, P.; Staels, B. Distinct but complementary contributions of PPAR isotypes to energy homeostasis. *J. Clin. Investig.* **2017**, *127*, 1202–1214. [CrossRef] [PubMed]
15. Neels, J.G.; Grimaldi, P.A. Physiological functions of peroxisome proliferator-activated receptor beta. *Physiol. Rev.* **2014**, *94*, 795–858. [CrossRef] [PubMed]

16. Cronet, P.; Petersen, J.F.; Folmer, R.; Blomberg, N.; Sjoblom, K.; Karlsson, U.; Lindstedt, E.L.; Bamberg, K. Structure of the PPARalpha and -gamma ligand binding domain in complex with az 242; ligand selectivity and agonist activation in the PPAR family. *Structure* **2001**, *9*, 699–706. [CrossRef]
17. Xu, H.E.; Lambert, M.H.; Montana, V.G.; Parks, D.J.; Blanchard, S.G.; Brown, P.J.; Sternbach, D.D.; Lehmann, J.M.; Wisely, G.B.; Willson, T.M.; et al. Molecular recognition of fatty acids by peroxisome proliferator-activated receptors. *Mol. Cell* **1999**, *3*, 397–403. [CrossRef]
18. Gampe, R.T., Jr.; Montana, V.G.; Lambert, M.H.; Miller, A.B.; Bledsoe, R.K.; Milburn, M.V.; Kliewer, S.A.; Willson, T.M.; Xu, H.E. Asymmetry in the PPARgamma/RXRalpha crystal structure reveals the molecular basis of heterodimerization among nuclear receptors. *Mol. Cell* **2000**, *5*, 545–555. [CrossRef]
19. Tan, C.K.; Zhuang, Y.; Wahli, W. Synthetic and natural peroxisome proliferator-activated receptor (PPAR) agonists as candidates for the therapy of the metabolic syndrome. *Expert Opin. Ther. Targets* **2017**, *21*, 333–348. [CrossRef] [PubMed]
20. Gross, B.; Pawlak, M.; Lefebvre, P.; Staels, B. PPARs in obesity-induced t2dm, dyslipidaemia and nafld. *Nat. Rev. Endocrinol.* **2017**, *13*, 36–49. [CrossRef] [PubMed]
21. Botta, M.; Audano, M.; Sahebkar, A.; Sirtori, C.R.; Mitro, N.; Ruscica, M. PPAR agonists and metabolic syndrome: An established role? *Int. J. Mol. Sci.* **2018**, *19*, 1197. [CrossRef] [PubMed]
22. DePaoli, A.M.; Higgins, L.S.; Henry, R.R.; Mantzoros, C.; Dunn, F.L.; Group, I.N.T.S. Can a selective PPARgamma modulator improve glycemic control in patients with type 2 diabetes with fewer side effects compared with pioglitazone? *Diabetes Care* **2014**, *37*, 1918–1923. [CrossRef] [PubMed]
23. Janani, C.; Ranjitha Kumari, B.D. PPAR gamma gene—A review. *Diabetes Metab. Syndr.* **2015**, *9*, 46–50. [CrossRef] [PubMed]
24. Koster, I.; Huppertz, E.; Hauner, H.; Schubert, I. Costs of diabetes mellitus (codim) in germany, direct per-capita costs of managing hyperglycaemia and diabetes complications in 2010 compared to 2001. *Exp. Clin. Endocrinol. Diabetes* **2014**, *122*, 510–516. [CrossRef] [PubMed]
25. Shah, A.D.; Langenberg, C.; Rapsomaniki, E.; Denaxas, S.; Pujades-Rodriguez, M.; Gale, C.P.; Deanfield, J.; Smeeth, L.; Timmis, A.; Hemingway, H. Type 2 diabetes and incidence of cardiovascular diseases: A cohort study in 1.9 million people. *Lancet Diabetes Endocrinol.* **2015**, *3*, 105–113. [CrossRef]
26. Chung, J.W.; Hartzler, M.L.; Smith, A.; Hatton, J.; Kelley, K. Pharmacological agents utilized in patients with type-2 diabetes: Beyond lowering a1c. *P & T* **2018**, *43*, 214–227.
27. Yasmin, S.; Jayaprakash, V. Thiazolidinediones and PPAR orchestra as antidiabetic agents: From past to present. *Eur. J. Med. Chem.* **2017**, *126*, 879–893. [CrossRef] [PubMed]
28. Investigators, D.T.; Gerstein, H.C.; Yusuf, S.; Bosch, J.; Pogue, J.; Sheridan, P.; Dinccag, N.; Hanefeld, M.; Hoogwerf, B.; Laakso, M.; et al. Effect of rosiglitazone on the frequency of diabetes in patients with impaired glucose tolerance or impaired fasting glucose: A randomised controlled trial. *Lancet* **2006**, *368*, 1096–1105.
29. Li, Y.; Zhang, Y.; Li, X.; Shi, L.; Tao, W.; Shi, L.; Yang, M.; Wang, X.; Yang, Y.; Yao, Y. Association study of polymorphisms in mirnas with t2dm in chinese population. *Int. J. Med. Sci.* **2015**, *12*, 875–880. [CrossRef] [PubMed]
30. Shaikh, S.; Muneera, M.S.; Thusleem, O.A.; Tahir, M.; Kondaguli, A.V.; Ruckmani, K. Development and validation of a selective online dissolution method for rosiglitazone maleate. *J. Chromatogr. Sci.* **2007**, *45*, 311–314. [CrossRef] [PubMed]
31. Kahn, B.B.; McGraw, T.E. Rosiglitazone, PPARgamma, and type 2 diabetes. *N. Engl. J. Med.* **2010**, *363*, 2667–2669. [CrossRef] [PubMed]
32. Mitka, M. Panel recommends easing restrictions on rosiglitazone despite concerns about cardiovascular safety. *JAMA* **2013**, *310*, 246–247. [CrossRef] [PubMed]
33. Tzanavaras, P.D.; Verdoukas, A.; Themelis, D.G. Development and validation of a flow-injection assay for dissolution studies of the anti-depressant drug venlafaxine. *Anal. Sci.* **2005**, *21*, 1515–1518. [CrossRef] [PubMed]
34. Aronoff, S.; Rosenblatt, S.; Braithwaite, S.; Egan, J.W.; Mathisen, A.L.; Schneider, R.L. Pioglitazone hydrochloride monotherapy improves glycemic control in the treatment of patients with type 2 diabetes: A 6-month randomized placebo-controlled dose-response study. The pioglitazone 001 study group. *Diabetes Care* **2000**, *23*, 1605–1611. [CrossRef] [PubMed]

35. Levin, D.; Bell, S.; Sund, R.; Hartikainen, S.A.; Tuomilehto, J.; Pukkala, E.; Keskimaki, I.; Badrick, E.; Renehan, A.G.; Buchan, I.E.; et al. Pioglitazone and bladder cancer risk: A multipopulation pooled, cumulative exposure analysis. *Diabetologia* **2015**, *58*, 493–504. [CrossRef] [PubMed]
36. Lewis, J.D.; Habel, L.A.; Quesenberry, C.P.; Strom, B.L.; Peng, T.; Hedderson, M.M.; Ehrlich, S.F.; Mamtani, R.; Bilker, W.; Vaughn, D.J.; et al. Pioglitazone use and risk of bladder cancer and other common cancers in persons with diabetes. *JAMA* **2015**, *314*, 265–277. [CrossRef] [PubMed]
37. Derosa, G.; Maffioli, P. Dipeptidyl peptidase-4 inhibitors: 3 years of experience. *Diabetes Technol. Ther.* **2012**, *14*, 350–364. [CrossRef] [PubMed]
38. Andukuri, R.; Drincic, A.; Rendell, M. Alogliptin: A new addition to the class of dpp-4 inhibitors. *Diabetes Metab. Syndr. Obes.* **2009**, *2*, 117–126. [PubMed]
39. Kaku, K.; Katou, M.; Igeta, M.; Ohira, T.; Sano, H. Efficacy and safety of pioglitazone added to alogliptin in japanese patients with type 2 diabetes mellitus: A multicentre, randomized, double-blind, parallel-group, comparative study. *Diabetes Obes. Metab.* **2015**, *17*, 1198–1201. [CrossRef] [PubMed]
40. Kim, S.G.; Kim, D.M.; Woo, J.T.; Jang, H.C.; Chung, C.H.; Ko, K.S.; Park, J.H.; Park, Y.S.; Kim, S.J.; Choi, D.S. Efficacy and safety of lobeglitazone monotherapy in patients with type 2 diabetes mellitus over 24-weeks: A multicenter, randomized, double-blind, parallel-group, placebo controlled trial. *PLoS ONE* **2014**, *9*, e92843. [CrossRef] [PubMed]
41. Jang, J.Y.; Bae, H.; Lee, Y.J.; Choi, Y.I.; Kim, H.J.; Park, S.B.; Suh, S.W.; Kim, S.W.; Han, B.W. Structural basis for the enhanced anti-diabetic efficacy of lobeglitazone on PPARgamma. *Sci. Rep.* **2018**, *8*, 31. [CrossRef] [PubMed]
42. Mittermayer, F.; Caveney, E.; De Oliveira, C.; Gourgiotis, L.; Puri, M.; Tai, L.J.; Turner, J.R. Addressing unmet medical needs in type 2 diabetes: A narrative review of drugs under development. *Curr. Diabetes Rev.* **2015**, *11*, 17–31. [CrossRef] [PubMed]
43. Lee, H.W.; Kim, B.Y.; Ahn, J.B.; Kang, S.K.; Lee, J.H.; Shin, J.S.; Ahn, S.K.; Lee, S.J.; Yoon, S.S. Molecular design, synthesis, and hypoglycemic and hypolipidemic activities of novel pyrimidine derivatives having thiazolidinedione. *Eur. J. Med. Chem.* **2005**, *40*, 862–874. [CrossRef] [PubMed]
44. Moon, K.S.; Lee, J.E.; Lee, H.S.; Hwang, I.C.; Kim, D.H.; Park, H.K.; Choi, H.J.; Jo, W.; Son, W.C.; Yun, H.I. Ckd-501, a novel selective PPARgamma agonist, shows no carcinogenic potential in icr mice following oral administration for 104 weeks. *J. Appl. Toxicol.* **2014**, *34*, 1271–1284. [CrossRef] [PubMed]
45. Lee, H.S.; Chang, M.; Lee, J.E.; Kim, W.; Hwang, I.C.; Kim, D.H.; Park, H.K.; Choi, H.J.; Jo, W.; Cha, S.W.; et al. Carcinogenicity study of ckd-501, a novel dual peroxisome proliferator-activated receptors alpha and gamma agonist, following oral administration to sprague dawley rats for 94-101 weeks. *Regul. Toxicol. Pharmacol.* **2014**, *69*, 207–216. [CrossRef] [PubMed]
46. Kim, S.H.; Kim, S.G.; Kim, D.M.; Woo, J.T.; Jang, H.C.; Chung, C.H.; Ko, K.S.; Park, J.H.; Park, Y.S.; Kim, S.J.; et al. Safety and efficacy of lobeglitazone monotherapy in patients with type 2 diabetes mellitus over 52 weeks: An open-label extension study. *Diabetes Res. Clin. Pract.* **2015**, *110*, e27–e30. [CrossRef] [PubMed]
47. Shin, D.; Kim, T.E.; Yoon, S.H.; Cho, J.Y.; Shin, S.G.; Jang, I.J.; Yu, K.S. Assessment of the pharmacokinetics of co-administered metformin and lobeglitazone, a thiazolidinedione antihyperglycemic agent, in healthy subjects. *Curr. Med. Res. Opin.* **2012**, *28*, 1213–1220. [CrossRef] [PubMed]
48. Choi, J.H.; Banks, A.S.; Estall, J.L.; Kajimura, S.; Bostrom, P.; Laznik, D.; Ruas, J.L.; Chalmers, M.J.; Kamenecka, T.M.; Bluher, M.; et al. Anti-diabetic drugs inhibit obesity-linked phosphorylation of PPARgamma by cdk5. *Nature* **2010**, *466*, 451–456. [CrossRef] [PubMed]
49. He, B.K.; Ning, Z.Q.; Li, Z.B.; Shan, S.; Pan, D.S.; Ko, B.C.; Li, P.P.; Shen, Z.F.; Dou, G.F.; Zhang, B.L.; et al. In vitro and in vivo characterizations of chiglitazar, a newly identified PPAR pan-agonist. *PPAR Res.* **2012**, *2012*, 546548. [CrossRef] [PubMed]
50. Konda, V.R.; Desai, A.; Darland, G.; Grayson, N.; Bland, J.S. Kdt501, a derivative from hops, normalizes glucose metabolism and body weight in rodent models of diabetes. *PLoS ONE* **2014**, *9*, e87848. [CrossRef] [PubMed]
51. Finlin, B.S.; Zhu, B.; Kok, B.P.; Godio, C.; Westgate, P.M.; Grayson, N.; Sims, R.; Bland, J.S.; Saez, E.; Kern, P.A. The influence of a kdt501, a novel isohumulone, on adipocyte function in humans. *Front. Endocrinol.* **2017**, *8*, 255. [CrossRef] [PubMed]

52. Kern, P.A.; Finlin, B.S.; Ross, D.; Boyechko, T.; Zhu, B.; Grayson, N.; Sims, R.; Bland, J.S. Effects of kdt501 on metabolic parameters in insulin-resistant prediabetic humans. *J. Endocr. Soc.* **2017**, *1*, 650–659. [CrossRef] [PubMed]
53. Raval, P.; Jain, M.; Goswami, A.; Basu, S.; Gite, A.; Godha, A.; Pingali, H.; Raval, S.; Giri, S.; Suthar, D.; et al. Revisiting glitazars: Thiophene substituted oxazole containing alpha-ethoxy phenylpropanoic acid derivatives as highly potent PPARalpha/gamma dual agonists devoid of adverse effects in rodents. *Bioorganic Med. Chem. Lett.* **2011**, *21*, 3103–3109. [CrossRef] [PubMed]
54. Dietz, M.; Mohr, P.; Kuhn, B.; Maerki, H.P.; Hartman, P.; Ruf, A.; Benz, J.; Grether, U.; Wright, M.B. Comparative molecular profiling of the PPARalpha/gamma activator aleglitazar: PPAR selectivity, activity and interaction with cofactors. *ChemMedChem* **2012**, *7*, 1101–1111. [CrossRef] [PubMed]
55. Henry, R.R.; Lincoff, A.M.; Mudaliar, S.; Rabbia, M.; Chognot, C.; Herz, M. Effect of the dual peroxisome proliferator-activated receptor-alpha/gamma agonist aleglitazar on risk of cardiovascular disease in patients with type 2 diabetes (synchrony): A phase ii, randomised, dose-ranging study. *Lancet* **2009**, *374*, 126–135. [CrossRef]
56. Lincoff, A.M.; Tardif, J.C.; Schwartz, G.G.; Nicholls, S.J.; Ryden, L.; Neal, B.; Malmberg, K.; Wedel, H.; Buse, J.B.; Henry, R.R.; et al. Effect of aleglitazar on cardiovascular outcomes after acute coronary syndrome in patients with type 2 diabetes mellitus: The alecardio randomized clinical trial. *JAMA* **2014**, *311*, 1515–1525. [CrossRef] [PubMed]
57. Oleksiewicz, M.B.; Southgate, J.; Iversen, L.; Egerod, F.L. Rat urinary bladder carcinogenesis by dual-acting PPARalpha + gamma agonists. *PPAR Res.* **2008**, *2008*, 103167. [CrossRef] [PubMed]
58. Sasarman, A.; Letowski, J.; Czaika, G.; Ramirez, V.; Nead, M.A.; Jacobs, J.M.; Morais, R. Nucleotide sequence of the hemg gene involved in the protoporphyrinogen oxidase activity of escherichia coli k12. *Can. J. Microbiol.* **1993**, *39*, 1155–1161. [CrossRef] [PubMed]
59. Stringer, F.; Scott, G.; Valbuena, M.; Kinley, J.; Nishihara, M.; Urquhart, R. The effect of genetic polymorphisms in ugt2b15 on the pharmacokinetic profile of sipoglitazar, a novel anti-diabetic agent. *Eur. J. Clin. Pharmacol.* **2013**, *69*, 423–430. [CrossRef] [PubMed]
60. Guo, L.; Zhang, L.; Sun, Y.; Muskhelishvili, L.; Blann, E.; Dial, S.; Shi, L.; Schroth, G.; Dragan, Y.P. Differences in hepatotoxicity and gene expression profiles by anti-diabetic PPAR gamma agonists on rat primary hepatocytes and human hepg2 cells. *Mol. Divers.* **2006**, *10*, 349–360. [CrossRef] [PubMed]
61. Henriksen, K.; Byrjalsen, I.; Qvist, P.; Beck-Nielsen, H.; Hansen, G.; Riis, B.J.; Perrild, H.; Svendsen, O.L.; Gram, J.; Karsdal, M.A.; et al. Efficacy and safety of the PPARgamma partial agonist balaglitazone compared with pioglitazone and placebo: A phase iii, randomized, parallel-group study in patients with type 2 diabetes on stable insulin therapy. *Diabetes Metab. Res. Rev.* **2011**, *27*, 392–401. [CrossRef] [PubMed]
62. Minoura, H.; Takeshita, S.; Kimura, C.; Hirosumi, J.; Takakura, S.; Kawamura, I.; Seki, J.; Manda, T.; Mutoh, S. Mechanism by which a novel non-thiazolidinedione peroxisome proliferator-activated receptor gamma agonist, fk614, ameliorates insulin resistance in zucker fatty rats. *Diabetes Obes. Metab.* **2007**, *9*, 369–378. [CrossRef] [PubMed]
63. Minoura, H.; Takeshita, S.; Yamamoto, T.; Mabuchi, M.; Hirosumi, J.; Takakura, S.; Kawamura, I.; Seki, J.; Manda, T.; Ita, M.; et al. Ameliorating effect of fk614, a novel nonthiazolidinedione peroxisome proliferator-activated receptor gamma agonist, on insulin resistance in zucker fatty rat. *Eur. J. Pharmacol.* **2005**, *519*, 182–190. [CrossRef] [PubMed]
64. Minoura, H.; Takeshita, S.; Ita, M.; Hirosumi, J.; Mabuchi, M.; Kawamura, I.; Nakajima, S.; Nakayama, O.; Kayakiri, H.; Oku, T.; et al. Pharmacological characteristics of a novel nonthiazolidinedione insulin sensitizer, fk614. *Eur. J. Pharmacol.* **2004**, *494*, 273–281. [CrossRef] [PubMed]
65. Colca, J.R. Discontinued drugs in 2005: Endocrine and metabolic. *Expert Opin. Investig. Drugs* **2007**, *16*, 129–136. [CrossRef] [PubMed]
66. Colca, J.R.; Tanis, S.P.; McDonald, W.G.; Kletzien, R.F. Insulin sensitizers in 2013: New insights for the development of novel therapeutic agents to treat metabolic diseases. *Expert Opin. Investig. Drugs* **2014**, *23*, 1–7. [CrossRef] [PubMed]
67. Diani, A.R.; Peterson, T.; Sawada, G.; Jodelis, K.; Wyse, B.M.; Gilchrist, B.J.; Hearron, A.E.; Chang, A.Y. Ciglitazone, a new hypoglycemic agent. 5. Effect on renal lesions in c57bl/ksj-db/db mice. *Nephron* **1986**, *42*, 72–77. [CrossRef] [PubMed]

68. Bortolini, M.; Wright, M.B.; Bopst, M.; Balas, B. Examining the safety of PPAR agonists—current trends and future prospects. *Expert Opin. Drug Saf.* **2013**, *12*, 65–79. [CrossRef] [PubMed]
69. Ansari, A.S.; de Lusignan, S.; Hinton, W.; Munro, N.; Taylor, S.; McGovern, A. Glycemic control is an important modifiable risk factor for uveitis in patients with diabetes: A retrospective cohort study establishing clinical risk and ophthalmic disease burden. *J. Diabetes Its Complicat.* **2018**, *32*, 602–608. [CrossRef] [PubMed]
70. Wanner, C.; Inzucchi, S.E.; Lachin, J.M.; Fitchett, D.; von Eynatten, M.; Mattheus, M.; Johansen, O.E.; Woerle, H.J.; Broedl, U.C.; Zinman, B.; et al. Empagliflozin and progression of kidney disease in type 2 diabetes. *N. Engl. J. Med.* **2016**, *375*, 323–334. [CrossRef] [PubMed]
71. Diabetes Prevention Program Research, G. Long-term effects of lifestyle intervention or metformin on diabetes development and microvascular complications over 15-year follow-up: The diabetes prevention program outcomes study. *Lancet Diabetes Endocrinol.* **2015**, *3*, 866–875.
72. Leiter, L.A.; Lundman, P.; da Silva, P.M.; Drexel, H.; Junger, C.; Gitt, A.K.; DYSIS investigators. Persistent lipid abnormalities in statin-treated patients with diabetes mellitus in europe and canada: Results of the dyslipidaemia international study. *Diabet. Med.* **2011**, *28*, 1343–1351. [CrossRef] [PubMed]
73. Feher, M.; Greener, M.; Munro, N. Persistent hypertriglyceridemia in statin-treated patients with type 2 diabetes mellitus. *Diabetes Metab. Syndr. Obes.* **2013**, *6*, 11–15. [PubMed]
74. Joshi, S.R. Saroglitazar for the treatment of dyslipidemia in diabetic patients. *Expert Opin. Pharmacother.* **2015**, *16*, 597–606. [CrossRef] [PubMed]
75. Bodkin, N.L.; Pill, J.; Meyer, K.; Hansen, B.C. The effects of k-111, a new insulin-sensitizer, on metabolic syndrome in obese prediabetic rhesus monkeys. *Horm. Metab. Res.* **2003**, *35*, 617–624. [CrossRef] [PubMed]
76. Ortmeyer, H.K.; Bodkin, N.L.; Haney, J.; Yoshioka, S.; Horikoshi, H.; Hansen, B.C. A thiazolidinedione improves in vivo insulin action on skeletal muscle glycogen synthase in insulin-resistant monkeys. *Int. J. Exp. Diabetes Res.* **2000**, *1*, 195–202. [CrossRef] [PubMed]
77. Hannah, J.S.; Bodkin, N.L.; Paidi, M.S.; Anh-Le, N.; Howard, B.V.; Hansen, B.C. Effects of acipimox on the metabolism of free fatty acids and very low lipoprotein triglyceride. *Acta Diabetol.* **1995**, *32*, 279–283. [CrossRef] [PubMed]
78. Bodkin, N.L.; Hansen, B.C. Antihypertensive effects of captopril without adverse effects on glucose tolerance in hyperinsulinemic rhesus monkeys. *J. Med. Primatol.* **1995**, *24*, 1–6. [CrossRef] [PubMed]
79. Dey, D.; Medicherla, S.; Neogi, P.; Gowri, M.; Cheng, J.; Gross, C.; Sharma, S.D.; Reaven, G.M.; Nag, B. A novel peroxisome proliferator-activated gamma (PPAR gamma) agonist, clx-0921, has potent antihyperglycemic activity with low adipogenic potential. *Metabolism* **2003**, *52*, 1012–1018. [CrossRef]
80. Medicherla, S.; Dey, D.; Neogi, P.; Lakner, F.J.; Nag, B. Clx-0921: A new PPAR-gamma agonist anti-diabetic thiazolidinedione compound. *Diabetes* **2000**, *49*, A117.
81. Soleymanian, T.; Hamid, G.; Arefi, M.; Najafi, I.; Ganji, M.R.; Amini, M.; Hakemi, M.; Tehrani, M.R.; Larijani, B. Non-diabetic renal disease with or without diabetic nephropathy in type 2 diabetes: Clinical predictors and outcome. *Ren. Fail.* **2015**, *37*, 572–575. [CrossRef] [PubMed]
82. Centers for Disease Control and Prevention. Incidence of end-stage renal disease attributed to diabetes among persons with diagnosed diabetes—united states and puerto rico, 1996–2007. *MMWR* **2010**, *59*, 1361–1366.
83. Weil, E.J.; Lemley, K.V.; Mason, C.C.; Yee, B.; Jones, L.I.; Blouch, K.; Lovato, T.; Richardson, M.; Myers, B.D.; Nelson, R.G. Podocyte detachment and reduced glomerular capillary endothelial fenestration promote kidney disease in type 2 diabetic nephropathy. *Kidney Int.* **2012**, *82*, 1010–1017. [CrossRef] [PubMed]
84. Henique, C.; Bollee, G.; Lenoir, O.; Dhaun, N.; Camus, M.; Chipont, A.; Flosseau, K.; Mandet, C.; Yamamoto, M.; Karras, A.; et al. Nuclear factor erythroid 2-related factor 2 drives podocyte-specific expression of peroxisome proliferator-activated receptor gamma essential for resistance to crescentic gn. *JASN* **2016**, *27*, 172–188. [CrossRef] [PubMed]
85. Zhang, J.; Villacorta, L.; Chang, L.; Fan, Z.; Hamblin, M.; Zhu, T.; Chen, C.S.; Cole, M.P.; Schopfer, F.J.; Deng, C.X.; et al. Nitro-oleic acid inhibits angiotensin ii-induced hypertension. *Circ. Res.* **2010**, *107*, 540–548. [CrossRef] [PubMed]
86. Cole, M.P.; Rudolph, T.K.; Khoo, N.K.; Motanya, U.N.; Golin-Bisello, F.; Wertz, J.W.; Schopfer, F.J.; Rudolph, V.; Woodcock, S.R.; Bolisetty, S.; et al. Nitro-fatty acid inhibition of neointima formation after endoluminal vessel injury. *Circ. Res.* **2009**, *105*, 965–972. [CrossRef] [PubMed]

87. Wang, H.; Liu, H.; Jia, Z.; Guan, G.; Yang, T. Effects of endogenous PPAR agonist nitro-oleic acid on metabolic syndrome in obese zucker rats. *PPAR Res.* **2010**, *2010*, 601562. [CrossRef] [PubMed]
88. Schopfer, F.J.; Cole, M.P.; Groeger, A.L.; Chen, C.S.; Khoo, N.K.; Woodcock, S.R.; Golin-Bisello, F.; Motanya, U.N.; Li, Y.; Zhang, J.; et al. Covalent peroxisome proliferator-activated receptor gamma adduction by nitro-fatty acids: Selective ligand activity and anti-diabetic signaling actions. *J. Biol. Chem.* **2010**, *285*, 12321–12333. [CrossRef] [PubMed]
89. Nie, H.; Xue, X.; Li, J.; Liu, X.; Lv, S.; Guan, G.; Liu, H.; Liu, G.; Liu, S.; Chen, Z. Nitro-oleic acid attenuates ogd/r-triggered apoptosis in renal tubular cells via inhibition of bax mitochondrial translocation in a PPAR-gamma-dependent manner. *Cell. Physiol. Biochem.* **2015**, *35*, 1201–1218. [CrossRef] [PubMed]
90. Taygerly, J.P.; McGee, L.R.; Rubenstein, S.M.; Houze, J.B.; Cushing, T.D.; Li, Y.; Motani, A.; Chen, J.L.; Frankmoelle, W.; Ye, G.; et al. Discovery of int131: A selective PPARgamma modulator that enhances insulin sensitivity. *Bioorganic Med. Chem.* **2013**, *21*, 979–992. [CrossRef] [PubMed]
91. Kintscher, U.; Goebel, M. Int-131, a PPARgamma agonist for the treatment of type 2 diabetes. *Curr. Opin. Investig. Drugs* **2009**, *10*, 381–387. [PubMed]
92. Berge, R.K.; Tronstad, K.J.; Berge, K.; Rost, T.H.; Wergedahl, H.; Gudbrandsen, O.A.; Skorve, J. The metabolic syndrome and the hepatic fatty acid drainage hypothesis. *Biochimie* **2005**, *87*, 15–20. [CrossRef] [PubMed]
93. Berge, R.K.; Hvattum, E. Impact of cytochrome p450 system on lipoprotein metabolism. Effect of abnormal fatty acids (3-thia fatty acids). *Pharmacol. Ther.* **1994**, *61*, 345–383. [CrossRef]
94. Vaagenes, H.; Madsen, L.; Asiedu, D.K.; Lillehaug, J.R.; Berge, R.K. Early modulation of genes encoding peroxisomal and mitochondrial beta-oxidation enzymes by 3-thia fatty acids. *Biochem. Pharmacol.* **1998**, *56*, 1571–1582. [CrossRef]
95. Wensaas, A.J.; Rustan, A.C.; Just, M.; Berge, R.K.; Drevon, C.A.; Gaster, M. Fatty acid incubation of myotubes from humans with type 2 diabetes leads to enhanced release of beta-oxidation products because of impaired fatty acid oxidation: Effects of tetradecylthioacetic acid and eicosapentaenoic acid. *Diabetes* **2009**, *58*, 527–535. [CrossRef] [PubMed]
96. Hafstad, A.D.; Khalid, A.M.; Hagve, M.; Lund, T.; Larsen, T.S.; Severson, D.L.; Clarke, K.; Berge, R.K.; Aasum, E. Cardiac peroxisome proliferator-activated receptor-alpha activation causes increased fatty acid oxidation, reducing efficiency and post-ischaemic functional loss. *Cardiovasc. Res.* **2009**, *83*, 519–526. [CrossRef] [PubMed]
97. Laurent, D.; Gounarides, J.S.; Gao, J.; Boettcher, B.R. Effects of cevoglitazar, a dual PPARalpha/gamma agonist, on ectopic fat deposition in fatty zucker rats. *Diabetes Obes. Metabol.* **2009**, *11*, 632–636.
98. Chen, H.; Dardik, B.; Qiu, L.; Ren, X.; Caplan, S.L.; Burkey, B.; Boettcher, B.R.; Gromada, J. Cevoglitazar, a novel peroxisome proliferator-activated receptor-alpha/gamma dual agonist, potently reduces food intake and body weight in obese mice and cynomolgus monkeys. *Endocrinology* **2010**, *151*, 3115–3124. [CrossRef] [PubMed]
99. Colca, J.R. Discontinued drugs in 2008: Endocrine and metabolic. *Expert Opin. Investig. Drugs* **2009**, *18*, 1243–1255. [CrossRef] [PubMed]
100. Murakami, K.; Tobe, K.; Ide, T.; Mochizuki, T.; Ohashi, M.; Akanuma, Y.; Yazaki, Y.; Kadowaki, T. A novel insulin sensitizer acts as a coligand for peroxisome proliferator-activated receptor-alpha (PPAR-alpha) and PPAR-gamma: Effect of ppar-alpha activation on abnormal lipid metabolism in liver of zucker fatty rats. *Diabetes* **1998**, *47*, 1841–1847. [CrossRef] [PubMed]
101. Nomura, M.; Kinoshita, S.; Satoh, H.; Maeda, T.; Murakami, K.; Tsunoda, M.; Miyachi, H.; Awano, K. (3-substituted benzyl)thiazolidine-2,4-diones as structurally new antihyperglycemic agents. *Bioorganic Med. Chem. Lett.* **1999**, *9*, 533–538. [CrossRef]
102. Doebber, T.W.; Kelly, L.J.; Zhou, G.; Meurer, R.; Biswas, C.; Li, Y.; Wu, M.S.; Ippolito, M.C.; Chao, Y.S.; Wang, P.R.; et al. Mk-0767, a novel dual PPARalpha/gamma agonist, displays robust antihyperglycemic and hypolipidemic activities. *Biochem. Biophys. Res. Commun.* **2004**, *318*, 323–328. [CrossRef] [PubMed]
103. Oleksiewicz, M.B.; Thorup, I.; Nielsen, H.S.; Andersen, H.V.; Hegelund, A.C.; Iversen, L.; Guldberg, T.S.; Brinck, P.R.; Sjogren, I.; Thinggaard, U.K.; et al. Generalized cellular hypertrophy is induced by a dual-acting PPAR agonist in rat urinary bladder urothelium in vivo. *Toxicol. Pathol.* **2005**, *33*, 552–560. [CrossRef] [PubMed]

104. Evans, J.L.; Lin, J.J.; Goldfine, I.D. Novel approach to treat insulin resistance, type 2 diabetes, and the metabolic syndrome: Simultaneous activation of PPARalpha, PPARgamma, and PPARdelta. *Curr. Diabetes Rev.* **2005**, *1*, 299–307. [CrossRef] [PubMed]
105. Cheang, W.S.; Tian, X.Y.; Wong, W.T.; Huang, Y. The peroxisome proliferator-activated receptors in cardiovascular diseases: Experimental benefits and clinical challenges. *Br. J. Pharmacol.* **2015**, *172*, 5512–5522. [CrossRef] [PubMed]
106. Goto, T.; Nakayama, R.; Yamanaka, M.; Takata, M.; Takazawa, T.; Watanabe, K.; Maruta, K.; Nagata, R.; Nagamine, J.; Tsuchida, A.; et al. Effects of dsp-8658, a novel selective peroxisome proliferator-activated receptors a/gamma modulator, on adipogenesis and glucose metabolism in diabetic obese mice. *Exp. Clin. Endocrinol. Diabetes* **2015**, *123*, 492–499. [PubMed]
107. Bray, G.A.; Fruhbeck, G.; Ryan, D.H.; Wilding, J.P. Management of obesity. *Lancet* **2016**, *387*, 1947–1956. [CrossRef]
108. Xu, P.; Dai, S.; Wang, J.; Zhang, J.; Liu, J.; Wang, F.; Zhai, Y. Preventive obesity agent montmorillonite adsorbs dietary lipids and enhances lipid excretion from the digestive tract. *Sci. Rep.* **2016**, *6*, 19659. [CrossRef] [PubMed]
109. Xu, P.; Wang, J.; Hong, F.; Wang, S.; Jin, X.; Xue, T.; Jia, L.; Zhai, Y. Melatonin prevents obesity through modulation of gut microbiota in mice. *J. Pineal Res.* **2017**, *62*. [CrossRef] [PubMed]
110. Vazquez-Carrera, M. Unraveling the effects of PPARbeta/delta on insulin resistance and cardiovascular disease. *Trends Endocrinol. Metab.* **2016**, *27*, 319–334. [CrossRef] [PubMed]
111. Vassilatou, E. Nonalcoholic fatty liver disease and polycystic ovary syndrome. *World J. Gastroenterol.* **2014**, *20*, 8351–8363. [CrossRef] [PubMed]
112. Xu, P.; Hong, F.; Wang, J.; Cong, Y.; Dai, S.; Wang, S.; Wang, J.; Jin, X.; Wang, F.; Liu, J.; et al. Microbiome remodeling via the montmorillonite adsorption-excretion axis prevents obesity-related metabolic disorders. *EBioMedicine* **2017**, *16*, 251–261. [CrossRef] [PubMed]
113. Lovren, F.; Teoh, H.; Verma, S. Obesity and atherosclerosis: Mechanistic insights. *Can. J. Cardiol.* **2015**, *31*, 177–183. [CrossRef] [PubMed]
114. Oda, N.; Imamura, S.; Fujita, T.; Uchida, Y.; Inagaki, K.; Kakizawa, H.; Hayakawa, N.; Suzuki, A.; Takeda, J.; Horikawa, Y.; et al. The ratio of leptin to adiponectin can be used as an index of insulin resistance. *Metabolism* **2008**, *57*, 268–273. [CrossRef] [PubMed]
115. Lalloyer, F.; Staels, B. Fibrates, glitazones, and peroxisome proliferator-activated receptors. *Arterioscler. Thromb. Vasc. Biol.* **2010**, *30*, 894–899. [CrossRef] [PubMed]
116. Robins, S.J.; Collins, D.; Wittes, J.T.; Papademetriou, V.; Deedwania, P.C.; Schaefer, E.J.; McNamara, J.R.; Kashyap, M.L.; Hershman, J.M.; Wexler, L.F.; et al. Relation of gemfibrozil treatment and lipid levels with major coronary events: Va-hit: A randomized controlled trial. *JAMA* **2001**, *285*, 1585–1591. [CrossRef] [PubMed]
117. Rodriguez-Cuenca, S.; Carobbio, S.; Barcelo-Coblijn, G.; Prieur, X.; Relat, J.; Amat, R.; Campbell, M.; Dias, A.R.; Bahri, M.; Gray, S.L.; et al. P465l PPARgamma mutation confers partial resistance to the hypolipidemic action of fibrates. *Diabetes Obes. Metab.* **2018**. [CrossRef] [PubMed]
118. Seiler, C.; Suter, T.M.; Hess, O.M. Exercise-induced vasomotion of angiographically normal and stenotic coronary arteries improves after cholesterol-lowering drug therapy with bezafibrate. *J. Am. Coll. Cardiol.* **1995**, *26*, 1615–1622. [CrossRef]
119. Khera, A.V.; Qamar, A.; Reilly, M.P.; Dunbar, R.L.; Rader, D.J. Effects of niacin, statin, and fenofibrate on circulating proprotein convertase subtilisin/kexin type 9 levels in patients with dyslipidemia. *Am. J. Cardiol.* **2015**, *115*, 178–182. [CrossRef] [PubMed]
120. Evans, M.; Anderson, R.A.; Graham, J.; Ellis, G.R.; Morris, K.; Davies, S.; Jackson, S.K.; Lewis, M.J.; Frenneaux, M.P.; Rees, A. Ciprofibrate therapy improves endothelial function and reduces postprandial lipemia and oxidative stress in type 2 diabetes mellitus. *Circulation* **2000**, *101*, 1773–1779. [CrossRef] [PubMed]
121. Song, D.; Chu, Z.; Min, L.; Zhen, T.; Li, P.; Han, L.; Bu, S.; Yang, J.; Gonzale, F.J.; Liu, A. Gemfibrozil not fenofibrate decreases systemic glucose level via PPARalpha. *Die Pharm.* **2016**, *71*, 205–212.
122. Parhofer, K.G. The treatment of disorders of lipid metabolism. *Deutsch. Arzteblatt Int.* **2016**, *113*, 261–268. [CrossRef] [PubMed]

123. Committee of Principal Investigators. WHO Cooperative trial on primary prevention of ischaemic heart disease using clofibrate to lower serum cholesterol: Mortality follow-up. Report of the committee of principal investigators. *Lancet* **1980**, *2*, 379–385.
124. Fazio, S.; Linton, M.F. The role of fibrates in managing hyperlipidemia: Mechanisms of action and clinical efficacy. *Curr. Atheroscler. Rep.* **2004**, *6*, 148–157. [CrossRef] [PubMed]
125. Keating, G.M.; Croom, K.F. Fenofibrate: A review of its use in primary dyslipidaemia, the metabolic syndrome and type 2 diabetes mellitus. *Drugs* **2007**, *67*, 121–153. [CrossRef] [PubMed]
126. Moutzouri, E.; Kei, A.; Elisaf, M.S.; Milionis, H.J. Management of dyslipidemias with fibrates, alone and in combination with statins: Role of delayed-release fenofibric acid. *Vasc. Health Risk Manag.* **2010**, *6*, 525–539. [PubMed]
127. Chachad, S.S.; Gole, M.; Malhotra, G.; Naidu, R. Comparison of pharmacokinetics of two fenofibrate tablet formulations in healthy human subjects. *Clin. Ther.* **2014**, *36*, 967–973. [CrossRef] [PubMed]
128. Zhang, X.; Chen, G.; Zhang, T.; Ma, Z.; Wu, B. Effects of pegylated lipid nanoparticles on the oral absorption of one bcs ii drug: A mechanistic investigation. *Int. J. Nanomed.* **2014**, *9*, 5503–5514.
129. Brown, W.V. Treatment of hypercholesterolaemia with fenofibrate: A review. *Curr. Med. Res. Opin.* **1989**, *11*, 321–330. [CrossRef] [PubMed]
130. Pellegrini, M.; Pallottini, V.; Marin, R.; Marino, M. Role of the sex hormone estrogen in the prevention of lipid disorder. *Curr. Med. Chem.* **2014**, *21*, 2734–2742. [CrossRef] [PubMed]
131. Tenenbaum, A.; Motro, M.; Fisman, E.Z. Dual and pan-peroxisome proliferator-activated receptors (PPAR) co-agonism: The bezafibrate lessons. *Cardiovasc. Diabetol.* **2005**, *4*, 14. [CrossRef] [PubMed]
132. Ericsson, C.G.; Hamsten, A.; Nilsson, J.; Grip, L.; Svane, B.; de Faire, U. Angiographic assessment of effects of bezafibrate on progression of coronary artery disease in young male postinfarction patients. *Lancet* **1996**, *347*, 849–853. [CrossRef]
133. Staels, B.; Dallongeville, J.; Auwerx, J.; Schoonjans, K.; Leitersdorf, E.; Fruchart, J.C. Mechanism of action of fibrates on lipid and lipoprotein metabolism. *Circulation* **1998**, *98*, 2088–2093. [CrossRef] [PubMed]
134. Saku, K.; Gartside, P.S.; Hynd, B.A.; Kashyap, M.L. Mechanism of action of gemfibrozil on lipoprotein metabolism. *J. Clin. Investig.* **1985**, *75*, 1702–1712. [CrossRef] [PubMed]
135. Mikhailidis, D.P.; Jagroon, I.A. Ciprofibrate versus gemfibrozil in the treatment of mixed hyperlipidemias: An open-label, multicenter study. *Metabolism* **2001**, *50*, 1385–1386. [CrossRef]
136. Rizos, E.; Bairaktari, E.; Ganotakis, E.; Tsimihodimos, V.; Mikhailidis, D.P.; Elisaf, M. Effect of ciprofibrate on lipoproteins, fibrinogen, renal function, and hepatic enzymes. *J. Cardiovasc. Pharmacol. Ther.* **2002**, *7*, 219–226. [CrossRef] [PubMed]
137. Fruchart, J.C. Selective peroxisome proliferator-activated receptor alpha modulators (sPPARmalpha): The next generation of peroxisome proliferator-activated receptor alpha-agonists. *Cardiovasc. Diabetol.* **2013**, *12*, 82. [CrossRef] [PubMed]
138. Yamazaki, Y.; Abe, K.; Toma, T.; Nishikawa, M.; Ozawa, H.; Okuda, A.; Araki, T.; Oda, S.; Inoue, K.; Shibuya, K.; et al. Design and synthesis of highly potent and selective human peroxisome proliferator-activated receptor alpha agonists. *Bioorganic Med. Chem. Lett.* **2007**, *17*, 4689–4693. [CrossRef] [PubMed]
139. Hennuyer, N.; Duplan, I.; Paquet, C.; Vanhoutte, J.; Woitrain, E.; Touche, V.; Colin, S.; Vallez, E.; Lestavel, S.; Lefebvre, P.; et al. The novel selective PPARalpha modulator (sPPARmalpha) pemafibrate improves dyslipidemia, enhances reverse cholesterol transport and decreases inflammation and atherosclerosis. *Atherosclerosis* **2016**, *249*, 200–208. [CrossRef] [PubMed]
140. Schima, S.M.; Maciejewski, S.R.; Hilleman, D.E.; Williams, M.A.; Mohiuddin, S.M. Fibrate therapy in the management of dyslipidemias, alone and in combination with statins: Role of delayed-release fenofibric acid. *Expert Opin. Pharmacother.* **2010**, *11*, 731–738. [CrossRef] [PubMed]
141. Athyros, V.G.; Mikhailidis, D.P.; Papageorgiou, A.A.; Didangelos, T.P.; Peletidou, A.; Kleta, D.; Karagiannis, A.; Kakafika, A.I.; Tziomalos, K.; Elisaf, M. Targeting vascular risk in patients with metabolic syndrome but without diabetes. *Metabolism* **2005**, *54*, 1065–1074. [CrossRef] [PubMed]
142. Mohiuddin, S.M.; Pepine, C.J.; Kelly, M.T.; Buttler, S.M.; Setze, C.M.; Sleep, D.J.; Stolzenbach, J.C. Efficacy and safety of abt-335 (fenofibric acid) in combination with simvastatin in patients with mixed dyslipidemia: A phase 3, randomized, controlled study. *Am. Heart J.* **2009**, *157*, 195–203. [CrossRef] [PubMed]

143. Shek, A.; Ferrill, M.J. Statin-fibrate combination therapy. *Ann. Pharmacother.* **2001**, *35*, 908–917. [CrossRef] [PubMed]
144. Backes, J.M.; Gibson, C.A.; Ruisinger, J.F.; Moriarty, P.M. Fibrates: What have we learned in the past 40 years? *Pharmacotherapy* **2007**, *27*, 412–424. [CrossRef] [PubMed]
145. Chalasani, N.; Younossi, Z.; Lavine, J.E.; Diehl, A.M.; Brunt, E.M.; Cusi, K.; Charlton, M.; Sanyal, A.J. The diagnosis and management of non-alcoholic fatty liver disease: Practice guideline by the american association for the study of liver diseases, american college of gastroenterology, and the american gastroenterological association. *Hepatology* **2012**, *55*, 2005–2023. [CrossRef] [PubMed]
146. Ratziu, V.; Bellentani, S.; Cortez-Pinto, H.; Day, C.; Marchesini, G. A position statement on nafld/nash based on the easl 2009 special conference. *J. Hepatol.* **2010**, *53*, 372–384. [CrossRef] [PubMed]
147. Staels, B.; Rubenstrunk, A.; Noel, B.; Rigou, G.; Delataille, P.; Millatt, L.J.; Baron, M.; Lucas, A.; Tailleux, A.; Hum, D.W.; et al. Hepatoprotective effects of the dual peroxisome proliferator-activated receptor alpha/delta agonist, gft505, in rodent models of nonalcoholic fatty liver disease/nonalcoholic steatohepatitis. *Hepatology* **2013**, *58*, 1941–1952. [CrossRef] [PubMed]
148. Cariou, B.; Hanf, R.; Lambert-Porcheron, S.; Zair, Y.; Sauvinet, V.; Noel, B.; Flet, L.; Vidal, H.; Staels, B.; Laville, M. Dual peroxisome proliferator-activated receptor alpha/delta agonist gft505 improves hepatic and peripheral insulin sensitivity in abdominally obese subjects. *Diabetes Care* **2013**, *36*, 2923–2930. [CrossRef] [PubMed]
149. Cariou, B.; Zair, Y.; Staels, B.; Bruckert, E. Effects of the new dual PPAR alpha/delta agonist gft505 on lipid and glucose homeostasis in abdominally obese patients with combined dyslipidemia or impaired glucose metabolism. *Diabetes Care* **2011**, *34*, 2008–2014. [CrossRef] [PubMed]
150. Bays, H.E.; Hallen, J.; Vige, R.; Fraser, D.; Zhou, R.; Hustvedt, S.O.; Orloff, D.G.; Kastelein, J.J. Icosabutate for the treatment of very high triglycerides: A placebo-controlled, randomized, double-blind, 12-week clinical trial. *J. Clin. Lipidol.* **2016**, *10*, 181–191. [CrossRef] [PubMed]
151. Fraser, D.A.; Skjaeret, T.; Qin, Y.; Larsen, L.N.; Husberg, C.; Hovland, R.; Pieterman, E.J.; van den Hoek, A.M.; Princen, H.M.; Hustvedt, S.O. Icosabutate, a novel structurally enhanced fatty-acid increases hepatic uptake of cholesterol and triglycerides in conjunction with increased hepatic LDL receptor expression. *Circulation* **2014**, *130*, A11889.
152. Billin, A.N. PPAR-beta/delta agonists for type 2 diabetes and dyslipidemia: An adopted orphan still looking for a home. *Expert Opin. Investig. Drugs* **2008**, *17*, 1465–1471. [CrossRef] [PubMed]
153. Strain, J.D.; Farver, D.K.; Clem, J.R. A review on the rationale and clinical use of concomitant rosuvastatin and fenofibrate/fenofibric acid therapy. *Clin. Pharmacol.* **2010**, *2*, 95–104. [CrossRef] [PubMed]
154. Harchaoui, K.E.; Visser, M.E.; Kastelein, J.J.; Stroes, E.S.; Dallinga-Thie, G.M. Triglycerides and cardiovascular risk. *Curr. Cardiol. Rev.* **2009**, *5*, 216–222. [CrossRef] [PubMed]
155. Stone, N.J.; Robinson, J.G.; Lichtenstein, A.H.; Bairey Merz, C.N.; Blum, C.B.; Eckel, R.H.; Goldberg, A.C.; Gordon, D.; Levy, D.; Lloyd-Jones, D.M.; et al. 2013 acc/aha guideline on the treatment of blood cholesterol to reduce atherosclerotic cardiovascular risk in adults: A report of the american college of cardiology/american heart association task force on practice guidelines. *J. Am. Coll. Cardiol.* **2014**, *63*, 2889–2934. [CrossRef] [PubMed]
156. Baigent, C.; Keech, A.; Kearney, P.M.; Blackwell, L.; Buck, G.; Pollicino, C.; Kirby, A.; Sourjina, T.; Peto, R.; Collins, R.; et al. Efficacy and safety of cholesterol-lowering treatment: Prospective meta-analysis of data from 90,056 participants in 14 randomised trials of statins. *Lancet* **2005**, *366*, 1267–1278. [PubMed]
157. Chapman, M.J.; Redfern, J.S.; McGovern, M.E.; Giral, P. Niacin and fibrates in atherogenic dyslipidemia: Pharmacotherapy to reduce cardiovascular risk. *Pharmacol. Ther.* **2010**, *126*, 314–345. [CrossRef] [PubMed]
158. Shitara, Y.; Sugiyama, Y. Pharmacokinetic and pharmacodynamic alterations of 3-hydroxy-3-methylglutaryl coenzyme a (hmg-coa) reductase inhibitors: Drug-drug interactions and interindividual differences in transporter and metabolic enzyme functions. *Pharmacol. Ther.* **2006**, *112*, 71–105. [CrossRef] [PubMed]
159. Aoki, T.; Nishimura, H.; Nakagawa, S.; Kojima, J.; Suzuki, H.; Tamaki, T.; Wada, Y.; Yokoo, N.; Sato, F.; Kimata, H.; et al. Pharmacological profile of a novel synthetic inhibitor of 3-hydroxy-3-methylglutaryl-coenzyme a reductase. *Arzneimittel-Forschung* **1997**, *47*, 904–909. [PubMed]
160. Wakida, Y.; Suzuki, S.; Nomura, H.; Isomura, T. Additional treatment with fenofibrate for patients treated with pitavastatin under ordinary medical practice for hypertriglyceridemia in japan (approach-j study). *Jpn. Clin. Med.* **2011**, *2*, 57–66. [CrossRef] [PubMed]

161. Tokuno, A.; Hirano, T.; Hayashi, T.; Mori, Y.; Yamamoto, T.; Nagashima, M.; Shiraishi, Y.; Ito, Y.; Adachi, M. The effects of statin and fibrate on lowering small dense LDL- cholesterol in hyperlipidemic patients with type 2 diabetes. *J. Atheroscler. Thromb.* **2007**, *14*, 128–132. [CrossRef] [PubMed]
162. Bisgaier, C.L.; Oniciu, D.C.; Srivastava, R.A.K. Comparative evaluation of gemcabene and PPAR ligands in transcriptional assays of peroxisome proliferator-activated receptors: Implication for the treatment of hyperlipidemia and cardiovascular disease. *J. Cardiovasc. Pharmacol.* **2018**, *72*, 3–10. [CrossRef] [PubMed]
163. Ridker, P.M.; MacFadyen, J.G.; Thuren, T.; Everett, B.M.; Libby, P.; Glynn, R.J.; Group, C.T. Effect of interleukin-1beta inhibition with canakinumab on incident lung cancer in patients with atherosclerosis: Exploratory results from a randomised, double-blind, placebo-controlled trial. *Lancet* **2017**, *390*, 1833–1842. [CrossRef]
164. Srivastava, R.A.K.; Cornicelli, J.A.; Markham, B.; Bisgaier, C.L. Gemcabene, a first-in-class lipid-lowering agent in late-stage development, down-regulates acute-phase c-reactive protein via c/ebp-delta-mediated transcriptional mechanism. *Mol. Cell. Biochem.* **2018**. [CrossRef] [PubMed]
165. Zhao, H.P.; Zhang, X.S.; Xiang, B.R. Discontinued drugs in 2010: Cardiovascular drugs. *Expert Opin. Investig. Drugs* **2011**, *20*, 1311–1325. [CrossRef] [PubMed]
166. Sprecher, D.L. Lipids, lipoproteins, and peroxisome proliferator activated receptor-delta. *Am. J. Cardiol.* **2007**, *100*, S20–S24. [CrossRef] [PubMed]
167. Sprecher, D.L.; Massien, C.; Pearce, G.; Billin, A.N.; Perlstein, I.; Willson, T.M.; Hassall, D.G.; Ancellin, N.; Patterson, S.D.; Lobe, D.C.; et al. Triglyceride:High-density lipoprotein cholesterol effects in healthy subjects administered a peroxisome proliferator activated receptor delta agonist. *Arterioscler. Thromb. Vasc. Biol.* **2007**, *27*, 359–365. [CrossRef] [PubMed]
168. Reilly, S.M.; Lee, C.H. PPAR delta as a therapeutic target in metabolic disease. *FEBS Lett.* **2008**, *582*, 26–31. [CrossRef] [PubMed]
169. Madrazo, J.A.; Kelly, D.P. The PPAR trio: Regulators of myocardial energy metabolism in health and disease. *J. Mol. Cell. Cardiol.* **2008**, *44*, 968–975. [CrossRef] [PubMed]
170. Schmuth, M.; Jiang, Y.J.; Dubrac, S.; Elias, P.M.; Feingold, K.R. Thematic review series: Skin lipids. Peroxisome proliferator-activated receptors and liver x receptors in epidermal biology. *J. Lipid Res.* **2008**, *49*, 499–509. [CrossRef] [PubMed]
171. Iwashita, A.; Muramatsu, Y.; Yamazaki, T.; Muramoto, M.; Kita, Y.; Yamazaki, S.; Mihara, K.; Moriguchi, A.; Matsuoka, N. Neuroprotective efficacy of the peroxisome proliferator-activated receptor delta-selective agonists in vitro and in vivo. *J. Pharmacol. Exp. Ther.* **2007**, *320*, 1087–1096. [CrossRef] [PubMed]
172. Peters, J.M.; Hollingshead, H.E.; Gonzalez, F.J. Role of peroxisome-proliferator-activated receptor beta/delta (PPARbeta/delta) in gastrointestinal tract function and disease. *Clin. Sci.* **2008**, *115*, 107–127. [CrossRef] [PubMed]
173. Hirschfield, G.M.; Gershwin, M.E. The immunobiology and pathophysiology of primary biliary cirrhosis. *Ann. Rev. Pathol.* **2013**, *8*, 303–330. [CrossRef] [PubMed]
174. Nevens, F.; Andreone, P.; Mazzella, G.; Strasser, S.I.; Bowlus, C.; Invernizzi, P.; Drenth, J.P.; Pockros, P.J.; Regula, J.; Beuers, U.; et al. A placebo-controlled trial of obeticholic acid in primary biliary cholangitis. *N. Engl. J. Med.* **2016**, *375*, 631–643. [CrossRef] [PubMed]
175. Corpechot, C.; Abenavoli, L.; Rabahi, N.; Chretien, Y.; Andreani, T.; Johanet, C.; Chazouilleres, O.; Poupon, R. Biochemical response to ursodeoxycholic acid and long-term prognosis in primary biliary cirrhosis. *Hepatology* **2008**, *48*, 871–877. [CrossRef] [PubMed]
176. Bays, H.E.; Schwartz, S.; Littlejohn, T., 3rd; Kerzner, B.; Krauss, R.M.; Karpf, D.B.; Choi, Y.J.; Wang, X.; Naim, S.; Roberts, B.K. Mbx-8025, a novel peroxisome proliferator receptor-delta agonist: Lipid and other metabolic effects in dyslipidemic overweight patients treated with and without atorvastatin. *J. Clin. Endocrinol. Metab.* **2011**, *96*, 2889–2897. [CrossRef] [PubMed]
177. Haczeyni, F.; Wang, H.; Barn, V.; Mridha, A.R.; Yeh, M.M.; Haigh, W.G.; Ioannou, G.N.; Choi, Y.J.; McWherter, C.A.; Teoh, N.C.; et al. The selective peroxisome proliferator-activated receptor-delta agonist seladelpar reverses nonalcoholic steatohepatitis pathology by abrogating lipotoxicity in diabetic obese mice. *Hepatol. Commun.* **2017**, *1*, 663–674. [CrossRef] [PubMed]
178. Jones, D.; Boudes, P.F.; Swain, M.G.; Bowlus, C.L.; Galambos, M.R.; Bacon, B.R.; Doerffel, Y.; Gitlin, N.; Gordon, S.C.; Odin, J.A.; et al. Seladelpar (mbx-8025), a selective PPAR-delta agonist, in patients with primary biliary cholangitis with an inadequate response to ursodeoxycholic acid: A double-blind, randomised,

placebo-controlled, phase 2, proof-of-concept study. *Lancet. Gastroenterol. Hepatol.* **2017**, *2*, 716–726. [CrossRef]

179. Wertheimer, A.; Morlock, R.; Becker, M.A. A revised estimate of the burden of illness of gout. *Curr. Ther. Res. Clin. Exp.* **2013**, *75*, 1–4. [CrossRef] [PubMed]
180. Edwards, N.L.; Sundy, J.S.; Forsythe, A.; Blume, S.; Pan, F.; Becker, M.A. Work productivity loss due to flares in patients with chronic gout refractory to conventional therapy. *J. Med. Econ.* **2011**, *14*, 10–15. [CrossRef] [PubMed]
181. Brook, R.A.; Forsythe, A.; Smeeding, J.E.; Lawrence Edwards, N. Chronic gout: Epidemiology, disease progression, treatment and disease burden. *Curr. Med. Res. Opin.* **2010**, *26*, 2813–2821. [CrossRef] [PubMed]
182. Aronow, W.S.; Harding, P.R.; Khursheed, M.; Vangrow, J.S.; Papageorge's, N.P. Effect of halofenate on serum uric acid. *Clin. Pharmacol. Ther.* **1973**, *14*, 371–373. [CrossRef] [PubMed]
183. Aronow, W.S.; Harding, P.R.; Khursheed, M.; Vangrow, J.S.; Papageorge's, N.P.; Mays, J. Effect of halofenate on serum lipids. *Clin. Pharmacol. Ther.* **1973**, *14*, 358–365. [CrossRef] [PubMed]
184. Poiley, J.; Steinberg, A.S.; Choi, Y.J.; Davis, C.S.; Martin, R.L.; McWherter, C.A.; Boudes, P.F.; Arhalofenate Flare Study, I. A randomized, double-blind, active- and placebo-controlled efficacy and safety study of arhalofenate for reducing flare in patients with gout. *Arthritis Rheumatol.* **2016**, *68*, 2027–2034. [CrossRef] [PubMed]
185. Tong, M.; Deochand, C.; Didsbury, J.; de la Monte, S.M. T3d-959: A multi-faceted disease remedial drug candidate for the treatment of alzheimer's disease. *J. Alzheimer's Dis.* **2016**, *51*, 123–138. [CrossRef] [PubMed]
186. Tong, M.; Dominguez, C.; Didsbury, J.; de la Monte, S.M. Targeting alzheimer's disease neuro-metabolic dysfunction with a small molecule nuclear receptor agonist (t3d-959) reverses disease pathologies. *J. Alzheimer's Dis. Parkinsonism* **2016**, *6*, pii:238. [CrossRef] [PubMed]
187. Jemal, A.; Bray, F.; Center, M.M.; Ferlay, J.; Ward, E.; Forman, D. Global cancer statistics. *CA Cancer J. Clin.* **2011**, *61*, 69–90. [CrossRef] [PubMed]
188. Koyi, H.; Hillerdal, G.; Branden, E. A prospective study of a total material of lung cancer from a county in sweden 1997–1999: Gender, symptoms, type, stage, and smoking habits. *Lung Cancer* **2002**, *36*, 9–14. [CrossRef]
189. Schiller, J.H.; Harrington, D.; Belani, C.P.; Langer, C.; Sandler, A.; Krook, J.; Zhu, J.; Johnson, D.H.; Eastern Cooperative Oncology, G. Comparison of four chemotherapy regimens for advanced non-small-cell lung cancer. *N. Engl. J. Med.* **2002**, *346*, 92–98. [CrossRef] [PubMed]
190. Blanquicett, C.; Roman, J.; Hart, C.M. Thiazolidinediones as anti-cancer agents. *Cancer Ther.* **2008**, *6*, 25–34. [PubMed]
191. Nemenoff, R.A.; Weiser-Evans, M.; Winn, R.A. Activation and molecular targets of peroxisome proliferator-activated receptor-gamma ligands in lung cancer. *PPAR Res.* **2008**, *2008*, 156875. [CrossRef] [PubMed]
192. Pishvaian, M.J.; Marshall, J.L.; Wagner, A.J.; Hwang, J.J.; Malik, S.; Cotarla, I.; Deeken, J.F.; He, A.R.; Daniel, H.; Halim, A.B.; et al. A phase 1 study of efatutazone, an oral peroxisome proliferator-activated receptor gamma agonist, administered to patients with advanced malignancies. *Cancer* **2012**, *118*, 5403–5413. [CrossRef] [PubMed]
193. Copland, J.A.; Marlow, L.A.; Kurakata, S.; Fujiwara, K.; Wong, A.K.; Kreinest, P.A.; Williams, S.F.; Haugen, B.R.; Klopper, J.P.; Smallridge, R.C. Novel high-affinity PPARgamma agonist alone and in combination with paclitaxel inhibits human anaplastic thyroid carcinoma tumor growth via p21waf1/cip1. *Oncogene* **2006**, *25*, 2304–2317. [CrossRef] [PubMed]
194. Shimazaki, N.; Togashi, N.; Hanai, M.; Isoyama, T.; Wada, K.; Fujita, T.; Fujiwara, K.; Kurakata, S. Anti-tumour activity of cs-7017, a selective peroxisome proliferator-activated receptor gamma agonist of thiazolidinedione class, in human tumour xenografts and a syngeneic tumour implant model. *Eur. J. Cancer* **2008**, *44*, 1734–1743. [CrossRef] [PubMed]
195. Budman, D.R.; Calabro, A. Studies of synergistic and antagonistic combinations of conventional cytotoxic agents with the multiple eicosanoid pathway modulator ly 293111. *Anti-Cancer Drugs* **2004**, *15*, 877–881. [CrossRef] [PubMed]

196. Schwartz, G.K.; Weitzman, A.; O'Reilly, E.; Brail, L.; de Alwis, D.P.; Cleverly, A.; Barile-Thiem, B.; Vinciguerra, V.; Budman, D.R. Phase i and pharmacokinetic study of ly293111, an orally bioavailable ltb4 receptor antagonist, in patients with advanced solid tumors. *J. Clin. Oncol.* **2005**, *23*, 5365–5373. [CrossRef] [PubMed]
197. Tong, W.G.; Ding, X.Z.; Talamonti, M.S.; Bell, R.H.; Adrian, T.E. Leukotriene b4 receptor antagonist ly293111 induces s-phase cell cycle arrest and apoptosis in human pancreatic cancer cells. *Anti-Cancer Drugs* **2007**, *18*, 535–541. [CrossRef] [PubMed]
198. Cleynen, I.; Boucher, G.; Jostins, L.; Schumm, L.P.; Zeissig, S.; Ahmad, T.; Andersen, V.; Andrews, J.M.; Annese, V.; Brand, S.; et al. Inherited determinants of crohn's disease and ulcerative colitis phenotypes: A genetic association study. *Lancet* **2016**, *387*, 156–167. [CrossRef]
199. Ricote, M.; Huang, J.; Fajas, L.; Li, A.; Welch, J.; Najib, J.; Witztum, J.L.; Auwerx, J.; Palinski, W.; Glass, C.K. Expression of the peroxisome proliferator-activated receptor gamma (PPARgamma) in human atherosclerosis and regulation in macrophages by colony stimulating factors and oxidized low density lipoprotein. *Proc. Natl. Acad. Sci. USA* **1998**, *95*, 7614–7619. [CrossRef] [PubMed]
200. Gosset, P.; Charbonnier, A.S.; Delerive, P.; Fontaine, J.; Staels, B.; Pestel, J.; Tonnel, A.B.; Trottein, F. Peroxisome proliferator-activated receptor gamma activators affect the maturation of human monocyte-derived dendritic cells. *Eur. J. Immunol.* **2001**, *31*, 2857–2865. [CrossRef]
201. Lewis, J.D.; Lichtenstein, G.R.; Deren, J.J.; Sands, B.E.; Hanauer, S.B.; Katz, J.A.; Lashner, B.; Present, D.H.; Chuai, S.; Ellenberg, J.H.; et al. Rosiglitazone for active ulcerative colitis: A randomized placebo-controlled trial. *Gastroenterology* **2008**, *134*, 688–695. [CrossRef] [PubMed]
202. Bertin, B.; Dubuquoy, L.; Colombel, J.F.; Desreumaux, P. PPAR-gamma in ulcerative colitis: A novel target for intervention. *Cur. Drug Targets* **2013**, *14*, 1501–1507. [CrossRef]
203. Pirat, C.; Farce, A.; Lebegue, N.; Renault, N.; Furman, C.; Millet, R.; Yous, S.; Speca, S.; Berthelot, P.; Desreumaux, P.; et al. Targeting peroxisome proliferator-activated receptors (PPARs): Development of modulators. *J. Med. Chem.* **2012**, *55*, 4027–4061. [CrossRef] [PubMed]
204. Avouac, J.; Konstantinova, I.; Guignabert, C.; Pezet, S.; Sadoine, J.; Guilbert, T.; Cauvet, A.; Tu, L.; Luccarini, J.M.; Junien, J.L.; et al. Pan-PPAR agonist iva337 is effective in experimental lung fibrosis and pulmonary hypertension. *Ann. Rheum. Dis.* **2017**, *76*, 1931–1940. [CrossRef] [PubMed]
205. Xu, P.; Hong, F.; Wang, J.; Wang, J.; Zhao, X.; Wang, S.; Xue, T.; Xu, J.; Zheng, X.; Zhai, Y. DBZ is a putative PPARgamma agonist that prevents high fat diet-induced obesity, insulin resistance and gut dysbiosis. *Biochim. Biophys. Acta* **2017**, *1861*, 2690–2701. [CrossRef] [PubMed]
206. Wang, J.; Xu, P.; Xie, X.; Li, J.; Zhang, J.; Wang, J.; Hong, F.; Li, J.; Zhang, Y.; Song, Y.; et al. DBZ (danshensu bingpian zhi), a novel natural compound derivative, attenuates atherosclerosis in apolipoprotein e-deficient mice. *J. Am. Heart Assoc.* **2017**, *6*, pii:e006297. [CrossRef] [PubMed]
207. Taniguchi, A.; Fukushima, M.; Sakai, M.; Tokuyama, K.; Nagata, I.; Fukunaga, A.; Kishimoto, H.; Doi, K.; Yamashita, Y.; Matsuura, T.; et al. Effects of bezafibrate on insulin sensitivity and insulin secretion in non-obese japanese type 2 diabetic patients. *Metabolism* **2001**, *50*, 477–480. [CrossRef] [PubMed]
208. Balakumar, P.; Rose, M.; Ganti, S.S.; Krishan, P.; Singh, M. PPAR dual agonists: Are they opening pandora's box? *Pharmacol. Res.* **2007**, *56*, 91–98. [CrossRef] [PubMed]
209. Pollock, C.B.; Rodriguez, O.; Martin, P.L.; Albanese, C.; Li, X.; Kopelovich, L.; Glazer, R.I. Induction of metastatic gastric cancer by peroxisome proliferator-activated receptordelta activation. *PPAR Res.* **2010**, *2010*, 571783. [CrossRef] [PubMed]
210. Fievet, C.; Fruchart, J.C.; Staels, B. PPARalpha and PPARgamma dual agonists for the treatment of type 2 diabetes and the metabolic syndrome. *Curr. Opin. Pharmacol.* **2006**, *6*, 606–614. [CrossRef] [PubMed]

International Journal of
Molecular Sciences

MDPI

Review

Functional Regulation of PPARs through Post-Translational Modifications

Reinhard Brunmeir [1] **and Feng Xu** [2,3,*]

1 Lee Kong Chian School of Medicine, Nanyang Technological University, 11 Mandalay Road, Singapore 308232, Singapore; Reinhard.Brunmeir@gmail.com
2 Institute of Molecular and Cell Biology, Agency for Science, Technology and Research (A*STAR), 61 Biopolis Drive, Singapore 138673, Singapore
3 Department of Biochemistry, Yong Loo Lin School of Medicine, National University of Singapore, 8 Medical Drive, Singapore 117596, Singapore
* Correspondence: fxu@imcb.a-star.edu.sg; Tel.: +65-6586-9678

Received: 3 May 2018; Accepted: 7 June 2018; Published: 12 June 2018

Abstract: Peroxisome proliferator-activated receptors (PPARs) belong to the nuclear receptor superfamily and they are essential regulators of cell differentiation, tissue development, and energy metabolism. Given their central roles in sensing the cellular metabolic state and controlling metabolic homeostasis, PPARs became important targets of drug development for the management of metabolic disorders. The function of PPARs is mainly regulated through ligand binding, which induces structural changes, further affecting the interactions with co-activators or co-repressors to stimulate or inhibit their functions. In addition, PPAR functions are also regulated by various Post-translational modifications (PTMs). These PTMs include phosphorylation, SUMOylation, ubiquitination, acetylation, and *O*-GlcNAcylation, which are found at numerous modification sites. The addition of these PTMs has a wide spectrum of consequences on protein stability, transactivation function, and co-factor interaction. Moreover, certain PTMs in PPAR proteins have been associated with the status of metabolic diseases. In this review, we summarize the PTMs found on the three PPAR isoforms PPARα, PPARβ/δ, and PPARγ, and their corresponding modifying enzymes. We also discuss the functional roles of these PTMs in regulating metabolic homeostasis and provide a perspective for future research in this intriguing field.

Keywords: nuclear receptors; PPARα; PPARγ; PPARβ/δ; post-translational modifications

1. Introduction

Nuclear receptors (NRs) are Transcription factors (TFs) capable of ligand binding, which modulates their activities to regulate gene expression. In this way, NRs directly process external signals to adapt relevant gene expression programs. Peroxisome proliferator-activated receptors (PPARs) are representative members of this large superfamily of NRs, which consist of three closely related isotypes: PPARα (NR1C1, encoded by the *Ppara* gene), PPARβ/δ (NR1C2, encoded by the *Ppard* gene), and PPARγ (NR1C3, encoded by the *Pparg* gene). The overall structure of PPAR proteins (and other NRs) is highly conserved and consists of six functional domains, A to F. The N-terminal portion of PPARs (domains A/B) is termed as the Activation-function 1 (AF-1) domain responsible for transcriptional activation. It provides constitutive activation function independent of ligand binding. The AF-1 domain is followed by a DNA-binding domain (DBD, domain C), containing two zinc-finger motifs involved in DNA recognition and protein-protein interaction. Finally, a more flexible hinge domain (domain D) is succeeded by the C-terminal Ligand-binding domain (LBD, domains E/F), which contains not only the ligand-binding pocket, but also regions important for dimerization, and the AF-2 domain. Ligand binding is thought to induce structural changes of the AF-2 domain, allowing the recruitment of co-activator proteins important for transcriptional activation, thereby serving as a switch to activate PPARs. To exert their biological functions, PPAR proteins form heterodimeric complexes with Retinoic acid receptor α (RXRα), another member of the NR family, through their dimerization domain. Binding to RXRα is a prerequisite for PPARs to bind to DNA, which usually occurs at regions known as PPAR response elements (PPREs) containing the conserved DNA sequence motif AGGTCANAGGTCA. PPAR:RXR heterodimers not bound to a ligand are thought to act as repressors through association with co-repressor complexes such as Nuclear receptor corepressor (NCoR) and the Silencing mediator of retinoid and thyroid hormone receptor (SMART). In contrast, ligand binding mediates the recruitment of co-activator complexes containing p300, CREB-binding protein (CBP), or Steroid receptor coactivator 1 (SRC1) to the heterodimers, leading to subsequent transcriptional activation of their target genes (Figure 1).

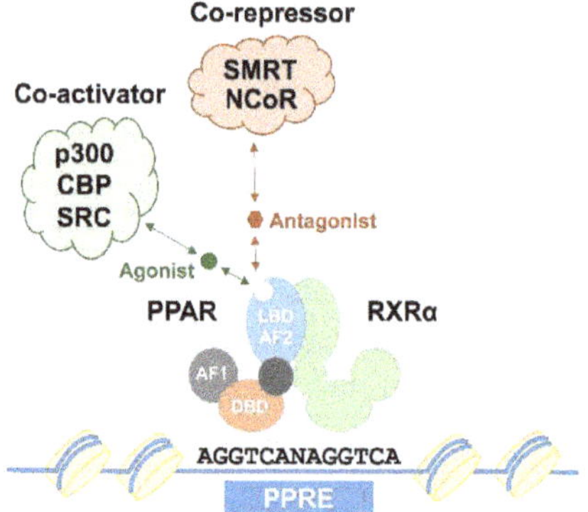

Figure 1. Transcriptional regulation by peroxisome proliferator-activated receptor (PPAR) proteins. PPARs form dimers with Retinoic acid receptor α (RXRα) proteins and subsequently bind to a DNA sequence known as peroxisome proliferator response elements (PPRE). Binding of agonists (green circle) or antagonists (red hexagon) lead to structural changes, enhancing co-activator (such as p300, CREB-binding protein (CBP), and Steroid receptor coactivator 1 (SRC1)) or co-repressor (such as Nuclear receptor corepressor (NCoR) and the Silencing mediator of retinoid and thyroid hormone receptor (SMART)) binding. AF1: activation function 1 domain; DBD: DNA-binding domain; LBD-AF2: ligand binding and activation function 2 domain.

A broad variety of natural compounds has been found to bind and activate PPAR proteins. Those natural ligands include fatty acids and their derivatives, coming either from external sources (diet) or arising as products of internal metabolic processes (de novo lipogenesis, lipolysis, etc.). Thus, via their sensitivity to intracellular levels of metabolites, PPARs act as sensors of the cellular metabolic states. Moreover, they have the ability to adjust gene regulatory networks according to fluctuating metabolic demands. Therefore, it is not surprising that PPARs have a central role in various cellular pathways linked to the energy homeostasis including glucose metabolism, lipid uptake and storage, insulin sensitivity, mitochondrial biogenesis, and thermogenesis. With the rise of metabolic disorders, commonly subsumed under the term "metabolic syndrome", over the last decades, PPAR proteins have emerged as interesting therapeutic targets to counter pathological conditions such as obesity, Type 2 diabetes (T2D), insulin resistance, Nonalcoholic fatty liver disease (NAFLD), Nonalcoholic steatosis (NASH), dyslipidema, and hypertension [1,2]. Numerous synthetic ligands targeting one, two, or all three PPARs have been developed and have entered various stages of (pre-)clinical trials, with several gaining admission. Currently, fibrates (synthetic PPARα agonists) are used to treat dyslipidemia, whereas the class of antidiabetic Thiazolidinediones (TZDs) targeting PPARγ had been widely prescribed for the management of T2D but are now partially withdrawn from clinical use due to their side effects [3–5].

The three different isoforms of PPAR have overlapping, but distinct roles, owing to their expression profiles in different tissues, sensitivities to agonists, and regulation of target genes (Reviewed in: [6]). PPARα is highly expressed in kidney, liver, Brown adipose tissue (BAT), heart, and skeletal muscle, the tissues with high capacities for Fatty acid oxidation (FAO). Accordingly, its main role seems to be the control of energy dissipation through the regulation of lipid metabolism in response to nutritional changes (such as fasting and feeding). PPARβ/δ shows a relatively broader expression pattern, with enriched levels in tissues associated with fatty acid metabolism, such as the gastrointestinal tract, heart, kidney, skeletal muscle, fat, and skin. Its physiological role in energy homeostasis is complex, as it not only controls plasma lipid levels through FAO in several tissues, but also modulates glucose handling in muscle and liver. The third member of the PPAR family, PPARγ, exists in two distinct protein forms: the shorter PPARγ1—lacking its first 30 amino acids due to alternative promoter usage—is expressed in a broad variety of cells including immune and brain cells, whereas the full length isoform PPARγ2 is highly abundant in BAT and White adipose tissue (WAT). PPARγ2 is considered the master regulator of adipocyte differentiation and stimulates energy storage by controlling fatty acid uptake and lipogenesis [7].

Many proteins undergo Post-translational modifications (PTMs), i.e., the covalent attachment of chemical groups to certain amino acid residues, at some points of their life-cycle. Those PTMs range from small entities such as methyl-, acetyl-, or phospho-groups to sizeable polypeptides such as ubiquitin chains with a size of several kDa. Their addition can have a wide spectrum of consequences on the chemical properties of targeted proteins, which further modulate protein functions. As expected, PTMs are important regulators of virtually every aspect of protein biology, including protein stability, cellular localization, enzyme function, and co-factor interaction. Several excellent recent reviews have covered various aspects of PPAR biology, including their roles in metabolic diseases [8], energy homeostasis [6], and as drug targets [9]. This review aims to give an overview of the current status of research on PTMs found in PPARα, PPARβ/δ, and PPARγ, and their functional roles.

2. Post-Translational Modifications of PPARα

2.1. Phosphorylation

It was reported as early as 1996 [10] that PPARα is a phosphoprotein. Its phosphorylation was shown to increase upon treatment with different stimuli such as insulin [10] and ciprofibrate, a PPARα agonist [11]. Specific serine residues in PPARα have emerged as important phosphorylation sites: serine 12 and 21, which are both targeted by either Mitogen-activated protein kinases (MAPKs) [12,13] or Cyclin-dependent kinase (CDK) 7 [14]. Functionally, phosphorylation of S12/S21 (S12ph/S21ph) correlates with increased transactivation of PPARα in hepatocytes and cardiac myocytes, potentially via decreased co-reperessor interaction (NCoR) or increased interaction with certain co-activators (Peroxisome proliferator-activated receptor gamma coactivator 1-alpha (PGC1α)). Lower S12ph/S21ph (together with decreased PPARγ phosphorylation, see below) is observed in Xeroderma pigmentosum group D (XPD) patients, which carry a mutation in the CDK7-containing Transcription factor II H (TFIIH) complex, and might partially explain their complex metabolic phenotypes, including reduced adipose mass and increased energy expenditure [14]. Another important phosphorylation event regulating PPARα function, S73ph, is mediated by Glycogen synthase kinase β (GSKβ), and leads to the degradation of PPARα [15]. Interestingly, in a mouse model of Gilbert's Syndrome, it was shown that the protective effect against hepatic steatosis might be mediated by increased PPARα protein levels and reduced S73ph [16]. A recent publication also reported increased S12ph in peripheral blood mononucleated cells of Gilbert's Syndrome patients [17]. The regulatory mechanism of S12ph/S21ph in PPARα is illustrated in Figure 2A.

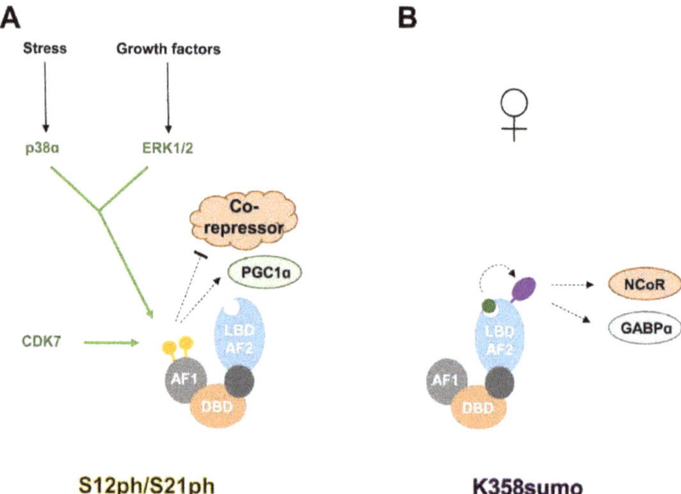

Figure 2. Regulatory mechanisms of S12ph/S21ph and K358sumo in PPARα. (**A**) Phosphorylation of serine 12 and 21 enhances PPARα activity, most likely via reduced co-repressor and/or increased Peroxisome proliferator-activated receptor gamma coactivator 1-alpha (PGC1α) recruitment. Both residues are targeted by Mitogen-activated protein kinase (MAPK) downstream kinases p38 and Extracellular signal–regulated kinase 1/2 (ERK1/2), as well as Cyclin-dependent kinase (CDK) 7. (**B**) Upon ligand binding, PPARα gets SUMOylated at K358 in female liver cells, leading to increased binding of NCoR and GA-binding protein α (GABPα), and silencing of androgen steroid genes. AF1: activation function 1 domain; DBD: DNA-binding domain; LBD-AF2: ligand binding and activation function 2 domain; enzymes depositing post-translational modifications (PTMs) are colored in green; green arrows indicate deposition of PTMs; green circle: PPARα-ligand; yellow circle: phosphorylated serine; purple oval: SUMOylated lysine; black arrow: activation; dotted arrow: increased interaction/stimulation; dotted T symbol: decreased interaction.

2.2. SUMOylation

SUMO (Small Ubiquitin-like MOdifier) polypeptides are roughly 12 kDa in size, which can be covalently attached to lysine residues via an enzymatic machinery analogous to that for protein ubiquitination. Its addition can have a wide range of effects on protein function [18]. Two lysine residues of PPARα have been reported to be subjected to this modification: K185 and K358 [19,20]. While SUMOylation of both residues increases the repressive ability of PPARα through enhanced co-repressor recruitment (NCoR, or GA-binding protein (GABP)), their regulation by PPARα agonists is marked different: K185sumo is blocked by the PPARα ligand GW7647, whereas agonist mediated conformational change of the LBD seems a prerequisite for efficient K358 SUMOylation. Functionally, K358 SUMOylation plays an interesting role in the establishment of sexual dimorphism of liver cells. The modification only occurs in female livers, where it helps to repress genes involved in the production of androgen steroids. The regulatory mechanism of K358sumo in PPARα is illustrated in Figure 2B.

2.3. Ubiquitination

There is a body of work showing that PPAR protein levels are regulated by the ubiquitin proteasome system [21]. Early findings implicated the E3 ligase function of Mouse double minute 2 homolog (MDM2) in the regulation of PPARα protein stability [22]. More recently, the addition of a single ubiquitin (mono-ubiquitination) has emerged as another way to regulate PPARα function in cardiomyocytes. Rodriguez et al. [23] found that the muscle-specific ubiquitin ligase Muscle-specific RING finger protein 1 (MuRF1) can modify PPARα, leading to the decreased activity of PPARα due to its export from the nucleus to the cytoplasm. Three lysine residues (K292, K310, and K358) located around a newly identified nuclear export signal in the LBD (aa300-308) were identified as putative mediators of this effect.

3. Post-Translational Modifications of PPARγ

3.1. Phosphorylation

PPARγ is by far the best studied member of the PPAR family, and phosphorylation of PPARγ has been reported as early as 1996 [24,25], shortly after its discovery as the master regulator of adipogenesis [7]. Numerous reports in quick succession showed that PPARγ gets phosphorylated upon stimulation of the MAPK activated pathway [24–28]. A variety of stimuli such as growth factors (Epidermal growth factor (EGF), Platelet-derived growth factor (PDGF), Transforming growth factor β (TGFβ) and insulin), Prostaglandin F2α (PGF2α), or cellular stress (UV, 12-*O*-tetradecanoyl-13-phorbol acetate (TPA) and anisomycin) were shown to trigger PPARγ phosphorylation through the activation of the downstream Extracellular signal-regulated kinases (ERKs) 1/2 or p38/c-Jun *N*-terminal kinase (JNK). The phosphorylation site was mapped to PPARγ2 serine 112 (corresponding to PPARγ1 S82), located in the AF1 region within a MAPK consensus site [24,28]. The functional role of S112ph was revealed through reporter assays, where the phosphorylation led to decreased transcriptional activity of PPARγ. Mutagenesis experiments further corroborated the notion that S112ph inhibits PPARγ function, as the expression of a nonphosphorylatable S112A led to increased transcriptional activity and enhanced adipogenic potential of fibroblasts [24,26–33]. On the flipside, the same mutation is detrimental for efficient osteoblast differentiation [34,35]. Another publication highlighted the role of the adaptor molecule Docking protein 1 (DOK1) as a modulator of this signaling cascade: DOK1 is induced by High fat diet (HFD) feeding and negatively regulates ERK1/2 mediated S112ph, thereby enhancing PPARγ activity even in a state of active insulin signaling [36]. Finally, our understanding of the mechanisms by which S112 gets dephosphorylated is also improved by the identification of Protein phosphatase 5 (PP5) [37], Protein phosphatase Mg^{2+}/Mn^{2+} dependent 1B (PPM1B) [38], and Wild-type p53-induced phosphatase 1 (WIP1) [39] as S112 phosphatases and PPARγ activators.

How is the repressive function of S112ph mediated mechanistically? Adams et al. showed that the phosphorylation event does not appear to impact PPARγ protein stability, or reduce its DNA binding activity. Instead, they proposed that S112ph might inhibit the transactivation function of PPARγ via co-repressor recruitment or co-activator release [26]. In another study, S112ph was shown to modulate PPARγ function by reducing ligand binding affinity, which involves the intramolecular communication between the AF1 and the ligand binding domain [30]. Finally, Grimaldi et al. described a mechanism by which S112ph regulates PPARγ-mediated transcription: phosphorylation of S112 enhances the interaction between PPARγ and the circadian clock protein Period circadian regulator 2 (PER2). PPARγ-PER2 interaction was shown to be detrimental to PPARγ recruitment to general adipogenic regulators as well as BAT-specific genes, such as *Ucp1*, *Elovl3*, and *Cidea*. Consequently, knockout of PER2 was found to cause increased BAT gene expression and oxidative capacity in WAT [40].

S112 is not exclusively targeted by the MAPK signaling pathway. Using the same xeroderma pigmentosum model mentioned earlier, Compe et al. [14] observed lower levels of PPARγ S112ph (together with decreased PPARα phosphorylation (see above)), which they attributed to the disruption of the CDK7 containing TFIIH complex. Indeed, they showed that CDK7 phosphorylates S112 in vitro. The authors also found reduced trans-activator function of PPARγ in their xeroderma pigmentosum system, and suggested a model where S112ph by CDK7 activates PPARγ function, in opposition to the repressive S112ph mediated by MAPK signaling. This result has been put into perspective by Helenius et al. [41], who found that MAT1, another THIIH complex member, and CDK7 itself, not only enhanced S112ph, but also inhibited adipocyte differentiation, which is in line with a generally repressive role of S112ph. Finally, another publication added the positive adipogenic regulator CDK9 to the list of S112ph kinases [42].

The physiological importance of S112ph has been highlighted by several lines of evidence: (1) In a (homozygous) S112A knock-in mouse model, Rangwala et al., found that the S112A mutation protects mice from obesity induced insulin resistance [43]; (2) A meta-analysis of Genome-wide association studies (GWAS) confirmed that the occurrence of the S112A allele is correlated with reduced type 2 diabetes risks [44]; and (3) subjects with the rare heterozygous variant P113Q, which renders the neighboring S112 nonphosphorylatable and increases its adipogenic potential [31], causes a range of metabolic symptoms ranging from obesity, type 2 diabetes, insulin resistance, and high fasting insulin levels [31,45]. This indicates that the phenotypic consequences are highly dependent on the genetic background, as well as the nutritional status. Additional studies will be necessary to untangle the complex relationship between genotype, PTM status, environmental cues, and disease risk.

In 2010, Choi et al. [46] revealed another phosphorylation event of PPARγ, S273ph, and since then this modification has attracted considerable interest. Serine 273 was found to be located within the consensus motif of CDK5, and readily get phosphorylated by the activated form of this kinase. Similar to S112, the loss of phosphorylation at S273 had activating effects on PPARγ, but the exact biological consequences were quite distinct: it did not increase the overall adipogenic activity of PPARγ, but upregulated a specific subset of target genes promoting insulin sensitivity. Mechanistically, this was caused by the loss of phosphorylation-dependent recruitment of the co-factor Thyroid hormone receptor associated protein 3 (THRAP3) [47]. Increased S273ph (which was induced by obesity) could be counteracted using PPARγ agonists, which led to improved metabolic profiles in HFD mice and patients with impaired glucose tolerance. Crucially, PPARγ binding compounds inhibiting S273ph with no or very low agonist activities elicited similar effects, without the side effects like weight gain, fluid retention, and bone loss, usually seen with PPARγ activation by full agonists such as TZDs [34,46,48]. Therefore, blocking S273ph seems to be an interesting avenue to treat metabolic disorders and a number of such compounds have been developed recently [47–50]. It will be intriguing to see their clinical potential in the future. In support of this notion, decreased S273ph was also detected in two genetic knockout models connected to an improved metabolic status in mice [51,52].

In a follow up paper to their work that identified CDK5 as a S273 kinase, Banks et al. generated adipocyte specific CDK5 knockout mice, and to their surprise found that S273ph levels were increased rather than decreased upon the ablation of this kinase [53]. This was explained by enhanced MEK/ERK (Extracellular signal–regulated kinase) signaling (caused by loss of CDK5), as ERK was subsequently identified as another potent S273ph kinase. In line with that notion, MEK inhibitor treatment produced beneficial metabolic effects [53]. In another publication, pharmacological inhibition of CDK5 via roscovitine evoked a somewhat different effect as genetic ablation, decreasing S273ph as well as S112ph, enhancing expression of BAT genes, increasing energy expenditure, and improving metabolic profiles [54]. This demonstrates that although the manipulation of signaling pathways connected to PPARγ phosphorylation is a highly promising approach to ameliorate metabolic disorders, more experimental work is needed to gain a comprehensive understanding of the underlying mechanisms.

Another important direction will be the identification and characterization of novel phosphorylation events in PPARγ. S112 and S273 are clearly not the only phosphorylated residues within PPARγ, as Banks et al. identified further phosphorylated sites (S133, T296) by Liquid chromatography-tandem mass spectrometry (LC-MS/MS) [53]. In addition, Choi et al. [55] recently described the phosphorylation of Y78, regulated by SRC proto-oncogene, nonreceptor tyrosine kinase (c-SRC), and Protein-tyrosine phosphatase 1B (PTP-1B), to be important for the regulation of genes involved in cytokine and chemokine expression. The regulatory mechanisms of S112ph and S273ph in PPARγ are illustrated in Figure 3A,B, respectively.

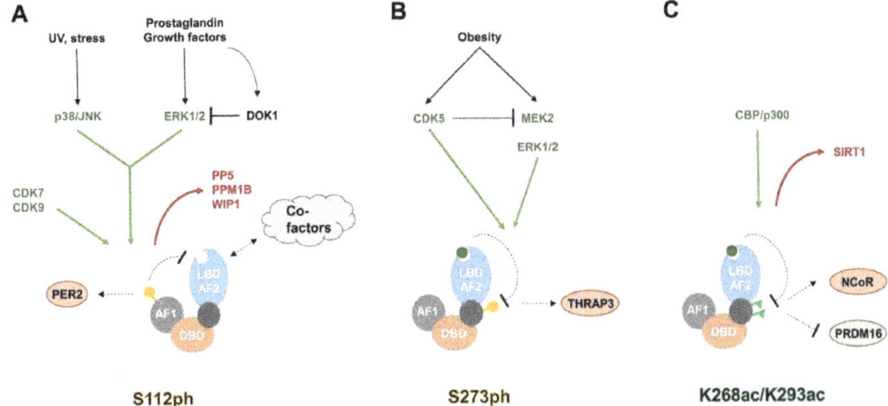

Figure 3. Regulatory mechanisms of selected modifications in PPARγ. (**A**) Activation of the MAPK pathway leads to the phosphorylation of serine 112 by p38/JNK or ERK1/2. S112ph decreases PPARγ activity, either through reducing its ligand binding affinity and co-activator binding, or by increasing Period circadian regulator 2 (PER2) binding, which leads to decreased recruitment to target genes. The adapter molecule Docking protein 1 (DOK1) modulates S112ph levels in response to nutritional inputs. S112ph is also targeted by CDK7 and CDK9. Phosphatases removing S112 phosphorylation from PPARγ are: Protein phosphatase 5 (PP5), Protein phosphatase Mg^{2+}/Mn^{2+} dependent 1B (PPM1B), and Wild-type p53-induced phosphatase 1 (WIP1). (**B**) Obesity-induced MAPK signaling leads to serine 273 phosphorylation, which enhances binding of the Thyroid hormone receptor associated protein 3 (THRAP3), and repression of certain PPARγ target genes. Phosphorylation levels are modulated by CDK5, either directly by CDK5-medatied S273 phosphorylation, or indirectly via phosphorylation of Dual specificity mitogen-activated protein kinase kinase 2 (MEK2) and suppression of MAPK signaling. Compounds with or without PPAR agonist activity can be used to block S273ph. (**C**) Acetylation of lysines 268 and 293 has been shown to increase NCoR co-repressor binding, whereas NAD (Nicotinamide adenine dinucleotide)-dependent deacetylase sirtuin-1 (SIRT1)-mediated deacetylation of K293 favours PR domain containing 16 (PRDM16) binding and expression of thermogenic genes. Ligand binding enhances SIRT1-PPARγ interaction and K268/K293 deacetylation. AF1: activation function 1 domain; DBD: DNA-binding domain; LBD-AF2: ligand binding and activation function 2 domain; enzymes depositing PTMs are colored in green, enzymes removing PTMs are shown in red; green circle: PPARγ-ligand; yellow circle: phosphorylated serine; green triangle: acetylated lysine; black arrow: activation; green arrow: PTM deposition; red arrow: PTM removal; black T symbol: inhibition; dotted arrow: increased interaction/stimulation; dotted T symbol: decreased interaction/inhibition.

3.2. SUMOylation

PPARγ SUMOylation with SUMO1 was first reported in 2004 [56–58]. The targeted lysine residue was identified as K107 on PPARγ2, located within a SUMOylation consensus motif (K77 in PPARγ1) [56–58]. Through analysis of cells expressing K107R mutant, it was found that the lack of PPARγ K107 SUMOylation correlated with transcriptional activation of PPARγ target genes [56–59], and enhanced adipogenesis [56]. These studies clearly defined K107 SUMOylation as a repressive mark for PPARγ, although the exact mechanism still remains to be elucidated. One proposed mechanism—the destabilization of PPARγ [58]—is most likely not the only important functional consequence of SUMO ligation. In support of this view, in the macrophage cell system, where PPARγ1 has a role in the repression of inflammatory genes, K77 SUMOylation was found to be important for the anti-inflammatory response triggered by apoptotic cells, possibly through stabilization of the co-repressor NCoR at target genes [60]. This is reminiscent of the effect of another SUMOylation event described earlier: also working in a macrophage cell system, Pascual et al. [61] showed that TZD-mediated SUMO1-modification of K365 (K395 in PPARγ2) is important for the repression of inflammatory response genes via PPARγ binding and stabilization of an NCoR-containing repressive complex. The precise biological roles of both modifications in the anti-inflammatory response, especially potential functional overlaps, remain to be determined.

A more recent publication reported that PPARγ can also be targeted by the SUMO2 modification and identified K33, K64, K68, and K77 (K63, K94, K98, and K107 in PPARγ2) as target sites, of which the first three sites are located within an inverted SUMOylation consensus motif. SUMOylation at either position was reported to be detrimental to PPARγ trans-activation [62].

The enzymatic machinery mediating PPARγ SUMOylation and de-SUMOylation has been identified earlier and consists of Ubiquitin conjugating enzyme 9 (UBC9, E2 ligase) [56,59,61], Protein inhibitor of activated STAT (PIAS1/PIASxβ, E3 ligase) [57,61,63,64], and SUMO-specific protease 2 (SENP2, protease) [65].

Interestingly, several reports have linked K107sumo to another PTM occurring in close proximity: S112ph. Initial reports showed that S112A, but not S112D phosphor-mimetic mutations, decreased PPARγ2 SUMOylation and transactivation function [56,59], supporting the model of a phospho-SUMOyl switch to regulate PPARγ function [66]. However, there might be additional mechanisms, allowing K107sumo regulation independent of S112ph. This notion is supported by two lines of evidence: (1) In Fibroblast growth factor 21 (FGF21) knockout mice, where PPARγ-dependent gene expression was reduced, increased K107sumo was not accompanied by elevated S112ph (and S273ph) [67]; and (2) Growth differentiation factor 11 (GDF11) treatment, which inhibits adipogenic differentiation and enhances osteoblastogenesis, increased PPARγ SUMOylation, again without concomitant changes of S112ph (and S273ph) [68].

While many details of the exact mechanisms and pathways governing SUMO-mediated PPARγ regulation remain open to future research, work from Mikkonen et al., has highlighted its physiological importance, as they showed that SUMO1 knockout mice exhibited a metabolic phenotype and decreased PPARγ target gene expression [69].

3.3. Acetylation

It was first noted in 2010 that PPARγ is a target for lysine acetylation [70], but only in 2012 another report gave a more detailed insight into its biological function [71]. Qiang and coworkers [71] identified five acetylated lysine residues at position K98, K107, K218, K268, and K293, of which two (K268ac and K293ac) could by blocked by administration of the TZD rosiglitazone, or by activation of the NAD (Nicotinamide adenine dinucleotide)-dependent deacetylase sirtuin-1 (SIRT1) deacetylase. It turned out that deacetylation of both residues, as seen in SIRT1 gain-of-function models, had beneficial metabolic effects, leading to browning of WAT and insulin sensitization. Mechanistically, this was achieved by modulation of co-factor recruitment. In detail, deacetylation of K293 favored the binding of the brown adipogenic activator PR domain containing 16 (PRDM16), whereas acetylation of K268 and K293 enhanced interaction with the co-repressor NCoR. Another mass spectrometric approach led to the identification of a total of nine putative acetylation sites on PPARγ1 (including the lysine residues corresponding to K218 and K268 on PPARγ2), of which K154 and K155 (K184 and K185 in PPARγ2) were further characterized [72]. K154/K155A and K154/K155Q mutants both showed severely diminished lipogenic potential compared to the WT protein. The regulatory mechanism of K268/K293ac in PPARγ is illustrated in Figure 3C.

3.4. Ubiquitination

Recently, two publications identified Seven in absentia homolog 2 (SIAH2) and Makorin RING finger protein 1 (MKRN1) as PPARγ E3 ligases, targeting PPARγ for proteasomal degradation [73,74]. MKRN1 activity was mainly directed towards K184 and K185. This work enhanced earlier work on PPARγ regulation through modulation of its stability (reviewed in [21]). A more unusual function for PPARγ ubiquitination was reported by two other publications: Watanabe et al. [75] and Li et al. [76]

showed that the E3 ligases Tripartite motif containing 23 (TRIM23) and Neural precursor cell expressed, developmentally downregulated 4 (NEDD4) confer atypical poly-ubiquitination to PPARγ (non-K48-mediated formation of poly-ubiquitin chains), which leads to reduced proteasomal degradation and stabilization of PPARγ.

3.5. *O-GlcNAcylation*

The addition of the single sugar modification β-*O*-linked *N*-acetylglucosamine (*O*-GlcNAc) to serine and threonine residues has been proposed to act as a nutrient sensor, linking signal transduction and gene expression to the metabolic status. Therefore it is interesting that PPARγ1 has been reported to get modified at T54 (corresponding to T84 of PPARγ2), leading to a decrease of its trans-activator function [77].

4. Post-Translational Modifications of PPARβ/δ

SUMOylation

PPARβ/δ is the least studied PPAR family member, and to our knowledge there is only one publication reporting a PTM in it: Koo et al. [78] show that PPARβ/δ SUMOylation at K104 is removed by SENP2, and (together with PPARγ, which is also targeted by SENP2, see above) this promotes the expression of FAO genes in muscle.

The PTMs in PPAR proteins and their corresponding modifying enzymes discussed above are summarized in Figure 4 and Table 1.

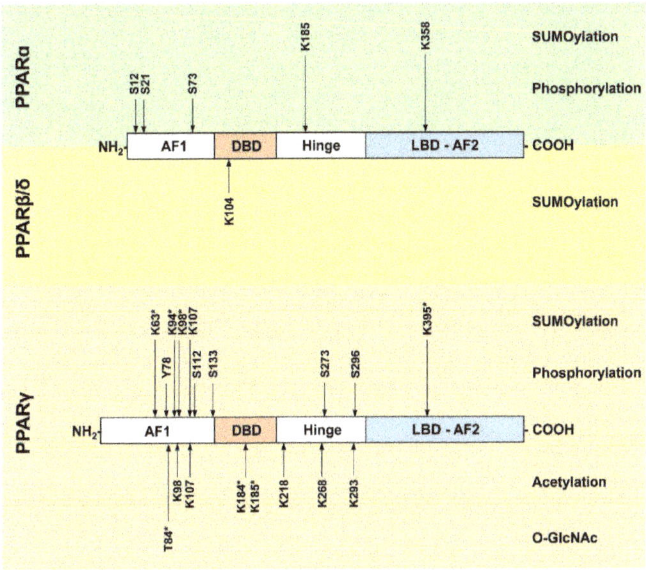

Figure 4. Post-translational modification sites in PPAR proteins. A schematic view of PPARα, PPARβ/δ, and PPARγ proteins and their functional domains is provided. The locations of PTM sites are indicated by arrows and the amino acid positions are given. Note that amino acids positions correspond to the murine proteins. For PPARγ, all amino acid positions refer to the PPARγ2 sequence; modifications which have so far only been described in PPARγ1 are highlighted with an asterisk. Ubiquitination events are not shown. AF1: activation function 1 domain; DBD: DNA-binding domain; Hinge domain; LBD-AF2: ligand binding and activation function 2 domain; K—lysine, S—serine, Y—tyrosine, T—threonine.

Table 1. Summary of PPAR modifying enzymes. Enzymes that deposit modifications are highlighted in green, while enzymes removing modifications are shown in red. For PPARγ, amino acid sequence positions refer to PPARγ2. If there is only experimental evidence for modification in PPARγ1, the corresponding amino acid position in PPARγ2 is given and highlighted with an asterisk. Question marks indicate undetermined target sites.

Modification	Enzyme	Target-Site	References
Phosphorylation	ERK1/2	PPARα S12, S21 PPARγ S112, S273, S133	[13,24,26,29,53,68]
	p38-α	PPARα S12, S21	[12]
	CDK7	PPARα S12, S21 PPARγ S112	[14,41]
	GSKβ	PPARα S73	[15]
	JNK	PPARγ S112	[26]
	CDK9	PPARγ S112	[42]
	CDK5	PPARγ S112, S273, S296	[46,53]
	MEK2	PPARγ S133	[53]
	c-SRC	PPARγ Y78	[55]
	PP5	PPARγ S112	[37]
	PPM1B	PPARγ S112	[38]
	WIP1	PPARγ S112	[39]
	PTB-1B	PPARγ Y78	[55]
Acetylation	CBP	PPARγ K98, K107, K218, K268, K293	[71]
	p300	PPARγ K?	[70]
	SIRT1	PPARγ K184/185 *, K268, K293	[70–72]
SUMOylation	PIAS1/PIASxβ	PPARα K358 PPARγ K107, K395 *	[20,57,61,63,64]
	PIASy	PPARα K185	[19]
	UBC9	PPARα K185 PPARγ K107, K395 *	[19,56,59,61]
	SENP2	PPARγ K107 PPARβ/δ K104	[65,78]
Ubiquitination	MKRN1	PPARγ K184/185	[74]
	SIAH2	PPARγ K?	[73]
	NEDD4	PPARγ K?	[76]
	TRIM23	PPARγ K?	[75]
	MDM2	PPARα K?	[22]
	MuRF	PPARα K?	[23]
O-GlcNAcylation	OGT	PPARγ T84 *	[77]

5. Outlook/Perspective

The last years have seen a wealth of information gathered on the role of PTMs on PPAR proteins. It is evident that PTMs are powerful modulators of PPAR function and we are getting an increasingly clearer picture of its complexity. They influence almost every aspect of PPAR biology, ranging from protein stability, localization, 3D structure, to ligand binding and co-factor interaction.

PTMs are the results of the action of signaling cascades, and therefore can be seen as representation of the physiological state of a cell. This is strikingly similar to the role of PPAR ligands, metabolites which are representing the metabolic status of a cell. PTMs and ligands can therefore be interpreted as two distinct, but related and partially overlapping, input signals and routes to modulate PPAR activity. It is not surprising that numerous instances of crosstalk between PTMs and agonist/antagonist action have been reported, but substantially more work is needed to dissect this complex network of relationships.

In the future, the use of high-throughput techniques will be instrumental to tackle questions related to the role of PTMs for target gene binding and genomic localization (via Chromatin immunoprecipitation-sequencing (ChIP-seq) using modification-specific antibodies), or the discovery of additional modifications (such as methylation) via mass-spectrometry based proteomic assays. The latter approach has already led to the identification of a fast growing number of new modification sites [53,71,72]. Due to increasing numbers of modifications, future studies will face the challenge that they will not only have to address their individual roles, but also take into consideration potential crosstalk between modifications. Introducing another layer of complexity is the fact that a growing number of amino acid residues has been shown to get targeted by more than one modification (e.g., PPARγ2 K98 and K107 can get SUMOylated as well as acetylated). This will make it necessary to revisit earlier results and critically re-evaluate some of the previous conclusions. Especially, assays based on the mutation of targeted residues might require careful reanalysis. Finally, it will be interesting to interrogate putative functional links between disease-risk connected Single-nucleotide polymorphisms (SNPs) and their potential effects on PTMs (as has been done for PPARγ2 S112ph and P113Q).

Importantly, some of those findings might lead to new approaches to tackle the prevalent epidemic of metabolic disorders. For example, the discovery that phosphorylation and acetylation events correlate with certain metabolic outcomes lends weight to the suggestions to specifically modulate responsible signaling pathways. A more directed approach, tackling not entire signaling pathways, but specifically blocking the modification of PPAR proteins itself via small molecules, seems to be an even more promising avenue that could decrease off-target/side effects. An example for the latter option is the use of small molecules to inhibit PPARγ2 S273ph [47–50]. It will be interesting to see if this approach can be successfully translated into the clinics and extended to other PTMs.

In summary, with an improving understanding of PPAR biology in general, and the role of PTMs specifically, PPARs remain promising targets for clinical interventions and will be in the focus of interest for years to come.

Acknowledgments: This work was supported by intramural funding from the Agency for Science, Technology and Research (A*STAR) of Singapore to Feng Xu.

Conflicts of Interest: The authors declare no conflict of interest.

Abbreviations

PPAR	Peroxisome Proliferator Activated Receptor
NR	Nuclear receptor
TF	Transcription factor
AF-1	Activation-function 1
DBD	DNA-binding domain
LBD	Ligand-binding domain
RXRα	Retinoic acid receptor α
PPRE	PPAR response element
NCoR	Nuclear receptor corepressor
SMART	Silencing mediator of retinoid and thyroid hormone receptor
CBP	CREB-binding protein
SRC1	Steroid receptor coactivator 1
T2D	Type 2 diabetes
NAFLD	Nonalcoholic fatty liver disease
NASH	Nonalcoholic steatosis
TZD	Thiazolidinedione
BAT	Brown adipose tissue
FAO	Fatty acid oxidation
WAT	White adipose tissue
PTM	Post-translational modification
MAPK	Mitogen-activated protein kinase
CDK	Cyclin-dependent kinase
PGC1α	Peroxisome proliferator-activated receptor gamma coactivator 1-alpha
XPD	Xeroderma pigmentosum group D
TFIIH	Transcription factor II H
GSKβ	Glycogen synthase kinase β
SUMO	Small ubiquitin-like modifier

GABP	GA-binding protein
MuRF1	Muscle-specific RING finger protein 1
EGF	Epidermal growth factor
PDGF	Platelet-derived growth factor
TGFβ	Transforming growth factor β
TPA	12-*O*-tetradecanoyl-13-phorbol acetate
PGF2α	Prostaglandin F2α
ERK	Extracellular signal–regulated kinase
MEK2	Dual specificity mitogen-activated protein kinase kinase 2
JNK	c-Jun *N*-terminal kinase
DOK1	Docking protein 1
HFD	High fat diet
PP5	Protein phosphatase 5
PPM1B	Protein phosphatase Mg^{2+}/Mn^{2+} dependent 1B
WIP1	Wild-type p53-induced phosphatase 1
PER2	Period circadian regulator 2
GWAS	Genome-wide association study
THRAP3	Thyroid hormone receptor associated protein 3
LC-MS/MS	Liquid chromatography-tandem mass spectrometry
c-SRC	SRC proto-oncogene, non-receptor tyrosine kinase
PTP-1B	Protein-tyrosine phosphatase 1B
UBC9	Ubiquitin conjugating enzyme 9
PIAS	Protein inhibitor of activated STAT
SENP2	SUMO-specific protease 2
FGF21	Fibroblast growth factor 21
GDF11	Growth differentiation factor 11
SIRT1	NAD-dependent deacetylase sirtuin-1
PRDM16	PR domain containing 16
SIAH2	Seven in absentia homolog 2
MKRN1	Makorin RING finger protein 1
TRIM23	Tripartite motif containing 23
NEDD4	Neural precursor cell expressed, developmentally down-regulated 4
O-GlcNAc	β-*O*-linked *N*-acetylglucosamine
ChIP-seq	Chromatin immunoprecipitation-sequencing
SNP	Single-nucleotide polymorphism

References

1. Monsalve, F.A.; Pyarasani, R.D.; Delgado-Lopez, F.; Moore-Carrasco, R. Peroxisome proliferator-activated receptor targets for the treatment of metabolic diseases. *Mediat. Inflamm.* **2013**, *2013*. [CrossRef] [PubMed]
2. Maccallini, C.; Mollica, A.; Amoroso, R. The positive regulation of enos signaling by PPAR agonists in cardiovascular diseases. *Am. J. Cardiovasc. Drugs* **2017**, *17*, 273–281. [CrossRef] [PubMed]
3. Tenenbaum, A.; Fisman, E.Z. Fibrates are an essential part of modern anti-dyslipidemic arsenal: Spotlight on atherogenic dyslipidemia and residual risk reduction. *Cardiovasc. Diabetol.* **2012**, *11*. [CrossRef] [PubMed]
4. Davidson, M.A.; Mattison, D.R.; Azoulay, L.; Krewski, D. Thiazolidinedione drugs in the treatment of type 2 diabetes mellitus: Past, present and future. *Crit. Rev. Toxicol.* **2018**, *48*, 52–108. [CrossRef] [PubMed]
5. Bortolini, M.; Wright, M.B.; Bopst, M.; Balas, B. Examining the safety of PPAR agonists—Current trends and future prospects. *Expert Opin. Drug Saf.* **2013**, *12*, 65–79. [CrossRef] [PubMed]
6. Dubois, V.; Eeckhoute, J.; Lefebvre, P.; Staels, B. Distinct but complementary contributions of PPAR isotypes to energy homeostasis. *J. Clin. Investig.* **2017**, *127*, 1202–1214. [CrossRef] [PubMed]
7. Tontonoz, P.; Hu, E.; Spiegelman, B.M. Stimulation of adipogenesis in fibroblasts by PPARγ 2, a lipid-activated transcription factor. *Cell* **1994**, *79*, 1147–1156. [CrossRef]
8. Gross, B.; Pawlak, M.; Lefebvre, P.; Staels, B. PPARs in obesity-induced T2dm, dyslipidaemia and NAFLD. *Nat. Rev. Endocrinol.* **2017**, *13*, 36–49. [CrossRef] [PubMed]

9. Tan, C.K.; Zhuang, Y.; Wahli, W. Synthetic and natural peroxisome proliferator-activated receptor (PPAR) agonists as candidates for the therapy of the metabolic syndrome. *Expert Opin. Ther. Targets* **2017**, *21*, 333–348. [CrossRef] [PubMed]
10. Shalev, A.; Siegrist-Kaiser, C.A.; Yen, P.M.; Wahli, W.; Burger, A.G.; Chin, W.W.; Meier, C.A. The peroxisome proliferator-activated receptor α is a phosphoprotein: Regulation by insulin. *Endocrinology* **1996**, *137*, 4499–4502. [CrossRef] [PubMed]
11. Passilly, P.; Schohn, H.; Jannin, B.; Cherkaoui Malki, M.; Boscoboinik, D.; Dauca, M.; Latruffe, N. Phosphorylation of peroxisome proliferator-activated receptor α in rat FAO cells and stimulation by ciprofibrate. *Biochem. Pharmacol.* **1999**, *58*, 1001–1008. [CrossRef]
12. Barger, P.M.; Browning, A.C.; Garner, A.N.; Kelly, D.P. P38 mitogen-activated protein kinase activates peroxisome proliferator-activated receptor α: A potential role in the cardiac metabolic stress response. *J. Biol. Chem.* **2001**, *276*, 44495–44501. [CrossRef] [PubMed]
13. Juge-Aubry, C.E.; Hammar, E.; Siegrist-Kaiser, C.; Pernin, A.; Takeshita, A.; Chin, W.W.; Burger, A.G.; Meier, C.A. Regulation of the transcriptional activity of the peroxisome proliferator-activated receptor α by phosphorylation of a ligand-independent trans-activating domain. *J. Biol. Chem.* **1999**, *274*, 10505–10510. [CrossRef] [PubMed]
14. Compe, E.; Drane, P.; Laurent, C.; Diderich, K.; Braun, C.; Hoeijmakers, J.H.; Egly, J.M. Dysregulation of the peroxisome proliferator-activated receptor target genes by xpd mutations. *Mol. Cell Biol.* **2005**, *25*, 6065–6076. [CrossRef] [PubMed]
15. Hinds, T.D., Jr.; Burns, K.A.; Hosick, P.A.; McBeth, L.; Nestor-Kalinoski, A.; Drummond, H.A.; AlAmodi, A.A.; Hankins, M.W.; Vanden Heuvel, J.P.; Stec, D.E. Biliverdin reductase a attenuates hepatic steatosis by inhibition of glycogen synthase kinase (GSK) 3β phosphorylation of serine 73 of peroxisome proliferator-activated receptor (PPAR) α. *J. Biol. Chem.* **2016**, *291*, 25179–25191. [CrossRef] [PubMed]
16. Hinds, T.D., Jr.; Hosick, P.A.; Chen, S.; Tukey, R.H.; Hankins, M.W.; Nestor-Kalinoski, A.; Stec, D.E. Mice with hyperbilirubinemia due to gilbert's syndrome polymorphism are resistant to hepatic steatosis by decreased serine 73 phosphorylation of PPARα. *Am. J. Physiol. Endocrinol. Metab.* **2017**, *312*, E244–E252. [CrossRef] [PubMed]
17. Molzer, C.; Wallner, M.; Kern, C.; Tosevska, A.; Schwarz, U.; Zadnikar, R.; Doberer, D.; Marculescu, R.; Wagner, K.H. Features of an altered AMPK metabolic pathway in gilbert's syndrome, and its role in metabolic health. *Sci. Rep.* **2016**, *6*. [CrossRef] [PubMed]
18. Flotho, A.; Melchior, F. SUMOylation: A regulatory protein modification in health and disease. *Annu. Rev. Biochem.* **2013**, *82*, 357–385. [CrossRef] [PubMed]
19. Pourcet, B.; Pineda-Torra, I.; Derudas, B.; Staels, B.; Glineur, C. SUMOylation of human peroxisome proliferator-activated receptor α inhibits its trans-activity through the recruitment of the nuclear corepressor ncor. *J. Biol. Chem.* **2010**, *285*, 5983–5992. [CrossRef] [PubMed]
20. Leuenberger, N.; Pradervand, S.; Wahli, W. Sumoylated PPARα mediates sex-specific gene repression and protects the liver from estrogen-induced toxicity in mice. *J. Clin. Investig.* **2009**, *119*, 3138–3148. [CrossRef] [PubMed]
21. Wadosky, K.M.; Willis, M.S. The story so far: Post-translational regulation of peroxisome proliferator-activated receptors by ubiquitination and SUMOylation. *Am. J. Physiol. Heart Circ. Physiol.* **2012**, *302*, H515–H526. [CrossRef] [PubMed]
22. Gopinathan, L.; Hannon, D.B.; Peters, J.M.; Vanden Heuvel, J.P. Regulation of peroxisome proliferator-activated receptor-α by MDM2. *Toxicol. Sci.* **2009**, *108*, 48–58. [CrossRef] [PubMed]
23. Rodriguez, J.E.; Liao, J.Y.; He, J.; Schisler, J.C.; Newgard, C.B.; Drujan, D.; Glass, D.J.; Frederick, C.B.; Yoder, B.C.; Lalush, D.S.; et al. The ubiquitin ligase MURF1 regulates PPARα activity in the heart by enhancing nuclear export via monoubiquitination. *Mol. Cell Endocrinol.* **2015**, *413*, 36–48. [CrossRef] [PubMed]
24. Hu, E.; Kim, J.B.; Sarraf, P.; Spiegelman, B.M. Inhibition of adipogenesis through MAP kinase-mediated phosphorylation of PPARγ. *Science* **1996**, *274*, 2100–2103. [CrossRef] [PubMed]
25. Zhang, B.; Berger, J.; Zhou, G.; Elbrecht, A.; Biswas, S.; White-Carrington, S.; Szalkowski, D.; Moller, D.E. Insulin- and mitogen-activated protein kinase-mediated phosphorylation and activation of peroxisome proliferator-activated receptor γ. *J. Biol. Chem.* **1996**, *271*, 31771–31774. [CrossRef] [PubMed]

26. Adams, M.; Reginato, M.J.; Shao, D.; Lazar, M.A.; Chatterjee, V.K. Transcriptional activation by peroxisome proliferator-activated receptor γ is inhibited by phosphorylation at a consensus mitogen-activated protein kinase site. *J. Biol. Chem.* **1997**, *272*, 5128–5132. [CrossRef] [PubMed]
27. Reginato, M.J.; Krakow, S.L.; Bailey, S.T.; Lazar, M.A. Prostaglandins promote and block adipogenesis through opposing effects on peroxisome proliferator-activated receptor γ. *J. Biol. Chem.* **1998**, *273*, 1855–1858. [CrossRef] [PubMed]
28. Camp, H.S.; Tafuri, S.R. Regulation of peroxisome proliferator-activated receptor γ activity by mitogen-activated protein kinase. *J. Biol. Chem.* **1997**, *272*, 10811–10816. [CrossRef] [PubMed]
29. Camp, H.S.; Tafuri, S.R.; Leff, T. c-Jun N-terminal kinase phosphorylates peroxisome proliferator-activated receptor-γ1 and negatively regulates its transcriptional activity. *Endocrinology* **1999**, *140*, 392–397. [CrossRef] [PubMed]
30. Shao, D.; Rangwala, S.M.; Bailey, S.T.; Krakow, S.L.; Reginato, M.J.; Lazar, M.A. Interdomain communication regulating ligand binding by PPAR-γ. *Nature* **1998**, *396*, 377–380. [CrossRef] [PubMed]
31. Ristow, M.; Muller-Wieland, D.; Pfeiffer, A.; Krone, W.; Kahn, C.R. Obesity associated with a mutation in a genetic regulator of adipocyte differentiation. *N. Engl. J. Med.* **1998**, *339*, 953–959. [CrossRef] [PubMed]
32. Iwata, M.; Haruta, T.; Usui, I.; Takata, Y.; Takano, A.; Uno, T.; Kawahara, J.; Ueno, E.; Sasaoka, T.; Ishibashi, O.; et al. Pioglitazone ameliorates tumor necrosis factor-α-induced insulin resistance by a mechanism independent of adipogenic activity of peroxisome proliferator—Activated receptor-γ. *Diabetes* **2001**, *50*, 1083–1092. [CrossRef] [PubMed]
33. Werman, A.; Hollenberg, A.; Solanes, G.; Bjorbaek, C.; Vidal-Puig, A.J.; Flier, J.S. Ligand-independent activation domain in the n terminus of peroxisome proliferator-activated receptor γ (PPARγ). Differential activity of PPARγ1 and -2 isoforms and influence of insulin. *J. Biol. Chem.* **1997**, *272*, 20230–20235. [CrossRef] [PubMed]
34. Stechschulte, L.A.; Czernik, P.J.; Rotter, Z.C.; Tausif, F.N.; Corzo, C.A.; Marciano, D.P.; Asteian, A.; Zheng, J.; Bruning, J.B.; Kamenecka, T.M.; et al. PPARg post-translational modifications regulate bone formation and bone resorption. *EBioMedicine* **2016**, *10*, 174–184. [CrossRef] [PubMed]
35. Ge, C.; Cawthorn, W.P.; Li, Y.; Zhao, G.; Macdougald, O.A.; Franceschi, R.T. Reciprocal control of osteogenic and adipogenic differentiation by ERK/MAP kinase phosphorylation of RUNX2 and PPARγ transcription factors. *J. Cell Physiol.* **2016**, *231*, 587–596. [CrossRef] [PubMed]
36. Hosooka, T.; Noguchi, T.; Kotani, K.; Nakamura, T.; Sakaue, H.; Inoue, H.; Ogawa, W.; Tobimatsu, K.; Takazawa, K.; Sakai, M.; et al. DOK1 mediates high-fat diet-induced adipocyte hypertrophy and obesity through modulation of PPAR-γ phosphorylation. *Nat. Med.* **2008**, *14*, 188–193. [CrossRef] [PubMed]
37. Hinds, T.D., Jr.; Stechschulte, L.A.; Cash, H.A.; Whisler, D.; Banerjee, A.; Yong, W.; Khuder, S.S.; Kaw, M.K.; Shou, W.; Najjar, S.M.; et al. Protein phosphatase 5 mediates lipid metabolism through reciprocal control of glucocorticoid receptor and peroxisome proliferator-activated receptor-γ (PPARγ). *J. Biol. Chem.* **2011**, *286*, 42911–42922. [CrossRef] [PubMed]
38. Tasdelen, I.; van Beekum, O.; Gorbenko, O.; Fleskens, V.; van den Broek, N.J.; Koppen, A.; Hamers, N.; Berger, R.; Coffer, P.J.; Brenkman, A.B.; et al. The serine/threonine phosphatase PPM1B (PP2Cβ) selectively modulates PPARγ activity. *Biochem. J.* **2013**, *451*, 45–53. [CrossRef] [PubMed]
39. Li, D.; Zhang, L.; Xu, L.; Liu, L.; He, Y.; Zhang, Y.; Huang, X.; Zhao, T.; Wu, L.; Zhao, Y.; et al. WIP1 phosphatase is a critical regulator of adipogenesis through dephosphorylating PPARγ serine 112. *Cell Mol. Life Sci.* **2017**, *74*, 2067–2079. [CrossRef] [PubMed]
40. Grimaldi, B.; Bellet, M.M.; Katada, S.; Astarita, G.; Hirayama, J.; Amin, R.H.; Granneman, J.G.; Piomelli, D.; Leff, T.; Sassone-Corsi, P. PER2 controls lipid metabolism by direct regulation of PPARγ. *Cell Metab.* **2010**, *12*, 509–520. [CrossRef] [PubMed]
41. Helenius, K.; Yang, Y.; Alasaari, J.; Makela, T.P. MAT1 inhibits peroxisome proliferator-activated receptor γ-mediated adipocyte differentiation. *Mol. Cell Biol.* **2009**, *29*, 315–323. [CrossRef] [PubMed]
42. Iankova, I.; Petersen, R.K.; Annicotte, J.S.; Chavey, C.; Hansen, J.B.; Kratchmarova, I.; Sarruf, D.; Benkirane, M.; Kristiansen, K.; Fajas, L. Peroxisome proliferator-activated receptor γ recruits the positive transcription elongation factor b complex to activate transcription and promote adipogenesis. *Mol. Endocrinol.* **2006**, *20*, 1494–1505. [CrossRef] [PubMed]

43. Rangwala, S.M.; Rhoades, B.; Shapiro, J.S.; Rich, A.S.; Kim, J.K.; Shulman, G.I.; Kaestner, K.H.; Lazar, M.A. Genetic modulation of PPARγ phosphorylation regulates insulin sensitivity. *Dev. Cell* **2003**, *5*, 657–663. [CrossRef]
44. Gouda, H.N.; Sagoo, G.S.; Harding, A.H.; Yates, J.; Sandhu, M.S.; Higgins, J.P. The association between the peroxisome proliferator-activated receptor-γ2 (PPARγ2) *Pro12Ala* gene variant and type 2 diabetes mellitus: A huge review and meta-analysis. *Am. J. Epidemiol.* **2010**, *171*, 645–655. [CrossRef] [PubMed]
45. Bluher, M.; Paschke, R. Analysis of the relationship between PPAR-γ 2 gene variants and severe insulin resistance in obese patients with impaired glucose tolerance. *Exp. Clin. Endocrinol. Diabetes* **2003**, *111*, 85–90. [CrossRef] [PubMed]
46. Choi, J.H.; Banks, A.S.; Estall, J.L.; Kajimura, S.; Bostrom, P.; Laznik, D.; Ruas, J.L.; Chalmers, M.J.; Kamenecka, T.M.; Bluher, M.; et al. Anti-diabetic drugs inhibit obesity-linked phosphorylation of PPARγ by CDK5. *Nature* **2010**, *466*, 451–456. [CrossRef] [PubMed]
47. Choi, J.H.; Choi, S.S.; Kim, E.S.; Jedrychowski, M.P.; Yang, Y.R.; Jang, H.J.; Suh, P.G.; Banks, A.S.; Gygi, S.P.; Spiegelman, B.M. THRAP3 docks on phosphoserine 273 of PPARγ and controls diabetic gene programming. *Genes Dev.* **2014**, *28*, 2361–2369. [CrossRef] [PubMed]
48. Choi, J.H.; Banks, A.S.; Kamenecka, T.M.; Busby, S.A.; Chalmers, M.J.; Kumar, N.; Kuruvilla, D.S.; Shin, Y.; He, Y.; Bruning, J.B.; et al. Antidiabetic actions of a non-agonist PPARγ ligand blocking CDK5-mediated phosphorylation. *Nature* **2011**, *477*, 477–481. [CrossRef] [PubMed]
49. Amato, A.A.; Rajagopalan, S.; Lin, J.Z.; Carvalho, B.M.; Figueira, A.C.; Lu, J.; Ayers, S.D.; Mottin, M.; Silveira, R.L.; Souza, P.C.; et al. Gq-16, a novel peroxisome proliferator-activated receptor γ (PPARγ) ligand, promotes insulin sensitization without weight gain. *J. Biol. Chem.* **2012**, *287*, 28169–28179. [CrossRef] [PubMed]
50. Zheng, W.; Qiu, L.; Wang, R.; Feng, X.; Han, Y.; Zhu, Y.; Chen, D.; Liu, Y.; Jin, L.; Li, Y. Selective targeting of PPARγ by the natural product chelerythrine with a unique binding mode and improved antidiabetic potency. *Sci. Rep.* **2015**, *5*. [CrossRef] [PubMed]
51. Li, P.; Fan, W.; Xu, J.; Lu, M.; Yamamoto, H.; Auwerx, J.; Sears, D.D.; Talukdar, S.; Oh, D.; Chen, A.; et al. Adipocyte NCOR knockout decreases PPARγ phosphorylation and enhances PPARγ activity and insulin sensitivity. *Cell* **2011**, *147*, 815–826. [CrossRef] [PubMed]
52. Mayoral, R.; Osborn, O.; McNelis, J.; Johnson, A.M.; Oh, D.Y.; Izquierdo, C.L.; Chung, H.; Li, P.; Traves, P.G.; Bandyopadhyay, G.; et al. Adipocyte SIRT1 knockout promotes PPARγ activity, adipogenesis and insulin sensitivity in chronic-HFD and obesity. *Mol. Metab.* **2015**, *4*, 378–391. [CrossRef] [PubMed]
53. Banks, A.S.; McAllister, F.E.; Camporez, J.P.; Zushin, P.J.; Jurczak, M.J.; Laznik-Bogoslavski, D.; Shulman, G.I.; Gygi, S.P.; Spiegelman, B.M. An ERK/CDK5 axis controls the diabetogenic actions of PPARγ. *Nature* **2015**, *517*, 391–395. [CrossRef] [PubMed]
54. Wang, H.; Liu, L.; Lin, J.Z.; Aprahamian, T.R.; Farmer, S.R. Browning of white adipose tissue with roscovitine induces a distinct population of $UCP1^+$ adipocytes. *Cell Metab.* **2016**, *24*, 835–847. [CrossRef] [PubMed]
55. Choi, S.; Jung, J.E.; Yang, Y.R.; Kim, E.S.; Jang, H.J.; Kim, E.K.; Kim, I.S.; Lee, J.Y.; Kim, J.K.; Seo, J.K.; et al. Novel phosphorylation of PPARγ ameliorates obesity-induced adipose tissue inflammation and improves insulin sensitivity. *Cell Signal* **2015**, *27*, 2488–2495. [CrossRef] [PubMed]
56. Yamashita, D.; Yamaguchi, T.; Shimizu, M.; Nakata, N.; Hirose, F.; Osumi, T. The transactivating function of peroxisome proliferator-activated receptor γ is negatively regulated by SUMO conjugation in the amino-terminal domain. *Genes Cells* **2004**, *9*, 1017–1029. [CrossRef] [PubMed]
57. Ohshima, T.; Koga, H.; Shimotohno, K. Transcriptional activity of peroxisome proliferator-activated receptor γ is modulated by SUMO-1 modification. *J. Biol. Chem.* **2004**, *279*, 29551–29557. [CrossRef] [PubMed]
58. Floyd, Z.E.; Stephens, J.M. Control of peroxisome proliferator-activated receptor γ2 stability and activity by SUMOylation. *Obes. Res.* **2004**, *12*, 921–928. [CrossRef] [PubMed]
59. Shimizu, M.; Yamashita, D.; Yamaguchi, T.; Hirose, F.; Osumi, T. Aspects of the regulatory mechanisms of PPAR functions: Analysis of a bidirectional response element and regulation by SUMOylation. *Mol. Cell Biochem.* **2006**, *286*, 33–42. [CrossRef] [PubMed]
60. Jennewein, C.; Kuhn, A.M.; Schmidt, M.V.; Meilladec-Jullig, V.; von Knethen, A.; Gonzalez, F.J.; Brune, B. SUMOylation of peroxisome proliferator-activated receptor γ by apoptotic cells prevents lipopolysaccharide-induced ncor removal from kappab binding sites mediating transrepression of proinflammatory cytokines. *J. Immunol.* **2008**, *181*, 5646–5652. [CrossRef] [PubMed]

61. Pascual, G.; Fong, A.L.; Ogawa, S.; Gamliel, A.; Li, A.C.; Perissi, V.; Rose, D.W.; Willson, T.M.; Rosenfeld, M.G.; Glass, C.K. A SUMOylation-dependent pathway mediates transrepression of inflammatory response genes by PPAR-γ. *Nature* **2005**, *437*, 759–763. [CrossRef] [PubMed]
62. Diezko, R.; Suske, G. Ligand binding reduces SUMOylation of the peroxisome proliferator-activated receptor γ (PPARγ) activation function 1 (AF1) domain. *PLoS ONE* **2013**, *8*. [CrossRef] [PubMed]
63. Ghisletti, S.; Huang, W.; Ogawa, S.; Pascual, G.; Lin, M.E.; Willson, T.M.; Rosenfeld, M.G.; Glass, C.K. Parallel SUMOylation-dependent pathways mediate gene- and signal-specific transrepression by lxrs and PPARγ. *Mol. Cell* **2007**, *25*, 57–70. [CrossRef] [PubMed]
64. Lu, Y.; Zhou, Q.; Shi, Y.; Liu, J.; Zhong, F.; Hao, X.; Li, C.; Chen, N.; Wang, W. SUMOylation of PPARγ by rosiglitazone prevents lps-induced ncor degradation mediating down regulation of chemokines expression in renal proximal tubular cells. *PLoS ONE* **2013**, *8*. [CrossRef] [PubMed]
65. Chung, S.S.; Ahn, B.Y.; Kim, M.; Kho, J.H.; Jung, H.S.; Park, K.S. Sumo modification selectively regulates transcriptional activity of peroxisome-proliferator-activated receptor γ in c2c12 myotubes. *Biochem. J.* **2011**, *433*, 155–161. [CrossRef] [PubMed]
66. Yang, X.J.; Gregoire, S. A recurrent phospho-Sumoyl switch in transcriptional repression and beyond. *Mol. Cell* **2006**, *23*, 779–786. [CrossRef] [PubMed]
67. Dutchak, P.A.; Katafuchi, T.; Bookout, A.L.; Choi, J.H.; Yu, R.T.; Mangelsdorf, D.J.; Kliewer, S.A. Fibroblast growth factor-21 regulates PPARγ activity and the antidibetic actions of thiazolidinediones. *Cell* **2012**, *148*, 556–567. [CrossRef] [PubMed]
68. Zhang, Y.; Shao, J.; Wang, Z.; Yang, T.; Liu, S.; Liu, Y.; Fan, X.; Ye, W. Growth differentiation factor 11 is a protective factor for osteoblastogenesis by targeting PPARγ. *Gene* **2015**, *557*, 209–214. [CrossRef] [PubMed]
69. Mikkonen, L.; Hirvonen, J.; Janne, O.A. Sumo-1 regulates body weight and adipogenesis via PPARγ in male and female mice. *Endocrinology* **2013**, *154*, 698–708. [CrossRef] [PubMed]
70. Han, L.; Zhou, R.; Niu, J.; McNutt, M.A.; Wang, P.; Tong, T. Sirt1 is regulated by a PPARγ-SIRT1 negative feedback loop associated with senescence. *Nucleic Acids Res.* **2010**, *38*, 7458–7471. [CrossRef] [PubMed]
71. Qiang, L.; Wang, L.; Kon, N.; Zhao, W.; Lee, S.; Zhang, Y.; Rosenbaum, M.; Zhao, Y.; Gu, W.; Farmer, S.R.; et al. Brown remodeling of white adipose tissue by SIRT1-dependent deacetylation of PPARγ. *Cell* **2012**, *150*, 620–632. [CrossRef] [PubMed]
72. Tian, L.; Wang, C.; Hagen, F.K.; Gormley, M.; Addya, S.; Soccio, R.; Casimiro, M.C.; Zhou, J.; Powell, M.J.; Xu, P.; et al. Acetylation-defective mutant of PPARγ is associated with decreased lipid synthesis in breast cancer cells. *Oncotarget* **2014**, *5*, 7303–7315. [CrossRef] [PubMed]
73. Kilroy, G.; Kirk-Ballard, H.; Carter, L.E.; Floyd, Z.E. The ubiquitin ligase SIAH2 regulates PPARγ activity in adipocytes. *Endocrinology* **2012**, *153*, 1206–1218. [CrossRef] [PubMed]
74. Kim, J.H.; Park, K.W.; Lee, E.W.; Jang, W.S.; Seo, J.; Shin, S.; Hwang, K.A.; Song, J. Suppression of PPARγ through MKRN1-mediated ubiquitination and degradation prevents adipocyte differentiation. *Cell Death Differ.* **2014**, *21*, 594–603. [CrossRef] [PubMed]
75. Watanabe, M.; Takahashi, H.; Saeki, Y.; Ozaki, T.; Itoh, S.; Suzuki, M.; Mizushima, W.; Tanaka, K.; Hatakeyama, S. The E3 ubiquitin ligase TRIM23 regulates adipocyte differentiation via stabilization of the adipogenic activator PPARγ. *ELife* **2015**, *4*. [CrossRef] [PubMed]
76. Li, J.J.; Wang, R.; Lama, R.; Wang, X.; Floyd, Z.E.; Park, E.A.; Liao, F.F. Ubiquitin ligase NEDD4 regulates PPARγ stability and adipocyte differentiation in 3t3-l1 cells. *Sci. Rep.* **2016**, *6*. [CrossRef] [PubMed]
77. Ji, S.; Park, S.Y.; Roth, J.; Kim, H.S.; Cho, J.W. *O*-Glcnac modification of PPARγ reduces its transcriptional activity. *Biochem. Biophys. Res. Commun.* **2012**, *417*, 1158–1163. [CrossRef] [PubMed]
78. Koo, Y.D.; Choi, J.W.; Kim, M.; Chae, S.; Ahn, B.Y.; Oh, B.C.; Hwang, D.; Seol, J.H.; Kim, Y.B.; Park, Y.J.; et al. Sumo-specific protease 2 (SENP2) is an important regulator of fatty acid metabolism in skeletal muscle. *Diabetes* **2015**, *64*, 2420–2431. [CrossRef] [PubMed]

International Journal of
Molecular Sciences

Review

The Role of Peroxisome Proliferator-Activated Receptors and Their Transcriptional Coactivators Gene Variations in Human Trainability: A Systematic Review

Miroslav Petr [1], Petr Stastny [1,*], Adam Zajac [2], James J. Tufano [1] and Agnieszka Maciejewska-Skrendo [3]

1. Faculty of Physical Education and Sport, Charles University, 162 52 Prague, Czech Republic; petr@ftvs.cuni.cz (M.P.); tufano@ftvs.cuni.cz (J.J.T.)
2. Department of Theory and Practice of Sport, The Jerzy Kukuczka Academy of Physical Education in Katowice, 40-065 Katowice, Poland; a.zajac@awf.katowice.pl
3. Faulty of Physical Education, Gdansk University of Physical Education and Sport, 80-336 Gdansk, Poland; maciejewska.us@wp.pl

* Correspondence: stastny@ftvs.cuni.cz; Tel.: +420-777-198-764

Received: 7 April 2018; Accepted: 12 May 2018; Published: 15 May 2018

Abstract: Background: The peroxisome proliferator-activated receptors (*PPARA, PPARG, PPARD*) and their transcriptional coactivators' (*PPARGC1A, PPARGC1B*) gene polymorphisms have been associated with muscle morphology, oxygen uptake, power output and endurance performance. The purpose of this review is to determine whether the PPARs and/or their coactivators' polymorphisms can predict the training response to specific training stimuli. Methods: In accordance with the Preferred Reporting Items for Systematic Reviews and Meta Analyses, a literature review has been run for a combination of PPARs and physical activity key words. Results: All ten of the included studies were performed using aerobic training in general, sedentary or elderly populations from 21 to 75 years of age. The non-responders for aerobic training (VO_2peak increase, slow muscle fiber increase and low-density lipoprotein decrease) are the carriers of *PPARGC1A* rs8192678 Ser/Ser. The negative responders for aerobic training (decrease in VO_2peak) are carriers of the *PPARD* rs2267668 G allele. The negative responders for aerobic training (decreased glucose tolerance and insulin response) are subjects with the *PPARG* rs1801282 Pro/Pro genotype. The best responders to aerobic training are *PPARGC1A* rs8192678 Gly/Gly, *PPARD* rs1053049 TT, *PPARD* rs2267668 AA and *PPARG* rs1801282 Ala carriers. Conclusions: The human response for aerobic training is significantly influenced by PPARs' gene polymorphism and their coactivators, where aerobic training can negatively influence glucose metabolism and VO_2peak in some genetically-predisposed individuals.

Keywords: human performance; aerobic training; genetic predisposition; anaerobic threshold; muscle fibers; glucose tolerance; insulin response; VO_2max; VO_2peak; mitochondria activity; cholesterol levels

---

1. Introduction

Although sport scientists strive to conceive of experiments that investigate the effects of specific diets or training strategies, their findings may not agree across general populations and specific groups of athletes. Even when an experiment is meticulously designed using homogeneous samples, the outcomes of a study can largely differ between individuals within a homogeneous group. As these inter-individual responses are often masked when reporting the mean values of dependent variables, some researchers have suggested that sport science should refrain from reporting mean data and should focus on individual data. Although this solution may seem logical, internal independent factors should also be considered when analyzing and discussing inter-individual differences in response to interventions.

As each individual might respond differently to the same external stimuli, the effectiveness of an intervention can likely be somewhat explained by an athlete's genetic make-up. For example, some individuals may exhibit minimal changes in a specific dependent variable, but others who undergo the exact same intervention may experience a massive improvement for the same variable. Using the famous HERITAGE study as an example, researchers demonstrated significant individual responses in VO_2max following 20 weeks of aerobic training [1] and, under same conditions, even negative metabolic responses manifested with common blood markers (systolic blood pressure, insulin, triacylglycerol, HDL cholesterol) in 8.4–13.3% of negative responders [2].

Peroxisome proliferator-activated receptor (PPAR) proteins belong to the steroid hormone receptor superfamily and combine with the retinoid X receptors to form heterodimers that regulate genes involved in lipid and glucose metabolism, adipocyte differentiation, fatty acid transport, carcinogenesis and inflammation [3,4]. PPARs exist in three different forms as PPAR-alpha (PPARα), PPAR-beta/delta (PPARβ/δ) and PPAR-gamma (PPARγ), which are encoded by the genes *PPARA*, *PPARD* and *PPARG*. PPARα and PPARβ/δ are present mainly in the liver and in tissues with high levels of fatty acid oxidation such as skeletal muscle, cardiac muscle and the kidneys. PPARγ are predominantly active in adipocytes affecting their differentiation and growth, and they are also an interesting target for pharmacotherapy of diabetes mellitus type 2 (DM2).

Peroxisome proliferator-activated receptor gamma, coactivator 1 alpha (PGC1α), encoded by the *PPARGC1A* gene, is a transcriptional coactivator of the PPAR superfamily. This protein interacts with PPARγ, which enables its interaction with many others transcriptional factors. PGC1α is involved in mitochondrial biogenesis, glucose utilization, fatty acid oxidation, thermogenesis, gluconeogenesis and insulin signaling [5]. Peroxisome proliferator-activated receptor gamma, coactivator 1 beta (PGC1β), encoded by the *PPARGC1B* gene, together with the *PPARGC1A* gene, encodes homologous proteins that, through nuclear transcription factor coactivation, regulate adipogenesis, insulin signaling, lipolysis, mitochondrial biogenesis, angiogenesis and hepatic gluconeogenesis [5].

Human performance is a multifactorial domain where genetic predisposition may act as a key intrinsic factor. Of the many genes that have been studied in relation to sport performance and exercise, *PPARA*, *PPARG*, *PPARD* and their transcriptional coactivators' *PPARGC1A* and *PPARGC1B* gene polymorphisms have been associated with elite athletic performance, which has been related to muscle morphology [6], oxygen uptake [7,8], power output [9] and endurance performance [6]. Therefore, identifying links between PPARs (and their coactivators) and human performance may shed light on the possibility of identifying athletes with specific genetic sporting potential, possibly also leading to genetically-specialized training methods for athletes who carry specific PPAR gene variants.

From various sets of single nucleotide polymorphisms (SNPs), the PPAR signaling pathway has been reported as one the most related to human cellular bioenergetics and VO_2max trainability in the genome-wide association study (GWAS) [10]. Previous reviews found that PPARs and/or their coactivators' genes were associated with endurance performance [11–13] or improvements in weight reduction following training programs [14]; however, their relation to training responses has not been determined yet.

Since there many studies showing the association between PPARs and/or their coactivators' gene polymorphisms and human performance, it is beneficial to also know the relationship between PPARs and different training responses. Therefore, the purpose of this review is to summarize whether the PPARs and/or their coactivators' polymorphisms can predict human response to specific training stimuli. We hypothesized that PPARs and/or their coactivators' gene polymorphisms can predict the response to aerobic and anaerobic training, that the PPARs and/or their coactivators' polymorphisms can predict the amount of appropriate training load and eventually can determine the responders to specific training methods (e.g., hypoxia-training).

2. Results

The literature search resulted in a total of 7389 articles, which was reduced to 4262 after removing duplicates. The number of eligible articles was further reduced to 64 after screening article titles and abstracts to include PPARs and/or their coactivators' gene polymorphisms in relation to physical activity (Figure 1). Of these studies, 53 were rejected following full-text screening, and one was rejected based on the methodological quality criteria. Finally, 10 studies (Table 1) were included in the analysis.

None of the 10 included studies were performed on elite athletes, using resistance training, using maximum training load or any other specific training method. One study [15] focusing on elite athlete response for resistance training and regarding the amount of training load had to be rejected due to a lack of a methodological approach, specifically a lack of reproducibility. Because this study showed a significant role of *ACE*, *ACTN3* and *PPARGC1A* genes with the volumes of specific training loads within the training process macrostructure of elite weightlifters [15], we suggest that PPARs' role in resistance training and elite sport should be studied in further research. On the other hand, all of the included studies were performed using aerobic training in general, sedentary or elderly populations from 21 to 75 years of age (Table 1), and one study did not find any association with *PPARGC1A* and trainability [16]. Therefore, our hypotheses have been confirmed only in the case of training responses to aerobic training in non-athletic populations. One study using a GWAS design [10] has been rejected by Exclusion Criterion 4.

The response to aerobic training is partly described in *PPARGC1A*, *PPARG* and *PPARD* in various aerobic training approaches referring to the improvement of training performance and the response of glucose metabolism and insulin sensitivity (Table 2). In the range of the population included in the study, we can state that PPARs and/or their coactivators' gene polymorphisms may be able to predict the human response to aerobic training at moderate intensities up to the lactate threshold. Specifically, the PPARs and their coactivators' gene polymorphisms can predict high response, no response or even negative response for aerobic training estimated by glucose tolerance, insulin response, body fat, VO_2peak, anaerobic threshold, mitochondria activity, cholesterol levels and slow muscle fibers' increase (Table 2).

The *PPARGC1A* rs8192678 Gly/Gly genotype has been associated with greater increases of an individual's anaerobic threshold [17], a greater increase of slow muscle fibers [18], greater mitochondria activity [18], a greater decrease of low-density and total lipoprotein cholesterol [19] and a greater VO_2peak increase after aerobic training than *PPARGC1A* rs8192678 Ser allele carriers. Moreover, *PPARGC1A* rs8192678 Ser allele carriers had no response in slow muscle fibers' changes, changes in low-density and total lipoprotein cholesterol and VO_2peak [20] after aerobic training.

PPARD rs1053049 TT homozygotes have been associated with greater increases in insulin sensitivity and greater decreases of fasting insulin levels than C allele carriers [17]. *PPARD* rs2267668 AA homozygotes have been found to have greater increases in insulin sensitivity, greater increases of the individual anaerobic threshold and greater increases in VO_2peak after aerobic training than G allele carriers [17]. Moreover, the *PPARD* rs2267668 G allele carriers have been found to have a negative response (decrease) of VO_2peak after aerobic training intervention [17]. The *PPARD* rs2016520 T allele carriers have been found to have a greater increase of VO_2max and maximum power output after aerobic training than CC homozygotes (only in black subjects), and the CC genotype had a higher increase of plasma HDL cholesterol (in white subjects) [21]. The *PPARD* rs2076167 GC genotype had a higher increase of plasma HDL cholesterol (only in white subjects).

PPARG rs1801282 Pro/Pro homozygotes have been found to have more decreased fasting insulin [22] than Ala/Pro heterozygotes and more decreased body fat than Ala allele carriers [23] after aerobic training. The Ala carriers have been found to have more increased glucose tolerance [24], more decreased fasting immunoreactive insulin and a more decreased insulin resistance index [25] after aerobic training than Pro/Pro homozygotes. *PPARG* Pro/Pro homozygotes have been found to have a negative response fasting immunoreactive insulin and a more decreased insulin resistance index after aerobic training.

Table 1. Basic description of included interventional studies. PPAR = peroxisome proliferator-activated receptor.

Study	Gene/Polymorphism	Population	Aim	Main Result
Stefan et al., 2007 [17]	*PPARGC1A* Gly482Ser (rs8192678) *PPARD* (rs2267668) (rs6902123) (rs2076167) (rs1053049)	German; n = 136 (men 63, women 73), Tuebingen Lifestyle Intervention Program. Age 45 ± 1 years, body mass 86.5 ± 1.5 kg	To investigate, whether selected SNPs predict the response of aerobic exercise training on changes in aerobic physical fitness and insulin sensitivity and whether they affect mitochondrial function in human myotubes in vitro.	Genetic variations in *PPARD* and *PPARGC1A* modulate mitochondrial function and changes in aerobic physical fitness and insulin sensitivity during lifestyle intervention.
Steinbacher et al., 2015 [18]	*PPARGC1A* Gly482Ser (rs8192678)	Austrian; n = 28 (men only), Salzburg Atherosclerosis Prevention Programme in Subjects at High Individual Risk. Age 59 ± 7 years (range 50–69), body mass 88 ± 2.2 kg	To investigate the myocellular responses in the vastus lateralis muscle of untrained male carriers of this SNP and of a control group after 10 weeks of endurance training.	The single nucleotide polymorphism Gly482Ser in the *PPARGC1A* gene impairs exercise-induced slow-twitch muscle fiber transformation in humans.
Tobina et al., 2017 [19]	*PPARGC1A* Gly482Ser (rs8192678)	Japanese; n = 119 (men 49, women 70), all participants >65 years of age. Age 71 ± 6 years, body mass 57.5 ± 9.8 kg	This study investigated the effects of *PPARGC1A* Gly482Ser polymorphisms on alterations in glucose and lipid metabolism induced by 12 weeks of exercise training.	The *PPARGC1A* Gly482Ser polymorphism is associated with the response of low-density lipoprotein cholesterol concentrations following exercise training in elderly Japanese.
Ring-Dimitriou, et al., 2014 [20]	*PPARGC1A* Gly482Ser (rs8192678)	Austrian; n = 24 (men only), untrained individuals selected from SAPHIR program. Age 58.3 ± 5.7 years, body mass 87.2 ± 7.6 kg	To test if untrained men who are homozygous or heterozygous carriers of the rare allele in *PPARGC1A* show a reduced change in oxygen uptake and work rate at the submaximal performance level compared to men characterized by the common genotype after 10 weeks of endurance exercise.	Investigated SNP affects the trainability of aerobic capacity measured as VO_2 or work rate at the respiratory compensation point of previously untrained middle-aged men. The highest responders were Gly/Gly genotypes compared to Gly/Ser and Ser/Ser genotypes.
He et al., 2008 [16]	*PPARGC1A* Thr394Thr (rs17847357) Gly482Ser (rs8192678) A2962G (rs6821591)	Chinese of Han origin; n = 102 (men only), soldiers from a local police army. Age 19 ± 1 years, height 171.7 ± 5.8 cm, body mass 60.3 ± 6.5 kg	To examine the possible association between *PPARGC1A* genotypes and both maximal (i.e., VO_2max) and submaximal endurance capacity (i.e., running economy in a pre-training state (baseline) and after endurance training.	None of the VO_2max and RE-related traits were associated with the Gly482Ser and Thr394Thr polymorphisms at baseline nor after training. The A2962G polymorphism was however associated with VO_2max at baseline, as carriers of the G allele (AG1GG genotypes; n = 49) had higher levels of VO_2max than the AA group (n = 53).
Weiss et al., 2005 [22]	*PPARG* Pro12Ala (rs1801282)	Caucasian; n = 73, (men 32, women 41), healthy sedentary subjects aged 50–75 years.	To investigate whether a common functional gene variant predicts insulin action and whether improvements in insulin action in response to endurance exercise training are associated with *PPARG* Pro12Ala.	Endurance training-induced changes in the insulin response to oral glucose are associated with the *PPARG* Pro12Ala genotype in men, but not in women.

Table 1. *Cont.*

Study	Gene/Polymorphism	Population	Aim	Main Result
Zarebska et al., 2014 [23]	*PPARG* Pro12Ala (rs1801282)	Polish; n = 201 (women only), no history of any metabolic or cardiovascular diseases. Age 21 ± 1 years	To examine the genotype distribution of the *PPARG* Pro12Ala allele in a group of Polish women measured for selected body mass and body composition variables before and after the completion of a 12-week training program.	The Pro12Ala polymorphism modifies the association of physical activity and body mass changes in Polish women.
Péruse et al., 2010 [24]	*PPARG* Pro12Ala (rs1801282)	White; n = 481 (men 233, women 248 from 98 nuclear families), sedentary non-diabetic subjects from the HERITAGE study who finished a 20-week endurance training program. Age 36 ± 0.67 years	To investigate whether variants either confirmed or newly identified as diabetes susceptibility variants through GWAS modulate changes in phenotypes derived from an intravenous glucose tolerance test (IVGTT) in response to an endurance training program.	Improvements in glucose homeostasis in response to regular exercise are influenced by the *PPARG* Pro12Ala variant.
Kahara et al., 2003 [25]	*PPARG* Pro12Ala	Japanese; n = 123, men, age 21–69 years. Age ± SD, 45.2 ± 11.6 years	To examine the association of *PPARG* gene polymorphism in Japanese healthy men with changes in insulin resistance after intervention with an exercise program.	The *PPARG* gene polymorphism may be a reliable indicator of whether exercise will have a beneficial effect as part of the treatment of insulin resistance syndrome.
Hautala et al., 2007 [21]	*PPARD* +15C/T (rs2016520) +65A/G (rs2076167)	American; n = 462 white subjects (223 males, 239 females) n = 256 black subjects (87 males, 169 females) from the HERITAGE study. Age 17–65 years	To test the hypothesis that *PPARD* gene polymorphisms are associated with cardiorespiratory fitness and plasma lipid responses to endurance training.	DNA sequence variation in the *PPARD* locus is a potential modifier of changes in cardiorespiratory fitness and plasma high-density lipoprotein cholesterol in healthy individuals in response to regular exercise.

Table 2. The summary of the trainability of different allele carriers. PPAR = peroxisome proliferator-activated receptor. w = week. ↑ = increase. ↓ = decrease. lbm = lean body mass. HR = heart rate.

Gene/Polymorphism	Intervention	Genotype/Allele Difference	Parameters	Study
PPARGC1A Gly482Ser rs8192678	9 months 3 h/w of moderate sports endurance exercise (e.g., walking, swimming)	Gly/Gly > Ser allele carriers	↑ individual anaerobic threshold (W)	Stefan et al., 2007 [17]
	10 w 3/w 60 min cycling training at a heart rate equaling 70–90% of peak oxygen uptake (VO_2peak)	Gly/Gly > Ser allele carriers (Ser allele-no response)	↑ slow muscle fibers' proportion	Steinbacher et al., 2015 [18]
		Gly/Gly > Ser allele carriers	↑ mitochondria activity–Complex II	Steinbacher et al., 2015 [18]

Table 2. *Cont.*

Gene/Polymorphism	Intervention	Genotype/Allele Difference	Parameters	Study
	12 w 140 min/w 20 cm bench-stepping exercise training at lactate threshold intensity	Gly/Gly > Ser allele carriers (Ser allele-no response)	↓ low-density and total lipoprotein cholesterol	Tobina et al., 2017 [19]
	10 w 3/w 45–60 min HR equaling 80–100% of the anaerobic threshold (ANT)	Gly/Gly > Ser allele carriers (Ser/Ser-no response)	↑ VO_2peak ($mL{\cdot}min^{-1}{\cdot}kg$)	Ring-Dimitriou, et al., 2014 [20]
PPARD rs1053049 (complete LD with rs2076167)	9 months 3 h/w of moderate sports endurance exercise (e.g., walking, swimming)	TT > C allele carriers	↑ insulin sensitivity	Stefan et al., 2007 [17]
		TT > C allele carriers	↓ fasting insulin levels	Stefan et al., 2007 [17]
PPARD rs2267668	9 months 3 h/w of moderate sports endurance exercise (e.g., walking, swimming)	AA > G allele carriers	↑ insulin sensitivity	Stefan et al., 2007 [17]
		AA > G allele carriers	↑ individual anaerobic threshold (W)	Stefan et al., 2007 [17]
		AA > G allele carriers (G allele-negative response)	↑ VO_2peak ($mL{\cdot}min^{-1}{\cdot}kg$ lbm)	Stefan et al., 2007 [17]
PPARD +15C/T (rs2016520) +65A/G (rs2076167)	20 w 3/w at 55–75% of baseline VO_2max for 30–50 min	T allele carriers > +15CC (only in black subjects)	↑ VO_2max ↑ maximum power output	Hautala et al., 2007 [21]
		+15CC > T allele carriers +65GG > A allele carriers (only in white subjects)	↑ plasma HDL cholesterol	Hautala et al., 2007 [21]
PPARG Pro12Ala rs1801282	10 w 3–4/w 40 min sessions of endurance treadmill walking and stationary cycling at 65–75% of heart rate reserve	Men: Pro/Pro < Ala/Pro	↓ fasting insulin and insulin AUC following intervention	Weiss et al., 2005 [22]
	12 w 3/w 60 min at 50–75% heart rate max. aerobic	Pro/Pro > Ala allele carriers	↓ body fat	Zarebska et al., 2014 [23]
	20 w 3/w at 55–75% of baseline VO_2max for 30–50 min	Ala carriers > Pro/Pro (Pro allele-negative response in some parameters)	↑ glucose tolerance (glucose effectiveness, acute insulin response to glucose, and disposition index)	Péruse et al., 2010 [24]
	3 months 2–3/w 2–3/day 20–60 min at 50% of the maximal heart rate of brisk walking	Ala allele carriers > Pro/Pro	↓ fasting immunoreactive insulin (IRI) ↓ homeostasis model assessment–insulin resistance index (HOMA-R)	Kahara et al., 2003 [25]

3. Discussion

The main finding of this review is that PPARs and their coactivators' gene polymorphisms may predict the human response to aerobic training at moderate intensities up to the lactate threshold, which might be expected. On the other hand, the lack of research in human training response to anaerobic training and specific training methods indicate that further research I needed. In this manner, there are significant cues that PPARs and their coactivators' gene polymorphisms can determine the anaerobic training effectiveness in response to training loads [15] (our finding includes also intensity at the anaerobic threshold). Although we had to exclude one study [15] for a lack of reproducibility, we have to highlight the importance of their findings (determination of resistance training load) as a significant suggestion for future research focus. A previous study on compound dinucleotide repeat polymorphism in *ALAS2* intron 7 in Han Chinese males determined that individuals with dinucleotide repeats $\leq$166 bp compared to individuals with dinucleotide repeats >166 bp were significantly better responders for high altitude training (measured as ΔVO_2max), especially to living-high exercise-high training-low (HiHiLo) training [26]. This specificity can be considered as the key information for creating a long-term endurance training program. Equally, women of multi-ethnicity origin from the FAMuSS cohort, homozygous for the mutant allele 577X in the *ACNT3* gene, demonstrated greater absolute and relative 1 repetition maximum gains of elbow flexors compared with the homozygous wild type (577RR) after resistance training when adjusted for body mass and age [27]. This review has to note that PPARs are not sufficiently analyzed for such specific training methods, although their connection to aerobic performance has been well known since the HERITAGE study results in 2001 [1].

This review summarized the best responders for aerobic training in relation to PPARs and their coactivators' genes polymorphisms (*PPARGC1A* rs8192678 Gly/Gly, *PPARD* rs1053049 TT, *PPARD* rs2267668 AA, *PPARD* rs2016520 T allele carriers and *PPARG* rs1801282 Ala allele carriers) in a common population [17,18,20,21]. On the other hand, the evaluation summary on the effects in *PPARD* rs2267668 G allele carriers and *PPARG* rs1801282 Pro/Pro homozygotes showed several negative responses to aerobic training. Most likely, this could be the most important information from exercise genomics studies, i.e., knowledge of genetic markers that can be beneficial for predicting the individual response to training in athletes and normal individuals, that is setting up the parameters of training protocols. On the other hand, the evaluation summary on effects in *PPARD* rs2267668 G allele carriers and *PPARG* rs1801282 Pro/Pro homozygotes showed several negative responses to aerobic training. Although the development of reliable tools for predicting exercise response based on one's genetic make-up is challenging and undoubtedly requires further research, the mentioned genetic variants seem to identify individuals who are not instructed to use classical aerobic training methods to improve their health or physical performance. Similarly, the non-responders for the *PPARGC1A* rs8192678 polymorphism who were Ser/Ser homozygotes might perform the aerobic training to improve metabolism functions such as mitochondria activity, but without a complex impact on improved health or endurance performance.

As was indicated earlier, post-training increase in aerobic fitness was found to be associated with the presence of a specific *PPARGC1A* rs8192678 Gly allele during a lifestyle intervention [17]. These observations led to the suggestion that the rs8192678 Gly allele may be a key element associated with the efficiency of aerobic metabolism; however, the question of how the rs8192678 Gly and Ser variants affect cardiorespiratory capacity remains. A general explanation is the engagement of the PGC-1α co-activator in the regulation of energy metabolism, as well as mitochondrial biogenesis and function, causing an upregulation of oxidative metabolism and parallel changes in muscle fiber types [28]. More detailed in vitro studies with the use of recombinant plasmids bearing Gly or Ser at position 482 in the PGC-1α protein showed that the PGC-1α 482Ser variant was less efficient as a co-activator of the MEF2C (myocyte enhancer factor 2C), which is a transcription factor especially important in the regulation of glucose transportation in skeletal muscle [29]. MEF2C, when coactivated by the PGC-1α, is particularly involved in the activation of GLUT4 (glucose transporter 4) via direct interaction with this gene promoter, which results in the facilitation of glucose uptake by the cell [30]. The Gly482Ser polymorphic site is located in the domain critical for the binding interaction between

MEF2C and PGC-1α proteins, and in this way, the rs8192678 Gly and Ser variants may influence the co-activation process, which may have consequences not only for glucose uptake, glycogen synthesis and the subsequent synthesis of fatty acids, but also for the transformation of muscle fiber type [28]. On the latter point, the expression of genes specific for type I slow-twitch fibers, such as MB (myoglobin) and TNNI1 (troponin I, slow skeletal muscle), is triggered by the calcineurin signaling pathway depending on PGC-1α/MEF2 coactivation [31].

The described structure of the *PPARD* gene differs from the classical eukaryotic gene model: it has been reported to encompass nine exons, of which exons 1–3, the 5′-end of exon 4 and the 3′-end of exon 9 are untranslated [32]. The rs2016520 polymorphic point is located precisely in the 5′UTR region of exon 4 of the *PPARD* gene. In this region, the recognition sites for Sp1 binding were found, raising the suggestion that rs2016520 may interfere with interaction between the *PPARD* gene and the Sp1 transcription factor, affecting in this way the *PPARD* expression level. Such an assumption was confirmed in the in vitro studies showing a higher transcriptional activity for the minor C allele compared with the major T allele of rs2016520 [33], which as a consequence may lead to impairment of PPARδ function and its ability to regulate the energy metabolism in skeletal muscles, in this manner influencing physical performance [34]. As was indicated in our metanalysis, during an intervention exercise training program performed in healthy (but previously sedentary) individuals of the HERITAGE Family Study, rs2016520 CC homozygotes were characterized by a smaller training-induced increase in maximal oxygen consumption and a lower training response in maximal power output compared with the CT and the TT genotypes, both in black and white subjects. Furthermore, CC homozygotes showed the greatest increases in HDL-C (white subjects) and Apo A-1 (black subjects) levels [21]. It was speculated that the greater promoter activity of *PPARD* rs2016520 CC homozygotes could result in higher PPARβ/δ levels. On the other hand, endurance training induces the elevated PPARβ/δ-specific agonists' availability, and the same ligands also increase the expression of the *ABCA1* gene, which is a key regulator of reverse cholesterol transport. All above-mentioned facts lead to the suggestion that the greatest increases in HDL-C levels observed in *PPARD* rs2016520 CC individuals might result from an increase in *ABCA1* gene expression [21].

Maintaining normal blood glucose levels is considered critical for preventing metabolic syndrome [35], and chronically impaired blood glucose responses comprise a significant risk factor for type II diabetes mellitus (DM2) [36]. Exercise in general has positive effects on glucose metabolism and DM2 prevention [37], thus encouraging individuals who are non-/poor responders to exercise is highly valuable. Our review shows that *PPARG* rs1801282 Ala allele carriers, *PPARD* rs2267668 AA homozygotes and *PPARD* rs1053049 TT homozygotes have, for some reason, more effectively improved glucose sensitivity and related parameters compared to their counterparts (Table 2). As regards the *PPARG* rs1801282 Ala allele, similar findings related to glycemic response to exercise have been found in diabetic patients [38] or in Japanese healthy men [25] who completed three months of supervised aerobic training. The *PPARG* Ala allele showed decreased binding affinity to the cognate promoter element and reduced ability to transactivate responsive promoters [39] and seems to be more responsive not only to exercise, but also to nutritional intervention; a significant decrease of waist circumference in diabetic patients was found following the swap from a normal to a Mediterranean diet [40]. The functional relevance of the Pro12Ala amino acid change in the PPARγ protein results from its localization within the molecule encoded by the *PPARG* gene. Pro12Ala substitution is a consequence of rs1801282 SNP located within the exon B sequence of the *PPARG* gene. This amino acid change is located within the AF-1 domain that controls the ligand-independent activation function of PPARγ. The presence of Ala at position 12 of the PPARγ protein may indirectly facilitate the chemical modification of some amino acid residues (phosphorylation and/or SUMOylation) responsible for decreasing the PPARγ activity as a transcriptional regulator involved in energy control and lipid/glucose homeostasis [41]. Different transcriptional activities of factors bearing Pro or Ala at position 12 in the PPARγ protein were confirmed in in vitro experiments, which recognized the Ala form as less active, characterized by a decreased ability to activate the transcription of appropriate constructs containing PPRE [42] or specific genes [39]. Moreover, analyses performed in vivo also revealed that expression of PPARγ target genes depends on the Pro12Ala genotypes [43]. Genetic association studies, as well as whole-body insulin sensitivity measurements documented that Ala allele carriers displayed a significantly improved insulin sensitivity [44], which may have the consequence of better glucose utilization in working skeletal muscles [45]. The studies investigating the effects of *PPARD* gene variants on glucose homeostasis are only marginal; only the effect of the contribution to the risk of DM2 of nine common variants in *PPARD* (including rs1053049 and rs2267668) in Chinese Hans was found in the rs6902123 polymorphism [46]. Another study also showed that *PPARD* polymorphisms (rs1053049, rs6902123 and rs2267668) could be involved in the development of insulin resistance and DM2 [47].

The combined effect of *PPARD, PPARG* and *PPARGC1A* gene polymorphisms on endurance exercise response and on health-related parameters is unclear, due to the amount of analyzed genes. Although, the results of studies included in our review seems to be promising in this manner, an evaluation demands larger cohorts with long-term supervised exercise programs to reach significance. At this moment, any life-style interventional program including exercise in normal people or a training regimen in athletes is not recommended according to the genomic data. On the other hand, the PPARs' relation to training methods' responses such as hypoxia [48] and resistance training [15] seems to have high potential to future research.

4. Materials and Methods

4.1. Review Process

The review was performed according to the Preferred Reporting Items for Systematic Reviews and Meta Analyses (PRISMA) [49] guidelines using the review protocol assigned in International Prospective Register of Systematic Reviews (PROSPERO) under Database No. CRD42018082236. The final articles' eligibility was assessed using the adapted "Standard Protocol Item Recommendation for Interventional Trials" (SPIRIT) checklist (Supplementary Material 1).

4.2. Literature Search

To find articles related to PPAR polymorphisms' role in physical activity, a systematic computerized literature search was conducted on 20 November 2017, in PubMed (1940 to the search date), Scopus (1823 to the search date) and Web of Science (1974 to the search date). A combination of the following search terms was used: (PPAR) OR (peroxisome AND proliferator AND activated AND receptor) AND (sport) OR (physical AND activity) OR (endurance) OR (exercise) OR (performance) OR (movement). The search did not include comments, proceedings, editorial letters, conference abstracts and dissertations. Reviews were included for a manual search of their reference lists. A manual search of the reference lists of included articles was also performed (Figure 1).

4.3. Literature Selection

After identifying potential articles, the titles and abstracts were reviewed by two independent reviewers (Petr Stastny, Miroslav Petr) to select relevant articles for full-text screening. The title and abstract screening focused on four related inclusion criteria:

- Sampling of genetic polymorphisms in the *PPARA, PPARG, PPARD, PPARGC1A* and *PPARGC1B*, genes.
- Analyses of genetic polymorphisms on sport phenotype (markers of sport phenotype) or related physical activity domains (e.g., body mass, fat mass, energy uptake, performance, physical fitness).
- Population of athletes and other healthy populations with a physical activity record.
- Cross-sectional, cohort, case control, intervention, control trials or GWAS.

When the inclusion of articles was questionable, the reviewers came to agreement after a personal discussion. The full texts of relevant articles were then analyzed to determine which were to be used in the final analysis. This full-text screening was performed by three independent reviewers (Petr Stastny, Miroslav Petr, Agnieszka Maciejewska-Skrendo), who also completed the data extraction form (Supplementary Material 2). Data collection was performed in interventional studies only. During the full-text screening, the following exclusion criteria were used:

(1) the full text was not available in English;
(2) the study did not contain an appropriate description of measuring devices, physical activity or genetic sampling procedures;
(3) the study did not include a specification of physical activity;
(4) the study did not report a quantitative performance outcome;
(5) the study did not perform the intervention of a physical training program;
(6) the study was not reproducible by the methodological quality criteria.

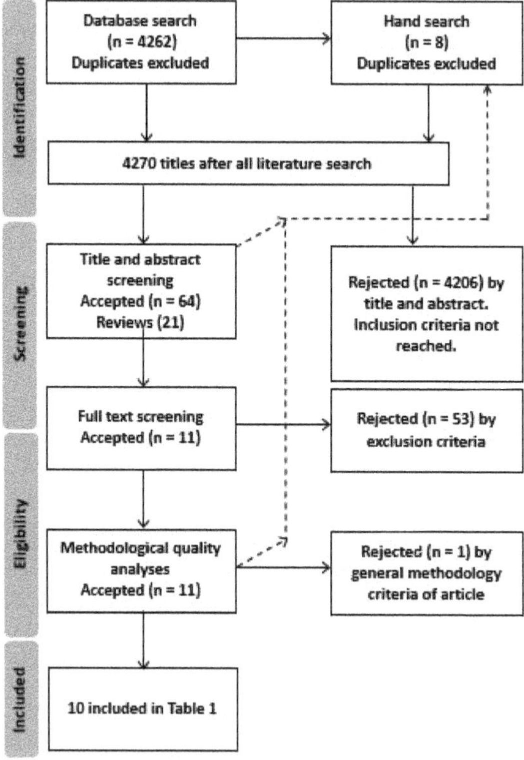

Figure 1. Review flow chart for articles included in tables.

5. Conclusions

PPARs and their coactivators' polymorphism genes can predict high response, no response or even negative response for aerobic training estimated by glucose tolerance, insulin response, body fat, VO_2max, anaerobic threshold, VO_2peak, mitochondria activity, cholesterol and slow muscle fibers' increase. Future studies should determine the role of PPARs and their coactivators in anaerobic training and more specific training methods (such as hypoxia) than moderate to lactate threshold aerobic training. The non-responders for aerobic training in VO_2peak, slow muscle fiber increase and low-density lipoprotein decrease are the carriers of *PPARGC1A* rs8192678 Ser/Ser. The negative responders for aerobic training in VO_2peak are carriers of the *PPARD* rs2267668 G allele. The negative responders for aerobic training in glucose tolerance and insulin response are carriers of the *PPARG* rs1801282 Pro/Pro genotype. The best responders to aerobic training are *PPARGC1A* rs8192678 Gly/Gly, rs1053049 TT, *PPARD* rs2267668 AA and *PPARG* rs1801282 Ala carriers.

Supplementary Materials: Supplementary materials can be found at http://www.mdpi.com/1422-0067/19/5/1472/s1.

Author Contributions: P.S., M.P. and A.Z. conceived of and designed the review. P.S., M.P., A.M.-S. performed the review. P.S., M.P., A.M.-S. analyzed the data. P.S., M.P., A.M.-S. and J.J.T. wrote the paper.

Acknowledgments: This study has been supported by the UNCE/HUM/032 grant at Charles University.

Conflicts of Interest: The authors declare no conflict of interest.

Abbreviations

GWAS	Genome-wide association study
PRISMA	Preferred Reporting Items for Systematic Reviews and Meta Analyses
SPIRIT	Standard Protocol Item Recommendation for Interventional Trials
PPAR	Peroxisome proliferator-activated receptor
IRI	Fasting immunoreactive insulin
HOMA-R	Homeostasis model assessment-insulin resistance index

References

1. Bouchard, C.; Rankinen, T. Individual differences in response to regular physical activity. *Med. Sci. Sports Exerc.* **2001**, *33*, S446–S451; discussion S52–S53. [CrossRef] [PubMed]
2. Bouchard, C.; Blair, S.N.; Church, T.S.; Earnest, C.P.; Hagberg, J.M.; Hakkinen, K.; Jenkins, N.T.; Karavirta, L.; Kraus, W.E.; Leon, A.S.; et al. Adverse metabolic response to regular exercise: Is it a rare or common occurrence? *PLoS ONE* **2012**, *7*, e37887. [CrossRef] [PubMed]
3. Dubuquoy, L.; Dharancy, S.; Nutten, S.; Pettersson, S.; Auwerx, J.; Desreumaux, P. Role of peroxisome proliferator-activated receptor γ and retinoid X receptor heterodimer in hepatogastroenterological diseases. *Lancet* **2002**, *360*, 1410–1418. [CrossRef]
4. Cabrero, A.; Laguna, J.; Vazquez, M. Peroxisome proliferator-activated receptors and the control of inflammation. *Curr. Drug Targets-Inflamm. Allergy* **2002**, *1*, 243–248. [CrossRef] [PubMed]
5. Franks, P.W.; Christophi, C.A.; Jablonski, K.A.; Billings, L.K.; Delahanty, L.M.; Horton, E.S.; Knowler, W.C.; Florez, J.C.; Diabetes Prevention Program Research Group. Common variation at PPARGC1A/B and change in body composition and metabolic traits following preventive interventions: The Diabetes Prevention Program. *Diabetologia* **2014**, *57*, 485–490. [CrossRef] [PubMed]
6. Ahmetov, I.I.; Williams, A.G.; Popov, D.V.; Lyubaeva, E.V.; Hakimullina, A.M.; Fedotovskaya, O.N.; Mozhayskaya, I.A.; Vinogradova, O.L.; Astratenkova, I.V.; Montgomery, H.E.; et al. The combined impact of metabolic gene polymorphisms on elite endurance athlete status and related phenotypes. *Hum. Genet.* **2009**, *126*, 751–761. [CrossRef] [PubMed]
7. Ahmetov, I.I.; Popov, D.V.; Mozhaiskaia, I.A.; Missina, S.S.; Astratenkova, I.V.; Vinogradova, O.L.; Rogozkin, V.A. Association of regulatory genes polymorphisms with aerobic and anaerobic performance of athletes. *Rossiskii Fiziologicheski Zhurnal Imeni IM Sechenova/Rossiskaia Akademiia Nauk* **2007**, *93*, 837–843.
8. Franks, P.W.; Barroso, I.; Luan, J.; Ekelund, U.; Crowley, V.E.F.; Brage, S.; Sandhu, M.S.; Jakes, R.; Middelberg, R.P.S.; Harding, A.-H.; et al. PGC-1α Genotype Modifies the Association of Volitional Energy Expenditure with VO_2 max. *Med. Sci. Sports Exerc.* **2003**, *35*, 1998–2004. [CrossRef] [PubMed]
9. Petr, M.; Šťastný, P.; Pecha, O.; Šteffl, M.; Šeda, O.; Kohlíková, E. PPARA intron polymorphism associated with power performance in 30-s anaerobic wingate test. *PLoS ONE* **2014**, *9*, e107171. [CrossRef] [PubMed]
10. Ghosh, S.; Vivar, J.C.; Sarzynski, M.A.; Sung, Y.J.; Timmons, J.A.; Bouchard, C.; Rankinen, T. Integrative pathway analysis of a genome-wide association study of VO_2 max response to exercise training. *J. Appl. Physiol.* **2013**, *115*, 1343–1359. [CrossRef] [PubMed]

11. Yang, R.Y.; Wang, Y.B.; Shen, X.Z.; Cai, G. Association of elite athlete performance and gene polymorphisms. *Chin. J. Tissue Eng. Res.* **2014**, *18*, 1121–1128.
12. Lopez-Leon, S.; Tuvblad, C.; Forero, D.A. Sports genetics: The PPARA gene and athletes' high ability in endurance sports. A systematic review and meta-analysis. *Biol. Sport* **2016**, *33*, 3–6. [PubMed]
13. Ahmetov, I.I.; Fedotovskaya, O.N. *Current Progress in Sports Genomics*; Advances in Clinical Chemistry; Academic Press Inc.: New York, NY, USA, 2015; pp. 247–314.
14. Leońska-Duniec, A.; Ahmetov, I.I.; Zmijewski, P. Genetic variants influencing effectiveness of exercise training programmes in obesity—An overview of human studies. *Biol. Sport* **2016**, *33*, 207–214. [CrossRef] [PubMed]
15. Aksenov, M.O.; Ilyin, A.B. Training process design in weightlifting sports customized to genetic predispositions. *Teoriya Praktika Fizicheskoy Kultury* **2017**, *2017*, 75–77.
16. He, Z.; Hu, Y.; Feng, L.; Bao, D.; Wang, L.; Li, Y.; Wang, J.; Liu, G.; Xi, Y.; Wen, L.; et al. Is there an association between PPARGC1A genotypes and endurance capacity in Chinese men? *Scand. J. Med. Sci. Sports* **2008**, *18*, 195–204. [CrossRef] [PubMed]
17. Stefan, N.; Thamer, C.; Staiger, H.; Machicao, F.; Machann, J.; Schick, F.; Venter, C.; Niess, A.; Laakso, M.; Fritsche, A.; et al. Genetic variations in PPARD and PPARGC1A determine mitochondrial function and change in aerobic physical fitness and insulin sensitivity during lifestyle intervention. *J. Clin. Endocrinol. Metab.* **2007**, *92*, 1827–1833. [CrossRef] [PubMed]
18. Steinbacher, P.; Feichtinger, R.G.; Kedenko, L.; Kedenko, I.; Reinhardt, S.; Schönauer, A.L.; Leitner, I.; Sänger, A.M.; Stoiber, W.; Kofler, B.; et al. The single nucleotide polymorphism Gly482Ser in the PGC-1α gene impairs exercise-induced slow-twitch muscle fibre transformation in humans. *PLoS ONE* **2015**, *10*, e0123881. [CrossRef] [PubMed]
19. Tobina, T.; Mori, Y.; Doi, Y.; Nakayama, F.; Kiyonaga, A.; Tanaka, H. Peroxisome proliferator-activated receptor gamma co-activator 1 gene Gly482Ser polymorphism is associated with the response of low-density lipoprotein cholesterol concentrations to exercise training in elderly Japanese. *J. Physiol. Sci.* **2017**, *67*, 595–602. [CrossRef] [PubMed]
20. Ring-Dimitriou, S.; Kedenko, L.; Kedenko, I.; Feichtinger, R.G.; Steinbacher, P.; Stoiber, W.; Förster, H.; Felder, T.K.; Müller, E.; Kofler, B.; et al. Does genetic variation in PPARGC1A affect exercise-induced changes in ventilatory thresholds and metabolic syndrome? *J. Exerc. Physiol. Online* **2014**, *17*, 1–18.
21. Hautala, A.J.; Leon, A.S.; Skinner, J.S.; Rao, D.C.; Bouchard, C.; Rankinen, T. Peroxisome proliferator-activated receptor-delta polymorphisms are associated with physical performance and plasma lipids: The HERITAGE Family Study. *Am. J. Physiol. Heart Circ. Physiol.* **2007**, *292*, H2498–H2505. [CrossRef] [PubMed]
22. Weiss, E.P.; Kulaputana, O.; Ghiu, I.A.; Brandauer, J.; Wohn, C.R.; Phares, D.A.; Shuldiner, A.R.; Hagberg, J.M. Endurance training-induced changes in the insulin response to oral glucose are associated with the peroxisome proliferator-activated receptor-γ2 Pro12Ala genotype in men but not in women. *Metabolism* **2005**, *54*, 97–102. [CrossRef] [PubMed]
23. Zarebska, A.; Jastrzebski, Z.; Cieszczyk, P.; Leonska-Duniec, A.; Kotarska, K.; Kaczmarczyk, M.; Sawczuk, M.; Maciejewska-Karlowska, A. The Pro12Ala polymorphism of the peroxisome proliferator-activated receptor gamma gene modifies the association of physical activity and body mass changes in Polish women. *PPAR Res.* **2014**, *2014*, 373782. [CrossRef] [PubMed]
24. Pérusse, L.; Ruchat, S.M.; Rankinen, T.; Weisnagel, S.J.; Rice, T.; Rao, D.C.; Bergman, R.N.; Bouchard, C.; Pérusse, L. Improvements in glucose homeostasis in response to regular exercise are influenced by the PPARG Pro12Ala variant: Results from the HERITAGE family study. *Diabetologia* **2010**, *53*, 679–689.
25. Kahara, T.; Takamura, T.; Hayakawa, T.; Nagai, Y.; Yamaguchi, H.; Katsuki, T.; Katsuki, K.; Katsuki, M.; Kobayashi, K. PPARγ gene polymorphism is associated with exercise-mediated changes of insulin resistance in healthy men. *Metabolism* **2003**, *52*, 209–212. [CrossRef] [PubMed]
26. Xu, Y.; Hu, Y.; Ren, Z.; Yi, L. Delta-aminolevulinate synthase 2 polymorphism is associated with maximal oxygen uptake after living-high exercise-high training-low in a male chinese population. *Int. J. Clin. Exp. Med.* **2015**, *8*, 21617. [PubMed]
27. Clarkson, P.M.; Devaney, J.M.; Gordish-Dressman, H.; Thompson, P.D.; Hubal, M.J.; Urso, M.; Price, T.B.; Angelopoulos, T.J.; Gordon, P.M.; Moyna, N.M.; et al. ACTN3 genotype is associated with increases in muscle strength in response to resistance training in women. *J. Appl. Physiol.* **2005**, *99*, 154–163. [CrossRef] [PubMed]

28. Maciejewska, A.; Sawczuk, M.; Cieszczyk, P.; Mozhayskaya, I.A.; Ahmetov, I. The PPARGC1A gene Gly482Ser in Polish and Russian athletes. *J. Sports Sci.* **2012**, *30*, 101–113. [CrossRef] [PubMed]
29. Zhang, S.-L.; Lu, W.-S.; Yan, L.; Wu, M.-C.; Xu, M.-T.; Chen, L.-H.; Cheng, H. Association between peroxisome proliferator-activated receptor-gamma coactivator-1alpha gene polymorphisms and type 2 diabetes in southern Chinese population: Role of altered interaction with myocyte enhancer factor 2C. *Chin. Med. J.* **2007**, *120*, 1878–1885. [PubMed]
30. Michael, L.F.; Wu, Z.; Cheatham, R.B.; Puigserver, P.; Adelmant, G.; Lehman, J.J.; Kelly, D.P.; Spiegelman, B.M. Restoration of insulin-sensitive glucose transporter (GLUT4) gene expression in muscle cells by the transcriptional coactivator PGC-1. *Proc. Natl. Acad. Sci. USA* **2001**, *98*, 3820–3825. [CrossRef] [PubMed]
31. Lin, J.; Wu, H.; Tarr, P.T.; Zhang, C.-Y.; Wu, Z.; Boss, O.; Michael, L.F.; Puigserver, P.; Isotani, E.; Olson, E.N.; et al. Transcriptional co-activator PGC-1α drives the formation of slow-twitch muscle fibres. *Nature* **2002**, *418*, 797–801. [CrossRef] [PubMed]
32. Skogsberg, J.; Kannisto, K.; Roshani, L.; Gagne, E.; Hamsten, A.; Larsson, C.; Ehrenborg, E. Characterization of the human peroxisome proliferator activated receptor delta gene and its expression. *Int. J. Mol. Med.* **2000**, *6*, 73–154. [CrossRef] [PubMed]
33. Skogsberg, J.; Kannisto, K.; Cassel, T.N.; Hamsten, A.; Eriksson, P.; Ehrenborg, E. Evidence that peroxisome proliferator–activated receptor delta influences cholesterol metabolism in men. *Arterioscler. Thromb. Vas. Biol.* **2003**, *23*, 637–643. [CrossRef] [PubMed]
34. Karpe, F.; Ehrenborg, E.E. PPARδ in humans: Genetic and pharmacological evidence for a significant metabolic function. *Curr. Opin. Lipidol.* **2009**, *20*, 333–336. [CrossRef] [PubMed]
35. Riccardi, G.; Rivellese, A. Dietary treatment of the metabolic syndrome—The optimal diet. *Br. J. Nutr.* **2000**, *83*, S143–S148. [CrossRef] [PubMed]
36. Nathan, D.M.; Davidson, M.B.; DeFronzo, R.A.; Heine, R.J.; Henry, R.R.; Pratley, R.; Zinman, B.; American Diabetes Association. Impaired fasting glucose and impaired glucose tolerance: Implications for care. *Diabetes Care* **2007**, *30*, 753–759. [CrossRef] [PubMed]
37. Kelley, D.E.; Goodpaster, B.H. Effects of physical activity on insulin action and glucose tolerance in obesity. *Med. Sci. Sports Exerc.* **1999**, *31* (Suppl. 11), S619–S623. [CrossRef] [PubMed]
38. Adamo, K.; Sigal, R.; Williams, K.; Kenny, G.; Prud'homme, D.; Tesson, F. Influence of Pro12Ala peroxisome proliferator-activated receptor γ2 polymorphism on glucose response to exercise training in type 2 diabetes. *Diabetologia* **2005**, *48*, 1503–1509. [CrossRef] [PubMed]
39. Deeb, S.S.; Fajas, L.; Nemoto, M.; Pihlajamäki, J.; Mykkänen, L.; Kuusisto, J.; Laakso, M.; Fujimoto, W.; Auwerx, J. A Pro12Ala substitution in PPARγ2 associated with decreased receptor activity, lower body mass index and improved insulin sensitivity. *Nat. Genet.* **1998**, *20*, 284–286. [CrossRef] [PubMed]
40. Razquin, C.; Martinez, J.A.; Martinez-Gonzalez, M.A.; Corella, D.; Santos, J.M.; Marti, A. The Mediterranean diet protects against waist circumference enlargement in 12Ala carriers for the PPARγ gene: 2 years' follow-up of 774 subjects at high cardiovascular risk. *Br. J. Nutr.* **2009**, *102*, 672–679. [CrossRef] [PubMed]
41. Yen, C.-J.; Beamer, B.A.; Negri, C.; Silver, K.; Brown, K.A.; Yarnall, D.P.; Burns, D.K.; Roth, J.; Shuldiner, A.R. Molecular scanning of the human peroxisome proliferator activated receptor γ (hPPARγ) gene in diabetic Caucasians: Identification of a Pro12Ala PPARγ2 missense mutation. *Biochem. Biophys. Res. Commun.* **1997**, *241*, 270–274. [CrossRef] [PubMed]
42. Masugi, J.; Tamori, Y.; Mori, H.; Koike, T.; Kasuga, M. Inhibitory effect of a proline-to-alanine substitution at codon 12 of peroxisome proliferator-activated receptor-γ2 on thiazolidinedione-induced adipogenesis. *Biochem. Biophys. Res. Commun.* **2000**, *268*, 178–182. [CrossRef] [PubMed]
43. Schneider, J.; Kreuzer, J.; Hamann, A.; Nawroth, P.P.; Dugi, K.A. The proline 12 alanine substitution in the peroxisome proliferator–Activated receptor-γ2 gene is associated with lower lipoprotein lipase activity in vivo. *Diabetes* **2002**, *51*, 867–870. [CrossRef] [PubMed]
44. Ek, J.; Andersen, G.; Urhammer, S.; Hansen, L.; Carstensen, B.; Borch-Johnsen, K.; Drivsholm, T.; Berglund, L.; Hansen, T.; Lithell, H.; et al. Studies of the Pro12Ala polymorphism of the peroxisome proliferator-activated receptor-γ2 (PPAR-γ2) gene in relation to insulin sensitivity among glucose tolerant Caucasians. *Diabetologia* **2001**, *44*, 1170–1176. [CrossRef] [PubMed]

45. Honka, M.-J.; Vänttinen, M.; Iozzo, P.; Virtanen, K.A.; Lautamäki, R.; Hällsten, K.; Borraa, R.J.H.; Takalaa, T.; Viljanena, A.P.M.; Kemppainen, J.; et al. The Pro12Ala polymorphism of the PPARγ2 gene is associated with hepatic glucose uptake during hyperinsulinemia in subjects with type 2 diabetes mellitus. *Metabolism* **2009**, *58*, 541–546. [CrossRef] [PubMed]
46. Lu, L.; Wu, Y.; Qi, Q.; Liu, C.; Gan, W.; Zhu, J.; Li, H.; Lin, X. Associations of type 2 diabetes with common variants in PPARD and the modifying effect of vitamin D among middle-aged and elderly Chinese. *PLoS ONE* **2012**, *7*, e34895. [CrossRef] [PubMed]
47. Thamer, C.; Machann, J.; Stefan, N.; Schäfer, S.A.; Machicao, F.; Staiger, H.; Laakso, M.; Böttcher, M.; Claussen, C.; Schick, F.; et al. Variations in PPARD determine the change in body composition during lifestyle intervention: A whole-body magnetic resonance study. *J. Clin. Endocrinol. Metab.* **2008**, *93*, 1497–1500. [CrossRef] [PubMed]
48. Masschelein, E.; Puype, J.; Broos, S.; Van Thienen, R.; Deldicque, L.; Lambrechts, D.; Hespel, P.; Thomis, M. A genetic predisposition score associates with reduced aerobic capacity in response to acute normobaric hypoxia in lowlanders. *High Alt. Med. Biol.* **2015**, *16*, 34–42. [CrossRef] [PubMed]
49. Moher, D.; Schulz, K.F.; Simera, I.; Altman, D.G. Guidance for developers of health research reporting guidelines. *PLoS Med.* **2010**, *7*, e1000217. [CrossRef] [PubMed]

International Journal of
Molecular Sciences

Article

The Contribution of EDF1 to PPARγ Transcriptional Activation in VEGF-Treated Human Endothelial Cells

Alessandra Cazzaniga †, Laura Locatelli †, Sara Castiglioni and Jeanette Maier *

Dipartimento di Scienze Biomediche e Cliniche L. Sacco, Università degli Studi di Milano, I-20157 Milan, Italy; alessandra.cazzaniga@unimi.it (A.C.); laura.locatelli@unimi.it (L.L.); sara.castiglioni@unimi.it (S.C.)

* Correspondence: Jeanette.maier@unimi.it; Tel.: +39-02-5031-9648

† The two authors contributed equally to this work.

Received: 31 May 2018; Accepted: 14 June 2018; Published: 21 June 2018

Abstract: Vascular endothelial growth factor (VEGF) is important for maintaining healthy endothelium, which is crucial for vascular integrity. In this paper, we show that VEGF stimulates the nuclear translocation of endothelial differentiation-related factor 1 (EDF1), a highly conserved intracellular protein implicated in molecular events that are pivotal to endothelial function. In the nucleus, EDF1 serves as a transcriptional coactivator of peroxisome proliferator-activated receptor gamma (PPARγ), which has a protective role in the vasculature. Indeed, silencing *EDF1* prevents VEGF induction of PPARγ activity as detected by gene reporter assay. Accordingly, silencing *EDF1* markedly inhibits the stimulatory effect of VEGF on the expression of *FABP4*, a PPARγ-inducible gene. As nitric oxide is a marker of endothelial function, it is noteworthy that we report a link between *EDF1* silencing, decreased levels of *FABP4*, and nitric oxide production. We conclude that EDF1 is required for VEGF-induced activation of the transcriptional activity of PPARγ.

Keywords: endothelial cells; vascular endothelial growth factor; Peroxisome proliferator-activated receptor γ; Endothelial Differentiation-related factor 1

---

1. Introduction

Peroxisome proliferator-activated receptor gamma (PPARγ) is a ubiquitous ligand-inducible transcription factor belonging to the nuclear receptor superfamily [1]. PPARγ, which is highly expressed in adipose tissue, is the master regulator of adipocyte differentiation and is fundamental for mature adipocyte function [2–4]. PPARγ is also implicated in glucose homeostasis as it upregulates genes involved in glucose uptake and controls the expression of adipokines, thereby having an effect on insulin sensitivity [3,5,6]. It is now clear that PPARγ plays an important protective role in the vasculature. Its activity has been proven both in smooth muscle cells, where it has a role in the regulation of vascular tone [7], and in endothelial cells (EC), where it exerts anti-inflammatory and antioxidant effects [8]. In endothelial-specific PPAR$\gamma^{-/-}$ mice, loss of PPARγ contributes to endothelial dysfunction associated with enhanced production of free radicals and exacerbated inflammation [9]. Accordingly, human and animal studies indicate that thiazolidinediones (TZD), which are largely utilized as antidiabetic drugs and PPARγ activators, attenuate vascular diseases including atherosclerosis [10,11]. Post-translational modifications—including phosphorylation, acetylation, and sumoylation—are important in carving PPARγ-driven gene expression [12]. Another layer of control over PPARγ activity depends on its interactions with coactivators and corepressors [12], which regulate transcriptional activity by reshaping chromatin structure via histone deacetylases and histone acetyltransferases [13]. Indeed, upon activation by small natural lipophilic ligands or synthetic agonists, the conformation of PPARγ changes—corepressors are released and coactivators are recruited [3]—thus resulting in transcriptional activation.

Endothelial differentiation-related factor 1 (EDF1), a highly conserved intracellular protein of 148 amino acids, has been identified as one of PPARγ's coactivators [14,15]. Initially, the role of EDF1 as a transcriptional coactivator was described in the silkworm *Bombyx mori* and in *Drosophila melanogaster* where EDF1 stimulates the activity of the FTZ-F1 nuclear receptor [16]. At the time, EDF1 was demonstrated to serve as a coactivator for several transcription factors [17,18]. This included some nuclear receptors implicated in lipid metabolism, such as steroidogenic factor 1, liver receptor homologue 1, liver X receptor α and, as mentioned above, PPARγ [14,15]. In particular, in 3T3-L1 preadipocytes, EDF1 is required for PPARγ-mediated differentiation and gene expression programs [15]. In human macrovascular EC EDF1 was described as a factor implicated in differentiation and spatial organization [19]. In these cells EDF1 is localized mainly in the cytosol where it binds calmodulin [20] under basal conditions. In response to various stimuli, it is translocated to the nucleus where it interacts with the TATA box-binding protein [20].

Apart from its pivotal role in vasculogenesis and angiogenesis, VEGF is essential for endothelial polarity and survival, thus contributing to the integrity of mature vessels [21]. This issue is relevant since ECs are key players in organogenesis as well as in promoting adult organ maintenance. To this purpose, it is noteworthy that VEGF is a critical component of the cross-talk between organs and tissues and the vessels [21].

For this study, we considered three known facts: (1) PPARγ ligands influence VEGF action [8]; (2) PPARγ contributes to maintain normal vascular function [7–11]; and (3) EDF1 modulates the activity of PPARγ [14,15]. We used these factors to investigate whether EDF1 acts as a regulator of PPARγ activity in human macrovascular EC under normal culture conditions and after treatment with VEGF.

2. Results

2.1. Translocation of EDF1 to the Nucleus in Response to VEGF

Initially, we evaluated whether VEGF modulates the total amounts of EDF1 and PPARγ in human umbilical vein endothelial cells (HUVEC). Confluent cells were treated with VEGF (50 ng/mL) for different times. We performed Real-Time PCR as well as western blot analysis and found no modulation in the levels of EDF1 and PPARγ after 8, 12, and 24 h exposure to VEGF (Figure 1 and Supplementary S1). Because EDF1 translocates to the nucleus when HUVEC are stimulated with the phorbol ester 12-O-Tetradecanoylphorbol-13-acetate (TPA) or with forskolin [20,22], we evaluated the subcellular localization of EDF1 in cells treated with VEGF (50 ng/mL) for different times. By immunofluorescence, EDF1 was detectable both in the cytosol and in the nucleus of unstimulated cells. After being treated with VEGF, EDF1 accumulated in the nuclei after 1 h and remained nuclear-associated for the following 24 h (Figure 2a and Supplementary S2a). Western blot on nuclear and cytosolic fractions isolated after 1 h treatment with VEGF confirmed these results (Figure 2b and Supplementary S2b).

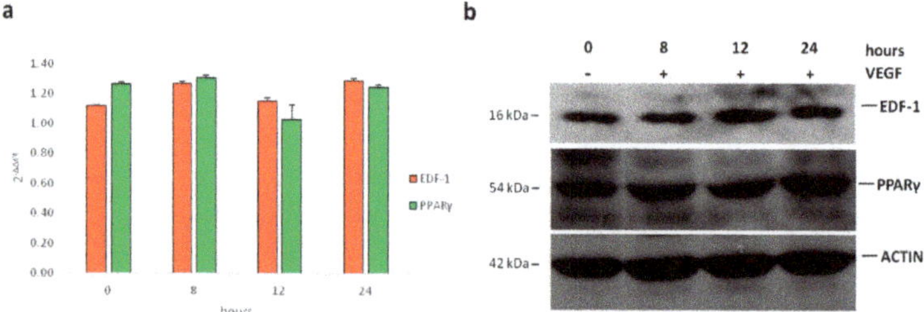

Figure 1. The total amounts of endothelial differentiation-related factor 1 (EDF1) and peroxisome proliferator-activated receptor gamma (PPARγ) in cells treated with VEGF. Human umbilical vein endothelial cells (HUVEC) were treated with 50 ng/mL of vascular endothelial growth factor (VEGF) for 0, 8, 12, and 24 h. (**a**) Real-Time PCR was performed on RNA samples. Two different experiments in triplicate were performed; (**b**) cell lysates were analyzed by western blot using antibodies against EDF1, PPARγ, and actin. A representative blot is shown.

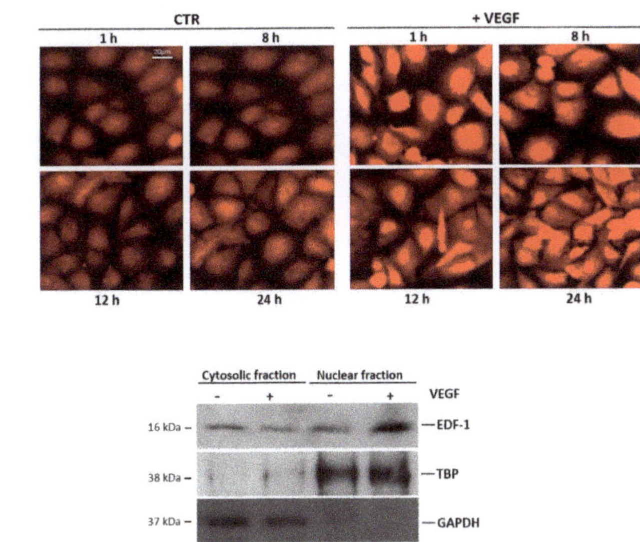

Figure 2. Subcellular localization of EDF1 in cells treated with VEGF. (**a**) HUVEC were treated with VEGF (50 ng/mL) for 1, 8, 12, and 24 h. Immunofluorescence was performed using anti-EDF1 immunopurified immunoglobulin G (IgGs) and rhodamine-conjugated anti-rabbit IgGs; (**b**) HUVEC were treated with VEGF (50 ng/mL) for 1 h. Western blot was performed on nuclear and cytosolic fractions using antibodies against EDF1. GAPDH and TBP were used as cytosolic and nuclear markers, respectively. A representative blot is shown.

2.2. Interaction between EDF1 and PPARγ in HUVEC

We evaluated the interaction between EDF1 and PPARγ in HUVEC treated with VEGF (50 ng/mL) for various times. Cell lysates were immunoprecipitated with antibodies against PPARγ. Western blot was performed on the immunoprecipitates to detect EDF1. EDF1 and PPARγ interacted in nonstimulated cells and VEGF did not significantly modulate this interaction at the time points tested (Figure 3 and Supplementary S3).

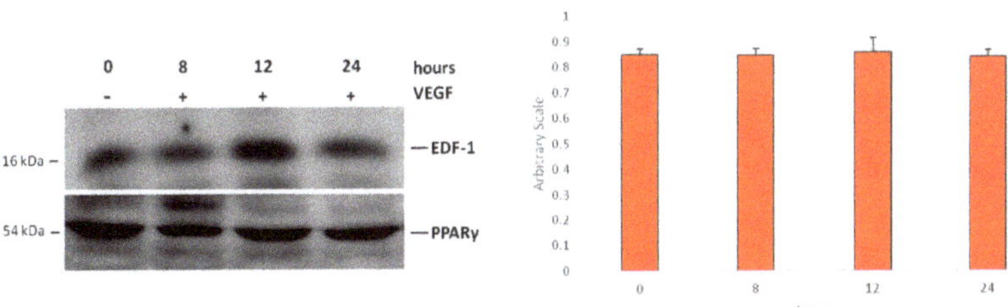

Figure 3. The interaction between EDF1 and PPARγ in HUVEC treated with VEGF. HUVEC were treated with VEGF (50 ng/mL) for different times. Cell lysates were immunoprecipitated with monoclonal antibodies against PPARγ and analyzed by western blot using rabbit antibodies against EDF1 (**upper panel**). The filter was then probed with rabbit anti-PPARγ antibodies to verify the equal amounts of immunoprecipitated proteins (**lower panel**). Densitometric analysis was performed using ImageJ software. EDF1/PPARγ ratio was calculated on three blots from separate experiments ± standard deviation.

2.3. Effect of Silencing EDF1 in VEGF-Induced PPARγ Activity

To study if EDF1 contributes to PPARγ transcriptional activity in VEGF-treated HUVEC, we utilized HUVEC with stably silenced EDF1, denominated αs1 cells [23]. We used HUVEC transfected with a nonsilencing sequence [23] as the control (CTR). It is noteworthy that PPARγ did not change in αs1 cells compared to their controls as demonstrated by western blot (Figure 4a and Supplementary S4).

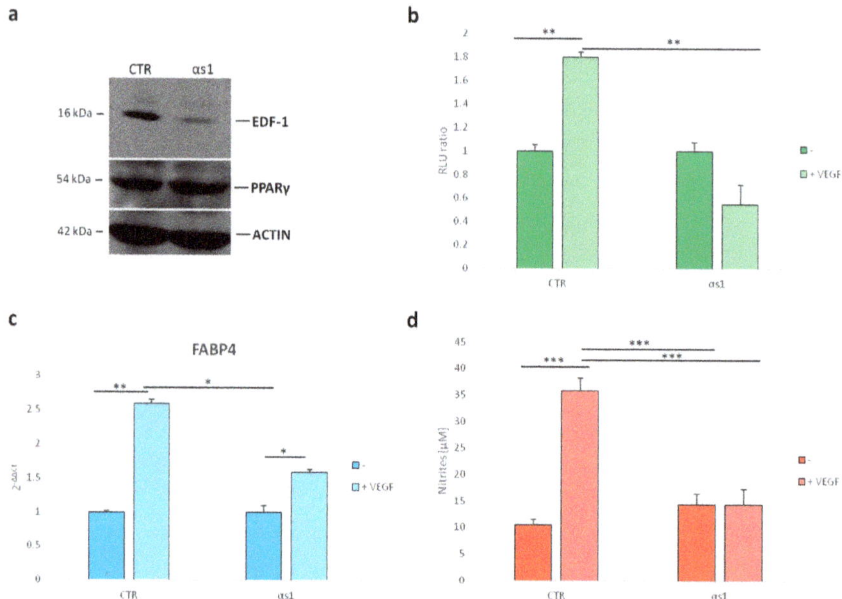

Figure 4. PPARγ transcriptional activity in HUVEC with silenced EDF1. (**a**) The modulation of PPARγ was evaluated in αs1 cells (αs1) (HUVEC with stably silenced EDF1) and compared to HUVEC transfected with a scrambled nonsilencing sequence (used as control) (CTR). Cell lysates were analyzed by western blot using antibodies against EDF1, PPARγ, and actin. A representative blot is shown; (**b**) PPARγ activity was evaluated by luciferase assay in αs1 cells and compared to the control HUVEC; (**c**) Real-Time PCR was performed on RNA samples from αs1 cells and relative control, treated or not with VEGF (50/ng/mL) for 24 h. Three different experiments in triplicate were performed; (**d**) Nitric oxide (NO) release was measured using the Griess method for nitrate quantification. The values were expressed as the mean of three different experiments in triplicate ± standard deviation. * $p < 0.05$, ** $p < 0.01$, *** $p < 0.001$.

We then transfected subconfluent αs1 cells and the control cells with a vector expressing luciferase under the control of a PPARγ responsive consensus (pDR1) [24]. After 4 h, the cells were treated with VEGF (50 ng/mL) and luciferase activity was measured after 24 h. While VEGF stimulated PPARγ transcriptional activation in control cells, this effect was prevented by silencing *EDF1* (Figure 4b). To reinforce this finding, we analyzed the expression of a PPARγ downstream target gene, i.e., fatty acid-binding protein 4 (FABP4), which is known to be upregulated in HUVEC after 24 h exposure to VEGF [25]. We cultured HUVEC in the presence of VEGF (50 ng/mL) for 24 h. Using Real-Time PCR, we confirmed the overexpression of *FABP4* RNA in control HUVEC treated with VEGF. In αs1 cells, which downregulate EDF1, the induction was significantly reduced (Figure 4c).

Because HUVEC with silenced FABP4 produce lower amounts of nitric oxide (NO) than controls and are insensitive to the stimulatory effect of VEGF [26], we measured the release of NO in αs1 cells and their controls that were treated or not with VEGF for 24 h. Figure 4d shows that while VEGF induced NO secretion in control cells, it did not exert any significant effect in αs1 cells.

3. Discussion

ECs line the inner face of blood vessels and their integrity is fundamental for vascular homeostasis and circulatory function [27]. Indeed, ECs are implicated in maintaining blood fluidity, governing leukocyte trafficking and vascular tone, and in regulating immune response. Consequently, it is not surprising that endothelial dysfunction, which is characterized by a pro-oxidant and pro-inflammatory phenotype, orchestrates events leading to cardiovascular diseases. Moreover, healthy endothelial cells are crucial for the maintenance of normal energy metabolism and, therefore, physiologic function of all tissues [27]. There is now evidence that PPARγ is a key regulator of endothelial function [21,27] and, accordingly, PPARγ activators inhibit the expression of proinflammatory molecules and the synthesis of free radicals [27]. Many factors contribute to the integrity of the endothelium including VEGF, which is critical for endothelial survival and barrier function in mature vessels [21]. On these bases, we investigated whether VEGF activates PPARγ in HUVEC by recruiting the transcriptional co-activator EDF1. We found that while VEGF does not change the total amounts of EDF1, it rapidly induces EDF1 nuclear translocation, which is maintained for 24 h. These results are in accordance with previous data showing EDF1 nuclear accumulation in HUVEC treated with the phorbol ester 12-O-Tetradecanoylphorbol-13-acetate (TPA) and the forskolin, which raise intracellular cAMP [20,23]. Interestingly, TPA, forskolin, and VEGF all induce the phosphorylation of EDF1 [20,22,23], and we hypothesize that phosphorylation has a role in increasing the nuclear accumulation of EDF1. It should be noted that EDF1 does not have a nuclear targeting sequence, and it is likely that a shuttle protein drives EDF1 to the nucleus. If this is the case, we postulate that VEGF enhances this shuttle mechanism. In the nucleus, VEGF stimulates the transcriptional activity of PPARγ in an EDF1-dependent manner. Indeed, silencing EDF1 prevents VEGF-induced PPARγ activity. Since VEGF increases the transcriptional activity of PPARγ in endothelial cells without altering its interaction with EDF1, it is possible that VEGF induces a specific PPARγ ligand or inhibits a corepressor, thus altering the balance between various transcriptional corepressors and coactivators. These results highlight an important difference between adipocytes and endothelial cells. In 3T3-L1 preadipocytes, silencing *EDF1* decreases the total amounts of PPARγ and, in parallel, its transcriptional activity [15]. By contrast, in HUVEC, silencing *EDF1* affects only its transcriptional activity.

We also evaluated the expression of a PPARγ-responsive gene in HUVEC with silenced EDF1. In particular, we focused on the modulation of *FABP4*, which encodes a cytoplasmic protein that has a role in endothelial fatty acid metabolism and free radical production. It also impairs proliferation and sprout elongation and impacts on nitric oxide synthesis [25,28]. Interestingly, pioglitazone—an insulin-sensitizing thiazolidinedione and a PPARγ agonist—increases FABP4 levels. In this study, we show that the activation of PPARγ by VEGF induces *FABP4* expression through the involvement of EDF1. Indeed, silencing *EDF1* markedly inhibits the stimulatory effect of VEGF on *FABP4* expression. It is known that VEGF induces *FABP4* through the Delta-like ligand (DLL) 4/NOTCH1 pathway [25]. Our results indicate that also PPARγ contributes to VEGF induction of *FABP4* in HUVEC.

NO is a pivotal mediator of VEGF-induced responses and is essential for vascular function [29]. The regulation of NO production is very complex and it is noteworthy that both PPARγ and FABP4 are involved. Indeed, in endothelial cells, the activation of PPARγ increases NO release [30] while the downregulation of *FABP4* reduces it [28]. To this purpose, we show that αs1 cells do not respond to VEGF by increasing NO release. We hypothesize that the downregulation of EDF1 impairs VEGF-induced activation of PPARγ with consequent reduction of *FABP4* and NO synthesis.

We conclude that VEGF induces EDF1 translocation to the nucleus where it acts as a transcriptional coactivator of PPARγ. The transcriptional activation of PPARγ increases the expression of FABP4, which is known to regulate NO production. Because NO is a marker of endothelial function, our results substantiate that PPARγ activation has a role in maintaining the integrity of vessels and highlight

EDF1 as a novel player in the complex regulation of PPARγ transcriptional activity in the endothelium. On these bases, we propose that EDF1 makes an important contribution to maintain endothelial integrity, and this may be crucial in the prevention of cardiovascular diseases.

4. Materials and Methods

4.1. Cell Culture

HUVEC—widely accepted as a model of macrovascular EC—were obtained from the American Type Culture Collection (ATCC) and cultured in M199 containing 10% fetal bovine serum, 1 mM glutamine, endothelial cell growth factor (150 μg/mL), 1 mM sodium pyruvate and heparin (5 units/mL) on 2% gelatin-coated dishes [20]. In some experiments, we utilized HUVEC stably transfected to silence *EDF1* (αs1), while their controls (CTR) were transfected with a scrambled, nonsilencing sequence as previously described [22].

4.2. Western Blot and Immunoprecipitation

Western blot was performed as described with antibodies against EDF1 (AVIVA Systems Biology Corporation, San Diego, CA, USA), rabbit anti-PPARγ, and anti-actin (Sigma Aldrich, St. Louis, MO, USA) [15]. Secondary antibodies were labeled with horseradish peroxidase (GE Healthcare, Milano, Italy). The immunoreactive proteins were visualized with the SuperSignal chemiluminescence kit (ThermoFisher Scientific, Waltham, MA, USA). To coimmunoprecipitate EDF1 and PPARγ, lysates were immunoprecipitated using monoclonal antibodies against PPARγ. Nonimmune immunoglobulin G (IgGs) were used as controls (Supplementary S3b). After binding to protein G-Sepharose, the samples were processed for western blot with rabbit anti-EDF1 antibodies. Nuclear and cytosolic fractions were obtained as described [20] and processed by western blot using antibodies against EDF1, anti-glyceraldehyde-3-phosphate dehydrogenase (GAPDH), and anti-TATA Binding Protein (TBP) (Santa Cruz Biotechnology-Tebu Bio, Huissen, The Netherlands). All the experiments were repeated at least three times. One representative blot is shown in the figures. Densitometry was performed using ImageJ software (1.50i, National Institutes of Health, Bethesda, MD, USA) on three blots and expressed using an arbitrary value scale. Results are shown as the mean ± standard deviation of three separate experiments.

4.3. Immunofluorescence Staining

Subconfluent HUVEC on gelatin-coated coverslips were treated or not treated with VEGF (50 ng/mL) (PeproTech, London, UK), fixed in phosphate-buffered saline containing 3% paraformaldehyde and 2% sucrose pH 7.6, permeabilized with HEPES-Triton 1%, incubated with anti-EDF1 immunopurified IgGs, and stained with rhodamine-conjugated anti-rabbit IgGs [20]. Staining with rabbit nonimmune IgGs did not yield any significant signal.

4.4. Reporter Gene Assay

To study PPARγ activity, subconfluent HUVEC with silenced *EDF1* and their controls were transfected with plasmids pDR1-Luc (0.2 μg/cm^2), using Arrest-in transfection reagent (Invitrogen) as described [24]. Luciferase activity was measured after 24 h of treatment with VEGF (50 ng/mL) using a luminometer. The transfection efficiency was normalized against a cotransfected reporter plasmid phRL-TK encoding Renilla luciferase (5 ng/cm^2), by dividing the firefly luciferase activity by the Renilla luciferase activity according to the Dual-Luciferase Reporter Assay kit manual (Promega, Madison, WI, USA). The experiments were performed in triplicate and the results are shown as the mean ± standard deviation of three separate experiments.

4.5. Real-Time-PCR

Total RNA was extracted using the PureLink RNA Mini kit (Ambion, Thermo Fisher Scientific, Waltham, MA, USA). After quantification, equivalent amounts of total RNA were assayed by first strand cDNA synthesis using SuperScript II RT (Invitrogen, Carlsbad, CA, USA). Real-time PCR was performed at least two times in triplicate on the 7500 FAST Real-Time PCR System instrument using TaqMan Gene Expression Assays (Life Technologies, Monza, Italy). We analyzed *FABP4* (Hs01086177_m1), *EDF1* (Hs00610152_m1), and *PPAR*γ (Hs01115513_m1) while the housekeeping gene *GAPDH* (Hs99999905_m1) was used as an internal reference gene. Relative changes in gene expression were analyzed by the $2^{-\Delta\Delta Ct}$ method.

4.6. NO Release

Griess assay was used to measure NO in cell culture media [23]. In particular, media were mixed 1:1 with fresh Griess solution and the absorbance was measured at 550 nm. The concentration of nitrites in the media were determined using calibration curve generated using known concentration of $NaNO_2$ solutions. The experiment was performed in triplicate and the results are shown as the mean $\pm$ standard deviation of three separate experiments.

4.7. Statistical Analysis

Statistical significance was determined using the Student's t test and set at p values less than 0.05. In the figures, * $p < 0.05$, ** $p < 0.01$, *** $p < 0.001$.

Supplementary Materials: Supplementary materials can be found at http://www.mdpi.com/1422-0067/19/7/1830/s1.

Author Contributions: A.C. and J.M. conceived and designed the experiments; A.C. and L.L. performed the experiments; A.C., L.L, and S.C. analyzed the data; J.M. wrote the paper.

Funding: This studied was sustained by intramural funds.

Conflicts of Interest: The authors declare no conflict of interest.

Abbreviations

VEGF	Vascular Endothelial Growth Factor
EDF	Endothelial Differentiation-related factor
PPARγ	Peroxisome proliferator-activated receptor γ
EC	Endothelial cells
NO	Nitric oxide

References

1. Duan, S.Z.; Usher, M.G.; Mortensen, R.M. Peroxisome Proliferator-Activated Receptor—Mediated Effects in the Vasculature. *Circ. Res.* **2008**, *102*, 283–294. [CrossRef] [PubMed]
2. Chawla, A.; Schwarz, E.J.; Dimaculangan, D.D.; Lazar, M.A. Peroxisome proliferator-activated receptor (PPAR) γ: Adipose predominant expression and induction early in adipocyte differentiation. *Endocrinology* **1994**, *135*, 798–800. [CrossRef] [PubMed]
3. Ahmadian, M.; Suh, J.M.; Hah, N.; Liddle, C.; Atkins, A.R.; Downes, M.; Evans, R.M. PPARgamma signaling and metabolism: The good, the bad and the future. *Nat. Med.* **2013**, *19*, 557–566. [CrossRef] [PubMed]
4. Imai, T.; Takakuwa, R.; Marchand, S.; Dentz, E.; Bornert, J.M.; Messaddeq, N.; Wendling, O.; Mark, M.; Desvergne, B.; Wahli, W.; et al. Peroxisome proliferator-activated receptor γ is required in mature white and brown adipocytes for their survival in the mouse. *Proc. Natl. Acad. Sci. USA* **2004**, *101*, 4543–4547. [CrossRef] [PubMed]
5. Tomaru, T.; Steger, D.J.; Lefterova, M.I.; Schupp, M.; Lazar, M.A. Adipocyte specific expression of murine resistin is mediated by synergism between peroxisome proliferator-activated receptor γ and CCAAT/enhancer-binding proteins. *J. Biol. Chem.* **2009**, *284*, 6116–6125. [CrossRef] [PubMed]

6. Iwaki, M.; Matsuda, M.; Maeda, N.; Funahashi, T.; Matsuzawa, Y.; Makishima, M.; Shimomura, I. Induction of adiponectin, a fat-derived antidiabetic and antiatherogenic factor, by nuclear receptors. *Diabetes* **2003**, *52*, 1655–1663. [CrossRef] [PubMed]
7. Halabi, C.M.; Beyer, A.M.; de Lange, W.J.; Keen, H.L.; Baumbach, G.L.; Faraci, F.M.; Sigmund, C.D. Interference with PPARγ Function in Smooth Muscle Causes Vascular Dysfunction and Hypertension. *Cell Metab.* **2008**, *7*, 215–226. [CrossRef] [PubMed]
8. Kotlinowski, J.; Jozkowicz, A. PPAR Gamma and Angiogenesis: Endothelial Cells Perspective. *J. Diabetes Res.* **2016**, *2016*, 8492353. [CrossRef] [PubMed]
9. Kleinhenz, J.M.; Kleinhenz, D.J.; You, S.; Ritzenthaler, J.D.; Hansen, J.M.; Archer, D.R.; Sutliff, R.L.; Hart, C.M. Disruption of endothelial peroxisome proliferator-activated receptor-γ reduces vascular nitric oxide production. *Am. J. Physiol. Heart Circ. Physiol.* **2009**, *297*, H1647–H1654. [CrossRef] [PubMed]
10. Chen, Z.; Ishibashi, S.; Perrey, S.; Osuga, J.; Gotoda, T.; Kitamine, T.; Tamura, Y.; Okazaki, H.; Yahagi, N.; Iizuka, Y.; et al. Troglitazone inhibits atherosclerosis in apolipoprotein E-knockout mice: Pleiotropic effects on CD36 expression and HDL. *Arterioscler. Thromb. Vasc. Biol.* **2001**, *21*, 372–377. [CrossRef] [PubMed]
11. Charbonnel, B.; Dormandy, J.; Erdmann, E.; Massi-Benedetti, M.; Skene, A.; PROactive Study Group. The prospective pioglitazone clinical trial in macrovascular events (PROactive): Can pioglitazone reduce cardiovascular events in diabetes? Study design and baseline characteristics of 5238 patients. *Diabetes Care* **2004**, *27*, 1647–1653. [CrossRef] [PubMed]
12. Wang, S.; Dougherty, E.J.; Danner, R.L. PPARγ signaling and emerging opportunities for improved therapeutics. *Pharmacol. Res.* **2016**, *111*, 76–85. [CrossRef] [PubMed]
13. Rosenfeld, M.G.; Lunyak, V.V.; Glass, C.K. Sensors and signals: A coactivator/corepressor/epigenetic code for integrating signal-dependent programs of transcriptional response. *Genes Dev.* **2005**, *20*, 1405–1428. [CrossRef] [PubMed]
14. Brendel, C.; Gelman, L.; Auwerx, J. Multiprotein bridging factor-1 (MBF-1) is a cofactor for nuclear receptors that regulate lipid metabolism. *Mol. Endocrinol.* **2002**, *16*, 1367–1377. [CrossRef] [PubMed]
15. Leidi, M.; Mariotti, M.; Maier, J.A. Transcriptional coactivator EDF-1 is required for PPARgamma-stimulated adipogenesis. *Cell. Mol. Life Sci.* **2009**, *66*, 2733–2742. [CrossRef] [PubMed]
16. Takemaru, K.; Li, F.Q.; Ueda, H.; Hirose, S. Multiprotein bridging factor 1 (MBF1) is an evolutionarily conserved transcriptional coactivator that connects a regulatory factor and TATA element-binding protein. *Proc. Natl. Acad. Sci. USA* **1997**, *94*, 7251–7256. [CrossRef] [PubMed]
17. Busk, P.K.; Wulf-Andersen, L.; Strøm, C.C.; Enevoldsen, M.; Thirstrup, K.; Haunsø, S.; Sheikh, S.P. Multiprotein bridging factor 1 cooperates with c-Jun and is necessary for cardiac hypertrophy in vitro. *Exp. Cell Res.* **2003**, *286*, 102–114. [CrossRef]
18. Liu, Q.X.; Jindra, M.; Ueda, H.; Hiromi, Y.; Hirose, S. Drosophila MBF1 is a co-activator for Tracheae Defective and contributes to the formation of tracheal and nervous systems. *Development* **2003**, *130*, 719–728. [CrossRef] [PubMed]
19. Dragoni, I.; Mariotti, M.; Consalez, G.G.; Soria, M.R.; Maier, J.A. EDF-1, a novel gene product down-regulated in human endothelial cell differentiation. *J. Biol. Chem.* **1998**, *273*, 31119–31124. [CrossRef] [PubMed]
20. Ballabio, E.; Mariotti, M.; De Benedictis, L.; Maier, J.A. The dual role of endothelial differentiation-related factor-1 in the cytosol and nucleus: Modulation by protein kinase A. *Cell. Mol. Life Sci.* **2004**, *61*, 1069–1074. [CrossRef] [PubMed]
21. Bautch, V.L. VEGF-Directed Blood Vessel Patterning: From Cells to Organism. *Cold Spring Harb. Perspect. Med.* **2012**, *2*, a006452. [CrossRef] [PubMed]
22. Mariotti, M.; De Benedictis, L.; Avon, E.; Maier, J.A.M. Interaction between endothelial differentiation-related factor-1 and calmodulin in vitro and in vivo. *J. Biol. Chem.* **2000**, *275*, 24047–24051. [CrossRef] [PubMed]
23. Leidi, M.; Mariotti, M.; Maier, J.A. EDF-1 contributes to the regulation of nitric oxide release in VEGF-treated human endothelial cells. *Eur. J. Cell Biol.* **2010**, *89*, 654–660. [CrossRef] [PubMed]
24. Castiglioni, S.; Cazzaniga, A.; Maier, J.A. Potential interplay between NFκB and PPARγ in human dermal microvascular endothelial cells cultured in low magnesium. *Magnes. Res.* **2014**, *27*, 86–93. [CrossRef] [PubMed]
25. Furuhashi, M.; Saitoh, S.; Shimamoto, K.; Miura, T. Fatty Acid-Binding Protein 4 (FABP4): Pathophysiological Insights and Potent Clinical Biomarker of Metabolic and Cardiovascular Diseases. *Clin. Med. Insights Cardiol.* **2015**, *8*, 23–33. [CrossRef] [PubMed]

26. Elmasri, H.; Ghelfi, E.; Yu, C.W.; Traphagen, S.; Cernadas, M.; Cao, H.; Shi, G.P.; Plutzky, J.; Sahin, M.; Hotamisligil, G.; et al. Endothelial cell-fatty acid binding protein 4 promotes angiogenesis: Role of stem cell factor/c-kit pathway. *Angiogenesis* **2012**, *15*, 457–468. [CrossRef] [PubMed]
27. Magri, C.J.; Gatt, N.; Xuereb, R.G.; Fava, S. Peroxisome proliferator-activated receptor-γ and the endothelium: Implications in cardiovascular disease. *Expert Rev. Cardiovasc. Ther.* **2011**, *9*, 1279–1294. [CrossRef] [PubMed]
28. Aragonès, G.; Saavedra, P.; Heras, M.; Cabré, A.; Girona, J.; Masana, L. Fatty acid-binding protein 4 impairs the insulin-dependent nitric oxide pathway in vascular endothelial cells. *Cardiovasc. Diabetol.* **2012**, *11*, 72. [CrossRef] [PubMed]
29. Gheibi, S.; Jeddi, S.; Kashfi, K.; Ghasemi, A. Regulation of vascular tone homeostasis by NO and H_2S: Implications in hypertension. *Biochem. Pharmacol.* **2018**, *149*, 42–59. [CrossRef] [PubMed]
30. Cho, D.H.; Choi, Y.J.; Jo, S.A.; Jo, I. Nitric oxide production and regulation of endothelial nitric-oxide synthase phosphorylation by prolonged treatment with troglitazone: Evidence for involvement of peroxisome proliferator-activated receptor (PPAR) gamma-dependent and PPARgamma-independent signaling pathways. *J. Biol. Chem.* **2004**, *279*, 2499–2506. [CrossRef] [PubMed]

Review

PGC-1α as a Pivotal Factor in Lipid and Metabolic Regulation

Ching-Feng Cheng [1,2,3], Hui-Chen Ku [1] and Heng Lin [4,5,*]

1. Department of Pediatrics, Taipei Tzu Chi Hospital, Buddhist Tzu Chi Medical Foundation, New Taipei City 23142, Taiwan; chengcf@mail.tcu.edu.tw (C.-F.C.); ku311@hotmail.com (H.-C.K.)
2. Institute of Biomedical Sciences, Academia Sinica, Taipei 11529, Taiwan
3. Department of Pediatrics, Tzu Chi University, Hualien 97004, Taiwan
4. Institute of Pharmacology, Taipei Medical University, 250 Wu-Hsing Street, Taipei 11031, Taiwan
5. Department of Physiology, School of Medicine, College of Medicine, Taipei Medical University, Taipei 11031, Taiwan

* Correspondence: linheng@tmu.edu.tw

Received: 31 August 2018; Accepted: 30 October 2018; Published: 2 November 2018

Abstract: Traditionally, peroxisome proliferator-activated receptor γ coactivator 1α (PGC-1α), a 91 kDa transcription factor, regulates lipid metabolism and long-chain fatty acid oxidation by upregulating the expression of several genes of the tricarboxylic acid cycle and the mitochondrial fatty acid oxidation pathway. In addition, PGC-1α regulates the expression of mitochondrial genes to control mitochondria DNA replication and cellular oxidative metabolism. Recently, new insights showed that several myokines such as irisin and myostatin are epigenetically regulated by PGC-1α in skeletal muscles, thereby modulating systemic energy balance, with marked expansion of mitochondrial volume density and oxidative capacity in healthy or diseased myocardia. In addition, in our studies evaluating whether PGC-1α overexpression in epicardial adipose tissue can act as a paracrine organ to improve or repair cardiac function, we found that overexpression of hepatic PGC-1α increased hepatic fatty acid oxidation and decreased triacylglycerol storage and secretion in vivo and in vitro. In this review, we discuss recent studies showing that PGC-1α may regulate mitochondrial fusion–fission homeostasis and affect the renal function in acute or chronic kidney injury. Furthermore, PGC-1α is an emerging protein with a biphasic role in cancer, acting both as a tumor suppressor and a tumor promoter and thus representing a new and unresolved topic for cancer biology studies. In summary, this review paper demonstrates that PGC-1α plays a central role in coordinating the gene expression of key components of mitochondrial biogenesis and as a critical metabolic regulator in many vital organs, including white and brown adipose tissue, skeletal muscle, heart, liver, and kidney.

Keywords: PGC-1α; metabolic homeostasis; adipose tissue; mitochondria

---

1. Introduction

The peroxisome proliferator-activated receptor gamma coactivator-1 (PGC-1) family includes ligands of multiple nuclear or non-nuclear receptors that control the expression of specific genes regulating cell metabolism. The first discovered member of the PGC-1 family, a 91 kDa nuclear protein [1] identified in brown adipose tissue (BAT) in mouse studies of cold-induced thermogenesis, was called peroxisome proliferator-activated receptor gamma (PPARγ) coactivator 1α (PGC-1α) [2]. The biological activity of PGC-1α is tightly controlled at several levels: by transcriptional control (of multiple promoter regions), alternative splicing of transcripts, and post-translational modification (e.g., phosphorylation, acetylation, or methylation). This activity results in several mRNA isoforms—PGC-1α-a, PGC-1α-b, PGC-1α-c, and NTPGC-1α—that enable cellular adaptation to various environmental conditions [3]. Studies have shown that PGC-1α can be used in different tissues with different coactivators to induce changes in lipid oxidation, energy homeostasis, mitochondrial mass, and insulin sensitivity. Here, we review these studies.

2. PGC-1α Can Regulate Lipid Metabolism

As a transcription factor, PGC-1α can bind to targets such as PPARα, PPARβ/δ, and PPARγ, which coordinate the expression of mitochondrial genes and indirectly contribute to fatty acid (FA) transport and utilization [4]. Furthermore, PGC-1α upregulates the expression of several genes of the tricarboxylic acid cycle [5] and the mitochondrial FA oxidation pathway [6]. PGC-1α also regulates the expression of nuclear and mitochondrial genes that encode components of the electron transport system and oxidative phosphorylation (OXPHOS) via nuclear respiratory factors 1 and 2 (NRF-1 and -2) and estrogen-related receptor α (ERRα) coactivation. These effects can increase the expression of mitochondrial transcription factor A (mtTFA), which is known to control mtDNA replication and transcription and therefore regulate cellular oxidative metabolism [7]. Accordingly, the augmented expression of cytochrome c, cytochrome-c-oxidase subunits II and IV, and adenosine triphosphate (ATP) synthase also result from PGC-1α activation [8–11].

Another noteworthy effect of PGC-1α is its ability to stimulate peroxisomal activity such as the oxidation of long-chain and very-long-chain FAs [12]. Briefly, PGC-1α level is positively correlated with the ability of cells to fully oxidize FA, an effect that may reduce intramuscular lipid deposition and improve tissue insulin sensitivity. Chromatin immunoprecipitation assays have shown that the mechanism of this effect includes the coactivation of liver X receptor α (LXRα), which stimulates PGC-1α binding to the LXR response element in the FAS promoter. In addition, muscle-specific PGC-1α expression in MPGC-1α transgenic mice exacerbated de novo free fatty acid (FFA) synthesis as well as FA esterification and triacylglycerol (TAG) accumulation [13]. Furthermore, PGC-1α is involved in lipid distribution and may upregulate FAT/CD36, FABPpm, and FATP1 mRNA and protein expression in mitochondrial fractions. The latter effect was confirmed solely in murine FAT/CD36 and FABP3 cells [14].

3. PGC-1α as a Coactivator for Metabolic Homeostasis in Skeletal Muscle

Muscle adjusts to endurance exercise by promoting mitochondrial biogenesis, angiogenesis, and changes of fiber composition [15–17]. Chinsomboona et al., had reported that mice lacking PGC-1α in skeletal muscle failed to increase capillary density in response to exercise. This study showed that β-adrenergic stimulation of a PGC-1 α/estrogen-related receptor alpha (ERRα)/vascular endothelial growth factor (VEGF) axis modulates exercise-induced angiogenesis in skeletal muscle [18] and truncated PGC-1α can lead to hypoxic induction of VEGF and angiogenesis in skeletal muscle [19]. In addition, PGC-1α activates transcription in cooperation with myocyte enhancer factor-2 (Mef2) and acts as a target for calcineurin signaling, which has been involved in slow fiber gene expression [20]. Skeletal muscle-specific PGC-1α knock-out mice demonstrate a shift from oxidative type I and IIa toward type IIx and IIb glycolytic muscle fibers [21]. Rasbach et al., reported that PGC-1α–mediated switch to slow, oxidative fibers in vitro is dependent on hypoxia-inducible factor 2 α (HIF2α), and mice lacking HIF2α in muscle increase the expression of genes and proteins related to a fast-twitch-fiber-type switch [22]. Transgenic mice with mildly elevated muscle levels of PGC1α are also resistant to age-related obesity and diabetes and show a prolonged lifespan [23]. These results strongly suggest that PGC1α expression in skeletal muscles can significantly contribute to regulating systemic energy balance. Recent studies also demonstrated that muscle contraction may induce the secretion of molecules called myokines, which enables the crosstalk between skeletal muscle and other organs such as adipose tissue, bone, liver, kidney, and brain; in this sense, skeletal muscle can be considered an endocrine organ. Indeed, several myokines discovered in the past decade via secretome analysis include interleukin-6, irisin/fibronectin type III domain-containing protein 5 (FNDC5), myostatin, interleukin-15, brain-derived neurotrophic factor (BDNF), β-aminoisobutyric acid, meteorin-like, leukemia inhibitory factor, and secreted protein acidic and rich in cysteine (SPARC).

Several myokines are regulated by PGC-1: irisin/FNDC5, myostatin, and *BDNF* [24]. (1) Irisin is a PGC-1α-dependent myokine. In mice with muscle-specific PGC-1α overexpression, PGC-1α induces the expression of a membrane protein, FNDC5, and exercise triggers the cleavage of FNDC5 to generate irisin and then secreted into the bloodstream, which elevates energy expenditure in subcutaneous adipose tissue via adipocyte browning [25]. This process implies that PGC-1α overexpression with exercise may increase the expression of uncoupling protein 1 (UCP-1) and eventually increase the browning of white fat cells [25]. Recently, mass spectrometry was used to measure circulating irisin levels in humans in an antibody-independent manner; irisin levels were increased by both short and prolonged period exercise [26,27]. Under physiological conditions, irisin stimulates glucose uptake and lipid metabolism via the activation of AMP-activated protein kinase (AMPK) [28–30] and is also involved in muscle growth by inducing insulin-like growth factor 1 and suppressing myostatin [31]. In addition to having effects on muscle, exogenous administration of irisin induces adipocyte browning in subcutaneous fat in mice via p38 mitogen-activated protein kinase (MAPK) and extracellular signal-regulated kinase 1/2 (ERK1/2) [32]. In the murine liver, irisin stimulates glycogenesis but reduces gluconeogenesis and lipogenesis by regulating GSK3, FOXO1, and SREBP2 [33–35]. (2) Myostatin is an autocrine and paracrine hormone secreted by muscle fibers and the only myokine with inhibited secretion during muscle contraction and exercise [36]. In addition to its local involvement in muscle atrophy [37], myostatin can also modulate metabolic homeostasis by regulating adipose tissue function [38–40]. The inhibition of myostatin was found to ameliorate the development of obesity and insulin resistance in mice fed a high-fat diet, presumably by mechanisms promoting lipolysis and mitochondrial lipid oxidation in adipose tissue and liver [41]. In addition, Dong et al., showed that inhibition of myostatin resulted in the conversion of white adipose tissue (WAT) to brown adipose tissue (BAT), while enhancing fatty acid oxidation and increasing energy expenditure. Inhibition of myostatin increased PGC-1α expression and irisin production in muscle. Irisin stimulated browning via mediating muscle-to-fat cross talk [42]. Myostatin knockout mice are characterized by increased expression and phosphorylation of AMPK in muscle, which subsequently activates PGC1α and Fndc5. This study demonstrated that Fndc5 is upregulated and secreted from muscle to induce browning of WAT in myostatin knockout mice [43]. (3) BDNF is known primarily as a molecule released by the hypothalamus and as a key element regulating neuronal development, plasticity, and energy homeostasis [44]. Cao et al., found that hypothalamic overexpression of BDNF via recombinant adeno-associated virus (rAAV) duplicated the enriched environment (EE)-associated activation of the brown fat program and lean phenotype. This study suggested that induction of hypothalamic BDNF expression in response to environmental stimuli results in selective sympathoneural regulation of white fat browning and increased energy dissipation [45]. Wrann et al., showed hippocampal BDNF gene expression [46]. PGC-1α knockout mice show decreased FNDC5 expression in the brain. Overexpression of FNDC5 increases BDNF expression in primary cortical neurons. Furthermore, peripheral delivery of FNDC5 to the liver leads to elevated blood irisin and increased BDNF expression in the hippocampus. Taken together, this study links endurance exercise and the significant metabolic mediators, PGC-1α and FNDC5, with BDNF expression in the brain [46] (Figure 1).

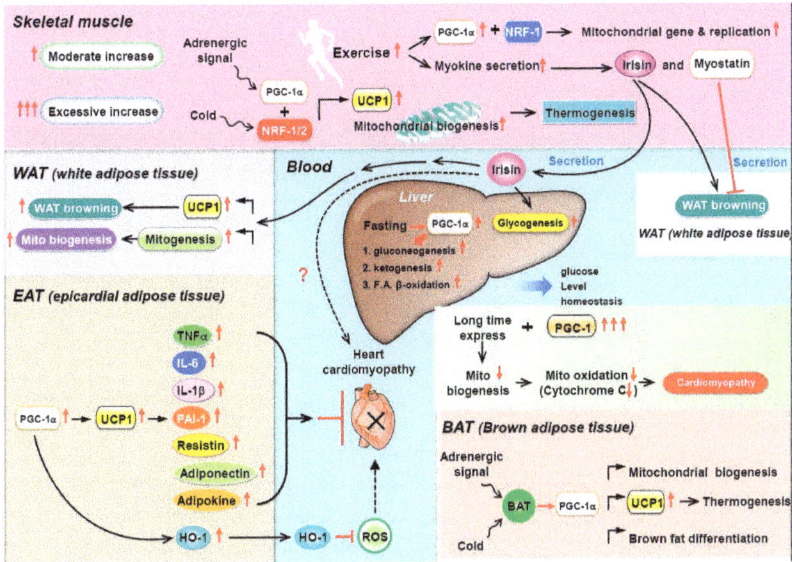

Figure 1. Schematic description of peroxisome proliferator-activated receptor gamma coactivator-1α (PGC-1α) function in various organs. Traditionally, in brown adipocytes (BAT) and white adipocytes (WAT), mitochondrial biogenesis and BAT gene expression are regulated by PGC-1α. Adrenergic stimulation and lower temperature trigger signaling cascades, including the upregulation of UCP-1 level, thereby resulting in body thermogenesis. In skeletal muscle and WAT, the transcriptional activity of PGC-1α is responsible for the expression of gene networks that control glucose uptake, glycolysis, fatty acid (FA) oxidation, tricarboxylic acid cycle, oxidative phosphorylation (OXPHOS), mitochondrial biogenesis, and protein uncoupling. Therefore, increasing exercise will increase mitochondrial gene biogenesis and secretion of myokines (such as irisin), which results in WAT browning and liver gluconeogenesis to prevent obesity and insulin resistance. In epicardial adipose tissue (EAT), increased heme oxygenase 1 (HO-1) expression depends on the PGC-1α–UCP-1 axis activity, which then decreases free radicals and reactive oxygen species (ROS) production, thus reducing cardiomyopathy. However, whether increased expression of cytokines such as TNF-α, IL-6, or adipokines by the PGC-1α–UCP-1 axis can reduce cardiomyopathy or not is still unclear.

4. PGC-1α as a Coactivator in WAT Browning, Thermogenesis, and Mitochondrial Biogenesis

Much of the adaptive thermogenesis in small mammals takes place in BAT. BAT is morphologically and metabolically different from WAT and partly exerts opposite physiological functions. Adipocytes from BAT contain multiple small triglyceride-filled droplets as well as a large number of mitochondria. In addition, their mitochondria contain a specific UCP-1, expressed only in brown adipocytes. Genetic studies with mice lacking UCP-1 or PGC-1α in adipocytes indicated that (1) PGC-1α is the only protein that can powerfully activate the UCP-1 enhancer in non-BAT cell lines and (2) when pharmacologically introduced into white adipocytes, PGC-1α induces mitochondrial gene expression and mitochondrial biogenesis. Finally, PGC-1α is a downstream target of adaptive thermogenesis in BAT via adrenergic receptor activation [47,48], the key mechanism in brown-fat differentiation in in vitro cell cultures and in vivo cellular responses to cold exposure [24]. Brown fat and skeletal muscle, in which PGC-1α is highly expressed and can be induced by cold or adrenergic stimuli with enhanced mitochondrial biogenesis, are the two main contributing tissues in adaptive thermogenesis via the adrenergic receptor PGC-1α–UCP-1 axis. Scarpulla and collaborators [49,50] identified and cloned two novel transcription factors, NRF-1 and -2, that bind to the promoter region of the mitochondrial genes β-ATP synthase, cytochrome-c, cytochrome-c-oxidase subunit IV, and mtTFA. PGC-1α has a major effect on the NRF system. When introduced into muscle cells in vitro, PGC-1α greatly induces the gene expression of NRF-1, NRF-2, and mtTFA. Furthermore, PGC-1α interacts directly with NRF-1 and co-activates its transcriptional activity [51] (Figure 1).

5. PGC-1 Controls Cardiac Energy Metabolism in Healthy or Diseased Myocardia

In mammalian embryos, proliferating cardiomyocyte precursor cells rely on glycolysis as their major energy source, and mitochondrial tissue and oxidative metabolism are poorly developed. Once precursor cells differentiate into mature cardiomyocytes, a shift occurs from glycolysis to FA metabolism as the main provider of the entry point for mitochondrial oxidative phosphorylation, which in mature heart cells yields most of the energy [52]. Therefore, during neonatal development, the healthy myocardium increases its rate of β-oxidation while simultaneously decreasing glycolytic activity. Eventually, adult heart muscle derives ~90% of its energy from oxidative phosphorylation in mitochondria, which occupy only ~30% of cardiomyocyte volume [53,54]. During various cardiac disease processes, such as hypertrophy or ischemia-induced cardiomyopathy, both the inhibition of mediators of mitochondrial oxidative phosphorylation (cytochrome-c-oxidase subunits) and the expression or activity of metabolic enzymes involved in oxidative phosphorylation [55,56] were noted [57]. These processes of cardiac remodeling result in a gradual decrease in mitochondrial biogenesis [58], and ATP is utilized for maintaining ion homeostasis rather than for force production during cardiomyocyte contraction; this process leads to irreversible hypertrophy or dilated cardiomyopathy. However, during prolonged periods of cardiac remodeling, cardiomyocyte energy metabolism is regulated by the actions of various transcription factors and their coactivators, such as the PGC-1 family. PGC-1α has been shown to interact with three families of transcription factors: (1) the PPAR family, which regulates the expression of genes involved in FA oxidation; (2) the ERR family; (3) NRF-1 [2,59–61], which controls genes that are involved in mitochondrial oxidative phosphorylation and the electron transport chain [62,63]. In cardiomyocytes, PGC-1α is considered a master regulator of metabolism because it co-activates PPARs, ERRs, and NRFs [4,64] and may thereby control the entire metabolic phenotype of cardiomyocytes [7].

In the heart, the interrelationship between PGC-1α and PPARα plays an important role in regulation of the expression of enzymes involved in FAO and uptake pathways [65] and may be involved in regulation of mitochondrial biogenesis [66]. PGC-1α loss of function in murine heart exhibited a damage to mitochondrial respiratory function and reduced expression of genes involved in several mitochondrial metabolic pathways [67–69]. Hearts from PGC-1α KO mice showed reductions in mitochondrial enzymatic activities and ATP levels [67]. Arany et al., had shown that PGC-1α KO mice are prone to develop of heart failure in response to transverse aortic constriction (TAC). Furthermore, induction of PGC-1α in cells via catecholamine treatment can reverse the mitochondrial genes inhibition, suggesting that PGC-1α may be a potential therapeutic target in heart failure [68]. In addition, PGC-1α deficient mice cause energy metabolic derangements in multiple systems [69]. Conversely, overexpression of PGC-1α in adult mice had shown a moderate mitochondrial proliferation, abnormal mitochondrial architecture and severe cardiac dysfunction [70], and constitutive overexpression of PGC-1α in murine heart resulted in unconstrained mitochondrial proliferation in cardiac myocytes leading to a dilated cardiomyopathy [10].

Cardiac energy substrate metabolism is disturbed in the hypertrophic and failing heart. The myocardium switches from dependence on fatty acid oxidation (FAO) to glucose utilization in the failing heart, mainly anaerobic glycolysis [71–74]. These alterations in energy substrate preference are regulated, at least partially, by the downregulation of the genes involved in OXPHOS and FAO and the PPARα–PGC-1α complex [56,73,75–78]. The expression levels of PPARα and PGC-1α are reduced in several mice models of pressure overload, hypertensive heart disease [68,75,79], ischemic heart disease [57,80–82], hypoxia [76], and genetically engineered mouse models of heart failure [83–85]. Additionally, under pathologic conditions, PPARα activity is inhibited by the lower levels of the heterodimeric partner, retinoid X receptor (RXR) [86] and by direct phosphorylation, dependent on the extracellular signal-related kinase and mitogen-activated protein kinase (ERK–MAPK) pathway [75]. These findings suggest that deactivation of the cardiac PPARα–PGC-1α axis is an important component of the switch in energy metabolism in the failing heart. It remains to be addressed whether the deactivation of the oxidative metabolism and the PPARα–PGC-1α complex in the hypertrophied and ischemic heart is adaptive or maladaptive. The increment of myocardial reliance on anaerobic glycolytic pathways for ATP production is likely an adaptive response to reduce oxygen consumption. Indeed, partial inhibitors of FAO exhibited a promising therapeutic effect for cardiac disease [87–89]. Liao et al., had reported that overexpression of the GLUT1 glucose transporter can prevent pressure overload-induced heart failure [90]. Moreover, overexpression of PGC-1α [83] and PPAR agonists [91–93] can prevent cardiac hypertrophy or improve cardiac myocyte contractility.

Cardiovascular disease is extraordinarily widespread in diabetic patients. Cardiomyopathy in diabetic subjects that occurs in the absence of known risk factors (hypertension, hyperlipidemia, etc.) is often referred to as "diabetic cardiomyopathy" [94–97]. Many studies have proposed that abnormalities in myocardial energy metabolism play an important role in the pathogenesis of diabetic cardiomyopathy. Indeed, the diabetic heart relies nearly exclusively on mitochondrial FAO for ATP requirements [98–101]. The expression levels of PPARα, PGC-1 α, and various target genes involved in FAO are increased in the murine insulin-resistant [66] and diabetic heart [102–104]. Moreover, transgenic mice that overexpress PPARα exclusively in the heart (MHC-PPARα mice) demonstrate a cardiac metabolic phenotype similar to that observed in the diabetic heart, including accelerated rates of FAO, reduction in glucose uptake and utilization, and repression of the mitochondrial biogenic response [66,102]. Mitochondria isolated from diabetic rodents showed reduced rates of OXPHOS [105,106] and decreased efficiency in ATP synthesis [107,108], likely due to increased uncoupled respiration [108]. The importance of PPARs and PGC-1α in the modulation of cardiac energy metabolism makes these regulatory pathways attractive therapeutic targets for diabetic cardiomyopathy.

In summary, increased PPARα and PGC-1α expression with the marked expansion of mitochondrial volume density and oxidative capacity accompany normal cardiac growth during postnatal maturation. Conversely, pathologic hypertrophy is associated with decreased PPARα–PGC-1α expression and/or activity and diminished reliance on oxidative mitochondrial metabolism, which leads to intramyocardial cell lipid accumulation. Finally, gain-of-function studies with PGC-1α overexpression in mice revealed that the extent of cardiomyopathy is primarily determined by the amount of PGC-1α that could be detected in the heart and, more importantly, the moment and duration of its emergence. Thus, both the synthesis and the moment of appearance of PGC-1α play important roles in the regulation of myocardial metabolism and mitochondrial biology (Figure 1).

6. Is PGC-1 a Paracrine Regulator in Epicardial Adipose Tissue?

Recently, a new type of adipose tissue, epicardial adipose tissue (EAT), was found in the heart of patients undergoing open-heart surgery. EAT is physically located next to the myocardium within the lateral wall of the right ventricle and the anterior wall of the left ventricle and surrounds the right coronary and left-anterior descending coronary arteries [109]. Similar to WAT, EAT shows high rates of lipogenesis but also high degrees of WAT lipolysis and thus serves as a local triacylglycerol (TAG) store [110]. EAT contains about five times more UCP-1 mRNA than WAT and also shows high expression of many genes of beige adipose tissue [111], that is, CD137, PRDM16, PGC-1α, C/EBPβ, and PPARα. The present understanding of the potential physiological roles of EAT includes: (1) the release of free FAs as energy to the myocardium under conditions associated with high metabolic demands, (2) the expression of the thermogenic protein UCP-1 in response to cold exposure, and (3) the expression and secretion of specific molecules for cardiovascular protection by vasocrine and paracrine pathways. EAT contributes to cardiovascular protection and vessel remodeling by secreting various paracrine factors. Several EAT-derived factors or cytokines, such as tumor necrosis factor alpha (TNF-α), monocyte chemoattractant protein-1 (MCP-1), interleukin-6 (IL-6), IL-1β, plasminogen activator inhibitor-1 (PAI-1), resistin, and adipokines, have both vasocrine and paracrine effects on the myocardium [112,113]. Other specific molecules secreted from EAT, such as adiponectin and adipocyte-derived relaxing factors called adipokines, can decrease contraction and vasoconstriction by increasing nitric oxide (NO) release or by reducing reactive oxygen species (ROS) production [114]. In addition, macrophages residing in EAT can release anti-inflammatory cytokines such as IL-10 [115]. EAT may contribute to cardioprotection by the local secretion of anti-inflammatory and anti-atherogenic adipokines such as adiponectin and adrenomedullin [116,117]. Both adiponectin and adrenomedullin are directly secreted from EAT into the coronary circulation, and their mRNA levels are correlated with their intracoronary levels [118,119]. Clinically, both adiponectin and adrenomedullin expression in EAT were significantly reduced in patients with coronary artery disease [118,119]. In a clear demonstration of the paracrine regulation of the cardiac function by PGC1 in mice, we found that chronic iron loading attenuated serum adiponectin concentration, thereby resulting in cardiomyopathy. In addition, adiponectin gene (ADIPOQ) overexpression in the heart after adeno-associated virus delivery (AAV-ADIPOQ) ameliorated cardiac iron deposition and restored the cardiac function in iron-overloaded mice; this occurred via the induced expression of heme oxygenase 1 (HO-1) through the PPARα–PGC-1 complex–dependent pathway in cardiomyocytes [120]. Craige et al., created mice with endothelial-specific loss of function (PGC-1α EC KO) that showed significantly reduced PGC-1α expression as well as decreased endothelial NO synthase (eNOS) expression and NO• bioactivity in response to angiotensin-II-induced hypertension [121]. The authors found that PGC-1α EC KO mice had significantly increased blood pressure with vascular dysfunction compared with Cre control mice. In summary, they showed that endothelial PGC-1α expression is required to exert vascular protection via increased bioactivity of NO• through ERRα-induced expression of eNOS, thus preventing cardiovascular disease.

7. Potential for PGC-1α to Modulate Paracrine Regulators in the Heart

PGC-1α is abundantly expressed in tissues with high energy requirements, such as the heart, skeletal muscle, kidney, and BAT [122,123]. In these tissues, PGC-1α controls the expression of genes involved in energy homeostasis, mitochondrial biogenesis, and free FA oxidation function [5,6]. In the heart, cardiomyopathy progression is determined by the amount and the time period of PGC-1α expression. However, the therapeutic window of PGC-1α in cardiomyocytes is relatively narrow because prolonged overexpression of this cofactor leads to uncontrolled mitochondrial proliferation, abnormal sarcomeric structure, and dilated cardiomyopathy [10,124]. Similar phenomena were found in kidney diseases. In the kidney, the basal expression of PGC-1α is stronger in the proximal than the distal tubules, whereas in the glomerulus it is low. A recent study showed aggravated glomerular cell injury when PGC-1 was chronically overexpressed, which is in contrast to the beneficial effects of PGC-1α expression in the proximal tubules promoting acute kidney injury recovery during systemic inflammation [125] or in cisplatin-induced acute renal injury [126]. The cardiac endothelium forms a continuous monolayer of cells that lines the cavity of the heart (endocardial endothelial cells (EECs)) and the luminal surface of the myocardial blood vessels (intramyocardial capillary endothelial cells (IMCEs)). Both EECs and IMCEs can master the contractility of cardiomyocytes by releasing various factors such as NO via endothelial NO-synthase (eNOS), angiotensin II, endothelin-1, peptide growth factors, prostaglandins, and neuregulin-1 (NRG-1) [127]. Craige et al., showed that PGC-1α expression protects the endothelium via increased eNOS expression and NO• bioactivity. ERRα is required for PGC-1α-mediated eNOS expression [121]. Chronic NRG-1 treatment increased oxidative metabolism and mitochondrial activity by enhancing the expression of PGC-1α and PPARδ [128]. Whether PGC-1α can modulate paracrine regulator in the heart needs further investigation. Besides, there are no reports confirming that EAT can act in a paracrine fashion to regulate PGC-1α expression in cardiomyocytes. However, prior reports have indicated that the expression of PGC1α in skeletal muscle may enable the production and release of myokines for the crosstalk between skeletal muscle and other organs. Therefore, future studies should focus on exploring whether PGC-1 in EAT stimulates the secretion of factors that regulate cardiac functions in a paracrine manner, in which cardiac muscle and skeletal muscle can act as endocrine organs.

8. PGC-1α Regulates Metabolic Homeostasis in the Liver

The expression of PGC-1α is induced in the liver at birth [129]. Starvation induces PGC-1α expression in the adult liver via glucagon and glucocorticoid (GR) signaling [130]. The fed-to-fasted transition cause metabolic changes in the liver to promote adaptation to nutrient deprivation. These metabolic changes consist in the activation of hepatic gluconeogenesis, FA β-oxidation, heme biosynthesis, bile acid homeostasis, and synthesis and secretion of ketone bodies [131]. In vitro studies in hepatocytes and in vivo studies have shown that PGC-1α is sufficient to activate the hepatic fasting responses, which include gluconeogenesis, ketogenesis, FA β-oxidation, and bile acid homeostasis [130,132,133]. PGC-1α regulates the metabolic adaption to fasting by coactivating key hepatic transcription factors such as HNF4α, PPARα, GR, FOXO1, LXR, and FXR [4]. PGC-1α-KO mice and RNAi-mediated liver-specific PGC-1α-knockdown mice showed defective gluconeogenic gene expression and hepatic glucose production [134,135]. These mice show a tendency for hypoglycemia and hepatic steatosis upon fasting [69]. In addition, PGC-1α stimulates the expression of genes involved in homocysteine metabolism in cultured primary hepatocytes and in the liver [136]. Hepatic PGC-1α protein expression and activation of mitochondrial biogenesis were reduced in a mouse model of hepatic steatosis [137]. PGC-1α plays an important role in exercise-induced hepatic mitochondrial adaptation [138]. PGC-1α expression was lower in the liver of obese, sedentary humans than lean humans [139]. Overexpression of hepatic PGC-1α increased hepatic FA oxidation with decreased TAG storage and secretion in vivo and in vitro [140]. In addition, PGC-1α integrates the mammalian clock and energy metabolism. PGC-1α stimulates the gene expression of the clock genes *Bmal1* (*Arntl*) and *Rev-erbα* (*Nr1d1*) by coactivation of the receptor tyrosine kinase-like orphan receptor family of orphan nuclear receptors. PGC-1α-null mice show abnormal diurnal rhythms of activity, body temperature, and metabolic rate [141]. Therefore, PGC-1α regulates both the fed-to-fasted energy transition and the diurnal rhythm in liver metabolic homeostasis.

9. PGC-1α Regulates Kidney Metabolism via Mitochondrial Homeostasis

As a bridge between homeostasis and mitochondrial function, PGC-1α activates NRF-1 and -2, which are nuclear-encoded transcription factors that promote the expression of multiple genes involved in mitochondrial DNA transcription and mitochondrial respiratory chains with anti-oxidative effects [142,143]. In the kidney, PGC-1α is predominantly expressed in the proximal tubules, and enforced expression of PGC-1α in cultured proximal tubular cells increased mitochondrial number, respiratory capacity, and mitochondrial protein level, which indicates the effectiveness of PGC-1α in proximal tubular homeostasis [122,144]. In the septic acute kidney injury (AKI) model, PGC-1α expression in tubular cells was proportionally decreased with an increasing degree of renal impairment. Although mice with PGC-1α gene deletion do not show altered kidney size [69,134], they exhibit increased serum blood urea nitrogen (BUN) and creatinine levels in these models [125], and patients and mouse models with acute and chronic kidney disease commonly show decreased PGC-1α expression accompanied by reduced FA oxidation. In addition, treatment with the PPARγ agonist rosiglitazone could induce PGC-1α expression in the nucleus of renal mesangial cells and significantly ameliorate renal fibrosis in mouse models of diabetic kidney disease. Furthermore, in vitro experiments with cultured renal mesangial cells demonstrated that PGC-1α knockdown increased glucose-induced ROS levels [145].

Studies from Rasbach et al., who used tertbutyl hydroperoxide (tBHP), an agent that profoundly depletes cellular glutathione, to induce oxidative stress in the rabbit proximal tubular cell culture system resulted in iron-dependent lipid peroxidation with extensive primary mitochondrial damage [146]. PGC-1α protein level was greatly increased after tBHP treatment, and the increase could be blocked by inhibiting the epidermal growth factor receptor–Src–p38 MAPK axis pathway. However, adenovirus-induced PGC-1α overexpression produced a 25% to 50% increase in mitochondrial number [147], which had a protective effect against tBHP-induced cell damage [144]. Choi et al., found that PGC1-α could attenuate ischemia-reperfusion-induced acute kidney injury by ameliorating the mitochondria dysfunction mediated by p38 signaling [148]. These consistent in vivo and in vitro findings indicate that PGC-1α expression may be increased in the early stage of acute and chronic kidney injury as a compensatory response and PGC-1α can regulate renal tubular mitochondrial biogenesis. Other kidney cells, such as podocytes and endothelial cells, are less metabolically active and have a narrow PGC-1α tolerance. Increasing PGC-1α levels in podocytes induce podocyte proliferation and collapsing glomerulopathy development, whereas increasing PGC-1α levels in endothelial cells alter the endothelial function and cause microangiopathy, thus highlighting the cell type-specific role of PGC-1α in the kidney (Figure 2A).

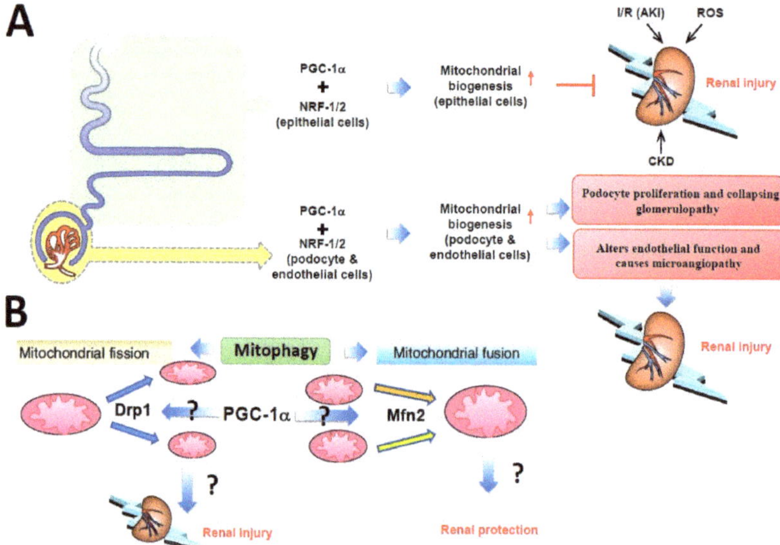

Figure 2. Schematic description of PGC-1α function in renal homeostasis. (**A**) PGC-1α associated with nuclear respiratory factors 1 and 2 (NRF-1/2) has a protective role in renal epithelial cells, including the proximal convoluted tubule, loop of Henle, and distal convoluted tubule, during renal injury by increasing mitochondrial biogenesis in epithelial cells. Bowman's capsule, podocytes, and endothelial cells have a narrow PGC-1α tolerance. Increased PGC-1α levels in podocytes induce podocyte proliferation and collapsing glomerulopathy development, whereas increased PGC1-α in endothelial cells alters endothelial function and causes microangiopathy, thereby resulting in renal injury. (**B**) The role of mitochondrial fusion and fission in mitophagy. Mitochondrial fusion is promoted by the *Mfn2* gene, whereas *Drp1* promotes mitochondrial fission. Increased *Drp1* and decreased *Mfn2* expression exacerbates tubular damage, thereby contributing to kidney disease; however, studies have shown opposite results and inconsistencies. Whether PGC-1α transcriptionally regulates *Drp1* and *Mfn2* requires further research.

Defects in mitochondrial fusion–fission homeostasis lead to altered mitochondrial morphology and impaired mitochondrial function and cause tubular damage in acute kidney injury. In addition, the balance between mitochondrial fusion and fission shifts to mitochondrial fission, resulting in mitochondrial fragmentation and then altered mitochondrial structure and renal tubular cell apoptosis [149]. Brooks et al. [149] observed mitochondrial fragmentation and *Drp1* mobilization to the outer mitochondrial membrane in injured tubular cells. *Drp1*, a mitochondrial fission mediator, is activated rapidly after ischemia-reperfusion-induced injury and induces mitochondrial fragmentation and subsequent renal tubular cell apoptosis [149]. By using dominant-negative mutants and RNA interference, Jiang et al., demonstrated that *Drp1* inhibition attenuated mitochondrial fragmentation, preserved mitochondrial integrity, limited renal cell apoptosis, and preserved kidney function. However, pharmacological inhibition or genetic deletion of autophagy-related genes worsened renal injury. These inconsistent results may imply that excessive mitochondrial fission during acute kidney injury is deleterious to organ function, and safe clearance of damaged mitochondria via mitophagy may be protective [150,151]. Meanwhile, primary cultured cells with tissue-specific knockout of *Mfn2*, a mitochondrial fusion mediator, in renal proximal tubular cells were highly sensitive to Bax activation and cytochrome c release, which led to cell apoptosis. However, *Mfn2* is also known to suppress cell proliferative effects via p21Ras, independently of mitochondrial dynamics [152,153]. Such *Mfn2*-mediated hyperplasia suppression may contribute greatly to renal recovery after stress; therefore, reducing *Mfn2* level in proximal tubular cells might actually accelerate organ recovery [154,155]. Gall et al., had shown that conditional knockout of proximal tubule *Mfn2* markedly boosts recovery of renal function and increased rodent survival after acute renal ischemia, partially by activating Ras and ERK1/2 signaling [156]. The above findings indicate, in general, that increased *Drp1* or decreased *Mfn2* levels exacerbate renal tubular damage via an imbalance in mitochondrial fission and fusion, with subsequent enhancement of mitochondrial fragmentation and aggravated acute kidney injury; however, studies with opposite results were also reported (Figure 2B). Further research is needed to investigate whether PGC-1α evokes the performance of the mitochondrial genes via the *Drp1–Mfn2* balance pathway and thus affects the function of the kidney in health or disease.

10. PGC-1α Regulates Cancer Metabolism

Metabolic reprogramming occurring in cancer cells refers to the ability to grow and survive under nutrient-starved or stressful microenvironments [157,158]. Increments in glycolysis, glutaminolytic flux, amino acid and lipid metabolism, and mitochondrial biogenesis have been observed in cancer development [159–162]. Deregulated metabolism is associated with oncogenesis, including the phenomenon of epithelial-to-mesenchymal transition (EMT), a complicated process that enables cancer cells to invade neighboring tissues and migrate to the vasculature [163–165]. Among the numerous regulators of cancer metabolism, PGC-1α has been shown to regulate many processes linked to oncogenesis by, for example, promoting the expression of antioxidant genes which protect cells from the detrimental effects of ROS, enhancing the catabolism of glucose and fatty acids, and promoting gluconeogenesis and lipogenesis which perform opposite anabolic functions [166–170]. No specific variant or isoform of PGC-1α has been reported in cancer studies. Some studies have shown that biphasic expression of PGC1α was observed in cancer biopsies or cells of breast cancer [171–174], melanoma [175–177], colon cancer [169,178], and ovarian cancer [179–181]. Low PGC-1α levels are associated with a worse outcome in breast and liver carcinomas [171,172,182]. The chemoresistant clear-cell subtype of ovarian carcinoma was identified by the lack of expression of both PGC-1α and mitochondrial transcription factor A (TFAM) [180]. In contrast, some studies showed that the plasma concentrations of PGC1α in breast cancer patients were higher than in healthy groups, and a multivariate analysis showed a correlation between high levels of PGC-1α and worse prognosis [183]. In a report of prostate cancer, androgens signaling via AMPK caused the increment of PGC1α mitobiogenesis, OXPHOS, and glycolysis. Furthermore, findings in mouse xenografts and patient samples suggested that AMPK–PGC1α function was associated with prostate cancer growth [184].

Even though many studies have investigated the role of PGC-1α in cancer by examining its expression via PGC-1α overexpression and siRNA knockdown experiments, the role of PGC-1α in cancer is still controversial. Several studies have shown that PGC-1α has tumor-suppressive effects.

PGC-1α overexpression in melanoma cells by adenovirus infection suppressed metastasis via the direct regulation of inhibitor of DNA binding protein 2 (ID2) and the inhibition of transcription factor 4 (TCF4)-mediated gene transcription [177]. Human ovarian cancer cell line Ho-8910 overexpressing PGC-1α has been shown to undergo apoptosis through downregulation of B-cell lymphoma 2 (Bcl-2) and upregulation of Bcl2-associated X protein (Bax) [185]. Wang et al., revealed that increased PGC-1α expression by a PPAR pan-agonist (bezafibrate) upregulated mitochondrial biogenesis, resulting in the inhibition of proliferation and invasion in HeLa, 143B, and MDA-MB-231 cancer cells [186]. Overexpression of PGC-1α by adenovirus infection in HepG2 human hepatoma cells upregulated E-cadherin expression and inhibited cell motility [187]. A study by Torrano et al., showed that PGC-1α inhibited the metastasization of prostate carcinoma via an estrogen-related receptor alpha (ERRα)-dependent transcriptional program [188]. PGC-1α overexpression in HT29 and HCT116 colorectal cancer cells induced apoptosis through ROS accumulation [178].

As opposed to the tumor-suppressive role of PGC-1α described above, many reports have shown that PGC-1α is a tumor promoter [169,170,176,184,189–191]. Bhalla et al., demonstrated that PGC-1α knockout mice had reduced chemical-induced liver and colon carcinogenesis, suggesting that PGC-1α may induce carcinogenesis [169]. This study reported that PGC-1α stimulates carcinogenesis and tumor growth via the induction of lipogenic enzymes (fatty acid synthase and acetyl-CoA carboxylase) in genetically modified PGC-1α mice [169]. In addition, knockdown PGC-1α significantly induced apoptosis in PGC-1α-positive melanoma cell lines, suggesting that PGC-1α regulates the survival of PGC-1α-positive melanoma cells [176]. PGC-1α promoted prostate cancer cell growth through the activation of androgen receptor [184,189]. It was shown that cell proliferation was inhibited in PGC-1α siRNA knockdown experiments in H1944 lung adenocarcinoma cells [191]. Similarly, overexpression of PGC-1α induced HEK293 cell proliferation and tumorigenesis through the upregulation of Specificity protein 1 (Sp1) and acyl-CoA-binding protein [190]. PGC-1α overexpression or ERRα activation conferred breast cancer cell growth ability, even under hypoxia conditions [170]. Despite the fact that PGC-1α can act as a tumor suppressor and a tumor promoter, there is no explicitly defined mechanism that can explain its dichotomous effects. However, its dual actions can be partially explained by its cell type-specific effects and varied interacting proteins.

11. Conclusions

PPARα and PGC-1α play a central role in metabolic flexibility by driving robust and coordinated changes in the expression of key components of mitochondrial biogenesis and by performing a critical metabolic regulation in many vital organs, including adipose tissue, skeletal muscle, heart, liver, and kidney (Figure 1). Traditionally, in BAT and WAT, mitochondrial biogenesis and BAT gene expression are regulated by PGC-1α. Adrenergic stimulation and reduced temperature trigger signaling cascades including the upregulation of UCP-1 level, thereby resulting in body thermogenesis. In skeletal muscle and WAT, the transcriptional activity of PGC-1α is responsible for the expression of gene networks that control glucose uptake, glycolysis, FA oxidation, the TCA cycle, OXPHOS, and mitochondrial biogenesis and uncoupling. Therefore, increased exercise will increase mitochondrial gene biogenesis and the secretion of myokines (such as irisin), resulting in WAT browning and liver gluconeogenesis and preventing obesity and insulin resistance. In EAT, increased HO-1 expression depends on the PGC-1α–UCP-1 axis, which subsequently decreases free radical and ROS production, thus reducing cardiomyopathy. However, whether long-term PGC-1α overexpression improves or impairs heart or kidney function under disease- or stress-induced remodeling is unclear. In cancer, the dichotomous effects of PGC-1α can be partially explained by its cell type-specific effects and diverse interacting proteins. Therefore, more details in vivo and pre-clinical work are required to assess the usefulness of PGC-1α-inducing drugs in cardiovascular, renal, and cancer therapy.

Funding: This research received no external funding.

Conflicts of Interest: The authors declare no conflict of interest.

References

1. Besseiche, A.; Riveline, J.P.; Gautier, J.F.; Breant, B.; Blondeau, B. Metabolic roles of PGC-1alpha and its implications for type 2 diabetes. *Diabetes Metab.* **2015**, *41*, 347–357. [CrossRef] [PubMed]
2. Puigserver, P.; Wu, Z.; Park, C.W.; Graves, R.; Wright, M.; Spiegelman, B.M. A cold-inducible coactivator of nuclear receptors linked to adaptive thermogenesis. *Cell* **1998**, *92*, 829–839. [CrossRef]
3. Popov, D.V.; Lysenko, E.A.; Kuzmin, I.V.; Vinogradova, V.; Grigoriev, A.I. Regulation of PGC-1alpha Isoform Expression in Skeletal Muscles. *Acta Nat.* **2015**, *7*, 48–59.
4. Lin, J.; Handschin, C.; Spiegelman, B.M. Metabolic control through the PGC-1 family of transcription coactivators. *Cell Metab.* **2005**, *1*, 361–370. [CrossRef] [PubMed]
5. Hatazawa, Y.; Senoo, N.; Tadaishi, M.; Ogawa, Y.; Ezaki, O.; Kamei, Y.; Miura, S. Metabolomic Analysis of the Skeletal Muscle of Mice Overexpressing PGC-1alpha. *PLoS ONE* **2015**, *10*, e0129084. [CrossRef] [PubMed]
6. Calvo, J.A.; Daniels, T.G.; Wang, X.; Paul, A.; Lin, J.; Spiegelman, B.M.; Stevenson, S.C.; Rangwala, S.M. Muscle-specific expression of PPARgamma coactivator-1alpha improves exercise performance and increases peak oxygen uptake. *J. Appl. Physiol.* **2008**, *104*, 1304–1312. [CrossRef] [PubMed]
7. Dillon, L.M.; Rebelo, A.P.; Moraes, C.T. The role of PGC-1 coactivators in aging skeletal muscle and heart. *IUBMB Life* **2012**, *64*, 231–241. [CrossRef] [PubMed]
8. Choi, C.S.; Befroy, D.E.; Codella, R.; Kim, S.; Reznick, R.M.; Hwang, Y.J.; Liu, Z.X.; Lee, H.Y.; Distefano, A.; Samuel, V.T.; et al. Paradoxical effects of increased expression of PGC-1alpha on muscle mitochondrial function and insulin-stimulated muscle glucose metabolism. *Proc. Natl. Acad. Sci. USA* **2008**, *105*, 19926–19931. [CrossRef] [PubMed]
9. Espinoza, D.O.; Boros, L.G.; Crunkhorn, S.; Gami, H.; Patti, M.E. Dual modulation of both lipid oxidation and synthesis by peroxisome proliferator-activated receptor-gamma coactivator-1alpha and -1beta in cultured myotubes. *FASEB J.* **2010**, *24*, 1003–1014. [CrossRef] [PubMed]
10. Lehman, J.J.; Barger, P.M.; Kovacs, A.; Saffitz, J.E.; Medeiros, D.M.; Kelly, D.P. Peroxisome proliferator-activated receptor gamma coactivator-1 promotes cardiac mitochondrial biogenesis. *J. Clin. Investig.* **2000**, *106*, 847–856. [CrossRef] [PubMed]
11. Smith, B.K.; Mukai, K.; Lally, J.S.; Maher, A.C.; Gurd, B.J.; Heigenhauser, G.J.; Spriet, L.L.; Holloway, G.P. AMP-activated protein kinase is required for exercise-induced peroxisome proliferator-activated receptor co-activator 1 translocation to subsarcolemmal mitochondria in skeletal muscle. *J. Physiol.* **2013**, *591*, 1551–1561. [CrossRef] [PubMed]
12. Huang, T.Y.; Zheng, D.; Houmard, J.A.; Brault, J.J.; Hickner, R.C.; Cortright, R.N. Overexpression of PGC-1alpha increases peroxisomal activity and mitochondrial fatty acid oxidation in human primary myotubes. *Am. J. Physiol. Endocrinol. Metab.* **2017**, *312*, E253–E263. [CrossRef] [PubMed]
13. Summermatter, S.; Baum, O.; Santos, G.; Hoppeler, H.; Handschin, C. Peroxisome proliferator-activated receptor {gamma} coactivator 1{alpha} (PGC-1{alpha}) promotes skeletal muscle lipid refueling in vivo by activating de novo lipogenesis and the pentose phosphate pathway. *J. Boil. Chem.* **2010**, *285*, 32793–32800. [CrossRef] [PubMed]
14. Supruniuk, E.; Miklosz, A.; Chabowski, A. The Implication of PGC-1alpha on Fatty Acid Transport across Plasma and Mitochondrial Membranes in the Insulin Sensitive Tissues. *Front. Physiol.* **2017**, *8*, 923. [CrossRef] [PubMed]
15. Booth, F.W.; Thomason, D.B. Molecular and cellular adaptation of muscle in response to exercise: Perspectives of various models. *Physiol. Rev.* **1991**, *71*, 541–585. [CrossRef] [PubMed]
16. Hood, D.A. Invited Review: Contractile activity-induced mitochondrial biogenesis in skeletal muscle. *J. Appl. Physiol.* **2001**, *90*, 1137–1157. [CrossRef] [PubMed]
17. Bassel-Duby, R.; Olson, E.N. Signaling pathways in skeletal muscle remodeling. *Annu. Rev. Biochem.* **2006**, *75*, 19–37. [CrossRef] [PubMed]
18. Chinsomboon, J.; Ruas, J.; Gupta, R.K.; Thom, R.; Shoag, J.; Rowe, G.C.; Sawada, N.; Raghuram, S.; Arany, Z. The transcriptional coactivator PGC-1alpha mediates exercise-induced angiogenesis in skeletal muscle. *Proc. Natl. Acad. Sci. USA* **2009**, *106*, 21401–21406. [CrossRef] [PubMed]
19. Thom, R.; Rowe, G.C.; Jang, C.; Safdar, A.; Arany, Z. Hypoxic induction of vascular endothelial growth factor (VEGF) and angiogenesis in muscle by truncated peroxisome proliferator-activated receptor gamma coactivator (PGC)-1alpha. *J. Boil. Chem.* **2014**, *289*, 8810–8817. [CrossRef] [PubMed]

20. Lin, J.; Wu, H.; Tarr, P.T.; Zhang, C.Y.; Wu, Z.; Boss, O.; Michael, L.F.; Puigserver, P.; Isotani, E.; Olson, E.N.; et al. Transcriptional co-activator PGC-1 alpha drives the formation of slow-twitch muscle fibres. *Nature* **2002**, *418*, 797–801. [CrossRef] [PubMed]
21. Handschin, C.; Chin, S.; Li, P.; Liu, F.; Maratos-Flier, E.; Lebrasseur, N.K.; Yan, Z.; Spiegelman, B.M. Skeletal muscle fiber-type switching, exercise intolerance, and myopathy in PGC-1alpha muscle-specific knock-out animals. *J. Boil. Chem.* **2007**, *282*, 30014–30021. [CrossRef] [PubMed]
22. Rasbach, K.A.; Gupta, R.K.; Ruas, J.L.; Wu, J.; Naseri, E.; Estall, J.L.; Spiegelman, B.M. PGC-1alpha regulates a HIF2alpha-dependent switch in skeletal muscle fiber types. *Proc. Natl. Acad. Sci. USA* **2010**, *107*, 21866–21871. [CrossRef] [PubMed]
23. Sandri, M.; Lin, J.; Handschin, C.; Yang, W.; Arany, Z.P.; Lecker, S.H.; Goldberg, A.L.; Spiegelman, B.M. PGC-1alpha protects skeletal muscle from atrophy by suppressing FoxO3 action and atrophy-specific gene transcription. *Proc. Natl. Acad. Sci. USA* **2006**, *103*, 16260–16265. [CrossRef] [PubMed]
24. Huh, J.Y. The role of exercise-induced myokines in regulating metabolism. *Arch. Pharm. Res.* **2018**, *41*, 14–29. [CrossRef] [PubMed]
25. Bostrom, P.; Wu, J.; Jedrychowski, M.P.; Korde, A.; Ye, L.; Lo, J.C.; Rasbach, K.A.; Bostrom, E.A.; Choi, J.H.; Long, J.Z.; et al. A PGC1-alpha-dependent myokine that drives brown-fat-like development of white fat and thermogenesis. *Nature* **2012**, *481*, 463–468. [CrossRef] [PubMed]
26. Daskalopoulou, S.S.; Cooke, A.B.; Gomez, Y.H.; Mutter, A.F.; Filippaios, A.; Mesfum, E.T.; Mantzoros, C.S. Plasma irisin levels progressively increase in response to increasing exercise workloads in young, healthy, active subjects. *Eur. J. Endocrinol.* **2014**, *171*, 343–352. [CrossRef] [PubMed]
27. Jedrychowski, M.P.; Wrann, C.D.; Paulo, J.A.; Gerber, K.K.; Szpyt, J.; Robinson, M.M.; Nair, K.S.; Gygi, S.P.; Spiegelman, B.M. Detection and Quantitation of Circulating Human Irisin by Tandem Mass Spectrometry. *Cell Metab.* **2015**, *22*, 734–740. [CrossRef] [PubMed]
28. Huh, J.Y.; Mougios, V.; Kabasakalis, A.; Fatouros, I.; Siopi, A.; Douroudos, I.I.; Filippaios, A.; Panagiotou, G.; Park, K.H.; Mantzoros, C.S. Exercise-induced irisin secretion is independent of age or fitness level and increased irisin may directly modulate muscle metabolism through AMPK activation. *J. Clin. Endocrinol. Metab.* **2014**, *99*, E2154–E2161. [CrossRef] [PubMed]
29. Lee, P.; Linderman, J.D.; Smith, S.; Brychta, R.J.; Wang, J.; Idelson, C.; Perron, R.M.; Werner, C.D.; Phan, G.Q.; Kammula, U.S.; et al. Irisin and FGF21 are cold-induced endocrine activators of brown fat function in humans. *Cell Metab.* **2014**, *19*, 302–309. [CrossRef] [PubMed]
30. Rodriguez, A.; Becerril, S.; Mendez-Gimenez, L.; Ramirez, B.; Sainz, N.; Catalan, V.; Gomez-Ambrosi, J.; Fruhbeck, G. Leptin administration activates irisin-induced myogenesis via nitric oxide-dependent mechanisms, but reduces its effect on subcutaneous fat browning in mice. *Int. J. Obes.* **2015**, *39*, 397–407. [CrossRef] [PubMed]
31. Huh, J.Y.; Dincer, F.; Mesfum, E.; Mantzoros, C.S. Irisin stimulates muscle growth-related genes and regulates adipocyte differentiation and metabolism in humans. *Int. J. Obes.* **2014**, *38*, 1538–1544. [CrossRef] [PubMed]
32. Zhang, Y.; Li, R.; Meng, Y.; Li, S.; Donelan, W.; Zhao, Y.; Qi, L.; Zhang, M.; Wang, X.; Cui, T.; et al. Irisin stimulates browning of white adipocytes through mitogen-activated protein kinase p38 MAP kinase and ERK MAP kinase signaling. *Diabetes* **2014**, *63*, 514–525. [CrossRef] [PubMed]
33. Liu, T.Y.; Shi, C.X.; Gao, R.; Sun, H.J.; Xiong, X.Q.; Ding, L.; Chen, Q.; Li, Y.H.; Wang, J.J.; Kang, Y.M.; et al. Irisin inhibits hepatic gluconeogenesis and increases glycogen synthesis via the PI3K/Akt pathway in type 2 diabetic mice and hepatocytes. *Clin. Sci.* **2015**, *129*, 839–850. [CrossRef] [PubMed]
34. Tang, S.; Zhang, R.; Jiang, F.; Wang, J.; Chen, M.; Peng, D.; Yan, J.; Wang, S.; Bao, Y.; Hu, C.; et al. Circulating irisin levels are associated with lipid and uric acid metabolism in a Chinese population. *Clin. Exp. Pharmacol. Physiol.* **2015**, *42*, 896–901. [CrossRef] [PubMed]
35. Xin, C.; Liu, J.; Zhang, J.; Zhu, D.; Wang, H.; Xiong, L.; Lee, Y.; Ye, J.; Lian, K.; Xu, C.; et al. Irisin improves fatty acid oxidation and glucose utilization in type 2 diabetes by regulating the AMPK signaling pathway. *Int. J. Obes.* **2016**, *40*, 443–451. [CrossRef] [PubMed]
36. McPherron, A.C.; Lee, S.J. Double muscling in cattle due to mutations in the myostatin gene. *Proc. Natl. Acad. Sci. USA* **1997**, *94*, 12457–12461. [CrossRef] [PubMed]
37. Allen, D.L.; Hittel, D.S.; McPherron, A.C. Expression and function of myostatin in obesity, diabetes, and exercise adaptation. *Med. Sci. Sports Exerc.* **2011**, *43*, 1828–1835. [CrossRef] [PubMed]

38. Feldman, B.J.; Streeper, R.S.; Farese, R.V., Jr.; Yamamoto, K.R. Myostatin modulates adipogenesis to generate adipocytes with favorable metabolic effects. *Proc. Natl. Acad. Sci. USA* **2006**, *103*, 15675–15680. [CrossRef] [PubMed]
39. Guo, T.; Jou, W.; Chanturiya, T.; Portas, J.; Gavrilova, O.; McPherron, A.C. Myostatin inhibition in muscle, but not adipose tissue, decreases fat mass and improves insulin sensitivity. *PLoS ONE* **2009**, *4*, e4937. [CrossRef] [PubMed]
40. Zhao, B.; Wall, R.J.; Yang, J. Transgenic expression of myostatin propeptide prevents diet-induced obesity and insulin resistance. *Biochem. Biophys. Res. Commun.* **2005**, *337*, 248–255. [CrossRef] [PubMed]
41. Zhang, C.; McFarlane, C.; Lokireddy, S.; Masuda, S.; Ge, X.; Gluckman, P.D.; Sharma, M.; Kambadur, R. Inhibition of myostatin protects against diet-induced obesity by enhancing fatty acid oxidation and promoting a brown adipose phenotype in mice. *Diabetologia* **2012**, *55*, 183–193. [CrossRef] [PubMed]
42. Dong, J.; Dong, Y.; Dong, Y.; Chen, F.; Mitch, W.E.; Zhang, L. Inhibition of myostatin in mice improves insulin sensitivity via irisin-mediated cross talk between muscle and adipose tissues. *Int. J. Obes.* **2016**, *40*, 434–442. [CrossRef] [PubMed]
43. Shan, T.; Liang, X.; Bi, P.; Kuang, S. Myostatin knockout drives browning of white adipose tissue through activating the AMPK-PGC1alpha-Fndc5 pathway in muscle. *FASEB J.* **2013**, *27*, 1981–1989. [CrossRef] [PubMed]
44. Lapchak, P.A.; Hefti, F. BDNF and NGF treatment in lesioned rats: Effects on cholinergic function and weight gain. *Neuroreport* **1992**, *3*, 405–408. [CrossRef] [PubMed]
45. Cao, L.; Choi, E.Y.; Liu, X.; Martin, A.; Wang, C.; Xu, X.; During, M.J. White to brown fat phenotypic switch induced by genetic and environmental activation of a hypothalamic-adipocyte axis. *Cell Metab.* **2011**, *14*, 324–338. [CrossRef] [PubMed]
46. Wrann, C.D.; White, J.P.; Salogiannnis, J.; Laznik-Bogoslavski, D.; Wu, J.; Ma, D.; Lin, J.D.; Greenberg, M.E.; Spiegelman, B.M. Exercise induces hippocampal BDNF through a PGC-1alpha/FNDC5 pathway. *Cell Metab.* **2013**, *18*, 649–659. [CrossRef] [PubMed]
47. Cannon, B.; Jacobsson, A.; Rehnmark, S.; Nedergaard, J. Signal transduction in brown adipose tissue recruitment: Noradrenaline and beyond. *Int. J. Obes. Relat. Metab. Disord. J. Int. Assoc. Study Obes.* **1996**, *20* (Suppl. 3), S36–S42.
48. Ricquier, D. Molecular biology of brown adipose tissue. *Proc. Nutr. Soc.* **1989**, *48*, 183–187. [CrossRef] [PubMed]
49. Gleyzer, N.; Vercauteren, K.; Scarpulla, R.C. Control of mitochondrial transcription specificity factors (TFB1M and TFB2M) by nuclear respiratory factors (NRF-1 and NRF-2) and PGC-1 family coactivators. *Mol. Cell. Boil.* **2005**, *25*, 1354–1366. [CrossRef] [PubMed]
50. Park, P.H.; Huang, H.; McMullen, M.R.; Mandal, P.; Sun, L.; Nagy, L.E. Suppression of lipopolysaccharide-stimulated tumor necrosis factor-alpha production by adiponectin is mediated by transcriptional and post-transcriptional mechanisms. *J. Boil. Chem.* **2008**, *283*, 26850–26858. [CrossRef] [PubMed]
51. Wu, Z.; Puigserver, P.; Andersson, U.; Zhang, C.; Adelmant, G.; Mootha, V.; Troy, A.; Cinti, S.; Lowell, B.; Scarpulla, R.C.; et al. Mechanisms controlling mitochondrial biogenesis and respiration through the thermogenic coactivator PGC-1. *Cell* **1999**, *98*, 115–124. [CrossRef]
52. Lopaschuk, G.D.; Jaswal, J.S. Energy metabolic phenotype of the cardiomyocyte during development, differentiation, and postnatal maturation. *J. Cardiovasc. Pharmacol.* **2010**, *56*, 130–140. [CrossRef] [PubMed]
53. Jafri, M.S.; Dudycha, S.J.; O'Rourke, B. Cardiac energy metabolism: Models of cellular respiration. *Annu. Rev. Biomed. Eng.* **2001**, *3*, 57–81. [CrossRef] [PubMed]
54. Ventura-Clapier, R.; Garnier, A.; Veksler, V. Energy metabolism in heart failure. *J. Physiol.* **2004**, *555 Pt 1*, 1–13. [CrossRef]
55. Razeghi, P.; Young, M.E.; Alcorn, J.L.; Moravec, C.S.; Frazier, O.H.; Taegtmeyer, H. Metabolic gene expression in fetal and failing human heart. *Circulation* **2001**, *104*, 2923–2931. [CrossRef] [PubMed]
56. Sack, M.N.; Rader, T.A.; Park, S.; Bastin, J.; McCune, S.A.; Kelly, D.P. Fatty acid oxidation enzyme gene expression is downregulated in the failing heart. *Circulation* **1996**, *94*, 2837–2842. [CrossRef] [PubMed]
57. Garnier, A.; Fortin, D.; Delomenie, C.; Momken, I.; Veksler, V.; Ventura-Clapier, R. Depressed mitochondrial transcription factors and oxidative capacity in rat failing cardiac and skeletal muscles. *J. Physiol.* **2003**, *551 Pt 2*, 491–501. [CrossRef]

58. Goffart, S.; von Kleist-Retzow, J.C.; Wiesner, R.J. Regulation of mitochondrial proliferation in the heart: Power-plant failure contributes to cardiac failure in hypertrophy. *Cardiovasc. Res.* **2004**, *64*, 198–207. [CrossRef] [PubMed]
59. Huss, J.M.; Torra, I.P.; Staels, B.; Giguere, V.; Kelly, D.P. Estrogen-related receptor alpha directs peroxisome proliferator-activated receptor alpha signaling in the transcriptional control of energy metabolism in cardiac and skeletal muscle. *Mol. Cell. Boil.* **2004**, *24*, 9079–9091. [CrossRef] [PubMed]
60. Scarpulla, R.C. Nuclear control of respiratory gene expression in mammalian cells. *J. Cell. Biochem.* **2006**, *97*, 673–683. [CrossRef] [PubMed]
61. Schreiber, S.N.; Emter, R.; Hock, M.B.; Knutti, D.; Cardenas, J.; Podvinec, M.; Oakeley, E.J.; Kralli, A. The estrogen-related receptor alpha (ERRalpha) functions in PPARgamma coactivator 1alpha (PGC-1alpha)-induced mitochondrial biogenesis. *Proc. Natl. Acad. Sci. USA* **2004**, *101*, 6472–6477. [CrossRef] [PubMed]
62. Puigserver, P.; Spiegelman, B.M. Peroxisome proliferator-activated receptor-gamma coactivator 1 alpha (PGC-1 alpha): Transcriptional coactivator and metabolic regulator. *Endocr. Rev.* **2003**, *24*, 78–90. [CrossRef] [PubMed]
63. Tuomainen, T.; Tavi, P. The role of cardiac energy metabolism in cardiac hypertrophy and failure. *Exp. Cell Res.* **2017**, *360*, 12–18. [CrossRef] [PubMed]
64. Rowe, G.C.; Jiang, A.; Arany, Z. PGC-1 coactivators in cardiac development and disease. *Circ. Res.* **2010**, *107*, 825–838. [CrossRef] [PubMed]
65. Vega, R.B.; Huss, J.M.; Kelly, D.P. The coactivator PGC-1 cooperates with peroxisome proliferator-activated receptor alpha in transcriptional control of nuclear genes encoding mitochondrial fatty acid oxidation enzymes. *Mol. Cell. Boil.* **2000**, *20*, 1868–1876. [CrossRef]
66. Duncan, J.G.; Fong, J.L.; Medeiros, D.M.; Finck, B.N.; Kelly, D.P. Insulin-resistant heart exhibits a mitochondrial biogenic response driven by the peroxisome proliferator-activated receptor-alpha/PGC-1alpha gene regulatory pathway. *Circulation* **2007**, *115*, 909–917. [CrossRef] [PubMed]
67. Arany, Z.; He, H.; Lin, J.; Hoyer, K.; Handschin, C.; Toka, O.; Ahmad, F.; Matsui, T.; Chin, S.; Wu, P.H.; et al. Transcriptional coactivator PGC-1 alpha controls the energy state and contractile function of cardiac muscle. *Cell Metab.* **2005**, *1*, 259–271. [CrossRef] [PubMed]
68. Arany, Z.; Novikov, M.; Chin, S.; Ma, Y.; Rosenzweig, A.; Spiegelman, B.M. Transverse aortic constriction leads to accelerated heart failure in mice lacking PPAR-gamma coactivator 1alpha. *Proc. Natl. Acad. Sci. USA* **2006**, *103*, 10086–10091. [CrossRef] [PubMed]
69. Leone, T.C.; Lehman, J.J.; Finck, B.N.; Schaeffer, P.J.; Wende, A.R.; Boudina, S.; Courtois, M.; Wozniak, D.F.; Sambandam, N.; Bernal-Mizrachi, C.; et al. PGC-1alpha deficiency causes multi-system energy metabolic derangements: Muscle dysfunction, abnormal weight control and hepatic steatosis. *PLoS Boil.* **2005**, *3*, e101.
70. Russell, L.K.; Mansfield, C.M.; Lehman, J.J.; Kovacs, A.; Courtois, M.; Saffitz, J.E.; Medeiros, D.M.; Valencik, M.L.; McDonald, J.A.; Kelly, D.P. Cardiac-specific induction of the transcriptional coactivator peroxisome proliferator-activated receptor gamma coactivator-1alpha promotes mitochondrial biogenesis and reversible cardiomyopathy in a developmental stage-dependent manner. *Circ. Res.* **2004**, *94*, 525–533. [CrossRef] [PubMed]
71. Allard, M.F.; Schonekess, B.O.; Henning, S.L.; English, D.R.; Lopaschuk, G.D. Contribution of oxidative metabolism and glycolysis to ATP production in hypertrophied hearts. *Am. J. Physiol.* **1994**, *267 Pt 2*, H742–H750. [CrossRef]
72. Christe, M.E.; Rodgers, R.L. Altered glucose and fatty acid oxidation in hearts of the spontaneously hypertensive rat. *J. Mol. Cell. Cardiol.* **1994**, *26*, 1371–1375. [CrossRef] [PubMed]
73. Taegtmeyer, H.; Overturf, M.L. Effects of moderate hypertension on cardiac function and metabolism in the rabbit. *Hypertension* **1988**, *11*, 416–426. [CrossRef] [PubMed]
74. Massie, B.M.; Schaefer, S.; Garcia, J.; McKirnan, M.D.; Schwartz, G.G.; Wisneski, J.A.; Weiner, M.W.; White, F.C. Myocardial high-energy phosphate and substrate metabolism in swine with moderate left ventricular hypertrophy. *Circulation* **1995**, *91*, 1814–1823. [CrossRef] [PubMed]
75. Barger, P.M.; Brandt, J.M.; Leone, T.C.; Weinheimer, C.J.; Kelly, D.P. Deactivation of peroxisome proliferator-activated receptor-alpha during cardiac hypertrophic growth. *J. Clin. Investig.* **2000**, *105*, 1723–1730. [CrossRef] [PubMed]

76. Razeghi, P.; Essop, M.F.; Huss, J.M.; Abbasi, S.; Manga, N.; Taegtmeyer, H. Hypoxia-induced switches of myosin heavy chain iso-gene expression in rat heart. *Biochem. Biophys. Res. Commun.* **2003**, *303*, 1024–1027. [CrossRef]
77. Van Bilsen, M.; Smeets, P.J.; Gilde, A.J.; van der Vusse, G.J. Metabolic remodelling of the failing heart: The cardiac burn-out syndrome? *Cardiovasc. Res.* **2004**, *61*, 218–226. [CrossRef] [PubMed]
78. Van Bilsen, M. "Energenetics" of heart failure. *Ann. N. Y. Acad. Sci.* **2004**, *1015*, 238–249. [CrossRef] [PubMed]
79. Young, M.E.; Laws, F.A.; Goodwin, G.W.; Taegtmeyer, H. Reactivation of peroxisome proliferator-activated receptor alpha is associated with contractile dysfunction in hypertrophied rat heart. *J. Boil. Chem.* **2001**, *276*, 44390–44395. [CrossRef] [PubMed]
80. Remondino, A.; Rosenblatt-Velin, N.; Montessuit, C.; Tardy, I.; Papageorgiou, I.; Dorsaz, P.A.; Jorge-Costa, M.; Lerch, R. Altered expression of proteins of metabolic regulation during remodeling of the left ventricle after myocardial infarction. *J. Mol. Cell. Cardiol.* **2000**, *32*, 2025–2034. [CrossRef] [PubMed]
81. Rosenblatt-Velin, N.; Montessuit, C.; Papageorgiou, I.; Terrand, J.; Lerch, R. Postinfarction heart failure in rats is associated with upregulation of GLUT-1 and downregulation of genes of fatty acid metabolism. *Cardiovasc. Res.* **2001**, *52*, 407–416. [CrossRef]
82. Dewald, O.; Sharma, S.; Adrogue, J.; Salazar, R.; Duerr, G.D.; Crapo, J.D.; Entman, M.L.; Taegtmeyer, H. Downregulation of peroxisome proliferator-activated receptor-alpha gene expression in a mouse model of ischemic cardiomyopathy is dependent on reactive oxygen species and prevents lipotoxicity. *Circulation* **2005**, *112*, 407–415. [CrossRef] [PubMed]
83. Sano, M.; Wang, S.C.; Shirai, M.; Scaglia, F.; Xie, M.; Sakai, S.; Tanaka, T.; Kulkarni, P.A.; Barger, P.M.; Youker, K.A.; et al. Activation of cardiac Cdk9 represses PGC-1 and confers a predisposition to heart failure. *EMBO J.* **2004**, *23*, 3559–3569. [CrossRef] [PubMed]
84. Sekiguchi, K.; Tian, Q.; Ishiyama, M.; Burchfield, J.; Gao, F.; Mann, D.L.; Barger, P.M. Inhibition of PPAR-alpha activity in mice with cardiac-restricted expression of tumor necrosis factor: Potential role of TGF-beta/Smad3. *Am. J. Physiol. Heart Circ. Physiol.* **2007**, *292*, H1443–H1451. [CrossRef] [PubMed]
85. Pellieux, C.; Aasum, E.; Larsen, T.S.; Montessuit, C.; Papageorgiou, I.; Pedrazzini, T.; Lerch, R. Overexpression of angiotensinogen in the myocardium induces downregulation of the fatty acid oxidation pathway. *J. Mol. Cell. Cardiol.* **2006**, *41*, 459–466. [CrossRef] [PubMed]
86. Huss, J.M.; Levy, F.H.; Kelly, D.P. Hypoxia inhibits the peroxisome proliferator-activated receptor alpha/retinoid X receptor gene regulatory pathway in cardiac myocytes: A mechanism for O_2-dependent modulation of mitochondrial fatty acid oxidation. *J. Boil. Chem.* **2001**, *276*, 27605–27612. [CrossRef] [PubMed]
87. Schofield, R.S.; Hill, J.A. The use of ranolazine in cardiovascular disease. *Expert Opin. Investig. Drugs* **2002**, *11*, 117–123. [PubMed]
88. Rupp, H.; Zarain-Herzberg, A.; Maisch, B. The use of partial fatty acid oxidation inhibitors for metabolic therapy of angina pectoris and heart failure. *Herz* **2002**, *27*, 621–636. [CrossRef] [PubMed]
89. Chandler, M.P.; Stanley, W.C.; Morita, H.; Suzuki, G.; Roth, B.A.; Blackburn, B.; Wolff, A.; Sabbah, H.N. Short-term treatment with ranolazine improves mechanical efficiency in dogs with chronic heart failure. *Circ. Res.* **2002**, *91*, 278–280. [CrossRef] [PubMed]
90. Liao, R.; Jain, M.; Cui, L.; D'Agostino, J.; Aiello, F.; Luptak, I.; Ngoy, S.; Mortensen, R.M.; Tian, R. Cardiac-specific overexpression of GLUT1 prevents the development of heart failure attributable to pressure overload in mice. *Circulation* **2002**, *106*, 2125–2131. [CrossRef] [PubMed]
91. Asakawa, M.; Takano, H.; Nagai, T.; Uozumi, H.; Hasegawa, H.; Kubota, N.; Saito, T.; Masuda, Y.; Kadowaki, T.; Komuro, I. Peroxisome proliferator-activated receptor gamma plays a critical role in inhibition of cardiac hypertrophy in vitro and in vivo. *Circulation* **2002**, *105*, 1240–1246. [CrossRef] [PubMed]
92. Yamamoto, K.; Ohki, R.; Lee, R.T.; Ikeda, U.; Shimada, K. Peroxisome proliferator-activated receptor gamma activators inhibit cardiac hypertrophy in cardiac myocytes. *Circulation* **2001**, *104*, 1670–1675. [CrossRef] [PubMed]
93. Planavila, A.; Rodriguez-Calvo, R.; Jove, M.; Michalik, L.; Wahli, W.; Laguna, J.C.; Vazquez-Carrera, M. Peroxisome proliferator-activated receptor beta/delta activation inhibits hypertrophy in neonatal rat cardiomyocytes. *Cardiovasc. Res.* **2005**, *65*, 832–841. [CrossRef] [PubMed]
94. Fein, F.S.; Sonnenblick, E.H. Diabetic cardiomyopathy. *Cardiovasc. Drugs Ther.* **1994**, *8*, 65–73. [CrossRef] [PubMed]

95. Rubler, S.; Dlugash, J.; Yuceoglu, Y.Z.; Kumral, T.; Branwood, A.W.; Grishman, A. New type of cardiomyopathy associated with diabetic glomerulosclerosis. *Am. J. Cardiol.* **1972**, *30*, 595–602. [PubMed]
96. Keen, H.; Jarrett, R.J. The WHO multinational study of vascular disease in diabetes: 2. Macrovascular disease prevalence. *Diabetes Care* **1979**, *2*, 187–195. [PubMed]
97. Fein, F.S.; Sonnenblick, E.H. Diabetic cardiomyopathy. *Prog. Cardiovasc. Dis.* **1985**, *27*, 255–270. [CrossRef]
98. Gamble, J.; Lopaschuk, G.D. Glycolysis and glucose oxidation during reperfusion of ischemic hearts from diabetic rats. *Biochim. Biophys. Acta* **1994**, *1225*, 191–199. [CrossRef]
99. Stanley, W.C.; Lopaschuk, G.D.; McCormack, J.G. Regulation of energy substrate metabolism in the diabetic heart. *Cardiovasc. Res.* **1997**, *34*, 25–33. [CrossRef]
100. Belke, D.D.; Larsen, T.S.; Gibbs, E.M.; Severson, D.L. Altered metabolism causes cardiac dysfunction in perfused hearts from diabetic (db/db) mice. *Am. J. Physiol. Endocrinol. Metab.* **2000**, *279*, E1104–E1113. [CrossRef] [PubMed]
101. Rodrigues, B.; McNeill, J.H. The diabetic heart: Metabolic causes for the development of a cardiomyopathy. *Cardiovasc. Res.* **1992**, *26*, 913–922. [CrossRef] [PubMed]
102. Finck, B.N.; Lehman, J.J.; Leone, T.C.; Welch, M.J.; Bennett, M.J.; Kovacs, A.; Han, X.; Gross, R.W.; Kozak, R.; Lopaschuk, G.D.; et al. The cardiac phenotype induced by PPARalpha overexpression mimics that caused by diabetes mellitus. *J. Clin. Investig.* **2002**, *109*, 121–130. [CrossRef] [PubMed]
103. Bernal-Mizrachi, C.; Weng, S.; Feng, C.; Finck, B.N.; Knutsen, R.H.; Leone, T.C.; Coleman, T.; Mecham, R.P.; Kelly, D.P.; Semenkovich, C.F. Dexamethasone induction of hypertension and diabetes is PPAR-alpha dependent in LDL receptor-null mice. *Nat. Med.* **2003**, *9*, 1069–1075. [CrossRef] [PubMed]
104. Buchanan, J.; Mazumder, P.K.; Hu, P.; Chakrabarti, G.; Roberts, M.W.; Yun, U.J.; Cooksey, R.C.; Litwin, S.E.; Abel, E.D. Reduced cardiac efficiency and altered substrate metabolism precedes the onset of hyperglycemia and contractile dysfunction in two mouse models of insulin resistance and obesity. *Endocrinology* **2005**, *146*, 5341–5349. [CrossRef] [PubMed]
105. Kuo, T.H.; Moore, K.H.; Giacomelli, F.; Wiener, J. Defective oxidative metabolism of heart mitochondria from genetically diabetic mice. *Diabetes* **1983**, *32*, 781–787. [CrossRef] [PubMed]
106. Tanaka, Y.; Konno, N.; Kako, K.J. Mitochondrial dysfunction observed in situ in cardiomyocytes of rats in experimental diabetes. *Cardiovasc. Res.* **1992**, *26*, 409–414. [CrossRef] [PubMed]
107. Boudina, S.; Sena, S.; O'Neill, B.T.; Tathireddy, P.; Young, M.E.; Abel, E.D. Reduced mitochondrial oxidative capacity and increased mitochondrial uncoupling impair myocardial energetics in obesity. *Circulation* **2005**, *112*, 2686–2695. [CrossRef] [PubMed]
108. Boudina, S.; Sena, S.; Theobald, H.; Sheng, X.; Wright, J.J.; Hu, X.X.; Aziz, S.; Johnson, J.I.; Bugger, H.; Zaha, V.G.; et al. Mitochondrial energetics in the heart in obesity-related diabetes: Direct evidence for increased uncoupled respiration and activation of uncoupling proteins. *Diabetes* **2007**, *56*, 2457–2466. [CrossRef] [PubMed]
109. Iacobellis, G. Epicardial and pericardial fat: Close, but very different. *Obesity* **2009**, *17*, 625–627. [CrossRef] [PubMed]
110. Fitzgibbons, T.P.; Czech, M.P. Epicardial and perivascular adipose tissues and their influence on cardiovascular disease: Basic mechanisms and clinical associations. *J. Am. Heart Assoc.* **2014**, *3*, e000582. [CrossRef] [PubMed]
111. Sacks, H.S.; Fain, J.N.; Holman, B.; Cheema, P.; Chary, A.; Parks, F.; Karas, J.; Optican, R.; Bahouth, S.W.; Garrett, E.; et al. Uncoupling protein-1 and related messenger ribonucleic acids in human epicardial and other adipose tissues: Epicardial fat functioning as brown fat. *J. Clin. Endocrinol. Metab.* **2009**, *94*, 3611–3615. [CrossRef] [PubMed]
112. Patel, V.B.; Mori, J.; McLean, B.A.; Basu, R.; Das, S.K.; Ramprasath, T.; Parajuli, N.; Penninger, J.M.; Grant, M.B.; Lopaschuk, G.D.; et al. ACE2 Deficiency Worsens Epicardial Adipose Tissue Inflammation and Cardiac Dysfunction in Response to Diet-Induced Obesity. *Diabetes* **2016**, *65*, 85–95. [CrossRef] [PubMed]
113. Cherian, S.; Lopaschuk, G.D.; Carvalho, E. Cellular cross-talk between epicardial adipose tissue and myocardium in relation to the pathogenesis of cardiovascular disease. *Am. J. Physiol. Endocrinol. Metab.* **2012**, *303*, E937–E949. [CrossRef] [PubMed]
114. Nour-Eldine, W.; Ghantous, C.M.; Zibara, K.; Dib, L.; Issaa, H.; Itani, H.A.; El-Zein, N.; Zeidan, A. Adiponectin Attenuates Angiotensin II-Induced Vascular Smooth Muscle Cell Remodeling through Nitric Oxide and the RhoA/ROCK Pathway. *Front. Pharmacol.* **2016**, *7*, 86. [CrossRef] [PubMed]

115. Bryan, N.; Ahswin, H.; Smart, N.; Bayon, Y.; Wohlert, S.; Hunt, J.A. Reactive oxygen species (ROS)—A family of fate deciding molecules pivotal in constructive inflammation and wound healing. *Eur. Cells Mater.* **2012**, *24*, 249–265. [CrossRef]
116. Iacobellis, G.; Pistilli, D.; Gucciardo, M.; Leonetti, F.; Miraldi, F.; Brancaccio, G.; Gallo, P.; di Gioia, C.R. Adiponectin expression in human epicardial adipose tissue in vivo is lower in patients with coronary artery disease. *Cytokine* **2005**, *29*, 251–255. [CrossRef] [PubMed]
117. Silaghi, A.; Achard, V.; Paulmyer-Lacroix, O.; Scridon, T.; Tassistro, V.; Duncea, I.; Clement, K.; Dutour, A.; Grino, M. Expression of adrenomedullin in human epicardial adipose tissue: Role of coronary status. *Am. J. Physiol. Endocrinol. Metab.* **2007**, *293*, E1443–E1450. [CrossRef] [PubMed]
118. Iacobellis, G.; Bianco, A.C. Epicardial adipose tissue: Emerging physiological, pathophysiological and clinical features. *Trends Endocrinol. Metab. TEM* **2011**, *22*, 450–457. [CrossRef] [PubMed]
119. Iacobellis, G.; di Gioia, C.R.; Cotesta, D.; Petramala, L.; Travaglini, C.; De Santis, V.; Vitale, D.; Tritapepe, L.; Letizia, C. Epicardial adipose tissue adiponectin expression is related to intracoronary adiponectin levels. *Horm. Metab. Res.* **2009**, *41*, 227–231. [CrossRef] [PubMed]
120. Lin, H.; Lian, W.S.; Chen, H.H.; Lai, P.F.; Cheng, C.F. Adiponectin ameliorates iron-overload cardiomyopathy through the PPARalpha-PGC-1-dependent signaling pathway. *Mol. Pharmacol.* **2013**, *84*, 275–285. [CrossRef] [PubMed]
121. Craige, S.M.; Kroller-Schon, S.; Li, C.; Kant, S.; Cai, S.; Chen, K.; Contractor, M.M.; Pei, Y.; Schulz, E.; Keaney, J.F., Jr. PGC-1alpha dictates endothelial function through regulation of eNOS expression. *Sci. Rep.* **2016**, *6*, 38210. [CrossRef] [PubMed]
122. Liang, H.; Ward, W.F. PGC-1alpha: A key regulator of energy metabolism. *Adv. Physiol. Educ.* **2006**, *30*, 145–151. [CrossRef] [PubMed]
123. Vega, R.B.; Horton, J.L.; Kelly, D.P. Maintaining ancient organelles: Mitochondrial biogenesis and maturation. *Circ. Res.* **2015**, *116*, 1820–1834. [CrossRef] [PubMed]
124. Lehman, J.J.; Kelly, D.P. Transcriptional activation of energy metabolic switches in the developing and hypertrophied heart. *Clin. Exp. Pharmacol. Physiol.* **2002**, *29*, 339–345. [CrossRef] [PubMed]
125. Tran, M.; Tam, D.; Bardia, A.; Bhasin, M.; Rowe, G.C.; Kher, A.; Zsengeller, Z.K.; Akhavan-Sharif, M.R.; Khankin, E.V.; Saintgeniez, M.; et al. PGC-1alpha promotes recovery after acute kidney injury during systemic inflammation in mice. *J. Clin. Investig.* **2011**, *121*, 4003–4014. [CrossRef] [PubMed]
126. Portilla, D.; Dai, G.; McClure, T.; Bates, L.; Kurten, R.; Megyesi, J.; Price, P.; Li, S. Alterations of PPARalpha and its coactivator PGC-1 in cisplatin-induced acute renal failure. *Kidney Int.* **2002**, *62*, 1208–1218. [CrossRef] [PubMed]
127. Noireaud, J.; Andriantsitohaina, R. Recent insights in the paracrine modulation of cardiomyocyte contractility by cardiac endothelial cells. *BioMed Res. Int.* **2014**, *2014*, 923805. [CrossRef] [PubMed]
128. Canto, C.; Pich, S.; Paz, J.C.; Sanches, R.; Martinez, V.; Orpinell, M.; Palacin, M.; Zorzano, A.; Guma, A. Neuregulins increase mitochondrial oxidative capacity and insulin sensitivity in skeletal muscle cells. *Diabetes* **2007**, *56*, 2185–2193. [CrossRef] [PubMed]
129. Lin, J.; Tarr, P.T.; Yang, R.; Rhee, J.; Puigserver, P.; Newgard, C.B.; Spiegelman, B.M. PGC-1beta in the regulation of hepatic glucose and energy metabolism. *J. Boil. Chem.* **2003**, *278*, 30843–30848. [CrossRef] [PubMed]
130. Yoon, J.C.; Puigserver, P.; Chen, G.; Donovan, J.; Wu, Z.; Rhee, J.; Adelmant, G.; Stafford, J.; Kahn, C.R.; Granner, D.K.; et al. Control of hepatic gluconeogenesis through the transcriptional coactivator PGC-1. *Nature* **2001**, *413*, 131–138. [PubMed]
131. Rui, L. Energy metabolism in the liver. *Compr. Physiol.* **2014**, *4*, 177–197. [PubMed]
132. Herzig, S.; Long, F.; Jhala, U.S.; Hedrick, S.; Quinn, R.; Bauer, A.; Rudolph, D.; Schutz, G.; Yoon, C.; Puigserver, P.; et al. CREB regulates hepatic gluconeogenesis through the coactivator PGC-1. *Nature* **2001**, *413*, 179–183. [CrossRef] [PubMed]
133. Rhee, J.; Inoue, Y.; Yoon, J.C.; Puigserver, P.; Fan, M.; Gonzalez, F.J.; Spiegelman, B.M. Regulation of hepatic fasting response by PPARgamma coactivator-1alpha (PGC-1): Requirement for hepatocyte nuclear factor 4alpha in gluconeogenesis. *Proc. Natl. Acad. Sci. USA* **2003**, *100*, 4012–4017. [CrossRef] [PubMed]
134. Lin, J.; Wu, P.H.; Tarr, P.T.; Lindenberg, K.S.; St-Pierre, J.; Zhang, C.Y.; Mootha, V.K.; Jager, S.; Vianna, C.R.; Reznick, R.M.; et al. Defects in adaptive energy metabolism with CNS-linked hyperactivity in PGC-1alpha null mice. *Cell* **2004**, *119*, 121–135. [CrossRef] [PubMed]

135. Koo, S.H.; Satoh, H.; Herzig, S.; Lee, C.H.; Hedrick, S.; Kulkarni, R.; Evans, R.M.; Olefsky, J.; Montminy, M. PGC-1 promotes insulin resistance in liver through PPAR-alpha-dependent induction of TRB-3. *Nat. Med.* **2004**, *10*, 530–534. [CrossRef] [PubMed]
136. Li, S.; Arning, E.; Liu, C.; Vitvitsky, V.; Hernandez, C.; Banerjee, R.; Bottiglieri, T.; Lin, J.D. Regulation of homocysteine homeostasis through the transcriptional coactivator PGC-1alpha. *Am. J. Physiol. Endocrinol. Metab.* **2009**, *296*, E543–E548. [CrossRef] [PubMed]
137. Aharoni-Simon, M.; Hann-Obercyger, M.; Pen, S.; Madar, Z.; Tirosh, O. Fatty liver is associated with impaired activity of PPARgamma-coactivator 1alpha (PGC1alpha) and mitochondrial biogenesis in mice. *Lab. Investig.* **2011**, *91*, 1018–1028. [CrossRef] [PubMed]
138. Haase, T.N.; Ringholm, S.; Leick, L.; Bienso, R.S.; Kiilerich, K.; Johansen, S.; Nielsen, M.M.; Wojtaszewski, J.F.; Hidalgo, J.; Pedersen, P.A.; et al. Role of PGC-1alpha in exercise and fasting-induced adaptations in mouse liver. *Am. J. Physiol. Regul. Integr. Comp. Physiol.* **2011**, *301*, R1501–R1509. [CrossRef] [PubMed]
139. Croce, M.A.; Eagon, J.C.; LaRiviere, L.L.; Korenblat, K.M.; Klein, S.; Finck, B.N. Hepatic lipin 1beta expression is diminished in insulin-resistant obese subjects and is reactivated by marked weight loss. *Diabetes* **2007**, *56*, 2395–2399. [CrossRef] [PubMed]
140. Morris, E.M.; Meers, G.M.; Booth, F.W.; Fritsche, K.L.; Hardin, C.D.; Thyfault, J.P.; Ibdah, J.A. PGC-1alpha overexpression results in increased hepatic fatty acid oxidation with reduced triacylglycerol accumulation and secretion. *Am. J. Physiol. Gastrointest. Liver Physiol.* **2012**, *303*, G979–G992. [CrossRef] [PubMed]
141. Liu, C.; Li, S.; Liu, T.; Borjigin, J.; Lin, J.D. Transcriptional coactivator PGC-1alpha integrates the mammalian clock and energy metabolism. *Nature* **2007**, *447*, 477–481. [CrossRef] [PubMed]
142. Weinberg, J.M. Mitochondrial biogenesis in kidney disease. *J. Am. Soc. Nephrol. JASN* **2011**, *22*, 431–436. [CrossRef] [PubMed]
143. Ishimoto, Y.; Inagi, R. Mitochondria: A therapeutic target in acute kidney injury. *Nephrol. Dial. Transplant.* **2016**, *31*, 1062–1069. [CrossRef] [PubMed]
144. Rasbach, K.A.; Schnellmann, R.G. PGC-1alpha over-expression promotes recovery from mitochondrial dysfunction and cell injury. *Biochem. Biophys. Res. Commun.* **2007**, *355*, 734–739. [CrossRef] [PubMed]
145. Zhang, L.; Liu, J.; Zhou, F.; Wang, W.; Chen, N. PGC-1alpha ameliorates kidney fibrosis in mice with diabetic kidney disease through an antioxidative mechanism. *Mol. Med. Rep.* **2018**, *17*, 4490–4498. [PubMed]
146. Sogabe, K.; Roeser, N.F.; Venkatachalam, M.A.; Weinberg, J.M. Differential cytoprotection by glycine against oxidant damage to proximal tubule cells. *Kidney Int.* **1996**, *50*, 845–854. [CrossRef] [PubMed]
147. Rasbach, K.A.; Schnellmann, R.G. Signaling of mitochondrial biogenesis following oxidant injury. *J. Boil. Chem.* **2007**, *282*, 2355–2362. [CrossRef] [PubMed]
148. Choi, H.I.; Kim, H.J.; Park, J.S.; Kim, I.J.; Bae, E.H.; Ma, S.K.; Kim, S.W. PGC-1alpha attenuates hydrogen peroxide-induced apoptotic cell death by upregulating Nrf-2 via GSK3beta inactivation mediated by activated p38 in HK-2 Cells. *Sci. Rep.* **2017**, *7*, 4319. [CrossRef] [PubMed]
149. Brooks, C.; Wei, Q.; Cho, S.G.; Dong, Z. Regulation of mitochondrial dynamics in acute kidney injury in cell culture and rodent models. *J. Clin. Investig.* **2009**, *119*, 1275–1285. [CrossRef] [PubMed]
150. Brooks, C.; Cho, S.G.; Wang, C.Y.; Yang, T.; Dong, Z. Fragmented mitochondria are sensitized to Bax insertion and activation during apoptosis. *Am. J. Physiol. Cell Physiol.* **2011**, *300*, C447–C455. [CrossRef] [PubMed]
151. Ishihara, M.; Urushido, M.; Hamada, K.; Matsumoto, T.; Shimamura, Y.; Ogata, K.; Inoue, K.; Taniguchi, Y.; Horino, T.; Fujieda, M.; et al. Sestrin-2 and BNIP3 regulate autophagy and mitophagy in renal tubular cells in acute kidney injury. *Am. J. Physiol. Ren. Physiol.* **2013**, *305*, F495–F509. [CrossRef] [PubMed]
152. Chen, K.H.; Guo, X.; Ma, D.; Guo, Y.; Li, Q.; Yang, D.; Li, P.; Qiu, X.; Wen, S.; Xiao, R.P.; et al. Dysregulation of HSG triggers vascular proliferative disorders. *Nat. Cell Boil.* **2004**, *6*, 872–883. [CrossRef] [PubMed]
153. Chen, K.H.; Dasgupta, A.; Ding, J.; Indig, F.E.; Ghosh, P.; Longo, D.L. Role of mitofusin 2 (Mfn2) in controlling cellular proliferation. *FASEB J.* **2014**, *28*, 382–394. [CrossRef] [PubMed]
154. Chen, J.; Chen, J.K.; Harris, R.C. Deletion of the epidermal growth factor receptor in renal proximal tubule epithelial cells delays recovery from acute kidney injury. *Kidney Int.* **2012**, *82*, 45–52. [CrossRef] [PubMed]
155. Jang, H.S.; Han, S.J.; Kim, J.I.; Lee, S.; Lipschutz, J.H.; Park, K.M. Activation of ERK accelerates repair of renal tubular epithelial cells, whereas it inhibits progression of fibrosis following ischemia/reperfusion injury. *Biochim. Biophys. Acta* **2013**, *1832*, 1998–2008. [CrossRef] [PubMed]

156. Gall, J.M.; Wang, Z.; Bonegio, R.G.; Havasi, A.; Liesa, M.; Vemula, P.; Borkan, S.C. Conditional knockout of proximal tubule mitofusin 2 accelerates recovery and improves survival after renal ischemia. *J. Am. Soc. Nephrol. JASN* **2015**, *26*, 1092–1102. [CrossRef] [PubMed]
157. Vander Heiden, M.G.; DeBerardinis, R.J. Understanding the Intersections between Metabolism and Cancer Biology. *Cell* **2017**, *168*, 657–669. [CrossRef] [PubMed]
158. Boroughs, L.K.; DeBerardinis, R.J. Metabolic pathways promoting cancer cell survival and growth. *Nat. Cell Boil.* **2015**, *17*, 351–359. [CrossRef] [PubMed]
159. Hensley, C.T.; Wasti, A.T.; DeBerardinis, R.J. Glutamine and cancer: Cell biology, physiology, and clinical opportunities. *J. Clin. Investig.* **2013**, *123*, 3678–3684. [CrossRef] [PubMed]
160. Currie, E.; Schulze, A.; Zechner, R.; Walther, T.C.; Farese, R.V., Jr. Cellular fatty acid metabolism and cancer. *Cell Metab.* **2013**, *18*, 153–161. [CrossRef] [PubMed]
161. Locasale, J.W. Serine, glycine and one-carbon units: Cancer metabolism in full circle. *Nat. Rev. Cancer* **2013**, *13*, 572–583. [CrossRef] [PubMed]
162. Seyfried, T.N.; Flores, R.E.; Poff, A.M.; D'Agostino, D.P. Cancer as a metabolic disease: Implications for novel therapeutics. *Carcinogenesis* **2014**, *35*, 515–527. [CrossRef] [PubMed]
163. Chaffer, C.L.; San Juan, B.P.; Lim, E.; Weinberg, R.A. EMT, cell plasticity and metastasis. *Cancer Metastasis Rev.* **2016**, *35*, 645–654. [CrossRef] [PubMed]
164. Sciacovelli, M.; Frezza, C. Metabolic reprogramming and epithelial-to-mesenchymal transition in cancer. *FEBS J.* **2017**, *284*, 3132–3144. [CrossRef] [PubMed]
165. Skrypek, N.; Goossens, S.; De Smedt, E.; Vandamme, N.; Berx, G. Epithelial-to-Mesenchymal Transition: Epigenetic Reprogramming Driving Cellular Plasticity. *Trends Genet. TIG* **2017**, *33*, 943–959. [CrossRef] [PubMed]
166. Valle, I.; Alvarez-Barrientos, A.; Arza, E.; Lamas, S.; Monsalve, M. PGC-1alpha regulates the mitochondrial antioxidant defense system in vascular endothelial cells. *Cardiovasc. Res.* **2005**, *66*, 562–573. [CrossRef] [PubMed]
167. St-Pierre, J.; Drori, S.; Uldry, M.; Silvaggi, J.M.; Rhee, J.; Jager, S.; Handschin, C.; Zheng, K.; Lin, J.; Yang, W.; et al. Suppression of reactive oxygen species and neurodegeneration by the PGC-1 transcriptional coactivators. *Cell* **2006**, *127*, 397–408. [CrossRef] [PubMed]
168. Li, X.; Monks, B.; Ge, Q.; Birnbaum, M.J. Akt/PKB regulates hepatic metabolism by directly inhibiting PGC-1alpha transcription coactivator. *Nature* **2007**, *447*, 1012–1016. [CrossRef] [PubMed]
169. Bhalla, K.; Hwang, B.J.; Dewi, R.E.; Ou, L.; Twaddel, W.; Fang, H.B.; Vafai, S.B.; Vazquez, F.; Puigserver, P.; Boros, L.; et al. PGC1alpha promotes tumor growth by inducing gene expression programs supporting lipogenesis. *Cancer Res.* **2011**, *71*, 6888–6898. [CrossRef] [PubMed]
170. McGuirk, S.; Gravel, S.P.; Deblois, G.; Papadopoli, D.J.; Faubert, B.; Wegner, A.; Hiller, K.; Avizonis, D.; Akavia, U.D.; Jones, R.G.; et al. PGC-1alpha supports glutamine metabolism in breast cancer. *Cancer Metab.* **2013**, *1*, 22. [CrossRef] [PubMed]
171. Watkins, G.; Douglas-Jones, A.; Mansel, R.E.; Jiang, W.G. The localisation and reduction of nuclear staining of PPARgamma and PGC-1 in human breast cancer. *Oncol. Rep.* **2004**, *12*, 483–488. [PubMed]
172. Jiang, W.G.; Douglas-Jones, A.; Mansel, R.E. Expression of peroxisome-proliferator activated receptor-gamma (PPARgamma) and the PPARgamma co-activator, PGC-1, in human breast cancer correlates with clinical outcomes. *Int. J. Cancer* **2003**, *106*, 752–757. [CrossRef] [PubMed]
173. LeBleu, V.S.; O'Connell, J.T.; Gonzalez Herrera, K.N.; Wikman, H.; Pantel, K.; Haigis, M.C.; de Carvalho, F.M.; Damascena, A.; Domingos Chinen, L.T.; Rocha, R.M.; et al. PGC-1alpha mediates mitochondrial biogenesis and oxidative phosphorylation in cancer cells to promote metastasis. *Nat. Cell Boil.* **2014**, *16*, 992–1003. [CrossRef] [PubMed]
174. Andrzejewski, S.; Klimcakova, E.; Johnson, R.M.; Tabaries, S.; Annis, M.G.; McGuirk, S.; Northey, J.J.; Chenard, V.; Sriram, U.; Papadopoli, D.J.; et al. PGC-1alpha Promotes Breast Cancer Metastasis and Confers Bioenergetic Flexibility against Metabolic Drugs. *Cell Metab.* **2017**, *26*, 778–787. [CrossRef] [PubMed]
175. Haq, R.; Shoag, J.; Andreu-Perez, P.; Yokoyama, S.; Edelman, H.; Rowe, G.C.; Frederick, D.T.; Hurley, A.D.; Nellore, A.; Kung, A.L.; et al. Oncogenic BRAF regulates oxidative metabolism via PGC1alpha and MITF. *Cancer Cell* **2013**, *23*, 302–315. [CrossRef] [PubMed]

176. Vazquez, F.; Lim, J.H.; Chim, H.; Bhalla, K.; Girnun, G.; Pierce, K.; Clish, C.B.; Granter, S.R.; Widlund, H.R.; Spiegelman, B.M.; et al. PGC1alpha expression defines a subset of human melanoma tumors with increased mitochondrial capacity and resistance to oxidative stress. *Cancer Cell* **2013**, *23*, 287–301. [CrossRef] [PubMed]
177. Luo, C.; Lim, J.H.; Lee, Y.; Granter, S.R.; Thomas, A.; Vazquez, F.; Widlund, H.R.; Puigserver, P. A PGC1alpha-mediated transcriptional axis suppresses melanoma metastasis. *Nature* **2016**, *537*, 422–426. [CrossRef] [PubMed]
178. D'Errico, I.; Salvatore, L.; Murzilli, S.; Lo Sasso, G.; Latorre, D.; Martelli, N.; Egorova, A.V.; Polishuck, R.; Madeyski-Bengtson, K.; Lelliott, C.; et al. Peroxisome proliferator-activated receptor-gamma coactivator 1-alpha (PGC1alpha) is a metabolic regulator of intestinal epithelial cell fate. *Proc. Natl. Acad. Sci. USA* **2011**, *108*, 6603–6608. [CrossRef] [PubMed]
179. Dar, S.; Chhina, J.; Mert, I.; Chitale, D.; Buekers, T.; Kaur, H.; Giri, S.; Munkarah, A.; Rattan, R. Bioenergetic Adaptations in Chemoresistant Ovarian Cancer Cells. *Sci. Rep.* **2017**, *7*, 8760. [CrossRef] [PubMed]
180. Gabrielson, M.; Bjorklund, M.; Carlson, J.; Shoshan, M. Expression of mitochondrial regulators PGC1alpha and TFAM as putative markers of subtype and chemoresistance in epithelial ovarian carcinoma. *PLoS ONE* **2014**, *9*, e107109. [CrossRef] [PubMed]
181. Kim, B.; Jung, J.W.; Jung, J.; Han, Y.; Suh, D.H.; Kim, H.S.; Dhanasekaran, D.N.; Song, Y.S. PGC1alpha induced by reactive oxygen species contributes to chemoresistance of ovarian cancer cells. *Oncotarget* **2017**, *8*, 60299–60311. [PubMed]
182. Liu, R.; Zhang, H.; Zhang, Y.; Li, S.; Wang, X.; Wang, X.; Wang, C.; Liu, B.; Zen, K.; Zhang, C.Y.; et al. Peroxisome proliferator-activated receptor gamma coactivator-1 alpha acts as a tumor suppressor in hepatocellular carcinoma. *Tumour Boil.* **2017**, *39*, 1010428317695031. [CrossRef] [PubMed]
183. Cai, F.F.; Xu, C.; Pan, X.; Cai, L.; Lin, X.Y.; Chen, S.; Biskup, E. Prognostic value of plasma levels of HIF-1a and PGC-1a in breast cancer. *Oncotarget* **2016**, *7*, 77793–77806. [CrossRef] [PubMed]
184. Tennakoon, J.B.; Shi, Y.; Han, J.J.; Tsouko, E.; White, M.A.; Burns, A.R.; Zhang, A.; Xia, X.; Ilkayeva, O.R.; Xin, L.; et al. Androgens regulate prostate cancer cell growth via an AMPK-PGC-1alpha-mediated metabolic switch. *Oncogene* **2014**, *33*, 5251–5261. [CrossRef] [PubMed]
185. Zhang, Y.; Ba, Y.; Liu, C.; Sun, G.; Ding, L.; Gao, S.; Hao, J.; Yu, Z.; Zhang, J.; Zen, K.; et al. PGC-1alpha induces apoptosis in human epithelial ovarian cancer cells through a PPARgamma-dependent pathway. *Cell Res.* **2007**, *17*, 363–373. [CrossRef] [PubMed]
186. Wang, X.; Moraes, C.T. Increases in mitochondrial biogenesis impair carcinogenesis at multiple levels. *Mol. Oncol.* **2011**, *5*, 399–409. [CrossRef] [PubMed]
187. Lee, H.J.; Su, Y.; Yin, P.H.; Lee, H.C.; Chi, C.W. PPAR(gamma)/PGC-1(alpha) pathway in E-cadherin expression and motility of HepG2 cells. *Anticancer. Res.* **2009**, *29*, 5057–5063. [PubMed]
188. Torrano, V.; Valcarcel-Jimenez, L.; Cortazar, A.R.; Liu, X.; Urosevic, J.; Castillo-Martin, M.; Fernandez-Ruiz, S.; Morciano, G.; Caro-Maldonado, A.; Guiu, M.; et al. The metabolic co-regulator PGC1alpha suppresses prostate cancer metastasis. *Nat. Cell Boil.* **2016**, *18*, 645–656. [CrossRef] [PubMed]
189. Shiota, M.; Yokomizo, A.; Tada, Y.; Inokuchi, J.; Tatsugami, K.; Kuroiwa, K.; Uchiumi, T.; Fujimoto, N.; Seki, N.; Naito, S. Peroxisome proliferator-activated receptor gamma coactivator-1alpha interacts with the androgen receptor (AR) and promotes prostate cancer cell growth by activating the AR. *Mol. Endocrinol.* **2010**, *24*, 114–127. [CrossRef] [PubMed]
190. Shin, S.W.; Yun, S.H.; Park, E.S.; Jeong, J.S.; Kwak, J.Y.; Park, J.I. Overexpression of PGC1alpha enhances cell proliferation and tumorigenesis of HEK293 cells through the upregulation of Sp1 and Acyl-CoA binding protein. *Int. J. Oncol.* **2015**, *46*, 1328–1342. [CrossRef] [PubMed]
191. Taguchi, A.; Delgado, O.; Celiktas, M.; Katayama, H.; Wang, H.; Gazdar, A.F.; Hanash, S.M. Proteomic signatures associated with p53 mutational status in lung adenocarcinoma. *Proteomics* **2014**, *14*, 2750–2759. [CrossRef] [PubMed]

International Journal of
Molecular Sciences

Article

Influence of Single-Nucleotide Polymorphisms in PPAR-δ, PPAR-γ, and PRKAA2 on the Changes in Anthropometric Indices and Blood Measurements through Exercise-Centered Lifestyle Intervention in Japanese Middle-Aged Men

Yuichiro Nishida [1,*], Minako Iyadomi [2], Hirotaka Tominaga [3], Hiroaki Taniguchi [4], Yasuki Higaki [5], Hiroaki Tanaka [5], Mikako Horita [1], Chisato Shimanoe [1], Megumi Hara [1] and Keitaro Tanaka [1]

1. Department of Preventive Medicine, Faculty of Medicine, Saga University, Saga 849-8501, Japan; horitam@cc.saga-u.ac.jp (M.H.); chisatos@cc.saga-u.ac.jp (C.S.); harameg@cc.saga-u.ac.jp (M.H.); tanakake@cc.saga-u.ac.jp (K.T.)
2. SUMCO Corporation, Saga 849-4271, Japan; miyadomi@sumcosi.com
3. Section of Clinical Cooperation System, Center for Comprehensive Community Medicine, Faculty of Medicine, Saga University, Saga 849-8501, Japan; hirotaka@cc.saga-u.ac.jp
4. Institute of Genetics and Animal Breeding of the Polish Academy of Sciences, 05-552 Jastrzebiec, Poland; h.taniguchi@ighz.pl
5. Laboratory of Exercise Physiology, Faculty of Health and Sports Science, Fukuoka University, Fukuoka 814-0180, Japan; higaki@fukuoka-u.ac.jp (Y.H.); htanaka@fukuoka-u.ac.jp (H.T.)

* Correspondence: ynishida@cc.saga-u.ac.jp; Tel.: +81-952-34-2065

Received: 9 February 2018; Accepted: 26 February 2018; Published: 1 March 2018

Abstract: The purpose of the current study was to examine the influence of single-nucleotide polymorphisms (SNPs) in the peroxisome proliferator-activated receptor-δ (PPAR-δ), PPAR-γ, and α2 isoforms of the catalytic subunit of AMP-activated protein kinase (PRKAA2) on the extent of changes in anthropometric indices and blood measurements through exercise-centered lifestyle intervention in middle-aged men. A total of 109 Japanese middle-aged male subjects (47.0 ± 0.4 years) participated in the baseline health checkup, 6-month exercise-centered lifestyle intervention, and second checkup conducted several months after the subject completed the intervention. The body mass index (BMI), waist circumference, and clinical measurements, including hemoglobin A_{1c} (HbA_{1c}), triglyceride (TG), alanine aminotransferase (ALT), and γ-glutamyl-transpeptidase (γ-GTP), were measured at the baseline and second checkup. The three SNPs of PPAR-δ A/G (rs2267668), PPAR-γ C/G (rs1801282), and PRKAA2 A/G (rs1418442) were determined. Blunted responses in the reduction in the BMI and waist circumference were observed in A/A carriers of PPAR-δ SNP compared with G allele carriers (all $p < 0.05$). The A/A carriers also displayed less-marked improvements in HbA_{1c}, TG, ALT, and γ-GTP (all $p < 0.05$). The current results suggest that A/A carriers of PPAR-δ SNP (rs2267668) may enjoy fewer beneficial effects of exercise-centered lifestyle intervention on anthropometric indices and blood measurements.

Keywords: exercise; PPAR; SNP; obesity; lipid; glucose; HbA_{1c}; liver enzyme

---

1. Introduction

Peroxisome proliferator-activated receptors (PPARs) are nuclear receptors that play an important role in obesity and metabolism [1,2]. There are three isoforms of PPARs—PPAR-α, PPAR-δ/β, and PPAR-γ—all of which are activated by dietary or endogenous fatty acids and their metabolic derivatives and regulate lipid and glucose metabolism by modulating the expression of their target genes [1,2]. While PPAR-α and PPAR-γ are preferentially expressed in liver and adipose tissue, respectively [1], PPAR-δ is ubiquitously expressed throughout the body, and its expression is especially high in skeletal muscle compared with the other two isoforms [3,4]. These three PPAR isoforms together control various aspects of fatty acid metabolism, energy balance, insulin sensitivity, and glucose homeostasis through their coordinated activities in adipose tissue, liver, and skeletal muscle [1]. Another key molecule in lipid and glucose metabolism as well as exercise adaptation is AMP-activated protein kinase (AMPK). AMPK is a heterotrimeric protein composed of a catalytic subunit (α) and regulatory subunits (β and γ), and the catalytic subunit α has two isoforms (α1 and α2) [5]. While no defects in glucose homoeostasis were observed in AMPK α1 knockout mice, AMPK α2 knockout mice were insulin-resistant [6].

There have been several reports regarding the influence of single-nucleotide polymorphisms (SNPs) in PPARs on changes in anthropometric and metabolic parameters following lifestyle interventions in Caucasians [7,8]. Thamer et al. [8] showed that G allele carriers of the A/G SNP in PPAR-δ gene (rs2267668) displayed reduced responses to a nine-month exercise and dietary lifestyle intervention, as they lost less adipose tissue mass, less hepatic lipid content, and had less improvement in insulin sensitivity following the lifestyle intervention than A/A genotype carriers. Similarly, it has also been reported that the G allele carriers of the PPAR-δ A/G enjoy fewer benefits than A/A genotype carriers regarding insulin sensitivity after nine-month exercise and dietary intervention [7]. Several studies have also shown associations of the Pro12Ala variant of PPAR-γ (rs1801282) with obesity, type 2 diabetes/insulin resistance, and metabolic syndrome [9–13], and the PRKAA2 SNP (rs1418442) has also been reported to be associated with serum cholesterol levels in Caucasians [14] and with type 2 diabetes in Japanese populations [2,15].

Gene-exercise interaction is influenced by many factors of exercise regimens (e.g., exercise type, frequency, duration) and subject characteristics (e.g., ethnicity, age, sex, energy intake, and baseline physical activity) [16]. Thus, the currently available evidence regarding the influence of gene polymorphisms on the responsiveness to exercise intervention is far from sufficient to develop tailored exercise interventions based on an individual's genetic information. In the current study, we selected the SNP PPAR-δ rs2267668 whose influence has previously been investigated in the aforementioned lifestyle intervention studies in Caucasians [7,8]. In addition, we also selected two other SNPs that are most well-known as metabolism-related SNPs (PPAR-γ rs1801282, PRKAA2 rs1418442) in Caucasians and Japanese [2,9–13,15]. Thus, the current study investigated the influence of these SNPs on the extent of changes in anthropometric indices and clinical blood measurements in response to six-month exercise-centered lifestyle intervention in Japanese middle-aged subjects.

2. Results

The genotype distributions of the three SNPs, PPAR-δ A/G (rs2267668), PPAR-γ C/G (rs1801282), and PRKAA2 A/G (rs1418442) in the current participants did not deviate from the Hardy-Weinberg equilibrium ($p > 0.05$). These genotype distributions were confirmed to be similar to those reported in a public database dbSNP (HapMap-JPT). The baseline levels of variables, such as the age, body mass index (BMI), waist circumference, systolic blood pressure (SBP), diastolic blood pressure (DBP), and clinical blood measurements (plasma glucose, hemoglobin A_{1c} (HbA_{1c}), triglyceride (TG), high-density lipoprotein cholesterol (HDL-C), low-density lipoprotein cholesterol (LDL-C), total cholesterol (Total-C)), were not significantly different among the genotypes of the three SNPs (Table 1).

Table 1. Baseline characteristics of the subjects according to genotypes of PPAR-δ, PPAR-γ, and PRKAA2 SNPs.

Variables	All Subjects (n = 109)	PPAR-δ rs2267668			PPAR-γ rs1801282			PRKAA2 rs1418442		
		A/A (n = 66)	A/G + G/G (n = 43)	p Value *	C/C (n = 99)	C/G + G/G (n = 10)	p Value *	A/A (n = 64)	A/G + G/G (n = 45)	p Value *
Age (years)	47.0 ± 0.4	46.8 ± 0.5	47.3 ± 0.6	0.61	47.1 ± 0.4	46.1 ± 1.3	0.49	47.1 ± 0.5	46.9 ± 0.7	0.78
Weight (kg)	75.3 ± 0.8	76.1 ± 1.0	74.2 ± 1.2	0.24	75.0 ± 0.8	79.1 ± 2.9	0.20	75.6 ± 1.0	74.9 ± 1.2	0.67
BMI (kg/m^2)	25.7 ± 0.2	25.8 ± 0.3	25.5 ± 0.4	0.64	25.5 ± 0.2	27.2 ± 0.6	0.02	25.8 ± 0.3	25.6 ± 0.3	0.67
Waist circumference (cm)	90.0 ± 0.5	90.3 ± 0.7	89.5 ± 0.8	0.47	89.7 ± 0.5	93.3 ± 2.2	0.15	90.5 ± 0.7	89.4 ± 0.7	0.29
SBP (mmHg)	125 ± 1	126 ± 1	122 ± 2	0.08	125 ± 1	125 ± 2	0.85	125 ± 1	124 ± 2	0.46
DBP (mmHg)	82 ± 1	82 ± 1	82 ± 1	0.95	82 ± 1	82 ± 2	0.85	82 ± 1	82 ± 1	0.84
Glucose (mg/dL)	99.9 ± 1.0	100.0 ± 1.3	99.8 ± 1.5	0.92	99.8 ± 1.1	100.9 ± 3.1	0.75	100.3 ± 1.4	99.4 ± 1.4	0.64
HbA$_{1c}$ (%)	5.48 ± 0.04	5.48 ± 0.06	5.47 ± 0.05	0.85	5.49 ± 0.04	5.37 ± 0.13	0.40	5.53 ± 0.06	5.40 ± 0.04	0.07
TG (mg/dL)	127 (93–171)	122 (92–169)	136 (102–179)	0.52	123 (92–171)	143 (121–166)	0.38	131 (94–166)	123 (92–189)	0.91
AST (IU/L)	23.0 (20.0–28.0)	22.0 (20.0–28.0)	24.0 (20.0–30.0)	0.35	22.0 (20.0–28.5)	25.5 (23.3–27.8)	0.49	23.5 (20.0–28.3)	23.0 (20.0–28.0)	1.00
ALT (IU/L)	28.0 (21.0–42.0)	27.0 (20.0–41.0)	30.0 (22.5–45.5)	0.32	28.0 (20.5–41.5)	29.0 (25.5–43.0)	0.53	28.0 (20.0–43.8)	27.0 (22.0–39.0)	0.87
γ-GTP (IU/L)	42.5 (31.8–58.9)	40.8 (30.1–55.1)	48.6 (34.4–77.6)	0.11	42.5 (32.3–59.2)	45.4 (26.7–53.8)	0.73	41.5 (32.5–57.7)	46.1 (29.6–61.5)	0.75
HDL-C (mg/dL)	50 ± 1	49 ± 1	52 ± 2	0.18	50 ± 1	47 ± 4	0.52	50 ± 1	50 ± 2	0.86
LDL-C (mg/dL)	133 ± 3	133 ± 3	134 ± 5	0.80	132 ± 3	144 ± 8	0.16	131 ± 3	137 ± 4	0.24
Total-C (mg/dL)	213 ± 3	210 ± 4	218 ± 4	0.18	212 ± 3	219 ± 8	0.47	211 ± 4	215 ± 4	0.48
Lactate threshold (mL/kg/min)	16.6 ± 0.2	16.7 ± 0.3	16.5 ± 0.3	0.68	16.6 ± 0.2	16.5 ± 0.8	0.94	16.5 ± 0.3	16.7 ± 0.3	0.56
Lactate threshold (METs)	4.74 ± 0.05	4.76 ± 0.07	4.72 ± 0.08	0.68	4.74 ± 0.05	4.73 ± 0.22	0.94	4.72 ± 0.08	4.78 ± 0.07	0.56

Values are the mean ± SE or the median (interquartile range). PPAR, peroxisome proliferator-activated receptor; PRKAA2, α2 isoform of catalytic subunit of AMP-activated protein kinase; SNP, single-nucleotide polymorphism; BMI, body mass index; SBP, systolic blood pressure; DBP, diastolic blood pressure; HbA$_{1c}$, hemoglobin A$_{1c}$; TG, triglyceride; AST, aspartate aminotransferase; ALT, alanine aminotransferase; GTP, glutamyl-transpeptidase, HDL-C, high-density lipoprotein cholesterol; LDL-C, low-density lipoprotein cholesterol; Total-C, total cholesterol; METs, metabolic equivalents. * p value for between-group comparison using Welch's t test (for age, weight, BMI, waist circumference, SBP, DBP, glucose, HbA$_{1c}$, HDL-C, LDL-C, Total-C, lactate threshold) or Wilcoxon's signed-rank test (for TG, AST, ALT, γ-GTP).

Table 2. Changes in the measured variables according to genotype groups of PPAR-δ, PPAR-γ, and PRKAA2 SNPs.

Variables	All Subjects (n = 109)	PPAR-δ rs2267668			PPAR-γ rs1801282			PRKAA2 rs1418442		
		A/A (n = 66)	A/G + G/G (n = 43)	p Value #	C/C (n = 99)	C/G + G/G (n = 10)	p Value #	A/A (n = 64)	A/G + G/G (n = 45)	p Value #
Weight (kg)	−1.1 ± 0.3 *	−0.1 ± 0.3	−2.5 ± 0.5 *	0.00	−1.2 ± 0.3 *	0.8 ± 1.2	0.13	−1.2 ± 0.4 *	−0.9 ± 0.5	0.58
BMI (kg/m^2)	−0.36 ± 0.10 *	−0.04 ± 0.10	−0.84 ± 0.18 *	0.00	−0.42 ± 0.10 *	0.24 ± 0.40	0.14	−0.40 ± 0.14 *	−0.30 ± 0.15	0.62
Waist circumference (cm)	−1.3 ± 0.3 *	−0.6 ± 0.3	−2.2 ± 0.6 *	0.02	−1.4 ± 0.3 *	−0.1 ± 1.2	0.33	−1.4 ± 0.4 *	−1.0 ± 0.4 *	0.56
Glucose (mg/dL)	−0.4 ± 0.8	0.2 ± 1.0	−1.4 ± 1.3	0.30	−0.1 ± 0.8	−3.6 ± 2.3	0.18	0.3 ± 1.0	−1.5 ± 1.3	0.27
HbA$_{1c}$ (%)	−0.03 ± 0.02	−0.002 ± 0.03	−0.08 ± 0.02 *	0.04	−0.03 ± 0.02	0.00 ± 0.05	0.51	−0.04 ± 0.03	−0.02 ± 0.03	0.79
TG (mg/dL)	−12.0 (−46.0–18.0) *	−2.5 (−33.5–21.8)	−33.0 (−59.0–−5.5) *	0.00	−12.0 (−48.5-16.0) *	−21.5 (−26.5–29.0)	0.63	−20.0 (−42.8–10.8)	−8.0 (−58.0–22.0)	0.54
AST (IU/L)	−1.0 (−5.0–3.0) *	0 (−5.8–3.8)	−3.0 (−5.0–2.0) *	0.15	−1.0 (−5.0–2.5) *	−1.0 (−4.3–3.8)	0.61	−1.0 (−5.0–2.3)	−1.0 (−5.0–3.0)	0.78
ALT (IU/L)	−3.0 (−12.0–4.0) *	−0.5 (−8.8–4.0)	−7.0 (−12.5–1.5) *	0.02	−3.0 (−12.0–3.5) *	−3.5 (−8.5–3.3)	0.82	−2.0 (−11.3–4.0) *	−5.0 (−12.0–3.0)	0.90
γ-GTP (IU/L)	−4.4 (−12.4–2.5)	−1.2 (−7.6–4.8)	−10.7 (−19.3–−2.5) *	0.00	−4.4 (−12.5–2.4)	−3.4 (−5.3–6.2)	0.52	−4.5 (−8.8–2.4)	−3.9 (−14.9–2.5)	0.35
HDL-C (mg/dL)	1.8 ± 0.7 *	1.2 ± 0.8	2.8 ± 1.1 *	0.26	2.0 ± 0.7 *	0.1 ± 2.8	0.52	1.5 ± 0.8	2.3 ± 1.1 *	0.57
LDL-C (mg/dL)	−4.1 ± 2.3	−4.5 ± 2.3	−3.5 ± 4.6	0.86	−4.3 ± 2.4	−1.8 ± 8.2	0.77	−5.5 ± 3.3	-2.1 ± 2.9	0.44
Total-C (mg/dL)	−6.7 ± 2.2 *	−4.0 ± 2.4	−10.8 ± 4.2 *	0.16	−7.3 ± 2.4 *	−0.6 ± 6.0	0.32	−9.5 ± 3.1 *	-2.7 ± 3.1	0.12

Values are the mean ± SE or the median (interquartile range). Note that post-intervention assessment (at the second checkup) was performed several months after the subject completed the six-month intervention. PPAR, peroxisome proliferator-activated receptor; PRKAA2, α2 isoform of catalytic subunit of AMP-activated protein kinase; SNP, single-nucleotide polymorphism; BMI, body mass index; SBP, systolic blood pressure; DBP, diastolic blood pressure; HbA$_{1c}$, hemoglobin A$_{1c}$; TG, triglyceride; AST, aspartate aminotransferase; ALT, alanine aminotransferase; GTP, glutamyl-transpeptidase; HDL-C, high-density lipoprotein cholesterol; LDL-C, low-density lipoprotein cholesterol; Total-C, total cholesterol. * $p < 0.05$ for within-group comparison between the baseline checkup and second checkup using Wilcoxon's signed-rank test. # p value for between-group comparison using Welch's t test (for weight, BMI, waist circumference, glucose, HbA$_{1c}$, HDL-C, LDL-C, Total-C) or Wilcoxon's signed-rank test (for TG, AST, ALT, γ-GTP).

In all 109 subjects, the anthropometric indices of the BMI and waist circumference, and blood measurements such as TG, aspartate aminotransferase (AST), alanine aminotransferase (ALT), HDL-C, and total-C were significantly improved after the exercise-centered lifestyle intervention (Table 2). The decreases in the body weight, BMI, and waist circumference were significantly less marked in the A/A genotype carriers of PPAR-δ A/G (rs2267668) than in the G allele carriers (A/G + G/G) (Table 2). Similarly, the PPAR-δ A/A carriers displayed significantly less-marked improvements in the HbA_{1c}, TG, ALT, and γ-glutamyl-transpeptidase (γ-GTP) (Table 2). However, no significant differences among genotype groups were observed in the other two tested SNPs (Table 2). The post-intervention values of variables measured at the second checkup are shown in Supplementary Table S1.

Univariate correlation analyses showed that, in the total subjects ($n = 109$), a greater reduction in the body weight was significantly correlated with greater improvements (decreases) in the HbA_{1c} ($r = 0.24$, $p < 0.05$), TG ($r = 0.42$, $p < 0.01$), ALT ($r = 0.44$, $p < 0.01$), and γ-GTP ($r = 0.40$, $p < 0.01$). The correlation analyses performed separately by the genotype groups of PPAR-δ SNP (A/A ($n = 66$) and A/G + G/G ($n = 43$)) showed a similar pattern of correlation (data not shown). Multiple linear regression analyses showed that the associations of the PPAR-δ SNP with the changes in HbA_{1c} ($p = 0.26$) and ALT ($p = 0.31$) were attenuated, being statistically insignificant, whereas the associations of the PPAR-δ SNP with the changes in TG (natural log-transformed) ($p = 0.02$) and γ-GTP ($p = 0.04$) remained significant even after adjustment for the change in body weight (Table 3).

Table 3. The multiple linear regression analyses on the associations of PPAR-δ *A/G* SNP (rs2267668) with the changes in the four clinical measures.

Outcome Variables	β	SE	*p* Value
Change in HbA_{1c}	−0.045	0.039	0.26
Change in log-transformed TG	−0.211	0.089	0.02
Change in ALT	−3.5	3.4	0.31
Change in γ-GTP	−9.5	4.6	0.04

PPAR, peroxisome proliferator-activated receptor; SNP, single-nucleotide polymorphism; HbA_{1c}, hemoglobin A_{1c}; TG, triglyceride; ALT, alanine aminotransferase; GTP, glutamyl-transpeptidase.

3. Discussion

In the current study, we investigated the influence of PPAR-δ A/G (rs2267668), PPAR-γ C/G (rs1801282), and PRKAA2 A/G (rs1418442) SNPs on the anthropometric and blood measurements in response to exercise-centered lifestyle intervention in Japanese middle-aged men. The major finding of the current study was that the A/A homozygotes of PPAR-δ rs2267668 had less marked weight loss after the intervention than the G allele carriers. Similarly, the A/A carriers also displayed less-marked improvements in clinical measurements, such as the values of HbA_{1c}, TG, ALT, and γ-GTP. Multiple linear regression analyses showed that the associations of the PPAR-δ SNP with the changes in HbA_{1c} and ALT were attenuated, being statistically insignificant, whereas the associations with TG and γ-GTP remained significant even after adjustment for changes in the body weight. These findings therefore suggest that the reduced improvements in TG and γ-GTP might be due to the A/A genotype itself, rather than the blunted weight loss response. In contrast, the other two tested SNPs in PPAR-γ and PRKAA2 did not influence the changes in the measured parameters.

PPAR-δ plays key roles in lipid and glucose metabolism and skeletal muscle adaptation to exercise [5]. In rodent studies, transgenic activation of PPAR-δ in adipose tissue has been shown to promote fatty acid oxidation in skeletal muscle and adipose tissue, preventing obesity [17], while PPAR-δ-deficient mice receiving a high-fat diet have reduced energy uncoupling and are prone to obesity [17]. Previous human studies have investigated the influence of the PPAR-δ A/G (rs2267668) SNP on the body fat reduction response and glucose metabolism indices after lifestyle intervention in Caucasians [7,8]. In these studies, lifestyle intervention-induced improvements (reductions) in adiposity and hepatic fat storage were blunted in G allele carriers of this SNP [8]. In contrast, in the current study A/A homozygous carriers of the PPAR-δ A/G (rs2267668) SNP displayed less-marked effects in anthropometric indices and clinical blood measurements than G allele carriers after six-month lifestyle intervention.

The reason for this discrepancy in findings between the previous reports and the present study is unclear. It may be due in part to differences in the exercise regimens (intensity, frequency, duration) and subject characteristics, such as the sex, age, energy intake, and baseline physical activity [16]. In addition, differences in the genetic backgrounds of European and Japanese subjects may be another possible reason, as the influence of gene polymorphisms on the degree of benefit from lifestyle intervention may not necessarily be the same (or may even be contrary) between different ethnic groups [16,18–20], although the reasons for the race-related differences in the influence of gene polymorphisms remain to be explored.

The physiological function of PPAR-δ A/G SNP (rs2267668) is unclear at present. However, one report evaluated the impact of this SNP (rs2267668) on the gene expression of PPAR-δ in human skeletal muscle [21]. This previous study showed that the A/A genotype carriers of PPAR-δ A/G SNP (rs2267668) had lower PPAR-δ mRNA expression in skeletal muscle than G allele carriers [21]. It is, therefore, biologically plausible that reduced PPAR-δ transcriptional activity in the muscle of A/A genotype carriers might be a reason for their blunt responses to exercise-centered health guidance intervention.

Less-marked decreases in the body weight and waist circumference observed in the A/A genotype of PPAR-δ A/G (rs2267668) SNP than in G allele carriers may have been due to insufficient increases in energy expenditure by physical activity and/or insufficient decreases in energy intake through the diet. We were unable to clarify which factors might have contributed most strongly to the blunted responses observed in the A/A genotype carriers. A rodent study found that treatment with a PPAR-δ agonist (GW50516) significantly retards weight gain but does not affect food consumption [1,22]; given those findings, we speculate that our subjects with the A/A genotype of PPAR-δ A/G (rs2267668) SNP may have expended less energy (rather than having a higher energy intake) than G allele carriers during the exercise-centered lifestyle intervention.

The major limitation of the current study was that we did not have precise data on the duration of exercise training performed by the subjects, although we instructed our subjects to regularly perform cycle ergometer exercise or brisk walking at the lactate threshold (LT) intensity, with a goal of >140 min/week. Second, we did not assess the effects of the exercise-centered lifestyle intervention immediately after completing the six-month intervention, instead evaluating the outcomes at the second checkup, which was conducted several months after the completion of the intervention. We were unable to clarify whether the reduced effects of the exercise-centered lifestyle intervention observed in the A/A genotype carriers of the PPAR-δ A/G SNP might have been due to less-marked changes in the measured parameters or an earlier return towards baseline values. We did not instruct the current subjects to maintain (not to change) their dietary habit during the exercise-centered lifestyle intervention period. We therefore cannot exclude the possibility that the subjects might have changed their dietary habits during the study period. We also did not collect any data on the eating habits of the study participants nor on their post-intervention habits. Although using bioelectric impedance or computed tomography to assess the body composition or abdominal/ectopic fat accumulation is ideal, we merely assessed the anthropometric indices (weight, BMI, waist circumference) included as inspection items for the annual specific health checkup. In addition, the current results were obtained in a Japanese population and cannot be generalized to other ethnic groups, such as Caucasians.

In conclusion, the current results suggest that the A/A genotype carriers of the PPAR-δ A/G SNP (rs2267668) may have experienced blunted effects in anthropometric measurements (weight, BMI, waist circumference), HbA_{1c}, TG, and serum liver enzymes (ALT, γ-GTP) through exercise-centered lifestyle intervention compared with G allele carriers among our population of middle-aged Japanese men. In addition, multiple linear regression analyses suggested that the less-marked improvements in TG and γ-GTP observed in the PPAR-δ SNP A/A carriers were not due to their attenuated weight reduction.

Studying the genetic background of multifactorial processes, such as weight loss, is challenging. In the present study, we were unable to clarify the extent to which the observed results depended on the genetic predisposition or difference in the subjects' compliance to the exercise-centered lifestyle intervention. Therefore, a further study on the influence of PPAR-δ A/G SNP (rs2267668) is needed, in which exercise intervention is standardized and its compliance is carefully monitored. The current results will aid in the development of tailored health guidance programs based on the individuals' genotypes for metabolism-related SNPs.

4. Materials and Methods

4.1. Subjects

The subjects were 109 Japanese middle-aged men (47.0 ± 0.4 years of age) employed at a silicon wafer manufacturer (Saga, Japan) who participated in the baseline (first) checkup (conducted as an annual specific health checkup), 6-month exercise-centered lifestyle intervention (conducted as a specific health guidance), and second checkup conducted 1 year after the baseline health checkup. The post-intervention assessments at the second checkup were conducted several months after each subject completed the six-month intervention. The specific health checkup (to identify individuals with metabolic syndrome) and the specific health guidance (to improve their metabolic syndrome) are currently conducted as part of a national effort against metabolic syndrome, in which medical insurers are obligated to provide insured middle-aged employees (40–74 years of age) with specific health checkups. Participants are recommended to participate in the specific health guidance if they meet certain conditions of metabolic syndrome (Available online: http://www.mhlw.go.jp/english/wp/wp-hw3/dl/2-007.pdf).

The inclusion criteria of the specific health guidance for men were as follows: abdominal obesity (waist circumference ≥ 85 cm) plus at least 1 of the following 3 components: (1) dyslipidemia (TG ≥ 150 mg/dL and/or HDL < 40 mg/dL); (2) high blood pressure (SBP/DBP ≥ 130/85 mm Hg); and (3) high blood glucose (fasting plasma glucose ≥ 100 mg/dL or HbA_{1c} expressed as National Glycohemoglobin Standardization Program (NGSP) value of 5.6%). In addition, even if the waist circumference was <85 cm, men who had a BMI ≥ 25 kg/m^2 plus at least 1 of the 3 abovementioned components met the requirements to participate in the specific health guidance. Subjects who were taking medications for type 2 diabetes, hypertension, or dyslipidemia were excluded.

4.2. Anthropometric and Blood Measurements at the Baseline (First) Checkup and Second Checkup (Conducted as Annual Specific Health Checkup)

The BMI was determined by dividing the body weight in kilograms by the square of the height in meters. The waist circumference was measured at the level of the umbilicus. Blood pressure (SBP and DBP) was measured in the sitting position after 5 min of rest using an automatic sphygmomanometer. For the blood biochemical test, blood samples were obtained from an antecubital vein after an overnight fast. Plasma glucose was measured via the standard method. HbA_{1c} was measured by a latex aggregation immunoassay (Japan Diabetes Society (JDS) value). The HbA_{1c} was estimated as a NGSP equivalent value calculated by the formula as follows: HbA_{1c} (NGSP (%)) = 1.02 × HbA_{1c} (JDS (%) + 0.25% [23]. Total-C and TG levels were measured enzymatically. HDL-C and LDL-C were measured via direct methods. Serum liver enzymes (AST, ALT, γ-GTP) were also measured by standard methods. The ectopic fat accumulation in the liver is an important factor for inducing insulin resistance [24], and serum liver enzymes, especially ALT, can be practical indices reflective of the liver fat accumulation [25]. All measurements and blood biochemical analyses were performed using similar methods at the baseline and second checkup.

4.3. *Six-Month Exercise-Centered Lifestyle Intervention (Conducted as Specific Health Guidance)*

First, an initial group-based health guidance (≤8 participants/group for 80 min) was conducted, in which the subjects received an explanation on the results of their baseline checkup from public health nurses and then set achievable goals related to lifestyle improvement with support or advice from public health nurses. The subjects were instructed to record information on whether their own-set goals were achieved on a recording sheet (made using the Excel software program). Throughout the six-month intervention, the subjects received personalized follow-up consultation through e-mail.

The main components of the 6-month exercise-centered intervention were two sessions of group-based (≤8 participants/group) exercise guidance (90 min/session). In the first exercise session, a submaximal graded exercise test using a cycle ergometer (Model EC-3600; Cateye Inc., Osaka, Japan) was performed to assess the lactate threshold (LT) as an index of aerobic capacity, and moderate exercise at the LT intensity was recommended to the subjects as an ideal exercise for their daily exercise training. For the graded exercise test, the work rate was initially set at a workload corresponding to 3 metabolic equivalents (METs) and then increased by 1 MET every 3 min (i.e., 3, 4, 5, 6, and 7 METs). Oxygen consumption was estimated based on the workload and subjects' body weight using the American College of Sports Medicine leg ergometer equation, as follows: the estimated oxygen consumption (mL/kg/min) = workload (watts) $\times$ 6.12 $\times$ 1.8/body weight (kg) + 7 [26]. The METs were calculated as the estimated oxygen consumption divided by 3.5. The end-point of the exercise test was determined based on either achieving a blood lactate concentration of 4 mM or the American College of Sports Medicine criteria [27]. The heart rate was measured in real time using a sensor (installed in the cycle ergometer) attached to the earlobe. The heart rate and rating of perceived exertion (RPE) [28] were recorded every 3 min during the tests. Blood samples (5 μL) were also obtained from the earlobe every 3 min to measure the blood lactate concentration using a portable blood lactate test meter (Lactate Pro; ARKRAY, Inc., Kyoto, Japan).

The blood lactate concentration (mM) was plotted against the exercise workload (watts) for each subject. The estimated oxygen consumption (or METs) at the first breakpoint of lactate concentration was used as the data of LT. The LT is a reliable indicator of aerobic fitness that is in no way inferior to VO_2 max [29,30], and this index can be simply and precisely measured by a graded cycle ergometer test using a portable lactate analyzer [20]. Immediately after the cycle ergometer exercise test, we provided an exercise prescription based on each subject's result for the cycle ergometer exercise test, and the subjects were instructed to perform cycle ergometer training at the LT intensity and/or brisk walking at a heart rate corresponding to the LT, with a goal of ≥140 min/week. The factories of the subjects' employer include training rooms (in which there were many cycle ergometers and treadmills) that were freely available to all subjects during their rest period or after working hours.

Two months after the first session of exercise guidance, the second session of exercise guidance (≤8 participants/group for 90 min) was conducted, mainly to revise the exercise intensity and to maintain or promote subjects' motivation to continue the cycle ergometer exercise and/or brisk walking at the LT intensity. In this second session of exercise guidance, the subjects performed a single bout of 15-min cycle ergometer exercise at the LT that was previously determined at the first session of exercise guidance, and the heart rate was monitored during the exercise to assess if the aerobic capacity had improved. The heart rate during the exercise was used to give feedback to the subjects. The subjects received individual advice on the exercise training based on the newly revised workload. Within one month after completing the six-month intervention, the body weight, waist circumference, and blood pressure were assessed, but these data were not collected for the current study, as this time point was not an endpoint of our study.

4.4. *Genotyping of Gene Variants*

Genomic DNA was extracted from saliva (2 mL) collected from subjects using Oragene DNA kits (DNA Genotek, Ottawa, ON, Canada). DNA was purified from 200-μL aliquots of Oragene DNA/saliva samples using an ethanol precipitation protocol supplied with the kits. Purified

DNA was redissolved in 200 μL of Tris ethylenediaminetetraacetic acid buffer (10 mM tris-HCl, 1 mM ethylenediaminetetraacetic acid, pH 8.0). The SNPs in PPAR-δ A/G rs2267668 (located in intron 2), PPAR-γ C/G rs1801282 (exon 1 (Pro12Ala), and PRKAA2 A/G rs1418442 (intron 4) were analyzed by a TaqMan® SNP Genotyping Assay using a StepOne Plus real-time PCR system (Applied Biosystems, Foster City, CA, USA). Their assay IDs were C__5872729_10 for PPAR-δ A/G (rs2267668), C__1129864_10 for PPAR-γ C/G (rs1801282)), and C__2821517_20 for PRKAA2 A/G (rs1418442).

4.5. Statistical Analyses

Values were shown as the mean ± standard error (SE) or the median (interquartile range). The chi-squared test was used to confirm whether the genotype frequencies were in Hardy-Weinberg equilibrium. Between-group (between genotype groups) comparisons were performed using Welch's *t*-test for the age, weight, BMI, waist circumference, SBP, DBP, glucose, HbA_{1c}, HDL-C, LDL-C, Total-C, and LT. Since the distributions of TG, AST, ALT, and γ-GTP may have been skewed, between-group comparisons of these four variables were performed using a nonparametric Wilcoxon's signed-rank test. Within-group comparisons between the baseline and second checkup were also performed using Wilcoxon's signed-rank test. The significance of correlations between two variables was assessed using Pearson's correlation coefficient. Multiple linear regression analyses with adjustment for change in body weight were performed to examine whether the observed significant differences in the changes in the clinical blood measurements between genotypes were independent of the change in the body weight. In these multiple linear regression analyses, PPAR-δ A/G SNP (rs2267668) was used as an independent variable, and the changes in four clinical measurements (HbA_{1c}, natural log-transformed TG, ALT, γ-GTP) were used as dependent variables, with the change in body weight used for adjustment. The statistical analyses were performed using the R software program (version 3.4.1). p Values of <0.05 were considered statistically significant.

Supplementary Materials: The following are available online at http://www.mdpi.com/1422-0067/19/3/703/s1.

Acknowledgments: This work was partially supported by grants from the Japanese Ministry of Education, Culture, Sports, Science, and Technology (No. 19200049, Strategic Research Infrastructure) and the Global FU Program, funded by Fukuoka University. We thank Hideaki Kumahara and Hiroyuki Higuchi for correcting the workload of the electric cycle ergometers.

Author Contributions: Yuichiro Nishida and Minako Iyadomi conceived and designed the experiments; Yuichiro Nishida and Mikako Horita performed the experiments; Hirotaka Tominaga, Chisato Shimanoe and Yuichiro Nishida analyzed the data; Yasuki Higaki, Hiroaki Tanaka, Megumi Hara and Keitaro Tanaka contributed reagents/materials/analysis tools; Yuichiro Nishida and Hiroaki Taniguchi wrote the paper.

Conflicts of Interest: The authors declare no conflicts of interest. The founding sponsors had no role in the design of the study; in the collection, analyses, or interpretation of data; in the writing of the manuscript; or in the decision to publish the results.

References

1. Evans, R.M.; Barish, G.D.; Wang, Y.X. PPARs and the complex journey to obesity. *Nat. Med.* **2004**, *10*, 355–361. [CrossRef] [PubMed]
2. Keshavarz, P.; Inoue, H.; Nakamura, N.; Yoshikawa, T.; Tanahashi, T.; Itakura, M. Single nucleotide polymorphisms in genes encoding LKB1 (STK11), TORC2 (CRTC2) and AMPK α2-subunit (PRKAA2) and risk of type 2 diabetes. *Mol. Genet. Metab.* **2008**, *93*, 200–209. [CrossRef] [PubMed]
3. Muoio, D.M.; MacLean, P.S.; Lang, D.B.; Li, S.; Houmard, J.A.; Way, J.M.; Winegar, D.A.; Corton, J.C.; Dohm, G.L.; Kraus, W.E. Fatty acid homeostasis and induction of lipid regulatory genes in skeletal muscles of peroxisome proliferator-activated receptor (PPAR) α knock-out mice. Evidence for compensatory regulation by *PPAR δ*. *J. Biol. Chem.* **2002**, *277*, 26089–26097. [CrossRef] [PubMed]
4. Barish, G.D.; Narkar, V.A.; Evans, R.M. *PPAR δ*: A dagger in the heart of the metabolic syndrome. *J. Clin. Investig.* **2006**, *116*, 590–597. [CrossRef] [PubMed]
5. Kjobsted, R.; Hingst, J.R.; Fentz, J.; Foretz, M.; Sanz, M.N.; Pehmoller, C.; Shum, M.; Marette, A.; Mounier, R.; Treebak, J.T.; et al. AMPK in skeletal muscle function and metabolism. *FASEB J.* **2018**. [CrossRef] [PubMed]

6. Viollet, B.; Andreelli, F.; Jorgensen, S.B.; Perrin, C.; Flamez, D.; Mu, J.; Wojtaszewski, J.F.; Schuit, F.C.; Birnbaum, M.; Richter, E.; et al. Physiological role of AMP-activated protein kinase (AMPK): Insights from knockout mouse models. *Biochem. Soc. Trans.* **2003**, *31*, 216–219. [CrossRef] [PubMed]
7. Stefan, N.; Thamer, C.; Staiger, H.; Machicao, F.; Machann, J.; Schick, F.; Venter, C.; Niess, A.; Laakso, M.; Fritsche, A.; et al. Genetic variations in PPARD and PPARGC1A determine mitochondrial function and change in aerobic physical fitness and insulin sensitivity during lifestyle intervention. *J. Clin. Endocrinol. Metab.* **2007**, *92*, 1827–1833. [CrossRef] [PubMed]
8. Thamer, C.; Machann, J.; Stefan, N.; Schafer, S.A.; Machicao, F.; Staiger, H.; Laakso, M.; Bottcher, M.; Claussen, C.; Schick, F.; et al. Variations in PPARD determine the change in body composition during lifestyle intervention: A whole-body magnetic resonance study. *J. Clin. Endocrinol. Metab.* **2008**, *93*, 1497–1500. [CrossRef] [PubMed]
9. Yen, C.J.; Beamer, B.A.; Negri, C.; Silver, K.; Brown, K.A.; Yarnall, D.P.; Burns, D.K.; Roth, J.; Shuldiner, A.R. Molecular scanning of the human peroxisome proliferator activated receptor γ (hPPAR γ) gene in diabetic Caucasians: Identification of a Pro12Ala *PPAR* γ 2 missense mutation. *Biochem. Biophys. Res. Commun.* **1997**, *241*, 270–274. [CrossRef] [PubMed]
10. Robitaille, J.; Despres, J.P.; Perusse, L.; Vohl, M.C. The PPAR-γ P12A polymorphism modulates the relationship between dietary fat intake and components of the metabolic syndrome: Results from the Quebec Family Study. *Clin. Genet.* **2003**, *63*, 109–116. [CrossRef] [PubMed]
11. Tellechea, M.L.; Aranguren, F.; Perez, M.S.; Cerrone, G.E.; Frechtel, G.D.; Taverna, M.J. Pro12Ala polymorphism of the peroxisome proliferatoractivated receptor-γ gene is associated with metabolic syndrome and surrogate measurements of insulin resistance in healthy men: Interaction with smoking status. *Circ. J.* **2009**, *73*, 2118–2124. [CrossRef] [PubMed]
12. Galbete, C.; Toledo, E.; Martinez-Gonzalez, M.A.; Martinez, J.A.; Guillen-Grima, F.; Marti, A. Pro12Ala variant of the PPARG2 gene increases body mass index: An updated meta-analysis encompassing 49,092 subjects. *Obesity* **2013**, *21*, 1486–1495. [CrossRef] [PubMed]
13. Wang, X.; Liu, J.; Ouyang, Y.; Fang, M.; Gao, H.; Liu, L. The association between the Pro12Ala variant in the *PPAR* γ 2 gene and type 2 diabetes mellitus and obesity in a Chinese population. *PLoS ONE* **2013**, *8*, e71985. [CrossRef]
14. Spencer-Jones, N.J.; Ge, D.; Snieder, H.; Perks, U.; Swaminathan, R.; Spector, T.D.; Carter, N.D.; O'Dell, S.D. AMP-kinase α2 subunit gene PRKAA2 variants are associated with total cholesterol, low-density lipoprotein-cholesterol and high-density lipoprotein-cholesterol in normal women. *J. Med. Genet.* **2006**, *43*, 936–942. [CrossRef] [PubMed]
15. Horikoshi, M.; Hara, K.; Ohashi, J.; Miyake, K.; Tokunaga, K.; Ito, C.; Kasuga, M.; Nagai, R.; Kadowaki, T. A polymorphism in the AMPKα2 subunit gene is associated with insulin resistance and type 2 diabetes in the Japanese population. *Diabetes* **2006**, *55*, 919–923. [CrossRef] [PubMed]
16. Mori, M.; Higuchi, K.; Sakurai, A.; Tabara, Y.; Miki, T.; Nose, H. Genetic basis of inter-individual variability in the effects of exercise on the alleviation of lifestyle-related diseases. *J. Physiol.* **2009**, *587*, 5577–5584. [CrossRef] [PubMed]
17. Wang, Y.X.; Lee, C.H.; Tiep, S.; Yu, R.T.; Ham, J.; Kang, H.; Evans, R.M. Peroxisome-proliferator-activated receptor delta activates fat metabolism to prevent obesity. *Cell* **2003**, *113*, 159–170. [CrossRef]
18. Lucia, A.; Gomez-Gallego, F.; Barroso, I.; Rabadan, M.; Bandres, F.; San Juan, A.F.; Chicharro, J.L.; Ekelund, U.; Brage, S.; Earnest, C.P.; et al. PPARGC1A genotype (Gly482Ser) predicts exceptional endurance capacity in European men. *J. Appl. Physiol.* **2005**, *99*, 344–348. [CrossRef] [PubMed]
19. Maciejewska, A.; Sawczuk, M.; Cieszczyk, P.; Mozhayskaya, I.A.; Ahmetov, I.I. The PPARGC1A gene Gly482Ser in Polish and Russian athletes. *J. Sports Sci.* **2012**, *30*, 101–113. [CrossRef] [PubMed]
20. Nishida, Y.; Iyadomi, M.; Higaki, Y.; Tanaka, H.; Kondo, Y.; Otsubo, H.; Horita, M.; Hara, M.; Tanaka, K. Association between the PPARGC1A polymorphism and aerobic capacity in Japanese middle-aged men. *Intern. Med.* **2015**, *54*, 359–366. [CrossRef] [PubMed]
21. Ordelheide, A.M.; Heni, M.; Gommer, N.; Gasse, L.; Haas, C.; Guirguis, A.; Machicao, F.; Haring, H.U.; Staiger, H. The myocyte expression of adiponectin receptors and *PPAR* δ is highly coordinated and reflects lipid metabolism of the human donors. *Exp. Diabetes Res.* **2011**, *2011*, 692536. [CrossRef] [PubMed]

22. Luquet, S.; Lopez-Soriano, J.; Holst, D.; Fredenrich, A.; Melki, J.; Rassoulzadegan, M.; Grimaldi, P.A. Peroxisome proliferator-activated receptor delta controls muscle development and oxidative capability. *FASEB J.* **2003**, *17*, 2299–2301. [CrossRef] [PubMed]
23. Kashiwagi, A.; Kasuga, M.; Araki, E.; Oka, Y.; Hanafusa, T.; Ito, H.; Tominaga, M.; Oikawa, S.; Noda, M.; Kawamura, T.; et al. International clinical harmonization of glycated hemoglobin in Japan: From Japan Diabetes Society to National Glycohemoglobin Standardization Program values. *J. Diabetes Investig.* **2012**, *3*, 39–40. [CrossRef] [PubMed]
24. Shulman, G.I. Cellular mechanisms of insulin resistance. *J. Clin. Investig.* **2000**, *106*, 171–176. [CrossRef] [PubMed]
25. Westerbacka, J.; Corner, A.; Tiikkainen, M.; Tamminen, M.; Vehkavaara, S.; Hakkinen, A.M.; Fredriksson, J.; Yki-Jarvinen, H. Women and men have similar amounts of liver and intra-abdominal fat, despite more subcutaneous fat in women: Implications for sex differences in markers of cardiovascular risk. *Diabetologia* **2004**, *47*, 1360–1369. [CrossRef] [PubMed]
26. American College of Sports Medicine. *ACSM's Metabolic Calculations Handbook*; Lippincott Williams & Wilkins: Philadelphia, PA, USA, 2006; pp. 52–65.
27. American College of Sports Medicine. *ACSM's Guideline for Exercise Testing and Prescription*, 8th ed.; Lippincott Williams & Wilkins: Philadelphia, PA, USA, 2011; pp. 114–141.
28. Borg, G.A. Psychophysical bases of perceived exertion. *Med. Sci. Sports Exerc.* **1982**, *14*, 377–381. [CrossRef] [PubMed]
29. Yoshida, T.; Chida, M.; Ichioka, M.; Suda, Y. Blood lactate parameters related to aerobic capacity and endurance performance. *Eur. J. Appl. Physiol. Occup. Physiol.* **1987**, *56*, 7–11. [CrossRef] [PubMed]
30. Tokmakidis, S.P.; Leger, L.A.; Pilianidis, T.C. Failure to obtain a unique threshold on the blood lactate concentration curve during exercise. *Eur. J. Appl. Physiol. Occup. Physiol.* **1998**, *77*, 333–342. [CrossRef] [PubMed]

International Journal of
Molecular Sciences

Review

PPARβ/δ: Linking Metabolism to Regeneration

Ajit Magadum [1,2] and Felix B. Engel [3,4,*]

1. Cardiovascular Research Center, Icahn School of Medicine at Mount Sinai, New York, NY 10029, USA; ajit23882@gmail.com
2. Department of Genetics and Genomic Sciences, Icahn School of Medicine at Mount Sinai, New York, NY 10029, USA
3. Department of Nephropathology, Experimental Renal and Cardiovascular Research, Institute of Pathology, Friedrich-Alexander-Universität Erlangen-Nürnberg, 91054 Erlangen, Germany
4. Muscle Research Center Erlangen (MURCE), 91054 Erlangen, Germany

* Correspondence: felix.engel@uk-erlangen.de; Tel.: +49-9131-85-25699

Received: 4 June 2018; Accepted: 5 July 2018; Published: 10 July 2018

Abstract: In contrast to the general belief that regeneration is a rare event, mainly occurring in simple organisms, the ability of regeneration is widely distributed in the animal kingdom. Yet, the efficiency and extent of regeneration varies greatly. Humans can recover from blood loss as well as damage to tissues like bone and liver. Yet damage to the heart and brain cannot be reversed, resulting in scaring. Thus, there is a great interest in understanding the molecular mechanisms of naturally occurring regeneration and to apply this knowledge to repair human organs. During regeneration, injury-activated immune cells induce wound healing, extracellular matrix remodeling, migration, dedifferentiation and/or proliferation with subsequent differentiation of somatic or stem cells. An anti-inflammatory response stops the regenerative process, which ends with tissue remodeling to achieve the original functional state. Notably, many of these processes are associated with enhanced glycolysis. Therefore, peroxisome proliferator-activated receptor (PPAR) β/δ—which is known to be involved for example in lipid catabolism, glucose homeostasis, inflammation, survival, proliferation, differentiation, as well as mammalian regeneration of the skin, bone and liver—appears to be a promising target to promote mammalian regeneration. This review summarizes our current knowledge of PPARβ/δ in processes associated with wound healing and regeneration.

Keywords: PPARβ/δ; regeneration; proliferation; differentiation; metabolism; Wnt signaling; PDK1; Akt; glycolysis

---

1. Introduction

Humankind has been fascinated by the phenomenon of regeneration since ancient history. While the phenomenon of regeneration is already mentioned in Greek mythology (punishment of Prometheus or Hercules' second labor—slaying the Lernean Hydra), the first written records date back to Empedocles (490–430 BCE) and Aristotle (384–322 BCE, lizard tail regeneration in his books *History of Animals* and *Generations of Animals*). The first known scientific publication appeared in 1712, in which René Antoine Ferchault de Réaumur described limb regeneration in crustaceans [1]. Considering the implications of regeneration, there was and is still great public interest. Initially, the observation of the possibility of the regeneration of entire animals (e.g., hydra (*Hydra vulgaris*) and planaria (*Schmidtea mediterranea*)) resulted in heated philosophical and religious discussions: can the soul be split? Where is its residence? Yet, there was also great optimism. For example, the great philosopher "Voltaire marveled briefly: he saw at once that the loss and replacement of one's head presented serious problems for those who saw that structure as the seat of a unique "spirit" or soul: and thought of the possible consequences of the experiment for man. Writing at this time to poor blind Madame du Deffand, he lamented that for snails but not for her the replacement of bad eyes by good was a possibility. Later he expressed confidence that men would one day so master the process of regeneration that they too would be able to replace their entire heads. There are many people, he implied, for whom the change could hardly be for the worse" [2]. Yet, while it appears soon possible to transplant a human head [3], we are far away from being able to induce the regeneration of damaged tissues/organs in humans.

In 1901, Thomas Morgan defined "regeneration" as "the replacement of missing structures following injury" [4]. It is often assumed that regeneration includes the restoration of structure and function of lost or damaged organs/tissues. Yet, during mammalian liver regeneration, upon resection of a liver lobe, the function is restored by increasing the size of the remaining lobes, not by re-growing a new lobe [5]. Thus, the main aim in regenerative medicine is to restore tissue/organ function.

The ability of regeneration is widely and randomly distributed in the animal kingdom [6]. Yet, the efficiency and extent of regeneration varies greatly. For example, *Hydra vulgaris* and *Schmidtea mediterranea* are considered immortal as they can reform from an individual, specialized cell type [6]. Amphibians and fish such as the newt *Notophthalmus viridescens* and the zebrafish *Danio rerio* can regenerate a large variety of organs including appendages, heart, lens, retina, and central nervous system [7]. Mammals are more restricted in their regenerative capacity, even though they can for example recover from blood loss as well as damage to the peripheral nervous system, skeletal muscle, and liver. Yet, as humans cannot recover from damage to essential organs such as heart and brain, there is a great interest in understanding the molecular mechanisms of natural occurring healing and regeneration and to apply this knowledge to repair human tissues/organs upon injury.

Peroxisome proliferator-activated receptor (PPAR) β/δ has been demonstrated, as described in detail below, to be involved in several key cellular processes relevant to regeneration: proliferation, differentiation, migration, and apoptosis. In addition, PPARβ/δ plays important roles in metabolism, angiogenesis, and inflammation that have been identified as important processes in regeneration. Thus, the aim of this review is to summarize the potential of PPARβ/δ as a therapeutic target for regenerative therapies.

2. PPARβ/δ

Three PPAR isoforms have so far been identified which are designated PPARα, PPARβ/δ, and PPARγ. They belong to the nuclear-receptor superfamily, meaning they act as transcription factors upon ligand activation. PPARβ/δ can be activated by endogenous ligands like polyunsaturated fatty acids and eicosanoid metabolites (e.g., prostacyclin and 15-hydroxyeicosatetraenoic acid (15-HETE)) as well as artificial agonists including GW501516, GW0742, L-165041, and carbacyclin [8,9]. In addition, the action of PPARβ/δ can be inhibited by several inverse agonists and antagonists [10]. Yet, there are currently neither agonistic nor antagonistic drugs clinically available [10,11].

PPARβ/δ is as a nuclear receptor characterized by classical domains: an N-terminal region containing a ligand-independent transactivation domain, often known as activation function 1 (AF-1), a DNA-binding domain (DBD), a flexible hinge region, and an AF-2 domain including a ligand-binding domain (LBD) and a ligand-dependent transactivation domain. The principle mode of action of PPARβ/δ is the heterodimerization with the 9-cis retinoic acid receptor (RXR or NR2B) and binding via two zinc-fingers in the DBD to peroxisome proliferator response elements (PPREs) located in the promoter regions of their target genes [12]. Chromatin immunoprecipitation sequencing has revealed three types of target genes [13]: (i) PPARβ/δ-RXR binds to PPREs as a repressor complex. Expression of such genes is induced upon siRNA-mediated depletion of PPARβ/δ but not by agonists; (ii) Type II genes are regulated as Type I genes but can be activated by agonists (canonical regulation); (iii) The third class of genes contains only PPRE-like motifs. They are bound by PPARβ/δ containing complexes which act as transcriptional activators. Expression of such genes is downregulated upon siRNA-mediated depletion of PPARβ/δ and respond weakly, if at all, to ligands. In addition, PPARβ/δ can regulate transcription independently of DNA binding by suppressing transcription factors via direct physical interaction, competition for limiting amounts of shared co-activators, and inhibition of mitogen-activated protein kinase (MAPK) signaling [12]. For instance, PPARβ/δ inhibits the nuclear factor κ-light-chain-enhancer of activated B cells (NF-κB) pathway by interacting with the NF-κB subunit p65 and thereby decreasing NF-κB binding to the DNA resulting in the inhibition of the transcription of NF-κB target genes [8]. In addition, it has been shown that PPARβ/δ interacts with β-catenin in colon cancer cells controlling the expression of vascular endothelial growth factor (VEGF) A [14]. In cardiomyocytes PPARβ/δ induces via β-catenin Cyclin D2 and c-MYC [9]. Another example is the interaction with the corepressor B-cell lymphoma 6 (BCL-6) [8,13]. It is important to note that PPARβ/δ exhibits different regulatory roles on the same gene dependent on its environment. This dependency on its environment explains the later described cell type-specific roles of PPARβ/δ. For example, it has been reported that several different signaling kinases can modulate the transcriptional activity of PPARβ/δ including protein kinase A and p38 mitogen-activated protein kinase [8]. In addition, PPARβ/δ is widely expressed with relative abundant expression in the brain, skeletal muscle, heart, gut, placenta, and skin [15–20]. It has been shown to play major roles in fatty acid metabolism and energy expenditure and thus also in skeletal and cardiac muscle homeostasis and disease as well as metabolic disorders [11,12,21]. In addition, PPARβ/δ is involved in a variety of other diseases such as cancer (including inflammation, cell survival and angiogenesis) [13,22,23], skin disease [24–26], atherosclerosis [27], retinopathy [28], and Alzheimer's disease [29]. Finally, more and more reports suggest that modulation of PPARβ/δ activity might provide an opportunity for wound healing and tissue regeneration [9,25,30–34].

3. PPARβ/δ Controls Basic Mechanisms of Wound Healing and Regeneration

Tissue protection, healing, and regeneration require a tight control of several processes including apoptosis (e.g., due to increased functional demand or lack of oxygen), proliferation and/or differentiation of stem cells to generate lost cell types, as well as extracellular matrix remodeling and breakdown (e.g., resolving scar tissue and restoration of a tissue support matrix). The analysis of natural occurring regeneration in model organisms such as zebrafish, newt, and the Murphy Roths Large (MRL) mouse have revealed that healing and regeneration depend on hypoxia-induced signaling, inflammation induced by inflammatory cytokines and eicosanoids produced during the first hours after injury, secretion of pro-angiogenic factors, and metabolic alteration [7,35–40]. In the following subchapters, we will highlight the roles of PPARβ/δ in these processes.

3.1. Energy Metabolism

In adult mammals, the vast majority of energy is produced by oxidative metabolism in mitochondria. In recent years, it has been proposed that this oxidative metabolism as basal metabolic state underlies the low regenerative capacity in mammals [37]. Accumulating evidence indicates that changes in metabolism, namely the activation of glycolysis, play an important role in regeneration. For example, increased glycolysis is correlated with planarian regeneration [41]. Moreover, inhibition of glycolysis impairs neonatal heart [42] as well as adult skeletal muscle [43] regeneration in mice. Intriguingly, the enhanced regenerative capacity of the MRL mouse has been attributed to increased glycolysis and reduced fatty acid oxidation as their basal metabolic state. Accordingly, enhancing fatty acid oxidation in MRL mice inhibited regeneration [37]. As PPAR plays an important role in energy metabolism [44,45] it is not surprising that PPARβ/δ activity/signaling is required for regeneration (see Section 4). However, even though regeneration is associated with glycolysis it remains elusive what advantage glycolysis has over an oxidative metabolism which generates 18 times as much ATP per mole of glucose. One issue might be that the utilization of oxygen results in reactive oxygen species which can oxidize lipids, nucleic acids, and proteins and thus might result in cellular dysfunction.

In high energy demand tissues (e.g., cardiac and skeletal muscle, brown adipose tissue) PPARβ/δ overexpression increases the expression of genes involved in fatty acid transport and beta-oxidation. Concordantly, PPARβ/δ deletion resulted in decreased expression of these genes [46–51]. Furthermore, PPARβ/δ overexpression increases in skeletal muscle the proportion of oxidative fibers [47] while PPARβ/δ deletion markedly increases glycolytic fibers with reduced fatty acid oxidation [52]. In contrast to muscle, PPARβ/δ overexpression in liver increases glucose utilization and lipogenesis [53]. Notably, PPARβ/δ overexpression in the heart does also increase glucose utilization via increased glucose transporter type 4 (Glut4) expression [54]. The principle regulatory mechanism is the protein-protein-interaction of PPARβ/δ with PPAR coactivator 1α (PGC1α) and nuclear receptor corepressor 1 (NCOR1) which are lacking DNA binding activity [45]. The resulting complexes regulate the transcription of genes encoding for example forkhead box protein O1 (FoxO1), pyruvate dehydrogenase kinase 4 (PDK4), cluster of differentiation 36 (CD36), lactate dehydrogenase B, and lipoprotein lipase [21,44].

3.2. Apoptosis

The link between apoptosis and metabolism has been known for a long time. One of the best examples of this link is the dual functionality of cytochrome c. On one hand, the metabolic role of cytochrome c is to pass an electron from respiratory complex III to complex IV in order to promote adenosine triphosphate (ATP) generation through oxidative phosphorylation. On the other hand, it is required for apoptosis in order to activate caspases. Notably, the apoptotic function of cytochrome c is inhibited by glucose-stimulated production of intracellular glutathione, a mechanism utilized by glycolytic cancer cells [55]. Considering the role of apoptosis during healing or regeneration usually results in the conclusion that inhibition of apoptosis is beneficial. However, apoptosis can also be beneficial as recently reviewed by Diwanji and Bergmann in the context of apoptosis-induced compensatory proliferation [56].

PPARβ/δ mediates retinoic acid-stimulated keratinocyte survival [57–59] and ligand activation of PPARβ/δ inhibited palmitate-induced apoptosis in neonatal cardiomyocytes by preventing an increase in interleukin (IL)-6 levels [60]. Concordantly with an anti-apoptotic function, inhibition of PPARβ/δ upon stimulation with 13-S-hydroxyoctadecadienoic acid induces apoptosis in colorectal cancer cells [61]. In addition, PPARβ/δ is required for the VEGF-mediated maintenance of endothelial cell (EC) survival [62]. In contrast, telmisartan (an angiotensin II receptor antagonist; used in the management of hypertension) stimulates in a PPARβ/δ-dependent manner apoptosis in prostate cancer cells [63]. Notably, the PPARβ/δ ligand retinoic acid (RA) can exhibit pro- as well as anti-apoptotic effects. The effect of RA depends on intracellular lipid binding proteins which transport RA to a specific nuclear receptor. RA is pro-apoptotic in cells in which cellular retinoic acid binding protein (CRABP)-II transports RA to the nucleus, mediating interaction with the RA receptor. In contrast, RA is anti-apoptotic in cells in which fatty acid binding protein (FABP) 5 transports RA to the nucleus mediating interaction with PPARβ/δ [57]. Yet, even though it is well known that PPARβ/δ regulates cell survival in several cell types, little is known about downstream signaling pathways. In keratinocytes PPARβ/δ prevented apoptosis by modulating Akt signaling via transcriptional upregulation of integrin-linked kinase (ILK) and PDK1 [58]. Experiments on ECs revealed that PPARβ/δ inhibits apoptosis by binding to the promoter of 14-3-3α resulting in increased 14-3-3α protein levels, reduction of Bad translocation to mitochondria via direct protein–protein interaction, and inhibition of Bad-triggered apoptosis [64]. This anti-apoptotic pathway is shared by all PPARs [65]. In addition, it has been shown that PPARβ/δ inhibits oxidative stress-induced apoptosis in H9c2 cells (rat cardiac myoblast) by a direct transcriptional activation of catalase gene expression [66] as well as in adult rat cardiomyocytes [67].

3.3. Inflammation: Fibrosis and/or Regeneration

Injury results in fibrosis/scaring or in healing/regeneration with or without transient fibrosis. Already, decades ago, it was assumed that fibrosis might inhibit endogenous repair mechanisms. Yet, initial attempts to enhance healing/regeneration by inhibiting fibrosis has been shown to be detrimental, for example resulting in heart wall rupture after myocardial infarction [68]. In recent years, accumulating evidence has been provided that fibrosis and regeneration are inversely correlated with each other [69,70]. Thus, great effort is invested to identify novel approaches to modulate scar formation and to promote healing/regeneration at the same time [71–73]. The major players in controlling the response to injury are inflammatory monocytes, tissue-resident macrophages, and fibroblasts [72]. Disturbances in macrophage function such as uncontrolled production of inflammatory macrophages or failed communication between macrophages and other cells such as tissue progenitor cells repress endogenous regenerative mechanisms. The importance of these processes has, for example, been demonstrated by utilizing the MRL as well as African spiny mouse. In these regenerative model systems, anti-inflammatory agents or macrophage depletion blocked ear hole closure [37]. Thus, it is assumed that modulations of inflammatory processes together with anti-fibrotic signals are required to promote regeneration.

Several studies have demonstrated that PPARβ/δ has direct anti-fibrotic effects. For example, it has been shown that genetic and/or pharmacological activation of PPARβ/δ decreases fibrosis in a model of corneal damage [74] as well as myocardial infarction [9]. On a cellular level, agonist treatment inhibits keratinocyte transdifferentiation into myofibroblasts and thus extracellular matrix (ECM) synthesis [74]. During cardiac fibrosis, PPARβ/δ is expressed both in cardiac fibroblasts as well as myofibroblasts. PPARβ/δ activation reduced the proliferation of both cell types, myofibroblast differentiation and collagen synthesis [75]. In addition, high-salt diet-induced fibrosis was associated with PPARβ/δ downregulation whereas capsaicin-inhibited fibrosis via the receptor transient receptor potential vanilloid type 1 (TRPV1) was associated with PPARβ/δ upregulation [76].

That inflammation and innate immunity are processes driven by aerobic glycolysis [77–79] indicates that modulation of PPARβ/δ activity might allow modulating these processes to promote regeneration. Yet, while PPARs have been identified as key regulators of inflammatory and immune responses, the role of PPARβ/δ in modulating inflammation during regeneration is poorly characterized. The anti-inflammatory properties of PPARβ/δ are mainly based on inhibiting NFκB signaling [8] as well as expression of pro-inflammatory cytokines (inducible nitric oxide synthases (iNOS), cyclooxygenase (COX) 2, tumor necrosis factor (TNF) α, and adhesion molecules (VCAM-1, ICAM-1, and E-selectin) in macrophages [8,12,27,80]. PPARβ/δ can for example inhibit NFκB signaling through direct binding to p65 or Akt-mediated inactivation of glycogen synthase kinase (GSK)-3β. In addition, PPARβ/δ can activate adenosine monophosphate-activated protein kinase (AMPK) through phosphorylation, resulting in the inactivation of p300 and activation of SIRT1 leading to a marked reduction in acetylation of p65 inhibiting the NFκB transcriptional activity [8]. Inflammation-related target genes of PPARβ/δ are for example TGFβ, 14-3-3α, superoxide dismutase (SOD), catalase, thioredoxin, and G protein signaling-4 and -5. However, the anti-inflammatory effect of PPARβ/δ is not only mediated by the induction of anti-inflammatory genes. An example of transrepression of pro-inflammatory genes is the inhibition of the anti-inflammatory corepressor BCL-6 by inactive PPARβ/δ [8].

HIF-1α is besides NF-κB another major gene that regulates inflammation [81]. Notably, Inoue and coworkers have reported the crosstalk of the PPARβ/δ and hypoxia-inducible factor HIF-1α signaling axes in ECs upon hypoxia-induced migration [82]. During osteoblast differentiation PPARβ/δ is regulated in a HIF-1α-dependent manner [83]. In addition, PPARβ/δ regulates HIF-1α expression via calcineurin promoter binding [84]. A possible connection between PPARβ/δ and HIF-1α is intriguing as the ancient HIF-1α pathway, operating through prolyl hydroxylase domain proteins, has been identified as a central player in mouse regeneration [37]. In future studies, it will be interesting to determine if PPARβ/δ-induced/enhanced healing or regeneration is mediated through HIF-1α or can be enhanced by modulating HIF-1α activity.

3.4. Proliferation

Recent years have revealed that proliferating cells such as stem cells and cancer cells exhibit high levels of glycolysis while differentiated, postmitotic cells utilize fatty acid oxidation [39,85].

Thus, it is not surprising that manipulation of PPARβ/δ activity affects both proliferation and differentiation of a large variety of somatic and cancer cell types. However, it is important to note that the effect of altered PPARβ/δ activity on proliferation and differentiation is cell type- and context-dependent. For example, it has been shown that increased PPARβ/δ activity promotes proliferation of endothelial progenitor cells as well as somatic ECs [86–88]. In addition, it promotes proliferation of cells like cardiomyocytes [9] and hepatocellular carcinoma cell lines [89]. In contrast, ligand activation of PPARβ/δ inhibits proliferation of vascular smooth muscle cells (VSMCs) [90–92], HaCaT keratinocytes [93], as well as breast cancer cell lines [94] and PPARβ/δ deletion promotes cancer EC proliferation [23].

How PPARβ/δ regulates proliferation remains unclear. The analysis of the available literature reveals that mainly the up- or downregulation of classical cell cycle promoting proteins has been reported such as Cyclin A (VSMC [95]), Cyclin D1 (VSMC [91]; embryonic stem cells (ESCs) [96]; Sertoli cells [97]), Cyclin D2 (Sertoli cells [97]), Cyclin E (ESC [96]; primary thyroid cells; mouse embryonic fibroblasts [98]), cdk2 (ESC [96]; VSMC [95]), and cdk4 (VSMC [90,91]; ESC [96]). In addition, few studies have determined the effect on the cell cycle inhibitors p21 (VSMC [90]; ESC, [96]), p27 (ESC [96]; Sertoli cells [97]), p53 (VSMC [90]), and p57 (cancer ECs [23]; VSMC [95]). In addition, PPARβ/δ regulates the transcription of growth factors that promote proliferation (heparin-binding epidermal growth factor-like growth factor (HB-EGF), adult primary epidermal keratinocytes, [99]). Moreover, several pathways have been suggested to mediate the effect of PPARβ/δ on proliferation including Akt (endothelial progenitor cells (EPCs) [86]; keratinocytes [100]), p38 MAPK (ESCs [101]), extracellular signal-regulated kinase (ERK) (Sertoli cells [97]; keratinocytes [102]), and Wnt/β-catenin signaling (ESCs [101]; cardiomyocytes [9]; epithelial cells [103]).

An example of cell type-specific regulation of proliferation is the binding of PPARβ/δ to the leptin promoter, resulting in decreased leptin expression and increased liposarcoma cell proliferation [104]. In addition, it has been demonstrated that silent mating type information regulation 2 homolog 1 (sirtuin 1) mediates the anti-proliferative effect of PPARβ/δ in VSMCs [105].

3.5. Differentiation

In agreement with the idea that differentiation is accompanied by a switch from glycolysis to fatty acid oxidation, ESC differentiation to cardiomyocytes involves upregulation of oxidative phosphorylation and downregulation of glycolysis [106]. In contrast, reprogramming fibroblasts to induced pluripotent stem cells is dependent on induction of glycolysis [107]. A metabolic switch from glycolysis to fatty acid oxidation occurs also during differentiation of immature somites to muscle progenitors [108] and during heart development when the mode of heart growth switches from hyperplasia (proliferation) to hypertrophy (increase in cell size) [102,109]. Moreover, multiple signaling pathways affecting differentiation control also cellular metabolism such as the phosphatidylinositol-4,5-bisphosphate 3-kinase (PI3K)/AKT/mammalian target of rapamycin (mTOR), the Ras, the liver kinase B (Lkb1)/AMPK, and the Hedgehog pathways [39]. Consequently, it could also be demonstrated that alterations of PPARβ/δ activity or signaling affect differentiation. For example, PPARβ/δ controls on a transcriptional level the endothelial differentiation gene (Edg)-2 and PPARβ/δ agonist stimulation enhances the vasculogenic potential of endothelial progenitor cells (EPCs) [110]. Moreover, it has been shown that PPARβ/δ can promote osteoblast differentiation via Wnt signaling [30], in a keratinocyte fatty acid binding protein (K-FABP)-dependent manner keratinocyte differentiation [111], early adipocyte differentiation via PPARγ [112,113], late sebocyte differentiation [114], oligodendrocyte [115,116] and neural [117] differentiation, as well as p53- and SOX2-mediated differentiation of neuroblastoma cells [118]. Finally, the PPARβ/δ target gene FoxO1 plays an important role as negative regulator of skeletal muscle differentiation [119,120].

3.6. Angiogenesis

Tissues and organs are vascularized to provide their cells with oxygen and nutrients as well as to remove metabolic waste products. Thus, new vessels have to be formed after an injury to maintain regenerated tissue. This process is called angiogenesis or neo-angiogenesis. As described under Section 3.4, PPARβ/δ is involved in the regulation of EPC, EC and VSMC proliferation. In addition, it has been shown that PPARβ/δ activation inhibits IL1β-stimulated VSMC migration via upregulation of IL-1 receptor antagonist and was associated with the down-regulation of matrix metalloproteinase (MMP)-2 and MMP-9 [90]. Moreover, oxidized low-density lipoprotein-induced VSMC migration was inhibited in a SIRT1-dependent manner by PPARβ/δ activation [105]. That PPARβ/δ directly regulates physiological angiogenesis has been demonstrated in skeletal [121] and cardiac [84] muscle by utilizing agonists and/or transgenic mice overexpressing PPARβ/δ in skeletal muscle cells. These data showed that PPARβ/δ bound directly to the calcineurin promoter inducing the expression of its target genes such as HIF-1α. Consequently, inhibition of calcineurin activity abolished the angiogenic response to PPARβ/δ agonist stimulation [84]. Notably, the effect of PPARβ/δ transgenic overexpression on angiogenesis was significantly lower than the effect of agonist treatment [121]. In a subsequent study, it has been revealed that PPARβ/δ agonist stimulation of EPCs resulted in MMP-9 expression by direct transcriptional activation which caused insulin-like growth factor-binding protein (IGFBP) 3 degradation and thus IGF-1 release. Conditioned medium of stimulated EPCs enhanced the number and functions of human umbilical vein ECs and C2C12 myoblasts via IGF-1 receptor activation. Importantly, PPARβ/δ agonist stimulation in vivo in a mouse hind limb ischemia model induced in an MMP-9-dependent manner IGF-1 receptor phosphorylation in ECs and skeletal muscle and promoted angiogenesis and skeletal muscle regeneration [122]. In the same study, the authors report that the same pro-angiogenic mechanism can be induced by PPARβ/δ agonist stimulation in a mouse skin punch wound model.

In addition to its pro-angiogenic effects, PPARβ/δ agonists have been shown to be vasoprotective by activating and or increasing the expression of endothelial nitric oxide synthase (eNOS) [123]. Importantly, injury to the endothelium (e.g., through angioplasty) results in inefficient regeneration as the regenerated endothelium cannot produce enough nitric oxide causing local nitric oxide deficiency which can lead to intravascular coagulations, vasospasm, and inflammation-mediated atherosclerosis [124]. Thus, PPARβ/δ agonists might be useful as anti-thrombotic and anti-atherosclerotic drugs.

Besides a physiological role of PPARβ/δ in angiogenesis, evidence is accumulating that modulation of PPARβ/δ can be used to control neo-angiogenesis during pathological conditions. For example, intravitreal injection of the PPARβ/δ antagonist GSK0660 inhibited neovascularization in a rat oxygen-induced retinopathy and reduced serum-induced in human retinal microvascular ECs proliferation and tube formation. Both cases were correlated with the reduced expression of the pro-angiogenic angiopoietin like (Angptl) 4. In contrast, the agonist PPARβ/δ GW0742 increased neovascularization and tube formation as well as Angptl4 expression [28]. In addition, tumor transplantation assays as well as Matrigel plug assays utilizing PPARβ/δ knockout mice indicate that PPARβ/δ is required for the formation of functional tumor microvessels [23].

4. PPARβ/δ in Wound Healing and Regeneration

It is essential for species to deal with injuries to survive. Notably, almost all species have some regenerative capacity, including humans. For example, they can regenerate liver [5] and bone [125,126]. The basic steps of regeneration are: (1) an inflammatory response [72] induced by injuries caused by infection, intoxication or mechanical infliction due to signal molecules released by dead or dying cells or invading organisms [127]; (2) wound healing [128] that can be accompanied by a transient scar [129]; (3) ECM remodeling to allow migration as well as induction of proliferation with subsequent differentiation to generate new tissue [6]; (4) an anti-inflammatory and anti-fibrotic response [72]; and (5) remodeling of the tissue to achieve a functional state [6,72,128]. As described above, PPARβ/δ is involved in all these mechanisms. In recent years, it has been shown that manipulation of PPARβ/δ activity is inhibiting or promoting healing as well as regeneration of a large variety of tissues/organs.

4.1. Skin

The skin is the largest organ of the human body consisting of epidermis and dermis. The epidermis consists of five layers, forming a protective outer barrier. The dermis consists of connective tissue and is separated from the epidermis by a thin sheet of fibers called the basement membrane. The dermis provides tensile strength and elasticity to the skin and serves as a location for the appendages of skin such as hair follicles, nails, and sweat glands. Skin injuries in adult mammals result usually in scar tissue that lack skin appendages. As long as the deepest layer of the epidermis, the basal layer containing stem cells, is not injured the mammalian skin can heal without forming a scar. Yet, deep injuries and third degree burns fail to regenerate and result into scarring or chronic wounds. As soon as a wound exceeds 4 cm in diameter, a tissue graft is needed. Yet, while enormous progress has been made in tissue engineering due to the clinical importance, to date there is no complete functional skin substitute available (reviewed in [130]).

Similar to other organs, wound healing is initiated by inflammation, followed by reepithialization due to proliferation and migration of keratinocytes. In parallel fibroblast proliferation is activated and angiogenesis is induced. In addition, fibroblasts produce collagens and other extracellular matrix proteins to aid in wound repair (reviewed in [131]). As PPARβ/δ has multiple functions in skin health and disease such as has pro-differentiating effects on keratinocytes, PPARβ/δ appears to be an ideal therapeutic target to enhance endogenous regenerative skin regeneration capacities [24].

That PPARβ/δ is involved in skin healing has been suggested based on the finding that its expression is strongly induced upon injury by inflammatory cytokines (e.g., TNF-α, [59]) and keratinocytes at the edges of wounds maintain high expression as long as the repair process has not been completed [132]. The analysis of wound healing in PPARβ/δ knockout mice revealed that PPARβ/δ is required during skin healing for keratinocyte proliferation resulting in a delay in healing by two to three days [132]. Activated PPARβ/δ signals via the PI3K/Akt1 pathway, which mediates cell survival via inactivation of BAD (BCL2-associated agonist of cell death) and adhesion as well as migration via inhibition of GSK3β [24,133]. During progression of the healing process, PPARβ/δ expression is decreasing mediated by transforming growth factor (TGF) β1-induced Smad3/Smad4 repressor complexes. Notably, keratinocyte proliferation is also regulated by dermal fibroblasts. An injury causes IL-1 secretion by keratinocytes, which activates in fibroblasts via IL-R1 the transforming growth factor beta-activated kinase 1 (TAK) 1/cJun/AP1 pathway resulting in growth factor and cytokine release promoting keratinocyte proliferation. In fibroblasts, however, activated PPARβ/δ induces the expression of the secretory IL-1 receptor antagonist sIL-1Ra. This attenuates the IL1 responsiveness of fibroblasts resulting in decreased secretion of pro-proliferative factors and thus reduced keratinocyte proliferation [134]. This regulatory mechanism demonstrates how important the local activation of PPARβ/δ is.

The available data on PPARβ/δ in regards to wound healing but also skin disorders has recently in detail been reviewed [24,25,34]. Yet, there appear to be no studies that attempted to utilize this knowledge to significantly enhance the regenerative capacity at least in mice or rat.

4.2. Corneal Epithelial Wound Healing

Nakamura and coworkers found that after surgical removal of corneal epithelium PPARβ/δ expression was temporally upregulated at the wound's edges like observed in skin wound healing. This phenomenon was additionally observed in a human corneal epithelial wound model ex vivo. PPARβ/δ activation enhanced healing of experimental corneal epithelial wounds in rats and wound closure in an in vitro system based on human corneal epithelial cells. Finally, PPARβ/δ activation was sufficient to inhibit TNFα–induced cell death of corneal epithelial cells [31]. If wound healing was impaired or the lesion was too large, activated keratocytes migrated, proliferated, and differentiated into fibroblasts and myofibroblasts leading to an altered ECM and corneal opacity. Gu and coworkers tested the effects of PPARβ/δ agonists in a model of corneal wound healing upon epithelial defects generated by laser ablation [74]. They observed that the agonist inhibited early stages of wound healing reepithelialization and promoted angiogenesis. Yet, during the remodeling phase agonist treatment decreased keratocyte transdifferentiation into myofibroblasts and thus also ECM synthesis/scaring and corneal opacity. These examples represent another good example for the need of a local and timed modulation of PPARβ/δ activity.

4.3. Reendothelialization

As described under Section 3, PPARβ/δ is involved in the regulation of EPC, EC, and VSMC proliferation and/or migration and can promote neo-angiogenesis. In addition, He and coworkers have shown in a mouse model of carotid artery injury that PPARβ/δ agonist treatment of human EPCs significantly enhanced the ability of transplanted EPCs to repair denuded endothelium. PPARβ/δ agonist treatment of human EPCs increased the production of tetrahydrobiopterin, an essential co-factor of eNOS, as well as expression and activity of GTP cyclohydrolase I, the rate-limiting enzyme responsible for de novo synthesis of tetrahydrobiopterin. These effects were dependent on PPARβ/δ agonist-induced suppression of the phosphatase and tensin homolog (PTEN) expression thereby promoting AKT signaling. Notably, PPARβ/δ agonist-induced EPC proliferation was primarily dependent on BH4 but independent of NO, while induced EPC migration was dependent on both [88].

4.4. Skeletal Muscle

Regeneration of skeletal muscle is among the best-understood regenerative processes in mammalians, including humans. Similar to bone, mammalian skeletal muscle can regenerate but the extent of regeneration is limited [135,136]. If an injury exceeds the endogenous regenerative capacity the skeletal muscle scars. This kind of injury does occur not only after an accident but are often also caused by surgical interventions such as total hip or knee arthroplasty [136]. Moreover, skeletal muscle loss occurs in a variety of congenital diseases (myofibrillar myopathies, [137]), cancer (cachexia, [138]), as well as aging (sarcopenia, [139]). As maintaining skeletal muscle function is essential for good health and independent living, there is a great interest in developing strategies to enhance the endogenous regenerative capacity of skeletal muscle or to generate muscle by stem cell-based therapies.

The mechanism of skeletal muscle regeneration has recently been reviewed in detail [135,136]. Briefly, the main cell type during skeletal muscle regeneration is the resident muscle stem (satellite) cell (MuSC). After an injury MuSCs are activated, enter the cell cycle, proliferate, differentiate into myoblasts, which finally fuse to damaged fibers or generate myofibers de novo. Maintenance of MuSCs is for example dependent on paired box proteinPax7, FoxO1/Notch signaling, and the ECM via β1-integrin signaling [119,135,140]. Yet, skeletal muscle regeneration is a complex process that involves several cell types (e.g., immune cells, adipogenic progenitors, fibroblasts, and pericytes) that interact with each other. The basic steps in skeletal muscle regeneration are: (1) bleeding triggering coagulation and hematoma formation; (2) induction of a pro-inflammatory reaction (e.g., MuSC-mediated recruitment of immune cells such as pro-inflammatory M1 macrophages); (3) induction of MuSC proliferation (e.g., by IL-6 secreted by M1 macrophages); (4) switching of immune cells such as macrophages to anaerobic glycolysis due to a hypoxic environment; (5) appearance of anti-inflammatory M2 macrophages, which inhibit myoblast proliferation and stimulate the subsequent differentiation and fusion of myofibers; and (6) initiation of re-vascularization. Notably, the cytokine pattern and mechanical tension decides, during the initial inflammatory reaction, whether fibroblasts differentiate into myofibroblasts promoting fibrosis or whether regeneration will occur [70].

The role of PPARβ/δ in skeletal muscle physiology and pathophysiology has recently been reviewed by Manickham and Wahli [21]. In 2009, Giordano and coworkers demonstrated that muscle-specific overexpression of PPARβ/δ as well as pharmacological activation promotes skeletal muscle fusion but not proliferation of MuSCs [141]. This is in agreement with the recent finding by

Lee and coworkers showing that PPARβ/δ agonist treatment, as well as PPARβ/δ overexpression, enhanced C2C12 myotube formation via p38 MAPK and Akt [142]. Conditional PPARβ/δ knockout mice utilizing Myf5-Cre deleter lines (affect MuSCs) exhibited no gross morphological phenotype. A detailed analysis revealed a reduced number of MuSCs (~40%) and a delayed regenerative response to cardiotoxin-induced injury. The number of small regenerating fibers was increased by ~30% while the number of large regenerating fibers was decreased by 20%. Moreover, MuSCs from conditional PPARβ/δ knockout mice displayed reduced in vitro and in vivo MuSCs proliferation but enhanced differentiation. These phenomena were associated with a downregulation of FoxO1, which is a negative regulator of skeletal muscle differentiation [120]. In 2015, Chandrashekar and coworkers reported that PPARβ/δ knockout mice exhibit reduced skeletal muscle weight and myofiber atrophy during postnatal development [143]. In agreement with the conditional knockout mice, the number of MuSCs was reduced (~25%). Yet, mass was affected in PPARβ/δ knockout mice while myofiber number was not significantly altered. Moreover, PPARβ/δ knockout mice contained significantly less myoblasts upon notexin-mediated injury (~50% at 28 dpi). In addition, the authors observed in the knockout animals increased necrosis (three days post injury (dpi)) and myofibers containing centrally located nuclei were smaller (7 dpi). While previous PPARβ/δ-related studies investigated mainly the effect on MuSCs, Chandrashekar and coworkers describe an increased infiltration of macrophages at 3 dpi. Yet, loss of PPARβ/δ did not significantly alter scar tissue formation or metabolic properties of regenerated muscle.

Even though several groups describe modulation of PPARβ/δ as affecting MuSCs as well as immune cells, the data obtained from mouse models utilizing overexpression and knockout strategies do not clarify if skeletal muscle regeneration can be enhanced by altering PPARβ/δ activity. In contrast, Haralampieva and coworkers aimed at manipulating MuSCs directly in order to enhance their regenerative capacity. For this purpose, they have overexpressed human peroxisome proliferator-activated receptor gamma coactivator 1-alpha (hPGC-1α) in hMuSCs and tested the effect in a crush-induced injury model [144]. The authors observed a decreased inflammatory response accompanied by enhanced expression of muscle markers in newly formed myotubes and increased muscle contraction force. Thus, injected hMuSCs overexpressing PGC-1α enhanced functional muscle regeneration after injury.

Collectively, these data indicate that PPARβ/δ is involved in MuSC proliferation but also in myoblast fusion [119]. It remains elusive if modulation of PPARβ/δ can be utilized to enhance regeneration even though it is involved in a large number of processes affecting skeletal muscle regeneration.

4.5. Bone

The bone is one of the few tissues/organs of the human body that can heal and regenerate [125,126]. Yet, the regenerative capacity is limited, does not occur in all cases and regeneration is complicated by comorbidities such as type 2 diabetes. That PPARβ/δ might play a role in bone regeneration has been suggested by the bone phenotype of PPARβ/δ knockout mice. These mice were characterized by increased myostatin expression, low bone formation, and increased resorption resulting in decreasing bone strength with age. In addition, they did not respond with bone formation upon exercise [145]. Conditional knockout mice utilizing a SOX2-Cre deleter line showed substantial osteopenia paralleled by lower serum concentrations of osteoprotegerin and osteocalcin, a higher RANKL-to-osteoprotegerin ratio, as well as a higher number of osteoclasts within the trabecular bones [30]. In contrast, activation of PPARβ/δ in vitro promoted osteogenic differentiation of osteoblasts and inhibited in co-cultures of osteoblasts and osteoclasts osteoclast differentiation and bone resorption. Moreover, pharmacological activation of PPARβ/δ in a mouse model of postmenopausal osteoporosis led to normalization of the altered RANKL-to-osteoprotegerin ratio and the restoration of normal bone density [30].

4.6. Liver

The mammalian liver can regenerate based on hepatocyte proliferation in contrast to skin, skeletal muscle, and bone after large injuries such as two-thirds partial hepatectomy [5]. Liu and coworkers demonstrated, utilizing PPARβ/δ knockout mice, that PPARβ/δ is required for the activation of hepatocyte proliferation upon injury to enable liver regeneration. A detailed analysis of their model revealed that PPARβ/δ deficiency blocked the induction of genes involved in glycolysis, the activation the PDK1/AKT pathway at 36 to 48 h after injury, as well as the proliferation associated transcription factors E2F1, 2, 7, and 8 resulting in delayed liver regeneration [33].

4.7. Cardiac Muscle

Significant effort is invested to develop novel regenerative therapies for the injured mammalian heart as heart failure represents a major socioeconomic burden [146]. Due to the fact that the embryonic heart growth during development is mediated by cardiomyocyte proliferation and as natural occurring heart regeneration in zebrafish and newt is based on the same cellular mechanism, one possible future approach appears to be the induction of adult mammalian cardiomyocyte proliferation [147]. Yet, it is poorly understood why mammalian cardiomyocytes stop proliferating shortly after birth. Recently, Magadum and coworkers wondered whether the metabolic shift in cardiomyocytes around birth from glycolysis to fatty acid oxidation to ensure ATP generation might be responsible for this phenomenon [9]. In the adult heart, about 70% of the cardiac energy metabolism relies on the oxidation of fatty acids and 30% on glucose, lactate, and ketone bodies. Notably, the heart can in contrast to other organs adapt its energy metabolism based on substrate availability [102,148]. Activation of PPARβ/δ in neonatal cardiomyocytes induced their proliferation via the PDK1/p308Akt/GSK3β/β-catenin pathway. This proliferative response could even be further enhanced by treatment with the GSK3β inhibitor 6-bromoindirubin-3′-oxime (BIO). Moreover, inhibition of PPARβ/δ reduced cardiomyocyte proliferation during zebrafish heart regeneration. Finally, genetic as well as pharmacological activation of PPARβ/δ in a myocardial infarct model induced cell cycle progression in cardiomyocytes, reduced scarring, and improved cardiac function [9]. While it has not been proven to what extent cardiomyocyte proliferation upon PPARβ/δ activation contributes to improved function, it appears likely that it is due to the pleiotropic effects of PPARβ/δ: (1) inhibiting apoptosis (see Section 3.2 and [54]); (2) modulating inflammation and inhibiting fibrosis (Section 3.3); (3) promoting cardiomyocyte proliferation (Section 3.4); (4) promoting angiogenesis (Section 3.6).

In addition to its role in healing and regeneration, PPARβ/δ-mediates, as recently reviewed, healing of metabolic diseases such as diabetes [11,149] and tissue protection [54,150–154].

5. Conclusions

Our literature analysis confirms that modulation of PPARβ/δ activity can regulate all cellular processes of regeneration (see Section 3 and Figure 1). However, it is important to consider that for example PPARβ/δ activation can have different outcomes not only in different but also the same cell type depending on intracellular and extracellular conditions (see Section 3). Importantly, our analysis also reveals that PPARβ/δ is involved in natural occurring regeneration of mammalian organs (Figure 1). However, very little information is available on the role of PPARβ/δ in model organisms characterized by extensive regenerative capacities such as zebrafish, the ability of enhancing natural occurring but limited regeneration (e.g., liver, bone, skin; see Section 4), and to induce regeneration of mammalian organs that have no significant endogenous regenerative capacity such as the brain and the heart. Our analysis also shows that the main signaling pathways controlled by PPARβ/δ have been demonstrated to be essential for regeneration (Figure 1). For example, Akt signaling is well known to be required for mammalian liver [155], skeletal muscle [156,157] and hair follicle [158], as well as planarian [159] regeneration and its activation promotes axonal regeneration in mice [160–162]. In addition, it has been shown that GSK3β or β-catenin are involved in regeneration of for example mature pancreatic acinar cells [163] and intestine [164] in mice; limb regeneration in the model organisms *axolotl*, *Xenopus*, and *zebrafish* [165]; and *zebrafish* heart regeneration [166]. Furthermore, GSK3β/β-catenin signaling also enhances skeletal muscle [167] and bone [168] regeneration and can induce mammalian cardiomyocyte proliferation [169] and central nervous system axon regeneration [170]. Finally, the class of FoxO transcription factors have been shown to play a role in stem cell aging [171]. Reduced FoxO1 expression accelerates skin wound healing [172] and skeletal muscle regeneration [173]. Moreover, it inhibits axon regeneration in *C. elegans* [174]. FoxO3 is known to inhibit oligodendrocyte progenitor cell and thus myelination [175]. In addition, it plays a role in mammalian spinal cord [176], skeletal muscle [177], and liver [178] regeneration. Collectively, modulation of PPARβ/δ has a great therapeutic potential to enhance or even promote regeneration and thus we suggest to intensify the analysis of PPARβ/δ signaling in regenerative model organisms in comparison to mammals.

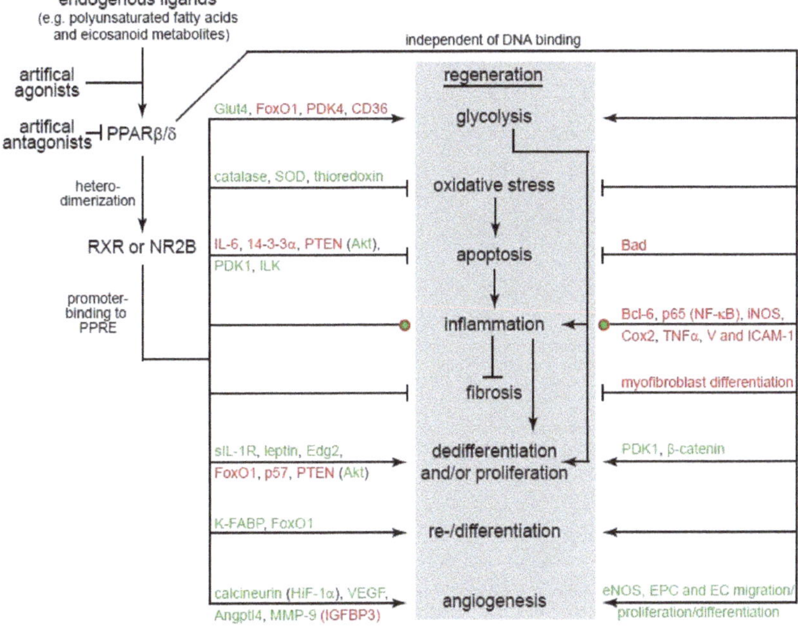

Figure 1. Examples of direct targets of PPARβ/δ involved in the different processes of regeneration. Red: inactivation. Green: activation. Affected pathways are indicated in brackets. Note: inflammation needs first to be activated and later inhibited during regeneration.

Acknowledgments: We thank the members of the Engel laboratory for critical reading of the manuscript. This work was supported by the Emerging Fields Initiative Cell "Cycle in Disease and Regeneration (CYDER)" (Friedrich-Alexander-University Erlangen-Nürnberg, to Felix B. Engel).

Conflicts of Interest: The authors declare no conflict of interest.

Abbreviations

AF	Activation function
AMPK	Adenosine monophosphate-activated protein kinase
Angptl4	Angiopoietin like 4

ATP	Adenosine triphosphate
BAD	BCL2-associated agonist of cell death
BCL-6	B-cell lymphoma 6
CD36	Cluster of differentiation 36
COX2	Cyclooxygenase 2
CRABP-II	Cellular retinoic acid binding protein-II
DBD	DNA-binding domain
EC	Endothelial cell
ECM	Extracellular matrix
Edg	Endothelial differentiation gene
eNOS	Endothelial nitric oxide synthase
EPC	Endothelial progenitor cell
ESC	Embryonic stem cell
FABP	Fatty acid binding protein
FoxO1	Forkhead box protein O1
Glut4	Glucose transporter type 4
GSK-3β	Glycogen synthase kinase 3β
GTP	Guanosine triphosphate
HB-EGF	Heparin-binding epidermal growth factor-like growth factor
HIF-1α	Hypoxia-inducible factor-1α
hPGC-1α	Human peroxisome proliferator-activated receptor gamma coactivator-1α
IGF-1	Insulin-like growth factor 1
IGFBP	Insulin-like growth factor-binding protein
IL-1	Interleukin
iNOS	Inducible nitric oxide synthase
K-FABP	Keratinocyte-fatty acid binding protein
LBD	Ligand-binding domain
LKB1	Liver kinase B1
MAPK	Mitogen-activated protein kinase
MMP	Matrix metalloproteinase
MRL	Murphy Roths Large
MuSC	Muscle stem (satellite) cell
mTOR	Mammalian target of rapamycin
NCOR1	Nuclear receptor corepressor 1
NF-κB	Nuclear factor κ-light-chain-enhancer of activated B cells
PDK4	Pyruvate dehydrogenase kinase 4
PGC1α	PPAR coactivator 1α
PI3K	Phosphatidylinositol-4,5-bisphosphate 3-kinase
PPAR	Peroxisome proliferator-activated receptor
PPRE	Putative PPAR response element
PTEN	Phosphatase and tensin homolog
RA	Retinoic acid
RANKL	Receptor activator of nuclear factor kappa-B ligand
RXR	Retinoic acid receptor
SIRT1	Silent mating type information regulation 2 homolog 1
SOD	Superoxide dismutase
SOX2	SRY (sex determining region Y)-box 2
TGFβ	Transforming growth factor β
TNFα	Tumor necrosis factor α
TRPV1	Transient receptor potential vanilloid type 1
VEGF	Vascular endothelial growth factor
VSMC	Vascular smooth muscle cells

References

1. De Réaumur, M. Sur les diverses reproductions. Qui se font dans les écrevisses, les homards, les crabes, etc. Et entre autres sur celles de leurs jambes et écailles. *Mem. Acad. R. Sci.* **1712**, 223–246.
2. Newth, D.R. *New (and Better?) Parts for Old*; Penguin Books: London, UK, 1958.
3. Furr, A.; Hardy, M.A.; Barret, J.P.; Barker, J.H. Surgical, ethical, and psychosocial considerations in human head transplantation. *Int. J. Surg.* **2017**, *41*, 190–195. [CrossRef] [PubMed]
4. Morgan, T.H. *Regeneration*; Macmillan: London, UK, 1901.
5. Michalopoulos, G.K.; DeFrances, M.C. Liver regeneration. *Science* **1997**, *276*, 60–66. [CrossRef] [PubMed]
6. Sanchez Alvarado, A.; Tsonis, P.A. Bridging the regeneration gap: Genetic insights from diverse animal models. *Nat. Rev. Genet.* **2006**, *7*, 873–884. [CrossRef] [PubMed]
7. Galliot, B.; Crescenzi, M.; Jacinto, A.; Tajbakhsh, S. Trends in tissue repair and regeneration. *Development* **2017**, *144*, 357–364. [CrossRef] [PubMed]
8. Neels, J.G.; Grimaldi, P.A. Physiological functions of peroxisome proliferator-activated receptor beta. *Physiol. Rev.* **2014**, *94*, 795–858. [CrossRef] [PubMed]
9. Magadum, A.; Ding, Y.; He, L.; Kim, T.; Vasudevarao, M.D.; Long, Q.; Yang, K.; Wickramasinghe, N.; Renikunta, H.V.; Dubois, N.; et al. Live cell screening platform identifies PPARdelta as a regulator of cardiomyocyte proliferation and cardiac repair. *Cell Res.* **2017**, *27*, 1002–1019. [CrossRef] [PubMed]
10. De Lellis, L.; Cimini, A.; Veschi, S.; Benedetti, E.; Amoroso, R.; Cama, A.; Ammazzalorso, A. The Anticancer Potential of Peroxisome Proliferator-Activated Receptor Antagonists. *ChemMedChem* **2018**, *13*, 209–219. [CrossRef] [PubMed]
11. Palomer, X.; Barroso, E.; Pizarro-Delgado, J.; Pena, L.; Botteri, G.; Zarei, M.; Aguilar, D.; Montori-Grau, M.; Vazquez-Carrera, M. PPARbeta/delta: A Key Therapeutic Target in Metabolic Disorders. *Int. J. Mol. Sci.* **2018**, *19*. [CrossRef] [PubMed]
12. Palomer, X.; Barroso, E.; Zarei, M.; Botteri, G.; Vazquez-Carrera, M. PPARbeta/delta and lipid metabolism in the heart. *Biochim. Biophys. Acta* **2016**, *1861*, 1569–1578. [CrossRef] [PubMed]
13. Muller, R. PPARbeta/delta in human cancer. *Biochimie* **2017**, *136*, 90–99. [CrossRef] [PubMed]
14. Hwang, I.; Kim, J.; Jeong, S. Beta-Catenin and peroxisome proliferator-activated receptor-delta coordinate dynamic chromatin loops for the transcription of vascular endothelial growth factor A gene in colon cancer cells. *J. Biol. Chem.* **2012**, *287*, 41364–41373. [CrossRef] [PubMed]
15. Kliewer, S.A.; Forman, B.M.; Blumberg, B.; Ong, E.S.; Borgmeyer, U.; Mangelsdorf, D.J.; Umesono, K.; Evans, R.M. Differential expression and activation of a family of murine peroxisome proliferator-activated receptors. *Proc. Natl. Acad. Sci. USA* **1994**, *91*, 7355–7359. [CrossRef] [PubMed]
16. Michalik, L.; Auwerx, J.; Berger, J.P.; Chatterjee, V.K.; Glass, C.K.; Gonzalez, F.J.; Grimaldi, P.A.; Kadowaki, T.; Lazar, M.A.; O'Rahilly, S.; et al. International Union of Pharmacology. LXI. Peroxisome proliferator-activated receptors. *Pharmacol. Rev.* **2006**, *58*, 726–741. [CrossRef] [PubMed]
17. Braissant, O.; Foufelle, F.; Scotto, C.; Dauca, M.; Wahli, W. Differential expression of peroxisome proliferator-activated receptors (PPARs): Tissue distribution of PPAR-alpha, -beta, and -gamma in the adult rat. *Endocrinology* **1996**, *137*, 354–366. [CrossRef] [PubMed]
18. Braissant, O.; Wahli, W. Differential expression of peroxisome proliferator-activated receptor-alpha, -beta, and -gamma during rat embryonic development. *Endocrinology* **1998**, *139*, 2748–2754. [CrossRef] [PubMed]
19. Abbott, B.D. Review of the expression of peroxisome proliferator-activated receptors alpha (PPAR alpha), beta (PPAR beta), and gamma (PPAR gamma) in rodent and human development. *Reprod. Toxicol.* **2009**, *27*, 246–257. [CrossRef] [PubMed]
20. Girroir, E.E.; Hollingshead, H.E.; He, P.; Zhu, B.; Perdew, G.H.; Peters, J.M. Quantitative expression patterns of peroxisome proliferator-activated receptor-beta/delta (PPARbeta/delta) protein in mice. *Biochem. Biophys. Res. Commun.* **2008**, *371*, 456–461. [CrossRef] [PubMed]
21. Manickam, R.; Wahli, W. Roles of Peroxisome Proliferator-Activated Receptor beta/delta in skeletal muscle physiology. *Biochimie* **2017**, *136*, 42–48. [CrossRef] [PubMed]
22. Peters, J.M.; Morales, J.L.; Gonzalez, F.J. Modulation of gastrointestinal inflammation and colorectal tumorigenesis by peroxisome proliferator-activated receptor-beta/delta (PPARbeta/delta). *Drug Discov. Today Dis. Mech.* **2011**, *8*, e85–e93. [CrossRef] [PubMed]

23. Muller-Brusselbach, S.; Komhoff, M.; Rieck, M.; Meissner, W.; Kaddatz, K.; Adamkiewicz, J.; Keil, B.; Klose, K.J.; Moll, R.; Burdick, A.D.; et al. Deregulation of tumor angiogenesis and blockade of tumor growth in PPARbeta-deficient mice. *EMBO J.* **2007**, *26*, 3686–3698. [CrossRef] [PubMed]
24. Montagner, A.; Wahli, W. Contributions of peroxisome proliferator-activated receptor beta/delta to skin health and disease. *Biomol. Concepts* **2013**, *4*, 53–64. [CrossRef] [PubMed]
25. Montagner, A.; Wahli, W.; Tan, N.S. Nuclear receptor peroxisome proliferator activated receptor (PPAR) beta/delta in skin wound healing and cancer. *Eur. J. Dermatol.* **2015**, *25* (Suppl. 1), 4–11. [PubMed]
26. Icre, G.; Wahli, W.; Michalik, L. Functions of the peroxisome proliferator-activated receptor (PPAR) alpha and beta in skin homeostasis, epithelial repair, and morphogenesis. *J. Investig. Dermatol. Symp. Proc.* **2006**, *11*, 30–35. [CrossRef] [PubMed]
27. Chinetti-Gbaguidi, G.; Staels, B. PPARbeta in macrophages and atherosclerosis. *Biochimie* **2017**, *136*, 59–64. [CrossRef] [PubMed]
28. Capozzi, M.E.; McCollum, G.W.; Savage, S.R.; Penn, J.S. Peroxisome proliferator-activated receptor-beta/delta regulates angiogenic cell behaviors and oxygen-induced retinopathy. *Investig. Ophthalmol. Vis. Sci.* **2013**, *54*, 4197–4207. [CrossRef] [PubMed]
29. Benedetti, E.; D'Angelo, B.; Cristiano, L.; Di Giacomo, E.; Fanelli, F.; Moreno, S.; Cecconi, F.; Fidoamore, A.; Antonosante, A.; Falcone, R.; et al. Involvement of peroxisome proliferator-activated receptor beta/delta (PPAR beta/delta) in BDNF signaling during aging and in Alzheimer disease: Possible role of 4-hydroxynonenal (4-HNE). *Cell Cycle* **2014**, *13*, 1335–1344. [CrossRef] [PubMed]
30. Scholtysek, C.; Katzenbeisser, J.; Fu, H.; Uderhardt, S.; Ipseiz, N.; Stoll, C.; Zaiss, M.M.; Stock, M.; Donhauser, L.; Bohm, C.; et al. PPARbeta/delta governs Wnt signaling and bone turnover. *Nat. Med.* **2013**, *19*, 608–613. [CrossRef] [PubMed]
31. Nakamura, Y.; Nakamura, T.; Tarui, T.; Inoue, J.; Kinoshita, S. Functional role of PPARdelta in corneal epithelial wound healing. *Am. J. Pathol.* **2012**, *180*, 583–598. [CrossRef] [PubMed]
32. Mothe-Satney, I.; Piquet, J.; Murdaca, J.; Sibille, B.; Grimaldi, P.A.; Neels, J.G.; Rousseau, A.S. Peroxisome Proliferator Activated Receptor Beta (PPARbeta) activity increases the immune response and shortens the early phases of skeletal muscle regeneration. *Biochimie* **2017**, *136*, 33–41. [CrossRef] [PubMed]
33. Liu, H.X.; Fang, Y.; Hu, Y.; Gonzalez, F.J.; Fang, J.; Wan, Y.J. PPARbeta Regulates Liver Regeneration by Modulating Akt and E2f Signaling. *PLoS ONE* **2013**, *8*, e65644.
34. Gupta, M.; Mahajan, V.K.; Mehta, K.S.; Chauhan, P.S.; Rawat, R. Peroxisome proliferator-activated receptors (PPARs) and PPAR agonists: The 'future' in dermatology therapeutics? *Arch. Dermatol. Res.* **2015**, *307*, 767–780. [CrossRef] [PubMed]
35. Karin, M.; Clevers, H. Reparative inflammation takes charge of tissue regeneration. *Nature* **2016**, *529*, 307–315. [CrossRef] [PubMed]
36. Gurtner, G.C.; Werner, S.; Barrandon, Y.; Longaker, M.T. Wound repair and regeneration. *Nature* **2008**, *453*, 314–321. [CrossRef] [PubMed]
37. Heber-Katz, E. Oxygen, Metabolism, and Regeneration: Lessons from Mice. *Trends Mol. Med.* **2017**, *23*, 1024–1036. [CrossRef] [PubMed]
38. Abnave, P.; Ghigo, E. Role of the immune system in regeneration and its dynamic interplay with adult stem cells. *Semin. Cell Dev. Biol.* **2018**, *10849*, 30200–30208. [CrossRef] [PubMed]
39. Agathocleous, M.; Harris, W.A. Metabolism in physiological cell proliferation and differentiation. *Trends Cell Biol.* **2013**, *23*, 484–492. [CrossRef] [PubMed]
40. Eming, S.A.; Wynn, T.A.; Martin, P. Inflammation and metabolism in tissue repair and regeneration. *Science* **2017**, *356*, 1026–1030. [CrossRef] [PubMed]
41. Osuma, E.A.; Riggs, D.W.; Gibb, A.A.; Hill, B.G. High throughput measurement of metabolism in planarians reveals activation of glycolysis during regeneration. *Regeneration* **2018**, *5*, 78–86. [CrossRef] [PubMed]
42. Wang, X.; Ha, T.; Liu, L.; Hu, Y.; Kao, R.; Kalbfleisch, J.; Williams, D.; Li, C. TLR3 Mediates Repair and Regeneration of Damaged Neonatal Heart through Glycolysis Dependent YAP1 Regulated miR-152 Expression. *Cell Death Differ.* **2018**, *25*, 966–982. [CrossRef] [PubMed]
43. Fu, X.; Zhu, M.J.; Dodson, M.V.; Du, M. AMP-activated protein kinase stimulates Warburg-like glycolysis and activation of satellite cells during muscle regeneration. *J. Biol. Chem.* **2015**, *290*, 26445–26456. [CrossRef] [PubMed]

44. Lamichane, S.; Dahal Lamichane, B.; Kwon, S.M. Pivotal Roles of Peroxisome Proliferator-Activated Receptors (PPARs) and Their Signal Cascade for Cellular and Whole-Body Energy Homeostasis. *Int. J. Mol. Sci.* **2018**, *19*, 949. [CrossRef] [PubMed]
45. Fan, W.; Evans, R. PPARs and ERRs: Molecular mediators of mitochondrial metabolism. *Curr. Opin. Cell Biol.* **2015**, *33*, 49–54. [CrossRef] [PubMed]
46. Gilde, A.J.; van der Lee, K.A.; Willemsen, P.H.; Chinetti, G.; van der Leij, F.R.; van der Vusse, G.J.; Staels, B.; van Bilsen, M. Peroxisome proliferator-activated receptor (PPAR) alpha and PPARbeta/delta, but not PPARgamma, modulate the expression of genes involved in cardiac lipid metabolism. *Circ. Res.* **2003**, *92*, 518–524. [CrossRef] [PubMed]
47. Gan, Z.; Rumsey, J.; Hazen, B.C.; Lai, L.; Leone, T.C.; Vega, R.B.; Xie, H.; Conley, K.E.; Auwerx, J.; Smith, S.R.; et al. Nuclear receptor/microRNA circuitry links muscle fiber type to energy metabolism. *J. Clin. Investig.* **2013**, *123*, 2564–2575. [CrossRef] [PubMed]
48. Cheng, L.; Ding, G.; Qin, Q.; Huang, Y.; Lewis, W.; He, N.; Evans, R.M.; Schneider, M.D.; Brako, F.A.; Xiao, Y.; et al. Cardiomyocyte-restricted peroxisome proliferator-activated receptor-delta deletion perturbs myocardial fatty acid oxidation and leads to cardiomyopathy. *Nat. Med.* **2004**, *10*, 1245–1250. [CrossRef] [PubMed]
49. Wang, Y.X.; Lee, C.H.; Tiep, S.; Yu, R.T.; Ham, J.; Kang, H.; Evans, R.M. Peroxisome-proliferator-activated receptor delta activates fat metabolism to prevent obesity. *Cell* **2003**, *113*, 159–170. [CrossRef]
50. Pan, D.; Fujimoto, M.; Lopes, A.; Wang, Y.X. Twist-1 is a PPARdelta-inducible, negative-feedback regulator of PGC-1alpha in brown fat metabolism. *Cell* **2009**, *137*, 73–86. [CrossRef] [PubMed]
51. Roberts, L.D.; Murray, A.J.; Menassa, D.; Ashmore, T.; Nicholls, A.W.; Griffin, J.L. The contrasting roles of PPARdelta and PPARgamma in regulating the metabolic switch between oxidation and storage of fats in white adipose tissue. *Genome Biol.* **2011**, *12*, R75. [CrossRef] [PubMed]
52. Schuler, M.; Ali, F.; Chambon, C.; Duteil, D.; Bornert, J.M.; Tardivel, A.; Desvergne, B.; Wahli, W.; Chambon, P.; Metzger, D. PGC1alpha expression is controlled in skeletal muscles by PPARbeta, whose ablation results in fiber-type switching, obesity, and type 2 diabetes. *Cell Metab.* **2006**, *4*, 407–414. [CrossRef] [PubMed]
53. Liu, S.; Hatano, B.; Zhao, M.; Yen, C.C.; Kang, K.; Reilly, S.M.; Gangl, M.R.; Gorgun, C.; Balschi, J.A.; Ntambi, J.M.; et al. Role of peroxisome proliferator-activated receptor {delta}/{beta} in hepatic metabolic regulation. *J. Biol. Chem.* **2011**, *286*, 1237–1247. [CrossRef] [PubMed]
54. Burkart, E.M.; Sambandam, N.; Han, X.; Gross, R.W.; Courtois, M.; Gierasch, C.M.; Shoghi, K.; Welch, M.J.; Kelly, D.P. Nuclear receptors PPARbeta/delta and PPARalpha direct distinct metabolic regulatory programs in the mouse heart. *J. Clin. Investig.* **2007**, *117*, 3930–3939. [PubMed]
55. Andersen, J.L.; Kornbluth, S. The tangled circuitry of metabolism and apoptosis. *Mol. Cell* **2013**, *49*, 399–410. [CrossRef] [PubMed]
56. Diwanji, N.; Bergmann, A. An unexpected friend-ROS in apoptosis-induced compensatory proliferation: Implications for regeneration and cancer. *Semin. Cell Dev. Biol.* **2018**, *80*, 74–82. [CrossRef] [PubMed]
57. Schug, T.T.; Berry, D.C.; Shaw, N.S.; Travis, S.N.; Noy, N. Opposing effects of retinoic acid on cell growth result from alternate activation of two different nuclear receptors. *Cell* **2007**, *129*, 723–733. [CrossRef] [PubMed]
58. Di-Poi, N.; Tan, N.S.; Michalik, L.; Wahli, W.; Desvergne, B. Antiapoptotic role of PPARbeta in keratinocytes via transcriptional control of the Akt1 signaling pathway. *Mol. Cell* **2002**, *10*, 721–733. [CrossRef]
59. Tan, N.S.; Michalik, L.; Noy, N.; Yasmin, R.; Pacot, C.; Heim, M.; Fluhmann, B.; Desvergne, B.; Wahli, W. Critical roles of PPAR beta/delta in keratinocyte response to inflammation. *Genes Dev.* **2001**, *15*, 3263–3277. [CrossRef] [PubMed]
60. Haffar, T.; Berube-Simard, F.A.; Bousette, N. Cardiomyocyte lipotoxicity is mediated by Il-6 and causes down-regulation of PPARs. *Biochem. Biophys. Res. Commun.* **2015**, *459*, 54–59. [CrossRef] [PubMed]
61. Shureiqi, I.; Jiang, W.; Zuo, X.; Wu, Y.; Stimmel, J.B.; Leesnitzer, L.M.; Morris, J.S.; Fan, H.Z.; Fischer, S.M.; Lippman, S.M. The 15-lipoxygenase-1 product 13-S-hydroxyoctadecadienoic acid down-regulates PPAR-delta to induce apoptosis in colorectal cancer cells. *Proc. Natl. Acad. Sci. USA* **2003**, *100*, 9968–9973. [CrossRef] [PubMed]
62. Domigan, C.K.; Warren, C.M.; Antanesian, V.; Happel, K.; Ziyad, S.; Lee, S.; Krall, A.; Duan, L.; Torres-Collado, A.X.; Castellani, L.W.; et al. Autocrine VEGF maintains endothelial survival through regulation of metabolism and autophagy. *J. Cell Sci.* **2015**, *128*, 2236–2248. [CrossRef] [PubMed]

63. Wu, T.T.; Niu, H.S.; Chen, L.J.; Cheng, J.T.; Tong, Y.C. Increase of human prostate cancer cell (DU145) apoptosis by telmisartan through PPAR-delta pathway. *Eur. J. Pharmacol.* **2016**, *775*, 35–42. [CrossRef] [PubMed]
64. Liou, J.Y.; Lee, S.; Ghelani, D.; Matijevic-Aleksic, N.; Wu, K.K. Protection of endothelial survival by peroxisome proliferator-activated receptor-delta mediated 14-3-3 upregulation. *Arterioscler. Thromb. Vasc. Biol.* **2006**, *26*, 1481–1487. [CrossRef] [PubMed]
65. Wu, K.K. Peroxisome Proliferator-Activated Receptors Protect against Apoptosis via 14-3-3. *PPAR Res.* **2010**, *2010*. [CrossRef] [PubMed]
66. Pesant, M.; Sueur, S.; Dutartre, P.; Tallandier, M.; Grimaldi, P.A.; Rochette, L.; Connat, J.L. Peroxisome proliferator-activated receptor delta (PPARdelta) activation protects H9c2 cardiomyoblasts from oxidative stress-induced apoptosis. *Cardiovasc. Res.* **2006**, *69*, 440–449. [CrossRef] [PubMed]
67. Barlaka, E.; Gorbe, A.; Gaspar, R.; Paloczi, J.; Ferdinandy, P.; Lazou, A. Activation of PPARbeta/delta protects cardiac myocytes from oxidative stress-induced apoptosis by suppressing generation of reactive oxygen/nitrogen species and expression of matrix metalloproteinases. *Pharmacol. Res.* **2015**, *95–96*, 102–110. [CrossRef] [PubMed]
68. Clarke, S.A.; Richardson, W.J.; Holmes, J.W. Modifying the mechanics of healing infarcts: Is better the enemy of good? *J. Mol. Cell. Cardiol.* **2016**, *93*, 115–124. [CrossRef] [PubMed]
69. Cordero-Espinoza, L.; Huch, M. The balancing act of the liver: Tissue regeneration versus fibrosis. *J. Clin. Investig.* **2018**, *128*, 85–96. [CrossRef] [PubMed]
70. Munoz-Canoves, P.; Serrano, A.L. Macrophages decide between regeneration and fibrosis in muscle. *Trends Endocrinol. Metab.* **2015**, *26*, 449–450. [CrossRef] [PubMed]
71. Li, J.; Tan, J.; Martino, M.M.; Lui, K.O. Regulatory T-Cells: Potential Regulator of Tissue Repair and Regeneration. *Front. Immunol.* **2018**, *9*, 585. [CrossRef] [PubMed]
72. Wynn, T.A.; Vannella, K.M. Macrophages in Tissue Repair, Regeneration, and Fibrosis. *Immunity* **2016**, *44*, 450–462. [CrossRef] [PubMed]
73. Chan, L.K.; Gerstenlauer, M.; Konukiewitz, B.; Steiger, K.; Weichert, W.; Wirth, T.; Maier, H.J. Epithelial NEMO/IKKgamma limits fibrosis and promotes regeneration during pancreatitis. *Gut* **2017**, *66*, 1995–2007. [CrossRef] [PubMed]
74. Gu, Y.; Li, X.; He, T.; Jiang, Z.; Hao, P.; Tang, X. The Antifibrosis Effects of Peroxisome Proliferator-Activated Receptor delta on Rat Corneal Wound Healing after Excimer Laser Keratectomy. *PPAR Res.* **2014**, *2014*, 464935. [CrossRef] [PubMed]
75. Teunissen, B.E.; Smeets, P.J.; Willemsen, P.H.; De Windt, L.J.; Van der Vusse, G.J.; Van Bilsen, M. Activation of PPARdelta inhibits cardiac fibroblast proliferation and the transdifferentiation into myofibroblasts. *Cardiovasc. Res.* **2007**, *75*, 519–529. [CrossRef] [PubMed]
76. Gao, F.; Liang, Y.; Wang, X.; Lu, Z.; Li, L.; Zhu, S.; Liu, D.; Yan, Z.; Zhu, Z. TRPV1 Activation Attenuates High-Salt Diet-Induced Cardiac Hypertrophy and Fibrosis through PPAR-delta Upregulation. *PPAR Res.* **2014**, *2014*, 491963. [CrossRef] [PubMed]
77. Nathan, C. Immunology: Oxygen and the inflammatory cell. *Nature* **2003**, *422*, 675–676. [CrossRef] [PubMed]
78. O'Neill, L.A.; Hardie, D.G. Metabolism of inflammation limited by AMPK and pseudo-starvation. *Nature* **2013**, *493*, 346–355. [CrossRef] [PubMed]
79. Mills, E.L.; Kelly, B.; Logan, A.; Costa, A.S.H.; Varma, M.; Bryant, C.E.; Tourlomousis, P.; Dabritz, J.H.M.; Gottlieb, E.; Latorre, I.; et al. Succinate Dehydrogenase Supports Metabolic Repurposing of Mitochondria to Drive Inflammatory Macrophages. *Cell* **2016**, *167*, 457–470. [CrossRef] [PubMed]
80. Djouad, F.; Ipseiz, N.; Luz-Crawford, P.; Scholtysek, C.; Kronke, G.; Jorgensen, C. PPARbeta/delta: A master regulator of mesenchymal stem cell functions. *Biochimie* **2017**, *136*, 55–58. [CrossRef] [PubMed]
81. Rius, J.; Guma, M.; Schachtrup, C.; Akassoglou, K.; Zinkernagel, A.S.; Nizet, V.; Johnson, R.S.; Haddad, G.G.; Karin, M. NF-kappaB links innate immunity to the hypoxic response through transcriptional regulation of HIF-1alpha. *Nature* **2008**, *453*, 807–811. [CrossRef] [PubMed]
82. Inoue, T.; Kohro, T.; Tanaka, T.; Kanki, Y.; Li, G.; Poh, H.M.; Mimura, I.; Kobayashi, M.; Taguchi, A.; Maejima, T.; et al. Cross-enhancement of ANGPTL4 transcription by HIF1 alpha and PPAR beta/delta is the result of the conformational proximity of two response elements. *Genome Biol.* **2014**, *15*, R63. [CrossRef] [PubMed]

83. Qu, B.; Hong, Z.; Gong, K.; Sheng, J.; Wu, H.H.; Deng, S.L.; Huang, G.; Ma, Z.H.; Pan, X.M. Inhibitors of Growth 1b Suppresses Peroxisome Proliferator-Activated Receptor-beta/delta Expression through Downregulation of Hypoxia-Inducible Factor 1alpha in Osteoblast Differentiation. *DNA Cell Biol.* **2016**, *35*, 184–191. [CrossRef] [PubMed]
84. Wagner, N.; Jehl-Pietri, C.; Lopez, P.; Murdaca, J.; Giordano, C.; Schwartz, C.; Gounon, P.; Hatem, S.N.; Grimaldi, P.; Wagner, K.D. Peroxisome proliferator-activated receptor beta stimulation induces rapid cardiac growth and angiogenesis via direct activation of calcineurin. *Cardiovasc. Res.* **2009**, *83*, 61–71. [CrossRef] [PubMed]
85. Yu, J.S.; Cui, W. Proliferation, survival and metabolism: The role of PI3K/AKT/mTOR signalling in pluripotency and cell fate determination. *Development* **2016**, *143*, 3050–3060. [CrossRef] [PubMed]
86. Han, J.K.; Lee, H.S.; Yang, H.M.; Hur, J.; Jun, S.I.; Kim, J.Y.; Cho, C.H.; Koh, G.Y.; Peters, J.M.; Park, K.W.; et al. Peroxisome proliferator-activated receptor-delta agonist enhances vasculogenesis by regulating endothelial progenitor cells through genomic and nongenomic activations of the phosphatidylinositol 3-kinase/Akt pathway. *Circulation* **2008**, *118*, 1021–1033. [CrossRef] [PubMed]
87. Piqueras, L.; Reynolds, A.R.; Hodivala-Dilke, K.M.; Alfranca, A.; Redondo, J.M.; Hatae, T.; Tanabe, T.; Warner, T.D.; Bishop-Bailey, D. Activation of PPARbeta/delta induces endothelial cell proliferation and angiogenesis. *Arterioscler. Thromb. Vasc. Biol.* **2007**, *27*, 63–69. [CrossRef] [PubMed]
88. He, T.; Smith, L.A.; Lu, T.; Joyner, M.J.; Katusic, Z.S. Activation of peroxisome proliferator-activated receptor-{delta} enhances regenerative capacity of human endothelial progenitor cells by stimulating biosynthesis of tetrahydrobiopterin. *Hypertension* **2011**, *58*, 287–294. [CrossRef] [PubMed]
89. Xu, L.; Han, C.; Lim, K.; Wu, T. Cross-talk between peroxisome proliferator-activated receptor delta and cytosolic phospholipase A(2)alpha/cyclooxygenase-2/prostaglandin E(2) signaling pathways in human hepatocellular carcinoma cells. *Cancer Res.* **2006**, *66*, 11859–11868. [CrossRef] [PubMed]
90. Kim, H.J.; Kim, M.Y.; Hwang, J.S.; Kim, H.J.; Lee, J.H.; Chang, K.C.; Kim, J.H.; Han, C.W.; Kim, J.H.; Seo, H.G. PPARdelta inhibits IL-1beta-stimulated proliferation and migration of vascular smooth muscle cells via up-regulation of IL-1Ra. *Cell. Mol. Life Sci.* **2010**, *67*, 2119–2130. [CrossRef] [PubMed]
91. Lim, H.J.; Lee, S.; Park, J.H.; Lee, K.S.; Choi, H.E.; Chung, K.S.; Lee, H.H.; Park, H.Y. PPAR delta agonist L-165041 inhibits rat vascular smooth muscle cell proliferation and migration via inhibition of cell cycle. *Atherosclerosis* **2009**, *202*, 446–454. [CrossRef] [PubMed]
92. Hytonen, J.; Leppanen, O.; Braesen, J.H.; Schunck, W.H.; Mueller, D.; Jung, F.; Mrowietz, C.; Jastroch, M.; von Bergwelt-Baildon, M.; Kappert, K.; et al. Activation of Peroxisome Proliferator-Activated Receptor-delta as Novel Therapeutic Strategy to Prevent In-Stent Restenosis and Stent Thrombosis. *Arterioscler. Thromb. Vasc. Biol.* **2016**, *36*, 1534–1548. [CrossRef] [PubMed]
93. Borland, M.G.; Foreman, J.E.; Girroir, E.E.; Zolfaghari, R.; Sharma, A.K.; Amin, S.; Gonzalez, F.J.; Ross, A.C.; Peters, J.M. Ligand activation of peroxisome proliferator-activated receptor-beta/delta inhibits cell proliferation in human HaCaT keratinocytes. *Mol. Pharmacol.* **2008**, *74*, 1429–1442. [CrossRef] [PubMed]
94. Yao, P.L.; Morales, J.L.; Zhu, B.; Kang, B.H.; Gonzalez, F.J.; Peters, J.M. Activation of peroxisome proliferator-activated receptor-beta/delta (PPAR-beta/delta) inhibits human breast cancer cell line tumorigenicity. *Mol. Cancer Ther.* **2014**, *13*, 1008–1017. [CrossRef] [PubMed]
95. Zhang, J.; Fu, M.; Zhu, X.; Xiao, Y.; Mou, Y.; Zheng, H.; Akinbami, M.A.; Wang, Q.; Chen, Y.E. Peroxisome proliferator-activated receptor delta is up-regulated during vascular lesion formation and promotes post-confluent cell proliferation in vascular smooth muscle cells. *J. Biol. Chem.* **2002**, *277*, 11505–11512. [CrossRef] [PubMed]
96. Kim, Y.H.; Han, H.J. High-glucose-induced prostaglandin E(2) and peroxisome proliferator-activated receptor delta promote mouse embryonic stem cell proliferation. *Stem Cells* **2008**, *26*, 745–755. [CrossRef] [PubMed]
97. Yao, P.L.; Chen, L.; Hess, R.A.; Muller, R.; Gonzalez, F.J.; Peters, J.M. Peroxisome Proliferator-activated Receptor-D (PPARD) Coordinates Mouse Spermatogenesis by Modulating Extracellular Signal-regulated Kinase (ERK)-dependent Signaling. *J. Biol. Chem.* **2015**, *290*, 23416–23431. [CrossRef] [PubMed]
98. Zeng, L.; Geng, Y.; Tretiakova, M.; Yu, X.; Sicinski, P.; Kroll, T.G. Peroxisome proliferator-activated receptor-delta induces cell proliferation by a cyclin E1-dependent mechanism and is up-regulated in thyroid tumors. *Cancer Res.* **2008**, *68*, 6578–6586. [CrossRef] [PubMed]

99. Romanowska, M.; al Yacoub, N.; Seidel, H.; Donandt, S.; Gerken, H.; Phillip, S.; Haritonova, N.; Artuc, M.; Schweiger, S.; Sterry, W.; et al. PPARdelta enhances keratinocyte proliferation in psoriasis and induces heparin-binding EGF-like growth factor. *J. Investig. Dermatol.* **2008**, *128*, 110–124. [CrossRef] [PubMed]
100. Montagner, A.; Rando, G.; Degueurce, G.; Leuenberger, N.; Michalik, L.; Wahli, W. New insights into the role of PPARs. *Prostag. Leukot. Essent. Fatty Acids* **2011**, *85*, 235–243. [CrossRef] [PubMed]
101. Jeong, A.Y.; Lee, M.Y.; Lee, S.H.; Park, J.H.; Han, H.J. PPARdelta agonist-mediated ROS stimulates mouse embryonic stem cell proliferation through cooperation of p38 MAPK and Wnt/beta-catenin. *Cell Cycle* **2009**, *8*, 611–619. [CrossRef] [PubMed]
102. Lopaschuk, G.D.; Jaswal, J.S. Energy metabolic phenotype of the cardiomyocyte during development, differentiation, and postnatal maturation. *J. Cardiovasc. Pharmacol.* **2010**, *56*, 130–140. [CrossRef] [PubMed]
103. Nagy, T.A.; Wroblewski, L.E.; Wang, D.; Piazuelo, M.B.; Delgado, A.; Romero-Gallo, J.; Noto, J.; Israel, D.A.; Ogden, S.R.; Correa, P.; et al. Beta-Catenin and p120 mediate PPARdelta-dependent proliferation induced by Helicobacter pylori in human and rodent epithelia. *Gastroenterology* **2011**, *141*, 553–564. [CrossRef] [PubMed]
104. Wagner, K.D.; Benchetrit, M.; Bianchini, L.; Michiels, J.F.; Wagner, N. Peroxisome proliferator-activated receptor beta/delta (PPARbeta/delta) is highly expressed in liposarcoma and promotes migration and proliferation. *J. Pathol.* **2011**, *224*, 575–588. [CrossRef] [PubMed]
105. Hwang, J.S.; Ham, S.A.; Yoo, T.; Lee, W.J.; Paek, K.S.; Lee, C.H.; Seo, H.G. Sirtuin 1 Mediates the Actions of Peroxisome Proliferator-Activated Receptor delta on the Oxidized Low-Density Lipoprotein-Triggered Migration and Proliferation of Vascular Smooth Muscle Cells. *Mol. Pharmacol.* **2016**, *90*, 522–529. [CrossRef] [PubMed]
106. Chung, S.; Dzeja, P.P.; Faustino, R.S.; Perez-Terzic, C.; Behfar, A.; Terzic, A. Mitochondrial oxidative metabolism is required for the cardiac differentiation of stem cells. *Nat. Clin. Pract. Cardiovasc. Med.* **2007**, *4* (Suppl. 1), S60. [CrossRef] [PubMed]
107. Folmes, C.D.; Nelson, T.J.; Martinez-Fernandez, A.; Arrell, D.K.; Lindor, J.Z.; Dzeja, P.P.; Ikeda, Y.; Perez-Terzic, C.; Terzic, A. Somatic oxidative bioenergetics transitions into pluripotency-dependent glycolysis to facilitate nuclear reprogramming. *Cell Metab.* **2011**, *14*, 264–271. [CrossRef] [PubMed]
108. Ozbudak, E.M.; Tassy, O.; Pourquie, O. Spatiotemporal compartmentalization of key physiological processes during muscle precursor differentiation. *Proc. Natl. Acad. Sci. USA* **2010**, *107*, 4224–4229. [CrossRef] [PubMed]
109. De Carvalho, A.; Bassaneze, V.; Forni, M.F.; Keusseyan, A.A.; Kowaltowski, A.J.; Krieger, J.E. Early Postnatal Cardiomyocyte Proliferation Requires High Oxidative Energy Metabolism. *Sci. Rep.* **2017**, *7*, 15434. [CrossRef] [PubMed]
110. Han, J.K.; Kim, B.K.; Won, J.Y.; Shin, Y.; Choi, S.B.; Hwang, I.; Kang, J.; Lee, H.J.; Koh, S.J.; Lee, J.; et al. Interaction between platelets and endothelial progenitor cells via LPA-Edg-2 axis is augmented by PPAR-delta activation. *J. Mol. Cell. Cardiol.* **2016**, *97*, 266–277. [CrossRef] [PubMed]
111. Tan, N.S.; Shaw, N.S.; Vinckenbosch, N.; Liu, P.; Yasmin, R.; Desvergne, B.; Wahli, W.; Noy, N. Selective cooperation between fatty acid binding proteins and peroxisome proliferator-activated receptors in regulating transcription. *Mol. Cell. Biol.* **2002**, *22*, 5114–5127. [CrossRef] [PubMed]
112. Bastie, C.; Holst, D.; Gaillard, D.; Jehl-Pietri, C.; Grimaldi, P.A. Expression of peroxisome proliferator-activated receptor PPARdelta promotes induction of PPARgamma and adipocyte differentiation in 3T3C2 fibroblasts. *J. Biol. Chem.* **1999**, *274*, 21920–21925. [CrossRef] [PubMed]
113. Grimaldi, P.A. The roles of PPARs in adipocyte differentiation. *Prog. Lipid Res.* **2001**, *40*, 269–281. [CrossRef]
114. Rosenfield, R.L.; Kentsis, A.; Deplewski, D.; Ciletti, N. Rat preputial sebocyte differentiation involves peroxisome proliferator-activated receptors. *J. Investig. Dermatol.* **1999**, *112*, 226–232. [CrossRef] [PubMed]
115. Peters, J.M.; Lee, S.S.; Li, W.; Ward, J.M.; Gavrilova, O.; Everett, C.; Reitman, M.L.; Hudson, L.D.; Gonzalez, F.J. Growth, adipose, brain, and skin alterations resulting from targeted disruption of the mouse peroxisome proliferator-activated receptor beta(delta). *Mol. Cell. Biol.* **2000**, *20*, 5119–5128. [CrossRef] [PubMed]
116. Saluja, I.; Granneman, J.G.; Skoff, R.P. PPAR delta agonists stimulate oligodendrocyte differentiation in tissue culture. *Glia* **2001**, *33*, 191–204. [CrossRef]
117. Mei, Y.Q.; Pan, Z.F.; Chen, W.T.; Xu, M.H.; Zhu, D.Y.; Yu, Y.P.; Lou, Y.J. A Flavonoid Compound Promotes Neuronal Differentiation of Embryonic Stem Cells via PPAR-beta Modulating Mitochondrial Energy Metabolism. *PLoS ONE* **2016**, *11*, e0157747. [CrossRef] [PubMed]

118. Yao, P.L.; Chen, L.; Dobrzanski, T.P.; Zhu, B.; Kang, B.H.; Muller, R.; Gonzalez, F.J.; Peters, J.M. Peroxisome proliferator-activated receptor-beta/delta inhibits human neuroblastoma cell tumorigenesis by inducing p53- and SOX2-mediated cell differentiation. *Mol. Carcinog.* **2017**, *56*, 1472–1483. [CrossRef] [PubMed]
119. Xu, M.; Chen, X.; Chen, D.; Yu, B.; Huang, Z. FoxO1: A novel insight into its molecular mechanisms in the regulation of skeletal muscle differentiation and fiber type specification. *Oncotarget* **2017**, *8*, 10662–10674. [CrossRef] [PubMed]
120. Angione, A.R.; Jiang, C.; Pan, D.; Wang, Y.X.; Kuang, S. PPARdelta regulates satellite cell proliferation and skeletal muscle regeneration. *Skelet. Muscle* **2011**, *1*, 33. [CrossRef] [PubMed]
121. Gaudel, C.; Schwartz, C.; Giordano, C.; Abumrad, N.A.; Grimaldi, P.A. Pharmacological activation of PPARbeta promotes rapid and calcineurin-dependent fiber remodeling and angiogenesis in mouse skeletal muscle. *Am. J. Physiol. Endocrinol. Metab.* **2008**, *295*, E297–E304. [CrossRef] [PubMed]
122. Han, J.K.; Kim, H.L.; Jeon, K.H.; Choi, Y.E.; Lee, H.S.; Kwon, Y.W.; Jang, J.J.; Cho, H.J.; Kang, H.J.; Oh, B.H.; et al. Peroxisome proliferator-activated receptor-delta activates endothelial progenitor cells to induce angio-myogenesis through matrix metallo-proteinase-9-mediated insulin-like growth factor-1 paracrine networks. *Eur. Heart J.* **2013**, *34*, 1755–1765. [CrossRef] [PubMed]
123. Maccallini, C.; Mollica, A.; Amoroso, R. The Positive Regulation of eNOS Signaling by PPAR Agonists in Cardiovascular Diseases. *Am. J. Cardiovasc. Drugs* **2017**, *17*, 273–281. [CrossRef] [PubMed]
124. Vanhoutte, P.M. Regenerated Endothelium and Its Senescent Response to Aggregating Platelets. *Circ. J.* **2016**, *80*, 783–790. [CrossRef] [PubMed]
125. Hankenson, K.D.; Gagne, K.; Shaughnessy, M. Extracellular signaling molecules to promote fracture healing and bone regeneration. *Adv. Drug Deliv. Rev.* **2015**, *94*, 3–12. [CrossRef] [PubMed]
126. Hankenson, K.D.; Zimmerman, G.; Marcucio, R. Biological perspectives of delayed fracture healing. *Injury* **2014**, *45* (Suppl. 2), S8–S15. [CrossRef] [PubMed]
127. Zhang, Q.; Raoof, M.; Chen, Y.; Sumi, Y.; Sursal, T.; Junger, W.; Brohi, K.; Itagaki, K.; Hauser, C.J. Circulating mitochondrial DAMPs cause inflammatory responses to injury. *Nature* **2010**, *464*, 104–107. [CrossRef] [PubMed]
128. Stoick-Cooper, C.L.; Moon, R.T.; Weidinger, G. Advances in signaling in vertebrate regeneration as a prelude to regenerative medicine. *Genes Dev.* **2007**, *21*, 1292–1315. [CrossRef] [PubMed]
129. Poss, K.D.; Wilson, L.G.; Keating, M.T. Heart regeneration in zebrafish. *Science* **2002**, *298*, 2188–2190. [CrossRef] [PubMed]
130. Vig, K.; Chaudhari, A.; Tripathi, S.; Dixit, S.; Sahu, R.; Pillai, S.; Dennis, V.A.; Singh, S.R. Advances in Skin Regeneration Using Tissue Engineering. *Int. J. Mol. Sci.* **2017**, *18*, 789. [CrossRef] [PubMed]
131. Takeo, M.; Lee, W.; Ito, M. Wound healing and skin regeneration. *Cold Spring Harb. Perspect. Med.* **2015**, *5*, a023267. [CrossRef] [PubMed]
132. Michalik, L.; Desvergne, B.; Tan, N.S.; Basu-Modak, S.; Escher, P.; Rieusset, J.; Peters, J.M.; Kaya, G.; Gonzalez, F.J.; Zakany, J.; et al. Impaired skin wound healing in peroxisome proliferator-activated receptor (PPAR)alpha and PPARbeta mutant mice. *J. Cell Biol.* **2001**, *154*, 799–814. [CrossRef] [PubMed]
133. Tan, N.S.; Icre, G.; Montagner, A.; Bordier-ten-Heggeler, B.; Wahli, W.; Michalik, L. The nuclear hormone receptor peroxisome proliferator-activated receptor beta/delta potentiates cell chemotactism, polarization, and migration. *Mol. Cell. Biol.* **2007**, *27*, 7161–7175. [CrossRef] [PubMed]
134. Chong, H.C.; Tan, M.J.; Philippe, V.; Tan, S.H.; Tan, C.K.; Ku, C.W.; Goh, Y.Y.; Wahli, W.; Michalik, L.; Tan, N.S. Regulation of epithelial-mesenchymal IL-1 signaling by PPARbeta/delta is essential for skin homeostasis and wound healing. *J. Cell Biol.* **2009**, *184*, 817–831. [CrossRef] [PubMed]
135. Baghdadi, M.B.; Tajbakhsh, S. Regulation and phylogeny of skeletal muscle regeneration. *Dev. Biol.* **2018**, *433*, 200–209. [CrossRef] [PubMed]
136. Sass, F.A.; Fuchs, M.; Pumberger, M.; Geissler, S.; Duda, G.N.; Perka, C.; Schmidt-Bleek, K. Immunology Guides Skeletal Muscle Regeneration. *Int. J. Mol. Sci.* **2018**, *19*, 835. [CrossRef] [PubMed]
137. Batonnet-Pichon, S.; Behin, A.; Cabet, E.; Delort, F.; Vicart, P.; Lilienbaum, A. Myofibrillar Myopathies: New Perspectives from Animal Models to Potential Therapeutic Approaches. *J. Neuromuscul. Dis.* **2017**, *4*, 1–15. [CrossRef] [PubMed]
138. Baracos, V.E.; Martin, L.; Korc, M.; Guttridge, D.C.; Fearon, K.C.H. Cancer-associated cachexia. *Nat. Rev. Dis. Primers* **2018**, *4*, 17105. [CrossRef] [PubMed]

139. Dalle, S.; Rossmeislova, L.; Koppo, K. The Role of Inflammation in Age-Related Sarcopenia. *Front. Physiol.* **2017**, *8*, 1045. [CrossRef] [PubMed]
140. Kitamura, T.; Kitamura, Y.I.; Funahashi, Y.; Shawber, C.J.; Castrillon, D.H.; Kollipara, R.; DePinho, R.A.; Kitajewski, J.; Accili, D. A Foxo/Notch pathway controls myogenic differentiation and fiber type specification. *J. Clin. Investig.* **2007**, *117*, 2477–2485. [CrossRef] [PubMed]
141. Giordano, C.; Rousseau, A.S.; Wagner, N.; Gaudel, C.; Murdaca, J.; Jehl-Pietri, C.; Sibille, B.; Grimaldi, P.A.; Lopez, P. Peroxisome proliferator-activated receptor beta activation promotes myonuclear accretion in skeletal muscle of adult and aged mice. *Pflugers Arch.* **2009**, *458*, 901–913. [CrossRef] [PubMed]
142. Lee, S.J.; Go, G.Y.; Yoo, M.; Kim, Y.K.; Seo, D.W.; Kang, J.S.; Bae, G.U. Peroxisome proliferator-activated receptor beta/delta (PPARbeta/delta) activates promyogenic signaling pathways, thereby promoting myoblast differentiation. *Biochem. Biophys. Res. Commun.* **2016**, *470*, 157–162. [CrossRef] [PubMed]
143. Chandrashekar, P.; Manickam, R.; Ge, X.; Bonala, S.; McFarlane, C.; Sharma, M.; Wahli, W.; Kambadur, R. Inactivation of PPARbeta/delta adversely affects satellite cells and reduces postnatal myogenesis. *Am. J. Physiol. Endocrinol. Metab.* **2015**, *309*, E122–E131. [CrossRef] [PubMed]
144. Haralampieva, D.; Salemi, S.; Betzel, T.; Dinulovic, I.; Kramer, S.D.; Schibli, R.; Sulser, T.; Handschin, C.; Ametamey, S.M.; Eberli, D. Injected Human Muscle Precursor Cells Overexpressing PGC-1alpha Enhance Functional Muscle Regeneration after Trauma. *Stem Cells Int.* **2018**, *2018*, 4658503. [CrossRef] [PubMed]
145. Fu, H.; Desvergne, B.; Ferrari, S.; Bonnet, N. Impaired musculoskeletal response to age and exercise in PPARbeta(−/−) diabetic mice. *Endocrinology* **2014**, *155*, 4686–4696. [CrossRef] [PubMed]
146. Zebrowski, D.C.; Becker, R.; Engel, F.B. Towards regenerating the mammalian heart: Challenges in evaluating experimentally induced adult mammalian cardiomyocyte proliferation. *Am. J. Physiol. Heart Circ. Physiol.* **2016**, *310*, H1045–H1054. [CrossRef] [PubMed]
147. Leone, M.; Magadum, A.; Engel, F.B. Cardiomyocyte proliferation in cardiac development and regeneration: A guide to methodologies and interpretations. *Am. J. Physiol. Heart Circ. Physiol.* **2015**, *309*, H1237–H1250. [CrossRef] [PubMed]
148. Lopaschuk, G.D.; Collins-Nakai, R.L.; Itoi, T. Developmental changes in energy substrate use by the heart. *Cardiovasc. Res.* **1992**, *26*, 1172–1180. [CrossRef] [PubMed]
149. Oikonomou, E.; Mourouzis, K.; Fountoulakis, P.; Papamikroulis, G.A.; Siasos, G.; Antonopoulos, A.; Vogiatzi, G.; Tsalamadris, S.; Vavuranakis, M.; Tousoulis, D. Interrelationship between diabetes mellitus and heart failure: The role of peroxisome proliferator-activated receptors in left ventricle performance. *Heart Fail. Rev.* **2018**, *23*, 389–408. [CrossRef] [PubMed]
150. Toral, M.; Romero, M.; Perez-Vizcaino, F.; Duarte, J.; Jimenez, R. Antihypertensive effects of peroxisome proliferator-activated receptor-beta/delta activation. *Am. J. Physiol. Heart Circ. Physiol.* **2017**, *312*, H189–H200. [CrossRef] [PubMed]
151. Esposito, E.; Paterniti, I.; Meli, R.; Bramanti, P.; Cuzzocrea, S. GW0742, a high-affinity PPAR-delta agonist, mediates protection in an organotypic model of spinal cord damage. *Spine* **2012**, *37*, E73–E78. [CrossRef] [PubMed]
152. Collino, M.; Patel, N.S.; Thiemermann, C. PPARs as new therapeutic targets for the treatment of cerebral ischemia/reperfusion injury. *Ther. Adv. Cardiovasc. Dis.* **2008**, *2*, 179–197. [CrossRef] [PubMed]
153. Miura, P.; Chakkalakal, J.V.; Boudreault, L.; Belanger, G.; Hebert, R.L.; Renaud, J.M.; Jasmin, B.J. Pharmacological activation of PPARbeta/delta stimulates utrophin A expression in skeletal muscle fibers and restores sarcolemmal integrity in mature mdx mice. *Hum. Mol. Genet.* **2009**, *18*, 4640–4649. [CrossRef] [PubMed]
154. Mandrekar-Colucci, S.; Sauerbeck, A.; Popovich, P.G.; McTigue, D.M. PPAR agonists as therapeutics for CNS trauma and neurological diseases. *ASN Neuro* **2013**, *5*, e00129. [CrossRef] [PubMed]
155. Pauta, M.; Rotllan, N.; Fernandez-Hernando, A.; Langhi, C.; Ribera, J.; Lu, M.; Boix, L.; Bruix, J.; Jimenez, W.; Suarez, Y.; et al. Akt-mediated foxo1 inhibition is required for liver regeneration. *Hepatology* **2016**, *63*, 1660–1674. [CrossRef] [PubMed]
156. Marshall, J.L.; Holmberg, J.; Chou, E.; Ocampo, A.C.; Oh, J.; Lee, J.; Peter, A.K.; Martin, P.T.; Crosbie-Watson, R.H. Sarcospan-dependent Akt activation is required for utrophin expression and muscle regeneration. *J. Cell Biol.* **2012**, *197*, 1009–1027. [CrossRef] [PubMed]

157. Kim, M.H.; Kay, D.I.; Rudra, R.T.; Chen, B.M.; Hsu, N.; Izumiya, Y.; Martinez, L.; Spencer, M.J.; Walsh, K.; Grinnell, A.D.; et al. Myogenic Akt signaling attenuates muscular degeneration, promotes myofiber regeneration and improves muscle function in dystrophin-deficient mdx mice. *Hum. Mol. Genet.* **2011**, *20*, 1324–1338. [CrossRef] [PubMed]
158. Wang, X.; Chen, H.; Tian, R.; Zhang, Y.; Drutskaya, M.S.; Wang, C.; Ge, J.; Fan, Z.; Kong, D.; Wang, X.; et al. Macrophages induce AKT/beta-catenin-dependent Lgr5(+) stem cell activation and hair follicle regeneration through TNF. *Nat. Commun.* **2017**, *8*, 14091. [CrossRef] [PubMed]
159. Peiris, T.H.; Ramirez, D.; Barghouth, P.G.; Oviedo, N.J. The Akt signaling pathway is required for tissue maintenance and regeneration in planarians. *BMC Dev. Biol.* **2016**, *16*, 7. [CrossRef] [PubMed]
160. Huang, Z.R.; Chen, H.Y.; Hu, Z.Z.; Xie, P.; Liu, Q.H. PTEN knockdown with the Y444F mutant AAV2 vector promotes axonal regeneration in the adult optic nerve. *Neural Regen. Res.* **2018**, *13*, 135–144. [PubMed]
161. Liu, K.; Lu, Y.; Lee, J.K.; Samara, R.; Willenberg, R.; Sears-Kraxberger, I.; Tedeschi, A.; Park, K.K.; Jin, D.; Cai, B.; et al. PTEN deletion enhances the regenerative ability of adult corticospinal neurons. *Nat. Neurosci.* **2010**, *13*, 1075–1081. [CrossRef] [PubMed]
162. Park, K.K.; Liu, K.; Hu, Y.; Smith, P.D.; Wang, C.; Cai, B.; Xu, B.; Connolly, L.; Kramvis, I.; Sahin, M.; et al. Promoting axon regeneration in the adult CNS by modulation of the PTEN/mTOR pathway. *Science* **2008**, *322*, 963–966. [CrossRef] [PubMed]
163. Keefe, M.D.; Wang, H.; De La, O.J.; Khan, A.; Firpo, M.A.; Murtaugh, L.C. Beta-catenin is selectively required for the expansion and regeneration of mature pancreatic acinar cells in mice. *Dis. Model Mech.* **2012**, *5*, 503–514. [CrossRef] [PubMed]
164. Suh, H.N.; Kim, M.J.; Jung, Y.S.; Lien, E.M.; Jun, S.; Park, J.I. Quiescence Exit of Tert(+) Stem Cells by Wnt/beta-Catenin Is Indispensable for Intestinal Regeneration. *Cell Rep.* **2017**, *21*, 2571–2584. [CrossRef] [PubMed]
165. Kawakami, Y.; Rodriguez Esteban, C.; Raya, M.; Kawakami, H.; Marti, M.; Dubova, I.; Izpisua Belmonte, J.C. Wnt/beta-catenin signaling regulates vertebrate limb regeneration. *Genes Dev.* **2006**, *20*, 3232–3237. [CrossRef] [PubMed]
166. Ozhan, G.; Weidinger, G. Wnt/beta-catenin signaling in heart regeneration. *Cell Regen. (Lond.)* **2015**, *4*, 3. [PubMed]
167. Pansters, N.A.; Schols, A.M.; Verhees, K.J.; de Theije, C.C.; Snepvangers, F.J.; Kelders, M.C.; Ubags, N.D.; Haegens, A.; Langen, R.C. Muscle-specific GSK-3beta ablation accelerates regeneration of disuse-atrophied skeletal muscle. *Biochim. Biophys. Acta* **2015**, *1852*, 490–506. [CrossRef] [PubMed]
168. Li, L.; Peng, X.; Qin, Y.; Wang, R.; Tang, J.; Cui, X.; Wang, T.; Liu, W.; Pan, H.; Li, B. Acceleration of bone regeneration by activating Wnt/beta-catenin signalling pathway via lithium released from lithium chloride/calcium phosphate cement in osteoporosis. *Sci. Rep.* **2017**, *7*, 45204. [CrossRef] [PubMed]
169. Tseng, A.S.; Engel, F.B.; Keating, M.T. The GSK-3 inhibitor BIO promotes proliferation in mammalian cardiomyocytes. *Chem. Biol.* **2006**, *13*, 957–963. [CrossRef] [PubMed]
170. Guo, X.; Snider, W.D.; Chen, B. GSK3beta regulates AKT-induced central nervous system axon regeneration via an eIF2Bepsilon-dependent, mTORC1-independent pathway. *eLife* **2016**, *5*, e11903. [CrossRef] [PubMed]
171. Artoni, F.; Kreipke, R.E.; Palmeira, O.; Dixon, C.; Goldberg, Z.; Ruohola-Baker, H. Loss of foxo rescues stem cell aging in Drosophila germ line. *eLife* **2017**, *6*, e27842. [CrossRef] [PubMed]
172. Mori, R.; Tanaka, K.; de Kerckhove, M.; Okamoto, M.; Kashiyama, K.; Tanaka, K.; Kim, S.; Kawata, T.; Komatsu, T.; Park, S.; et al. Reduced FOXO1 expression accelerates skin wound healing and attenuates scarring. *Am. J. Pathol.* **2014**, *184*, 2465–2479. [CrossRef] [PubMed]
173. Lara-Pezzi, E.; Winn, N.; Paul, A.; McCullagh, K.; Slominsky, E.; Santini, M.P.; Mourkioti, F.; Sarathchandra, P.; Fukushima, S.; Suzuki, K.; et al. A naturally occurring calcineurin variant inhibits FoxO activity and enhances skeletal muscle regeneration. *J. Cell Biol.* **2007**, *179*, 1205–1218. [CrossRef] [PubMed]
174. Byrne, A.B.; Walradt, T.; Gardner, K.E.; Hubbert, A.; Reinke, V.; Hammarlund, M. Insulin/IGF1 signaling inhibits age-dependent axon regeneration. *Neuron* **2014**, *81*, 561–573. [CrossRef] [PubMed]
175. Srivastava, T.; Diba, P.; Dean, J.M.; Banine, F.; Shaver, D.; Hagen, M.; Gong, X.; Su, W.; Emery, B.; Marks, D.L.; et al. A TLR/AKT/FoxO3 immune tolerance-like pathway disrupts the repair capacity of oligodendrocyte progenitors. *J. Clin. Investig.* **2018**, *128*, 2025–2041. [CrossRef] [PubMed]

176. Lu, H.; Zhang, L.H.; Yang, L.; Tang, P.F. The PI3K/Akt/FOXO3a pathway regulates regeneration following spinal cord injury in adult rats through TNF-alpha and p27kip1 expression. *Int. J. Mol. Med.* **2018**, *41*, 2832–2838. [PubMed]
177. Dentice, M.; Marsili, A.; Ambrosio, R.; Guardiola, O.; Sibilio, A.; Paik, J.H.; Minchiotti, G.; DePinho, R.A.; Fenzi, G.; Larsen, P.R.; et al. The FoxO3/type 2 deiodinase pathway is required for normal mouse myogenesis and muscle regeneration. *J. Clin. Investig.* **2010**, *120*, 4021–4030. [CrossRef] [PubMed]
178. Kurinna, S.; Stratton, S.A.; Tsai, W.W.; Akdemir, K.C.; Gu, W.; Singh, P.; Goode, T.; Darlington, G.J.; Barton, M.C. Direct activation of forkhead box O3 by tumor suppressors p53 and p73 is disrupted during liver regeneration in mice. *Hepatology* **2010**, *52*, 1023–1032. [CrossRef] [PubMed]

International Journal of
Molecular Sciences

Article

PPARγ Controls Ectopic Adipogenesis and Cross-Talks with Myogenesis During Skeletal Muscle Regeneration

Gabriele Dammone [1,2,†], Sonia Karaz [1,†], Laura Lukjanenko [1,3], Carine Winkler [2], Federico Sizzano [1], Guillaume Jacot [1], Eugenia Migliavacca [1], Alessio Palini [1], Béatrice Desvergne [2], Federica Gilardi [2] and Jerome N. Feige [1,3,*]

1 Nestle Institute of Health Sciences, EPFL Innovation Park, Building H, EPFL Campus, 1015 Lausanne, Switzerland

2 Center for Integrative Genomics, Faculty of Biology and Medicine, University of Lausanne, 1015 Lausanne, Switzerland; beatrice.desvergne@unil.ch (B.D.); Federica.Gilardi@hcuge.ch (F.G.)

3 School of Life Sciences, Ecole Polytechnique Federale de Lausanne (EPFL), 1015 Lausanne, Switzerland

* Correspondence: Jerome.Feige@rd.nestle.com

† These authors contribute equally to this work.

Received: 8 May 2018; Accepted: 9 July 2018; Published: 13 July 2018

Abstract: Skeletal muscle is a regenerative tissue which can repair damaged myofibers through the activation of tissue-resident muscle stem cells (MuSCs). Many muscle diseases with impaired regeneration cause excessive adipose tissue accumulation in muscle, alter the myogenic fate of MuSCs, and deregulate the cross-talk between MuSCs and fibro/adipogenic progenitors (FAPs), a bi-potent cell population which supports myogenesis and controls intra-muscular fibrosis and adipocyte formation. In order to better characterize the interaction between adipogenesis and myogenesis, we studied muscle regeneration and MuSC function in whole body *Pparg* null mice generated by epiblast-specific Cre/lox deletion (*Pparg*$^{\Delta/\Delta}$). We demonstrate that deletion of PPARγ completely abolishes ectopic muscle adipogenesis during regeneration and impairs MuSC expansion and myogenesis after injury. Ex vivo assays revealed that perturbed myogenesis in *Pparg*$^{\Delta/\Delta}$ mice does not primarily result from intrinsic defects of MuSCs or from perturbed myogenic support from FAPs. The immune transition from a pro- to anti-inflammatory MuSC niche during regeneration is perturbed in *Pparg*$^{\Delta/\Delta}$ mice and suggests that PPARγ signaling in macrophages can interact with ectopic adipogenesis and influence muscle regeneration. Altogether, our study demonstrates that a PPARγ-dependent adipogenic response regulates muscle fat infiltration during regeneration and that PPARγ is required for MuSC function and efficient muscle repair.

Keywords: Skeletal muscle; regeneration; myogenesis; adipogenesis; muscle stem cells; satellite cells; inflammation; *PPARg*

1. Introduction

Skeletal muscle is a highly plastic tissue, which can adapt to different metabolic or physical challenges, such as physical activity, disuse, or injuries. When muscle fibers are damaged as a consequence of intense exercise, trauma, surgery, or genetic diseases, complementary cellular mechanisms allow skeletal muscle to regenerate and recover tissue architecture and functionality. This regenerative capacity is primarily driven by muscle stem cells (MuSCs), also known as satellite cells, which express the transcription factor paired box 7 (Pax7) [1,2]. As a consequence of injury, MuSCs break quiescence and enter the cell cycle, undertake myogenic commitment after expansion, and reconstitute the contractile architecture by forming fusing to damaged fibers or forming de novo myofibers [3,4]. Successful muscle regeneration is tightly dependent on a temporally controlled inflammatory response. A first wave of pro-inflammatory macrophages clears cellular debris [5] and promotes MuSC expansion while inhibiting their premature differentiation [6]. Pro-inflammatory macrophages are then replaced by a second wave of anti-inflammatory and pro-regenerative macrophages, which switches the muscle stem cell niche to a state permissive to fusion and differentiation to allow regeneration to complete [5,6].

Infiltration of fat in skeletal muscle has been reported in various physiological and pathological settings. In particular, intra-muscular adipose tissue (IMAT; [7,8]) accumulates during muscle diseases with impaired regeneration such as sarcopenia, muscular dystrophies, and rotator cuff tears [9–13]. Genetic experiments in mice with depletion of $Pax7^+$ MuSCs have demonstrated that MuSC failure and impaired regeneration directly causes ectopic muscle adipogenesis and fibrosis [14,15], and other studies have associated unsuccessful or delayed muscle regeneration with the development of muscle ectopic fat and fibrosis [16–18]. Fibro/adipogenic progenitors (FAPs) are specific progenitor cells residing in the stem cell niche that are distinct from MuSCs and can differentiate to adipocytes in vitro and in vivo [3,19,20]. In response to injury, uncommitted FAPs activate and proliferate to support the myogenic capacity of MuSCs [19–22]. The fate of FAPs is then precisely regulated during regeneration, in particular through a cross-talk with immune cells where tumor necrosis factor α (TNFα) initially promotes FAP amplification and survival during MuSC expansion, and transforming growth factor β1 (TGF-β1) subsequently induces FAP apoptosis in later phases of regeneration to limit uncontrolled expansion of FAPs [23].

We and others have shown that various models of muscle injury induced by intra-muscular injection of cardiotoxin (CTX) or glycerol lead to ectopic adipogenesis in muscle with accumulation of fat containing cells between muscle fibers [24–27]. Under physiological conditions, this response is transient and cleared during the terminal phases of regeneration. Given this observation, we hypothesized that ectopic adipogenesis could be important for efficient regeneration and set-out to test this possibility by interfering with adipogenesis through genetic modulation of the peroxisome proliferator-activated receptor γ (PPARγ). PPARγ is a nuclear receptor activated by fatty acids and eicosanoids and is the master regulator of the adipogenesis [28,29]. Pre-adipocytes cannot differentiate in the absence of PPARγ, and PPARγ activation during adipocyte differentiation induces several target genes required for cell cycle exit, adipogenic commitment, and triglyceride metabolism and storage in lipid droplets [30–32]. During skeletal muscle regeneration, PPARγ is strongly upregulated three days after injury [24], at the time where FAPs expand to support myogenesis. PPARγ is also expressed in macrophages where it indirectly modulates MuSC function and regeneration via a GDF-3 mediated crosstalk in the muscle stem cell niche [33]. Total invalidation of PPARγ in mice is lethal, and genetic studies in mice have largely focused on tissue-specific conditional knock-outs. Deletion of PPARγ from adipose tissue caused lipodystrophy and confirmed the prominent role of PPARγ in adipocyte differentiation and metabolism, while PPARγ deletion from skeletal muscle caused insulin resistance [34,35]. PPARγ is also important in macrophages where it controls inflammation and foam cell formation during atherosclerosis [36–38]. Recent work has demonstrated that the lethality of PPARγ knock-out mice results from placental defects that can be overcome by using an epiblast-specific Cre deleter which removes PPARγ in the entire embryo while maintaining its expression in the placenta during pregnancy [39,40]. In the present study, we used this genetic model of whole-body PPARγ deletion to study muscle regeneration. By analyzing histological and molecular readouts in vivo and the cross-talk between FAPs and MuSCs in vitro, we demonstrate that the loss of PPARγ totally blunts ectopic adipogenesis during muscle regeneration and alters MuSC function and regeneration.

2. Results

2.1. PPARγ Deletion Completely Abolishes Ectopic Adipogenesis During Skeletal Muscle Regeneration

Several studies have reported an upregulation of PPARγ during skeletal muscle regeneration, which precedes the infiltration of lipid-droplet positive cells [24,26]. One open question is whether the induction of PPARγ is the causal mechanism leading to fat infiltration, and whether this infiltration is triggered by the adipogenic differentiation of progenitors in mature adipocytes. In order to answer the question, we used whole-body PPARγ null mice (*Pparg*$^{\Delta/\Delta}$ mice) which were recently generated by epiblast-specific Cre deletion which deletes PPARγ in the entire embryo but not in the placenta [40]. Using this strategy, *Pparg*$^{\Delta/\Delta}$ mice are viable and totally lack mature adipocytes throughout the body [40], but have a compensatory fat accumulation in non-adipose tissues leading to increased body weight (Figure S1A). Muscle weight is also lower in *Pparg*$^{\Delta/\Delta}$ mice than in control littermates (Figure S1B), most likely because of impaired metabolic homeostasis and steatosis in these mice, as these conditions have been shown to alter insulin and anabolic signaling in other settings [41,42]. Tibialis anterior muscle of *Pparg*$^{\Delta/\Delta}$ mice and control littermates was injected intramuscularly with glycerol to cause fatty muscle degeneration as previously reported [19,20,24,26] and intramuscular fat infiltration was quantified by Oil Red O staining throughout the time course of regeneration (3, 7, and 14 days post-injury (dpi)). As expected, intramuscular fat infiltration peaked at 14 dpi in control mice. In contrast, *Pparg*$^{\Delta/\Delta}$ mice failed to induce ectopic adipogenesis as *Pparg*$^{\Delta/\Delta}$ muscles were completely devoid of any intra-muscular lipid accumulation (Figure 1A). The level of fat infiltration in *Pparg*$^{\Delta/\Delta}$ injured muscle was blunted by 98% (Figure 1B), and did not differ from non-injured contralateral muscles used as control. Thus, muscle ectopic adipogenesis, which develops transiently during regeneration, is due to bona fide adipocyte formation and fully relies on the presence of PPARγ.

To confirm the impaired adipogenic signature of *Pparg*$^{\Delta/\Delta}$ mice, we freshly isolated FAPs by flow cytometry and induced their differentiation into adipocytes ex vivo. As expected, approximately 45% of control FAPs differentiated in adipocytes after induction. In contrast, ex vivo adipogenesis was completely abolished in FAPs isolated from *Pparg*$^{\Delta/\Delta}$ muscles (reported hereafter as *Pparg*$^{\Delta/\Delta}$ FAPs) (Figure 1C,D). The loss of adipogenic fate of *Pparg*$^{\Delta/\Delta}$ FAPs did not primarily result from cell death as a large number of undifferentiated KO FAPs remained in culture after differentiation (Figure 1E) and *Pparg*$^{\Delta/\Delta}$ FAPs could efficiently expand and proliferate ex vivo. Nevertheless, we detected approximately 25% less *Pparg*$^{\Delta/\Delta}$ than control FAPs after differentiation (Figure 1E), likely because PPARγ is known to regulate pre-adipocyte clonal expansion, a process where adipogenic cells undergo a final cell division to exit the cell cycle and trigger adipogenic commitment and differentiation [43]. Altogether, our results confirmed the role of PPARγ as a master regulator of adipogenesis, and extended this observation to the context of intra-muscular adipogenesis. The complete absence of adipogenesis and ectopic fat infiltration in *Pparg*$^{\Delta/\Delta}$ muscle prompted us to use this model to investigate the influence of ectopic adipogenesis on muscle stem cell function, myogenesis, and muscle regeneration.

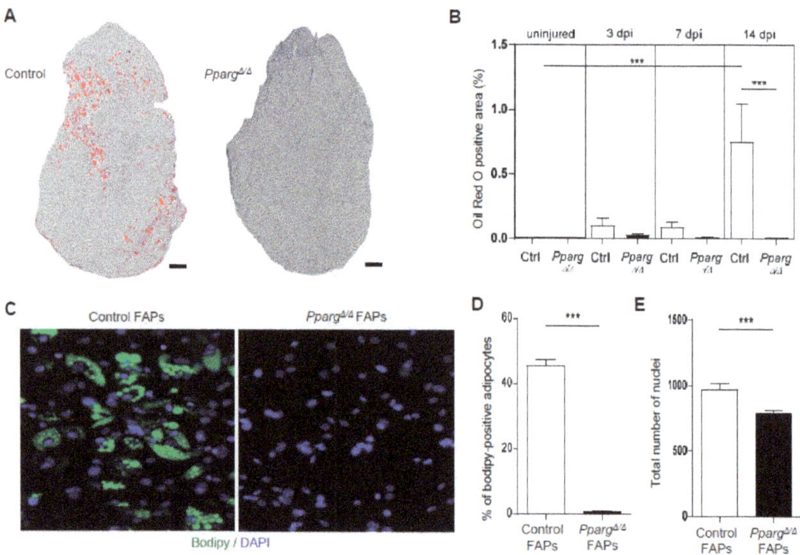

Figure 1. Ectopic adipogenesis during muscle regeneration is abolished in *Pparg*$^{\Delta/\Delta}$ mice. Tibialis anterior muscles from *Pparg*$^{\Delta/\Delta}$ and control littermates mice were injured with 50% glycerol intramuscular injection and collected at 3, 7, and 14 days post injury (dpi), $n \geq 7$ mice per group. (**A**) Representative image of tibialis anterior (TA) section of control (Ctl) and *Pparg*$^{\Delta/\Delta}$ (KO) mice stained by Oil Red O at 14 dpi. Scale bars: 200 μm. (**B**) Quantification of the area covered by Oil Red O in uninjured muscle and at 3, 7, and 14 dpi in control and *Pparg*$^{\Delta/\Delta}$. (**C–E**) FAPs were isolated from control and *Pparg*$^{\Delta/\Delta}$ muscle by flow cytometry and cultured in adipogenic conditions, with $n \geq 24$ cell culture replicates pooled from at least three mice per genotype and replicated twice on different days. (**C**) Representative staining of lipid droplets by bodipy (green) and DNA (blue). Magnification: 20× objective. (**D**) Percentage of differentiated Bodipy-positive FAPs quantified by high content imaging. (**E**) Total number of FAPs quantified by high content imaging via the DAPI staining. *** $p < 0.001$ by one-way analysis of variance (ANOVA) followed by Bonferroni's test (**A**) and by Mann–Whitney test (**D**,**E**).

2.2. Loss of PPARγ Impairs Muscle Stem Cell Function and Cross-Talks with Myogenesis In Vivo

To interrogate whether MuSC function and muscle repair are affected in *Pparg*$^{\Delta/\Delta}$ mice, we first quantified the expression of Pax7 and MyoD, which are expressed by regenerating cells, during the time course of muscle repair. As expected, we observed a strong upregulation of Pax7 and MyoD at 3 dpi, as a consequence of MuSC activation in response to injury (Figure 2A). Interestingly, the induction of Pax7 and MyoD was impaired in *Pparg*$^{\Delta/\Delta}$ mice during the time course of regeneration. Similar analyses in contra-lateral (CL) muscles demonstrated that these myogenic markers were not deregulated in non-injured *Pparg*$^{\Delta/\Delta}$ muscles, demonstrating that lack of PPARγ specifically impairs the myogenic activity of MuSCs after their activation in response to injury. In order to confirm the evidence obtained at a molecular level, we next analyzed the in vivo activity of MuSCs by immunofluorescence against Pax7. The total number of Pax7-positive MuSCs was lower in *Pparg*$^{\Delta/\Delta}$ than control muscle after injury (Figure 2B,C), confirming that *Pparg*$^{\Delta/\Delta}$ mice fail to efficiently amplify myogenic precursors during muscle regeneration. Altered amplification of *Pparg*$^{\Delta/\Delta}$ MuSCs resulted from impaired proliferation of MuSCs as the number of Pax7$^+$/Ki-67$^+$ proliferating MuSCs was significantly reduced by approximately 45% in *Pparg*$^{\Delta/\Delta}$ muscle (Figure 2B–D).

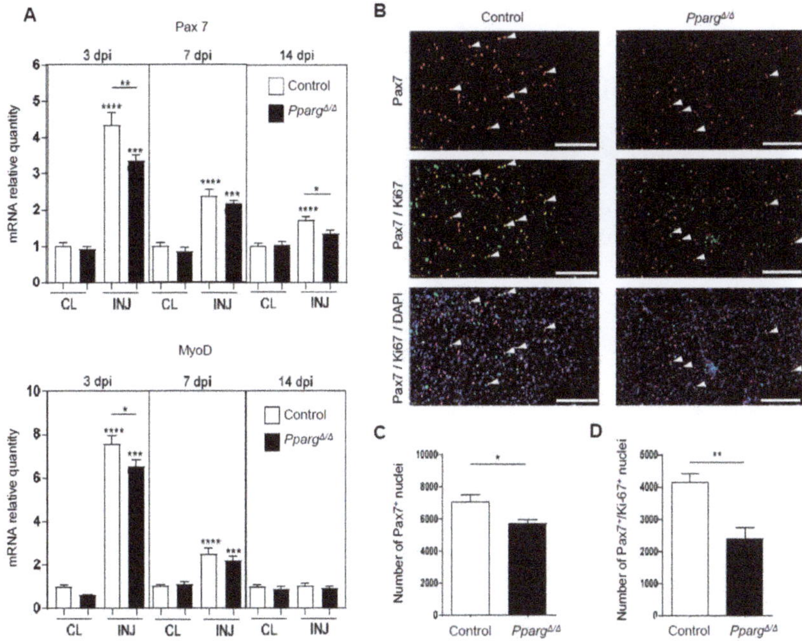

Figure 2. MuSC amplification and commitment are impaired in *Pparg*$^{\Delta/\Delta}$ mice. Tibialis anterior (TA) muscles from *Pparg*$^{\Delta/\Delta}$ and control littermates mice were injured with 50% glycerol intramuscular injection and collected at 3, 7, and 14 days post injury (dpi). (**A**) Pax7 and MoyD mRNA quantification by qPCR in injured (INJ) and non-injured contralateral (CL) muscles. Data are normalized to CL muscles of control mice per timepoint, $n \geq 7$ mice per group. (**B**) Representative images of TA sections stained for Pax7, Ki-67, and DAPI at 7dpi. White arrows indicate Pax7$^+$/Ki-67$^+$ nuclei. Scale bar: 200 μm. (**C**,**D**) In vivo quantification of total Pax7$^+$ nuclei (**C**) and proliferating Pax7$^+$/Ki-67$^+$ nuclei (**D**) in injured muscles at 7dpi, with n = 4–5 mice per group. **** $p < 0.0001$, *** $p = 0.001$, ** $p = 0.01$, * $p < 0.05$ by one-way ANOVA (**A**) or Mann–Whitney test (**C**,**D**). Unless otherwise indicated, statistical comparisons relate to the CL muscle of control mice.

To investigate how the MuSC phenotype of *Pparg*$^{\Delta/\Delta}$ mice influences myogenesis and muscle repair, we quantified the amount of regenerating fibers with centralized nuclei as well as well their cross-sectional area during the time course of muscle regeneration. Loss of PPARγ did not affect the size of the injured area as the percentage of fibers with centralized nuclei was equal in control and *Pparg*$^{\Delta/\Delta}$ mice during the early steps of regeneration (Figure 3B). However, the muscle of *Pparg*$^{\Delta/\Delta}$ mice failed to efficiently regenerate as they had more fibers with centralized nuclei at late stages of regeneration (Figure 3A,B), indicating that their recovery to fully matured fibers with peripherally located nuclei was delayed. At 7dpi when MuSCs start to fuse with damaged or newly formed fibers, fiber size distribution was significantly impaired in *Pparg*$^{\Delta/\Delta}$ mice in a bimodal fashion (Figure 3C; gray arrows). The smallest fibers, which encompass the newly formed fibers through de novo MuSC fusion, were smaller in *Pparg*$^{\Delta/\Delta}$ mice, indicating that the defective MuSC amplification in *Pparg*$^{\Delta/\Delta}$ mice translates to perturbed myogenesis and altered MuSC fusion. The largest fibers, which likely correspond to the repair of pre-existing fibers that were only mildly damaged were larger in *Pparg*$^{\Delta/\Delta}$ mice and this phenotype persisted at later time points of regeneration (Figure 3C), likely indicating that PPARγ can influence myofiber growth through additional mechanisms which are MuSC-independent. Altogether, these results demonstrate that loss of PPARγ impairs MuSC expansion and myogenesis in vivo and cross-talks with muscle repair.

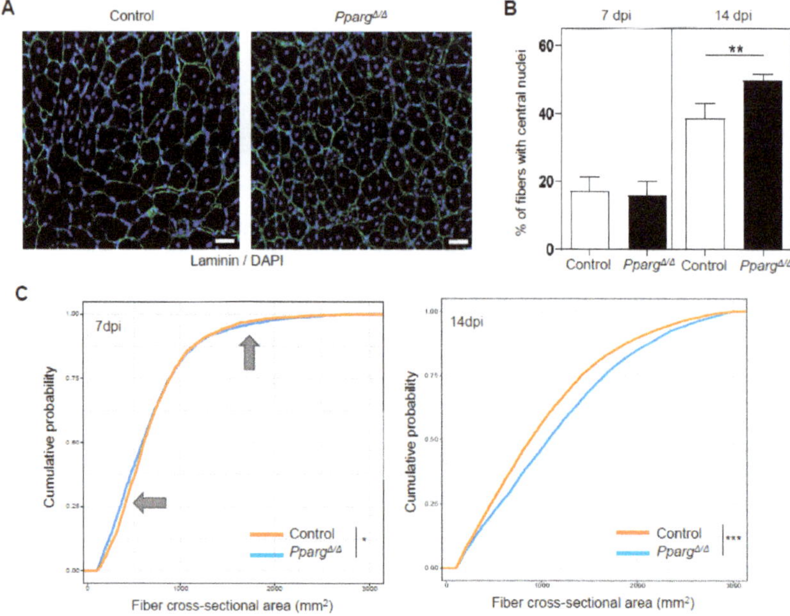

Figure 3. Muscle regeneration is impaired in *Pparg*$^{\Delta/\Delta}$ mice. (**A**) Representative images of *Pparg*$^{\Delta/\Delta}$ and control muscle sections stained with Laminin and DAPI at 14dpi. Scale bar: 50 µm. (**B**) Quantification of the percentage of fibers possessing centralized nuclei at 7 and 14dpi, with n=7 mice per group. (**C**) Cumulative distribution of cross sectional area of regenerating fibers with centralized nuclei in regenerating muscles of *Pparg*$^{\Delta/\Delta}$ and control littermates mice at 7dpi (left panel) and 14dpi (right panel). Gray arrows depict the bimodal differences of *Pparg*$^{\Delta/\Delta}$ and control cross sectional area distributions. $n = 7$ mice per group. *** $p = 0.001$, ** $p = 0.01$, * $p < 0.05$ by Mann–Whitney test (**B**) or Kolmogorov–Smirnov test (**C**).

2.3. *PPARγ KO Does Not Directly Impair the Intrinsic Function of MuSC Function and the FAP/MuSC Cross-Talk Ex Vivo*

Since muscle regeneration is regulated by an integrated cooperation between many residing and recruited cell types in the MuSC niche, we asked whether the in vivo MuSC phenotype of *Pparg*$^{\Delta/\Delta}$ mice is either due to a direct intrinsic role of PPARγ in MuSCs, or indirectly influenced by a PPARγ-dependent paracrine signal from other cell types of the MuSC niche. As a first step, PPARγ expression in freshly isolated MuSCs and FAPs was compared to its expression in white adipose tissue (Figure 4A). PPARγ expression was detected in MuSCs, but at lower levels than in FAPs and much lower than in adipose tissue (Figure 4A), thus making it unlikely that PPARγ could drive MuSC fate in a cell-autonomous fashion. This was confirmed by demonstrating that the lower activation and proliferation of MuSCs detected in vivo in *Pparg*$^{\Delta/\Delta}$ mice was not reproduced in an ex vivo assay with Edu incorporation 1.5 and 3 days after MuSCs isolation from control and *Pparg*$^{\Delta/\Delta}$ mice (Figure 4B,C). Surprisingly, the ex vivo activation during the first cellular division after exit of quiescence was even greater in *Pparg*$^{\Delta/\Delta}$ cells than in control cells (Figure 4B). Given the low expression of PPARγ in MuSCs and the fact that the proliferation of *Pparg*$^{\Delta/\Delta}$ MuSCs was normal and steady after 3 days of culture, higher ex vivo activation of *Pparg*$^{\Delta/\Delta}$ MuSCs is likely caused by indirect mechanisms. This was further confirmed by assessing the terminal differentiation of *Pparg*$^{\Delta/\Delta}$ MuSCs into myofibers after 6 days of ex vivo culture. Neither the number of differentiated myosin-heavy chain (MHC)-positive cells nor the fusion index of nuclei in multi-nucleated myofibers differed between *Pparg*$^{\Delta/\Delta}$ and control MuSCs (Figure 4D).

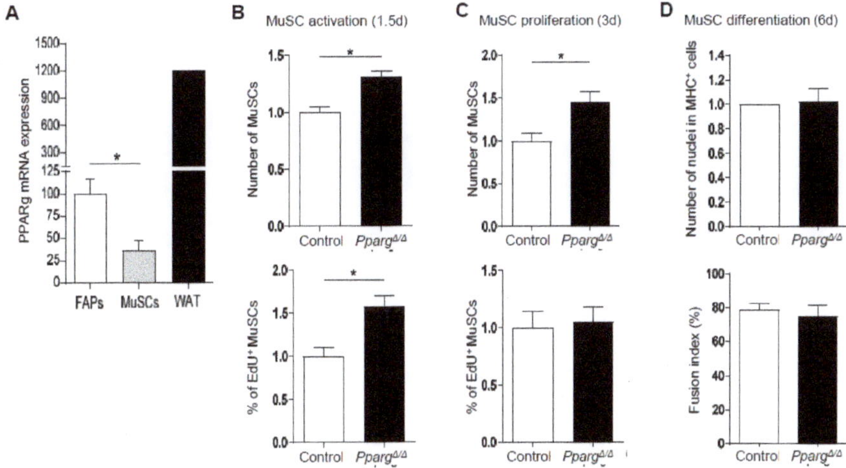

Figure 4. *Pparg*$^{\Delta/\Delta}$ MuSCs activate faster than control MuSCs ex vivo. (**A**) PPARγ mRNA quantified by qPCR in freshly isolated MuSCs, FAPs and mouse white adipose tissue (WAT). n = 4 mice (**B**,**C**) Relative number (top panel) and percentage of EdU$^+$ (lower panel) MuSCs cultured for 36 h (**B**) or 72 h (**C**) after FACS isolation. (**D**) Myogenic differentiation of *Pparg*$^{\Delta/\Delta}$ and control MuSCs cultured for six days after FACS isolation. Relative quantification of the total number of nuclei having fused in MHC + myofibers is shown in the top panel, and fusion index (% nuclei in MHC + myofibers) is shown in the lower panel. (**B**,**D**) $n \geq 24$ cell culture replicates pooled from at least three mice per genotype and replicated twice on different days. * $p < 0.05$; by Mann–Whitney test.

We then tested whether PPARγ could indirectly regulate MuSC function by regulating the cellular cross-talk between MuSCs and other PPARγ-expressing accessory cells in the muscle stem cell niche. FAPs appeared as good candidates as they support MuSC activity as progenitors [19] and possess an adipogenic fate. We first evaluated how loss of PPARγ influences FAP expansion ex vivo using total cell counts and EdU incorporation to reveal activated and proliferating cells 2 and 6 days after isolation, respectively. After 2 days in culture, we did not detect any difference in the total FAP number, highlighting that both control and *Pparg*$^{\Delta/\Delta}$ FAPs had similar adhesion and survival (Figure 5A). Cell cycle entry assessed by Edu incorporation was faster in *Pparg*$^{\Delta/\Delta}$ FAPs 2 days post-isolation (Figure 5A), but did not result in more FAPs or higher proliferation 6 days post-isolation (Figure 5B). Thus, while PPARγ is essential for FAP differentiation to adipocytes, it does not play an intrinsic role in the regulation of FAP expansion. We next interrogated the role of PPARγ in the cross talk between FAPs and MuSCs through direct ex vivo co-cultures. As expected, co-culture of MuSCs with control FAPs was beneficial for their myogenic differentiation in myosin-heavy chain positive myotubes (Figure 5C). However, *Pparg*$^{\Delta/\Delta}$ FAPs also supported myogenic differentiation of MuSCs to a similar extent (Figure 5C). These results therefore revealed that in their undifferentiated precursor state, FAPs efficiently sustain the myogenic differentiation of MuSCs through PPARγ-independent mechanisms.

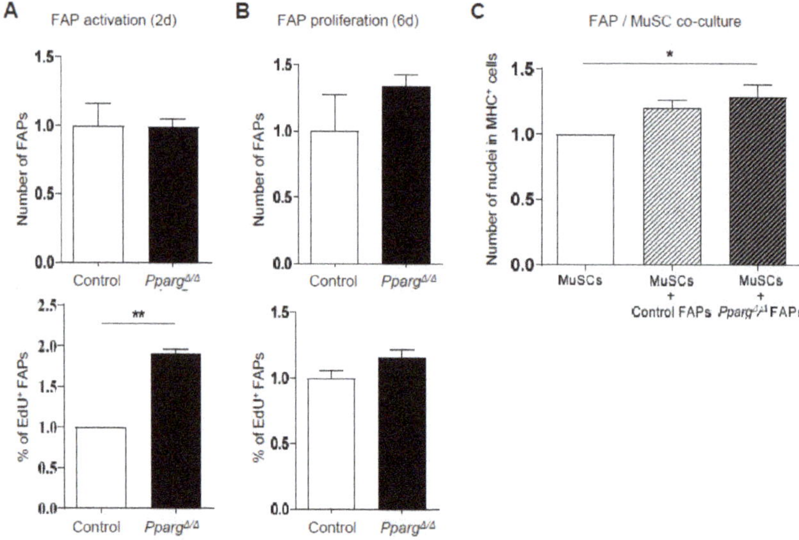

Figure 5. PPARγ deficiency does not affect the myogenic support of undifferentiated FAPs to MuSCs. (**A**,**B**) Relative number (top panels) and percentage of Edu-positive FAPs from $Pparg^{\Delta/\Delta}$ and control littermates mice cultured for 48 h (**A**) and six days after isolation (**B**,**C**) Relative number of differentiated MHC⁺ MuSCs, when cultured alone or co-cultured with $Pparg^{\Delta/\Delta}$ and control FAPs. Data are normalized to MuSCs monoculture. $n \geq 24$ cell culture replicates pooled from at least three mice per genotype and replicated twice on different days ** $p < 0.01$, * $p < 0.05$ by Mann–Whitney test.

2.4. Altered Inflammatory Signatures in PPARγ-KO Mice During Muscle Regeneration

Our results demonstrating that the intrinsic loss of PPARγ in MuSCs and FAPs does not directly cause the MuSC activation and regeneration defects of $Pparg^{\Delta/\Delta}$ mice prompted us to evaluate for a role of PPARγ in other cell types that remodel the stem cell niche during regeneration. In particular, the fact that MuSCs are normal in uninjured muscle of $Pparg^{\Delta/\Delta}$ mice and only affected upon muscle regeneration in a dynamic fashion during the early steps of muscle repair suggested a potential role of PPARγ in a cell type recruited to damaged muscle after injury. We thus quantified the expression level of macrophage surface markers and chemokines in control and $Pparg^{\Delta/\Delta}$ mice. General inflammatory markers such as F4/80, CD11b, and CD11c were differentially regulated in injured muscles of $Pparg^{\Delta/\Delta}$ *vs* control mice at various time points (Figure 6A), while the induction of IL-1, TNF and IL-6 mRNA in response to injury did not change significantly at 3dpi between $Pparg^{\Delta/\Delta}$ and control muscles (Figure 6B). The analysis of anti-inflammatory macrophage cell surface markers revealed a defect in regenerating muscle of $Pparg^{\Delta/\Delta}$ mice, as deletion of PPARγ severely blunted the induction of the mannose receptor (MR) and the macrophage scavenger receptor (MRS1) at 3dpi (Figure 6C), when the regenerating niche shifts to anti-inflammatory conditions to support myogenic commitment. Resolution of the IL-6 inflammatory cytokine production at 7dpi was also altered in $Pparg^{\Delta/\Delta}$ (Figure 6B). These observations highlight that PPARγ influences the temporal profiles of the inflammatory response during muscle regeneration. Altogether, our study demonstrates that whole body loss of PPARγ influences muscle regeneration by acting on MuSC function indirectly through various complementary mechanisms in various cell types of the stem cell niche.

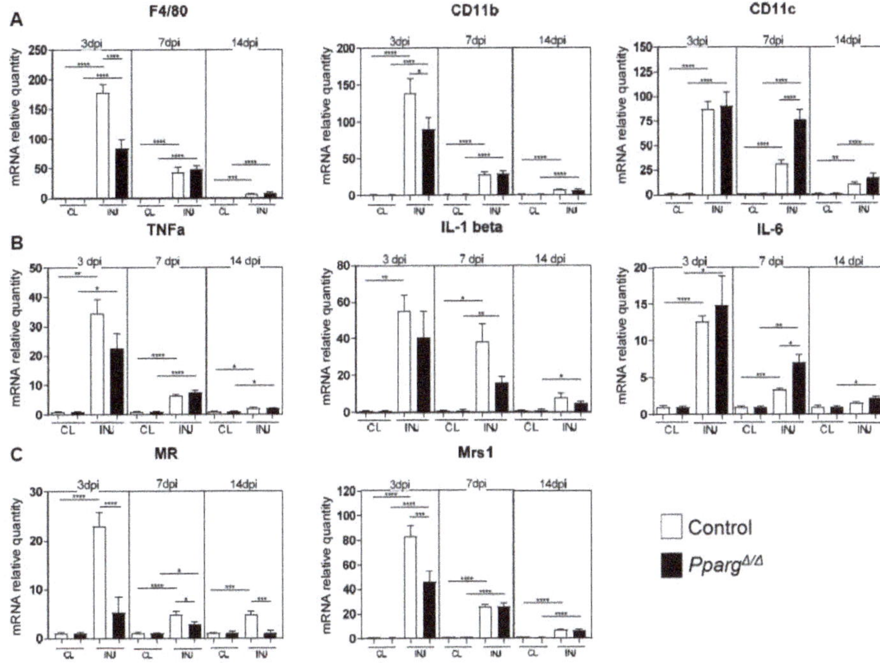

Figure 6. The inflammatory response during muscle regeneration is altered in *Pparg*$^{\Delta/\Delta}$ mice. mRNA quantification by qPCR of immune cell marker in injured (INJ) and non-injured contra-lateral (CL) muscles at 3, 7 and 14 dpi. Results are normalized to the non-injured CL muscles. (**A**) Cellular surface macrophages markers: *F4/80* = *EMR1* (EGF-like module-containing mucin-like hormone receptor-like 1), *CD11b* = Integrin alpha M, *CD11c* = Integrin alpha X. (**B**) Cytokines: *TNFα* = Tumor necrosis factor alpha, *IL-1 β* = Interleukin 1 beta, *IL-6* = Interleukin 6. (**C**) Anti-inflammatory cellular macrophages markers: *MR* = mannose receptor, *Msr1* = Macrophage scavenger receptor 1. $n \geq 7$ mice. **** $p < 0.0001$, *** $p < 0.001$, ** $p < 0.01$ * $p < 0.05$ by one-way analysis of variance (ANOVA). Unless otherwise indicated, statistical comparisons relate to the CL condition of the same genotype.

3. Discussion

Ectopic infiltration of adipocytes in skeletal muscle has been widely associated with altered muscle function, mainly because adipose infiltration is observed in many muscle degenerative conditions such as dystrophies or rotator cuff tears [9–12]. In contrast, recent studies have suggested that muscle adipogenesis may actually be a physiological process required for muscle plasticity, which gets deregulated and over-amplified during pathology. First, ectopic muscle adipogenesis is detected in several models of efficient muscle regeneration [24–27], and this response is transient and cleared during the terminal phases of muscle repair when myofibers return to homeostasis. In addition, the FAP lineage can both support myogenesis and differentiate to intra-muscular adipocytes, with different states of permissivity according to the patho-physiological status [19,20]. Finally, muscle adipogenesis is blunted in response to unloading and lack of muscle contraction [25]. In the present study, we have demonstrated that the ectopic accumulation of fat during muscle regeneration fully relies on the adipogenic transcription factor PPARγ. *Pparg*$^{\Delta/\Delta}$ mice are totally devoid of intra-muscular adipocytes after muscle injury and *Pparg*$^{\Delta/\Delta}$ FAPs completely lose the ability to differentiate to adipocytes. On top of confirming the dominant role of PPARγ on adipogenesis [40] in skeletal muscle, this result also highlights that muscle ectopic fat formation after injury is an active cellular process tightly regulated at the molecular level, and not just the aspecific accumulation of cellular lipids from damaged myofibers.

Interestingly, our results consistently demonstrate that the absence of PPARγ and ectopic adipocyte formation alter the activation and myogenic commitment of MuSCs in vivo. The induction of Pax7 and MyoD as well as the proliferation of Pax7-positive MuSCs are impaired in the regenerating muscle of *Pparg*$^{\Delta/\Delta}$ mice. PPARβ/δ another member of the PPAR nuclear receptor family, has previously been demonstrated to control MuSC function and muscle regeneration [44–46], most likely through a direct effect on the myogenic lineage as Myf5-Cre conditional deletion of PPARγ is sufficient to alter MuSCs and regeneration [44]. In contrast, the alterations of muscle regeneration in *Pparg*$^{\Delta/\Delta}$ mice in vivo did not result from cell autonomous defects linked to the absence of PPARγ in MuSCs as PPARγ expression is low in MuSCs and the activation and proliferation of *Pparg*$^{\Delta/\Delta}$ MuSCs was not impaired ex vivo. Thus, the perturbed regenerative response of *Pparg*$^{\Delta/\Delta}$ mice support the notion that a functional adipogenic response is required to efficiently sustain myogenesis during physiological muscle healing. In particular, we can hypothesize that cellular and membrane lipids from damaged myofibers are released in the muscle stem niche a few days after injury and need to be stored intracellularly to avoid damage to regenerating fibers. Interestingly, the adipogenic lineage has also been demonstrated to support self-renewal and regeneration in other organs. For example, adipocyte precursors have been involved in skin epithelial stem cell activation in the hair follicle [47,48], and the adipogenic lineage was shown to recruit fibroblasts during skin wound healing [49]. A cross-talk between adipose-derived stem cells and chondrocytes has also recently been found in vitro, potentially revealing similar cross-communications during cartilage repair [50].

Our study has analyzed muscle regeneration in a mouse model where PPARγ is deleted in all cells and organs [39,40]. At the undifferentiated progenitor level, we demonstrated that control and *Pparg*$^{\Delta/\Delta}$ FAPs had a similar ability to support myogenesis. We also further showed that the function of isolated *Pparg*$^{\Delta/\Delta}$ MuSCs is not lower ex vivo. Surprisingly, *Pparg*$^{\Delta/\Delta}$ MuSCs even activated faster than control cells ex vivo, possibly because the lipodystrophy and metabolic phenotype of *Pparg*$^{\Delta/\Delta}$ mice alter MuSC quiescence in vivo and trigger earlier activation ex vivo. Given these observations, it is therefore likely that impaired in vivo MuSC function in *Pparg*$^{\Delta/\Delta}$ mice is indirect, through a cross-talk with other cell types of the muscle stem cell niche or via systemic signals from other tissues. Along this line, two mechanisms most likely cross-talk during regeneration. First, the absence of adipogenic differentiation of *Pparg*$^{\Delta/\Delta}$ FAPs can contribute to the in vivo phenotype by controlling adipokine signals during adipogenic and myogenic differentiation. Second, the immune signature is altered in the regenerating niche of *Pparg*$^{\Delta/\Delta}$ mice, where anti-inflammatory macrophage markers fail to get efficiently induced. Recent work has demonstrated that PPARγ in macrophages regulates myogenesis and muscle repair via secreted paracrine signals [33]. Our results are consistent with the indirect modulation of myogenesis via PPARγ in immune cells and fully supports the model where macrophage PPARγ cross-talks with MuSCs through secreted cytokines. Interestingly, the fate of FAPs is tightly controlled by inflammatory cytokines, which dynamically influence their expansion, differentiation and apoptosis during the resolution of inflammation and the different phases of muscle repair [23,51–53]. Thus, PPARγ most likely influences MuSCs and myogenesis indirectly via a complex cross-talk between macrophages and FAPs.

Altogether, our work has demonstrated that PPARγ is required for ectopic adipocyte infiltration during muscle regeneration and alters MuSC proliferation and myogenic commitment via indirect cellular cross-talks in the stem cell niche. Genetic PPARγ invalidation in macrophages and bone marrow transplant from PPARγ KO mice have already demonstrated that macrophages are required for this cross-talk [33]. Further studies with cell-type specific invalidation of PPARγ, in particular in FAPs, will be important to further refine the roles of the distinct cellular compartments of the niche and the interactions between adipogenesis and myogenesis during muscle regeneration and muscle diseases.

4. Materials and Methods

4.1. Animals

Animals were housed under standard conditions and allowed to access to food and water ad libitum. Since the whole-body *Pparg* null mice die during embryonic development as a result of placental defects, we used mice obtained through an epiblast-specific Cre recombinase expression ($Sox2\text{-}Cre^{tg/+}$:$Pparg^{\Delta/em\Delta}$ mice, hereafter called $Pparg^{\Delta/\Delta}$ mice) as it has been demonstrated that PPARγ is necessary for placental function but not for embryo development [39]. The generation and breeding of mice for these experiments was performed as previously described [40]. Briefly, $Sox2\text{-}Cre^{tg/+}$ male mice (Jackson Laboratory, Bar Harbor, ME, USA) were mated with female $Pparg^{\Delta/+}$ mice [54] to obtain male $Sox2\text{-}Cre^{tg/+}$:$Pparg^{\Delta/+}$ mice that were next mated with $Pparg^{fl/fl}$ females to obtain $Sox2\text{-}Cre^{tg/+}$:$Pparg^{\Delta/em\Delta}$ mice ($Pparg^{\Delta/\Delta}$) and $Sox2\text{-}Cre^{+/+}$:$Pparg^{fl/+}$ (control). All experimental animals were breeding littermates of a mixed C57BL6/SV129 genetic background, and control animals were $PPARg^{fl/+}$ that do not express $Sox2^{Cre}$. Only female mice were used in this study since adult male $Pparg^{\Delta/\Delta}$ mice present a higher mortality. For muscle regeneration experiments, 11–15 week old mice were anaesthetized with 2% isoflurane and 50 uL of 50% v/v glycerol was injected into the tibialis anterior (TA) muscle. Mice received buprenorphine (Temgesic) 0.1 mg/kd/day i.p. as analgesic during the 48h following glycerol injection. Mice were sacrificed by CO_2 inhalation followed by cervical dislocation 3, 7, 14 days post-injury (dpi), and regenerating and contra-lateral TA muscles were cut in half and snap-frozen in liquid nitrogen for molecular analyses or frozen in liquid nitrogen cooled isopentane for histology. All in vivo experiments were performed following the regulations of the Swiss Animal Experimentation Ordinance and approved by the ethical committee of the canton de Vaud under license VD2818, 01 Mar 2014.

4.2. Muscle Progenitor Cell Isolation

For the isolation of muscle stem cells, uninjured hindlimb muscles representing various fiber types (gastrocnemius/soleus/plantaris complex, tibialis anterior/EDL complex and quadriceps) were collected and from $Pparg^{\Delta/\Delta}$ mice and their control littermates and freshly digested with dispase II (2.5 U/mL) (Roche, Basel, Switzerland), collagenase B (0.2%) (Roche), and $MgCl_2$ (5 mM). Cells were then incubated at 4 °C for 30 min with fluorescently-coupled antibodies against CD45 (Invitrogen, Carlsbad, CA, USA) MCD4501 or MCD4528; dilution for both 1/25), CD31 (Invitrogen, RM5201 or RM5228; dilution for both 1/25), CD11b (Invitrogen, RM2801 or RM2828; dilution for both 1/25), CD34 (BD Biosciences, 560230 or 560238; dilution for both 1/60), Ly-6A–Ly-6E (Sca1) (BD Biosciences, Franklin Lakes, NJ, USA) 561021; dilution 1/150), α7-integrin (R&D, FAB3518N; dilution 1/30) and CD140a (eBioscience, San Diego, CA, USA, 12-1401-81 or 17-1401-81; dilution for both 1/30). Cell isolation was performed on a Beckman–Coulter Astrios Cell sorter; Beckman-Coulter, Brea, CA, USA. MuSCs isolated by flow-cytometry were $CD45^-CD31^-CD11b^-Sca1^-CD34^+Integrin\alpha7^+$; fibro/adipogenic progenitors (FAPs) were $CD45^-CD31^-CD11b^-Sca1^+CD34^+PDGFR\alpha^+$.

4.3. Cell Culture

MuSCs and FAPs were sorted in 96 well plates at a density of 600 cells per well. Freshly sorted cells were grown in growth medium: 20 mM glucose DMEM, 20% heat-inactivated FBS, 10% inactivated horse serum, 2.5 ng/mL bFGF (Invitrogen), 1% P/S + 1% L+-Glutamine, 1% Na-pyruvate (Invitrogen). For MuSC activation 1µM EdU (5-Ethynyl-2′-deoxyuridine) was added to the growth medium after cell sorting. Cells were then fixed 1.5 days after with 4% paraformaldehyde (PFA) for 15 min. For MuSC proliferation, 1 µM EdU was added to the growth medium 3 days after cell sorting, for 2.5 h. Cells were then fixed 15 min in 4% PFA. MuSC differentiation was induced after 4 days in culture by switching to myogenic differentiation medium (20 mM glucose DMEM, 5% inactivated horse serum, 1% P/S). Cells were cultured in these conditions for two days and then fixed in 4% PFA for 5 min. For FAP activation 1 µM EdU was added to the growth medium after cell sorting. Cells were then fixed two days after with 4% PFA for 15 min. For FAP proliferation, 1 μM EdU was added six days after cell sorting. Cells were then incubated for 6 h and fixed in 4% PFA for 15 min. For FAP differentiation, after six days of culture in growth medium, the medium was replaced by adipogenic differentiation medium for an additional seven days (20 mM glucose DMEM, 20% heat-inactivated FBS, 1% P/S, 0.25 μM dexamethasone, 1 μg/mL insulin, 5 μM troglitazone, 0.5 mM isobutylmethylxanthine). For MuCS and FAP co-cultures, we cultured PPARγ-KO MuSCs alone, or with WT or PPARγ-KO FAPs.

4.4. *Immunocytochemistry for In Vitro Studies*

EdU incorporation was revealed by the Click-iT assay (Molecular Probes, Eugene, OR, USA) according to manufacturer's instructions. After fixation, cells were permeabilized during 20 min in PBTX 0.5%, stained with the Click-iT reaction mix, and counterstained with DAPI. For myosin heavy vhain (MHC) staining, after fixation, cells were permeabilized using EtOH/MetOH (*v/v*) for 5 min, incubated for 1 h with the primary antibody anti-MHC 1/200 (Millipore clone A4.1025) in PBS, 1% Horse Serum at room temperature, and incubated during 30 min with the secondary antibody Alexa488 anti-mouse IgG diluted at 1/1000 (Life Tech (Carlsbad, CA, USA) A-10680) and Hoeschst33342 in PBS, 1% horse serum at room temperature. Fusion index was determined as the percentage of nuclei located within MHC positive fibers. Image acquisition was performed using the ImageXpress (Molecular Devices, San Jose, CA, USA) platform. Quantifications were done using the MetaXpress software (Molecular Devices).

4.5. *Immunohistochemistry for In Vivo Studies*

TA muscles were frozen in isopentane and cooled with liquid nitrogen. Sections of 10 μm were obtained after sectioning on a cryostat. For Pax7-Ki67 staining, sections were allowed to dry for 10 min and then fixed 10 min in 4% PFA. Tissue sections were permeabilized by applying cold methanol (−20 °C) for 6 min. For the antigen revealing, slides were immersed in a glass container filled with hot acid citric (0.01 M), and then in boiling water bath for 5 min, repeated twice. Slides were then blocked 3 h in blocking solution (IgG-free Bovine Serum Albumin (BSA) 4% (Jackson #001-000-162), followed by 30 min incubation with Goat-anti-mouse FAB (Jackson #115-007-003) diluted 1/50 in PBS. Rabbit anti-Ki67 polyclonal (Abcam, Cambridge, United Kingdom, #ab15580) was used (1/200) and chicken anti-laminin (Life-Span Bioscience, Seattle, WA, USA, #LS-C96142) antibodies were incubated 3 h in blocking solution in1/200 and 1/300 dilution, respectively. Mouse anti-Pax7 (DSHB, purified) IgG1 (2.5 μg/mL) was then incubated overnight at 4 °C. Secondary antibodies (Alexa488-Goat anti-rabbit (1/1000) and Alexa647-Goat anti-chicken (1/1000) were incubated for 1 h in blocking solution. Pax7 signal was further amplified using a goat-anti mouse IgM1-biotin (Jackson), followed by a streptavidin (1/2000) treatment, together with Hoechst.

For Oil-Red-O staining slides were air-dried and then incubated in 50% ethanol during 30 min, followed by 15 min incubation in 2.5 g/L oil-red-O solution in 70% ethanol and 1 min washes in 50% ethanol and water. Finally, slides were counterstained with Mayer's hematoxylin. Stained tissues were imaged using an Olympus VS120 virtual microscopy slide scanning system and analyzed using the VS-ASW FL software measurement tools. The Pax7-Ki67 double positive nuclei (DAPI positive) were counted by randomly selecting four or more areas within the injured region. Fiber size and Oil-Red-O area analyses were performed on an automatized software developed in house.

For BODIPY staining, BODIPY 493/503 (4,4-difluro-1,3,5,7-tetramethyl-4-bora-3a,4a-diaza-s-indacene, Thermo Fisher Scientific, Waltham, MA, USA) was used according to manufacturer's instructions. Images were acquired with a Leica microscope (DMI4000B Leica widefield inverted microscope; Leica, Wetzlar, Germany), at 20× magnification.

4.6. *Quantitative PCR*

RNA was extracted from sorted cells or muscles and white adipose tissue using RNeasy Micro Kit (Qiagen, Hilden, Germany) or miRNeasy Mini Kit (Qiagen), respectively. According to the manufacturer's instruction, cells were stored in RTL buffer and frozen after the FACS sorting. RNA samples were processed by reverse transcription (high capacity cDNA reverse transcription kits, Invitrogen) using random primers

(High Capacity cDNA Reverse Transcription Kit, Applied Biosystems, Foster City, CA, USA. All of the quantitative PCRs were performed using the SybR Green I master kit (Roche) on a LightCycler 480. For sorted cells and adipose tissue, reference gene HPRT was selected. PPARγ primers were designed on the gene region deleted in the KO mice:

Forward: AAGAGCTGACCCAATGGTTG, reverse: GCATCCTTCACAAGCATGAA. For muscles, three reference genes (ATP5b, EIF2a, and PSMB4) were selected from previous micro-array data on the base of their stability through the different time points of regeneration.

qPCR probes were designed as follows:

ATP5b forward: ACCTCGGTGCAGGCTATCTA

ATP5b reverse: AATAGCCCGGGACAACACAG;

EIF2a forward: CACGGTGCTTCCCAGAGAAT

EIF2a reverse: TGCAGTAGTCCCTTGTTAGCG;

PSMB4 forward: GCGAGTCAACGACAGCACTA

PSMB4 reverse: TCATCAATCACCATCTGGCCG;

Pax7 forward: AAGTTCGGGAAGAAAGAGGACGAC

Pax7 reverse: GAGGTCGGGTTCTGATTCCACATC;

MyoD forward: GCAGATGCACCACCAGAGTC

MyoD reverse: GCACCTGATAAATCGCATTGG;

F4/80 forward: CTCTTCTGGGGCTTCAGTGG

F4/80 reverse: TGTCAGTGCAGGTGGCATAA;

ITGAM (CD11b) forward: GCCTGTGAAGTACGCCATCT

ITGAM (CD11b) reverse: GCCCAGGTTGTTGAACTGGT;

ITGAX (CD11c) forward: ACACTGAGTGATGCCACTGT

ITGAX (CD11c) reverse: TTCGGAGGTCACCTAGTTGGG;

Msr1 forward and reverse: [24]

MR forward: ATGCCAAGTGGGAAAATCTG

MR reverse: TGTAGCAGTGGCCTGCATAG;

TNFα forward: AGCCGATGGGTTGTACCTTG

TNFα reverse: ATAGCAAATCGGCTGACGGT;

IL1beta forward: TGCCACCTTTTGACAGTGAATGA

IL1beta reverse: TGCCTGCCTGAAGCTCTTGT;

IL-6 forward: CCAGAAACCGCTATGAAGTTCC

IL-6 reverse: TTGTCACCAGCATCAGTCCC;

4.7. Statistical Analysis

In vitro experiments were performed in at least three independent replicates from different mice, using several cell culture replicates from the same isolation. All of the data are expressed as mean value + the standard error of the mean (s.e.m.). For statistical analysis, GraphPad Prism (GraphPad Software Inc., La Jolla, CA, USA) was used. Statistical significance of two-group comparisons was assessed by a non-parametric Mann–Whitney test which does not assume normal distribution, using variance correction when required. For comparison of more than two groups, one-way ANOVA followed by Bonferroni multiple-comparison test was used.

Supplementary Materials: Supplementary materials can be found at http://www.mdpi.com/1422-0067/19/7/2044/s1.

Author Contributions: These authors have contributed as follow: Conceptualization, L.L., B.D. and J.N.F.; Methodology, S.K., L.L. and J.N.F.; Software, G.J.; Validation, C.W.; F.S.; A.P.; Investigation, G.D., S.K.; L.L.; Data Curation and Analysis, E.M.; Writing-Original Draft Preparation, G.D.; J.N.F.; Writing-Review & Editing, G.D., L.L.; F.G.; B.D.; J.N.F.; Supervision, F.G.; B.D.; J.N.F.; Project Administration, X.X.; Funding Acquisition, Y.Y.

Funding: This research received external funding from the Etat de Vaud and FNRS to B.D. and F.G.

Acknowledgments: We thank members of the CIG and NIHS communities for fruitful discussions.

Conflicts of Interest: All authors except C.W., F.G., and B.D. are full-time employees of the Nestlé Institute of Health Sciences SA.

References

1. Mauro, A. Satellite cell of skeletal muscle fibers. *J. Biophys. Biochem. Cytol.* **1961**, 9, 493–495. [CrossRef] [PubMed]
2. Seale, P.; Sabourin, L.A.; Girgis-Gabardo, A.; Mansouri, A.; Gruss, P.; Rudnicki, M.A. Pax7 is required for the specification of myogenic satellite cells. *Cell* **2000**, 102, 777–786. [CrossRef]
3. Bentzinger, C.F.; Wang, Y.X.; Dumont, N.A.; Rudnicki, M.A. Cellular dynamics in the muscle satellite cell niche. *EMBO Rep.* **2013**, 14, 1062–1072. [CrossRef] [PubMed]
4. Abmayr, S.M.; Pavlath, G.K. Myoblast fusion: Lessons from flies and mice. *Development* **2012**, 139, 641–656. [CrossRef] [PubMed]
5. Varga, T.; Mounier, R.; Horvath, A.; Cuvellier, S.; Dumont, F.; Poliska, S.; Ardjoune, H.; Juban, G.; Nagy, L.; Chazaud, B. Highly dynamic transcriptional signature of distinct macrophage subsets during sterile inflammation, resolution, and tissue repair. *J. Immunol.* **2016**, 196, 4771–4782. [CrossRef] [PubMed]
6. Saclier, M.; Cuvellier, S.; Magnan, M.; Mounier, R.; Chazaud, B. Monocyte/macrophage interactions with myogenic precursor cells during skeletal muscle regeneration. *FEBS J.* **2013**, 280, 4118–4130. [CrossRef] [PubMed]
7. Gallagher, D.; Kuznia, P.; Heshka, S.; Albu, J.; Heymsfield, S.B.; Goodpaster, B.; Visser, M.; Harris, T.B. Adipose tissue in muscle: A novel depot similar in size to visceral adipose tissue. *Am. J. Clin. Nutr.* **2005**, 81, 903–910. [CrossRef] [PubMed]
8. Vettor, R.; Milan, G.; Franzin, C.; Sanna, M.; De Coppi, P.; Rizzuto, R.; Federspil, G. The origin of intermuscular adipose tissue and its pathophysiological implications. *Am. J. Physiol. Endocrinol. Metab.* **2009**, 297, E987–E998. [CrossRef] [PubMed]
9. Gladstone, J.N.; Bishop, J.Y.; Lo, I.K.; Flatow, E.L. Fatty infiltration and atrophy of the rotator cuff do not improve after rotator cuff repair and correlate with poor functional outcome. *Am. J. Sports Med.* **2007**, 35, 719–728. [CrossRef] [PubMed]
10. Mankodi, A.; Bishop, C.A.; Auh, S.; Newbould, R.D.; Fischbeck, K.H.; Janiczek, R.L. Quantifying disease activity in fatty-infiltrated skeletal muscle by ideal-cpmg in duchenne muscular dystrophy. *Neuromuscul. Disord.* **2016**, 26, 650–658. [CrossRef] [PubMed]
11. Samagh, S.P.; Kramer, E.J.; Melkus, G.; Laron, D.; Bodendorfer, B.M.; Natsuhara, K.; Kim, H.T.; Liu, X.; Feeley, B.T. MRI quantification of fatty infiltration and muscle atrophy in a mouse model of rotator cuff tears. *J. Orthop. Res.* **2013**, 31, 421–426. [CrossRef] [PubMed]

12. Sambasivan, R.; Yao, R.; Kissenpfennig, A.; Van Wittenberghe, L.; Paldi, A.; Gayraud-Morel, B.; Guenou, H.; Malissen, B.; Tajbakhsh, S.; Galy, A. Pax7-expressing satellite cells are indispensable for adult skeletal muscle regeneration. *Development* **2011**, *138*, 3647–3656. [CrossRef] [PubMed]
13. Brioche, T.; Pagano, A.F.; Py, G.; Chopard, A. Muscle wasting and aging: Experimental models, fatty infiltrations, and prevention. *Mol. Asp. Med.* **2016**, *50*, 56–87. [CrossRef] [PubMed]
14. Von Maltzahn, J.; Jones, A.E.; Parks, R.J.; Rudnicki, M.A. Pax7 is critical for the normal function of satellite cells in adult skeletal muscle. *Proc. Natl. Acad. Sci.* **2013**, *110*, 16474–16479. [CrossRef] [PubMed]
15. Murphy, M.M.; Lawson, J.A.; Mathew, S.J.; Hutcheson, D.A.; Kardon, G. Satellite cells, connective tissue fibroblasts and their interactions are crucial for muscle regeneration. *Development* **2011**, *138*, 3625–3637. [CrossRef] [PubMed]
16. Arnold, L.; Henry, A.; Poron, F.; Baba-Amer, Y.; van Rooijen, N.; Plonquet, A.; Gherardi, R.K.; Chazaud, B. Inflammatory monocytes recruited after skeletal muscle injury switch into antiinflammatory macrophages to support myogenesis. *J. Exp. Med.* **2007**, *204*, 1057–1069. [CrossRef] [PubMed]
17. Vidal, B.; Serrano, A.L.; Tjwa, M.; Suelves, M.; Ardite, E.; De Mori, R.; Baeza-Raja, B.; Martinez de Lagran, M.; Lafuste, P.; Ruiz-Bonilla, V.; et al. Fibrinogen drives dystrophic muscle fibrosis via a tgfbeta/alternative macrophage activation pathway. *Genes Dev.* **2008**, *22*, 1747–1752. [CrossRef] [PubMed]
18. Perdiguero, E.; Sousa-Victor, P.; Ruiz-Bonilla, V.; Jardi, M.; Caelles, C.; Serrano, A.L.; Munoz-Canoves, P. P38/MKP-1-regulated AKT coordinates macrophage transitions and resolution of inflammation during tissue repair. *J. Cell Biol.* **2011**, *195*, 307–322. [CrossRef] [PubMed]
19. Joe, A.W.; Yi, L.; Natarajan, A.; Le Grand, F.; So, L.; Wang, J.; Rudnicki, M.A.; Rossi, F.M. Muscle injury activates resident fibro/adipogenic progenitors that facilitate myogenesis. *Nat. Cell Biol.* **2010**, *12*, 153–163. [CrossRef] [PubMed]
20. Uezumi, A.; Fukada, S.; Yamamoto, N.; Takeda, S.; Tsuchida, K. Mesenchymal progenitors distinct from satellite cells contribute to ectopic fat cell formation in skeletal muscle. *Nat. Cell Biol.* **2010**, *12*, 143–152. [CrossRef] [PubMed]
21. Uezumi, A.; Fukada, S.; Yamamoto, N.; Ikemoto-Uezumi, M.; Nakatani, M.; Morita, M.; Yamaguchi, A.; Yamada, H.; Nishino, I.; Hamada, Y.; et al. Identification and characterization of pdgfralpha+ mesenchymal progenitors in human skeletal muscle. *Cell Death Dis.* **2014**, *5*, e1186. [CrossRef] [PubMed]
22. Uezumi, A.; Nakatani, M.; Ikemoto-Uezumi, M.; Yamamoto, N.; Morita, M.; Yamaguchi, A.; Yamada, H.; Kasai, T.; Masuda, S.; Narita, A.; et al. Cell-surface protein profiling identifies distinctive markers of progenitor cells in human skeletal muscle. *Stem Cell Rep.* **2016**, *7*, 263–278. [CrossRef] [PubMed]
23. Lemos, D.R.; Babaeijandaghi, F.; Low, M.; Chang, C.K.; Lee, S.T.; Fiore, D.; Zhang, R.H.; Natarajan, A.; Nedospasov, S.A.; Rossi, F.M. Nilotinib reduces muscle fibrosis in chronic muscle injury by promoting tnf-mediated apoptosis of fibro/adipogenic progenitors. *Nat. Med.* **2015**, *21*, 786–794. [CrossRef] [PubMed]
24. Lukjanenko, L.; Brachat, S.; Pierrel, E.; Lach-Trifilieff, E.; Feige, J.N. Genomic profiling reveals that transient adipogenic activation is a hallmark of mouse models of skeletal muscle regeneration. *PLoS ONE* **2013**, *8*, e71084. [CrossRef] [PubMed]
25. Pagano, A.F.; Demangel, R.; Brioche, T.; Jublanc, E.; Bertrand-Gaday, C.; Candau, R.; Dechesne, C.A.; Dani, C.; Bonnieu, A.; Py, G.; et al. Muscle regeneration with intermuscular adipose tissue (IMAT) accumulation is modulated by mechanical constraints. *PLoS ONE* **2015**, *10*, e0144230. [CrossRef] [PubMed]
26. Pisani, D.F.; Bottema, C.D.; Butori, C.; Dani, C.; Dechesne, C.A. Mouse model of skeletal muscle adiposity: A glycerol treatment approach. *Biochem. Biophys. Res. Commun.* **2010**, *396*, 767–773. [CrossRef] [PubMed]
27. Wagatsuma, A. Adipogenic potential can be activated during muscle regeneration. *Mol. Cell. Biochem.* **2007**, *304*, 25–33. [CrossRef] [PubMed]
28. Tontonoz, P.; Nagy, L.; Alvarez, J.G.; Thomazy, V.A.; Evans, R.M. PPARgamma promotes monocyte/macrophage differentiation and uptake of oxidized LDL. *Cell* **1998**, *93*, 241–252. [CrossRef]
29. Rosen, E.D.; Walkey, C.J.; Puigserver, P.; Spiegelman, B.M. Transcriptional regulation of adipogenesis. *Genes Dev.* **2000**, *14*, 1293–1307. [PubMed]
30. Feige, J.N.; Gelman, L.; Michalik, L.; Desvergne, B.; Wahli, W. From molecular action to physiological outputs: Peroxisome proliferator-activated receptors are nuclear receptors at the crossroads of key cellular functions. *Prog. Lipid Res.* **2006**, *45*, 120–159. [CrossRef] [PubMed]
31. Spiegelman, B.M. PPAR-gamma: Adipogenic regulator and thiazolidinedione receptor. *Diabetes* **1998**, *47*, 507–514. [CrossRef] [PubMed]

32. Tontonoz, P.; Spiegelman, B.M. Fat and beyond: The diverse biology of PPARgamma. *Annu. Rev. Biochem.* **2008**, *77*, 289–312. [CrossRef] [PubMed]
33. Varga, T.; Mounier, R.; Patsalos, A.; Gogolak, P.; Peloquin, M.; Horvath, A.; Pap, A.; Daniel, B.; Nagy, G.; Pintye, E.; et al. Macrophage PPARgamma, a lipid activated transcription factor controls the growth factor GDF3 and skeletal muscle regeneration. *Immunity* **2016**, *45*, 1038–1051. [CrossRef] [PubMed]
34. Norris, A.W.; Chen, L.; Fisher, S.J.; Szanto, I.; Ristow, M.; Jozsi, A.C.; Hirshman, M.F.; Rosen, E.D.; Goodyear, L.J.; Gonzalez, F.J.; et al. Muscle-specific PPARγ-deficient mice develop increased adiposity and insulin resistance but respond to thiazolidinediones. *J. Clin. Investig.* **2003**, *112*, 608–618. [CrossRef] [PubMed]
35. Hevener, A.L.; He, W.; Barak, Y.; Le, J.; Bandyopadhyay, G.; Olson, P.; Wilkes, J.; Evans, R.M.; Olefsky, J. Muscle-specific Pparg deletion causes insulin resistance. *Nat. Med.* **2003**, *9*, 1491–1497. [CrossRef] [PubMed]
36. Chawla, A. Control of macrophage activation and function by PPARS. *Circ. Res.* **2010**, *106*, 1559–1569. [CrossRef] [PubMed]
37. Li, A.C.; Binder, C.J.; Gutierrez, A.; Brown, K.K.; Plotkin, C.R.; Pattison, J.W.; Valledor, A.F.; Davis, R.A.; Willson, T.M.; Witztum, J.L.; et al. Differential inhibition of macrophage foam-cell formation and atherosclerosis in mice by PPARalpha, beta/delta, and gamma. *J. Clin. Investig.* **2004**, *114*, 1564–1576. [CrossRef] [PubMed]
38. Odegaard, J.I.; Ricardo-Gonzalez, R.R.; Goforth, M.H.; Morel, C.R.; Subramanian, V.; Mukundan, L.; Red Eagle, A.; Vats, D.; Brombacher, F.; Ferrante, A.W.; et al. Macrophage-specific PPARgamma controls alternative activation and improves insulin resistance. *Nature* **2007**, *447*, 1116–1120. [CrossRef] [PubMed]
39. Nadra, K.; Quignodon, L.; Sardella, C.; Joye, E.; Mucciolo, A.; Chrast, R.; Desvergne, B. PPARgamma in placental angiogenesis. *Endocrinology* **2010**, *151*, 4969–4981. [CrossRef] [PubMed]
40. Sardella, C.; Winkler, C.; Quignodon, L.; Hardman, J.A.; Toffoli, B.; Giordano Attianese, G.M.P.; Hundt, J.E.; Michalik, L.; Vinson, C.R.; Paus, R.; et al. Delayed hair follicle morphogenesis and hair follicle dystrophy in a lipoatrophy mouse model of pparg total deletion. *J. Investig. Dermatol.* **2018**, *138*, 500–510. [CrossRef] [PubMed]
41. Bassil, M.S.; Gougeon, R. Muscle protein anabolism in type 2 diabetes. *Curr. Opin. Clin. Nutr. Metab. Care* **2013**, *16*, 83–88. [CrossRef] [PubMed]
42. Savage, D.B. Mouse models of inherited lipodystrophy. *Dis. Model Mech.* **2009**, *2*, 554–562. [CrossRef] [PubMed]
43. Fajas, L.; Landsberg, R.L.; Huss-Garcia, Y.; Sardet, C.; Lees, J.A.; Auwerx, J. E2Fs regulate adipocyte differentiation. *Dev. Cell* **2002**, *3*, 39–49. [PubMed]
44. Angione, A.R.; Jiang, C.; Pan, D.; Wang, Y.X.; Kuang, S. PPARdelta regulates satellite cell proliferation and skeletal muscle regeneration. *Skelet. Muscle* **2011**, *1*, 33. [CrossRef] [PubMed]
45. Chandrashekar, P.; Manickam, R.; Ge, X.; Bonala, S.; McFarlane, C.; Sharma, M.; Wahli, W.; Kambadur, R. Inactivation of PPARbeta/delta adversely affects satellite cells and reduces postnatal myogenesis. *Am. J. Physiol. Endocrinol. Metab.* **2015**, *309*, E122–E131. [CrossRef] [PubMed]
46. Mothe-Satney, I.; Piquet, J.; Murdaca, J.; Sibille, B.; Grimaldi, P.A.; Neels, J.G.; Rousseau, A.S. Peroxisome proliferator activated receptor beta (PPARbeta) activity increases the immune response and shortens the early phases of skeletal muscle regeneration. *Biochimie* **2017**, *136*, 33–41. [CrossRef] [PubMed]
47. Donati, G.; Proserpio, V.; Lichtenberger, B.M.; Natsuga, K.; Sinclair, R.; Fujiwara, H.; Watt, F.M. Epidermal wnt/beta-catenin signaling regulates adipocyte differentiation via secretion of adipogenic factors. *Proc. Natl. Acad. Sci. USA* **2014**, *111*, E1501–E1509. [CrossRef] [PubMed]
48. Festa, E.; Fretz, J.; Berry, R.; Schmidt, B.; Rodeheffer, M.; Horowitz, M.; Horsley, V. Adipocyte lineage cells contribute to the skin stem cell niche to drive hair cycling. *Cell* **2011**, *146*, 761–771. [CrossRef] [PubMed]
49. Schmidt, B.A.; Horsley, V. Intradermal adipocytes mediate fibroblast recruitment during skin wound healing. *Development* **2013**, *140*, 1517–1527. [CrossRef] [PubMed]
50. Zhong, J.; Guo, B.; Xie, J.; Deng, S.; Fu, N.; Lin, S.; Li, G.; Lin, Y.; Cai, X. Crosstalk between adipose-derived stem cells and chondrocytes: When growth factors matter. *Bone Res.* **2016**, *4*, 15036. [CrossRef] [PubMed]
51. Heredia, J.E.; Mukundan, L.; Chen, F.M.; Mueller, A.A.; Deo, R.C.; Locksley, R.M.; Rando, T.A.; Chawla, A. Type 2 innate signals stimulate fibro/adipogenic progenitors to facilitate muscle regeneration. *Cell* **2013**, *153*, 376–388. [CrossRef] [PubMed]

52. Kuswanto, W.; Burzyn, D.; Panduro, M.; Wang, K.K.; Jang, Y.C.; Wagers, A.J.; Benoist, C.; Mathis, D. Poor repair of skeletal muscle in aging mice reflects a defect in local, interleukin-33-dependent accumulation of regulatory t cells. *Immunity* **2016**, *44*, 355–367. [CrossRef] [PubMed]
53. Fiore, D.; Judson, R.N.; Low, M.; Lee, S.; Zhang, E.; Hopkins, C.; Xu, P.; Lenzi, A.; Rossi, F.M.; Lemos, D.R. Pharmacological blockage of fibro/adipogenic progenitor expansion and suppression of regenerative fibrogenesis is associated with impaired skeletal muscle regeneration. *Stem Cell Res.* **2016**, *17*, 161–169. [CrossRef] [PubMed]
54. Imai, T.; Takakuwa, R.; Marchand, S.; Dentz, E.; Bornert, J.M.; Messaddeq, N.; Wendling, O.; Mark, M.; Desvergne, B.; Wahli, W.; et al. Peroxisome proliferator-activated receptor gamma is required in mature white and brown adipocytes for their survival in the mouse. *Proc. Natl. Acad. Sci. USA* **2004**, *101*, 4543–4547. [CrossRef] [PubMed]

International Journal of
Molecular Sciences

Review

An aPPARent Functional Consequence in Skeletal Muscle Physiology via Peroxisome Proliferator-Activated Receptors

Wendy Wen Ting Phua [1,2,3], Melissa Xin Yu Wong [1], Zehuan Liao [1] and Nguan Soon Tan [1,2,4,5,*]

1 School of Biological Sciences, Nanyang Technological University 60 Nanyang Drive, Singapore 637551, Singapore; wphua003@e.ntu.edu.sg (W.W.T.P.); mwong018@e.ntu.edu.sg (M.X.Y.W.); liao0058@e.ntu.edu.sg (Z.L.)
2 Lee Kong Chian School of Medicine, Nanyang Technological University, 50 Nanyang Avenue, Singapore 639798, Singapore
3 NTU Institute for Health Technologies, Interdisciplinary Graduate School, Nanyang Technological University, 50 Nanyang Drive, Singapore 637553, Singapore
4 Institute of Molecular and Cell Biology, A*STAR, 61 Biopolis Drive, Proteos, Singapore 138673, Singapore
5 KK Women's and Children's Hospital, 100 Bukit Timah Road, Singapore 229899, Singapore
* Correspondence: nstan@ntu.edu.sg; Tel.: +65-6316-2941; Fax: +65-67913-856

Received: 27 March 2018; Accepted: 8 May 2018; Published: 10 May 2018

Abstract: Skeletal muscle comprises 30–40% of the total body mass and plays a central role in energy homeostasis in the body. The deregulation of energy homeostasis is a common underlying characteristic of metabolic syndrome. Over the past decades, peroxisome proliferator-activated receptors (PPARs) have been shown to play critical regulatory roles in skeletal muscle. The three family members of PPAR have overlapping roles that contribute to the myriad of processes in skeletal muscle. This review aims to provide an overview of the functions of different PPAR members in energy homeostasis as well as during skeletal muscle metabolic disorders, with a particular focus on human and relevant mouse model studies.

Keywords: peroxisome proliferator-activated receptor; skeletal muscle; lipid metabolism; insulin resistance; aging; physical exercise; type 2 diabetes; muscle regeneration

---

1. Skeletal Muscle

Skeletal muscle is the largest metabolic organ in the human body, and it contributes ~40% of the total human body mass in healthy non-obese adults. Beyond its well-recognized role in physical movement and postural stabilization, the importance of skeletal muscle in the whole-body metabolism has been increasingly acknowledged, as it can impact overall health and quality of life [1]. Skeletal muscle is a heterogeneous tissue composed of different fiber types, and it exhibits high metabolic flexibility when adapting to metabolic or energy demands, as well as prevailing conditions and activities. Skeletal muscle can withstand massive and sudden changes, both mechanically and bioenergetically, from rest to rapid contractile activity, because it has effective mechanisms for coping with ATP consumption and re-synthesis. While skeletal muscle is anatomically fixed at birth in mammals, postnatal muscle growth can undergo cellular changes, such as increases in length and girth, and some myofibers can experience changes in contractile activity and humoral factors in response to the nutrient availability [2]. The mammalian skeletal muscle can be classified across a spectrum, according to its contractile and metabolic properties, but it is broadly classified into two categories, namely, slow-twitch type I fibers and fast-twitch type II fibers. Slow-twitch type I fibers are rich in mitochondria and have a higher insulin sensitivity and glucose transporter 4 (GLUT4) expression levels than the fast-twitch type II fibers [3]. The type I fibers are rich in myoglobin, surrounded by many capillaries, and contain relatively abundant intracellular lipid levels for oxidative metabolism. These characteristics support long-duration contractile activities, such as walking and postural stabilization. In contrast, the fast-twitch type II fibers are large fibers with vast glycogen reserves that support their role in glycolytic metabolism. Type II fibers produce rapid contractions that are used for intense activities, but these fibers are easily fatigued. In mammals, type II muscle fibers can be further categorized into type IIa (fast-twitch oxidative), type IIb (fast-twitch glycolytic), and type IIx (an intermediate type between IIa and IIb). However, type IIb fibers are not detectable in the human skeletal muscle [4]. Muscle fiber type switching and tissue remodeling can occur on demand during exercise or during obesity and metabolic-related diseases. In response to exercise training, the metabolic phenotype of the muscle that is used changes along with the increase in size and strength. At rest, a trained muscle uses more energy from fat and less from carbohydrates than the untrained muscle [5]. Skeletal muscle is the predominant site of the insulin-mediated glucose uptake. The deregulation of skeletal muscle energy homeostasis plays a major role in the pathogenesis of peripheral insulin resistance and type 2 diabetes mellitus (T2DM). T2DM is characterized by chronic hyperglycemia, as a result of inefficient pancreatic beta-cell insulin secretion compensation. T2DM is also characterized by a chronic increase in plasma free fatty acid (FFA) levels and dyslipidemia. Excessive triglyceride accumulation in skeletal muscle, both the intramuscular and intramyocellular deposition, induces lipotoxicity, reduces glucose uptake, and ultimately leads to insulin resistance and T2DM [6]. Physiologically, the deregulation of the metabolic homeostasis in skeletal muscle causes muscle fiber type switching, from the slow-twitch to fast-twitch, as the disease worsens over time [7]. Understanding the changes in skeletal muscle during obesity and T2DM development is thus crucial for elucidating the underlying causes of insulin resistance.

The peroxisome proliferator-activated receptors (PPARs) have emerged as the master regulators of both lipid and glucose homeostasis, and are considered as valuable pharmaceutical targets for treating metabolic dysfunctions and T2DM. PPARs are ligand-activated transcription factors that belong to the nuclear hormone receptor superfamily, and they are activated by a variety of synthetic ligands and endogenous ligands, such as the naturally occurring FFAs and their metabolites, arachidonic acid and eicosanoids. The synthetic ligands of PPARs have been used successfully to treat T2DM and dyslipidemia. Specifically, thiazolidinediones (TZDs), such as rosiglitazone and pioglitazone, are specific PPARγ activators and are used as insulin sensitizers in order to improve insulin resistance in T2DM patients. Fibrates include fenofibrate, clofibrate, and ciprofibrate, which exhibit a predominant PPARα activity and induce lipid uptake and oxidation. The PPARα agonist clofibrate has been used to treat dyslipidemia. Insulin-sensitizing effects can also occur as a consequence of PPARα and PPARβ/δ activation. Physiologically, the members of the PPAR family also modulate basic processes, such as proliferation, differentiation, and postnatal development [8,9]. In this review, we will focus on the metabolic regulatory roles of PPARs in the skeletal muscle during healthy and diseased states, primarily with studies that have used human and mouse models.

2. Transcription Regulation by PPARs

Three related PPAR members, each encoded by distinct genes, have been identified and designated as PPARα, PPARβ/δ, and PPARγ. PPARγ has two distinct isoforms, PPARγ1 and PPARγ2. PPARγ2 is predominantly expressed in adipose tissue and is 30 amino acids longer than the PPARγ1 at the N-terminal [10]. As with most nuclear receptors, PPARs share modular structural characteristics. The N-terminal A/B domains encode the activation function 1 (AF-1), the C-domain consisting of the DNA binding domain (DBD), the D-domain, or the hinge domain that provides structural flexibility, and the E-domain containing the ligand binding domain (LBD) and the ligand-dependent activation function 2 (AF-2). Of the PPAR members, the LBDs of PPARα and PPARγ are the most similar in shape and size, whereas the LBD of PPARβ/δ is significantly smaller [11,12]. The differences in amino acid sequences among the PPAR members also indicate that the LBD pocket of PPARα is more lipophilic than that of the two others. These structural differences among the PPAR LBDs suggest the influences of the structurally distinct ligands with varying binding affinities that contribute to ligand selectivity [13]. All members of PPAR form obligate heterodimers with retinoid X receptors (RXRs) and bind as a complex to the consensus sequences, known as peroxisome proliferator response elements (PPREs), located in the regulatory region of their target genes.

In addition to ligand binding, the activity of PPARs is also affected by post-translational modifications, such as phosphorylation, SUMOylation, and ubiquitination, as well as through regulatory proteins, such as AMP-activated protein kinase (AMPK) and cryptochrome (CRY1). Regulation by insulin and insulin-induced PPAR phosphorylation has been reported to enhance the PPAR transcriptional activity [14]. Post-translational modification by ubiquitination has been shown to be affected by the presence of the PPAR ligand. In the absence of the ligand, PPARα and PPARβ/δ are poly-ubiquitinated and targeted for subsequent degradation [15,16]. The presence of PPARγ agonists, on the other hand, enhances the PPARγ polyubiquitination, which promotes its degradation. The monoSUMOlyation of PPARα and PPARγ has been reported, in which the transcriptional activities of both PPARs are inhibited [17]. The role of energy metabolism and circadian regulation in skeletal muscle has recently been understood through the modulation of the PAR protein. Recent studies by Jordan et al. (2017), on skeletal muscle circadian rhythm, have shown that the circadian transcriptional repressors CRY1 and CRY2 function as co-repressors of PPARβ/δ, possibly via an AMPK-dependent signaling pathway [18]. Collectively, the post-translational regulation of the PPAR protein has a direct impact on the cellular metabolism and energy production.

PPARs are diverse regulators that fundamentally regulate the energy metabolism at the transcription level. Each member displays distinct tissue distribution patterns and pharmacological profiles. PPARα is highly expressed in active metabolic tissues, such as the liver, kidney, heart, and skeletal muscle [19], whereas PPARγ is expressed in primarily the white and brown adipose tissue, where most of the free fatty acids are deposited [20]. PPARβ/δ is ubiquitously expressed because of its importance in the systemic and basic cellular functions, which include the energy modulation in metabolically active tissues, inflammation, wound healing, and keratinocyte and intestinal cell differentiation [7,9]. The *PPARD* gene ablation in mice results in a high embryonic lethality [21], and the PPARγ-deficient mice exhibit an embryonic lethality by E10 [22]. These findings highlight the importance and complex physiological roles of PPARs. All three of the PPARs are expressed in the skeletal muscle at different amounts, as follows: PPARβ/δ has the highest expression levels, followed by PPARα and PPARγ [23–25].

3. Nutrient Sensing by PPARs

Members of the PPAR family modulate metabolic responses through sensing and responding to fluctuations in the nutrient availability. Major dietary constituents, such as fatty acids and carbohydrates, can regulate the gene expression of several metabolic pathways via hormones and PPARs and, in turn, induce their utilization [26]. In a post-prandial state, the availability of metabolic precursors promotes the synthesis of natural PPAR ligands and induces PPAR trans-regulation so as to promote anabolism and storage. Upon nutrient scarcity, PPARs are directly activated by the release of FFAs from lipid reserves, and they stimulate the transcription of genes that are involved in FFA uptake and fatty acid oxidation in the skeletal muscle, as well as glycogenolysis, gluconeogenesis, and ketone body synthesis in the liver, reviewed in [27].

Nutrient intake and energy metabolism are closely associated and are subject to hormonal regulation. Insulin, one of the main hormones that regulates whole-body metabolism, promotes glucose uptake in the metabolically active tissues, such as the liver, fat, and skeletal muscle. During post-prandial state, insulin is secreted from the pancreatic beta cells into the bloodstream in response to increased blood glucose levels. At the peripheral tissues, such as skeletal muscle, insulin binds to the insulin receptors at the plasma membrane in order to trigger the insulin signaling cascade via insulin receptor substrate 1 (IRS1) phosphorylation, protein kinase B (AKT/PKB) activation, and glucose transporter type 4 (GLUT4) translocation to the plasma membrane [28]. These actions promote an extracellular glucose clearance [29]. Skeletal muscle accounts for over 80% of the insulin-dependent glucose uptake [30]. Glucose serves as an immediate source of energy and is subsequently converted into acetyl-coenzyme A (acetyl-CoA), by the pyruvate dehydrogenase complex (PDC). Then, it is channeled into the tricarboxylic acid (TCA) cycle and undergoes oxidative phosphorylation in the mitochondria [31]. In skeletal muscle, the excess glucose is stored as glycogen or used as a precursor for lipid synthesis [31]. As blood glucose levels drop over time, the body transits from a fed to fasted state, triggering the change from glucose to free fatty acids (FFAs) as the preferred fuel substrates of the skeletal muscle and liver. This dynamic glucose-FFA cycle, also known as the Randle cycle, provides metabolic flexibility and survival adaptation so as to conserve the whole-body glucose supply and is of major quantitative importance in the skeletal muscle, as reviewed in [32,33].

During fasting, both PPARα and PPARβ/δ are upregulated in the skeletal muscle in rodents [34], but only PPARβ/δ is upregulated in the human skeletal muscle [7,35]. Upon the increased FFA influx, the FFAs are hydrolyzed into acyl-CoA complexes, which are then channeled into the mitochondria by the carnitine palmitoyltransferase I (CPT1) for fatty acid oxidation. One of the key genes that regulates the glucose-FFA shuttle is the pyruvate dehydrogenase kinase (PDK), which is a classical PPAR target gene. PDK inactivates PDC, via phosphorylation, and reduces oxidation of the glycolysis-derived pyruvate. These effects decrease the glucose utilization in order to conserve glucose. In human skeletal muscle, all four of the PDK isozymes are PPARβ/δ target genes, and PDK2 and PDK4 are the most abundantly expressed [35,36]. In the skeletal muscle of PPARβ/δ knockout mice, PDK4 expression is markedly blunted [37]. Interestingly, the PDK4 expression is unaffected in the skeletal muscle of the fasted PPARα knockout mice [25]. These findings thus suggest that PPARβ/δ is the primary PPAR member that regulates the skeletal muscle substrate utilization.

4. Regulation of Lipid Metabolism in Skeletal Muscle by PPARs

Fat and excess calories from the diet are converted into the concentrated form of triglycerides to store metabolic energy over extended periods of time. Triglycerides are typically stored in three main organs (ranked in order, from the greatest to least amount stored), namely, adipose tissue, skeletal muscle, and liver [38]. During fasting or increased energy demands, triglycerides in adipose tissue are hydrolyzed into FFAs and delivered to tissues through the action of lipoprotein lipase (LPL), and can be used either for fatty acid β-oxidation in the energy-converting mitochondria or as building blocks for cellular functions and signaling.

Over the past decade of research, PPARs have emerged as master regulators of the lipid metabolism. In humans, skeletal muscle accounts for more than 30% of the total energy expenditure, and up to 70% of this energy is derived from FFAs in resting muscle. Of the three PPAR members, PPARα and PPARβ/δ play central roles in regulating lipid homeostasis [25]. PPARγ promotes glucose uptake in skeletal muscle, in order to play a role in insulin-stimulated glucose metabolism [39]. In vivo, PPARα and PPARβ/δ regulate the genes that are involved in FFA uptake, such as cluster of differentiation 36/SR-B2 (CD36) and LPL; FFA intracellular transport, such as fatty acid binding protein 3 (FABP3); and fatty acid oxidation, such as CPT1 and stearoyl-CoA desaturase (SCD). The genes that are involved in fatty acid oxidation and utilization are similarly regulated by PPARα and PPARβ/δ in skeletal muscle, as shown by overexpression studies [40–42]. Skeletal muscle-specific PPARβ/δ overexpression also induces characteristic shifts towards oxidative fibers and increased oxidative capacity [7]. Conversely, selective PPARβ/δ ablation in skeletal muscle leads to lower oxidative capacity in the fibers, resulting in obesity and T2DM [43]. In contrast to PPARβ/δ, PPARα overexpression promotes fiber type shifts towards glycolytic type II fibers, and these fibers are protected from diet-induced obesity. Interestingly, when fed a high-fat diet, PPARα-overexpressing mice have significantly higher intramuscular triglyceride concentrations than control mice, and they develop glucose intolerance [41]. In PPARα knockout mice, however, fatty acid oxidation is reduced during starvation despite an increase in oxidative fibers [25].

4.1. Regulation of Lipid Transport in Skeletal Muscle by PPARs

Unlike glucose, which is water soluble, circulating FFAs are usually associated with albumin or exist as fatty esters and phospholipids in lipoproteins. On the plasma membrane surface, LPL mediates the hydrolysis of triglyceride-rich lipoproteins. This hydrolysis releases the encapsulated lipids and is thus considered a rate-limiting step for lipid uptake. The cellular lipid uptake was initially thought to occur via passive diffusion because of the hydrophobic nature of the plasma membrane. However, it is now widely recognized that FFA uptake requires a highly regulated, protein-mediated action by the transporter proteins. In humans and rodents, CD36, FABPs, and fatty acid transport proteins (FATPs) are co-expressed in the skeletal muscle, which is key in facilitating FA transport, and their expression levels are regulated predominantly by PPARβ/δ [7] (Figure 1). Approximately 70% of total FFA uptake is mediated by CD36 [44], although the mechanisms of FFA transmembrane movement and the binding specificity of CD36 are not understood [45]. It has been suggested that CD36 promotes fatty acid partitioning at the outer leaflet for translocation through the lipid bilayer and that it provides a docking site for FABPs and other enzymes at the intracellular side of the membrane, so as to facilitate the transport of the incoming FFAs [45]. Cytoplasmic FABP (FABPc) serves as an acceptor for FFAs, shuttles them through the cellular compartments, and protects against lipotoxic accumulation and aggregation within the cell [46]. The fatty acid transporters in skeletal muscle exhibit different capacities for FFA transport and metabolism. An in vivo study of CD36, plasma membrane FABP (FABPpm), FATP1 or FATP4 overexpression in the anterior tibialis muscle of rats showed the differential effects on FFA transport and utilization in skeletal muscle [47]. The authors have reported that CD36 and FATP4 are quantitatively the most effective in FFA transport. Interestingly, the transporter overexpression did not alter the rates of FFA esterification into triglycerides, but it increased fatty acid oxidation that was observed with CD36 and FABPpm overexpression [47,48] (Figure 1).

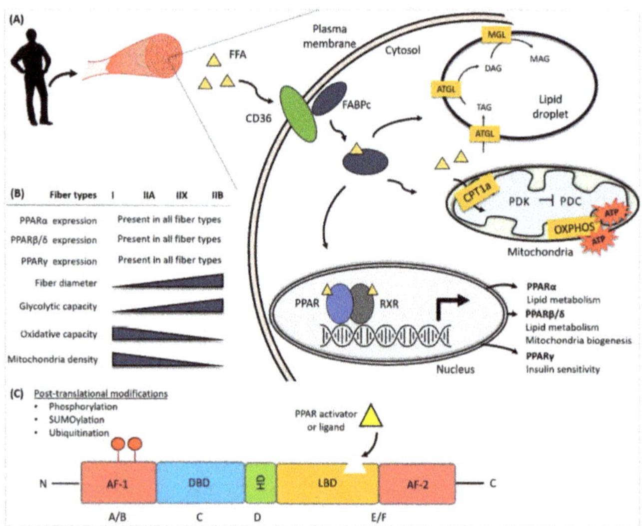

Figure 1. Schematic diagram of skeletal muscle fiber and its fatty acid handling. (**A**) The fate of free fatty acid (FFA) in skeletal muscle. FFA uptake is mediated by receptors, such as CD36, at the plasma membrane. Within the cell, FFA is transported throughout the cellular compartments, via the lipid transporter, FABPc. FFAs can either be targeted to the lipid droplet for storage, translocated to the mitochondria for fatty acid oxidation, or serve as a ligand for peroxisome proliferator-activated receptors (PPARs) within the nucleus. In the skeletal muscle, PPARα and PPARβ/δ are mainly involved in lipid metabolism regulation. PPARβ/δ is also involved in regulating mitochondria biogenesis while PPARγ is involved in skeletal muscle insulin sensitivity and glucose regulation. (**B**) The spectrum of skeletal muscle fiber type characteristics. All three of the PPAR isotypes are expressed regardless of the fiber types. Slow-twitch type I fibers are smaller in fiber diameter, with high oxidative capacity and mitochondria density, while fast-twitch type II fibers have a range in their fiber diameters, typically higher glycolytic capacity with lower mitochondria density, and oxidative capacity in comparison to type I fibers. (**C**) Schematic diagram of PPAR protein structure. PPARs are regulated by post-translational modifications, such as phosphorylation, SUMOylation, and ubiquitination in the presence or absence of ligand. Activation function, AF; DNA-binding domain, DBD; Hinge domain, HD; ligand binding domain, LBD.

Mammalian FABPs are small skeletal muscle proteins of approximately 15 kDa and are involved in the reversible binding of FFAs, in order to facilitate trafficking to various cellular compartments, such as peroxisomes, mitochondria, and nuclei. In humans, there are nine FABP isoforms (FABP1-9) that are differentially expressed in metabolically active tissues [49]. In adult skeletal muscle, FABP3 is predominantly expressed and is more abundant in type 1 oxidative fibers than in type 2 glycolytic fibers. FABP3 is responsible for FFA shuttling to the outer mitochondrial membrane, where FFAs are converted into their acyl-CoA derivatives by acyl-CoA synthetase, and are subsequently channeled for mitochondrial β-oxidation [50,51]. A small amount of acyl-CoA is converted into lipid intermediates, such as ceramide, diacylglycerol, and phospholipids, which can function as lipid secondary messengers or modulate membrane structures [52,53]. FABPs have been shown to interact with PPARs in the nucleus, so as to mediate transcriptional activities. Recently, the ligand-sensitive nuclear localization signal in FABP5 protein conformation has been described. In this conformation, FABP5 facilitates PPARβ/δ transcriptional activation through nuclear trafficking of linoleic acid and arachidonic acid [54]. Similar reports have shown that FABP1, FABP2, and FABP3 can increase FFA shuttling to the nucleus and enhance ligand-mediated PPARα transactivation [51,55,56], whereas PPARγ agonists can induce the nuclear localization of FABP4 [57,58]. However, the specificity of the lipid species with FABP chaperone activity and the significant impact of each FABP isoform on the transcriptional regulation in the skeletal muscle remains unclear.

4.2. Regulation of Muscle Lipolysis by PPARs

Lipolysis is the process through which FFAs are sequentially hydrolyzed. Lipolysis is first mediated by the rate-limiting enzyme adipose triglyceride lipase (ATGL), which hydrolyzes triglyceride to diacylglycerol and one fatty acid molecule. Diacylglycerol is then converted into monoacylglycerol, by hormone-sensitive lipase (HSL). The final step of FFA hydrolysis by monoacylglycerol lipase (MGL) produces glycerol and the third fatty acid molecule. In the mouse skeletal muscle, lipolysis can occur without stimulation (basal conditions) or with hormone stimulation [59]. Under either condition, ATGL and HSL collectively account for most of the hydrolysis activity [60]. ATGL is an evolutionarily conserved enzyme for fat storage lipolysis [61]. It is highly expressed in type I fibers in both mice and humans and is a reported transcriptional target of PPARα in rodents [62]. ATGL knockout mice have a shorter life-span and exhibit defective lipolysis and increased triglyceride accumulation in non-adipose tissues, including skeletal muscle [63,64]. These mice also show a concomitant decrease in muscle relaxation and have an increased reliance on carbohydrates as the major fuel source at rest [65]. Interestingly, pharmacological treatment of ATGL knockout mice with PPARα agonists reversed the excessive systemic lipid accumulation, improved metabolic flexibility in substrate switching from glucose to fatty acids, and prevented premature death [64]. ATGL overexpression in different muscles has varied effects on skeletal muscle fatty acid oxidation in mice. These varied effects are most likely due to the differential expression levels of ATGL among fiber types. Increased fatty acid oxidation was reported upon ATGL overexpression in the soleus muscle via electroporation [62]. However, adenovirus-mediated ATGL overexpression in the tibialis anterior muscle was not sufficient to alter fatty acid oxidation rates [66]. Similarly, mutations in the *PNPLA2* gene, which encodes ATGL in humans, can lead to neutral lipid storage diseases with myopathy. In humans, ATGL is exclusively expressed in type I muscle fibers and plays an important role in skeletal muscle FA turnover [67]. ATGL deficiency in young adults resulted in increased lipid accumulation in primarily type I skeletal muscle [68]. However, treatment with a PPARα agonist was less successful in humans than in rodents [69].

4.3. Regulation of Muscle Lipid Storage by PPARs

Skeletal muscles stockpile excess FFAs in lipid droplets as an energy reservoir. These FFA stores are commonly referred to as intramuscular triglycerides. Excess FFAs are converted in the endoplasmic reticulum (ER) and stored in lipid droplets (LDs), which are surrounded by a phospholipid monolayer and LD-associated surface proteins. These LDs are dynamic structures that function as more than temporary fuel storage. In fact, they serve as a reserve pool of intracellular signaling mediators for ligands, such as PPAR, and are thought to have a protective mechanism against possible lipid aggregation that leads to lipotoxicity and ER stress after the excess uptake of FFAs and sterols. In skeletal muscle, lipid droplets are distributed between myofibrils (intermyofibrillar LDs) and beneath the plasma membrane (subsarcolemmal LDs). These LDs serve as transport organelles between cellular compartments and as a readily available energy pool for short-term or long-term muscular contractions. PPAR agonists have been reported to regulate LD-associated proteins, such as perilipins (PLIN1-5), in various organs. Perilipins, except PLIN1, are expressed in skeletal muscle in humans and rodents [70]. PLIN2, one of the most abundantly expressed LD-coating proteins in skeletal muscle, is thought to maintain insulin sensitivity in skeletal muscle and promote the storage of FFAs in the form of triglycerides [71]. PLIN2 is induced upon PPARβ/δ activation by GW501516 in both human primary myocytes and mouse skeletal muscle [72,73]. In PPARα knockout mice, PLIN2 and PLIN5 expression levels are decreased in the soleus, whereas PLIN3 and PLIN4 expression levels seem to be unaffected [74]. Interestingly, immunofluorescent staining of human and rodent skeletal muscle sections have shown that PLIN2 is abundantly expressed in type I fibers, which contain more intramuscular triglyceride contents than type II fibers [75,76]. Similarly, the direct regulation of PLIN5 by PPARβ/δ in the soleus and gastrocnemius of wild-type mice has been observed. In this study, a conserved PPRE in humans and mice had been found in the first intron of PLIN5 [74]. However, PLIN5 protein levels in the skeletal muscle did not seem to be altered in PPARβ/δ knockout mice.

PLIN5 has been suggested to regulate FFAs storage and to be involved in skeletal muscle adaptation in type II fibers, in response to exercise and fasting [77]. Similarly to PLIN2, the PLIN5 expression levels are higher in the oxidative fibers than in glycolytic fibers [78], and its protein levels are associated with intramuscular triglyceride levels in both rodents and humans [75]. In glucose-intolerant human subjects, it has been reported that PPARγ agonists can induce PLIN5 mRNA expression, and PLIN5 mRNA expression is negatively correlated with the body mass index (BMI) in non-diabetic subjects [79]. The role of PPAR regulation and its effects on perilipin functions in skeletal muscle physiology, however, need further investigation, as most of the studies on PLIN5 have been performed in vitro [46].

5. Regulation of Mitochondrial Biogenesis and Function by PPARs

The members of the PPARγ-coactivator 1 (PGC-1) family, such as PGC-1α and PGC-1β, regulate mitochondrial oxidative metabolism and biogenesis, and activate gene transcription through coordination with PPARα, PPARβ/δ, and other nuclear receptors. PGC-1α is reported to be a direct target of PPARβ/δ, but not PPARα, in the skeletal muscle, via agonism [43,80] and during conditions of increased energy demands, such as cold, exercise, and fasting [43,81]. Moreover, in vivo PPARβ/δ overexpression, via electroporation in adult rat muscle, caused an increase in PGC-1α protein levels [82]. PGC-1α thus mediates a positive feed-forward transcriptional control of the PPAR-regulated genes that are involved in fatty acid oxidation and carbohydrate metabolism, as well as an auto-regulatory loop, in which PGC-1α regulates its own gene expression [83]. Gene manipulation of PGC-1α and PGC-1β in skeletal muscle produces phenotypes similar to those of PPARβ/δ transgenic mice. Conversely, PGC-1α or PGC-1β overexpression in mouse skeletal muscle induces oxidative fiber development, promotes fatty acid oxidation and increases the capacity to sustain physical activity in mice [84,85]. However, PPARβ/δ overexpression in mice does not increase PGC-1α mRNA levels and does not affect mitochondrial function [42,86]. Additionally, transcription factors such as mitochondrial transcription factor A (TFAM) and mitochondrial transcription factors B1 (TFB1M) and B2 (TFB2M), which directly regulate mitochondrial biogenesis via nuclear respiratory factors (NRF1 and NRF2) are not known to be classic PPARβ/δ target genes [87]. Thus, the precise regulation of PPARβ/δ and PGC-1α in mitochondrial function and biogenesis has been a long-standing question. Recently, Koh et al. [88] used an electroporation-mediated PPARβ/δ overexpression in mouse muscles to demonstrate that PPARβ/δ modulates mitochondrial biogenesis and PGC-1α expression, in both a transcriptional manner and a posttranslational manner. PPARβ/δ overexpression in adult mice increases NRF1 and mitochondrial electron transport chain enzyme protein levels, before increasing PGC-1α protein levels. Moreover, PPARβ/δ decreased PGC-1α degradation via ubiquitin-proteasome system, through binding and blocking its ubiquitin-binding site. These actions led to the gradual accumulation of the PGC-1α protein [88]. The authors also reported the auto-regulation of PPARβ/δ, suggesting a feed-forward mechanism that is important in the mitochondrial oxidative metabolism and biogenesis.

6. Dysregulation of Lipid Metabolism and PPAR during Insulin Resistance and T2DM

Insulin resistance is the key pathophysiological feature of obesity and T2DM, and is caused by imbalances in insulin action in peripheral tissues, insulin secretion, or both. In skeletal muscle, the major causes of insulin resistance are thought to be the excess accumulation of intramyocellular lipid (IMCL) and the inhibition of one or several steps in the insulin signaling cascade [89]. IMCL includes all types of lipids within the myocytes. Myocytes are composed of mostly triglycerides, but also include the lipid intermediates of lipid metabolism, ceramides, diacylglycerol, phospholipids, and sphingolipids [90]. The most common cause of lipid accumulation is overnutrition, which leads to an increase in FFA uptake that exceeds the rates of fatty acid oxidation and storage [91]. High IMCL concentrations have also been negatively associated with insulin sensitivity in non-obese adults [92], high-fat diet rodent models [93], and lean offspring of T2DM patients [94]. Similarly, acute lipid overload in skeletal muscle decreases peripheral insulin sensitivity in healthy individuals [5,95]. Paradoxically, it has been reported that endurance athletes are highly insulin-sensitive, despite possessing higher IMCL concentrations than normal healthy individuals. This phenomenon is thus called the 'athlete's paradox' [96]. These trained athletes, however, have a high capacity for fat oxidation and have high glucose disposal rates, but are not totally immune to lipid-induced insulin resistance [5,96]. Unlike obese individuals and T2DM patients, the turnover rates of IMCL in trained athletes is high, and this turnover is an adaptive physiological response rather than a pathological condition [5]. Thus, endurance athletes do not bear the ascribed toxic effects on insulin signaling.

Ceramide and diacylglycerol accumulation interfere with the insulin signaling cascade through the direct interaction with and activation of protein kinase C (PKC) isoforms, so as to reduce glucose uptake [91,97,98]. In the skeletal muscle, a 50% increase in endogenous ceramide levels, induced by treatment with a high concentration of saturated FFAs, is sufficient to inhibit AKT/PKB activity [99]. In obese insulin-resistant human subjects, ceramide concentrations were found to be nearly two-fold higher in muscle compared with lean insulin-sensitive human subjects [100]. In contrast, overexpressing acid ceramidase, which converts ceramide into sphingosine, fully negates the inhibitory effects of high FFA treatment on insulin signaling [101]. Additionally, ceramide has also been shown to stimulate protein phosphatase 2A (PP2A), a phosphatase long known to negatively regulate AKT/PKB [102]. The inverse relationship between ceramide and insulin sensitivity has been reviewed [97]. Furthermore, PKCθ activation by diacylglycerol, induces insulin resistance through inhibiting IRS1-associated phosphatidylinositol-3 kinase (PI3K) activity [103,104]. Diacylglycerol acyltransferase 1 (DGAT1), a downstream PPARβ/δ target gene, catalyzes the conversion of diacylglycerol and fatty acyl-CoA to triglyceride [105]. The skeletal muscle-specific DGAT1-overexpressing mice have low diacylglycerol concentrations and are protected from diet-induced insulin resistance, despite the increased FFAs accumulation in their skeletal muscle [106].

PPAR agonists have been of clinical interest since the discovery of fibrates and the TZDs for treating metabolic-related diseases. Below, we describe the impact of PPAR regulation in skeletal muscle, during insulin resistance and T2DM.

6.1. *PPARγ Agonists and Insulin Resistance and T2DM Treatment*

PPARγ ligands, including TZDs, have hypoglycemic effects, reduce insulin resistance, and improve insulin sensitivity. In the early 1980s, TZDs were reported as insulin sensitizers. Currently, pioglitazone is the only FDA-approved TZD for treating T2DM. This drug has lipid-modifying benefits and can reduce adverse cardiovascular outcomes. The insulin-sensitizing effects of TZDs can be attributed to the activation of skeletal muscle PPARγ. This activation maintains insulin signaling activity, even though PPARγ is expressed at low levels. Given the whole-body skeletal muscle mass, the regulation of the skeletal muscle PPARγ remains physiologically relevant. The direct action of TZDs on non-adipose tissues has been indicated in adipose tissue-specific PPARγ-silenced mice, in which TZD treatment improved insulin sensitivity in the skeletal muscle and the liver, despite an increase in triglyceride deposition [107]. In obese Zucker rats, short-term treatment with rosiglitazone increases the skeletal muscle tyrosine phosphorylation of insulin receptor and IRS-1, and induces AKT/PKB activation [108]. Similarly, muscle biopsies that were obtained from T2DM patients that were treated with either rosiglitazone or pioglitazone showed increased insulin-stimulated IRS-1 tyrosine phosphorylation, IRS-1-associated PI3-kinase activity, and AKT/PKB activity [109,110]. The TZD administration, however, has been reported to stimulate skeletal muscle glucose uptake acutely and improve glucose handling through a PPARγ-independent mechanism [111,112]. Moreover, the PPARγ-sparing TZD analogs have similar insulin-sensitizing pharmacological effects to rosiglitazone and pioglitazone in rodent models [113]. These results suggest that the insulin-sensitizing effects of TZDs may be independent of PPARγ regulation, to some degree. Despite the varied pharmacological actions of TZDs via PPARγ regulation, the role of PPARγ in the skeletal muscle in glucose homeostasis and insulin sensitivity remains physiologically and clinically relevant. In the human skeletal muscle, PPARγ expression is acutely regulated and increased by insulin [114]. PPARγ activation directly regulates the expression of the glucose transporters GLUT1 and GLUT4, and promotes their translocation to the cell surface so as to increase the cellular glucose uptake. In addition, GLUT4 regulation by PPARγ is remarkably conserved across the vertebrate evolution, from fish to mammals [115]. In L6 muscle cells, PPARγ agonists, but not PPARα agonist WY14643, have been shown to increase IRS1 protein expression directly [116]. Moreover, constitutive PPARγ activation in the mouse skeletal muscle decreases intramuscular lipid accumulation, induces a shift towards the oxidative fiber type, and protects against susceptibility to diet-induced insulin resistance [117]. Conversely, skeletal muscle-specific PPARγ knockout mice have an increased adiposity and are glucose intolerant and insulin resistant [118,119]. However, the young skeletal muscle of PPARγ-deficient mice remained responsive to the TZD treatment, despite a high-fat diet-induced hepatic insulin resistance and excess adiposity [119]. These findings led to the suggestion of age-dependent differences in TZD insulin-sensitizing effects and the potential role of tissue crosstalk in the regulation of whole-body insulin sensitivity [120]. In humans, dominant negative PPARγ mutations are associated with obesity [121], dyslipidemia, and severe insulin resistance [122], whereas a common polymorphism (Pro12Ala) has been shown to decrease PPARγ receptor activity, improve insulin sensitivity, and decrease T2DM risk [123,124].

6.2. PPARα Agonists and Insulin Resistance and T2DM Treatment

PPARα plays a pivotal role in the liver during the nutritional transitions and intricately controls hepatic lipid metabolism and whole-body glucose homeostasis [27]. The role of skeletal muscle PPARα in regulating the insulin signaling pathway is, however, less clear. Though PPARα has metabolic regulatory roles, its expression in skeletal muscle remains unchanged during fasting [35]. The clinical use of fibrates for treating hyperlipidemia in obese individuals and T2DM patients was first approved in the late 1960s [125]. The fibrates that are commonly used for clinical treatment are bezafibrate, fenofibrate, and gemfibrozil. Fenofibrate treatment in patients with metabolic syndrome improves lipid profiles and increases insulin sensitivity [126,127]. Recently, bezafibrate has been reported to increase skeletal muscle AKT/PKB phosphorylation and improve the insulin sensitivity in insulin-deficient streptozotocin-treated mice [128]. However, bezafibrate and fenofibrate exhibit weak PPARβ/δ and/or PPARγ agonist activity [125,129]. Therefore, the direct pharmacological activity of PPARα on human skeletal muscle insulin sensitivity requires further investigation.

6.3. Evidence for PPARβ/δ Agonist Treatment of Insulin Resistance and T2DM

PPARβ/δ agonists may be insulin sensitizers and have been suggested as a therapeutic approach for treating metabolic dysfunction and T2DM. Currently, there are no PPARβ/δ agonists that are approved for clinical treatment, but several are in the development and clinical study phases [8]. One prominent PPARβ/δ-selective agonist is seladelpar (MBX-8025), which is currently in clinical phase 2/3 for primary biliary cirrhosis, and has previously been shown to improve the insulin sensitivity and dyslipidemia in overweight subjects [130]. The well-known GW501516, though its development was halted in 2007, has since served as an important PPARβ/δ-specific agonist in the elucidation for PPARβ/δ physiological and pathophysiological functions. In animal models of obesity and T2DM, PPARβ/δ activation, through specific agonists or genetic manipulation, ameliorates hyperglycemia, insulin resistance, and dyslipidemia. PPARβ/δ silencing renders mice glucose intolerant and less metabolically active [131]. Similarly, the skeletal muscle-specific PPARβ/δ knockout mice exhibit insulin insensitivity and impaired glucose tolerance [43]. PPARβ/δ agonist treatment improves whole-body insulin sensitivity through complementary actions in the liver and skeletal muscle. In insulin-resistant ob/ob mice, activating PPARβ/δ through GW501516 ameliorates hyperglycemia-mediated glycolysis, and lipogenesis increases in the liver so as to reduce hepatic glucose output. Simultaneously, GW501516 promotes FAO in the skeletal muscle to enhance insulin sensitivity [131]. In addition, long-term GW501516 treatment in wild-type mice reduces body weight and circulating triglyceride levels [42].

7. Regulation of PPARs during Physical Exercise

Adopting and maintaining physical activity is by far the best intervention and prevention for obesity and T2DM. Short-term aerobic exercise can increase glucose uptake by muscles during exercise and can increase insulin-mediated glucose storage in muscles after exercise [132]. In addition, both short-term exercise and endurance training have been reported to increase PPARβ/δ expression levels in both human and rodent muscles [73,133]. In obese and overweight humans, PPARβ/δ expression levels increase with exercise and are associated with the transcription of oxidative and lipoprotein metabolism genes, as well as PGC-1α [133] (Figure 2). In mice, endogenous PPARβ/δ activation with GW501516 treatment can enhance physical performance and upregulate oxidative genes, mitochondrial biogenesis, and fiber type switching [42]. A recent study showed that GW501516 promotes running endurance by preserving glucose. Activation of muscle PPARβ/δ coordinately reduces glucose catabolism to prevent hypoglycemia and facilitate a progressively longer running time [105]. Similarly, the authors also showed that overexpressing constitutively active PPARβ/δ in rodent skeletal muscle increased the running endurance of these transgenic mice [42]. Furthermore, in the mouse model of ischemic cardiomyopathy, the impaired exercise endurance following myocardial infarction could be reversed by the PPARβ/δ agonist GW501516 [134]. The pharmaceutical activation of PPARβ/δ has attracted much interest as an exercise mimetic to promote oxidative myofibers and running endurance without exercise. Despite a lack of evidence for its clinical safety, GW501516 has become an interest in endurance athletes because of its ability to influence energy expenditure and improve adaptations to training. Unfortunately, this drug has added complexity to the doping dilemma in competitive sports, which has culminated in the suspension of many athletes from the Olympics. The clinical development of PPARβ/δ agonists has been unsuccessful to date, and GW501516 remains a banned metabolic modulator by the World Anti-Doping Agency. Pharmaco-equivalents with better safety profiles, however, are still heavily researched [135,136].

Similarly to PPARβ/δ in humans, the expression levels of PPARα and its downstream target genes increase upon endurance training [133,137]. In skeletal muscle biopsies from spinal cord-injured subjects, the fiber type switching from type 1 oxidative fibers to type II glycolytic fibers often occurs as a result of muscle disuse, and PPARα expression is reduced [138]. In rodents, PPARα knockout mice are less tolerant of endurance exercise, although their skeletal muscle glycogen depletion rate is similar to their wild-type counterparts [25]. Interestingly, genetic variations in PPARα and PPARγ appear to play a role in athleticism. A recent study has found that *PPARA* gene intron 7 G/C polymorphism correlates to an endurance ability. Athletes with high levels of performance in endurance sports have a higher frequency of the GG genotype and G allele [139]. This genotype has also been associated with an increased skeletal muscle fatty acid β-oxidation rate and an increased proportion of type I slow-twitch fibers [140]. The *PPARG* Pro12Ala polymorphism, which is associated with an improved glucose utilization in skeletal muscles, is prevalent in Polish athletes who are involved in sports that involve short-term and intense exercises, such as power-lifters, weight-lifters, and throwers [141].

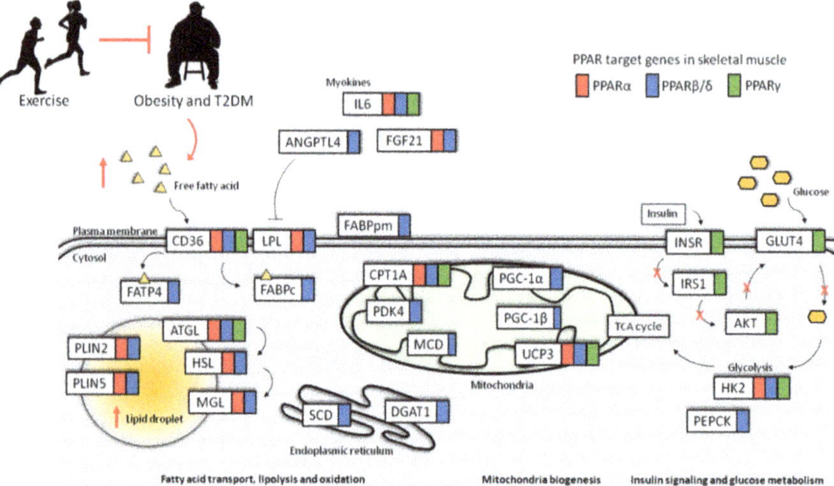

Figure 2. List of genes regulated by PPARα (red box), PPARβ/δ (blue box), and PPARγ (green box) in skeletal muscle. During obesity (red arrows), increased free fatty acid flux leads to excess lipid droplet accumulation, lipid dysregulation, and deregulation of insulin signaling and glucose uptake. Physical exercise can prevent obesity-related disorders and T2DM. Fibroblast growth factor 21, FGF21; malonyl-CoA decarboxylase, MCD; uncoupling protein 1, UCP1; insulin receptor, INSR; hexokinase 2, HK2; and phosphoenolpyruvate, PEPCK.

8. Regulation of Skeletal Muscle Regeneration by PPARs

Skeletal muscle injuries are among the most common soft tissue injuries [142,143], which occur not only during sports traumas and daily activities, but they are also a major concern of diabetic complications, such as muscle ischemia and peripheral vascular disease—the major risk factor of limb amputation in diabetic patients [144,145].

Skeletal muscle regeneration is initiated shortly upon injury and undergoes three main coordinated phases of healing—destruct, repair, and remodel [142]. Upon injury, ruptured myofibres first undergo necrosis, which induces an inflammatory reaction. The damaged tissues are then cleared by infiltrated immune cells, such as macrophages and neutrophils, through phagocytosis [146]. The activation and infiltration of the immune cells further promote the activation of myogenic-reserve stem cells (satellite cells), which then proliferate and differentiate to form new myofibers that orchestrate the muscle reparation [147–149]. During the remodeling phase, angiogenesis of skeletal muscle capillaries and the maturation of regenerated myofibres occur, restoring muscle metabolism and contraction functions [149–152].

8.1. Roles of PPARβ/δ Regulation in Satellite Cells during Muscle Regeneration

After an injury, satellite cells, as the main adult muscle stem cells, get activated and provide an indispensable role during muscle regeneration [153,154]. The satellite cells and their progeny expand as myogenic precursor cells, where most commit towards terminal differentiation and fuse with existing myofibres, so as to regenerate and restore functional myofibers [154]. A small percentage of these myogenic precursor cells, which do not commit into terminal differentiation, return to a quiescent state, providing a pool of satellite cells so as to sustain the muscle's capacity for future regeneration [155]. Satellite cells are notoriously difficult to study, because of their low abundance under the basal lamina of skeletal muscle. Currently, knowledge of human satellite cells is limited, and most of the studies of satellite cells are performed using mice models [155].

PPARβ/δ has been shown to be important for the proper maintenance of satellite cells, as well as postnatal muscle myogenesis, and it is better studied among the PPAR proteins, because of its abundant expression in skeletal muscle. The specific ablation of PPARβ/δ in the mouse satellite cells has been reported, with approximately 40% fewer satellite cells than their wild-type littermates [156]. A similar observation was also reported in total PPARβ/δ-knock out mice [157]. Mice with PPARβ/δ-deficient muscle progenitor cells exhibited impaired muscle regeneration after cardiotoxin-induced injury and exhibited reduced growth kinetics and proliferation in primary cultures [156]. Furthermore, these mice developed metabolic syndrome upon aging, similar to the PPARβ/δ knockout mice [43,156,157]. The authors found reduced foxhead box protein (FOXO1) expression in quiescent PPARβ/δ-deficient satellite cells, which impaired the proliferation and differentiation ability of these satellite cells during muscle regeneration, thus suggesting that PPARβ/δ regulates the regenerative capability of skeletal muscle through FOXO1 [156]. In addition, CPT1β expression was also found to be reduced during quiescence , but the differences were abolished on day 5 of muscle regeneration [156], suggesting a possible PPARβ/δ-regulated metabolic role during quiescence [156].

Recent findings on the role of the lipid and glucose metabolism in stem cell cellular homeostasis have been increasingly postulated to be vital in stem cell maintenance and their proliferative activity [158,159]. Delineation of cellular metabolism in satellite cell fate could potentially offer pharmacological strategies in the treatment of degenerative muscle diseases, such as Duchenne muscular dystrophy (DMD). PPARβ/δ has been suggested as a direct transcriptional regulator of utrophin A, a key member of the dystrophin-associated protein complex [160,161]. The expression of utrophin A, stimulated by the PPARβ/δ agonist, GW501516, in the mdx mouse model of DMD has been shown to improve sarcolemma integrity, protect muscles from contraction-induced damage, and help to alleviate muscle wasting, which ultimately slowed down the disease progression [161]. Therefore, understanding the function of PPARβ/δ, and potentially the two other PPAR members, in skeletal muscle progenitor cells has important implications for muscle regeneration and the treatment of degenerative muscle diseases.

8.2. PPAR-Regulated Paracrine Networks between Muscle and Other Cell Types

Inflammation, specifically the infiltration of macrophages during early phases of muscle regeneration, is a major component for efficient healing and repair. Varga et al. [162] showed that myeloid-specific conditional PPARγ knockout mice exhibited a pronounced delay in muscle regeneration following a toxin-induced injury, compared with their wild-type counterparts. The injured muscle in these mice displayed a reduced muscle differentiation without differences in macrophage infiltration and phagocytic activity. They determined that the macrophage secretion of growth differentiation factor 3 (GDF3), through a direct PPARγ regulation, is a potent inducer of myotube formation, demonstrating the role of PPARγ-dependent paracrine signaling between the infiltrated macrophages and regenerating muscle [162].

Skeletal muscle is known to be highly vascularised, and numerous studies have demonstrated the importance of myogenesis and angiogenesis during skeletal muscle regeneration [163–165]. Recent findings on the PPARβ/δ-modulated paracrine network between the endothelial progenitor cells and regenerating myofibers, have been reported to promote both myogenesis and capillary angiogenesis [165]. PPARβ/δ activation in endothelial progenitor cells promotes insulin-like growth factor 1 (IGF1) signaling pathway in both the skeletal muscle and endothelial cells, via a direct PPARβ/δ induced transcriptional activation of matrix metalloproteinase 9 (MMP9) [165]. Matrix metalloproteinases are well known for their proteolytic activities in the extracellular matrix and they promote angiogenesis [166]. The increased MMP9 secretion from PPARβ/δ agonist-treated endothelial progenitor cells, promotes the (MMP9)-mediated insulin-like growth factor-binding protein 3 (IGFBP3) proteolysis, and thereby modulates the IGF1 activity [165,167]. The MMP9-dependent increase in IGF1 signaling was further demonstrated via the transplantation of PPARβ/δ-activated endothelial progenitor cells to a hindlimb ischaemic mice model. These mice showed an increase in regenerating the myofiber numbers and an enhanced capillary-to-myocyte ratio. The enhanced muscle regeneration and increased angiogenesis promoted a better muscle architecture with reduced fibrosis, and thereby protected the ischaemic limb from hypoxic damage [165].

Interestingly, recent reports on adiponectin produced by skeletal muscle as a myokine, exert anti-diabetic metabolic effects similar to PPAR activation [168]. The skeletal muscle-derived adiponectin has been demonstrated to regulate the fatty acid metabolism, increase glucose uptake, and induce mitochondrial biogenesis, through human skeletal muscle primary culture, muscle biopsies, and gain/loss function studies in rodent models [169–172]. Adiponectin promotes fatty acid uptake and oxidation through a series of sequential activation, involving AMPK, p38 mitogen-activated protein kinase (MAPK), and PPARα. In skeletal muscle, the activation of AMPK has been known to inhibit lipid biosynthesis through the phosphorylation of acetyl-CoA carboxylase (ACC) [173,174]. Indeed, adiponectin treatment in mouse myotube inhibited ACC phosphorylation in a time-dependent manner [170]. The PPARγ agonist, rosiglitazone, has been shown to induce adiponectin production and secretion directly [175], and is directly correlated with the rosiglitazone-mediated improvement in insulin sensitivity [176]. The overexpression of PPARγ in the mouse skeletal muscle also increased adiponectin expression, which protected these mice from high-fat diet induced insulin resistance [117].

9. Regulation of PPARs during Aging

Both physical exercise and aging are two physiological situations that have marked, but opposite, effects on muscle mass. Aging is a complex and multifactorial process that is characterized by progressive, endogenous, and irreversible alterations in cellular signaling, and it is associated with the slow and concerted decline of physiological functions [177]. Moreover, age is the single most significant risk factor for metabolic disorders, such as obesity, T2DM, and other major debilitating and life-threatening conditions [178]. In humans, aging leads to a loss of muscle mass, though the magnitude of loss varies substantially among individuals [179]. Age-related muscle loss is also accompanied by fiber type transformation, metabolic changes, and ectopic fat accumulation over time [180]. In aged muscles, type II glycolytic fibers, particularly type IIx, are susceptible to both atrophy and fiber type switching [180]. Compared to the percentage of glycolytic fibers, an increased percentage of oxidative fibers has been reported in the elderly [181]. Although type I muscle fiber size is largely unaffected [182,183], lower maximal force generation by type I and type IIa fibers was observed in older men, in comparison to that of the similar fibers in younger men [184].

Evidence for the Involvement of PPARs during Aging

In aged muscles, all three PPAR expression levels are decreased and contribute to carbohydrate-lipid metabolism dysregulation, reduced muscle regeneration, and fiber remodeling [185–187]. In addition, the PGC-1α expression levels, as well as both the oxidative and glycolytic enzymatic capacity, are compromised in the aged skeletal muscle. The age-related decreases in fat oxidation have been consistently associated with reductions in both the quantity and the oxidative capacity to metabolize fats [188]. Lipid metabolism may be further impaired because of the increased lipid accumulation in aged muscle [189]. The decrease in both myonuclear density and mitochondria numbers in aged muscle has been associated with PPARβ/δ deficiency [43,186]. In rodents, PPARβ/δ overexpression and pharmacological activation stimulate nuclei accretion through the fusion of pre-existing muscle precursor cells to myofibers [186,190]. PPARβ/δ agonist treatment in aged mice restores the muscle fiber distribution profile and the oxidative capacity of the fast-twitch fibers, similar to those of the young untreated counterparts [186].

PPARα may play a role in glucose utilization in aged muscle. In PPARα knockout mice, an age-dependent reduction in glycolysis has been observed in the soleus muscle, which comprises mainly of slow-twitch type I fibers [185]. In addition, decreased muscle glycogen concentrations have been detected in aged PPARα-deficient mice. This suggests a role for PPARα in modulating metabolic changes during the normal aging process. Interestingly, the clinical use of fibrates may cause muscle weakness and pain (myopathy), or rhabdomyolysis in rare cases [191]. The exact mechanism of PPARα activation in diseased and aged skeletal muscle remains unclear. However, the mechanism may be partly mediated by the increased oxidative stress and tissue damage associated with PPARα-induced activity [192,193].

Aging is associated with progressive declines in both insulin sensitivity and glucose tolerance [194,195]. These effects are partly caused by decreased insulin production by the pancreatic islets and deregulated insulin signaling in muscle [196]. The PPARγ and GLUT4 expression levels are reduced in the skeletal muscle of aged rodents and humans [187,197,198]. In middle-aged adults with both diabetic and non-diabetic histories, insulin-sensitizing TZD compounds improve insulin sensitivity and glucose tolerance, and increase the likelihood of regression from pre-diabetes to normal glucose regulation [199,200]. In aged rodents, rosiglitazone treatment reverses age-related alterations in plasma triglyceride and glucose levels [201]. Paradoxically, in aged animals, mice that were heterogenous for PPARγ displayed greater insulin sensitivity than their wild-type counterparts [202]. This increased insulin sensitivity was lost upon TZD treatment or high-fat diet administration [203]. The authors suggest that PPARγ deficiency partially protects from normal physiological age-induced decreases in insulin sensitivity. In short, the physiological impact and role of diminished PPARγ expression in insulin resistance during the aging process are not clearly understood.

Although PPAR activation has beneficial effects on various metabolic dysfunctions, its beneficial effects on the aging process are not fully understood. More importantly, given the complexity of aging, there are other factors that contribute to aging that have not been discussed here. However, increasing evidence demonstrates that countermeasures can improve age-related metabolic syndromes and muscle loss, partially through modulating endogenous PPAR expression. In addition to pharmacological PPAR activation, interventions such as exercise have been shown to preserve muscle integrity in both aging humans and rodent models. The molecular changes in both lipid and glucose metabolism, after a single bout of exercise in aged humans, have been reported to increase skeletal muscle insulin action [204]. The loss of muscle mass not only reduces mobility and functional capacities which affect the quality of life, but also increases the risks associated with falls and age-related diseases. Developing treatments for age-related and disease-related muscle loss may improve the active life expectancy of older adults, thus leading to substantial health-care savings and an improved quality of life.

10. Concluding Remarks and Perspectives

Numerous studies have provided compelling evidence for important roles of PPAR in skeletal muscle physiology. The capacity to modulate PPAR activity with appropriate agonists or antagonist, further underscores their potential as therapeutic targets. However, the widespread use of these ligands is plagued by their accompanying side effects. Beside myopathy, fibrates are also known to increase the risk for gallstones formation [205] and renal failure [206]. The safety reputation of TZDs suffered as well when the extended use of rosiglitazone and pioglitazone were associated with an increased risk of heart attack/stroke and bladder cancer [207,208]. Although drugs for PPARβ/δ have not been clinically approved, the selective agonist GW501516 has been sold illegally as an endurance booster by its online supplement name, endurobol. GW501516 has been included in the banned substance list since 2009 by the World Anti-Doping Agency, and was re-categorized as a 'hormone and metabolic modulator' drug in 2012. The clinical development of GW501516 was halted in 2007 after increased incidences of several cancer types were observed in rodents [209]. Recent developments in dual- and pan-PPAR agonists displayed therapeutic benefits for the complex and wide-range metabolic disorders [8]. One example is saroglitazar, a dual PPARα/γ agonist, currently approved in India for the treatment of T2DM and dyslipidemia. Thus, the pharmacological effort in the development of combined PPARs therapeutic effects, with reduced side effects, will be crucial for next-generation drug candidates for metabolic disorders.

Skeletal muscle has been identified as an endocrine organ that expresses and releases myokines as messengers among different organs, as well as within the muscle itself. There are limited studies on the effect of PPAR on the expression of myokines, and even fewer studies on the reciprocal effect of myokines on PPAR expression and activity. For example, the expression of angiopoietin-like 4 (ANGPTL4) is an exercise-responsive myokine and is regulated by PPARs [210,211]. ANGPTL4 may regulate the lipoprotein lipase-dependent plasma clearance of triglyceride from the skeletal muscle during exercise. Another prominent PPAR-regulated myokine is interleukin-6 (IL6), whose expression can be paradoxically exercised-induced or increased during obesity and T2DM [212]. The exact mechanistic involvement of muscle-derived IL6 in health and disease, however, remains elusive, and almost nothing for the IL6 autocrine feedback regulation on PPAR. It is conceivable that pharmacological compounds that mimic the benefits of exercises will also be helpful for elderly adults, as well as for individuals with poignant mobility impairment [213].

The impact of gut microbiota on the whole-body physiology is beginning to be recognized. The bidirectional signaling between the gut microbiota and the brain has been shown to influence neurotransmission and alter behavioral responses through the changes of microbiota-derived metabolites composition. One of the dominant gut-derived metabolites are the short chains fatty acids, such as acetate and propionate, which have been shown to strongly exhibit anti-lipolysis activity in the adipose tissue [214,215].

The gut microbiota and their metabolites or components can modulate the immune system, based on their translocation into tissues and the circulatory system [216]. In recent years, the gut microbiota has been implicated in altered skeletal muscle fiber type proportions in obese porcine, offering a new perspective on the development of dietary supplements for muscle maintenance and regeneration [217]. However, the biological impact, as well as the cause and effect of this gut-muscle connection, remains to be fully understood.

In conclusion, it is clear that PPARs play an essential role in regulating energy homeostasis in skeletal muscle. It is foreseeable that, with a new development in drug design and a better understanding of PPAR's relationship with myokines, among others, PPARs remain important pharmaceutical targets for the therapeutic strategies in order to combat different facets of metabolic syndrome.

Acknowledgments: This research is supported by the Singapore Ministry of Education under its Singapore Ministry of Education Academic Research Fund Tier 1 (2014-T1-002-138-04) to N.S.T., and W.W.T.P is a recipient of the scholarship from the Interdisciplinary Graduate School, NTU, Singapore.We apologize to all fellow scientists whose works were not cited in this review.

Conflicts of Interest: The authors declare no conflict of interest.

Abbreviations

acetyl-CoA	acetyl-coenzyme A
AF1	activation function 1
AF2	activation function 2
AKT/PKB	protein kinase B
AMPK	AMP-activated protein kinase
ANGPTL4	angiopoietin-like 4
ATGL	adipose triglyceride lipase
BMI	body mass index

CD36	cluster of differentiation 36/SR-B2
CPT1	carnitine palmitoyltransferase I
CRY1	Cryptochrome 1
DBD	DNA binding domain
DGAT1	diacylglycerol acyltransferase 1
ER	endoplasmic reticulum
FABP3	fatty acid binding protein 3
FATP	fatty acid transport protein
FATPc	cytoplasmic FABP
FATPpm	plasma membrane FABP
GLUT4	glucose transporter 4
HSL	hormone-sensitive lipase
IGFBP3	insulin-like growth factor-binding protein 3
IL6	interleukin-6
IMCL	intramyocellular lipid
IRS1	insulin receptor substrate 1
LBD	ligand binding domain
LD	lipid droplets
LPL	lipoprotein lipase
MAG	monoacylglycerol
MAPK	mitogen-activated protein kinase
MGL	monoacylglycerol lipase
NRF	nuclear respiratory factor
PDC	pyruvate dehydrogenase complex
PDK	pyruvate dehydrogenase kinase
PGC-1	PPARγ-coactivator 1
PI3K	phosphatidylinositol-3 kinase
PKC	protein kinase C
PLIN	perilipin
PP2A	protein phosphatase 2A
PPAR	peroxisome proliferator-activated receptor
PPRE	peroxisome proliferator response element
RXR	retinoid X receptors
SCD	stearoyl-CoA desaturase
T2DM	type 2 diabetes mellitus
TA	tibialis anterior
TCA	tricarboxylic acid
TFAM	mitochondrial transcription factor A
TFB1M	mitochondrial transcription factors B1
TFB2M	mitochondrial transcription factors B2
TZD	Thiazolidinediones

References

1. McLeod, M.; Breen, L.; Hamilton, D.L.; Philp, A. Live strong and prosper: The importance of skeletal muscle strength for healthy ageing. *Biogerontology* **2016**, *17*, 497–510. [CrossRef] [PubMed]
2. Li, M.; Zhou, X.; Chen, Y.; Nie, Y.; Huang, H.; Chen, H.; Mo, D. Not all the number of skeletal muscle fibers is determined prenatally. *BMC Dev. Biol.* **2015**, *15*, 42. [CrossRef] [PubMed]
3. Albers, P.H.; Pedersen, A.J.; Birk, J.B.; Kristensen, D.E.; Vind, B.F.; Baba, O.; Nohr, J.; Hojlund, K.; Wojtaszewski, J.F. Human muscle fiber type-specific insulin signaling: Impact of obesity and type 2 diabetes. *Diabetes* **2015**, *64*, 485–497. [CrossRef] [PubMed]
4. Schiaffino, S.; Reggiani, C. Fiber types in mammalian skeletal muscles. *Physiol. Rev.* **2011**, *91*, 1447–1531. [CrossRef] [PubMed]

5. Dube, J.J.; Coen, P.M.; DiStefano, G.; Chacon, A.C.; Helbling, N.L.; Desimone, M.E.; Stafanovic-Racic, M.; Hames, K.C.; Despines, A.A.; Toledo, F.G.; et al. Effects of acute lipid overload on skeletal muscle insulin resistance, metabolic flexibility, and mitochondrial performance. *Am. J. Physiol. Endocrinol. Metab.* **2014**, *307*, E1117–E1124. [CrossRef] [PubMed]
6. Stump, C.S.; Henriksen, E.J.; Wei, Y.; Sowers, J.R. The metabolic syndrome: Role of skeletal muscle metabolism. *Ann. Med.* **2006**, *38*, 389–402. [CrossRef] [PubMed]
7. Ehrenborg, E.; Krook, A. Regulation of skeletal muscle physiology and metabolism by peroxisome proliferator-activated receptor delta. *Pharmacol. Rev.* **2009**, *61*, 373–393. [CrossRef] [PubMed]
8. Tan, C.K.; Zhuang, Y.; Wahli, W. Synthetic and natural peroxisome proliferator-activated receptor (PPAR) agonists as candidates for the therapy of the metabolic syndrome. *Expert Opin. Ther. Targets* **2017**, *21*, 333–348. [CrossRef] [PubMed]
9. Tan, N.S.; Vazquez-Carrera, M.; Montagner, A.; Sng, M.K.; Guillou, H.; Wahli, W. Transcriptional control of physiological and pathological processes by the nuclear receptor pparbeta/delta. *Prog. Lipid Res.* **2016**, *64*, 98–122. [CrossRef] [PubMed]
10. Fajas, L.; Auboeuf, D.; Raspe, E.; Schoonjans, K.; Lefebvre, A.M.; Saladin, R.; Najib, J.; Laville, M.; Fruchart, J.C.; Deeb, S.; et al. The organization, promoter analysis, and expression of the human ppargamma gene. *J. Biol. Chem.* **1997**, *272*, 18779–18789. [CrossRef] [PubMed]
11. Xu, H.E.; Lambert, M.H.; Montana, V.G.; Plunket, K.D.; Moore, L.B.; Collins, J.L.; Oplinger, J.A.; Kliewer, S.A.; Gampe, R.T., Jr.; McKee, D.D.; et al. Structural determinants of ligand binding selectivity between the peroxisome proliferator-activated receptors. *Proc. Natl. Acad. Sci. USA* **2001**, *98*, 13919–13924. [CrossRef] [PubMed]
12. Zoete, V.; Grosdidier, A.; Michielin, O. Peroxisome proliferator-activated receptor structures: Ligand specificity, molecular switch and interactions with regulators. *Biochim. Biophys. Acta* **2007**, *1771*, 915–925. [CrossRef] [PubMed]
13. Escher, P.; Wahli, W. Peroxisome proliferator-activated receptors: Insight into multiple cellular functions. *Mutat. Res.* **2000**, *448*, 121–138. [CrossRef]
14. Shalev, A.; Siegrist-Kaiser, C.A.; Yen, P.M.; Wahli, W.; Burger, A.G.; Chin, W.W.; Meier, C.A. The peroxisome proliferator-activated receptor alpha is a phosphoprotein: Regulation by insulin. *Endocrinology* **1996**, *137*, 4499–4502. [CrossRef] [PubMed]
15. Blanquart, C.; Barbier, O.; Fruchart, J.C.; Staels, B.; Glineur, C. Peroxisome proliferator-activated receptor alpha (PPARα) turnover by the ubiquitin-proteasome system controls the ligand-induced expression level of its target genes. *J. Biol. Chem.* **2002**, *277*, 37254–37259. [CrossRef] [PubMed]
16. Genini, D.; Catapano, C.V. Block of nuclear receptor ubiquitination. A mechanism of ligand-dependent control of peroxisome proliferator-activated receptor delta activity. *J. Biol. Chem.* **2007**, *282*, 11776–11785. [CrossRef] [PubMed]
17. Wadosky, K.M.; Willis, M.S. The story so far: Post-translational regulation of peroxisome proliferator-activated receptors by ubiquitination and sumoylation. *Am. J. Physiol. Heart Circ. Physiol.* **2012**, *302*, H515–H526. [CrossRef] [PubMed]
18. Jordan, S.D.; Kriebs, A.; Vaughan, M.; Duglan, D.; Fan, W.; Henriksson, E.; Huber, A.L.; Papp, S.J.; Nguyen, M.; Afetian, M.; et al. Cry1/2 selectively repress PPARδ and limit exercise capacity. *Cell Metab.* **2017**, *26*, 243–255.e6. [CrossRef] [PubMed]
19. Abbott, B.D. Review of the expression of peroxisome proliferator-activated receptors alpha (PPARα), beta (PPARβ), and gamma (PPARγ) in rodent and human development. *Reprod. Toxicol.* **2009**, *27*, 246–257. [CrossRef] [PubMed]
20. Lehrke, M.; Lazar, M.A. The many faces of ppargamma. *Cell* **2005**, *123*, 993–999. [CrossRef] [PubMed]
21. Giordano, A.G.M.; Desvergne, B. Integrative and systemic approaches for evaluating PPARβ/δ (PPARD) function. *Nucl. Recept. Signal.* **2015**, *13*, e001.
22. Barak, Y.; Nelson, M.C.; Ong, E.S.; Jones, Y.Z.; Ruiz-Lozano, P.; Chien, K.R.; Koder, A.; Evans, R.M. PPARγ is required for placental, cardiac, and adipose tissue development. *Mol. Cell* **1999**, *4*, 585–595. [CrossRef]
23. Braissant, O.; Foufelle, F.; Scotto, C.; Dauca, M.; Wahli, W. Differential expression of peroxisome proliferator-activated receptors (PPARs): Tissue distribution of PPAR-α, -β, and -γ in the adult rat. *Endocrinology* **1996**, *137*, 354–366. [CrossRef] [PubMed]

24. De Lange, P.; Ragni, M.; Silvestri, E.; Moreno, M.; Schiavo, L.; Lombardi, A.; Farina, P.; Feola, A.; Goglia, F.; Lanni, A. Combined cdna array/RT-PCR analysis of gene expression profile in rat gastrocnemius muscle: Relation to its adaptive function in energy metabolism during fasting. *FASEB J.* **2004**, *18*, 350–352. [CrossRef] [PubMed]
25. Muoio, D.M.; MacLean, P.S.; Lang, D.B.; Li, S.; Houmard, J.A.; Way, J.M.; Winegar, D.A.; Corton, J.C.; Dohm, G.L.; Kraus, W.E. Fatty acid homeostasis and induction of lipid regulatory genes in skeletal muscles of peroxisome proliferator-activated receptor (PPAR) α knock-out mice. Evidence for compensatory regulation by ppar delta. *J. Biol. Chem.* **2002**, *277*, 26089–26097. [CrossRef] [PubMed]
26. Grygiel-Gorniak, B. Peroxisome proliferator-activated receptors and their ligands: Nutritional and clinical implications—A review. *Nutr. J.* **2014**, *13*, 17. [CrossRef] [PubMed]
27. Peeters, A.; Baes, M. Role of PPARα in hepatic carbohydrate metabolism. *PPAR Res.* **2010**, *2010*, 572405. [CrossRef] [PubMed]
28. Kido, Y.; Nakae, J.; Accili, D. Clinical review 125: The insulin receptor and its cellular targets. *J. Clin. Endocrinol. Metab.* **2001**, *86*, 972–979. [PubMed]
29. Foley, K.; Boguslavsky, S.; Klip, A. Endocytosis, recycling, and regulated exocytosis of glucose transporter 4. *Biochemistry* **2011**, *50*, 3048–3061. [CrossRef] [PubMed]
30. Thiebaud, D.; Jacot, E.; DeFronzo, R.A.; Maeder, E.; Jequier, E.; Felber, J.P. The effect of graded doses of insulin on total glucose uptake, glucose oxidation, and glucose storage in man. *Diabetes* **1982**, *31*, 957–963. [CrossRef] [PubMed]
31. Dashty, M. A quick look at biochemistry: Carbohydrate metabolism. *Clin. Biochem.* **2013**, *46*, 1339–1352. [CrossRef] [PubMed]
32. Hue, L.; Taegtmeyer, H. The randle cycle revisited: A new head for an old hat. *Am. J. Physiol. Endocrinol. Metab.* **2009**, *297*, E578–E591. [CrossRef] [PubMed]
33. Randle, P.J. Regulatory interactions between lipids and carbohydrates: The glucose fatty acid cycle after 35 years. *Diabetes Metab. Rev.* **1998**, *14*, 263–283. [CrossRef]
34. Peters, S.J.; Harris, R.A.; Heigenhauser, G.J.; Spriet, L.L. Muscle fiber type comparison of PDH kinase activity and isoform expression in fed and fasted rats. *Am. J. Physiol. Regul. Integr. Comp. Physiol.* **2001**, *280*, R661–R668. [CrossRef] [PubMed]
35. Spriet, L.L.; Tunstall, R.J.; Watt, M.J.; Mehan, K.A.; Hargreaves, M.; Cameron-Smith, D. Pyruvate dehydrogenase activation and kinase expression in human skeletal muscle during fasting. *J. Appl. Physiol. (1985)* **2004**, *96*, 2082–2087. [CrossRef] [PubMed]
36. Degenhardt, T.; Saramaki, A.; Malinen, M.; Rieck, M.; Vaisanen, S.; Huotari, A.; Herzig, K.H.; Muller, R.; Carlberg, C. Three members of the human pyruvate dehydrogenase kinase gene family are direct targets of the peroxisome proliferator-activated receptor β/δ. *J. Mol. Biol.* **2007**, *372*, 341–355. [CrossRef] [PubMed]
37. Nahle, Z.; Hsieh, M.; Pietka, T.; Coburn, C.T.; Grimaldi, P.A.; Zhang, M.Q.; Das, D.; Abumrad, N.A. CD36-dependent regulation of muscle foxo1 and PDK4 in the PPAR β/δ -mediated adaptation to metabolic stress. *J. Biol. Chem.* **2008**, *283*, 14317–14326. [CrossRef] [PubMed]
38. Frayn, K.N.; Arner, P.; Yki-Jarvinen, H. Fatty acid metabolism in adipose tissue, muscle and liver in health and disease. *Essays Biochem.* **2006**, *42*, 89–103. [CrossRef] [PubMed]
39. Ciaraldi, T.P.; Cha, B.S.; Park, K.S.; Carter, L.; Mudaliar, S.R.; Henry, R.R. Free fatty acid metabolism in human skeletal muscle is regulated by PPARγ and RXR agonists. *Ann. N. Y. Acad. Sci.* **2002**, *967*, 66–70. [CrossRef] [PubMed]
40. Fan, W.; Evans, R. Ppars and errs: Molecular mediators of mitochondrial metabolism. *Curr. Opin. Cell Biol.* **2015**, *33*, 49–54. [CrossRef] [PubMed]
41. Finck, B.N.; Bernal-Mizrachi, C.; Han, D.H.; Coleman, T.; Sambandam, N.; LaRiviere, L.L.; Holloszy, J.O.; Semenkovich, C.F.; Kelly, D.P. A potential link between muscle peroxisome proliferator-activated receptor-alpha signaling and obesity-related diabetes. *Cell Metab.* **2005**, *1*, 133–144. [CrossRef] [PubMed]
42. Wang, Y.-X.; Zhang, C.-L.; Yu, R.T.; Cho, H.K.; Nelson, M.C.; Bayuga-Ocampo, C.R.; Ham, J.; Kang, H.; Evans, R.M. Regulation of muscle fiber type and running endurance by PPARδ. *PLoS Biol.* **2004**, *2*, e294. [CrossRef] [PubMed]
43. Schuler, M.; Ali, F.; Chambon, C.; Duteil, D.; Bornert, J.M.; Tardivel, A.; Desvergne, B.; Wahli, W.; Chambon, P.; Metzger, D. Pgc1alpha expression is controlled in skeletal muscles by PPARβ, whose ablation results in fiber-type switching, obesity, and type 2 diabetes. *Cell Metab.* **2006**, *4*, 407–414. [CrossRef] [PubMed]

44. Harmon, C.M.; Luce, P.; Beth, A.H.; Abumrad, N.A. Labeling of adipocyte membranes by sulfo-*n*-succinimidyl derivatives of long-chain fatty acids: Inhibition of fatty acid transport. *J. Membr. Biol.* **1991**, *121*, 261–268. [CrossRef] [PubMed]
45. Glatz, J.F.; Luiken, J.J. From fat to fat (CD36/SR-B2): Understanding the regulation of cellular fatty acid uptake. *Biochimie* **2017**, *136*, 21–26. [CrossRef] [PubMed]
46. Watt, M.J.; Hoy, A.J. Lipid metabolism in skeletal muscle: Generation of adaptive and maladaptive intracellular signals for cellular function. *Am. J. Physiol. Endocrinol. Metab.* **2012**, *302*, E1315–E1328. [CrossRef] [PubMed]
47. Nickerson, J.G.; Alkhateeb, H.; Benton, C.R.; Lally, J.; Nickerson, J.; Han, X.X.; Wilson, M.H.; Jain, S.S.; Snook, L.A.; Glatz, J.F.; et al. Greater transport efficiencies of the membrane fatty acid transporters FAT/CD36 and FATP4 compared with FABPpm and FATP1 and differential effects on fatty acid esterification and oxidation in rat skeletal muscle. *J. Biol. Chem.* **2009**, *284*, 16522–16530. [CrossRef] [PubMed]
48. Holloway, G.P.; Lally, J.; Nickerson, J.G.; Alkhateeb, H.; Snook, L.A.; Heigenhauser, G.J.; Calles-Escandon, J.; Glatz, J.F.; Luiken, J.J.; Spriet, L.L.; et al. Fatty acid binding protein facilitates sarcolemmal fatty acid transport but not mitochondrial oxidation in rat and human skeletal muscle. *J. Physiol.* **2007**, *582*, 393–405. [CrossRef] [PubMed]
49. Storch, J.; Thumser, A.E. Tissue-specific functions in the fatty acid-binding protein family. *J. Biol. Chem.* **2010**, *285*, 32679–32683. [CrossRef] [PubMed]
50. Koonen, D.P.; Glatz, J.F.; Bonen, A.; Luiken, J.J. Long-chain fatty acid uptake and FAT/CD36 translocation in heart and skeletal muscle. *Biochim. Biophys. Acta* **2005**, *1736*, 163–180. [CrossRef] [PubMed]
51. Schaap, F.G.; Binas, B.; Danneberg, H.; van der Vusse, G.J.; Glatz, J.F. Impaired long-chain fatty acid utilization by cardiac myocytes isolated from mice lacking the heart-type fatty acid binding protein gene. *Circ. Res.* **1999**, *85*, 329–337. [CrossRef] [PubMed]
52. Escriba, P.V.; Gonzalez-Ros, J.M.; Goni, F.M.; Kinnunen, P.K.; Vigh, L.; Sanchez-Magraner, L.; Fernandez, A.M.; Busquets, X.; Horvath, I.; Barcelo-Coblijn, G. Membranes: A meeting point for lipids, proteins and therapies. *J. Cell. Mol. Med.* **2008**, *12*, 829–875. [CrossRef] [PubMed]
53. Van Blitterswijk, W.J.; van der Luit, A.H.; Veldman, R.J.; Verheij, M.; Borst, J. Ceramide: Second messenger or modulator of membrane structure and dynamics? *Biochem. J.* **2003**, *369*, 199–211. [CrossRef] [PubMed]
54. Armstrong, E.H.; Goswami, D.; Griffin, P.R.; Noy, N.; Ortlund, E.A. Structural basis for ligand regulation of the fatty acid-binding protein 5, peroxisome proliferator-activated receptor β/δ (FABP5-PPARβ/δ) signaling pathway. *J. Biol. Chem.* **2014**, *289*, 14941–14954. [CrossRef] [PubMed]
55. Schug, T.T.; Berry, D.C.; Shaw, N.S.; Travis, S.N.; Noy, N. Opposing effects of retinoic acid on cell growth result from alternate activation of two different nuclear receptors. *Cell* **2007**, *129*, 723–733. [CrossRef] [PubMed]
56. Wolfrum, C.; Borrmann, C.M.; Borchers, T.; Spener, F. Fatty acids and hypolipidemic drugs regulate peroxisome proliferator-activated receptors α—and γ-mediated gene expression via liver fatty acid binding protein: A signaling path to the nucleus. *Proc. Natl. Acad. Sci. USA* **2001**, *98*, 2323–2328. [CrossRef] [PubMed]
57. Ayers, S.D.; Nedrow, K.L.; Gillilan, R.E.; Noy, N. Continuous nucleocytoplasmic shuttling underlies transcriptional activation of PPARγ by FABP4. *Biochemistry* **2007**, *46*, 6744–6752. [CrossRef] [PubMed]
58. Tan, N.S.; Shaw, N.S.; Vinckenbosch, N.; Liu, P.; Yasmin, R.; Desvergne, B.; Wahli, W.; Noy, N. Selective cooperation between fatty acid binding proteins and peroxisome proliferator-activated receptors in regulating transcription. *Mol. Cell. Biol.* **2002**, *22*, 5114–5127. [CrossRef] [PubMed]
59. Bezaire, V.; Langin, D. Regulation of adipose tissue lipolysis revisited. *Proc. Nutr. Soc.* **2009**, *68*, 350–360. [CrossRef] [PubMed]
60. Alsted, T.J.; Ploug, T.; Prats, C.; Serup, A.K.; Hoeg, L.; Schjerling, P.; Holm, C.; Zimmermann, R.; Fledelius, C.; Galbo, H.; et al. Contraction-induced lipolysis is not impaired by inhibition of hormone-sensitive lipase in skeletal muscle. *J. Physiol.* **2013**, *591*, 5141–5155. [CrossRef] [PubMed]
61. Gronke, S.; Mildner, A.; Fellert, S.; Tennagels, N.; Petry, S.; Muller, G.; Jackle, H.; Kuhnlein, R.P. Brummer lipase is an evolutionary conserved fat storage regulator in drosophila. *Cell Metab.* **2005**, *1*, 323–330. [CrossRef] [PubMed]
62. Biswas, D.; Ghosh, M.; Kumar, S.; Chakrabarti, P. PPARα-ATGL pathway improves muscle mitochondrial metabolism: Implication in aging. *FASEB J.* **2016**, *30*, 3822–3834. [CrossRef] [PubMed]

63. Haemmerle, G.; Lass, A.; Zimmermann, R.; Gorkiewicz, G.; Meyer, C.; Rozman, J.; Heldmaier, G.; Maier, R.; Theussl, C.; Eder, S.; et al. Defective lipolysis and altered energy metabolism in mice lacking adipose triglyceride lipase. *Science* **2006**, *312*, 734–737. [CrossRef] [PubMed]
64. Haemmerle, G.; Moustafa, T.; Woelkart, G.; Buttner, S.; Schmidt, A.; van de Weijer, T.; Hesselink, M.; Jaeger, D.; Kienesberger, P.C.; Zierler, K.; et al. ATGL-mediated fat catabolism regulates cardiac mitochondrial function via ppar-alpha and pgc-1. *Nat. Med.* **2011**, *17*, 1076–1085. [CrossRef] [PubMed]
65. Huijsman, E.; van de Par, C.; Economou, C.; van der Poel, C.; Lynch, G.S.; Schoiswohl, G.; Haemmerle, G.; Zechner, R.; Watt, M.J. Adipose triacylglycerol lipase deletion alters whole body energy metabolism and impairs exercise performance in mice. *Am. J. Physiol. Endocrinol. Metab.* **2009**, *297*, E505–E513. [CrossRef] [PubMed]
66. Meex, R.C.; Hoy, A.J.; Mason, R.M.; Martin, S.D.; McGee, S.L.; Bruce, C.R.; Watt, M.J. ATGL-mediated triglyceride turnover and the regulation of mitochondrial capacity in skeletal muscle. *Am. J. Physiol. Endocrinol. Metab.* **2015**, *308*, E960–E970. [CrossRef] [PubMed]
67. Jocken, J.W.; Smit, E.; Goossens, G.H.; Essers, Y.P.; van Baak, M.A.; Mensink, M.; Saris, W.H.; Blaak, E.E. Adipose triglyceride lipase (ATGL) expression in human skeletal muscle is type I (oxidative) fiber specific. *Histochem. Cell Biol.* **2008**, *129*, 535–538. [CrossRef] [PubMed]
68. Wu, J.W.; Yang, H.; Wang, S.P.; Soni, K.G.; Brunel-Guitton, C.; Mitchell, G.A. Inborn errors of cytoplasmic triglyceride metabolism. *J. Inherit. Metab. Dis.* **2015**, *38*, 85–98. [CrossRef] [PubMed]
69. Van de Weijer, T.; Havekes, B.; Bilet, L.; Hoeks, J.; Sparks, L.; Bosma, M.; Paglialunga, S.; Jorgensen, J.; Janssen, M.C.; Schaart, G.; et al. Effects of bezafibrate treatment in a patient and a carrier with mutations in the PNPLA2 gene, causing neutral lipid storage disease with myopathy. *Circ. Res.* **2013**, *112*, e51–e54. [CrossRef] [PubMed]
70. Peters, S.J.; Samjoo, I.A.; Devries, M.C.; Stevic, I.; Robertshaw, H.A.; Tarnopolsky, M.A. Perilipin family (PLIN) proteins in human skeletal muscle: The effect of sex, obesity, and endurance training. *Appl. Physiol. Nutr. Metab.* **2012**, *37*, 724–735. [CrossRef] [PubMed]
71. De Wilde, J.; Smit, E.; Snepvangers, F.J.; de Wit, N.W.; Mohren, R.; Hulshof, M.F.; Mariman, E.C. Adipophilin protein expression in muscle—A possible protective role against insulin resistance. *FEBS J.* **2010**, *277*, 761–773. [CrossRef] [PubMed]
72. Feng, Y.Z.; Nikolic, N.; Bakke, S.S.; Boekschoten, M.V.; Kersten, S.; Kase, E.T.; Rustan, A.C.; Thoresen, G.H. PPARδ activation in human myotubes increases mitochondrial fatty acid oxidative capacity and reduces glucose utilization by a switch in substrate preference. *Arch. Physiol. Biochem.* **2014**, *120*, 12–21. [CrossRef] [PubMed]
73. Narkar, V.A.; Downes, M.; Yu, R.T.; Embler, E.; Wang, Y.X.; Banayo, E.; Mihaylova, M.M.; Nelson, M.C.; Zou, Y.; Juguilon, H.; et al. AMPK and PPARδ agonists are exercise mimetics. *Cell* **2008**, *134*, 405–415. [CrossRef] [PubMed]
74. Bindesboll, C.; Berg, O.; Arntsen, B.; Nebb, H.I.; Dalen, K.T. Fatty acids regulate perilipin5 in muscle by activating ppardelta. *J. Lipid Res.* **2013**, *54*, 1949–1963. [CrossRef] [PubMed]
75. Minnaard, R.; Schrauwen, P.; Schaart, G.; Jorgensen, J.A.; Lenaers, E.; Mensink, M.; Hesselink, M.K. Adipocyte differentiation-related protein and OXPAT in rat and human skeletal muscle: Involvement in lipid accumulation and type 2 diabetes mellitus. *J. Clin. Endocrinol. Metab.* **2009**, *94*, 4077–4085. [CrossRef] [PubMed]
76. Shaw, C.S.; Sherlock, M.; Stewart, P.M.; Wagenmakers, A.J. Adipophilin distribution and colocalization with lipid droplets in skeletal muscle. *Histochem. Cell Biol.* **2009**, *131*, 575–581. [CrossRef] [PubMed]
77. Harris, L.A.; Skinner, J.R.; Shew, T.M.; Pietka, T.A.; Abumrad, N.A.; Wolins, N.E. Perilipin 5-driven lipid droplet accumulation in skeletal muscle stimulates the expression of fibroblast growth factor 21. *Diabetes* **2015**, *64*, 2757–2768. [CrossRef] [PubMed]
78. Dalen, K.T.; Dahl, T.; Holter, E.; Arntsen, B.; Londos, C.; Sztalryd, C.; Nebb, H.I. LSDP5 is a pat protein specifically expressed in fatty acid oxidizing tissues. *Biochim. Biophys. Acta* **2007**, *1771*, 210–227. [CrossRef] [PubMed]
79. Wolins, N.E.; Quaynor, B.K.; Skinner, J.R.; Tzekov, A.; Croce, M.A.; Gropler, M.C.; Varma, V.; Yao-Borengasser, A.; Rasouli, N.; Kern, P.A.; et al. OXPAT/PAT-1 is a PPAR-induced lipid droplet protein that promotes fatty acid utilization. *Diabetes* **2006**, *55*, 3418–3428. [CrossRef] [PubMed]

80. Kleiner, S.; Nguyen-Tran, V.; Bare, O.; Huang, X.; Spiegelman, B.; Wu, Z. PPARδ agonism activates fatty acid oxidation via PGC-1α but does not increase mitochondrial gene expression and function. *J. Biol. Chem.* **2009**, *284*, 18624–18633. [CrossRef] [PubMed]
81. Hondares, E.; Pineda-Torra, I.; Iglesias, R.; Staels, B.; Villarroya, F.; Giralt, M. Ppardelta, but not PPARα, activates PGC-1alpha gene transcription in muscle. *Biochem. Biophys. Res. Commun.* **2007**, *354*, 1021–1027. [CrossRef] [PubMed]
82. Hancock, C.R.; Han, D.H.; Chen, M.; Terada, S.; Yasuda, T.; Wright, D.C.; Holloszy, J.O. High-fat diets cause insulin resistance despite an increase in muscle mitochondria. *Proc. Natl. Acad. Sci. USA* **2008**, *105*, 7815–7820. [CrossRef] [PubMed]
83. Handschin, C.; Rhee, J.; Lin, J.; Tarr, P.T.; Spiegelman, B.M. An autoregulatory loop controls peroxisome proliferator-activated receptor gamma coactivator 1 α expression in muscle. *Proc. Natl. Acad. Sci. USA* **2003**, *100*, 7111–7116. [CrossRef] [PubMed]
84. Arany, Z.; Lebrasseur, N.; Morris, C.; Smith, E.; Yang, W.; Ma, Y.; Chin, S.; Spiegelman, B.M. The transcriptional coactivator PGC-1β drives the formation of oxidative type IIx fibers in skeletal muscle. *Cell Metab.* **2007**, *5*, 35–46. [CrossRef] [PubMed]
85. Lin, J.; Wu, H.; Tarr, P.T.; Zhang, C.Y.; Wu, Z.; Boss, O.; Michael, L.F.; Puigserver, P.; Isotani, E.; Olson, E.N.; et al. Transcriptional co-activator PGC-1 α drives the formation of slow-twitch muscle fibres. *Nature* **2002**, *418*, 797–801. [CrossRef] [PubMed]
86. Luquet, S.; Lopez-Soriano, J.; Holst, D.; Fredenrich, A.; Melki, J.; Rassoulzadegan, M.; Grimaldi, P.A. Peroxisome proliferator-activated receptor δ controls muscle development and oxidative capability. *FASEB J.* **2003**, *17*, 2299–2301. [CrossRef] [PubMed]
87. Scarpulla, R.C. Transcriptional paradigms in mammalian mitochondrial biogenesis and function. *Physiol. Rev.* **2008**, *88*, 611–638. [CrossRef] [PubMed]
88. Koh, J.H.; Hancock, C.R.; Terada, S.; Higashida, K.; Holloszy, J.O.; Han, D.H. PPARβ is essential for maintaining normal levels of PGC-1alpha and mitochondria and for the increase in muscle mitochondria induced by exercise. *Cell Metab.* **2017**, *25*, 1176–1185.e5. [CrossRef] [PubMed]
89. Gemmink, A.; Goodpaster, B.H.; Schrauwen, P.; Hesselink, M.K.C. Intramyocellular lipid droplets and insulin sensitivity, the human perspective. *Biochim. Biophys. Acta* **2017**, *1862*, 1242–1249. [CrossRef] [PubMed]
90. Li, Y.; Xu, S.; Zhang, X.; Yi, Z.; Cichello, S. Skeletal intramyocellular lipid metabolism and insulin resistance. *Biophys. Rep.* **2015**, *1*, 90–98. [CrossRef] [PubMed]
91. Erion, D.M.; Shulman, G.I. Diacylglycerol-mediated insulin resistance. *Nat. Med.* **2010**, *16*, 400–402. [CrossRef] [PubMed]
92. Krssak, M.; Falk Petersen, K.; Dresner, A.; DiPietro, L.; Vogel, S.M.; Rothman, D.L.; Roden, M.; Shulman, G.I. Intramyocellular lipid concentrations are correlated with insulin sensitivity in humans: A 1 h NMR spectroscopy study. *Diabetologia* **1999**, *42*, 113–116. [CrossRef] [PubMed]
93. Dobbins, R.L.; Szczepaniak, L.S.; Bentley, B.; Esser, V.; Myhill, J.; McGarry, J.D. Prolonged inhibition of muscle carnitine palmitoyltransferase-1 promotes intramyocellular lipid accumulation and insulin resistance in rats. *Diabetes* **2001**, *50*, 123–130. [CrossRef] [PubMed]
94. Jacob, S.; Machann, J.; Rett, K.; Brechtel, K.; Volk, A.; Renn, W.; Maerker, E.; Matthaei, S.; Schick, F.; Claussen, C.D.; et al. Association of increased intramyocellular lipid content with insulin resistance in lean nondiabetic offspring of type 2 diabetic subjects. *Diabetes* **1999**, *48*, 1113–1119. [CrossRef] [PubMed]
95. Boden, G.; Lebed, B.; Schatz, M.; Homko, C.; Lemieux, S. Effects of acute changes of plasma free fatty acids on intramyocellular fat content and insulin resistance in healthy subjects. *Diabetes* **2001**, *50*, 1612–1617. [CrossRef] [PubMed]
96. Goodpaster, B.H.; He, J.; Watkins, S.; Kelley, D.E. Skeletal muscle lipid content and insulin resistance: Evidence for a paradox in endurance-trained athletes. *J. Clin. Endocrinol. Metab.* **2001**, *86*, 5755–5761. [CrossRef] [PubMed]
97. Chavez, J.A.; Summers, S.A. A ceramide-centric view of insulin resistance. *Cell Metab.* **2012**, *15*, 585–594. [CrossRef] [PubMed]
98. Itani, S.I.; Ruderman, N.B.; Schmieder, F.; Boden, G. Lipid-induced insulin resistance in human muscle is associated with changes in diacylglycerol, protein kinase C, and IκB-α. *Diabetes* **2002**, *51*, 2005–2011. [CrossRef] [PubMed]

99. Schmitz-Peiffer, C.; Craig, D.L.; Biden, T.J. Ceramide generation is sufficient to account for the inhibition of the insulin-stimulated pkb pathway in C2C12 skeletal muscle cells pretreated with palmitate. *J. Biol. Chem.* **1999**, *274*, 24202–24210. [CrossRef] [PubMed]
100. Adams, J.M., 2nd; Pratipanawatr, T.; Berria, R.; Wang, E.; DeFronzo, R.A.; Sullards, M.C.; Mandarino, L.J. Ceramide content is increased in skeletal muscle from obese insulin-resistant humans. *Diabetes* **2004**, *53*, 25–31. [CrossRef] [PubMed]
101. Chavez, J.A.; Knotts, T.A.; Wang, L.P.; Li, G.; Dobrowsky, R.T.; Florant, G.L.; Summers, S.A. A role for ceramide, but not diacylglycerol, in the antagonism of insulin signal transduction by saturated fatty acids. *J. Biol. Chem.* **2003**, *278*, 10297–10303. [CrossRef] [PubMed]
102. Kuo, Y.C.; Huang, K.Y.; Yang, C.H.; Yang, Y.S.; Lee, W.Y.; Chiang, C.W. Regulation of phosphorylation of Thr-308 of Akt, cell proliferation, and survival by the B55α regulatory subunit targeting of the protein phosphatase 2A holoenzyme to Akt. *J. Biol. Chem.* **2008**, *283*, 1882–1892. [CrossRef] [PubMed]
103. Griffin, M.E.; Marcucci, M.J.; Cline, G.W.; Bell, K.; Barucci, N.; Lee, D.; Goodyear, L.J.; Kraegen, E.W.; White, M.F.; Shulman, G.I. Free fatty acid-induced insulin resistance is associated with activation of protein kinase C theta and alterations in the insulin signaling cascade. *Diabetes* **1999**, *48*, 1270–1274. [CrossRef] [PubMed]
104. Yu, C.; Chen, Y.; Cline, G.W.; Zhang, D.; Zong, H.; Wang, Y.; Bergeron, R.; Kim, J.K.; Cushman, S.W.; Cooney, G.J.; et al. Mechanism by which fatty acids inhibit insulin activation of insulin receptor substrate-1 (IRS-1)-associated phosphatidylinositol 3-kinase activity in muscle. *J. Biol. Chem.* **2002**, *277*, 50230–50236. [CrossRef] [PubMed]
105. Fan, W.; Waizenegger, W.; Lin, C.S.; Sorrentino, V.; He, M.X.; Wall, C.E.; Li, H.; Liddle, C.; Yu, R.T.; Atkins, A.R.; et al. PPARδ promotes running endurance by preserving glucose. *Cell Metab.* **2017**, *25*, 1186–1193.e4. [CrossRef] [PubMed]
106. Liu, L.; Zhang, Y.; Chen, N.; Shi, X.; Tsang, B.; Yu, Y.H. Upregulation of myocellular DGAT1 augments triglyceride synthesis in skeletal muscle and protects against fat-induced insulin resistance. *J. Clin. Investig.* **2007**, *117*, 1679–1689. [CrossRef] [PubMed]
107. He, W.; Barak, Y.; Hevener, A.; Olson, P.; Liao, D.; Le, J.; Nelson, M.; Ong, E.; Olefsky, J.M.; Evans, R.M. Adipose-specific peroxisome proliferator-activated receptor γ knockout causes insulin resistance in fat and liver but not in muscle. *Proc. Natl. Acad. Sci. USA* **2003**, *100*, 15712–15717. [CrossRef] [PubMed]
108. Jiang, G.; Dallas-Yang, Q.; Li, Z.; Szalkowski, D.; Liu, F.; Shen, X.; Wu, M.; Zhou, G.; Doebber, T.; Berger, J.; et al. Potentiation of insulin signaling in tissues of zucker obese rats after acute and long-term treatment with PPARγ agonists. *Diabetes* **2002**, *51*, 2412–2419. [CrossRef] [PubMed]
109. Kim, Y.B.; Ciaraldi, T.P.; Kong, A.; Kim, D.; Chu, N.; Mohideen, P.; Mudaliar, S.; Henry, R.R.; Kahn, B.B. Troglitazone but not metformin restores insulin-stimulated phosphoinositide 3-kinase activity and increases P110β protein levels in skeletal muscle of type 2 diabetic subjects. *Diabetes* **2002**, *51*, 443–448. [CrossRef] [PubMed]
110. Miyazaki, Y.; He, H.; Mandarino, L.J.; DeFronzo, R.A. Rosiglitazone improves downstream insulin receptor signaling in type 2 diabetic patients. *Diabetes* **2003**, *52*, 1943–1950. [CrossRef] [PubMed]
111. Divakaruni, A.S.; Wiley, S.E.; Rogers, G.W.; Andreyev, A.Y.; Petrosyan, S.; Loviscach, M.; Wall, E.A.; Yadava, N.; Heuck, A.P.; Ferrick, D.A.; et al. Thiazolidinediones are acute, specific inhibitors of the mitochondrial pyruvate carrier. *Proc. Natl. Acad. Sci. USA* **2013**, *110*, 5422–5427. [CrossRef] [PubMed]
112. LeBrasseur, N.K.; Kelly, M.; Tsao, T.S.; Farmer, S.R.; Saha, A.K.; Ruderman, N.B.; Tomas, E. Thiazolidinediones can rapidly activate AMP-activated protein kinase in mammalian tissues. *Am. J. Physiol. Endocrinol. Metab.* **2006**, *291*, E175–E181. [CrossRef] [PubMed]
113. Chen, Z.; Vigueira, P.A.; Chambers, K.T.; Hall, A.M.; Mitra, M.S.; Qi, N.; McDonald, W.G.; Colca, J.R.; Kletzien, R.F.; Finck, B.N. Insulin resistance and metabolic derangements in obese mice are ameliorated by a novel peroxisome proliferator-activated receptor γ-sparing thiazolidinedione. *J. Biol. Chem.* **2012**, *287*, 23537–23548. [CrossRef] [PubMed]
114. Park, K.S.; Ciaraldi, T.P.; Abrams-Carter, L.; Mudaliar, S.; Nikoulina, S.E.; Henry, R.R. PPAR-γ gene expression is elevated in skeletal muscle of obese and type II diabetic subjects. *Diabetes* **1997**, *46*, 1230–1234. [CrossRef] [PubMed]
115. Marin-Juez, R.; Diaz, M.; Morata, J.; Planas, J.V. Mechanisms regulating GLUT4 transcription in skeletal muscle cells are highly conserved across vertebrates. *PLoS ONE* **2013**, *8*, e80628. [CrossRef] [PubMed]

116. Hammarstedt, A.; Smith, U. Thiazolidinediones (PPARγ ligands) increase IRS-1, UCP-2 and C/EBPα expression, but not transdifferentiation, in l6 muscle cells. *Diabetologia* **2003**, *46*, 48–52. [CrossRef] [PubMed]
117. Amin, R.H.; Mathews, S.T.; Camp, H.S.; Ding, L.; Leff, T. Selective activation of PPARγ in skeletal muscle induces endogenous production of adiponectin and protects mice from diet-induced insulin resistance. *Am. J. Physiol. Endocrinol. Metab.* **2010**, *298*, E28–E37. [CrossRef] [PubMed]
118. Hevener, A.L.; He, W.; Barak, Y.; Le, J.; Bandyopadhyay, G.; Olson, P.; Wilkes, J.; Evans, R.M.; Olefsky, J. Muscle-specific PPARG deletion causes insulin resistance. *Nat. Med.* **2003**, *9*, 1491–1497. [CrossRef] [PubMed]
119. Norris, A.W.; Chen, L.; Fisher, S.J.; Szanto, I.; Ristow, M.; Jozsi, A.C.; Hirshman, M.F.; Rosen, E.D.; Goodyear, L.J.; Gonzalez, F.J.; et al. Muscle-specific PPARγ-deficient mice develop increased adiposity and insulin resistance but respond to thiazolidinediones. *J. Clin. Investig.* **2003**, *112*, 608–618. [CrossRef] [PubMed]
120. Ahmadian, M.; Suh, J.M.; Hah, N.; Liddle, C.; Atkins, A.R.; Downes, M.; Evans, R.M. PPARγ signaling and metabolism: The good, the bad and the future. *Nat. Med.* **2013**, *19*, 557–566. [CrossRef] [PubMed]
121. Ristow, M.; Muller-Wieland, D.; Pfeiffer, A.; Krone, W.; Kahn, C.R. Obesity associated with a mutation in a genetic regulator of adipocyte differentiation. *N. Engl. J. Med.* **1998**, *339*, 953–959. [CrossRef] [PubMed]
122. Barroso, I.; Gurnell, M.; Crowley, V.E.; Agostini, M.; Schwabe, J.W.; Soos, M.A.; Maslen, G.L.; Williams, T.D.; Lewis, H.; Schafer, A.J.; et al. Dominant negative mutations in human PPARγ associated with severe insulin resistance, diabetes mellitus and hypertension. *Nature* **1999**, *402*, 880–883. [CrossRef] [PubMed]
123. Deeb, S.S.; Fajas, L.; Nemoto, M.; Pihlajamaki, J.; Mykkanen, L.; Kuusisto, J.; Laakso, M.; Fujimoto, W.; Auwerx, J. A PRO12ALA substitution in PPARγ 2 associated with decreased receptor activity, lower body mass index and improved insulin sensitivity. *Nat. Genet.* **1998**, *20*, 284–287. [CrossRef] [PubMed]
124. Altshuler, D.; Hirschhorn, J.N.; Klannemark, M.; Lindgren, C.M.; Vohl, M.C.; Nemesh, J.; Lane, C.R.; Schaffner, S.F.; Bolk, S.; Brewer, C.; et al. The common PPARγ PRO12ALA polymorphism is associated with decreased risk of type 2 diabetes. *Nat. Genet.* **2000**, *26*, 76–80. [PubMed]
125. Jones, P.H. Chapter 26—fibrates A2—Ballantyne, christie m. In *Clinical Lipidology*; W.B. Saunders: Philadelphia, PA, USA, 2009; pp. 315–325.
126. Koh, K.K.; Han, S.H.; Quon, M.J.; Yeal Ahn, J.; Shin, E.K. Beneficial effects of fenofibrate to improve endothelial dysfunction and raise adiponectin levels in patients with primary hypertriglyceridemia. *Diabetes Care* **2005**, *28*, 1419–1424. [CrossRef] [PubMed]
127. Ueno, H.; Saitoh, Y.; Mizuta, M.; Shiiya, T.; Noma, K.; Mashiba, S.; Kojima, S.; Nakazato, M. Fenofibrate ameliorates insulin resistance, hypertension and novel oxidative stress markers in patients with metabolic syndrome. *Obes. Res. Clin. Pract.* **2011**, *5*, e267–e360. [CrossRef] [PubMed]
128. Franko, A.; Huypens, P.; Neschen, S.; Irmler, M.; Rozman, J.; Rathkolb, B.; Neff, F.; Prehn, C.; Dubois, G.; Baumann, M.; et al. Bezafibrate improves insulin sensitivity and metabolic flexibility in STZ-induced diabetic mice. *Diabetes* **2016**, *65*, 2540–2552. [CrossRef] [PubMed]
129. Tenenbaum, A.; Fisman, E.Z. Balanced pan-PPAR activator bezafibrate in combination with statin: Comprehensive lipids control and diabetes prevention? *Cardiovasc. Diabetol.* **2012**, *11*, 140. [CrossRef] [PubMed]
130. Bays, H.E.; Schwartz, S.; Littlejohn, T., 3rd; Kerzner, B.; Krauss, R.M.; Karpf, D.B.; Choi, Y.J.; Wang, X.; Naim, S.; Roberts, B.K. Mbx-8025, a novel peroxisome proliferator receptor-δ agonist: Lipid and other metabolic effects in dyslipidemic overweight patients treated with and without atorvastatin. *J. Clin. Endocrinol. Metab.* **2011**, *96*, 2889–2897. [CrossRef] [PubMed]
131. Lee, C.H.; Olson, P.; Hevener, A.; Mehl, I.; Chong, L.W.; Olefsky, J.M.; Gonzalez, F.J.; Ham, J.; Kang, H.; Peters, J.M.; et al. PPARδ regulates glucose metabolism and insulin sensitivity. *Proc. Natl. Acad. Sci. USA* **2006**, *103*, 3444–3449. [CrossRef] [PubMed]
132. Turcotte, L.P.; Fisher, J.S. Skeletal muscle insulin resistance: Roles of fatty acid metabolism and exercise. *Phys. Ther.* **2008**, *88*, 1279–1296. [CrossRef] [PubMed]
133. Greene, N.P.; Fluckey, J.D.; Lambert, B.S.; Greene, E.S.; Riechman, S.E.; Crouse, S.F. Regulators of blood lipids and lipoproteins? PPARδ and AMPK, induced by exercise, are correlated with lipids and lipoproteins in overweight/obese men and women. *Am. J. Physiol. Endocrinol. Metab.* **2012**, *303*, E1212–E1221. [CrossRef] [PubMed]
134. Myers, R.B.; Yoshioka, J. Regulating PPARδ signaling as a potential therapeutic strategy for skeletal muscle disorders in heart failure. *Am. J. Physiol. Heart Circ. Physiol.* **2015**, *308*, H967–H969. [CrossRef] [PubMed]

135. Lagu, B.; Kluge, A.F.; Fredenburg, R.A.; Tozzo, E.; Senaiar, R.S.; Jaleel, M.; Panigrahi, S.K.; Tiwari, N.K.; Krishnamurthy, N.R.; Takahashi, T.; et al. Novel highly selective peroxisome proliferator-activated receptor delta (PPARδ) modulators with pharmacokinetic properties suitable for once-daily oral dosing. *Bioorg. Med. Chem. Lett.* **2017**, *27*, 5230–5234. [CrossRef] [PubMed]
136. Lagu, B.; Kluge, A.F.; Goddeeris, M.M.; Tozzo, E.; Fredenburg, R.A.; Chellur, S.; Senaiar, R.S.; Jaleel, M.; Babu, D.R.K.; Tiwari, N.K.; et al. Highly selective peroxisome proliferator-activated receptor delta (PPARδ) modulator demonstrates improved safety profile compared to GW501516. *Bioorg. Med. Chem. Lett.* **2018**, *28*, 533–536. [CrossRef] [PubMed]
137. Russell, A.P.; Feilchenfeldt, J.; Schreiber, S.; Praz, M.; Crettenand, A.; Gobelet, C.; Meier, C.A.; Bell, D.R.; Kralli, A.; Giacobino, J.P.; et al. Endurance training in humans leads to fiber type-specific increases in levels of peroxisome proliferator-activated receptor-γ coactivator-1 and peroxisome proliferator-activated receptor-α in skeletal muscle. *Diabetes* **2003**, *52*, 2874–2881. [CrossRef] [PubMed]
138. Kramer, D.K.; Ahlsen, M.; Norrbom, J.; Jansson, E.; Hjeltnes, N.; Gustafsson, T.; Krook, A. Human skeletal muscle fibre type variations correlate with PPARα, PPARδ and PGC-1α MRNA. *Acta Physiol. (Oxf.)* **2006**, *188*, 207–216. [CrossRef] [PubMed]
139. Lopez-Leon, S.; Tuvblad, C.; Forero, D.A. Sports genetics: The PPARA gene and athletes' high ability in endurance sports. A systematic review and meta-analysis. *Biol. Sport* **2016**, *33*, 3–6. [PubMed]
140. Ahmetov, I.I.; Mozhayskaya, I.A.; Flavell, D.M.; Astratenkova, I.V.; Komkova, A.I.; Lyubaeva, E.V.; Tarakin, P.P.; Shenkman, B.S.; Vdovina, A.B.; Netreba, A.I.; et al. PPARα gene variation and physical performance in russian athletes. *Eur. J. Appl. Physiol.* **2006**, *97*, 103–108. [CrossRef] [PubMed]
141. Maciejewska-Karlowska, A.; Sawczuk, M.; Cieszczyk, P.; Zarebska, A.; Sawczyn, S. Association between the PRO12ALA polymorphism of the peroxisome proliferator-activated receptor gamma gene and strength athlete status. *PLoS ONE* **2013**, *8*, e67172. [CrossRef] [PubMed]
142. Huard, J.; Li, Y.; Fu, F.H. Muscle injuries and repair: Current trends in research. *J. Bone Jt. Surg. Am.* **2002**, *84-A*, 822–832. [CrossRef]
143. Jarvinen, T.A.; Jarvinen, T.L.; Kaariainen, M.; Kalimo, H.; Jarvinen, M. Muscle injuries: Biology and treatment. *Am. J. Sports Med.* **2005**, *33*, 745–764. [CrossRef] [PubMed]
144. Delos, D.; Maak, T.G.; Rodeo, S.A. Muscle injuries in athletes: Enhancing recovery through scientific understanding and novel therapies. *Sports Health* **2013**, *5*, 346–352. [CrossRef] [PubMed]
145. Jude, E.B.; Eleftheriadou, I.; Tentolouris, N. Peripheral arterial disease in diabetes—A review. *Diabet. Med.* **2010**, *27*, 4–14. [CrossRef] [PubMed]
146. St Pierre, B.A.; Tidball, J.G. Differential response of macrophage subpopulations to soleus muscle reloading after rat hindlimb suspension. *J. Appl. Physiol. (1985)* **1994**, *77*, 290–297. [CrossRef] [PubMed]
147. Mauro, A. Satellite cell of skeletal muscle fibers. *J. Biophys. Biochem. Cytol.* **1961**, *9*, 493–495. [CrossRef] [PubMed]
148. Chazaud, B.; Sonnet, C.; Lafuste, P.; Bassez, G.; Rimaniol, A.C.; Poron, F.; Authier, F.J.; Dreyfus, P.A.; Gherardi, R.K. Satellite cells attract monocytes and use macrophages as a support to escape apoptosis and enhance muscle growth. *J. Cell Biol.* **2003**, *163*, 1133–1143. [CrossRef] [PubMed]
149. Yang, W.; Hu, P. Skeletal muscle regeneration is modulated by inflammation. *J. Orthop. Transl.* **2018**, *13*, 25–32. [CrossRef] [PubMed]
150. Potthoff, M.J.; Olson, E.N.; Bassel-Duby, R. Skeletal muscle remodeling. *Curr. Opin. Rheumatol.* **2007**, *19*, 542–549. [CrossRef] [PubMed]
151. Hoier, B.; Hellsten, Y. Exercise-induced capillary growth in human skeletal muscle and the dynamics of VEGF. *Microcirculation* **2014**, *21*, 301–314. [CrossRef] [PubMed]
152. Poole, D.C.; Copp, S.W.; Ferguson, S.K.; Musch, T.I. Skeletal muscle capillary function: Contemporary observations and novel hypotheses. *Exp. Physiol.* **2013**, *98*, 1645–1658. [CrossRef] [PubMed]
153. Collins, C.A.; Olsen, I.; Zammit, P.S.; Heslop, L.; Petrie, A.; Partridge, T.A.; Morgan, J.E. Stem cell function, self-renewal, and behavioral heterogeneity of cells from the adult muscle satellite cell niche. *Cell* **2005**, *122*, 289–301. [CrossRef] [PubMed]
154. Yin, H.; Price, F.; Rudnicki, M.A. Satellite cells and the muscle stem cell niche. *Physiol. Rev.* **2013**, *93*, 23–67. [CrossRef] [PubMed]
155. Zammit, P.S. All muscle satellite cells are equal, but are some more equal than others? *J. Cell Sci.* **2008**, *121*, 2975–2982. [CrossRef] [PubMed]

156. Angione, A.R.; Jiang, C.; Pan, D.; Wang, Y.X.; Kuang, S. PPARδ regulates satellite cell proliferation and skeletal muscle regeneration. *Skelet Muscle* **2011**, *1*, 33. [CrossRef] [PubMed]
157. Chandrashekar, P.; Manickam, R.; Ge, X.; Bonala, S.; McFarlane, C.; Sharma, M.; Wahli, W.; Kambadur, R. Inactivation of PPARβ/δ adversely affects satellite cells and reduces postnatal myogenesis. *Am. J. Physiol. Endocrinol. Metab.* **2015**, *309*, E122–E131. [CrossRef] [PubMed]
158. Knobloch, M.; Braun, S.M.; Zurkirchen, L.; von Schoultz, C.; Zamboni, N.; Arauzo-Bravo, M.J.; Kovacs, W.J.; Karalay, O.; Suter, U.; Machado, R.A.; et al. Metabolic control of adult neural stem cell activity by FASN-dependent lipogenesis. *Nature* **2013**, *493*, 226–230. [CrossRef] [PubMed]
159. Folmes, C.D.; Dzeja, P.P.; Nelson, T.J.; Terzic, A. Metabolic plasticity in stem cell homeostasis and differentiation. *Cell Stem Cell* **2012**, *11*, 596–606. [CrossRef] [PubMed]
160. Wagner, K.D.; Wagner, N. Peroxisome proliferator-activated receptor β/δ (PPARβ/δ) acts as regulator of metabolism linked to multiple cellular functions. *Pharmacol. Ther.* **2010**, *125*, 423–435. [CrossRef] [PubMed]
161. Miura, P.; Chakkalakal, J.V.; Boudreault, L.; Belanger, G.; Hebert, R.L.; Renaud, J.M.; Jasmin, B.J. Pharmacological activation of PPARβ/δ stimulates utrophin a expression in skeletal muscle fibers and restores sarcolemmal integrity in mature mdx mice. *Hum. Mol. Genet.* **2009**, *18*, 4640–4649. [CrossRef] [PubMed]
162. Varga, T.; Mounier, R.; Patsalos, A.; Gogolak, P.; Peloquin, M.; Horvath, A.; Pap, A.; Daniel, B.; Nagy, G.; Pintye, E.; et al. Macrophage PPARγ, a lipid activated transcription factor controls the growth factor GDF3 and skeletal muscle regeneration. *Immunity* **2016**, *45*, 1038–1051. [CrossRef] [PubMed]
163. Latroche, C.; Weiss-Gayet, M.; Muller, L.; Gitiaux, C.; Leblanc, P.; Liot, S.; Ben-Larbi, S.; Abou-Khalil, R.; Verger, N.; Bardot, P.; et al. Coupling between myogenesis and angiogenesis during skeletal muscle regeneration is stimulated by restorative macrophages. *Stem Cell Rep.* **2017**, *9*, 2018–2033. [CrossRef] [PubMed]
164. Latroche, C.; Gitiaux, C.; Chretien, F.; Desguerre, I.; Mounier, R.; Chazaud, B. Skeletal muscle microvasculature: A highly dynamic lifeline. *Physiology (Bethesda)* **2015**, *30*, 417–427. [CrossRef] [PubMed]
165. Han, J.K.; Kim, H.L.; Jeon, K.H.; Choi, Y.E.; Lee, H.S.; Kwon, Y.W.; Jang, J.J.; Cho, H.J.; Kang, H.J.; Oh, B.H.; et al. Peroxisome proliferator-activated receptor-delta activates endothelial progenitor cells to induce angio-myogenesis through matrix metallo-proteinase-9-mediated insulin-like growth factor-1 paracrine networks. *Eur. Heart J.* **2013**, *34*, 1755–1765. [CrossRef] [PubMed]
166. Haas, T.L.; Nwadozi, E. Regulation of skeletal muscle capillary growth in exercise and disease. *Appl. Physiol. Nutr. Metab.* **2015**, *40*, 1221–1232. [CrossRef] [PubMed]
167. Hwa, V.; Oh, Y.; Rosenfeld, R.G. The insulin-like growth factor-binding protein (IGFBP) superfamily. *Endocr. Rev.* **1999**, *20*, 761–787. [CrossRef] [PubMed]
168. Liu, Y.; Sweeney, G. Adiponectin action in skeletal muscle. *Best Pract. Res. Clin. Endocrinol. Metab.* **2014**, *28*, 33–41. [CrossRef] [PubMed]
169. Ceddia, R.B.; Somwar, R.; Maida, A.; Fang, X.; Bikopoulos, G.; Sweeney, G. Globular adiponectin increases GLUT4 translocation and glucose uptake but reduces glycogen synthesis in rat skeletal muscle cells. *Diabetologia* **2005**, *48*, 132–139. [CrossRef] [PubMed]
170. Yoon, M.J.; Lee, G.Y.; Chung, J.J.; Ahn, Y.H.; Hong, S.H.; Kim, J.B. Adiponectin increases fatty acid oxidation in skeletal muscle cells by sequential activation of AMP-activated protein kinase, p38 mitogen-activated protein kinase, and peroxisome proliferator-activated receptor alpha. *Diabetes* **2006**, *55*, 2562–2570. [CrossRef] [PubMed]
171. Chen, M.B.; McAinch, A.J.; Macaulay, S.L.; Castelli, L.A.; O'Brien P, E.; Dixon, J.B.; Cameron-Smith, D.; Kemp, B.E.; Steinberg, G.R. Impaired activation of AMP-kinase and fatty acid oxidation by globular adiponectin in cultured human skeletal muscle of obese type 2 diabetics. *J. Clin. Endocrinol. Metab.* **2005**, *90*, 3665–3672. [CrossRef] [PubMed]
172. Qiao, L.; Kinney, B.; Yoo, H.S.; Lee, B.; Schaack, J.; Shao, J. Adiponectin increases skeletal muscle mitochondrial biogenesis by suppressing mitogen-activated protein kinase phosphatase-1. *Diabetes* **2012**, *61*, 1463–1470. [CrossRef] [PubMed]
173. Vavvas, D.; Apazidis, A.; Saha, A.K.; Gamble, J.; Patel, A.; Kemp, B.E.; Witters, L.A.; Ruderman, N.B. Contraction-induced changes in acetyl-CoA carboxylase and 5′-amp-activated kinase in skeletal muscle. *J. Biol. Chem.* **1997**, *272*, 13255–13261. [CrossRef] [PubMed]

174. Saha, A.K.; Schwarsin, A.J.; Roduit, R.; Masse, F.; Kaushik, V.; Tornheim, K.; Prentki, M.; Ruderman, N.B. Activation of malonyl-CoA decarboxylase in rat skeletal muscle by contraction and the AMP-activated protein kinase activator 5-aminoimidazole-4-carboxamide-1-β-D-ribofuranoside. *J. Biol. Chem.* **2000**, *275*, 24279–24283. [CrossRef] [PubMed]
175. Maeda, N.; Takahashi, M.; Funahashi, T.; Kihara, S.; Nishizawa, H.; Kishida, K.; Nagaretani, H.; Matsuda, M.; Komuro, R.; Ouchi, N.; et al. PPARγ ligands increase expression and plasma concentrations of adiponectin, an adipose-derived protein. *Diabetes* **2001**, *50*, 2094–2099. [CrossRef] [PubMed]
176. Liu, Y.; Chewchuk, S.; Lavigne, C.; Brule, S.; Pilon, G.; Houde, V.; Xu, A.; Marette, A.; Sweeney, G. Functional significance of skeletal muscle adiponectin production, changes in animal models of obesity and diabetes, and regulation by rosiglitazone treatment. *Am. J. Physiol. Endocrinol. Metab.* **2009**, *297*, E657–E664. [CrossRef] [PubMed]
177. Erol, A. The functions of PPARs in aging and longevity. *PPAR Res.* **2007**, *2007*, 39654. [CrossRef] [PubMed]
178. Niccoli, T.; Partridge, L. Ageing as a risk factor for disease. *Curr. Biol.* **2012**, *22*, R741–R752. [CrossRef] [PubMed]
179. Keller, K.; Engelhardt, M. Strength and muscle mass loss with aging process. Age and strength loss. *Muscles Ligaments Tendons J.* **2013**, *3*, 346–350. [PubMed]
180. Miljkovic, N.; Lim, J.Y.; Miljkovic, I.; Frontera, W.R. Aging of skeletal muscle fibers. *Ann. Rehabil. Med.* **2015**, *39*, 155–162. [CrossRef] [PubMed]
181. Larsson, L.; Karlsson, J. Isometric and dynamic endurance as a function of age and skeletal muscle characteristics. *Acta Physiol. Scand.* **1978**, *104*, 129–136. [CrossRef] [PubMed]
182. Clark, B.C.; Taylor, J.L. Age-related changes in motor cortical properties and voluntary activation of skeletal muscle. *Curr. Aging Sci.* **2011**, *4*, 192–199. [CrossRef] [PubMed]
183. Verdijk, L.B.; Koopman, R.; Schaart, G.; Meijer, K.; Savelberg, H.H.; van Loon, L.J. Satellite cell content is specifically reduced in type II skeletal muscle fibers in the elderly. *Am. J. Physiol. Endocrinol. Metab.* **2007**, *292*, E151–E157. [CrossRef] [PubMed]
184. Frontera, W.R.; Suh, D.; Krivickas, L.S.; Hughes, V.A.; Goldstein, R.; Roubenoff, R. Skeletal muscle fiber quality in older men and women. *Am. J. Physiol. Cell Physiol.* **2000**, *279*, C611–C618. [CrossRef] [PubMed]
185. Atherton, H.J.; Gulston, M.K.; Bailey, N.J.; Cheng, K.K.; Zhang, W.; Clarke, K.; Griffin, J.L. Metabolomics of the interaction between PPAR-α and age in the PPAR-α-null mouse. *Mol. Syst. Biol.* **2009**, *5*, 259. [CrossRef] [PubMed]
186. Giordano, C.; Rousseau, A.S.; Wagner, N.; Gaudel, C.; Murdaca, J.; Jehl-Pietri, C.; Sibille, B.; Grimaldi, P.A.; Lopez, P. Peroxisome proliferator-activated receptor beta activation promotes myonuclear accretion in skeletal muscle of adult and aged mice. *Pflugers Arch.* **2009**, *458*, 901–913. [CrossRef] [PubMed]
187. Ye, P.; Zhang, X.J.; Wang, Z.J.; Zhang, C. Effect of aging on the expression of peroxisome proliferator-activated receptor γ and the possible relation to insulin resistance. *Gerontology* **2006**, *52*, 69–75. [CrossRef] [PubMed]
188. Toth, M.J.; Tchernof, A. Lipid metabolism in the elderly. *Eur. J. Clin. Nutr.* **2000**, *54* (Suppl. 3), 121S–S125. [CrossRef]
189. Johannsen, D.L.; Conley, K.E.; Bajpeyi, S.; Punyanitya, M.; Gallagher, D.; Zhang, Z.; Covington, J.; Smith, S.R.; Ravussin, E. Ectopic lipid accumulation and reduced glucose tolerance in elderly adults are accompanied by altered skeletal muscle mitochondrial activity. *J. Clin. Endocrinol. Metab.* **2012**, *97*, 242–250. [CrossRef] [PubMed]
190. Gaudel, C.; Schwartz, C.; Giordano, C.; Abumrad, N.A.; Grimaldi, P.A. Pharmacological activation of PPARβ promotes rapid and calcineurin-dependent fiber remodeling and angiogenesis in mouse skeletal muscle. *Am. J. Physiol. Endocrinol. Metab.* **2008**, *295*, E297–E304. [CrossRef] [PubMed]
191. Hodel, C. Myopathy and rhabdomyolysis with lipid-lowering drugs. *Toxicol. Lett.* **2002**, *128*, 159–168. [CrossRef]
192. Burri, L.; Thoresen, G.H.; Berge, R.K. The role of pparalpha activation in liver and muscle. *PPAR Res.* **2010**, *2010*. [CrossRef] [PubMed]
193. Faiola, B.; Falls, J.G.; Peterson, R.A.; Bordelon, N.R.; Brodie, T.A.; Cummings, C.A.; Romach, E.H.; Miller, R.T. PPARα, more than PPARδ, mediates the hepatic and skeletal muscle alterations induced by the PPAR agonist GW0742. *Toxicol. Sci.* **2008**, *105*, 384–394. [CrossRef] [PubMed]
194. Finkel, T. The metabolic regulation of aging. *Nat. Med.* **2015**, *21*, 1416–1423. [CrossRef] [PubMed]

195. Guillet, C.; Boirie, Y. Insulin resistance: A contributing factor to age-related muscle mass loss? *Diabetes Metab.* **2005**, *31*, 5S20–5S26. [CrossRef]
196. Morley, J.E. Hormones and the aging process. *J. Am. Geriatr. Soc.* **2003**, *51*, S333–S337. [CrossRef]
197. Houmard, J.A.; Weidner, M.D.; Dolan, P.L.; Leggett-Frazier, N.; Gavigan, K.E.; Hickey, M.S.; Tyndall, G.L.; Zheng, D.; Alshami, A.; Dohm, G.L. Skeletal muscle GLUT4 protein concentration and aging in humans. *Diabetes* **1995**, *44*, 555–560. [CrossRef] [PubMed]
198. Ulrich-Lai, Y.M.; Ryan, K.K. Ppargamma and stress: Implications for aging. *Exp. Gerontol.* **2013**, *48*, 671–676. [CrossRef] [PubMed]
199. Investigators, D.T.; Gerstein, H.C.; Yusuf, S.; Bosch, J.; Pogue, J.; Sheridan, P.; Dinccag, N.; Hanefeld, M.; Hoogwerf, B.; Laakso, M.; et al. Effect of rosiglitazone on the frequency of diabetes in patients with impaired glucose tolerance or impaired fasting glucose: A randomised controlled trial. *Lancet* **2006**, *368*, 1096–1105.
200. Nolan, J.J.; Ludvik, B.; Beerdsen, P.; Joyce, M.; Olefsky, J. Improvement in glucose tolerance and insulin resistance in obese subjects treated with troglitazone. *N. Engl. J. Med.* **1994**, *331*, 1188–1193. [CrossRef] [PubMed]
201. Sanguino, E.; Roglans, N.; Alegret, M.; Sanchez, R.M.; Vazquez-Carrera, M.; Laguna, J.C. Different response of senescent female Sprague-Dawley rats to gemfibrozil and rosiglitazone administration. *Exp. Gerontol.* **2005**, *40*, 588–598. [CrossRef] [PubMed]
202. Miles, P.D.; Barak, Y.; He, W.; Evans, R.M.; Olefsky, J.M. Improved insulin-sensitivity in mice heterozygous for PPAR-γ deficiency. *J. Clin. Investig.* **2000**, *105*, 287–292. [CrossRef] [PubMed]
203. Miles, P.D.; Barak, Y.; Evans, R.M.; Olefsky, J.M. Effect of heterozygous PPARγ deficiency and TZD treatment on insulin resistance associated with age and high-fat feeding. *Am. J. Physiol. Endocrinol. Metab.* **2003**, *284*, E618–E626. [CrossRef] [PubMed]
204. Stephens, F.B.; Tsintzas, K. Metabolic and molecular changes associated with the increased skeletal muscle insulin action 24–48 h after exercise in young and old humans. *Biochem. Soc. Trans.* **2018**, *46*, 111–118. [CrossRef] [PubMed]
205. Post, S.M.; Duez, H.; Gervois, P.P.; Staels, B.; Kuipers, F.; Princen, H.M. Fibrates suppress bile acid synthesis via peroxisome proliferator-activated receptor-α-mediated downregulation of cholesterol 7α-hydroxylase and sterol 27-hydroxylase expression. *Arterioscler. Thromb. Vasc. Biol.* **2001**, *21*, 1840–1845. [CrossRef] [PubMed]
206. Kostapanos, M.S.; Florentin, M.; Elisaf, M.S. Fenofibrate and the kidney: An overview. *Eur J. Clin Investig.* **2013**, *43*, 522–531. [CrossRef] [PubMed]
207. Turner, R.M.; Kwok, C.S.; Chen-Turner, C.; Maduakor, C.A.; Singh, S.; Loke, Y.K. Thiazolidinediones and associated risk of bladder cancer: A systematic review and meta-analysis. *Br. J. Clin. Pharmacol.* **2014**, *78*, 258–273. [CrossRef] [PubMed]
208. Singh, S.; Loke, Y.K.; Furberg, C.D. Thiazolidinediones and heart failure: A teleo-analysis. *Diabetes Care* **2007**, *30*, 2148–2153. [CrossRef] [PubMed]
209. Geiger, L.N.; Dunsford, W.S.; Lewis, D.J.; Brennan, C.; Liu, K.C.; Newsholme, S.J. *Rat Carcinogenicity Study with gw501516*, A Ppar Delta Agonist. *Toxicol. Sci.* **2009**, *108*, 895.
210. Catoire, M.; Alex, S.; Paraskevopulos, N.; Mattijssen, F.; Evers-van Gogh, I.; Schaart, G.; Jeppesen, J.; Kneppers, A.; Mensink, M.; Voshol, P.J.; et al. Fatty acid-inducible anGPTL4 governs lipid metabolic response to exercise. *Proc. Natl. Acad. Sci. USA* **2014**, *111*, E1043–E1052. [CrossRef] [PubMed]
211. Kersten, S.; Lichtenstein, L.; Steenbergen, E.; Mudde, K.; Hendriks, H.F.; Hesselink, M.K.; Schrauwen, P.; Muller, M. Caloric restriction and exercise increase plasma ANGPTl4 levels in humans via elevated free fatty acids. *Arterioscler. Thromb. Vasc. Biol.* **2009**, *29*, 969–974. [CrossRef] [PubMed]
212. Munoz-Canoves, P.; Scheele, C.; Pedersen, B.K.; Serrano, A.L. Interleukin-6 myokine signaling in skeletal muscle: A double-edged sword? *FEBS J.* **2013**, *280*, 4131–4148. [CrossRef] [PubMed]
213. Weihrauch, M.; Handschin, C. Pharmacological targeting of exercise adaptations in skeletal muscle: Benefits and pitfalls. *Biochem. Pharmacol.* **2018**, *147*, 211–220. [CrossRef] [PubMed]
214. Hong, Y.H.; Nishimura, Y.; Hishikawa, D.; Tsuzuki, H.; Miyahara, H.; Gotoh, C.; Choi, K.C.; Feng, D.D.; Chen, C.; Lee, H.G.; et al. Acetate and propionate short chain fatty acids stimulate adipogenesis via GPCR43. *Endocrinology* **2005**, *146*, 5092–5099. [CrossRef] [PubMed]

215. Al-Lahham, S.H.; Peppelenbosch, M.P.; Roelofsen, H.; Vonk, R.J.; Venema, K. Biological effects of propionic acid in humans; metabolism, potential applications and underlying mechanisms. *Biochim. Biophys. Acta* **2010**, *1801*, 1175–1183. [CrossRef] [PubMed]
216. Bengmark, S. Gut microbiota, immune development and function. *Pharmacol. Res.* **2013**, *69*, 87–113. [CrossRef] [PubMed]
217. Yan, H.; Diao, H.; Xiao, Y.; Li, W.; Yu, B.; He, J.; Yu, J.; Zheng, P.; Mao, X.; Luo, Y.; et al. Gut microbiota can transfer fiber characteristics and lipid metabolic profiles of skeletal muscle from pigs to germ-free mice. *Sci. Rep.* **2016**, *6*, 31786. [CrossRef] [PubMed]

International Journal of
Molecular Sciences

Article

PIMT/NCOA6IP Deletion in the Mouse Heart Causes Delayed Cardiomyopathy Attributable to Perturbation in Energy Metabolism

Yuzhi Jia [1,†], Ning Liu [1,†], Navin Viswakarma [2], Ruya Sun [1], Mathew J. Schipma [3], Meng Shang [4], Edward B. Thorp [1], Yashpal S. Kanwar [1], Bayar Thimmapaya [5,*] and Janardan K. Reddy [1,*]

1 Department of Pathology, Feinberg School of Medicine, Northwestern University, Chicago, IL 60611, USA; y-jia@northwestern.edu (Y.J.); nliu2224@163.com (N.L.); ruya.sun@northwestern.edu (R.S.); ebthorp@northwestern.edu (E.B.T.); y-kanwar@northwestern.edu (Y.S.K.)
2 Department of Surgery, Division of Surgical Oncology, University of Illinois at Chicago, Chicago, IL 60612, USA; navinv@uic.edu
3 Next Generation Sequencing Core Facility, Feinberg School of Medicine, Northwestern University, Chicago, IL 60611, USA; m-schipma@northwestern.edu
4 Feinberg Cardiovascular Research Institute and Department of Medicine, Feinberg School of Medicine, Northwestern University, Chicago, IL 60611, USA; meng.shang@northwestern.edu
5 Department of Microbiology and Immunology, Feinberg School of Medicine, Northwestern University, Chicago, IL 60611, USA
* Correspondence: b-thimmapaya@northwestern.edu (B.T.); jkreddy@northwestern.edu (J.K.R.)
† These authors contributed equally to this work.

Received: 3 May 2018; Accepted: 9 May 2018; Published: 16 May 2018

Abstract: PIMT/NCOA6IP, a transcriptional coactivator PRIP/NCOA6 binding protein, enhances nuclear receptor transcriptional activity. Germline disruption of PIMT results in early embryonic lethality due to impairment of development around blastocyst and uterine implantation stages. We now generated mice with Cre-mediated cardiac-specific deletion of PIMT (csPIMT$^{-/-}$) in adult mice. These mice manifest enlargement of heart, with nearly 100% mortality by 7.5 months of age due to dilated cardiomyopathy. Significant reductions in the expression of genes (i) pertaining to mitochondrial respiratory chain complexes I to IV; (ii) calcium cycling cardiac muscle contraction (*Atp2a1, Atp2a2, Ryr2*); and (iii) nuclear receptor PPAR- regulated genes involved in glucose and fatty acid energy metabolism were found in csPIMT$^{-/-}$ mouse heart. Elevated levels of *Nppa* and *Nppb* mRNAs were noted in csPIMT$^{-/-}$ heart indicative of myocardial damage. These hearts revealed increased reparative fibrosis associated with enhanced expression of *Tgfβ2* and *Ctgf*. Furthermore, cardiac-specific deletion of PIMT in adult mice, using tamoxifen-inducible Cre-approach (TmcsPIMT$^{-/-}$), results in the development of cardiomyopathy. Thus, cumulative evidence suggests that PIMT functions in cardiac energy metabolism by interacting with nuclear receptor coactivators and this property could be useful in the management of heart failure.

Keywords: PIMT/NCOA6IP; PRIP/NCOA6; PPARα; dilated cardiomyopathy; cardiac fibrosis; energy metabolism

---

1. Introduction

The nuclear receptor coactivators, as exemplified by some components of Mediator complex and others such as PRIP/NCOA6, (proliferator-activated receptor (PPAR) interacting protein (PRIP)/Nuclear receptor coactivator 6) participate in the transcriptional activation of specific genes regulated by nuclear receptors and other transcription factors [1–5]. In an effort to understand the role of coactivator PRIP (NCOA6), we previously isolated a PRIP-interacting protein, designated PIMT/NCOA6IP/TGS1 (PRIP-interacting protein with methyltransferase domain (PIMT)/NCOA6-interacting protein (NCOA6IP)/Trimethylguanosine Synthase1 (TGS1)) an RNA binding protein with RNA methyltransferase activity [1]. PIMT is expressed ubiquitously including in liver, kidney and skeletal muscle. The methyltransferase activity of PIMT hypermethylates 2,2,7-trimethylguanosine cap structures of small nuclear RNA (snRNA), and small nucleolar RNA (snoRNA), that are important in RNA splicing [6,7]. PIMT binds to PRIP under in vivo and in vitro conditions and may serve as a bridge to transduce signals from upstream transcription factor-coactivator complex to the Mediator complex to drive RNA polymerase II mediated gene transcription [8]. Thus, available evidence suggests that coactivators PIMT, PRIP and Med1 are important in nuclear receptor PPARα controlled fatty acid β-oxidation [3].

Because heart derives the bulk of its functional energy from fatty acid β-oxidation, we asked whether PIMT is essential for normal cardiac functions and if cardiac-specific ablation of this gene causes dilated cardiomyopathy (DCM) similar to that noted with ablation of coactivators Med1 and PRIP [9,10]. First, we used a mouse model in which cardiomyocyte-specific deletion of PIMT (csPIMT$^{-/-}$) was carried out during late gestational and early postnatal development by intercrossing PIMT$^{fl/fl}$ mice with α-MyHC-Cre transgenic mice [11]. In this csPIMT$^{-/-}$ mouse model, hearts develop lethal DCM between four to eight months after birth. The csPIMT$^{-/-}$ mouse heart showed severe mitochondrial damage, reduced expression of several genes related to energy metabolism, and calcium signaling related cardiac muscle contraction. Some of the essential findings noted in csPIMT$^{-/-}$ mouse were independently confirmed using another mouse model in which cardiac-specific deletion of PIMT in adult mice was accomplished by using tamoxifen-inducible Cre-approach [9]. Many of the cardiac-specific changes noted in csPIMT$^{-/-}$ mice we report here bear resemblance to that reported recently for the cardiac-specific ablation of coactivators *Med1* and *Ncoa6* [9,10]. Collectively, these observations lead us to propose a model in which a protein complex consisting of PIMT, NCOA6, and MED1 (Mediator1) interact with other chromatin modifiers such as p300/CBP to target a common set of transcription factors to regulate metabolic pathways critical for cardiac functions.

2. Results

2.1. Generation of Cardiomyocyte-Specific PIMT Heart Knockout Mice

Previously, we reported that global disruption of PIMT gene in mice results in early embryonic lethality by affecting development around blastocyst and uterine implantation stages [11]. To evaluate the heart-specific function of PIMT, we generated mice with cardiomyocyte-specific disruption of PIMT gene (csPIMT$^{-/-}$). We crossed mice with a loxP flanked allele targeting exons 3–4 of PIMT gene (PIMT$^{fl/fl}$) with α-MyHC-Cre recombinase transgenic mice that express Cre-recombinase in cardiomyocytes under the control of α-myosin heavy chain (α-MyHC) gene promoter to yield csPIMT$^{-/-}$ mice following protocols as described in our recent paper [9]. Disruption of the PIMT gene in cardiomyocytes was confirmed by PCR genotyping and by q-PCR analysis of RNA from mouse heart (Figure 1A). PIMT mRNA levels greatly decreased in csPIMT$^{-/-}$ hearts but not in the liver, kidney or skeletal muscle, confirming heart-specific PIMT deletion (Figure 1A). Immunohistochemical localization of PIMT revealed prominent cardiomyocyte nuclear staining in PIMT$^{fl/fl}$ mouse heart but not in the myocardium of csPIMT$^{-/-}$ mouse littermates (Figure 1B). Furthermore, on Western blot analysis, PIMT was barely detectable at the protein level in csPIMT$^{-/-}$ mouse hearts (Figure 1C).

2.2. Cardiomyocyte-Specific Disruption of PIMT Causes Dilated Cardiomyopathy in Mice

csPIMT$^{-/-}$ mice are viable at birth with no grossly appreciable morphological abnormalities. There was no significant change in the heart size at two months of age in these csPIMT$^{-/-}$ mice but sectioning revealed mild degree of heart dilation as evidenced by thinning of the walls of left ventricular chamber (Figure 1D,E). The csPIMT$^{-/-}$ mice continued to show myocardial damage, with dilated heart associated with thinning of heart walls. At age six months, csPIMT$^{-/-}$ mice showed significant increase in heart size and they were increasingly flaccid when compared to that of littermate controls (Figure 1E). To further assess the heart damage, we assayed the mRNA levels of heart failure indicators atrial natriuretic peptide (ANP, gene *Nppa*) and brain natriuretic peptide (BNP, gene *Nppb*) [12]. Both *Nppa* and *Nppb* RNA levels increase in heart failure as ventricular cells are recruited to secrete both these peptides in response to left ventricular dysfunction [12]. Both *Nppa* and *Nppb* RNA levels increased dramatically in csPIMT$^{-/-}$ hearts at two months and the levels continued to remain high until six months (Figure 1G,F). Nearly 100% of csPIMT$^{-/-}$ mice died within 7.5 months after weaning due to dilated cardiomyopathy-related atrial and ventricular dilatation and heart failure (Figure 1H).

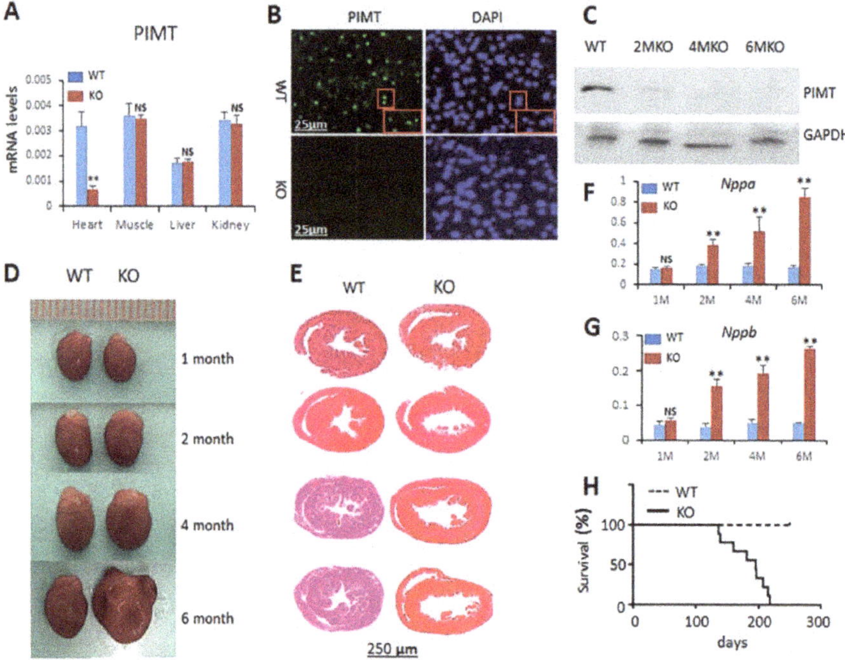

Figure 1. Cardiac-specific ablation of PIMT expression causes dilated cardiomyopathy. (**A**) Quantification of PIMT mRNA relative to 18S ribosomal RNA by RT-qPCR in PIMT$^{fl/fl}$ (WT) and csPIMT$^{-/-}$ (KO) mouse heart, muscle, liver and kidney;(**B**) Immunohistochemical localization of PIMT in 2-month-old PIMT$^{fl/fl}$ (WT) and csPIMT$^{-/-}$ (KO) mouse hearts. Nuclear localization of PIMT is evident in WT but not in KO hearts; compare DAPI stained images shown in right; (**C**) Western blot analysis for detecting PIMT protein level in PIMT$^{fl/fl}$ and csPIMT$^{-/-}$ mouse heart homogenates; (**D**) Representative photographs of heart of 1-, 2-, 4-, and 6-month-old csPIMT$^{-/-}$ mice and their PIMT$^{fl/fl}$ littermate controls. Six-month-old csPIMT$^{-/-}$ mouse hearts were flaccid and flabby; (**E**) Cross sections of hearts shown in Figure 1D were stained with H&E to reveal thinning of ventricular walls and dilation of chambers in csPIMT$^{-/-}$ mouse hearts (right panel); (**F**,**G**) *Nppa* and *Nppb* mRNA levels, respectively, in PIMT$^{fl/fl}$ and csPIMT$^{-/-}$ mouse hearts obtained at indicated ages. Each group was analyzed using 5 different mice (each mouse was assayed separately) and the values were expressed as the mean ± SD. * $p < 0.05$, ** $p < 0.01$, NS: not significant; (**H**) Survival curve showing lethality of mice with csPIMT$^{-/-}$ hearts. 36 mice for each group of PIMT$^{fl/fl}$ and csPIMT$^{-/-}$ were used for the generation of survival curve. Kaplan-Meier method was used to determine the survival rates and data were compared using log rank test. Each group was analyzed using 5 different mice and the values were expressed as the mean ± SD. * $p < 0.05$, ** $p < 0.01$, NS: not significant.

2.3. Echocardiographic Observations of csPIMT$^{-/-}$ Mouse Heart Indicate Poor Contractility

The effects of PIMT deletion on cardiac function were evaluated by obtaining the 2D and M-mode echocardiographic images (Figure 2A). Echocardiographic analysis of two-, four-, and six-month-old csPIMT$^{-/-}$ mice revealed increased left ventricular end-diastolic internal dimension (LVID-d), decreased fractional shortening and also decreased ejection fraction (see Figure 2B for quantification of these changes). At six months of age, the contractility of csPIMT$^{-/-}$ mouse heart was diminished with a fractional shortening of 18% vs. 67% for littermate controls. Likewise, the ejection fraction in four- and six-month-old csPIMT$^{-/-}$ mouse was 57% and 41%, respectively vs. 79% and 73%, respectively, for the floxed littermate controls. These values suggest poor contractility of PIMT null hearts and support the conclusion that PIMT deficient mice die of heart failure.

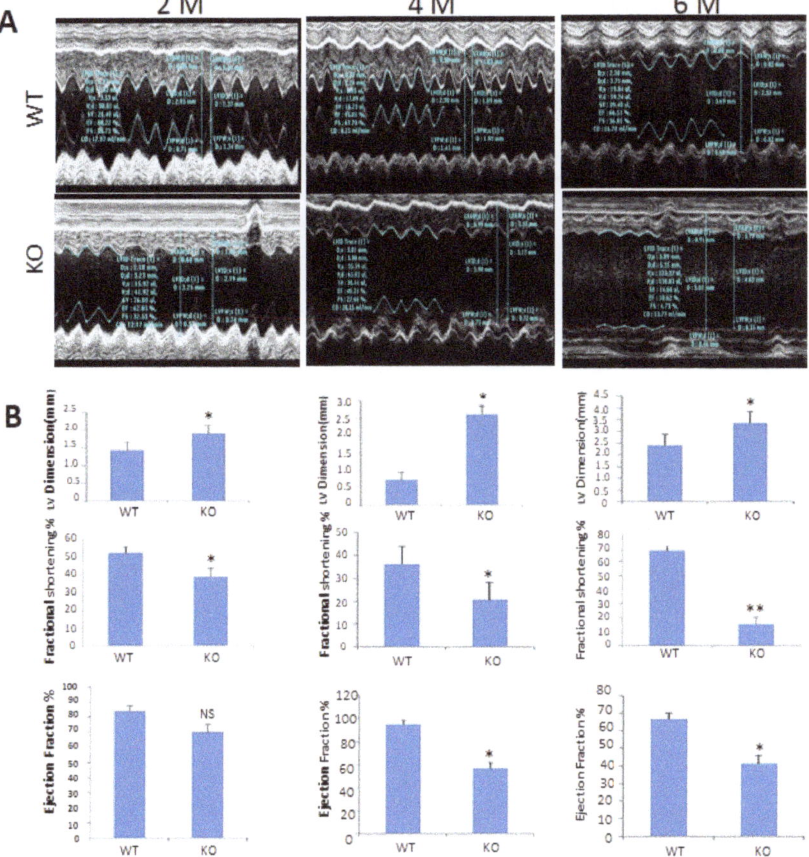

Figure 2. Echocardiographic results showing poor contractility of csPIMT$^{-/-}$ hearts. (**A**) Representative profiles of M-mode echocardiographic analyses of 2-, 4-, and 6-month-old PIMT$^{fl/fl}$ (WT, **upper** panel) and csPIMT$^{-/-}$ (KO, **lower** panel) mice; (**B**) Quantification of left ventricular dimension (upper panel), and fractional shortening (middle panel) and ejection fraction (lower panel) are shown below for 2-, 4-, and 6-month PIMT$^{fl/fl}$ and csPIMT$^{-/-}$ echocardiographic images. Values were expressed as the mean ± SD. * $p < 0.05$, ** $p < 0.01$, NS: not significant.

2.4. Global RNA Sequence Analysis of csPIMT$^{-/-}$ Hearts Suggests that Loss of PIMT Affects Multiple Pathways That Are Critical for Heart Function

The structural and functional changes observed so far in the heart of csPIMT$^{-/-}$ mice prompted us to evaluate the alterations in myocardial gene expression. First, we carried out expression profile analysis of PIMT$^{-/-}$ heart tissue for two- and six-month-old mice using the RNA-seq approach to obtain a global view of changes in gene expression. Heart RNA samples from five controls and five csPIMT$^{-/-}$ mice (two- and six-month-old) were pooled then subjected to RNA-seq protocol [9]. For the two-month-old mice, the RNA analysis identified a total of 708 genes with greater than two-fold expression difference between control and PIMT$^{-/-}$ heart RNA samples. Of these, 635 genes showed decreased expression, whereas expression of the rest of the genes was elevated. The down- and up-regulated genes at two months time point are presented in Supplemental section (Tables S1 and S2, respectively). Similarly, for the six-month-old mice, 600 genes showed a greater than two-fold expression difference between control and csPIMT$^{-/-}$ heart RNA samples. These include 417 downregulated genes and 183 upregulated genes; shown in Tables S3 and S4, respectively, in Supplemental section. Some of the down- and upregulated genes classified according to KEGG pathway and their role in some of the pathways related to heart function are shown in Table 1.

Table 1. Gene expression changes in csPIMT$^{-/-}$ relative to PIMT$^{fl/fl}$ mouse heart.

Function	Gene	KO/WT	KO/WT	KO/WT	KO/WT
		2 M, qPCR	2 M RNA-Seq	6 M, qPCR	6 M, RNA-Seq
OXPHOS					
	Ndufaf4	0.35 *	0.21	0.18 **	0.20
	Ndufaf5	0.29 *	0.33	0.21 **	0.17
	Ndufs4	0.51 *	0.42	0.27 **	0.24
	Cox7b	0.32 *	0.27	0.36 *	0.41
	Cox10	0.59	0.27	0.25 **	0.28
	Sdha	0.63	0.52	0.41 *	0.37
	Sucla2	0.24 *	0.12	0.12 **	0.08
Energy metabolism/fatty acid					
	Pparα	0.54	0.66	0.49	0.57
	Ppargc1a	0.51	0.36	0.55	0.45
	Acadm	0.36 *	0.39	0.12 **	0.27
	Ucp3	0.63	0.61	0.15 **	0.24
	Abcc9	0.61	0.67	0.16 *	0.49
Glucose metabolism					
	Gck(GK)	3.93 *	1.97	2.46 *	3.87
	Pck1	1.88	4.23	0.51	0.63
	Pdk4	0.38 *	0.37	0.24 *	0.99
	HK2	0.74	0.97	0.45 *	0.83
	Glut4	0.59	0.92	0.36 *	0.78
Transcription factor, coactivator					
	Med1	0.64	0.51	0.66	0.48
	NcoA6	0.69	0.66	0.57	0.62
	Tfam	0.47	0.25	0.19 **	0.48
Mitophagy/mitochondria fission					
	Pink1	2.16 *	1.93	1.57	1.65
	Drp1	0.58	0.41	0.39 *	0.46
Cardiomyopathy/Fibrosis					
	Tgfb2	0.66	0.91	9.41 **	5.23
	Ctgf	1.16	2.83	6.42 **	7.71
	Col9a2	3.24 *	6.42	8.33 **	26.06
	Fgf6	1.86	2.37	10.92 **	17.65
	Fgf21	2.35 *	3.08	5.18 **	6.22
	Mmp3	3.92 *	2.37	10.86 **	7.61
	Timp1	4.58 **	4.97	4.63 **	6.18

Table 1. *Cont.*

Function	Gene	KO/WT	KO/WT	KO/WT	KO/WT
		2 M, qPCR	2 M RNA-Seq	6 M, qPCR	6 M, RNA-Seq
Hypertrophy/dilation					
	Atf3	1.19	1.57	3.91 *	3.02
	Ace	2.69 *	2.23	3.78 *	6.48
	Wisp2	4.72 **	3.62	11.61 **	8.74
	Thbs4	5.91 **	6.68	9.34 **	8.56
Calcium homeostasis and signaling pathway					
	Atp2b1	0.34 *	0.22	0.31 *	0.25
	Atp2a1	0.53	0.52	0.23 *	0.43
	Ryr2	0.65	0.53	0.14 **	0.54
	Map3k6	2.14	3.07	3.01 *	8.01
	Cacnb1	2.49 *	1.98	3.72 *	5.12
	Pde1c	0.38	0.21	0.21 **	0.14
	Cacna1h	0.29 *	0.23	0.19 **	0.01
	Mapk8	0.27 *	0.21	0.33 *	0.18

* $p < 0.05$, ** $p < 0.01$. KO, csPIMT$^{-/-}$; WT, PIMT $^{fl/fl}$; M, months.

The entire list of genes analyzed by RNA-seq that showed significant difference in expression levels at two months and six months has been deposited in Gene Expression Omnibus (GEO number is GSE111862). Overall, these results indicate significant changes in the expression levels of several important genes that would impact on multiple pathways critical for heart function. These include mitochondrial oxidative phosphorylation, energy metabolism, mitophagy, calcium signaling, cardiac muscle contraction, cardiac hypertrophy and myocardial fibrosis. Changes in expression of several of these genes were also confirmed by RT-qPCR. As expected, RT-qPCR analysis validated the changes of gene expression levels observed by RNA-seq analysis (Table 1).

2.5. Reduced Expression of Genes Related to Mitochondrial Functions in csPIMT$^{-/-}$ Hearts

Genes involved in mitochondrial gene expression and mitochondrial biogenesis are downregulated: Mitochondrial transcription factor A (*Tfam*), is a nuclear encoded gene whose function is to transcribe mitochondrial DNA, and maintain mitochondrial genome copy number [13]. *Tfam* is also necessary for energy generation from oxidative phosphorylation [14]. A 60 to 70% decrease in *Tfam* gene expression was noted in csPIMT$^{-/-}$ mouse heart (Table 1 and Figure 3A) which could explain the reduced population of mitochondria in cardiomyocytes.

Genes involved in oxidative phosphorylation and respiratory chain complexes and fatty acid β-oxidation pathway are expressed at lower levels: In mitochondria, ATP is generated in inner mitochondrial membrane by five respiratory complexes (Complexes I, II, III, IV and V) through coupled electron transport and oxidative phosphorylation [15,16]. Reduced expression of any one of the genes related to these subunits would affect oxidative phosphorylation and ATP production. Examples of genes whose expression levels decreased include Ndufs4, Ndufaf4, Ndufaf5, Cox7b and Cox10, Sucla2 and Sdha [17], (Table 1). Expression of the mitochondrial genes were also assayed using Western blots which showed a significant reduction in the protein levels for Complex II, which catalyzes three out of the four steps in β-oxidation [17] (Figures 3 and 4).

Expression of mitochondrial calcium homeostasis related genes is reduced: RNA-seq data of csPIMT$^{-/-}$ hearts, which, in several cases were confirmed by RT-qPCR show significant changes in the expression levels of key genes, namely *Atp2b1*, *Atp2a1*, *Ryr2*, *Cacnb1*, and *Pde1c* that are involved in calcium signaling pathway and cardiac muscle contraction (Table 1). These changes in gene expression could contribute to the development of DCM and are consistent with the echocardiographic observations shown in Figure 2, which indicated poor contractility of PIMT null hearts.

Electron microscopic analysis of csPIMT$^{-/-}$ heart reveal structural damage to mitochondria: Gene expression data suggest dramatic changes in mitochondrial functions in csPIMT$^{-/-}$ mouse heart. To further analyze the damage occurred in mitochondria of csPIMT$^{-/-}$ myocardial cells, we carried out electron microscopic analysis of six-month-old csPIMT$^{-/-}$ heart tissue. Results shown in Figure 3D indicate the presence of lipid vacuoles of differing sizes in cardiomyocytes (Figure 3D). Some mitochondria contained lipid droplets and membranous swirls (*yellow arrows*). Irregularities in Z band pattern were also noted (*red arrows*). These observations combined with the changes in gene expressions described above strongly argues that loss of PIMT leads to damage in mitochondria.

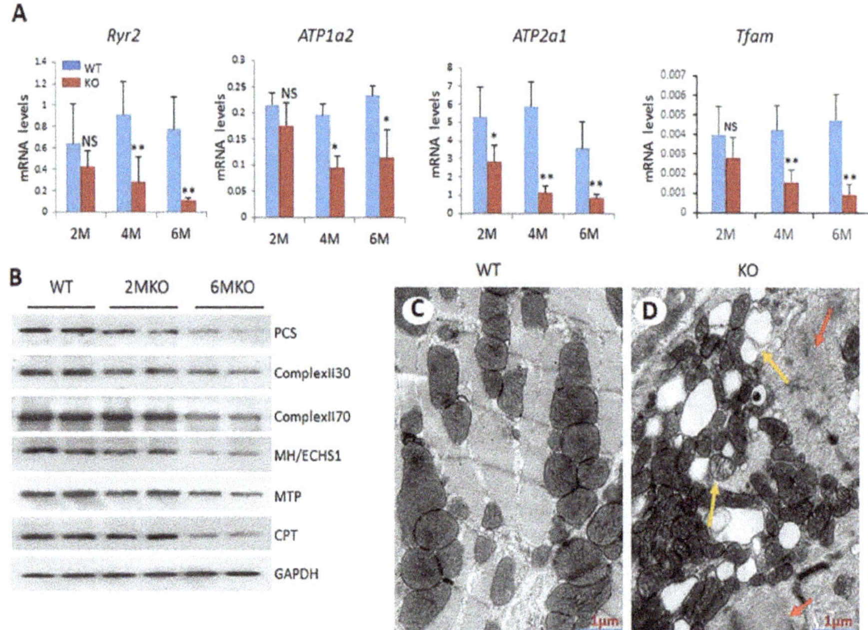

Figure 3. csPIMT$^{-/-}$ hearts show significant mitochondrial damage. (**A**) Quantification of mRNA levels of Atp1a2, Atp2a1, Ryr2 and Tfam genes. Each group was analyzed using 5 different mice (assayed individually) and the values are expressed as the mean ± SD. * $p < 0.05$, ** $p <$ 001, NS: not significant; (**B**) Western blot showing protein levels for PCS (palmitoyl-CoA synthetase; 62 Kda), complex II30 and 70, MH/ECHS1 (mitochondrial enoyl-CoA hydratase; 31 Kda), MTP (mitochondrial trifunctional protein; 100 Kda) and CPT1α (carnitine palmitoyltransferase; 88 Kda)). The protein extracts were prepared from 5 hearts pooled together. The protein expression of each gene was normalized to GAPDH. Percent decrease as compared WT controls were as follows: PCS, 67%; complex II30 and 70, 28% and 38%; MTP, 26%, and CPT, 42%; (**C**,**D**) display the electron micrographs of 6-month csPIMT$^{fl/fl}$ and csPIMT$^{-/-}$ mouse hearts. Red arrows in **D** indicate abnormal sarcomeres and H zone absent. Yellow arrows point to lipid droplets and damage in mitochondria.

csPIMT$^{-/-}$ cardiomyocytes undergo increased mitophagy. Because we observed severe mitochondrial damage in csPIMT$^{-/-}$ cardiomyocytes, we ascertained whether csPIMT$^{-/-}$ cardiomyocytes display increased mitophagy. Examination of RNA-seq data revealed changes in the expression of several key genes related to mitophagy including *Pink1* (PTEN-induced putative kinase 1), and *Drp1* (dynamin-related protein 1) [18–21]. These genes play important roles in maintaining mitochondrial homeostasis through complex mechanisms. As shown in Table 1, *Pink1* RNA levels increased between two- to three-fold and Drp1 RNA levels decreased two-fold. The changes in the expression of these genes were also confirmed by Western blots (Figure 4C). In agreement with RNA data, PINK levels increased at least two-fold whereas DRP1 levels decreased two-fold. To sum up, these results suggest that csPIMT$^{-/-}$ heart cells undergo increased mitophagy, and also possibly decreased fission due to increased mitochondrial damage.

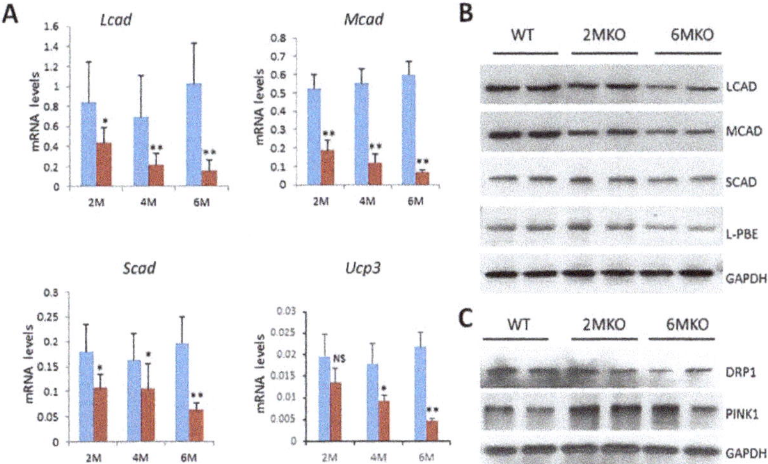

Figure 4. Expression of fatty acid metabolism genes is decreased in csPIMT$^{-/-}$ hearts. (**A**) Quantification of mRNA levels of *Lcad* (Long chain acly-CoA dehydrogenase), *Mcad* (medium-chain acyl-CoA dehydrogenase), *Scad* (short-chain acyl-CoA dehydrogenase) and *Ucp3* (uncoupling protein 3) genes. * $p < 0.05$, ** $p < 0.01$, NS: not significant; (**B**) Western blot showing protein levels of LCAD (47 Kda), MCAD (46 Kda), SCAD (44 Kda) and L-PBE (78 Kda; Enoyl-CoA hydratase /L-3-hydroxyacyl-CoA dehydrogenase). Percent reduction (KO vs. WT) for LCAD, 40%; MCAD, 39%; SCAD, 45%, and L-PBE, 27% for 6-month time point. Each group was analyzed using 5 different mice and the values were expressed as the mean ± SD; (**C**) Western blot showing the protein levels of DRP1 (78 Kda) and PINK1 (60 Kda). Details of same as in (**A**). See Materials and Methods for antibody sources and dilutions.

Genes involved in the β-oxidation process are expressed at lower levels: Heart muscle cells contract constantly in a coordinated fashion. Therefore, to maintain its contractile function, heart cells must receive constant supply of metabolic substrates to generate ATP. The major source (about 70%) of the energy for cardiac muscle cells come from fatty acids, especially long chain fatty acids [22]. The remainder of the energy in myocardial cells is derived from glucose and lactose [22]. β-Oxidation, a catabolic process by which fatty acid molecules are oxidized, is primarily facilitated by an enzyme complex (mitochondrial trifunctional protein, MTP) that is associated with the inner mitochondrial membrane. Therefore, we assessed the expression levels of genes related to fatty acid oxidation. Data presented in Table 1 show a reduced expression of several genes involved in fatty acid oxidation including *Pparα*, *Pgc1α*, *Mtp*, *Mcad*, *Ucp3* and *Abcc9* [23,24]. Decreased expression of several of these genes was also confirmed at the protein level by analyzing key mitochondrial and peroxisomal fatty acid β-oxidation enzymes. Western blots from total cell extracts derived from two- and six-month-old csPIMT$^{-/-}$ hearts along with that of csPIMT$^{fl/fl}$ hearts reveal that protein levels for mitochondrial enoyl-CoA hydratase (ECHS1), MTP, MCAD, SCAD and peroxisomal EHHADH/L-PBE are decreased two- to five-fold as compared to that of control heart extracts (Figures 3 and 4). These data are consistent with the decreased RNA levels shown in Table 1.

2.6. csPIMT$^{-/-}$ Mice Develop Cardiac Fibrosis

Cardiac fibrosis is an important complication in all types of heart diseases including DCM and it is associated with excessive accumulation of extra cellular matrix. To determine whether DCM in csPIMT$^{-/-}$ heart is associated with cardiac fibrosis, we examined the heart tissue for fibrosis by staining paraffin sections of heart with Masson's trichrome staining. Figure 5A shows significant fibrosis in csPIMT$^{-/-}$ heart as compared to the normal hearts at six months of age. Gene expression analysis supported this observation. At six months of age both RNA-seq and RT-qPCR data showed elevated levels of *Tgfβ2*, *Ctgf*, *Col9a2*, *Mmp3* and *Timp1* RNAs that stimulate signaling mechanisms involved in the regulation of extracellular matrix and promote fibrosis [25,26] (Table 1 and Figure 5B).

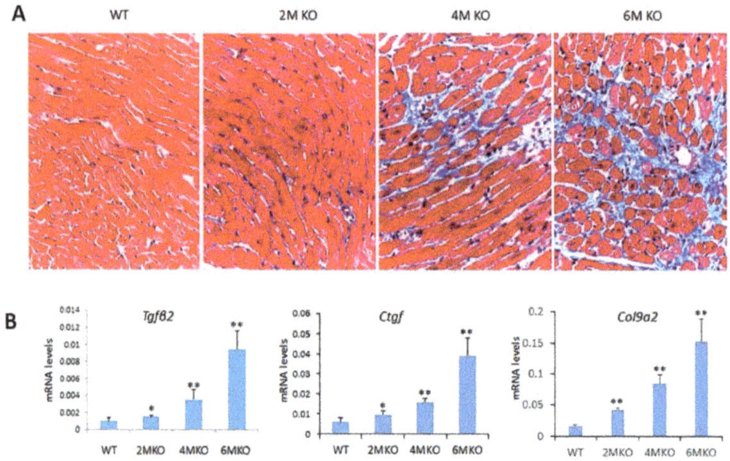

Figure 5. Myocardial fibrosis in csPIMT$^{-/-}$ mouse hearts. (**A**) Images of Masson trichrome staining patterns of a representative PIMT$^{fl/fl}$ and csPIMT$^{-/-}$ hearts of 2, 4 and 6 months are shown (magnification 400×). Note the intensely stained (blue color) interstitial fibrous strands in 6-month-old csPIMT$^{-/-}$ hearts; (**B**) Quantification of mRNA levels for *Tgfβ2*, *Ctgf* and *Col9a2* in csPIMT$^{fl/fl}$ and csPIMT$^{-/-}$ hearts of 2, 4 and 6 months of age. mRNA levels were quantified by RT-qPCR assays. Each group was analyzed using 5 different mice and the values were expressed as the mean ± SD. * $p < 0.05$, ** $p < 0.01$.

2.7. Genes Related to Glucose Metabolism Are Downregulated in csPIMT$^{-/-}$ Heart Leading to Glycogen Storage

As stated above, glucose and lactose also serve as significant energy source for myocardial cells. Glucose transporters (GLUT) are a family of proteins which mediate entry of glucose into cells [27,28]. Of these, GLUT4 is the most abundant glucose transporter in heart [29]. We observed a three-fold reduced expression of *Glut4* and hexokinase 2 (*Hk2*, ref [30]) in csPIMT$^{-/-}$ heart as compared to that of PIMT$^{fl/fl}$ heart (Table 1) that potentially could reduce uptake of glucose by myocardial cells (*Glut4*), phosphorylation of glucose (*Hk2*) and curtail the energy source from glucose and lactose for heart cells (Table 1). A likely consequence of reduced expression of these genes is that glucose is not properly utilized in csPIMT$^{-/-}$ myocardial cells and stored as glycogen. We also noted significant reduction in the *Pdk4* mRNA level in the myocardium of csPIMT$^{-/-}$ mice (Figure S1). This enzyme plays a key role in the regulation of glucose and fatty acid metabolism via phosphorylation [31]. Glucokinase (*Gck1,2*) provides G6P for the synthesis of glycogen [32]. GK mRNA level is increased in PIMT$^{-/-}$ heart, as compared to that of PIMT$^{fl/fl}$ heart (Figure S1).

2.8. Tamoxifen-Inducible Heart-Specific Cre-Recombinase to Disrupt PIMT Gene (TmcsPIMT$^{-/-}$) in Adult Mouse

To further validate the findings that lack of PIMT expression is solely responsible for the heart abnormalities and associated heart failure observed in csPIMT$^{-/-}$ mice, we used tamoxifen-inducible heart-specific Cre (Myh6-MCM)/PIMT$^{fl/fl}$ mouse model (TmcsPIMT$^{-/-}$). The tamoxifen-inducible gene knockout approach has clear advantages in that expression of a selected gene can be ablated in adult mice, as necessary, in a tissue-specific manner [33]. The Myh6-MCM/PIMT$^{fl/f}$ mice were given daily intraperitoneal injection of tamoxifen for five days. By 14 days after the first tamoxifen injection, the size of the heart increased dramatically as compared to that of littermate controls (Figure 6A, upper panel). Figure 6A also shows (lower panel) dilatation of the left ventricular chambers with thinning of the walls. PIMT RNA levels become almost non-detectable in heart (Figure 6B). PIMT expression was also evaluated using Western blot method. Figure 6C shows that PIMT expression was negligible in TmcsPIMT$^{-/-}$ hearts as compared to control hearts. Immunostaining of TmcsPIMT$^{-/-}$ heart tissue confirmed the absence of PIMT in nuclei of TmcsPIMT$^{-/-}$ mouse cardiomyocytes Echocardiographic analysis of TmcsPIMT$^{-/-}$ mice heart revealed increased left ventricular end-diastolic internal dimension (LVID-d), decreased fractional shortening and also decreased ejection fraction (Figure 6D,E). The contractility of TmcsPIMT$^{-/-}$ mouse heart was diminished with the ejection fraction in TmcsPIMT$^{-/-}$ mouse was 45% vs. 78% for floxed littermate controls (Figure 6F). Likewise, a fractional shortening of 21% vs. 43% for littermate controls was also observed (Figure 6G). These values suggest poor contractility of PIMT null hearts and support the conclusion that PIMT deficient mice die of DCM. Accordingly, the mRNA levels of heart failure indicator BNP were significantly elevated in TmcsPIMT$^{-/-}$ mouse heart (Figure 6H).

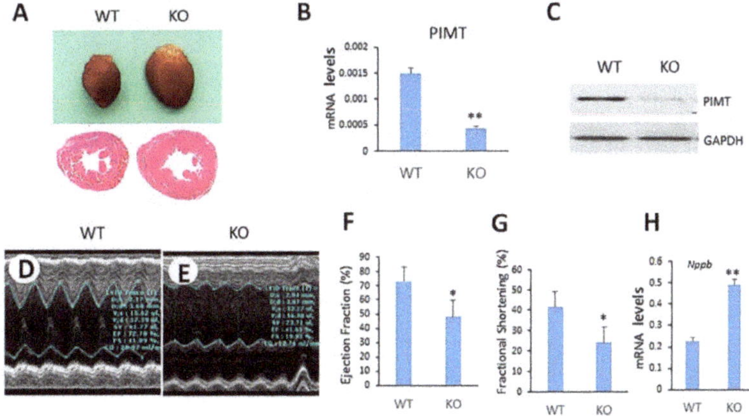

Figure 6. Tamoxifen-inducible cardiac-specific disruption of PIMT (TmcsPIMT$^{-/-}$) in adult mice causes dilated cardiomyopathy. Mice were killed 14 days after first tamoxifen injection in the experiments. (**A**) Representative photographs of adult hearts after tamoxifen-inducible heart-specific Cre mediated PIMT deletion. It is evident that TmcsPIMT$^{-/-}$ mouse heart is bigger than that of TmcsPIMT$^{fl/fl}$ mouse. Lower panel in (**A**) shows H&E cross sections of TmcsPIMT$^{-/-}$ and TmcsPIMT$^{fl/fl}$ hearts; (**B**) Relative PIMT mRNA expression in TmcsPIMT$^{fl/fl}$ (WT) and TmcsPIMT$^{-/-}$ (KO) mouse heart. (**C**) Western blot analysis of PIMT in TmcsPIMT$^{fl/fl}$ and TmcsPIMT$^{-/-}$ hearts. Total proteins from the heart tissues of appropriate mice were prepared as described (see Materials and Methods). They were then Western immunoblotted and probed with an anti-PIMT antibody (Bethyl IHC-00467, 1:1000); (**D**,**E**) Representative profiles of M-mode echocardiographic analyses of TmcsPIMT$^{-/-}$ and littermate control mice; (**F**,**G**) represent ejection fraction and fractional shortening respectively. Data were derived from (**D**,**E**); (**H**) Relative mRNA levels of BNP (*Nppb*) in TmcsPIMT$^{fl/fl}$ and TmcsPIMT$^{-/-}$ mouse hearts. The day of initial injection of Tamoxifen was counted as day 1. Results are expressed as the mean ± SD. * $p < 0.05$, ** $p < 0.01$.

We also examined whether TmcsPIMT$^{-/-}$ hearts develop cardiac fibrosis. As shown in Figure S2, Masson Trichrome staining indicated significant cardiac fibrosis in PIMT null hearts in Tamoxifen inducible model (shown by arrows) which is in agreement with the development of cardiac fibrosis in csPIMT$^{-/-}$ hearts. There was also significant increase in the RNA levels of *Ctgf* and Tgfβ2 genes (Figure S2C,D).

3. Discussion

The molecular mechanisms that lead to the development of DCM are not well understood. We now report the role played by PIMT in mouse heart functions. PIMT, an RNA binding protein with methyl transferase activity, participates in the formation of 2,2,7-trimethylguanosine cap structures of small nuclear and nucleolar RNAs that are important in RNA splicing [1]. Furthermore, PIMT interacts with nuclear receptor coactivators NCOA6, p300/CBP histone acetyltransferases, and the MED1 subunit of the Mediator complex and these interactions appear to influence energy metabolism in heart [8]. Mice with germ line deletion of *Pimt* gene, manifest early embryonic lethality by affecting development during preimplantation stage [11]. These and other results suggest that PIMT has the potential to control metabolic pathways at the chromatin level by influencing fatty acid oxidation and gluconeogenesis-related genes on its own merit and in concert with other transcription factors.

The results presented in this paper clearly demonstrate that PIMT is an essential gene for normal heart function and that heart-specific ablation of PIMT results in DCM (Figures 2 and 6). Cardiomyocyte-specific conditional PIMT deleted mice (csPIMT$^{-/-}$ mice) died by 7.5 months of age (this manuscript) whereas csMed1$^{-/-}$ mice died at age of one month [9]. It is also worth noting that deletion of the Mediator subunit genes including *Med1*, *Med12* or *Med30* in heart is more damaging in causing DCM [9,34–36]. It is not surprising because Med1 is necessary for the completion of transcriptional signaling of PPAR subfamily nuclear receptors [3].

The heart sections of two-month-old csPIMT$^{-/-}$ mice showed detectable thinning of the ventricular walls with considerable ventricular enlargement (Figure 1). That the two-month-old csPIMT$^{-/-}$ hearts suffer with DCM is supported by the data showing elevated levels of mRNAs coding for the BNP and ANP proteins. Both BNP and ANP levels increase during heart failure as ventricular cells secrete both these peptides in response to left ventricular dysfunction [12]. Left ventricular dilation increases progressively in csPIMT$^{-/-}$ hearts as evidenced by the images of H&E stained heart cross sections and sustained increase of ANP and BNP RNA levels (Figure 1). Other evidence including diminished contractility of csPIMT$^{-/-}$ hearts, loss of structural integrity of mitochondria and reduced expression of most of the genes related to oxidative phosphorylation and the fatty acid β-oxidation. Evidence supports the assertion that mitochondrial damage significantly contributes to the development of DCM and myocardial dysfunction. Cardiac myocytes are a type of muscle cells that contract and expand continuously, and this contractility is dependent on the constant supply of ATP generated by mitochondria through a complex interaction between oxidative phosphorylation and electron transport chain systems and both mitochondrial and peroxisomal fatty acid β-oxidation systems. Our gene expression analysis showed that the majority of the genes related to both these β-oxidation systems are expressed at lower levels in csPIMT$^{-/-}$ hearts at four and six months of age. For example, expression of several genes related to mitochondrial complexes I to IV and the mitochondrial uncoupling protein UCP3 decreased by about two- to five-fold. Similarly, expression of genes such as *Pparα*, *Pargc1a*, *Acadm* that are involved in energy homeostasis is also reduced about two- to six-fold. Another important gene *Tfam*, involved in mitochondrial DNA replication and transcription is also expressed at six-fold reduced levels that correlates with decreased mitochondrial copy number in csPIMT$^{-/-}$ hearts. TFAM is a multifunctional transcriptional factor that is critical for the mitochondrial DNA transcription and maintenance of mitochondrial genome copy number [13]. The mitochondria of csPIMT$^{-/-}$ heart cells also suffer structural damage as evidenced by the presence of lipid droplets and membranous swirls and irregularities in Z band pattern (Figure 3). In addition, evidence indicates that csPIMT$^{-/-}$ cardiomyocytes display increased mitophagy. In summary, mitochondria of csPIMT$^{-/-}$ hearts are unable to perform their normal functions and thus contribute to DCM and heart failure.

The role played by the peroxisomal and mitochondrial β-oxidation in the pathogenesis of cardiomyocyte and mitochondrial damage leading to the development of DCM is that both inhibition and profound elevation of fatty acid β oxidation can be pathogenic [37]. Diminished fatty acid β-oxidation can occur in the absence of PPARα and this causes toxic lipid injury due to un-metabolized very long chain fatty acids and fatty acyl CoAs and other intermediate metabolic products [37]. Likewise, excess fatty acid β-oxidation resulting from PPARα activation can also be deleterious [37,38].

Proper transport of calcium in and out of cardiac myocytes and coordination between calcium channel function and ATP production are critical for normal heart functions. Defects in calcium regulation and energy production are hallmarks of heart failure [39]. Global RNA analysis of csPIMT$^{-/-}$ heart tissue showed downregulation of several genes related to calcium channel structure and function, cardiac muscle contraction and calcium homeostasis. For example, *Atp2a2* (also known as *Serca-2a*) which encodes Ca2β ATPase isoform 2a protein and *Cacna1h* encodes a structural protein of voltage gated calcium channel are involved in calcium mediated changes in cardiomyocyte contractility [39]. Similarly, ryanodine receptor 2 gene (*Ryr2*), which encodes Ryr2 protein initiates cardiac muscle contraction by calcium channeling [29]. *Ryr2* regulates mitochondrial Ca^{2+} and ATP levels as well as a cascade of transcription factors that modulate metabolism and survival [39,40]. Overall, these results suggest that calcium regulation is defective in myocardial cells of csPIMT$^{-/-}$ hearts owing to reduced expression of the relevant genes described above. Thus, PIMT along with Med1 and Med12 contribute to the regulation of calcium handling genes [9,36].

We also observed significant myocardial fibrosis in csPIMT$^{-/-}$ heart at six months of age. In normal heart, the fibroblasts form a network throughout myocardium and contribute in part to the mechanical and structural maintenance of heart [41]. When there is cardiac injury, myocardial fibroblasts are activated due to cytokine and neurohumoral factors released by the heart tissue which deregulate the extracellular matrix leading to the development of fibrosis [26]. The fibrosis process involves activation of a number of genes related to formation of extracellular matrix including *Tgfβ*, connective tissue growth factor (*Ctgf*), matrix metalloprotease *Mmp3*, alpha 2 type IX collagen (*Col9a2*), and the tissue inhibitor of metalloproteinase-1 Timp1 [26]. Our RNAseq and RT-qPCR data confirmed the upregulation of these genes beginning two months after birth, with the expression levels ranging from 1.5- to eight-fold depending on the type of the gene. Sustained upregulation of these genes was observed at the age of six months that could explain the development of the cardiac fibrosis in csPIMT$^{-/-}$ hearts.

At present, we do not know the molecular mechanisms by which PIMT affects gene expression except that it occurs at the chromatin level. We showed earlier that PIMT physically interacts with NCOA6, p300/CBP, and MED1 to transcriptionally stimulate reporter genes [8]. Based on these observations, we propose a model in which PIMT forms a complex with NCOA6 and cooperates with histone acetyl transferases p300/CBP, MED1 and perhaps with other unknown chromatin factors to affect transcription of a group of genes specific for cardiac functions [8] (see Figure 7). This model at least in part can explain the broad effects we observed about PIMT deletion in PIMT$^{-/-}$ hearts. However, additional studies are needed to address many of the issues raised in this study.

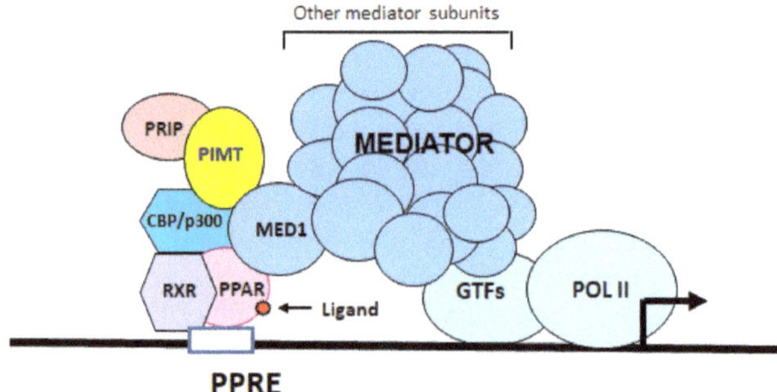

Figure 7. A model showing the interactions of PIMT with PRIP/NCOA6, histone acetyl transferases p300/CBP, and MED1 of the Mediator complex in the regulation of transcription of PPAR regulated genes. Note that in addition to Med1 subunit of the Mediator complex MED30 and MED12 also modulate genes involved in energy metabolism. See Discussion for further details. See Discussion for further details.

In summary, we have for the first time, reported here an essential function for PIMT gene in heart function and consequences of ablation of PIMT gene in heart including the development of DCM and heart failure. Thus, PIMT is a member of a growing list of essential genes that are critical for heart function and understanding the PIMT functions in heart will aid in the efforts to develop novel drugs and other therapeutic strategies in the management of heart failure.

4. Materials and Methods

4.1. Animals

PIMT conditional knock-out mice were generated using the two-loxP, two-frt recombination system [42]. PIMT$^{fl/fl}$ mice were crossbred with cardiac α-myosin heavy chain promoter driven Cre (α-MyHC-Cre) transgenic mice [11] to generate cardiomyocyte-specific PIMT null mice (csPIMT$^{-/-}$) with deletion of exons 3 and 4 of *PIMT* gene commencing during late embryonic period. PCR genotyping was performed using the primers P4: 5′-CTGCATGTATGAATCTTGGGAG-3′, P5: 5′-GCATCAAGAATATACAGAACAGAGA CTC-3′ and P6: 5′-CTCCTTCCTTCTGTACCTCTGTAGC-3′. Primers P6/P5 yielded a 376 bp Wild-type PIMT allele in PIMT$^{+/+}$ mice; primers P4/P5 yielded a 298 bp PIMT$^{fl/fl}$ allele. *Cre*-specific primers used included: 5′-AGGTGTAGAGAAGGCACTCAGC-3′ and 5′-CTAATCGCCATCTTCCAGCAGG-3′.

To generate mice with tamoxifen-inducible heart-specific PIMT deletion (TmcsPIMT$^{-/-}$), PIMT$^{fl/fl}$ mice were cross-bred with Myh6-MCM (tamoxifen-inducible heart-specific Cre) transgenic mice purchased from the Jackson Laboratory [33]. TmcsPIMT$^{-/-}$ mice and their littermate controls were administered tamoxifen intraperitoneally at seven weeks of age at a daily dose of 65 mg/kg body weight for five days and then killed at selected intervals. Survival curves were obtained by following 36 csPIMT$^{-/-}$ mice and the same number of csPIMT$^{fl/fl}$ genotype. The criteria used for animal euthanasia were as listed previously [5], and included absence of food and water consumption, diminished or absence of mobility, absence of heart beat and respiratory movement. Pentobarbital was injected intraperitoneally at the dose of 150 mg/kg body weight to minimize suffering. Animals had access to food and water ad libitum and maintained on a 12-h light-dark cycle. All procedures were performed in accordance with the National Institutes of Health Guide for Care and Use of Laboratory Animals. The animal protocols were reviewed and approved by the Institutional Animal Care and Use Committee of Northwestern University (protocol number 2013–3198, 1 July 2013).

4.2. Echocardiography

Echocardiography was performed as described previously using a VisualSonics Vevo 770 high-resolution noninvasive transthoracic imaging system with a 30 MHz scanhead [5]. Short- and long-axis parasternal views were used to obtain 2D and M-mode images which facilitated examination of the septum, posterior wall and left ventricular outflow tract. We recorded at least eight independent cardiac cycles per experiment.

4.3. Histological Analysis

Heart tissues from csPIMT$^{-/-}$ and PIMT$^{fl/fl}$ and also from TmcsPIMT$^{-/-}$ and the corresponding control mice were fixed in 4% paraformaldehyde for 48 h and processed for embedding in paraffin. Sections, 4-μm thick, were stained with hematoxylin and eosin (H&E). Immunohistochemical localization of PIMT was carried out using anti-PIMT antibody (catalog number IHC-00467, Bethyl, Montgomery, TX, USA). Masson's trichrome staining was used for the visualization of cardiac fibrosis. Heart specimens were also embedded in O.C.T. compound (Tissue-TeK, Torrance City, CA, USA), and 6-um thick sections were stained with Oil Red O for the visualization of neutral lipid.

4.4. Electron Microscopy

Heart tissue samples obtained from the left ventricle were fixed overnight at 4 °C with 3% glutaraldehyde in sodium cacodylate buffer. The tissue was then post-fixed in 1% osmium tetroxide in cacodylate buffer (pH 7.4) for 2 h at 4 °C and embedded in Epon [9]. Ultra-thin sections were cut with a Leica UC6 ultramicrotome and examined with a FEI Tecnai Spirit transmission electron microscope (FEI, Hilsboro City, OR, USA).

4.5. Library Construction and Sequencing

Library construction and sequencing were carried out at the Genomics Core facility of University of Chicago. To generate single-end 50 bp (SR50) RNA sequencing libraries, RNA quality and quantity were determined with Agilent Bioanalyzer 2100, selecting RNA integrity numbers (RIN) of >7 and quantities of 100 nanograms or more per sample. Directional mRNA libraries were generated using Illumina TruSeq mRNA Sample Preparation Kits (Illumina, San Diego, CA, USA). Briefly, polyadenylated mRNAs were captured from total RNA using oligo-dT selection and then converted to cDNA by reverse transcription. They were then ligated to Illumina sequencing adapters containing unique barcode sequences. These were then amplified by PCR and the resulting cDNA libraries quantified using RT-RT-qPCR. Finally, equimolar concentrations of ach cDNA library were pooled and sequenced on the Illumina HiSeq2500 (Illumina, San Diego, CA, USA).

4.6. Transcriptome Analysis

The quality of DNA reads, in fastq format, was evaluated using FastQC. Adapters were trimmed and reads of poor quality or aligning to rRNA sequences were filtered. The cleaned reads were aligned to the *Mus musculus* genome (mm10) using STAR [43]. Read counts for each gene were calculated using htseq-count [44] in conjunction with a gene annotation file for mm10 obtained from UCSC (University of California Santa Cruz; http://genome.ucsc.edu). Differential expression was determined using DESeq2 [45]. The cutoff for determining significantly differentially expressed genes was an FDR-adjusted *p*-value less than 0.05. A pathway analysis was performed on both upregulated and downregulated gene lists using GeneCoDis [46,47].

4.7. Quantitative Real-Time PCR

Total RNA was extracted from the csPIMT$^{-/-}$ and TmcPIMT$^{-/-}$ and the corresponding control mice using TRIzol® reagent (Life Technology, Carlsbad, CA, USA). RNA was further purified using Qiagen RNeasy columns (Life Technology, Carlsbad, CA, USA). cDNA was prepared with 2 μg of total RNA using SuperScript VILO First-Strand Synthesis System (Life Technology, Carlsbad, CA, USA). Expression of specific genes was verified using SYBR Green (Life Technologies) in triplicates and normalized with 18S ribosomal RNA. Each PCR reaction contained 1 μL (100 pmol) of forward and reverse primers and 10 μL of 2× SYBR Green PCR Master Mix to make a final volume of 20 μL. The reaction was run by using an ABI 7300 (Applied Biosystems, Foster City, CA, USA)). The relative gene expression changes were measured using the comparative C_t method, $X = 2^{-\Delta\Delta Ct}$. Sequences of all primers are shown in Table S5.

4.8. Western Blot Analysis

Total proteins were extracted from the heart tissue of csPIMT$^{-/-}$ mice and corresponding littermates and subjected to 4–20% SDS-PAGE. Samples were analyzed in duplicates for each time point. Protein extracts were prepared from pooled samples using five animals. Same pooled hearts were used for protein extracts and for RT-qPCR assays, as well as for RNA-seq analysis. They were then transferred to a nitrocellulose membrane (Invitrogen). Immunoblotting was performed using relevant antibodies as described with GAPDH as loading control. The protein bands were developed with an enhanced chemiluminescence substrate. Quantification of blots was performed using ImageJ software (NIH). Sources of antibodies and dilutions: Complex II30, Invitrogen cat# 459230; ComplexII70, Invitrogen cat#459200; DRP1, Cell Signaling, cat#8570; Pink1, Cell Signaling cat#6946. All antibodies mentioned above were diluted 1:1000. GAPDH, Cell Signaling cat# 5174, dilution 1:1500. PCS, MH, MTP, CPT, LCAD, MCAD, SCAD, L-PBE antibodies (dilution 1:2000) are rabbit polyclonal antibodies, kind gifts of Dr. T. Hashimoto, Department of Pediatrics, Gifu University School of Medicine, Japan.

4.9. Mitochondrial DNA Content

To determine the mitochondrial DNA copy number, total DNA from heart tissue was first isolated. The quantity of nuclear-encoded 18S ribosomal RNA (rRNA) and the mitochondrial encoded gene cytochrome c oxidase subunit 1 (CO1) were estimated by RT-qPCR. Mitochondrial DNA copy number was expressed as the ratio of CO1 to 18S rRNA as described [9,48].

4.10. Statistical Analysis

Student's *t* test was used to determine whether the sample was significantly different from the control. Differences were considered statistically significant at $p < 0.05$, while $p < 0.01$ represented more significant change.

Supplementary Materials: Supplementary materials can be found at http://www.mdpi.com/1422-0067/19/5/1485/s1.

Author Contributions: Y.J. designed and performed the experiments. N.L. and R.S. participated in the animal breeding and genotyping. N.V. designed and analyzed the heart-specific PIMT null studies. M.J.S. performed RNA-seq experiments. M.S. conducted echocardiographic studies. E.B.T., and Y.S.K. contributed expert advice and helped analyzing the data. B.T. and J.K.R. supervised the study and wrote the manuscript. All authors corrected and approved the final manuscript.

Acknowledgments: This research was supported by grants NIH R01 DK083163 awarded to J.K.R., NIH R21 AI1094296 awarded to B.T, and NIH RO1 DK60635 awarded to YSK.

Conflicts of Interest: The authors declare no conflict of interests.

References

1. Zhu, Y.; Qi, C.; Cao, W.Q.; Yeldandi, A.V.; Rao, M.S.; Reddy, J.K. Cloning and characterization of PIMT, a protein with a methyltransferase domain, which interacts with and enhances nuclear receptor coactivator PRIP function. *Proc. Natl. Acad. Sci. USA* **2001**, *98*, 10380–10385. [CrossRef] [PubMed]
2. Carlsten, J.O.; Zhu, X.; Gustafsson, C.M. The multitalented Mediator complex. *Trends Biochem. Sci.* **2013**, *38*, 531–537. [CrossRef] [PubMed]

3. Jia, Y.; Viswakarma, N.; Reddy, J.K. Med1 subunit of the mediator complex in nuclear receptor-regulated energy metabolism, liver regeneration, and hepatocarcinogenesis. *Gene Expr.* **2014**, *16*, 63–75. [CrossRef] [PubMed]
4. Burkart, E.M.; Sambandam, N.; Han, X.; Gross, R.W.; Courtois, M.; Gierasch, C.M.; Shoghi, K.; Welch, M.J.; Kelly, D.P. Nuclear receptors PPARβ/δ and PPARalpha direct distinct metabolic regulatory programs in the mouse heart. *J. Clin. Investig.* **2007**, *117*, 3930–3939. [PubMed]
5. Cheng, L.; Ding, G.; Qin, Q.; Huang, Y.; Lewis, W.; He, N.; Evans, R.M.; Schneider, M.D.; Brako, F.A.; Xiao, Y.; et al. Cardiomyocyte-restricted peroxisome proliferator-activated receptor-δ deletion perturbs myocardial fatty acid oxidation and leads to cardiomyopathy. *Nat. Med.* **2004**, *10*, 1245–1250. [CrossRef] [PubMed]
6. Mouaikel, J.; Bujnicki, J.M.; Tazi, J.; Bordonne, R. Sequence-structure-function relationships of Tgs1, the yeast snRNA/snoRNA cap hypermethylase. *Nucleic Acids Res.* **2003**, *31*, 4899–4909. [CrossRef] [PubMed]
7. Mouaikel, J.; Verheggen, C.; Bertrand, E.; Tazi, J.; Bordonne, R. Hypermethylation of the cap structure of both yeast snRNAs and snoRNAs requires a conserved methyltransferase that is localized to the nucleolus. *Mol. Cell* **2002**, *9*, 891–901. [CrossRef]
8. Misra, P.; Qi, C.; Yu, S.; Shah, S.H.; Cao, W.Q.; Rao, M.S.; Thimmapaya, B.; Zhu, Y.; Reddy, J.K. Interaction of PIMT with transcriptional coactivators CBP, p300, and PBP differential role in transcriptional regulation. *J. Biol. Chem.* **2002**, *277*, 20011–20019. [CrossRef] [PubMed]
9. Jia, Y.; Chang, H.C.; Schipma, M.J.; Liu, J.; Shete, V.; Liu, N.; Sato, T.; Thorp, E.B.; Barger, P.M.; Zhu, Y.J.; et al. Cardiomyocyte-Specific Ablation of Med1 Subunit of the Mediator Complex Causes Lethal Dilated Cardiomyopathy in Mice. *PLoS ONE* **2016**, *11*, e0160755.
10. Roh, J.I.; Cheong, C.; Sung, Y.H.; Lee, J.; Oh, J.; Lee, B.S.; Lee, J.E.; Gho, Y.S.; Kim, D.K.; Park, C.B.; et al. Perturbation of NCOA6 leads to dilated cardiomyopathy. *Cell Rep.* **2014**, *8*, 991–998. [CrossRef] [PubMed]
11. Jia, Y.; Viswakarma, N.; Crawford, S.E.; Sarkar, J.; Sambasiva Rao, M.; Karpus, W.J.; Kanwar, Y.S.; Zhu, Y.J.; Reddy, J.K. Early embryonic lethality of mice with disrupted transcription cofactor PIMT/NCOA6IP/Tgs1 gene. *Mech. Dev.* **2012**, *129*, 193–207. [CrossRef] [PubMed]
12. Kuwahara, K.; Nakao, K. Regulation and significance of atrial and brain natriuretic peptides as cardiac hormones. *Endocr. J.* **2010**, *57*, 555–565. [CrossRef] [PubMed]
13. Campbell, C.T.; Kolesar, J.E.; Kaufman, B.A. Mitochondrial transcription factor A regulates mitochondrial transcription initiation, DNA packaging, and genome copy number. *Biochim. Biophys. Acta* **2012**, *1819*, 921–929. [CrossRef] [PubMed]
14. Bar-Yaacov, D.; Blumberg, A.; Mishmar, D. Mitochondrial-nuclear co-evolution and its effects on OXPHOS activity and regulation. *Biochim. Biophys. Acta* **2012**, *1819*, 1107–1111. [CrossRef] [PubMed]
15. Lobo-Jarne, T.; Ugalde, C. Respiratory chain supercomplexes: Structures, function and biogenesis. *Semin. Cell. Dev. Biol* **2017**, *76*, 179–190. [CrossRef] [PubMed]
16. Vonck, J.; Schafer, E. Supramolecular organization of protein complexes in the mitochondrial inner membrane. *Biochim. Biophys. Acta* **2009**, *1793*, 117–124. [CrossRef] [PubMed]
17. Papa, S.; Martino, P.L.; Capitanio, G.; Gaballo, A.; De Rasmo, D.; Signorile, A.; Petruzzella, V. The oxidative phosphorylation system in mammalian mitochondria. *Adv. Exp. Med. Biol.* **2012**, *942*, 3–37. [PubMed]
18. Ikeda, Y.; Shirakabe, A.; Brady, C.; Zablocki, D.; Ohishi, M.; Sadoshima, J. Molecular mechanisms mediating mitochondrial dynamics and mitophagy and their functional roles in the cardiovascular system. *J. Mol. Cell. Cardiol.* **2015**, *78*, 116–122. [CrossRef] [PubMed]
19. Mukherjee, U.A.; Ong, S.B.; Ong, S.G.; Hausenloy, D.J. Parkinson's disease proteins: Novel mitochondrial targets for cardioprotection. *Pharmacol. Ther.* **2015**, *156*, 34–43. [CrossRef] [PubMed]
20. Shirihai, O.S.; Song, M.; Dorn, G.W., 2nd. How mitochondrial dynamism orchestrates mitophagy. *Circ. Res.* **2015**, *116*, 1835–1849. [CrossRef] [PubMed]
21. Dorn, G.W., 2nd. Parkin-dependent mitophagy in the heart. *J. Mol. Cell. Cardiol.* **2016**, *95*, 42–49. [CrossRef] [PubMed]
22. Lopaschuk, G.D.; Ussher, J.R.; Folmes, C.D.; Jaswal, J.S.; Stanley, W.C. Myocardial fatty acid metabolism in health and disease. *Physiol. Rev.* **2010**, *90*, 207–258. [CrossRef] [PubMed]
23. Houten, S.M.; Wanders, R.J. A general introduction to the biochemistry of mitochondrial fatty acid β-oxidation. *J. Inherit. Metab. Dis.* **2010**, *33*, 469–477. [CrossRef] [PubMed]

24. Settembre, C.; De Cegli, R.; Mansueto, G.; Saha, P.K.; Vetrini, F.; Visvikis, O.; Huynh, T.; Carissimo, A.; Palmer, D.; Klisch, T.J.; et al. TFEB controls cellular lipid metabolism through a starvation-induced autoregulatory loop. *Nat. Cell Biol.* **2013**, *15*, 647–658. [CrossRef] [PubMed]
25. Teekakirikul, P.; Eminaga, S.; Toka, O.; Alcalai, R.; Wang, L.; Wakimoto, H.; Nayor, M.; Konno, T.; Gorham, J.M.; Wolf, C.M.; et al. Cardiac fibrosis in mice with hypertrophic cardiomyopathy is mediated by non-myocyte proliferation and requires TGF-β. *J. Clin. Investig.* **2010**, *120*, 3520–3529. [CrossRef] [PubMed]
26. Leask, A. Potential therapeutic targets for cardiac fibrosis: TGFβ, angiotensin, endothelin, CCN2, and PDGF, partners in fibroblast activation. *Circ. Res.* **2010**, *106*, 1675–1680. [CrossRef] [PubMed]
27. Byers, M.S.; Howard, C.; Wang, X. Avian and Mammalian Facilitative Glucose Transporters. *Microarrays* **2017**, *6*, 7. [CrossRef] [PubMed]
28. Kain, V.; Kapadia, B.; Viswakarma, N.; Seshadri, S.; Prajapati, B.; Jena, P.K.; Teja Meda, C.L.; Subramanian, M.; Kaimal Suraj, S.; Kumar, S.T.; et al. Co-activator binding protein PIMT mediates TNF-alpha induced insulin resistance in skeletal muscle via the transcriptional down-regulation of MEF2A and GLUT4. *Sci. Rep.* **2015**, *5*, 15197. [CrossRef] [PubMed]
29. Szablewski, L. Glucose transporters in healthy heart and in cardiac disease. *Int. J. Cardiol.* **2017**, *230*, 70–75. [CrossRef] [PubMed]
30. McCommis, K.S.; Douglas, D.L.; Krenz, M.; Baines, C.P. Cardiac-specific hexokinase 2 overexpression attenuates hypertrophy by increasing pentose phosphate pathway flux. *J. Am. Heart Assoc.* **2013**, *2*, e000355. [CrossRef] [PubMed]
31. Depre, C.; Rider, M.H.; Hue, L. Mechanisms of control of heart glycolysis. *Eur. J. Biochem.* **1998**, *258*, 277–290. [CrossRef] [PubMed]
32. Petersen, M.C.; Vatner, D.F.; Shulman, G.I. Regulation of hepatic glucose metabolism in health and disease. *Nat. Rev. Endocrinol.* **2017**, *13*, 572–587. [CrossRef] [PubMed]
33. Sohal, D.S.; Nghiem, M.; Crackower, M.A.; Witt, S.A.; Kimball, T.R.; Tymitz, K.M.; Penninger, J.M.; Molkentin, J.D. Temporally regulated and tissue-specific gene manipulations in the adult and embryonic heart using a tamoxifen-inducible Cre protein. *Circ. Res.* **2001**, *89*, 20–25. [CrossRef] [PubMed]
34. Spitler, K.M.; Ponce, J.M.; Oudit, G.Y.; Hall, D.D.; Grueter, C.E. Cardiac Med1 deletion promotes early lethality, cardiac remodeling, and transcriptional reprogramming. *Am. J. Physiol. Heart Circ. Physiol.* **2017**, *312*, H768–H780. [CrossRef] [PubMed]
35. Krebs, P.; Fan, W.; Chen, Y.H.; Tobita, K.; Downes, M.R.; Wood, M.R.; Sun, L.; Li, X.; Xia, Y.; Ding, N.; et al. Lethal mitochondrial cardiomyopathy in a hypomorphic Med30 mouse mutant is ameliorated by ketogenic diet. *Proc. Natl. Acad. Sci. USA* **2011**, *108*, 19678–19682. [CrossRef] [PubMed]
36. Baskin, K.K.; Makarewich, C.A.; DeLeon, S.M.; Ye, W.; Chen, B.; Beetz, N.; Schrewe, H.; Bassel-Duby, R.; Olson, E.N. MED12 regulates a transcriptional network of calcium-handling genes in the heart. *JCI Insight* **2017**, *2*, e91920. [CrossRef] [PubMed]
37. Drosatos, K.; Schulze, P.C. Cardiac lipotoxicity: Molecular pathways and therapeutic implications. *Curr. Heart Fail. Rep.* **2013**, *10*, 109–121. [CrossRef] [PubMed]
38. Xiaoli; Yang, F. Mediating lipid biosynthesis: Implications for cardiovascular disease. *Trends Cardiovasc. Med.* **2013**, *23*, 269–273.
39. Marks, A.R. Calcium cycling proteins and heart failure: Mechanisms and therapeutics. *J. Clin. Investig.* **2013**, *123*, 46–52. [CrossRef] [PubMed]
40. Kushnir, A.; Marks, A.R. The ryanodine receptor in cardiac physiology and disease. *Adv. Pharmacol.* **2010**, *59*, 1–30. [PubMed]
41. Li, L.; Zhao, Q.; Kong, W. Extracellular matrix remodeling and cardiac fibrosis. *Matrix Biol.* **2018**. [CrossRef] [PubMed]
42. Agah, R.; Frenkel, P.A.; French, B.A.; Michael, L.H.; Overbeek, P.A.; Schneider, M.D. Gene recombination in postmitotic cells. Targeted expression of Cre recombinase provokes cardiac-restricted, site-specific rearrangement in adult ventricular muscle in vivo. *J. Clin. Investig.* **1997**, *100*, 169–179. [CrossRef] [PubMed]
43. Dobin, A.; Davis, C.A.; Schlesinger, F.; Drenkow, J.; Zaleski, C.; Jha, S.; Batut, P.; Chaisson, M.; Gingeras, T.R. STAR: Ultrafast universal RNA-seq aligner. *Bioinformatics* **2013**, *29*, 15–21. [CrossRef] [PubMed]
44. Anders, S.; Pyl, P.T.; Huber, W. HTSeq—A Python framework to work with high-throughput sequencing data. *Bioinformatics* **2015**, *31*, 166–169. [CrossRef] [PubMed]

45. Love, M.I.; Huber, W.; Anders, S. Moderated estimation of fold change and dispersion for RNA-seq data with DESeq2. *Genome Biol.* **2014**, *15*, 550. [CrossRef] [PubMed]
46. Tabas-Madrid, D.; Nogales-Cadenas, R.; Pascual-Montano, A. GeneCodis3: A non-redundant and modular enrichment analysis tool for functional genomics. *Nucleic Acids Res.* **2012**, *40*, W478–W483. [CrossRef] [PubMed]
47. Nogales-Cadenas, R.; Carmona-Saez, P.; Vazquez, M.; Vicente, C.; Yang, X.; Tirado, F.; Carazo, J.M.; Pascual-Montano, A. GeneCodis: Interpreting gene lists through enrichment analysis and integration of diverse biological information. *Nucleic Acids Res.* **2009**, *37*, W317–W322. [CrossRef] [PubMed]
48. Wu, R.; Chang, H.C.; Khechaduri, A.; Chawla, K.; Tran, M.; Chai, X.; Wagg, C.; Ghanefar, M.; Jiang, X.; Bayeva, M.; et al. Cardiac-specific ablation of ARNT leads to lipotoxicity and cardiomyopathy. *J. Clin. Investig.* **2014**, *124*, 4795–4806. [CrossRef] [PubMed]

International Journal of
Molecular Sciences

Review

Elucidating the Beneficial Role of PPAR Agonists in Cardiac Diseases

Zaza Khuchua [1,2,3,*], Aleksandr I. Glukhov [2,4], Arnold W. Strauss [1] and Sabzali Javadov [5,*]

1 The Heart Institute, Cincinnati Children's Hospital Medical Center, Cincinnati, OH 45229-7020, USA; arnold.strauss@cchmc.org
2 Department of Biochemistry, Sechenov University, 119991 Moscow, Russia; aiglukhov1958@gmail.com
3 Department of Biochemistry, Ilia University, Tbilisi 0162, Georgia
4 Department of Biology, Lomonosov Moscow State University, 119991 Moscow, Russia
5 Department of Physiology, University of Puerto Rico School of Medicine, San Juan, PR 00936-5067, USA
* Correspondence: zkhuchua@gmail.com (Z.K.); sabzali.javadov@upr.edu (S.J.); Tel.: +1-513-490-2206 (Z.K.); +1-787-758-2525 (ext. 2909) (S.J.)

Received: 20 September 2018; Accepted: 2 November 2018; Published: 4 November 2018

Abstract: Peroxisome proliferator-activated receptors (PPARs) are nuclear hormone receptors that bind to DNA and regulate transcription of genes involved in lipid and glucose metabolism. A growing number of studies provide strong evidence that PPARs are the promising pharmacological targets for therapeutic intervention in various diseases including cardiovascular disorders caused by compromised energy metabolism. PPAR agonists have been widely used for decades as lipid-lowering and anti-inflammatory drugs. Existing studies are mainly focused on the anti-atherosclerotic effects of PPAR agonists; however, their role in the maintenance of cellular bioenergetics remains unclear. Recent studies on animal models and patients suggest that PPAR agonists can normalize lipid metabolism by stimulating fatty acid oxidation. These studies indicate the importance of elucidation of PPAR agonists as potential pharmacological agents for protection of the heart from energy deprivation. Here, we summarize and provide a comprehensive analysis of previous studies on the role of PPARs in the heart under normal and pathological conditions. In addition, the review discusses the PPARs as a therapeutic target and the beneficial effects of PPAR agonists, particularly bezafibrate, to attenuate cardiomyopathy and heart failure in patients and animal models.

Keywords: PPAR agonists; bezafibrate; heart; cardiomyopathy; heart failure; lipids; fatty acid oxidation; energy metabolism; mitochondria

---

1. Introduction

Peroxisome proliferator-activated receptors (PPARs) play an important role in the regulation of carbohydrate and lipid metabolism in the cell. They are involved in the transcriptional regulation of multiple processes and play a central role in the pathogenesis of metabolic disorders, cardiovascular diseases, diabetes, cancer, inflammation, and other diseases. PPARs are members of the nuclear hormone receptor superfamily and act as ligand-activated transcription factors. They were first discovered in the early 1990s as transcription factors that mediate proliferation of peroxisomes in the cell [1–3]. Interestingly, biological effects of bezafibrate (BF), a potent pan-specific activator of PPARs, were demonstrated before cloning and discovery of PPARs [4–6]. Currently, there are three PPAR isoforms, PPARα, PPARβ/δ, and PPARγ that are encoded by separate genes. All three isoforms possess a high degree of inter-species sequence homology, particularly in the DNA-binding domain (DBD) and ligand-binding domain (LBD) [7,8] (Figure 1). The central role of PPARs in heart metabolism, particularly fatty acid oxidation (FAO) and mitochondrial bioenergetics, makes them a promising therapeutic target for the treatment of cardiac diseases, such as myocardial infarction (MI) and heart failure (HF). A growing number of studies using experimental animal models and patients often provide controversial data on the beneficial effects of PPAR agonists in cardiac diseases. In this review, we summarize and discuss the role of PPARs, particularly PPARα, in the healthy heart and cardiac diseases. In addition, we provide a comprehensive discussion of PPAR agonists in the treatment of cardiac diseases, particularly cardiomyopathy and HF.

```
Q07869_PPARA_HUMAN    1    -----------------------------MVDTESPLCPLSPLEAGDLESPLSEEF-LQ    29
Q03181_PPARD_HUMAN    1    ------------------------------------------------MEQPQEEAPEVR    12
P37231_PPARG_HUMAN    1    MGETLGDSPIDPESDSFTDTLSANISQEMTMVDTEMPFWPTNFGI--------------    45

Q07869_PPARA_HUMAN    30   EMGNIQEISQSIGEDSSGSFGFT-------------EYQ---YLGSCPG-SD--------    64
Q03181_PPARD_HUMAN    13   EEEKEEVAEAEG--------AP-------------E------LNGGPQHAL--------    37
P37231_PPARG_HUMAN    46   -----SSVDLSVMEDHSHSFDIKPFTTVDFSSISTPHYEDIPFTRTDPVVADYKYDLKLQ    100
                                ..:  :                    .           *    :
                                                                          ZFD
Q07869_PPARA_HUMAN    65   GSVITDTLSPASSPSSVTYPV-VPGSVDESPSGALNIECRICGDKASGYHYGVHACEGCK    123
Q03181_PPARD_HUMAN    38   PSSSYTDLSRSSSPPSLLD-Q-LQMGCDGASCGSLNMECRVCGDKASGFHYGVHACEGCK    95
P37231_PPARG_HUMAN    101  EYQSAIKVEPASPPYYSEKTQLYNKPHEEPSNSLMAIECRVCGDKASGFHYGVHACEGCK    160
                              :. :* *            :       . : :***:*******:**********
                                DBD             ZFD
Q07869_PPARA_HUMAN    124  GFFRRTIRLKLVYDKCDRSCKIQKKNRNKCQYCRFHKCLSVGMSHNAIRFGRMPRSEKAK    183
Q03181_PPARD_HUMAN    96   GFFRRTIRMKLEYEKCERSCKIQKKNRNKCQYCRFQKCLALGMSHNAIRFGRMPEAEKRK    155
P37231_PPARG_HUMAN    161  GFFRRTIRLKLIYDRCDLNCRIHKKSRNKCQYCRFQKCLAVGMSHNAIRFGRMPQAEKEK    220
                           ********:** *::*: .*:*:**.*********:***::*************.:** *

Q07869_PPARA_HUMAN    184  LKAEILTCEHDIEDSETADLKSLAKRIYEAYLKNFNMNKVKARVILSGKASNNPPFVIHD    243
Q03181_PPARD_HUMAN    156  LVAGLTANEGSQYNPQVADLKAFSKHIYNAYLKNFNMTKKKARSILTGKASHTAPFVIHD    215
P37231_PPARG_HUMAN    221  LLAEISS-DIDQLNPESADLRALAKHLYDSYIKSFPLTKAKARAILTGKTTDKSPFVIYD    279
                           * * : : : .  : : ***::::*::*::*:*.* :.* *** **:**::.. ****:*

Q07869_PPARA_HUMAN    244  METLCMAEKTLVAKLVANGIQ-NKEAEVRIFHCCQCTSVETVTELTEFAKAIPGFANLDL    302
Q03181_PPARD_HUMAN    216  IETLWQAEKGLVWKQLVNGLPPYKEISVHVFYRCQCTTVETVRELTEFAKSIPSFSSLFL    275
P37231_PPARG_HUMAN    280  MNSLMMGEDKIKFKHITPLQEQSKEVAIRIFQGCQFRSVEAVQEITEYAKSIPGFVNLDL    339
                           :::*  .*. :  * :.        **  :::*  **  :**:* *:**:**:**.* .* *
                                                                LBD
Q07869_PPARA_HUMAN    303  NDQVTLLKYGVYEAIFAMLSSVMNKDGMLVAYGNGFITREFLKSLRKPFCDIMEPKFDFA    362
Q03181_PPARD_HUMAN    276  NDQVTLLKYGVHEAIFAMLASIVNKDGLLVANGSGFVTREFLRSLRKPFSDIIEPKFEFA    335
P37231_PPARG_HUMAN    340  NDQVTLLKYGVHEIIYTMLASLMNKDGVLISEGQGFMTREFLKSLRKPFGDFMEPKFEFA    399
                           ***********:* *::**:*::****:*:: *.**:*****:****** *::****:**

Q07869_PPARA_HUMAN    363  MKFNALELDDSDISLFVAAIICCGDRPGLLNVGHIEKMQEGIVHVLRLHLQSNHPDDIFL    422
Q03181_PPARD_HUMAN    336  VKFNALELDDSDLALFIAAIILCGDRPGLMNVPRVEAIQDTILRALEFHLQANHPDAQYL    395
P37231_PPARG_HUMAN    400  VKFNALELDDSDLAIFIAVIILSGDRPGLLNVKPIEDIQDNLLQALELQLKLNHPESSQL    459
                           :***********:::*:*.** .******:**  :* :*: :::.*.::*: ***:    *

Q07869_PPARA_HUMAN    423  FPKLLQKMADLRQLVTEHAQLVQIIKKTESDAALHPLLQEIYRDMY                  468
Q03181_PPARD_HUMAN    396  FPKLLQKMADLRQLVTEHAQMMQRIKKTETETSLHPLLQEIYKDMY                  441
P37231_PPARG_HUMAN    460  FAKLLQKMTDLRQIVTEHVQLLQVIKKTETDMSLHPLLQEIYKDLY                  505
                           * ******:****:****.*::* *****:: :*********:*:*
```

Figure 1. Amino acid sequence alignments of human peroxisome proliferator-activated receptors (PPAR) isoforms. DNA-binding domain (DBD, purple), zinc-finger domains (ZFD, purple), and ligand binding domain (LBD, yellow) are highlighted. All three isoforms of PPAR possess a high degree of inter-species sequence homology, particularly in the DBD and LBD. The sequence positions that are conserved within PPAR isoforms are important for identification of the structural dynamics, ligand affinity, and DNA binding specificity. Amino acid residues, which participate in ligand binding, are boxed. Alignment was performed with CLUSTALO (https://www.uniprot.org/align/). (*)—fully conserved residues; (:)—residues with strongly similar properties; (.)—residues with weakly similar properties.

2. Biological Role and Tissue-Specific Expression of PPARs

PPARs form heterodimers with the retinoid X receptor in the nucleus. These heterodimers recruit coactivators and corepressors and bind to specific peroxisome proliferator response elements (PPRE) in regulatory regions of PPAR target genes (Figure 2). Ligand binding releases the corepressor from the complex and allows activation of coactivator leading to changes in target gene expression [9]. DNA-pull down of PPARγ with subsequent MS-based proteomics identification of binding partners revealed highly complex patterns of interaction of PPARγ with other proteins in the cytoplasm and nucleus [10]. In addition, this study revealed that interactions of PPARγ with its binding partners are highly ligand- and DNA-dependent. In silico analysis of protein–protein interactions between PPARα and PPARβ/δ predicted the interaction of PPARs and retinoid X receptors (RXRs) with chromatin state modifiers, such as histone deacetylases (HDACs) that can play a role in epigenetic modifications of the diseases [11].

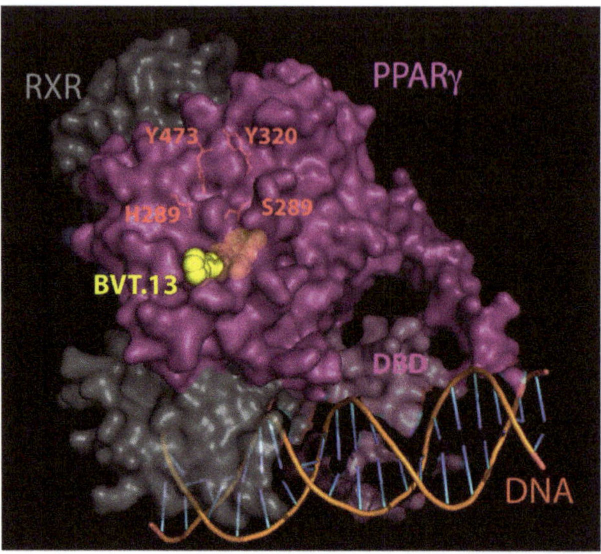

Figure 2. X-ray crystal structure of the complex of PPARγ (magenta) and retinoid X receptors (RXR) (grey) at 3.2 Å resolution. The BVT.13 agonist ligand is displayed as yellow balls. The amino acids residues, which form a ligand binding pocket, are shown in *red*. DNA-binding domain (DBD, light magenta) and DNA fragment are shown. The structure is derived from Protein Data Bank (PDB: 3DZU) [12] and visualized using PyMol software (v. 2.0.7).

PPARα plays a crucial role in the regulation of FAO, a major source of ATP in high energy-consuming organs and tissues. Hence, PPARα is highly expressed in skeletal muscle, heart, liver, and brown adipose tissue [13–15]. PPARγ is mainly expressed in adipose tissue, large intestine, and spleen. It regulates adipogenesis, lipid and glucose metabolism, and inflammatory pathways. The least studied PPARβ/δ is expressed ubiquitously with the highest levels found in the liver, intestine, kidney, adipose tissue, and skeletal muscle thereby, suggesting its fundamental role in cellular biology (reviewed in References [16–19]).

Transcriptional activities of PPARs are regulated, in part, by the PPARγ coactivator 1α (PGC-1α). PGC-1α is an integrator of the transcriptional network regulating mitochondrial biogenesis and, like PPARα, highly expressed in high energy-consuming cells. In addition to PPARs, PGC-1α mediates its effects through other downstream transcriptional regulatory circuits such as estrogen-related receptors (ERRs), and nuclear respiratory factors (NRF) 1 and 2. The nuclear respiratory factors, in turn, regulate downstream genes, including mitochondrial transcription factor A (TFAM), which is responsible for the maintenance as well as replication and transcription of mitochondrial DNA (reviewed in References [20–22]). Thus, PGC-1α is an inducible co-activator that coordinately regulates mitochondrial biogenesis through the network of transcription factors PPARs/NRF/ERRs. Mitochondrial biogenesis via the PGC-1α/NRF pathway is apparently regulated by AMP kinase (AMPK) [23,24]. Indeed, direct phosphorylation of PGC-1α by AMPK in vitro and in cultured cells has been shown recently [25].

Various natural fatty acids and eicosanoids act as natural ligands for PPARs. Generally, polyunsaturated fatty acids (PUFAs) display a higher affinity to PPARγ and PPARβ/δ, while both saturated and unsaturated fatty acids interact with PPARα equally efficiently (reviewed in Reference [8]). Therefore, PPARs represent attractive molecular targets for the development of pharmacological agents and treatment of metabolic disorders, including obesity, type 2 diabetes, dyslipidemia, and cardiovascular diseases.

3. The Role of PPARs in Cardiac Diseases

PPARα is highly expressed in cardiomyocytes, and genetic studies demonstrated the importance of PPARα in fatty acid metabolism in the heart [26,27]. PPARα knockout mice demonstrated normal [27] or reduced [26,28] cardiac function. Cardiac dysfunction in PPARα$^{-/-}$ mice was associated with structural abnormalities in mitochondria [26] and increased oxidative stress due to downregulation of the antioxidant capacity in the heart [29]. High workload decreased cardiac performance in PPARα knockout mice associated with lower levels of ATP in the myocardium [30]. In response to transverse aortic constriction (TAC), PPARα-null mice showed pronounced cardiac hypertrophy [31]. On the other hand, overexpression of PPARα increased mild cardiac hypertrophy, ventricular dysfunction, and lipotoxicity associated with reciprocal repression of glucose uptake and oxidation in the mouse heart [32]. These mice developed a phenotype strikingly similar to diabetic cardiomyopathy [33]. The contrasting metabolic phenotypes induced by genetic upregulation or downregulation of PPARα in mice indicate the central role of the receptors in regulating glucose and lipid metabolism in the heart.

Recent studies demonstrated that the expression of PPARα significantly decreases in cardiomyocytes in a pressure–overload mouse model of HF induced by TAC. Expression of PPARα target genes, carnitine palmitoyltransferase 1 (CPT-1) and fatty acid transport protein 1 (FATP1) were also significantly reduced in the HF hearts. Activation of PPARα either by cardiac-specific overexpression of *PPARα* gene or by treating mice with the specific PPARα agonist, WY-1463 improved cardiac function, attenuated cardiac fibrosis, and preserved FAO and high-energy phosphates in a mouse model of HF induced by TAC [34]. The energy substrate switch from FAO to glucose oxidation and other metabolic changes in hearts with hypertrophy and HF is mediated, at least in part, through downregulation of genes encoding FAO and oxidative phosphorylation enzymes due to deactivation of the PGC-1α/PPARα pathway [21,22]. Reduced PGC-1α and PPARα expression occurs in animal models of HF [35,36] and in failing human hearts [37,38], suggesting that deactivation of the PGC-1α/PPARα pathway in the failing heart plays a critical role in coincident mitochondrial dysfunction.

The role of post-translational modifications (PTMs) in activation/inactivation of PPARα is debated. PPARs have been shown to undergo several types of PTMs including phosphorylation, acetylation, sumoylation, and ubiquitination, among others. Several protein kinases, including extracellular signal-regulated protein kinases 1 and 2 and c-Jun N-terminal kinase, AMPK, protein kinase A, and glycogen synthase kinase 3 phosphorylate PPARα and PPARγ [39]. Protein kinase A [40] and p38 [41] phosphorylated PPARα that resulted in a ligand-dependent increase of PPARα activity in neonatal rat cardiomyocytes and HEK-293 cells. On the other hand, ventricular pressure overload in mice and PPARα overexpression in cardiomyocytes revealed that downregulation of cardiac PPARα and alteration of its activity during hypertrophic growth occur at the posttranscriptional level via activation of extracellular signal-regulated protein kinases 1 and 2 [42]. Phosphorylation increased transcriptional activation of PPARα [43] but decreased that of PPARγ [44].

PPARα seems to be a downstream target for AMPK and mediates its beneficial effects by improving mitochondrial metabolism. AMPK is the main cellular energy sensor that initiates ATP generating processes while blocking ATP consuming processes. It is also involved in the regulation of mitochondrial metabolism and the redox state in the cell (reviewed in References [45,46]). Pharmacological activation of AMPK stimulates FAO through increased expression of PPARα target genes in skeletal muscle cells [47]. Furthermore, the PPARα inhibitor, GW6471, prevented the cardioprotective effects of metformin, an AMPK activator, against ischemia-reperfusion in rat hearts [48]. Oxidative stress-induced phosphorylation of PGC-1α and PPARα in cardiac cells. However, the protective effects of the AMPK activators metformin and A-769662 on hydrogen peroxide-treated H9c2 cells and in vivo cardiac ischemia-reperfusion in rats were not associated with phosphorylation of PPARα [49,50]. These studies suggest that PTMs of PPARα during cardiac oxidative stress and hypertrophic growth can occur at several levels.

In addition to regulation of the mitochondrial transcriptional network, PPARα can translocate to mitochondria and affect metabolism and function of mitochondria. Hydrogen peroxide-induced oxidative stress in H9c2 cells [49] and ischemia-reperfusion in the rat heart [50] stimulated protein-protein interactions between PPARα and cyclophilin D (CyP-D), a major regulator of the mitochondrial permeability transition pore. The interaction provoked the opening of the mitochondrial permeability transition pores. Conversely, activation of AMPK with metformin or A-769662 prevented PPARα-CyP-D interaction leading to inhibition of mitochondrial permeability transition pore opening, and improved cell survival and post-infarction recovery [49,50]. These studies indicate the role of PPARα in mediating the beneficial effects of AMPK in cardiac ischemia-reperfusion.

Similar to PPARα, heart-specific PPARγ knockout mice developed cardiac hypertrophy with preserved normal cardiac metabolism and function [51,52]. Decreased expression of genes encoding FAO enzymes and impaired fatty acid utilization with unchanged glucose oxidation were found in inducible cardiomyocyte-specific PPARγ$^{-/-}$ mice [53]. It should be noted that cardiac hypertrophy in heart-specific PPARγ knockout mice associated with oxidative damage and mitochondrial dysfunction progresses with age and leads to dilated cardiomyopathy and premature death [54]. Like PPARα$^{-/-}$ mice, antioxidant therapy attenuated cardiac dysfunction in the PPARγ$^{-/-}$ mice. Heart-specific PPARγ overexpression induced a dilated cardiomyopathy associated with increased expression of FAO genes, lipotoxicity, and mitochondrial structural abnormalities such as cristae disruption in the heart [55]. Interestingly, glucose uptake was not decreased in these hearts.

Overexpression of PPARβ/δ in mouse hearts enhanced mitochondrial biogenesis, myocardial oxidative metabolism, improved cardiac performance, and reduced cardiac fibrosis [56]. These effects of PPARβ/δ overexpression were not affected by TAC-induced cardiac hypertrophy. In rats with congestive HF, the PPARβ/δ-specific agonist, GW610742X, normalized cardiac substrate metabolism in a dose-dependent manner, dramatically reduced right ventricular hypertrophy, and decreased the level of the arterial natriuretic peptide in the right ventricle. However, GW610742X had no beneficial effect on the left ventricular function [57,58].

The activity of a large number of proteins is regulated through acetylation/deacetylation of lysine residues. Four classes of HDACs play a central role in cell metabolism, including energy metabolism in the heart. Mitochondrial bioenergetics including fatty acid metabolism, electron transfer chain, and oxidative phosphorylation are regulated by the class III HDACs sirtuins, particularly SIRT3 [59]. Interestingly, the interaction of HDAC3 with PPARγ induced deacetylation of the protein and reduced its activity [60]. Inhibition of HDAC3 stimulated ligand-independent activation of PPARγ by protein acetylation suggesting that acetylation of PPARγ induces its activation through a ligand-independent mechanism. Cardiac-specific HDAC3 knockout mice demonstrated a modest increase in expression of FAO genes with no changes in gene expression of PPARs [61]. Oxidative stress induced by hydrogen peroxide did not increase acetylation of PGC-1α and PPARα in H9c2 cardioblasts [49]. Further studies are needed to establish a cause–effect relationship between acetylation and activity of PPARs in the healthy heart and cardiac diseases.

Other forms of PTM, such as sumoylation [62] and ubiquitination [63], have been shown to affect the PPAR activity (reviewed in Reference [58]). These studies were conducted mostly using various cell lines, and there are few, if any, studies on the PPAR sumoylation and ubiquitination in the heart. For instance, upregulation of the ubiquitin ligase, muscle ring finger-1 increased its interaction and ubiquitination of PPARα in neonatal cardiomyocytes [64]. The ubiquitination reduced PPARα activity and FAO suggesting a critical role of ubiquitination in regulating cardiac PPARα and fatty acid metabolism in the heart.

Polymorphisms in PPARs are significantly associated with cardiac disorders. Intronic rs4253778 polymorphism and common L162V (rs1800206) polymorphism in *PPARα* are significantly associated with coronary heart disease (CHD) risk [65]. L162V variant at the DBD region of PPARα affects the transactivation activity of PPAR ligands [66,67]. A Rs135551 intronic variant in *PPARα* showed significant association with CHD [68]. T allele carriers of C161T polymorphism in *PPARγ* (rs3856806) have lower CHD risk, but higher risk of acute coronary syndrome (ACS). +294T/C polymorphism at *PPARγ/δ* (rs2016520) is significantly associated with ACS [65].

Thus, PPARs play an important role in fatty acid metabolism in the heart and are involved in the pathogenesis of cardiac hypertrophy, cardiomyopathy, and HF. Apparently, beneficial or detrimental effects of PTMs of PPARα depend on the severity and timing of oxidative and energy stresses that are associated with the diminished capacity of the myocardium to maintain lipid and glucose metabolism.

4. Therapeutic Potential of PPAR Agonists in Cardiac Diseases

4.1. Studies in Animal Models

Lower rates of FAO are associated with cardiomyopathies and HF [69–73]. Due to high expression and the beneficial role of PPARα and PPARβ/δ in the heart, numerous studies have been conducted to study the efficiency of PPARα and PPARβ/δ agonists on various animal models with HF.

Fibrates have been used for many years as PPAR agonists for treatment heart attacks and strokes. The fibrates are a family of hypolipidemic drugs that are structural derivatives of the parent compound, clofibrate (ethyl 2-(4-chlorophenoxy)-2-methylpropionate (Figure 3). They lower serum triglycerides, raise high-density lipoprotein cholesterol (HDL-C) and lower low-density lipoprotein cholesterol (LDL-C) levels. Therefore, long-term therapy with fibrates could help to prevent cardiovascular disease events. However, fibrates demonstrate a high risk for developing rhabdomyolysis and renal failure (reviewed in Reference [74]). The main list of fibrates currently used in experimental studies and clinical trials include gemfibrozil, fenofibrate, BF, etofibrate, and ciprofibrate (Figure 3). The clofibrate previously used in studies is no longer in use due to safety concerns.

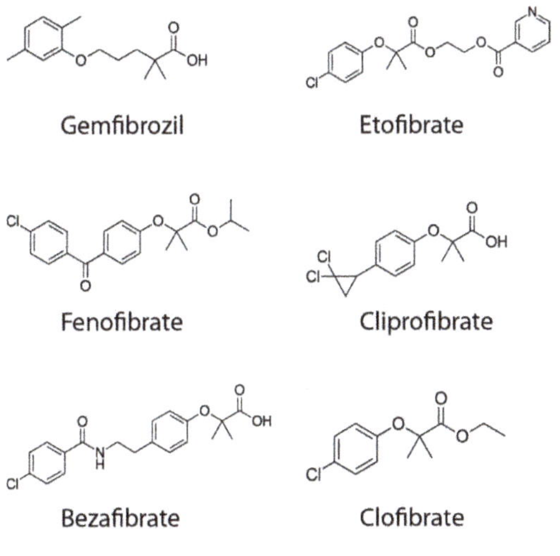

Figure 3. Chemical structures of fibrates.

Fenofibrate is a dual activator of PPARα and PPARγ, with 10-fold selectivity for PPARα [18]. Oral intake of fenofibrate (100 mg/kg body weight) significantly attenuated end-diastolic and end-systolic left ventricular dimensions and cardiac fibrosis in aldosterone-induced hypertrophy model independently of an effect of the drug on blood pressure [75]. Similar effects were observed in porcine and canine tachycardia-induced cardiomyopathy models [76,77]. Fenofibrate attenuated hypertrophy, inhibited the inflammatory response, improved the survival of Dahl salt-sensitive rats [78] and decreased fibrosis in the rat model with the pressure overload-induced HF [79]. While fenofibrate had clear beneficial effects in wild-type mice, this drug had deleterious consequences on cardiac hypertrophy and fibrosis in PPARα$^{-/-}$ mice [80]. The detrimental effects of fenofibrate in PPARα knockout mice might be a result of anomalous activation of PPARγ and PPARβ/δ in cardiomyocytes in the absence of PPARα. Administration of another PPARα agonist, tetradecylthioacetic acid, elevated expression of PPARα target genes, myocardial oxygen consumption, and FAO with concomitant reduction of glucose oxidation in the heart. However, this drug had a negative impact on the post-ischemic recovery of cardiac function in an isolated perfused heart model [81].

Fibrates require micromolar concentrations to activate PPARα. The half maximal effective concentration (EC_{50}) for fenofibrate is approximately 30 μM for human PPARα [18] that requires high doses of the drug (>100 mg/kg) to achieve a clinical effect. Attempts to discover more potent and more selective PPARα agonists resulted in the synthesis of several more potent compounds that work in nanomolar ranges. The PPARα agonist AVE8134 has a high affinity for PPARα (EC_{50} 0.01 and 0.03 μM for human and rodent PPARα, respectively). AVE8134 at the daily dose of 3–30 mg/kg improved lipid profile and augmented glucose metabolism; prevented post-MI hypertrophy, fibrosis and cardiac dysfunction, and reduced mortality in rats [82,83]. Other selective PPARα agonists, WY-14643 (pirinixic acid) and GW7647, have EC_{50} in the micromolar range. WY-14643 (0.01% *w*/*w* in rodent food, or ~20 mg/kg) significantly attenuated cardiac dysfunction and remodeling induced by pressure–overload HF in mice [84].

Treatment with GW7647 did not prevent the development of hypertrophy but preserved the left ventricular ejection fraction during pressure–overload cardiac hypertrophy in rabbits [85]. The effects of GW7647 were associated with an increased cardiac FAO and overall ATP production that resulted in an improved post-ischemic recovery of cardiac function. In addition, GW7647 treatment resulted in relived endoplasmic reticulum stress, preserved mitochondrial membrane potential, and activated sarcoplasmic reticulum Ca-ATPase (ATP2A2) [85]. Additionally, GW2331 and GW9578, dual PPARα/PPARγ agonists that work in the nanomolar range, have been synthesized [86,87].

Bezafibrate was introduced as a lipid-lowering drug by Boehringer Mannheim in 1970s [6,88]. It reduces heart rate, blood pressure, insulin level, and free fatty acids in patients with hypertriglyceridemia [89]. Bezafibrate also is a widely used pan-PPAR agonist in animal trials. It activates all three PPAR subtypes with the highest affinity for PPARα and PPARβ/δ isoforms [90]. The EC_{50}s for BF are 50, 60, and 20 μM for human PPARα, PPARγ, and PPARβ/δ, respectively; and 90, 55, and 110 μM for mouse PPARα, PPARγ, and PPARβ/δ, respectively [18]. There are significant inconsistencies in BF studies in humans and rodents. In clinical practice, BF is typically prescribed at a daily dose of 10–25 mg/kg [6,91,92]. Conversely, in animal studies BF is usually administrated *per os* with diet in the amount of 0.5% *w*/*w*, corresponding to a daily dose of 600–800 mg per kg [93–95]. It is conceivable that the relatively low affinity of BF for murine PPARβ/δ requires a higher dose of the drug to achieve a biological response in mice, particularly in skeletal muscle, where the BF effects are predominantly mediated by PPARβ/δ rather than PPARα [96].

Experiments with PPARα knockout mice suggested that more clinically relevant, low-dose BF decreases serum and liver triglycerides in a PPAR-independent manner by suppressing the expression of sterol regulatory element-binding protein 1c (*SPREBP1*) affecting hepatic lipogenesis and triglyceride secretion [97]. Several groups have reported on the use of BF in mouse models with varying success. Two mouse models of cytochrome *c*-oxidase deficiency, systemic *Surf1*$^{-/-}$ and muscle-specific *Cox15*$^{-/-}$, were given BF at the dose of 0.5% in rodent diet (Table 1). At this dose, BF was highly toxic, causing massive apoptosis in skeletal muscles. In these models, BF induced expression of the genes encoding proteins that are involved in FAO but not oxidative phosphorylation in mitochondria.

PPARs and PGC-1α regulate mitochondrial aerobic metabolism, acting on different, though partially overlapping sets of genes. PPARs regulate expression of FAO genes, including *CD36/FAT*, *ACOX*, *SCAD*, while PGC-1α controls the expression of genes involved in oxidative phosphorylation. Treatment with BF of *Surf1*$^{-/-}$ mice increased expression of the PPAR isoforms present in the skeletal muscle, PPARα and PPARβ/δ. However, no increase in PGC-1α was observed in the skeletal muscles of BF-treated mice [98]. On the contrary, BF increased mitochondrial biogenesis and significantly increased expression of PGC-1α and battery of downstream its targets cytochrome c, TFAM, and subunits of ATP synthase in muscle, brown adipose tissue (BAT), and brain in mice with Huntington disease [99]. Treatment with BF rescued neuropathological features in the brain, increased motor activity, and muscle strength, prevented fiber–type switching in muscles, attenuated vacuolization in BAT and increased survival rate in Huntington mice. BF also was found to interact with hemoglobin and lower its affinity to oxygen. However, it is not clear whether the pharmacological doses of BF can achieve a concentration of the drug in erythrocytes sufficient to benefit the oxygen transport capacity [100,101].

Table 1. Bezafibrate trials using mouse genetic models with mitochondrial defects.

Disease Model	Tissue Studied	BF Dose	Effects	Ref.
OXPHOS defect: *Surf1*$^{-/-}$	Muscle	0.5% (0.6–0.8 g/kg)	Weight loss, hepatomegaly. Increased expression of FAO genes, PPARα and PPARβ/δ.	[98]
OXPHOS defect: *Cox15*$^{-/-}$	Muscle	0.5% (0.6–0.8 g/kg)	Toxic, mitochondrial myopathy, excessive apoptosis.	[98]
Huntington disease: *Htt*-ex1 (R6/2)	Brain, Muscle, BAT	0.5% (0.6–0.8 g/kg)	Attenuated neurodegeneration in brain, prevented muscle–type switching. Increased exercise capacity and muscle strength, increased vacuolization in BAT, and extend survival.	[99]
Premature aging: mtDNA polymerase $\gamma^{-/-}$	Skin, Spleen	0.5% (0.6–0.8 g/kg)	Delayed hair loss and restored skin structure. Improved spleen size and structure.	[93]
BTHS: TAZ knockdown	Heart	0.5% (0.6–0.8 g/kg)	Preserved cardiac systolic function. Reduced cardiolipin level in mitochondria.	[95]
BTHS: TAZ knockdown	Heart, Muscle	0.05% (0.06–0.08 g/kg)	Restored cardiac systolic function. Ameliorated exercise intolerance phenotype when treatment was combined with everyday voluntary exercise.	[102]

Recent studies demonstrated the therapeutic effectiveness of BF to attenuate left-ventricular defects in the mouse model of Barth syndrome (BTHS). Barth syndrome is an X-linked rare genetic disease that is manifested by dilated cardiomyopathy, muscle weakness, and exercise intolerance. Causative gene is *TAZ* that encodes mitochondrial cardiolipin transacylase, tafazzin, and mutations in *TAZ* results in a deficiency of the essential mitochondrial phospholipid cardiolipin. Intake of BF with diet during the 4-month period in daily doses of 60–80 mg/kg (0.05% in rodent diet) or 600–800 mg/kg (0.5% in rodent diet) effectively prevented the development of systolic dysfunction and cardiomyopathy in *TAZ* knockdown mice [95,102]. Surprisingly, improvement of systolic function in mice treated with a high-dose (0.5%) of BF was accompanied by a simultaneous reduction of cardiolipin in the heart that can be explained by an increased number of mitochondria with a reduced content of cardiolipin [95].

Differential transcriptomic analysis of hearts demonstrated that treatment with a low dose (0.05%) of BF resulted in robust activation of genes involved in a wide-spectrum of biological processes that included metabolism of fatty acids, ketone bodies, amino acids and glucose, metabolism of proteins, mitochondrial protein transport, RNA metabolism, gene expression, DNA repair, chromatin organization, immune system, and organelle biogenesis and maintenance [102]. Bezafibrate failed to ameliorate the exercise intolerance phenotype in BTHS mice. However, when treatment with BF was combined with voluntary daily exercise on the running wheel, BF's effect on exercise capacity in BTHS mice was significantly potentiated. The mechanisms underlying this synergistic effect of BF with everyday voluntary exercise are unclear. Apparently, exercise alters the epigenetic landscape in skeletal muscles and facilitates transcription of PPAR target genes thereby enhancing cellular metabolic plasticity.

4.2. Clinical Studies

Currently, 41 clinical trials are registered to investigate the therapeutic efficiency of numerous PPAR agonists in various diseases worldwide (www.clinicaltrials.gov). Among those, eight trials have been completed, and no serious adverse effects have been reported. Among 41 trials, 13 are aimed to elucidate the efficacy of PPAR agonists in cardiac diseases. To date only one phase-2 clinical trial investigating the therapeutic potential of PPARγ agonist rosiglitazone in patients with congestive HF has been completed (NCT00064727), however, findings have not yet been reported.

Recently, comprehensive systemic research of major cardiovascular disease prevention trials with fibrates was performed [103,104]. In these trials, data associated with the effects of clofibrate, which is no longer in use, were excluded from the analysis. Moderate-quality evidence from six primary prevention trials with 16,135 participants (8087 in the intervention group and 8048 in the placebo group) suggested that fibrate therapy reduced the combined outcome of death due to cardiovascular disease, heart attack, or stroke by 16% [103].

The Bezafibrate Infarction Prevention (BIP) study was initiated in 1998 with a total of 3090 enrolled participants (1548 in the intervention group, 1542 in the placebo group). The goal of this trial was to evaluate whether treatment with BF was effective in preventing MI injury and death in coronary artery disease patients. At 8.2 years of follow-up, there was an 18% risk reduction of major cardiac events (occurrence of cardiac death or nonfatal MI) ($p = 0.02$) [105]. Prolonged, 16 years of follow-up, showed that there was an 11% reduction ($p = 0.06$) in mortality in patients that were treated with BF [106–108]. BF had no therapeutic outcome on the risk of coronary heart disease and stroke in men with lower extremity arterial disease, although reduced the incidence of non-fatal coronary events in men aged 65 years or older [109].

Patients, who develop metabolic syndrome are at particularly increased risk of myocardial infarction (MI). The efficacy of BF to prevent MI was analyzed on the subgroup of patients from BIP study, who had developed metabolic syndrome (740 patients from BF group and 730 patients from the placebo group). The rate of nonfatal MI was significantly lower in patients in the BF group (9.5% and 13.8% in BF and placebo groups, respectively; $p = 0.009$). The decrease in MI incidence in patients taking BF was reflected in a trend to a 26% reduction of cardiac mortality rate ($p = 0.056$) [110].

Existing trials are mainly directed towards the studies of lipid-lowering and anti-atherosclerotic effects of PPAR agonists, whereas the bioenergetics actions of PPAR agonists on energy metabolism in cardiomyocytes remain less investigated. Although no clinical trial has prospectively studied the effects of PPAR agonists in patients with HF, there are several compelling evidences that PPAR agonists can improve clinical outcomes in HF.

Treatment with BF in combination with ursodeoxycholic acid had a beneficial effect in patients with primary biliary cholangitis, a progressive liver disease, compared to a control group that was treated with ursodeoxycholic acid alone [111].

Studies including the cohort of six patients with the myopathic form of CPT-2 deficiency showed that six-month-long treatment with BF (200 mg three times a day) markedly upregulated CPT-2, increased oxidation rates of the long-chain fatty acids, decreased muscle pain and increased physical activity in all BF-treated patients [112]. BF failed to improve FAO in skeletal muscles and exercise tolerance in patients with CPT-2 and very long-chain acyl-CoA dehydrogenase deficiencies [92]. The authors ascribed the lack of effect FAO to the suppression of lipolysis by BF. However, an alternative explanation is that high plasma insulin in BF-treated patients had markedly inhibited lipolysis, hence hindering any increase of FAO and masking the effects of BF [113]. In vitro studies of patient cells with very long-chain acyl-CoA dehydrogenase and CPT-2 deficiencies revealed that treatment with BF is beneficial in cells of mildly affected patients that retain residual FAO capacities. In contrast, no increase in FAO capacities is expected in response to BF if the gene mutations impact the catalytic site or lead to highly unstable or severely misfolded proteins [96,114]. A phase 2 clinical trial of BF in BTHS patients (CARDIOMAN) is underway at University Hospital in Bristol, England.

5. Conclusions

PPARs play a central role in the pathogenesis of cardiac hypertrophy and HF and thereby, represent a potentially attractive therapeutic target for the treatment of these diseases. PPAR agonists, particularly PPARα and PPARβ/δ agonists appear to stimulate FAO and energy metabolism in cardiomyocytes. Subsequently, improved cellular energy homeostasis in the heart attenuates systolic dysfunction in HF patients as well as in animal models of cardiomyopathy and HF. Skeletal muscle appears to be more resistant to the treatment with PPARα and PPARβ/δ agonists. Full understanding of the therapeutic potential of PPAR agonists requires more detailed studies using various animal models of cardiac diseases. The effects of PPAR agonists on cellular transcriptional and epigenetic landscapes as well as activation/inhibition of individual PPAR isoforms on cellular metabolic and signaling systems need to be evaluated in detail using systems biology approaches.

Author Contributions: Z.K. supervised the project, edited and compiled the final version of the manuscript. Z.K. and S.J. reviewed and analyzed the literature, prepared a draft of the manuscript and proofed it. A.I.G. and A.W.S. reviewed and analyzed the literature and participated in writing the manuscript.

Funding: Studies of the authors discussed in this article were supported by the National Heart, Lung, and Blood Institute (Grants R01HL108867 to Z.K. and SC1HL118669 to S.J.), and National Institute of General Medical Sciences (Grant SC1GM128210 to S.J.) of the National Institutes of Health.

Acknowledgments: The authors apologize to all colleagues whose important studies were not cited due to space restrictions.

Conflicts of Interest: The authors declare no conflict of interest.

Abbreviations

AMPK	AMP kinase
BAT	brown adipose tissue
BF	bezafibrate
BTHS	Barth syndrome
CHD	coronary heart disease
CPT	carnitine palmitoyltransferase
CyP-D	cyclophilin D
FAO	fatty acid oxidation
HF	heart failure
MI NRF	myocardial infarction nuclear respiratory factors
PGC-1α	proliferator-activated receptor gamma coactivator-1 alpha
PPAR	peroxisome proliferator-activated receptor
TAC	transverse aortic constriction

References

1. Schmidt, A.; Endo, N.; Rutledge, S.J.; Vogel, R.; Shinar, D.; Rodan, G.A. Identification of a new member of the steroid hormone receptor superfamily that is activated by a peroxisome proliferator and fatty acids. *Mol. Endocrinol.* **1992**, *6*, 1634–1641. [CrossRef] [PubMed]
2. Dreyer, C.; Krey, G.; Keller, H.; Givel, F.; Helftenbein, G.; Wahli, W. Control of the peroxisomal beta-oxidation pathway by a novel family of nuclear hormone receptors. *Cell* **1992**, *68*, 879–887. [CrossRef]
3. Issemann, I.; Green, S. Activation of a member of the steroid hormone receptor superfamily by peroxisome proliferators. *Nature* **1990**, *347*, 645–650. [CrossRef] [PubMed]
4. Eleff, S.; Kennaway, N.G.; Buist, N.R.; Darley-Usmar, V.M.; Capaldi, R.A.; Bank, W.J.; Chance, B. ^{31}P NMR study of improvement in oxidative phosphorylation by vitamins K3 and C in a patient with a defect in

electron transport at complex III in skeletal muscle. *Proc. Natl. Acad. Sci. USA* **1984**, *81*, 3529–3533. [CrossRef] [PubMed]

5. Ledermann, H.; Kaufmann, B. Comparative pharmacokinetics of 400 mg bezafibrate after a single oral administration of a new slow-release preparation and the currently available commercial form. *J. Int. Med. Res.* **1981**, *9*, 516–520. [CrossRef] [PubMed]
6. Olsson, A.G.; Lang, P.D. Dose-response study of bezafibrate on serum lipoprotein concentrations in hyperlipoproteinanemia. *Atherosclerosis* **1978**, *31*, 421–428. [CrossRef]
7. Nolte, R.T.; Wisely, G.B.; Westin, S.; Cobb, J.E.; Lambert, M.H.; Kurokawa, R.; Rosenfeld, M.G.; Willson, T.M.; Glass, C.K.; Milburn, M.V. Ligand binding and co-activator assembly of the peroxisome proliferator-activated receptor-gamma. *Nature* **1998**, *395*, 137–143. [CrossRef] [PubMed]
8. Xu, H.E.; Lambert, M.H.; Montana, V.G.; Parks, D.J.; Blanchard, S.G.; Brown, P.J.; Sternbach, D.D.; Lehmann, J.M.; Wisely, G.B.; Willson, T.M.; et al. Molecular recognition of fatty acids by peroxisome proliferator-activated receptors. *Mol. Cell* **1999**, *3*, 397–403. [CrossRef]
9. Komen, J.C.; Thorburn, D.R. Turn up the power—Pharmacological activation of mitochondrial biogenesis in mouse models. *Br. J. Pharmacol.* **2014**, *171*, 1818–1836. [CrossRef] [PubMed]
10. Lam, V.Q.; Zheng, J.; Griffin, P.R. Unique Interactome Network Signatures for Peroxisome Proliferator-activated Receptor Gamma (PPARgamma) Modulation by Functional Selective Ligands. *Mol. Cell. Proteom.* **2017**, *16*, 2098–2110. [CrossRef] [PubMed]
11. Sookoian, S.; Pirola, C.J. Elafibranor for the treatment of NAFLD: One pill, two molecular targets and multiple effects in a complex phenotype. *Ann. Hepatol.* **2016**, *15*, 604–609. [PubMed]
12. Chandra, V.; Huang, P.; Hamuro, Y.; Raghuram, S.; Wang, Y.; Burris, T.P.; Rastinejad, F. Structure of the intact PPAR-gamma-RXR-nuclear receptor complex on DNA. *Nature* **2008**, *456*, 350–356. [CrossRef] [PubMed]
13. Braissant, O.; Foufelle, F.; Scotto, C.; Dauca, M.; Wahli, W. Differential expression of peroxisome proliferator-activated receptors (PPARs): Tissue distribution of PPAR-alpha, -beta, and -gamma in the adult rat. *Endocrinology* **1996**, *137*, 354–366. [CrossRef] [PubMed]
14. Auboeuf, D.; Rieusset, J.; Fajas, L.; Vallier, P.; Frering, V.; Riou, J.P.; Staels, B.; Auwerx, J.; Laville, M.; Vidal, H. Tissue distribution and quantification of the expression of mRNAs of peroxisome proliferator-activated receptors and liver X receptor-alpha in humans: No alteration in adipose tissue of obese and NIDDM patients. *Diabetes* **1997**, *46*, 1319–1327. [CrossRef] [PubMed]
15. Abbott, B.D. Review of the expression of peroxisome proliferator-activated receptors alpha (PPAR alpha), beta (PPAR beta), and gamma (PPAR gamma) in rodent and human development. *Reprod. Toxicol.* **2009**, *27*, 246–257. [CrossRef] [PubMed]
16. Grygiel-Gorniak, B. Peroxisome proliferator-activated receptors and their ligands: Nutritional and clinical implications—A review. *Nutr. J.* **2014**, *13*, 17. [CrossRef] [PubMed]
17. Sertznig, P.; Seifert, M.; Tilgen, W.; Reichrath, J. Present concepts and future outlook: Function of peroxisome proliferator-activated receptors (PPARs) for pathogenesis, progression, and therapy of cancer. *J. Cell. Physiol.* **2007**, *212*, 1–12. [CrossRef] [PubMed]
18. Willson, T.M.; Brown, P.J.; Sternbach, D.D.; Henke, B.R. The PPARs: From orphan receptors to drug discovery. *J. Med. Chem.* **2000**, *43*, 527–550. [CrossRef] [PubMed]
19. Tenenbaum, A.; Fisman, E.Z. Balanced pan-PPAR activator bezafibrate in combination with statin: Comprehensive lipids control and diabetes prevention? *Cardiovasc. Diabetol.* **2012**, *11*, 140. [CrossRef] [PubMed]
20. Kelly, D.P.; Scarpulla, R.C. Transcriptional regulatory circuits controlling mitochondrial biogenesis and function. *Genes Dev.* **2004**, *18*, 357–368. [CrossRef] [PubMed]
21. Huss, J.M.; Kelly, D.P. Mitochondrial energy metabolism in heart failure: A question of balance. *J. Clin. Investig.* **2005**, *115*, 547–555. [CrossRef] [PubMed]
22. Ventura-Clapier, R.; Garnier, A.; Veksler, V. Transcriptional control of mitochondrial biogenesis: The central role of PGC-1alpha. *Cardiovasc. Res.* **2008**, *79*, 208–217. [CrossRef] [PubMed]
23. Zong, H.; Ren, J.M.; Young, L.H.; Pypaert, M.; Mu, J.; Birnbaum, M.J.; Shulman, G.I. AMP kinase is required for mitochondrial biogenesis in skeletal muscle in response to chronic energy deprivation. *Proc. Natl. Acad. Sci. USA* **2002**, *99*, 15983–15987. [CrossRef] [PubMed]
24. Reznick, R.M.; Shulman, G.I. The role of AMP-activated protein kinase in mitochondrial biogenesis. *J. Physiol.* **2006**, *574*, 33–39. [CrossRef] [PubMed]

25. Jager, S.; Handschin, C.; St-Pierre, J.; Spiegelman, B.M. AMP-activated protein kinase (AMPK) action in skeletal muscle via direct phosphorylation of PGC-1alpha. *Proc. Natl. Acad. Sci. USA* **2007**, *104*, 12017–12022. [CrossRef] [PubMed]
26. Watanabe, K.; Fujii, H.; Takahashi, T.; Kodama, M.; Aizawa, Y.; Ohta, Y.; Ono, T.; Hasegawa, G.; Naito, M.; Nakajima, T.; et al. Constitutive regulation of cardiac fatty acid metabolism through peroxisome proliferator-activated receptor alpha associated with age-dependent cardiac toxicity. *J. Biol. Chem.* **2000**, *275*, 22293–22299. [CrossRef] [PubMed]
27. Campbell, F.M.; Kozak, R.; Wagner, A.; Altarejos, J.Y.; Dyck, J.R.; Belke, D.D.; Severson, D.L.; Kelly, D.P.; Lopaschuk, G.D. A role for peroxisome proliferator-activated receptor alpha (PPARalpha) in the control of cardiac malonyl-CoA levels: Reduced fatty acid oxidation rates and increased glucose oxidation rates in the hearts of mice lacking PPARalpha are associated with higher concentrations of malonyl-CoA and reduced expression of malonyl-CoA decarboxylase. *J. Biol. Chem.* **2002**, *277*, 4098–4103. [CrossRef] [PubMed]
28. Loichot, C.; Jesel, L.; Tesse, A.; Tabernero, A.; Schoonjans, K.; Roul, G.; Carpusca, I.; Auwerx, J.; Andriantsitohaina, R. Deletion of peroxisome proliferator-activated receptor-alpha induces an alteration of cardiac functions. *Am. J. Physiol. Heart Circ. Physiol.* **2006**, *291*, H161–H166. [CrossRef] [PubMed]
29. Guellich, A.; Damy, T.; Conti, M.; Claes, V.; Samuel, J.L.; Pineau, T.; Lecarpentier, Y.; Coirault, C. Tempol prevents cardiac oxidative damage and left ventricular dysfunction in the PPAR-alpha KO mouse. *Am. J. Physiol. Heart Circ. Physiol.* **2013**, *304*, H1505–H1512. [CrossRef] [PubMed]
30. Luptak, I.; Balschi, J.A.; Xing, Y.; Leone, T.C.; Kelly, D.P.; Tian, R. Decreased contractile and metabolic reserve in peroxisome proliferator-activated receptor-alpha-null hearts can be rescued by increasing glucose transport and utilization. *Circulation* **2005**, *112*, 2339–2346. [CrossRef] [PubMed]
31. Smeets, P.J.; Teunissen, B.E.; Willemsen, P.H.; van Nieuwenhoven, F.A.; Brouns, A.E.; Janssen, B.J.; Cleutjens, J.P.; Staels, B.; van der Vusse, G.J.; van Bilsen, M. Cardiac hypertrophy is enhanced in PPAR alpha-/- mice in response to chronic pressure overload. *Cardiovasc. Res.* **2008**, *78*, 79–89. [CrossRef] [PubMed]
32. Oka, S.; Alcendor, R.; Zhai, P.; Park, J.Y.; Shao, D.; Cho, J.; Yamamoto, T.; Tian, B.; Sadoshima, J. PPARalpha-Sirt1 complex mediates cardiac hypertrophy and failure through suppression of the ERR transcriptional pathway. *Cell Metab.* **2011**, *14*, 598–611. [CrossRef] [PubMed]
33. Finck, B.N.; Lehman, J.J.; Leone, T.C.; Welch, M.J.; Bennett, M.J.; Kovacs, A.; Han, X.; Gross, R.W.; Kozak, R.; Lopaschuk, G.D.; et al. The cardiac phenotype induced by PPARalpha overexpression mimics that caused by diabetes mellitus. *J. Clin. Investig.* **2002**, *109*, 121–130. [CrossRef] [PubMed]
34. Kaimoto, S.; Hoshino, A.; Ariyoshi, M.; Okawa, Y.; Tateishi, S.; Ono, K.; Uchihashi, M.; Fukai, K.; Iwai-Kanai, E.; Matoba, S. Activation of PPAR-alpha in the early stage of heart failure maintained myocardial function and energetics in pressure-overload heart failure. *Am. J. Physiol. Heart Circ. Physiol.* **2017**, *312*, H305–H313. [CrossRef] [PubMed]
35. Garnier, A.; Fortin, D.; Delomenie, C.; Momken, I.; Veksler, V.; Ventura-Clapier, R. Depressed mitochondrial transcription factors and oxidative capacity in rat failing cardiac and skeletal muscles. *J. Physiol.* **2003**, *551*, 491–501. [CrossRef] [PubMed]
36. Javadov, S.; Purdham, D.M.; Zeidan, A.; Karmazyn, M. NHE-1 inhibition improves cardiac mitochondrial function through regulation of mitochondrial biogenesis during postinfarction remodeling. *Am. J. Physiol. Heart Circ. Physiol.* **2006**, *291*, H1722–H1730. [CrossRef] [PubMed]
37. Sebastiani, M.; Giordano, C.; Nediani, C.; Travaglini, C.; Borchi, E.; Zani, M.; Feccia, M.; Mancini, M.; Petrozza, V.; Cossarizza, A.; et al. Induction of mitochondrial biogenesis is a maladaptive mechanism in mitochondrial cardiomyopathies. *J. Am. Coll. Cardiol.* **2007**, *50*, 1362–1369. [CrossRef] [PubMed]
38. Sack, M.N.; Rader, T.A.; Park, S.; Bastin, J.; McCune, S.A.; Kelly, D.P. Fatty acid oxidation enzyme gene expression is downregulated in the failing heart. *Circulation* **1996**, *94*, 2837–2842. [CrossRef] [PubMed]
39. Burns, K.A.; Vanden Heuvel, J.P. Modulation of PPAR activity via phosphorylation. *Biochim. Biophys. Acta* **2007**, *1771*, 952–960. [CrossRef] [PubMed]
40. Lazennec, G.; Canaple, L.; Saugy, D.; Wahli, W. Activation of peroxisome proliferator-activated receptors (PPARs) by their ligands and protein kinase A activators. *Mol. Endocrinol.* **2000**, *14*, 1962–1975. [CrossRef] [PubMed]
41. Barger, P.M.; Browning, A.C.; Garner, A.N.; Kelly, D.P. p38 mitogen-activated protein kinase activates peroxisome proliferator-activated receptor alpha: A potential role in the cardiac metabolic stress response. *J. Biol. Chem.* **2001**, *276*, 44495–44501. [CrossRef] [PubMed]

42. Barger, P.M.; Brandt, J.M.; Leone, T.C.; Weinheimer, C.J.; Kelly, D.P. Deactivation of peroxisome proliferator-activated receptor-alpha during cardiac hypertrophic growth. *J. Clin. Investig.* **2000**, *105*, 1723–1730. [CrossRef] [PubMed]
43. Shalev, A.; Siegrist-Kaiser, C.A.; Yen, P.M.; Wahli, W.; Burger, A.G.; Chin, W.W.; Meier, C.A. The peroxisome proliferator-activated receptor alpha is a phosphoprotein: Regulation by insulin. *Endocrinology* **1996**, *137*, 4499–4502. [CrossRef] [PubMed]
44. Adams, M.; Reginato, M.J.; Shao, D.; Lazar, M.A.; Chatterjee, V.K. Transcriptional activation by peroxisome proliferator-activated receptor gamma is inhibited by phosphorylation at a consensus mitogen-activated protein kinase site. *J. Biol. Chem.* **1997**, *272*, 5128–5132. [CrossRef] [PubMed]
45. Lopaschuk, G.D.; Ussher, J.R.; Folmes, C.D.; Jaswal, J.S.; Stanley, W.C. Myocardial fatty acid metabolism in health and disease. *Physiol. Rev.* **2010**, *90*, 207–258. [CrossRef] [PubMed]
46. Arad, M.; Seidman, C.E.; Seidman, J.G. AMP-activated protein kinase in the heart: Role during health and disease. *Circ. Res.* **2007**, *100*, 474–488. [CrossRef] [PubMed]
47. Yoon, M.J.; Lee, G.Y.; Chung, J.J.; Ahn, Y.H.; Hong, S.H.; Kim, J.B. Adiponectin increases fatty acid oxidation in skeletal muscle cells by sequential activation of AMP-activated protein kinase, p38 mitogen-activated protein kinase, and peroxisome proliferator-activated receptor alpha. *Diabetes* **2006**, *55*, 2562–2570. [CrossRef] [PubMed]
48. Barreto-Torres, G.; Parodi-Rullan, R.; Javadov, S. The role of PPARalpha in metformin-induced attenuation of mitochondrial dysfunction in acute cardiac ischemia/reperfusion in rats. *Int. J. Mol. Sci.* **2012**, *13*, 7694–7709. [CrossRef] [PubMed]
49. Barreto-Torres, G.; Hernandez, J.S.; Jang, S.; Rodriguez-Munoz, A.R.; Torres-Ramos, C.A.; Basnakian, A.G.; Javadov, S. The beneficial effects of AMP kinase activation against oxidative stress are associated with prevention of PPARalpha-cyclophilin D interaction in cardiomyocytes. *Am. J. Physiol. Heart Circ. Physiol.* **2015**, *308*, H749–H758. [CrossRef] [PubMed]
50. Barreto-Torres, G.; Javadov, S. Possible Role of Interaction between PPARalpha and Cyclophilin D in Cardioprotection of AMPK against In Vivo Ischemia-Reperfusion in Rats. *PPAR Res.* **2016**, *2016*, 9282087. [CrossRef] [PubMed]
51. Duan, S.Z.; Ivashchenko, C.Y.; Russell, M.W.; Milstone, D.S.; Mortensen, R.M. Cardiomyocyte-specific knockout and agonist of peroxisome proliferator-activated receptor-gamma both induce cardiac hypertrophy in mice. *Circ. Res.* **2005**, *97*, 372–379. [CrossRef] [PubMed]
52. Barbieri, M.; Di Filippo, C.; Esposito, A.; Marfella, R.; Rizzo, M.R.; D'Amico, M.; Ferraraccio, F.; Di Ronza, C.; Duan, S.Z.; Mortensen, R.M.; et al. Effects of PPARs agonists on cardiac metabolism in littermate and cardiomyocyte-specific PPAR-gamma-knockout (CM-PGKO) mice. *PLoS ONE* **2012**, *7*, e35999. [CrossRef] [PubMed]
53. Luo, J.; Wu, S.; Liu, J.; Li, Y.; Yang, H.; Kim, T.; Zhelyabovska, O.; Ding, G.; Zhou, Y.; Yang, Y.; et al. Conditional PPARgamma knockout from cardiomyocytes of adult mice impairs myocardial fatty acid utilization and cardiac function. *Am. J. Transl. Res.* **2010**, *3*, 61–72. [PubMed]
54. Ding, G.; Fu, M.; Qin, Q.; Lewis, W.; Kim, H.W.; Fukai, T.; Bacanamwo, M.; Chen, Y.E.; Schneider, M.D.; Mangelsdorf, D.J.; et al. Cardiac peroxisome proliferator-activated receptor gamma is essential in protecting cardiomyocytes from oxidative damage. *Cardiovasc. Res.* **2007**, *76*, 269–279. [CrossRef] [PubMed]
55. Son, N.H.; Park, T.S.; Yamashita, H.; Yokoyama, M.; Huggins, L.A.; Okajima, K.; Homma, S.; Szabolcs, M.J.; Huang, L.S.; Goldberg, I.J. Cardiomyocyte expression of PPARgamma leads to cardiac dysfunction in mice. *J. Clin. Investig.* **2007**, *117*, 2791–2801. [CrossRef] [PubMed]
56. Liu, J.; Wang, P.; Luo, J.; Huang, Y.; He, L.; Yang, H.; Li, Q.; Wu, S.; Zhelyabovska, O.; Yang, Q. Peroxisome proliferator-activated receptor beta/delta activation in adult hearts facilitates mitochondrial function and cardiac performance under pressure-overload condition. *Hypertension* **2011**, *57*, 223–230. [CrossRef] [PubMed]

57. Jucker, B.M.; Doe, C.P.; Schnackenberg, C.G.; Olzinski, A.R.; Maniscalco, K.; Williams, C.; Hu, T.C.; Lenhard, S.C.; Costell, M.; Bernard, R.; et al. PPARdelta activation normalizes cardiac substrate metabolism and reduces right ventricular hypertrophy in congestive heart failure. *J. Cardiovasc. Pharmacol.* **2007**, *50*, 25–34. [CrossRef] [PubMed]
58. Pol, C.J.; Lieu, M.; Drosatos, K. PPARs: Protectors or Opponents of Myocardial Function? *PPAR Res.* **2015**, *2015*, 835985. [CrossRef] [PubMed]
59. Parodi-Rullan, R.M.; Chapa-Dubocq, X.R.; Javadov, S. Acetylation of Mitochondrial Proteins in the Heart: The Role of SIRT3. *Front. Physiol.* **2018**, *9*, 1094. [CrossRef] [PubMed]
60. Jiang, X.; Ye, X.; Guo, W.; Lu, H.; Gao, Z. Inhibition of HDAC3 promotes ligand-independent PPARgamma activation by protein acetylation. *J. Mol. Endocrinol.* **2014**, *53*, 191–200. [CrossRef] [PubMed]
61. Montgomery, R.L.; Potthoff, M.J.; Haberland, M.; Qi, X.; Matsuzaki, S.; Humphries, K.M.; Richardson, J.A.; Bassel-Duby, R.; Olson, E.N. Maintenance of cardiac energy metabolism by histone deacetylase 3 in mice. *J. Clin. Investig.* **2008**, *118*, 3588–3597. [CrossRef] [PubMed]
62. Diezko, R.; Suske, G. Ligand binding reduces SUMOylation of the peroxisome proliferator-activated receptor gamma (PPARgamma) activation function 1 (AF1) domain. *PLoS ONE* **2013**, *8*, e66947. [CrossRef] [PubMed]
63. Kim, J.H.; Park, K.W.; Lee, E.W.; Jang, W.S.; Seo, J.; Shin, S.; Hwang, K.A.; Song, J. Suppression of PPARgamma through MKRN1-mediated ubiquitination and degradation prevents adipocyte differentiation. *Cell Death Differ.* **2014**, *21*, 594–603. [CrossRef] [PubMed]
64. Rodriguez, J.E.; Liao, J.Y.; He, J.; Schisler, J.C.; Newgard, C.B.; Drujan, D.; Glass, D.J.; Frederick, C.B.; Yoder, B.C.; Lalush, D.S.; et al. The ubiquitin ligase MuRF1 regulates PPARalpha activity in the heart by enhancing nuclear export via monoubiquitination. *Mol. Cell. Endocrinol.* **2015**, *413*, 36–48. [CrossRef] [PubMed]
65. Qian, Y.; Li, P.; Zhang, J.; Shi, Y.; Chen, K.; Yang, J.; Wu, Y.; Ye, X. Association between peroxisome proliferator-activated receptor-alpha, delta, and gamma polymorphisms and risk of coronary heart disease: A case-control study and meta-analysis. *Medicine (Baltimore)* **2016**, *95*, e4299. [CrossRef] [PubMed]
66. Rudkowska, I.; Verreault, M.; Barbier, O.; Vohl, M.C. Differences in transcriptional activation by the two allelic (L162V Polymorphic) variants of PPARalpha after Omega-3 fatty acids treatment. *PPAR Res.* **2009**, *2009*, 369602. [CrossRef] [PubMed]
67. Tai, E.S.; Corella, D.; Demissie, S.; Cupples, L.A.; Coltell, O.; Schaefer, E.J.; Tucker, K.L.; Ordovas, J.M. Polyunsaturated fatty acids interact with the PPARA-L162V polymorphism to affect plasma triglyceride and apolipoprotein C-III concentrations in the Framingham Heart Study. *J. Nutr.* **2005**, *135*, 397–403. [CrossRef] [PubMed]
68. Reinhard, W.; Stark, K.; Sedlacek, K.; Fischer, M.; Baessler, A.; Neureuther, K.; Weber, S.; Kaess, B.; Wiedmann, S.; Mitsching, S.; et al. Association between PPARalpha gene polymorphisms and myocardial infarction. *Clin. Sci. (Lond.)* **2008**, *115*, 301–308. [CrossRef] [PubMed]
69. Ingwall, J.S. Energy metabolism in heart failure and remodelling. *Cardiovasc. Res.* **2009**, *81*, 412–419. [CrossRef] [PubMed]
70. Neubauer, S. The failing heart—An engine out of fuel. *N. Engl. J. Med.* **2007**, *356*, 1140–1151. [CrossRef] [PubMed]
71. Stanley, W.C.; Recchia, F.A.; Lopaschuk, G.D. Myocardial substrate metabolism in the normal and failing heart. *Physiol. Rev.* **2005**, *85*, 1093–1129. [CrossRef] [PubMed]
72. Xiong, D.; He, H.; James, J.; Tokunaga, C.; Powers, C.; Huang, Y.; Osinska, H.; Towbin, J.A.; Purevjav, E.; Balschi, J.A.; et al. Cardiac-specific VLCAD deficiency induces dilated cardiomyopathy and cold intolerance. *Am. J. Physiol. Heart Circ. Physiol.* **2014**, *306*, H326–H338. [CrossRef] [PubMed]
73. Exil, V.J.; Gardner, C.D.; Rottman, J.N.; Sims, H.; Bartelds, B.; Khuchua, Z.; Sindhal, R.; Ni, G.; Strauss, A.W. Abnormal mitochondrial bioenergetics and heart rate dysfunction in mice lacking very-long-chain acyl-CoA dehydrogenase. *Am. J. Physiol. Heart Circ. Physiol.* **2006**, *290*, H1289–H1297. [CrossRef] [PubMed]
74. Wu, J.; Song, Y.; Li, H.; Chen, J. Rhabdomyolysis associated with fibrate therapy: Review of 76 published cases and a new case report. *Eur. J. Clin. Pharmacol.* **2009**, *65*, 1169–1174. [CrossRef] [PubMed]
75. Lebrasseur, N.K.; Duhaney, T.A.; De Silva, D.S.; Cui, L.; Ip, P.C.; Joseph, L.; Sam, F. Effects of fenofibrate on cardiac remodeling in aldosterone-induced hypertension. *Hypertension* **2007**, *50*, 489–496. [CrossRef] [PubMed]

76. Brigadeau, F.; Gele, P.; Wibaux, M.; Marquie, C.; Martin-Nizard, F.; Torpier, G.; Fruchart, J.C.; Staels, B.; Duriez, P.; Lacroix, D. The PPARalpha activator fenofibrate slows down the progression of the left ventricular dysfunction in porcine tachycardia-induced cardiomyopathy. *J. Cardiovasc. Pharmacol.* **2007**, *49*, 408–415. [CrossRef] [PubMed]
77. Labinskyy, V.; Bellomo, M.; Chandler, M.P.; Young, M.E.; Lionetti, V.; Qanud, K.; Bigazzi, F.; Sampietro, T.; Stanley, W.C.; Recchia, F.A. Chronic activation of peroxisome proliferator-activated receptor-alpha with fenofibrate prevents alterations in cardiac metabolic phenotype without changing the onset of decompensation in pacing-induced heart failure. *J. Pharmacol. Exp. Ther.* **2007**, *321*, 165–171. [CrossRef] [PubMed]
78. Ichihara, S.; Obata, K.; Yamada, Y.; Nagata, K.; Noda, A.; Ichihara, G.; Yamada, A.; Kato, T.; Izawa, H.; Murohara, T.; et al. Attenuation of cardiac dysfunction by a PPAR-alpha agonist is associated with down-regulation of redox-regulated transcription factors. *J. Mol. Cell. Cardiol.* **2006**, *41*, 318–329. [CrossRef] [PubMed]
79. Ogata, T.; Miyauchi, T.; Sakai, S.; Irukayama-Tomobe, Y.; Goto, K.; Yamaguchi, I. Stimulation of peroxisome-proliferator-activated receptor alpha (PPAR alpha) attenuates cardiac fibrosis and endothelin-1 production in pressure-overloaded rat hearts. *Clin. Sci. (Lond.)* **2002**, *103* (Suppl. 48), 284S–288S. [CrossRef] [PubMed]
80. Duhaney, T.A.; Cui, L.; Rude, M.K.; Lebrasseur, N.K.; Ngoy, S.; De Silva, D.S.; Siwik, D.A.; Liao, R.; Sam, F. Peroxisome proliferator-activated receptor alpha-independent actions of fenofibrate exacerbates left ventricular dilation and fibrosis in chronic pressure overload. *Hypertension* **2007**, *49*, 1084–1094. [CrossRef] [PubMed]
81. Hafstad, A.D.; Khalid, A.M.; Hagve, M.; Lund, T.; Larsen, T.S.; Severson, D.L.; Clarke, K.; Berge, R.K.; Aasum, E. Cardiac peroxisome proliferator-activated receptor-alpha activation causes increased fatty acid oxidation, reducing efficiency and post-ischaemic functional loss. *Cardiovasc. Res.* **2009**, *83*, 519–526. [CrossRef] [PubMed]
82. Linz, W.; Wohlfart, P.; Baader, M.; Breitschopf, K.; Falk, E.; Schafer, H.L.; Gerl, M.; Kramer, W.; Rutten, H. The peroxisome proliferator-activated receptor-alpha (PPAR-alpha) agonist, AVE8134, attenuates the progression of heart failure and increases survival in rats. *Acta Pharmacol. Sin.* **2009**, *30*, 935–946. [CrossRef] [PubMed]
83. Schafer, H.L.; Linz, W.; Falk, E.; Glien, M.; Glombik, H.; Korn, M.; Wendler, W.; Herling, A.W.; Rutten, H. AVE8134, a novel potent PPARalpha agonist, improves lipid profile and glucose metabolism in dyslipidemic mice and type 2 diabetic rats. *Acta Pharmacol. Sin.* **2012**, *33*, 82–90. [CrossRef] [PubMed]
84. Forman, B.M.; Chen, J.; Evans, R.M. Hypolipidemic drugs, polyunsaturated fatty acids, and eicosanoids are ligands for peroxisome proliferator-activated receptors alpha and delta. *Proc. Natl. Acad. Sci. USA* **1997**, *94*, 4312–4317. [CrossRef] [PubMed]
85. Lam, V.H.; Zhang, L.; Huqi, A.; Fukushima, A.; Tanner, B.A.; Onay-Besikci, A.; Keung, W.; Kantor, P.F.; Jaswal, J.S.; Rebeyka, I.M.; et al. Activating PPARalpha Prevents Post-Ischemic Contractile Dysfunction in Hypertrophied Neonatal Hearts. *Circ. Res.* **2015**. [CrossRef] [PubMed]
86. Kliewer, S.A.; Sundseth, S.S.; Jones, S.A.; Brown, P.J.; Wisely, G.B.; Koble, C.S.; Devchand, P.; Wahli, W.; Willson, T.M.; Lenhard, J.M.; et al. Fatty acids and eicosanoids regulate gene expression through direct interactions with peroxisome proliferator-activated receptors alpha and gamma. *Proc. Natl. Acad. Sci. USA* **1997**, *94*, 4318–4323. [CrossRef] [PubMed]
87. Brown, P.J.; Winegar, D.A.; Plunket, K.D.; Moore, L.B.; Lewis, M.C.; Wilson, J.G.; Sundseth, S.S.; Koble, C.S.; Wu, Z.; Chapman, J.M.; et al. A ureido-thioisobutyric acid (GW9578) is a subtype-selective PPARalpha agonist with potent lipid-lowering activity. *J. Med. Chem.* **1999**, *42*, 3785–3788. [CrossRef] [PubMed]
88. Olsson, A.G.; Lang, P.D. One-year study of the effect of bezafibrate on serum lipoprotein concentrations in hyperlipoproteinaemia. *Atherosclerosis* **1978**, *31*, 429–433. [CrossRef]
89. Jonkers, I.J.; de Man, F.H.; van der Laarse, A.; Frolich, M.; Gevers Leuven, J.A.; Kamper, A.M.; Blauw, G.J.; Smelt, A.H. Bezafibrate reduces heart rate and blood pressure in patients with hypertriglyceridemia. *J. Hypertens.* **2001**, *19*, 749–755. [CrossRef] [PubMed]
90. Peters, J.M.; Aoyama, T.; Burns, A.M.; Gonzalez, F.J. Bezafibrate is a dual ligand for PPARalpha and PPARbeta: Studies using null mice. *Biochim. Biophys. Acta* **2003**, *1632*, 80–89. [CrossRef]

91. Yamaguchi, S.; Li, H.; Purevsuren, J.; Yamada, K.; Furui, M.; Takahashi, T.; Mushimoto, Y.; Kobayashi, H.; Hasegawa, Y.; Taketani, T.; et al. Bezafibrate can be a new treatment option for mitochondrial fatty acid oxidation disorders: Evaluation by in vitro probe acylcarnitine assay. *Mol. Genet. Metab.* **2012**, *107*, 87–91. [CrossRef] [PubMed]
92. Orngreen, M.C.; Madsen, K.L.; Preisler, N.; Andersen, G.; Vissing, J.; Laforet, P. Bezafibrate in skeletal muscle fatty acid oxidation disorders: A randomized clinical trial. *Neurology* **2014**, *82*, 607–613. [CrossRef] [PubMed]
93. Dillon, L.M.; Hida, A.; Garcia, S.; Prolla, T.A.; Moraes, C.T. Long-term bezafibrate treatment improves skin and spleen phenotypes of the mtDNA mutator mouse. *PLoS ONE* **2012**, *7*, e44335. [CrossRef] [PubMed]
94. Dumont, M.; Stack, C.; Elipenahli, C.; Jainuddin, S.; Gerges, M.; Starkova, N.; Calingasan, N.Y.; Yang, L.; Tampellini, D.; Starkov, A.A.; et al. Bezafibrate administration improves behavioral deficits and tau pathology in P301S mice. *Hum. Mol. Genet.* **2012**, *21*, 5091–5105. [CrossRef] [PubMed]
95. Huang, Y.; Powers, C.; Moore, V.; Schafer, C.; Ren, M.; Phoon, C.K.; James, J.F.; Glukhov, A.V.; Javadov, S.; Vaz, F.M.; et al. The PPAR pan-agonist bezafibrate ameliorates cardiomyopathy in a mouse model of Barth syndrome. *Orphanet J. Rare Dis.* **2017**, *12*, 49. [CrossRef] [PubMed]
96. Djouadi, F.; Bastin, J. PPARs as therapeutic targets for correction of inborn mitochondrial fatty acid oxidation disorders. *J. Inherit. Metab. Dis.* **2008**, *31*, 217–225. [CrossRef] [PubMed]
97. Nakajima, T.; Tanaka, N.; Kanbe, H.; Hara, A.; Kamijo, Y.; Zhang, X.; Gonzalez, F.J.; Aoyama, T. Bezafibrate at clinically relevant doses decreases serum/liver triglycerides via down-regulation of sterol regulatory element-binding protein-1c in mice: A novel peroxisome proliferator-activated receptor alpha-independent mechanism. *Mol. Pharmacol.* **2009**, *75*, 782–792. [CrossRef] [PubMed]
98. Viscomi, C.; Bottani, E.; Civiletto, G.; Cerutti, R.; Moggio, M.; Fagiolari, G.; Schon, E.A.; Lamperti, C.; Zeviani, M. In vivo correction of COX deficiency by activation of the AMPK/PGC-1alpha axis. *Cell Metab.* **2011**, *14*, 80–90. [CrossRef] [PubMed]
99. Johri, A.; Calingasan, N.Y.; Hennessey, T.M.; Sharma, A.; Yang, L.; Wille, E.; Chandra, A.; Beal, M.F. Pharmacologic activation of mitochondrial biogenesis exerts widespread beneficial effects in a transgenic mouse model of Huntington's disease. *Hum. Mol. Genet.* **2012**, *21*, 1124–1137. [CrossRef] [PubMed]
100. Sugihara, J.; Imamura, T.; Nagafuchi, S.; Bonaventura, J.; Bonaventura, C.; Cashon, R. Hemoglobin Rahere, a human hemoglobin variant with amino acid substitution at the 2,3-diphosphoglycerate binding site. Functional consequences of the alteration and effects of bezafibrate on the oxygen bindings. *J. Clin. Investig.* **1985**, *76*, 1169–1173. [CrossRef] [PubMed]
101. Shibayama, N.; Miura, S.; Tame, J.R.; Yonetani, T.; Park, S.Y. Crystal structure of horse carbonmonoxyhemoglobin-bezafibrate complex at 1.55-A resolution. A novel allosteric binding site in R-state hemoglobin. *J. Biol. Chem.* **2002**, *277*, 38791–38796. [CrossRef] [PubMed]
102. Schafer, C.; Moore, V.; Dasgupta, N.; Javadov, S.; James, J.F.; Glukhov, A.I.; Strauss, A.W.; Khuchua, Z. The Effects of PPAR Stimulation on Cardiac Metabolic Pathways in Barth Syndrome Mice. *Front. Pharmacol.* **2018**, *9*, 318. [CrossRef] [PubMed]
103. Jakob, T.; Nordmann, A.J.; Schandelmaier, S.; Ferreira-Gonzalez, I.; Briel, M. Fibrates for primary prevention of cardiovascular disease events. *Cochrane Database Syst. Rev.* **2016**, *11*, CD009753. [CrossRef] [PubMed]
104. Wang, D.; Liu, B.; Tao, W.; Hao, Z.; Liu, M. Fibrates for secondary prevention of cardiovascular disease and stroke. *Cochrane Database Syst. Rev.* **2015**, CD009580. [CrossRef] [PubMed]
105. Goldenberg, I.; Benderly, M.; Goldbourt, U. Secondary prevention with bezafibrate therapy for the treatment of dyslipidemia: An extended follow-up of the BIP trial. *J. Am. Coll. Cardiol.* **2008**, *51*, 459–465. [CrossRef] [PubMed]
106. Goldenberg, I.; Boyko, V.; Tennenbaum, A.; Tanne, D.; Behar, S.; Guetta, V. Long-term benefit of high-density lipoprotein cholesterol-raising therapy with bezafibrate: 16-year mortality follow-up of the bezafibrate infarction prevention trial. *Arch. Intern. Med.* **2009**, *169*, 508–514. [CrossRef] [PubMed]
107. Goldenberg, I.; Benderly, M.; Sidi, R.; Boyko, V.; Tenenbaum, A.; Tanne, D.; Behar, S. Relation of clinical benefit of raising high-density lipoprotein cholesterol to serum levels of low-density lipoprotein cholesterol in patients with coronary heart disease (from the Bezafibrate Infarction Prevention Trial). *Am. J. Cardiol.* **2009**, *103*, 41–45. [CrossRef] [PubMed]
108. Arbel, Y.; Klempfner, R.; Erez, A.; Goldenberg, I.; Benzekry, S.; Shlomo, N.; Fisman, E.Z.; Tenenbaum, A. Bezafibrate for the treatment of dyslipidemia in patients with coronary artery disease: 20-year mortality follow-up of the BIP randomized control trial. *Cardiovasc. Diabetol.* **2016**, *15*, 11. [CrossRef] [PubMed]

109. Meade, T.; Zuhrie, R.; Cook, C.; Cooper, J. Bezafibrate in men with lower extremity arterial disease: Randomised controlled trial. *BMJ* **2002**, *325*, 1139. [CrossRef] [PubMed]
110. Tenenbaum, A.; Motro, M.; Fisman, E.Z.; Tanne, D.; Boyko, V.; Behar, S. Bezafibrate for the secondary prevention of myocardial infarction in patients with metabolic syndrome. *Arch. Intern. Med.* **2005**, *165*, 1154–1160. [CrossRef] [PubMed]
111. Corpechot, C.; Chazouilleres, O.; Rousseau, A.; Le Gruyer, A.; Habersetzer, F.; Mathurin, P.; Goria, O.; Potier, P.; Minello, A.; Silvain, C.; et al. A Placebo-Controlled Trial of Bezafibrate in Primary Biliary Cholangitis. *N. Engl. J. Med.* **2018**, *378*, 2171–2181. [CrossRef] [PubMed]
112. Bonnefont, J.P.; Bastin, J.; Laforet, P.; Aubey, F.; Mogenet, A.; Romano, S.; Ricquier, D.; Gobin-Limballe, S.; Vassault, A.; Behin, A.; et al. Long-term follow-up of bezafibrate treatment in patients with the myopathic form of carnitine palmitoyltransferase 2 deficiency. *Clin. Pharmacol. Ther.* **2010**, *88*, 101–108. [CrossRef] [PubMed]
113. Bastin, J.; Bonnefont, J.P.; Djouadi, F.; Bresson, J.L. Should the beneficial impact of bezafibrate on fatty acid oxidation disorders be questioned? *J. Inherit. Metab. Dis.* **2014**, *38*, 371–372. [CrossRef] [PubMed]
114. Gobin-Limballe, S.; McAndrew, R.P.; Djouadi, F.; Kim, J.J.; Bastin, J. Compared effects of missense mutations in Very-Long-Chain Acyl-CoA Dehydrogenase deficiency: Combined analysis by structural, functional and pharmacological approaches. *Biochim. Biophys. Acta* **2010**, *1802*, 478–484. [CrossRef] [PubMed]

International Journal of
Molecular Sciences

Review

Maintenance of Kidney Metabolic Homeostasis by PPAR Gamma

Patricia Corrales [1], Adriana Izquierdo-Lahuerta [1] and Gema Medina-Gómez [1,2,*]

[1] Área de Bioquímica y Biología Molecular, Departamento de Ciencias Básicas de la Salud, Facultad de Ciencias de la Salud, Universidad Rey Juan Carlos. Avda. de Atenas s/n. Alcorcón, 28922 Madrid, Spain; patricia.corrales@urjc.es (P.C.); adriana.zquierdo@urjc.es (A.I.-L.)

[2] MEMORISM Research Unit of University Rey Juan Carlos-Institute of Biomedical Research "Alberto Sols" (CSIC), 28029 Madrid, Spain

* Correspondence: gema.medina@urjc.es; Tel.: +34-914-888-632

Received: 29 June 2018; Accepted: 11 July 2018; Published: 16 July 2018

Abstract: Peroxisome proliferator-activated receptors (PPARs) are a family of nuclear hormone receptors that control the transcription of specific genes by binding to regulatory DNA sequences. Among the three subtypes of PPARs, PPARγ modulates a broad range of physiopathological processes, including lipid metabolism, insulin sensitization, cellular differentiation, and cancer. Although predominantly expressed in adipose tissue, PPARγ expression is also found in different regions of the kidney and, upon activation, can redirect metabolism. Recent studies have highlighted important roles for PPARγ in kidney metabolism, such as lipid and glucose metabolism and renal mineral control. PPARγ is also implicated in the renin-angiotensin-aldosterone system and, consequently, in the control of systemic blood pressure. Accordingly, synthetic agonists of PPARγ have reno-protective effects both in diabetic and nondiabetic patients. This review focuses on the role of PPARγ in renal metabolism as a likely key factor in the maintenance of systemic homeostasis.

Keywords: PPARγ; metabolism; lipid; RAAS; nuclear receptors; kidney

1. Introduction

The peroxisome proliferator-activated receptors (PPARs) are ligand-activated nuclear hormone receptors that participate in the transactivation or transrepression of networks of target genes, resulting in complex biological effects. PPARs are class 2 receptors that, upon ligand binding, heterodimerize with retinoid X receptors (RXRs) and translocate to the nucleus, whereupon the PPAR:RXR heterodimer binds to the PPAR response element (PPRE) generally in the promoter region of target genes, to control their expression [1,2]. The affinity of PPARs for ligands, and hence their transcriptional response, is determined by the conformational changes induced by ligand binding within a complex pocket with multiple interaction points [3].

There are three known subtypes of PPARs that have distinct physiological roles in energy metabolism in different tissues: PPARα, PPARδ, and PPARγ. Overall, PPARs function as lipid sensors to govern metabolic homeostasis through binding to dietary metabolites: PPARα regulates catabolism, mainly in the liver and the heart, PPARγ regulates anabolism in adipose tissue, and PPARδ is involved in fatty acid transport and oxidation in skeletal muscle. PPARs not only serve critical roles in the control of lipid metabolism, but they are also implicated in the regulation of vascular diseases, cellular differentiation, insulin sensitization, and cancer [4]. In the kidney, PPARα plays an important role in the metabolic control of renal energy homeostasis and is expressed in the proximal tubules and medullary thick ascending limb, with lower expression in glomerular mesangial cells [5]). Likewise, kidney PPARδ has a role in renal metabolic adaptation to fasting and refeeding [6], and is expressed in the renal cortex and medulla [7].

In humans, the *PPARγ* gene is located on chromosome 3 (3p25.2) and contains nine exons spanning more than 100 kb [8]. Four *PPARγ* splice variants(γ1, γ2, γ3, and γ4), generated by alternative splicing and differential promoter usage, are found in human [9]; however, only two protein isoforms (γ1 and γ2) are encoded [10]. *PPARγ* is also present in other animals (rodents, chicken, lizard, *Xenopus*, and Zebrafish) [11,12], although, they do not appear to display significant functional differences. Mutations in *PPARγ* lead to dysfunctional lipid and glucose homeostasis and have been directly related to type 2 diabetes mellitus and obesity [13], familial partial lipodystrophy type 3 (FPLD3) [14], and also cancer [15,16].

PPARγ is expressed predominantly in the adipose tissue where, together with the coexpression of *C/EBP alpha* and other proteins involved in lipid and glucose metabolism, serves as a key regulator both of adipocyte differentiation and triglyceride energy stores, and has pleiotropic vascular effects that are independent of its glucose blood-lowering effect, protecting against the progression of hypertension and atherosclerosis [17,18]. Specifically, *PPARγ1* is expressed in many tissues and cell types, including white and brown adipose tissue, skeletal muscle, liver, pancreatic β-cells, macrophages, colon, bone, and placenta. By contrast, *PPARγ2* has a more restricted pattern of expression with significant amounts found only in white and brown adipose tissue under physiological conditions, although it is induced in liver and skeletal muscle in response to overnutrition or genetic obesity. *PPARγ3* mRNA expression appears to be limited to human white adipocytes, but it is also found in a variety of cell lines with different origins, including liver hepatocellular cells (HepG2), human intestinal cells (Caco-2), and cervical cancer (HeLa) cells. *PPARγ4* is expressed in human adipose tissue [19] and mutations in his promotor are associated to FPLD, familial partial lipodystrophy [20].

At the protein level, PPARγ is subject to several post-translational modifications, including glycosylation and phosphorylation, which function to modify its activity [21]. For example, *O*-GlcNAcylation at Thr-84 has been found to reduce PPARγ transcriptional activity in adipocytes cultured in vitro. Moreover, PPARγ2 is phosphorylated at Ser-112 by MAPK in response to different mitogenic growth factors that inhibit fat cell differentiation. The cdk5-dependent phosphorylation of PPARγ at Ser-273 occurs in inflamed obese white adipose tissue and decreases PPARγ activity. By contrast, PKA phosphorylation has been reported to positively affect the activity of PPARγ [22]. Furthermore, CK-II-dependent PPARγ1 phosphorylation at Ser-16 and Ser-21 is necessary for its nuclear translocation [23,24]. Overall, these events illustrate the complex regulation of this nuclear receptor. Interestingly, *PPARγ* expression can be inhibited by miRNAs in specific settings [25], constituting another layer of regulation.

In kidney, PPARγ is expressed in different regions of the renal collecting system under physiological conditions, including connective renal tubules and collecting ducts [26] (Figure 1). PPARγ is also abundant in the inner renal medulla and is localized to the epithelial layer, from the medullary collecting ducts to the urothelium of the ureter and the bladder. It is additionally expressed in renal medullary interstitial cells and in the juxtaglomerular apparatus and the glomeruli, including podocytes, mesangial cells, and renal microvascular endothelial cells [27]. Given that multiple renal cell types have endogenous PPARγ expression and activity, its activation in kidney may be critical for governing renal function. Indeed, as we describe later, synthetic PPARγ ligand agonists have been shown to have reno-protective effects both in diabetic and nondiabetic patients [28].

In this review, we focus on several key observations that illustrate the central role of PPARγ in renal metabolism and maintenance of systemic homeostasis. We will examine the involvement of PPARγ in renal lipid, glucose, and mineral metabolism, and also blood pressure control. Against this background, we will also address the potential use of PPARγ agonists in the clinical setting as therapeutic agents for renal pathologies.

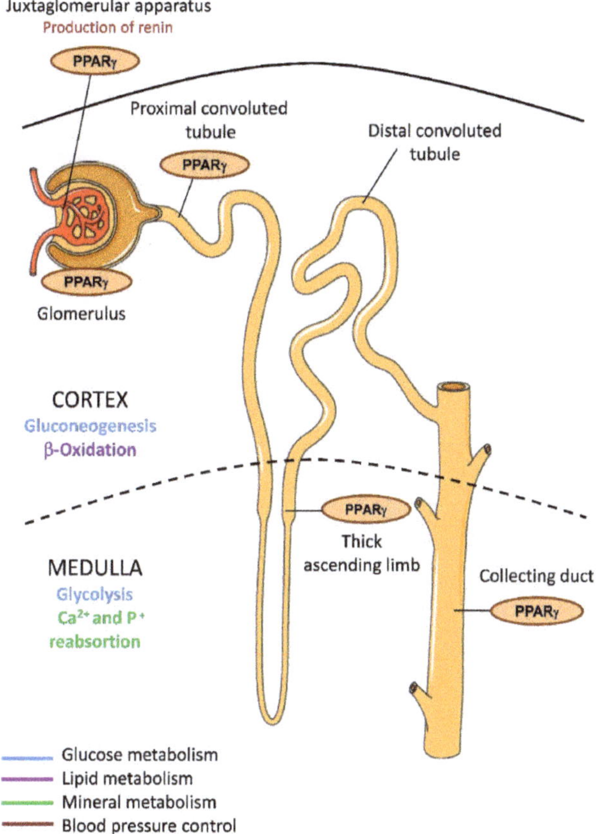

Figure 1. *PPARγ* expression in the kidney and implications for metabolism. PPARγ is expressed in different areas of kidney including the cortex and the medulla, which have different metabolic specializations. Glucose metabolism (blue); Lipid metabolism (purple); Mineral metabolism (green); Blood pressure control (red).

2. PPARγ in Renal Lipid Metabolism

Human kidney contains about 3% fat, although this varies greatly among individuals. Much of this (~50%) is in the form of phospholipids that form cell membranes, 15% is in the form of triglycerides, and around 0.3% is nonesterified (free) fatty acids (FFAs) [29]. Under physiological conditions, the kidney can metabolize a variety of substrates, including FFAs, lactate, glutamine, 3-hydroxybutyrate, citrate, pyruvate, α-ketoglutarate, glycerol, proline, and other amino acids. Proximal tubule reabsorption is responsible for about 70% of these substrates, the metabolic fate of which depends on the extracellular medium, hormonal influences, and metabolic conditions (acid-base). Once reabsorbed, FFAs are metabolized within mitochondria of proximal tubule cells by β-oxidation (as their glycolytic capacity is poor), constituting the largest source of Adenosine triphosphate (ATP). An increase in the availability of intracellular fatty acids for mitochondria results in competition with other oxidizable substrates, triggering a decrease in the use of glutamine, with the consequent reduction in ammoniagenesis [30].

Moorhead et al. in 1982 were the first to establish a link between alterations in lipid metabolism and kidney disease [31]. Mutations in enzymes responsible for lipid metabolism or an increase in serum lipids promotes a decline in renal function [32]. Accordingly, mice deficient for PPARγ present alterations in renal lipid metabolism, and the extent of renal damage induced by a high-fat diet is lower in PPARγ heterozygous knockout mice than in wild-type mice concomitant with a decrease in lipid accumulation and lipogenesis and an attenuation of lipid-mediated kidney damage [33]. We recently demonstrated that accelerated kidney damage manifests in the POKO mouse—A model of the metabolic syndrome generated by ablation of the *PPARγ2* isoform in a leptin-deficient obese (*ob/ob*) background [34,35]. Despite having similar body weight and blood pressure to *ob/ob* littermates at an early age, POKO mice present renal hypertrophy and dyslipidemia, and with alterations in some proliferation markers. Moreover, they develop incipient insulin resistance associated with a decrease in the expression of renal adiponectin. POKO mice also exhibit faster progression of kidney disease compared with *ob/ob* mice, accompanied by an increase in the expression of transforming growth factor beta (TGFβ) and also inflammatory and profibrotic markers in glomeruli, which associates with lipotoxicity and insulin resistance. This model has confirmed the key role of PPARγ2 in regulating renal lipid metabolism and insulin sensitivity, which is closely associated with glomerular filtration rate and albuminuria [34]. Furthermore, we showed that podocytes treated with saturated palmitic acid present a tendency for decreased *PPARγ1* expression that correlates with a proinflammatory and proapoptotic state, and with changes in the gene expression of enzymes involved in fatty acid synthesis including a decrease in acyl-CoA carboxylase and fatty acid synthase [36].

In addition to lipid metabolism, PPARγ is involved in adipokine expression from adipose tissue, including adiponectin [37], a circulating plasma protein produced by white adipose tissue that negatively correlates with obesity. Adiponectin stimulates fatty acid oxidation, suppresses hepatic gluconeogenesis, increases insulin sensitivity, and acts to counter the effects of tumor necrosis factor, an inflammatory cytokine [38]. Especially relevant to kidney function, low levels of adiponectin correlate with albuminuria both in mice and humans, and adiponectin is postulated as a renoprotective protein after podocyte injury [39].

Heterozygous mutations in *PPARγ* cause FPLD3 (OMIM 604367), clinically characterized by loss of subcutaneous limb and gluteal fat with preservation of visceral and subcutaneous abdominal fat, fatty infiltration of the liver, and hyperuricemia [14]. All patients develop type II diabetes mellitus and hypertension at an unusually early age. Interestingly, a single case study showed that the adipokine leptin was effective in treating metabolic complications in a patient with FPLD3 [40]. It is known that the kidney expresses a leptin receptor, and that leptin is cleared from the bloodstream principally by the kidney [41].

Systemic lupus erythematosus is an autoimmune connective tissue disease marked by immune complex-mediated lesions in small blood vessels of different organs, particularly the kidneys. Mice lacking macrophage-specific expression of PPARγ or RXRα develop glomerulonephritis and autoantibodies to nuclear antigens, resembling the nephritis seen in systemic lupus erythematosus [42]. Moreover, these mice exhibit tubule-interstitial lipid deposition that leads to lipid-mediated tissue damage in all areas of the nephron [43].

Overall, these findings illustrate the crucial role of PPARγ as a master regulator of systemic lipid metabolism in the kidney.

3. PPARγ in Renal Glucose Metabolism

The kidney contributes to glucose homeostasis not only through the processes of utilization (i.e., glucose filtration, reabsorption, and consumption), but it is also increasingly recognized as having a significant role in gluconeogenesis (~20% of all glucose production), and uniquely contributes to plasma glucose regulation by controlling glucose reabsorption from renal tubules following glomerular filtration [44]. Whereas the poorly vascularized, and consequently relatively hypoxic, medulla is a site of considerable glycolysis, the cortex is the renal site of gluconeogenesis. Moreover, the proximal tubule is the only region of the kidney with the appropriate enzymes necessary for gluconeogenesis.

Blood glucose is freely filtered at the glomerulus and is then reabsorbed predominantly in the proximal tubule. This process is performed by two sodium-dependent glucose cotransporter (SGLT) proteins situated in different segments of epithelial cells of the proximal tubule (SGLT2 in the S1 segment and SGLT1 in the S3 segment). Once glucose has been concentrated in the epithelial cells, it diffuses out to the interstitium via specific facilitative glucose transporters (GLUTs) localized at the basolateral membrane [44]. Hyperglycemic conditions lead to a dysfunction of SGLT-mediated glucose transport in proximal tubular cells and promote epithelial–mesenchymal transition (EMT), leading to renal fibrosis. PPARγ agonists have been shown to reverse this hyperglycemia-induced EMT and to restore functional SGLT-mediated glucose uptake [45]. Moreover, PPARγ agonists have significant renoprotective properties in experimental models of nephropathy. Correspondingly, thiazolidinedione compounds (TZDs), a class of insulin-sensitizing drug used in the treatment of type 2 diabetes, and a synthetic PPARγ ligand improve glucose tolerance, which may indirectly ameliorate the progression of chronic kidney disease (CKD). Importantly, it has been suggested that the hypoglycemic action of TZDs relates to the inhibition of gluconeogenesis that is confined to proximal tubule cells, with minimal hepatic consequences [44,46]. For instance, administration of the TZD rosiglitazone to rabbit renal tubules resulted in a ~70% decrease in the rate of gluconeogenesis, accompanied by a ~75% decrease in alanine utilization, and a ~35% increase in lactate synthesis [46]. PPARγ agonists have also been tested in humans, and seem to improve glucose tolerance and reduce the urinary albumin excretion rate, thus indirectly delaying renal disease progression. Unfortunately, two PPARγ agonists, rosiglitazone and pioglitazone, had to be withdrawn from the US and European markets because of cardiovascular disease and bladder cancer as associated side effects [21,47]. Thus, more efforts should be made to discover new PPARγ agonists with beneficial effects on CKD.

A recent review has addressed the association of metabolic traits, specifically glucose metabolism, with PPARγ genetic polymorphisms in humans [48], the most common of which leads to the replacement of alanine for proline at codon 12 (Pro12Ala) in PPARγ2. This polymorphism is associated with a risk of diabetic nephropathy in Caucasians, but no similar association is observed in Asians. Additionally, the Ala-12 polymorphism is associated with a decreased risk of albuminuria [49]. Also, whereas heterozygosity for the frameshift/premature stop codon mutation [A553ΔAAAiT]fs.185[stop186] in the DNA-binding domain of PPARγ is not associated with insulin resistance, individuals doubly heterozygous, with an additional defect in an unrelated gene encoding the muscle-specific regulatory subunit of protein phosphatase 1 (PPP1R3A), exhibit severe insulin resistance [50]. Along this line, Dyment et al. described a woman with biallelic mutations in PPARγ who presented with congenital generalized lipodystrophy, hypertriglyceridemia, hepatosplenomegaly, insulin resistance, and renal failure since birth [51].

The discovery of PPARγ as a target for TZDs prompted the screening of a cohort of subjects with severe insulin resistance for mutations in *PPARγ*. Through this analysis, two new heterozygous mutations in PPARγ, Pro467Leu, and Val290Met, were identified in three subjects [14]. In addition to severe insulin resistance, all three patients developed liver steatosis, type 2 diabetes, and also hypertension at a very early age [52]. Similarly, rare mutations in the ligand-binding domain of PPARγ (Arg425Cys and Phe388Leu) have been found in patients with insulin resistance [53]. Moreover, the novel PPARγ mutations Arg165Thr and Leu339X, which are linked to familial partial lipodystrophies, are associated with a defective transrepression of RAS, leading to cellular dysfunction and contributing to the specific FPLD3-linked severe hypertension [54]. In addition to these findings, other mutations such as Pro115Glu have been identified in nonlipodystrophic subjects, ascertained based upon obesity and diabetes [13].

4. PPARγ in Renal Mineral Metabolism

Chronic kidney disease is directly related to the development of abnormalities in bone mineral metabolism. The kidney is known to regulate the levels of vitamin D3 by producing its most active form, 1,25 dihydroxyvitamin D3 (calcitriol), which participates in calcium and phosphate metabolism. The key players in this hormonal bone-parathyroid-kidney axis include fibroblast growth factor 23

(FGF23), Klotho, parathyroid hormone (PTH), and calcitriol, and recent work has also implicated the involvement of PPARγ [55].

Osteoblasts and adipocytes share a common multipotent mesenchymal stem cell progenitor. Akune et al. [56] showed that homozygous *PPARγ*-deficient embryonic stem cells failed to differentiate into adipocytes, but spontaneously differentiated into osteoblasts. They further showed that adipogenesis was restored by reintroduction of the *PPARγ*, indicating that PPARγ insufficiency stimulates osteoblastogenesis in vivo. The canonical Wnt/β-catenin-PPARγ system determines the molecular switch between osteoblastogenesis and adipogenesis [57]. Activation of this pathway leads to osteogenesis, not adipogenesis, and its inhibition leads to an increase in transcription of *PPARγ*. The osteogenic pathway is linked to the stimulation of Wnt signaling, leading to the final transcriptional activation of early osteogenic markers such as runt-related transcription factor 2 (*RUNX-2*) and alkaline phosphatase (*ALP*), which is mediated by β-catenin. Conversely, the adipogenic pathway involves inhibition of the Wnt pathway leading to ubiquitination/degradation of β-catenin, which results in the transcription of *PPARγ*, a pivotal initiator of adipogenesis [58].

It has been recently shown that the murine *klotho* gene is a target of PPARγ [59]. *Klotho* was initially discovered as an antiaging gene and encodes a single-pass transmembrane protein that forms a complex with the FGF receptor (FGFR) to create a de novo high-affinity binding site for FGF23. In addition to membrane-anchored Klotho, a secreted form of the protein is directly released into the extracellular compartment and is present in body fluids [55]. *Klotho* is expressed in multiple tissues but is particularly high in the kidney in distal convoluted tubules, proximal convoluted tubules, and also in inner medullary collecting duct-derived cell lines. Recently, two *Klotho*-related genes were identified based on sequence similarity, *βKlotho* and *γKlotho*. The originally described *Klotho* is now referred to as *αKlotho* to distinguish it from the β and γ forms. *βKlotho* is expressed in various tissues, most notably in the liver and white adipose tissue, whereas *γKlotho* is expressed in the eye [60].

Klotho is involved in the regulation of calcitriol production and modulates urinary phosphate, calcium and potassium excretion, and is downregulated in conditions related to kidney injury [61]. Consistent with its activation by PPARγ, oral administration of the PPARγ agonist troglitazone augmented renal *klotho* mRNA expression in Otsuka Long–Evans Tokushima Fatty rats and protected against the endothelial dysfunction induced in this model of atherogenesis [62]. Unfortunately, TZD treatment is associated with an increased risk of hip fractures and is linked to the formation of excessive calcium phosphate precipitates in the urinary bladder [63,64]. Likely, this effect is caused by excess PPARγ activity, which increases adipogenesis and downregulates the β-catenin pathway and osteoblastogenesis. Overall, these data point to PPARγ as an important player in maintaining mineral metabolism, both at the renal and the systemic level.

5. PPARγ in Systemic Blood Pressure Control

In line with a major role in the regulation of vascular tone and blood pressure, mutations in *PPARγ* induce severe hypertension and type 2 diabetes. Mice with mutations in *PPARγ* in smooth muscle present vascular dysfunction and severe systolic hypertension [65]. Cells of the juxtaglomerular apparatus express *PPARγ* (Figure 1) and produce renin, a protease that cleaves angiotensinogen to generate angiotensin I. Renin is a key component of the renin-angiotensin-aldosterone system (RAAS) that mediates extracellular volume. The human renin gene is activated by endogenous and pharmacological PPARγ agonists and is a direct target of PPARγ, containing two PPARγ binding sequences that control its transcription [66].

PPARγ also acts as a negative regulator of angiotensin II receptor 1 transcription, another important component of RAAS. In 2004, the angiotensin II receptor 1 antagonist telmisartan, which is to treat hypertension and diabetes, was identified as a partial agonist of PPARγ [67,68]. By activating PPARγ, telmisartan exerts beneficial effects on the kidney by decreasing proteinuria and inflammation, and consequently confers renoprotection. Since telmisartan can bind to the ligand-binding domain of PPARγ at a site that is different to that used by TZDs, it is not surprising that telmisartan possesses unique properties unrelated to that of conventional TZDs. Accordingly, telmisartan has the capacity to reverse the progression of EMT induced by TGFβ1 in cultured human kidney proximal tubule epithelial cells and can counteract EMT-related pathological changes such as renal fibrosis [69]. The combined use of TZDs and angiotensin receptor blockers, however, fails to provide synergistic protective action, and it has been shown that co-administration of RAAS inhibitors and PPARγ agonists promotes anemia in uncomplicated diabetic patients [70]. The development of new therapies (combined or not) that exploit the beneficial effects of PPARγ activation in the treatment of renal disease are therefore warranted.

It has been also observed that troglitazone has vasodilating effects on efferent and afferent arterioles from rabbit kidney, consistent with PPARγ expression in intima/media renal vasculature, thereby decreasing glomerular capillary pressure and hence excretion of urinary protein [71].

The beneficial effects of PPARγ agonists in the control of blood pressure underscore the pivotal role of PPARγ at this level. PPARγ also has a critical role in systemic fluid retention through the regulation of renal sodium transport in the collecting duct, as the adverse effect of TZD use in the treatment of diabetes is PPARγ-dependent. Thus, the *Scnn1g* gene, encoding the gamma subunit of the epithelial Na^+ channel, was identified as a critical PPARγ target gene in the control of edema [72,73]. The TZD-induced fluid retention effects are attenuated in patients by the combination treatment with diuretics (Figure 2).

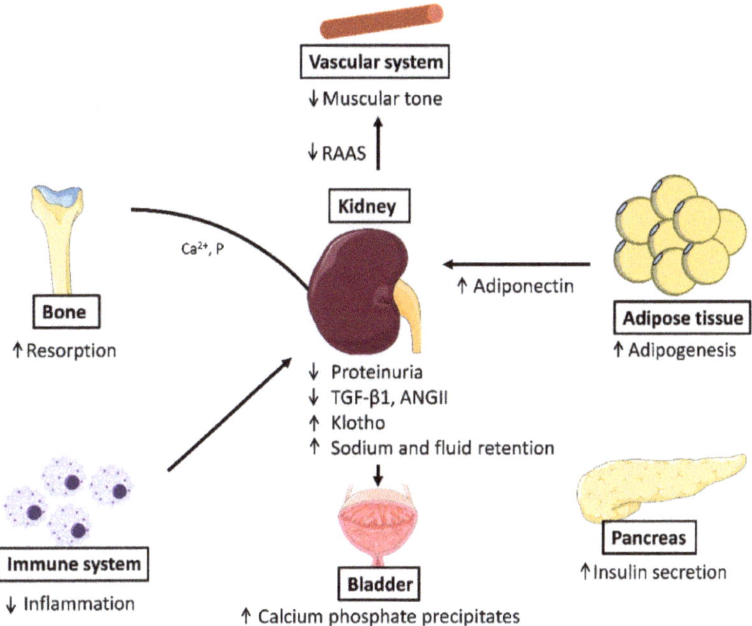

Figure 2. Effects of PPARγ activation in different systems: In the vascular system, PPARγ decreases arterial pressure, relaxes muscular tone, and downregulates RAAS; in adipose tissue, the activation of PPARγ promotes adipogenesis; in β-cells it increases insulin secretion; PPARγ is also anti-inflammatory and from the bone resorption is produced accompanied of release of calcium and phosphate that form stones in the bladder. ANG-II: Angiotensin II; RAAS, Renin Angiotensin Aldosterone System; TGF-β1, transforming growth factor beta 1.

6. PPARγ and Circadian Rhythm

All PPAR isoforms are known to be rhythmically expressed [25]. Specifically, PPARγ has direct interactions with the core clock genes [26–28], suggesting that it may act as a molecular link between circadian rhythm and energy metabolism. Moreover, PPARγ exhibits variations in its diurnal expression in mouse fat, liver, and blood vessels [74], which is exacerbated by consumption of a high-fat diet [75]. In addition, deletion of *PPARγ* in mouse suppresses or diminishes diurnal rhythms [76]. In this regard, *nocturnin*, a circadian-regulated gene that encodes a deadenylase thought to be involved in the removal of polyA tail from mRNAs, binds to PPARγ and enhances its transcriptional activity. Enhanced activity of PPARγ by nocturnin may result in increased bone marrow adiposity and bone loss [77].

The circadian expression of PPARγ has not yet been established in kidney. An evaluation of this phenomenon could be especially interesting to associate the regulation of renal metabolism with central and peripheral circadian networks.

7. Conclusions

PPARγ is a nuclear receptor that regulates systemic glucose and lipid homeostasis and participates also in immunity and vascular health. In the kidney, PPARγ is expressed in many different types of cells with diverse metabolic specializations, underscoring important roles for this nuclear receptor in renal lipid metabolism, glucose management, mineral metabolism, and the control of systemic blood pressure. PPARγ activation improves insulin sensitivity and reduces cardiovascular complications and renal injury in clinical practice. Nevertheless, the complex regulation of *PPARγ* together with the adverse effects of PPARγ agonists hinders efforts to develop safe clinical treatments. The new challenges with regard to future applications include a comprehensive analysis of PPARγ cell-specific actions in the kidney and the manipulation of PPARγ expression. Finally, a better understanding of the molecular mechanisms of action of PPARγ in specific pathways together with its systemic implications may allow the development of new agonists and modulators to improve the management of kidney disease and consequently global homeostasis.

Author Contributions: P.C., A.I.-L., and G.M.-G. wrote the paper.

Funding: This study was funded by Ministerio de Economíay Competitividad, Spain (BFU2012-33594, BFU2013-47384-R and BFU2016-78951-R) and Comunidad de Madrid (B2017/BMD-3684).

Acknowledgments: We thank Ismael Velasco for technical assistance.

Conflicts of Interest: The authors declare no conflict of interest.

Abbreviations

CKD	chronic kidney disease
EMT	epithelial-mesenchymal transition
FFA	free fatty acids
FGF23	fibroblast growth factor 23
GLUT	glucose transporter
PPAR	Peroxisome proliferator-activated receptors
PPRE	PPAR response element
RAAS	Renin Angiotensin Aldosterone System
RXR	retinoid X receptors
SGLT	sodium-dependent glucose cotransporter
TGFβ	transforming growth factor beta
TZD	thiazolidinedione

References

1. Khurana, S.; Bruggeman, L.A.; Kao, H.-Y. Nuclear hormone receptors in podocytes. *Cell Biosci.* **2012**, *2*, 33. [CrossRef] [PubMed]
2. Thomas, M.C.; Jandeleit-Dahm, K.A.; Tikellis, C. The renoprotective actions of peroxisome proliferator-activated receptors agonists in diabetes. *PPAR Res.* **2012**, *2012*, 456529. [CrossRef] [PubMed]

3. Higgins, L.S.; Depaoli, A.M. Selective peroxisome proliferator-activated receptor gamma (PPAR gamma) modulation as a strategy for safer therapeutic PPAR gamma activation 1–3. *Am. J. Clin. Nutr.* **2010**, *91*, 267–272. [CrossRef] [PubMed]
4. Vitale, S.G.; Laganà, A.S.; Nigro, A.; La Rosa, V.L.; Rossetti, P.; Rapisarda, A.M.C.; La Vignera, S.; Condorelli, R.A.; Corrado, F.; Buscema, M.; et al. Peroxisome Proliferator-Activated Receptor Modulation during Metabolic Diseases and Cancers: Master Minions. *PPAR Res.* **2016**, *2016*, 6517313. [CrossRef] [PubMed]
5. Tovar-Palacio, C.; Torres, N.; Diaz-Villaseñor, A.; Tovar, A.R. The role of nuclear receptors in the kidney in obesity and metabolic syndrome. *Genes Nutr.* **2012**, *7*, 483–498. [CrossRef] [PubMed]
6. Escher, P.; Braissant, O.; Basu-Modak, S.; Michalik, L.; Wahli, W.; Desvergne, B. Rat PPARs: Quantitative Analysis in Adult Rat Tissues and Regulation in Fasting and Refeeding. *Endocrinology* **2001**, *142*, 4195–4202. [CrossRef] [PubMed]
7. Guan, Y.; Zhang, Y.; Davis, L.; Breyer, M.D. Expression of peroxisome proliferator-activated receptors in urinary tract of rabbits and humans. *Am. J. Physiol.* **1997**, *273*, F1013–F1022. [CrossRef] [PubMed]
8. Beamer, B.A.; Negri, C.; Yen, C.J.; Gavrilova, O.; Rumberger, J.M.; Durcan, M.J.; Yarnall, D.P.; Hawkins, A.L.; Griffin, C.A.; Burns, D.K.; et al. Chromosomal localization and partial genomic structure of the human peroxisome proliferator activated receptor-gamma (hPPAR gamma) gene. *Biochem. Biophys. Res. Commun.* **1997**, *233*, 756–759. [CrossRef] [PubMed]
9. Martin, G.; Schoonjans, K.; Staels, B.; Auwerx, J. PPARγ activators improve glucose-homeostasis by stimulating fatty-acid uptake in the adipocytes. *Atherosclerosis* **1998**, *137*, S75–S80. [CrossRef]
10. Astarci, E.; Banerjee, S. PPARG (peroxisome proliferator-activated receptor gamma). *Atlas Genet. Cytogenet. Oncol. Heamatol.* **2009**, *13*, 417–421. [CrossRef]
11. Navidshad, B.; Royan, M. Ligands and Regulatory Modes of Peroxisome Proliferator-Activated Receptor Gamma (PPARγ) in Avians. *Crit. Rev. Eukaryot. Gene Expr.* **2015**, *25*, 287–292. [CrossRef] [PubMed]
12. Zhu, J.; Janesick, A.; Wu, L.; Hu, L.; Tang, W.; Blumberg, B.; Shi, H. The unexpected teratogenicity of RXR antagonist UVI3003 via activation of PPARγ in Xenopus tropicalis. *Toxicol. Appl. Pharmacol.* **2017**, *314*, 91–97. [CrossRef] [PubMed]
13. Ristow, M.; Muller-Wieland, D.; Pfeiffer, A.; Krone, W.; Kahn, C.R. Obesity associated with a mutation in a genetic regulator of adipocite diferentiation. *Engl. J. Med.* **1998**, *339*, 953. [CrossRef] [PubMed]
14. Barroso, I.; Gurnell, M.; Crowley, V.E.F.; Agostini, M.; Schwabe, J.W.; Soos, M.A.; Maslen, G.L.; Williams, T.D.M.; Lewis, H.; Schafer, A.J.; et al. Dominant negative mutations in human PPARgamma associated with severe insulin resistance, diabetes mellitus and hypertension. *Nature* **1999**, *402*, 880–883. [CrossRef] [PubMed]
15. Nikiforova, M.N.; Lynch, R.A.; Biddinger, P.W.; Alexander, E.K.; Dorn, G.W.; Tallini, G.; Kroll, T.G.; Nikiforov, Y.E. RAS point mutations and PAX8-PPARγ Rearrangement in thyroid tumors: Evidence for distinct molecular pathways in thyroid follicular carcinoma. *J. Clin. Endocrinol. Metab.* **2003**, *88*, 2318–2326. [CrossRef] [PubMed]
16. Marques, A.R.; Espadinha, C.; Catarino, A.L.; Moniz, S.; Pereira, T.; Sobrinho, L.G.; Leite, V. Expression of PAX8-PPAR gamma 1 rearrangements in both follicular thyroid carcinomas and adenomas. *J. Clin. Endocrinol. Metab.* **2002**, *87*, 3947–3952. [CrossRef] [PubMed]
17. Rosen, E.D.; Spiegelman, B.M. PPARgamma: A nuclear regulator of metabolism, differentiation, and cell growth. *J. Biol. Chem.* **2001**, *276*, 37731–37734. [CrossRef] [PubMed]
18. Sugawara, A.; Uruno, A.; Kudo, M.; Matsuda, K.; Yang, C.W.; Ito, S. Effects of PPARγ on hypertension, atherosclerosis, and chronic kidney disease. *Endocr. J.* **2010**, *57*, 847–852. [CrossRef] [PubMed]
19. Iwanishi, M.; Ebihara, K.; Kusakabe, T.; Chen, W.; Ito, J.; Masuzaki, H.; Hosoda, K.; Nakao, K. Clinical characteristics and efficacy of pioglitazone in a Japanese diabetic patient with an unusual type of familial partial lipodystrophy. *Metabolism* **2009**, *58*, 1681–1687. [CrossRef] [PubMed]
20. Al-Shali, K.; Cao, H.; Knoers, N.; Hermus, A.R.; Tack, C.J.; Hegele, R.A. A single-base mutation in the peroxisome proliferator-activated receptor gamma4 promoter associated with altered in vitro expression and partial lipodystrophy. *J. Clin. Endocrinol. Metab.* **2004**, *89*, 5655–5660. [CrossRef] [PubMed]
21. Ahmadian, M.; Suh, J.M.; Hah, N.; Liddle, C.; Atkins, A.R.; Downes, M.; Evans, R.M. PPARγ signaling and metabolism: The good, the bad and the future. *Nat. Med.* **2013**, *19*, 557–566. [CrossRef] [PubMed]

22. Burns, K.A.; Vanden Heuvel, J.P. Modulation of PPAR activity via phosphorylation. *Biochim. Biophys. Acta* **2007**, *1771*, 952–960. [CrossRef] [PubMed]
23. Berrabah, W.; Aumercier, P.; Lefebvre, P.; Staels, B. Control of nuclear receptor activities in metabolism by post-translational modifications. *FEBS Lett.* **2011**, *585*, 1640–1650. [CrossRef] [PubMed]
24. Choi, J.H.; Banks, A.S.; Estall, J.L.; Kajimura, S.; Boström, P.; Laznik, D.; Ruas, J.L.; Chalmers, M.J.; Kamenecka, T.M.; Blüher, M.; et al. Anti-diabetic drugs inhibit obesity-linked phosphorylation of PPARgamma by Cdk5. *Nature* **2010**, *466*, 451–456. [CrossRef] [PubMed]
25. Povero, D.; Panera, N.; Eguchi, A.; Johnson, C.D.; Papouchado, B.G.; de Araujo Horcel, L.; Pinatel, E.M.; Alisi, A.; Nobili, V.; Feldstein, A.E. Lipid-Induced Hepatocyte-Derived Extracellular Vesicles Regulate Hepatic Stellate Cells via MicroRNA Targeting Peroxisome Proliferator-Activated Receptor-γ. *Cell. Mol. Gastroenterol. Hepatol.* **2015**, *1*, 646–663. [CrossRef] [PubMed]
26. Yang, T.; Michele, D.E.; Park, J.; Smart, A.M.; Lin, Z.; Brosius, F.C.; Schnermann, J.B.; Briggs, J.P. Expression of peroxisomal proliferator-activated receptors and retinoid X receptors in the kidney. *Am. J. Physiol.* **1999**, *277*, F966–F973. [CrossRef] [PubMed]
27. Kiss-Tóth, É.; Röszer, T. PPAR γ in kidney physiology and pathophysiology. *PPAR Res.* **2008**, *2008*, 183108. [CrossRef] [PubMed]
28. Sarafidis, P.A.; Bakris, G.L. Protection of the kidney by thiazolidinediones: An assessment from bench to bedside. *Kidney Int.* **2006**, *70*, 1223–1233. [CrossRef] [PubMed]
29. Druilhet, R.E.; Overturf, M.L.; Kirkendall, W.M. Cortical and medullary lipids of normal and nephrosclerotic human kidney. *Int. J. Biochem.* **1978**, *9*, 729–734. [CrossRef]
30. Bobulescu, I.A.; Lotan, Y.; Zhang, J.; Rosenthal, T.R.; Rogers, J.T.; Adams-Huet, B.; Sakhaee, K.; Moe, O.W. Triglycerides in the human kidney cortex: Relationship with body size. *PLoS ONE* **2014**, *9*, e101285. [CrossRef] [PubMed]
31. Moorhead, J.F.; Chan, M.K.; El-Nahas, M.; Varghese, Z. Lipid nephrotoxicity in chronic progressive glomerular and tubulo-interstitial disease. *Lancet* **1982**, *2*, 1309–1311. [CrossRef]
32. Izquierdo-Lahuerta, A.; Martínez-García, C.; Medina-Gómez, G. Lipotoxicity as a trigger factor of renal disease. *J. Nephrol.* **2016**, *29*, 603–610. [CrossRef] [PubMed]
33. Kume, S.; Uzu, T.; Araki, S.I.; Sugimoto, T.; Isshiki, K.; Chin-Kanasaki, M.; Sakaguchi, M.; Kubota, N.; Terauchi, Y.; Kadowaki, T.; et al. Role of Altered Renal Lipid Metabolism in the Development of Renal Injury Induced by a High-Fat Diet. *J. Am. Soc. Nephrol.* **2007**, *18*, 2715–2723. [CrossRef] [PubMed]
34. Martínez-García, C.; Izquierdo, A.; Velagapudi, V.; Vivas, Y.; Velasco, I.; Campbell, M.; Burling, K.; Cava, F.; Ros, M.; Oresič, M.; et al. Accelerated renal disease is associated with the development of metabolic syndrome in a glucolipotoxic mouse model. *Dis. Models Mech.* **2012**, *5*, 636–648. [CrossRef] [PubMed]
35. Medina-Gomez, G.; Yetukuri, L.; Velagapudi, V.; Campbell, M.; Blount, M.; Jimenez-Linan, M.; Ros, M.; Orešič, M.; Vidal-Puig, A. Adaptation and failure of pancreatic cells in murine models with different degrees of metabolic syndrome. *Dis. Models Mech.* **2009**, *2*, 582–592. [CrossRef] [PubMed]
36. Martínez-García, C.; Izquierdo-Lahuerta, A.; Vivas, Y.; Velasco, I.; Yeo, T.K.; Chen, S.; Medina-Gomez, G. Renal Lipotoxicity-Associated Inflammation and Insulin Resistance Affects Actin Cytoskeleton Organization in Podocytes. *PLoS ONE* **2015**, *10*, e0142291. [CrossRef] [PubMed]
37. Berg, A.H.; Combs, T.P.; Scherer, P.E. ACRP30/adiponectin: An adipokine regulating glucose and lipid metabolism. *Trends Endocrinol. Metab.* **2002**, *13*, 84–89. [CrossRef]
38. Long, Q.; Lei, T.; Feng, B.; Yin, C.; Jin, D.; Wu, Y.; Zhu, X.; Chen, X.; Gan, L.; Yang, Z. Peroxisome proliferator-activated receptor-γ increases adiponectin secretion via transcriptional repression of endoplasmic reticulum chaperone protein ERp44. *Endocrinology* **2010**, *151*, 3195–3203. [CrossRef] [PubMed]
39. Rutkowski, J.M.; Wang, Z.V.; Park, A.S.D.; Zhang, J.; Zhang, D.; Hu, M.C.; Moe, O.W.; Susztak, K.; Scherer, P.E. Adiponectin Promotes Functional Recovery after Podocyte Ablation. *J. Am. Soc. Nephrol.* **2013**, *24*, 268–282. [CrossRef] [PubMed]
40. Guettier, J.M.; Park, J.Y.; Cochran, E.K.; Poitou, C.; Basdevant, A.; Meier, M.; Clément, K.; Magré, J.; Gorden, P. Leptin therapy for partial lipodystrophy linked to a PPAR-γ mutation. *Clin. Endocrinol.* **2008**, *68*, 547–554. [CrossRef] [PubMed]

41. Simonds, S.E.; Pryor, J.T.; Ravussin, E.; Greenway, F.L.; Dileone, R.; Allen, A.M.; Bassi, J.; Elmquist, J.K.; Keogh, J.M.; Henning, E.; et al. Leptin mediates the increase in blood pressure associated with obesity. *Cell* **2014**, *159*, 1404–1416. [CrossRef] [PubMed]
42. Rőszer, T.; Menéndez-Gutiérrez, M.P.; Lefterova, M.I.; Alameda, D.; Núñez, V.; Lazar, M.A.; Fischer, T.; Ricote, M. Autoimmune Kidney Disease and Impaired Engulfment of Apoptotic Cells in Mice with Macrophage Peroxisome Proliferator-Activated Receptor or Retinoid X Receptor Deficiency. *J. Immunol.* **2011**, *186*, 621–631. [CrossRef] [PubMed]
43. Luzar, B.; Ferluga, D. Role of lipids in the progression of renal disease in systemic lupus erythematosus patients. *Wiener Klinische Wochenschrift* **2000**, *112*, 716–721. [PubMed]
44. Mather, A.; Pollock, C. Glucose handling by the kidney. *Kidney Int. Suppl.* **2011**, *79*, S1–S6. [CrossRef] [PubMed]
45. Lee, Y.J.; Han, H.J. Troglitazone ameliorates high glucose-induced EMT and dysfunction of SGLTs through PI3K/Akt, GSK-3, Snail1, and -catenin in renal proximal tubule cells. *AJP Renal Physiol.* **2010**, *298*, F1263–F1275. [CrossRef] [PubMed]
46. Derlacz, R.A.; Hyc, K.; Usarek, M.; Jagielski, A.K.; Drozak, J.; Jarzyna, R. PPAR-gamma-independent inhibitory effect of rosiglitazone on glucose synthesis in primary cultured rabbit kidney-cortex tubules. *Biochem. Cell. Biol.* **2008**, *86*, 396–404. [CrossRef] [PubMed]
47. Kung, J.; Henry, R.R. Thiazolidinedione safety. *Expert Opin. Drug Saf.* **2012**, *11*, 565–579. [CrossRef] [PubMed]
48. Pap, A.; Cuaranta-Monroy, I.; Peloquin, M.; Nagy, L. Is the mouse a good model of human PPAR??-related metabolic diseases? *Int. J. Mol. Sci.* **2016**, *17*, 1236. [CrossRef] [PubMed]
49. Zhang, H.; Zhu, S.; Chen, J.; Tang, Y.; Hu, H.; Mohan, V.; Venkatesan, R.; Wang, J.; Chen, H. Peroxisome proliferator-activated receptor γ polymorphism Pro12Ala is associated with nephropathy in type 2 diabetes: Evidence from meta-analysis of 18 studies. *Diabetes Care* **2012**, *35*, 1388–1393. [CrossRef] [PubMed]
50. Agostini, M.; Schoenmakers, E.; Mitchell, C.; Szatmari, I.; Savage, D.; Smith, A.; Rajanayagam, O.; Semple, R.; Luan, J.A.; Bath, L.; et al. Non-DNA binding, dominant-negative, human PPARγ mutations cause lipodystrophic insulin resistance. *Cell Metab.* **2006**, *4*, 303–311. [CrossRef] [PubMed]
51. Dyment, D.A.; Gibson, W.T.; Huang, L.; Bassyouni, H.; Hegele, R.A.; Innes, A.M. Biallelic mutations at PPARG cause a congenital, generalized lipodystrophy similar to the Berardinelli–Seip syndrome. *Eur. J. Genet.* **2014**, *57*, 524–526. [CrossRef] [PubMed]
52. Savage, D.B.; Tan, G.D.; Acerini, C.L.; Jebb, S.A.; Agostini, M.; Gurnell, M.; Williams, R.L.; Umpleby, A.M.; Thomas, E.L.; Bell, J.D.; et al. Human metabolic syndrome resulting from dominant-negative mutations in the nuclear receptor peroxisome proliferator-activated receptor-gamma. *Diabetes* **2003**, *52*, 910–917. [CrossRef] [PubMed]
53. Hegele, R.A.; Cao, H.; Frankowski, C.; Mathews, S.T.; Leff, T. PPARG F388L, a transactivation-deficient mutant, in familial partial lipodystrophy. *Diabetes* **2002**, *51*, 3586–3590. [CrossRef] [PubMed]
54. Auclair, M.; Vigouroux, C.; Boccara, F.; Capel, E.; Vigeral, C.; Guerci, B.; Lascols, O.; Capeau, J.; Caron-Debarle, M. Peroxisome proliferator-activated receptor-γ mutations responsible for lipodystrophy with severe hypertension activate the cellular renin-angiotensin system. *Arterioscler. Thromb. Vasc. Biol.* **2013**, *33*, 829–838. [CrossRef] [PubMed]
55. Hu, M.C.; Kuro-o, M.; Moe, O.W. Klotho and Chronic Kidney Disease. *Contrib. Nephrol.* **2013**, *180*, 47–63. [CrossRef] [PubMed]
56. Akune, T.; Ohba, S.; Kamekura, S.; Yamaguchi, M.; Chung, U.I.; Kubota, N.; Terauchi, Y.; Harada, Y.; Azuma, Y.; Nakamura, K.; et al. PPARγ insufficiency enhances osteogenesis through osteoblast formation from bone marrow progenitors. *J. Clin. Investig.* **2004**, *113*, 846–855. [CrossRef] [PubMed]
57. Takada, I.; Kouzmenko, A.P.; Kato, S. Wnt and PPARγ signaling in osteoblastogenesis and adipogenesis. *Nat. Rev. Rheumatol.* **2009**, *5*, 442–447. [CrossRef] [PubMed]
58. Lecarpentier, Y.; Claes, V.; Duthoit, G.; Hébert, J.-L. Circadian rhythms, Wnt/beta-catenin pathway and PPAR alpha/gamma profiles in diseases with primary or secondary cardiac dysfunction. *Front. Physiol.* **2014**, *5*, 429. [CrossRef] [PubMed]
59. Zhang, H.; Li, Y.; Fan, Y.; Wu, J.; Zhao, B.; Guan, Y.; Chien, S.; Wang, N. Klotho is a target gene of PPAR-gamma. *Kidney Int.* **2008**, *74*, 732–739. [CrossRef] [PubMed]
60. Kuro-o, M. Klotho and the Aging Process. *Korean J. Intern. Med.* **2011**, *26*, 113. [CrossRef] [PubMed]

61. Izquierdo, M.C.; Perez-Gomez, M.V.; Sanchez-Niño, M.D.; Sanz, A.B.; Ruiz-Andres, O.; Poveda, J.; Moreno, J.A.; Egido, J.; Ortiz, A. Klotho, phosphate and inflammation/ageing in chronic kidney disease. *Nephrol. Dial. Transplant.* **2012**, *27*, 6–10. [CrossRef] [PubMed]
62. Yamagishi, T.; Saito, Y.; Nakamura, T.; Takeda, S.I.; Kanai, H.; Sumino, H.; Kuro-o, M.; Nabeshima, Y.I.; Kurabayashi, M.; Nagai, R. Troglitazone improves endothelial function and augments renal klotho mRNA expression in Otsuka Long-Evans Tokushima Fatty (OLETF) rats with multiple atherogenic risk factors. *Hypertens. Res.* **2001**, *24*, 705–709. [CrossRef] [PubMed]
63. Dominick, M.A.; White, M.R.; Sanderson, T.P.; Van Vleet, T.; Cohen, S.M.; Arnold, L.E.; Cano, M.; Tannehill-Gregg, S.; Moehlenkamp, J.D.; Waites, C.R.; et al. Urothelial carcinogenesis in the urinary bladder of male rats treated with muraglitazar, a PPAR alpha/gamma agonist: Evidence for urolithiasis as the inciting event in the mode of action. *Toxicol. Pathol.* **2006**, *34*, 903–920. [CrossRef] [PubMed]
64. Monami, M.; Dicembrini, I.; Mannucci, E. Thiazolidinediones and cancer: Results of a meta-analysis of randomized clinical trials. *Acta Diabetol.* **2014**, *51*, 91–101. [CrossRef] [PubMed]
65. Halabi, C.M.; Beyer, A.M.; de Lange, W.J.; Keen, H.L.; Baumbach, G.L.; Faraci, F.M.; Sigmund, C.D. Interference with PPARγ Function in Smooth Muscle Causes Vascular Dysfunction and Hypertension. *Cell Metab.* **2008**, *7*, 215–226. [CrossRef] [PubMed]
66. Todorov, V.T.; Desch, M.; Schmitt-Nilson, N.; Todorova, A.; Kurtz, A. Peroxisome proliferator-activated receptor-gamma is involved in the control of renin gene expression. *Hypertension* **2007**, *50*, 939–944. [CrossRef] [PubMed]
67. Benson, S.C.; Pershadsingh, H.A.; Ho, C.I.; Chittiboyina, A.; Desai, P.; Pravenec, M.; Qi, N.; Wang, J.; Avery, M.A.; Kurtz, T.W. Identification of Telmisartan as a Unique Angiotensin II Receptor Antagonist with Selective PPARγ-Modulating Activity. *Hypertension* **2004**, *43*, 993–1002. [CrossRef] [PubMed]
68. Schupp, M.; Janke, J.; Clasen, R.; Unger, T.; Kintscher, U. Angiotensin Type 1 Receptor Blockers Induce Peroxisome Proliferator-Activated Receptor-g Activity. *Circulation* **2004**, *109*, 2054–2057. [CrossRef] [PubMed]
69. Chen, Y.; Luo, Q.; Xiong, Z.; Liang, W.; Chen, L.; Xiong, Z. Telmisartan counteracts TGF-β1 induced epithelial-to-mesenchymal transition via PPAR-γ in human proximal tubule epithelial cells. *Int. J. Clin. Exp. Pathol.* **2012**, *5*, 522–529. [PubMed]
70. Raptis, A.E.; Bacharaki, D.; Mazioti, M.; Marathias, K.P.; Markakis, K.P.; Raptis, S.A.; Dimitriadis, G.D.; Vlahakos, D.V. Anemia due to coadministration of renin-angiotensin-system inhibitors and PPARγ agonists in uncomplicated diabetic patients. *Exp. Clin. Endocrinol. Diabetes* **2012**, *120*, 416–419. [CrossRef] [PubMed]
71. Arima, S.; Kohagura, K.; Takeuchi, K.; Taniyama, Y.; Sugawara, A.; Ikeda, Y.; Abe, M.; Omata, K.; Ito, S. Biphasic vasodilator action of troglitazone on the renal microcirculation. *J. Am. Soc. Nephrol.* **2002**, *13*, 342–349. [PubMed]
72. Guan, Y.; Hao, C.; Cha, D.R.; Rao, R.; Lu, W.; Kohan, D.E.; Magnuson, M.A.; Redha, R.; Zhang, Y.; Breyer, M.D. Thiazolidinediones expand body fluid volume through PPARgamma stimulation of ENaC-mediated renal salt absorption. *Nat. Med.* **2005**, *11*, 861–866. [CrossRef] [PubMed]
73. Zhang, H.; Zhang, A.; Kohan, D.E.; Nelson, R.D.; Gonzalez, F.J.; Yang, T. Collecting duct-specific deletion of peroxisome proliferator-activated receptor blocks thiazolidinedione-induced fluid retention. *Proc. Natl. Acad. Sci. USA* **2005**, *102*, 9406–9411. [CrossRef] [PubMed]
74. Wang, N.; Yang, G.; Jia, Z.; Zhang, H.; Aoyagi, T.; Soodvilai, S.; Symons, J.D.; Schnermann, J.B.; Gonzalez, F.J.; Litwin, S.E. Vascular PPARγ Controls Circadian Variation in Blood Pressure and Heart Rate through Bmal1. *Cell Metab.* **2008**, *8*, 482–491. [CrossRef] [PubMed]
75. Kawai, M.; Rosen, C.J. PPARγ: A circadian transcription factor in adipogenesis and osteogenesis. *Nat. Rev. Endocrinol.* **2010**, *6*, 629–636. [CrossRef] [PubMed]
76. Yang, G.; Jia, Z.; Aoyagi, T.; McClain, D.; Mortensen, R.M.; Yang, T. Systemic PPARγ Deletion Impairs Circadian Rhythms of Behavior and Metabolism. *PLoS ONE* **2012**, *7*, e38117. [CrossRef] [PubMed]
77. Stubblefield, J.J.; Terrien, J.; Green, C.B. Nocturnin: At the crossroads of clocks and metabolism. *Trends Endocrinol. Metab.* **2012**, *23*, 326–333. [CrossRef] [PubMed]

International Journal of
Molecular Sciences

Review

PPARs and Metabolic Disorders Associated with Challenged Adipose Tissue Plasticity

Patricia Corrales [1,*], Antonio Vidal-Puig [2,3] and Gema Medina-Gómez [1,*]

1 Área de Bioquímica y Biología Molecular, Departamento de Ciencias Básicas de la Salud, Facultad de Ciencias de la Salud, Universidad Rey Juan Carlos, Avda. de Atenas s/n. Alcorcón, 28922 Madrid, Spain
2 Metabolic Research Laboratories, Wellcome Trust MRC Institute of Metabolic Science, Addenbrooke's Hospital, University of Cambridge, Cambridge CB2 0QQ, UK; ajv22@medschl.cam.ac.uk
3 Wellcome Trust Sanger Institute, Hinxton, Cambridgeshire CB10 1SA, UK
* Correspondence: patricia.corrales@urjc.es (P.C.); gema.medina@urjc.es (G.M.-G.); Tel.: +34-91-4888632 (G.M.-G.)

Received: 19 June 2018; Accepted: 18 July 2018; Published: 21 July 2018

Abstract: Peroxisome proliferator-activated receptors (PPARs) are members of a family of nuclear hormone receptors that exert their transcriptional control on genes harboring PPAR-responsive regulatory elements (PPRE) in partnership with retinoid X receptors (RXR). The activation of PPARs coordinated by specific coactivators/repressors regulate networks of genes controlling diverse homeostatic processes involving inflammation, adipogenesis, lipid metabolism, glucose homeostasis, and insulin resistance. Defects in PPARs have been linked to lipodystrophy, obesity, and insulin resistance as a result of the impairment of adipose tissue expandability and functionality. PPARs can act as lipid sensors, and when optimally activated, can rewire many of the metabolic pathways typically disrupted in obesity leading to an improvement of metabolic homeostasis. PPARs also contribute to the homeostasis of adipose tissue under challenging physiological circumstances, such as pregnancy and aging. Given their potential pathogenic role and their therapeutic potential, the benefits of PPARs activation should not only be considered relevant in the context of energy balance-associated pathologies and insulin resistance but also as potential relevant targets in the context of diabetic pregnancy and changes in body composition and metabolic stress associated with aging. Here, we review the rationale for the optimization of PPAR activation under these conditions.

Keywords: PPAR; metabolism; adipose tissue; obesity; pregnancy; aging; caloric restriction

1. Introduction

Approximately 39% of the world's adult population is overweight and no less than 13% is obese. Obesity is currently the most prevalent chronic metabolic disorder and its current prevalence is predicted to triple by 2030 according to the World Health Organization (WHO). Beyond the obvious physical constraints and associated psychological stress, the main cause of morbimortality associated with obesity is its associated cardiometabolic metabolic pathologies, namely insulin resistance, dyslipidemia, and type 2 diabetes (T2D), a cluster of pathological entities globally designated as metabolic syndrome (MetS). Under normal circumstances, an excess of calories is considered advantageous for the organism as long as it is efficiently stored in the adipose tissue in the form of fat. However, excessive amounts of fat, beyond the available storing capacity of the adipose tissue (AT), or when accreted at a relatively fast pace, may overwhelm the functional capacity of the organ. When that happens, the excess of nutrients can, to a certain extent, be burnt, and/or alternatively be accumulated ectopically in other metabolically relevant organs, such as the liver, skeletal muscle, kidney, and pancreas—organs not purposely designed to be a main storage compartment. In these organs, as in the white AT (WAT), the excessive nutrient load induces metabolic stress causing lipid-related toxicity, a known cause for insulin resistance and inflammation [1,2].

Given these gloomy prospects, it has become increasingly necessary to identify pathogenic molecular mechanisms and diagnostic and prognostic biomarkers that can predict evolution and potential outcomes, as well as suitable therapeutic targets. Although metabolic syndrome has by definition different potential pathogenic entrances, we believe that given the relevance of its association with obesity, it is quite likely that in a high percentage of these predominantly obese patients, the dysfunction of their adipose tissue becomes a main contributor to subsequent associated complications. Peroxisome proliferator-activated receptors (PPARs) play important regulatory roles that control the homeostasis of the adipose tissue through the regulation of the balance between anabolic and oxidative processes. In this regard, we think that the PPARs associated with specific processes could be targeted, given their objective to beneficially improve insulin sensitivity, and that their agonists could be suitable candidates in the therapeutic arsenal to treat MetS.

PPARs are a group of ligand-activated nuclear hormone receptors. These transcription factors exist within a protein superfamily, which includes the receptors for retinoids, vitamin D, steroids, and thyroid hormones. These nuclear receptors bind to PPAR-responsive regulatory elements (PPRE) and heterodimerize with the retinoid X receptors (RXR), translocating to the nucleus where they contribute to transactivate and/or transrepress specific genes. In some respect, the PPARs are well placed to connect the environment represented by nutritional inputs [3,4] to specific genetic programs controlling genes involved in inflammation, adipogenesis, lipid metabolism, and glucose homeostasis [5].

There are three different isoforms of PPARs in mammals: PPARα, PPARβ/δ, and PPARγ. The three PPARs isoforms show structural similarities. However, despite their similarities, the isoforms exhibit differences in tissue distribution, ligand specificities, and functions. Recently, PPARs have been suggested to relate to the crossroads of obesity, diabetes, inflammation, and cancer [6]. Their topographic distribution and context-dependent regulation may be more important than the specific repertoire of genes they regulate, and collectively, they play an essential role in the maintenance of metabolic homeostasis [7–9].

PPARα is predominantly expressed in the liver and to a lesser extent in muscle, heart, bone, and brown adipose tissue (BAT), all of which are eminently prooxidative tissues rich in mitochondria content. In the liver, PPARα is activated under energy deprivation conditions. It is part of the adaptive response to fasting, and its main net contribution is to increase ATP production from β-oxidative phosphorylation, a process that requires coupling to the ancillary systems related to fatty acid transport and ketogenesis [10,11]. Moreover, the role of PPARα in controlling the expression of genes involved in lipid metabolism goes beyond its immediate effect of increasing energy availability in the liver, by also providing energy for supply to the peripheral tissues according to energetic demands in the heart, muscle, kidney, and brown AT during fasting. Through its prooxidative anti lipotoxic effects, PPARα ligands are successfully used therapeutically to treat primary and secondary forms of hypertriglyceridemia particularly associated with MetS [12,13].

PPARβ/δ is ubiquitously expressed but is particularly active in skeletal muscle, where it contributes to sustain the energy requirements for physical exercise by upregulating fatty acid β-oxidation specifically during fasting [14]. This PPAR isoform is also expressed in adipocytes and macrophages, where it reduces the expression of proinflammatory markers, such as nuclear factor kappa B (NF-κB), conferring the systemic anti-inflammatory activity of this isoform [15,16].

PPARγ is predominantly expressed in adipose tissues, both white and brown, where it plays an important anabolic role in facilitating fat storage, adipogenesis, and thermogenesis [17,18]. There are two main isoforms, PPARγ1 and PPARγ2, differentiated by an extra exon of 90 nucleotides in the end terminus of the γ2 isoform. PPARγ1 has a widespread distribution and seems to support a sort of housekeeping metabolic role, which is particularly relevant in the intestine, macrophages, the liver, muscle, pancreatic β-cells, bone, placenta, and adipose tissue. Conversely, the expression of the PPARγ2 isoform is restricted under physiological conditions to adipose tissues. However, under conditions such as long-term overnutrition or obesity, PPARγ2 is induced de novo in the liver and skeletal muscle, in parallel with the development of ectopic accumulation of lipids in these and other organs [19]. Both PPARγ isoforms contribute to the uptake of glucose and lipids, and when expressed ectopically, they promote safe deposition of lipids in peripheral tissues, such as the liver, muscle, and adipose tissue. When interpreting the role of PPARγ isoforms on the maintenance of energy homeostasis, it is important to consider the effect that the ectopic induction of PPARγ2 contributes to facilitating the reorganization of the inter-organ communication of nutrients and energy fluxes, which will help to understand how, when defective, it may lead to insulin resistance [20]. PPARγ has been heavily studied in part because the availability of its pharmacological agonists (TZDs) ligands, such as rosiglitazone and pioglitazone, both known to improve insulin resistance and exert anti-inflammatory effects in the adipose tissue [21–23] and on a systemic level. Such effects are potentially important in the treatment of obesity and T2D but also could have therapeutic value in physiological states, such as pregnancy and aging, characterized by insulin resistance and changes in body composition.

In this review, we summarize the contribution of PPARs to the maintenance of the adipose tissue physiology and discuss the pathogenic role mediated by dysfunctional PPARs in different contexts, characterized by defective adipose tissue expandability or functional failure associated with the development of insulin resistance and T2D, such as obesity, pregnancy, and aging.

2. Adipose Tissue Physiology and Lipotoxicity

Two main types of adipose tissue—white and brown adipose tissue (WAT and BAT, respectively)—exist. WAT is an endocrine organ that stores and mobilizes energy reserves as fat, whereas BAT uses lipids to produce heat by promoting uncoupled fatty acid oxidation converting nutrients in heat upon β-adrenergic stimulation or cold exposure. Both white and brown are necessary and contribute to maintain whole-body energy homeostasis [24].

Beyond its storage function, the WAT is an important endocrine organ responsible for synthesizing hormones, chemokines, and cytokines that modulate food intake, insulin sensitivity, or inflammation, which contribute to the maintenance of whole metabolism functionality [25,26]. In healthy conditions, the main function of the subcutaneous WAT is lipid storage of free fatty acids (FFAs) as triglycerides (TGs) in large unilocular droplets. However, in the context of chronic energy surplus leading to weight gain, the subcutaneous WAT adapts by increasing the cell size of the existing adipocytes (hypertrophy) and/or increasing the number through differentiation of new adipocytes (hyperplasia). Initially, this adaptation is sufficient to store lipids in the WAT, preventing them from ectopically accumulating in the liver or muscle. But, once adipose tissue storage capacity is exceeded above an individualized threshold, where the subcutaneous WAT cannot accommodate the excess of lipids, then these lipids are ectopically deposited in the liver, pancreas, muscle, kidney, and other important peripheral tissues. As obesity progresses the adipose tissue becomes inflamed and fibrotic, further contributing to the dysfunction of the AT. Both the failure to take upon lipids and to appropriately mobilize them decreases the metabolic flexibility of the WAT and exerts a knock-on effect on other organs leading to the development of metabolic abnormalities, such as dyslipidemia and peripheral insulin resistance [24,27].

In contrast to the unique, large lipid droplets of white adipocytes, BAT stores TGs in multilocular lipid droplets. This distribution of lipids in small vesicles helps to titrate the release of lipids destined to be oxidized by the mitochondria to produce heat through "uncoupling" of oxidative phosphorylation of FFAs stimulated by β-adrenergic sympathetic nervous system (SNS) typically observed in cold exposure. By oxidizing nutrients, BAT activation counteracts obesity, reduces TGs in the plasma, and reduces atherosclerosis development [28,29]. Furthermore, another type of thermogenic adipocytes can also be found interspersed in white fat depots, such as cells known as 'brite/beige' adipocytes [30]. This type of adipocytes appears to respond to thermogenic stimulation and, in principle, are expected to contribute to the regulation of body weight and the improvement of insulin resistance [31]; however, their functional relevance has not been clearly demonstrated so far.

Although there is a correlation between fat mass and insulin resistance, the adipose tissue mass by itself is not the main determinant factor linking obesity and insulin resistance. The relative mismatch between storage capacity and lipid load supply could be considered to be more relevant. By upregulating the capacity of adipose tissue to expand and store and/or by decreasing the supply of lipids, we may tweak this balance, preventing the mismatch. PPARs are key transcriptional regulators of this balance by contributing to increase the storage capacity of the adipose tissue and also, through their prooxidative effects, to decrease the demand for storage or the supply of lipids. A second layer of complexity comes from the knock-on positive effect, that by restoring the balance of the AT, it will exert an influence on the function of other metabolic tissues as a result of removing the toxic effects determined by ectopic fat deposition.

3. PPARs and Fat Mass Expansion and Function

The binding of ligands to PPARγ results in molecular changes, including the dissociation of co-repressors and the recruitment of co-activators, ultimately leading to changes in the coordinated expression of networks of genes functionally linked to adipogenesis, lipid metabolism, inflammation, thermogenesis, and body glucose homeostasis. PPARγ activation facilitates fat accretion and retains the functionality of adipose tissue by coordinating adipogenesis, fat transport, and lipolysis upon reaching an individualized threshold of adipose tissue mass. Defects in PPARγ, either in the form of mutants or secondary-to-decreased expression, compromises adipose tissue function, plasticity, and lipotoxicity. This is accompanied by the development of peripheral insulin resistance and ultimately global metabolic disruption. Restoration/maintenance of PPARγ functionality senses the lipid load and enables the recovery of the homeostasis of essential metabolic pathways (Figure 1).

PPARα is preferentially expressed in liver, where its activation is essential to generate energy, particularly under conditions of energy deprivation paradigms by promoting fatty acid uptake and oxidation [10]. This isoform exerts a pleiotropic effect controlling liver glucose metabolism, as observed in mice where PPARα activation by fibrates decreases expression levels of glucokinase [32] and suppresses pyruvate transformation to acetyl-CoA [33]. PPARα is also expressed in the WAT, where its activation elicits systemic effects in rodents by decreasing adiposity and ameliorating the insulin resistance in obese mouse models [34]. Moreover, it has been reported that treatment with PPAR-α agonists also increases the expression of adiponectin by the WAT [34,35] and decreases the tumor necrosis factor α (TNF-α) levels [36,37]. This anti-inflammatory effect in the WAT suggests PPARα activation has the capacity to improve insulin resistance and ameliorate obesity. In the BAT, PPARα exerts a thermogenic effect, cooperating with the peroxisome proliferator-activated receptor gamma coactivator 1-alpha (PGC1α) to control lipid oxidation and thermogenesis in response to β-adrenergic stimulation in response to cold exposure [38]. Moreover, activation of PPARα in obese mice increases energy expenditure and activates thermogenic pathways that facilitate weight loss [39]. Because of PPARα agonists' prooxidative actions, activators of this nuclear hormone receptor may be used to improve obesity-induced insulin resistance.

PPAR-β/δ in the BAT regulates the fatty acid oxidation [14,40] and the thermogenic response contributing to the induction of the uncoupling protein 1 (Ucp1) expression and leading to the reduction of the WAT mass [40]. Moreover, this isoform may also have an anti-inflammatory effect when activated [41,42]. The metabolic function of PPAR-β/δ in the WAT has been much less studied, although it is known that this isoform facilitates preadipocyte differentiation [43]. Nowadays, PPAR-β/δ activators are under study for their clinical advantages in treating obesity.

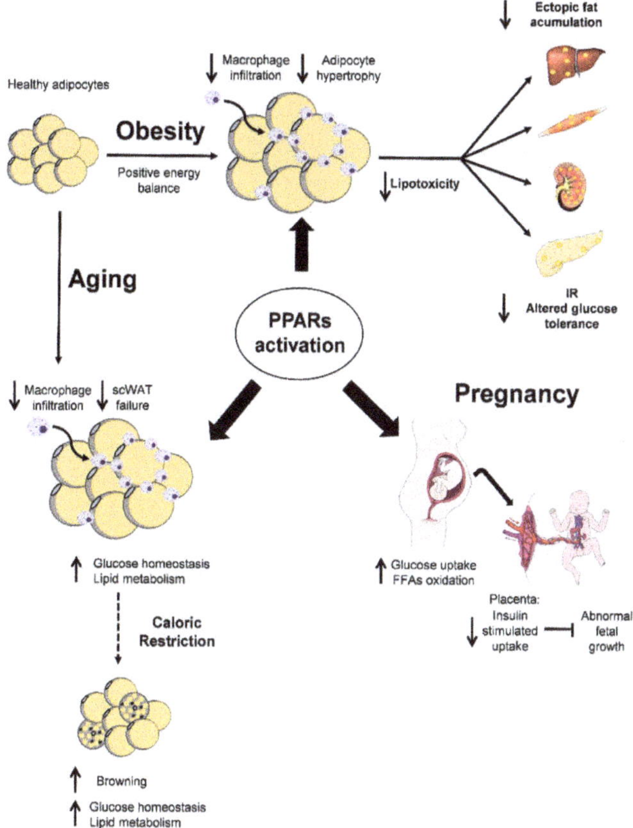

Figure 1. Overview of the effects of peroxisome proliferator-activated receptors (PPARs) activation in obesity, pregnancy, and aging. In obesity, PPARs activation decreases fibro-inflammation and ectopic fat accumulation in the adipose tissue (AT). In pregnancy, PPARs activation stimulates glucose uptake and fatty acid oxidation in the mother, while the placenta decreases insulin uptake to ameliorate abnormal fetal growth. In aging, this PPARs activation increases the glucose uptake and lipid metabolism in the subcutaneous white AT (scWAT). Moreover, both PPARs activation and caloric restriction (CR, dotted arrow) in aging promote browning in the AT, which improves the whole-body metabolism. The final effect of PPARs activation in all situations is the improvement of insulin resistance (IR). ➡ PPARs activation; ⟶. effects or PPARs activation; - - -▸ CR effect; ⊣ inhibition; ↑ increase or ↓ decrease effect.

PPARγ is expressed predominantly in the adipose tissues, where it acts as sensor of lipids, hormones, vitamins, and endogenous metabolites. This isoform is an important regulator of adipocyte differentiation, fat storage as triglyceride, and energy homeostasis. Both isoforms of PPARγ, PPARγ1 and PPARγ2, are necessary for the adipogenic function, and alteration in their expression increases susceptibility to lipodystrophy, insulin resistance, and T2D [44]. PPARγ is necessary for fat cell differentiation in all adipose depots and contributes to define the maximum threshold of expansion of the WAT. This is supported by studies showing that ectopic presence of PPARγ in non-adipogenic cells trans-differentiates them into mature adipocytes [45,46]. Moreover, PPARγ-deficient mice cannot develop adipose tissue [47,48]. Mice with PPARγ knockout in mature cells also develop insulin resistance and hyperlipidemia through dysregulation of molecular pathways of insulin signaling,

FFAs uptake, and lipolysis [49]. Specific knockdown of the PPARγ2 isoform in mice results in adipose tissue dysfunction and insulin resistance [2]. Moreover, when the adipose tissue of this model is challenged with increased lipid supply, as characteristically observed in a leptin-deficient obese (*ob/ob*) background (POKO mouse [50]), these mice are precociously more insulin resistant, as young as 4 weeks, an age where the differences in fat mass in comparison with an *ob/ob* mice are not well established. In addition, the POKO mice became diabetic and hyperlipidaemic at 16 weeks of age, despite weighing less and having less fat than an *ob/ob* mouse at that age, with increased toxic reactive lipid species in different tissues, behaving like a mouse model of lipotoxicity and metabolic syndrome [51]. This reinforces the concept that it is not the absolute amount of fat mass, but the mismatch between nutrient supply and storage capacity that results in dysfunctional adipose tissue and metabolic stress. Furthermore, in states of obesity, the expression of PPARγ decreases with the consequent induction of a high grade of inflammation, angiogenesis, and fibrosis in the WAT [52] and low levels of adiponectin, which limits the adipose tissue expansion. Consistent with this, patients with mutations of PPARγ develop lipodystrophy and insulin resistance [53]. Conversely, increased expression of PPARγ protects from the insulin resistance associated with obesity [54].

TZDs, the pharmacological agonists of PPARγ, have been used clinically as antidiabetic agents, and their beneficial effects are well documented in relation with insulin resistance and obesity. The activation of PPARγ by TZDs in the WAT improves WAT expansion, alleviates peripheral lipotoxicity and normalizes adipokine secretion [24]. This activation improves the WAT's ability to store lipids and reduces lipotoxicity in the liver and muscle by the activation of metabolic pathways implicated in FFA oxidation. The metabolic effects include lower levels of TGs in circulation, in the liver, and muscle, coupled with increased TGs in the adipose tissue [52]. The expression of TNF-α is also inhibited using TZDs [55]. Furthermore, PPARγ stimulates adiponectin production in the WAT, which contributes to further stimulating FFA oxidation, reducing hepatic glucose, and increasing the use of glucose by muscle [56]. Recently, it has been shown that PPARγ activated by TZDs promotes the expression of the fibroblast growth factor family (FGF1 and FGF21) showing the key role of the PPARγ–FGF axis, which contributes to the remodeling of the adipose tissue and the maintenance of metabolic homeostasis [57,58]. Thus, the result of pharmacological intervention in obesity with TZDs is the improvement of insulin sensitivity derived from the effects of TZDs improving adipose tissue function despite the associated increase in fat mass.

TZDs can promote browning in the WAT. Activating PPARγ [59] increases the expression of BAT-specific genes, such as *Ucp-1* and *Prdm16* [60], via *Sirt1* [61], priming the oxidative capacity of the adipose tissue through its transformation into brown-like adipocytes. These effects confer thermogenic properties by promoting mitochondrial biogenesis in the WAT, which can help in the remodeling of the adipose tissue and insulin resistance improvement under obesity conditions.

Therefore, the role of PPARγ improving glucose metabolism and insulin sensitivity is well established and provides insights into the molecular regulation of adipose tissue expansion in normal and obese/lipodystrophy pathological states but also in other situations in which the adipose tissue is physiologically stressed, such as pregnancy and aging.

4. PPARs and Pregnancy

Pregnancy involves hormonal and metabolic adaptations that directly affect maternal adiposity. In the early stages of pregnancy, the adipose tissue mass expands due to an increase in the lipid accumulation (known as a primarily anabolic phase). There is an increase in lipid synthesis and fat storage that prepare the mother's metabolism for the prospective increase in fetal energy needs at a later phase. This increase in lipid/energy supply is enabled by maternal hyperphagia and improved insulin sensitivity, which stimulates FFAs synthesis in adipocytes and the uptake of FFAs from circulating TGs for preferential accumulation in the adipose tissue. Moreover, the production of hormones, such as progesterone, cortisol, and leptin, also contribute to facilitated fat storage and adipocyte hypertrophy. However, in contrast to this early anabolic phase of gestation, the adipose mass decreases in a later phase (known as a net catabolic phase). During the late phase of gestation, IR and a low grade of inflammation, especially in the adipose tissue, are developed, which should be considered as a physiological adaptation. Moreover, the decrease in insulin sensitivity enhances lipolysis, helping to mobilize the stored TGs. The human placental lactogen (Hpl) also stimulates lipolysis in adipocytes, coupled with the decrease in FFAs uptake from TGs in the plasma. The net result of these changes is a reduction in the adipose tissue mass and an increased glucose flux from mother to fetus. Although the physiological IR developed in the late phase of pregnancy is well documented, the mechanisms causing the changes in the adipose tissue and insulin resistance during pregnancy are still unclear.

Together with the exacerbated insulin resistance, insulin secretion may also become inadequate to meet the increased demands in the late stage of pregnancy, leading to gestational diabetes mellitus (GDM). Moreover, their offspring have an increased risk of perinatal complications, obesity, and diabetes in adulthood [62]. GDM is defined as glucose intolerance on first recognition during pregnancy [63] and characteristically shows altered plasma adipokine levels, inflammation, deregulation of the insulin signaling pathway, and oxidative stress [64–66]. The mechanisms underlying the GDM are not fully understood; however, it is known the association between inappropriate PPARγ function/levels and GDM through its function in both the adipose tissue and the placenta [67,68].

During early pregnancy, as in obesity, the mechanisms leading to energy storage and adipose tissue expansion are activated. It has also been reported that in advanced pregnancy, PPARγ declines, thus accelerating adipose tissue insulin resistance and facilitating lipolysis in the subcutaneous adipose tissue of obese pregnant women with GDM [67]. Moreover, PPARα and PPARβ/δ expression also decreases in adipose tissue from obese pregnant women and/or women with GDM [69]. In pregnant mouse models, the association between a decrease in PPARγ expression, exacerbated lipolysis in the AT [70], and the subcutaneous AT dysfunction has been reported. In agreement with this, we have shown that genetic ablation of PPARγ2 in pregnant mice is associated with poor AT expandability and the worsening of insulin resistance [71]. The contribution of PPARγ2 is also important for the process of pancreatic β-cell mass expansion and adaptation in murine models of MetS [50,71,72]. A missing study is the PPARγ deleted specifically in the pancreatic β-cell in order to study the mechanisms implicated in its adaptation when pregnancy occurs. Furthermore, it has been reported that the use of PPARγ agonists reverse the insulin resistance associated with late pregnancy in murine models [65]. For this reason, more studies are necessary to elucidate the potential of PPARγ agonism to overcome defects in pregnancy related to insulin resistance and GDM.

The role of PPARs in the placenta is potentially important. The placenta is an endocrine gland that synthesizes the peptides and steroid hormones during pregnancy that are essential for the maintenance of mother and fetus. Regarding their roles, PPARα null female mice become diabetic during pregnancy and have an increased risk of spontaneous abortion [73]. PPARδ also has a relevant role in embryonic, decidual, and placental function [74]. Moreover, PPARδ and PPARγ null mice are not viable and exhibit a failure in the development of the placenta [75]. PPARγ is downregulated in the placenta in human patients with GDM. PPARγ has anti-inflammatory effects in the placenta and modulates embryogenesis, implantation, trophoblast invasion, and maternal spiral artery transformation [76]. Moreover, it has been demonstrated that reduced expression of PPARγ in placental tissues and serum is contributing to the development of preeclampsia, a specific pregnancy disorder in humans that contributes to maternal mortality [77]. However, in a mice study, PPARγ expression was upregulated in the placentas of diabetic pregnant mice [78]. These contradictory effects may lead to specie-specific effects and would need to be further elucidated to be used to improve the GDM in pregnancy.

5. PPARs and Aging

Aging is a complex and multifactorial progressive physiological decay process, associated with an increased risk of metabolic disorders, such as obesity, insulin resistance, and other manifestations related to MetS, which are exacerbated by age. Moreover, aging is associated with changes in body composition characterized by increased total adiposity and topographical redistribution of adipose tissue defined by preferential loss of the subcutaneous WAT coupled with expansion of adipose in the visceral compartment [79,80]. Accretion of visceral, rather than subcutaneous WAT has been associated with the development of insulin resistance. The expansion of the intraabdominal adipose tissue may also be considered another example of peripheral lipotoxicity determined by a primary defect in the subcutaneous WAT. An important concept is the fact that as we age, the adipose tissue ages as other tissues, such as muscle or the liver, do. When muscle becomes frail/sarcopenic, with decreased oxidative capacity, lipids are redirected to the adipose tissue for storage precisely at a time when the adipose tissue itself has aged and is less functional and competent to deal with increasing metabolic challenges. In this regard the adipose tissue of the aged individual is subjected to even more lipid load, increasing the chances of AT dysfunction and peripheral lipotoxicity. In fact, ectopic fat accumulation in muscle is observed in lean elder individuals.

This conflict between increased storage demand and age-related decay results in increased adiposity coupled with macrophage infiltration and inflammation that interferes with insulin signaling. Thus, a key factor determining the shift of the AT towards the inflammatory state is the mismatch between the demand for storage and capacity [81]. Moreover, as inflammatory cells can be high-level producers of selective fibrotic molecules [82], the old adipose tissue characteristically shows age-related fibro-inflammatory changes. Fibrosis is a disease process that deposits collagen-rich extracellular matrix (ECM) in an attempt to remodel and repair the tissue morphology and organ functionality of a failing organ. Transforming growth factor β (TGF-β) plays a key role in fibrosis, modulating the balance between the rate of synthesis and the degradation of matrix collagen proteins. It has been reported that a role for PPARγ is as a potent antifibrotic factor in the kidney [83]. In the AT, it has also been reported that TGF-β, apart from increasing collagen deposition, also increases mechanical stress on the adipocyte membrane and the rigidity of AT, compromising its further expansion. This rigid matrix can result in cell death by necrosis. In this situation, PPARγ agonists decrease collagen levels in the AT and confers a more flexible environment for the adipocyte growth and remodeling [84]. Thus, we could speculate that age-related defects in the adipose tissue remodeling may also contribute, particularly in the context of obesity, to exacerbated pathological conditions linked to insulin resistance [85].

Based on this evidence it is conceivable that age-related changes in PPARs may contribute to some of these pathological changes. However, at present, there is a paucity of information about how PPARs activation may contribute to delay or ameliorate these pathological changes.

Given the importance of PPARγ in adipose tissue biology, it is important to determine the contribution of PPARγ dysfunction in aging-associated metabolic decline. It is well documented, the role of PPARγ in coordinating gene expression programs of adipocyte differentiation, lipid storage, and lipolysis. Previous reports also suggest that PPARγ deficiency selectively in the subcutaneous AT during aging is associated with increased AT expansion that is associated with the development of insulin resistance [86]. Moreover, other studies have shown that oxidative stress and reactive oxygen species (ROS) production (characteristically observed in aging and linked to insulin resistance in adipocytes) modulate other proinflammatory pathways, linking PPARγ dysfunction with inflammation [87], such as NF-κB [88], likely to contribute to altered tissue expansion and inflammation and associated age-related insulin resistance. In addition, mitochondria, as the core organelle required to maintain cellular functionality and glucose and lipid homeostasis [89], have been suggested as key contributors to adipocyte formation through ROS signaling [90]. In aging, as in obesity, mitochondrial enzyme expression is reduced in the AT leading to decreased oxygen consumption and oxidative phosphorylation [91,92] in response to lipid overload, usually coupled with decreased AT insulin sensitivity [93]. Interventions with TZDs induce mitochondrial biogenesis, ROS, and remodeling in the AT, enhancing fatty acid oxidation and oxygen consumption, which seems to contribute to changes in the whole-body energy metabolism and insulin sensitivity [94]. Moreover, insulin resistance in elderly patient populations has been associated with decreases in mitochondrial oxidative phosphorylation [95]; however, further work is required to identify the mediators of nuclear-encoded mitochondrial genes induced by PPARγ ligand-dependent mechanisms that could be helpful during aging.

Furthermore, both PPARα and PPARγ decrease in the kidney with aging. This is correlated with accelerated oxidative stress and counteracted by the antioxidative action of caloric restriction [96]. All these studies suggest that defective PPARs are important for the defects in energy expenditure and lipogenic function leading to lipid accumulation in the whole body during aging. From this, it is conceivable that targeting these isoforms could be a helpful approach to reduce or prevent age-associated metabolic decline and protect from lipid accumulation and lipotoxicity. It could be speculated that TZDs treatment may help to reduce the side effects of weight gain in the elderly by minimal/function specific PPARγ activation stimulating insulin sensitivity without promoting adipogenesis [97]. But, in any case, it would be important to overcome some of the negative effects observed when using TZDs in the elderly. Amongst them is the increased bone marrow adiposity and reduced bone formation, resulting in osteopenia, bone fracture, and other complications [98]. Moreover, the use of some PPAR agonists had to be withdrawn from the United States (US) and European markets because of associated complications, such as edema, weight gain, macular oedema, heart failure, and bladder cancer, that have been associated side effects [99]. These are important drawbacks but potentially addressable with increased knowledge of the specific PPARα dependent pathways mediating them.

An added value of the PPARα agonists occurs in the muscle. Old age is associated with dyslipidemia, which together with the increase and dysfunction of the AT can lead to preferential deposition of TGs in skeletal muscle, ultimately leading to IR. Agonists of PPARα and PPARβ/δ have been used to treat dyslipidemia by increasing oxidative capacity in muscle fibers and improving insulin sensitivity [100]. Furthermore, TZDs as specific activators of PPARγ are used as insulin sensitizers and as regulators of FFAs storage, which may prevent intramuscular lipid accumulation [80] and maintain skeletal muscle insulin action [101].

Caloric restriction (CR) is another therapeutic paradigm representing a non-pharmacological intervention to efficiently delay the deleterious effects of age-related metabolic diseases [102]. Previous studies in animal models have shown that CR exerts physiological effects leading to reduced body weight and glucose and insulin serum levels [103]. Moreover, the reduction of adiposity by CR [104] or fat removal [105] have been demonstrated to ameliorate age-associated insulin resistance. Of note, CR alters the expression of genes that are regulated by PPARs and that are involved in lipid metabolism and insulin signaling. In some ways, the beneficial effects derived from fasting may be mediated, at least in part, by these nuclear receptors [106]. The effects of CR in the AT on PPARα and PPAR-β/δ have not been shown yet, but it is known that CR and PPARγ agonists can improve the reduced mitochondrial function in the WAT due to aging and obesity [107]. Moreover, it has been reported that CR induces BAT functionality [102], and it has been speculated that the induction of a brown fat phenotype in the WAT by CR or PPARγ agonists would result in an increased mitochondrial functionality with beneficial effects on aging and metabolism. Due to the improvement in aging conditions by using PPARs agonists, more studies are needed to document the role of PPARs on adipose tissue plasticity during aging.

6. Conclusions

Defective adipose tissue synergizes with the age-related metabolic defects to exacerbate metabolic diseases. Thus, the understanding of cellular mechanisms governing the plasticity of adipose tissue should help to understand and provide therapeutic rationale to address metabolic disorders. Understanding the molecular alterations that determine the impaired adipose tissue plasticity may identify therapeutic targets to optimize AT expandability and function. Thus, PPARs should be considered candidates to improve age-related metabolism through their influence in the balance between anabolic and catabolic processes and by limiting unwanted inflammatory reactions that may compromise lipid and glucose homeostasis. Thus, PPARs may have the clue to restore, delay, and improve the metabolic balance in those conditions that render someone particularly susceptible to developing insulin resistance, such as obesity, pregnancy and aging.

Author Contributions: P.C., A.V.-P., and G.M.-G. wrote the paper.

Acknowledgments: Research conducted for this publication was supported by Ministerio de Economía y Competitividad (BFU2013-47384-R and BFU2016-78951-R) and Comunidad de Madrid (S2010/BMD-2423 and B2017/BMD-3684).

Conflicts of Interest: The authors declare no conflict of interest.

Abbreviations

T2D	Type 2 diabetes
MetS	Metabolic Syndrome
AT	Adipose tissue
PPARs	Peroxisome proliferator-activated receptors
PPRE	Ppar response element
RXR	Retinoid X receptor
NF-κB	Nuclear factor kappa B
TZDs	Thiazolidenidiones
WAT	White adipose tissue
BAT	Brown adipose tissue
FFAs	Free fatty acids
TGs	Triglycerides
TNF-α	Tumor necrosis factor α
PGC1α	Peroxisome proliferator-activated receptor gamma coactivator 1-alpha
Ucp-1	Uncoupling protein 1
GDM	Gestational diabetes mellitus
CR	Caloric restriction

References

1. Virtue, S.; Vidal-Puig, A. Adipose tissue expandability, lipotoxicity and the Metabolic Syndrome—An allostatic perspective. *Biochim. Biophys. Acta Mol. Cell Biol. Lipids* **2010**, *1801*, 338–349. [CrossRef] [PubMed]
2. Medina-Gomez, G.; Gray, S.L.; Yetukuri, L.; Shimomura, K.; Virtue, S.; Campbell, M.; Curtis, R.K.; Jimenez-Linan, M.; Blount, M.; Yeo, G.S.H.; et al. PPAR gamma 2 Prevents Lipotoxicity by Controlling Adipose Tissue Expandability and Peripheral Lipid Metabolism. *PLoS Genet.* **2007**, *3*, e64. [CrossRef] [PubMed]
3. Venteclef, N.; Jakobsson, T.; Steffensen, K.R.; Treuter, E. Metabolic nuclear receptor signaling and the inflammatory acute phase response. *Trends Endocrinol. Metab.* **2011**, *22*, 333–343. [CrossRef] [PubMed]
4. Michalik, L.; Wahli, W. Involvement of PPAR nuclear receptors in tissue injury and wound repair. *J. Clin. Investig.* **2006**, *116*, 598–606. [CrossRef] [PubMed]
5. Evans, R.M.; Barish, G.D.; Wang, Y.X. PPARs and the complex journey to obesity. *Nat. Med.* **2004**, *10*, 355–361. [CrossRef] [PubMed]
6. Vitale, S.G.; Laganà, A.S.; Nigro, A.; La Rosa, V.L.; Rossetti, P.; Rapisarda, A.M.C.; La Vignera, S.; Condorelli, R.A.; Corrado, F.; Buscema, M.; et al. Peroxisome Proliferator-Activated Receptor Modulation during Metabolic Diseases and Cancers: Master and Minions. *PPAR Res.* **2016**, *2016*, 6517313. [CrossRef] [PubMed]
7. Dubrac, S.; Stoitzner, P.; Pirkebner, D.; Elentner, A.; Schoonjans, K.; Auwerx, J.; Saeland, S.; Hengster, P.; Fritsch, P.; Romani, N.; et al. Peroxisome proliferator-activated receptor-alpha activation inhibits Langerhans cell function. *J. Immunol.* **2007**, *178*, 4362–4372. [CrossRef] [PubMed]
8. Dalen, K.T.; Schoonjans, K.; Ulven, S.M.; Weedon-Fekjaer, M.S.; Bentzen, T.G.; Kontnikova, H.; Auwerx, J.; Nebb, H.I. Adipose Tissue Expression of the Lipid Droplet-Associating Proteins31 S–S2 and Perilipin Is Controlled by Peroxisome Proliferator-Activated Receptor-γ. *Diabetes* **2004**, *53*, 1243–1252. [CrossRef] [PubMed]

9. Moore, K.J.; Rosen, E.D.; Fitzgerald, M.L.; Randow, F.; Andersson, L.P.; Altshuler, D.; Milstone, D.S.; Mortensen, R.M.; Spiegelman, B.M.; Freeman, M.W. The role of PPAR-gamma in macrophage differentiation and cholesterol uptake. *Nat. Med.* **2001**, *7*, 41–47. [CrossRef] [PubMed]
10. Kersten, S.; Seydoux, J.; Peters, J.M.; Gonzalez, F.J.; Desvergne, B.; Wrahli, W. Peroxisome proliferator-activated receptor alpha mediates the adaptive response to fasting. *J. Clin. Investig.* **1999**, *103*, 1489–1498. [CrossRef] [PubMed]
11. Gross, B.; Pawlak, M.; Lefebvre, P.; Staels, B. PPARs in obesity-induced T2DM, dyslipidaemia and NAFLD. *Nat. Rev. Endocrinol.* **2017**, *13*, 36–49. [CrossRef] [PubMed]
12. Lalloyer, F.; Wouters, K.; Baron, M.; Caron, S.; Vallez, E.; Vanhoutte, J.; Baugé, E.; Shiri-Sverdlov, R.; Hofker, M.; Staels, B.; et al. Peroxisome proliferator-activated receptor-α gene level differently affects lipid metabolism and inflammation in apolipoprotein E2 knock-in mice. *Arterioscler. Thromb. Vasc. Biol.* **2011**, *31*, 1573–1579. [CrossRef] [PubMed]
13. Staels, B.; Maes, M.; Zambon, A. Fibrates and future PPARalpha agonists in the treatment of cardiovascular disease. *Nat. Clin. Pract. Cardiovasc. Med.* **2008**, *5*, 542–553. [CrossRef] [PubMed]
14. Holst, D.; Luquet, S.; Nogueira, V.; Kristiansen, K.; Leverve, X.; Grimaldi, P.A. Nutritional regulation and role of peroxisome proliferator-activated receptor δ in fatty acid catabolism in skeletal muscle. *Biochim. Biophys. Acta Mol. Cell Biol. Lipids* **2003**, *1633*, 43–50. [CrossRef]
15. Zoete, V.; Grosdidier, A.; Michielin, O. Peroxisome proliferator-activated receptor structures: Ligand specificity, molecular switch and interactions with regulators. *Biochim. Biophys. Acta Mol. Cell Biol. Lipids* **2007**, *1771*, 915–925. [CrossRef] [PubMed]
16. Zingarelli, B.; Piraino, G.; Hake, P.W.; O'Connor, M.; Denenberg, A.; Fan, H.; Cook, J.A. Peroxisome Proliferator-Activated Receptor δ Regulates Inflammation via NF-κB Signaling in Polymicrobial Sepsis. *Am. J. Pathol.* **2010**, *177*, 1834–1847. [CrossRef] [PubMed]
17. Siersbaek, R.; Nielsen, R.; Mandrup, S. PPARgamma in adipocyte differentiation and metabolism—Novel insights from genome-wide studies. *FEBS Lett.* **2010**, *584*, 3242–3249. [CrossRef] [PubMed]
18. Ferre, P. The Biology of Peroxisome Proliferator-Activated Receptors: Relationship With Lipid Metabolism and Insulin Sensitivity. *Diabetes* **2004**, *53*, S43–S50. [CrossRef] [PubMed]
19. Medina-Gomez, G.; Virtue, S.; Lelliott, C.; Boiani, R.; Campbell, M.; Christodoulides, C.; Perrin, C.; Jimenez-Linan, M.; Blount, M.; Dixon, J.; et al. The link between nutritional status and insulin sensitivity is dependent on the adipocyte-specific peroxisome proliferator-activated receptor-gamma2 isoform. *Diabetes* **2005**, *54*, 1706–1716. [CrossRef] [PubMed]
20. Tontonoz, P.; Spiegelman, B.M. Fat and Beyond: The Diverse Biology of PPARγ. *Annu. Rev. Biochem.* **2008**, *77*, 289–312. [CrossRef] [PubMed]
21. Kubota, N.; Terauchi, Y.; Kubota, T.; Kumagai, H.; Itoh, S.; Satoh, H.; Yano, W.; Ogata, H.; Tokuyama, K.; Takamoto, I.; et al. Pioglitazone ameliorates insulin resistance and diabetes by both adiponectin-dependent and -independent pathways. *J. Biol. Chem.* **2006**, *281*, 8748–8755. [CrossRef] [PubMed]
22. Olefsky, J.M. Treatment of insulin resistance with peroxisome proliferator–activated receptor γ agonists. *J. Clin. Investig.* **2000**, *106*, 467–472. [CrossRef] [PubMed]
23. Odegaard, J.I.; Ricardo-Gonzalez, R.R.; Goforth, M.H.; Morel, C.R.; Subramanian, V.; Mukundan, L.; Red Eagle, A.; Vats, D.; Brombacher, F.; Ferrante, A.W.; et al. Macrophage-specific PPARgamma controls alternative activation and improves insulin resistance. *Nature* **2007**, *447*, 1116–1120. [CrossRef] [PubMed]
24. Pellegrinelli, V.; Carobbio, S.; Vidal-Puig, A. Adipose tissue plasticity: How fat depots respond differently to pathophysiological cues. *Diabetologia* **2016**, *59*, 1075–1088. [CrossRef] [PubMed]
25. Coelho, M.; Oliveira, T.; Fernandes, R. Biochemistry of adipose tissue: An endocrine organ. *Arch. Med. Sci.* **2013**, *9*, 191–200. [CrossRef] [PubMed]
26. Kershaw, E.E.; Flier, J.S. Adipose tissue as an endocrine organ. *J. Clin. Endocrinol. Metab.* **2004**, *89*, 2548–2556. [CrossRef] [PubMed]
27. Brestoff, J.R.; Artis, D. Immune regulation of metabolic homeostasis in health and disease. *Cell* **2015**, *161*, 146–160. [CrossRef] [PubMed]
28. Yoneshiro, T.; Aita, S.; Matsushita, M.; Kayahara, T.; Kameya, T.; Kawai, Y.; Iwanaga, T.; Saito, M. Recruited brown adipose tissue as an antiobesity agent in humans. *J. Clin. Investig.* **2013**, *123*, 3404–3408. [CrossRef] [PubMed]

29. Berbée, J.F.P.; Boon, M.R.; Khedoe, P.P.S.J.; Bartelt, A.; Schlein, C.; Worthmann, A.; Kooijman, S.; Hoeke, G.; Mol, I.M.; John, C.; et al. Brown fat activation reduces hypercholesterolaemia and protects from atherosclerosis development. *Nat. Commun.* **2015**, *6*, 6356. [CrossRef] [PubMed]
30. Pisani, D.F.; Barquissau, V.; Chambard, J.-C.; Beuzelin, D.; Ghandour, R.A.; Giroud, M.; Mairal, A.; Pagnotta, S.; Cinti, S.; Langin, D.; et al. Mitochondrial fission is associated with UCP1 activity in human brite/beige adipocytes. *Mol. Metab.* **2018**, *7*, 35–44. [CrossRef] [PubMed]
31. Bartelt, A.; Bruns, O.T.; Reimer, R.; Hohenberg, H.; Ittrich, H.; Peldschus, K.; Kaul, M.G.; Tromsdorf, U.I.; Weller, H.; Waurisch, C.; et al. Brown adipose tissue activity controls triglyceride clearance. *Nat. Med.* **2011**, *17*, 200–205. [CrossRef] [PubMed]
32. Oosterveer, M.H.; Grefhorst, A.; van Dijk, T.H.; Havinga, R.; Staels, B.; Kuipers, F.; Groen, A.K.; Reijngoud, D.-J. Fenofibrate simultaneously induces hepatic fatty acid oxidation, synthesis, and elongation in mice. *J. Biol. Chem.* **2009**, *284*, 34036–34044. [CrossRef] [PubMed]
33. Wu, P.; Peters, J.M.; Harris, R.A. Adaptive increase in pyruvate dehydrogenase kinase 4 during starvation is mediated by peroxisome proliferator-activated receptor alpha. *Biochem. Biophys. Res. Commun.* **2001**, *287*, 391–396. [CrossRef] [PubMed]
34. Tsuchida, A.; Yamauchi, T.; Takekawa, S.; Hada, Y.; Ito, Y.; Maki, T.; Kadowaki, T. Peroxisome proliferator-activated receptor (PPAR)alpha activation increases adiponectin receptors and reduces obesity-related inflammation in adipose tissue: Comparison of activation of PPARalpha, PPARgamma, and their combination. *Diabetes* **2005**, *54*, 3358–3370. [CrossRef] [PubMed]
35. Veiga, F.M.S.; Graus-Nunes, F.; Rachid, T.L.; Barreto, A.B.; Mandarim-de-Lacerda, C.A.; Souza-Mello, V. Anti-obesogenic effects of WY14643 (PPAR-alpha agonist): Hepatic mitochondrial enhancement and suppressed lipogenic pathway in diet-induced obese mice. *Biochimie* **2017**, *140*, 106–116. [CrossRef] [PubMed]
36. Kleemann, R.; Gervois, P.P.; Verschuren, L.; Staels, B.; Princen, H.M.G.; Kooistra, T. Fibrates down-regulate IL-1-stimulated C-reactive protein gene expression in hepatocytes by reducing nuclear p50-NFkappa B-C/EBP-beta complex formation. *Blood* **2003**, *101*, 545–551. [CrossRef] [PubMed]
37. Wang, W.; Lin, Q.; Lin, R.; Zhang, J.; Ren, F.; Zhang, J.; Ji, M.; Li, Y. PPARα agonist fenofibrate attenuates TNF-α-induced CD40 expression in 3T3-L1 adipocytes via the SIRT1-dependent signaling pathway. *Exp. Cell Res.* **2013**, *319*, 1523–1533. [CrossRef] [PubMed]
38. Hondares, E.; Rosell, M.; Díaz-Delfín, J.; Olmos, Y.; Monsalve, M.; Iglesias, R.; Villarroya, F.; Giralt, M. Peroxisome proliferator-activated receptor α (PPARα) induces PPARγ coactivator 1α (PGC-1α) gene expression and contributes to thermogenic activation of brown fat: Involvement of PRDM16. *J. Biol. Chem.* **2011**, *286*, 43112–43122. [CrossRef] [PubMed]
39. Rachid, T.L.; Penna-de-Carvalho, A.; Bringhenti, I.; Aguila, M.B.; Mandarim-de-Lacerda, C.A.; Souza-Mello, V. PPAR-α agonist elicits metabolically active brown adipocytes and weight loss in diet-induced obese mice. *Cell Biochem. Funct.* **2015**, *33*, 249–256. [CrossRef] [PubMed]
40. Wang, Y.-X.; Lee, C.-H.; Tiep, S.; Yu, R.T.; Ham, J.; Kang, H.; Evans, R.M. Peroxisome-proliferator-activated receptor delta activates fat metabolism to prevent obesity. *Cell* **2003**, *113*, 159–170. [CrossRef]
41. Shearer, B.G.; Steger, D.J.; Way, J.M.; Stanley, T.B.; Lobe, D.C.; Grillot, D.A.; Iannone, M.A.; Lazar, M.A.; Willson, T.M.; Billin, A.N. Identification and characterization of a selective peroxisome proliferator-activated receptor beta/delta (NR1C2) antagonist. *Mol. Endocrinol.* **2008**, *22*, 523–529. [CrossRef] [PubMed]
42. Kang, K.; Reilly, S.M.; Karabacak, V.; Gangl, M.R.; Fitzgerald, K.; Hatano, B.; Lee, C.-H. Adipocyte-derived Th2 cytokines and myeloid PPARdelta regulate macrophage polarization and insulin sensitivity. *Cell Metab.* **2008**, *7*, 485–495. [CrossRef] [PubMed]
43. Hansen, J.B.; Zhang, H.; Rasmussen, T.H.; Petersen, R.K.; Flindt, E.N.; Kristiansen, K. Peroxisome proliferator-activated receptor delta (PPARdelta)-mediated regulation of preadipocyte proliferation and gene expression is dependent on cAMP signaling. *J. Biol. Chem.* **2001**, *276*, 3175–3182. [CrossRef] [PubMed]
44. Majithia, A.R.; Tsuda, B.; Agostini, M.; Gnanapradeepan, K.; Rice, R.; Peloso, G.; Patel, K.A.; Zhang, X.; Broekema, M.F.; Patterson, N.; et al. Prospective functional classification of all possible missense variants in PPARG. *Nat. Genet.* **2016**, *48*, 1570–1575. [CrossRef] [PubMed]
45. Tang, W.; Zeve, D.; Suh, J.M.; Bosnakovski, D.; Kyba, M.; Hammer, R.E.; Tallquist, M.D.; Graff, J.M. White Fat Progenitor Cells Reside in the Adipose Vasculature. *Science* **2008**, *322*, 583–586. [CrossRef] [PubMed]
46. Tontonoz, P.; Hu, E.; Spiegelman, B.M. Stimulation of adipogenesis in fibroblasts by PPAR gamma 2, a lipid-activated transcription factor. *Cell* **1994**, *79*, 1147–1156. [CrossRef]

47. Rosen, E.D.; Sarraf, P.; Troy, A.E.; Bradwin, G.; Moore, K.; Milstone, D.S.; Spiegelman, B.M.; Mortensen, R.M. PPAR gamma is required for the differentiation of adipose tissue in vivo and in vitro. *Mol. Cell* **1999**, *4*, 611–617. [CrossRef]
48. Semple, R.K.; Meirhaeghe, A.; Vidal-Puig, A.J.; Schwabe, J.W.R.; Wiggins, D.; Gibbons, G.F.; Gurnell, M.; Chatterjee, V.K.K.; O'Rahilly, S. A dominant negative human peroxisome proliferator-activated receptor (PPAR){alpha} is a constitutive transcriptional corepressor and inhibits signaling through all PPAR isoforms. *Endocrinology* **2005**, *146*, 1871–1882. [CrossRef] [PubMed]
49. Gray, S.L.; Nora, E.D.; Grosse, J.; Manieri, M.; Stoeger, T.; Medina-Gomez, G.; Burling, K.; Wattler, S.; Russ, A.; Yeo, G.S.H.; et al. Leptin deficiency unmasks the deleterious effects of impaired peroxisome proliferator-activated receptor gamma function (P465L PPARgamma) in mice. *Diabetes* **2006**, *55*, 2669–2677. [CrossRef] [PubMed]
50. Medina-Gomez, G.; Yetukuri, L.; Velagapudi, V.; Campbell, M.; Blount, M.; Jimenez-Linan, M.; Ros, M.; Oresic, M.; Vidal-Puig, A. Adaptation and failure of pancreatic beta cells in murine models with different degrees of metabolic syndrome. *Dis. Model. Mech.* **2009**, *2*, 582–592. [CrossRef] [PubMed]
51. Savage, D.B.; Tan, G.D.; Acerini, C.L.; Jebb, S.A.; Agostini, M.; Gurnell, M.; Williams, R.L.; Umpleby, A.M.; Thomas, E.L.; Bell, J.D.; et al. Human metabolic syndrome resulting from dominant-negative mutations in the nuclear receptor peroxisome proliferator-activated receptor-gamma. *Diabetes* **2003**, *52*, 910–917. [CrossRef] [PubMed]
52. Yamauchi, T.; Kamon, J.; Waki, H.; Murakami, K.; Motojima, K.; Komeda, K.; Ide, T.; Kubota, N.; Terauchi, Y.; Tobe, K.; et al. The mechanisms by which both heterozygous peroxisome proliferator-activated receptor gamma (PPARgamma) deficiency and PPARgamma agonist improve insulin resistance. *J. Biol. Chem.* **2001**, *276*, 41245–41254. [CrossRef] [PubMed]
53. Rangwala, S.M.; Rhoades, B.; Shapiro, J.S.; Rich, A.S.; Kim, J.K.; Shulman, G.I.; Kaestner, K.H.; Lazar, M.A. Genetic modulation of PPARgamma phosphorylation regulates insulin sensitivity. *Dev. Cell* **2003**, *5*, 657–663. [CrossRef]
54. Lefebvre, B.; Benomar, Y.; Guédin, A.; Langlois, A.; Hennuyer, N.; Dumont, J.; Bouchaert, E.; Dacquet, C.; Pénicaud, L.; Casteilla, L.; et al. Proteasomal degradation of retinoid X receptor alpha reprograms transcriptional activity of PPARgamma in obese mice and humans. *J. Clin. Investig.* **2010**, *120*, 1454–1468. [CrossRef] [PubMed]
55. Peraldi, P.; Xu, M.; Spiegelman, B.M. Thiazolidinediones block tumor necrosis factor-alpha-induced inhibition of insulin signaling. *J. Clin. Investig.* **1997**, *100*, 1863–1869. [CrossRef] [PubMed]
56. Yamauchi, T.; Kamon, J.; Minokoshi, Y.; Ito, Y.; Waki, H.; Uchida, S.; Yamashita, S.; Noda, M.; Kita, S.; Ueki, K.; et al. Adiponectin stimulates glucose utilization and fatty-acid oxidation by activating AMP-activated protein kinase. *Nat. Med.* **2002**, *8*, 1288–1295. [CrossRef] [PubMed]
57. Jonker, J.W.; Suh, J.M.; Atkins, A.R.; Ahmadian, M.; Li, P.; Whyte, J.; He, M.; Juguilon, H.; Yin, Y.Q.; Phillips, C.T.; et al. A PPARγ-FGF1 axis is required for adaptive adipose remodelling and metabolic homeostasis. *Nature* **2012**, *485*, 391–394. [CrossRef] [PubMed]
58. Dutchak, P.A.; Katafuchi, T.; Bookout, A.L.; Choi, J.H.; Yu, R.T.; Mangelsdorf, D.J.; Kliewer, S.A. Fibroblast Growth Factor-21 Regulates PPARγ Activity and the Antidiabetic Actions of Thiazolidinediones. *Cell* **2012**, *148*, 556–567. [CrossRef] [PubMed]
59. Ahmadian, M.; Suh, J.M.; Hah, N.; Liddle, C.; Atkins, A.R.; Downes, M.; Evans, R.M. PPARγ signaling and metabolism: The good, the bad and the future. *Nat. Med.* **2013**, *19*, 557–566. [CrossRef] [PubMed]
60. Seale, P.; Kajimura, S.; Yang, W.; Chin, S.; Rohas, L.M.; Uldry, M.; Tavernier, G.; Langin, D.; Spiegelman, B.M. Transcriptional Control of Brown Fat Determination by PRDM16. *Cell Metab.* **2007**, *6*, 38–54. [CrossRef] [PubMed]
61. Qiang, L.; Wang, L.; Kon, N.; Zhao, W.; Lee, S.; Zhang, Y.; Rosenbaum, M.; Zhao, Y.; Gu, W.; Farmer, S.R.; et al. Brown remodeling of white adipose tissue by SirT1-dependent deacetylation of Pparγ. *Cell* **2012**, *150*, 620–632. [CrossRef] [PubMed]
62. Metzger, B.E. Long-term Outcomes in Mothers Diagnosed With Gestational Diabetes Mellitus and Their Offspring. *Clin. Obstet. Gynecol.* **2007**, *50*, 972–979. [CrossRef] [PubMed]
63. Mericq, V.; Martinez-Aguayo, A.; Uauy, R.; Iñiguez, G.; Van der Steen, M.; Hokken-Koelega, A. Long-term metabolic risk among children born premature or small for gestational age. *Nat. Rev. Endocrinol.* **2017**, *13*, 50–62. [CrossRef] [PubMed]

64. Worda, C.; Leipold, H.; Gruber, C.; Kautzky-Willer, A.; Knöfler, M.; Bancher-Todesca, D. Decreased plasma adiponectin concentrations in women with gestational diabetes mellitus. *Am. J. Obstet. Gynecol.* **2004**, *191*, 2120–2124. [CrossRef] [PubMed]
65. Resi, V.; Basu, S.; Haghiac, M.; Presley, L.; Minium, J.; Kaufman, B.; Bernard, S.; Catalano, P.; Hauguel-de Mouzon, S. Molecular inflammation and adipose tissue matrix remodeling precede physiological adaptations to pregnancy. *Am. J. Physiol. Endocrinol. Metab.* **2012**, *303*, E832–E840. [CrossRef] [PubMed]
66. Sevillano, J.; de Castro, J.; Bocos, C.; Herrera, E.; Ramos, M.P. Role of insulin receptor substrate-1 serine 307 phosphorylation and adiponectin in adipose tissue insulin resistance in late pregnancy. *Endocrinology* **2007**, *148*, 5933–5942. [CrossRef] [PubMed]
67. Catalano, P.M.; Nizielski, S.E.; Shao, J.; Preston, L.; Qiao, L.; Friedman, J.E. Downregulated IRS-1 and PPARgamma in obese women with gestational diabetes: Relationship to FFA during pregnancy. *Am. J. Physiol. Endocrinol. Metab.* **2002**, *282*, E522–E533. [CrossRef] [PubMed]
68. Capobianco, E.; Martínez, N.; Fornes, D.; Higa, R.; Di Marco, I.; Basualdo, M.N.; Faingold, M.C.; Jawerbaum, A. PPAR activation as a regulator of lipid metabolism, nitric oxide production and lipid peroxidation in the placenta from type 2 diabetic patients. *Mol. Cell. Endocrinol.* **2013**, *377*, 7–15. [CrossRef] [PubMed]
69. Lappas, M. Effect of pre-existing maternal obesity, gestational diabetes and adipokines on the expression of genes involved in lipid metabolism in adipose tissue. *Metabolism* **2014**, *63*, 250–262. [CrossRef] [PubMed]
70. Rodriguez-Cuenca, S.; Carobbio, S.; Velagapudi, V.R.; Barbarroja, N.; Moreno-Navarrete, J.M.; Tinahones, F.J.; Fernandez-Real, J.M.; Orešic, M.; Vidal-Puig, A. Peroxisome proliferator-activated receptor γ-dependent regulation of lipolytic nodes and metabolic flexibility. *Mol. Cell. Biol.* **2012**, *32*, 1555–1565. [CrossRef] [PubMed]
71. Vivas, Y.; Díez-Hochleitner, M.; Izquierdo-Lahuerta, A.; Corrales, P.; Horrillo, D.; Velasco, I.; Martínez-García, C.; Campbell, M.; Sevillano, J.; Ricote, M.; et al. Peroxisome Proliferator-Activated Receptor γ 2 Modulates Late-Pregnancy Homeostatic Metabolic Adaptations. *Mol. Med.* **2016**, *22*, 1. [CrossRef] [PubMed]
72. Vivas, Y.; Martínez-García, C.; Izquierdo, A.; Garcia-Garcia, F.; Callejas, S.; Velasco, I.; Campbell, M.; Ros, M.; Dopazo, A.; Dopazo, J.; et al. Early peroxisome proliferator-activated receptor gamma regulated genes involved in expansion of pancreatic beta cell mass. *BMC Med. Genom.* **2011**, *4*, 86. [CrossRef] [PubMed]
73. Yessoufou, A.; Hichami, A.; Besnard, P.; Moutairou, K.; Khan, N.A. Peroxisome Proliferator-Activated Receptor α Deficiency Increases the Risk of Maternal Abortion and Neonatal Mortality in Murine Pregnancy with or without Diabetes Mellitus: Modulation of T Cell Differentiation. *Endocrinology* **2006**, *147*, 4410–4418. [CrossRef] [PubMed]
74. Kurtz, M.; Capobianco, E.; Martínez, N.; Fernández, J.; Higa, R.; White, V.; Jawerbaum, A. Carbaprostacyclin, a PPARδ agonist, ameliorates excess lipid accumulation in diabetic rat placentas. *Life Sci.* **2010**, *86*, 781–790. [CrossRef] [PubMed]
75. Barak, Y.; Sadovsky, Y.; Shalom-Barak, T. PPAR Signaling in Placental Development and Function. *PPAR Res.* **2008**, *2008*, 1–11. [CrossRef] [PubMed]
76. Gao, Y.; She, R.; Sha, W. Gestational diabetes mellitus is associated with decreased adipose and placenta peroxisome proliferator-activator receptor γ expression in a Chinese population. *Oncotarget* **2017**, *8*, 113928–113937. [CrossRef] [PubMed]
77. Liu, L.; Zhuang, X.; Jiang, M.; Guan, F.; Fu, Q.; Lin, J. ANGPTL4 mediates the protective role of PPARγ activators in the pathogenesis of preeclampsia. *Cell Death Dis.* **2017**, *8*, e3054. [CrossRef] [PubMed]
78. Suwaki, N.; Masuyama, H.; Masumoto, A.; Takamoto, N.; Hiramatsu, Y. Expression and potential role of peroxisome proliferator-activated receptor gamma in the placenta of diabetic pregnancy. *Placenta* **2007**, *28*, 315–323. [CrossRef] [PubMed]
79. Carrascosa, J.M.; Andrés, A.; Ros, M.; Bogónez, E.; Arribas, C.; Fernández-Agulló, T.; De Solís, A.J.; Gallardo, N.; Martínez, C. Development of insulin resistance during aging: Involvement of central processes and role of adipokines. *Curr. Protein Pept. Sci.* **2011**, *12*, 305–315. [CrossRef] [PubMed]
80. Redman, L.M.; Smith, S.R.; Burton, J.H.; Martin, C.K.; Il'yasova, D.; Ravussin, E. Metabolic Slowing and Reduced Oxidative Damage with Sustained Caloric Restriction Support the Rate of Living and Oxidative Damage Theories of Aging. *Cell Metab.* **2018**, *27*, 805. [CrossRef] [PubMed]

81. Carobbio, S.; Pellegrinelli, V.; Vidal-Puig, A. Adipose Tissue Function and Expandability as Determinants of Lipotoxicity and the Metabolic Syndrome. *Adv. Exp. Med. Biol.* **2017**, *960*, 161–196. [PubMed]
82. Sun, K.; Tordjman, J.; Clément, K.; Scherer, P.E. Fibrosis and adipose tissue dysfunction. *Cell Metab.* **2013**, *18*, 470–477. [CrossRef] [PubMed]
83. Cha, D.R.; Zhang, X.; Zhang, Y.; Wu, J.; Su, D.; Han, J.Y.; Fang, X.; Yu, B.; Breyer, M.D.; Guan, Y. Peroxisome proliferator activated receptor alpha/gamma dual agonist tesaglitazar attenuates diabetic nephropathy in db/db mice. *Diabetes* **2007**, *56*, 2036–2045. [CrossRef] [PubMed]
84. Khan, T.; Muise, E.S.; Iyengar, P.; Wang, Z.V.; Chandalia, M.; Abate, N.; Zhang, B.B.; Bonaldo, P.; Chua, S.; Scherer, P.E. Metabolic Dysregulation and Adipose Tissue Fibrosis: Role of Collagen VI. *Mol. Cell. Biol.* **2009**, *29*, 1575–1591. [CrossRef] [PubMed]
85. Huh, J.Y.; Park, Y.J.; Ham, M.; Kim, J.B. Crosstalk between adipocytes and immune cells in adipose tissue inflammation and metabolic dysregulation in obesity. *Mol. Cells* **2014**, *37*, 365–371. [CrossRef] [PubMed]
86. Xu, L.; Ma, X.; Verma, N.K.; Wang, D.; Gavrilova, O.; Proia, R.L.; Finkel, T.; Mueller, E. Ablation of PPARγ in subcutaneous fat exacerbates age-associated obesity and metabolic decline. *Aging Cell* **2018**, *17*, e12721. [CrossRef]
87. Miard, S.; Dombrowski, L.; Carter, S.; Boivin, L.; Picard, F. Aging alters PPARgamma in rodent and human adipose tissue by modulating the balance in steroid receptor coactivator-1. *Aging Cell* **2009**, *8*, 449–459. [CrossRef] [PubMed]
88. Park, J.; Choe, S.S.; Choi, A.H.; Kim, K.H.; Yoon, M.J.; Suganami, T.; Ogawa, Y.; Kim, J.B. Increase in Glucose-6-Phosphate Dehydrogenase in Adipocytes Stimulates Oxidative Stress and Inflammatory Signals. *Diabetes* **2006**, *55*, 2939–2949. [CrossRef] [PubMed]
89. De Pauw, A.; Tejerina, S.; Raes, M.; Keijer, J.; Arnould, T. Mitochondrial (dys)function in adipocyte (de)differentiation and systemic metabolic alterations. *Am. J. Pathol.* **2009**, *175*, 927–939. [CrossRef] [PubMed]
90. Tormos, K.V.; Anso, E.; Hamanaka, R.B.; Eisenbart, J.; Joseph, J.; Kalyanaraman, B.; Chandel, N.S. Mitochondrial complex III ROS regulate adipocyte differentiation. *Cell Metab.* **2011**, *14*, 537–544. [CrossRef] [PubMed]
91. Mennes, E.; Dungan, C.M.; Frendo-Cumbo, S.; Williamson, D.L.; Wright, D.C. Aging-associated reductions in lipolytic and mitochondrial proteins in mouse adipose tissue are not rescued by metformin treatment. *J. Gerontol. A Biol. Sci. Med. Sci.* **2014**, *69*, 1060–1068. [CrossRef] [PubMed]
92. Hallgren, P.; Sjöström, L.; Hedlund, H.; Lundell, L.; Olbe, L. Influence of age, fat cell weight, and obesity on O_2 consumption of human adipose tissue. *Am. J. Physiol.* **1989**, *256*, E467–E474. [CrossRef] [PubMed]
93. Graier, W.F.; Malli, R.; Kostner, G.M. Mitochondrial protein phosphorylation: Instigator or target of lipotoxicity? *Trends Endocrinol. Metab.* **2009**, *20*, 186–193. [CrossRef] [PubMed]
94. Schosserer, M.; Grillari, J.; Wolfrum, C.; Scheideler, M. Age-Induced Changes in White, Brite, and Brown Adipose Depots: A Mini-Review. *Gerontology* **2018**, *64*, 229–236. [CrossRef] [PubMed]
95. Shin, W.; Okamatsu-Ogura, Y.; Machida, K.; Tsubota, A.; Nio-Kobayashi, J.; Kimura, K. Impaired adrenergic agonist-dependent beige adipocyte induction in aged mice. *Obesity* **2017**, *25*, 417–423. [CrossRef] [PubMed]
96. Sung, B.; Park, S.; Yu, B.P.; Chung, H.Y. Modulation of PPAR in aging, inflammation, and calorie restriction. *J. Gerontol. A Biol. Sci. Med. Sci.* **2004**, *59*, 997–1006. [CrossRef] [PubMed]
97. Mukherjee, R.; Hoener, P.A.; Jow, L.; Bilakovics, J.; Klausing, K.; Mais, D.E.; Faulkner, A.; Croston, G.E.; Paterniti, J.R. A selective peroxisome proliferator-activated receptor-gamma (PPARgamma) modulator blocks adipocyte differentiation but stimulates glucose uptake in 3T3-L1 adipocytes. *Mol. Endocrinol.* **2000**, *14*, 1425–1433. [CrossRef] [PubMed]
98. Schwartz, A.V. Diabetes, TZDs, and Bone: A Review of the Clinical Evidence. *PPAR Res.* **2006**, *2006*, 24502. [CrossRef] [PubMed]
99. Kung, J.; Henry, R.R. Thiazolidinedione safety. *Expert Opin. Drug Saf.* **2012**, *11*, 565–579. [CrossRef] [PubMed]
100. Fan, W.; Evans, R. PPARs and ERRs: Molecular mediators of mitochondrial metabolism. *Curr. Opin. Cell Biol.* **2015**, *33*, 49–54. [CrossRef] [PubMed]
101. Hevener, A.L.; He, W.; Barak, Y.; Le, J.; Bandyopadhyay, G.; Olson, P.; Wilkes, J.; Evans, R.M.; Olefsky, J. Muscle-specific Pparg deletion causes insulin resistance. *Nat. Med.* **2003**, *9*, 1491–1497. [CrossRef] [PubMed]

102. Fabbiano, S.; Suá Rez-Zamorano, N.; Rigo, D.E.; Veyrat-Durebex, C.; Stevanovic Dokic, A.; Colin, D.J.; Trajkovski, M. Caloric Restriction Leads to Browning of White Adipose Tissue through Type 2 Immune Signaling. *Cell Metab.* **2016**, *24*, 1–13. [CrossRef] [PubMed]
103. Mitchell, S.J.; Madrigal-Matute, J.; Scheibye-Knudsen, M.; Fang, E.; Aon, M.; González-Reyes, J.A.; Cortassa, S.; Kaushik, S.; Gonzalez-Freire, M.; Patel, B.; et al. Effects of Sex, Strain, and Energy Intake on Hallmarks of Aging in Mice. *Cell Metab.* **2016**, *23*, 1093–1112. [CrossRef] [PubMed]
104. Escrivá, F.; Gavete, M.L.; Fermín, Y.; Pérez, C.; Gallardo, N.; Alvarez, C.; Andrés, A.; Ros, M.; Carrascosa, J.M. Effect of age and moderate food restriction on insulin sensitivity in Wistar rats: Role of adiposity. *J. Endocrinol.* **2007**, *194*, 131–141. [CrossRef] [PubMed]
105. Gabriely, I.; Ma, X.H.; Yang, X.M.; Atzmon, G.; Rajala, M.W.; Berg, A.H.; Scherer, P.; Rossetti, L.; Barzilai, N. Removal of visceral fat prevents insulin resistance and glucose intolerance of aging: An adipokine-mediated process? *Diabetes* **2002**, *51*, 2951–2958. [CrossRef] [PubMed]
106. Masternak, M.M.; Bartke, A. PPARs in Calorie Restricted and Genetically Long-Lived Mice. *PPAR Res.* **2007**, *2007*, 28436. [CrossRef] [PubMed]
107. Wilson-Fritch, L.; Nicoloro, S.; Chouinard, M.; Lazar, M.A.; Chui, P.C.; Leszyk, J.; Straubhaar, J.; Czech, M.P.; Corvera, S. Mitochondrial remodeling in adipose tissue associated with obesity and treatment with rosiglitazone. *J. Clin. Investig.* **2004**, *114*, 1281–1289. [CrossRef] [PubMed]

International Journal of
Molecular Sciences

Review

PPARs and Energy Metabolism Adaptation during Neurogenesis and Neuronal Maturation

Michele D'Angelo, Andrea Antonosante, Vanessa Castelli, Mariano Catanesi, NandhaKumar Moorthy, Dalila Iannotta, Annamaria Cimini and Elisabetta Benedetti *

Department of Life, Health and Environmental Sciences, University of L'Aquila, 67100 L'Aquila, Italy; dangelomiche@gmail.com (M.D.); andrea.antonosante@gmail.com (A.A.); vanessa.castelli@graduate.univaq.it (V.C.); mariano.catanesi86@gmail.com (M.C.); nandhabinfo@gmail.com (N.M.); iannottadalila@gmail.com (D.I.); annamaria.cimini@univaq.it (A.C.)

* Correspondence: elisabetta.benedetti@univaq.it; Tel.: +39-0862-433267

Received: 7 May 2018; Accepted: 24 June 2018; Published: 26 June 2018

Abstract: Peroxisome proliferator activated receptors (PPARs) are a class of ligand-activated transcription factors, belonging to the superfamily of receptors for steroid and thyroid hormones, retinoids, and vitamin D. PPARs control the expression of several genes connected with carbohydrate and lipid metabolism, and it has been demonstrated that PPARs play important roles in determining neural stem cell (NSC) fate. Lipogenesis and aerobic glycolysis support the rapid proliferation during neurogenesis, and specific roles for PPARs in the control of different phases of neurogenesis have been demonstrated. Understanding the changes in metabolism during neuronal differentiation is important in the context of stem cell research, neurodegenerative diseases, and regenerative medicine. In this review, we will discuss pivotal evidence that supports the role of PPARs in energy metabolism alterations during neuronal maturation and neurodegenerative disorders.

Keywords: stem cells; metabolism; PPARs

---

1. Introduction

Neurogenesis, the process of generating neurons, occurs during embryonic and perinatal stages in mammals. It occurs also in the adult mammalian brain in two principal neurogenic niches, the subventricular zone (SVZ) of the lateral ventricles, and the subgranular zone (SGZ) of the dentate gyrus (DG) in the hippocampus [1]. Similarly to other adult stem cells, neural stem cells (NSCs) participate in tissue repair after brain damage. Consequently, it has been reported that neurogenesis follows different types of central nervous system (CNS) injury, including ischemic injury, seizure, and mechanical and excitotoxic injury. In line with the role of neurogenesis in the normal turnover of neuronal populations, recently through ^{14}C, it has been demonstrated that about one third of the human adult hippocampal neurons is replaced with 700 new neurons per day [2]. Although, many transcription factors, participating in regulating adult neurogenesis, have been shown to control cell metabolism outside the brain [3]. Metabolism was, for a long time, considered to occur secondary to cell fate switch during neurogenesis. Nowadays, as recently reviewed by Lorenz and Prigione 2017, the emerging picture is that metabolism can be fine-tuned at different levels during neural commitment [4].

Glucose and lipid metabolism are regulated by transcriptional control exerted by peroxisome proliferator activated receptors (PPAR) α, β/δ, and γ, type II nuclear receptors that are particularly active in the brain [5]. In fact, PPAR isotypes are all expressed in the CNS (central nervous system) of rodents during embryonic development, as well as in adults. PPARβ/δ is broadly distributed in the brain, while PPARα and PPARγ are located in more restricted regions [6–8]. Although it has been demonstrated that PPARs can directly regulate neural cell differentiation [9–14] and play important roles in determining NSC fate [15–18]; less is known about their function in regulating NSC metabolism during differentiation. In this review, we will discuss some recent important evidence that supports the role of PPARs on adaptation of energy metabolism during neurogenesis, neuronal development, and neurodegenerative disorders.

2. Metabolic States in Neural Stem Cells Lineage

NSCs are multipotent stem cells, which generate neurons and glial cells. NSCs use symmetrical division for a quick expansion of the progenitor pool; subsequently to the beginning of neurogenesis, they undergo an asymmetric division, by which a stem cell makes another stem cell and an intermediate progenitor committed to neurogenesis. The passage to gliogenesis involves a return to the symmetric division of progenitors [19]. During embryonic development, the choice between neuronal and glial fates is fine-regulated, particularly in vertebrates, in which different cell types are generated in a precise sequence: first neurons, followed by oligodendrocytes and astrocytes [20]. The specification of neuronal and glial cell types, consequently, may help to understand the complex interactions between multiple signaling pathways, transcription factors, and epigenetic mechanisms in the control of fate decision.

Metabolism can be fine-tuned at different levels during neural commitment, and it can play an important role in the specification of neuronal and glial cell types [4]. Neurons and glial cells have different metabolic programs; in fact, neurons are dependent on mitochondrial-based oxidative phosphorylation (OXPHOS), while glia stand on glycolysis [21,22]. NSCs, like glia cells, show a glycolytic nature, and this kind of metabolism is proposed to be an effect of cells' elevated rate of proliferation, because it produces the precursor molecules for biomass generation via the pentose phosphate pathway (PPP) that results from the upstream branches of glycolysis [23]. In agreement with this concept, low oxygen typical of stem cell niches (<1–6%) [24] may influence cell metabolism, inducing anaerobic glycolysis. Hence, hypoxia induces stem cells self-renewal with respect to differentiation, and in concert, the hypoxia-inducible factors (HIFs) control the expression of genes involved in glycolysis and fructose metabolism [25]. Accordingly, in vivo evidence revealed that the modulation of blood vessel function in stem cell niches of the developing mouse cerebral cortex influenced neurogenesis in an oxygen-dependent manner [26]. The NSC state seems correlated with glycolytic metabolism coupled to non-fused mitochondrial morphology [27], while OXPHOS metabolism is commonly associated with differentiated neurons [22,28], which showed a typical tubular mitochondrial network. Recently, these concepts have been confirmed in several works investigating the mitochondrial state of neurons derived in vitro from human pluripotent stem cells (PSCs) [29–31]. Mitochondrial biogenesis and dynamics have a pivotal role in neuronal functions, since they regulate mitochondrial number, location, morphology, and function [32]. It is important to underline that these processes need synchronization refinement in the metabolic enzymes of fatty acid oxidation and oxidative phosphorylation [33], and PPARs are important regulators of these processes. Moreover, Mitofusin2 (Mfn2), a selective target of PPAR β/δ, [34], regulates mitochondrial fusion [35] and seems to be crucial for the efficiency of mitochondrial uptake of Ca^{2+} ions [36,37]. Although NSCs in vivo can rapidly divide during development, becoming quiescent in adult age [38], however, they still maintain glycolytic metabolism. One hypothesis to explain this behavior is that glycolytic metabolism also regulates redox metabolism; particularly, the use of glycolysis may reduce the intracellular levels of reactive oxygen species (ROS) [39]. Glycolysis produces reducing equivalents by means of the pentose cycle and, by reduced mitochondrial activity, promptly limits the generation of ROS. In fact, emerging evidence suggests that ROS can function as second messengers, playing a crucial role in the self-renewal of NSCs [40]. The correct intracellular ROS levels regulation may help to neurogenesis induction, suggesting that low ROS levels are beneficial for NSCs, while committed neural progenitor stem cells (NPCs) increase ROS production to promote differentiation [4]. However, also in NSCs, a determined amount of oxidative metabolism might even be necessary to prevent oncologic transformation of NSCs, as has been recently suggested that inhibition of mitochondrial metabolism in NSCs led to a switch towards more glycolysis with higher proliferation and less inducible differentiation [41]. A significant role in this control seems to be explained by de novo lipogenesis, in fact, an increase of fatty acid oxidation (FAO) was found to be high in adult NSCs in the SVZ, and pharmacological inhibition of FAO resulted in reduced proliferation [42]. In addition, de novo lipogenesis is crucial for adult stem cell behavior, as demonstrated by an interesting experiment of Knobloch et al., 2013, in which they showed a decrease of stem cell proliferation upon genetic deletion or pharmacological inhibition of the key enzyme fatty acid synthase [43]. Meanwhile, an elevated lipogenesis seems to be associated with an increase of NSC proliferation, and in quiescent NSCs, FAO appears, instead, to be favored. Data from single-cell RNA experiment demonstrate that a low rate oxidative metabolism, because of FAO in quiescent NSCs, may correspond to an alternative energy fuel to glucose [44]. Furthermore, congenital defects in mitochondrial FAO in NSCs, leads to differentiation with the loss of NSC self-renewal in the developing mouse brain [45]. In addition, silencing of promyelocytic leukemia gene (PML), which it is known to regulate FAO and is involved in modulation of PPAR β/δ signaling, reduces the hematopoietic stem cell pool in mice [46].

In the brain, during pathological conditions, an alteration in metabolic status occurs; in fact, recent studies showed an impaired NSCs function in metabolic disease underlying the role of lipid metabolism in neurogenesis. In example, high fat diet (HFD) decreases hippocampal neurogenesis in male rats. These mice exhibit reduced hippocampal neurogenesis and neuronal precursor cells proliferation paralleled with increased lipid peroxidation and decreased expression of trophic and pro-neurogenic BDNF (brain derived neurotrophic factor). Moreover, young mice treated with HFD exhibited decreased hippocampal neurogenesis respect adult mice under the same diet [2]. It has been demonstrated that lipid accumulation perturbs niche microenvironment and inhibits neurogenesis in unhealthy brains, thus supporting evidence for a novel FA-mediated mechanism suppressing NSC activity.

In this context, it is important to underline recent evidence suggesting that sporadic Alzheimer's disease (AD) etiopathogenesis could also involve dysfunctional brain insulin signaling, with subsequent glucose dysmetabolism and metabolic shift to alternative energy sources, also known as type 3 diabetes [47].

3. Roles of PPARs in the Energetic Metabolic Switch Occurring during Neurogenesis and Neuronal Maturation

PPARs are ligand-activated transcription factors included into nuclear receptor superfamily, three isotypes have been determined, encoded by separate genes (α, *NR1C1*; β/δ, *NR1C2*; and γ, *NR1C3*). PPARs, once activated by the ligand, form a heterodimer with the 9-cis retinoic acid receptor (RXR) and modulate the transcription of their target genes by binding to the putative PPRE (AGGTCAAAGGTCA) in the promoter regions of them. Regarding their protein structure, in the N-terminal there is the A/B domain (AF-1), which holds a ligand-independent function, while the C-terminal domain, that holds the DNA binding domain (DBD), is composed of two zinc finger-like motifs that can bind the PPARs response element (PPRE). The D domain is a hinge region important for the cofactor interaction, and consequently, for DNA binding. The E/F (LBD) domain is involved in the dimerization with RXR and a ligand-dependent transcriptional activating function (AF-2) [38,48]. PPARs transcriptional activity and stability can be modified covalently by phosphorylation, ubiquitylation, and SUMOylation [49,50]. PPARα, the first PPAR to be identified, is expressed mainly in the liver, heart, and brown adipose tissue, in which it regulates the ketogenesis, lipid storage, and fatty acid oxidation pathways. PPARβ/δ is ubiquitously expressed, and it has a leading role in glucose and fatty acid oxidation in key metabolic tissues, such as liver, skeletal muscle, and heart. Finally, PPARγ is expressed in white adipose tissue, where it is a master regulator of adipogenesis, as well as a potent modulator of whole-body lipid metabolism and insulin sensitivity [51].

Regarding PPAR ligands, some of them, such as fibrates (PPARα ligands), are currently used as treatment of dyslipidemia; while, glitazones (PPARγ ligands) are antidiabetic and insulin-sensitizing agents, otherwise, PPARβ/δ ligands have only confirmations obtained from animal models [52]. Moreover, PPARα/γ dual agonists, (glitazar) PPAR α/δ dual agonists (elafibranor), and pan-PPAR agonists have been recently become available [52].

Regarding their expression in the brain, all PPAR isotypes are expressed in CNS, both during embryonic development and in the adult. PPARα and PPARγ are located in more restricted regions, while PPARβ/δ is widely distributed in the brain [6–8]. PPARs are implicated in the regulation of the proliferation, migration, and differentiation of NSCs by signaling pathways, such as STAT3, NFkB, and Wnt [15–17], and it has been demonstrated that in neurospheres, grown in vitro from adult mouse SVZ, all three PPAR isotypes are expressed [18,53]. PPARβ/δ resulted the most abundant isotype; it is not surprising due to its early expression and its abundance during brain development [6]. Moreover, the concurrent expression of the three isotypes in the NSC nucleus does not mean that they are all transcriptionally active; in fact, it has been suggested that unliganded PPARβ/δ may act as potent inhibitor of the transcriptional activity of α and γ isotypes [54]. In the astroglial differentiating NSCs, PPARs undergo quantitative modifications. A strong decrease of PPARβ/δ was observed, in this context, it might be considered as inhibitor of astroglial differentiation. PPARγ did not change, both at mRNA and protein levels, while PPARα was significantly increased in agreement with our previous findings on astrocytes in vitro differentiation [14], suggesting a role for this transcription factor in astroglial differentiation, confirmed by the results achieved when NSCs were treated with a specific PPARα agonist [18]. Finally, in the cytoplasm of neural stem cells, large lipid droplets were found in SVZ adult NSCs, in accordance with de novo lipogenesis [42]. Moreover, lipid droplet withdrawal, during astroglial differentiation, agrees with the view that differentiated astrocytes develop catabolic lipid metabolism, rather than anabolic, needing PPARα activity.

In Figure 1, is shown a scheme summarizing the effects of PPARs on energy metabolism adaptation during neural stem cell differentiation in neurons and astrocytes.

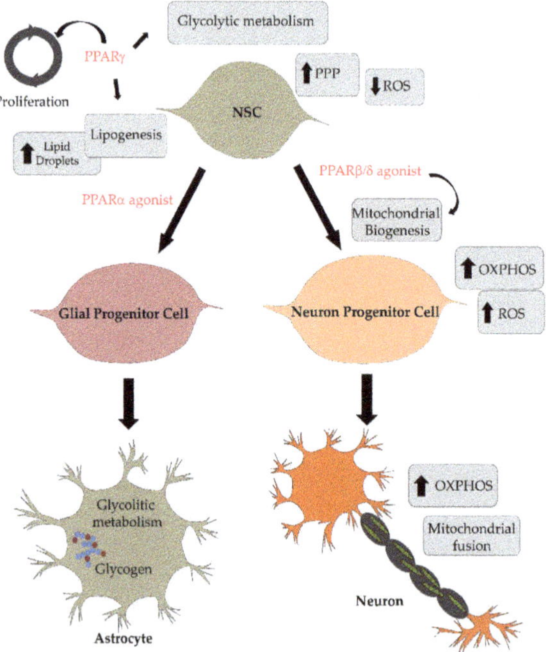

Figure 1. Scheme summarizing the effects of peroxisome proliferator activated receptors (PPARs) on energy metabolism adaptation during neural stem cells differentiation in neurons and astrocytes.

4. Roles of PPARβ/δ in Neurogenesis and Neuronal Maturation

The PPARβ/δ isotype is highly expressed in the brain [55], and its deletion in mice is associated with brain developmental defects [56]. In fact, PPARβ/δ has important roles in neuronal function; it has been demonstrated that PPARβ/δ-deficient mice are viable, but they show several defects in CNS such as altered myelination [56] and bad performance in memory tests, paralleled with an increase in inflammatory markers, astrogliosis, and tau hyperphosphorylation [57]. The presence and modulation of PPARβ/δ in embryonic rat cortical neurons during their in vitro maturation were observed by us [9], suggesting a potential role of PPARβ/δ in neuronal maturation. In addition, we demonstrated in human neuroblastoma cell line, SH-SY5Y, a neuronal differentiating effect of PPARβ/δ [58,59]. The signal transduction pathways activated by PPARβ/δ during neuronal differentiation were studied on this in vitro model. In particular, it has been demonstrated that the PPARβ/δ activation was able to determine the activation of MAPK-ERK1/2 and to increase the expression of BDNF and p75 receptor, in parallel to a decrease in BDNF TrkB receptor, suggesting that activation of PPARβ/δ was involved, directly or indirectly in neuritogenesis and neuronal maturation. Finally, these results were further confirmed by the use of a specific agonist and antagonist of PPAR β/δ in primary neuronal cultures [11], in which we also observed a specific effect of PPARβ/δ activation on cholesterol biosynthesis during neuronal maturation. Furthermore, it has been demonstrated that retinoic acid (RA) promotes neurogenesis by activating both retinoic acid receptors (RARs) and PPAR β/δ in P19 mouse embryonal carcinoma cell line [10]. Recently, Mei and Coll, in 2016, have been reported that, by modulating mitochondrial energy metabolism via Mfn2 and mitochondrial Ca^{2+}, PPAR β/δ plays a key role in neuronal differentiation. This study provides novel insights for the role of PPARβ/δ and energy metabolism adaptation during neurogenesis and neuronal maturation [33]. In particular, the authors have been shown that flavonoid compound 4a facilitated embryonic stem cells (ESC) to differentiate into neurons morphologically as well as functionally, and that the PPAR β/δ gene silencing blocked compound 4a-induced neurogenesis of ES cells, demonstrating the important role of PPARβ/δ in neuronal differentiation. In this kind of model, mitochondrial biogenesis was upregulated by compound 4a treatment, and was altered by sh-PPAR β/δ knockdown, suggesting a key role of PPAR β/δ in mitochondrial biogenesis during neuronal differentiation. Moreover, they showed that the compound 4a was able to increase the protein expression of Mfn2, which was abolished by PPARβ/δ knockdown, and that sh-PPAR β/δ reduced mitochondrial Ca^{2+} concentration. Thus, PPARβ/δ seems strongly implicated in the induction of neuronal lineage, increasing mitochondrial fusion, modulating BDNF expression, cholesterol biosynthesis, and mitochondrial FAO. Finally, it should be emphasized that a natural ligand of this receptor, the 4-hydroxynonenal (4-HNE) [60], is a product of oxidative stress and, thus, it should be possible that the increased ROS levels in committed neuroblast could trigger the activation of PPAR β/δ.

5. Roles of PPARγ in Neurogenesis and Neuronal Maturation

PPARγ activation induces the transcription of genes associated with lipid uptake and storage, playing critical roles in lipid homeostasis [61]. PPARγ controls murine NSC proliferation and survival [27]; particularly, when activated by low concentrations of specific agonists, PPARγ stimulates proliferation concurrently constraining neuronal differentiation, while activation by high concentrations of agonists leads to NSC death. This dual role suggests that PPARγ controls the expansion of NSC population in a concentration-dependent manner, and it shows that precise concentrations of its agonists are critical for the survival and proliferation of NSCs in vivo.

Regarding metabolism, in order to examine the mechanisms of PPARγ in the control of energy balance in CNS, Stump and colleagues 2016 used a Cre-recombinase dependent ($Nestin^{Cre}$), conditionally activatable transgene expressing either wildtype (WT) or dominant-negative (P467L) PPARγ. What they found is that Nes^{Cre}/PPARγ-WT mice displayed severe microcephaly and brain malformation, indicating that PPARγ can control brain development. On the contrary, global interference with PPARγ function caused impaired growth, resistance to diet induced obesity, decreased lean mass, redistribution of adipose tissue, GH resistance, and abnormalities in glucose and insulin [62].

Recently, we have shown, in vitro, the energetic metabolism pathways controlled by PPARγ [63] in neuroblast differentiation. We used the human neuroblastoma cell lines SH-SY5Y, as a model of neuroblast induced to differentiate neuron. During the early phases of neuronal differentiation, a significant downregulation of PPARγ was observed, concomitant with a change in its cellular localization, in fact, it came to be cytoplasmic after the differentiation challenge. In addition, we observed that the decrease of PPARγ was paralleled by a strong decrease of glycogen and lipid droplets content in differentiating cells. PPARγ knockdown showed a strong decrease of glycogen content, concomitant with a significant increase of phosphorylase glycogen brain (PYGB), indicating that PPARγ is critical for NPCs maintenance and energetic storage.

6. Energy Metabolism Imbalance in Neurodegenerative Disorders

During aging, there is an increase of circulating glucose due to the cellular inability to increase glucose uptake in response to insulin, and this peripheral insulin resistance has been related with poorer cognitive function [64]. Insulin signaling pathway results in phosphorylation of the insulin receptor-interacting protein (IRS-1), particularly, a decrease in IRS-1 phosphorylation may induce insulin resistance, while an increased phosphorylation on serine 312 of IRS-1 has opposite effects. Studies on post mortem brain tissue from elderly subjects showed an increased IRS-1 phosphorylation on serine 312, suggesting neuronal insulin resistance [65,66]. Concomitant with insulin resistance, also, the neuronal glucose transporter GLUT3 is susceptible to aging factors [67,68]. During aging, the metabolism of several lipid species is altered, such as long-chain ceramides [69] and omega-3 fatty acids [70]. Dyslipidemia is often associated with dementia, and it may increase the risk of AD [71]. Moreover, individuals having the ε4 allele of the gene encoding apolipoprotein E, the protein that transports cholesterol and lipoproteins, have an increased risk of developing sporadic AD [72].

Accordingly, age-related neurodegenerative disorders, such as AD and PD, share common pathogenic pathway with metabolic syndromes like obesity and type 2 diabetes, such as deregulation of brain insulin signaling and insulin growth factor-1 (IGF-1) signaling. This signaling induces insulin resistance, and energy and lipid metabolism imbalance, that have a direct negative impact on the CNS [47]. Moreover, neurodegenerative disorders, such as metabolic syndromes, are characterized also by mitochondrial and peroxisomal dysfunction, and alterations in energy metabolism [73,74].

Alzheimer's disease is the most common form of dementia, characterized by age-related cognitive decline that starts as mild short-term memory impairment, and then progresses to severe deficits in essentially all cognitive domains. The hallmarks of this disease are amyloid β plaques (Aβ) and hyperphosphorylated tau tangles [75].

Parkinson's disease (PD), like AD, is a long-term degenerative disorder of the CNS, characterized by degeneration of dopaminergic neurons in the substantia nigra that innervate the striatum [76]. The hallmarks of PD are "Lewy bodies", large accumulations of α-synuclein in the cytoplasm [77]; experimental evidence suggests that the accumulation of α-synuclein aggregates induces mitochondrial dysfunction in neurons, and these are pivotal events in the pathogenesis of PD.

As reviewed by Agarwal and colleagues 2017, it is becoming increasingly evident that mitochondrial abnormalities play an import role in the onset, progression, and neuronal cell death in age-related neurodegenerative disorders [73].

Recently, in neurodegenerative disorders, it has been demonstrated that functional and structural changes in mitochondria are early features that conduce to neuronal death, paralleled by cognitive and neurobehavioral abnormalities [78]. In age-related neurodegenerative disorders, the mitochondrial population is decreased, due to dysregulation of mitochondrial biogenesis [79]. The mitochondrial dysfunction observed in neurodegenerative disorders leads to the damage in mitochondrial electron transport chain, in the mitochondrial DNA, and calcium buffering [79]. Mitochondria is the second major intracellular Ca^{2+} store after endoplasmic reticulum, and Ca^{2+} deregulation plays a critical role in the pathogenesis of several neurodegenerative disorders [80]. In fact, mitochondrial Ca^{2+} plays an important role in preserving cellular physiology, activating the respiratory chain [81]. When mitochondria accumulate excessive Ca^{2+} ions, this causes mitochondrial swelling, injury of mitochondrial membrane potential, and finally, it induces apoptosis in neurons [82].

Mitochondrial dynamics/biogenesis helps to maintain the characteristic morphology of mitochondria and a healthy mitochondrial pool in neurons; it is a tightly controlled balance between three important phenomena: mitochondria fission, fusion, and degradation. [78]. Mitochondrial fission consists of replacement of damaged mitochondria, and it plays a main role in the appropriate function and assembly of mitochondrial electron transport chain complex [78]; the main protein mediators of mitochondrial fission are Fis-1 and Drp-1 [78]. Fusion is related with the improvement of mitochondrial functions, and is regulated by three main proteins: mitofusin 1 (Mfn-1), mitofusin 2 (Mfn-2), and optic atrophy protein 1 (OPA-1) [78]. The expression and protein levels of Drp-1, Opa-1, Mfn-1, and Mfn-2 are decreased in numerous neurodegenerative disorders. Moreover, mutations in several PD-linked genes, like *PINK-1*, *Parkin*, *DJ-1*, *LRRK2*, and *VPS35*, are directly or indirectly, linked to mitochondrial dysfunction [83,84]. In particular, PINK/parkin pathway promotes mitochondrial fission or inhibits mitochondrial fusion in drosophila [85]. A key factor for mitochondria biogenesis is the PGC-1 α; any loss or impairment in PGC-1α activity may result in metabolic defects and mitochondrial dysfunctions in most neurodegenerative disease [78]. PPARs bind this transcriptional co-activator, modulating the expression of the gene encoding for mitochondrial fatty acid oxidation and glucose metabolism enzymes [86], but also the genes encoding for antioxidant enzymes such as catalase, glutathione peroxidase, and MnSOD, thus reducing oxidative damage [87,88].

The role of peroxisomal dysfunction in aging has been largely undervalued; however, accumulating evidence suggests that peroxisomal function declines with aging and in age-related neurological disorders, such as AD and PD [89]. Interestingly, not only mitochondria, but also peroxisomes, are organelles involved in the response to the redox unbalance, characterizing the earliest phases of Aβ pathology [90–92].

Peroxisomal dysfunction was also linked to disease, principally through ROS metabolism [93,94], in fact, peroxisome-mediated ROS production may have also a deeper effect on mitochondrial integrity, as demonstrated by the induction of intraperoxisomal ROS, using a peroxisome-localized photosensitizer [95]. Interestingly, genetic inactivation of catalase, a PPAR target gene, perturbs mitochondrial redox potential in mice [96]. Reflecting the intimate link between the two organelles, these studies suggest that peroxisomal dysfunction may be a precursor for mitochondrial impairment. Moreover, proteins involved in peroxisomal fatty acid oxidation, ether lipid synthesis, and other peroxisomal processes, were also decreased in in age-related neurological disorders [93], suggesting that peroxisomal dysfunction extends beyond dysregulated ROS metabolism. Remarkably, increased very long chain fatty acids (VLCFAs) and reduced plasmalogen levels are observed in the brain of AD patients, suggesting a possible defect in peroxisomal beta oxidation and peroxisomal lipid synthesis [97]. Peroxisomal dysfunction is present also in PD, particularly, plasmalogen levels are significantly reduced in PD post mortem human frontal cortex lipid rafts [98].

7. Roles of PPARs in Neurodegenerative Disorders

The most studied PPAR in neurodegenerative disease is the γ isotype. Combs and colleagues [99] were the first to report the relationship between PPARγ activation and neurodegeneration, and this evidence was supported by several lines of evidence in animal and cellular models of Alzheimer's disease (AD), Parkinson's disease (PD), amyotrophic lateral sclerosis (ALS), Huntington's disease (HD), stroke, and traumatic injuries [100].

In numerous mouse models of AD, it has been indicated that administration of PPARγ agonists can ameliorate memory and cognition performance, reduce inflammation, and decrease amyloid levels. Searcy and colleagues [101] have been demonstrated that PPAR agonists are able to ameliorate synaptic function in AD mouse models.

Since it is known that PPARγ agonists decrease insulin resistance in type II diabetes, the beneficial effects of PPARγ agonists in AD mice indicate that they can act in the same manner in CNS [102]. Escribano and his research group demonstrated that rosiglitazone, a high-affinity PPARγ agonist, rescues memory impairment in a mouse model of AD [103]. Specifically, these authors indicated that rosiglitazone promotes Aβ clearance, by promoting microglial phagocytic ability and decreasing the expression of proinflammatory markers.

Moreover, an interesting meta-analysis compared the efficacy of glitazones (antidiabetic and insulin-sensitizing agents) for Alzheimer's disease (AD) and mild cognitive impairment (MCI). In particular, this analysis included 20 comparisons from 4855 individuals randomly assigned to 6 different antidiabetic drugs with various doses. The results have shown that pioglitazone and rosiglitazone had the major pro-cognitive effects in subjects with AD/MCI [104].

Recently a role for PPARγ has been recognized in regional transcriptional regulation of chr19q13.32; this region contains genes such as *TOMM40* and *APOE*, implicated in AD. Mostly, this region holds a number of PPARγ binding sites, and understanding how those sites regulate the expression of genes in the region could help in the development of more efficient therapies [105].

In a recent study, Cheng and collaborators (2015) studied the effects of PPARα activation on neuronal degeneration by inducing Aβ42 cytotoxicity in an in vitro model. They established that the mitochondrial-associated AIF/Endo G-dependent pathway could be prevented by activation of the receptor in this model [106]. Recently, Fidaleo et al. [107] reported that PPARα ligands, such as palmitoylethanolamide (PEA), are able to protect neurons from degeneration, leading to a reduction in oxidative stress, inflammation, and neurogenesis, and glial cell proliferation/differentiation, thus further suggesting the use of PPARα as a potential therapeutic agent for neurodegeneration.

In 2003, Brune and colleagues [108] screened for polymorphisms in the PPARα gene, and they detected two known polymorphisms located in exon 5 and intron 7. They studied the possible association of these polymorphisms with AD and its effect in carriers of an insulin gene (*INS*) polymorphism. They showed that carriers of a *PPARαL162V* allele and an *INS-1* allele presented an increased risk for AD. These authors also found an increased level of βamyloid in cerebrospinal fluid in PPAR-α L162V genotype carriers. These results suggested that PPARα polymorphism may be considered a risk factor for AD. Moreover, since altered glucose metabolism has been indicated in AD, the interaction of the insulin and the PPARα genes in AD risk in the Epistasis Project, have been assayed. The authors proposed that dysregulation of glucose metabolism leads to the development of AD, and might be due, in part, to genetic variations in *INS* and *PPARα*, and their interaction especially in Northern Europeans [109]. Recently, it has been reported that statins serve as ligands of PPARα, and that Leu331 and Tyr 334 residues of PPARα are important for statin binding [110]. Upon binding, statins induce upregulation of neurotrophins through PPARα-mediated transcriptional activation of cAMP-response element binding protein (CREB). Consequently, simvastatin increases CREB and also BDNF in the hippocampus of PPARα null mice receiving full-length lentiviral PPARα, but not L331M/Y334D statin-binding domain mutated lentiviral PPARα. This study identifies statins as ligands of PPARα analyzing the importance of PPARα in the therapeutic success of simvastatin in an animal model of Alzheimer's disease. Limited studies indicated a protective role for PPARα agonists in models of PD: treatment with the PPARα agonist fenofibrate [111] protected nigral dopaminergic neurons in the 1-methyl-4-phenyl-1,2,3,6-tetrahydropyridine (MPTP) mouse model of PD. The role of PPARβ/δ in neurodegeneration is less studied than PPARγ and α, and more controversial. PPARβ/δ agonists, acting through PPARβ/δ activation, induce protection in many pathological CNS states, such as a transgenic mouse model of Alzheimer's disease, MPTP model of Parkinson's disease, stroke, EAE, spinal cord injury and in a streptozotocin-induced experimental type 3 diabetes [100]; in all these cases, the effect has been mainly attributed to reduction of inflammation and oxidative stress. However, the main question regarding this nuclear receptor is that further studies are needed in order to better characterize this receptor in a more systemic manner, to support the possibility that PPARβ/δ might be used as a therapeutic target [112].

Regarding mitochondrial biogenesis, PPAR agonists can increase the functionality of mitochondrial, and they enhance Ca^{2+} buffering ability of mitochondria. Therefore, it seems attractive to examine the cellular and molecular mechanisms by which PPARs determine changes in cytosolic Ca^{2+} concentration to develop new strategies in the field of drug development for neurodegenerative disorders [73]. Moreover, PPAR agonists are able to induce mitochondrial biogenesis through PGC-1α, preventing mitochondrial dysfunction caused by oxidative insults [113]. In Table 1, are shown the references on energy metabolism imbalance in neurodegenerative disorders, and about PPARs ligands.

Table 1. Table summarizing the references on energy metabolism imbalance in neurodegenerative disorders and on PPARs and PPAR ligands.

AD Ref. Energy Metabolism Imbalance	Neurodegenerative Diseases i.e. and	PD Ref. PPARs and Their Ligands
[47,64–68]	Insulin Resistance	[102–104,109,110]
[78–85]	Mitochondrial Dysregulation	[73,105,106,113]
[89–98]	Peroxisomal Dysregulation	[90–92]

8. Conclusions

The data summarized here underlines the significant role of PPARs in energy metabolism adaptation during brain development. However, we still need to better elucidate the molecular networks driven by these nuclear receptors in regulating NSC metabolism during self-renewal and differentiation. In the brain, during pathological conditions, an alteration in metabolic status occurs, whereby elucidate the crucial steps in energetic metabolism and the involvement of PPARs in NSCs neuronal fate (lineage) may be useful for the future design of preventive and/or therapeutic interventions. However, the future use of PPAR ligands as therapeutic agent is related to an important problem of design of drugs: the new molecules have to be able to pass the BBB (blood–brain barrier) and they have to be projected in order to avoid the classical pharmacokinetic problems related to the drugs active on CNS.

Funding: This research was supported by the RIA fund E.B.

Conflicts of Interest: The authors declare no conflicts of interest.

References

1. Bond, A.M.; Ming, G.L.; Song, H. Adult mammalian neural stem cells and neurogenesis: Five decades later. *Cell Stem Cell* **2015**, *17*, 385–395. [CrossRef] [PubMed]
2. Fidaleo, M.; Cavallucci, V.; Pani, G. Nutrients, neurogenesis and brain ageing: From disease mechanisms to therapeutic opportunities. *Biochem. Pharmacol.* **2017**, *141*, 63–76. [CrossRef] [PubMed]
3. Gonçalves, J.T.; Schafer, S.T.; Gage, F.H. Adult Neurogenesis in the Hippocampus: From Stem Cells to Behavior. *Cell* **2016**, *167*, 897–914. [CrossRef] [PubMed]
4. Lorenz, C.; Prigione, A. Mitochondrial metabolism in early neural fate and its relevance for neuronal disease modeling. *Curr. Opin. Cell Biol.* **2017**, *49*, 71–76. [CrossRef] [PubMed]
5. Skerrett, R.; Malm, T.; Landreth, G. Nuclear receptors in neurodegenerative diseases. *Neurobiol. Dis.* **2014**, *72*, 104–116. [CrossRef] [PubMed]
6. Moreno, S.; Farioli-vecchioli, S.; Cerù, M.P. Immunolocalization of peroxisome proliferator-activated receptors and retinoid X receptors in the adult rat CNS. *Neuroscience* **2004**, *123*, 131–145. [CrossRef] [PubMed]
7. Woods, J.W.; Tanen, M.; Figueroa, D.J.; Biswas, C.; Zycband, E.; Moller, D.E.; Austin, C.P.; Berger, J.P. Localization of PPARδ; in murine central nervous system: Expression in oligodendrocytes and neurons. *Brain Res.* **2003**, *975*, 10–21. [CrossRef]
8. Cullingford, T.E.; Bhakoo, K.; Peuchen, S.; Dolphin, C.T.; Patel, R.; Clark, J.B. Distribution of mRNAs encoding the peroxisome proliferator-activated receptor α, β, and γ and the retinoid X receptor α, β, and γ in rat central nervous system. *J. Neurochem.* **1998**, *70*, 1366–1375. [CrossRef] [PubMed]

9. Cimini, A.; Benedetti, E.; Cristiano, L.; Sebastiani, P.; D'Amico, M.A.; D'Angelo, B.; Di Loreto, S. Expression of peroxisome proliferator-activated receptors (PPARs) and retinoic acid receptors (RXRs) in rat cortical neurons. *Neuroscience* **2005**, *130*, 325–337. [CrossRef] [PubMed]
10. Yu, S.; Levi, L.; Siegel, R.; Noy, N. Retinoic acid induces neurogenesis by activating both retinoic acid receptors (RARs) and peroxisome proliferator-activated receptor β/δ (PPARβ/δ). *J. Biol. Chem.* **2012**, *287*, 195–205. [CrossRef] [PubMed]
11. Benedetti, E.; Di Loreto, S.; D'Angelo, B.; Cristiano, L.; d'Angelo, M.; Antonosante, A.; Fidoamore, A.; Golini, R.; Cinque, B.; Cifone, M.G.; et al. The PPARβ/δ Agonist GW0742 Induces Early Neuronal Maturation of Cortical Post-Mitotic Neurons: Role of PPARβ/δ in Neuronal Maturation. *J. Cell. Physiol.* **2015**, *23*, 597–606. [CrossRef] [PubMed]
12. Saluja, I.; Granneman, J.G.; Skoff, R.P. PPAR delta agonists stimulate oligodendrocyte differentiation in tissue culture. *Glia* **2001**, *33*, 191–204. [CrossRef]
13. Cimini, A.; Bernardo, A.; Cifone, G.; Di Muzio, L.; Di Loreto, S. TNFα downregulates PPARδ expression in oligodendrocyte progenitor cells: Implications for demyelinating diseases. *Glia* **2003**, *41*, 3–14. [CrossRef] [PubMed]
14. Cristiano, L.; Cimini, A.; Moreno, S.; Ragnelli, A.M.; Cerù, M.P. Peroxisome proliferator-activated receptors (PPARs) and related transcription factors in differentiating astrocyte cultures. *Neuroscience* **2005**, *131*, 577–587. [CrossRef] [PubMed]
15. Wada, K.; Nakajima, A.; Katayama, K.; Kudo, C.; Shibuya, A.; Kubota, N.; Terauchi, Y.; Tachibana, M.; Miyoshi, H.; Kamisaki, Y.; et al. Peroxisome proliferator-activated receptor γ-mediated regulation of neural stem cell proliferation and differentiation. *J. Biol. Chem.* **2006**, *81*, 12673–12681. [CrossRef] [PubMed]
16. Mulholland, D.J.; Dedhar, S.; Coetzee, G.A.; Nelson, C.C. Interaction of nuclear receptors with the Wnt/β-catenin/Tcf signaling axis: Wnt you like to know? *Endocr. Rev.* **2008**, *26*, 898–915. [CrossRef] [PubMed]
17. Jung, Y.; Song, S.; Choi, C. Peroxisome proliferator activated receptor γ agonists suppress TNFα-induced ICAM-1 expression by endothelial cells in a manner potentially dependent on inhibition of reactive oxygen species. *Immunol. Lett.* **2008**, *117*, 63–69. [CrossRef] [PubMed]
18. Cimini, A.; Cristiano, L.; Benedetti, E.; D'Angelo, B.; Cerù, M.P. PPAR expression in adult mouse neural stem cells (NSC). Modulation of PPARs during astroglial differentiation. *PPAR Res.* **2007**, *2007*, 48242. [CrossRef] [PubMed]
19. Temple, S. The development of neural stem cells. *Nature* **2001**, *414*, 112–117. [CrossRef] [PubMed]
20. Bayer, S.A.; Altman, J. *Neocortical Development*, 1st ed.; Raven Press: New York, NY, USA, 1999.
21. Magistretti, P.J.; Allaman, I. A cellular perspective on brain energy metabolism and functional imaging. *Neuron* **2015**, *86*, 883–901. [CrossRef] [PubMed]
22. Hall, C.N.; Klein-Flugge, M.C.; Howarth, C.; Attwell, D. Oxidative phosphorylation, not glycolysis, powers presynaptic and postsynaptic mechanisms underlying brain information processing. *J. Neurosci.* **2012**, *32*, 8940–8951. [CrossRef] [PubMed]
23. Vander Heiden, M.G.; Cantley, L.C.; Thompson, C.B. Understanding the Warburg effect: The metabolic requirements of cell proliferation. *Science* **2009**, *324*, 1029–1033. [CrossRef] [PubMed]
24. Ochocki, J.D.; Simon, M.C. Nutrient-sensing pathways and metabolic regulation in stem cells. *J. Cell Biol.* **2013**, *203*, 23–33. [CrossRef] [PubMed]
25. Majmundar, A.J.; Wong, W.J.; Simon, M.C. Hypoxia-inducible factors and the response to hypoxic stress. *Mol. Cell* **2010**, *40*, 294–309. [CrossRef] [PubMed]
26. Lange, C.; Turrero Garcia, M.; Decimo, I.; Bifari, F.; Eelen, G.; Quaegebeur, A.; Boon, R.; Zhao, H.; Boeckx, B.; Chang, J.; et al. Relief of hypoxia by angiogenesis promotes neural stem cell differentiation by targeting glycolysis. *EMBO J.* **2016**, *35*, 924–941. [CrossRef] [PubMed]
27. Chen, H.; Chan, D.C. Mitochondrial Dynamics in Regulating the Unique Phenotypes of Cancer and Stem Cells. *Cell Metab.* **2017**, *26*, 39–48. [CrossRef] [PubMed]
28. Rafalski, V.A.; Brunet, A. Energy metabolism in adult neural stem cell fate. *Prog. Neurobiol.* **2011**, *93*, 182–203. [CrossRef] [PubMed]
29. Zheng, X.; Boyer, L.; Jin, M.; Mertens, J.; Kim, Y.; Ma, L.; Ma, L.; Hamm, M.; Gage, F.H.; Hunter, T. Metabolic reprogramming during neuronal differentiation from aerobic glycolysis to neuronal oxidative phosphorylation. *eLIFE* **2016**, *10*, e13374. [CrossRef] [PubMed]

30. O'Brien, L.C.; Keeney, P.M.; Bennett, J.P., Jr. Differentiation of Human Neural Stem Cells into Motor Neurons Stimulates Mitochondrial Biogenesis and Decreases Glycolytic Flux. *Stem Cells Dev.* **2015**, *24*, 1984–1994. [CrossRef] [PubMed]
31. Fang, D.; Qing, Y.; Yan, S.; Chen, D.; Yan, S.S. Development and Dynamic Regulation of Mitochondrial Network in Human Midbrain Dopaminergic Neurons Differentiated from iPSCs. *Stem Cell Rep.* **2016**, *7*, 678–692. [CrossRef] [PubMed]
32. Cheng, A.; Hou, Y.; Mattson, M.P. Mitochondria and neuroplasticity. *ASN Neuro* **2010**, *2*, e00045. [CrossRef] [PubMed]
33. Mei, Y.Q.; Pan, Z.F.; Chen, W.T.; Xu, M.H.; Zhu, D.Y.; Yu, Y.P.; Lou, Y.J. A Flavonoid Compound Promotes Neuronal Differentiation of Embryonic Stem Cells via PPAR-β Modulating Mitochondrial Energy Metabolism. *PLoS ONE* **2016**, *11*, e0157747. [CrossRef] [PubMed]
34. Li, Y.; Yin, R.; Liu, J.; Wang, P.; Wu, S.; Luo, J.; Zhelyabovska, O.; Yang, Q. Peroxisome proliferator-activated receptor delta regulates mitofusin 2 expression in the heart. *J. Mol. Cell. Cardiol.* **2009**, *46*, 876–882. [CrossRef] [PubMed]
35. Huang, P.; Galloway, C.A.; Yoon, Y. Control of mitochondrial morphology through differential interactions of mitochondrial fusion and fission proteins. *PLoS ONE* **2011**, *6*, e20655. [CrossRef] [PubMed]
36. De Brito, O.M.; Scorrano, L. Mitofusin 2 tethers endoplasmic reticulum to mitochondria. *Nature* **2008**, *456*, 605–610. [CrossRef] [PubMed]
37. Merkwirth, C.; Langer, T. Mitofusin 2 builds a bridge between ER and mitochondria. *Cell* **2008**, *135*, 1165–1167. [CrossRef] [PubMed]
38. Furutachi, S.; Miya, H.; Watanabe, T.; Kawai, H.; Yamasaki, N.; Harada, Y.; Imayoshi, I.; Nelson, M.; Nakayama, K.I.; Hirabayashi, Y.; et al. Slowly dividing neural progenitors are an embryonic origin of adult neural stem cells. *Nat. Neurosci.* **2015**, *18*, 657–665. [CrossRef] [PubMed]
39. Stincone, A.; Prigione, A.; Cramer, T.; Wamelink, M.M.; Campbell, K.; Cheung, E.; Olin-Sandoval, V.; Grüning, N.M.; Krüger, A.; Tauqeer Alam, M.; et al. The return of metabolism: Biochemistry and physiology of the pentose phosphate pathway. *Biol. Rev. Camb. Philos. Soc.* **2015**, *90*, 927–963. [CrossRef] [PubMed]
40. Prozorovski, T.; Schneider, R.; Berndt, C.; Hartung, H.P.; Aktas, O. Redox-regulated fate of neural stem progenitor cells. *Biochim. Biophys. Acta* **2015**, *1850*, 1543–1554. [CrossRef] [PubMed]
41. Bartesaghi, S.; Graziano, V.; Galavotti, S.; Henriquez, N.V.; Betts, J.; Saxena, J.; Minieri, V.A.D.; Karlsson, A.; Martins, L.M.; Capasso, M.; et al. Inhibition of oxidative metabolism leads to p53 genetic inactivation and transformation in neural stem cells. *Proc. Natl. Acad. Sci. USA* **2015**, *112*, 1059–1064. [CrossRef] [PubMed]
42. Stoll, E.A.; Makin, R.; Sweet, I.R.; Trevelyan, A.J.; Miwa, S.; Horner, P.J.; Turnbull, D.M. Neural Stem Cells in the Adult Subventricular Zone Oxidize Fatty Acids to Produce Energy and Support Neurogenic Activity. *Stem Cells* **2015**, *33*, 2306–2319. [CrossRef] [PubMed]
43. Knobloch, M.; Braun, S.M.; Zurkirchen, L.; von Schoultz, C.; Zamboni, N.; Araúzo-Bravo, M.J.; Kovacs, W.J.; Karalay, O.; Suter, U.; Machado, R.A.; et al. Metabolic control of adult neural stem cell activity by Fasn-dependent lipogenesis. *Nature* **2013**, *493*, 226–230. [CrossRef] [PubMed]
44. Shin, J.; Berg, D.A.; Zhu, Y.; Shin, J.Y.; Song, J.; Bonaguidi, M.A.; Enikolopov, G.; Nauen, D.W.; Christian, K.M.; Ming, G.L.; et al. Single-Cell RNA-Seq with Waterfall Reveals Molecular Cascades underlying Adult Neurogenesis. *Cell Stem Cell* **2015**, *17*, 360–372. [CrossRef] [PubMed]
45. Xie, Z.; Jones, A.; Deeney, J.T.; Hur, S.K.; Bankaitis, V.A. Inborn Errors of Long-Chain Fatty Acid β-Oxidation Link Neural Stem Cell Self-Renewal to Autism. *Cell Rep.* **2016**, *14*, 991–999. [CrossRef] [PubMed]
46. Ito, K.; Carracedo, A.; Weiss, D.; Arai, F.; Ala, U.; Avigan, D.E.; Schafer, Z.T.; Evans, R.M.; Suda, T.; Lee, C.H.; et al. A PML–PPAR-δ pathway for fatty acid oxidation regulates hematopoietic stem cell maintenance. *Nat. Med.* **2012**, *18*, 1350–1358. [CrossRef] [PubMed]
47. Duarte, A.I.; Santos, M.S.; Oliveira, C.R.; Moreira, P.I. Brain insulin signalling, glucose metabolism and females' reproductive aging: A dangerous triad in Alzheimer's disease. *Neuropharmacology* **2018**. [CrossRef] [PubMed]
48. Hihi, A.K.; Michalik, L.; Wahli, W. PPARs: Transcriptional effectors of fatty acids and their derivatives. *Cell. Mol. Life Sci.* **2002**, *59*, 790–798. [CrossRef] [PubMed]
49. Diradourian, C.; Girard, J.; Pégorier, J.P. Phosphorylation of PPARs: From molecular characterization to physiological relevance. *Biochimie* **2005**, *87*, 33–38. [CrossRef] [PubMed]

50. Anbalagan, M.; Huderson, B.; Murphy, L.; Rowan, B.G. Post-translational modifications of nuclear receptors and human disease. *Nucl. Recept. Signal.* **2012**, *10*, e001. [CrossRef] [PubMed]
51. Poulsen, L.; Siersbæk, M.; Mandrup, S. PPARs: Fatty acid sensors controlling metabolism. *Semin. Cell Dev. Biol.* **2012**, *23*, 631–639. [CrossRef] [PubMed]
52. Derosa, G.; Sahebkar, A.; Maffioli, P. The role of various peroxisome proliferator-activated receptors and their ligands in clinical practice. *J. Cell. Physiol.* **2018**, *233*, 153–161. [CrossRef] [PubMed]
53. Cimini, A.; Cerù, M.P. Emerging roles of peroxisome proliferator-activated receptors (PPARs) in the regulation of neural stem cells proliferation and differentiation. *Stem Cell Rev.* **2008**, *4*, 293–303. [CrossRef] [PubMed]
54. Shi, Y.; Hon, M.; Evans, R.M. The peroxisome proliferator-activated receptor delta, an integrator of transcriptional repression and nuclear receptor signaling. *Proc. Natl. Acad. Sci. USA* **2002**, *99*, 2613–2618. [CrossRef] [PubMed]
55. Braissant, O.; Foufelle, F.; Scotto, C.; Dauça, M.; Wahli, W. Differential expression of peroxisome proliferator-activated receptors (PPARs): Tissue distribution of PPAR-α, -β, and -γ in the adult rat. *Endocrinology* **1996**, *137*, 354–366. [CrossRef] [PubMed]
56. Peters, J.M.; Lee, S.S.; Li, W.; Ward, J.M.; Gavrilova, O.; Everett, C.; Reitman, M.L.; Hudson, L.D.; Gonzalez, F.J. Growth, adipose, brain, and skin alterations resulting from targeted disruption of the mouse peroxisome proliferator-activated receptor β(δ). *Mol. Cell. Biol.* **2000**, *20*, 5119–5128. [CrossRef] [PubMed]
57. Barroso, E.; del Valle, J.; Porquet, D.; Vieira Santos, A.M.; Salvadó, L.; Rodríguez-Rodríguez, R.; Gutiérrez, P.; Anglada-Huguet, M.; Alberch, J.; Camins, A.; et al. Tau hyperphosphorylation and increased BACE1 and RAGE levels in the cortex of PPARβ/δ-null mice. *Biochim. Biophys. Acta* **2013**, *1832*, 1241–1248. [CrossRef] [PubMed]
58. Di Loreto, S.; D'Angelo, B.; D'Amico, M.A.; Benedetti, E.; Cristiano, L.; Cinque, B.; Cifone, M.G.; Cerù, M.P.; Festuccia, C.; Cimini, A. PPARbeta agonists trigger neuronal differentiation in the human neuroblastoma cell line SH-SY5Y. *J. Cell. Physiol.* **2007**, *211*, 837–847. [CrossRef] [PubMed]
59. D'Angelo, B.; Benedetti, E.; Di Loreto, S.; Cristiano, L.; Laurenti, G.; Cerù, M.P.; Cimini, A. Signal transduction pathways involved in PPARβ/δ-induced neuronal differentiation. *J. Cell. Physiol.* **2011**, *226*, 2170–2180. [CrossRef] [PubMed]
60. Beaven, S.W.; Tontonoz, P. Nuclear receptors in lipid metabolism: Targeting the heart of dyslipidemia. *Annu. Rev. Med.* **2006**, *57*, 313–329. [CrossRef] [PubMed]
61. Coleman, J.D.; Prabhu, K.S.; Thompson, J.T.; Reddy, P.S.; Peters, J.M.; Peterson, B.R.; Reddy, C.C.; Vanden Heuvel, J.P. The oxidative stress mediator 4-hydroxynonenal is anintracellular agonist of the nuclear receptor peroxisome proliferator-activatedreceptor-β/δ (PPARβ/δ). *Free Radic. Biol. Med.* **2007**, *42*, 1155–1164. [CrossRef] [PubMed]
62. Stump, M.; Guo, D.F.; Lu, K.T.; Mukohda, M.; Cassell, M.D.; Norris, A.W.; Rahmouni, K.; Sigmund, C.D. Nervous System Expression of PPARγ and Mutant PPARγ Has Profound Effects on Metabolic Regulation and Brain Development. *Endocrinology* **2016**, *157*, 4266–4275. [CrossRef] [PubMed]
63. Di Giacomo, E.; Benedetti, E.; Cristiano, L.; Antonosante, A.; d'Angelo, M.; Fidoamore, A.; Barone, D.; Moreno, S.; Ippoliti, R.; Cerù, M.P.; et al. Roles of PPAR transcription factors in the energetic metabolic switch occurring during adult neurogenesis. *Cell Cycle* **2017**, *16*, 59–72. [CrossRef] [PubMed]
64. Thambisetty, M.; Beason-Held, L.L.; An, Y.; Kraut, M.; Metter, J.; Egan, J.; Ferrucci, L.; O'Brien, R.; Resnick, S.M. Impaired glucose tolerance in midlife and longitudinal changes in brain function during aging. *Neurobiol. Aging* **2013**, *34*, 2271–2276. [CrossRef] [PubMed]
65. Moloney, A.M.; Griffin, R.J.; Timmons, S.; O'Connor, R.; Ravid, R.; O'Neill, C. Defects in IGF-1 receptor, insulin receptor and IRS-1/2 in Alzheimer's disease indicate possible resistance to IGF-1 and insulin signalling. *Neurobiol. Aging* **2010**, *31*, 224–243. [CrossRef] [PubMed]
66. Yarchoan, M.; Toledo, J.B.; Lee, E.B.; Arvanitakis, Z.; Kazi, H.; Han, L.Y.; Louneva, N.; Lee, V.M.; Kim, S.F.; Trojanowski, J.Q.; et al. Abnormal serine phosphorylation of insulin receptor substrate 1 is associated with tau pathology in Alzheimer's disease and tauopathies. *Acta Neuropathol.* **2014**, *128*, 679–689. [CrossRef] [PubMed]
67. Mark, R.J.; Pang, Z.; Geddes, J.W.; Uchida, K.; Mattson, M.P. Amyloid beta-peptide impairs glucose transport in hippocampal and cortical neurons: Involvement of membrane lipid peroxidation. *J. Neurosci.* **1997**, *17*, 1046–1054. [CrossRef] [PubMed]

68. Mattson, M.P. Roles of the lipid peroxidation product 4-hydroxynonenal in obesity, the metabolic syndrome, and associated vascular and neurodegenerative disorders. *Exp. Gerontol.* **2009**, *44*, 625–633. [CrossRef] [PubMed]
69. Cutler, R.G.; Kelly, J.; Storie, K.; Pedersen, W.A.; Tammara, A.; Hatanpaa, K.; Troncoso, J.C.; Mattson, M.P. Involvement of oxidative stress induced abnormalities in ceramide and cholesterol metabolism in brain aging and Alzheimer's disease. *Proc. Natl. Acad. Sci. USA* **2004**, *101*, 2070–2075. [CrossRef] [PubMed]
70. Denis, I.; Potier, B.; Vancassel, S.; Heberden, C.; Lavialle, M. Omega-3 fatty acids and brain resistance to ageing and stress: Body of evidence and possible mechanisms. *Ageing Res. Rev.* **2013**, *12*, 579–594. [CrossRef] [PubMed]
71. Appleton, J.P.; Scutt, P.; Sprigg, N.; Bath, P.M. Hypercholesterolaemia and vascular dementia. *Clin. Sci.* **2017**, *131*, 1561–1578. [CrossRef] [PubMed]
72. Lane-Donovan, C.; Philips, G.T.; Herz, J. More than cholesterol transporters: Lipoprotein receptors in CNS function and neurodegeneration. *Neuron* **2014**, *83*, 771–787. [CrossRef] [PubMed]
73. Agarwal, S.; Yadav, A.; Chaturvedi, R.K. Peroxisome proliferator-activated receptors (PPARs) as therapeutic target in neurodegenerative disorders. *Biochem. Biophys. Res. Commun.* **2017**, *483*, 1166–1177. [CrossRef] [PubMed]
74. Cipolla, C.M.; Lodhi, I.J. Peroxisomal Dysfunction in Age-Related Diseases. *Trends Endocrinol. Metab.* **2017**, *28*, 297–308. [CrossRef] [PubMed]
75. Geddes, J.W.; Tekirian, T.L.; Soultanian, N.S.; Ashford, J.W.; Davis, D.G.; Markesbery, W.R. Comparison of neuropathologic criteria for the diagnosis of Alzheimer's disease. *Neurobiol. Aging* **1997**, *18*, 99–105. [CrossRef]
76. Rodriguez-Oroz, M.C.; Jahanshahi, M.; Krack, P.; Litvan, I.; Macias, R.; Bezard, E.; Obeso, J.A. Initial clinical manifestations of Parkinson's disease: Features and pathophysiological mechanisms. *Lancet Neurol.* **2009**, *8*, 1128–1139. [CrossRef]
77. Klingelhoefer, L.; Reichmann, H. Pathogenesis of Parkinson disease–the gut-brain axis and environmental factors. *Nat. Rev. Neurol.* **2015**, *11*, 625–636. [CrossRef] [PubMed]
78. Morsci, N.S.; Hall, D.H.; Driscoll, M.; Sheng, Z.H. Age-Related Phasic Patterns of Mitochondrial Maintenance in Adult Caenorhabditis elegans Neurons. *J. Neurosci.* **2016**, *36*, 1373–1385. [CrossRef] [PubMed]
79. Chaturvedi, R.K.; Flint Beal, M. Mitochondrial diseases of the brain. *Free Radic. Biol. Med.* **2013**, *63*, 1–29. [CrossRef] [PubMed]
80. Marambaud, P.; Dreses-Werringloer, U.; Vingtdeux, V. Calcium signaling in neurodegeneration. *Mol. Neurodegener.* **2009**, *47*, 140–149. [CrossRef] [PubMed]
81. Rcom-H'cheo-Gauthier, A.; Goodwin, J.; Pountney, D.L. Interactions between calcium and alpha synuclein in neurodegeneration. *Biomolecules* **2014**, *4*, 795–811. [CrossRef] [PubMed]
82. Denton, R.M.; Rutter, G.A.; Midgley, P.J.; McCormack, J.G. Effects of Ca^{2+} on the activities of the calcium-sensitive dehydrogenases within the mitochondria of mammalian tissues. *J. Cardiovasc. Pharmacol.* **1988**, *12*, S69–S72. [CrossRef] [PubMed]
83. Perier, C.; Vila, M. Mitochondrial biology and Parkinson's disease. *Cold Spring Harb. Perspect. Med.* **2012**, *2*, a009332. [CrossRef] [PubMed]
84. Cookson, M.R. Parkinsonism due to mutations in PINK1, parkin, and DJ-1 and oxidative stress and mitochondrial pathways. *Cold Spring Harb. Perspect. Med.* **2012**, *2*, a009415. [CrossRef] [PubMed]
85. Deng, H.; Dodson, M.W.; Huang, H.; Guo, M. The Parkinson's disease genes pink1 and parkin promote mitochondrial fission and/or inhibit fusion in Drosophila. *Proc. Natl. Acad. Sci. USA* **2008**, *105*, 14503–14508. [CrossRef] [PubMed]
86. Vega, R.B.; Huss, J.M.; Kelly, D.P. The coactivator PGC-1 cooperates with peroxisome proliferator-activated receptor alpha in transcriptional control of nuclear gene encoding mitochondrial fatty acid oxidation enzymes. *Mol. Cell. Biol.* **2000**, *20*, 1868–1876. [CrossRef] [PubMed]
87. St-Pierre, J.; Drori, S.; Uldry, M.; Silvaggi, J.M.; Rhee, J.; Jäger, S.; Handschin, C.; Zheng, K.; Lin, J.; Yang, W.; et al. Suppression of reactive oxygen species and neurodegeneration by the PGC-1 transcriptional coactivators. *Cell* **2006**, *127*, 397–408. [CrossRef] [PubMed]
88. Uldry, M.; Yang, W.; St-Pierre, J.; Lin, J.; Seale, P.; Spiegelman, B.M. Complementary action of the PGC-1 coactivators in mitochondrial biogenesis and brown fat differentiation. *Cell Metab.* **2006**, *3*, 333–341. [CrossRef] [PubMed]

89. Nunnari, J.; Suomalainen, A. Mitochondria: In sickness and in health. *Cell* **2012**, *148*, 1145–1159. [CrossRef] [PubMed]
90. Porcellotti, S.; Fanelli, F.; Fracassi, A.; Sepe, S.; Cecconi, F.; Bernardi, C.; Cimini, A.; Cerù, M.P.; Moreno, S. Oxidative Stress during the Progression of β-Amyloid Pathology in the Neocortex of the Tg2576 Mouse Model of Alzheimer's Disease. *Oxidative Med. Cell. Longev.* **2015**, *2015*, 967203. [CrossRef] [PubMed]
91. Fanelli, F.; Sepe, S.; D'Amelio, M.; Bernardi, C.; Cristiano, L.; Cimini, A.; Cecconi, F.; Ceru, M.P.; Moreno, S. Age-dependent roles of peroxisomes in the hippocampus of a transgenic mouse model of Alzheimer's disease. *Mol. Neurodegener.* **2013**, *8*, 8. [CrossRef] [PubMed]
92. Cimini, A.; Moreno, S.; D'Amelio, M.; Cristiano, L.; D'Angelo, B.; Falone, S.; Benedetti, E.; Carrara, P.; Fanelli, F.; Cecconi, F.; et al. Early biochemical and morphological modifications in the brain of a transgenic mouse model of Alzheimer's disease: A role for peroxisomes. *J. Alzheimers Dis.* **2009**, *18*, 935–952. [CrossRef] [PubMed]
93. Fransen, M.; Nordgren, M.; Wang, B.; Apanasets, O.; Van Veldhoven, P.P. Aging, age-related diseases and peroxisomes. *Subcell. Biochem.* **2013**, *69*, 45–65. [PubMed]
94. Terlecky, S.R.; Terlecky, L.J.; Giordano, C.R. Peroxisomes, oxidative stress, and inflammation. *World J. Biol. Chem.* **2012**, *3*, 93–97. [CrossRef] [PubMed]
95. Ivashchenko, O.; Van Veldhoven, P.P.; Brees, C.; Ho, Y.S.; Terlecky, S.R.; Fransen, M. Intraperoxisomal redox balance in mammalian cells: Oxidative stress and interorganellar cross-talk. *Mol. Biol. Cell* **2011**, *22*, 1440–1451. [CrossRef] [PubMed]
96. Hwang, I.; Lee, J.; Huh, J.Y.; Park, J.; Lee, H.B.; Ho, Y.S.; Ha, H. Catalase deficiency accelerates diabetic renal injury through peroxisomal dysfunction. *Diabetes* **2012**, *61*, 728–738. [CrossRef] [PubMed]
97. Fabelo, N.; Martín, V.; Santpere, G.; Marín, R.; Torrent, L.; Ferrer, I.; Díaz, M. Severe alterations in lipid composition of frontal cortex lipid rafts from Parkinson's disease and incidental Parkinson's disease. *Mol. Med.* **2011**, *17*, 1107–1118. [CrossRef] [PubMed]
98. Kou, J.; Kovacs, G.G.; Höftberger, R.; Kulik, W.; Brodde, A.; Forss-Petter, S.; Hönigschnabl, S.; Gleiss, A.; Brügger, B.; Wanders, R.; et al. Peroxisomal alterations in Alzheimer's disease. *Acta Neuropathol.* **2011**, *122*, 271–283. [CrossRef] [PubMed]
99. Combs, C.K.; Johnson, D.E.; Karlo, J.C.; Cannady, S.B.; Landreth, G.E. Inflammatory mechanisms in Alzheimer's disease: Inhibition of beta-amyloid-stimulated proinflammatory responses and neurotoxicity by PPARgamma agonists. *J. Neurosci.* **2000**, *20*, 558–567. [CrossRef] [PubMed]
100. Benedetti, E.; Cristiano, L.; Antonosante, A.; d'Angelo, M.; D'Angelo, B.; Selli, S.; Castelli, V.; Ippoliti, R.; Giordano, A.; Cimini, A. PPARs in Neurodegenerative and Neuroinflammatory Pathways. *Curr. Alzheimer Res.* **2018**, *15*, 336–344. [CrossRef] [PubMed]
101. Searcy, J.L.; Phelps, J.T.; Pancani, T.; Kadish, I.; Popovic, J.; Anderson, K.L.; Beckett, T.L.; Murphy, M.P.; Chen, K.C.; Blalock, E.M.; et al. Long-term pioglitazone treatment improves learning and attenuates pathological markers in a mouse model of Alzheimer's disease. *J. Alzheimers Dis.* **2012**, *30*, 943–961. [CrossRef] [PubMed]
102. Craft, S.; Cholerton, B.; Baker, L.D. Insulin and Alzheimer's disease: Untangling the web. *J. Alzheimers Dis.* **2013**, *33*, S263–S275. [CrossRef] [PubMed]
103. Escribano, L.; Simón, A.M.; Gimeno, E.; Cuadrado-Tejedor, M.; López de Maturana, R.; García-Osta, A.; Ricobaraza, A.; Pérez-Mediavilla, A.; Del Río, J.; Frechilla, D. Rosiglitazone rescues memory impairment in Alzheimer's transgenic mice: Mechanisms involving a reduced amyloid and tau pathology. *Neuropsychopharmacology* **2010**, *35*, 1593–1604. [CrossRef] [PubMed]
104. Cao, B.; Rosenblat, J.D.; Brietzke, E.; Park, C.; Lee, Y.; Musial, N.; Pan, Z.; Mansur, R.B.; McIntyre, R.S. Comparative Efficacy and Acceptability of Anti-Diabetic Agents for Alzheimer's Disease and Mild Cognitive Impairment: A Systematic Review and Network Meta-analysis. *Diabetes Obes. Metab.* **2018**. [CrossRef] [PubMed]
105. Subramanian, S.; Gottschalk, W.K.; Kim, S.Y.; Roses, A.D.; Chiba-Falek, O. The effects of PPARγ on the regulation of the TOMM40-APOE-C1 genes cluster. *Biochim. Biophys. Acta* **2017**, *1863*, 810–816. [CrossRef] [PubMed]
106. Cheng, Y.H.; Lai, S.W.; Chen, P.Y.; Chang, J.H.; Chang, N.W. PPARα activation attenuates amyloid-β-dependent neurodegeneration by modulating Endo G and AIF translocation. *Neurotoxic. Res.* **2015**, *27*, 55–68. [CrossRef] [PubMed]

107. Fidaleo, M.; Fanelli, F.; Cerù, M.P.; Moreno, S. Neuroprotective properties of peroxisome proliferator-activated receptor alpha (PPARα) and its lipid ligands. *Curr. Med. Chem.* **2014**, *21*, 2803–2821. [CrossRef] [PubMed]
108. Brune, S.; Kölsch, H.; Ptok, U.; Majores, M.; Schulz, A.; Schlosser, R.; Rao, M.L.; Maier, W.; Heun, R. Polymorphism in the peroxisome proliferator-activated receptor alpha gene influences the risk for Alzheimer's disease. *J. Neural Transm.* **2003**, *110*, 1041–1050. [CrossRef] [PubMed]
109. Kölsch, H.; Lehmann, D.J.; Ibrahim-Verbaas, C.A.; Combarros, O.; van Duijn, C.M.; Hammond, N.; Belbin, O.; Cortina-Borja, M.; Lehmann, M.G.; Aulchenko, Y.S.; et al. Interaction of insulin and PPAR-α genes in Alzheimer's disease: The Epistasis Project. *J. Neural Transm.* **2012**, *119*, 473–479. [CrossRef] [PubMed]
110. Roy, A.; Jana, M.; Kundu, M.; Corbett, G.T.; Rangaswamy, S.B.; Mishra, R.K.; Luan, C.H.; Gonzalez, F.J.; Pahan, K. HMG-CoA Reductase Inhibitors Bind to PPARα to Upregulate Neurotrophin Expression in the Brain and Improve Memory in Mice. *Cell Metab.* **2015**, *22*, 253–265. [CrossRef] [PubMed]
111. Kreisler, A.; Gelé, P.; Wiart, J.F.; Lhermitte, M.; Destée, A.; Bordet, R. Lipid-lowering drugs in the MPTP mouse model of Parkinson's disease: Fenofibrate has a neuroprotective effect, whereas bezafibrate and HMG-CoA reductase inhibitors do not. *Brain Res.* **2007**, *1135*, 77–84. [CrossRef] [PubMed]
112. Giordano Attianese, G.M.; Desvergne, B. Integrative and systemic approaches for evaluating PPARβ/δ (PPARD) function. *Nucl. Recept. Signal.* **2015**, *13*, e001. [CrossRef] [PubMed]
113. Zolezzi, J.M.; Bastías-Candia, S.; Santos, M.J.; Inestrosa, N.C. Alzheimer's disease: Relevant molecular and physiopathological events affecting amyloid-β brain balance and the putative role of PPARs. *Front. Aging Neurosci.* **2014**, *6*, 176. [CrossRef] [PubMed]

International Journal of
Molecular Sciences

Review

Demyelination in Multiple Sclerosis: Reprogramming Energy Metabolism and Potential PPARγ Agonist Treatment Approaches

Alexandre Vallée [1,*], Yves Lecarpentier [2], Rémy Guillevin [3] and Jean-Noël Vallée [4,5]

1. Délégation à la Recherche Clinique et à l'Innovation (DRCI), Hôpital Foch, 92150 Suresnes, France
2. Centre de Recherche Clinique, Grand Hôpital de l'Est Francilien (GHEF), 77100 Meaux, France; yves.c.lecarpentier@gmail.com
3. Data Analysis and Computations Through Imaging Modeling-Mathématiques (DACTIM), Unité mixte de recherche (UMR), Centre National de la Recherche Scientifique (CNRS) 7348 (Laboratoire de Mathématiques et Application), University of Poitiers, Centre Hospitalier Universitaire (CHU) de Poitiers, 86000 Poitiers, France; remy.guillevin@chu-poitiers.fr
4. Centre Hospitalier Universitaire (CHU) Amiens Picardie, University of Picardie Jules Verne (UPJV), 80000 Amiens, France; valleejn@gmail.com
5. LMA (Laboratoire de Mathématiques et Applications), Unité mixte de recherche (UMR), Centre National de la Recherche Scientifique (CNRS) 7348, Université de Poitiers, 86000 Poitiers, France

* Correspondence: alexandre.g.vallee@gmail.com; Tel.: +33-629-303-240

Received: 21 March 2018; Accepted: 11 April 2018; Published: 16 April 2018

Abstract: Demyelination in multiple sclerosis (MS) cells is the site of several energy metabolic abnormalities driven by dysregulation between the opposed interplay of peroxisome proliferator-activated receptor γ (PPARγ) and WNT/β-catenin pathways. We focus our review on the opposing interactions observed in demyelinating processes in MS between the canonical WNT/β-catenin pathway and PPARγ and their reprogramming energy metabolism implications. Demyelination in MS is associated with chronic inflammation, which is itself associated with the release of cytokines by $CD4^+$ Th17 cells, and downregulation of PPARγ expression leading to the upregulation of the WNT/β-catenin pathway. Upregulation of WNT/β-catenin signaling induces activation of glycolytic enzymes that modify their energy metabolic behavior. Then, in MS cells, a large portion of cytosolic pyruvate is converted into lactate. This phenomenon is called the Warburg effect, despite the availability of oxygen. The Warburg effect is the shift of an energy transfer production from mitochondrial oxidative phosphorylation to aerobic glycolysis. Lactate production is correlated with increased WNT/β-catenin signaling and demyelinating processes by inducing dysfunction of $CD4^+$ T cells leading to axonal and neuronal damage. In MS, downregulation of PPARγ decreases insulin sensitivity and increases neuroinflammation. PPARγ agonists inhibit Th17 differentiation in $CD4^+$ T cells and then diminish release of cytokines. In MS, abnormalities in the regulation of circadian rhythms stimulate the WNT pathway to initiate the demyelination process. Moreover, PPARγ contributes to the regulation of some key circadian genes. Thus, PPARγ agonists interfere with reprogramming energy metabolism by directly inhibiting the WNT/β-catenin pathway and circadian rhythms and could appear as promising treatments in MS due to these interactions.

Keywords: WNT/β-catenin pathway; PPARγ; multiple sclerosis; energy metabolism; aerobic glycolysis; demyelination; Warburg effect; circadian rhythms; clock genes

1. Introduction

Multiple sclerosis (MS) presents chronic inflammation, immune responses, blood–brain barrier (BBB) breakdown, and demyelination in the white matter of the central nervous system (CNS) [1,2].

In brain and spinal cord areas, chronic inflammation leads to axonal myelin sheath destruction and the progressive loss of neurological functions with neuronal death. The inflammatory process in MS is initiated by the microglia in association with the release of players CD4$^+$ helper (Th) (Th1 and Th17), the markers of the chronic inflammation [3]. Pro-inflammatory mediators, such as cytokines (interleukin (IL-6, IL-17, IL-22), tumor necrosis factor α (TNF-α)), are synthetized by Th17 cells, which are the main immune actors in the pathogenesis of MS [4]. MS can be considered as an autoimmune disease which presents neurological disability and many genetic and environmental determinant etiologies [5].

Glial cells, called oligodendrocytes (OLs), synthetize myelin sheaths in CNS by wrapping axons with multi-lamellar sheets of plasma membrane which are composed of specific lipids and proteins. Loss of myelinating OLs is considered as the origin of MS pathogenesis [6–9]. In white matter lesions of MS, oligodendrocyte precursor cells (OPC) present a stop state and a non-differentiation into myelinating OLs [6,10–14].

Altered cells in MS are derived from exergonic processes and emit heat that flows to the surrounding environment. Several irreversible processes occur by changing reprogramming energy metabolism [15,16].

Peroxisome proliferator-activated receptor γ (PPARγ) and the WNT/β-catenin pathway act in an opposite manner in many diseases, including MS [17,18]. Numerous autoimmune disorders present this opposed interplay, such as type 1 diabetes [19,20], thyroid autoimmunity [21,22] and rheumatoid arthritis [23,24].

In MS, the dysregulation of both PPARγ [25] and the WNT/β-catenin pathway [26] influence several statistical mechanisms by modifying energy metabolism leading to aerobic glycolysis, called the Warburg effect [27,28].

PPARγ is a member of the nuclear superfamily of ligand-activated transcription factors which regulates glucose metabolism and cellular homeostasis. WNT ligands belong to the family of glycoproteins participating in the control of cell cycle, cell life and embryogenesis.

The Warburg effect is the shift of an energy transfer production from mitochondrial oxidative phosphorylation to aerobic glycolysis. The Warburg effect was discovered by Otto Warburg in 1930 in cancer processes [28]. This energy shift is partly due to injury of mitochondrial respiration, leading to an increase of adenosine triphosphate (ATP) production by glycolysis. Indeed, although aerobic glycolysis is less efficient in producing ATP molecules than oxidative phosphorylation, its production cycles are much faster than those of oxidation phosphorylation [29], which results in higher ATP molecule production than oxidative phosphorylation [30]. Recent studies have shown that this phenomenon is not specific to cancers but is also observed in non-tumor diseases, such as MS [31].

In parallel, dysregulation of circadian rhythms (CRs) has been observed in MS [32]. This dysfunction leads to upregulation of the canonical WNT/β-catenin pathway that contributes to MS pathogenesis. PPARγ can control CRs by regulating some key circadian genes, like Bmal1 (brain and muscle aryl-hydrocarbon receptor nuclear translocator-like 1) [33] and can directly target the WNT pathway [34] and energy balance in CNS [35]. By acting on these systems, PPARγ appears as an interesting therapeutic pathway. In MS, the opposed interplay between PPARγ and the WNT/β-catenin pathway has a major role in the dysregulation of energy metabolism and the disruption of CRs. Several energy balance abnormalities found in MS are induced by several cellular processes involved in both of these. We focus this review on the opposing interactions observed in MS between PPARγ and the canonical WNT/β-catenin pathway and their reprogramming energy metabolism implications.

2. PPARγ

Peroxisome proliferator-activated receptor γ (PPARγ) is an orphan nuclear receptor which is a member of the nuclear superfamily of ligand-activated transcription factors [36,37]. PPARγ is composed of a ligand binding domain which is hydrophobic and a type II zinc finger DNA-binding domain [38].

PPARγ ligands form a heterodimer with the retinoic X receptor (RXR). RXR is a 9-*cis* retinoic acid receptor. The heterodimer binds to peroxisome proliferator response element (PPRE) to activate several target genes [39]. PPARγ is highly expressed in adipose tissues [40] and in cardiac and skeletal muscle, pancreatic β-cells, kidney, macrophages [41], and other vascular cells, like endothelial cells [42,43].

PPARγ expression is implicated in numerous homeostasis pathways such as glucose and lipid metabolism. Likewise, PPARγ expression is implicated in migration, apoptosis, cell growth, antioxidant and inflammatory responses [39,44,45]. PPARγ is normally little expressed in CNS [46], but its expression is found in neurons, OLs, astrocytes, microglia/macrophages [47], T and B lymphocytes, dendritic cells [48] and brain endothelial cells [49]. PPARγ can repress inflammation by decreasing nuclear factor-κB (NF-κB) activity [50].

Synthetic ligands of PPARγ are prostaglandins like 15-deoxy-Δ, 14 prostaglandin J2 [51], hydroxyl octadecadienoic acid with derivatives of fatty acid oxidation [52] and lysophosphatidic acid (LPA) [53]. Pioglitazone and rosiglitazone are thiazolidinediones (TZD) which are synthetic PPARγ ligands [52].

3. Canonical WNT/β-Catenin Pathway (Figure 1)

Canonical WNT/β-catenin pathway is named as the discovery of the cascade gene "W"ingless in drosophila and its homologue in mice "INT"(Integration site) [54] (Figure 1). The WNT/β-catenin pathway is involved in numerous life cycles, such as embryogenesis in migration, proliferation, differentiation, apoptosis and cell polarity [55]. Deregulation of the WNT/β-catenin pathway is observed in several pathologies, such as cancers, fibrosis, neurodegenerative diseases, and atherosclerosis, and its targeting appears as an emerging therapeutic pathway [56].

WNT ligands are glycoproteins, which activate the canonical WNT/β-catenin pathway [57]. Extracellular WNT ligands bind the receptor Frizzled (FZD) and then stimulate the co-receptor Low-Density Lipoprotein (LDL) receptor-related proteins 5 and 6 (LRP 5/6) [58].

β-catenin is considered as the main molecule of the canonical WNT pathway. Its major function is transcriptional activity. In physiologic conditions, cytoplasmic β-catenin is in constant turnover between synthetized and destroyed intracellular cycles.

Cytosolic β-catenin is maintained at a minimal level through the activation of the β-catenin destruction complex, which is formed by a combination of AXIN (a cytoplasmic protein regulating G-protein signaling), glycogen synthase kinase-3β (GSK-3β, a serine-theronine kinase), adenomatous polyposis coli (APC, a tumor suppressor gene), and casein kinase 1 (CK-1, a serine/threonine-selective enzyme) [59]. CK-1 and GSK-3β target β-catenin by phosphorylating the serine and threonine residues located in the amino acid terminus [60–62]. CK-1 phosphorylates an N-terminus of β-catenin and GSK-3β phosphorylates a threonine 41 (Th41), Ser33 and Ser37 sites of β-catenin [55,63]. These phosphorylations result in the recruiting of APC in the destruction complex. APC modulates the degradation of the cytosolic β-catenin into the proteasome through its tumor suppressor properties [59,64].

Activation of the WNT/β-catenin pathway is characterized by the initiation of WNT ligands and their interactions with FZD and LRP 5/6 co-receptors [65]. This binding stimulates Disheveled (DSH) to inhibit the destruction complex and to permit cytosolic β-catenin accumulation. Nuclear β-catenin binds T-cell factor/lymphoid enhancer factor (TCF/LEF) to activate several WNT target genes, such as c-Myc and cyclin D1 [66,67].

Demyelinating events present an upregulation of the WNT/β-catenin pathway correlated with a release of pro-inflammatory cytokines [68]. Moreover, PPARγ stimulation has a beneficial role in MS [69,70] through the decrease of neuroinflammation [71] and the downregulation of the WNT/β-catenin pathway in MS [17,18]. PPARγ agonists are considered as potential therapeutic perspectives against neuroinflammation and neurodegeneration [72]. In MS, these two pathways operate in an opposed interplay [18] and their dysregulations lead to energy metabolism reprogramming. The objectives of this review are to describe this opposed crosstalk with circadian rhythms regulation, and to better understand the energy remodeling aspect, called the Warburg effect, observed in MS and the potential therapeutic benefits of targeting these two pathways to improve MS-related symptoms.

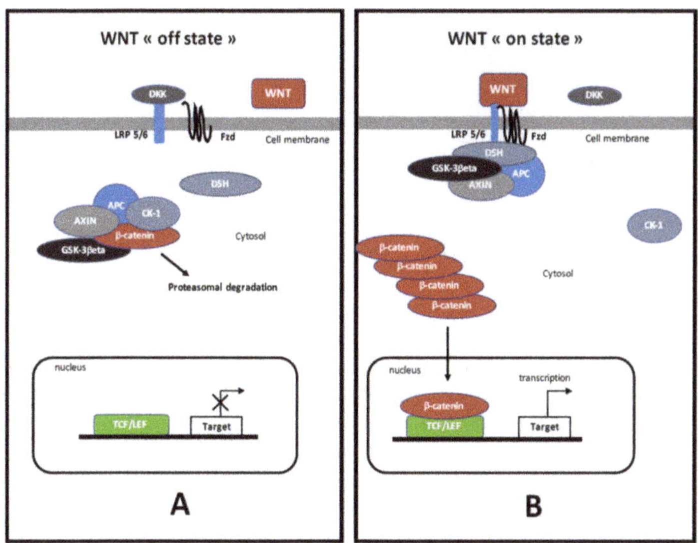

Figure 1. The canonical WNT/β-catenin pathway. (**A**) Under physiological circumstances, the WNT "off state", the cytosolic β-catenin is bound to its destruction complex, consisting of adenomatous polyposis coli (APC), AXIN and glycogen synthase kinase-3β (GSK-3β). After CK-1 phosphorylates on Ser45 residue, β-catenin is further phosphorylated on Thr41, Ser37, and Ser33 residues by GSK-3β. Then, phosphorylated β-catenin is degraded into the proteasome. Therefore, the cytosolic level of β-catenin is kept low in the absence of WNT ligands. If β-catenin is not present in the nucleus, the T-cell factor/lymphoid enhancer factor (TCF/LEF) complex cannot activate the target genes. Dickkopf (DKK) can inhibit the WNT/β-catenin pathway by binding to WNT ligands or low-density lipoprotein receptor-related protein 5/6 (LRP 5/6). (**B**) When WNT ligands bind to both Frizzled (FZD) and LRP 5/6, the WNT "on state", Disheveled (DSH) is recruited and phosphorylated by FZD. Phosphorylated DSH in turn recruits AXIN, which dissociates the β-catenin destruction complex. Therefore, β-catenin escapes from phosphorylation and subsequently accumulates in the cytosol. The accumulated cytosolic β-catenin goes into the nucleus, where it binds to TCF/LEF and activates the transcription of target genes.

4. Crosstalk between PPARγ and Canonical WNT/β-Catenin Pathway in Diseases

The opposed interplay between the canonical WNT/β-catenin pathway and PPARγ has been observed in numerous pathologies. Cancers, such as gliomas [73–75] and colon cancer [76], present an upregulation of the canonical WNT/β-catenin pathway associated with a decrease of PPARγ expression [77]. The process of fibrosis exhibits the same mechanism [78–80]. Neurodegenerative diseases are classified in two categories [34], i.e., diseases that present a downregulation of the canonical WNT/β-catenin pathway and an upregulation of PPARγ, such as Alzheimer's disease [81–83], and diseases with an upregulation of the canonical WNT/β-catenin pathway whereas PPARγ is decreased, such as exudative age related macular degeneration [84,85], amyotrophic lateral sclerosis [86], and multiple sclerosis [18].

Numerous studies have suggested that PPARγ may be considered as a negative β-catenin target [87,88]. The β-catenin pathway can decrease PPARγ expression [89–98]. Indeed, PPARγ and WNT/β-catenin pathway interact via a catenin-binding domain within PPARγ and a TCF/LEF β-catenin domain [99–102].

The decrease of the WNT/β-catenin pathway stimulates the expression of PPARγ [103], while the increase of PPARγ expression inhibits β-catenin levels in numerous cellular systems [104–106]. Troglitazone, a PPARγ agonist, can downregulate c-Myc expression, a WNT target gene [107]. PPARγ agonists, can activate WNT inhibitors, such as Dicckopf-1 (DKK1) [108] and GSK-3β [109] to decrease β-catenin levels. In parallel, the WNT target COUP II can decrease PPARγ [110]. Inflammatory cytokines and cellular pathways, such as WNT/β-catenin pathway, interleukin 1 (IL-1) and TNF-α, can inhibit PPARγ expression [111–113].

5. PPARγ and the Canonical WNT/β-Catenin Pathway in MS

5.1. PPARγ in MS

Several studies have shown that PPARγ agonists can reduce the clinical expression of experimental autoimmune encephalomyelitis (EAE) models of MS (Table 1). In EAE models, the PPARγ agonist 15-deoxy-Δ(12,14)-prostaglandin acts by inhibiting NF-κB activity [114–116]. In addition, PPARγ deficiency has been shown to exacerbate the clinical symptoms of EAE models [117]. The downregulation of PPARγ during demyelination in MS is well-described in previous studies [18]. However, the stimulation of PPARγ [118,119] leads to decreased inflammation and permits the remyelination in oligodendrocytes (OLs) models of MS [120]. The overexpression of PPARγ is correlated with neuroprotection in both OLs and neurons [121–128]. Th17 differentiation is decreased by PPARγ agonists in both murine $CD4^+$ T cells and in human models [129]. In CNS-infiltrating $CD4^+$ T cells, IL-17 expression is diminished by PPARγ overexpression [130]. The anti-inflammatory role of PPARγ is responsible for both the decreased release of inflammatory cytokines [41,131,132], and the expansion of encephalitogenic Th1 [117], Th17 cells [129] and B lymphocytes [133]. Lovastatin induces the expression of PPARγ in the central nervous system (CNS) of EAE models [134]. However, simvastatin impedes the remyelination mechanism in cuprizone-CNS demyelinating models (non-EAE-models) [135,136].

5.2. Demyelination and Activation of WNT/β-Catenin Pathway

Several studies have shown that the WNT/β-catenin pathway is overexpressed during the demyelination process (for review, see [18]) (Table 1). The expression of WNT/β-catenin pathway is overexpressed in the spinal cord dorsal horn (SCDH) in EAE models of mice [68]. The increase of the β-catenin inhibitor indomethacin is known to decrease mechanical allodynia in EAE mice [68]. In EAE models, over-activation of the WNT/β-catenin pathway impairs and delays OPC differentiation [137]. The WNT/β-catenin pathway, by stimulating pro-inflammatory cytokines, has a major role in neuropathic pain pathogenesis [138]. β-catenin accumulation and nuclear transcription are associated with alteration of endothelial adherens in experimental models [139,140] and in MS brain tissue [141].

5.3. Opposed Interaction between PPARγ and WNT Pathway in MS

In MS models, moringin (4-[α-L-rhamnopyranosyloxy]-benzyl isothiocyanate) can modulate neuroinflammation through both decreased β-catenin signaling and increased PPARγ expression [142]. Moringin can also repress inflammatory factors, such as IL-1, IL-6 and cyclo-oxygenase-2 (COX2) in EAE mice by increasing PPARγ levels [142]. In MS, moringin is known to protect against neurodegenerative disorders [143,144].

6. Reprogramming Energy Metabolism in Demyelination

6.1. Aerobic Glycolysis

Aerobic glycolysis, called the Warburg effect, is the conversion of glucose to lactate in the presence of oxygen sufficient to support glucose catabolism via the tricarboxylic acid (TCA) cycle with oxidative phosphorylation [28] (Figure 2). Numerous studies have shown that the canonical WNT/β-catenin pathway stimulates aerobic glycolysis and glycolytic enzymes such as glucose transporter (Glut), hexokinase (HK), pyruvate kinase M2 (PKM2), lactate dehydrogenase A (LDH-A), monocarboxylate transporter 1 (MCT-1) [27,73,77,85,145,146]. An increased rate of glucose metabolism is correlated with activation of the PI3K/Akt pathway [147]. The WNT/β-catenin pathway directly stimulates PI3K/Akt signaling [148,149]. Activation of the PI3K/Akt pathway leads to HIF-1α stimulation (hypoxia-inducible factor 1-α) [150] to induce overexpression of glycolytic enzymes such as Glut, LDH-A, pyruvate dehydrogenase kinase 1 (PDK1) and PKM2 [150,151]. The allosteric enzyme Phosphofructokinase (PFK) catalyzes the conversion between β-D-fructose-6-phosphate and β-D-fructose-1,6-biphosphate. This reaction, by using ATP, leads to glycolytic oscillations and can be organized in time and space driven by PFK with a positive feedback responsible for periodic behavior [152].

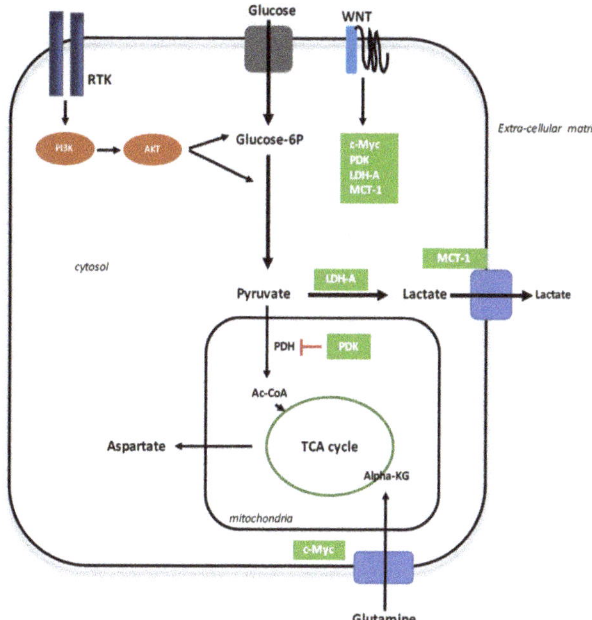

Figure 2. Aerobic glycolysis stimulation by activated canonical WNT/β-catenin pathway. Activation of the receptor tyrosine kinase (RTK) is required to take up enough glucose to cell survival. PI3K/Akt pathway is stimulated to maintain a sufficient ATP production through the metabolism of glucose. Glucose is transformed into pyruvate into the mitochondria for the oxidative phosphorylation process. However, during WNT activation, WNT signal transduction results in activation of c-Myc, lactate dehydrogenase A (LDH-A), pyruvate dehydrogenase kinase (PDK) and monocarboxylate transporter 1 (MCT-1). The WNT target genes cooperate to divert glycolytically derived pyruvate into lactate which is expelled out the cell by MCT-1. Moreover, c-Myc induces glutamine uptake and glutaminolysis to support mitochondrial integrity and aspartate production. Accumulation of cytosolic lactate involves several pathways such as nucleotide synthesis, lipid synthesis and cell division.

6.2. Aerobic Glycolysis in MS

An imbalance between energy production and consumption has been observed in MS [153–155]. Decrease of oxidative phosphorylation and mRNA deletions observed in MS neuronal cell bodies indicate a mitochondrial dysregulation [156,157]. Indeed, alteration of energy metabolism is observed in urine [158] and in serum of MS patients [159]. Activation of aerobic glycolysis and decrease of oxidative phosphorylation aggravate MS pathogenesis by inducing dysfunction of $CD4^+$ T cell [160]. $CD4^+$ T cell dysregulation has a major role in MS pathogenesis by aggravating axonal and neuronal damage [1,161].

Inhibition of aerobic glycolysis in MS by copaxone restores mitochondrial activity and then diminish $CD4^+$ T cell dysregulation [162]. Glycolytic metabolism reduces ROS (reactive oxygen species) production, oxidative damage and promotes the production of lipids and fatty acid required by OLs for myelin production [163–165].

Neuronal cell death and astrocytic inflammation processes are associated with the increase of glycolytic activity [166,167]. Shunt of TCA cycle by decrease of pyruvate dehydrogenase (PDH) activity is associated with neurodegeneration [168,169]. Lactate metabolism is upregulated upon the increase of aerobic glycolysis in MS [170,171]. The increase in lactate levels is correlated with the progression of MS [172,173]. Reduction of oxidative phosphorylation, shunt of the TCA cycle and activation of aerobic glycolysis inducing lactate production are observed in MS lesions [31,159]. Recently, magnetic resonance spectroscopy and positron emission tomography (PET) have shown that lactate levels are increased in MS lesions [170,174] and that lactate concentration in the cerebral spinal fluid is associated with the number of inflammatory plaques and mitochondrial dysregulation in MS [175–177]. Modulation of aerobic glycolysis appears as a potential treatment for myelin maintenance in MS lesions [26].

Table 1. WNT pathway, peroxisome proliferator-activated receptor γ (PPARγ) and aerobic glycolysis in multiple sclerosis (MS) models.

Pathway	Expression	Actions	Model	References
PPARγ	Agonists	Inhibition of NF-κB	EAE models	[114–116]
		Decrease inflammation, permits remyelination	OLs models	[118,119]
		Neuroprotection	EAE models	[121–128]
		Th17 differentiation	Murine $CD4^+$ T cells	[129]
		Decrease IL-17 expression	EAE models	[130]
		Decrease IL-1, IL-6 and COX2	EAE models	[142]
		Decrease β-catenin	EAE models	[142]
WNT	Overexpression	Chronic pain	EAE models	[68]
		Impairs OPC differentiation	EAE models	[137]
		Alteration of endothelial adherens	EAE models	[139,140]
		Alteration of endothelial adherens	MS brain tissue	[141]
Aerobic Glycolysis	Activation	Neuronal cell death and astrocytic inflammation	EAE models	[166,167]
		MS progression	Human models	[172,173]
		Increased lactate production	Human models	[31,159]
		Mitochondrial dysregulation	Human models	[175–177]

NF-κB: nuclear factor-κB; EAE: experimental autoimmune encephalomyelitis; OLs: oligodendrocytes; OPC: oligodendrocyte precursor cells; MS: multiple sclerosis.

7. Circadian Rhythms in MS

7.1. Circadian Rhythms, Definition

Several biologic mechanisms in the body are controlled by the circadian "clock" (circadian locomotors output cycles kaput). The circadian clock is located in the hypothalamic suprachiasmatic nucleus (SCN). CRs are endogenous and entrainable free-running periods that last approximately

24 h. Numerous transcription determinants are responsible for the regulation of CRs. They are called circadian locomotor output cycles kaput (Clock), brain and muscle aryl-hydrocarbon receptor nuclear translocator-like 1 (Bmal1), Period 1 (Per1), Period 2 (Per2), Period 3 (Per3), and Cryptochrome (Cry 1 and Cry 2) [178,179] (Figure 3). These transcription factors are controlled by positive and negative feedbacks mediated by CRs [180,181]. Clock and Bmal1 heterodimerize and then initiate transcription of Per1, Per2, Cry1 and Cry2 [182]. The Per/Cry heterodimer can inhibit its activation through negative feedback. It translocates back to the nucleus to directly inhibit the Clock/Bmal1 complex and then inhibits its own transcription [182].

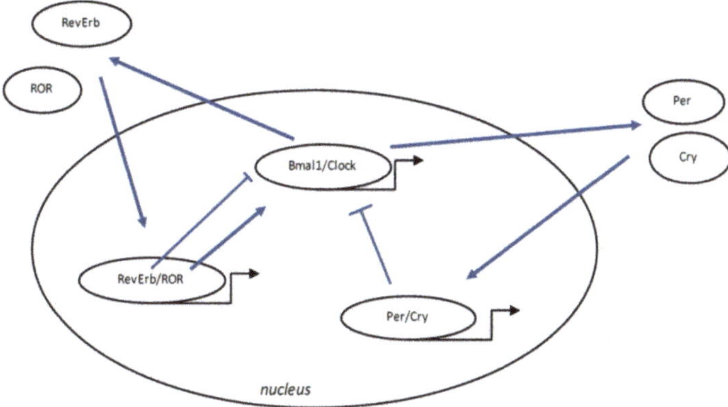

Figure 3. Circadian clock genes process. The clock is considered as a stimulatory loop, with the Bmal1/Clock heterodimer activating the transcription of Period (Per) and Cryptochrome (Cry) genes, and then a negative feedback loop with the Per/Cry heterodimer which translocates to the nucleus and then represses the transcription of the Clock and Bmal1 genes. An additional loop implicates the RORs and Rev Erbs factors with a positive feedback by retinoid-related orphan receptors (ROR) and a negative feedback by Rev Erbs. Arrows: activation; T bar: inhibition.

The Clock/Bmal1 heterodimer activates the transcription of retinoic acid-related orphan nuclear receptors, Rev-Erbs and retinoid-related orphan receptors (RORs). By a positive self-regulation RORs can activate Bmal1 transcription, whereas Rev-Erbs can repress their transcription through negative feedback [182].

7.2. Circadian Rhythm Disruption in MS

Several studies have shown that circadian rhythms have a main role in MS [183]. Late-night shift work in MS patients is associated with disruption of circadian rhythms and sleep [184,185]. Indeed, sleep dysregulation worsens EAE symptoms [186] by increasing the infiltration of inflammatory cells in the CNS, such as $CD4^+$ T cells [186]. EAE severity is associated with both sleep disruption and degree of sleep fragmentation [187]. Few studies have examined CRs dysregulation in MS. However, MS is associated with many symptoms such as hypertension, heart disease, anxiety, depression and sleep disturbances [188,189]. Sleep disorders observed in MS patients [190] are characterized by prolonged sleep latency and frequent nocturnal awakening [191]. In MS, fatigue symptom is associated with CRs abnormalities [192,193], such as sleep phase syndrome and irregular sleep wake pattern. In MS, dorsolateral hypothalamic neurons secrete less neuropeptide hypocretin-1/orexin-A [194]. The Orexin-A system is responsible for the modulation of sleep-wake cycle [195]. Decreased orexin-A levels lead to the promotion of consolidated night sleep [196].

Hypersomnia observed in MS patients is associated with low cerebrospinal fluid (CSF) orexin-A and hypothalamic lesions [197]. Inflammation may suppress the orexin-A system [198] through the overexpression of cytokines, TNF-α and interferon γ (IFN-γ) leading to fatigue syndrome in MS patients [199].

The orexin-A system is influenced by seasonal fluctuations and day length [200]. Demyelination process may put MS patients at risk for CR disorders [201]. Seasonal fluctuations observed in MS may be due to variations of melatonin levels which increase in winter and decrease in summer [202]. Moreover, this seasonal variation could also act through the birth month in susceptibility to developing MS [203,204]. Cytokine and chemokine expression in lymphoid tissues present some seasonal variation in EAE mice [204,205].

By inducing Rev-Erbs, CRs can regulate the balance of Th17/Th1/Treg in EAE mice [206]. The number of Th17 cells decreases during the acute phase of MS and is associated with melatonin levels [206].

7.3. Interaction between WNT/β-Catenin Pathway and Circadian Rhythms

The WNT/β-catenin pathway is downstream of the RORs regulation factors and contains diverse putative Bmal1 clock-binding sites within its promoter [207]. By these interactions, circadian genes can regulate cell cycle progression through the WNT pathway [208] (Figure 4). Expression of WNT pathway can be downregulated by Bmal1 knockdown [209]. Expression levels of WNT-related genes in wild-type mice are higher than levels of WNT-related genes with Bmal1 knockdown mice [210,211]. Bmal1 appears to be upregulated in MS [212]. Cell proliferation and cell cycle progression are regulated by Bmal1 via stimulation of the canonical WNT/β-catenin pathway [213]. Bmal1 activation increases β-catenin transcription and decreases both β-catenin degradation and GSK-3β activation [214]. Per2 degradation induced by β-catenin involves the dysregulation of circadian genes in intestinal mucosa of ApcMin/+ mice [215].

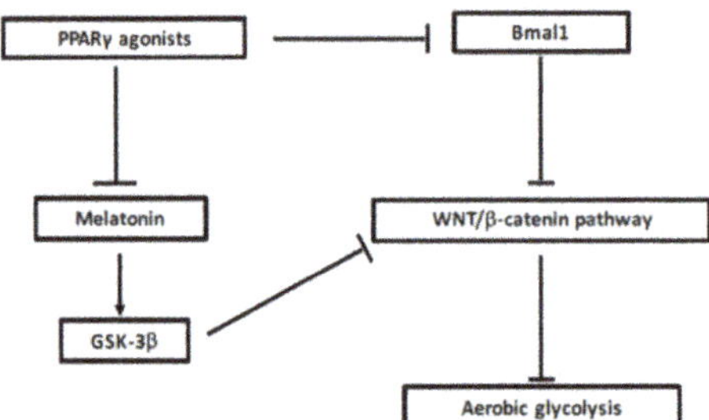

Figure 4. Schematic interaction between WNT, peroxisome proliferator-activated receptor γ (PPARγ) and circadian rhythms. PPARγ agonists can decrease Bmal1 expression. The knockout of Bmal1 leads to decrease WNT/β-catenin pathway activity and then in absence of initiation of aerobic glycolysis. In parallel, PPARγ agonists can decrease melatonin levels leading to activate GSK-3β activity, the main inhibitor of WNT pathway.

In normal circumstances, the core circadian genes work in accurate feedback loops and keep the molecular clockworks in the SCN. They permit regulation of peripheral clocks [180,181].

Per1 and Per2 control CRs cells and modulate cell-related genes expression, such as c-Myc, to sustain the normal cell cycle [216,217]. mRNAs and proteins levels of circadian genes oscillate throughout the 24-hour period [180].

7.4. Action of PPARγ on Circadian Rhythms

PPARγ directly acts with the core clock genes and presents diurnal fluctuations in liver and blood vessels [33,218]. In mice, impaired diurnal rhythms are induced by a knockdown of PPARγ [219]. PPARγ agonists can regulate Bmal1 and the constitution of the heterodimer Clock/Bmal1 [33,220] and can then target Rev-Erb [221] (Figure 4). Decrease of the clock-controlled gene Nocturin inhibits

PPARγ oscillations in the liver of mice fed on high-fat diet. In normal circumstances, nocturin binds to PPARγ to enhance its transcriptional activity [222]. The inhibtion of PPARγ expression prevents circadian function of 15-Deoxy-D 12,14-prostaglandin J2 (15-PGJ2) [219]. The partner of PPARγ, RXR, interacts with Clock protein in a ligand-dependent manner and then decreases the formation and transcriptional activity of the Clock/Bmal1 heterodimer [223]. PPARγ acts on the mammalian clock and energy metabolism [223]. Circadian metabolism is directly regulated by PPARγ [219]. Retinoic acid receptor-related orphan receptor γ t (RORγt) is considered as a key transcriptional factor for Th17 differentiation [224,225]. PPARγ can influence the function of Th cell clones [226]. PPARγ agonists inhibit Th17 differentiation through the inhibition of RORγt induction [129,227,228]. CD4$^+$ T cells fail to express RORγt under the action of PPARγ agonists [129].

7.5. Interest of Cortisol in MS

Cortisol production by the HPA axis (hypothalamic-pituitary-adrenal axis) during the acute phase of MS leads to suppression of T cell secretion of inflammatory factors, such as cytokines [229]. The peak of inflammatory factor production occurs in association with low levels of cortisol. TNF-α and IL-6 production during active phases is consistent with reduction of night-time cortisol production [230]. Cortisol production is regulated by circadian rhythms that present an elevated morning level, but normalizes by the evening in MS patients [231]. However, the role of cortisol in MS remains unclear [183]. HPA axis seems to be activated in relapsing-remitting MS patients but this phenomenon has not been shown in all studies [232,233]. Nevertheless, a normal cortisol level is associated with a more severe disease course [234]. Rat strains with low HPA axis activity present worse EAE in comparison to rat strains with low HPA axis activity [235]. These results suggest that elevated levels of cortisol suppress inflammation in MS even if other studies have shown that over-active HPA axis in association with high serum cortisol contributed to worse forms of MS [236]. Cortisol seems not to act alone, and corticosteroids present in the CNS and blood may have an impact on the immune response. Corticosteroid concentration in cerebrospinal fluid present high levels in MS patients with stable disease and low levels with worse forms despite similar serum cortisol level [237]. High serum cortisol level can inhibit inflammation process through an over-active HPA axis leading to protection in MS whereas disruption of HPA axis is associated with worse forms of the disease. Cortisol is known to have immunosuppressive effects by affecting cytokine secretion and T cell activation [238]. Glucocorticoids are not limited to the inhibition of T cell response but also affect the decrease of macrophages, B cells and dendritic cells [239]. Glucocorticoids in MS present many beneficial effects but the resistance observed in humans may complicate its use [240]. Nevertheless, cortisol can affect CRs by glucocorticoid receptors [by activating the transcription of Per1 and Per2 [241,242]. In MS, the circadian oscillations of cortisol levels show that cortisol may have a key role in the regulation of peripheral clocks. In MS, no study has shown a link between PPARγ expression and cortisol level. Few studies have observed that high PPARγ agonists can increase cortisol levels in cancers [243].

7.6. Interest of Melatonin in MS

Melatonin (also named 5-methoxy-*N*-acetyltryptamine) is a secreted by the pineal gland [244]. Melatonin is released during darkness and thereby regulates the circadian regulation of sleep [245,246]. An inverse correlation is observed between melatonin levels and MS progression [247,248]. Melatonin has anti-inflammatory, anti-oxidant and neuroprotective effects [245,249–253]. Administration of melatonin reduces EAE severity through the suppression of Th17 cell number [202,254]. Moreover, CRs could be related to inflammation by affecting immunization [255]. TNF-α and IL-1β overexpression can inhibit the melatonin synthesis pathway [256–258]. TNF-α directly inhibits melatonin expression [259]. Melatonin ameliorates EAE development by suppressing Th17 cells generation [202,254,260–262]. Melatonin also ameliorates symptoms in EAE mouse models [202,254,260–262] through the inhibition of Rev-Erb and ROR expressions, and by limiting Th17 cell differentiation and function [206,260]. Melatonin decreases phosphorylation of GSK-3β [263,264]. PPARγ agonists can upregulate melatonin levels to restore mitochondrial membrane potential, stimulate the biogenesis of mitochondria [265] and enhance mitochondrial function [266].

8. Conclusions

Demyelination during MS lesions is associated with reprogramming energy metabolism through the dysregulation of the opposed interplay of PPARγ and the WNT/β-catenin pathway (Table 1). The canonical WNT/β-catenin pathway is upregulated by chronic neuroinflammation, whereas PPARγ is downregulated during demyelinating processes. These two systems act in an opposed and reverse manner. Demyelinating processes are associated with the increase of the WNT/β-catenin pathway and dysregulation of the circadian clock genes. In MS, over-activation of Bmal1 leads to stimulation of the canonical WNT/β-catenin pathway. Then, activation of the WNT/β-catenin pathway results in stimulation of glycolytic enzymes leading to activation of aerobic glycolysis. Lactate production induces dysfunction of CD4$^+$ T cells leading to axonal and neuronal damage during the MS demyelinating processes. PPARγ agonists can inhibit Th17 differentiation in CD4$^+$ T cells and can diminish cytokine release. In parallel, PPARγ agonists can interfere with reprogramming energy metabolism by directly inhibiting the WNT/β-catenin pathway and interacting with clock genes and thus, could be a promising therapeutic pathway in MS due to their interactions (Figure 5). These findings support the possibility of targeting these pathways with the goal of improving the symptoms of MS. Clinical trials and studies are needed to confirm this hypothesis in MS pathogenesis.

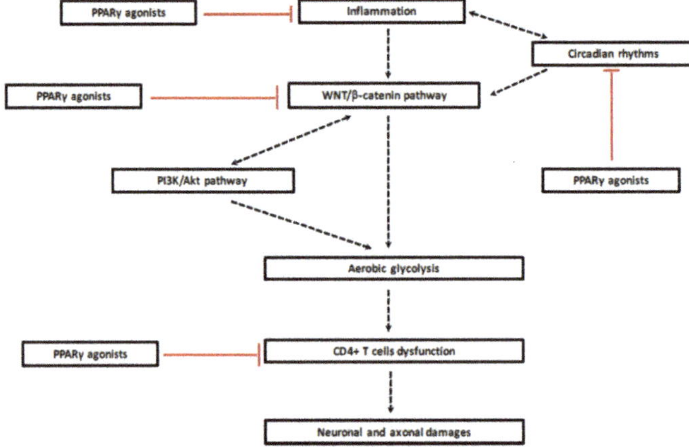

Figure 5. Potential PPARγ agonists treatment approach in demyelination. During acute phase, inflammation processes, activated by disruption of circadian rhythms, lead to release of several cytokines and pro-inflammatory factors which stimulate the canonical WNT/β-catenin pathway. Activation of the WNT ligands involves WNT target genes that are responsible for the initiation of the shunt of the tricarboxylic acid (TCA) resulting in aerobic glycolysis instead of oxidative phosphorylation. Lactate production, the main factor of energy metabolism alteration, and its release out the cells enhance CD4$^+$ T cells dysfunction which aggravates MS pathogenesis, neuronal and axonal damages. Using PPARγ agonists could be interesting because of their four interactions in the demyelination cascade. First, PPARγ agonists directly inhibits neuroinflammation by inhibiting cytokines and inflammatory factors release. Secondly, PPARγ agonists can regulate circadian clocks, such as Bmal1, to decrease inflammatory factors release and to target WNT ligands. Third, their opposed interaction with the canonical WNT/β-catenin pathway can prevent the initiation of aerobic glycolysis process and then the energy metabolism reprogramming enable MS. At last, PPARγ agonists have neuroprotective effects by targeting CD4$^+$ T cells to prevent neuronal and axonal damages. Arrow: activation; T bar: inhibition.

Author Contributions: Alexandre Vallée, Yves Lecarpentier, Rémy Guillevin and Jean-Noël Vallée have contributed to this review. Alexandre Vallée, Yves Lecarpentier, Rémy Guillevin and Jean-Noël Vallée read and approved the final manuscript.

Conflicts of Interest: The authors declare that they have no competing interests.

Abbreviations

Acetyl-coA	Acetyl-coenzyme A
APC	Adenomatous polyposis coli
Bmal1	Brain and muscle aryl-hydrocarbon receptor nuclear translocator-like 1
Clock	Circadian locomotor output cycles kaput
Cry	Cryptochrome
CRs	Circadian rhythms
DSH	Disheveled
FZD	Frizzled
Glut	Glucose transporter
GSK-3β	Glycogen synthase kinase-3β
LDH	Lactate dehydrogenase
LRP 5/6	Low-density lipoprotein receptor-related protein 5/6
MCT-1	Monocarboxylate lactate transporter-1
Per	Period
PPARγ	Peroxisome proliferator-activated receptor γ
PI3K-Akt	Phosphatidylinositol 3-kinase-protein kinase B
PDH	Pyruvate dehydrogenase complex
PDK	Pyruvate dehydrogenase kinase
RORs	Retinoid-related orphan receptors
TCF/LEF	T-cell factor/lymphoid enhancer factor
TCA	Tricarboxylic acid

References

1. Goverman, J. Autoimmune T cell responses in the central nervous system. *Nat. Rev. Immunol.* **2009**, *9*, 393–407. [CrossRef] [PubMed]
2. Trapp, B.D.; Nave, K.-A. Multiple sclerosis: An immune or neurodegenerative disorder? *Annu. Rev. Neurosci.* **2008**, *31*, 247–269. [CrossRef] [PubMed]
3. Rasmussen, S.; Wang, Y.; Kivisäkk, P.; Bronson, R.T.; Meyer, M.; Imitola, J.; Khoury, S.J. Persistent activation of microglia is associated with neuronal dysfunction of callosal projecting pathways and multiple sclerosis-like lesions in relapsing—Remitting experimental autoimmune encephalomyelitis. *Brain J. Neurol.* **2007**, *130*, 2816–2829. [CrossRef] [PubMed]
4. Jadidi-Niaragh, F.; Mirshafiey, A. Th17 cell, the new player of neuroinflammatory process in multiple sclerosis. *Scand. J. Immunol.* **2011**, *74*, 1–13. [CrossRef] [PubMed]
5. Didonna, A.; Oksenberg, J.R. Genetic determinants of risk and progression in multiple sclerosis. *Clin. Chim. Acta Int. J. Clin. Chem.* **2015**, *449*, 16–22. [CrossRef] [PubMed]
6. Chang, A.; Tourtellotte, W.W.; Rudick, R.; Trapp, B.D. Premyelinating oligodendrocytes in chronic lesions of multiple sclerosis. *N. Engl. J. Med.* **2002**, *346*, 165–173. [CrossRef] [PubMed]
7. Franklin, R.J.M. Why does remyelination fail in multiple sclerosis? *Nat. Rev. Neurosci.* **2002**, *3*, 705–714. [CrossRef] [PubMed]
8. Khwaja, O.; Volpe, J.J. Pathogenesis of cerebral white matter injury of prematurity. *Arch. Dis. Child. Fetal Neonatal Ed.* **2008**, *93*, F153–F161. [CrossRef] [PubMed]
9. Woodward, L.J.; Anderson, P.J.; Austin, N.C.; Howard, K.; Inder, T.E. Neonatal MRI to predict neurodevelopmental outcomes in preterm infants. *N. Engl. J. Med.* **2006**, *355*, 685–694. [CrossRef] [PubMed]
10. Back, S.A.; Rosenberg, P.A. Pathophysiology of glia in perinatal white matter injury. *Glia* **2014**, *62*, 1790–1815. [CrossRef] [PubMed]

11. Billiards, S.S.; Haynes, R.L.; Folkerth, R.D.; Borenstein, N.S.; Trachtenberg, F.L.; Rowitch, D.H.; Ligon, K.L.; Volpe, J.J.; Kinney, H.C. Myelin abnormalities without oligodendrocyte loss in periventricular leukomalacia. *Brain Pathol.* **2008**, *18*, 153–163. [CrossRef] [PubMed]
12. Buser, J.R.; Maire, J.; Riddle, A.; Gong, X.; Nguyen, T.; Nelson, K.; Luo, N.L.; Ren, J.; Struve, J.; Sherman, L.S.; et al. Arrested preoligodendrocyte maturation contributes to myelination failure in premature infants. *Ann. Neurol.* **2012**, *71*, 93–109. [CrossRef] [PubMed]
13. Fancy, S.P.J.; Kotter, M.R.; Harrington, E.P.; Huang, J.K.; Zhao, C.; Rowitch, D.H.; Franklin, R.J.M. Overcoming remyelination failure in multiple sclerosis and other myelin disorders. *Exp. Neurol.* **2010**, *225*, 18–23. [CrossRef] [PubMed]
14. Kuhlmann, T.; Miron, V.; Cui, Q.; Cuo, Q.; Wegner, C.; Antel, J.; Brück, W. Differentiation block of oligodendroglial progenitor cells as a cause for remyelination failure in chronic multiple sclerosis. *Brain J. Neurol.* **2008**, *131*, 1749–1758. [CrossRef] [PubMed]
15. Sandler, S. *Chemical and Engineering Thermodynamics*, 4th ed.; Wiely: New York, NY, USA, 2006.
16. Garcia, H.G.; Kondev, J.; Orme, N.; Theriot, J.A.; Phillips, R. Thermodynamics of biological processes. *Methods Enzymol.* **2011**, *492*, 27–59. [CrossRef] [PubMed]
17. Lecarpentier, Y.; Claes, V.; Duthoit, G.; Hébert, J.-L. Circadian rhythms, Wnt/beta-catenin pathway and PPAR alpha/gamma profiles in diseases with primary or secondary cardiac dysfunction. *Front. Physiol.* **2014**, *5*, 429. [CrossRef] [PubMed]
18. Vallée, A.; Vallée, J.-N.; Guillevin, R.; Lecarpentier, Y. Interactions between the Canonical WNT/Beta-Catenin Pathway and PPAR Gamma on Neuroinflammation, Demyelination, and Remyelination in Multiple Sclerosis. *Cell. Mol. Neurobiol.* **2017**, *38*, 783–795. [CrossRef] [PubMed]
19. Tsentidis, C.; Gourgiotis, D.; Kossiva, L.; Marmarinos, A.; Doulgeraki, A.; Karavanaki, K. Increased levels of Dickkopf-1 are indicative of Wnt/β-catenin downregulation and lower osteoblast signaling in children and adolescents with type 1 diabetes mellitus, contributing to lower bone mineral density. *Osteoporos. Int.* **2017**, *28*, 945–953. [CrossRef] [PubMed]
20. Pörksen, S.; Nielsen, L.B.; Mortensen, H.B.; Danne, T.; Kocova, M.; Castaño, L.; Pociot, F.; Hougaard, P.; Ekstrøm, C.T.; Gammeltoft, S.; et al. Hvidøre Study Group on Childhood Diabetes Variation within the PPARG gene is associated with residual beta-cell function and glycemic control in children and adolescents during the first year of clinical type 1 diabetes. *Pediatr. Diabetes* **2008**, *9*, 297–302. [CrossRef] [PubMed]
21. Ferrari, S.M.; Fallahi, P.; Vita, R.; Antonelli, A.; Benvenga, S. Peroxisome Proliferator-Activated Receptor-γ in Thyroid Autoimmunity. *PPAR Res.* **2015**, *2015*, 232818. [CrossRef] [PubMed]
22. Tao, W.; Ayala-Haedo, J.A.; Field, M.G.; Pelaez, D.; Wester, S.T. RNA-Sequencing Gene Expression Profiling of Orbital Adipose-Derived Stem Cell Population Implicate HOX Genes and WNT Signaling Dysregulation in the Pathogenesis of Thyroid-Associated Orbitopathy. *Investig. Ophthalmol. Vis. Sci.* **2017**, *58*, 6146–6158. [CrossRef] [PubMed]
23. Xiao, C.Y.; Pan, Y.F.; Guo, X.H.; Wu, Y.Q.; Gu, J.R.; Cai, D.Z. Expression of β-catenin in rheumatoid arthritis fibroblast-like synoviocytes. *Scand. J. Rheumatol.* **2011**, *40*, 26–33. [CrossRef] [PubMed]
24. Marder, W.; Khalatbari, S.; Myles, J.D.; Hench, R.; Lustig, S.; Yalavarthi, S.; Parameswaran, A.; Brook, R.D.; Kaplan, M.J. The peroxisome proliferator activated receptor-γ pioglitazone improves vascular function and decreases disease activity in patients with rheumatoid arthritis. *J. Am. Heart Assoc.* **2013**, *2*, e000441. [CrossRef] [PubMed]
25. Lecarpentier, Y.; Krokidis, X.; Martin, P.; Pineau, T.; Hébert, J.-L.; Quillard, J.; Cortes-Morichetti, M.; Coirault, C. Increased entropy production in diaphragm muscle of PPAR alpha knockout mice. *J. Theor. Biol.* **2008**, *250*, 92–102. [CrossRef] [PubMed]
26. Rone, M.B.; Cui, Q.-L.; Fang, J.; Wang, L.-C.; Zhang, J.; Khan, D.; Bedard, M.; Almazan, G.; Ludwin, S.K.; Jones, R.; et al. Oligodendrogliopathy in Multiple Sclerosis: Low Glycolytic Metabolic Rate Promotes Oligodendrocyte Survival. *J. Neurosci. Off. J. Soc. Neurosci.* **2016**, *36*, 4698–4707. [CrossRef] [PubMed]
27. Thompson, C.B. Wnt meets Warburg: Another piece in the puzzle? *EMBO J.* **2014**, *33*, 1420–1422. [CrossRef] [PubMed]
28. Warburg, O. On the origin of cancer cells. *Science* **1956**, *123*, 309–314. [CrossRef] [PubMed]
29. Yang, D.; Wang, M.-T.; Tang, Y.; Chen, Y.; Jiang, H.; Jones, T.T.; Rao, K.; Brewer, G.J.; Singh, K.K.; Nie, D. Impairment of mitochondrial respiration in mouse fibroblasts by oncogenic H-RAS(Q61L). *Cancer Biol. Ther.* **2010**, *9*, 122–133. [CrossRef] [PubMed]

30. Pfeiffer, T.; Schuster, S.; Bonhoeffer, S. Cooperation and competition in the evolution of ATP-producing pathways. *Science* **2001**, *292*, 504–507. [CrossRef] [PubMed]
31. Chen, Z.; Liu, M.; Li, L.; Chen, L. Involvement of the Warburg effect in non-tumor diseases processes. *J. Cell. Physiol.* **2018**, *233*, 2839–2849. [CrossRef] [PubMed]
32. Newland, P.; Starkweather, A.; Sorenson, M. Central fatigue in multiple sclerosis: A review of the literature. *J. Spinal Cord Med.* **2016**, *39*, 386–399. [CrossRef] [PubMed]
33. Wang, N.; Yang, G.; Jia, Z.; Zhang, H.; Aoyagi, T.; Soodvilai, S.; Symons, J.D.; Schnermann, J.B.; Gonzalez, F.J.; Litwin, S.E.; et al. Vascular PPARgamma controls circadian variation in blood pressure and heart rate through Bmal1. *Cell Metab.* **2008**, *8*, 482–491. [CrossRef] [PubMed]
34. Vallée, A.; Lecarpentier, Y.; Guillevin, R.; Vallée, J.-N. Thermodynamics in Neurodegenerative Diseases: Interplay between Canonical WNT/Beta-Catenin Pathway-PPAR Gamma, Energy Metabolism and Circadian Rhythms. *Neuromolecular Med.* **2018**. [CrossRef] [PubMed]
35. Ryan, K.K.; Li, B.; Grayson, B.E.; Matter, E.K.; Woods, S.C.; Seeley, R.J. A role for central nervous system PPAR-γ in the regulation of energy balance. *Nat. Med.* **2011**, *17*, 623–626. [CrossRef] [PubMed]
36. Bookout, A.L.; Jeong, Y.; Downes, M.; Yu, R.T.; Evans, R.M.; Mangelsdorf, D.J. Anatomical profiling of nuclear receptor expression reveals a hierarchical transcriptional network. *Cell* **2006**, *126*, 789–799. [CrossRef] [PubMed]
37. Michalik, L.; Auwerx, J.; Berger, J.P.; Chatterjee, V.K.; Glass, C.K.; Gonzalez, F.J.; Grimaldi, P.A.; Kadowaki, T.; Lazar, M.A.; O'Rahilly, S.; et al. International Union of Pharmacology. LXI. Peroxisome proliferator-activated receptors. *Pharmacol. Rev.* **2006**, *58*, 726–741. [CrossRef] [PubMed]
38. Abbas, A.; Blandon, J.; Rude, J.; Elfar, A.; Mukherjee, D. PPAR-γ agonist in treatment of diabetes: Cardiovascular safety considerations. *Cardiovasc. Hematol. Agents Med. Chem.* **2012**, *10*, 124–134. [CrossRef] [PubMed]
39. Oyekan, A. PPARs and their effects on the cardiovascular system. *Clin. Exp. Hypertens.* **2011**, *33*, 287–293. [CrossRef] [PubMed]
40. Adams, M.; Montague, C.T.; Prins, J.B.; Holder, J.C.; Smith, S.A.; Sanders, L.; Digby, J.E.; Sewter, C.P.; Lazar, M.A.; Chatterjee, V.K.; et al. Activators of peroxisome proliferator-activated receptor gamma have depot-specific effects on human preadipocyte differentiation. *J. Clin. Investig.* **1997**, *100*, 3149–3153. [CrossRef] [PubMed]
41. Ricote, M.; Li, A.C.; Willson, T.M.; Kelly, C.J.; Glass, C.K. The peroxisome proliferator-activated receptor-gamma is a negative regulator of macrophage activation. *Nature* **1998**, *391*, 79–82. [CrossRef] [PubMed]
42. Asakawa, M.; Takano, H.; Nagai, T.; Uozumi, H.; Hasegawa, H.; Kubota, N.; Saito, T.; Masuda, Y.; Kadowaki, T.; Komuro, I. Peroxisome proliferator-activated receptor gamma plays a critical role in inhibition of cardiac hypertrophy in vitro and in vivo. *Circulation* **2002**, *105*, 1240–1246. [CrossRef] [PubMed]
43. Chandra, M.; Miriyala, S.; Panchatcharam, M. PPARγ and Its Role in Cardiovascular Diseases. *PPAR Res.* **2017**, *2017*, 6404638. [CrossRef] [PubMed]
44. Chen, R.; Liang, F.; Moriya, J.; Yamakawa, J.; Takahashi, T.; Shen, L.; Kanda, T. Peroxisome proliferator-activated receptors (PPARs) and their agonists for hypertension and heart failure: Are the reagents beneficial or harmful? *Int. J. Cardiol.* **2008**, *130*, 131–139. [CrossRef] [PubMed]
45. Polvani, S.; Tarocchi, M.; Galli, A. PPARγ and Oxidative Stress: Con(β) Catenating NRF2 and FOXO. *PPAR Res.* **2012**, *2012*, 641087. [CrossRef] [PubMed]
46. Braissant, O.; Foufelle, F.; Scotto, C.; Dauça, M.; Wahli, W. Differential expression of peroxisome proliferator-activated receptors (PPARs): Tissue distribution of PPAR-alpha, -beta, and -gamma in the adult rat. *Endocrinology* **1996**, *137*, 354–366. [CrossRef] [PubMed]
47. Bernardo, A.; Minghetti, L. Regulation of Glial Cell Functions by PPAR-gamma Natural and Synthetic Agonists. *PPAR Res.* **2008**, *2008*, 864140. [CrossRef] [PubMed]
48. Yang, Y.; Lovett-Racke, A.E.; Racke, M.K. Regulation of Immune Responses and Autoimmune Encephalomyelitis by PPARs. *PPAR Res.* **2010**, *2010*, 104705. [CrossRef] [PubMed]
49. Klotz, L.; Diehl, L.; Dani, I.; Neumann, H.; von Oppen, N.; Dolf, A.; Endl, E.; Klockgether, T.; Engelhardt, B.; Knolle, P. Brain endothelial PPARgamma controls inflammation-induced $CD4^+$ T cell adhesion and transmigration in vitro. *J. Neuroimmunol.* **2007**, *190*, 34–43. [CrossRef] [PubMed]

50. Ricote, M.; Glass, C.K. PPARs and molecular mechanisms of transrepression. *Biochim. Biophys. Acta* **2007**, *1771*, 926–935. [CrossRef] [PubMed]
51. Touyz, R.M.; Schiffrin, E.L. Ang II-stimulated superoxide production is mediated via phospholipase D in human vascular smooth muscle cells. *Hypertens.* **1999**, *34*, 976–982. [CrossRef]
52. Széles, L.; Töröcsik, D.; Nagy, L. PPARgamma in immunity and inflammation: Cell types and diseases. *Biochim. Biophys. Acta* **2007**, *1771*, 1014–1030. [CrossRef] [PubMed]
53. McIntyre, T.M.; Pontsler, A.V.; Silva, A.R.; St Hilaire, A.; Xu, Y.; Hinshaw, J.C.; Zimmerman, G.A.; Hama, K.; Aoki, J.; Arai, H.; et al. Identification of an intracellular receptor for lysophosphatidic acid (LPA): LPA is a transcellular PPARgamma agonist. *Proc. Natl. Acad. Sci. USA* **2003**, *100*, 131–136. [CrossRef] [PubMed]
54. Nusse, R.; Brown, A.; Papkoff, J.; Scambler, P.; Shackleford, G.; McMahon, A.; Moon, R.; Varmus, H. A new nomenclature for int-1 and related genes: The Wnt gene family. *Cell* **1991**, *64*, 231. [CrossRef]
55. MacDonald, B.T.; Tamai, K.; He, X. Wnt/beta-catenin signaling: Components, mechanisms, and diseases. *Dev. Cell* **2009**, *17*, 9–26. [CrossRef] [PubMed]
56. Nusse, R.; Clevers, H. Wnt/β-Catenin Signaling, Disease, and Emerging Therapeutic Modalities. *Cell* **2017**, *169*, 985–999. [CrossRef] [PubMed]
57. Nusse, R. Wnt signaling. *Cold Spring Harb. Perspect. Biol.* **2012**, *4*. [CrossRef] [PubMed]
58. He, X.; Semenov, M.; Tamai, K.; Zeng, X. LDL receptor-related proteins 5 and 6 in Wnt/beta-catenin signaling: Arrows point the way. *Dev. Camb. Engl.* **2004**, *131*, 1663–1677. [CrossRef]
59. Miller, J.R.; Hocking, A.M.; Brown, J.D.; Moon, R.T. Mechanism and function of signal transduction by the Wnt/beta-catenin and Wnt/Ca^{2+} pathways. *Oncogene* **1999**, *18*, 7860–7872. [CrossRef] [PubMed]
60. Reya, T.; Clevers, H. Wnt signalling in stem cells and cancer. *Nature* **2005**, *434*, 843–850. [CrossRef] [PubMed]
61. Anastas, J.N.; Moon, R.T. WNT signalling pathways as therapeutic targets in cancer. *Nat. Rev. Cancer* **2013**, *13*, 11–26. [CrossRef] [PubMed]
62. Kim, W.; Kim, M.; Jho, E. Wnt/β-catenin signalling: From plasma membrane to nucleus. *Biochem. J.* **2013**, *450*, 9–21. [CrossRef] [PubMed]
63. Rao, T.P.; Kühl, M. An updated overview on Wnt signaling pathways: A prelude for more. *Circ. Res.* **2010**, *106*, 1798–1806. [CrossRef] [PubMed]
64. Aoki, K.; Taketo, M.M. Adenomatous polyposis coli (APC): A multi-functional tumor suppressor gene. *J. Cell Sci.* **2007**, *120*, 3327–3335. [CrossRef] [PubMed]
65. Komiya, Y.; Habas, R. Wnt signal transduction pathways. *Organogenesis* **2008**, *4*, 68–75. [CrossRef] [PubMed]
66. Singh, R.; De Aguiar, R.B.; Naik, S.; Mani, S.; Ostadsharif, K.; Wencker, D.; Sotoudeh, M.; Malekzadeh, R.; Sherwin, R.S.; Mani, A. LRP6 enhances glucose metabolism by promoting TCF7L2-dependent insulin receptor expression and IGF receptor stabilization in humans. *Cell Metab.* **2013**, *17*, 197–209. [CrossRef] [PubMed]
67. Wang, S.; Song, K.; Srivastava, R.; Dong, C.; Go, G.-W.; Li, N.; Iwakiri, Y.; Mani, A. Nonalcoholic fatty liver disease induced by noncanonical Wnt and its rescue by Wnt3a. *FASEB J.* **2015**, *29*, 3436–3445. [CrossRef] [PubMed]
68. Yuan, S.; Shi, Y.; Tang, S.-J. Wnt signaling in the pathogenesis of multiple sclerosis-associated chronic pain. *J. Neuroimmune Pharmacol.* **2012**, *7*, 904–913. [CrossRef] [PubMed]
69. Kaiser, C.C.; Shukla, D.K.; Stebbins, G.T.; Skias, D.D.; Jeffery, D.R.; Stefoski, D.; Katsamakis, G.; Feinstein, D.L. A pilot test of pioglitazone as an add-on in patients with relapsing remitting multiple sclerosis. *J. Neuroimmunol.* **2009**, *211*, 124–130. [CrossRef] [PubMed]
70. Shukla, D.K.; Kaiser, C.C.; Stebbins, G.T.; Feinstein, D.L. Effects of pioglitazone on diffusion tensor imaging indices in multiple sclerosis patients. *Neurosci. Lett.* **2010**, *472*, 153–156. [CrossRef] [PubMed]
71. Klotz, L.; Schmidt, M.; Giese, T.; Sastre, M.; Knolle, P.; Klockgether, T.; Heneka, M.T. Proinflammatory stimulation and pioglitazone treatment regulate peroxisome proliferator-activated receptor gamma levels in peripheral blood mononuclear cells from healthy controls and multiple sclerosis patients. *J. Immunol. 1950* **2005**, *175*, 4948–4955. [CrossRef]
72. Esmaeili, M.A.; Yadav, S.; Gupta, R.K.; Waggoner, G.R.; Deloach, A.; Calingasan, N.Y.; Beal, M.F.; Kiaei, M. Preferential PPAR-α activation reduces neuroinflammation, and blocks neurodegeneration in vivo. *Hum. Mol. Genet.* **2016**, *25*, 317–327. [CrossRef] [PubMed]
73. Vallée, A.; Lecarpentier, Y.; Guillevin, R.; Vallée, J.-N. Thermodynamics in Gliomas: Interactions between the Canonical WNT/Beta-Catenin Pathway and PPAR Gamma. *Front. Physiol.* **2017**, *8*, 352. [CrossRef] [PubMed]

74. Vallée, A.; Guillevin, R.; Vallée, J.-N. Vasculogenesis and angiogenesis initiation under normoxic conditions through Wnt/β-catenin pathway in gliomas. *Rev. Neurosci.* **2018**, *29*, 71–91. [CrossRef] [PubMed]
75. Vallée, A.; Lecarpentier, Y.; Guillevin, R.; Vallée, J.-N. Opposite Interplay between the Canonical WNT/β-Catenin Pathway and PPAR Gamma: A Potential Therapeutic Target in Gliomas. *Neurosci. Bull.* **2018**. [CrossRef] [PubMed]
76. Lecarpentier, Y.; Claes, V.; Vallée, A.; Hébert, J.-L. Interactions between PPAR Gamma and the Canonical Wnt/Beta-Catenin Pathway in Type 2 Diabetes and Colon Cancer. *PPAR Res.* **2017**, *2017*, 5879090. [CrossRef] [PubMed]
77. Lecarpentier, Y.; Claes, V.; Vallée, A.; Hébert, J.-L. Thermodynamics in cancers: Opposing interactions between PPAR gamma and the canonical WNT/beta-catenin pathway. *Clin. Transl. Med.* **2017**, *6*, 14. [CrossRef] [PubMed]
78. Vallée, A.; Lecarpentier, Y.; Vallée, J.-N. Thermodynamic Aspects and Reprogramming Cellular Energy Metabolism during the Fibrosis Process. *Int. J. Mol. Sci.* **2017**, *18*, 2537. [CrossRef] [PubMed]
79. Vallée, A.; Lecarpentier, Y.; Guillevin, R.; Vallée, J.-N. Interactions between TGF-β1, canonical WNT/β-catenin pathway and PPAR γ in radiation-induced fibrosis. *Oncotarget* **2017**, *8*, 90579–90604. [CrossRef] [PubMed]
80. Lecarpentier, Y.; Schussler, O.; Claes, V.; Vallée, A. The Myofibroblast: TGFβ-1, A Conductor which Plays a Key Role in Fibrosis by Regulating the Balance between PPARγ and the Canonical WNT Pathway. *Nucl. Recept. Res.* **2017**, *23*. [CrossRef] [PubMed]
81. Vallée, A.; Lecarpentier, Y.; Guillevin, R.; Vallée, J.-N. Effects of cannabidiol interactions with Wnt/β-catenin pathway and PPARγ on oxidative stress and neuroinflammation in Alzheimer's disease. *Acta Biochim. Biophys. Sin.* **2017**, *49*, 853–866. [CrossRef] [PubMed]
82. Vallée, A.; Lecarpentier, Y.; Guillevin, R.; Vallée, J.-N. Reprogramming energetic metabolism in Alzheimer's disease. *Life Sci.* **2018**, *193*, 141–152. [CrossRef] [PubMed]
83. Vallée, A.; Lecarpentier, Y. Alzheimer Disease: Crosstalk between the Canonical Wnt/Beta-Catenin Pathway and PPARs Alpha and Gamma. *Front. Neurosci.* **2016**, *10*, 459. [CrossRef] [PubMed]
84. Vallée, A.; Lecarpentier, Y.; Guillevin, R.; Vallée, J.-N. Aerobic Glycolysis Hypothesis Through WNT/Beta-Catenin Pathway in Exudative Age-Related Macular Degeneration. *J. Mol. Neurosci.* **2017**, *62*, 368–379. [CrossRef] [PubMed]
85. Vallée, A.; Lecarpentier, Y.; Guillevin, R.; Vallée, J.-N. PPARγ agonists: Potential treatments for exudative age-related macular degeneration. *Life Sci.* **2017**, *188*, 123–130. [CrossRef] [PubMed]
86. Lecarpentier, Y.; Vallée, A. Opposite Interplay between PPAR Gamma and Canonical Wnt/Beta-Catenin Pathway in Amyotrophic Lateral Sclerosis. *Front. Neurol.* **2016**, *7*, 100. [CrossRef] [PubMed]
87. Ajmone-Cat, M.A.; D'Urso, M.C.; di Blasio, G.; Brignone, M.S.; De Simone, R.; Minghetti, L. Glycogen synthase kinase 3 is part of the molecular machinery regulating the adaptive response to LPS stimulation in microglial cells. *Brain Behav. Immun.* **2016**, *55*, 225–235. [CrossRef] [PubMed]
88. Jansson, E.A.; Are, A.; Greicius, G.; Kuo, I.-C.; Kelly, D.; Arulampalam, V.; Pettersson, S. The Wnt/beta-catenin signaling pathway targets PPARgamma activity in colon cancer cells. *Proc. Natl. Acad. Sci. USA* **2005**, *102*, 1460–1465. [CrossRef] [PubMed]
89. Drygiannakis, I.; Valatas, V.; Sfakianaki, O.; Bourikas, L.; Manousou, P.; Kambas, K.; Ritis, K.; Kolios, G.; Kouroumalis, E. Proinflammatory cytokines induce crosstalk between colonic epithelial cells and subepithelial myofibroblasts: Implication in intestinal fibrosis. *J. Crohns Colitis* **2013**, *7*, 286–300. [CrossRef] [PubMed]
90. Farmer, S.R. Regulation of PPARgamma activity during adipogenesis. *Int. J. Obes.* **2005**, *29* (Suppl. 1), S13–S16. [CrossRef] [PubMed]
91. Jeon, K.-I.; Kulkarni, A.; Woeller, C.F.; Phipps, R.P.; Sime, P.J.; Hindman, H.B.; Huxlin, K.R. Inhibitory effects of PPARγ ligands on TGF-β1-induced corneal myofibroblast transformation. *Am. J. Pathol.* **2014**, *184*, 1429–1445. [CrossRef] [PubMed]
92. Kumar, V.; Mundra, V.; Mahato, R.I. Nanomedicines of Hedgehog inhibitor and PPAR-γ agonist for treating liver fibrosis. *Pharm. Res.* **2014**, *31*, 1158–1169. [CrossRef] [PubMed]
93. Lee, Y.; Kim, S.H.; Lee, Y.J.; Kang, E.S.; Lee, B.-W.; Cha, B.S.; Kim, J.W.; Song, D.H.; Lee, H.C. Transcription factor Snail is a novel regulator of adipocyte differentiation via inhibiting the expression of peroxisome proliferator-activated receptor γ. *Cell. Mol. Life Sci.* **2013**, *70*, 3959–3971. [CrossRef] [PubMed]

94. Li, Q.; Yan, Z.; Li, F.; Lu, W.; Wang, J.; Guo, C. The improving effects on hepatic fibrosis of interferon-γ liposomes targeted to hepatic stellate cells. *Nanotechnology* **2012**, *23*, 265101. [CrossRef] [PubMed]
95. Liu, J.; Farmer, S.R. Regulating the balance between peroxisome proliferator-activated receptor gamma and beta-catenin signaling during adipogenesis. A glycogen synthase kinase 3beta phosphorylation-defective mutant of beta-catenin inhibits expression of a subset of adipogenic genes. *J. Biol. Chem.* **2004**, *279*, 45020–45027. [CrossRef] [PubMed]
96. Qian, J.; Niu, M.; Zhai, X.; Zhou, Q.; Zhou, Y. β-Catenin pathway is required for TGF-β1 inhibition of PPARγ expression in cultured hepatic stellate cells. *Pharmacol. Res.* **2012**, *66*, 219–225. [CrossRef] [PubMed]
97. Segel, M.J.; Izbicki, G.; Cohen, P.Y.; Or, R.; Christensen, T.G.; Wallach-Dayan, S.B.; Breuer, R. Role of interferon-gamma in the evolution of murine bleomycin lung fibrosis. *Am. J. Physiol. Lung Cell. Mol. Physiol.* **2003**, *285*, L1255–L1262. [CrossRef] [PubMed]
98. Shim, C.Y.; Song, B.-W.; Cha, M.-J.; Hwang, K.-C.; Park, S.; Hong, G.-R.; Kang, S.-M.; Lee, J.E.; Ha, J.-W.; Chung, N. Combination of a peroxisome proliferator-activated receptor-gamma agonist and an angiotensin II receptor blocker attenuates myocardial fibrosis and dysfunction in type 2 diabetic rats. *J. Diabetes Investig.* **2014**, *5*, 362–371. [CrossRef] [PubMed]
99. Sharma, C.; Pradeep, A.; Wong, L.; Rana, A.; Rana, B. Peroxisome proliferator-activated receptor gamma activation can regulate beta-catenin levels via a proteasome-mediated and adenomatous polyposis coli-independent pathway. *J. Biol. Chem.* **2004**, *279*, 35583–35594. [CrossRef] [PubMed]
100. Li, X.-F.; Sun, Y.-Y.; Bao, J.; Chen, X.; Li, Y.-H.; Yang, Y.; Zhang, L.; Huang, C.; Wu, B.-M.; Meng, X.-M.; et al. Functional role of PPAR-γ on the proliferation and migration of fibroblast-like synoviocytes in rheumatoid arthritis. *Sci. Rep.* **2017**, *7*, 12671. [CrossRef] [PubMed]
101. Lu, D.; Carson, D.A. Repression of beta-catenin signaling by PPAR gamma ligands. *Eur. J. Pharmacol.* **2010**, *636*, 198–202. [CrossRef] [PubMed]
102. Takada, I.; Kouzmenko, A.P.; Kato, S. Wnt and PPARgamma signaling in osteoblastogenesis and adipogenesis. *Nat. Rev. Rheumatol.* **2009**, *5*, 442–447. [CrossRef] [PubMed]
103. Garcia-Gras, E.; Lombardi, R.; Giocondo, M.J.; Willerson, J.T.; Schneider, M.D.; Khoury, D.S.; Marian, A.J. Suppression of canonical Wnt/beta-catenin signaling by nuclear plakoglobin recapitulates phenotype of arrhythmogenic right ventricular cardiomyopathy. *J. Clin. Investig.* **2006**, *116*, 2012–2021. [CrossRef] [PubMed]
104. Elbrecht, A.; Chen, Y.; Cullinan, C.A.; Hayes, N.; Leibowitz, M.D.; Moller, D.E.; Berger, J. Molecular cloning, expression and characterization of human peroxisome proliferator activated receptors gamma 1 and gamma 2. *Biochem. Biophys. Res. Commun.* **1996**, *224*, 431–437. [CrossRef] [PubMed]
105. Fajas, L.; Auboeuf, D.; Raspé, E.; Schoonjans, K.; Lefebvre, A.M.; Saladin, R.; Najib, J.; Laville, M.; Fruchart, J.C.; Deeb, S.; et al. The organization, promoter analysis, and expression of the human PPARgamma gene. *J. Biol. Chem.* **1997**, *272*, 18779–18789. [CrossRef] [PubMed]
106. Moldes, M.; Zuo, Y.; Morrison, R.F.; Silva, D.; Park, B.-H.; Liu, J.; Farmer, S.R. Peroxisome-proliferator-activated receptor gamma suppresses Wnt/beta-catenin signalling during adipogenesis. *Biochem. J.* **2003**, *376*, 607–613. [CrossRef] [PubMed]
107. Akinyeke, T.O.; Stewart, L.V. Troglitazone suppresses c-Myc levels in human prostate cancer cells via a PPARγ-independent mechanism. *Cancer Biol. Ther.* **2011**, *11*, 1046–1058. [CrossRef] [PubMed]
108. Gustafson, B.; Eliasson, B.; Smith, U. Thiazolidinediones increase the wingless-type MMTV integration site family (WNT) inhibitor Dickkopf-1 in adipocytes: A link with osteogenesis. *Diabetologia* **2010**, *53*, 536–540. [CrossRef] [PubMed]
109. Jeon, M.; Rahman, N.; Kim, Y.-S. Wnt/β-catenin signaling plays a distinct role in methyl gallate-mediated inhibition of adipogenesis. *Biochem. Biophys. Res. Commun.* **2016**, *479*, 22–27. [CrossRef] [PubMed]
110. Okamura, M.; Kudo, H.; Wakabayashi, K.; Tanaka, T.; Nonaka, A.; Uchida, A.; Tsutsumi, S.; Sakakibara, I.; Naito, M.; Osborne, T.F.; et al. COUP-TFII acts downstream of Wnt/beta-catenin signal to silence PPARgamma gene expression and repress adipogenesis. *Proc. Natl. Acad. Sci. USA* **2009**, *106*, 5819–5824. [CrossRef] [PubMed]
111. Simon, M.F.; Daviaud, D.; Pradère, J.P.; Grès, S.; Guigné, C.; Wabitsch, M.; Chun, J.; Valet, P.; Saulnier-Blache, J.S. Lysophosphatidic acid inhibits adipocyte differentiation via lysophosphatidic acid 1 receptor-dependent down-regulation of peroxisome proliferator-activated receptor gamma2. *J. Biol. Chem.* **2005**, *280*, 14656–14662. [CrossRef] [PubMed]

112. Tan, J.T.M.; McLennan, S.V.; Song, W.W.; Lo, L.W.-Y.; Bonner, J.G.; Williams, P.F.; Twigg, S.M. Connective tissue growth factor inhibits adipocyte differentiation. *Am. J. Physiol. Cell Physiol.* **2008**, *295*, C740–C751. [CrossRef] [PubMed]
113. Yamasaki, S.; Nakashima, T.; Kawakami, A.; Miyashita, T.; Tanaka, F.; Ida, H.; Migita, K.; Origuchi, T.; Eguchi, K. Cytokines regulate fibroblast-like synovial cell differentiation to adipocyte-like cells. *Rheumatology* **2004**, *43*, 448–452. [CrossRef] [PubMed]
114. Niino, M.; Iwabuchi, K.; Kikuchi, S.; Ato, M.; Morohashi, T.; Ogata, A.; Tashiro, K.; Onoé, K. Amelioration of experimental autoimmune encephalomyelitis in C57BL/6 mice by an agonist of peroxisome proliferator-activated receptor-gamma. *J. Neuroimmunol.* **2001**, *116*, 40–48. [CrossRef]
115. Diab, A.; Deng, C.; Smith, J.D.; Hussain, R.Z.; Phanavanh, B.; Lovett-Racke, A.E.; Drew, P.D.; Racke, M.K. Peroxisome proliferator-activated receptor-gamma agonist 15-deoxy-Delta(12,14)-prostaglandin J(2) ameliorates experimental autoimmune encephalomyelitis. *J. Immunol.* **2002**, *168*, 2508–2515. [CrossRef] [PubMed]
116. Feinstein, D.L.; Galea, E.; Gavrilyuk, V.; Brosnan, C.F.; Whitacre, C.C.; Dumitrescu-Ozimek, L.; Landreth, G.E.; Pershadsingh, H.A.; Weinberg, G.; Heneka, M.T. Peroxisome proliferator-activated receptor-gamma agonists prevent experimental autoimmune encephalomyelitis. *Ann. Neurol.* **2002**, *51*, 694–702. [CrossRef] [PubMed]
117. Natarajan, C.; Bright, J.J. Peroxisome proliferator-activated receptor-gamma agonists inhibit experimental allergic encephalomyelitis by blocking IL-12 production, IL-12 signaling and Th1 differentiation. *Genes Immun.* **2002**, *3*, 59–70. [CrossRef] [PubMed]
118. Schulman, I.G.; Shao, G.; Heyman, R.A. Transactivation by retinoid X receptor-peroxisome proliferator-activated receptor gamma (PPARgamma) heterodimers: Intermolecular synergy requires only the PPARgamma hormone-dependent activation function. *Mol. Cell. Biol.* **1998**, *18*, 3483–3494. [CrossRef] [PubMed]
119. Westin, S.; Kurokawa, R.; Nolte, R.T.; Wisely, G.B.; McInerney, E.M.; Rose, D.W.; Milburn, M.V.; Rosenfeld, M.G.; Glass, C.K. Interactions controlling the assembly of nuclear-receptor heterodimers and co-activators. *Nature* **1998**, *395*, 199–202. [CrossRef] [PubMed]
120. Huang, J.K.; Jarjour, A.A.; Oumesmar, B.N.; Kerninon, C.; Williams, A.; Krezel, W.; Kagechika, H.; Bauer, J.; Zhao, C.; Baron-Van Evercooren, A.; et al. Retinoid X receptor gamma signaling accelerates CNS remyelination. *Nat. Neurosci.* **2011**, *14*, 45–53. [CrossRef] [PubMed]
121. Benedusi, V.; Martorana, F.; Brambilla, L.; Maggi, A.; Rossi, D. The peroxisome proliferator-activated receptor γ (PPARγ) controls natural protective mechanisms against lipid peroxidation in amyotrophic lateral sclerosis. *J. Biol. Chem.* **2012**, *287*, 35899–35911. [CrossRef] [PubMed]
122. Drew, P.D.; Storer, P.D.; Xu, J.; Chavis, J.A. Hormone regulation of microglial cell activation: Relevance to multiple sclerosis. *Brain Res. Brain Res. Rev.* **2005**, *48*, 322–327. [CrossRef] [PubMed]
123. Drew, P.D.; Xu, J.; Racke, M.K. PPAR-gamma: Therapeutic Potential for Multiple Sclerosis. *PPAR Res.* **2008**, *2008*, 627463. [CrossRef] [PubMed]
124. Duvanel, C.B.; Honegger, P.; Pershadsingh, H.; Feinstein, D.; Matthieu, J.-M. Inhibition of glial cell proinflammatory activities by peroxisome proliferator-activated receptor gamma agonist confers partial protection during antimyelin oligodendrocyte glycoprotein demyelination in vitro. *J. Neurosci. Res.* **2003**, *71*, 246–255. [CrossRef] [PubMed]
125. Luna-Medina, R.; Cortes-Canteli, M.; Alonso, M.; Santos, A.; Martínez, A.; Perez-Castillo, A. Regulation of inflammatory response in neural cells in vitro by thiadiazolidinones derivatives through peroxisome proliferator-activated receptor gamma activation. *J. Biol. Chem.* **2005**, *280*, 21453–21462. [CrossRef] [PubMed]
126. Paintlia, A.S.; Paintlia, M.K.; Singh, I.; Singh, A.K. IL-4-induced peroxisome proliferator-activated receptor gamma activation inhibits NF-kappaB trans activation in central nervous system (CNS) glial cells and protects oligodendrocyte progenitors under neuroinflammatory disease conditions: Implication for CNS-demyelinating diseases. *J. Immunol.* **2006**, *176*, 4385–4398. [PubMed]
127. Swanson, C.R.; Joers, V.; Bondarenko, V.; Brunner, K.; Simmons, H.A.; Ziegler, T.E.; Kemnitz, J.W.; Johnson, J.A.; Emborg, M.E. The PPAR-γ agonist pioglitazone modulates inflammation and induces neuroprotection in parkinsonian monkeys. *J. Neuroinflamm.* **2011**, *8*, 91. [CrossRef] [PubMed]
128. Xing, B.; Xin, T.; Hunter, R.L.; Bing, G. Pioglitazone inhibition of lipopolysaccharide-induced nitric oxide synthase is associated with altered activity of p38 MAP kinase and PI3K/Akt. *J. Neuroinflamm.* **2008**, *5*, 4. [CrossRef] [PubMed]

129. Klotz, L.; Burgdorf, S.; Dani, I.; Saijo, K.; Flossdorf, J.; Hucke, S.; Alferink, J.; Nowak, N.; Novak, N.; Beyer, M.; et al. The nuclear receptor PPAR gamma selectively inhibits Th17 differentiation in a T cell-intrinsic fashion and suppresses CNS autoimmunity. *J. Exp. Med.* **2009**, *206*, 2079–2089. [CrossRef] [PubMed]
130. Unoda, K.; Doi, Y.; Nakajima, H.; Yamane, K.; Hosokawa, T.; Ishida, S.; Kimura, F.; Hanafusa, T. Eicosapentaenoic acid (EPA) induces peroxisome proliferator-activated receptors and ameliorates experimental autoimmune encephalomyelitis. *J. Neuroimmunol.* **2013**, *256*, 7–12. [CrossRef] [PubMed]
131. Storer, P.D.; Xu, J.; Chavis, J.; Drew, P.D. Peroxisome proliferator-activated receptor-gamma agonists inhibit the activation of microglia and astrocytes: Implications for multiple sclerosis. *J. Neuroimmunol.* **2005**, *161*, 113–122. [CrossRef] [PubMed]
132. Xu, J.; Drew, P.D. Peroxisome proliferator-activated receptor-gamma agonists suppress the production of IL-12 family cytokines by activated glia. *J. Immunol.* **2007**, *178*, 1904–1913. [CrossRef] [PubMed]
133. Padilla, J.; Leung, E.; Phipps, R.P. Human B lymphocytes and B lymphomas express PPAR-gamma and are killed by PPAR-gamma agonists. *Clin. Immunol.* **2002**, *103*, 22–33. [CrossRef] [PubMed]
134. Paintlia, A.S.; Paintlia, M.K.; Singh, A.K.; Stanislaus, R.; Gilg, A.G.; Barbosa, E.; Singh, I. Regulation of gene expression associated with acute experimental autoimmune encephalomyelitis by Lovastatin. *J. Neurosci. Res.* **2004**, *77*, 63–81. [CrossRef] [PubMed]
135. Klopfleisch, S.; Merkler, D.; Schmitz, M.; Klöppner, S.; Schedensack, M.; Jeserich, G.; Althaus, H.H.; Brück, W. Negative impact of statins on oligodendrocytes and myelin formation in vitro and in vivo. *J. Neurosci.* **2008**, *28*, 13609–13614. [CrossRef] [PubMed]
136. Miron, V.E.; Zehntner, S.P.; Kuhlmann, T.; Ludwin, S.K.; Owens, T.; Kennedy, T.E.; Bedell, B.J.; Antel, J.P. Statin therapy inhibits remyelination in the central nervous system. *Am. J. Pathol.* **2009**, *174*, 1880–1890. [CrossRef] [PubMed]
137. Fancy, S.P.J.; Baranzini, S.E.; Zhao, C.; Yuk, D.-I.; Irvine, K.-A.; Kaing, S.; Sanai, N.; Franklin, R.J.M.; Rowitch, D.H. Dysregulation of the Wnt pathway inhibits timely myelination and remyelination in the mammalian CNS. *Genes Dev.* **2009**, *23*, 1571–1585. [CrossRef] [PubMed]
138. Kiguchi, N.; Kobayashi, Y.; Kishioka, S. Chemokines and cytokines in neuroinflammation leading to neuropathic pain. *Curr. Opin. Pharmacol.* **2012**, *12*, 55–61. [CrossRef] [PubMed]
139. Kam, Y.; Quaranta, V. Cadherin-bound beta-catenin feeds into the Wnt pathway upon adherens junctions dissociation: Evidence for an intersection between beta-catenin pools. *PLoS ONE* **2009**, *4*, e4580. [CrossRef] [PubMed]
140. Lock, C.; Hermans, G.; Pedotti, R.; Brendolan, A.; Schadt, E.; Garren, H.; Langer-Gould, A.; Strober, S.; Cannella, B.; Allard, J.; et al. Gene-microarray analysis of multiple sclerosis lesions yields new targets validated in autoimmune encephalomyelitis. *Nat. Med.* **2002**, *8*, 500–508. [CrossRef] [PubMed]
141. Padden, M.; Leech, S.; Craig, B.; Kirk, J.; Brankin, B.; McQuaid, S. Differences in expression of junctional adhesion molecule-A and beta-catenin in multiple sclerosis brain tissue: Increasing evidence for the role of tight junction pathology. *Acta Neuropathol.* **2007**, *113*, 177–186. [CrossRef] [PubMed]
142. Giacoppo, S.; Soundara Rajan, T.; De Nicola, G.R.; Iori, R.; Bramanti, P.; Mazzon, E. Moringin activates Wnt canonical pathway by inhibiting GSK3β in a mouse model of experimental autoimmune encephalomyelitis. *Drug Des. Dev. Ther.* **2016**, *10*, 3291–3304. [CrossRef] [PubMed]
143. Galuppo, M.; Giacoppo, S.; De Nicola, G.R.; Iori, R.; Navarra, M.; Lombardo, G.E.; Bramanti, P.; Mazzon, E. Antiinflammatory activity of glucomoringin isothiocyanate in a mouse model of experimental autoimmune encephalomyelitis. *Fitoterapia* **2014**, *95*, 160–174. [CrossRef] [PubMed]
144. Giacoppo, S.; Galuppo, M.; Montaut, S.; Iori, R.; Rollin, P.; Bramanti, P.; Mazzon, E. An overview on neuroprotective effects of isothiocyanates for the treatment of neurodegenerative diseases. *Fitoterapia* **2015**, *106*, 12–21. [CrossRef] [PubMed]
145. Pate, K.T.; Stringari, C.; Sprowl-Tanio, S.; Wang, K.; TeSlaa, T.; Hoverter, N.P.; McQuade, M.M.; Garner, C.; Digman, M.A.; Teitell, M.A.; et al. Wnt signaling directs a metabolic program of glycolysis and angiogenesis in colon cancer. *EMBO J.* **2014**, *33*, 1454–1473. [CrossRef] [PubMed]
146. Vallée, A.; Vallée, J.-N. Warburg effect hypothesis in autism Spectrum disorders. *Mol. Brain* **2018**, *11*, 1. [CrossRef] [PubMed]
147. Reuter, S.; Gupta, S.C.; Chaturvedi, M.M.; Aggarwal, B.B. Oxidative stress, inflammation, and cancer: How are they linked? *Free Radic. Biol. Med.* **2010**, *49*, 1603–1616. [CrossRef] [PubMed]

148. Park, K.S.; Lee, R.D.; Kang, S.-K.; Han, S.Y.; Park, K.L.; Yang, K.H.; Song, Y.S.; Park, H.J.; Lee, Y.M.; Yun, Y.P.; et al. Neuronal differentiation of embryonic midbrain cells by upregulation of peroxisome proliferator-activated receptor-gamma via the JNK-dependent pathway. *Exp. Cell Res.* **2004**, *297*, 424–433. [CrossRef] [PubMed]
149. Yue, X.; Lan, F.; Yang, W.; Yang, Y.; Han, L.; Zhang, A.; Liu, J.; Zeng, H.; Jiang, T.; Pu, P.; et al. Interruption of β-catenin suppresses the EGFR pathway by blocking multiple oncogenic targets in human glioma cells. *Brain Res.* **2010**, *1366*, 27–37. [CrossRef] [PubMed]
150. Sun, Q.; Chen, X.; Ma, J.; Peng, H.; Wang, F.; Zha, X.; Wang, Y.; Jing, Y.; Yang, H.; Chen, R.; et al. Mammalian target of rapamycin up-regulation of pyruvate kinase isoenzyme type M2 is critical for aerobic glycolysis and tumor growth. *Proc. Natl. Acad. Sci. USA* **2011**, *108*, 4129–4134. [CrossRef] [PubMed]
151. Semenza, G.L. HIF-1: Upstream and downstream of cancer metabolism. *Curr. Opin. Genet. Dev.* **2010**, *20*, 51–56. [CrossRef] [PubMed]
152. Goldbeter, A. Patterns of spatiotemporal organization in an allosteric enzyme model. *Proc. Natl. Acad. Sci. USA* **1973**, *70*, 3255–3259. [CrossRef] [PubMed]
153. Cambron, M.; D'Haeseleer, M.; Laureys, G.; Clinckers, R.; Debruyne, J.; De Keyser, J. White-matter astrocytes, axonal energy metabolism, and axonal degeneration in multiple sclerosis. *J. Cereb. Blood Flow Metab.* **2012**, *32*, 413–424. [CrossRef] [PubMed]
154. Hattingen, E.; Magerkurth, J.; Pilatus, U.; Hübers, A.; Wahl, M.; Ziemann, U. Combined (1)H and (31)P spectroscopy provides new insights into the pathobiochemistry of brain damage in multiple sclerosis. *NMR Biomed.* **2011**, *24*, 536–546. [CrossRef] [PubMed]
155. Trapp, B.D.; Stys, P.K. Virtual hypoxia and chronic necrosis of demyelinated axons in multiple sclerosis. *Lancet Neurol.* **2009**, *8*, 280–291. [CrossRef]
156. Campbell, G.R.; Ziabreva, I.; Reeve, A.K.; Krishnan, K.J.; Reynolds, R.; Howell, O.; Lassmann, H.; Turnbull, D.M.; Mahad, D.J. Mitochondrial DNA deletions and neurodegeneration in multiple sclerosis. *Ann. Neurol.* **2011**, *69*, 481–492. [CrossRef] [PubMed]
157. Witte, M.E.; Bø, L.; Rodenburg, R.J.; Belien, J.A.; Musters, R.; Hazes, T.; Wintjes, L.T.; Smeitink, J.A.; Geurts, J.J.G.; De Vries, H.E.; et al. Enhanced number and activity of mitochondria in multiple sclerosis lesions. *J. Pathol.* **2009**, *219*, 193–204. [CrossRef] [PubMed]
158. Gebregiworgis, T.; Nielsen, H.H.; Massilamany, C.; Gangaplara, A.; Reddy, J.; Illes, Z.; Powers, R. A Urinary Metabolic Signature for Multiple Sclerosis and Neuromyelitis Optica. *J. Proteome Res.* **2016**, *15*, 659–666. [CrossRef] [PubMed]
159. Nijland, P.G.; Molenaar, R.J.; van der Pol, S.M.A.; van der Valk, P.; van Noorden, C.J.F.; de Vries, H.E.; van Horssen, J. Differential expression of glucose-metabolizing enzymes in multiple sclerosis lesions. *Acta Neuropathol. Commun.* **2015**, *3*, 79. [CrossRef] [PubMed]
160. De Riccardis, L.; Rizzello, A.; Ferramosca, A.; Urso, E.; De Robertis, F.; Danieli, A.; Giudetti, A.M.; Trianni, G.; Zara, V.; Maffia, M. Bioenergetics profile of CD4(+) T cells in relapsing remitting multiple sclerosis subjects. *J. Biotechnol.* **2015**, *202*, 31–39. [CrossRef] [PubMed]
161. Kebir, H.; Kreymborg, K.; Ifergan, I.; Dodelet-Devillers, A.; Cayrol, R.; Bernard, M.; Giuliani, F.; Arbour, N.; Becher, B.; Prat, A. Human TH17 lymphocytes promote blood-brain barrier disruption and central nervous system inflammation. *Nat. Med.* **2007**, *13*, 1173–1175. [CrossRef] [PubMed]
162. De Riccardis, L.; Ferramosca, A.; Danieli, A.; Trianni, G.; Zara, V.; De Robertis, F.; Maffia, M. Metabolic response to glatiramer acetate therapy in multiple sclerosis patients. *BBA Clin.* **2016**, *6*, 131–137. [CrossRef] [PubMed]
163. Bauernfeind, A.L.; Barks, S.K.; Duka, T.; Grossman, L.I.; Hof, P.R.; Sherwood, C.C. Aerobic glycolysis in the primate brain: Reconsidering the implications for growth and maintenance. *Brain Struct. Funct.* **2014**, *219*, 1149–1167. [CrossRef] [PubMed]
164. Bongarzone, E.R.; Pasquini, J.M.; Soto, E.F. Oxidative damage to proteins and lipids of CNS myelin produced by in vitro generated reactive oxygen species. *J. Neurosci. Res.* **1995**, *41*, 213–221. [CrossRef] [PubMed]
165. Brand, K.A.; Hermfisse, U. Aerobic glycolysis by proliferating cells: A protective strategy against reactive oxygen species. *FASEB J.* **1997**, *11*, 388–395. [CrossRef] [PubMed]
166. Rodriguez-Rodriguez, P.; Fernandez, E.; Almeida, A.; Bolaños, J.P. Excitotoxic stimulus stabilizes PFKFB3 causing pentose-phosphate pathway to glycolysis switch and neurodegeneration. *Cell Death Differ.* **2012**, *19*, 1582–1589. [CrossRef] [PubMed]

167. Wang, J.; Li, G.; Wang, Z.; Zhang, X.; Yao, L.; Wang, F.; Liu, S.; Yin, J.; Ling, E.-A.; Wang, L.; et al. High glucose-induced expression of inflammatory cytokines and reactive oxygen species in cultured astrocytes. *Neuroscience* **2012**, *202*, 58–68. [CrossRef] [PubMed]
168. Patel, K.P.; O'Brien, T.W.; Subramony, S.H.; Shuster, J.; Stacpoole, P.W. The spectrum of pyruvate dehydrogenase complex deficiency: Clinical, biochemical and genetic features in 371 patients. *Mol. Genet. Metab.* **2012**, *106*, 385–394. [CrossRef] [PubMed]
169. Trofimova, L.K.; Araújo, W.L.; Strokina, A.A.; Fernie, A.R.; Bettendorff, L.; Bunik, V.I. Consequences of the α-ketoglutarate dehydrogenase inhibition for neuronal metabolism and survival: Implications for neurodegenerative diseases. *Curr. Med. Chem.* **2012**, *19*, 5895–5906. [CrossRef] [PubMed]
170. Schiepers, C.; Van Hecke, P.; Vandenberghe, R.; Van Oostende, S.; Dupont, P.; Demaerel, P.; Bormans, G.; Carton, H. Positron emission tomography, magnetic resonance imaging and proton NMR spectroscopy of white matter in multiple sclerosis. *Mult. Scler. J.* **1997**, *3*, 8–17. [CrossRef] [PubMed]
171. Simone, I.L.; Tortorella, C.; Federico, F.; Liguori, M.; Lucivero, V.; Giannini, P.; Carrara, D.; Bellacosa, A.; Livrea, P. Axonal damage in multiple sclerosis plaques: A combined magnetic resonance imaging and 1H-magnetic resonance spectroscopy study. *J. Neurol. Sci.* **2001**, *182*, 143–150. [CrossRef]
172. Amorini, A.M.; Nociti, V.; Petzold, A.; Gasperini, C.; Quartuccio, E.; Lazzarino, G.; Di Pietro, V.; Belli, A.; Signoretti, S.; Vagnozzi, R.; et al. Serum lactate as a novel potential biomarker in multiple sclerosis. *Biochim. Biophys. Acta* **2014**, *1842*, 1137–1143. [CrossRef] [PubMed]
173. Petzold, A.; Nijland, P.G.; Balk, L.J.; Amorini, A.M.; Lazzarino, G.; Wattjes, M.P.; Gasperini, C.; van der Valk, P.; Tavazzi, B.; Lazzarino, G.; et al. Visual pathway neurodegeneration winged by mitochondrial dysfunction. *Ann. Clin. Transl. Neurol.* **2015**, *2*, 140–150. [CrossRef] [PubMed]
174. Schocke, M.F.H.; Berger, T.; Felber, S.R.; Wolf, C.; Deisenhammer, F.; Kremser, C.; Seppi, K.; Aichner, F.T. Serial contrast-enhanced magnetic resonance imaging and spectroscopic imaging of acute multiple sclerosis lesions under high-dose methylprednisolone therapy. *NeuroImage* **2003**, *20*, 1253–1263. [CrossRef]
175. Lutz, N.W.; Viola, A.; Malikova, I.; Confort-Gouny, S.; Audoin, B.; Ranjeva, J.-P.; Pelletier, J.; Cozzone, P.J. Inflammatory multiple-sclerosis plaques generate characteristic metabolic profiles in cerebrospinal fluid. *PLoS ONE* **2007**, *2*, e595. [CrossRef] [PubMed]
176. Regenold, W.T.; Phatak, P.; Makley, M.J.; Stone, R.D.; Kling, M.A. Cerebrospinal fluid evidence of increased extra-mitochondrial glucose metabolism implicates mitochondrial dysfunction in multiple sclerosis disease progression. *J. Neurol. Sci.* **2008**, *275*, 106–112. [CrossRef] [PubMed]
177. Simone, I.L.; Federico, F.; Trojano, M.; Tortorella, C.; Liguori, M.; Giannini, P.; Picciola, E.; Natile, G.; Livrea, P. High resolution proton MR spectroscopy of cerebrospinal fluid in MS patients. Comparison with biochemical changes in demyelinating plaques. *J. Neurol. Sci.* **1996**, *144*, 182–190. [CrossRef]
178. Hogenesch, J.B.; Gu, Y.Z.; Jain, S.; Bradfield, C.A. The basic-helix-loop-helix-PAS orphan MOP3 forms transcriptionally active complexes with circadian and hypoxia factors. *Proc. Natl. Acad. Sci. USA* **1998**, *95*, 5474–5479. [CrossRef] [PubMed]
179. Gekakis, N.; Staknis, D.; Nguyen, H.B.; Davis, F.C.; Wilsbacher, L.D.; King, D.P.; Takahashi, J.S.; Weitz, C.J. Role of the CLOCK protein in the mammalian circadian mechanism. *Science* **1998**, *280*, 1564–1569. [CrossRef] [PubMed]
180. Reppert, S.M.; Weaver, D.R. Coordination of circadian timing in mammals. *Nature* **2002**, *418*, 935–941. [CrossRef] [PubMed]
181. Schibler, U.; Sassone-Corsi, P. A web of circadian pacemakers. *Cell* **2002**, *111*, 919–922. [CrossRef]
182. Ko, C.H.; Takahashi, J.S. Molecular components of the mammalian circadian clock. *Hum. Mol. Genet.* **2006**, *15*, R271–R277. [CrossRef] [PubMed]
183. De Somma, E.; Jain, R.W.; Poon, K.W.C.; Tresidder, K.A.; Segal, J.P.; Ghasemlou, N. Chronobiological regulation of psychosocial and physiological outcomes in multiple sclerosis. *Neurosci. Biobehav. Rev.* **2018**. [CrossRef] [PubMed]
184. Hedström, A.K.; Åkerstedt, T.; Hillert, J.; Olsson, T.; Alfredsson, L. Shift work at young age is associated with increased risk for multiple sclerosis. *Ann. Neurol.* **2011**, *70*, 733–741. [CrossRef] [PubMed]
185. Hedström, A.K.; Åkerstedt, T.; Olsson, T.; Alfredsson, L. Shift work influences multiple sclerosis risk. *Mult. Scler. J.* **2015**, *21*, 1195–1199. [CrossRef] [PubMed]

186. Haspel, J.A.; Chettimada, S.; Shaik, R.S.; Chu, J.-H.; Raby, B.A.; Cernadas, M.; Carey, V.; Process, V.; Hunninghake, G.M.; Ifedigbo, E.; et al. Circadian rhythm reprogramming during lung inflammation. *Nat. Commun.* **2014**, *5*, 4753. [CrossRef] [PubMed]
187. He, J.; Hsuchou, H.; He, Y.; Kastin, A.J.; Wang, Y.; Pan, W. Sleep restriction impairs blood-brain barrier function. *J. Neurosci.* **2014**, *34*, 14697–14706. [CrossRef] [PubMed]
188. Akpinar, Z.; Tokgöz, S.; Gökbel, H.; Okudan, N.; Uğuz, F.; Yilmaz, G. The association of nocturnal serum melatonin levels with major depression in patients with acute multiple sclerosis. *Psychiatry Res.* **2008**, *161*, 253–257. [CrossRef] [PubMed]
189. Marck, C.H.; Neate, S.L.; Taylor, K.L.; Weiland, T.J.; Jelinek, G.A. Prevalence of Comorbidities, Overweight and Obesity in an International Sample of People with Multiple Sclerosis and Associations with Modifiable Lifestyle Factors. *PLoS ONE* **2016**, *11*, e0148573. [CrossRef] [PubMed]
190. Lunde, H.M.B.; Bjorvatn, B.; Myhr, K.-M.; Bø, L. Clinical assessment and management of sleep disorders in multiple sclerosis: A literature review. *Acta Neurol. Scand.* **2013**, *127*, 24–30. [CrossRef] [PubMed]
191. Turek, F.W.; Dugovic, C.; Zee, P.C. Current understanding of the circadian clock and the clinical implications for neurological disorders. *Arch. Neurol.* **2001**, *58*, 1781–1787. [CrossRef] [PubMed]
192. Attarian, H.P.; Brown, K.M.; Duntley, S.P.; Carter, J.D.; Cross, A.H. The relationship of sleep disturbances and fatigue in multiple sclerosis. *Arch. Neurol.* **2004**, *61*, 525–528. [CrossRef] [PubMed]
193. Najafi, M.R.; Toghianifar, N.; Etemadifar, M.; Haghighi, S.; Maghzi, A.H.; Akbari, M. Circadian rhythm sleep disorders in patients with multiple sclerosis and its association with fatigue: A case-control study. *J. Res. Med. Sci.* **2013**, *18*, S71–S73. [PubMed]
194. Ayache, S.S.; Chalah, M.A. Fatigue in multiple sclerosis—Insights into evaluation and management. *Neurophysiol. Clin. Clin. Neurophysiol.* **2017**, *47*, 139–171. [CrossRef] [PubMed]
195. Tsujino, N.; Sakurai, T. Role of orexin in modulating arousal, feeding, and motivation. *Front. Behav. Neurosci.* **2013**, *7*, 28. [CrossRef] [PubMed]
196. Kiyashchenko, L.I.; Mileykovskiy, B.Y.; Maidment, N.; Lam, H.A.; Wu, M.-F.; John, J.; Peever, J.; Siegel, J.M. Release of hypocretin (orexin) during waking and sleep states. *J. Neurosci.* **2002**, *22*, 5282–5286. [CrossRef] [PubMed]
197. Oka, Y.; Kanbayashi, T.; Mezaki, T.; Iseki, K.; Matsubayashi, J.; Murakami, G.; Matsui, M.; Shimizu, T.; Shibasaki, H. Low CSF hypocretin-1/orexin-A associated with hypersomnia secondary to hypothalamic lesion in a case of multiple sclerosis. *J. Neurol.* **2004**, *251*, 885–886. [CrossRef] [PubMed]
198. Grossberg, A.J.; Zhu, X.; Leinninger, G.M.; Levasseur, P.R.; Braun, T.P.; Myers, M.G.; Marks, D.L. Inflammation-induced lethargy is mediated by suppression of orexin neuron activity. *J. Neurosci.* **2011**, *31*, 11376–11386. [CrossRef] [PubMed]
199. Pokryszko-Dragan, A.; Frydecka, I.; Kosmaczewska, A.; Ciszak, L.; Bilińska, M.; Gruszka, E.; Podemski, R.; Frydecka, D. Stimulated peripheral production of interferon-gamma is related to fatigue and depression in multiple sclerosis. *Clin. Neurol. Neurosurg.* **2012**, *114*, 1153–1158. [CrossRef] [PubMed]
200. Boddum, K.; Hansen, M.H.; Jennum, P.J.; Kornum, B.R. Cerebrospinal Fluid Hypocretin-1 (Orexin-A) Level Fluctuates with Season and Correlates with Day Length. *PLoS ONE* **2016**, *11*, e0151288. [CrossRef] [PubMed]
201. Taphoorn, M.J.; van Someren, E.; Snoek, F.J.; Strijers, R.L.; Swaab, D.F.; Visscher, F.; de Waal, L.P.; Polman, C.H. Fatigue, sleep disturbances and circadian rhythm in multiple sclerosis. *J. Neurol.* **1993**, *240*, 446–448. [CrossRef] [PubMed]
202. Farez, M.F.; Mascanfroni, I.D.; Méndez-Huergo, S.P.; Yeste, A.; Murugaiyan, G.; Garo, L.P.; Balbuena Aguirre, M.E.; Patel, B.; Ysrraelit, M.C.; Zhu, C.; et al. Melatonin Contributes to the Seasonality of Multiple Sclerosis Relapses. *Cell* **2015**, *162*, 1338–1352. [CrossRef] [PubMed]
203. Torkildsen, O.; Aarseth, J.; Benjaminsen, E.; Celius, E.; Holmøy, T.; Kampman, M.T.; Løken-Amsrud, K.; Midgard, R.; Myhr, K.-M.; Riise, T.; et al. Month of birth and risk of multiple sclerosis: Confounding and adjustments. *Ann. Clin. Transl. Neurol.* **2014**, *1*, 141–144. [CrossRef] [PubMed]
204. Reynolds, J.D.; Case, L.K.; Krementsov, D.N.; Raza, A.; Bartiss, R.; Teuscher, C. Modeling month-season of birth as a risk factor in mouse models of chronic disease: From multiple sclerosis to autoimmune encephalomyelitis. *FASEB J.* **2017**, *31*, 2709–2719. [CrossRef] [PubMed]
205. Dang, A.K.; Tesfagiorgis, Y.; Jain, R.W.; Craig, H.C.; Kerfoot, S.M. Meningeal Infiltration of the Spinal Cord by Non-Classically Activated B Cells is Associated with Chronic Disease Course in a Spontaneous B Cell-Dependent Model of CNS Autoimmune Disease. *Front. Immunol.* **2015**, *6*, 470. [CrossRef] [PubMed]

206. Yu, X.; Rollins, D.; Ruhn, K.A.; Stubblefield, J.J.; Green, C.B.; Kashiwada, M.; Rothman, P.B.; Takahashi, J.S.; Hooper, L.V. TH17 cell differentiation is regulated by the circadian clock. *Science* **2013**, *342*, 727–730. [CrossRef] [PubMed]
207. Chen, T.L. Inhibition of growth and differentiation of osteoprogenitors in mouse bone marrow stromal cell cultures by increased donor age and glucocorticoid treatment. *Bone* **2004**, *35*, 83–95. [CrossRef] [PubMed]
208. Soták, M.; Sumová, A.; Pácha, J. Cross-talk between the circadian clock and the cell cycle in cancer. *Ann. Med.* **2014**, *46*, 221–232. [CrossRef] [PubMed]
209. Guo, F.; He, D.; Zhang, W.; Walton, R.G. Trends in prevalence, awareness, management, and control of hypertension among United States adults, 1999 to 2010. *J. Am. Coll. Cardiol.* **2012**, *60*, 599–606. [CrossRef] [PubMed]
210. Yasuniwa, Y.; Izumi, H.; Wang, K.-Y.; Shimajiri, S.; Sasaguri, Y.; Kawai, K.; Kasai, H.; Shimada, T.; Miyake, K.; Kashiwagi, E.; et al. Circadian disruption accelerates tumor growth and angio/stromagenesis through a Wnt signaling pathway. *PLoS ONE* **2010**, *5*, e15330. [CrossRef] [PubMed]
211. Janich, P.; Pascual, G.; Merlos-Suárez, A.; Batlle, E.; Ripperger, J.; Albrecht, U.; Cheng, H.-Y.M.; Obrietan, K.; Di Croce, L.; Benitah, S.A. The circadian molecular clock creates epidermal stem cell heterogeneity. *Nature* **2011**, *480*, 209–214. [CrossRef] [PubMed]
212. Lavtar, P.; Rudolf, G.; Maver, A.; Hodžić, A.; Starčević Čizmarević, N.; Živković, M.; Šega Jazbec, S.; Klemenc Ketiš, Z.; Kapović, M.; Dinčić, E.; et al. Association of circadian rhythm genes ARNTL/BMAL1 and CLOCK with multiple sclerosis. *PLoS ONE* **2018**, *13*, e0190601. [CrossRef] [PubMed]
213. Lin, F.; Chen, Y.; Li, X.; Zhao, Q.; Tan, Z. Over-expression of circadian clock gene Bmal1 affects proliferation and the canonical Wnt pathway in NIH-3T3 cells. *Cell Biochem. Funct.* **2013**, *31*, 166–172. [CrossRef] [PubMed]
214. Sahar, S.; Sassone-Corsi, P. Metabolism and cancer: The circadian clock connection. *Nat. Rev. Cancer* **2009**, *9*, 886–896. [CrossRef] [PubMed]
215. Yang, X.; Wood, P.A.; Ansell, C.M.; Ohmori, M.; Oh, E.-Y.; Xiong, Y.; Berger, F.G.; Peña, M.M.O.; Hrushesky, W.J.M. Beta-catenin induces beta-TrCP-mediated PER2 degradation altering circadian clock gene expression in intestinal mucosa of ApcMin/+ mice. *J. Biochem.* **2009**, *145*, 289–297. [CrossRef] [PubMed]
216. Duffield, G.E.; Best, J.D.; Meurers, B.H.; Bittner, A.; Loros, J.J.; Dunlap, J.C. Circadian programs of transcriptional activation, signaling, and protein turnover revealed by microarray analysis of mammalian cells. *Curr. Biol.* **2002**, *12*, 551–557. [CrossRef]
217. Sancar, A.; Lindsey-Boltz, L.A.; Unsal-Kaçmaz, K.; Linn, S. Molecular mechanisms of mammalian DNA repair and the DNA damage checkpoints. *Annu. Rev. Biochem.* **2004**, *73*, 39–85. [CrossRef] [PubMed]
218. Yang, X.; Downes, M.; Yu, R.T.; Bookout, A.L.; He, W.; Straume, M.; Mangelsdorf, D.J.; Evans, R.M. Nuclear receptor expression links the circadian clock to metabolism. *Cell* **2006**, *126*, 801–810. [CrossRef] [PubMed]
219. Yang, G.; Jia, Z.; Aoyagi, T.; McClain, D.; Mortensen, R.M.; Yang, T. Systemic PPARγ deletion impairs circadian rhythms of behavior and metabolism. *PLoS ONE* **2012**, *7*, e38117. [CrossRef] [PubMed]
220. Wang, H.-M.; Zhao, Y.-X.; Zhang, S.; Liu, G.-D.; Kang, W.-Y.; Tang, H.-D.; Ding, J.-Q.; Chen, S.-D. PPARgamma agonist curcumin reduces the amyloid-beta-stimulated inflammatory responses in primary astrocytes. *J. Alzheimers Dis.* **2010**, *20*, 1189–1199. [CrossRef] [PubMed]
221. Fontaine, C.; Dubois, G.; Duguay, Y.; Helledie, T.; Vu-Dac, N.; Gervois, P.; Soncin, F.; Mandrup, S.; Fruchart, J.-C.; Fruchart-Najib, J.; et al. The orphan nuclear receptor Rev-Erbalpha is a peroxisome proliferator-activated receptor (PPAR) gamma target gene and promotes PPARgamma-induced adipocyte differentiation. *J. Biol. Chem.* **2003**, *278*, 37672–37680. [CrossRef] [PubMed]
222. Green, C.B.; Douris, N.; Kojima, S.; Strayer, C.A.; Fogerty, J.; Lourim, D.; Keller, S.R.; Besharse, J.C. Loss of Nocturnin, a circadian deadenylase, confers resistance to hepatic steatosis and diet-induced obesity. *Proc. Natl. Acad. Sci. USA* **2007**, *104*, 9888–9893. [CrossRef] [PubMed]
223. Chen, L.; Yang, G. PPARs Integrate the Mammalian Clock and Energy Metabolism. *PPAR Res.* **2014**, *2014*, 653017. [CrossRef] [PubMed]
224. Ivanov, I.I.; McKenzie, B.S.; Zhou, L.; Tadokoro, C.E.; Lepelley, A.; Lafaille, J.J.; Cua, D.J.; Littman, D.R. The orphan nuclear receptor RORgammat directs the differentiation program of proinflammatory IL-17+ T helper cells. *Cell* **2006**, *126*, 1121–1133. [CrossRef] [PubMed]

225. Manel, N.; Unutmaz, D.; Littman, D.R. The differentiation of human T(H)-17 cells requires transforming growth factor-beta and induction of the nuclear receptor RORgammat. *Nat. Immunol.* **2008**, *9*, 641–649. [CrossRef] [PubMed]
226. Clark, R.B.; Bishop-Bailey, D.; Estrada-Hernandez, T.; Hla, T.; Puddington, L.; Padula, S.J. The nuclear receptor PPAR gamma and immunoregulation: PPAR gamma mediates inhibition of helper T cell responses. *J. Immunol.* **2000**, *164*, 1364–1371. [CrossRef] [PubMed]
227. Li, W.; Zhang, Z.; Zhang, K.; Xue, Z.; Li, Y.; Zhang, Z.; Zhang, L.; Gu, C.; Zhang, Q.; Hao, J.; et al. Arctigenin Suppress Th17 Cells and Ameliorates Experimental Autoimmune Encephalomyelitis Through AMPK and PPAR-γ/ROR-γt Signaling. *Mol. Neurobiol.* **2016**, *53*, 5356–5366. [CrossRef] [PubMed]
228. Lochner, M.; Peduto, L.; Cherrier, M.; Sawa, S.; Langa, F.; Varona, R.; Riethmacher, D.; Si-Tahar, M.; Di Santo, J.P.; Eberl, G. In vivo equilibrium of proinflammatory IL-17+ and regulatory IL-10+ Foxp3+ RORgamma t+ T cells. *J. Exp. Med.* **2008**, *205*, 1381–1393. [CrossRef] [PubMed]
229. Cermakian, N.; Lange, T.; Golombek, D.; Sarkar, D.; Nakao, A.; Shibata, S.; Mazzoccoli, G. Crosstalk between the circadian clock circuitry and the immune system. *Chronobiol. Int.* **2013**, *30*, 870–888. [CrossRef] [PubMed]
230. Cutolo, M.; Sulli, A.; Pincus, T. Circadian use of glucocorticoids in rheumatoid arthritis. *Neuroimmunomodulation* **2015**, *22*, 33–39. [CrossRef] [PubMed]
231. Kern, S.; Schultheiss, T.; Schneider, H.; Schrempf, W.; Reichmann, H.; Ziemssen, T. Circadian cortisol, depressive symptoms and neurological impairment in early multiple sclerosis. *Psychoneuroendocrinology* **2011**, *36*, 1505–1512. [CrossRef] [PubMed]
232. Wipfler, P.; Heikkinen, A.; Harrer, A.; Pilz, G.; Kunz, A.; Golaszewski, S.M.; Reuss, R.; Oschmann, P.; Kraus, J. Circadian rhythmicity of inflammatory serum parameters: A neglected issue in the search of biomarkers in multiple sclerosis. *J. Neurol.* **2013**, *260*, 221–227. [CrossRef] [PubMed]
233. Powell, D.J.H.; Moss-Morris, R.; Liossi, C.; Schlotz, W. Circadian cortisol and fatigue severity in relapsing-remitting multiple sclerosis. *Psychoneuroendocrinology* **2015**, *56*, 120–131. [CrossRef] [PubMed]
234. Melief, J.; de Wit, S.J.; van Eden, C.G.; Teunissen, C.; Hamann, J.; Uitdehaag, B.M.; Swaab, D.; Huitinga, I. HPA axis activity in multiple sclerosis correlates with disease severity, lesion type and gene expression in normal-appearing white matter. *Acta Neuropathol.* **2013**, *126*, 237–249. [CrossRef] [PubMed]
235. Mason, D.; MacPhee, I.; Antoni, F. The role of the neuroendocrine system in determining genetic susceptibility to experimental allergic encephalomyelitis in the rat. *Immunology* **1990**, *70*, 1–5. [PubMed]
236. Gold, S.M.; Raji, A.; Huitinga, I.; Wiedemann, K.; Schulz, K.-H.; Heesen, C. Hypothalamo-pituitary-adrenal axis activity predicts disease progression in multiple sclerosis. *J. Neuroimmunol.* **2005**, *165*, 186–191. [CrossRef] [PubMed]
237. Heidbrink, C.; Häusler, S.F.M.; Buttmann, M.; Ossadnik, M.; Strik, H.M.; Keller, A.; Buck, D.; Verbraak, E.; van Meurs, M.; Krockenberger, M.; et al. Reduced cortisol levels in cerebrospinal fluid and differential distribution of 11beta-hydroxysteroid dehydrogenases in multiple sclerosis: Implications for lesion pathogenesis. *Brain Behav. Immun.* **2010**, *24*, 975–984. [CrossRef] [PubMed]
238. Van den Brandt, J.; Lühder, F.; McPherson, K.G.; de Graaf, K.L.; Tischner, D.; Wiehr, S.; Herrmann, T.; Weissert, R.; Gold, R.; Reichardt, H.M. Enhanced glucocorticoid receptor signaling in T cells impacts thymocyte apoptosis and adaptive immune responses. *Am. J. Pathol.* **2007**, *170*, 1041–1053. [CrossRef] [PubMed]
239. Bellavance, M.-A.; Rivest, S. The HPA—Immune Axis and the Immunomodulatory Actions of Glucocorticoids in the Brain. *Front. Immunol.* **2014**, *5*, 136. [CrossRef] [PubMed]
240. Van Winsen, L.M.L.; Muris, D.F.R.; Polman, C.H.; Dijkstra, C.D.; van den Berg, T.K.; Uitdehaag, B.M.J. Sensitivity to glucocorticoids is decreased in relapsing remitting multiple sclerosis. *J. Clin. Endocrinol. Metab.* **2005**, *90*, 734–740. [CrossRef] [PubMed]
241. Cheon, S.; Park, N.; Cho, S.; Kim, K. Glucocorticoid-mediated Period2 induction delays the phase of circadian rhythm. *Nucleic Acids Res.* **2013**, *41*, 6161–6174. [CrossRef] [PubMed]
242. So, A.Y.-L.; Bernal, T.U.; Pillsbury, M.L.; Yamamoto, K.R.; Feldman, B.J. Glucocorticoid regulation of the circadian clock modulates glucose homeostasis. *Proc. Natl. Acad. Sci. USA* **2009**, *106*, 17582–17587. [CrossRef] [PubMed]
243. Pan, Z.; Xie, D.; Choudhary, V.; Seremwe, M.; Tsai, Y.-Y.; Olala, L.; Chen, X.; Bollag, W.B. The effect of pioglitazone on aldosterone and cortisol production in HAC15 human adrenocortical carcinoma cells. *Mol. Cell. Endocrinol.* **2014**, *394*, 119–128. [CrossRef] [PubMed]

244. Csernus, V.; Mess, B. Biorhythms and pineal gland. *Neuro Endocrinol. Lett.* **2003**, *24*, 404–411. [PubMed]
245. Mauriz, J.L.; Collado, P.S.; Veneroso, C.; Reiter, R.J.; González-Gallego, J. A review of the molecular aspects of melatonin's anti-inflammatory actions: Recent insights and new perspectives. *J. Pineal Res.* **2013**, *54*, 1–14. [CrossRef] [PubMed]
246. Crowley, S.J.; Eastman, C.I. Melatonin in the afternoons of a gradually advancing sleep schedule enhances the circadian rhythm phase advance. *Psychopharmacology* **2013**, *225*, 825–837. [CrossRef] [PubMed]
247. Farez, M.F.; Calandri, I.L.; Correale, J.; Quintana, F.J. Anti-inflammatory effects of melatonin in multiple sclerosis. *BioEssays* **2016**, *38*, 1016–1026. [CrossRef] [PubMed]
248. Sandyk, R.; Awerbuch, G.I. Nocturnal plasma melatonin and alpha-melanocyte stimulating hormone levels during exacerbation of multiple sclerosis. *Int. J. Neurosci.* **1992**, *67*, 173–186. [CrossRef] [PubMed]
249. Wang, X.; Sirianni, A.; Pei, Z.; Cormier, K.; Smith, K.; Jiang, J.; Zhou, S.; Wang, H.; Zhao, R.; Yano, H.; et al. The melatonin MT1 receptor axis modulates mutant Huntingtin-mediated toxicity. *J. Neurosci.* **2011**, *31*, 14496–14507. [CrossRef] [PubMed]
250. Rosales-Corral, S.A.; Acuña-Castroviejo, D.; Coto-Montes, A.; Boga, J.A.; Manchester, L.C.; Fuentes-Broto, L.; Korkmaz, A.; Ma, S.; Tan, D.-X.; Reiter, R.J. Alzheimer's disease: Pathological mechanisms and the beneficial role of melatonin. *J. Pineal Res.* **2012**, *52*, 167–202. [CrossRef] [PubMed]
251. Galano, A.; Tan, D.X.; Reiter, R.J. On the free radical scavenging activities of melatonin's metabolites, AFMK and AMK. *J. Pineal Res.* **2013**, *54*, 245–257. [CrossRef] [PubMed]
252. Calvo, J.R.; González-Yanes, C.; Maldonado, M.D. The role of melatonin in the cells of the innate immunity: A review. *J. Pineal Res.* **2013**, *55*, 103–120. [CrossRef] [PubMed]
253. Zhang, H.-M.; Zhang, Y. Melatonin: A well-documented antioxidant with conditional pro-oxidant actions. *J. Pineal Res.* **2014**, *57*, 131–146. [CrossRef] [PubMed]
254. Chen, S.-J.; Huang, S.-H.; Chen, J.-W.; Wang, K.-C.; Yang, Y.-R.; Liu, P.-F.; Lin, G.-J.; Sytwu, H.-K. Melatonin enhances interleukin-10 expression and suppresses chemotaxis to inhibit inflammation in situ and reduce the severity of experimental autoimmune encephalomyelitis. *Int. Immunopharmacol.* **2016**, *31*, 169–177. [CrossRef] [PubMed]
255. Markowska, M.; Bialecka, B.; Ciechanowska, M.; Koter, Z.; Laskowska, H.; Karkucinska-Wieckowska, A.; Skwarlo-Sonta, K. Effect of immunization on nocturnal NAT activity in chicken pineal gland. *Neuro Endocrinol. Lett.* **2000**, *21*, 367–373. [PubMed]
256. Fernandes, P.A.C.M.; Cecon, E.; Markus, R.P.; Ferreira, Z.S. Effect of TNF-alpha on the melatonin synthetic pathway in the rat pineal gland: Basis for a "feedback" of the immune response on circadian timing. *J. Pineal Res.* **2006**, *41*, 344–350. [CrossRef] [PubMed]
257. Herman, A.P.; Bochenek, J.; Król, K.; Krawczyńska, A.; Antushevich, H.; Pawlina, B.; Herman, A.; Romanowicz, K.; Tomaszewska-Zaremba, D. Central Interleukin-1β Suppresses the Nocturnal Secretion of Melatonin. *Mediators Inflamm.* **2016**, *2016*, 2589483. [CrossRef] [PubMed]
258. Pontes, G.N.; Cardoso, E.C.; Carneiro-Sampaio, M.M.S.; Markus, R.P. Pineal melatonin and the innate immune response: The TNF-alpha increase after cesarean section suppresses nocturnal melatonin production. *J. Pineal Res.* **2007**, *43*, 365–371. [CrossRef] [PubMed]
259. Kallaur, A.P.; Oliveira, S.R.; Simão, A.N.C.; Alfieri, D.F.; Flauzino, T.; Lopes, J.; de Carvalho Jennings Pereira, W.L.; de Meleck Proença, C.; Borelli, S.D.; Kaimen-Maciel, D.R.; et al. Cytokine Profile in Patients with Progressive Multiple Sclerosis and Its Association with Disease Progression and Disability. *Mol. Neurobiol.* **2017**, *54*, 2950–2960. [CrossRef] [PubMed]
260. Álvarez-Sánchez, N.; Cruz-Chamorro, I.; López-González, A.; Utrilla, J.C.; Fernández-Santos, J.M.; Martínez-López, A.; Lardone, P.J.; Guerrero, J.M.; Carrillo-Vico, A. Melatonin controls experimental autoimmune encephalomyelitis by altering the T effector/regulatory balance. *Brain Behav. Immun.* **2015**, *50*, 101–114. [CrossRef] [PubMed]
261. Constantinescu, C.S.; Hilliard, B.; Ventura, E.; Rostami, A. Luzindole, a melatonin receptor antagonist, suppresses experimental autoimmune encephalomyelitis. *Pathobiology* **1997**, *65*, 190–194. [CrossRef] [PubMed]
262. Kang, J.C.; Ahn, M.; Kim, Y.S.; Moon, C.; Lee, Y.; Wie, M.B.; Lee, Y.J.; Shin, T. Melatonin ameliorates autoimmune encephalomyelitis through suppression of intercellular adhesion molecule-1. *J. Vet. Sci.* **2001**, *2*, 85–89. [PubMed]

263. Giese, K.P. GSK-3: A key player in neurodegeneration and memory. *IUBMB Life* **2009**, *61*, 516–521. [CrossRef] [PubMed]
264. Hoppe, J.B.; Frozza, R.L.; Horn, A.P.; Comiran, R.A.; Bernardi, A.; Campos, M.M.; Battastini, A.M.O.; Salbego, C. Amyloid-beta neurotoxicity in organotypic culture is attenuated by melatonin: Involvement of GSK-3beta, tau and neuroinflammation. *J. Pineal Res.* **2010**, *48*, 230–238. [CrossRef] [PubMed]
265. Guven, C.; Taskin, E.; Akcakaya, H. Melatonin Prevents Mitochondrial Damage Induced by Doxorubicin in Mouse Fibroblasts Through Ampk-Ppar Gamma-Dependent Mechanisms. *Med. Sci. Monit. Int. Med. J. Exp. Clin. Res.* **2016**, *22*, 438–446. [CrossRef]
266. Kato, H.; Tanaka, G.; Masuda, S.; Ogasawara, J.; Sakurai, T.; Kizaki, T.; Ohno, H.; Izawa, T. Melatonin promotes adipogenesis and mitochondrial biogenesis in 3T3-L1 preadipocytes. *J. Pineal Res.* **2015**, *59*, 267–275. [CrossRef] [PubMed]

International Journal of
Molecular Sciences

Review

Regulation of Immune Cell Function by PPARs and the Connection with Metabolic and Neurodegenerative Diseases

Gwenaëlle Le Menn and Jaap G. Neels *

Université Côte d'Azur, Inserm, C3M Nice, France; gwenaelle.le-menn@unice.fr
* Correspondence: Jaap.Neels@unice.fr; Tel.: +33-(0)48-906-4266

Received: 25 April 2018; Accepted: 23 May 2018; Published: 25 May 2018

Abstract: Increasing evidence points towards the existence of a bidirectional interconnection between metabolic disease and neurodegenerative disorders, in which inflammation is linking both together. Activation of members of the peroxisome proliferator-activated receptor (PPAR) family has been shown to have beneficial effects in these interlinked pathologies, and these improvements are often attributed to anti-inflammatory effects of PPAR activation. In this review, we summarize the role of PPARs in immune cell function, with a focus on macrophages and T cells, and how this was shown to contribute to obesity-associated inflammation and insulin resistance, atherosclerosis, and neurodegenerative disorders. We address gender differences as a potential explanation in observed contradictory results, and we highlight PPAR-induced metabolic changes as a potential mechanism of regulation of immune cell function through these nuclear receptors. Together, immune cell-specific activation of PPARs present a promising therapeutic approach to treat both metabolic and neurodegenerative diseases.

Keywords: obesity; type 2 diabetes; atherosclerosis; neurodegenerative disease; inflammation; macrophages; T cells; PPARs; metabolism; gender

---

1. The Interrelationship between Metabolism, Inflammation, and Neurodegenerative Disease

1.1. *Inflammation and Metabolic Disease*

Although inflammation is a vital response to infection and tissue injury, non-resolved chronic inflammation is associated with many pathological processes. Several of these pathologies, in which inflammation is a common denominator, are grouped under metabolic syndrome, including obesity, type 2 diabetes, cardiovascular disease, and fatty liver disease [1].

Over the past two decades, a clear link has been established between obesity-associated inflammation and the development of insulin resistance, which eventually leads to type 2 diabetes [1]. As a result of insulin resistance, the body needs higher levels of insulin to help glucose enter cells. The β cells in the pancreas try to keep up with this increased demand for insulin by producing more. Over time, however, insulin resistance can lead to type 2 diabetes and prediabetes, because the β cells fail to keep up with the body's increased need for insulin.

Initially, studies showed that adipose tissue expansion in obesity is accompanied by an increase in cytokine and chemokine expression, such as tumor necrosis factor (TNF)-α, interleukin (IL)-6, monocyte chemoattractant protein (MCP)-1, and interferon (IFN)-γ. Some of these cytokines/chemokines were shown to impair insulin action in normally insulin-sensitive tissues, leading to insulin resistance. Later, it was demonstrated that this obesity-induced adipose tissue inflammation was largely the result of a shift in the balance of anti-inflammatory towards pro-inflammatory immune cells [2]. In lean adipose tissue, regulatory B cells (Bregs), regulatory T cells (Tregs), T helper 2 (Th2) cells, eosinophils, and type 2 innate lymphoid cells (ILC2s) maintain an anti-inflammatory environment through the production of IL-10, IL-4, IL-5, and IL-13. These anti-inflammatory cytokines promote anti-inflammatory M2 polarized macrophages in adipose tissue. By contrast, obesity-associated adipose tissue expansion is accompanied by an increase in elastase-secreting neutrophils, mast cells, and IFNγ-secreting $CD8^+$ T cells, Th1 cells, and natural killer (NK) cells. Inflammatory mediators secreted by these cells promote pro-inflammatory M1 macrophage polarization and their release of IL-1β, IL-6, and TNF-α cytokines [2].

Likewise, atherosclerosis is also associated with a chronic and non-resolving immune response. The accumulation of lipoproteins in the arterial wall, characteristic of atherosclerosis, triggers first an innate immune response, dominated by monocyte/macrophages, followed by an adaptive immune response involving primarily Th1, but also Th17 and Th2 cells and B cells, alongside a progressive decrease in Tregs [3]. As in adipose tissue, atherosclerotic plaques can contain both inflammatory and resolving macrophages. The pro-inflammatory macrophages secrete cytokines, proteases, and other factors that can cause plaque morphological changes and progression that can eventually trigger plaque rupture, whereas resolving macrophages carry out functions that can suppress plaque progression and promote plaque regression and/or stabilization [3].

1.2. *Inflammation as a Link between Metabolic Disease and Neurodegenerative Disorders*

Both human studies and animal models concur to suggest an interrelationship between metabolic disease and neurodegenerative disorders (NDDs), such as Alzheimer's disease, Huntington's disease, Parkinson's disease, and multiple sclerosis [4–9]. Higher body mass index represents a risk factor for the development of these NDDs [4–9]. Inflammation might be linking metabolic disease to NDDs, since a growing body of observational and experimental data shows that inflammatory processes, termed neuroinflammation, contribute to the onset and progression of neuronal degeneration [10]. Furthermore, this link between metabolic disease and neuroinflammation goes both ways, since hypothalamic inflammation has been linked to the development and progression of obesity and its sequelae [11,12]. Hypothalamic inflammation induced by obesogenic diets occurs before significant body weight gain, and precedes inflammation in peripheral tissues. This results in the uncoupling of caloric intake and energy expenditure, not only leading to overeating and weight gain, but also contributes to obesity-associated insulin resistance via altered neurocircuit functions. For example, hypothalamic inflammation modulates insulin secretion by pancreatic β cells, adipose tissue lipolysis, and hepatic glucose production [13,14]. Microglia cells, the brain counterpart of macrophages, play a major role in the neuroinflammation observed in both NDDs and the obesity-associated hypothalamic inflammation [10,11]. The aggregates of amyloid β-peptide (Aβ) and α-synuclein, that respectively characterize Alzheimer's and Parkinson's disease, have been shown to induce microglia activation, which augments the level of neuroinflammatory mediators, that in turn worsen these NDDs [10]. Likewise, an obesogenic diet leads to an accumulation of activated microglia within the hypothalamus that produce a variety of proinflammatory cytokines [11]. Furthermore, high fat feeding is associated with the accumulation and activation of astrocytes in the hypothalamus, which also produce a variety of inflammatory factors [11]. In Huntington's disease, expression of mutant Huntingtin (HTT) protein results in a cell-autonomous pro-inflammatory state of activation of microglia and, to a certain extent, of astrocytes [15]. Multiple sclerosis is characterized by the progressive destruction of axon myelin sheaths by the action of autoreactive immune cells (including T cells and macrophages) [10].

Taken together, both animal models and human studies strongly suggest that there is a close interconnection between metabolism, inflammation, and neurodegeneration (see Figure 1). With inflammation as a link between metabolic disease and NDDs, therapies targeting inflammation might both re-establish metabolic homeostasis and have efficacy in counteracting cognitive decline.

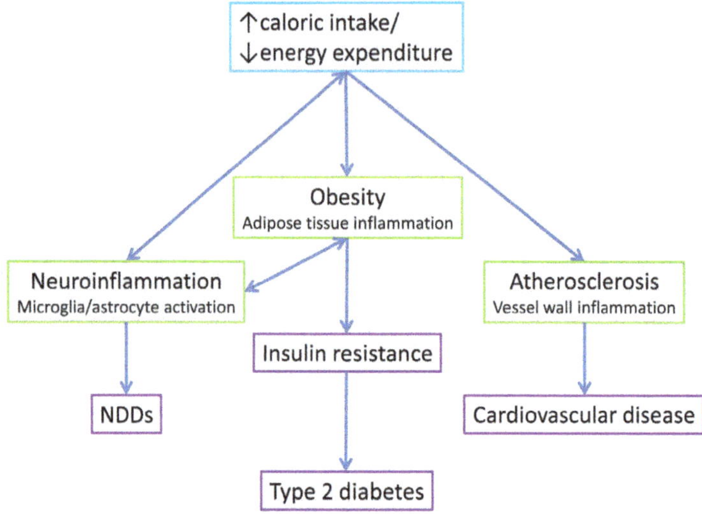

Figure 1. Interconnection between metabolism, inflammation, and neurodegeneration. An imbalance between caloric intake and energy expenditure has been linked to both metabolic disease (obesity and atherosclerosis) and neurodegenerative disorders. These pathologies all have a state of unresolved chronic inflammation in common. The link between neuroinflammation and obesity and associated sequelae is bidirectional, since hypothalamic inflammation leads to uncoupling of caloric intake and energy expenditure, leading to obesity, but also contributes to obesity-induced insulin resistance (and subsequent type 2 diabetes) via altered neurocircuit functions.

2. The Role of Metabolism in Immune Cell Function

Glycolysis, oxidative phosphorylation (OXPHOS), glutaminolysis, and/or fatty acid oxidation (FAO) are metabolic pathways that generate energy needed to satisfy basic cellular functions. Regarding immune cells, it was shown over the years that these cells can adapt their metabolism, from one pathway to another, to support the bioenergetically demanding processes of growth and effector function during an immune response.

2.1. Adaptive Immune Cells

The first metabolic change encountered by lymphocytes appears upon activation when shifting from quiescent cells with a relatively low metabolism to activated and proliferating cells, that have high metabolic needs. This shift is supported by a switch from an oxidative metabolism towards anaerobic glycolysis (Warburg effect) following antigen recognition by both T and B cells [16,17]. Indeed, lymphocyte activation is accompanied by an elevated glucose uptake through increased translocation of glucose transporter 1 (GLUT1) to the cellular membrane [18,19]. Increase in glutaminolysis in also observed in both cell types as glutamine is an essential substrate for the tricarboxylic acid cycle [20,21]. For B cells, activation is also accompanied by an increased OXPHOS, but data on the metabolic profile of distinct B cell subsets is still lacking [17]. As for T cells, activated $CD4^+$ T cells will polarize into different subpopulations with their own inflammatory and metabolic phenotype (Th1, Th2, Th17, and Tregs). Anti-inflammatory Tregs are poorly proliferative, whereas pro-inflammatory T cell subsets can be highly proliferative. In this regard, studies showed that Th1, Th2, and Th17 cells use glycolysis to meet their energy demands, whereas Tregs have high lipid oxidation rates [22,23]. Furthermore, it was demonstrated that by directly manipulating cell metabolism one can regulate $CD4^+$ T cell fate; for example, inhibition of glycolysis blocks Th17 development and promotes T cell polarization towards Treg cells [23]. CD8$^+$ memory T cells largely depend on FAO for their metabolic needs, and in line with this, carnitine palmitoyltransferase Ia (CPT1a) expression (rate-limiting enzyme of FAO pathway) was found to promote the differentiation into this subpopulation [24].

2.2. Innate Immune Cells

Granulocytes, dendritic cells (DC), and M1 type macrophages rely on glucose metabolism upon activation, while M2 macrophages depend on FAO. Unlike lymphocytes, activated myeloid cells tend to be non-proliferative, but still mostly exhibit an increased glycolytic metabolism upon activation, which is essential to acquire their effector function.

Indeed, neutrophil effector functions, such as neutrophil extracellular trap formation, tissue infiltration and phagocytosis, were decreased in the presence of the 2-deoxy-glucose, an inhibitor of glycolysis [25,26]. In a recent study on mast cells, seahorse experiment results showed an increase of glycolysis, as well as OXPHOS, following their activation. The latter was particularly implicated in the degranulation process and cytokine production [27]. As for eosinophil and basophil metabolism, evidence suggests a glycolytic metabolism after their activation, but this needs to be investigated further [28]. DCs shift from naïve DCs, using mainly FAO and OXPHOS metabolism, to glycolysis, upon activation. Increase of glucose metabolism is then mainly implicated in the increase in de novo fatty acid synthesis that seems to correlate with the immunogenic phenotype of DCs [29]. Similar to T cells, macrophage activation can give rise to the polarization into pro-inflammatory M1 or anti-inflammatory M2 macrophages that exhibit metabolic differences. While M1 macrophages preferentially use glycolysis to support the production of inflammatory cytokines, such as IL-1β and TNF-α via the activation of nuclear factor-κB (NF-κB) and activator protein-1 (AP-1) signaling, M2 macrophages use lipid oxidation as energy source [30]. In this case, lipid oxidation is supported by an increase in the expression of fatty acid translocase (FAT)/CD36 and CPT1a, that favors lipid import into cells and mitochondria, respectively [30,31].

It is clear from these findings that metabolism plays an important role in the immune cell fate and inflammatory phenotype. Overall, a distinction can be made between pro-inflammatory cells, that require a rapid burst of energy and macromolecule synthesis via glycolysis to produce cytokines, and quiescent or anti-inflammatory cells, that use mostly oxidation (FAO and OXPHOS) for their survival and longevity. As a consequence, manipulating immune cell metabolism has become an interesting approach to control the immune response.

3. Role of PPARs in Immune Cell Function

3.1. PPARs and Their Mode of Action

The peroxisome proliferator-activated receptor (PPAR) subfamily of nuclear hormone receptors consist of three different isoforms; PPARα, PPARβ, and PPARγ, that are each expressed in various tissues and cell types, and regulate the transcription of a large variety of genes implicated in metabolism, cell proliferation/differentiation, and inflammation [32]. These different PPAR members have a conserved structure that includes an N-terminal ligand-independent transactivation domain, a DNA binding domain, and a C-terminal ligand-binding domain and ligand-dependent activation domain [33]. This C-terminal region is implicated in receptor heterodimerization with the obligatory transcriptional partner, the retinoid X receptor (RXR). These heterodimers bind to specific DNA sequence elements called peroxisome proliferator response elements (PPREs) in the regulatory region of their target genes. Binding of synthetic or endogenous ligands (fatty acids and their derivatives) induces a conformational switch in the receptors, leading to dissociation of co-repressor proteins and recruitment of co-activator proteins to enhance the transcription of target genes [33]. This direct transcriptional regulation of PPARs through binding to PPREs largely concerns target genes involved in transport, synthesis, storage, mobilization, activation, and oxidation of fatty acids. However, the regulation of immune cell function by PPARs, the topic of this review, is thought to mostly implicate transcription regulation of target genes through indirect mechanisms. The best-known mechanism by which PPARs regulate inflammation is through transrepression [34]. This activity involves indirect association (tethering) of the PPARs with target genes. There are many mechanisms by which PPARs can transrepress inflammatory responses, including competition for a limiting pool of coactivators, direct interaction with the p65 subunit of NF-κB and c-Jun subunit of AP-1, modulation of p38 mitogen-activated protein kinase (MAPK) activity, and partitioning the corepressor B-cell lymphoma 6 (BCL-6) [34].

3.2. Role of PPARs in Immune Cells

There is a vast amount of literature (including many excellent reviews) on the anti-inflammatory roles of the different PPARs in a multitude of inflammatory diseases (for selection of reviews, see [32,35–49]). Many of these studies were performed in global knockout models and/or PPAR agonists/antagonists were administered systemically. The global/systemic nature of these latter studies often does not allow for the interpretation of the role of PPARs in specific immune cells, since the effects observed could be due to numerous PPAR actions unrelated to their function in immune cells. Furthermore, several studies treated immune cells with endogenous PPAR ligands that are also known to have PPAR-independent effects, so again, this complicates the interpretation of the results obtained. As a consequence, we limit this review to studies that (1) use mouse models that are deficient for, or overexpress, PPARs specifically in certain immune cells, (2) performed in vitro studies on immune cells deficient for, or overexpressing PPARs, and/or (3) used PPAR-specific (ant)agonists directly on (mouse or human) immune cells. In particular, we focus on studies concerning PPAR actions in macrophages and T cells, and how that impacts inflammatory disease (with a focus on metabolic and neurodegenerative diseases).

3.2.1. Role of PPARs in Macrophages

All three PPAR family members have been shown to play a role in mouse macrophage polarization. PPARα, β, or γ activation was demonstrated to potentiate the polarization of mouse macrophages towards the anti-inflammatory M2 phenotype, while M2-type responses are compromised in the absence of PPARγ or β expression (effect of PPARα absence has not been studied) [50–66]. In human macrophages results are less clear-cut; while PPARγ activation has been shown to stimulate M2 polarization, PPARα or β activation did not seem to have any effect [67–71]. These anti-inflammatory actions of PPARs in macrophages have often been described to involve transrepression mechanisms involving NF-κB and AP-1 [51,53,60,61]. However, in line with the importance of metabolism in macrophage polarization (see Section 2.2 above), deletion of PPARγ in macrophages leads to reduced rates of β-oxidation of fatty acids, and consequently, these PPARγ-deficient macrophages are unable to clear the metabolic checkpoint required for full conversion to the alternative phenotype [50]. One mechanism through which PPARβ activation was proposed to exert its anti-inflammatory actions in macrophages involves the repressor BCL-6; unliganded PPARβ binds and sequesters BCL-6, and upon ligand binding, BCL-6 is released, and can repress transcription of pro-inflammatory target genes, including IL-1β, MCP-1, and matrix metalloproteinase 9 (MMP9) [72]. Based on this mechanism, PPARβ-deficient macrophages should exhibit an anti-inflammatory phenotype (BCL-6 would be free to repress pro-inflammatory genes). However, this is contradicted by two different studies that show that absence of PPARβ does not suppress pro-inflammatory responses during alternative activation of macrophages [66,73].

3.2.2. Role of PPARs in T Cells

In T cells, PPARs have been shown to regulate survival, activation, and $CD4^+$ T cell differentiation into the Th1, Th2, Th17, and Treg lineages [39]. PPARβ activation was shown to inhibit Th1 and Th17 polarization, and augment Th2 polarization, and the opposite was seen when PPARβ was deleted [74–76]. We have recently shown that activation or overexpression of PPARβ increases

FAO in T cells [77]. Furthermore, using both in vivo and in vitro models, we demonstrated that PPARβ activation/overexpression inhibits thymic T cell development by decreasing proliferation of $CD4^{-}CD8^{-}$ double-negative stage 4 (DN4) thymocytes [77]. These results support a model where PPARβ activation/overexpression favors oxidation of fatty acids, instead of glucose, in developing T cells, thereby hampering the proliferative burst normally occurring at the DN4 stage of T cell development. As a consequence, the αβ T cells that are derived from DN4 thymocytes were dramatically decreased in peripheral lymphoid tissues, while the γδ T cell population remained untouched [77].

PPARγ activation was shown to impair T cell proliferation through an IL-2 dependent mechanism involving repression of nuclear factor of activated T cells (NFAT) [78,79]. Deletion of PPARγ in $CD4^{+}$ T cells resulted in increased antigen-specific proliferation and overproduction of IFN-γ in response to IL-12, highlighting the importance of PPARγ expression in downregulating excessive Th1 responses [80]. Furthermore, PPARγ is highly expressed in both mouse and human Th2 cells, as opposed to other Th subsets, and although having a minor direct role in regulating Th2 differentiation, controls Th2 sensitivity to IL-33 and thus, has an impact on Th2 effector function [81]. However, PPARγ activation was reported to downregulate IL-4 production in T cells (through downregulation of NFAT) and expression of other Th2 cytokines (IL-5 and IL-13) was also reported to be decreased, as well as c-Maf, a Th2-specific transcription factor [82,83]. Together, these studies indicate that the effect of PPARγ activation on Th2 differentiation remains unclear.

Loss of PPARγ in Tregs has been shown to impair their ability to control effector $CD4^{+}$ T cell responses while PPARγ activation in naïve $CD4^{+}$ T cells enhanced induction of forkhead box P3 $(FoxP3)^{+}$ inducible regulatory T cells [80,84,85]. Moreover, a recent study demonstrated that T cell-specific deletion of PPARγ leads to a specific reduction in GATA binding protein 3 (GATA3)-expressing Tregs [81]. In addition, a population of Tregs that highly expresses PPARγ has been identified in visceral adipose tissue, and Treg-specific deletion of PPARγ prevents accumulation of Tregs in visceral adipose tissue [86]. Furthermore, phosphorylation of serine 273 of PPARγ in Tregs changes the characteristic transcriptional signature of these Tregs [87]. Together, these studies suggest that PPARγ may contribute to the quality and quantity of Tregs.

In regard to Th17 differentiation, PPARγ activation was shown to have inhibitory effects while PPARγ deficiency led to increased Th17 differentiation [88]. Th17 differentiation depends on the transcription factor retinoic acid receptor (RAR)-related orphan receptor (ROR) γt, and the latter study by Klotz et al. demonstrated that under physiological conditions, the co-repressor silencing mediator of retinoid and thyroid hormone receptors (SMRT) is bound to the RORγt promoter and inhibits its transcription, and that PPARγ activation prevents removal of this corepressor complex, thereby suppressing RORγt expression and Th17 differentiation. It should also be mentioned that Klotz et al. did not observe an effect of PPARγ activation on Th1, Th2, or Treg T cell subsets, contradicting the above-mentioned studies.

3.2.3. Gender-Specific Differences in the Role of PPARs in T Cells

One explanation for these contradicting results could be sex-specific roles of PPARs in T cells [89]. One of the first observations of gender differences in the role of PPARs in T cells was that T cells from male mice have increased expression of PPARα, compared to their female counterparts, and that the male sex hormone androgen has been suggested to regulate PPARα expression [90,91]. In the same study it was shown that PPARα-deficient T cells were predisposed to a Th1 response at the expense of Th2 function, and this was mediated by PPARα modulation of NF-κB and c-Jun activity. These results were recently confirmed by using a PPARα antagonist [92]. While PPARα expression is high in male T cells, PPARγ expression is high in female T cells [91], and the female sex hormone estrogen seems to influence expression of PPARγ [93]. As a result, the inhibitory role of PPARγ in T cell activation (see Section 3.2.2 above) is observed in female PPARγ-deficient T cells, but not in male T cells [94]. Similarly, PPARγ activation inhibits the differentiation of female Th1, Th2, and Th17 cells, whereas it specifically reduces only Th17-cell differentiation in males [95]. This provides a strong argument that, indeed, gender-specific differences in PPARγ expression in T cells could explain the contradictory results regarding the role of PPARγ in Th differentiation. PPARβ expression did not differ much when comparing male and female naïve and activated T cells [90].

Taken together, these studies demonstrate that the differential regulation of PPAR expression by sex hormones has an impact on the roles these receptors play in T cell biology. Furthermore, it cannot be excluded that contradictions in studies on the role of PPARs in macrophages, specifically the differences between mice and humans, could also potentially be the consequence of gender differences. Based on the importance of metabolism in immune cells (see Section 2 above), and the fact that most of the directly regulated PPAR target genes are involved in different aspects of fatty acid metabolism, it would seem obvious that the observed effects of PPARs on macrophage and T cell polarization/proliferation can be mechanistically explained by PPAR-induced changes in metabolism. However, this possibility was only rarely explored in the studies described above (and below).

4. Consequences of PPAR Actions in Immune Cells for Metabolic and Neurodegenerative Diseases

4.1. Metabolic Diseases

We focus here on the role of PPARs in immune cells in the context of atherosclerosis and obesity-associated inflammation and insulin resistance. Again, for reasons mentioned above (Section 3.2), studies using global knockouts or systemic treatments with agonists will not be discussed. Transplantation of PPARβ$^{-/-}$ bone marrow into atherogenic diet-fed low-density lipoprotein receptor (LDLR)-deficient mice resulted in a reduction of aortic valve lesion surface compared to mice transplanted with wild type bone marrow [72]. Similarly, transplantation of bone marrow cells infected with lentivirus expressing selective microRNA (miRNA) targeting PPARβ into recipient LDLR$^{-/-}$ mice resulted in reduction of atherosclerotic lesions, accompanied by a reduced presence of macrophages and expression of MCP-1 and MMP9 in the plaque [96]. This reduction of inflammation in absence of PPARβ in bone marrow cells is in line with the BCL-6 mechanistic model of PPARβ regulation of macrophage function. By contrast, transplantation of PPARγ$^{-/-}$ bone marrow cells or conditional knockout of macrophage PPARγ increases atherosclerosis in both wild type and LDLR$^{-/-}$ mice fed an atherogenic diet [97,98].

Two studies showed that macrophage-specific deletion of PPARγ predisposes mice to development of diet-induced obesity and insulin resistance [50,99]. Similar results were obtained when the effect of PPARβ-deficient bone marrow or macrophage-specific PPARβ$^{-/-}$ on HFD-induced obesity and insulin resistance was studied [65,66]. However, one study found preserved glucose tolerance in mice transplanted with PPARγ$^{-/-}$ or PPARβ$^{-/-}$ bone marrow [100]. Since bone marrow-derived cells include T cells, some of the results outlined above could also be due to PPAR actions in T cells, even though the cited studies often interpreted them as macrophage specific. T cell-specific actions of PPARs, in the context of atherosclerosis or obesity-associated inflammation and insulin resistance, have largely been unexplored, with the exception of the role of PPARγ in adipose tissue Tregs in the latter. As mentioned already above (Section 3.2.2), PPARγ has been shown to be a crucial molecular orchestrator of visceral adipose tissue Treg accumulation, phenotype, and function [86,87]. Another area of PPAR research that deserves further exploration, not counting global knockout studies and systemic agonist treatment, is the specific role of PPARα in immune cells in the context of atherosclerosis and obesity-associated inflammation and insulin resistance.

4.2. Neurodegenerative Diseases

Even though neuroinflammation plays an important role in NDDs (outlined above in Section 1.2), and numerous studies have demonstrated beneficial effects of treatment with PPAR agonists in those pathologies, few studies have investigated how much PPAR actions in immune cells contribute to these positive effects observed [101,102]. In the context of Alzheimer's disease, in vitro studies demonstrated that PPARγ agonists stimulated Aβ phagocytosis by rat primary microglia through induction of CD36 expression [103]. A similar study showed that PPARγ activation stimulated Aβ degradation by both primary mouse microglia and astrocytes, and that this involved a M1 to M2 shift for microglia [104]. Other in vitro studies revealed that pharmacological activation of PPARα attenuates the inflammatory responses of both primary mouse astrocytes and microglia [105,106]. The same group showed that PPARα activation in lipopolysaccharide (LPS)-treated microglia suppressed secretion of IL-12 family cytokines that are known to stimulate Th1 and Th17 differentiation [107]. Furthermore, they showed a similar decrease in IL-12 family cytokines in both microglia and astrocytes treated with PPARγ agonists [108,109], and PPARγ agonist inhibited the inflammatory response of those central nervous system (CNS) cells [110,111]. PPARγ activation in neuron–microglia co-cultures protected the neurons from damage induced by LPS-induced insults, by inhibiting microglia activation through interference with the NF-κB and AP-1 pathways [112]. In addition, PPARβ activation was shown to reduce LPS-stimulated nitric oxide (NO) production in enriched microglia and astrocyte cultures [113]. Likewise, PPARβ activation can also modulate radiation-induced oxidative stress and pro-inflammatory responses in microglia [114]. The latter was shown to occur through PPARβ interaction with the p65 subunit NF-κB.

Taken together, these in vitro cell culture studies demonstrate that PPAR activation reduces inflammation in both microglia and astrocytes and it is therefore likely that some (or most) of the beneficial effects observed with PPAR activation in NDDs are the consequence of anti-inflammatory PPAR actions in these cells. However, to study the specific role of microglial and astrocyte PPARs in NDDs in an in vivo context, it would be of great interest to overexpress or knockout PPARs in a cell-type specific fashion using CX3C chemokine receptor 1 (CX3CR1)-Cre or glial fibrillary acid protein (GFAP)-Cre mice, respectively. Even though the CX3CR1-Cre approach will also affect other CX3CR1-expressing myeloid cell populations, these types of studies would still be very informative.

5. Conclusions

In summary, inflammation has been shown to be a common denominator in both metabolic syndrome and NDDs, and targeting this inflammation from a therapeutic standpoint could potentially have beneficial consequences for both pathologies. Based on the anti-inflammatory effects that have been attributed to PPARs, and the roles that have been described for these receptors in regard to immune cell functions, activating these receptors, specifically in immune cells, could be considered as such a therapeutic approach (see Figure 2). This immune cell-specific approach could circumvent certain adverse effects that have been observed in the past with systemic treatments with PPAR agonists. However, before pursuing such an ambitious goal, several insufficiently explored questions in PPAR research should be further addressed. While many studies strongly suggest that beneficial effects of PPAR activation in the context of metabolic syndrome and NDDs can be explained by anti-inflammatory effects, direct proof of an important role for PPAR-induced changes in immune cell function is often lacking. This missing proof could be supplied by studying the effects of immune cell-specific deficiency or overexpression of PPARs in the context of metabolic disease and NDD mouse models. It is important that potential gender-specific differences should be taken into account while conducting these types of studies. Lastly, PPAR-induced metabolic changes should be more often considered/explored as a mechanistic explanation of the regulatory functions that are attributed to these nuclear receptors in immune cells.

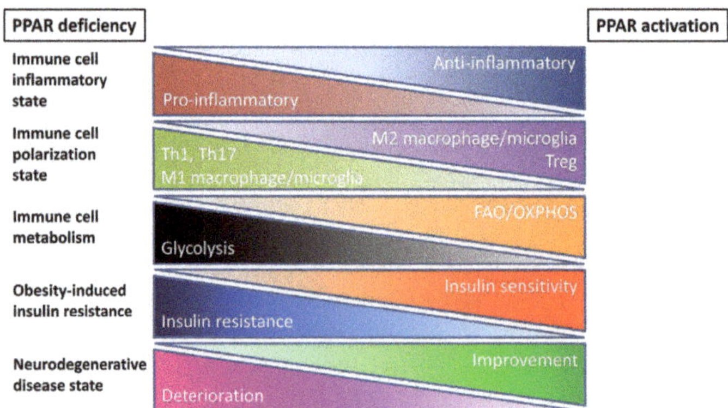

Figure 2. Effects of peroxisome proliferator-activated receptor (PPAR) deficiency or activation on immune cell properties and metabolic and neurodegenerative disease states. Despite some contradictory results (perhaps due to gender differences), the overall impression we deduce from the literature is that PPAR activation has anti-inflammatory effects on immune cells by stimulating the polarization of these cells towards more anti-inflammatory subsets. Perhaps the switch towards FAO/OXPHOS (fatty acid oxidation/oxidative phosphorylation) metabolism induced by PPAR activation plays an important role in this shift towards anti-inflammatory immune cell subsets. By contrast, PPAR deficiency has often been shown to have the opposite effects. Together, these PPAR-regulated properties of immune cells might contribute to the severity of the disease state both in metabolic diseases (e.g., obesity-induced insulin resistance) and neurodegenerative disorders NDDs.

Acknowledgments: This work was financially supported by INSERM, the Université Côte d'Azur, the Fondation pour la Recherche Médicale (FRM, Grant DRM20101220437), the French National Research Agency (ANR, N° ANR-14-CE12-0008-02) and the Agence Française de Lutte contre le Dopage (AFLD).

Conflicts of Interest: The authors declare no conflict of interest.

Abbreviations

AP-1	activator protein-1
Aβ	amyloid β-peptide
BCL-6	B-cell lymphoma 6
Bregs	regulatory B cells 6
CNS	central nervous system
CPT1a	carnitine palmitoyltransferase Ia
CX3CR1	CX3C chemokine receptor 1
DC	dendritic cells
FAO	fatty acid oxidation
FAT	fatty acid translocase
FoxP3	forkhead box P3
GATA3	GATA binding protein 3
GFAP	glial fibrillary acid protein
GLUT1	glucose transporter 1
HTT	huntingtin
IFNγ	interferon γ
IL	interleukin
ILC2s	type 2 innate lymphoid cells
LDLR	low-density lipoprotein receptor
LPS	lipopolysaccharide
MAPK	mitogen-activated protein kinase

MCP-1	monocyte chemoattractant protein 1
miRNA	microRNA
MMP9	matrix metalloproteinase-9
NDDs	neurodegenerative disorders
NF-κB	nuclear factor-κB
NFAT	nuclear factor of activated T cells
NK	natural killer cells
NO	nitric oxide
OXPHOS	oxidative phosphorylation
PPAR	peroxisome proliferator-activated receptor
PPREs	peroxisome proliferator response elements
RAR	retinoic acid receptor
ROR	related orphan receptor
RXR	retinoid X receptor
SMRT	silencing mediator of retinoid and thyroid hormone receptors
Th	T helper
TNFα	tumor necrosis factor alpha
Tregs	regulatory T cells

References

1. Hotamisligil, G.S. Inflammation and metabolic disorders. *Nature* **2006**, *444*, 860–867. [CrossRef] [PubMed]
2. McLaughlin, T.; Ackerman, S.E.; Shen, L.; Engleman, E. Role of innate and adaptive immunity in obesity-associated metabolic disease. *J. Clin. Investig.* **2017**, *127*, 5–13. [CrossRef] [PubMed]
3. Tabas, I.; Lichtman, A.H. Monocyte-Macrophages and T Cells in Atherosclerosis. *Immunity* **2017**, *47*, 621–634. [CrossRef] [PubMed]
4. De Candia, P.; Matarese, G. Leptin and ghrelin: Sewing metabolism onto neurodegeneration. *Neuropharmacology* **2017**. [CrossRef] [PubMed]
5. Abbott, R.D.; Ross, G.W.; White, L.R.; Nelson, J.S.; Masaki, K.H.; Tanner, C.M.; Curb, J.D.; Blanchette, P.L.; Popper, J.S.; Petrovitch, H. Midlife adiposity and the future risk of Parkinson's disease. *Neurology* **2002**, *59*, 1051–1057. [CrossRef] [PubMed]
6. Kivipelto, M.; Ngandu, T.; Fratiglioni, L.; Viitanen, M.; Kareholt, I.; Winblad, B.; Helkala, E.L.; Tuomilehto, J.; Soininen, H.; Nissinen, A. Obesity and vascular risk factors at midlife and the risk of dementia and Alzheimer disease. *Arch. Neurol.* **2005**, *62*, 1556–1560. [CrossRef] [PubMed]
7. Xu, W.L.; Atti, A.R.; Gatz, M.; Pedersen, N.L.; Johansson, B.; Fratiglioni, L. Midlife overweight and obesity increase late-life dementia risk: A population-based twin study. *Neurology* **2011**, *76*, 1568–1574. [CrossRef] [PubMed]
8. Colman, R.J.; Anderson, R.M.; Johnson, S.C.; Kastman, E.K.; Kosmatka, K.J.; Beasley, T.M.; Allison, D.B.; Cruzen, C.; Simmons, H.A.; Kemnitz, J.W.; et al. Caloric restriction delays disease onset and mortality in rhesus monkeys. *Science* **2009**, *325*, 201–204. [CrossRef] [PubMed]
9. Guerrero-Garcia, J.J.; Carrera-Quintanar, L.; Lopez-Roa, R.I.; Marquez-Aguirre, A.L.; Rojas-Mayorquin, A.E.; Ortuno-Sahagun, D. Multiple Sclerosis and Obesity: Possible Roles of Adipokines. *Mediat. Inflamm.* **2016**, *2016*, 4036232. [CrossRef] [PubMed]
10. Ransohoff, R.M. How neuroinflammation contributes to neurodegeneration. *Science* **2016**, *353*, 777–783. [CrossRef] [PubMed]
11. Jais, A.; Bruning, J.C. Hypothalamic inflammation in obesity and metabolic disease. *J. Clin. Investig.* **2017**, *127*, 24–32. [CrossRef] [PubMed]
12. Maldonado-Ruiz, R.; Montalvo-Martinez, L.; Fuentes-Mera, L.; Camacho, A. Microglia activation due to obesity programs metabolic failure leading to type two diabetes. *Nutr. Diabetes* **2017**, *7*, e254. [CrossRef] [PubMed]
13. Calegari, V.C.; Torsoni, A.S.; Vanzela, E.C.; Araujo, E.P.; Morari, J.; Zoppi, C.C.; Sbragia, L.; Boschero, A.C.; Velloso, L.A. Inflammation of the hypothalamus leads to defective pancreatic islet function. *J. Biol. Chem.* **2016**, *291*, 26935. [CrossRef] [PubMed]

14. Scherer, T.; Lindtner, C.; Zielinski, E.; O'Hare, J.; Filatova, N.; Buettner, C. Short term voluntary overfeeding disrupts brain insulin control of adipose tissue lipolysis. *J. Biol. Chem.* **2012**, *287*, 33061–33069. [CrossRef] [PubMed]
15. Crotti, A.; Glass, C.K. The choreography of neuroinflammation in Huntington's disease. *Trends Immunol.* **2015**, *36*, 364–373. [CrossRef] [PubMed]
16. Frauwirth, K.A.; Riley, J.L.; Harris, M.H.; Parry, R.V.; Rathmell, J.C.; Plas, D.R.; Elstrom, R.L.; June, C.H.; Thompson, C.B. The CD28 signaling pathway regulates glucose metabolism. *Immunity* **2002**, *16*, 769–777. [CrossRef]
17. Caro-Maldonado, A.; Wang, R.; Nichols, A.G.; Kuraoka, M.; Milasta, S.; Sun, L.D.; Gavin, A.L.; Abel, E.D.; Kelsoe, G.; Green, D.R.; et al. Metabolic reprogramming is required for antibody production that is suppressed in anergic but exaggerated in chronically BAFF-exposed B cells. *J. Immunol.* **2014**, *192*, 3626–3636. [CrossRef] [PubMed]
18. Macintyre, A.N.; Gerriets, V.A.; Nichols, A.G.; Michalek, R.D.; Rudolph, M.C.; Deoliveira, D.; Anderson, S.M.; Abel, E.D.; Chen, B.J.; Hale, L.P.; et al. The glucose transporter Glut1 is selectively essential for CD4 T cell activation and effector function. *Cell Metab.* **2014**, *20*, 61–72. [CrossRef] [PubMed]
19. Doughty, C.A.; Bleiman, B.F.; Wagner, D.J.; Dufort, F.J.; Mataraza, J.M.; Roberts, M.F.; Chiles, T.C. Antigen receptor-mediated changes in glucose metabolism in B lymphocytes: Role of phosphatidylinositol 3-kinase signaling in the glycolytic control of growth. *Blood* **2006**, *107*, 4458–4465. [CrossRef] [PubMed]
20. Wang, R.; Dillon, C.P.; Shi, L.Z.; Milasta, S.; Carter, R.; Finkelstein, D.; McCormick, L.L.; Fitzgerald, P.; Chi, H.; Munger, J.; et al. The transcription factor Myc controls metabolic reprogramming upon T lymphocyte activation. *Immunity* **2011**, *35*, 871–882. [CrossRef] [PubMed]
21. Le, A.; Lane, A.N.; Hamaker, M.; Bose, S.; Gouw, A.; Barbi, J.; Tsukamoto, T.; Rojas, C.J.; Slusher, B.S.; Zhang, H.; et al. Glucose-independent glutamine metabolism via TCA cycling for proliferation and survival in B cells. *Cell Metab.* **2012**, *15*, 110–121. [CrossRef] [PubMed]
22. Michalek, R.D.; Gerriets, V.A.; Jacobs, S.R.; Macintyre, A.N.; MacIver, N.J.; Mason, E.F.; Sullivan, S.A.; Nichols, A.G.; Rathmell, J.C. Cutting edge: Distinct glycolytic and lipid oxidative metabolic programs are essential for effector and regulatory $CD4^+$ T cell subsets. *J. Immunol.* **2011**, *186*, 3299–3303. [CrossRef] [PubMed]
23. Shi, L.Z.; Wang, R.; Huang, G.; Vogel, P.; Neale, G.; Green, D.R.; Chi, H. HIF1α-dependent glycolytic pathway orchestrates a metabolic checkpoint for the differentiation of TH17 and Treg cells. *J. Exp. Med.* **2011**, *208*, 1367–1376. [CrossRef] [PubMed]
24. van der Windt, G.J.; Everts, B.; Chang, C.H.; Curtis, J.D.; Freitas, T.C.; Amiel, E.; Pearce, E.J.; Pearce, E.L. Mitochondrial respiratory capacity is a critical regulator of $CD8^+$ T cell memory development. *Immunity* **2012**, *36*, 68–78. [CrossRef] [PubMed]
25. Rodriguez-Espinosa, O.; Rojas-Espinosa, O.; Moreno-Altamirano, M.M.; Lopez-Villegas, E.O.; Sanchez-Garcia, F.J. Metabolic requirements for neutrophil extracellular traps formation. *Immunology* **2015**, *145*, 213–224. [CrossRef] [PubMed]
26. Lane, T.A.; Lamkin, G.E. A reassessment of the energy requirements for neutrophil migration: Adenosine triphosphate depletion enhances chemotaxis. *Blood* **1984**, *64*, 986–993. [PubMed]
27. Phong, B.; Avery, L.; Menk, A.V.; Delgoffe, G.M.; Kane, L.P. Cutting Edge: Murine Mast Cells Rapidly Modulate Metabolic Pathways Essential for Distinct Effector Functions. *J. Immunol.* **2017**, *198*, 640–644. [CrossRef] [PubMed]
28. Sumbayev, V.V.; Nicholas, S.A.; Streatfield, C.L.; Gibbs, B.F. Involvement of hypoxia-inducible factor-1 HiF(1α) in IgE-mediated primary human basophil responses. *Eur. J. Immunol.* **2009**, *39*, 3511–3519. [CrossRef] [PubMed]
29. Everts, B.; Amiel, E.; Huang, S.C.; Smith, A.M.; Chang, C.H.; Lam, W.Y.; Redmann, V.; Freitas, T.C.; Blagih, J.; van der Windt, G.J.; et al. TLR-driven early glycolytic reprogramming via the kinases TBK1-IKKvarepsilon supports the anabolic demands of dendritic cell activation. *Nat. Immunol.* **2014**, *15*, 323–332. [CrossRef] [PubMed]
30. Huang, S.C.; Everts, B.; Ivanova, Y.; O'Sullivan, D.; Nascimento, M.; Smith, A.M.; Beatty, W.; Love-Gregory, L.; Lam, W.Y.; O'Neill, C.M.; et al. Cell-intrinsic lysosomal lipolysis is essential for alternative activation of macrophages. *Nat. Immunol.* **2014**, *15*, 846–855. [CrossRef] [PubMed]

31. Namgaladze, D.; Lips, S.; Leiker, T.J.; Murphy, R.C.; Ekroos, K.; Ferreiros, N.; Geisslinger, G.; Brune, B. Inhibition of macrophage fatty acid β-oxidation exacerbates palmitate-induced inflammatory and endoplasmic reticulum stress responses. *Diabetologia* **2014**, *57*, 1067–1077. [CrossRef] [PubMed]
32. Varga, T.; Czimmerer, Z.; Nagy, L. PPARs are a unique set of fatty acid regulated transcription factors controlling both lipid metabolism and inflammation. *Biochim. Biophys. Acta* **2011**, *1812*, 1007–1022. [CrossRef] [PubMed]
33. Zoete, V.; Grosdidier, A.; Michielin, O. Peroxisome proliferator-activated receptor structures: Ligand specificity, molecular switch and interactions with regulators. *Biochim. Biophys. Acta* **2007**, *1771*, 915–925. [CrossRef] [PubMed]
34. Ricote, M.; Glass, C.K. PPARs and molecular mechanisms of transrepression. *Biochim. Biophys. Acta* **2007**, *1771*, 926–935. [CrossRef] [PubMed]
35. Neels, J.G.; Grimaldi, P.A. Physiological functions of peroxisome proliferator-activated receptor β. *Physiol. Rev.* **2014**, *94*, 795–858. [CrossRef] [PubMed]
36. Fuentes, E.; Guzman-Jofre, L.; Moore-Carrasco, R.; Palomo, I. Role of PPARs in inflammatory processes associated with metabolic syndrome. *Mol. Med. Rep.* **2013**, *8*, 1611–1616. [CrossRef] [PubMed]
37. Gervois, P.; Mansouri, R.M. PPARα as a therapeutic target in inflammation-associated diseases. *Expert Opin. Ther. Targets* **2012**, *16*, 1113–1125. [CrossRef] [PubMed]
38. Wahli, W.; Michalik, L. PPARs at the crossroads of lipid signaling and inflammation. *Trends Endocrinol. Metab. TEM* **2012**, *23*, 351–363. [CrossRef] [PubMed]
39. Choi, J.M.; Bothwell, A.L. The nuclear receptor PPARs as important regulators of T-cell functions and autoimmune diseases. *Mol. Cells* **2012**, *33*, 217–222. [CrossRef] [PubMed]
40. Bishop-Bailey, D.; Bystrom, J. Emerging roles of peroxisome proliferator-activated receptor-β/δ in inflammation. *Pharmacol. Ther.* **2009**, *124*, 141–150. [CrossRef] [PubMed]
41. Hong, C.; Tontonoz, P. Coordination of inflammation and metabolism by PPAR and LXR nuclear receptors. *Curr. Opin. Genet. Dev.* **2008**, *18*, 461–467. [CrossRef] [PubMed]
42. Straus, D.S.; Glass, C.K. Anti-inflammatory actions of PPAR ligands: New insights on cellular and molecular mechanisms. *Trends Immunol.* **2007**, *28*, 551–558. [CrossRef] [PubMed]
43. Szeles, L.; Torocsik, D.; Nagy, L. PPARγ in immunity and inflammation: Cell types and diseases. *Biochim. Biophys. Acta* **2007**, *1771*, 1014–1030. [CrossRef] [PubMed]
44. Rizzo, G.; Fiorucci, S. PPARs and other nuclear receptors in inflammation. *Curr. Opin. Pharmacol.* **2006**, *6*, 421–427. [CrossRef] [PubMed]
45. Kostadinova, R.; Wahli, W.; Michalik, L. PPARs in diseases: Control mechanisms of inflammation. *Curr. Med. Chem.* **2005**, *12*, 2995–3009. [CrossRef] [PubMed]
46. Moraes, L.A.; Piqueras, L.; Bishop-Bailey, D. Peroxisome proliferator-activated receptors and inflammation. *Pharmacol. Ther.* **2006**, *110*, 371–385. [CrossRef] [PubMed]
47. Chinetti, G.; Fruchart, J.C.; Staels, B. Peroxisome proliferator-activated receptors and inflammation: From basic science to clinical applications. *Int. J. Obes.* **2003**, *27* (Suppl. 3), S41–S45. [CrossRef] [PubMed]
48. Cabrero, A.; Laguna, J.C.; Vazquez, M. Peroxisome proliferator-activated receptors and the control of inflammation. *Curr. Drug Targets Inflamm. Allergy* **2002**, *1*, 243–248. [CrossRef] [PubMed]
49. Chinetti, G.; Fruchart, J.C.; Staels, B. Peroxisome proliferator-activated receptors (PPARs): Nuclear receptors at the crossroads between lipid metabolism and inflammation. *Inflamm. Res.* **2000**, *49*, 497–505. [CrossRef] [PubMed]
50. Odegaard, J.I.; Ricardo-Gonzalez, R.R.; Goforth, M.H.; Morel, C.R.; Subramanian, V.; Mukundan, L.; Red Eagle, A.; Vats, D.; Brombacher, F.; Ferrante, A.W.; et al. Macrophage-specific PPARγ controls alternative activation and improves insulin resistance. *Nature* **2007**, *447*, 1116–1120. [CrossRef] [PubMed]
51. Penas, F.; Mirkin, G.A.; Vera, M.; Cevey, A.; Gonzalez, C.D.; Gomez, M.I.; Sales, M.E.; Goren, N.B. Treatment in vitro with PPARα and PPARγ ligands drives M1-to-M2 polarization of macrophages from T. cruzi-infected mice. *Biochim. Biophys. Acta* **2015**, *1852*, 893–904. [CrossRef] [PubMed]
52. Gallardo-Soler, A.; Gomez-Nieto, C.; Campo, M.L.; Marathe, C.; Tontonoz, P.; Castrillo, A.; Corraliza, I. Arginase I induction by modified lipoproteins in macrophages: A peroxisome proliferator-activated receptor-γ/δ-mediated effect that links lipid metabolism and immunity. *Mol. Endocrinol.* **2008**, *22*, 1394–1402. [CrossRef] [PubMed]

53. Luo, W.; Xu, Q.; Wang, Q.; Wu, H.; Hua, J. Effect of modulation of PPAR-γ activity on Kupffer cells M1/M2 polarization in the development of non-alcoholic fatty liver disease. *Sci. Rep.* **2017**, *7*, 44612. [CrossRef] [PubMed]
54. Zhong, X.; Liu, H. Honokiol attenuates diet-induced non-alcoholic steatohepatitis by regulating macrophage polarization through activating peroxisome proliferator-activated receptor γ. *J. Gastroenterol. Hepatol.* **2018**, *33*, 524–532. [CrossRef] [PubMed]
55. Li, C.; Ying, W.; Huang, Z.; Brehm, T.; Morin, A.; Vella, A.T.; Zhou, B. IRF6 Regulates Alternative Activation by Suppressing PPARγ in Male Murine Macrophages. *Endocrinology* **2017**, *158*, 2837–2847. [CrossRef] [PubMed]
56. Bermudez, B.; Dahl, T.B.; Medina, I.; Groeneweg, M.; Holm, S.; Montserrat-de la Paz, S.; Rousch, M.; Otten, J.; Herias, V.; Varela, L.M.; et al. Leukocyte Overexpression of Intracellular NAMPT Attenuates Atherosclerosis by Regulating PPARγ-Dependent Monocyte Differentiation and Function. *Arterioscler. Thromb. Vasc. Biol.* **2017**, *37*, 1157–1167. [CrossRef] [PubMed]
57. Tikhanovich, I.; Zhao, J.; Olson, J.; Adams, A.; Taylor, R.; Bridges, B.; Marshall, L.; Roberts, B.; Weinman, S.A. Protein arginine methyltransferase 1 modulates innate immune responses through regulation of peroxisome proliferator-activated receptor γ-dependent macrophage differentiation. *J. Biol. Chem.* **2017**, *292*, 6882–6894. [CrossRef] [PubMed]
58. Assuncao, L.S.; Magalhaes, K.G.; Carneiro, A.B.; Molinaro, R.; Almeida, P.E.; Atella, G.C.; Castro-Faria-Neto, H.C.; Bozza, P.T. Schistosomal-derived lysophosphatidylcholine triggers M2 polarization of macrophages through PPARγ dependent mechanisms. *Biochim. Biophys. Acta* **2017**, *1862*, 246–254. [CrossRef] [PubMed]
59. Zhang, M.; Zhou, Z.; Wang, J.; Li, S. MiR-130b promotes obesity associated adipose tissue inflammation and insulin resistance in diabetes mice through alleviating M2 macrophage polarization via repression of PPAR-γ. *Immunol. Lett.* **2016**, *180*, 1–8. [CrossRef] [PubMed]
60. Feng, X.; Weng, D.; Zhou, F.; Owen, Y.D.; Qin, H.; Zhao, J.; WenYu; Huang, Y.; Chen, J.; Fu, H.; Yang, N.; et al. Activation of PPARγ by a natural flavonoid modulator, apigenin ameliorates obesity-related inflammation via regulation of macrophage polarization. *EBioMedicine* **2016**, *9*, 61–76. [CrossRef] [PubMed]
61. Deng, X.; Zhang, P.; Liang, T.; Deng, S.; Chen, X.; Zhu, L. Ovarian cancer stem cells induce the M2 polarization of macrophages through the PPAγ and NF-κB pathways. *Int. J. Mol. Med.* **2015**, *36*, 449–454. [CrossRef] [PubMed]
62. Zhang, X.; Zhou, M.; Guo, Y.; Song, Z.; Liu, B. 1,25-Dihydroxyvitamin D_3 Promotes High Glucose-Induced M1 Macrophage Switching to M2 via the VDR-PPARγ Signaling Pathway. *Biomed. Res. Int.* **2015**, *2015*, 157834. [PubMed]
63. Chang, H.Y.; Lee, H.N.; Kim, W.; Surh, Y.J. Docosahexaenoic acid induces M2 macrophage polarization through peroxisome proliferator-activated receptor γ activation. *Life Sci.* **2015**, *120*, 39–47. [CrossRef] [PubMed]
64. Feng, X.; Qin, H.; Shi, Q.; Zhang, Y.; Zhou, F.; Wu, H.; Ding, S.; Niu, Z.; Lu, Y.; Shen, P. Chrysin attenuates inflammation by regulating M1/M2 status via activating PPARγ. *Biochem. Pharmacol.* **2014**, *89*, 503–514. [CrossRef] [PubMed]
65. Kang, K.; Reilly, S.M.; Karabacak, V.; Gangl, M.R.; Fitzgerald, K.; Hatano, B.; Lee, C.H. Adipocyte-derived Th2 cytokines and myeloid PPARδ regulate macrophage polarization and insulin sensitivity. *Cell Metab.* **2008**, *7*, 485–495. [CrossRef] [PubMed]
66. Odegaard, J.I.; Ricardo-Gonzalez, R.R.; Red Eagle, A.; Vats, D.; Morel, C.R.; Goforth, M.H.; Subramanian, V.; Mukundan, L.; Ferrante, A.W.; Chawla, A. Alternative M2 activation of Kupffer cells by PPARδ ameliorates obesity-induced insulin resistance. *Cell Metab.* **2008**, *7*, 496–507. [CrossRef] [PubMed]
67. Bouhlel, M.A.; Brozek, J.; Derudas, B.; Zawadzki, C.; Jude, B.; Staels, B.; Chinetti-Gbaguidi, G. Unlike PPARγ, PPARα or PPARβ/δ activation does not promote human monocyte differentiation toward alternative macrophages. *Biochem. Biophys. Res. Commun.* **2009**, *386*, 459–462. [CrossRef] [PubMed]
68. Bouhlel, M.A.; Derudas, B.; Rigamonti, E.; Dievart, R.; Brozek, J.; Haulon, S.; Zawadzki, C.; Jude, B.; Torpier, G.; Marx, N.; et al. PPARγ activation primes human monocytes into alternative M2 macrophages with anti-inflammatory properties. *Cell Metab.* **2007**, *6*, 137–143. [CrossRef] [PubMed]
69. Zhang, T.; Shao, B.; Liu, G.A. Rosuvastatin promotes the differentiation of peripheral blood monocytes into M2 macrophages in patients with atherosclerosis by activating PPAR-γ. *Eur. Rev. Med. Pharmacol. Sci.* **2017**, *21*, 4464–4471. [PubMed]

70. Zizzo, G.; Cohen, P.L. The PPAR-γ antagonist GW9662 elicits differentiation of M2c-like cells and upregulation of the MerTK/Gas6 axis: A key role for PPAR-γ in human macrophage polarization. *J. Inflamm.* **2015**, *12*, 36. [CrossRef] [PubMed]
71. Zhang, O.; Zhang, J. Atorvastatin promotes human monocyte differentiation toward alternative M2 macrophages through p38 mitogen-activated protein kinase-dependent peroxisome proliferator-activated receptor γ activation. *Int. Immunopharmacol.* **2015**, *26*, 58–64. [CrossRef] [PubMed]
72. Lee, C.H.; Chawla, A.; Urbiztondo, N.; Liao, D.; Boisvert, W.A.; Evans, R.M.; Curtiss, L.K. Transcriptional repression of atherogenic inflammation: Modulation by PPARδ. *Science* **2003**, *302*, 453–457. [CrossRef] [PubMed]
73. Mukundan, L.; Odegaard, J.I.; Morel, C.R.; Heredia, J.E.; Mwangi, J.W.; Ricardo-Gonzalez, R.R.; Goh, Y.P.; Eagle, A.R.; Dunn, S.E.; Awakuni, J.U.; et al. PPAR-δ senses and orchestrates clearance of apoptotic cells to promote tolerance. *Nat. Med.* **2009**, *15*, 1266–1272. [CrossRef] [PubMed]
74. Kanakasabai, S.; Chearwae, W.; Walline, C.C.; Iams, W.; Adams, S.M.; Bright, J.J. Peroxisome proliferator-activated receptor δ agonists inhibit T helper type 1 (Th1) and Th17 responses in experimental allergic encephalomyelitis. *Immunology* **2010**, *130*, 572–588. [CrossRef] [PubMed]
75. Kanakasabai, S.; Walline, C.C.; Chakraborty, S.; Bright, J.J. PPARδ deficient mice develop elevated Th1/Th17 responses and prolonged experimental autoimmune encephalomyelitis. *Brain Res.* **2011**, *1376*, 101–112. [CrossRef] [PubMed]
76. Dunn, S.E.; Bhat, R.; Straus, D.S.; Sobel, R.A.; Axtell, R.; Johnson, A.; Nguyen, K.; Mukundan, L.; Moshkova, M.; Dugas, J.C.; et al. Peroxisome proliferator-activated receptor δ limits the expansion of pathogenic Th cells during central nervous system autoimmunity. *J. Exp. Med.* **2010**, *207*, 1599–1608. [CrossRef] [PubMed]
77. Mothe-Satney, I.; Murdaca, J.; Sibille, B.; Rousseau, A.S.; Squillace, R.; Le Menn, G.; Rekima, A.; Larbret, F.; Pele, J.; Verhasselt, V.; et al. A role for Peroxisome Proliferator-Activated Receptor Beta in T cell development. *Sci. Rep.* **2016**, *6*, 34317. [CrossRef] [PubMed]
78. Clark, R.B.; Bishop-Bailey, D.; Estrada-Hernandez, T.; Hla, T.; Puddington, L.; Padula, S.J. The nuclear receptor PPAR γ and immunoregulation: PPAR γ mediates inhibition of helper T cell responses. *J. Immunol.* **2000**, *164*, 1364–1371. [CrossRef] [PubMed]
79. Yang, X.Y.; Wang, L.H.; Chen, T.; Hodge, D.R.; Resau, J.H.; DaSilva, L.; Farrar, W.L. Activation of human T lymphocytes is inhibited by peroxisome proliferator-activated receptor γ (PPARγ) agonists. PPARγ co-association with transcription factor NFAT. *J. Biol. Chem.* **2000**, *275*, 4541–4544. [CrossRef] [PubMed]
80. Hontecillas, R.; Bassaganya-Riera, J. Peroxisome proliferator-activated receptor γ is required for regulatory $CD4^+$ T cell-mediated protection against colitis. *J. Immunol.* **2007**, *178*, 2940–2949. [CrossRef] [PubMed]
81. Nobs, S.P.; Natali, S.; Pohlmeier, L.; Okreglicka, K.; Schneider, C.; Kurrer, M.; Sallusto, F.; Kopf, M. PPARγ in dendritic cells and T cells drives pathogenic type-2 effector responses in lung inflammation. *J. Exp. Med.* **2017**, *214*, 3015–3035. [CrossRef] [PubMed]
82. Chung, S.W.; Kang, B.Y.; Kim, T.S. Inhibition of interleukin-4 production in $CD4^+$ T cells by peroxisome proliferator-activated receptor-γ (PPAR-γ) ligands: Involvement of physical association between PPAR-γ and the nuclear factor of activated T cells transcription factor. *Mol. Pharmacol.* **2003**, *64*, 1169–1179. [CrossRef] [PubMed]
83. Won, H.Y.; Min, H.J.; Ahn, J.H.; Yoo, S.E.; Bae, M.A.; Hong, J.H.; Hwang, E.S. Anti-allergic function and regulatory mechanisms of KR62980 in allergen-induced airway inflammation. *Biochem. Pharmacol.* **2010**, *79*, 888–896. [CrossRef] [PubMed]
84. Guri, A.J.; Mohapatra, S.K.; Horne, W.T., II; Hontecillas, R.; Bassaganya-Riera, J. The role of T cell PPAR γ in mice with experimental inflammatory bowel disease. *BMC Gastroenterol.* **2010**, *10*, 60. [CrossRef] [PubMed]
85. Wohlfert, E.A.; Nichols, F.C.; Nevius, E.; Clark, R.B. Peroxisome proliferator-activated receptor γ (PPARγ) and immunoregulation: Enhancement of regulatory T cells through PPARγ-dependent and -independent mechanisms. *J. Immunol.* **2007**, *178*, 4129–4135. [CrossRef] [PubMed]
86. Cipolletta, D.; Feuerer, M.; Li, A.; Kamei, N.; Lee, J.; Shoelson, S.E.; Benoist, C.; Mathis, D. PPAR-γ is a major driver of the accumulation and phenotype of adipose tissue Treg cells. *Nature* **2012**, *486*, 549–553. [CrossRef] [PubMed]

87. Cipolletta, D.; Cohen, P.; Spiegelman, B.M.; Benoist, C.; Mathis, D. Appearance and disappearance of the mRNA signature characteristic of Treg cells in visceral adipose tissue: Age, diet, and PPARγ effects. *Proc. Natl. Acad. Sci. USA* **2015**, *112*, 482–487. [CrossRef] [PubMed]
88. Klotz, L.; Burgdorf, S.; Dani, I.; Saijo, K.; Flossdorf, J.; Hucke, S.; Alferink, J.; Nowak, N.; Beyer, M.; Mayer, G.; et al. The nuclear receptor PPAR γ selectively inhibits Th17 differentiation in a T cell-intrinsic fashion and suppresses CNS autoimmunity. *J. Exp. Med.* **2009**, *206*, 2079–2089. [CrossRef] [PubMed]
89. Park, H.J.; Choi, J.M. Sex-specific regulation of immune responses by PPARs. *Exp. Mol. Med.* **2017**, *49*, e364. [CrossRef] [PubMed]
90. Dunn, S.E.; Ousman, S.S.; Sobel, R.A.; Zuniga, L.; Baranzini, S.E.; Youssef, S.; Crowell, A.; Loh, J.; Oksenberg, J.; Steinman, L. Peroxisome proliferator-activated receptor (PPAR)α expression in T cells mediates gender differences in development of T cell-mediated autoimmunity. *J. Exp. Med.* **2007**, *204*, 321–330. [CrossRef] [PubMed]
91. Zhang, M.A.; Rego, D.; Moshkova, M.; Kebir, H.; Chruscinski, A.; Nguyen, H.; Akkermann, R.; Stanczyk, F.Z.; Prat, A.; Steinman, L.; et al. Peroxisome proliferator-activated receptor (PPAR)α and -γ regulate IFNγ and IL-17A production by human T cells in a sex-specific way. *Proc. Natl. Acad. Sci. USA* **2012**, *109*, 9505–9510. [CrossRef] [PubMed]
92. Zhang, M.A.; Ahn, J.J.; Zhao, F.L.; Selvanantham, T.; Mallevaey, T.; Stock, N.; Correa, L.; Clark, R.; Spaner, D.; Dunn, S.E. Antagonizing Peroxisome Proliferator-Activated Receptor α Activity Selectively Enhances Th1 Immunity in Male Mice. *J. Immunol.* **2015**, *195*, 5189–5202. [CrossRef] [PubMed]
93. Park, H.J.; Park, H.S.; Lee, J.U.; Bothwell, A.L.; Choi, J.M. Gender-specific differences in PPARγ regulation of follicular helper T cell responses with estrogen. *Sci. Rep.* **2016**, *6*, 28495. [CrossRef] [PubMed]
94. Park, H.J.; Kim, D.H.; Choi, J.Y.; Kim, W.J.; Kim, J.Y.; Senejani, A.G.; Hwang, S.S.; Kim, L.K.; Tobiasova, Z.; Lee, G.R.; et al. PPARγ negatively regulates T cell activation to prevent follicular helper T cells and germinal center formation. *PLoS ONE* **2014**, *9*, e99127. [CrossRef] [PubMed]
95. Park, H.J.; Park, H.S.; Lee, J.U.; Bothwell, A.L.; Choi, J.M. Sex-Based Selectivity of PPARγ Regulation in Th1, Th2, and Th17 Differentiation. *Int. J. Mol. Sci.* **2016**, *17*, 1347. [CrossRef] [PubMed]
96. Li, G.; Chen, C.; Laing, S.D.; Ballard, C.; Biju, K.C.; Reddick, R.L.; Clark, R.A.; Li, S. Hematopoietic knockdown of PPARδ reduces atherosclerosis in LDLR$^{-/-}$ mice. *Gene Ther.* **2016**, *23*, 78–85. [CrossRef] [PubMed]
97. Babaev, V.R.; Yancey, P.G.; Ryzhov, S.V.; Kon, V.; Breyer, M.D.; Magnuson, M.A.; Fazio, S.; Linton, M.F. Conditional knockout of macrophage PPARγ increases atherosclerosis in C57BL/6 and low-density lipoprotein receptor-deficient mice. *Arterioscler. Thromb. Vasc. Boil.* **2005**, *25*, 1647–1653. [CrossRef] [PubMed]
98. Chawla, A.; Boisvert, W.A.; Lee, C.H.; Laffitte, B.A.; Barak, Y.; Joseph, S.B.; Liao, D.; Nagy, L.; Edwards, P.A.; Curtiss, L.K.; et al. A PPAR γ-LXR-ABCA1 pathway in macrophages is involved in cholesterol efflux and atherogenesis. *Mol. Cell* **2001**, *7*, 161–171. [CrossRef]
99. Hevener, A.L.; Olefsky, J.M.; Reichart, D.; Nguyen, M.T.; Bandyopadyhay, G.; Leung, H.Y.; Watt, M.J.; Benner, C.; Febbraio, M.A.; Nguyen, A.K.; et al. Macrophage PPAR γ is required for normal skeletal muscle and hepatic insulin sensitivity and full antidiabetic effects of thiazolidinediones. *J. Clin. Investig.* **2007**, *117*, 1658–1669. [CrossRef] [PubMed]
100. Marathe, C.; Bradley, M.N.; Hong, C.; Chao, L.; Wilpitz, D.; Salazar, J.; Tontonoz, P. Preserved glucose tolerance in high-fat-fed C57BL/6 mice transplanted with PPARγ$^{-/-}$, PPARδ$^{-/-}$, PPARγδ$^{-/-}$, or LXRαβ$^{-/-}$ bone marrow. *J. Lipid Res.* **2009**, *50*, 214–224. [CrossRef] [PubMed]
101. Skerrett, R.; Malm, T.; Landreth, G. Nuclear receptors in neurodegenerative diseases. *Neurobiol. Dis.* **2014**, *72*, 104–116. [CrossRef] [PubMed]
102. Iglesias, J.; Morales, L.; Barreto, G.E. Metabolic and Inflammatory Adaptation of Reactive Astrocytes: Role of PPARs. *Mol. Neurobiol.* **2017**, *54*, 2518–2538. [CrossRef] [PubMed]
103. Yamanaka, M.; Ishikawa, T.; Griep, A.; Axt, D.; Kummer, M.P.; Heneka, M.T. PPARγ/RXRα-induced and CD36-mediated microglial amyloid-β phagocytosis results in cognitive improvement in amyloid precursor protein/presenilin 1 mice. *J. Neurosci.* **2012**, *32*, 17321–17331. [CrossRef] [PubMed]
104. Mandrekar-Colucci, S.; Karlo, J.C.; Landreth, G.E. Mechanisms underlying the rapid peroxisome proliferator-activated receptor-γ-mediated amyloid clearance and reversal of cognitive deficits in a murine model of Alzheimer's disease. *J. Neurosci.* **2012**, *32*, 10117–10128. [CrossRef] [PubMed]

105. Xu, J.; Chavis, J.A.; Racke, M.K.; Drew, P.D. Peroxisome proliferator-activated receptor-α and retinoid X receptor agonists inhibit inflammatory responses of astrocytes. *J. Neuroimmunol.* **2006**, *176*, 95–105. [CrossRef] [PubMed]

106. Xu, J.; Storer, P.D.; Chavis, J.A.; Racke, M.K.; Drew, P.D. Agonists for the peroxisome proliferator-activated receptor-α and the retinoid X receptor inhibit inflammatory responses of microglia. *J. Neurosci. Res.* **2005**, *81*, 403–411. [CrossRef] [PubMed]

107. Xu, J.; Racke, M.K.; Drew, P.D. Peroxisome proliferator-activated receptor-α agonist fenofibrate regulates IL-12 family cytokine expression in the CNS: Relevance to multiple sclerosis. *J. Neurochem.* **2007**, *103*, 1801–1810. [CrossRef] [PubMed]

108. Xu, J.; Drew, P.D. Peroxisome proliferator-activated receptor-γ agonists suppress the production of IL-12 family cytokines by activated glia. *J. Immunol.* **2007**, *178*, 1904–1913. [CrossRef] [PubMed]

109. Xu, J.; Barger, S.W.; Drew, P.D. The PPAR-γ Agonist 15-Deoxy-Delta-Prostaglandin J(2) Attenuates Microglial Production of IL-12 Family Cytokines: Potential Relevance to Alzheimer's Disease. *PPAR Res.* **2008**, *2008*, 349185. [CrossRef] [PubMed]

110. Storer, P.D.; Xu, J.; Chavis, J.; Drew, P.D. Peroxisome proliferator-activated receptor-γ agonists inhibit the activation of microglia and astrocytes: Implications for multiple sclerosis. *J. Neuroimmunol.* **2005**, *161*, 113–122. [CrossRef] [PubMed]

111. Storer, P.D.; Xu, J.; Chavis, J.A.; Drew, P.D. Cyclopentenone prostaglandins PGA2 and 15-deoxy-δ12,14 PGJ2 suppress activation of murine microglia and astrocytes: Implications for multiple sclerosis. *J. Neurosci. Res.* **2005**, *80*, 66–74. [CrossRef] [PubMed]

112. Xing, B.; Liu, M.; Bing, G. Neuroprotection with pioglitazone against LPS insult on dopaminergic neurons may be associated with its inhibition of NF-κB and JNK activation and suppression of COX-2 activity. *J. Neuroimmunol.* **2007**, *192*, 89–98. [CrossRef] [PubMed]

113. Polak, P.E.; Kalinin, S.; Dello Russo, C.; Gavrilyuk, V.; Sharp, A.; Peters, J.M.; Richardson, J.; Willson, T.M.; Weinberg, G.; Feinstein, D.L. Protective effects of a peroxisome proliferator-activated receptor-β/δ agonist in experimental autoimmune encephalomyelitis. *J. Neuroimmunol.* **2005**, *168*, 65–75. [CrossRef] [PubMed]

114. Schnegg, C.I.; Kooshki, M.; Hsu, F.C.; Sui, G.; Robbins, M.E. PPARδ prevents radiation-induced proinflammatory responses in microglia via transrepression of NF-κB and inhibition of the PKCα/MEK1/2/ERK1/2/AP-1 pathway. *Free Radic. Biol. Med.* **2012**, *52*, 1734–1743. [CrossRef] [PubMed]

International Journal of
Molecular Sciences

Review

Metabolic Dysfunction and Peroxisome Proliferator-Activated Receptors (PPAR) in Multiple Sclerosis

Véronique Ferret-Sena [1], Carlos Capela [2] and Armando Sena [1,*]

[1] Centro de Investigação Interdisciplinar Egas Moniz (CiiEM), Instituto Universitário Egas Moniz, Campus Universitário, Quinta da Granja, Monte de Caparica, 2819-511 Caparica, Portugal; versena@egasmoniz.edu.pt

[2] Centro Hospitalar de Lisboa Central, EPE, Hospital de Santo António dos Capuchos, Departamento de Neurociências, Alameda de Santo António dos Capuchos, 1169-050 Lisboa, Portugal; carlos.capela.trabalho@gmail.com

* Correspondence: asena@egasmoniz.edu.pt

Received: 28 April 2018; Accepted: 28 May 2018; Published: 1 June 2018

Abstract: Multiple sclerosis (MS) is an inflammatory and neurodegenerative disease of the central nervous system (CNS) probably caused, in most cases, by the interaction of genetic and environmental factors. This review first summarizes some clinical, epidemiological and pathological characteristics of MS. Then, the involvement of biochemical pathways is discussed in the development and repair of the CNS lesions and the immune dysfunction in the disease. Finally, the potential roles of peroxisome proliferator-activated receptors (PPAR) in MS are discussed. It is suggested that metabolic mechanisms modulated by PPAR provide a window to integrate the systemic and neurological events underlying the pathogenesis of the disease. In conclusion, the reviewed data highlight molecular avenues of understanding MS that may open new targets for improved therapies and preventive strategies for the disease.

Keywords: multiple sclerosis; metabolism; peroxisome proliferator-activated receptors; immune system; neuroinflammation; neurodegeneration

---

1. An Overview of Multiple Sclerosis (MS)

Multiple sclerosis (MS) afflicts about 2.5 million patients worldwide and is one of the most common causes of permanent disability in young adults. MS is a disease of the central nervous system (CNS) with the most frequent onset in young adulthood and very different clinical courses in individual patients. About 85% of patients develop the relapsing-remitting type of the disease (RRMS), which is more common in women and starts with episodes of clinical symptoms (relapses) with variable recovery. Most of these patients develop secondary progressive MS (SPMS), a stage of progressive disability not associated with relapses. Patients with primary progressive MS (PPMS) have a progressive course from disease onset regardless of eventual relapse episodes. Clinically isolated syndromes (CIS) categorize patients who experienced a first clinical presentation suggestive of MS. Because many of these patients convert to clinically definite MS, some authors include CIS as another element of the MS phenotype spectrum [1].

Histopathological hallmarks of the disease include multifocal lesions of demyelination, inflammation, gliosis and axon loss or damage (the MS plaque). Typically, these lesions are associated with breakdown of the blood-brain barrier (BBB) and are thought to be mainly mediated by type 1 helper T cells (Th1), Th17, and $CD8^+$ cells, despite increasing evidence for an early involvement of B cells and the innate immune system (including macrophages, microglia, dendritic cells and astrocytes) [2,3]. Four patterns of active lesions were described, characterized by different immunological involvement and glial and neuronal injury. Interestingly, these features were found to differ between patients but to be identical in the same individual. These findings support the important concept that different pathogenic mechanisms and targets may underlie the development of MS lesions in different patients [4,5].

During the last few decades, the concept of MS pathogenesis was profoundly influenced by research conducted in experimental autoimmune encephalomyelitis (EAE). These animal models mimic many pathological aspects and all clinical forms of MS [6–8]. This has led many authors to suggest that MS has a primary autoimmune etiology. Supporting this hypothesis, studies on EAE were crucial for the development of the majority of current MS treatments. These drugs have the adaptive immune system as the main target of mechanism of action and reduce the rate of relapses in most patients. However, their efficacies in preventing disability associated with progressive MS (SPMS and PPMS) have been disappointing. It is possible that the newer and most powerful approved therapies may benefit some patients with these forms of the disease but considerable concern exists regarding their toxicity and serious adverse reactions [9–11]. In short, current available therapies confirm previous clinical evidence suggesting that disability accumulation is largely independent of inflammatory relapse activity [12]. This scenario is interpreted as supporting the so-called two-stage hypothesis of MS immunopathogenesis. This framework suggests that the adaptive immune system drives autoimmune lesions in an early stage of the disease and clinical relapses, whereas an innate immune system dysfunction predominates in a second progressive and neurodegenerative stage independently of relapses [13].

MS is not just a white matter disease. It has been known for a long time that grey matter, which, besides myelin, is enriched in neuronal cell bodies and synaptic structures, is also affected [14,15]. Recent studies indicate the presence of inflammatory grey matter lesions of demyelination and neuronal damage during the earliest phases of the disease, which are, at least in part, independent of white matter lesions. Most importantly, the severity of grey matter atrophy and neuronal damage is the major correlate of disability progression [15]. These findings are consistent with the view that MS could be a primary neurodegenerative disorder [14]. This scenario is also in agreement with clinical data suggesting that MS, instead of being two-staged, is a one-stage disorder of progressive neurodegeneration from onset [12]. However, if MS is a chronic neurodegenerative disease from onset, myeloid cells should have a more prominent role in its pathogenesis than suggested by the two-stage hypothesis. The innate immune system could have a critical role in the regulation of autoimmunity mechanisms from the earliest states of the disease [2,16].

Immune-mediated processes are essential in the pathophysiology of MS and their qualities probably differ in relapsing and progressive phenotypes of the disease. Nevertheless, a critical issue, for which an old disagreement persists, concerns the mechanisms that initiate the inflammatory nature of the disorder [15,17]. Some authors believe that MS is caused by a primary peripheral immune dysfunction leading to autoimmune mechanisms of CNS damage (the "outside-in" model) [3,13]. An alternative hypothesis suggests that the inflammatory reaction is secondary to an abnormal development or degenerative damage of oligodendrocyte-myelin complex or/and neurons (the "inside-out" model) [14,18–20]. Two points should be emphasized in this context. First, there is a physiological cross-talk between the immune system and the CNS [21]. Neurons and glial cells and the peripheral immune system share the expression of many molecules, including human leukocyte antigens (HLA), complement proteins, cytokines and neurotransmitters [15,21]. Myeloid cells (macrophages and microglia) and astrocytes are involved in maintaining neuronal homeostasis, synaptic development and plasticity and myelin remodelling [2,16,22]. Regulatory T cells (Treg) are impaired in MS [3] and also possess neurotrophic proprieties and promote myelin production independently of immunomodulatory functions and overt inflammation [23]. Consequently, it would not be unexpected to find a physiological disturbance concomitantly expressed in the peripheral immune system and the CNS. Indeed, many neuropsychiatric and classic primary neurodegenerative disorders, such as Alzheimer's disease, are associated with an abnormal systemic and brain inflammatory reactivity. Secondly, the "outside-in" and "inside-in" models are not necessarily in conflict. As outlined above, MS is clinically and pathologically a heterogeneous disease. Therefore, the pathways inducing or maintaining the abnormal immune reactivity in MS do not need to be the same in all individuals or forms/stages of the disease. These points are further discussed in this manuscript in the context of recent studies that suggest an abnormal metabolism and an involvement of peroxisome proliferator-activated receptors (PPAR) in the pathophysiology of MS.

2. Genes and the Environment in MS

As in many other complex diseases, the cause of MS in most patients is thought to be due to an interaction of multiple potential genetic and environmental factors. Large genome-associated studies (GWAS) identified more than 200 genetic variants associated with MS susceptibility. The HLADRB1*1501 haplotype is the most significant genetic risk factor for the MS and possibly promotes a more severe course of the disease [3,5,15]. The genes encoding the HLA molecules are only a minority of the over 250 expressed genes in the extended major histocompatibility complex (xMHC), which are mainly related to immune functions. However, disease-associations have been found in more than 100 loci outside the xMHC. A recent study identified 60 genes shared by MS and cardiovascular disease (CVD) risk factors [24]. Multiple loci in the xMHC region were overlapping between MS and triglyceride and high density lipoprotein (HDL) cholesterol while the polygenic overlap between MS and low-density lipoprotein (LDL) cholesterol and some other CVD factors was not dependent on the xMHC region. On one hand, these results are in line with the involvement of immune-mediated processes in vascular diseases; and, on the other hand, they suggest genetic influences on lipid metabolism shared by the pathogenesis of CVD and MS. Supporting this scenario, a recent large study concluded that genetically increased body mass index (BMI) is also associated with the risk of MS [25]. These results are in accordance with previous studies linking obesity in childhood or adolescence with increased risk for the disease. Multiple environmental or life style factors have been associated with the risk and severity of MS [26,27]. However, the impact of such environmental factors for the risk of MS seems greater if exposure occurs before 15 years of age; and interactions with individual and ethnic genetic backgrounds could contribute to explain the disparity of some studies regarding the influences, for example, of Epstein–Barr virus, tobacco use or salt intake [28–30]. While regions of higher latitude and decreased levels of sunlight exposure have higher prevalence of MS, genetic influences on the vitamin D level or action could be important [31]. Some genetic and environment factors have been found to be protective of MS [3,27]. In short, although the development of MS requires in most cases an exposure to environmental factors, even in most of these cases the disease probably occurs only in genetically susceptible individuals [32]. It should be emphasized that neither the genetic nor the environmental factors associated with the risk and clinical course of the disease need to be the same in different patients.

In a remarkable paper published in 1965, R.H.S. Thompson reviewed the large biochemical and epidemiological data implicating a disturbed phospholipid metabolism and essential polyunsaturated fatty acid (PUFA) deficiency in MS pathogenesis [33]. Notably, these metabolic alterations were interpreted to support the contribution of an abnormal brain chemical composition and vascular/anoxia mechanisms to the genesis of MS lesions. Later, Goldberg (1974) suggested a link between vitamin D and calcium deficiency and an abnormal lipid metabolism in the aetiology of the disease, which could affect the development and stability of myelin [34,35]. In the 1970s, Swank [36] reported the benefit of treating MS patients with a low fat diet for more than 20 years and epidemiological studies confirmed a correlation between high saturated animal-fat and low PUFA intake and the prevalence of MS [37,38]. More recently, a large cross-sectional study found strong associations of plant-based ω-3 supplementation (but not fish oil intake) with lower disability and relapse rate [39]. These findings are in agreement with prospective studies supporting a link of lower PUFA intake (especially α-linolenic acid) with increased risk of MS [40] and high saturated fat and low vegetable intakes with relapse risk in paediatric MS [41]. The protective effects of fatty fish intake described by some studies could be confined to individuals exposed to low ultraviolet radiation and vitamin D deficiency [42]. A very recent large cross-sectional survey in 6989 patients concluded that a high intake of fruits, vegetables and legumes and whole grains and a low intake of sugar and red meat were associated with lower levels of disability in people with MS [43].

In summary, wide clinical and epidemiological data support the view that an interplay between genetic and environmental or life-style factors affecting lipid and energetic metabolism is implicated in the development and clinical course of MS. These data suggest pathogenic links between MS and vascular diseases.

3. Genesis and Repair of MS Lesions

3.1. Demyelination Lesions

The landmark work by Shore et al. [44] in EAE concluded that "major changes in ApoE-containing lipoproteins are undoubtedly significant in the altered immune function in EAE". This paper was followed by increased research on the role of lipids in the genesis of lesions and clinical course of EAE and MS. Newcombe et al. [45] suggested that, in MS patients, plasma low density lipoprotein (LDL) enters the CNS parenchyma as result of BBB increased permeability and is then largely oxidized and taken up by infiltrating macrophages and microglia. This mechanism is thought to contribute to the activation of these cells and phagocytosis of myelin. Oxidized phospholipids were identified in myelin, oligodendrocytes and neurons, and may be involved not only in active demyelination but also in neurodegenerative lesions [46]. However, ingestion of myelin by foam myeloid cells change their pro-inflammatory (M1) to an anti-inflammatory (M2) phenotype, which could downregulate the development of lesions and promote repair [47]. The low-density lipoprotein receptor-related protein1 (LRP1) is implicated in myelin phagocytosis and also controls BBB permeability and immune activation [48,49]. Myelin is especially enriched in sphingomyelin and other ceramide-derived compounds that are liberated during the destructive process and can be found in the CSF of MS patients [50]. However, as in other neuroinflammatory processes, astrocytes are the main origin of the increased levels of ceramide found in MS active lesions, which promotes leukocyte migration across the BBB [51]. Higher levels of palmitic acid-containing hexosylceramide (Cer16:0) in CSF were found in CIS patients who converted to MS within three years from sampling [52]. In contrast, the ceramide metabolite shingosine-1-phopsphate (S1P), among other effects, can signal endothelial cells and astrocytes to reduce leukocyte transmigration and CNS inflammatory activity. Fingolimod, a modulator of the S1P receptor, attenuates the BBB dysfunction induced by ceramide, which might contribute to its beneficial effects in MS patients [51,53]. As in plasma, S1P in the brain is mainly associated with high density lipoproteins (HDL) and may mediate some of its protective and anti-inflammatory effects [54]. Interestingly, S1P was found to be increased in the CSF of RRMS patients who were still not treated with immunomodulatory agents, supporting an involvement of ceramide metabolism from the earliest stages of the disease [55]. Myelin proteins and lipids are recognized antigenic targets of the adaptive immune system. For example, intrathecal synthesis of lipid-specific antibodies were correlated with increased relapses and a more aggressive disease [56]. However, it should be emphasized that brain phospholipids and glycolipids may also resolve inflammatory reactivity, including by suppressing activation and inducing apoptosis of autoreactive T cells [57,58].

In adult CNS, neurons and oligodendrocytes are largely dependent on cholesterol delivered by astrocytes. Apolipoprotein E (ApoE) in the brain is mainly expressed by astrocytes and necessary for the transport of cholesterol to neurons and oligodendrocytes through HDL-like particles [59]. Recently, a decrease in expression of cholesterol synthesis genes was found in chronic EAE and MS lesions and related to inflammatory infiltrates [60]. Lavrnja et al. [61] observed an altered expression of cholesterol metabolism-related genes during the development of demyelinating lesions in EAE. This work suggests that an increased cholesterol synthesis and expression of ApoE occurs in later stages of the destructive process, contributing to the regeneration of myelin and neurons. ApoE has immunosuppressive effects, inducing the differentiation of macrophages into an anti-inflammatory phenotype [62]. However, phagocytosis of myelin cholesterol and its oxygenated metabolites (oxysterols) may induce regenerative and protective mechanisms as well. Mailleux et al. [63] have shown that oxysterols are also present in myelin and that foam phagocytes generate 27-hydroxycholesterol in MS lesion. These authors observed that uptake of oxysterols by foam cells induces the expression of ApoE and other liver X receptor (LXR) genes, and stimulates anti-inflammatory mechanisms. LXR activation decreases disease severity, Th17 polarization and IL-17 secretion in EAE, though certain oxysterols may have pro-inflammatory effects in these models [64]. The Mailleux group also reported that in EAE, LDL receptor deficiency attenuates the severity of the disease in females (not in males) mice, through the induction of ApoE [65]. Interestingly, other authors have found that, in ApoE knock-out mice, ApoE deficiency increases EAE severity only in female animals [8]. These findings suggest interactions between sex steroids and cholesterol metabolism in the development of CNS lesions and severity of the disease. Sex steroids have anti-inflammatory and neuroprotective effects and the CNS is not only a target of these bloodstream hormones because it also produces sex steroids (neurosteroids), mainly by astrocytes. Luchetti et al. [66] observed that oestrogen signalling is induced in male and progesterone signalling is increased in female brain lesions of MS patients. Importantly, different alterations in sex steroid metabolism were detected even in normal-appearing white matter (NAWM). These data indicate that sex steroids could mediate gender differences associated with the genesis and repair of MS lesions.

3.2. Neurodegeneration and Progressive MS

R.H.S. Thompson has suggested that "... chemical differences may exist in apparently unaffected areas of brain tissue in multiple sclerosis. Such differences, possibly inborn in nature, might render the central nervous system more sensitive to other potentially damaging factors" [33]. Several studies have detected an abnormal lipid composition of NAWM that could precede myelin and neuronal inflammatory damage [18,33,35]. The Moscarello group observed that an abnormal maturation of myelin basic protein (MBP) could also contribute to the activation of autoimmune mechanisms [18]. Recent spectrometry analysis of NAWM and normal and normal-appearing grey matter (NAGM) from MS patients support the view that an abnormal brain chemical composition could precede the development of immune-mediated lesions [67]. The last study found an increased lipid peroxidation in white matter and a pattern of composition in NAWM and NAGM, suggesting a metabolic disturbance leading to a decrease of sphingolipids and increase of phospholipids content that could destabilize myelin. In this line of thought, Vidaurre et al. [68] found increased levels of ceramide C16:0 and C24:0 in the CSF of MS patients, without changes of cytokines levels. Ceramide compounds were sufficient to induce mitochondrial dysfunction, increased expression of genes involved in oxidative damage and glutamate excitotoxicity and decreased expression of neuroprotective genes. Other authors found that high levels of hexosylceramide C16:0 in the CSF correlated with disability scores only in progressive patients [52]. Increasingly data support major roles of mitochondrial dysfunction, reactive oxygen and nitrogen species, glutamate excitotoxicity and ion channel dysfunction (mainly of calcium homeostasis) in the neurodegenerative process and progression of the disease [69–71]. Accordingly, *N*-acetyl aspartate, a metabolite only produced by neuronal mitochondria and required for myelin synthesis is decreased in NAWM and NAGM and correlated with disability [72,73]. Reduced oxygen consumption (hypoxia) and age-dependent brain iron accumulation may amplify oxidative damage and the neurodegenerative process. Local cyclooxygenase-dependent lipid oxidation metabolites were also suggested to contribute to the mechanisms of disease leading to disability progression [74]. The transcriptional factor (erythroid-derived2)-like 2 (Nrf2), a target of dimethyl fumarate therapy in MS, is activated by excessive free radical production, induces the expression of many antioxidant defenses and is upregulated in inflammatory lesions of the disease [75]. A diminished expression of Nrf2 was correlated with reduced levels of glutathione in EAE [76] and a marked reduction of glutathione was recently observed in brains of progressive MS (SPMS and PPMS) in comparison to RRMS patients [77]. It should be noted that neuron–astrocyte interactions may have major roles in the regulation Nrf2 signalling in the CNS [78]. In a chronic pro-inflammatory milieu, diminished synthesis of cholesterol by astrocytes critically compromise myelin remodelling and maintenance of neuronal structural integrity [59,60]. Several studies have shown a correlation between neuronal damage and brain atrophy and decreased synthesis of 24S-hydroxycholesterol (24OHC), which mainly occurs in neurons [79]. An additional element of complexity concerns the complement proteins, which in the adult human brain are mainly synthesized in neurons and have physiological roles in synaptic elimination and remodelling. As in the immune system, these proteins may opsonize cellular components for clearance by activated macrophages and microglia and could drive synaptic loss from the earliest stages of MS [80]. Recent work observed a widespread and pronounced synaptic loss in the cerebral cortex of MS patients independent of cortical demyelination and axonal loss [81]. Neuronal dysfunction could be a source of local complement production independent of systemic circulation and genesis of acute demyelination lesions and related with the progression of the disease [82]. Again, astrocytes constitute important players in this scenario. Cytokines and complement from activated microglia change astrocytes to a "reactive" and toxic phenotype, inducing synaptic loss and death of neurons and oligodendrocytes [83]. An increased synthesis and release of lactosylceramide by astrocytes was found in EAE and MS lesions, promoting the recruitment and activation of monocytes and microglia and neurodegeneration [84]. S1P receptors in astrocytes control the development of acute lesions and nuclear factor-kappa B (NF-κB) activity associated with CNS inflammation in chronic progressive EAE and MS [85]. Kynurenine acid (KA) is also produced in the brain mainly by astrocytes, has anti-inflammatory and antioxidant effects and protects neurons against glutamate toxicity. Abnormalities in the kynurenine pathway of tryptophan metabolism possibly have important roles in the neurodegenerative mechanisms of MS (71) (see below). In addition, cortical and meningeal infiltrates of lymphocytes are frequently present in progressive disease and may contribute to grey matter and neurodegenerative pathology as well [15,70].

In summary, it is generally accepted that white matter focal demyelination and axonal lesions are mainly driven by immune cell infiltration from the periphery, whereas a compartmentalized diffuse microglial and astrocyte activation mainly drives grey matter and synaptic loss pathology and progressive MS [3,15]. Nevertheless, myeloid cells and astrocytes are critical players in all pathogenic processes of MS [2]. Furthermore, alterations in lipid, oxidative and other metabolic pathways are present in the CNS from the earliest stages of the disease and are involved in mechanisms modulating immune activation and the development and repair of demyelinating and neurodegenerative lesions.

4. Systemic Metabolism in MS

4.1. Plasma Lipids

Giubilei et al. [86] were the first to report a correlation between the number of new brain lesions evaluated by magnetic resonance imaging (MRI) and the mean plasma level of total and LDL cholesterol in patients with the first clinical episode suggestive of MS (CIS). Recent prospective studies have variably found associations between total cholesterol (TC), LDL, non-HDL, TC/HDL, triglycerides, apolipoprotein B and the risk of new lesions accumulation and disability progression in patients with RRMS and/or CIS. Some studies also reported correlations between higher plasma ApoE levels and severity of EAE, higher disability in RRMS, and deep grey matter atrophy in CIS patients. In contrast, higher levels of HDL and of its major apolipoprotein, ApoA1, were associated with lower blood–brain barrier permeability and protection to the development inflammatory lesions [87]. RRMS subjects have smaller LDL in comparison to healthy controls and in some cases increased levels of small HDL with impaired anti-inflammatory activity [88]. Interestingly, this study observed some differences between male and female patients, suggesting gender differences in lipid metabolism associated with MS [65]. Supporting this hypothesis, recent research from our group suggests that sex steroids modify the serum lipid profile associated with disability in these patients (unpublished results). Plasma oxysterols levels are increased in RRMS and mainly associated with small dense LDL, supporting a link with mechanisms promoting atherogenesis [89]. These data disclose many evident similarities between MS and the mechanisms involved in atherosclerotic plaque development and progression, as mentioned by Ludewig and Laman [90]. In their paper, these authors concluded that ... "Systematic comparison of these two diseases involving foam cells in chronic lesions may prove fruitful". Besides hypercholesterolemia, the coexistence in the MS patient of other related vascular pathology (such as diabetes and hypertension) may indeed be associated with more rapid disability progression [91] and increased risk of relapses [92]. Taken together, the results from these studies and those summarized in the above sections strongly suggest that CVD and MS may share certain pathophysiological mechanisms [93]. Statins have well-known immunosuppressive proprieties and decrease inflammatory activity and clinical signs in EAE. In 2003, our group published a pilot study suggesting potential benefits of lovastatin monotherapy in RRMS [94] and similar results were obtained by Vollmer et al. [95] using simvastatin. More recently, simvastatin was shown to reduce brain atrophy and to improve cognitive and physical quality of life measures in SPMS patients [96,97]. Interferon beta therapy changes the associations between serum lipoprotein levels and the clinical activity and neurodegenerative process of the disease [98,99]. Current approved drugs for MS may induce specific alterations in systemic lipid metabolism. In one study, interferon beta was shown to increase subspecies of ceramides and natalizumab to increase S1P and sphinganine-1 phosphate, whereas fingolimod did not affect the levels of these lipids in plasma [100].

Systemic metabolic alterations in MS must have characteristics unique to the disease and to its different pathological and clinical phenotypes. Metabolomic investigations have found distinctive serum phospholipid and sphingolipid patterns in MS patients in comparison to healthy controls subjects or patients with other neurological disorders [101–103]. In one study, an increase of phospholipids and a decrease of sphingolipids were observed and certain phospholipids, glutamic acid, tryptophan and arachidonic acid metabolites levels were correlated with a more severe disease [102]. In addition, Quintana et al. [104] reported different patterns of serum antibodies to lipids and other CNS antigens associated with RR, SP and PP forms of MS and Bakshi et al. [105] found that serum lipid antibodies associated with atrophy differed from those associated with brain focal lesions. Some authors believe that this profile of alterations in the serum of MS patients is mainly due to the chronic activation of the immune system [102]. However, an underlying abnormal metabolism could also affect immune reactivity. Obviously, these two mechanisms are not mutually exclusive and their role could differ, depending on the individual patient and activity/type of the disease. We have found that in patients displaying similar clinical activity and disability scores, lower serum ApoE levels were associated with the increased risk for development of neutralising antibodies to interferon beta [106]. These results are consistent with anti-inflammatory effects of ApoE [62,63,107] and suggest that individual differences in lipoprotein metabolism could influence the reactivity of the immune system in these patients [106].

4.2. Metabolism and Immune Dysfunction

It is presently indisputable that specific metabolic processes are needed to support the different functions of immune cells. Distinct metabolic programs are required for differentiation and function of effector T cells (Th1, Th2, Th17) and inducible regulatory T cells (Treg) [108]. Activation and proliferation of $CD4^+$ and $CD8^+$ T effector cells depend on glycolysis, whereas T memory cells depend on fatty acid oxidation for ATP production [108,109]. A reduction in proliferation and suppressive functions of Treg cells is one of the main characteristics of the immune dysfunction in RRMS patients [3,110]. Although Treg cells also rely on lipid oxidation, an engagement on glycolysis may be necessary to generate the suppressive functions of these cells [111]. Interestingly, an impairment of glycolysis and mitochondrial respiration was recently observed during T cell activation in RRMS patients, which was reversed by interferon beta treatment [112]. One of the main signalling pathways that trigger glycolysis during inflammatory immune activations involves the transcriptional factor hypoxia-inducible factor 1α (HIF 1α), which is suppressed by dimethyl fumarate [113,114]. Different types of fatty acids have distinctive effects on immunity. Medium- and long- chain saturated fatty acids (LCFA) promote the differentiation of CD4$^+$ T cells towards Th1 and Th17 cells and pro-inflammatory M1 macrophages, while suppressing the differentiation and function of Treg cells [114]. ω-6 PUFA may induce inflammatory reactivity, whereas ω-3 PUFA suppress innate and adaptive immune reactivity, in line with protective effects in MS [115]. In contrast, short chain saturated fatty acids (SCFA) promote Treg functions and suppress Th17 inflammatory activity (see below). Proliferation of immune cells require the generation of nucleotides, cholesterol and specific fatty acids, lipids and proteins. The glycolytic derived pentose phosphate pathway allows the production of nucleotides and the reduced form of nicotinamide adenine dinucleotide phosphate (NADPH) which is used for fatty acid synthesis and to generate glutathione and other antioxidants. Fatty acid synthesis is especially needed for differentiation and inflammatory functions of M1 macrophage, dendritic cells and effector T cells [114]. Certain sphingolipids, apolipoproteins and amino acids (such as glutamate and tryptophan) also have specific roles in the signalling or transduction mechanisms of immune cells [64,114,116] . In particular, tryptophan derived metabolites of the L-kynurenine pathway, among many other compounds in circulation, regulate inflammation through the aryl hydrocarbon receptor (AHR) in immune cells and astrocytes [117]. RRMS patients have lower serum levels of AHR agonists in comparison to healthy controls. However, increased AHR agonists levels were detected during acute CNS inflammation, probably reflecting an attempt to restrict immune activation [118]. In addition, an abnormal L-kynurenine metabolism was linked to the development of progressive forms of MS [71,119].

Steroid hormones, insulin, leptin and adiponectin all have distinctive modulatory roles on immune cells functions [120]. An impairment of insulin signalling following CD4$^+$ and CD8$^+$ effector T cells activation attenuates the symptoms in EAE [121] and alterations in peripheral insulin sensibility have been described in RRMS patients [88,92,122]. In MS patients who developed metabolic syndrome, metformin treatment decreases serum leptin and increases adiponectin levels. A decreased secretion of pro-inflammatory cytokines by peripheral blood mononuclear cells (PBMC) and increased Treg number and function was also observed after metformin treatment [122]. In EAE, a diet mimicking fasting promotes remyelination and a lower pro-inflammatory state and increases in plasma levels of corticosterone and adiponectin [123]. In contrast to leptin, adiponectin has anti-inflammatory protective roles in MS possibly mediated by ceramide metabolites, such as S1P [116,120]. Leptin-deficient genetically obese mice (ob/ob) present an altered systemic and brain development, including in the amount and fatty acid composition of myelin [124]. Interestingly, although these mice are congenitally resistant to EAE induction, exogenous leptin replacement render these animals susceptible to the disease [120]. These findings support the view that alterations induced in systemic metabolism may have critical roles in promoting the development of MS, even in individuals not genetically susceptible to the disease.

The hypothesis that an "enteropathy" [33] or a defect in the metabolism of "some component of westernized diet" [17] could contribute to the development of MS were discussed many years ago. During the last decade, the evidence for important physiological roles of gut microbiota (GM) in regulating the immune functions has revisited and illuminated those old, intriguing hypotheses. These regulatory effects are mediated through microbiota-dependent alterations in hormonal levels and metabolism of carbohydrates, lipids and amino acids [125–128]. Sex differences in the GM result in altered serum levels of testosterone and of glycerophospholipids and sphingolipids, which could contribute to the increased susceptibility of females to MS [129] . On one hand, GM is required for the induction of autoimmune T and B cells demyelination in EAE [130]. On the other hand, GM produces SCFA (acetate, butyrate, and propionate) from dietary polysaccharides, inducing the generation of Treg cells and protection against the disease [131]. Butyrate was also shown to downregulate innate response receptors in human monocytes [132]. Dietary LCFA enhances differentiation and proliferation of effector T cells and exacerbates the symptoms of EAE, whereas administration of SCFA reduces Th1 and increase regulatory T cells and is protective [133,134]. Alterations in GM composition promoting a pro-inflammatory milieu have been observed in MS patients regardless of immunomodulatory treatments and associated with clinical relapses, suggesting a decrease in fatty acid metabolism and engagements in defence pathways linked to oestrogen signalling and production of bile acid metabolites and glutathione [135,136]. Notably, transplanted GM from MS patients promotes the induction or exacerbates EAE symptoms in association with reduced proportions of interleulin-10 (IL-10) secreting Treg cells [137,138]. High frequency of intestinal Th17 cells in MS patients was correlated with decreased abundance of Prevotella strains and increased activity of the disease [139]. Treatment with interferon beta or glatiramer acetate raises the quantity of Prevotella and other current approved drugs for the disease and vitamin D supplementation could change microbial intestinal composition as well [125–128]. GM may regulate BBB permeability [140], brain myelination [141], microglia and astrocyte activities [117,142]. In humans, dietary ω-3 fatty acids change microbiota composition and induce the production of anti-inflammatory compounds like butyrate [143]. Pilot trials in MS suggest that a ketogenic diet normalizes the mass and diversity of the colonic microbiome [144] and that the modulation of dysbiosis by a high-vegetable/low protein diet is associated with an increase of Treg differentiation and IL-10 production and improvement in clinical courses [145]. Several strategies to manipulate GM are presently being considered for therapeutic proposes in MS [127]. Recent studies suggest that alterations in GM could mediate the possible influences of salt intake on the risk and progression of the disease. A high salt diet affects the composition of mice GM, inducing Th17 cells, the development of EAE, hypertension [146] and cognitive dysfunction [147].

In short, the studies reviewed here and in the sections above indicate that the abnormal immune reactivity in MS can be modulated by genetic and environmental factors shaping the metabolic and hormonal milieu of the organism. These metabolic processes may either result in loss of immune tolerance to self and drive the induction the disease or be protective of its development and progression. In this context, Peroxisome proliferator-activated receptors (PPAR) may comprise important players in MS pathogenesis, as discussed below.

5. Peroxisome Proliferator-Activated Receptors (PPAR) in MS

PPAR are transcriptional factors involved in the regulation of lipid and glucose metabolism, cell differentiation and proliferation. The PPAR subfamily of nuclear receptors comprise the members PPARα (NR1C1), PPARβ/δ (NR1C2) and PPARγ (NRC1C3), which, after ligand activation, regulate gene transcription by dimerizing with the retinoid X receptor and acting in specific DNA sequences. In addition, PPAR can repress gene expression in a DNA-binding independent way by interfering with other transcription factors. PPAR can be activated by synthetic agonists and specific endogenous ligands, mainly PUFA and eicosanoid metabolites [148–150]. All PPAR subtypes are variably expressed in immune cells and have important roles in the control of innate and adaptive immune functions [151–153]. Several studies have shown protective effects of PPAR agonists in EAE [154]. Current data suggest that each PPAR isoform is able to control the development of T cell- mediated autoimmunity in EAE by distinctive mechanisms. PPARα promotes Th2 cells differentiation and cytokines production [155], whereas PPARγ strongly restricts Th17 differentiation [156] and PPARβ/δ expression inhibits the production of interferon-γ (INF-γ)and IL-12 [157]. PPARγ could be a major driver of Treg cell accumulation and functioning [158]. Accordingly, administration of eicosapentaenoic acid in the diet ameliorates the clinical symptoms of EAE. Importantly, although this treatment slightly increased only PPARγ in the periphery, all PPAR isotypes were induced in CNS-infiltrating CD4 T cells [159]. In the CNS, PPAR activation inhibits the NF-κB and Janus kinase-signal transducer and activator of transcription (JAK-STAT) pathways of a complex network involving adaptive immune cells, myeloid cells and astrocytes [151,160]. All subtypes of PPAR may regulate astrocyte Toll-like receptors stimulation in inflammatory responses [161]. PPAR activation is in general inhibitory of innate immune cells activation and foam cell formation, promoting macrophage polarization to an anti-inflammatory phenotype [149,152]. PPARγ activation in the CNS during EAE reduces the production of inflammatory and neurotoxic mediators by macrophages and astrocytes and re-stimulation of infiltrating autoreactive T cells, supporting a major role in the control of disease progression. These findings suggested that endogenous PPARγ ligands, namely induced by IL-4, could inhibit the formation of foam cells and CNS autoimmunity [162]. In EAE animals and active MS lesions, specific activation PPARβ/δ was also able to promote an anti-inflammatory phenotype in macrophages after uptake of myelin lipids and foam cell formation, decreasing immune cell infiltration in the CNS [163].

MS patients present an impairment of oligodendrocyte precursor cells (OPC) differentiation compromising remyelination, which is increasingly deficient with the progression of the disease. Selective PPARβ/δ agonists stimulate oligodendrocyte differentiation [164]. Gemfibrozil, an activator of PPARα often prescribed to lower triglycerides levels, was shown to increase the expression of myelin genes (including myelin basic protein, MBP and myelin oligodendrocyte glycoprotein, MOG) in human oligodendrocytes via PPARβ/δ (not via PPARα or PPARγ) [165]. PPARγ agonists induce astrocyte and prominently OPC differentiation genes of neural stem cells,, though not MBP and MOG [166]. PPARγ activators further promote OPC maturation toward myelin-forming oligodendrocytes, increasing MBP expression and membranes enriched in cholesterol and plasmalogens [167–169]. Promotion of myelinogenesis by PPARγ signalling involves an increase of antioxidant defenses and mitochondrial respiratory activity [168,169]. Importantly, docosahexaenoic acid and others' endogenous and synthetic PPARγ activators may protect against the impairment of oligodendrocyte maturation and myelination associated with inflammatory conditions [170,171]. The mechanisms promoting remyelination in MS by PPAR could involve downregulation of NF-κB/β-catenin and activation of PI3K/Akt pathways [172].

PPAR are expressed in all cells of adult mouse and human brains, PPARβ/δ being the most abundant [173,174]. Curiously, all PPAR isotypes are mainly expressed in neurons. In normal conditions, it was observed that PPARβ/δ is only expressed in neurons, whereas PPARα colocalizes in neurons, astrocytes and microglia. PPARγ colocalizes in neurons and astrocytes, but its expression could be induced in microglia after LPS treatment [174]. These recent observations highlight the importance of interactions in the PPAR triad for the CNS physiology and pathology accordingly with distinctive roles of these three isomorphs in immune function and myelin formation. In the CNS, PPARβ/δ probably has a major integrative role and was suggested as the main target to be considered in therapeutic strategies for neurodegenerative diseases [175]. PPARβ/δ agonist GW0742 promotes neuronal maturation in parallel with increased expression of brain-derived neurotrophic factor (BDNF) and activation of cholesterol synthesis [176]. Nevertheless, selective agonists of other PPAR subtypes also induce neuroprotective effects in vitro and in animal models of neurodegenerative disorders [177]. Recent research supports an essential role of these nuclear receptors in brain development and suggest that individual variability in PPAR signalling may be associated with susceptibility to various neurological insults and diseases [178]. In parallel with anti-inflammatory and remyelination effects, many studies have demonstrated protective effects of PPAR agonists against central players in the neurodegenerative process in MS and other disorders, including mitochondrial dysfunction, calcium dysregulation, glutamate toxicity and oxidative injury [175,177]. For example, PPARγ agonists have protective effects on glutamate excitotoxicity by increasing the expression of the major transporter (GLT-1) of this neurotransmitter in astrocytes [179]. As mentioned above, Nrf2 signalling and glutathione availability are compromised very early in the disease. Deficiency in glutathione induces macrophage CD36 expression at a translational level, independently of PPARγ activation, resulting in enhancement uptake of oxidized LDL and foam formation [180]. PPARγ and Nrf2 are linked by a positive feedback loop that sustains the expression of both transcriptional factors and antioxidant defences [181]. PPARγ is a critical mediator on the effects of galangin (a polyphenolic abundant in honey and certain vegetables) in decreasing NF-κB activation and increasing Nrf2 signalling in LPS-stimulated microglia [182]. PPARγ agonists have protective effects on mouse cochlea against gentamicin toxicity through anti-apoptotic effects and upregulation of glutathione and other antioxidant defences [183]. Interestingly, this work revealed some protective effects of fenofibric acid (a PPARα-specific agonist) by affecting different antioxidant mechanisms. Decreased neuronal concentrations of glutathione stimulates the generation of 12/15-lipoxygenase metabolites and the inhibition of this pathway by baicalein (a type of flavonoid) induces the expression of PPARβ/δ in microglia and attenuates inflammation in EAE [184]. Recent studies suggest protective effects of α-lipoic acid (an endogenous antioxidant) therapy on the neurodegenerative process of EAE and MS [185], which could be mediated by PPARγ activity [186]. In brief, a huge body of studies in vitro and in animal models indicate that all PPAR subtypes could be involved in the control of immune activity and mechanisms promoting myelination and neuronal protection in MS by affecting different but complementary cellular and molecular pathways.

The Heneka group was the first to investigate PPAR signalling in MS patients. In one study, it was shown that peripheral blood mononuclear cells (PBMC) from these patients expressed decreased PPARγ levels correlated with disease activity and preceding the development of clinical relapses. Moreover, pre-treatment with the PPARγ agonist pioglitazone was able to increase PPARγ DNA-binding and decrease NF-κB DNA-binding activity in patients in a stable stage of the disease. These results indicate a suppression of PPARγ by inflammatory stimuli and that PPARγ agonists are protective through increased expression of their own receptors [187]. Preliminary results suggest an influence of PPARγ gene polymorphism in modulating the onset of MS [188], and beneficial effects of pioglitazone treatment in reducing the development of brain lesions and grey matter atrophy and clinical progression of the disease [189–191]. A pronounced elevation of PPARγ levels was observed in the CSF of MS patients free of therapy, which were correlated with intrathecal inflammatory parameters and clinical disability [192]. In accordance with the results reviewed above, these findings could reflect an attempt of PPARγ-mediated processes of brain cells in restricting immune activity. A crucial and very early event in MS pathogenesis concerns the abnormal leucocyte transmigration to the CNS across the BBB, a process promoted by inflammatory activation of endothelium and involving integrin β1-mediated mechanisms. PPARγ was shown to inhibit leucocyte migration across activated brain endothelial cells [193]. Interestingly, ω-3 PUFA increases PPARγ expression and decreases the expression of integrin β1 in glomerular mesangial cells treated with LPS [194]. PPARγ is a well-known inductor of CD36 receptor expression. This scavenger receptor requires interactions with integrins for many of its functions [195], and may regulate BBB permeability [196]. Natalizumab (NTZ) treatment blocks α4/β1 integrin resulting in peripheral sequestration of activated T cells and increased production of proinflammatory cytokines. In female patients, we have found an increase of PPARβ/δ mRNA and a decrease of PPARγ and CD36 mRNA expression in PBMC at three months of NTZ therapy. In contrast, PPARα was unchanged [197]. Interestingly, inflammatory activity of monocyte derived human macrophages is associated with lower PPARγ/CD36 and higher PPARβ/δ expression levels [198]. In addition, PPARβ/δ agonists may stimulate pathways enhancing immune reactivity under hypoxic stress, suggesting context-dependent functions [199]. These selective responses support the concept that PPAR subtypes may have reciprocal or complementary roles in immune regulation. Fingolimod (FTY720) is an S1P mimetic that down-modulates S1P signalling, retaining central memory T cells and B lymphocytes in lymph nodes and preventing autoreactive cell infiltration in the CNS [200]. In contrast to NTZ, fingolimod treatment decreases the number of reactive lymphocytes in circulation. Preliminary results from our group have found that, at six months of therapy, fingolimod augmented PPARγ and CD36 mRNA expression in total blood leucocytes from female patients, whereas PPARβ/δ and PPARα were unaffected. Importantly, this selective alteration in PPARγ/CD36 mRNA expression was associated with a significant increase of plasma HDL levels and decrease of total cholesterol/HDL ratio [201]. As mentioned above, S1P is mainly associated with HDL particles and is involved in the control of BBB dysfunction and brain lesions in MS. Therefore, the diminished systemic immune reactivity induced by fingolimod could be related to protective effects of HDL. In fact, the antioxidant and anti-inflammatory proprieties of HDL are thought to be regulated, at least in part, by upregulation of PPARγ/CD36 pathway [198,202] As already mentioned, MS patients display an adverse lipoprotein metabolism that may be associated with a dysfunctional pro-inflammatory HDL particle affecting PPARγ/CD36/Nrf2 signalling [203]. Further work by our group is now on course to evaluate whether upregulation of PPARγ/CD36 pathway is associated with the clinical benefits of fingolimod treatment in the disease. The fact that PPARα expression was unchanged in female patients treated with

NTZ and fingolimod is of considerable interest. The Steinman group has shown that PPARα is expressed at higher levels in males in comparison to female naïve T cells through androgen-mediated effects. Higher PPARα expression in males was associated with protection to developing EAE [155]. In contrast, 17β-estradiol increases PPARγ expression, supporting evidence for an increased efficacy of pioglitazone in females [204]. However, the neuroprotective effects of the phytoestrogen daidzein were shown to be due to an increase of PPARγ activity not mediated by receptor binding and not additive with rosiglitazone [205]. PPARγ signalling has more wide and profound anti-inflammatory effects in females, selectively reducing only Th17-cell differentiation in males [206]. These data indicate gender differences associated with the roles of PPAR in the control of immune functions and neurological homeostasis [207], possibly contributing to the modulatory effects of sex steroids in susceptibility, genesis/repair of CNS lesions, and clinical courses of MS [66,208,209].

PPAR-mediated processes could be implicated in other pathogenic pathways suggested for the disease as mentioned above. Vitamin D and PPARγ may interact in the mechanisms leading to differentiation of immune cells restricting pro-inflammatory activity [210]. In MS patients also developing metabolic syndrome, pioglitazone treatment decreases leptin and increases adiponectin serum levels in association with reduced secretion of pro-inflammatory cytokines by PBMC [122]. As quoted above, oxysterols' uptake by macrophages and LXR activation is probably an important driver in repair mechanisms of MS lesions. PPAR and LXR are well known to cooperate in suppressing innate and adaptive reactivity and cell foam formation [149,152,153]. However, LXR activation may affect PPARγ-dependent expression of adiponectin, promoting inflammation and insulin resistance [211]. Therefore, LXR activation is probably not a good therapeutic approach for MS. PPAR activation may induce LXR transcription and both PPARγ and LXR agonists upregulate ApoE expression. Interestingly, PPARγ-mediated neuroprotective effects of daidzein were shown to be critically dependent of ApoE induction [212]. Finally, the increasing evidence for an involvement of gut dysbiosis in MS further supports a potential role of PPAR in its pathogenesis. The anti-inflammatory effects of microbiota derived SCFA (butyrate and propionate) or α-linolenic acid are mediated through PPARγ activation [213,214]. Microbiota-activated PPARγ signalling prevents gut dysbiotic expansion [215], and a high fat diet promotes intestinal dysbiosis, at least in part by interfering with PPARγ activation [216]. PPARα deficiency also results in microbiota-dependent increase in intestinal inflammation [217]. Intestinal dysbiosis may affect the generation of tryptophan-derived ligands of AHR in immune cells and astrocytes [118]. Butyrate and propionate are able to control AHR gene expression and activity [213]. Therefore, PPAR- dependent pathways may be involved in the regulation of systemic and brain inflammatory reactivity by AHR ligands. The main pathophysiological pathways in MS potentially modulated by PPAR-mediated mechanisms are summarized in Figure 1.

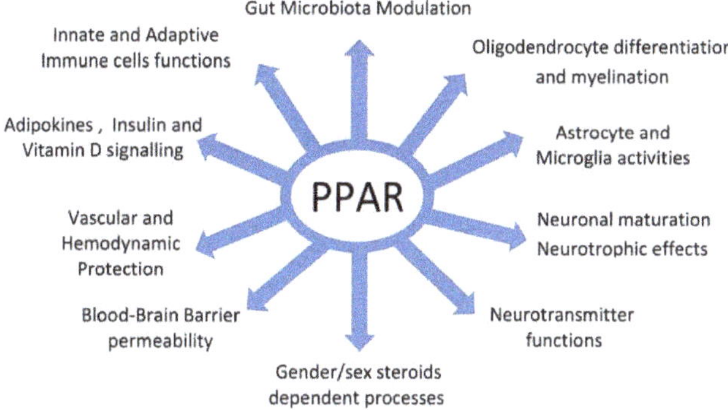

Figure 1. Pathophysiological pathways in multiple sclerosis potentially modulated by Peroxisome Proliferator-Activated Receptors (PPAR)-mediated metabolic processes.

In 2011, Angelique Corthals advocated a major role of lipid dysregulation in the pathogenesis of MS, possibly involving PPAR signalling [218]. The present review emphasized that the last decade of research has revisited this topic with new insights, with the very old evidence for the importance of metabolic mechanisms in MS. Nevertheless, these advances do not allow for concluding that … "multiple sclerosis is not a disease of the immune system" [218]. The development and progression of MS only occurs in a framework of immune dysfunction. If PPAR are involved in the mechanisms of disease, it will be mainly due to the fact, that, among other roles, they are important modulators of the innate and adaptive immune functions. Notably, metabolic pathways modulated by PPAR could integrate the concomitant development of many systemic and neurological pathological processes in the disease. In addition, this scenario is coherent with much clinical and epidemiological data. However, most evidence to date is based on studies in vitro and in animal models of MS and much more research needs to be performed in human patients. As MS is pathological and clinically heterogeneous, and is associated with multiple potential genetic and environmental risk factors, PPAR-dependent processes may not necessarily be implicated in every individual and type/stage of the disease. As in other areas of human health, much work in the field is presently aimed to search for useful biomarkers for personalized preventive and treatment strategies of the disease, the so-called "precision medicine" [3,5,128]. In this context, PPAR deserve to be explored as potential targets to monitor the disease and to discover improved and individualized therapeutic measures. In a real-world setting, clinical evidence indicates that even the most powerful current strategies to modulate or suppress immune functions are of doubtful benefit or not tolerated in some patients, and may carry the risk of serious adverse events. Furthermore, very probably, they will never be sufficient to delay the evolution of progressive forms and the neurodegenerative process of the disease. In the future, much effort is still needed to trace the metabolic roads that may light and nourish "the fire" of immune activation in MS. Perhaps we should recognize that … "we have mistaken the smoke for the fire" [17].

6. Conclusions

During the last few decades, much progress has been made in clarifying the role of metabolic mechanisms in the pathogenesis of MS. In this context, PPAR-mediated processes have emerged as potential players able to integrate many systemic and CNS events occurring in the disease. These nuclear receptors could be implicated in pathophysiological pathways involving lipid, glucose, amino acid and antioxidant metabolism and the regulatory roles of insulin, adipokines, steroid hormones and gut microbiota. Importantly, metabolic roads modulated by PPAR seem to provide strategies to control the immune dysfunction and the genesis of CNS lesions of MS, and to promote neuroprotection and myelin formation. PPAR seem to constitute research targets that may lead to further therapeutic and preventive measures for this disorder.

Author Contributions: V.F.S., C.C. and A.S. contribute equally to this work.

Acknowledgments: The author´s research work reported in this manuscript was supported by Searle Pharma (LovMS-2000); Bayer (Pillbeta-2003; Lipesbeta-2005); Biogen (Reconvert-2000; Natalippar-2012); Teva (Litericop-2004); Merck (PPAR-2007) and Novartis (Fingoppar-2013).

Conflicts of Interest: The authors declare no conflict of interest.

References

1. Lublin, F.D.; Reingold, S.C.; Cohen, J.A.; Cutter, G.R.; Sørensen, P.S.; Thompson, A.J.; Wolinsky, J.S.; Balcer, L.J.; Banwell, B.; Barkhof, F.; et al. Defining the clinical course of multiple sclerosis: The 2013 revisions. *Neurology* **2014**, *83*, 278–286. [CrossRef] [PubMed]
2. Mishra, M.K.; Yong, V.W. Myeloid cells—Targets of medication in multiple sclerosis. *Nat. Rev. Neurol.* **2016**, *12*, 539–551. [CrossRef] [PubMed]
3. Baecher-Allan, C.; Kaskow, B.J.; Weiner, H.L. Multiple Sclerosis: Mechanisms and Immunotherapy. *Neuron* **2018**, *97*, 742–768. [CrossRef] [PubMed]

4. Metz, I.; Weigand, S.D.; Popescu, B.F.G.; Frischer, J.M.; Parisi, J.E.; Guo, Y.; Lassmann, H.; Brück, W.; Lucchinetti, C.F. Pathologic heterogeneity persists in early active multiple sclerosis lesions. *Ann. Neurol.* **2014**, *75*, 728–738. [CrossRef] [PubMed]
5. Reich, D.S.; Lucchinetti, C.F.; Calabresi, P.A. Multiple Sclerosis. *N. Engl. J. Med.* **2018**, *378*, 169–180. [CrossRef] [PubMed]
6. Steinman, L.; Zamvil, S.S. Virtues and pitfalls of EAE for the development of therapies for multiple sclerosis. *Trends Immunol.* **2005**, *26*, 565–571. [CrossRef] [PubMed]
7. Schuh, C.; Wimmer, I.; Hametner, S.; Haider, L.; Van Dam, A.-M.; Liblau, R.S.; Smith, K.J.; Probert, L.; Binder, C.J.; Bauer, J.; et al. Oxidative tissue injury in multiple sclerosis is only partly reflected in experimental disease models. *Acta Neuropathol.* **2014**, *128*, 247–266. [CrossRef] [PubMed]
8. Schrewe, L.; Lill, C.M.; Liu, T.; Salmen, A.; Gerdes, L.A.; Guillot-Noel, L.; Akkad, D.A.; Blaschke, P.; Graetz, C.; Hoffjan, S.; et al. Investigation of sex-specific effects of apolipoprotein E on severity of EAE and MS. *J. Neuroinflamm.* **2015**, *12*, 234. [CrossRef] [PubMed]
9. Giovannoni, G.; Cohen, J.A.; Coles, A.J.; Hartung, H.-P.; Havrdova, E.; Selmaj, K.W.; Margolin, D.H.; Lake, S.L.; Kaup, S.M.; Panzara, M.A.; et al. CARE-MS II Investigators Alemtuzumab improves preexisting disability in active relapsing-remitting MS patients. *Neurology* **2016**, *87*, 1985–1992. [CrossRef] [PubMed]
10. Lorscheider, J.; Jokubaitis, V.G.; Spelman, T.; Izquierdo, G.; Lugaresi, A.; Havrdova, E.; Horakova, D.; Trojano, M.; Duquette, P.; Girard, M.; et al. MSBase Study Group Anti-inflammatory disease-modifying treatment and short-term disability progression in SPMS. *Neurology* **2017**, *89*, 1050–1059. [CrossRef] [PubMed]
11. Westad, A.; Venugopal, A.; Snyder, E. The multiple sclerosis market. *Nat. Rev. Drug Discov.* **2017**, *16*, 675–676. [CrossRef] [PubMed]
12. Confavreux, C.; Vukusic, S. Natural history of multiple sclerosis: A unifying concept. *Brain* **2006**, *129*, 606–616. [CrossRef] [PubMed]
13. Weiner, H.L. The challenge of multiple sclerosis: How do we cure a chronic heterogeneous disease? *Ann. Neurol.* **2009**, *65*, 239–248. [CrossRef] [PubMed]
14. Trapp, B.D.; Nave, K.-A. Multiple Sclerosis: An Immune or Neurodegenerative Disorder? *Annu. Rev. Neurosci.* **2008**, *31*, 247–269. [CrossRef] [PubMed]
15. Calabrese, M.; Magliozzi, R.; Ciccarelli, O.; Geurts, J.J.G.; Reynolds, R.; Martin, R. Exploring the origins of grey matter damage in multiple sclerosis. *Nat. Rev. Neurosci.* **2015**, *16*, 147–158. [CrossRef] [PubMed]
16. Hossain, M.J.; Tanasescu, R.; Gran, B. Innate immune regulation of autoimmunity in multiple sclerosis: Focus on the role of Toll-like receptor 2. *J. Neuroimmunol.* **2017**, *304*, 11–20. [CrossRef] [PubMed]
17. Wolfgram, F. What if multiple sclerosis isn't an immunological or a viral disease? The case for a circulating toxin. *Neurochem. Res.* **1979**, *4*, 1–14. [CrossRef] [PubMed]
18. Mastronardi, F.G.; Moscarello, M.A. Molecules affecting myelin stability: A novel hypothesis regarding the pathogenesis of multiple sclerosis. *J. Neurosci. Res.* **2005**, *80*, 301–308. [CrossRef] [PubMed]
19. Barnett, M.H.; Henderson, A.P.D.; Prineas, J.W. The macrophage in MS: Just a scavenger after all? Pathology and pathogenesis of the acute MS lesion. *Mult. Scler.* **2006**, *12*, 121–132. [CrossRef] [PubMed]
20. Stys, P.K.; Zamponi, G.W.; Van Minnen, J.; Geurts, J.J.G. Will the real multiple sclerosis please stand up? *Nat. Rev. Neurosci.* **2012**, *13*, 507–514. [CrossRef] [PubMed]
21. Qureshi, I.A.; Mehler, M.F. Towards a "systems"-level understanding of the nervous system and its disorders. *Trends Neurosci.* **2013**, *36*, 674–684. [CrossRef] [PubMed]
22. McMurran, C.E.; Jones, C.A.; Fitzgerald, D.C.; Franklin, R.J.M. CNS Remyelination and the Innate Immune System. *Front. Cell Dev. Biol.* **2016**, *4*, 38. [CrossRef] [PubMed]
23. Dombrowski, Y.; O'Hagan, T.; Dittmer, M.; Penalva, R.; Mayoral, S.R.; Bankhead, P.; Fleville, S.; Eleftheriadis, G.; Zhao, C.; Naughton, M.; et al. Regulatory T cells promote myelin regeneration in the central nervous system. *Nat. Neurosci.* **2017**, *20*, 674–680. [CrossRef] [PubMed]
24. Wang, Y.; Bos, S.D.; Harbo, H.F.; Thompson, W.K.; Schork, A.J.; Bettella, F.; Witoelar, A.; Lie, B.A.; Li, W.; McEvoy, L.K.; et al. Genetic overlap between multiple sclerosis and several cardiovascular disease risk factors. *Mult. Scler. J.* **2016**, *22*, 1783–1793. [CrossRef] [PubMed]
25. Mokry, L.E.; Ross, S.; Timpson, N.J.; Sawcer, S.; Davey Smith, G.; Richards, J.B. Obesity and Multiple Sclerosis: A Mendelian Randomization Study. *PLoS Med.* **2016**, *13*, e1002053. [CrossRef] [PubMed]
26. O'Gorman, C.; Lucas, R.; Taylor, B. Environmental Risk Factors for Multiple Sclerosis: A Review with a Focus on Molecular Mechanisms. *Int. J. Mol. Sci.* **2012**, *13*, 11718–11752. [CrossRef] [PubMed]

27. Olsson, T.; Barcellos, L.F.; Alfredsson, L. Interactions between genetic, lifestyle and environmental risk factors for multiple sclerosis. *Nat. Rev. Neurol.* **2016**, *13*, 26–36. [CrossRef] [PubMed]
28. Sena, A.; Couderc, R.; Ferret-Sena, V.; Pedrosa, R.; Andrade, M.L.; Araujo, C.; Roque, R.; Cascais, M.J.; Morais, M.G. Apolipoprotein E polymorphism interacts with cigarette smoking in progression of multiple sclerosis. *Eur. J. Neurol.* **2009**, *16*, 832–837. [CrossRef] [PubMed]
29. Munger, K.L.; Fitzgerald, K.C.; Freedman, M.S.; Hartung, H.-P.; Miller, D.H.; Montalbán, X.; Edan, G.; Barkhof, F.; Suarez, G.; Radue, E.-W.; et al. No association of multiple sclerosis activity and progression with EBV or tobacco use in BENEFIT. *Neurology* **2015**, *85*, 1694–1701. [CrossRef] [PubMed]
30. Cortese, M.; Yuan, C.; Chitnis, T.; Ascherio, A.; Munger, K.L. No association between dietary sodium intake and the risk of multiple sclerosis. *Neurology* **2017**, *89*, 1322–1329. [CrossRef] [PubMed]
31. Mokry, L.E.; Ross, S.; Ahmad, O.S.; Forgetta, V.; Smith, G.D.; Leong, A.; Greenwood, C.M.T.; Thanassoulis, G.; Richards, J.B. Vitamin D and Risk of Multiple Sclerosis: A Mendelian Randomization Study. *PLoS Med.* **2015**, *12*, e1001866. [CrossRef] [PubMed]
32. Goodin, D.S. The nature of genetic susceptibility to multiple sclerosis: Constraining the possibilities. *BMC Neurol.* **2016**, *16*, 56. [CrossRef] [PubMed]
33. Thompson, R.H. A biochemical approach to the problem of multiple sclerosis. *Proc. R. Soc. Med.* **1966**, *59*, 269–276. [PubMed]
34. Goldberg, P. Multiple sclerosis: Vitamin D and calcium as environmental determinants of prevalence. (A Viewpoint) Part I: Sunlight, Dietary Factors and Epidemiology. *Intern. J. Environ. Stud.* **1974**, *6*, 19–27. [CrossRef]
35. Goldberg, P. Multiple sclerosis: Vitamin D and calcium as environmental determinants of prevalence. (A viewpoint). Part II. Biochemical and genetic factores. *Intern. J. Environ. Stud.* **1974**, *6*, 121–129. [CrossRef]
36. Swank, R.L. Multiple sclerosis: Twenty years on low fat diet. *Arch. Neurol.* **1970**, *23*, 460–474. [CrossRef] [PubMed]
37. Alter, M.; Yamoor, M.; Harshe, M. Multiple sclerosis and nutrition. *Arch. Neurol.* **1974**, *31*, 267–272. [CrossRef] [PubMed]
38. Agranoff, B.W.; Goldberg, D. Diet and the geographical distribution of multiple sclerosis. *Lancet* **1974**, *2*, 1061–1066. [CrossRef]
39. Jelinek, G.A.; De Livera, A.M.; Marck, C.H.; Brown, C.R.; Neate, S.L.; Taylor, K.L.; Weiland, T.J. Associations of Lifestyle, Medication, and Socio-Demographic Factors with Disability in People with Multiple Sclerosis: An International Cross-Sectional Study. *PLoS ONE* **2016**, *11*, e0161701. [CrossRef] [PubMed]
40. Bjørnevik, K.; Chitnis, T.; Ascherio, A.; Munger, K.L. Polyunsaturated fatty acids and the risk of multiple sclerosis. *Mult. Scler. J.* **2017**, *23*, 1830–1838. [CrossRef] [PubMed]
41. Azary, S.; Schreiner, T.; Graves, J.; Waldman, A.; Belman, A.; Guttman, B.W.; Aaen, G.; Tillema, J.-M.; Mar, S.; Hart, J.; et al. Contribution of dietary intake to relapse rate in early paediatric multiple sclerosis. *J. Neurol. Neurosurg. Psychiatry* **2018**, *89*, 28–33. [CrossRef] [PubMed]
42. Bäärnhielm, M.; Olsson, T.; Alfredsson, L. Fatty fish intake is associated with decreased occurrence of multiple sclerosis. *Mult. Scler. J.* **2014**, *20*, 726–732. [CrossRef] [PubMed]
43. Fitzgerald, K.C.; Tyry, T.; Salter, A.; Cofield, S.S.; Cutter, G.; Fox, R.; Marrie, R.A. Diet quality is associated with disability and symptom severity in multiple sclerosis. *Neurology* **2018**, *90*, e1–e11. [CrossRef] [PubMed]
44. Shore, V.G.; Smith, M.E.; Perret, V.; Laskaris, M.A. Alterations in plasma lipoproteins and apolipoproteins in experimental allergic encephalomyelitis. *J. Lipid Res.* **1987**, *28*, 119–129. [PubMed]
45. Newcombe, J.; Li, H.; Cuzner, M.L. Low density lipoprotein uptake by macrophages in multiple sclerosis plaques: Implications for pathogenesis. *Neuropathol. Appl. Neurobiol.* **1994**, *20*, 152–162. [CrossRef] [PubMed]
46. Haider, L.; Fischer, M.T.; Frischer, J.M.; Bauer, J.; Hoftberger, R.; Botond, G.; Esterbauer, H.; Binder, C.J.; Witztum, J.L.; Lassmann, H. Oxidative damage in multiple sclerosis lesions. *Brain* **2011**, *134*, 1914–1924. [CrossRef] [PubMed]
47. Boven, L.A.; Van Meurs, M.; Van Zwam, M.; Wierenga-Wolf, A.; Hintzen, R.Q.; Boot, R.G.; Aerts, J.M.; Amor, S.; Nieuwenhuis, E.E.; Laman, J.D. Myelin-laden macrophages are anti-inflammatory, consistent with foam cells in multiple sclerosis. *Brain* **2006**, *129*, 517–526. [CrossRef] [PubMed]
48. Yepes, M.; Sandkvist, M.; Moore, E.G.; Bugge, T.H.; Strickland, D.K.; Lawrence, D.A. Tissue-type plasminogen activator induces opening of the blood-brain barrier via the LDL receptor–related protein. *J. Clin. Investig.* **2003**, *112*, 1533–1540. [CrossRef] [PubMed]

49. Gaultier, A.; Wu, X.; Le Moan, N.; Takimoto, S.; Mukandala, G.; Akassoglou, K.; Campana, W.M.; Gonias, S.L. Low-density lipoprotein receptor-related protein 1 is an essential receptor for myelin phagocytosis. *J. Cell Sci.* **2009**, *122*, 1155–1162. [CrossRef] [PubMed]
50. Podbielska, M.; Hogan, E. Molecular and immunogenic features of myelin lipids: Incitants or modulators of multiple sclerosis? *Mult. Scler. J.* **2009**, *15*, 1011–1029. [CrossRef] [PubMed]
51. Van Doorn, R.; Nijland, P.G.; Dekker, N.; Witte, M.E.; Lopes-Pinheiro, M.A.; van het Hof, B.; Kooij, G.; Reijerkerk, A.; Dijkstra, C.; van van der Valk, P.; et al. Fingolimod attenuates ceramide-induced blood–brain barrier dysfunction in multiple sclerosis by targeting reactive astrocytes. *Acta Neuropathol.* **2012**, *124*, 397–410. [CrossRef] [PubMed]
52. Checa, A.; Khademi, M.; Sar, D.G.; Haeggström, J.Z.; Lundberg, J.O.; Piehl, F.; Olsson, T.; Wheelock, C.E. Hexosylceramides as intrathecal markers of worsening disability in multiple sclerosis. *Mult. Scler.* **2015**, *21*, 1271–1279. [CrossRef] [PubMed]
53. Spampinato, S.F.; Obermeier, B.; Cotleur, A.; Love, A.; Takeshita, Y.; Sano, Y.; Kanda, T.; Ransohoff, R.M. Sphingosine 1 Phosphate at the Blood Brain Barrier: Can the Modulation of S1P Receptor 1 Influence the Response of Endothelial Cells and Astrocytes to Inflammatory Stimuli? *PLoS ONE* **2015**, *10*, e0133392. [CrossRef] [PubMed]
54. Sato, K.; Malchinkhuu, E.; Horiuchi, Y.; Mogi, C.; Tomura, H.; Tosaka, M.; Yoshimoto, Y.; Kuwabara, A.; Okajima, F. HDL-like lipoproteins in cerebrospinal fluid affect neural cell activity through lipoprotein-associated sphingosine 1-phosphate. *Biochem. Biophys. Res. Commun.* **2007**, *359*, 649–654. [CrossRef] [PubMed]
55. Kułakowska, A.; Żendzian-Piotrowska, M.; Baranowski, M.; Konończuk, T.; Drozdowski, W.; Górski, J.; Bucki, R. Intrathecal increase of sphingosine 1-phosphate at early stage multiple sclerosis. *Neurosci. Lett.* **2010**, *477*, 149–152. [CrossRef] [PubMed]
56. Villar, L.M.; Sádaba, M.C.; Roldán, E.; Masjuan, J.; González-Porqué, P.; Villarrubia, N.; Espiño, M.; García-Trujillo, J.A.; Bootello, A.; Alvarez-Cermeño, J.C. Intrathecal synthesis of oligoclonal IgM against myelin lipids predicts an aggressive disease course in MS. *J. Clin. Investig.* **2005**, *115*, 187–194. [CrossRef] [PubMed]
57. Ho, P.P.; Kanter, J.L.; Johnson, A.M.; Srinagesh, H.K.; Chang, E.-J.; Purdy, T.M.; van Haren, K.; Wikoff, W.R.; Kind, T.; et al. Identification of naturally occurring fatty acids of the myelin sheath that resolve neuroinflammation. *Sci. Transl. Med.* **2012**, *4*, 137ra73. [CrossRef] [PubMed]
58. Mycko, M.P.; Sliwinska, B.; Cichalewska, M.; Cwiklinska, H.; Raine, C.S.; Selmaj, K.W. Brain glycolipids suppress T helper cells and inhibit autoimmune demyelination. *J. Neurosci.* **2014**, *34*, 8646–8658. [CrossRef] [PubMed]
59. Saher, G.; Stumpf, S.K. Cholesterol in myelin biogenesis and hypomyelinating disorders. *Biochim. Biophys. Acta* **2015**, *1851*, 1083–1094. [CrossRef] [PubMed]
60. Itoh, N.; Itoh, Y.; Tassoni, A.; Ren, E.; Kaito, M.; Ohno, A.; Ao, Y.; Farkhondeh, V.; Johnsonbaugh, H.; Burda, J.; et al. Cell-specific and region-specific transcriptomics in the multiple sclerosis model: Focus on astrocytes. *Proc. Natl. Acad. Sci. USA* **2017**, *115*, E302–E309. [CrossRef] [PubMed]
61. Lavrnja, I.; Smiljanic, K.; Savic, D.; Mladenovic-Djordjevic, A.; Tesovic, K.; Kanazir, S.; Pekovic, S. Expression profiles of cholesterol metabolism-related genes are altered during development of experimental autoimmune encephalomyelitis in the rat spinal cord. *Sci. Rep.* **2017**, *7*, 2702. [CrossRef] [PubMed]
62. Baitsch, D.; Bock, H.H.; Engel, T.; Telgmann, R.; Muller-Tidow, C.; Varga, G.; Bot, M.; Herz, J.; Robenek, H.; von Eckardstein, A.; et al. Apolipoprotein E Induces Antiinflammatory Phenotype in Macrophages. *Arterioscler. Thromb. Vasc. Biol.* **2011**, *31*, 1160–1168. [CrossRef] [PubMed]
63. Mailleux, J.; Vanmierlo, T.; Bogie, J.F.; Wouters, E.; Lütjohann, D.; Hendriks, J.J.; van Horssen, J. Active liver X receptor signaling in phagocytes in multiple sclerosis lesions. *Mult. Scler. J.* **2018**, *24*, 279–289. [CrossRef] [PubMed]
64. Howie, D.; Bokum, A. Ten; Necula, A.S.; Cobbold, S.P.; Waldmann, H. The role of lipid metabolism in T lymphocyte differentiation and survival. *Front. Immunol.* **2018**, *8*, 1949. [CrossRef] [PubMed]
65. Mailleux, J.; Timmermans, S.; Nelissen, K.; Vanmol, J.; Vanmierlo, T.; van Horssen, J.; Bogie, J.F.J.; Hendriks, J.J.A. Low-Density Lipoprotein Receptor Deficiency Attenuates Neuroinflammation through the Induction of Apolipoprotein E. *Front. Immunol.* **2017**, *8*, 1701. [CrossRef] [PubMed]

66. Luchetti, S.; van Eden, C.G.; Schuurman, K.; van Strien, M.E.; Swaab, D.F.; Huitinga, I. Gender Differences in Multiple Sclerosis. *J. Neuropathol. Exp. Neurol.* **2014**, *73*, 123–135. [CrossRef] [PubMed]
67. Wheeler, D.; Bandaru, V.V.R.; Calabresi, P.A.; Nath, A.; Haughey, N.J. A defect of sphingolipid metabolism modifies the properties of normal appearing white matter in multiple sclerosis. *Brain* **2008**, *131*, 3092–3102. [CrossRef] [PubMed]
68. Vidaurre, O.G.; Haines, J.D.; Katz Sand, I.; Adula, K.P.; Huynh, J.L.; McGraw, C.A.; Zhang, F.; Varghese, M.; Sotirchos, E.; Bhargava, P.; et al. Cerebrospinal fluid ceramides from patients with multiple sclerosis impair neuronal bioenergetics. *Brain* **2014**, *137*, 2271–2286. [CrossRef] [PubMed]
69. Friese, M.A.; Schattling, B.; Fugger, L. Mechanisms of neurodegeneration and axonal dysfunction in multiple sclerosis. *Nat. Rev. Neurol.* **2014**, *10*, 225–238. [CrossRef] [PubMed]
70. Mahad, D.H.; Trapp, B.D.; Lassmann, H. Pathological mechanisms in progressive multiple sclerosis. *Lancet Neurol.* **2015**, *14*, 183–193. [CrossRef]
71. Rajda, C.; Pukoli, D.; Bende, Z.; Majláth, Z.; Vécsei, L. Excitotoxins, Mitochondrial and Redox Disturbances in Multiple Sclerosis. *Int. J. Mol. Sci.* **2017**, *18*, 353. [CrossRef] [PubMed]
72. De Stefano, N.; Matthews, P.M.; Antel, J.P.; Preul, M.; Francis, G.; Arnold, D.L. Chemical pathology of acute demyelinating lesions and its correlation with disability. *Ann. Neurol.* **1995**, *38*, 901–909. [CrossRef] [PubMed]
73. Paling, D.; Golay, X.; Wheeler-Kingshott, C.; Kapoor, R.; Miller, D. Energy failure in multiple sclerosis and its investigation using MR techniques. *J. Neurol.* **2011**, *258*, 2113–2127. [CrossRef] [PubMed]
74. Lam, M.A.; Maghzal, G.J.; Khademi, M.; Piehl, F.; Ratzer, R.; Romme Christensen, J.; Sellebjerg, F.T.; Olsson, T.; Stocker, R. Absence of systemic oxidative stress and increased CSF prostaglandin $F_{2\alpha}$ in progressive MS. *Neurol. Neuroimmunol. Neuroinflamm.* **2016**, *3*, e256. [CrossRef] [PubMed]
75. Van Horssen, J.; Drexhage, J.A.R.; Flor, T.; Gerritsen, W.; van der Valk, P.; de Vries, H.E. Nrf2 and DJ1 are consistently upregulated in inflammatory multiple sclerosis lesions. *Free Radic. Biol. Med.* **2010**, *49*, 1283–1289. [CrossRef] [PubMed]
76. Morales Pantoja, I.E.; Hu, C.; Perrone-Bizzozero, N.I.; Zheng, J.; Bizzozero, O.A. Nrf2-dysregulation correlates with reduced synthesis and low glutathione levels in experimental autoimmune encephalomyelitis. *J. Neurochem.* **2016**, *139*, 640–650. [CrossRef] [PubMed]
77. Choi, I.-Y.; Lee, P.; Adany, P.; Hughes, A.J.; Belliston, S.; Denney, D.R.; Lynch, S.G. In vivo evidence of oxidative stress in brains of patients with progressive multiple sclerosis. *Mult. Scler. J.* **2017**, 135245851771156. [CrossRef] [PubMed]
78. Habas, A.; Hahn, J.; Wang, X.; Margeta, M. Neuronal activity regulates astrocytic Nrf2 signaling. *Proc. Natl. Acad. Sci. USA* **2013**, *110*, 18291–18296. [CrossRef] [PubMed]
79. Van de Kraats, C.; Killestein, J.; Popescu, V.; Rijkers, E.; Vrenken, H.; Lütjohann, D.; Barkhof, F.; Polman, C.H.; Teunissen, C.E. Oxysterols and cholesterol precursors correlate to magnetic resonance imaging measures of neurodegeneration in multiple sclerosis. *Mult. Scler.* **2014**, *20*, 412–417. [CrossRef] [PubMed]
80. Friese, M.A. Widespread synaptic loss in multiple sclerosis. *Brain* **2016**, *139*, 2–4. [CrossRef] [PubMed]
81. Jürgens, T.; Jafari, M.; Kreutzfeldt, M.; Bahn, E.; Brück, W.; Kerschensteiner, M.; Merkler, D. Reconstruction of single cortical projection neurons reveals primary spine loss in multiple sclerosis. *Brain* **2016**, *139*, 39–46. [CrossRef] [PubMed]
82. Michailidou, I.; Naessens, D.M.P.; Hametner, S.; Guldenaar, W.; Kooi, E.-J.; Geurts, J.J.G.; Baas, F.; Lassmann, H.; Ramaglia, V. Complement C3 on microglial clusters in multiple sclerosis occur in chronic but not acute disease: Implication for disease pathogenesis. *Glia* **2017**, *65*, 264–277. [CrossRef] [PubMed]
83. Liddelow, S.A.; Guttenplan, K.A.; Clarke, L.E.; Bennett, F.C.; Bohlen, C.J.; Schirmer, L.; Bennett, M.L.; Münch, A.E.; Chung, W.-S.; Peterson, T.C.; et al. Neurotoxic reactive astrocytes are induced by activated microglia. *Nature* **2017**, *541*, 481–487. [CrossRef] [PubMed]
84. Mayo, L.; Trauger, S.A.; Blain, M.; Nadeau, M.; Patel, B.; Alvarez, J.I.; Mascanfroni, I.D.; Yeste, A.; Kivisäkk, P.; Kallas, K.; et al. Regulation of astrocyte activation by glycolipids drives chronic CNS inflammation. *Nat. Med.* **2014**, *20*, 1147–1156. [CrossRef] [PubMed]
85. Rothhammer, V.; Kenison, J.E.; Tjon, E.; Takenaka, M.C.; de Lima, K.A.; Borucki, D.M.; Chao, C.-C.; Wilz, A.; Blain, M.; Healy, L.; et al. Sphingosine 1-phosphate receptor modulation suppresses pathogenic astrocyte activation and chronic progressive CNS inflammation. *Proc. Natl. Acad. Sci. USA* **2017**, *114*, 2012–2017. [CrossRef] [PubMed]

86. Giubilei, F.; Antonini, G.; Di Legge, S.; Sormani, M.P.; Pantano, P.; Antonini, R.; Sepe-Monti, M.; Caramia, F.; Pozzilli, C. Blood cholesterol and MRI activity in first clinical episode suggestive of multiple sclerosis. *Acta Neurol. Scand.* **2002**, *106*, 109–112. [CrossRef] [PubMed]
87. Zhornitsky, S.; McKay, K.A.; Metz, L.M.; Teunissen, C.E.; Rangachari, M. Cholesterol and markers of cholesterol turnover in multiple sclerosis: Relationship with disease outcomes. *Mult. Scler. Relat. Disord.* **2016**, *5*, 53–65. [CrossRef] [PubMed]
88. Jorissen, W.; Wouters, E.; Bogie, J.F.; Vanmierlo, T.; Noben, J.-P.; Sviridov, D.; Hellings, N.; Somers, V.; Valcke, R.; Vanwijmeersch, B.; et al. Relapsing-remitting multiple sclerosis patients display an altered lipoprotein profile with dysfunctional HDL. *Sci. Rep.* **2017**, *7*, 43410. [CrossRef] [PubMed]
89. Mukhopadhyay, S.; Fellows, K.; Browne, R.W.; Khare, P.; Krishnan Radhakrishnan, S.; Hagemeier, J.; Weinstock-Guttman, B.; Zivadinov, R.; Ramanathan, M. Interdependence of oxysterols with cholesterol profiles in multiple sclerosis. *Mult. Scler.* **2017**, *23*, 792–801. [CrossRef] [PubMed]
90. Ludewig, B.; Laman, J.D. The in and out of monocytes in atherosclerotic plaques: Balancing inflammation through migration. *Proc. Natl. Acad. Sci. USA* **2004**, *101*, 11529–11530. [CrossRef] [PubMed]
91. Marrie, R.A.; Rudick, R.; Horwitz, R.; Cutter, G.; Tyry, T.; Campagnolo, D.; Vollmer, T. Vascular comorbidity is associated with more rapid disability progression in multiple sclerosis. *Neurology* **2010**, *74*, 1041–1047. [CrossRef] [PubMed]
92. Kowalec, K.; McKay, K.A.; Patten, S.B.; Fisk, J.D.; Evans, C.; Tremlett, H.; Marrie, R.A. CIHR Team in Epidemiology and Impact of Comorbidity on Multiple Sclerosis (ECoMS) Comorbidity increases the risk of relapse in multiple sclerosis. *Neurology* **2017**, *89*, 2455–2461. [CrossRef] [PubMed]
93. Palavra, F.; Reis, F.; Marado, D.; Sena, A. (Eds.) *Biomarkers of Cardiometabolic Risk, Inflammation and Disease*; Springer International Publishing: Cham, Switzerland, 2015; ISBN 9783319160184.
94. Sena, A.; Pedrosa, R.; Graça Morais, M. Therapeutic potential of lovastatin in multiple sclerosis. *J. Neurol.* **2003**, *250*, 754–755. [CrossRef] [PubMed]
95. Vollmer, T.; Key, L.; Durkalski, V.; Tyor, W.; Corboy, J.; Markovic-Plese, S.; Preiningerova, J.; Rizzo, M.; Singh, I. Oral simvastatin treatment in relapsing-remitting multiple sclerosis. *Lancet* **2004**, *363*, 1607–1608. [CrossRef]
96. Chataway, J.; Schuerer, N.; Alsanousi, A.; Chan, D.; MacManus, D.; Hunter, K.; Anderson, V.; Bangham, C.R.M.; Clegg, S.; Nielsen, C.; et al. Effect of high-dose simvastatin on brain atrophy and disability in secondary progressive multiple sclerosis (MS-STAT): A randomised, placebo-controlled, phase 2 trial. *Lancet* **2014**, *383*, 2213–2221. [CrossRef]
97. Chan, D.; Binks, S.; Nicholas, J.M.; Frost, C.; Cardoso, M.J.; Ourselin, S.; Wilkie, D.; Nicholas, R.; Chataway, J. Effect of high-dose simvastatin on cognitive, neuropsychiatric, and health-related quality-of-life measures in secondary progressive multiple sclerosis: Secondary analyses from the MS-STAT randomised, placebo-controlled trial. *Lancet Neurol.* **2017**, *16*, 591–600. [CrossRef]
98. Sena, A.; Pedrosa, R.; Ferret-Sena, V.; Almeida, R.; Andrade, M.L.; Morais, M.G.; Couderc, R. Interferon β1a therapy changes lipoprotein metabolism in patients with multiple sclerosis. *Clin. Chem. Lab. Med.* **2000**, *38*, 209–213. [CrossRef] [PubMed]
99. Uher, T.; Fellows, K.; Horakova, D.; Zivadinov, R.; Vaneckova, M.; Sobisek, L.; Tyblova, M.; Seidl, Z.; Krasensky, J.; Bergsland, N.; et al. Serum lipid profile changes predict neurodegeneration in interferon-β1a-treated multiple sclerosis patients. *J. Lipid Res.* **2017**, *58*, 403–411. [CrossRef] [PubMed]
100. Ottenlinger, F.M.; Mayer, C.A.; Ferreirós, N.; Schreiber, Y.; Schwiebs, A.; Schmidt, K.G.; Ackermann, H.; Pfeilschifter, J.M.; Radeke, H.H. Interferon-β Increases Plasma Ceramides of Specific Chain Length in Multiple Sclerosis Patients, Unlike Fingolimod or Natalizumab. *Front. Pharmacol.* **2016**, *7*, 412. [CrossRef] [PubMed]
101. Del Boccio, P.; Pieragostino, D.; Di Ioia, M.; Petrucci, F.; Lugaresi, A.; De Luca, G.; Gambi, D.; Onofrj, M.; Di Ilio, C.; Sacchetta, P.; et al. Lipidomic investigations for the characterization of circulating serum lipids in multiple sclerosis. *J. Proteom.* **2011**, *74*, 2826–2836. [CrossRef] [PubMed]
102. Villoslada, P.; Alonso, C.; Agirrezabal, I.; Kotelnikova, E.; Zubizarreta, I.; Pulido-Valdeolivas, I.; Saiz, A.; Comabella, M.; Montalban, X.; Villar, L.; et al. Metabolomic signatures associated with disease severity in multiple sclerosis. *Neurol. Neuroimmunol. Neuroinflamm.* **2017**, *4*, e321. [CrossRef] [PubMed]

103. Lötsch, J.; Thrun, M.; Lerch, F.; Brunkhorst, R.; Schiffmann, S.; Thomas, D.; Tegder, I.; Geisslinger, G.; Ultsch, A. Machine-Learned Data Structures of Lipid Marker Serum Concentrations in Multiple Sclerosis Patients Differ from Those in Healthy Subjects. *Int. J. Mol. Sci.* **2017**, *18*, 1217. [CrossRef] [PubMed]
104. Quintana, F.J.; Farez, M.F.; Viglietta, V.; Iglesias, A.H.; Merbl, Y.; Izquierdo, G.; Lucas, M.; Basso, A.S.; Khoury, S.J.; Lucchinetti, C.F.; et al. Antigen microarrays identify unique serum autoantibody signatures in clinical and pathologic subtypes of multiple sclerosis. *Proc. Natl. Acad. Sci. USA* **2008**, *105*, 18889–18894. [CrossRef] [PubMed]
105. Bakshi, R.; Yeste, A.; Patel, B.; Tauhid, S.; Tummala, S.; Rahbari, R.; Chu, R.; Regev, K.; Kivisäkk, P.; Weiner, H.L.; et al. Serum lipid antibodies are associated with cerebral tissue damage in multiple sclerosis. *Neurol. Neuroimmunol. Neuroinflamm.* **2016**, *3*, e200. [CrossRef] [PubMed]
106. Sena, A.; Bendtzen, K.; Cascais, M.J.; Pedrosa, R.; Ferret-Sena, V.; Campos, E. Influence of apolipoprotein E plasma levels and tobacco smoking on the induction of neutralising antibodies to interferon-β. *J. Neurol.* **2010**, *257*, 1703–1707. [CrossRef] [PubMed]
107. Li, F.-Q.; Sempowski, G.D.; McKenna, S.E.; Laskowitz, D.T.; Colton, C.A.; Vitek, M.P. Apolipoprotein E-derived peptides ameliorate clinical disability and inflammatory infiltrates into the spinal cord in a murine model of multiple sclerosis. *J. Pharmacol. Exp. Ther.* **2006**, *318*, 956–965. [CrossRef] [PubMed]
108. Michalek, R.D.; Gerriets, V.A.; Jacobs, S.R.; Macintyre, A.N.; MacIver, N.J.; Mason, E.F.; Sullivan, S.A.; Nichols, A.G.; Rathmell, J.C. Cutting Edge: Distinct Glycolytic and Lipid Oxidative Metabolic Programs Are Essential for Effector and Regulatory CD4+ T Cell Subsets. *J. Immunol.* **2011**, *186*, 3299–3303. [CrossRef] [PubMed]
109. Pearce, E.L.; Poffenberger, M.C.; Chang, C.-H.; Jones, R.G. Fueling Immunity: Insights into Metabolism and Lymphocyte Function. *Science* **2013**, *342*, 1242454. [CrossRef] [PubMed]
110. Carbone, F.; De Rosa, V.; Carrieri, P.B.; Montella, S.; Bruzzese, D.; Porcellini, A.; Procaccini, C.; La Cava, A.; Matarese, G. Regulatory T cell proliferative potential is impaired in human autoimmune disease. *Nat. Med.* **2014**, *20*, 69–74. [CrossRef] [PubMed]
111. Gerriets, V.A.; Kishton, R.J.; Johnson, M.O.; Cohen, S.; Siska, P.J.; Nichols, A.G.; Warmoes, M.O.; de Cubas, A.A.; MacIver, N.J.; Locasale, J.W.; et al. Foxp3 and Toll-like receptor signaling balance Tregcell anabolic metabolism for suppression. *Nat. Immunol.* **2016**, *17*, 1459–1466. [CrossRef] [PubMed]
112. La Rocca, C.; Carbone, F.; De Rosa, V.; Colamatteo, A.; Galgani, M.; Perna, F.; Lanzillo, R.; Brescia Morra, V.; Orefice, G.; et al. Immunometabolic profiling of T cells from patients with relapsing-remitting multiple sclerosis reveals an impairment in glycolysis and mitochondrial respiration. *Metabolism* **2017**, *77*, 39–46. [CrossRef] [PubMed]
113. Zhao, G.; Liu, Y.; Fang, J.; Chen, Y.; Li, H.; Gao, K. Dimethyl fumarate inhibits the expression and function of hypoxia-inducible factor-1α (HIF-1α). *Biochem. Biophys. Res. Commun.* **2014**, *448*, 303–307. [CrossRef] [PubMed]
114. O'Neill, L.A.J.; Kishton, R.J.; Rathmell, J. A guide to immunometabolism for immunologists. *Nat. Rev. Immunol.* **2016**, *16*, 553–565. [CrossRef] [PubMed]
115. Haase, S.; Haghikia, A.; Gold, R.; Linker, R.A. Dietary fatty acids and susceptibility to multiple sclerosis. *Mult. Scler. J.* **2018**, *24*, 12–16. [CrossRef] [PubMed]
116. Maceyka, M.; Spiegel, S. Sphingolipid metabolites in inflammatory disease. *Nature* **2014**, *510*, 58–67. [CrossRef] [PubMed]
117. Rothhammer, V.; Mascanfroni, I.D.; Bunse, L.; Takenaka, M.C.; Kenison, J.E.; Mayo, L.; Chao, C.-C.; Patel, B.; Yan, R.; Blain, M.; et al. Type I interferons and microbial metabolites of tryptophan modulate astrocyte activity and central nervous system inflammation via the aryl hydrocarbon receptor. *Nat. Med.* **2016**, *22*, 586–597. [CrossRef] [PubMed]
118. Rothhammer, V.; Borucki, D.M.; Garcia Sanchez, M.I.; Mazzola, M.A.; Hemond, C.C.; Regev, K.; Paul, A.; Kivisäkk, P.; Bakshi, R.; Izquierdo, G.; et al. Dynamic regulation of serum aryl hydrocarbon receptor agonists in MS. *Neurol. Neuroimmunol. Neuroinflamm.* **2017**, *4*, e359. [CrossRef] [PubMed]
119. Lim, C.K.; Bilgin, A.; Lovejoy, D.B.; Tan, V.; Bustamante, S.; Taylor, B.V.; Bessede, A.; Brew, B.J.; Guillemin, G.J. Kynurenine pathway metabolomics predicts and provides mechanistic insight into multiple sclerosis progression. *Sci. Rep.* **2017**, *7*, 41473. [CrossRef] [PubMed]

120. Procaccini, C.; Santopaolo, M.; Faicchia, D.; Colamatteo, A.; Formisano, L.; De Candia, P.; Galgani, M.; De Rosa, V.; Matarese, G. Role of metabolism in neurodegenerative disorders. *Metabolism* **2016**, *65*, 1376–1390. [CrossRef] [PubMed]
121. Fischer, H.J.; Sie, C.; Schumann, E.; Witte, A.-K.; Dressel, R.; van den Brandt, J.; Reichardt, H.M. The Insulin Receptor Plays a Critical Role in T Cell Function and Adaptive Immunity. *J. Immunol.* **2017**, *198*, 1910–1920. [CrossRef] [PubMed]
122. Negrotto, L.; Farez, M.F.; Correale, J. Immunologic Effects of Metformin and Pioglitazone Treatment on Metabolic Syndrome and Multiple Sclerosis. *JAMA Neurol.* **2016**, *73*, 520–528. [CrossRef] [PubMed]
123. Choi, I.Y.; Piccio, L.; Childress, P.; Bollman, B.; Ghosh, A.; Brandhorst, S.; Suarez, J.; Michalsen, A.; Cross, A.H.; Morgan, T.E.; et al. A Diet Mimicking Fasting Promotes Regeneration and Reduces Autoimmunity and Multiple Sclerosis Symptoms. *Cell Rep.* **2016**, *15*, 2136–2146. [CrossRef] [PubMed]
124. Sena, A.; Sarliève, L.L.; Rebel, G. Brain myelin of genetically obese mice. *J. Neurol. Sci.* **1985**, *68*, 233–243. [CrossRef]
125. Fleck, A.K.; Schuppan, D.; Wiendl, H.; Klotz, L. Gut–CNS-axis as possibility to modulate inflammatory disease activity—Implications for multiple sclerosis. *Int. J. Mol. Sci.* **2017**, *18*, 1526. [CrossRef] [PubMed]
126. Adamczyk-Sowa, M.; Medrek, A.; Madej, P.; Michlicka, W.; Dobrakowski, P. Does the Gut Microbiota Influence Immunity and Inflammation in Multiple Sclerosis Pathophysiology? *J. Immunol. Res.* **2017**, *2017*, 7904821. [CrossRef] [PubMed]
127. Van den Hoogen, W.J.; Laman, J.D.; 't Hart, B.A. Modulation of Multiple Sclerosis and Its Animal Model Experimental Autoimmune Encephalomyelitis by Food and Gut Microbiota. *Front. Immunol.* **2017**, *8*, 1081. [CrossRef] [PubMed]
128. Trott, S.; King, I.L. An introduction to the microbiome and MS. *Mult. Scler. J.* **2018**, *24*, 53–57. [CrossRef] [PubMed]
129. Markle, J.G.M.; Frank, D.N.; Mortin-Toth, S.; Robertson, C.E.; Feazel, L.M.; Rolle-Kampczyk, U.; von Bergen, M.; McCoy, K.D.; Macpherson, A.J.; Danska, J.S. Sex differences in the gut microbiome drive hormone-dependent regulation of autoimmunity. *Science* **2013**, *339*, 1084–1088. [CrossRef] [PubMed]
130. Berer, K.; Mues, M.; Koutrolos, M.; AlRasbi, Z.; Boziki, M.; Johner, C.; Wekerle, H.; Krishnamoorthy, G. Commensal microbiota and myelin autoantigen cooperate to trigger autoimmune demyelination. *Nature* **2011**, *479*, 538–541. [CrossRef] [PubMed]
131. Arpaia, N.; Campbell, C.; Fan, X.; Dikiy, S.; van der Veeken, J.; deRoos, P.; Liu, H.; Cross, J.R.; Pfeffer, K.; Coffer, P.J.; et al. Metabolites produced by commensal bacteria promote peripheral regulatory T-cell generation. *Nature* **2013**, *504*, 451–455. [CrossRef] [PubMed]
132. Lasitschka, F.; Giese, T.; Paparella, M.; Kurzhals, S.R.; Wabnitz, G.; Jacob, K.; Gras, J.; Bode, K.A.; Heninger, A.-K.; Sziskzai, T.; et al. Human monocytes downregulate innate response receptors following exposure to the microbial metabolite n-butyrate. *Immunity Inflamm. Dis.* **2017**, *5*, 480–492. [CrossRef] [PubMed]
133. Haghikia, A.; Jörg, S.; Duscha, A.; Berg, J.; Manzel, A.; Waschbisch, A.; Hammer, A.; Lee, D.-H.; May, C.; Wilck, N.; et al. Dietary Fatty Acids Directly Impact Central Nervous System Autoimmunity via the Small Intestine. *Immunity* **2016**, *44*, 951–953. [CrossRef] [PubMed]
134. Mizuno, M.; Noto, D.; Kaga, N.; Chiba, A.; Miyake, S. The dual role of short fatty acid chains in the pathogenesis of autoimmune disease models. *PLoS ONE* **2017**, *12*, e0173032. [CrossRef] [PubMed]
135. Chen, J.; Chia, N.; Kalari, K.R.; Yao, J.Z.; Novotna, M.; Soldan, M.M.P.; Luckey, D.H.; Marietta, E.V.; Jeraldo, P.R.; Chen, X.; et al. Multiple sclerosis patients have a distinct gut microbiota compared to healthy controls. *Sci. Rep.* **2016**, *6*, 28484. [CrossRef] [PubMed]
136. Tremlett, H.; Fadrosh, D.W.; Faruqi, A.A.; Zhu, F.; Hart, J.; Roalstad, S.; Graves, J.; Lynch, S.; Waubant, E. US Network of Pediatric MS Centers Gut microbiota in early pediatric multiple sclerosis: A case-control study. *Eur. J. Neurol.* **2016**, *23*, 1308–1321. [CrossRef] [PubMed]
137. Cekanaviciute, E.; Yoo, B.B.; Runia, T.F.; Debelius, J.W.; Singh, S.; Nelson, C.A.; Kanner, R.; Bencosme, Y.; Lee, Y.K.; Hauser, S.L.; et al. Gut bacteria from multiple sclerosis patients modulate human T cells and exacerbate symptoms in mouse models. *Proc. Natl. Acad. Sci. USA* **2017**, *114*, 10713–10718. [CrossRef] [PubMed]

138. Berer, K.; Gerdes, L.A.; Cekanaviciute, E.; Jia, X.; Xiao, L.; Xia, Z.; Liu, C.; Klotz, L.; Stauffer, U.; Baranzini, S.E.; et al. Gut microbiota from multiple sclerosis patients enables spontaneous autoimmune encephalomyelitis in mice. *Proc. Natl. Acad. Sci. USA* **2017**, *114*, 10719–10724. [CrossRef] [PubMed]
139. Cosorich, I.; Dalla-Costa, G.; Sorini, C.; Ferrarese, R.; Messina, M.J.; Dolpady, J.; Radice, E.; Mariani, A.; Testoni, P.A.; Canducci, F.; et al. High frequency of intestinal TH17 cells correlates with microbiota alterations and disease activity in multiple sclerosis. *Sci. Adv.* **2017**, *3*, e1700492. [CrossRef] [PubMed]
140. Braniste, V.; Al-Asmakh, M.; Kowal, C.; Anuar, F.; Abbaspour, A.; Toth, M.; Korecka, A.; Bakocevic, N.; Ng, L.G.; Kundu, P.; et al. The gut microbiota influences blood-brain barrier permeability in mice. *Sci. Transl. Med.* **2014**, *6*, 263ra158. [CrossRef] [PubMed]
141. Hoban, A.E.; Stilling, R.M.; Ryan, F.J.; Shanahan, F.; Dinan, T.G.; Claesson, M.J.; Clarke, G.; Cryan, J.F. Regulation of prefrontal cortex myelination by the microbiota. *Transl. Psychiatry* **2016**, *6*, e774. [CrossRef] [PubMed]
142. Erny, D.; Hrabě de Angelis, A.L.; Prinz, M. Communicating systems in the body: How microbiota and microglia cooperate. *Immunology* **2017**, *150*, 7–15. [CrossRef] [PubMed]
143. Costantini, L.; Molinari, R.; Farinon, B.; Merendino, N. Impact of Omega-3 Fatty Acids on the Gut Microbiota. *Int. J. Mol. Sci.* **2017**, *18*, 2645. [CrossRef] [PubMed]
144. Swidsinski, A.; Dörffel, Y.; Loening-Baucke, V.; Gille, C.; Göktas, Ö.; Reißhauer, A.; Neuhaus, J.; Weylandt, K.-H.; Guschin, A.; Bock, M. Reduced Mass and Diversity of the Colonic Microbiome in Patients with Multiple Sclerosis and Their Improvement with Ketogenic Diet. *Front. Microbiol.* **2017**, *8*, 1141. [CrossRef] [PubMed]
145. Saresella, M.; Mendozzi, L.; Rossi, V.; Mazzali, F.; Piancone, F.; LaRosa, F.; Marventano, I.; Caputo, D.; Felis, G.E.; Clerici, M. Immunological and Clinical Effect of Diet Modulation of the Gut Microbiome in Multiple Sclerosis Patients: A Pilot Study. *Front. Immunol.* **2017**, *8*, 1391. [CrossRef] [PubMed]
146. Wilck, N.; Matus, M.G.; Kearney, S.M.; Olesen, S.W.; Forslund, K.; Bartolomaeus, H.; Haase, S.; Mähler, A.; Balogh, A.; Markó, L.; et al. Salt-responsive gut commensal modulates TH17 axis and disease. *Nature* **2017**, *551*, 585–589. [CrossRef] [PubMed]
147. Faraco, G.; Brea, D.; Garcia-Bonilla, L.; Wang, G.; Racchumi, G.; Chang, H.; Buendia, I.; Santisteban, M.M.; Segarra, S.G.; et al. Dietary salt promotes neurovascular and cognitive dysfunction through a gut-initiated TH17 response. *Nat. Neurosci.* **2018**, *21*, 240–249. [CrossRef] [PubMed]
148. Barbier, O.; Torra, I.P.; Duguay, Y.; Blanquart, C.; Fruchart, J.-C.; Glineur, C.; Staels, B. Pleiotropic actions of peroxisome proliferator-activated receptors in lipid metabolism and atherosclerosis. *Arterioscler. Thromb. Vasc. Biol.* **2002**, *22*, 717–726. [CrossRef] [PubMed]
149. Rigamonti, E.; Chinetti-Gbaguidi, G.; Staels, B. Regulation of macrophage functions by PPAR-α, PPAR-γ, and LXRs in mice and men. *Arterioscler. Thromb. Vasc. Biol.* **2008**, *28*, 1050–1059. [CrossRef] [PubMed]
150. Grygiel-Górniak, B. Peroxisome proliferator-activated receptors and their ligands: Nutritional and clinical implications—A review. *Nutr. J.* **2014**, *13*, 17. [CrossRef] [PubMed]
151. Daynes, R.A.; Jones, D.C. Emerging roles of PPARS in inflammation and immunity. *Nat. Rev. Immunol.* **2002**, *2*, 748–759. [CrossRef] [PubMed]
152. Ricote, M.; Valledor, A.F.; Glass, C.K. Decoding Transcriptional Programs Regulated by PPARs and LXRs in the Macrophage: Effects on Lipid Homeostasis, Inflammation, and Atherosclerosis. *Arterioscler. Thromb. Vasc. Biol.* **2004**, *24*, 230–239. [CrossRef] [PubMed]
153. Kidani, Y.; Bensinger, S.J. Liver X receptor and peroxisome proliferator-activated receptor as integrators of lipid homeostasis and immunity. *Immunol. Rev.* **2012**, *249*, 72–83. [CrossRef] [PubMed]
154. Racke, M.K.; Gocke, A.R.; Muir, M.; Diab, A.; Drew, P.D.; Lovett-Racke, A.E. Nuclear receptors and autoimmune disease: The potential of PPAR agonists to treat multiple sclerosis. *J. Nutr.* **2006**, *136*, 700–703. [CrossRef] [PubMed]
155. Dunn, S.E.; Ousman, S.S.; Sobel, R.A.; Zuniga, L.; Baranzini, S.E.; Youssef, S.; Crowell, A.; Loh, J.; Oksenberg, J.; Steinman, L. Peroxisome proliferator–activated receptor (PPAR)α expression in T cells mediates gender differences in development of T cell–mediated autoimmunity. *J. Exp. Med.* **2007**, *204*, 321–330. [CrossRef] [PubMed]
156. Klotz, L.; Burgdorf, S.; Dani, I.; Saijo, K.; Flossdorf, J.; Hucke, S.; Alferink, J.; Nowak, N.; Novak, N.; Beyer, M.; et al. The nuclear receptor PPAR γ selectively inhibits Th17 differentiation in a T cell-intrinsic fashion and suppresses CNS autoimmunity. *J. Exp. Med.* **2009**, *206*, 2079–2089. [CrossRef] [PubMed]

157. Dunn, S.E.; Bhat, R.; Straus, D.S.; Sobel, R.A.; Axtell, R.; Johnson, A.; Nguyen, K.; Mukundan, L.; Moshkova, M.; Dugas, J.C.; et al. Peroxisome proliferator–activated receptor δ limits the expansion of pathogenic Th cells during central nervous system autoimmunity. *J. Exp. Med.* **2010**, *207*, 1599–1608. [CrossRef] [PubMed]
158. Cipolletta, D.; Feuerer, M.; Li, A.; Kamei, N.; Lee, J.; Shoelson, S.E.; Benoist, C.; Mathis, D. PPAR-γ is a major driver of the accumulation and phenotype of adipose tissue Treg cells. *Nature* **2012**, *486*, 549–553. [CrossRef] [PubMed]
159. Unoda, K.; Doi, Y.; Nakajima, H.; Yamane, K.; Hosokawa, T.; Ishida, S.; Kimura, F.; Hanafusa, T. Eicosapentaenoic acid (EPA) induces peroxisome proliferator-activated receptors and ameliorates experimental autoimmune encephalomyelitis. *J. Neuroimmunol.* **2013**, *256*, 7–12. [CrossRef] [PubMed]
160. Bright, J.J.; Kanakasabai, S.; Chearwae, W.; Chakraborty, S. PPAR Regulation of Inflammatory Signaling in CNS Diseases. *PPAR Res.* **2008**, *2008*, 658520. [CrossRef] [PubMed]
161. Chistyakov, D.V.; Aleshin, S.E.; Astakhova, A.A.; Sergeeva, M.G.; Reiser, G. Regulation of peroxisome proliferator-activated receptors (PPAR) α and -γ of rat brain astrocytes in the course of activation by toll-like receptor agonists. *J. Neurochem.* **2015**, *134*, 113–124. [CrossRef] [PubMed]
162. Hucke, S.; Floßdorf, J.; Grützke, B.; Dunay, I.R.; Frenzel, K.; Jungverdorben, J.; Linnartz, B.; Mack, M.; Peitz, M.; Brüstle, O.; et al. Licensing of myeloid cells promotes central nervous system autoimmunity and is controlled by peroxisome proliferator-activated receptor γ. *Brain* **2012**, *135*, 1586–1605. [CrossRef] [PubMed]
163. Bogie, J.F.; Jorissen, W.; Mailleux, J.; Nijland, P.G.; Zelcer, N.; Vanmierlo, T.; Van Horssen, J.; Stinissen, P.; Hellings, N.; Hendriks, J.J. Myelin alters the inflammatory phenotype of macrophages by activating PPARs. *Acta Neuropathol. Commun.* **2013**, *1*, 43. [CrossRef] [PubMed]
164. Sakuma, S.; Endo, T.; Kanda, T.; Nakamura, H.; Yamasaki, S.; Yamakawa, T. Synthesis of a novel human PPARδ selective agonist and its stimulatory effect on oligodendrocyte differentiation. *Bioorg. Med. Chem. Lett.* **2011**, *21*, 240–244. [CrossRef] [PubMed]
165. Jana, M.; Mondal, S.; Gonzalez, F.J.; Pahan, K. Gemfibrozil, a Lipid-lowering Drug, Increases Myelin Genes in Human Oligodendrocytes via Peroxisome Proliferator-activated Receptor-β. *J. Biol. Chem.* **2012**, *287*, 34134–34148. [CrossRef] [PubMed]
166. Kanakasabai, S.; Pestereva, E.; Chearwae, W.; Gupta, S.K.; Ansari, S.; Bright, J.J. PPARγ Agonists Promote Oligodendrocyte Differentiation of Neural Stem Cells by Modulating Stemness and Differentiation Genes. *PLoS ONE* **2012**, *7*, e50500. [CrossRef] [PubMed]
167. Roth, A.D.; Leisewitz, A.V.; Jung, J.E.; Cassina, P.; Barbeito, L.; Inestrosa, N.C.; Bronfman, M. PPARγ activators induce growth arrest and process extension in B12 oligodendrocyte-like cells and terminal differentiation of cultured oligodendrocytes. *J. Neurosci. Res.* **2003**, *72*, 425–435. [CrossRef] [PubMed]
168. Bernardo, A.; Bianchi, D.; Magnaghi, V.; Minghetti, L. Peroxisome proliferator-activated receptor-gamma agonists promote differentiation and antioxidant defenses of oligodendrocyte progenitor cells. *J. Neuropathol. Exp. Neurol.* **2009**, *68*, 797–808. [CrossRef] [PubMed]
169. De Nuccio, C.; Bernardo, A.; De Simone, R.; Mancuso, E.; Magnaghi, V.; Visentin, S.; Minghetti, L. Peroxisome Proliferator-Activated Receptor γ Agonists Accelerate Oligodendrocyte Maturation and Influence Mitochondrial Functions and Oscillatory Ca^{2+} Waves. *J. Neuropathol. Exp. Neurol.* **2011**, *70*, 900–912. [CrossRef] [PubMed]
170. De Nuccio, C.; Bernardo, A.; Cruciani, C.; De Simone, R.; Visentin, S.; Minghetti, L. Peroxisome proliferator activated receptor-γ agonists protect oligodendrocyte progenitors against tumor necrosis factor-alpha-induced damage: Effects on mitochondrial functions and differentiation. *Exp. Neurol.* **2015**, *271*, 506–514. [CrossRef] [PubMed]
171. Bernardo, A.; Giammarco, M.L.; De Nuccio, C.; Ajmone-Cat, M.A.; Visentin, S.; De Simone, R.; Minghetti, L. Docosahexaenoic acid promotes oligodendrocyte differentiation via PPAR-γ signalling and prevents tumor necrosis factor-α-dependent maturational arrest. *Biochim. Biophys. Acta Mol. Cell Biol. Lipids* **2017**, *1862*, 1013–1023. [CrossRef] [PubMed]
172. Vallée, A.; Vallée, J.-N.; Guillevin, R.; Lecarpentier, Y. Interactions Between the Canonical WNT/Beta-Catenin Pathway and PPAR γ on Neuroinflammation, Demyelination, and Remyelination in Multiple Sclerosis. *Cell Mol. Neurobiol.* **2017**. [CrossRef] [PubMed]
173. Moreno, S.; Farioli-Vecchioli, S.; Cerù, M.P. Immunolocalization of peroxisome proliferator-activated receptors and retinoid X receptors in the adult rat CNS. *Neuroscience* **2004**, *123*, 131–145. [CrossRef] [PubMed]

174. Warden, A.; Truitt, J.; Merriman, M.; Ponomareva, O.; Jameson, K.; Ferguson, L.B.; Mayfield, R.D.; Harris, R.A. Localization of PPAR isotypes in the adult mouse and human brain. *Sci. Rep.* **2016**, *6*, 27618. [CrossRef] [PubMed]
175. Aleshin, S.; Strokin, M.; Sergeeva, M.; Reiser, G. Peroxisome proliferator-activated receptor (PPAR)β/δ, a possible nexus of PPARα- and PPARγ-dependent molecular pathways in neurodegenerative diseases: Review and novel hypotheses. *Neurochem. Int.* **2013**, *63*, 322–330. [CrossRef] [PubMed]
176. Benedetti, E.; Di Loreto, S.; D'Angelo, B.; Cristiano, L.; d'Angelo, M.; Antonosante, A.; Fidoamore, A.; Golini, R.; Cinque, B.; Cifone, M.G.; et al. The PPARβ/δ Agonist GW0742 Induces Early Neuronal Maturation of Cortical Post-Mitotic Neurons: Role of PPARβ/δ in Neuronal Maturation. *J. Cell Physiol.* **2016**, *231*, 597–606. [CrossRef] [PubMed]
177. Agarwal, S.; Yadav, A.; Chaturvedi, R.K. Peroxisome proliferator-activated receptors (PPARs) as therapeutic target in neurodegenerative disorders. *Biochem. Biophys. Res. Commun.* **2017**, *483*, 1166–1177. [CrossRef] [PubMed]
178. Krishnan, M.L.; Wang, Z.; Aljabar, P.; Ball, G.; Mirza, G.; Saxena, A.; Counsell, S.J.; Hajnal, J.V.; Montana, G.; Edwards, A.D. Machine learning shows association between genetic variability inPPARGand cerebral connectivity in preterm infants. *Proc. Natl. Acad. Sci. USA* **2017**, *114*, 13744–13749. [CrossRef] [PubMed]
179. Omeragic, A.; Hoque, M.T.; Choi, U.-Y.; Bendayan, R. Peroxisome proliferator-activated receptor-gamma: Potential molecular therapeutic target for HIV-1-associated brain inflammation. *J. Neuroinflamm.* **2017**, *14*, 183. [CrossRef] [PubMed]
180. Yang, X.; Yao, H.; Chen, Y.; Sun, L.; Li, Y.; Ma, X.; Duan, S.; Li, X.; Xiang, R.; Han, J.; et al. Inhibition of Glutathione Production Induces Macrophage CD36 Expression and Enhances Cellular-oxidized Low Density Lipoprotein (oxLDL) Uptake. *J. Biol. Chem.* **2015**, *290*, 21788–21799. [CrossRef] [PubMed]
181. Polvani, S.; Tarocchi, M.; Galli, A. PPARγ and Oxidative Stress: Con(β) Catenating NRF2 and FOXO. *PPAR Res.* **2012**, *2012*, 641087. [CrossRef] [PubMed]
182. Choi, M.-J.; Lee, E.-J.; Park, J.-S.; Kim, S.-N.; Park, E.-M.; Kim, H.-S. Anti-inflammatory mechanism of galangin in lipopolysaccharide-stimulated microglia: Critical role of PPAR-γ signaling pathway. *Biochem. Pharmacol.* **2017**, *144*, 120–131. [CrossRef] [PubMed]
183. Sekulic-Jablanovic, M.; Petkovic, V.; Wright, M.B.; Kucharava, K.; Huerzeler, N.; Levano, S.; Brand, Y.; Leitmeyer, K.; Glutz, A.; Bausch, A.; et al. Effects of peroxisome proliferator activated receptors (PPAR)-γ and -α agonists on cochlear protection from oxidative stress. *PLoS ONE* **2017**, *12*, e0188596. [CrossRef] [PubMed]
184. Xu, J.; Zhang, Y.; Xiao, Y.; Ma, S.; Liu, Q.; Dang, S.; Jin, M.; Shi, Y.; Wan, B.; Zhang, Y. Inhibition of 12/15-lipoxygenase by baicalein induces microglia PPARβ/δ: A potential therapeutic role for CNS autoimmune disease. *Cell. Death Dis.* **2013**, *4*, e569. [CrossRef] [PubMed]
185. Dietrich, M.; Helling, N.; Hilla, A.; Heskamp, A.; Issberner, A.; Hildebrandt, T.; Kohne, Z.; Küry, P.; Berndt, C.; Aktas, O.; et al. Early alpha-lipoic acid therapy protects from degeneration of the inner retinal layers and vision loss in an experimental autoimmune encephalomyelitis-optic neuritis model. *J. Neuroinflamm.* **2018**, *15*, 71. [CrossRef] [PubMed]
186. Wang, K.-C.; Tsai, C.-P.; Lee, C.-L.; Chen, S.-Y.; Lin, G.-J.; Yen, M.-H.; Sytwu, H.-K.; Chen, S.-J. α-Lipoic acid enhances endogenous peroxisome-proliferator-activated receptor-γ to ameliorate experimental autoimmune encephalomyelitis in mice. *Clin. Sci.* **2013**, *125*, 329–340. [CrossRef] [PubMed]
187. Klotz, L.; Schmidt, M.; Giese, T.; Sastre, M.; Knolle, P.; Klockgether, T.; Heneka, M.T. Proinflammatory stimulation and pioglitazone treatment regulate peroxisome proliferator-activated receptor gamma levels in peripheral blood mononuclear cells from healthy controls and multiple sclerosis patients. *J. Immunol.* **2005**, *175*, 4948–4955. [CrossRef] [PubMed]
188. Klotz, L.; Schmidt, S.; Heun, R.; Klockgether, T.; Kölsch, H. Association of the PPARγ gene polymorphism Pro12Ala with delayed onset of multiple sclerosis. *Neurosci. Lett.* **2009**, *449*, 81–83. [CrossRef] [PubMed]
189. Pershadsingh, H.A.; Heneka, M.T.; Saini, R.; Amin, N.M.; Broeske, D.J.; Feinstein, D.L. Effect of pioglitazone treatment in a patient with secondary multiple sclerosis. *J. Neuroinflamm.* **2004**, *1*, 3. [CrossRef] [PubMed]
190. Kaiser, C.C.; Shukla, D.K.; Stebbins, G.T.; Skias, D.D.; Jeffery, D.R.; Stefoski, D.; Katsamakis, G.; Feinstein, D.L. A pilot test of pioglitazone as an add-on in patients with relapsing remitting multiple sclerosis. *J. Neuroimmunol.* **2009**, *211*, 124–130. [CrossRef] [PubMed]

191. Shukla, D.K.; Kaiser, C.C.; Stebbins, G.T.; Feinstein, D.L. Effects of pioglitazone on diffusion tensor imaging indices in multiple sclerosis patients. *Neurosci. Lett.* **2010**, *472*, 153–156. [CrossRef] [PubMed]
192. Szalardy, L.; Zadori, D.; Tanczos, E.; Simu, M.; Bencsik, K.; Vecsei, L.; Klivenyi, P. Elevated levels of PPAR-γ in the cerebrospinal fluid of patients with multiple sclerosis. *Neurosci. Lett.* **2013**, *554*, 131–134. [CrossRef] [PubMed]
193. Huang, W.; Rha, G.B.; Han, M.-J.; Eum, S.Y.; András, I.E.; Zhong, Y.; Hennig, B.; Toborek, M. PPARα and PPARgγ effectively protect against HIV-induced inflammatory responses in brain endothelial cells. *J. Neurochem.* **2008**, *107*, 497–509. [CrossRef] [PubMed]
194. Han, W.; Zhao, H.; Jiao, B.; Liu, F. EPA and DHA increased PPARγ expression and deceased integrin-linked kinase and integrin β1 expression in rat glomerular mesangial cells treated with lipopolysaccharide. *Biosci. Trends* **2014**, *8*, 120–125. [CrossRef] [PubMed]
195. Silverstein, R.L.; Febbraio, M. CD36, a Scavenger Receptor Involved in Immunity, Metabolism, Angiogenesis, and Behavior. *Sci. Signal.* **2009**, *2*, re3. [CrossRef] [PubMed]
196. Park, L.; Zhou, J.; Zhou, P.; Pistick, R.; El Jamal, S.; Younkin, L.; Pierce, J.; Arreguin, A.; Anrather, J.; Younkin, S.G.; et al. Innate immunity receptor CD36 promotes cerebral amyloid angiopathy. *Proc. Natl. Acad. Sci. USA* **2013**, *110*, 3089–3094. [CrossRef] [PubMed]
197. Ferret-Sena, V.; Maia E Silva, A.; Sena, A.; Cavaleiro, I.; Vale, J.; Derudas, B.; Chinetti-Gbaguidi, G.; Staels, B. Natalizumab Treatment Modulates Peroxisome Proliferator-Activated Receptors Expression in Women with Multiple Sclerosis. *PPAR Res.* **2016**, *2016*, 5716415. [CrossRef] [PubMed]
198. Sarov-Blat, L.; Kiss, R.S.; Haidar, B.; Kavaslar, N.; Jaye, M.; Bertiaux, M.; Steplewski, K.; Hurle, M.R.; Sprecher, D.; McPherson, R.; et al. Predominance of a Proinflammatory Phenotype in Monocyte-Derived Macrophages From Subjects With Low Plasma HDL-Cholesterol. *Arterioscler. Thromb. Vasc. Biol.* **2007**, *27*, 1115–1122. [CrossRef] [PubMed]
199. Adhikary, T.; Wortmann, A.; Schumann, T.; Finkernagel, F.; Lieber, S.; Roth, K.; Toth, P.M.; Diederich, W.E.; Nist, A.; et al. The transcriptional PPARβ/δ network in human macrophages defines a unique agonist-induced activation state. *Nucleic Acids Res.* **2015**, *43*, 5033–5051. [CrossRef] [PubMed]
200. Hla, T.; Brinkmann, V. Sphingosine 1-phosphate (S1P): Physiology and the effects of S1P receptor modulation. *Neurology* **2011**, *76*, S3–S8. [CrossRef] [PubMed]
201. Ferret-Sena, V.; Capela, C.; Pedrosa, R.; Salgado, V.; Derudas, B.; Staels, B.; Sena, A. Fingolimod treatment increase peroxisome proliferator-activated receptor (PPAR) γ and CD36 receptor gene expression in blood leukocytes of multiple sclerosis patients. *Mult. Scler. J.* **2017**, *23*, 427–679.
202. Zhong, Q.; Zhao, S.; Yu, B.; Wang, X.; Matyal, R.; Li, Y.; Jiang, Z. High-density Lipoprotein Increases the Uptake of Oxidized Low Density Lipoprotein *via* PPARγ/CD36 Pathway in Inflammatory Adipocytes. *Int. J. Biol. Sci.* **2015**, *11*, 256–265. [CrossRef] [PubMed]
203. Sini, S.; Deepa, D.; Harikrishnan, S.; Jayakumari, N. High-density lipoprotein from subjects with coronary artery disease promotes macrophage foam cell formation: Role of scavenger receptor CD36 and ERK/MAPK signaling. *Mol. Cell Biochem.* **2017**, *427*, 23–34. [CrossRef] [PubMed]
204. Sato, H.; Sugai, H.; Kurosaki, H.; Ishikawa, M.; Funaki, A.; Kimura, Y.; Ueno, K. The effect of sex hormones on peroxisome proliferator-activated receptor gamma expression and activity in mature adipocytes. *Biol. Pharm. Bull.* **2013**, *36*, 564–573. [CrossRef] [PubMed]
205. Hurtado, O.; Ballesteros, I.; Cuartero, M.I.; Moraga, A.; Pradillo, J.M.; Ramírez-Franco, J.; Bartolomé-Martín, D.; Pascual, D.; Torres, M.; Sánchez-Prieto, J.; et al. Daidzein has neuroprotective effects through ligand-binding-independent PPARγ activation. *Neurochem. Int.* **2012**, *61*, 119–127. [CrossRef] [PubMed]
206. Park, H.J.; Park, H.S.; Lee, J.U.; Bothwell, A.L.M.; Choi, J.M. Sex-based selectivity of PPARγ regulation in Th1, Th2, and Th17 differentiation. *Int. J. Mol. Sci.* **2016**, *17*, 1347. [CrossRef] [PubMed]
207. Park, H.-J.; Choi, J.-M. Sex-specific regulation of immune responses by PPARs. *Exp. Mol. Med.* **2017**, *49*, e364. [CrossRef] [PubMed]
208. Sena, A.; Couderc, R.; Vasconcelos, J.C.; Ferret-Sena, V.; Pedrosa, R. Oral contraceptive use and clinical outcomes in patients with multiple sclerosis. *J. Neurol. Sci.* **2012**, *317*, 47–51. [CrossRef] [PubMed]
209. Bove, R.; Chitnis, T. The role of gender and sex hormones in determining the onset and outcome of multiple sclerosis. *Mult. Scler. J.* **2014**, *20*, 520–526. [CrossRef] [PubMed]

210. Zhang, X.; Zhou, M.; Guo, Y.; Song, Z.; Liu, B. 1,25-Dihydroxyvitamin D_3 Promotes High Glucose-Induced M1 Macrophage Switching to M2 via the VDR-PPARγ Signaling Pathway. *Biomed. Res. Int.* **2015**, *2015*, 157834 1–157834 14. [CrossRef] [PubMed]
211. Zheng, F.; Zhang, S.; Lu, W.; Wu, F.; Yin, X.; Yu, D.; Pan, Q.; Li, H. Regulation of Insulin Resistance and Adiponectin Signaling in Adipose Tissue by Liver X Receptor Activation Highlights a Cross-Talk with PPARγ. *PLoS ONE* **2014**, *9*, e101269. [CrossRef] [PubMed]
212. Kim, E.; Woo, M.-S.; Qin, L.; Ma, T.; Beltran, C.D.; Bao, Y.; Bailey, J.A.; Corbett, D.; Ratan, R.R.; Lahiri, D.K.; et al. Daidzein Augments Cholesterol Homeostasis via ApoE to Promote Functional Recovery in Chronic Stroke. *J. Neurosci.* **2015**, *35*, 15113–15126. [CrossRef] [PubMed]
213. Ranhotra, H.S. Gut Microbiota and Host Nuclear Receptors Signalling. *Nucl. Recept. Res.* **2017**, *4*, 10136. [CrossRef] [PubMed]
214. Ohue-Kitano, R.; Yasuoka, Y.; Goto, T.; Kitamura, N.; Park, S.-B.; Kishino, S.; Kimura, I.; Kasubuchi, M.; Takahashi, H.; Li, Y.; et al. α-Linolenic acid-derived metabolites from gut lactic acid bacteria induce differentiation of anti-inflammatory M2 macrophages through G protein-coupled receptor 40. *FASEB J.* **2018**, *32*, 304–318. [CrossRef] [PubMed]
215. Byndloss, M.X.; Olsan, E.E.; Rivera-Chávez, F.; Tiffany, C.R.; Cevallos, S.A.; Lokken, K.L.; Torres, T.P.; Byndloss, A.J.; Faber, F.; Gao, Y.; et al. Microbiota-activated PPAR-γ signaling inhibits dysbiotic Enterobacteriaceae expansion. *Science* **2017**, *357*, 570–575. [CrossRef] [PubMed]
216. Tomas, J.; Mulet, C.; Saffarian, A.; Cavin, J.-B.; Ducroc, R.; Regnault, B.; Kun Tan, C.; Duszka, K.; Burcelin, R.; Wahli, W.; et al. High-fat diet modifies the PPAR-γ pathway leading to disruption of microbial and physiological ecosystem in murine small intestine. *Proc. Natl. Acad. Sci. USA* **2016**, *113*, E5934–E5943. [CrossRef] [PubMed]
217. Manoharan, I.; Suryawanshi, A.; Hong, Y.; Ranganathan, P.; Shanmugam, A.; Ahmad, S.; Swafford, D.; Manicassamy, B.; Ramesh, G.; Koni, P.A.; et al. Homeostatic PPARα Signaling Limits Inflammatory Responses to Commensal Microbiota in the Intestine. *J. Immunol.* **2016**, *196*, 4739–4749. [CrossRef] [PubMed]
218. Corthals, A.P. Multiple sclerosis is not a disease of the immune system. *Q. Rev. Biol.* **2011**, *86*, 287–321. [CrossRef] [PubMed]

International Journal of
Molecular Sciences

Article

PPARγ Modulates Long Chain Fatty Acid Processing in the Intestinal Epithelium

Kalina Duszka [1,2,3], Matej Oresic [4], Cedric Le May [5], Jürgen König [3,6] and Walter Wahli [1,2,7,*]

1. Lee Kong Chian School of Medicine, Nanyang Technological University, 11 Mandalay Road, Singapore 308232, Singapore; Kalina.duszka@univie.ac.at
2. Center for Integrative Genomics, University of Lausanne, Génopode, CH-1015 Lausanne, Switzerland
3. Department of Nutritional Sciences, University of Vienna, Althanstrasse 14, 1090 Vienna, Austria; juergen.koenig@univie.ac.at
4. Turku Centre for Biotechnology, University of Turku and Åbo Akademi University, Tykisokatu 6, 20520 Turku, Finland; matej.oresic@utu.fi
5. Institut du Thorax, INSERM, CNRS, UNIV Nantes, 44007 Nantes, France; cedric.lemay@univ-nantes.fr
6. Vienna Metabolomics Center (VIME), University of Vienna, Althanstrasse 14, 1090 Vienna, Austria
7. ToxAlim, Research Center in Food Toxicology, National Institute for Agricultural Research (INRA), 180 Chemin de Tournefeuille, 31300 Toulouse, France

* Correspondence: Walter.Wahli@ntu.edu.sg; Tel.: +65-6592-3927 or +65-9026-6430

Received: 7 November 2017; Accepted: 27 November 2017; Published: 28 November 2017

Abstract: Nuclear receptor PPARγ affects lipid metabolism in several tissues, but its role in intestinal lipid metabolism has not been explored. As alterations have been observed in the plasma lipid profile of ad libitum fed intestinal epithelium-specific PPARγ knockout mice (iePPARγKO), we submitted these mice to lipid gavage challenges. Within hours after gavage with long chain unsaturated fatty acid (FA)-rich canola oil, the iePPARγKO mice had higher plasma free FA levels and lower gastric inhibitory polypeptide levels than their wild-type (WT) littermates, and altered expression of incretin genes and lipid metabolism-associated genes in the intestinal epithelium. Gavage with the medium chain saturated FA-rich coconut oil did not result in differences between the two genotypes. Furthermore, the iePPARγKO mice did not exhibit defective lipid uptake and stomach emptying; however, their intestinal transit was more rapid than in WT mice. When fed a canola oil-rich diet for 4.5 months, iePPARγKO mice had higher body lean mass than the WT mice. We conclude that intestinal epithelium PPARγ is activated preferentially by long chain unsaturated FAs compared to medium chain saturated FAs. Furthermore, we hypothesize that the iePPARγKO phenotype originates from altered lipid metabolism and release in epithelial cells, as well as changes in intestinal motility.

Keywords: PPARγ; intestine; lipid metabolism

1. Introduction

The digestion of lipids starts in the oral cavity and involves lipases secreted by the lingual glands. The process continues in the stomach, where fats become emulsified and enter the duodenum as fine lipid droplets. There, they are further emulsified, micellized, and processed by bile acids and the pancreatic juice, eventually resulting in the formation of monoglycerides, free glycerol, and free fatty acids (FFAs) [1,2]. CD36 and various Fatty acids biding proteins (FABPs) facilitate long chain fatty acid (LCFA) transport across the apical membrane of enterocytes [3,4]. After entering enterocytes, FFAs and glycerol arrive at the crossroads of several pathways; they can be metabolized within mitochondria or be transported to the endoplasmic reticulum, where several enzymes, including GPAT, AGPAT, Lipin, and DGAT, catalyze the formation of triglycerides (TGs) [5–7]. The resulting

TGs bind to the microsomal triglyceride transport protein (MTTP), which assists in the generation of chylomicrons in the endoplasmic reticulum [8,9]. Depending on the cellular lipid load, TGs can also be temporarily stored in cytosolic lipid droplets (CLDs) within enterocytes [10], from which they can be released by lipases, such as ATGL and HSL, and further trafficked to chylomicrons [11,12]. Afterwards, chylomicrons are transported to the Golgi complex and are secreted from enterocytes to the lymph [2]. As long chain FAs (LCFAs) go through these complex absorption, rebuilding, and secretion steps, medium chain fatty acids (MCFAs) are processed faster and more easily. Given their lower mass, MCFAs are hydrolyzed rapidly and more completely by pancreatic lipase than LCFAs, and do not form micelles. In addition, their short carbon chain makes them weak electrolytes that are highly ionized at neutral pH, which increases their solubility and accelerates their transporter-free absorption. Due to the bias of TG-assembling enzymes in enterocytes towards FAs with chains >12 carbons, MCFAs are not incorporated into TGs. Therefore, 95% of MCFAs are not integrated into chylomicrons, but are directly shed into the portal vein and travel quickly to the liver as FFAs. Therefore, MCFAs reach this organ much faster than LCFAs [13,14].

Canola oil is much appreciated by nutritionists due to its high unsaturated FA content. The oil is composed of 71% monounsaturated fatty acids (MUFAs), 21% polyunsaturated fatty acids (PUFAs), and only 6.3% saturated FAs [15]. Because of its plant sterol and tocopherol content, canola oil is thought to be cardioprotective [16], and canola oil-based diets reduce plasma TG and low-density lipoprotein cholesterol (LDL-C) levels, as well as biomarkers of coronary heart disease [17]. In contrast, coconut oil consists mainly of saturated FAs (92%) with a high lauric acid content (47%), and also other MCFAs (17%) [13]. Lauric acid with its 12-carbon atom chain, shares only some of the properties of MCFAs; however, during the digestion process, it can be released faster and absorbed more rapidly than LCFAs [13]. As a plant-derived oil, coconut oil is considered as a healthier alternative to animal fat, but it increases total cholesterol, high-density lipoprotein cholesterol (HDL-C), and LDL-C levels in the blood [13].

PPARs form a subfamily of the nuclear receptor family, which consists of PPARα, PPARβ/δ, and PPARγ [18]. They are expressed in various tissues at varying levels, and the individual roles of these receptors remain distinct. In the gastrointestinal tract, PPARα regulates the expression of genes that are associated with FA, cholesterol, glucose, and amino acid metabolism, transport, and intestinal motility in response to dietary lipids [19,20]. PPARβ/δ in the intestine regulates multiple processes, including cell proliferation, differentiation [21], and lipid uptake [22]. Compared to the small intestine, PPARγ is expressed at higher levels in the colon, where it inhibits dysbiotic Enterobacteriaceae expansion [23]. However, in both of these sections of the intestine, PPARγ is present at relatively higher levels in the proximal regions and its expression decreases towards the distal regions [24–27]. In the small intestine, PPARγ is directly exposed to ligands that naturally occur in food, with its activity being regulated by FAs, glutamine, curcumin, capsaicin, and vitamin E [28,29]. Thus, dietary composition impacts PPARγ functions. Though much attention has been paid to the anti-inflammatory [30–37] and anticarcinogenic role of PPARγ in the colon [38–41], and intestines in general [42], very little is known about its function in the small intestine. In this section of the intestine, its expression and nuclear translocation is activated during inflammation and injury [43]. We previously reported that intestinal PPARγ regulates adipocyte energy mobilization via the sympathetic nervous system during caloric restriction (CR) [44].

When considering that PPARγ is under-investigated in the small intestine and its importance in lipid metabolism in other tissues [45], we assessed its role in small intestine lipid metabolism using intestinal epithelium-specific PPARγ knockout mice. This approach is superior to antagonist treatment, which does not allow for tissue-specific inhibition of receptor activity. Here, we present evidence that PPARγ is preferentially involved in the metabolism of LCFAs in the small intestine.

2. Results

2.1. PPARγ Regulates Lipid Transit but Not Uptake in Small Intestine

As we reported previously, the intestinal epithelium-specific PPARγ knockout mouse (PPARγ$^{\Delta/\Delta}$ VillinCre$^{+/-}$, or iePPARγKO) does not demonstrate any easily apparent phenotype in basic ad libitum conditions with respect to body weight, internal organ size, or plasma markers (TGs, FFAs, glucose, and cholesterol) [44]. However, advanced lipidomics analysis showed that plasma levels of several lipids differ in ad libitum iePPARγKO compared to their wild-type (PPARγ$^{fl/fl}$VillinCre$^{-/-}$, WT) littermates. Sphingomyelins (SMs, d18:1/18:0) and phosphatidylethanolamines (PEs, 36:0) were underrepresented in plasma from iePPARγKO as compared to WT mice (Figure 1a), whereas the concentrations of TGs (53:0) and several types of phosphatidylcholines (PCs) were higher in plasma from iePPARγKO mice than WT mice. In general, saturated FAs containing lipids were less abundant and unsaturated FAs occurred at higher concentrations in iePPARγKO when compared to WT mice. Choline is an essential component of PC; it is also essential for bile acid homeostasis and plays a role in the lipid uptake process. These observations suggested the involvement of PPARγ from the intestinal epithelium in lipid uptake and/or metabolism in this tissue.

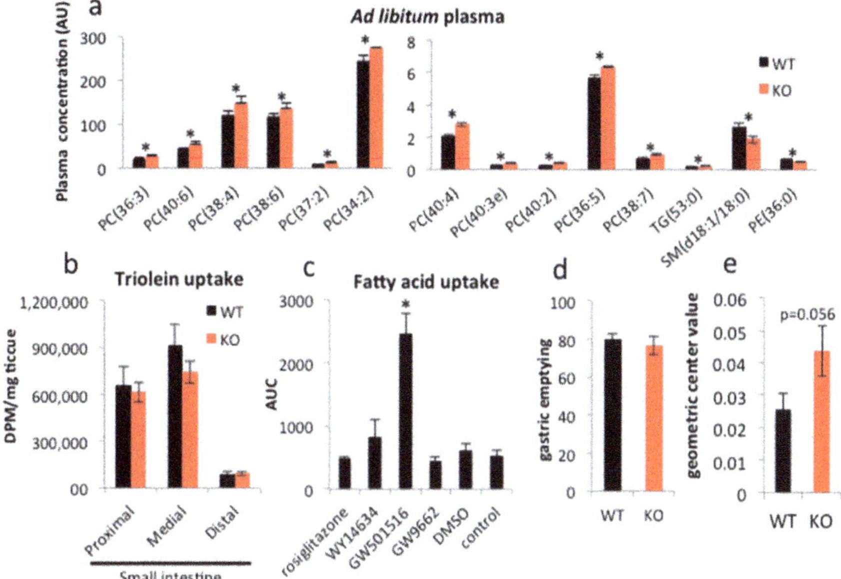

Figure 1. PPARγ does not affect lipid uptake, but regulates intestinal transit. (**a**) Blood was collected from mice fed ad libitum and plasma lipid composition analyzed (n = 5). (**b**) Lipid uptake was quantified by recording the radioactive tracer uptake (^{3}H-triolein) in the duodenal epithelium 30 min after labeled oil gavage (wild-type (WT) n = 8, KO n = 12). (**c**) Following 24 h incubation with the indicated compounds, fluorescent fatty acids (FAs) were added to each well and uptake by Caco-2 cells measured over 2 h (n = 3). (**d**) WT and iePPARγKO mice were gavaged with FITC-dextran in canola oil and the fluorescence in the stomach and (**e**) small intestine measured. For the intestine, the geometric center was quantified 30 min after gavage (WT n = 12, KO n = 10). The Student's t-test was performed for (**a**,**b**,**d**,**e**). For (**c**), one-way ANOVA with a Bonferroni post-hoc test was applied. Data are presented as means ± SEM (standard error of mean). * $p < 0.05$.

To characterize the role of the intestinal PPARγ, we performed a lipid uptake test in the small intestine in iePPARγKO mice. In animals that were gavaged with a mix of canola oil and ^{3}H-triolein, the amount of ^{3}H-tracer taken up by the intestinal epithelium within 30 min after gavage did not differ between iePPARγKO and WT mice (Figure 1b). In analogous in vitro experiments, Caco-2 cells were treated with various agonists and antagonists of the different PPAR isotypes and were incubated with fluorescently labeled FAs. Only the agonist specific for PPARβ/δ, GW501516, clearly increased FA uptake (Figure 1c). In contrast, rosiglitazone, an agonist of PPARγ, and WY14634, a strong agonist of PPARα, which also weakly activates PPARβ/δ and PPARγ, did not significantly affect FA uptake. Furthermore, GW9662, an antagonist of all three PPAR isotypes, had no significant effect on FA uptake.

Next, we performed a gastrointestinal transit assay using fluorescently labeled FAs. Stomach emptying activity was not affected by the absence of PPARγ (Figure 1d). However, the assay revealed an increased intestinal transit speed in iePPARγKO compared to WT mice (Figure 1e). We conclude that although PPARγ does not affect lipid uptake in the intestinal epithelium, it regulates intestinal transit.

2.2. Long-Term Canola Oil Diet Results in Modest Body Composition Changes in iePPARγKO Compared to WT Mice

In order to disclose an iePPARγKO phenotype, we submitted the iePPARγKO mice to an 18-week feeding with different high-lipid diets. The animals were fed a standard high-fat diet (HFD; 60% energy from mixed fat sources) and two fat-rich diets with 45% energy from lard or canola oil. The latter two were set at 45% energy from fat because it was the maximum possible percentage at which the use of liquid canola oil still allowed for the production of solid food pellets. Control groups were fed standard chow containing 4.5% energy from fat, mostly of soy and sunflower origin. The samples were collected in the late morning following a 2 h fast during the resting/non-eating phase to avoid acute fat/oil effects as studied below. The animals fed HFD presented with the highest body weight gain, followed by the canola and lard diet groups when compared to the control animals (Figure 2a). However, only the weight increase of iePPARγKO mice fed the HFD was significant. All mice fed fat diets (HFD, lard, and canola) consumed less food than the control mice (Figure S1a). Interestingly, there were no significant differences in final body weight and food intake between iePPARγKO and WT mice in any of the four groups (Figure 2a and Figure S1a). Mice fed a HFD or lard diet presented elevated VO_2 when compared to control mice (Figure S1b). Canola oil-fed mice exhibited a similar trend, but did not reach significance. Respiratory exchange ratio (RER) was reduced in all of the mice that were consuming fat diets (Figure S1c). No differences were noted for VO_2, VCO_2, RER, or heat release between iePPARγKO and WT mice (Figure S1b–e).

Gene expression analysis revealed that PPARγ expression in the intestinal epithelium of WT mice was not significantly modified by the fatty diets as compared to the control diet (Figure 2b). Among the genes whose expression in the intestine was affected by canola oil gavage, only *Fxr* was downregulated in iePPARγKO compared to WT mice after the 18-week canola oil diet (Figure 2c). Further perusal of the expression of FXR target genes, such as *Fabp6*, *Nr0b2*, and *Fgf15*, did not reveal significant changes, not least because of relatively high variability in expression. A trend of downregulation in the iePPARγKO epithelium was observed for the PPAR-regulated genes *Atgl*, *Dgat2*, and *Tip47*. Furthermore, canola oil diet did not trigger differences between iePPARγKO and WT mice in regards to plasma TG, FFA, cholesterol, and glucose levels (Figure 2d). In the oral glucose tolerance test (OGTT), mice that were fed fatty diets had significantly higher glucose plasma levels than control mice (Figure 2e and Figure S1f). Interestingly, mice that were fed a canola diet had higher glucose levels than HFD-fed mice. No differences in plasma glucose levels were found between iePPARγKO and WT mice at any of the time points of the OGTT for any diet. Similarly, liver size was comparable between the two genotypes (Figure S1g).

As expected, the mice that were fed fat diets had increased relative epididymal, subcutaneous abdominal, and dorsal white adipose tissue (WAT) weight when compared to control mice, with HFD mice having the highest amount of WAT (Figure 2f) and the canola oil diet-fed WT mice the lowest. The iePPARγKO mice fed a canola diet exhibited a trend towards heavier fat pads for all three of the diets tested and increased total body fat mass compared to their WT littermates, but the difference was not significant. EchoMRI confirmed the trend towards increased total body fat mass in canola oil-fed animals

(Figure 2g) and revealed a significant decrease in the lean mass of iePPARγKO vs. WT canola diet mice (Figure 2h). Expression of *Acc* and *Fas* in the WAT of mice consuming the canola diet was decreased as compared to control mice (Figure 2i), but no difference was detected between iePPARγKO and WT mice. Thus, a canola oil diet increased the fat mass in iePPARγKO mice when compared to WT mice and resulted in a difference in body mass composition between the two genotypes. Furthermore, the effect on gene expression in the duodenum of these animals fed canola oil for 18 weeks was less than that observed with acute canola oil gavage (see below).

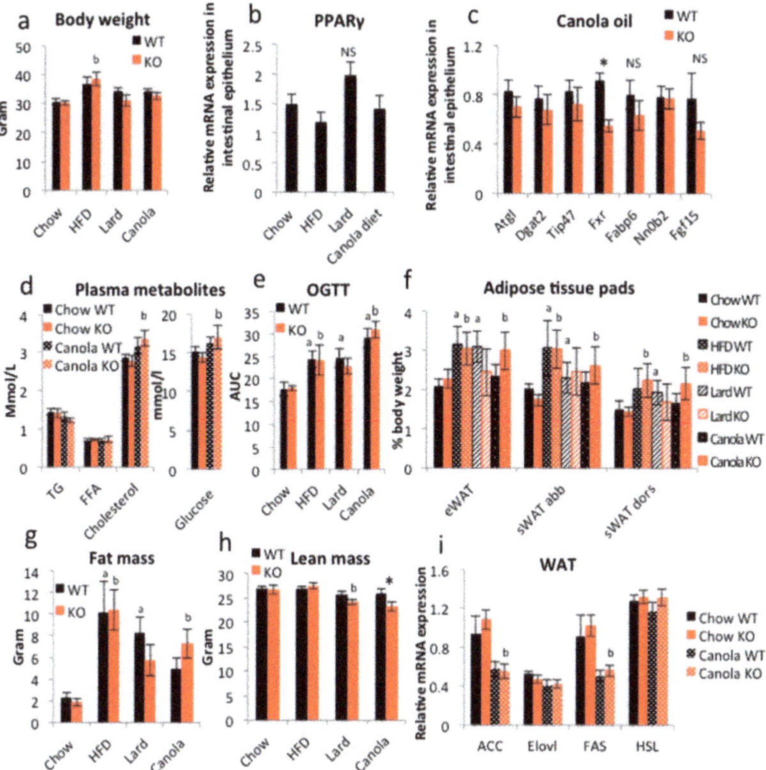

Figure 2. Long-term exposure to canola oil triggers mild body composition changes in iePPARγKO vs. WT mice. (**a**) Body weight of mice fed chow, high fat diet (HFD), lard diet, or canola oil diet (n = 7–10). (**b**) The relative mRNA expression levels of PPARγ and (**c**) lipid metabolism-associated genes in the duodenal epithelium were assayed by RT-qPCR (n = 9–18). (**d**) Concentration of TGs, FFAs, cholesterol, and glucose in plasma of mice fed chow and canola diet (n = 7–12). (**e**) Mice were submitted to oral glucose tolerance test (OGTT) and their plasma glucose levels monitored over 2 h (n = 5–9). (**f**) Weight of epididymal white adipose tissue (eWAT), subcutaneous abdominal WAT (sWAT abb), and subcutaneous dorsal WAT (sWAT dors) presented as % of total body weight (n = 8–10). (**g**) Total body fat and (**h**) lean mass were estimated using EchoMRI (n = 8–10). (**i**) Relative mRNA expression levels in epididymal WAT from chow and canola fed mice were measured using RT-qPCR (n = 8–10). One-way ANOVA followed by the Bonferroni post-hoc test was used to compare the experimental groups in (**b**,**d**,**f**,**i**). The two-tail Student's t-test was applied to verify significance ($p < 0.05$) in (**a**,**c**,**e**,**g**,**h**). * $p < 0.05$ for canola iePPARγKO vs. canolaWT, ^{a} significantly differ from Chow WT, ^{b} significantly differ from Chow KO. Data are presented as mean ± SEM.

2.3. PPARγ Affects Lipid Metabolism in Duodenal Enterocytes

As the long-term lipid challenge with canola oil disclosed a mild iePPARγKO phenotype, we challenged iePPARγKO and WT mice with acute lipid loads via a single gavage of canola oil (5 μL/g body weight), which is very rich in long chain unsaturated FAs, following an overnight fast. In WT animals, plasma TG and FFA levels were maximal after 2 h, with lower levels already at 3 h. In iePPARγKO mice, TG levels were still increased and FFAs remained higher at 3 h with significant differences from WT mice (Figure 3a,b). These results suggest that PPARγ in the intestinal epithelium impacts the processing of these molecules in the small intestine. WT and iePPARγKO mice that were gavaged with the same volume of coconut oil, which is very rich in saturated MCFAs, did not exhibit differences in plasma TGs or FFAs, which were highest at 3 h (Figure 3a,b), but TGs were significantly lower than after gavage with canola oil (Figure 3a). Plasma levels of total cholesterol, HDL, and glucose were similar in iePPARγKO and WT mice after either of the two oil gavages (Figure S2a,b). These results suggest that PPARγ selectively affected the intestinal processing of unsaturated LCFAs, but did not impact that of saturated MCFAs.

Because PPARγ is a transcription factor, we assessed whether the above observations result from changes in gene expression in the intestinal epithelium due to oil gavage, and whether tissue-specific deletion of PPARγ affects them. As PPARγ is expressed at higher levels in the proximal parts of the small intestine [24,27], we measured the mRNA levels in the duodenum. It noteworthy that the deletion of PPARγ did not significantly affect the expression of PPARα and PPARβ/δ, which could have impacted the results (Figure S2e). In WT mice, canola oil gavage stimulated the expression of *Gip* and *Secretin* after 2 and 3 h, respectively, whereas the expression of *Cck* (cholecystokinin) and *Dpp4* (dipeptidyl peptidase-4) was not affected (Figure 3c, Table S1). When compared to WT, *Cck*, *Dpp4*, and *Secretin*, expression was reduced in the iePPARγKO duodenum (Figure 3c, Table S1). Plasma gastric inhibitory polypeptide (GIP) protein levels were also significantly diminished in iePPARγKO when compared to WT mice at 3 h, and glucagon-like peptide-1 (GLP-1) at 4 h, after canola oil gavage (Figure 3d).

Canola oil gavage resulted in the stimulation of several genes of lipid metabolism in the duodenum of WT mice 2 and/or 3 h after gavage (Figure 3e and Figure S2c,e and Table S1). Importantly, the gene expression profiles differed between the WT and iePPARγKO duodenum (Figure 3e and Figure S2c,d, Table S1). In iePPARγKO mice, the genes stimulated in WT mice were expressed at lower levels 2 and/or 3 h after gavage (Figure 3e and Figure S2d, Table S1). Moreover, in iePPARγKO mice some of the genes were initially downregulated at 2 h compared to 0 h. Among the altered transcripts were those encoded by genes associated with lipid uptake (*Cd36*), TG synthesis (*Dgat2*, *Agpat9*), FA metabolism (*Acot11*, *Fasn*, *Mlysd*), FA transport to mitochondria (*Cact*), lipid droplet formation (*Hsl*, *Atgl*, *Tip47*), and chylomicron production (*Mttp*) (Figure 3e). Notably, the mRNA levels of *Fxr* were affected, suggesting a possible impact of PPARγ on bile acid signaling. Among the genes that were not influenced by the absence of intestinal PPARγ were those associated with cholesterol and lipid absorption (*Abca1*, *Abcg5*, *Ppap2a*), lipid metabolism (*Lcad*, *Cpt-1*), lipoprotein composition (*ApoAIV*, *ApoB*, *Vti1A*), mitochondrial ATP production (*Atp5e*), and mitochondrial respiratory chain (*Uqcr2*). Notably, the *Pparγ* mRNA level was not altered in WT mice after canola oil gavage (Figure S2f). Importantly, the level of mRNA of *Pparα* was downregulated in iePPARγKO 2 h after the gavage (Figure S2e). However, the expression pattern of *Pparα* and *β/δ* did not differ between iePPARγKO and WT mice at any time point following the gavage (Figure S2e), which indicates that their action in lipid metabolism is independent of *Pparγ*.

In contrast to the above results, coconut oil gavage did not trigger differences in intestinal epithelium gene expression between iePPARγKO and WT mice, with the exception of *Tip47*, which is involved in the biogenesis of lipid droplets (Figure 3f) and shares a significant homology with the other members of this family, including perilipin and adipophilin [46].

Interestingly, *Npy* ($p = 0.03$), which is associated with the regulation of metabolism and behavior, and *Mchr1* ($p = 0.05$) whose product is thought to have a number of functions, including the regulation of appetite [47,48], were down- and slightly up-regulated, respectively, in the hypothalami of iePPARγKO compared to WT mice 3 h after canola oil gavage (Figure 3g). Meanwhile, the expression of other hypothalamic appetite-related genes (*Hpmr, Hcrtr1, Mc4r, Npbw1*) was not affected in iePPARγKO mice after canola oil gavage (Figure S2g).

Collectively, these results show that PPARγ in enterocytes is activated by canola oil to specifically control pathways that are connected with FA metabolism and mitochondrial function, and possibly affect some hypothalamic functions. In contrast, PPARγ activity appears to not be influenced much by saturated MCFAs. This finding is in line with the previously reported preference of PPARγ for PUFAs as ligands [49].

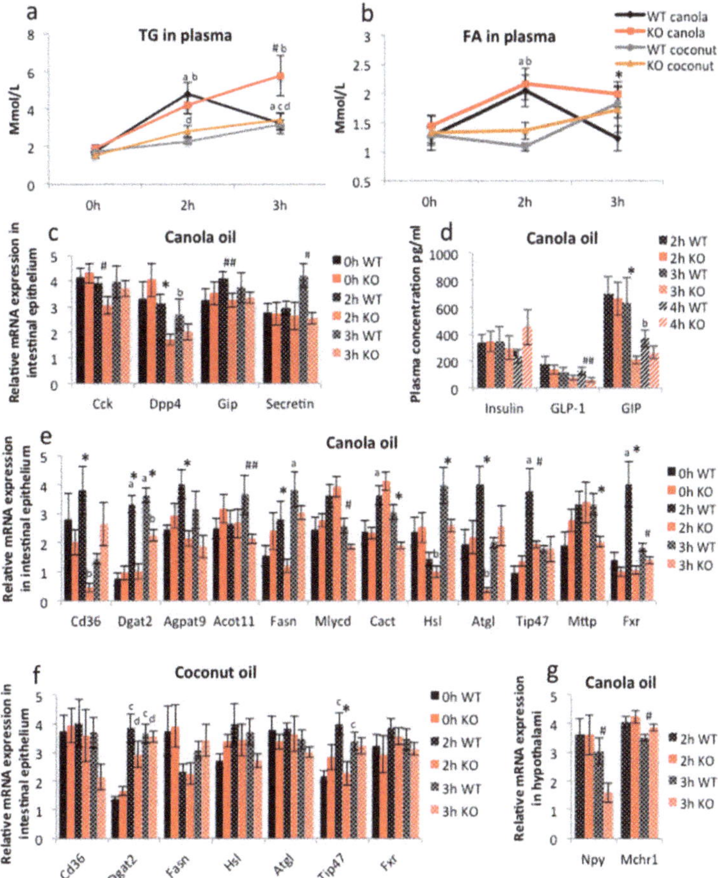

Figure 3. Canola oil gavage triggers differences in lipid metabolism signaling between iePPARγKO and WT mice. (**a**) Triglyceride (TG) and (**b**) free fatty acid (FFA) levels were measured in plasma after canola ($n = 6$) and coconut ($n = 5$–6) oil gavage. (**c**) Applying RT-qPCR, the relative mRNA expression levels in the duodenal epithelium were analyzed for intestinal hormones. (**d**) Plasma concentrations of insulin, GLP-1, and GIP were measured for WT and iePPARγKO mice gavaged with canola oil ($n = 6$–7). (**e**) The relative mRNA expression levels were quantified for lipid metabolism-associated genes in the duodenal epithelium of animals gavaged with canola oil and (**f**) coconut oil ($n = 5$–6) and (**g**) for hunger-related genes in the hypothalami of canola oil gavaged WT and iePPARγKO mice ($n = 6$–10). * Significant differences between iePPARγKO and WT mice. # $p < 0.05$; ## $p < 0.08$. [a] Significant differences between the labeled group and 0 h WT canola, [b] 0 h KO canola, [c] 0 h WT coconut, and [d] 0 h KO coconut. One-way ANOVA with a Bonferroni post-hoc test was applied for statistical analysis. Error bars depict the standard error.

3. Discussion

A previous investigation of iePPARγKO mice fed a chow diet when compared to WT mice did not reveal an easily recognized phenotype [44]. Here, a more in-depth plasma analysis revealed differences in circulating lipids, particularly the PC fraction. We also found that, after long-term exposure to a canola oil-rich diet (18 weeks), the iePPARγKO mice had reduced relative lean mass compared to WT animals, which correlated with a trend of higher fat mass, in line with the previously reported adipose tissue dysregulation in these animals under CR [44]. Furthermore, after canola oil gavage, we observed changes in plasma TG and FFA levels between iePPARγKO and WT mice. These modifications in circulating lipids were not due to faulty lipid uptake, but were correlated with increased intestinal transit in iePPARγKO mice, and, importantly, iePPARγ-dependent changes in enterocyte gene expression. The modulated genes are associated with lipid metabolism, mitochondrial functions, and gut hormones.

As mentioned above, the plasma levels of several PCs were increased in iePPARγKO mice. In humans, PCs are derived mostly from bile acids (10–20 g/day), but also from the diet (1–2 g/day) [2]. If this also prevails in rodents, the level of PCs in plasma may reflect changes in bile acid metabolism. The loss of *Fxr* upregulation after canola oil gavage and canola diet in iePPARγKO mice also suggests that bile acid metabolism may be affected by the absence of PPARγ in the intestinal epithelium. Although there was a trend for a lower expression of several FXR target genes in iePPARγKO mice, the difference from WT did not reach significance. Therefore, a possible alteration of the role of FXR in the iePPARγKO phenotype remains to be investigated more in-depth in the future. Choline and its metabolites are needed for the structural integrity of cell membranes and their signaling roles, cholinergic neurotransmission, and participation in the *S*-adenosylmethionine (SAMe) synthesis pathways. As PCs are the predominant type of phospholipids in the intestinal lumen and were increased in iePPARγKO mice, we evaluated whether intestinal lipid uptake was perturbed in iePPARγKO mice. Although iePPARγ did not modify the amount of lipid that was taken up, canola oil gavage led to differences in plasma lipid levels between WT and iePPARγKO mice, which is in line with alterations in epithelial gene expression in the latter. The persistence of high plasma TG and FFA levels 3 h after gavage may suggest a modified intestinal transit time, release from the epithelium, or clearance from the bloodstream. The expression of several genes in the intestinal epithelium was reduced in the absence of iePPARγ. Together, these genes are implicated in all of the processes of lipid metabolism in enterocytes (Figure 4), including lipid transport (*Cd36* [50–53]), lipolysis (*Hsl* [50,52,54,55], and *Atgl* [51,55,56]), and various lipid metabolism pathways (*Cact* [57], *Fasn* [58,59], *Mlycd* [60], *Dgat2* [50,52,55,59], and *Agpat9* [51,55,61]). Interestingly, *Acot 11* (hydrolysis of various coenzyme A esters), *Tip 47* (lipid droplet formation), and *Mttp* (chylomicron assembly) were previously not associated with PPARγ in any tissue. In addition, *Tip 47* was differentially expressed between the two phenotypes after both canola oil and coconut oil gavage, suggesting that coconut oil contains some FAs that may moderately affect some PPARγ pathways. In the future, an investigation at the protein level (expression, posttranslational modifications) will further the present study.

Previously, we demonstrated that the intestinal PPARγ negatively affects the expression of incretins and their plasma levels during CR [44]. Here we showed that, after canola oil gavage, the mRNA and plasma levels of incretins are reduced in iePPARγKO compared to WT mice, which demonstrates the different roles of PPARγ in intestinal hormone synthesis in different nutritional contexts. Interestingly, based on previously published findings by others [62,63], this downregulation of GIP, CCK, or secretin levels in iePPARγKO mice may explain the difference in intestinal passage time, which was increased in these mice. When considering that the lipid load increases gut motility [64], we hypothesize that fat may act through PPARγ to regulate intestinal transit, a function that has also been attributed to PPARα [19]. Such a putative role of PPARγ remains to be studied, as we have observed slightly accelerated transit in the iePPARγKO mice. Interestingly, gavage with saturated fat-rich coconut oil had much weaker effects than canola oil. This is very much in line with PPARγ having a preference for PUFAs as ligands, which are enriched in canola oil, over saturated FAs as ligands, which are abundant in coconut oil [28,29,49].

Interestingly, an 18-week-long canola oil feeding with sampling after 2 h fast during the resting non-feeding time did not produce the same clear effects as acute gavage. These observations suggest that the feeding time and the type of fats in the food directly regulate PPARγ activity in the intestinal epithelium. Alternatively, long-term fat feeding may change the lipid uptake and processing in the intestine [65], and, thus, the iePPARγKO phenotype may be attenuated under this condition. Nonetheless, we observed an effect of long-term canola oil feeding with a change in the ratio between lean and fat body mass in iePPARγKO mice when compared to WT mice. This difference in body composition may originate from a faulty metabolism of lipids in the intestine, as discussed above. Alternatively, canola-activated PPARγ could also lead to a similar effect on lipid release from WAT via PPARγ-dependent sympathetic nervous system signaling, as described previously [44]. The absence of this signal would result in fat retention in the adipose tissue, which is in line with our present observations.

Oils with different FA composition causing different phenotypes in iePPARγKO implies that intestinal PPARγ specifically regulates complex pathways under the influence of LCFAs, which are enriched in canola oil as naturally occurring agonists of PPARγ [28,29]. Our results suggest that consumption of oils rich in PPARγ agonists may improve the efficiency of lipid metabolism in the intestine and also impact the lean/fat mass ratio. In conclusion, we hypothesize that intestinal epithelium PPARγ affects lipid processing and/or the storage in enterocytes and adipose tissue, and its deletion would result in delayed trafficking in enterocytes and, possibly as described for CR [44], fat retention in adipose tissue.

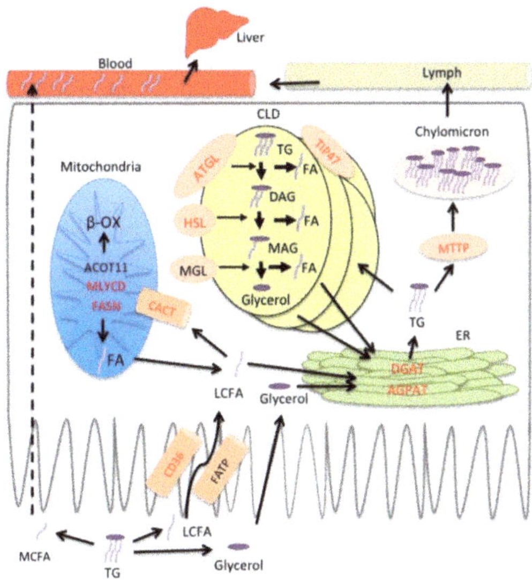

Figure 4. Model of lipid processing regulation by PPARγ in enterocytes. Red font indicates genes whose upregulation is lost or expression is reduced in enterocytes of iePPARγKO mice after canola oil gavage. Following intestinal digestion, FAs and glycerol are absorbed by enterocytes. Medium chain FAs (MCFA) travel through the enterocyte directly to blood (portal vein) (dashed arrow) and are transported to the liver as free FAs. Long chain FAs (LCFAs) are taken up by the enterocytes with the assistance of transporter proteins (CD36 and FATP). FAs are trafficked to mitochondria, where they are catabolized, or to the endoplasmic reticulum (ER), where there serve as substrates for TG assembly. Depending on the lipid load, TGs can be temporarily stored in cytoplasmic lipid droplets (CLD) or incorporated into chylomicrons and secreted into the lymph.

4. Materials and Methods

4.1. Mouse Handling

All of the animal experiment protocols were approved by the Vaud Cantonal Authority (SCAV 24735; authorization: VD 2440.3; 01 April 2015), Switzerland. As described previously [44], the intestinal epithelium-specific PPARγ knockout mouse was obtained by crossing floxed *Pparγ* (PPARγ$^{fl/fl}$) mice with mice expressing the Cre recombinase transgene under control of the villin promoter (VillinCre$^{+/-}$). The offspring PPARγ$^{\Delta/\Delta}$ VillinCre$^{+/-}$ mice with targeted disruption of PPARγ in the intestinal epithelium were named iePPARγKO mice and were used in parallel with littermate control PPARγ$^{fl/fl}$ (WT) mice with the same genetic background. Male mice were kept under a 12-h light/12-h dark cycle in standard housing cages. The animals were fed a standard laboratory diet, unless otherwise stated, and housed with free water access. For the oil gavage experiments, 10 to 12-week-old mice were fasted overnight. The next morning, the mice received 5 µL canola or coconut oil (Sigma-Aldrich, Buchs, Switzerland) per gram of body weight via gavage. The animals were dissected directly after overnight fasting or 2 and 3 h after oil gavage. The mice were euthanized using CO_2 and blood was drawn by cardiac puncture. The blood was mixed with 2% aprotinin-EDTA (Sigma, Mendota Heights, MN, USA) and DPPIV inhibitor (Merck, Kenilworth, NJ, USA), centrifuged for 10 min at 8000× *g*, and plasma frozen. Epididymal WAT, subcutaneous abdominal WAT, dorsal WAT, and liver weight were recorded. Duodenum scrapings and hypothalami were collected. All tissues were frozen in liquid nitrogen and stored at −80 °C until use.

For the diet experiments, five-week-old mice were randomly assigned to one of the diets: chow containing 4.5% energy from fat, mostly of soy and sunflower origin (Diet 3436, Provimi Kliba AG, Penthalaz, Switzerland); HFD with 60% kcal from fat in which the main fat source was lard (D12492 OpenSource Diets, Research Diets, New Brunswick, NJ, USA); or, HFD with 45% kcal fat from canola oil or lard (custom made modified D12451 diets, Research Diets). Body weight and food intake were measured weekly. After 15 weeks feeding with these diets, metabolic parameters (VO_2, VCO_2, heat production) and locomotor activity were monitored for three days using the Comprehensive Lab Animal Monitoring System (CLAMS, Columbus Instruments, Columbus, OH, USA). After 16 weeks of feeding with the control and HFDs, mice were submitted to the OGTT. Briefly, mice were fasted overnight, placed in single cages, and the first blood samples drawn from the tail. Next, the mice were gavaged a glucose solution and received the equivalent of 3 mg of glucose per gram body weight. Blood glucose levels were monitored after 15, 30, 60, 90, and 120 min. After 17 weeks of the diet, bedding maintained in the cage for 24 h was collected. Feces were separated from the bedding, dried, and fecal energy load measured using direct calorimetry (IKA-Kalorimeter C2000; IKA®-Werke GmbH & Co. KG; Staufen, Germany). Afterwards, the mouse body composition was measured under anesthesia using an EchoMRI whole-body composition analyzer (EchoMRI, Huston, TX, USA). After the EchoMRI, the mice were given 1 week to recover and then dissected following the procedure described above between 9 a.m. and 11 a.m. following 2 h fasting.

4.2. Intestinal ^{3}H-Triolein

After overnight fasting, mice received 200 µL canola oil containing 15 µCi 3H-triolein by gavage and sacrificed 30 min later. Blood was removed by perfusing the heart for 3 min with PBS. The intestinal lumina was flushed four times with 5 mM taurocholate, the small intestine divided into three equal segments (proximal, medial, and distal), and the segments were dissolved in Solvable™ (Perkin Elmer, Courtaboeuf, Villejust, France) overnight at 60 °C and incubated in scintillation fluid (Betaplate Scint, Perkin Elmer, Waltham, MA, USA). The radioactivity in each intestinal segment was measured by a liquid scintillation analyzer.

4.3. Gastric Emptying and Intestinal Motility

Overnight-fasted mice were gavaged with 200 µL of 5 mmol/L FITC-dextran (70 kDa FITC-dextran, Sigma) in canola oil and sacrificed 30 min later. Animals' small intestines were divided into 10 equal parts. The stomach and each part of the intestine was opened longitudinally, vortexed thoroughly with PBS, and centrifuged at 1200 rpm for 5 min. The intensity of fluorescence in the supernatant was measured. The geometric center used as an index of intestinal transit was calculated as the sum of the % fluorescence per segment × segment number [66].

4.4. RT-qPCR

RNA was isolated from intestinal scrapings using the RNeasy mini kit (Qiagen, Hombrechtikon, Switzerland). Samples were thawed in lysis buffer, disrupted using a syringe and needle, and processed following the manufacturer's recommendations. RNA was extracted from adipose tissue and the hypothalamus using the RNeasy Lipid Tissue mini kit (Qiagen). SuperScript® II Reverse Transcriptase (Thermo Fisher Scientific, Lausanne, Switzerland) and random primers (Promega, Madison, WI, USA) were used for the reverse transcription step for all of the samples. Quantitative real-time PCR (qRT-PCR) reactions were carried out using the Applied Biosystems 7900HT (Thermo Fisher Scientific) with the SYBR green PCR Master Mix (Applied Biosystems, Thermo Fisher Scientific). Primers used for qRT-PCR are listed in Table S2.

4.5. Plasma Analysis

Serum (10 µL) samples were diluted with 0.9% NaCl (10 µL) buffer. All of the samples were spiked with an internal standard (10 µL). Subsequently, the samples were extracted with chloroform: methanol (2:1) solvent (100 µL), homogenized with a glass rod (serum) at 4 °C by adding two zirconium oxide grinding balls, vortexed (1 min), incubated at room temperature (1 h), and centrifuged at 5590× g for 3 min. An aliquot of the separated lower phase (60 µL) was mixed with a labeled standard mixture (three stable isotope-labeled reference compounds; 10 µL) and 0.5–1.0 µL injection used for the analysis. The sample order for analysis was established by randomization. Lipid extracts were analyzed on a Q-ToF Premier mass spectrometer (Waters, Milford, MA, USA), and combined with an Acquity ultra performance liquid chromatograph (UPLC/MS).

Plasma glucose, lipid, and cholesterol levels were measured using a Hitachi chemistry analyzer (Roche Diagnostics, Basel, Switzerland), according to the manufacturer's instructions.

Plasma insulin, GLP-1, and GIP concentrations were estimated using Bio-Plex® (Luminex Corporation, Austin, TX, USA).

4.6. Cell Culture

Caco-2 cells were maintained in high glucose DMEM supplemented with 10% fetal bovine serum, 100 U/mL penicillin and 100 U/mL streptomycin (all from Sigma-Aldrich) in a humidified atmosphere of 5% CO_2 at 37 °C. Cells were cultured for 10 days after reaching confluence. Rosiglitazone, WY14634, GW501516, and GW9662 (all from Sigma) were added to the culture at final concentrations of 10 µM for 24 h. Control cells received the DMSO (Sigma) vehicle or no treatment. Afterwards, BODIPY-labeled fatty acids (QBT Fatty Acid Uptake Assay Kit, Molecular Devices, Wokingham, Berkshire, UK) were added to the culture and fluorescence measured over 2 h. The results are presented as area under the curve.

Supplementary Materials: Supplementary materials can be found at www.mdpi.com/1422-0067/18/12/2559/s1.

Acknowledgments: The authors would like to acknowledge the staff at the Metabolic Evaluation Facility at the Center for Integrative Genomics (University of Lausanne) for help with the plasma analysis, and direct and indirect calorimetry and Hervé Guillou for useful comments on the manuscript. This study was funded by the Swiss National Science Foundation (Walter Wahli); the 7th EU program TORNADO (Walter Wahli, Matej Oresic); the Bonizzi-Theler-Stiftung (Walter Wahli); the Etat de Vaud (Walter Wahli) and a start-up grant from the Lee Kong Chian School of Medicine, Nanyang Technological University, Singapore (Walter Wahli).

Author Contributions: Kalina Duszka designed and performed the experiments. Matej Oresic analyzed plasma lipid profiles. Cedric Le May assisted with the intestinal ^{3}H-triolein assay. Jürgen König contributed expert advice and helped write the manuscript. Walter Wahli supervised the study and wrote the manuscript. All authors corrected and approved the final manuscript.

Conflicts of Interest: The authors declare no conflict of interests.

References

1. Hofmann, A.F.; Borgstrom, B. Hydrolysis of long-chain monoglycerides in micellar solution by pancreatic lipase. *Biochim. Biophys. Acta* **1963**, *70*, 317–731. [CrossRef]
2. Iqbal, J.; Hussain, M.M. Intestinal lipid absorption. *Am. J. Physiol. Endocrinol. Metab.* **2009**, *296*, E1183–E1194. [CrossRef] [PubMed]
3. Chabowski, A.; Gorski, J.; Luiken, J.J.; Glatz, J.F.; Bonen, A. Evidence for concerted action of FAT/CD36 and FABPpm to increase fatty acid transport across the plasma membrane. *Prostaglandins Leukot. Essent. Fatty Acids* **2007**, *77*, 345–353. [CrossRef] [PubMed]
4. Schaffer, J.E.; Lodish, H.F. Expression cloning and characterization of a novel adipocyte long chain fatty acid transport protein. *Cell* **1994**, *79*, 427–436. [CrossRef]
5. Coleman, R.A.; Haynes, E.B. Monoacylglycerol acyltransferase. Evidence that the activities from rat intestine and suckling liver are tissue-specific isoenzymes. *J. Biol. Chem.* **1986**, *261*, 224–228. [PubMed]
6. Yen, C.L.; Stone, S.J.; Koliwad, S.; Harris, C.; Farese, R.V., Jr. Thematic review series: Glycerolipids. DGAT enzymes and triacylglycerol biosynthesis. *J. Lipid Res.* **2008**, *49*, 2283–2301. [PubMed]
7. Takeuchi, K.; Reue, K. Biochemistry, physiology, and genetics of GPAT, AGPAT, and lipin enzymes in triglyceride synthesis. *Am. J. Physiol. Endocrinol. Metab.* **2009**, *296*, E1195–E1209. [CrossRef] [PubMed]
8. Black, D.D. Development and physiological regulation of intestinal lipid absorption. I. Development of intestinal lipid absorption: Cellular events in chylomicron assembly and secretion. *Am. J. Physiol. Gastrointest. Liver Physiol.* **2007**, *293*, G519–G524. [PubMed]
9. Mansbach, C.M., 2nd; Gorelick, F. Development and physiological regulation of intestinal lipid absorption. II. Dietary lipid absorption, complex lipid synthesis, and the intracellular packaging and secretion of chylomicrons. *Am. J. Physiol. Gastrointest. Liver Physiol.* **2007**, *293*, G645–G650.
10. Zhu, J.; Lee, B.; Buhman, K.K.; Cheng, J.X. A dynamic, cytoplasmic triacylglycerol pool in enterocytes revealed by ex vivo and in vivo coherent anti-Stokes Raman scattering imaging. *J. Lipid Res.* **2009**, *50*, 1080–1089. [CrossRef] [PubMed]
11. Grober, J.; Lucas, S.; Sorhede-Winzell, M.; Zaghini, I.; Mairal, A.; Contreras, J.A.; Besnard, P.; Holm, C.; Langin, D. Hormone-sensitive lipase is a cholesterol esterase of the intestinal mucosa. *J. Biol. Chem.* **2003**, *278*, 6510–6515. [CrossRef] [PubMed]
12. Haemmerle, G.; Lass, A.; Zimmermann, R.; Gorkiewicz, G.; Meyer, C.; Rozman, J.; Heldmaier, G.; Maier, R.; Theussl, C.; Eder, S.; et al. Defective lipolysis and altered energy metabolism in mice lacking adipose triglyceride lipase. *Science* **2006**, *312*, 734–737. [CrossRef] [PubMed]
13. Eyres, L.; Eyres, M.F.; Chisholm, A.; Brown, R.C. Coconut oil consumption and cardiovascular risk factors in humans. *Nutr. Rev.* **2016**, *74*, 267–280. [CrossRef] [PubMed]
14. Bach, A.C.; Babayan, V.K. Medium-chain triglycerides: An update. *Am. J. Clin. Nutr.* **1982**, *36*, 950–962. [PubMed]
15. Orsavova, J.; Misurcova, L.; Ambrozova, J.V.; Vicha, R.; Mlcek, J. Fatty Acids Composition of Vegetable Oils and Its Contribution to Dietary Energy Intake and Dependence of Cardiovascular Mortality on Dietary Intake of Fatty Acids. *Int. J. Mol. Sci.* **2015**, *16*, 12871–12890. [CrossRef] [PubMed]
16. Schwartz, H.; Ollilainen, V.; Piironen, V.; Lampi, A.M. Tocopherol, tocotrienol and plant sterol contents of vegetable oils and industrial fats. *J. Food Compos. Anal.* **2008**, *21*, 152–161. [CrossRef]
17. Lin, L.; Allemekinders, H.; Dansby, A.; Campbell, L.; Durance-Tod, S.; Berger, A.; Jones, P.J. Evidence of health benefits of canola oil. *Nutr. Rev.* **2013**, *71*, 370–385. [CrossRef] [PubMed]
18. Michalik, L.; Auwerx, J.; Berger, J.P.; Chatterjee, V.K.; Glass, C.K.; Gonzalez, F.J.; Grimaldi, P.A.; Kadowaki, T.; Lazar, M.A.; O'Rahilly, S.; et al. International Union of Pharmacology. LXI. Peroxisome proliferator-activated receptors. *Pharmacol. Rev.* **2006**, *58*, 726–741. [PubMed]

19. De Vogel-van den Bosch, H.M.; Bunger, M.; de Groot, P.J.; Bosch-Vermeulen, H.; Hooiveld, G.J.; Muller, M. PPARα-mediated effects of dietary lipids on intestinal barrier gene expression. *BMC Genom.* **2008**, *9*. [CrossRef] [PubMed]
20. Bunger, M.; van den Bosch, H.M.; van der Meijde, J.; Kersten, S.; Hooiveld, G.J.; Muller, M. Genome-wide analysis of PPARα activation in murine small intestine. *Physiol. Genom.* **2007**, *30*, 192–204. [CrossRef] [PubMed]
21. Varnat, F.; Heggeler, B.B.; Grisel, P.; Boucard, N.; Corthesy-Theulaz, I.; Wahli, W.; Desvergne, B. PPARβ/delta regulates paneth cell differentiation via controlling the hedgehog signaling pathway. *Gastroenterology* **2006**, *131*, 538–553. [CrossRef] [PubMed]
22. Poirier, H.; Niot, I.; Monnot, M.C.; Braissant, O.; Meunier-Durmort, C.; Costet, P.; Pineau, T.; Wahli, W.; Willson, T.M.; Besnard, P. Differential involvement of peroxisome-proliferator-activated receptors α and delta in fibrate and fatty-acid-mediated inductions of the gene encoding liver fatty-acid-binding protein in the liver and the small intestine. *Biochem. J.* **2001**, *355*, 481–488. [CrossRef] [PubMed]
23. Byndloss, M.X.; Olsan, E.E.; Rivera-Chavez, F.; Tiffany, C.R.; Cevallos, S.A.; Lokken, K.L.; Torres, T.P.; Byndloss, A.J.; Faber, F.; Gao, Y.; et al. Microbiota-activated PPAR-γ signaling inhibits dysbiotic Enterobacteriaceae expansion. *Science* **2017**, *357*, 570–575. [CrossRef] [PubMed]
24. Escher, P.; Braissant, O.; Basu-Modak, S.; Michalik, L.; Wahli, W.; Desvergne, B. Rat PPARs: Quantitative analysis in adult rat tissues and regulation in fasting and refeeding. *Endocrinology* **2001**, *142*, 4195–4202. [CrossRef] [PubMed]
25. Harmon, G.S.; Dumlao, D.S.; Ng, D.T.; Barrett, K.E.; Dennis, E.A.; Dong, H.; Glass, C.K. Pharmacological correction of a defect in PPAR-γ signaling ameliorates disease severity in *Cftr*-deficient mice. *Nat. Med.* **2010**, *16*, 313–318. [CrossRef] [PubMed]
26. Mansen, A.; Guardiola-Diaz, H.; Rafter, J.; Branting, C.; Gustafsson, J.A. Expression of the peroxisome proliferator-activated receptor (PPAR) in the mouse colonic mucosa. *Biochem. Biophys. Res. Commun.* **1996**, *222*, 844–851. [CrossRef] [PubMed]
27. Braissant, O.; Foufelle, F.; Scotto, C.; Dauca, M.; Wahli, W. Differential expression of peroxisome proliferator-activated receptors (PPARs): Tissue distribution of PPAR-α, -β, and -γ in the adult rat. *Endocrinology* **1996**, *137*, 354–366. [CrossRef] [PubMed]
28. Marion-Letellier, R.; Dechelotte, P.; Iacucci, M.; Ghosh, S. Dietary modulation of peroxisome proliferator-activated receptor γ. *Gut* **2009**, *58*, 586–593. [CrossRef] [PubMed]
29. Willson, T.M.; Wahli, W. Peroxisome proliferator-activated receptor agonists. *Curr. Opin. Chem. Biol.* **1997**, *1*, 235–241. [CrossRef]
30. Bassaganya-Riera, J.; Hontecillas, R. CLA and n-3 PUFA differentially modulate clinical activity and colonic PPAR-responsive gene expression in a pig model of experimental IBD. *Clin. Nutr.* **2006**, *25*, 454–465. [CrossRef] [PubMed]
31. Lewis, J.D.; Lichtenstein, G.R.; Deren, J.J.; Sands, B.E.; Hanauer, S.B.; Katz, J.A.; Lashner, B.; Present, D.H.; Chuai, S.; Ellenberg, J.H.; et al. Rosiglitazone for active ulcerative colitis: A randomized placebo-controlled trial. *Gastroenterology* **2008**, *134*, 688–695. [CrossRef] [PubMed]
32. Lewis, J.D.; Lichtenstein, G.R.; Stein, R.B.; Deren, J.J.; Judge, T.A.; Fogt, F.; Furth, E.E.; Demissie, E.J.; Hurd, L.B.; Su, C.G.; et al. An open-label trial of the PPAR-γ ligand rosiglitazone for active ulcerative colitis. *Am. J. Gastroenterol.* **2001**, *96*, 3323–3328. [PubMed]
33. Sanchez-Hidalgo, M.; Martin, A.R.; Villegas, I.; de la Lastra, C.A. Rosiglitazone, a PPARγ ligand, modulates signal transduction pathways during the development of acute TNBS-induced colitis in rats. *Eur. J. Pharmacol.* **2007**, *562*, 247–258. [CrossRef] [PubMed]
34. Shah, Y.M.; Morimura, K.; Gonzalez, F.J. Expression of peroxisome proliferator-activated receptor-γ in macrophage suppresses experimentally induced colitis. *Am. J. Physiol. Gastrointest. Liver Physiol.* **2007**, *292*, G657–G666. [CrossRef] [PubMed]
35. Su, C.G.; Wen, X.; Bailey, S.T.; Jiang, W.; Rangwala, S.M.; Keilbaugh, S.A.; Flanigan, A.; Murthy, S.; Lazar, M.A.; Wu, G.D. A novel therapy for colitis utilizing PPAR-γ ligands to inhibit the epithelial inflammatory response. *J. Clin Investig.* **1999**, *104*, 383–389. [CrossRef] [PubMed]
36. Rousseaux, C.; Lefebvre, B.; Dubuquoy, L.; Lefebvre, P.; Romano, O.; Auwerx, J.; Metzger, D.; Wahli, W.; Desvergne, B.; Naccari, G.C.; et al. Intestinal antiinflammatory effect of 5-aminosalicylic acid is dependent on peroxisome proliferator-activated receptor-γ. *J. Exp. Med.* **2005**, *201*, 1205–1215. [CrossRef] [PubMed]

37. Wahli, W. A gut feeling of the PXR, PPAR and NF-κB connection. *J. Intern. Med.* **2008**, *263*, 613–619. [CrossRef] [PubMed]
38. Cerbone, A.; Toaldo, C.; Laurora, S.; Briatore, F.; Pizzimenti, S.; Dianzani, M.U.; Ferretti, C.; Barrera, G. 4-Hydroxynonenal and PPARγ ligands affect proliferation, differentiation, and apoptosis in colon cancer cells. *Free Radic. Biol. Med.* **2007**, *42*, 1661–1670. [CrossRef] [PubMed]
39. Martinasso, G.; Oraldi, M.; Trombetta, A.; Maggiora, M.; Bertetto, O.; Canuto, R.A.; Muzio, G. Involvement of PPARs in Cell Proliferation and Apoptosis in Human Colon Cancer Specimens and in Normal and Cancer Cell Lines. *PPAR Res.* **2007**, *2007*. [CrossRef] [PubMed]
40. Sharma, C.; Pradeep, A.; Wong, L.; Rana, A.; Rana, B. Peroxisome proliferator-activated receptor γ activation can regulate β-catenin levels via a proteasome-mediated and adenomatous polyposis coli-independent pathway. *J. Biol. Chem.* **2004**, *279*, 35583–35594. [CrossRef] [PubMed]
41. Xu, W.P.; Zhang, X.; Xie, W.F. Differentiation therapy for solid tumors. *J. Dig. Dis.* **2014**, *15*, 159–165. [CrossRef] [PubMed]
42. Shao, J.; Sheng, H.; DuBois, R.N. Peroxisome proliferator-activated receptors modulate K-Ras-mediated transformation of intestinal epithelial cells. *Cancer Res.* **2002**, *62*, 3282–3288. [PubMed]
43. Sato, N.; Kozar, R.A.; Zou, L.; Weatherall, J.M.; Attuwaybi, B.; Moore-Olufemi, S.D.; Weisbrodt, N.W.; Moore, F.A. Peroxisome proliferator-activated receptor γ mediates protection against cyclooxygenase-2-induced gut dysfunction in a rodent model of mesenteric ischemia/reperfusion. *Shock* **2005**, *24*, 462–469. [CrossRef] [PubMed]
44. Duszka, K.; Picard, A.; Ellero-Simatos, S.; Chen, J.; Defernez, M.; Paramalingam, E.; Pigram, A.; Vanoaica, L.; Canlet, C.; Parini, P.; et al. Intestinal PPARγ signalling is required for sympathetic nervous system activation in response to caloric restriction. *Sci. Rep.* **2016**, *6*. [CrossRef] [PubMed]
45. Anghel, S.I.; Wahli, W. Fat poetry: A kingdom for PPAR γ. *Cell Res.* **2007**, *17*, 486–511. [CrossRef] [PubMed]
46. Brasaemle, D.L. Thematic review series: Adipocyte biology. The perilipin family of structural lipid droplet proteins: Stabilization of lipid droplets and control of lipolysis. *J. Lipid Res.* **2007**, *48*, 2547–2559. [PubMed]
47. Macneil, D.J. The role of melanin-concentrating hormone and its receptors in energy homeostasis. *Front. Endocrinol.* **2013**, *4*. [CrossRef] [PubMed]
48. Shearman, L.P.; Camacho, R.E.; Sloan Stribling, D.; Zhou, D.; Bednarek, M.A.; Hreniuk, D.L.; Feighner, S.D.; Tan, C.P.; Howard, A.D.; van der Ploeg, L.H.; et al. Chronic MCH-1 receptor modulation alters appetite, body weight and adiposity in rats. *Eur. J. Pharmacol.* **2003**, *475*, 37–47. [CrossRef]
49. Krey, G.; Braissant, O.; L'Horset, F.; Kalkhoven, E.; Perroud, M.; Parker, M.G.; Wahli, W. Fatty acids, eicosanoids, and hypolipidemic agents identified as ligands of peroxisome proliferator-activated receptors by coactivator-dependent receptor ligand assay. *Mol. Endocrinol.* **1997**, *11*, 779–791. [CrossRef] [PubMed]
50. Yu, S.; Viswakarma, N.; Batra, S.K.; Sambasiva Rao, M.; Reddy, J.K. Identification of promethin and PGLP as two novel up-regulated genes in PPARγ1-induced adipogenic mouse liver. *Biochimie* **2004**, *86*, 743–761. [CrossRef] [PubMed]
51. Madsen, M.S.; Siersbaek, R.; Boergesen, M.; Nielsen, R.; Mandrup, S. Peroxisome proliferator-activated receptor γ and C/EBPα synergistically activate key metabolic adipocyte genes by assisted loading. *Mol. Cell. Biol.* **2014**, *34*, 939–954. [CrossRef] [PubMed]
52. Yu, S.; Matsusue, K.; Kashireddy, P.; Cao, W.Q.; Yeldandi, V.; Yeldandi, A.V.; Rao, M.S.; Gonzalez, F.J.; Reddy, J.K. Adipocyte-specific gene expression and adipogenic steatosis in the mouse liver due to peroxisome proliferator-activated receptor γ1 (PPARγ1) overexpression. *J. Biol. Chem.* **2003**, *278*, 498–505. [CrossRef] [PubMed]
53. Berry, A.; Balard, P.; Coste, A.; Olagnier, D.; Lagane, C.; Authier, H.; Benoit-Vical, F.; Lepert, J.C.; Seguela, J.P.; Magnaval, J.F.; et al. IL-13 induces expression of CD36 in human monocytes through PPARγ activation. *Eur. J. Immunol.* **2007**, *37*, 1642–1652. [CrossRef] [PubMed]
54. Deng, T.; Shan, S.; Li, P.P.; Shen, Z.F.; Lu, X.P.; Cheng, J.; Ning, Z.Q. Peroxisome proliferator-activated receptor-γ transcriptionally up-regulates hormone-sensitive lipase via the involvement of specificity protein-1. *Endocrinology* **2006**, *147*, 875–884. [CrossRef] [PubMed]
55. Nielsen, R.; Pedersen, T.A.; Hagenbeek, D.; Moulos, P.; Siersbaek, R.; Megens, E.; Denissov, S.; Borgesen, M.; Francoijs, K.J.; Mandrup, S.; et al. Genome-wide profiling of PPARγ: RXR and RNA polymerase II occupancy reveals temporal activation of distinct metabolic pathways and changes in RXR dimer composition during adipogenesis. *Genes Dev.* **2008**, *22*, 2953–2967. [CrossRef] [PubMed]

56. Kershaw, E.E.; Schupp, M.; Guan, H.P.; Gardner, N.P.; Lazar, M.A.; Flier, J.S. PPARγ regulates adipose triglyceride lipase in adipocytes in vitro and in vivo. *Am. J. Physiol. Endocrinol. Metab.* **2007**, *293*, E1736–E1745. [CrossRef] [PubMed]
57. Lapsys, N.M.; Kriketos, A.D.; Lim-Fraser, M.; Poynten, A.M.; Lowy, A.; Furler, S.M.; Chisholm, D.J.; Cooney, G.J. Expression of genes involved in lipid metabolism correlate with peroxisome proliferator-activated receptor γ expression in human skeletal muscle. *J. Clin. Endocrinol. Metab.* **2000**, *85*, 4293–4297. [CrossRef] [PubMed]
58. Matsusue, K.; Haluzik, M.; Lambert, G.; Yim, S.H.; Gavrilova, O.; Ward, J.M.; Brewer, B., Jr.; Reitman, M.L.; Gonzalez, F.J. Liver-specific disruption of PPARγ in leptin-deficient mice improves fatty liver but aggravates diabetic phenotypes. *J. Clin. Investig.* **2003**, *111*, 737–747. [CrossRef] [PubMed]
59. Graugnard, D.E.; Piantoni, P.; Bionaz, M.; Berger, L.L.; Faulkner, D.B.; Loor, J.J. Adipogenic and energy metabolism gene networks in longissimus lumborum during rapid post-weaning growth in Angus and Angus x Simmental cattle fed high-starch or low-starch diets. *BMC Genom.* **2009**, *10*. [CrossRef] [PubMed]
60. Young, M.E.; Goodwin, G.W.; Ying, J.; Guthrie, P.; Wilson, C.R.; Laws, F.A.; Taegtmeyer, H. Regulation of cardiac and skeletal muscle malonyl-CoA decarboxylase by fatty acids. *Am. J. Physiol. Endocrinol. Metab.* **2001**, *280*, E471–E479. [PubMed]
61. Cao, J.; Li, J.L.; Li, D.; Tobin, J.F.; Gimeno, R.E. Molecular identification of microsomal acyl-CoA: Glycerol-3-phosphate acyltransferase, a key enzyme in de novo triacylglycerol synthesis. *Proc. Natl. Acad. Sci. USA* **2006**, *103*, 19695–19700. [CrossRef] [PubMed]
62. Meyer, B.M.; Werth, B.A.; Beglinger, C.; Hildebrand, P.; Jansen, J.B.; Zach, D.; Rovati, L.C.; Stalder, G.A. Role of cholecystokinin in regulation of gastrointestinal motor functions. *Lancet* **1989**, *2*, 12–15. [CrossRef]
63. Harvey, R.F. Hormonal control of gastrointestinal motility. *Am. J. Dig. Dis.* **1975**, *20*, 523–539. [CrossRef] [PubMed]
64. Hammer, J.; Hammer, K.; Kletter, K. Lipids infused into the jejunum accelerate small intestinal transit but delay ileocolonic transit of solids and liquids. *Gut* **1998**, *43*, 111–116. [CrossRef] [PubMed]
65. Petit, V.; Arnould, L.; Martin, P.; Monnot, M.C.; Pineau, T.; Besnard, P.; Niot, I. Chronic high-fat diet affects intestinal fat absorption and postprandial triglyceride levels in the mouse. *J. Lipid Res.* **2007**, *48*, 278–287. [CrossRef] [PubMed]
66. Miller, M.S.; Galligan, J.J.; Burks, T.F. Accurate measurement of intestinal transit in the rat. *J. Pharmacol. Methods* **1981**, *6*, 211–217. [CrossRef]

International Journal of
Molecular Sciences

Article

Activation of PPARα by Oral Clofibrate Increases Renal Fatty Acid Oxidation in Developing Pigs

Yonghui He, Imad Khan, Xiumei Bai, Jack Odle and Lin Xi *

Laboratory of Developmental Nutrition, Department of Animal Sciences, North Carolina State University, Raleigh, NC 27695, USA; hyonghui@163.com (Y.H.); ikhan@ncsu.edu (I.K.); bxm8302@126.com (X.B.); jodle@ncsu.edu (J.O.)

* Correspondence: lin_xi@ncsu.edu; Tel.: +1-919-515-4014; Fax: +1-919-515-6884

Received: 15 November 2017; Accepted: 5 December 2017; Published: 8 December 2017

Abstract: The objective of this study was to evaluate the effects of peroxisome proliferator-activated receptor α (PPARα) activation by clofibrate on both mitochondrial and peroxisomal fatty acid oxidation in the developing kidney. Ten newborn pigs from 5 litters were randomly assigned to two groups and fed either 5 mL of a control vehicle (2% Tween 80) or a vehicle containing clofibrate (75 mg/kg body weight, treatment). The pigs received oral gavage daily for three days. In vitro fatty acid oxidation was then measured in kidneys with and without mitochondria inhibitors (antimycin A and rotenone) using [1-^{14}C]-labeled oleic acid (C18:1) and erucic acid (C22:1) as substrates. Clofibrate significantly stimulated C18:1 and C22:1 oxidation in mitochondria ($p < 0.001$) but not in peroxisomes. In addition, the oxidation rate of C18:1 was greater in mitochondria than peroxisomes, while the oxidation of C22:1 was higher in peroxisomes than mitochondria ($p < 0.001$). Consistent with the increase in fatty acid oxidation, the mRNA abundance and enzyme activity of carnitine palmitoyltransferase I (CPT I) in mitochondria were increased. Although mRNA of mitochondrial 3-hydroxy-3-methylglutaryl-coenzyme A synthase (mHMGCS) was increased, the β-hydroxybutyrate concentration measured in kidneys did not increase in pigs treated with clofibrate. These findings indicate that PPARα activation stimulates renal fatty acid oxidation but not ketogenesis.

Keywords: peroxisome proliferator-activated receptor α (PPARα); clofibrate; fatty acid β-oxidation; pigs

---

1. Introduction

The kidney is an organ with a high energy requirement due to its central role in the elimination of water-soluble metabolic waste products. Thus, energy metabolism is very active and important for renal physiology. In support of the high energy metabolism, renal fatty acid oxidation and carnitine biosynthesis are very active, generating ketone bodies when fatty acids are catabolized and in maintaining carnitine homeostasis, respectively [1]. Recently, a strong link between impaired renal energy metabolism and chronic kidney disease has been highly identified [2,3].

Peroxisome proliferator-activated receptor α (PPARα), a member of a large nuclear receptor superfamily, is expressed primarily in the liver, the intestine, and the kidney [4,5]. The critical role of PPARα activation in regulation of hepatic fatty acid oxidation, lipid metabolism, and inflammatory and vascular responses has been well studied [6]. In contrast with the liver, however, the data on the role of PPARα activation in the regulation of renal fatty acid oxidation and metabolism is scant, especially for developing animals. By comparison, both mitochondrial and peroxisomal β-oxidation enzymes are expressed in the liver and the kidney, but the enzymes in peroxisomes are less abundant in the kidney than in the liver. The response of mitochondrial and peroxisomal β-oxidation enzymes to PPARα activation in the kidney is also moderate [7]. Despite all this, the importance of peroxisomal β-oxidation in short-, long-, and very long-chain fatty acids has been well recognized. Moreover, the essential role of PPARα-induction of fatty acid metabolism in the prevention of renal ischemia and renal damage induced by drugs has been observed in rodent species [8–10].

Potential ligands for the PPARα transcription factor include fatty acids, eicosanoids, and pharmacological drugs such as the fibrates. Clofibrate is a potent PPARα activator that stimulates peroxisome proliferation and increases fatty acid oxidation in rodent species. The target genes of PPARα encode enzymes involved in peroxisomal and mitochondrial β-oxidation and ketone body synthesis. The peroxisome proliferation elicited by fibrates has drawn much attention because peroxisome proliferation has been associated with oxidative stress and hepatocellular carcinoma [11]. However, less is known about the impacts of the agonist in the kidney. Fatty acids are the preferred energy substrate for the kidney, and defects in fatty acid oxidation and mitochondrial and peroxisomal dysfunction are involved in acute renal injury and chronic disease. Indeed, PPARα signaling may play a protective role in acute free fatty acid-associated renal tubule toxicity [12]. PPARα activation has been recognized as essential for kidney function under both healthy and pathophysiological states [7].

Data regarding inborn errors in the kidney such as neonatal urea cycle defects and disorders of long-chain fatty acid oxidation associated with energy deficiency in infants is very limited in the literature. Understanding the renal kinetics and adaptation of energy metabolism is very important for human infant health. The domestic neonatal pig (*Sus scrofa*) ranks among the most prominent research models for the study of pediatric nutrition and metabolism due to the similarity of human infant and piglet physiology [13]. Unlike rodent species, the peroxisome proliferation and hepatocarcinogenic potencies of clofibrate are not observed in the livers of humans or pigs [14,15]. Peroxisomal β-oxidation (enzymes) increase with the age in the renal cortex of suckling rat pups, and this might be involved in PPARα-mediated mechanisms [16]. Similarly, previous work from our laboratory showed that fatty acid β-oxidation capacity was increased with age in the kidney of pigs as well, and the capacity was higher during the preweaning period than in adults [17]. The enzymatic responses to PPARα activation also were compared in the heart, kidney, and liver of pigs in our previous work, but effects of the activation on fatty acid oxidative metabolism were not determined. Promoting energy supply and thermogenesis after birth are critical for the survivor of neonatal piglets [17]. Therefore, to provide basic knowledge on the regulation of energy metabolism in the developing kidney, the present study assessed changes in peroxisomal and mitochondrial long-chain fatty acid oxidation in the kidney during early development in response to the activation of PPAR by clofibrate.

2. Results

2.1. β-Hydroxybutyrate Concentration

No differences were detected in β-hydroxybutyrate concentration measured in plasma and kidney tissues between control and clofibrate-treated pigs ($p > 0.05$). The concentration of β-hydroxybutyrate was on average 8-fold higher in kidney tissue compared with plasma (Figure 1).

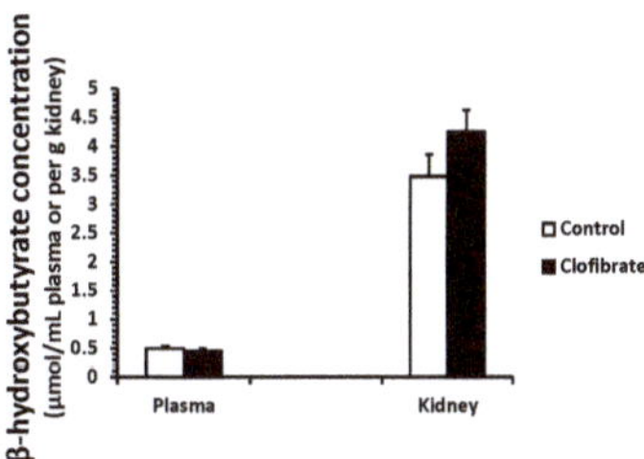

Figure 1. β-Hydroxybutyrate concentration in plasma and kidneys of neonatal piglets. Values are means ± SE ($n = 5$).

2.2. Fatty Acid Oxidation In Vitro

Clofibrate had no effects on the accumulation of $^{14}CO_2$ in peroxisomes from either [1-^{14}C]-C18:1 or C22:1 oxidation ($p > 0.05$), but the accumulation in mitochondria and in homogenates (a total of peroxisomes and mitochondria) from [1-^{14}C]-C18:1 was significantly higher in clofibrate-treated than control piglets ($p < 0.05$; Figure 2A). The $^{14}CO_2$ accumulation from [1-^{14}C]-C18:1 and C22:1 oxidation were on average 133- and 25-fold higher in mitochondria than peroxisomes ($p < 0.0025$). In addition, the $^{14}CO_2$ accumulation in mitochondria or homogenates were 2.5-fold greater from [1-^{14}C]-C18:1 than C22:1 ($p < 0.0009$).

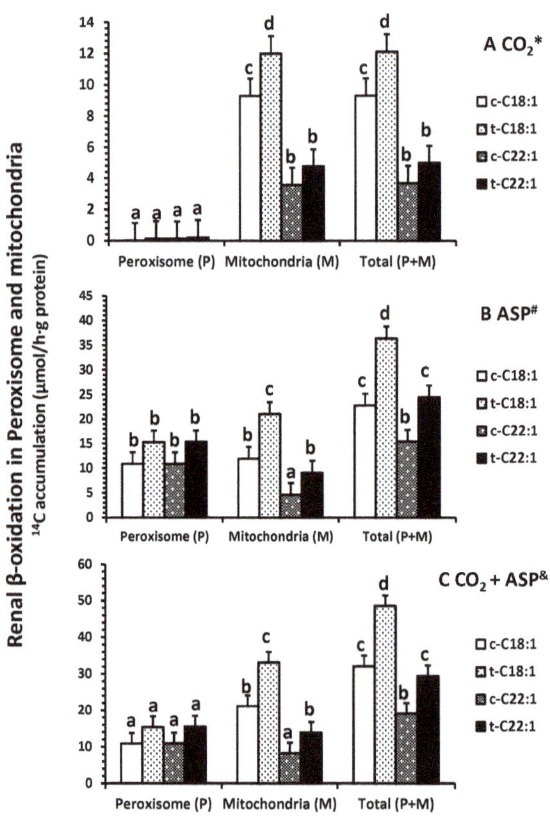

Figure 2. Effects of oral clofibrate on renal β-oxidation (^{14}C accumulation in CO_2 * (**A**); ASP # (**B**) and CO_2 + ASP & (**C**)) in peroxisomes and mitochondria of neonatal pigs. Values are least square means ± SEM ($n = 5$). Abbreviations: ASP: acid soluble product; c-C18:1: control with oleate; t-C18:1: treatment with oleate; c-C22:1: control with erucate; t-C22:1: treatment with erucate. * Clofibrate effect ($p < 0.037$) and fatty acid effect ($p < 0.0001$); # Clofibrate effect ($p < 0.0001$) and fatty acid effect ($p < 0.0001$); & Clofibrate effect ($p < 0.0001$) and fatty acid effect ($p < 0.0001$). Column with different letters differ ($p < 0.05$).

Clofibrate tended to increase the accumulation of ^{14}C in acid-soluble product (ASP) in peroxisomes from both [1-^{14}C]-C18:1 and C22:1 oxidation ($p = 0.06$), but the accumulation of ^{14}C in ASP from C18:1 and C22:1 in mitochondria and in homogenate were increased in clofibrate-treated compared to the control pigs ($p < 0.006$; Figure 2B). There was no difference between C18:1 and C22:1 in ^{14}C-ASP accumulation in peroxisomes, but the ^{14}C-ASP accumulation from C18:1 was greater than that from

C22:1 in mitochondria. The accumulations of ^{14}C-ASP in the homogenates also were 1.5-fold higher from [1-^{14}C]-C18:1 compared with C22:1 ($p < 0.001$).

By combining both $^{14}CO_2$ and ^{14}C-ASP (Figure 2C), the total oxidation (CO_2 + ASP) from either C18:1 or C22:1 was not affected by clofibrate in peroxisomes ($p > 0.05$). However, clofibrate increased mitochondrial oxidation of C18:1 by 56% and C22:1 by 70%. Thus, the total oxidation in homogenates was significantly higher from clofibrate-treated than control piglets ($p < 0.001$). The oxidation from C18:1 was on average 1.7-fold greater than that from C22:1 ($p < 0.03$).

No difference was observed in the percentage of ^{14}C accumulation in CO_2 (less than 2%) in peroxisomes ($p = 0.9$), but clofibrate reduced the percentage of accumulation of C22:1 in CO_2 in mitochondria ($p < 0.01$) (Figure 3A). Over 98% of the oxidative metabolites were ASP in peroxisomes, while only about 60% (54–64%) of the ASP was detected in mitochondria (Figure 3B). Clofibrate administration did not affect the percentage of ASP from C18:1 ($p = 0.13$) but increased the ASP from C22:1 significantly ($p < 0.04$) (Figure 3B). The percentage of total oxidation (CO_2 + ASP) from C22:1 in peroxisomes was 1.5-fold higher than that from C18:1, and the percentage of total oxidation from C18:1 in mitochondria was 1.5-fold higher than that from C22:1 (Figure 3C).

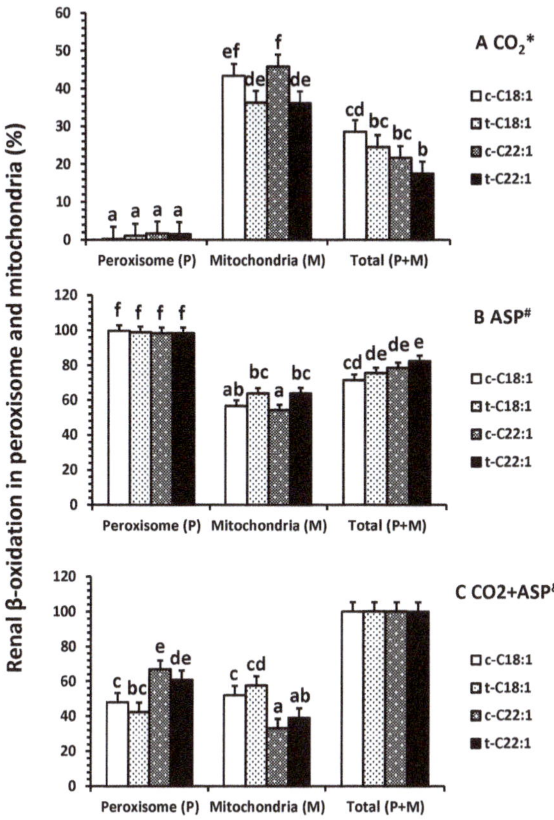

Figure 3. Percentage distribution of renal β-oxidation (% of ^{14}C accumulation in CO_2 * (**A**), ASP # (**B**), and CO_2 + ASP & (**C**)) in peroxisomes and mitochondria of neonatal pigs. Values are least square means ± SEM ($n = 5$). Abbreviations: ASP: acid soluble product; c-C18:1: control with oleate; t-C18:1: treatment with oleate; c-C22:1: control with erucate; t-C22:1: treatment with erucate. * Clofibrate effect ($p < 0.040$) and fatty acid effect ($p = 0.39$); # Clofibrate effect ($p < 0.0001$) and fatty acid effect ($p = 0.39$); & Clofibrate effect ($p = 1.0$) and fatty acid effect ($p = 1.0$). Column with different letters differ ($p < 0.05$).

2.3. Renal Enzyme Activity

The activity of carnitine palmitoyltransferase I (CPT I) was increased 25% by clofibrate ($p < 0.05$), but no effect on the activity of CPT II was detected ($p > 0.05$; Figure 4A). The activity of acyl-CoA oxidase (ACO) was increased 2.2-fold in clofibrate-treated pigs ($p < 0.05$; Figure 4B).

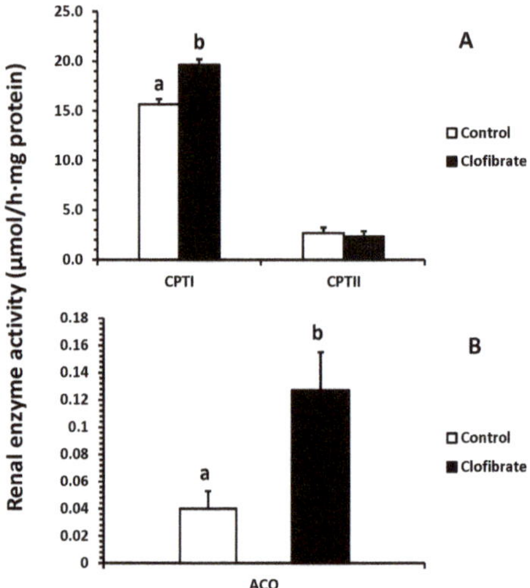

Figure 4. Effects of oral clofibrate on renal enzyme activity in neonatal pigs. Values are least square means ± SEM (n = 5). CPT I and CPT II, carnitine palmitoyltransferase I and II (**A**), and ACO, acyl-CoA oxidase (**B**). Columns with different letters are different ($p < 0.05$).

2.4. Renal mRNA Enrichment

Clofibrate administration had a great impact on the relative mRNA abundance of CPT I, CPT II, and mHMG-CoA, but had no effects on PPARα and ACO (Figure 5). The mRNA enrichments of CPT I, CPTII, and mitochondrial 3-hydroxy-3-methylglutaryl-CoA synthase (mHMG-CoA) were 3.5-, 2.6-, and 9.7-fold greater from clofibrate-treated pigs than control pigs ($p < 0.001$).

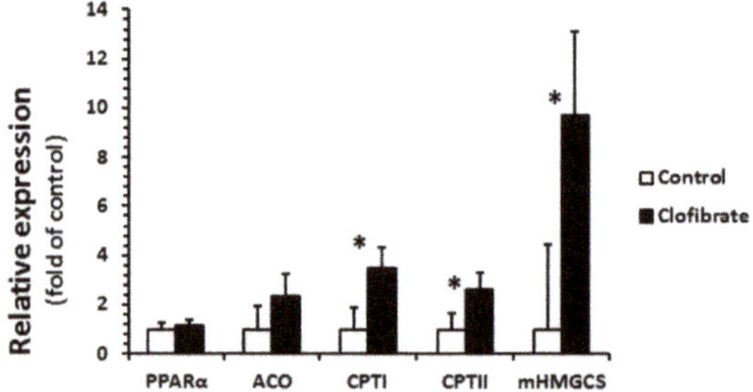

Figure 5. Renal mRNA abundance in pigs with and without oral clofibrate. Values are least square means ± SEM (n = 5). PPARα: peroxisome proliferator-activated receptor α; ACO: acyl-CoA oxidase; CPT I and CPT II: carnitine palmitoyltransferase I and II; mHMGCS: mitochondrial 3-hydroxy-3-methylglutaryl-coenzyme A synthase. * Significant difference between control and treatment groups ($p < 0.05$).

3. Discussion

Activation of PPARα by oral clofibrate administration to newborn piglets resulted in a significant increase in renal fatty acid β-oxidation. Similar observations were reported in humans and rats [18]. Fatty acid β-oxidation is the primary pathway of ATP production for the kidney to meet its daily function requirement. Therefore, this result implied that PPARα could play an important regulatory role in ATP production and energy metabolism in the developing kidney. We also noticed that the induction profiles were different in mitochondria and peroxisomes for the long- and very long-chain fatty acids, suggesting that the response of renal fatty acid β-oxidation to PPARα activation depends on the subcellular and substrates.

The activation had no significant impact on the fatty acid β-oxidation (^{14}C accumulation in CO_2 or/and ASP) in renal peroxisomes, although the ACO activity increased 2.2-fold in clofibrate-treated piglets. Only a tendency of increase in ASP (p = 0.06) was observed, and the mild response of peroxisomal β-oxidation to the PPARα agonist was similar to that reported in adult rats [19]. As in mitochondria, fatty acid β-oxidation in peroxisomes involves multiple enzymes that ultimately yield acetyl-CoA [20]. However, the peroxisomal fatty acid β-oxidation is not coupled with ATP synthesis and catalase is required for H_2O_2 produced in peroxisomes by transferring electrons to O_2. It was reported that the activation of PPARα had no influence on catalase activity in 14-day-old piglets [21], and catalase increases fast after birth [22]. This result could be related to the catalase or other enzymes in β-oxidation system of peroxisomes such as the bifunctional protein and 3-ketoacyl-CoA thiolase during development. In addition, we did not find any difference in renal PPARα and ACO mRNA enrichments between control and clofibrate-treated piglets. The low response of PPARα and ACO mRNA to clofibrate induction was observed in the livers of newborn, 24-hour-old, and 4-day-old fasted neonatal piglets [21,23,24]. Besides, the ACO activity measured in kidneys of 14-day-old control pigs was not different from pigs treated with clofibrate [21]. Because the rates of mitochondrial and peroxisomal β-oxidation of palmitate change during postnatal development and food deprivation in pig kidneys [22], age or physiological status and even species could contribute to these differences.

A similar ^{14}C-accumulation rate in CO_2 or/and ASP from both C18:1 and C22:1 was detected in peroxisomes, indicating that the chain-length of these two fatty acids had no effects on peroxismal fatty acid β-oxidation. However, the percentage of peroxisomal fatty acid β-oxidation increased with the increase in the fatty acid chain-length. The percentage of β-oxidation of C22:1 was on average 40% higher than that of C18:1, although the total fatty acid oxidation rate had no difference. A similar result was detected in the liver [23], demonstrating that C22:1 has a preference to be oxidized in peroxisomes. The preference for C22:1 appeared to be associated with the affinity of fatty acid activation systems for long-chain fatty acid and very-long-chain fatty acid identified in rat [25]. It was very interesting that a high percentage (about 42–67%) of the fatty acids were oxidized in renal peroxisomes with 98–99% as ASP and 1–2% as CO_2, and the activation of PPARα had no influence on the percentage distribution of fatty acid oxidation. The contribution of peroxisomal fatty acid β-oxidation to the total fatty acid β-oxidation in the kidney was similar to that measured in the liver (40–47) and 2-fold higher than that in rats (20–35% [26]).

Mitochondrial fatty acid oxidation was increased significantly by the activation of PPARα induced by clofibrate administration. Consistent with the increase in fatty acid β-oxidation, the CPT I activity was increased by 25% and mRNA expression was increased 3.5-fold. In addition, the chain-length of fatty acid significantly affected mitochondrial β-oxidation, and the ^{14}C-accumulations were much greater from C18:1 than C22:1 in both of CO_2 (2.6-fold) and ASP (2.3-fold). Similar results were observed in livers of PPARα-activated neonatal pigs with clofibrate administration [23]. Swine milk fat is known to be composed of mainly long chain fatty acids (LCFAs) and very long chain fatty acids (VLCFAs). These results indicate that mitochondrial oxidation of LCFAs provides an important source of energy for kidneys, and activation of PPARα could promote the utilization of LCFAs and VLCFAs in developing kidneys.

The $^{14}CO_2$ accumulation rates from C18:1 and C22:1 (μmol/h·g protein) were on average 64% and 50% higher in the kidney (10.7 and 4.3) than in the liver (3.9 and 2.2; [23]), while the ^{14}C accumulations in ASP from C18:1 and C22:1 were 52% and 55% greater in the liver (44.9 and 30.8; [23]) than in the kidney (29.6 and 19.9). It was recently demonstrated that, in rat kidneys, proximal tubules do not generate energy via glycolysis and are completely dependent on oxidative phosphorylation for ATP production, although energy production is primarily from fuels such as lactate, glutamine, and free fatty acids [27]. On the other hand, fatty acid elongation can occur in both livers and kidneys, but it was reported that the specific activity of the fatty acid elongation in the kidney is about 30% compared to the liver. Different incorporation rates [1-^{14}C] acetate into fatty acids were observed in the mitochondria elongation system between livers and kidneys in the presence of nicotinamide adenine dinucleotide + hydrogen (NADH), nicotinamide adenine dinucleotide phosphate + hydrogen (NADPH), or both NADH and NADPH as the hydrogen donor [28]. Thus, the results demonstrated that fatty acid catabolic metabolism in mitochondria and citric acid cycle is the primary emergent source in developing kidneys and that activation of PPARα might have a benefit to kidney development via improving fatty acid utilization. Compared with kidneys, the liver may need to produce more ASP in which acetate was found to be one of the primary product in piglets [29].

The mitochondrial 3-hydroxy-3-methylglutaryl-CoA synthase (mHMGCS) mRNA increased 9.7-fold in clofibrate-treated pigs, but the induction of mHMGCS had no influence on plasma and renal β-hydroxybutyrate concentrations. Although the activity of mHMGCS was not measured in this study, available evidence confirms that the enzyme activity in the liver remains low until the weaned age of pigs [30]. Ketone bodies are transferred in and out of cells by monocarboxylate transporter 1. In wild-type mice, treatment with WY 14,643 increased mRNA concentrations of monocarboxylate transporter 1 in the liver, the small intestine, and the kidney, but no upregulation was observed in PPARα-null mice [31]. This suggested that activation of PPARα could potentially promote ketone body production and transfer from organs to plasma. However, we found that β-hydroxybutyrate concentration was 8-fold higher in the kidney tissue than plasma, suggesting that the contribution of the kidney to plasma ketone bodies is minimal in this species. It has been well known that suckling pigs are hypoketonemic despite elevated dietary fat after birth [30].

4. Materials and Methods

4.1. Experiment Design and Animal Model

All experimental procedures were approved by the North Carolina State University Animal Care and Use Committee. Ten male newborn pigs (Landrace × Yorkshire × Duroc), 2 from each of 5 L, were used in this experiment. The selected newborn piglets (Body weight (BW) = 1.61 ± 0.06 kg) were allocated randomly into two treatments: control and clofibrate. The control piglets were orogastrically gavaged with 2 mL of 2% Tween 80, and the clofibrate-treated piglets were orogastrically gavaged to 2 mL of 2% Tween 80 containing clofibrate (75 mg/kg BW; Cayman Chemicals, Ann Arbor, MI, USA) at 8:00 a.m. of each day for 4 days as described previously [23]. All piglets were kept with their dams and siblings at the North Carolina State University Swine Educational Unit in Raleigh, North Carolina during the experiment. The piglets were euthanized by AVMA-approved electrocution on Day 4 after gavaging and feeding, and kidney and blood samples were collected. Fresh kidney samples were collected and stored in a homogenate buffer, and extra kidney samples were immersed in liquid nitrogen and stored at −80 °C. The blood was sampled with vacutainer containing sodium heparin and centrifuged at 2500 rpm × 10 min. The plasma was collected and stored at −20 °C.

4.2. β-Hydroxybutyrate Concentration

A BioVision β-hydroxybutyrate assay kit (K632-100; BioVision, Milpitas, CA, USA) was used to measure the β-hydroxybutyrate concentration in the plasma and kidney samples. The standard curve and samples were prepared according to the BioVision assay procedure and allowed to develop at room temperature for 30 min. The samples were measured with a BioTek reader (Synergy HT, Winooski, VT, USA) at an absorbance of 450 nm.

4.3. Fatty Acid Oxidation In Vitro

Fresh kidney homogenates (~5 mg) were incubated in 3 mL of reverse Krebs–Henseleit bicarbonate medium with or without rotenone and antimycin A (10 + 50 μmol/L), blockers of mitochondrial respiratory system. Mitochondrial and peroxisomal fatty acid oxidations were measured in the medium using either [1-^{14}C]-labeled oleic acid (C18:1) or erucic acid (C22:1) purchased from American Radiolabeled Chemicals (ARC; Saint Louis, MO, USA) as substrate. The biochemical and radio-chemical purities of both C18:1 and C22:1 were greater than 99% based on TLC and HPLC analyses. The fatty acids were bound to fatty acid-free BSA (5:1, molar ratio) and dissolved in the reaction medium. The measurements were performed in 25 mL Erlenmeyer flasks containing 2 mL of the reaction medium. The medium was incubated with 2 μmol [1-^{14}C]-C18:1 (0.98 kBq/μmol) or [1-^{14}C]-C22:1 (1.37 kBq/μmol). The incubation was stopped after 30 min by the addition of 0.5 mL of 35% $HClO_4$. The ^{14}C accumulation in CO_2 and acid-soluble products (ASP) were collected, processed, and analyzed by liquid scintillation spectrometry (Beckman LS 6000IC, Fullerton, CA, USA) according to the procedures by Lin et al. [24].

4.4. CPTI Activity

Kidney mitochondria were isolated from fresh samples. The samples were homogenized in an isolation buffer and centrifuged with a gradient centrifugation [32]. The mitochondria pellet was collected, and the protein concentration was determined using the biuret method as previously described [32]. The CPTI activity was assayed in the mitochondria at 30 °C with 80 μmol/L palmitoyl-CoA following the method used previously [32]. The assays were performed with or without supplementation of 4.7 μg/mL of malonyl-CoA. The assay was initiated by the addition of 20 μL of ^{3}H-carnitine (166.5 kBq/μmol) purchased from ARC and terminated with the addition of 4 mL of 6% $HClO_4$ after 6 min incubation. The activity was determined by measuring the ^{3}H-labeled palmitoyl-carnitine generated from the reactions. The radioactivity was determined using the Beckman liquid scintillation spectrometry (Beckman LS 6000IC, Fullerton, CA, USA).

4.5. ACO Activity

The fatty acyl-CoA oxidase (ACO) activity was measured by using a fluorometric procedure with scopoletin, a fluorescing compound as described previously [24]. The reduction of the ACO produced H_2O_2 was coupled to the oxidation of scopoletin to its non-fluorescing product. The control and treatment kidney samples were prepared as described previously [32] and were incubated at 37 °C for 20 min. A standard curve was generated consisting of (0–0.1 μm) concentrations of H_2O_2. The samples were measured with a BioTek reader (Synergy HT, Winooski, VT, USA) with an emission at 460 nm and an excitation at 360 nm.

4.6. mRNA Expression

Total mRNA was extracted using guanidine isothiocynate and phenol, and was quantified using NanoDrop spectrometer (Thermo Scientific, Wilmington, DE, USA). The mRNA was treated with Turbo DNase (Ambion, Austin, TX, USA) and transcribed using iScripTM Select cDNA synthesis kit (Bio-Rad Laboratories, Hercules, CA, USA). Primers were designed with the use of GenBank as described previously [32]. The mRNA abundances were measured with MyiQ Single Color RT-PCR (Bio-Rad Laboratories, Hercules, CA, USA).

4.7. Statistical Analysis

Data from plasma β-hydroxybutyrate, tissue enzyme activity and mRNA enrichment assays, were analyzed using the GLM procedure of SAS (Proprietary Software 9.3 (TS1M1), SAS Institute Inc., Cary, NC, USA) according to a randomized complete block design with 2 treatments (control and clofibrate), blocked by litter. Data from in vitro fatty acid oxidation measurements were analyzed with a split-plot design, including a main plot (control vs. clofibrate) in randomized blocks and a subplot modeling fatty acid chain length (C18:1 vs. C22:1) effects, subcellular (mitochondria vs. peroxisomes) differences, and interactions. Multiple comparisons between treatments were performed using Tukey's test, with significance declared when $p \leq 0.05$ and tendencies noted when $0.05 \leq p \leq 0.1$.

5. Conclusions

Activation of PPARα by clofibrate resulted in a greater increase in mitochondrial long-chain fatty acid oxidation in developing kidneys. The increase was elicited with induced enzyme activity and mRNA expression implies that PPARα activation could improve renal energy utilization during development. More than 40% of the catabolic metabolism occurred in mitochondria and citric acid cycle, suggesting that mitochondrial fatty acid oxidation plays a primary role in energy generation in developing kidneys. However, the activation did not alter the β-hydroxybutyrate concentration in plasma or kidneys.

Acknowledgments: This project was supported by a National Research Initiative Competitive Grant No. 2007-35206-17897 and 2015-67015-23245 from the United States Department of Agriculture (USDA) National Institute of Food and Agriculture and by the North Carolina Agricultural Research Service. Funds received from the grants cover the costs to publish this article in open access.

Author Contributions: Yonghui He conducted sample analyses, summarized data analyses, and wrote the manuscript. Imad Khan performed tissue processing and fatty acid oxidation assays in vitro and conducted sample analyses. Xiumei Bai conducted the animal experiment and collected the tissue samples. Jack Odle participated in the experimental design, result discussion, and manuscript review. Lin Xi designed the experiment, organized the experiment procedures and sample analyses, and participated in result discussion and manuscript writing.

Conflicts of Interest: The authors declare no conflicts of interest.

Abbreviations

C18:1	oleic acid
C20:1	erucic acid
PPARα	peroxisome proliferator-activated receptor α
mHMGCS	mitochondrial 3-hydroxy-3-methylglutaryl-coenzyme A synthase
CPT	carnitine palmitoyltransferase

References

1. Broderick, T.L.; Cusimano, F.A.; Carlson, C.; Tamura, L.K. Acute exercise stimulates carnitine biosynthesis and OCTN2 expression in mouse kidney. *Kidney Blood Press Res.* **2017**, *42*, 398–405. [CrossRef] [PubMed]
2. Mount, P.F.; Power, D.A. Balancing the energy equation for healthy kidneys. *J. Pathol.* **2015**, *237*, 407–410. [CrossRef] [PubMed]
3. Vasko, R. Peroxisomes and kidney injury. *Antioxid. Redox Signal.* **2016**, *25*, 217–231. [CrossRef] [PubMed]
4. Latruffe, N.; Cherkaoui Malki, M.; Nicolas-Frances, V.; Clemencet, M.C.; Jannin, B.; Berlot, J.P. Regulation of the peroxisomal beta-oxidation-dependent pathway by peroxisome proliferator-activated receptor alpha and kinases. *Biochem. Pharmacol.* **2000**, *60*, 1027–1032. [CrossRef]
5. Latruffe, N.; Cherkaoui Malki, M.; Nicolas-Frances, V.; Jannin, B.; Clemencet, M.C.; Hansmannel, F.; Passilly-Degrace, P.; Berlot, J.P. Peroxisome-proliferator-activated receptors as physiological sensors of fatty acid metabolism: Molecular regulation in peroxisomes. *Biochem. Soc. Trans.* **2001**, *29*, 305–309. [CrossRef] [PubMed]
6. Kouroumichakis, I.; Papanas, N.; Zarogoulidis, P.; Liakopoulos, V.; Maltezos, E.; Mikhailidis, D.P. Fibrates: Therapeutic potential for diabetic nephropathy? *Eur. J. Intern. Med.* **2012**, *23*, 309–316. [CrossRef] [PubMed]

7. Cook, W.S.; Yeldandi, A.V.; Rao, M.S.; Hashimoto, T.; Reddy, J.K. Less extrahepatic induction of fatty acid beta-oxidation enzymes by PPAR alpha. *Biochem. Biophys. Res. Commun.* **2000**, *278*, 250–257. [CrossRef] [PubMed]
8. Sugden, M.C.; Bulmer, K.; Gibbons, G.F.; Holness, M.J. Role of peroxisome proliferator-activated receptor-alpha in the mechanism underlying changes in renal pyruvate dehydrogenase kinase isoform 4 protein expression in starvation and after refeeding. *Arch. Biochem. Biophys.* **2001**, *395*, 246–252. [CrossRef] [PubMed]
9. Lopez-Hernandez, F.J.; Lopez-Novoa, J.M. Potential utility of PPARalpha activation in the prevention of ischemic and drug-induced acute renal damage. *Kidney Int.* **2009**, *76*, 1022–1024. [CrossRef] [PubMed]
10. Sivarajah, A.; Chatterjee, P.K.; Hattori, Y.; Brown, P.A.; Stewart, K.N.; Todorovic, Z.; Mota-Filipe, H.; Thiemermann, C. Agonists of peroxisome-proliferator activated receptor-alpha (clofibrate and WY14643) reduce renal ischemia/reperfusion injury in the rat. *Med. Sci. Monit.* **2002**, *8*, BR532–BR539. [PubMed]
11. Reddy, J.K.; Warren, J.R.; Reddy, M.K.; Lalwani, N.D. Hepatic and renal effects of peroxisome proliferators: Biological implications. *Ann. N. Y. Acad. Sci.* **1982**, *386*, 81–110. [CrossRef] [PubMed]
12. Takahashi, K.; Kamijo, Y.; Hora, K.; Hashimoto, K.; Higuchi, M.; Nakajima, T.; Ehara, T.; Shigematsu, H.; Gonzalez, F.J.; Aoyama, T. Pretreatment by low-dose fibrates protects against acute free fatty acid-induced renal tubule toxicity by counteracting PPARα deterioration. *Toxicol. Appl. Pharmacol.* **2011**, *252*, 237–249. [CrossRef] [PubMed]
13. Odle, J.; Lin, X.; Jacobi, S.K.; Kim, S.W.; Stahl, C.H. The suckling piglet as an agrimedical model for the study of pediatric nutrition and metabolism. *Annu. Rev. Anim. Biosci.* **2014**, *2*, 419–444. [CrossRef] [PubMed]
14. Vamecq, J.; Draye, J.P. Pathophysiology of peroxisomal beta-oxidation. *Essays Biochem.* **1989**, *24*, 115–225. [PubMed]
15. Palmer, C.N.; Hsu, M.H.; Griffin, K.J.; Raucy, J.L.; Johnson, E.F. Peroxisome proliferator activated receptor-alpha expression in human liver. *Mol. Pharmacol.* **1998**, *53*, 14–22. [PubMed]
16. Ouali, F.; Djouadi, F.; Merlet-Bénichou, C.; Bastin, J. Dietary lipids regulate beta-oxidation enzyme gene expression in the developing rat kidney. *Am. J. Physiol.* **1998**, *275*, F777–F784. [PubMed]
17. Yu, X.X.; Drackley, J.K.; Odle, J. Rates of mitochondrial and peroxisomal beta-oxidation of palmitate change during postnatal development and food deprivation in liver, kidney and heart of pigs. *J. Nutr.* **1997**, *127*, 1814–1821. [PubMed]
18. Veerkamp, J.H.; van Moerkerk, H.T. Peroxisomal fatty acid oxidation in rat and human tissues. Effect of nutritional state, clofibrate treatment and postnatal development in the rat. *Biochim. Biophys. Acta* **1986**, *875*, 301–310. [CrossRef]
19. Vamecq, J.; Draye, J.P. Peroxisomal and mitochondrial beta-oxidation of monocarboxylyl-CoA, omega-hydroxymonocarboxylyl-CoA and dicarboxylyl-CoA esters in tissues from untreated and clofibrate-treated rats. *J. Biochem.* **1989**, *106*, 216–222. [CrossRef] [PubMed]
20. Poirier, Y.; Antonenkov, V.D.; Glumoff, T.; Hiltunen, J.K. Peroxisomal beta-oxidation—A metabolic pathway with multiple functions. *Biochim. Biophys. Acta Mol. Cell Res.* **2006**, *1763*, 1413–1426. [CrossRef] [PubMed]
21. Yu, X.X.; Odle, J.; Drackley, J.K. Differential induction of peroxisomal beta-oxidation enzymes by clofibric acid and aspirin in piglet tissues. *Am. J. Physiol. Regul. Integr. Comp. Physiol.* **2001**, *281*, R1553–R1561. [PubMed]
22. Yu, X.X.; Drackley, J.K.; Odle, J. Food deprivation changes peroxisomal beta-oxidation activity but not catalase activity during postnatal development in pig tissues. *J. Nutr.* **1998**, *128*, 1114–1121. [PubMed]
23. Bai, X.; Lin, X.; Drayton, J.; Liu, Y.; Ji, C.; Odle, J. Clofibrate increases long-chain fatty acid oxidation by neonatal pigs. *J. Nutr.* **2014**, *144*, 1688–1693. [CrossRef] [PubMed]
24. Lin, X.; Jacobi, S.; Odle, J. Transplacental induction of fatty acid oxidation in term fetal pigs by the peroxisome proliferator-activated receptor alpha agonist clofibrate. *J. Anim. Sci. Biotechnol.* **2015**, *6*, 11–22. [CrossRef] [PubMed]
25. Wanders, R.J.; van Roermund, C.W.; van Wijland, M.J.; Schutgens, R.B.; Heikoop, J.; van den Bosch, H.; Schram, A.W.; Tager, J.M. Peroxisomal fatty acid beta-oxidation in relation to the accumulation of very long chain fatty acids in cultured skin fibroblasts from patients with Zellweger syndrome and other peroxisomal disorders. *J. Clin. Investig.* **1987**, *80*, 1778–1783. [CrossRef] [PubMed]
26. Guzmán, M.; Geelen, M.J.H. Regulation of fatty acid oxidation in mammalian liver. *Biochim. Biophys. Acta Lipids Lipid Metab.* **1993**, *1167*, 227–241. [CrossRef]

27. Hall, A.M.; Unwin, R.J. The not so 'mighty chondrion': Emergence of renal diseases due to mitochondrial dysfunction. *Nephron Physiol.* **2007**, *105*, 1–10. [CrossRef] [PubMed]
28. Hinsch, W.; Seubert, W. On the mechanism of malonyl-CoA-independent fatty-acid synthesis. Characterization of the mitochondrial chain-elongating system of rat liver and pig-kidney cortex. *Eur. J. Biochem.* **1975**, *53*, 437–447. [CrossRef] [PubMed]
29. Lin, X.; Adams, S.H.; Odle, J. Acetate represents a major product of heptanoate and octanoate beta-oxidation in hepatocytes isolated from neonatal piglets. *Biochem. J.* **1996**, *318*, 235–240. [CrossRef] [PubMed]
30. Adams, S.H.; Alho, C.S.; Asins, G.; Hegardt, F.G.; Marrero, P.F. Gene expression of mitochondrial 3-hydroxy-3-methylglutaryl-CoA synthase in a poorly ketogenic mammal: Effect of starvation during the neonatal period of the piglet. *Biochem. J.* **1997**, *324*, 65–73. [CrossRef] [PubMed]
31. König, B.; Koch, A.; Giggel, K.; Dordschbal, B.; Eder, K.; Stangl, G.I. Monocarboxylate transporter (MCT)-1 is up-regulated by PPARalpha. *Biochim. Biophys. Acta Gen. Subj.* **2008**, *1780*, 899–904. [CrossRef] [PubMed]
32. Lin, X.; Shim, K.; Odle, J. Carnitine palmitoyltransferase I control of acetogenesis, the major pathway of fatty acid {beta}-oxidation in liver of neonatal swine. *Am. J. Physiol. Regul. Integr. Comp. Physiol.* **2010**, *298*, R1435–R1443. [CrossRef] [PubMed]

International Journal of
Molecular Sciences

Review

Lysophospholipid-Related Diseases and PPARγ Signaling Pathway

Tamotsu Tsukahara [1,*], Yoshikazu Matsuda [2] and Hisao Haniu [3]

1 Department of Pharmacology and Therapeutic Innovation, Nagasaki University Graduate School of Biomedical Sciences, 1-14 Bunkyo-machi, Nagasaki 852-8521, Japan
2 Clinical Pharmacology Educational Center, Nihon Pharmaceutical University, Ina-machi, Saitama 362-0806, Japan; yomatsuda@nichiyaku.ac.jp
3 Institute for Biomedical Sciences, Interdisciplinary Cluster for Cutting Edge Research, Shinshu University, 3-1-1 Asahi, Matsumoto, Nagano 390-8621, Japan; hhaniu@shinshu-u.ac.jp
* Correspondence: ttamotsu@nagasaki-u.ac.jp

Received: 18 November 2017; Accepted: 15 December 2017; Published: 16 December 2017

Abstract: The nuclear receptor superfamily includes ligand-inducible transcription factors that play diverse roles in cell metabolism and are associated with pathologies such as cardiovascular diseases. Lysophosphatidic acid (LPA) belongs to a family of lipid mediators. LPA and its naturally occurring analogues interact with G protein-coupled receptors on the cell surface and an intracellular nuclear hormone receptor. In addition, several enzymes that utilize LPA as a substrate or generate it as a product are under its regulatory control. Recent studies have demonstrated that the endogenously produced peroxisome proliferator-activated receptor gamma (PPARγ) antagonist cyclic phosphatidic acid (cPA), which is structurally similar to LPA, inhibits cancer cell invasion and metastasis in vitro and in vivo. We recently observed that cPA negatively regulates PPARγ function by stabilizing the binding of the co-repressor protein, a silencing mediator of retinoic acid, and the thyroid hormone receptor. We also showed that cPA prevents neointima formation, adipocyte differentiation, lipid accumulation, and upregulation of PPARγ target gene transcription. The present review discusses the arbitrary aspects of the physiological and pathophysiological actions of lysophospholipids in vascular and nervous system biology.

Keywords: lysophospholipids; PPARγ; vascular diseases; dementia; spinal cord injury

---

1. PPARγ and Lysophospholipids

Phospholipids are hydrolyzed by phospholipase A_2 (PLA2) to produce lysophospholipids and free fatty acids. One of the most attractive targets of PLA2 is lysophosphatidic acid (LPA), a naturally occurring phospholipid that functions as a bioactive lipid mediator and a second messenger [1]. It consists of a glycerol backbone with a hydroxyl group, a phosphate group, and a long-chain saturated or unsaturated fatty acid. LPA has been detected in biological fluids, and it performs a wide range of biological functions in cell proliferation, migration, and survival [2,3]. LPA is produced by platelet activation after activation of multiple biochemical pathways [4,5]. The plasma contains nanomolar quantities of LPA, whereas LPA concentration can reach physiological levels in the serum during blood clotting [6,7]. LPA has attracted considerable interest because of its multiple roles in physiological and pathological conditions. Recent studies suggest that LPA receptor (LPAR) antagonists abolish platelet aggregation elicited by mildly oxidized low-density lipoprotein (LDL) (mox-LDL), indicating that an LPA-like lipid plays an essential role in the thrombogenic effects of mox-LDL [8]. LDL oxidation generates peroxisome proliferator-activated receptor (PPAR)γ agonists [9], including alkyl glycerophosphate (AGP) [10]. AGP is also formed enzymatically from alkyl dihydroxyacetone phosphate [11]. AGP concentration in the brain is 0.44 nmol/g, which is

15% that of acyl-LPA [12]. Here, we provide evidence that AGP is a PPARγ ligand, with potency similar to that of the thiazolidinedione rosiglitazone, but with only 40% efficacy. Computational and mutational analysis of the AGP-PPARγ complex indicates differential interaction with key residues in the ligand binding and activation domains that explains the partial activation elicited by AGP. Several reports have identified putative intracellular agonists of PPARγ. For example, selected forms of LPA, which accumulate as oxidatively modified LDL, also activate PPARγ [13]. LPA exerts growth-like effects in almost every mammalian cell type. Although LPA is the known ligand for G-coupled cell surface LPARs, some of the effects of LPA are also mediated by PPARγ activation [8]. PPARγ plays key roles in regulating lipid and glucose homeostasis, cell proliferation, apoptosis, and inflammation. In contrast, cPA, which is structurally similar to LPA, is generated by phospholipase D2 (PLD2) and negatively regulate PPARγ functions [14]. cPA shows several unique functions compared to LPA [15]. Unlike LPA, cPA inhibits cell proliferation [16]. Reports show that cPA attenuates neointima formation, which is an early step in the development of atherosclerotic plaques [17]. cPA is a second messenger and a physiological inhibitor of PPARγ, revealing that PPARγ is regulated by both agonists and antagonists.

2. Lysophospholipid and Vascular Pathologies

LPA has been identified as a platelet-activating lipid of mox-LDL in human atherosclerotic lesions [8]. Relatively few intracellular binding partners for LPA are known. Previous studies have identified some candidate proteins, including C-terminal-binding protein/brefeldin A-dependent ADP ribosylated substrate [18], liver fatty-acid-binding protein [19], and gelsolin [20]. Recently, we reported that the isolation and purification of heart-type fatty-acid-binding protein (FABP3) from human coronary artery endothelial cells (HCAECs) were coupled to their identification by proteomics techniques [21]. FABP3, a small cytoplasmic protein with a molecular mass of about 15 kDa, transports fatty acids and other lipophilic substances from the cytoplasm to the nucleus, where these lipids are released to a group of nuclear receptors such as PPARs [21]. FABP3 did not bind LPC or activate PPARγ in HCAEC, showing that FABP3 distributes from the cytosol to the nucleus in response to LPA-mediated PPARγ activation. Recent reports showed that AGP plays an important role in the vascular system [22]. Our group reported that AGP activates PPARγ-mediated transcription more than LPA [10]. Activation of biochemical pathways linked to platelet activation induces AGP production in the serum [8]. Binding studies using the PPARγ ligand-binding domain (LBD) showed that the binding affinity of AGP to PPARγ was similar to that of the synthetic agonist, rosiglitazone [10]. AGP has been detected in several biological fluids and tissues, including the human brain, ascitic fluid, and saliva [23–25]. Recently, we identified that AGP and rosiglitazone induce neointima formation when applied topically within the carotid artery [14]. Neointimal lesions are characterized by the accumulation of cells within the arterial wall and are a prelude to atherosclerotic disease [8]. Recent reports showed that the knockdown of the gene encoding 1-acyl-sn-glycerol-3-phosphate acyltransferase β (AGPAT2) increased cPA levels [26]. AGPAT2 is located the endoplasmic reticulum membrane and converts LPA to phosphatidic acid (PA). Mutations in *AGPAT2* have been associated with congenital generalized lipodystrophy (CGL) [26,27]. Lipodystrophies, including CGL, are heterogeneous acquired or inherited disorders characterized by the selective loss of adipose tissue and development of severe insulin resistance. Histone deacetylases (HDACs), which have been shown to activate PPARγ and enhance the expression of its target genes, regulate chromatin structure and gene transcription via interactions with nuclear receptor corepressors, such as SMRT and nuclear receptor corepressor (NCoR) [28]. HDAC3 inhibits PPARγ and nuclear transcription factor-κB (NF-κB) [29], and HDAC3 inhibition restores PPARγ function in obesity [30]. Additionally, HDAC2-containing complexes are involved in the regulation of nuclear receptor-dependent gene transcription [31]. A previous study demonstrated that topical application of AGP onto uninjured carotid arteries of rats induces arterial wall remodeling in a PPARγ-dependent manner [14]. Our current study also identified increased AGP levels in the carotid artery of $apoE^{-/-}$

mice [32]. These results suggest that AGP in the circulatory system may be a risk factor for development of diabetes-mediated atherosclerosis.

3. Lysophospholipids and Vascular Dementia

The brain is a lipid-rich organ, the structure and function of which are influenced by diet and nutrients [33]. Bioactive lipids within the brain are shown to be pivotal for central nervous system homeostasis by modulating neurotransmission, synaptic plasticity, enzyme function, ion channel activities, gene expression, and inflammation [34]. Lysophospholipids are also involved in a variety of important processes, including vascular dementia. Vascular dementia is a progressive disease caused by reduced blood flow to the brain, and it affects cognitive abilities especially executive function [35,36]. Vascular dementia is poorly understood, and the dearth of suitable animal models limits the understanding of the molecular basis of the disease and development of suitable therapies [36]. On the basis of their chemical structures, different bioactive lysophospholipids can be assigned either to the group of lysophospholipids, LPA, and lysophosphatidylcholine (LPC), or the group of lysosphingolipids, lysosphingomyelin (SPC), and sphingosine 1-phosphate (S1P). LPA is present in the embryonic brain, neural tube, spinal cord, and cerebrospinal fluid at nanomolar to micromolar concentrations and plays several significant roles in the nervous system during development and injury [34]. In the adult brain, LPA receptors are differentially expressed in various neural cell types; for example, the LPA_1 receptor affects cerebral cortical neuron growth, growth cone and process retraction, survival, migration, adhesion, and proliferation [37]. Our recent study suggested that LPA treatment profoundly induced the expression of Kruppel-like factor 9 (KLF9) in human induced pluripotent stem cell-derived neurons [38]. Furthermore, we observed that the effects of LPA on neurite outgrowth and proliferation were also mediated through the PPARγ pathway [38]. Studies show that KLF9, a member of the KLF family of evolutionarily conserved zinc finger transcription factors [39], has been implicated in mediating a diverse range of biological processes including neural stem cell maintenance [40]. *KLF9* expression is induced by neuronal activity as dentate granule neurons functionally integrate in the developing and adult dentate gyrus (DG). During brain development, dentate granule neurons lacking KLF9 show delayed maturation as reflected by the altered expression of early-phase markers and dendritic spine formation [41,42]. Adult *KLF9*-null mice exhibit normal stem cell proliferation and cell fate specification in the DG but show impaired differentiation of adult-born neurons and decreased neurogenesis-dependent synaptic plasticity [41]. Although further investigations will be needed to ascertain the underlying mechanism, these reports highlight that the KLF9-LPC axis is essential for neuronal development. The presence of PPARs has been extensively studied in nervous tissue [43]; PPARs are present in astrocytes, oligodendrocytes, microglia, and neural stem cells (NSCs) [44–47], where it inhibits proinflammatory gene and protein expression. For example, PPARγ inhibits proinflammatory transcription factors, nuclear factor-κB (NF-κB) [48], and activator protein 1 (AP-1) [49]. Owing to the anti-inflammatory and potentially neuroprotective effects of PPARγ, PPARγ agonists are increasingly being used for the treatment of neurodegenerative diseases [50]. Since PPARγ does not colocalize significantly within microglia, several studies indicated a reduction in microglial activity after PPARγ agonist administration [51]. A recent study suggested that LPC, a precursor of LPA, exerts direct biological effects, especially on vascular dementia [52,53]. Plasma LPC is produced by lecithin-cholesterol acyltransferase, hepatic secretion, or by the action of phospholipase A2 (PLA2) [54]. PLA2 are enzymes that catalyze the cleavage of fatty acids from the sn-2 position of phospholipids, producing free fatty acids and LPC. However, abundant evidence exists regarding the capacity of free LPC to increase cytosolic Ca^{2+} and activate inflammatory signaling pathways [55]. In a study of the plasma metabolic profile of Alzheimer's disease (AD), a decrease in LPC 16:0 and 18:2 was reported [56]. Furthermore, previous studies have suggested that oxidative stress is related to AD [57]. These stimulations can activate PC metabolism and downregulate LPC [58]. Therefore, it is important to further evaluate the significance of targeting these bioactive lipids.

4. Lysophospholipids and Spinal Cord Injury (SCI)

A recent estimate shows that the annual incidence of spinal cord injury (SCI) is approximately 54 cases per one million people in the United States, or about 17,500 new SCI cases per year [59]. SCI results in serious damage at the site of injury in the initial stages of neurotrauma, and is complicated by the inflammatory response, which prevents neuronal regeneration and recovery by the central nervous system (CNS) [60]. In addition, a considerable extent of the post-traumatic degeneration of the spinal cord is due to a multifactorial secondary injury [61]. Currently, therapeutic research is focused on two main areas—neuroprotection and neuroregeneration. Several therapeutic strategies have been developed to potentially intervene in these progressive neurodegenerative events and minimize secondary damage to the spinal cord. A variety of promising drugs have been tested in animal models, but few can be applied on human patients with SCI. Neuroprotective drugs target secondary injury effects, including inflammation, oxidative stress-mediated damage, glutamate excitotoxicity, and programmed cell death. Several potentially neuroprotective agents that target the above pathways are under investigation in human clinical trials [62]. Reports show that blocking of LPA signaling is a useful and novel therapeutic strategy for SCI [63]. In the murine SCI model, the use of a specific anti-LPA monoclonal antibody indicated that LPA produced endogenously after neurotrauma inhibits SCI regeneration [63]. In the normal spinal cord, six different LPA receptors (LPA_1-LPA_6) were expressed constitutively, and LPA_1 was the most highly expressed [64]. LPA leads to demyelination via activation of microglia LPA_1. Moreover, we demonstrate that selective blockade of LPA_1 after SCI reduces functional deficits and demyelination, altogether revealing important contributions of LPA–LPA_1 signaling in secondary damage after SCI [64]. In addition, FTY720, an orally available sphingosine-1-phosphate (S1P) receptor modulator known clinically as fingolimod [65], protects an animal model of ischemia-reperfusion after cerebral ischemia and improves functional outcomes in a rat model of SCI. FTY720 is a first-in-class S1P receptor modulator that was highly effective in phase II clinical trials for multiple sclerosis. S1P is a bioactive lysophospholipid mediator that produces a variety of cellular responses, including proliferation, survival, and motility via association of the receptor with G protein-coupled receptor (GPCR) [66]. The efficacy of FTY720 in SCI is possibly because of its role in immune modulation. These studies suggest that lysophospholipids are key modulators of nervous system disorders, including SCI. Furthermore, PPARγ can potentially minimize or prevent dysfunction after SCI [67]. Increased intracellular calcium levels, mitochondrial dysfunction, arachidonic acid breakdown, and activation of nitric oxide synthase (NOS) immediately after SCI results in the formation of reactive oxygen (ROS) and nitrogen species (RNS) [67]. Treatment with the PPARγ agonist pioglitazone increased the number of motor neurons after SCI, which might partially reduce post-SCI oxidative damage [67]. However, none of the agents tested until now have demonstrated strong clinical beneficial outcomes in patients with SCI. Thus, the search for pharmacological drugs capable of improving neurological function is still on. Strategies targeted at modulating lysophospholipid levels in the injured CNS may lead to new therapeutic approaches toward repairing various CNS disorders.

5. Conclusions

In this review, we have focused on recent developments that elucidate the role of lysophospholipids in vascular and nervous system biology. Our proposed mechanism of action for lysophospholipid-related diseases is summarized in Figure 1. Lysophospholipids act as mediators via the activation of cell surface GPCRs, and as intracellular second messengers through PPARγ activation and inhibition in diseases such as atherosclerosis, dementia, and spinal cord injury. However, the physiological role of lysophospholipids in PPARγ signaling is still unclear; further understanding would promote the synthesis of novel medicines that modulate lysophospholipid-mediated PPARγ regulation.

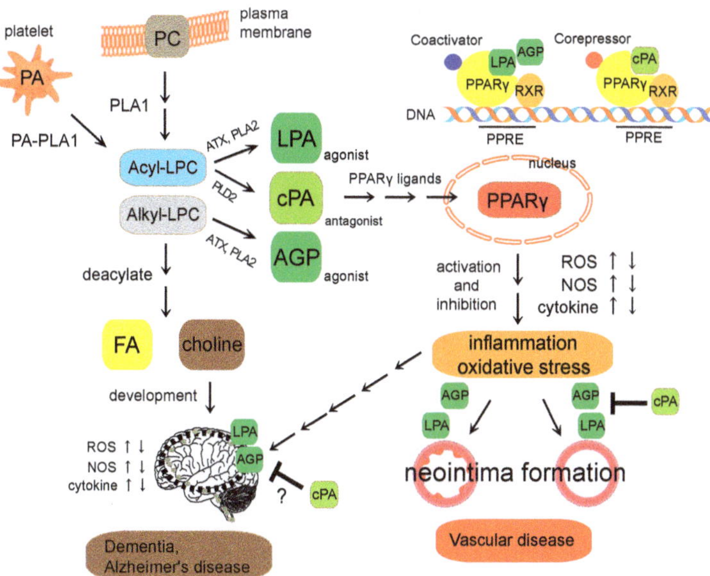

Figure 1. Schematic diagram of lysophospholipid-mediated PPARγ signaling. Lysophosphatidylcholine (LPC) is a bioactive phospholipid generated primarily by the action of phospholipase A2 (PLA2) enzymes on the plasma membrane. After cellular uptake, free LPC is reacylated yielding PC or deacylated yielding FA and choline. LPA and AGP are generated intracellularly in a stimulus-coupled manner by the ATX or PLA2 enzyme. cPA is generated intracellularly in a stimulus-coupled manner by the PLD2 enzyme. LPA and AGP induced neointima formation through the activation of PPARγ, whereas cPA inhibited PPARγ-mediated arterial wall remodeling in a noninjury infusion model. However, the physiological context of cPA in PPARγ signaling in brain is still unclear. Imbalance of the PPARγ agonist-antagonist equilibrium is involved in changes in cellular functions, including ROS generation, NOS and cytokine expression. These endogenous lysophospholipids regulate PPARγ function required for vascular wall pathologies, and metabolic-related diseases. PPRE (PPAR response element); RXR (retinoid X receptor); ATX (autotaxin).

Acknowledgments: This work was supported by the Japan Society for the Promotion of Science KAKENHI (grant number 16K15660) to Hisao Haniu, and by SENSHIN Medical Research Foundation (grant number 2-06) to Tamotsu Tsukahara.

Conflicts of Interest: The authors declare no conflicts of interest.

Abbreviations

PPAR	Peroxisome proliferator-activated receptor
LPA	Lysophosphatidic acid
AGP	1-*O*-octadecenyl-2-hydroxy-sn-glycero-3-phosphate
LPC	Lysophosphatidylcholine
cPA	Cyclic phosphatidic acid
S1P	Sphingosine 1-phosphate
SPC	Lysosphingomyelin
KLF9	Kruppel-like factor 9
mox-LDL	Mildly oxidized low-density lipoprotein
NF-κB	Nuclear factor-κB

AP-1	Activator protein 1
PLA2	Phospholipase A2
AD	Alzheimer's disease
SCI	Spinal cord injury
ROS	Reactive oxygen species
RNS	Reactive nitrogen species
NOS	Nitric oxide synthase
CNS	Central nervous system

References

1. Schneider, G.; Sellers, Z.P.; Abdel-Latif, A.; Morris, A.J.; Ratajczak, M.Z. Bioactive lipids, LPC and LPA, are novel prometastatic factors and their tissue levels increase in response to radio/chemotherapy. *Mol. Cancer Res.* **2014**, *12*, 1560–1573. [CrossRef] [PubMed]
2. Tigyi, G. Aiming drug discovery at lysophosphatidic acid targets. *Br. J. Pharmacol.* **2010**, *161*, 241–270. [CrossRef] [PubMed]
3. Choi, J.W.; Herr, D.R.; Noguchi, K.; Yung, Y.C.; Lee, C.W.; Mutoh, T.; Lin, M.E.; Teo, S.T.; Park, K.E.; Mosley, A.N.; et al. LPA receptors: Subtypes and biological actions. *Annu. Rev. Pharmacol. Toxicol.* **2010**, *50*, 157–186. [CrossRef] [PubMed]
4. Siess, W.; Zangl, K.J.; Essler, M.; Bauer, M.; Brandl, R.; Corrinth, C.; Bittman, R.; Tigyi, G.; Aepfelbacher, M. Lysophosphatidic acid mediates the rapid activation of platelets and endothelial cells by mildly oxidized low density lipoprotein and accumulates in human atherosclerotic lesions. *Proc. Natl. Acad. Sci. USA* **1999**, *96*, 6931–6936. [CrossRef] [PubMed]
5. Tigyi, G.; Parrill, A.L. Molecular mechanisms of lysophosphatidic acid action. *Prog. Lipid Res.* **2003**, *42*, 498–526. [CrossRef]
6. Tigyi, G. Selective ligands for lysophosphatidic acid receptor subtypes: Gaining control over the endothelial differentiation gene family. *Mol. Pharmacol.* **2001**, *60*, 1161–1164. [PubMed]
7. Sano, T.; Baker, D.; Virag, T.; Wada, A.; Yatomi, Y.; Kobayashi, T.; Igarashi, Y.; Tigyi, G. Multiple mechanisms linked to platelet activation result in lysophosphatidic acid and sphingosine 1-phosphate generation in blood. *J. Biol. Chem.* **2002**, *277*, 21197–21206. [CrossRef] [PubMed]
8. Zhang, C.; Baker, D.L.; Yasuda, S.; Makarova, N.; Balazs, L.; Johnson, L.R.; Marathe, G.K.; McIntyre, T.M.; Xu, Y.; Prestwich, G.D.; et al. Lysophosphatidic acid induces neointima formation through PPARγ activation. *J. Exp. Med.* **2004**, *199*, 763–774. [CrossRef] [PubMed]
9. Chui, P.C.; Guan, H.P.; Lehrke, M.; Lazar, M.A. PPARγ regulates adipocyte cholesterol metabolism via oxidized LDL receptor 1. *J. Clin. Investig.* **2005**, *115*, 2244–2256. [CrossRef] [PubMed]
10. Tsukahara, T.; Tsukahara, R.; Yasuda, S.; Makarova, N.; Valentine, W.J.; Allison, P.; Yuan, H.; Baker, D.L.; Li, Z.; Bittman, R.; et al. Different residues mediate recognition of 1-*O*-oleyllysophosphatidic acid and rosiglitazone in the ligand binding domain of peroxisome proliferator-activated receptor γ. *J. Biol. Chem.* **2006**, *281*, 3398–3407. [CrossRef] [PubMed]
11. Gaits, F.; Fourcade, O.; Le Balle, F.; Gueguen, G.; Gaige, B.; Gassama-Diagne, A.; Fauvel, J.; Salles, J.P.; Mauco, G.; Simon, M.F.; et al. Lysophosphatidic acid as a phospholipid mediator: Pathways of synthesis. *FEBS Lett.* **1997**, *410*, 54–58. [CrossRef]
12. Sugiura, T.; Nakane, S.; Kishimoto, S.; Waku, K.; Yoshioka, Y.; Tokumura, A.; Hanahan, D.J. Occurrence of lysophosphatidic acid and its alkyl ether-linked analog in rat brain and comparison of their biological activities toward cultured neural cells. *Biochim. Biophys. Acta* **1999**, *1440*, 194–204. [CrossRef]
13. McIntyre, T.M.; Pontsler, A.V.; Silva, A.R.; St Hilaire, A.; Xu, Y.; Hinshaw, J.C.; Zimmerman, G.A.; Hama, K.; Aoki, J.; Arai, H.; et al. Identification of an intracellular receptor for lysophosphatidic acid (LPA): LPA is a transcellular PPARγ agonist. *Proc. Natl. Acad. Sci. USA* **2003**, *100*, 131–136. [CrossRef] [PubMed]
14. Tsukahara, T.; Tsukahara, R.; Fujiwara, Y.; Yue, J.; Cheng, Y.; Guo, H.; Bolen, A.; Zhang, C.; Balazs, L.; Re, F.; et al. Phospholipase D2-dependent inhibition of the nuclear hormone receptor PPARγ by cyclic phosphatidic acid. *Mol. Cell* **2010**, *39*, 421–432. [CrossRef] [PubMed]
15. Tsukahara, T. PPARγ Networks in Cell Signaling: Update and Impact of Cyclic Phosphatidic Acid. *J. Lipids* **2013**, *2013*, 246597. [CrossRef] [PubMed]

16. Fujiwara, Y. Cyclic phosphatidic acid—A unique bioactive phospholipid. *Biochim. Biophys. Acta* **2008**, *1781*, 519–524. [CrossRef] [PubMed]
17. Steinberg, D. Atherogenesis in perspective: Hypercholesterolemia and inflammation as partners in crime. *Nat. Med.* **2002**, *8*, 1211–1217. [CrossRef] [PubMed]
18. Nardini, M.; Spano, S.; Cericola, C.; Pesce, A.; Massaro, A.; Millo, E.; Luini, A.; Corda, D.; Bolognesi, M. CtBP/BARS: A dual-function protein involved in transcription co-repression and Golgi membrane fission. *EMBO J.* **2003**, *22*, 3122–3130. [CrossRef] [PubMed]
19. Thumser, A.E.; Voysey, J.E.; Wilton, D.C. The binding of lysophospholipids to rat liver fatty acid-binding protein and albumin. *Biochem. J.* **1994**, *301*, 801–806. [CrossRef] [PubMed]
20. Goetzl, E.J.; Lee, H.; Azuma, T.; Stossel, T.P.; Turck, C.W.; Karliner, J.S. Gelsolin binding and cellular presentation of lysophosphatidic acid. *J. Biol. Chem.* **2000**, *275*, 14573–14578. [CrossRef] [PubMed]
21. Tsukahara, R.; Haniu, H.; Matsuda, Y.; Tsukahara, T. Heart-type fatty-acid-binding protein (FABP3) is a lysophosphatidic acid-binding protein in human coronary artery endothelial cells. *FEBS Open Bio* **2014**, *4*, 947–951. [CrossRef] [PubMed]
22. Tsukahara, T.; Tsukahara, R.; Haniu, H.; Matsuda, Y.; Murakami-Murofushi, K. Cyclic phosphatidic acid inhibits the secretion of vascular endothelial growth factor from diabetic human coronary artery endothelial cells through peroxisome proliferator-activated receptor γ. *Mol. Cell. Endocrinol.* **2015**, *412*, 320–329. [CrossRef] [PubMed]
23. Xiao, Y.J.; Schwartz, B.; Washington, M.; Kennedy, A.; Webster, K.; Belinson, J.; Xu, Y. Electrospray ionization mass spectrometry analysis of lysophospholipids in human ascitic fluids: Comparison of the lysophospholipid contents in malignant vs nonmalignant ascitic fluids. *Anal. Biochem.* **2001**, *290*, 302–313. [CrossRef] [PubMed]
24. Sugiura, T.; Nakane, S.; Kishimoto, S.; Waku, K.; Yoshioka, Y.; Tokumura, A. Lysophosphatidic acid, a growth factor-like lipid, in the saliva. *J. Lipid Res.* **2002**, *43*, 2049–2055. [CrossRef] [PubMed]
25. Nakane, S.; Tokumura, A.; Waku, K.; Sugiura, T. Hen egg yolk and white contain high amounts of lysophosphatidic acids, growth factor-like lipids: Distinct molecular species compositions. *Lipids* **2001**, *36*, 413–419. [CrossRef] [PubMed]
26. Subauste, A.R.; Das, A.K.; Li, X.; Elliott, B.G.; Evans, C.; El Azzouny, M.; Treutelaar, M.; Oral, E.; Leff, T.; Burant, C.F. Alterations in lipid signaling underlie lipodystrophy secondary to AGPAT2 mutations. *Diabetes* **2012**, *61*, 2922–2931. [CrossRef] [PubMed]
27. Agarwal, A.K.; Arioglu, E.; De Almeida, S.; Akkoc, N.; Taylor, S.I.; Bowcock, A.M.; Barnes, R.I.; Garg, A. AGPAT2 is mutated in congenital generalized lipodystrophy linked to chromosome 9q34. *Nat. Genet.* **2002**, *31*, 21–23. [CrossRef] [PubMed]
28. Yu, C.; Markan, K.; Temple, K.A.; Deplewski, D.; Brady, M.J.; Cohen, R.N. The nuclear receptor corepressors NCoR and SMRT decrease peroxisome proliferator-activated receptor γ transcriptional activity and repress 3T3-L1 adipogenesis. *J. Biol. Chem.* **2005**, *280*, 13600–13605. [CrossRef] [PubMed]
29. Gao, Z.; He, Q.; Peng, B.; Chiao, P.J.; Ye, J. Regulation of nuclear translocation of HDAC3 by IκBα is required for tumor necrosis factor inhibition of peroxisome proliferator-activated receptor γ function. *J. Biol. Chem.* **2006**, *281*, 4540–4547. [CrossRef] [PubMed]
30. Ferrari, A.; Longo, R.; Fiorino, E.; Silva, R.; Mitro, N.; Cermenati, G.; Gilardi, F.; Desvergne, B.; Andolfo, A.; Magagnotti, C.; et al. HDAC3 is a molecular brake of the metabolic switch supporting white adipose tissue browning. *Nat. Commun.* **2017**, *8*, 93. [CrossRef] [PubMed]
31. Tsukahara, T.; Haniu, H.; Matsuda, Y. Cyclic phosphatidic acid inhibits alkyl-glycerophosphate-induced downregulation of histone deacetylase 2 expression and suppresses the inflammatory response in human coronary artery endothelial cells. *Int. J. Med. Sci.* **2014**, *11*, 955–961. [CrossRef] [PubMed]
32. Tsukahara, T.; Haniu, H.; Matsuda, Y.; Murakmi-Murofushi, K. Short-term treatment with a 2-carba analog of cyclic phosphatidic acid induces lowering of plasma cholesterol levels in ApoE-deficient mice. *Biochem. Biophys. Res. Commun.* **2016**, *473*, 107–113. [CrossRef] [PubMed]
33. Dietschy, J.M.; Turley, S.D. Thematic review series: Brain Lipids. Cholesterol metabolism in the central nervous system during early development and in the mature animal. *J. Lipid Res.* **2004**, *45*, 1375–1397. [CrossRef] [PubMed]
34. Yung, Y.C.; Stoddard, N.C.; Mirendil, H.; Chun, J. Lysophosphatidic Acid signaling in the nervous system. *Neuron* **2015**, *85*, 669–682. [CrossRef] [PubMed]

35. Iadecola, C. The pathobiology of vascular dementia. *Neuron* **2013**, *80*, 844–866. [CrossRef] [PubMed]
36. Venkat, P.; Chopp, M.; Chen, J. Models and mechanisms of vascular dementia. *Exp. Neurol.* **2015**, *272*, 97–108. [CrossRef] [PubMed]
37. Yung, Y.C.; Stoddard, N.C.; Chun, J. LPA receptor signaling: Pharmacology, physiology, and pathophysiology. *J. Lipid Res.* **2014**, *55*, 1192–1214. [CrossRef] [PubMed]
38. Tsukahara, T.; Yamagishi, S.; Matsuda, Y.; Haniu, H. Lysophosphatidic acid signaling regulates the KLF9-PPARγ axis in human induced pluripotent stem cell-derived neurons. *Biochem. Biophys. Res. Commun.* **2017**, *491*, 223–227. [CrossRef] [PubMed]
39. Kaczynski, J.; Cook, T.; Urrutia, R. Sp1- and Kruppel-like transcription factors. *Genome Biol.* **2003**, *4*, 206. [CrossRef] [PubMed]
40. Jiang, J.; Chan, Y.S.; Loh, Y.H.; Cai, J.; Tong, G.Q.; Lim, C.A.; Robson, P.; Zhong, S.; Ng, H.H. A core Klf circuitry regulates self-renewal of embryonic stem cells. *Nat. Cell Biol.* **2008**, *10*, 353–360. [CrossRef] [PubMed]
41. Scobie, K.N.; Hall, B.J.; Wilke, S.A.; Klemenhagen, K.C.; Fujii-Kuriyama, Y.; Ghosh, A.; Hen, R.; Sahay, A. Kruppel-like factor 9 is necessary for late-phase neuronal maturation in the developing dentate gyrus and during adult hippocampal neurogenesis. *J. Neurosci.* **2009**, *29*, 9875–9887. [CrossRef] [PubMed]
42. Bialkowska, A.B.; Yang, V.W.; Mallipattu, S.K. Kruppel-like factors in mammalian stem cells and development. *Development* **2017**, *144*, 737–754. [CrossRef] [PubMed]
43. Moreno, S.; Ceru, M.P. In search for novel strategies towards neuroprotection and neuroregeneration: Is PPARalpha a promising therapeutic target? *Neural. Regen. Res.* **2015**, *10*, 1409–1412. [CrossRef] [PubMed]
44. Bernardo, A.; Minghetti, L. Regulation of Glial Cell Functions by PPARγ Natural and Synthetic Agonists. *PPAR Res.* **2008**, *2008*, 864140. [CrossRef] [PubMed]
45. Cristiano, L.; Bernardo, A.; Ceru, M.P. Peroxisome proliferator-activated receptors (PPARs) and peroxisomes in rat cortical and cerebellar astrocytes. *J. Neurocytol.* **2001**, *30*, 671–683. [CrossRef] [PubMed]
46. Bernardo, A.; Bianchi, D.; Magnaghi, V.; Minghetti, L. Peroxisome proliferator-activated receptor γ agonists promote differentiation and antioxidant defenses of oligodendrocyte progenitor cells. *J. Neuropathol. Exp. Neurol.* **2009**, *68*, 797–808. [CrossRef] [PubMed]
47. Cimini, A.; Cristiano, L.; Benedetti, E.; D'Angelo, B.; Ceru, M.P. PPARs Expression in Adult Mouse Neural Stem Cells: Modulation of PPARs during Astroglial Differentiaton of NSC. *PPAR Res.* **2007**, *2007*, 48242. [CrossRef] [PubMed]
48. Gupta, R.A.; Polk, D.B.; Krishna, U.; Israel, D.A.; Yan, F.; DuBois, R.N.; Peek, R.M., Jr. Activation of peroxisome proliferator-activated receptor γ suppresses nuclear factor κ B-mediated apoptosis induced by Helicobacter pylori in gastric epithelial cells. *J. Biol. Chem.* **2001**, *276*, 31059–31066. [CrossRef] [PubMed]
49. Fu, M.; Zhang, J.; Lin, Y.; Zhu, X.; Zhao, L.; Ahmad, M.; Ehrengruber, M.U.; Chen, Y.E. Early stimulation and late inhibition of peroxisome proliferator-activated receptor γ (PPARγ) gene expression by transforming growth factor β in human aortic smooth muscle cells: Role of early growth-response factor-1 (EGR-1), activator protein 1 (AP1) and Smads. *Biochem. J.* **2003**, *370*, 1019–1025. [PubMed]
50. Chen, Y.C.; Wu, J.S.; Tsai, H.D.; Huang, C.Y.; Chen, J.J.; Sun, G.Y.; Lin, T.N. Peroxisome proliferator-activated receptor γ (PPARγ) and neurodegenerative disorders. *Mol. Neurobiol.* **2012**, *46*, 114–124. [CrossRef] [PubMed]
51. Bernardo, A.; Minghetti, L. PPARγ agonists as regulators of microglial activation and brain inflammation. *Curr. Pharm. Des.* **2006**, *12*, 93–109. [CrossRef] [PubMed]
52. Wood, P.L. Lipidomics of Alzheimer's disease: Current status. *Alzheimers Res. Ther.* **2012**, *4*, 5. [CrossRef] [PubMed]
53. Lam, S.M.; Wang, Y.; Duan, X.; Wenk, M.R.; Kalaria, R.N.; Chen, C.P.; Lai, M.K.; Shui, G. Brain lipidomes of subcortical ischemic vascular dementia and mixed dementia. *Neurobiol. Aging* **2014**, *35*, 2369–2381. [CrossRef] [PubMed]
54. Liebisch, G.; Drobnik, W.; Lieser, B.; Schmitz, G. High-throughput quantification of lysophosphatidylcholine by electrospray ionization tandem mass spectrometry. *Clin. Chem.* **2002**, *48*, 2217–2224. [PubMed]
55. Kume, N.; Cybulsky, M.I.; Gimbrone, M.A., Jr. Lysophosphatidylcholine, a component of atherogenic lipoproteins, induces mononuclear leukocyte adhesion molecules in cultured human and rabbit arterial endothelial cells. *J. Clin. Investig.* **1992**, *90*, 1138–1144. [CrossRef] [PubMed]

56. Li, N.J.; Liu, W.T.; Li, W.; Li, S.Q.; Chen, X.H.; Bi, K.S.; He, P. Plasma metabolic profiling of Alzheimer's disease by liquid chromatography/mass spectrometry. *Clin. Biochem.* **2010**, *43*, 992–997. [CrossRef] [PubMed]
57. Huang, W.J.; Zhang, X.; Chen, W.W. Role of oxidative stress in Alzheimer's disease. *Biomed. Rep.* **2016**, *4*, 519–522. [CrossRef] [PubMed]
58. Xu, Y.; Xiao, Y.J.; Zhu, K.; Baudhuin, L.M.; Lu, J.; Hong, G.; Kim, K.S.; Cristina, K.L.; Song, L.; Williams, F.S.; et al. Unfolding the pathophysiological role of bioactive lysophospholipids. *Curr. Drug Targets Immune Endocr. Metabol. Disord.* **2003**, *3*, 23–32. [CrossRef] [PubMed]
59. Spinal Cord Injury (SCI) 2016 Facts and Figures at a Glance. *J. Spinal Cord Med.* **2016**, *39*, 493–494.
60. Silva, N.A.; Sousa, N.; Reis, R.L.; Salgado, A.J. From basics to clinical: A comprehensive review on spinal cord injury. *Prog. Neurobiol.* **2014**, *114*, 25–57. [CrossRef] [PubMed]
61. Bareyre, F.M. Neuronal repair and replacement in spinal cord injury. *J. Neurol. Sci.* **2008**, *265*, 63–72. [CrossRef] [PubMed]
62. Kim, Y.H.; Ha, K.Y.; Kim, S.I. Spinal Cord Injury and Related Clinical Trials. *Clin. Orthop. Surg.* **2017**, *9*, 1–9. [CrossRef] [PubMed]
63. Goldshmit, Y.; Matteo, R.; Sztal, T.; Ellett, F.; Frisca, F.; Moreno, K.; Crombie, D.; Lieschke, G.J.; Currie, P.D.; Sabbadini, R.A.; et al. Blockage of lysophosphatidic acid signaling improves spinal cord injury outcomes. *Am. J. Pathol.* **2012**, *181*, 978–992. [CrossRef] [PubMed]
64. Santos-Nogueira, E.; Lopez-Serrano, C.; Hernandez, J.; Lago, N.; Astudillo, A.M.; Balsinde, J.; Estivill-Torrus, G.; de Fonseca, F.R.; Chun, J.; Lopez-Vales, R. Activation of Lysophosphatidic Acid Receptor Type 1 Contributes to Pathophysiology of Spinal Cord Injury. *J. Neurosci.* **2015**, *35*, 10224–10235. [CrossRef] [PubMed]
65. Brinkmann, V. FTY720 (fingolimod) in Multiple Sclerosis: Therapeutic effects in the immune and the central nervous system. *Br. J. Pharmacol.* **2009**, *158*, 1173–1182. [CrossRef] [PubMed]
66. Sadahira, Y.; Ruan, F.; Hakomori, S.; Igarashi, Y. Sphingosine 1-phosphate, a specific endogenous signaling molecule controlling cell motility and tumor cell invasiveness. *Proc. Natl. Acad. Sci. USA* **1992**, *89*, 9686–9690. [CrossRef] [PubMed]
67. McTigue, D.M. Potential Therapeutic Targets for PPARgamma after Spinal Cord Injury. *PPAR Res.* **2008**, *2008*, 517162. [CrossRef] [PubMed]

International Journal of
Molecular Sciences

Review

Pivotal Roles of Peroxisome Proliferator-Activated Receptors (PPARs) and Their Signal Cascade for Cellular and Whole-Body Energy Homeostasis

Shreekrishna Lamichane [1,2], Babita Dahal Lamichane [1,2] and Sang-Mo Kwon [1,2,3,*]

1 Laboratory for Vascular Medicine and Stem Cell Biology, Medical Research Institute, Department of Physiology, School of Medicine, Pusan National University, Yangsan 50612, Korea; lamichaneshreekrishna@gmail.com (S.L.); dahalbabita2@gmail.com (B.D.L.)
2 Convergence Stem Cell Research Center, Pusan National University, Yangsan 50612, Korea
3 Research Institute of Convergence Biomedical Science and Technology, Pusan National University Yangsan Hospital, Yangsan 50612, Korea
* Correspondence: smkwon323@pusan.ac.kr; Tel.: +82-51-510-8070

Received: 27 February 2018; Accepted: 20 March 2018; Published: 22 March 2018

Abstract: Peroxisome proliferator-activated receptors (PPARs), members of the nuclear receptor superfamily, are important in whole-body energy metabolism. PPARs are classified into three isoforms, namely, PPARα, β/δ, and γ. They are collectively involved in fatty acid oxidation, as well as glucose and lipid metabolism throughout the body. Importantly, the three isoforms of PPARs have complementary and distinct metabolic activities for energy balance at a cellular and whole-body level. PPARs also act with other co-regulators to maintain energy homeostasis. When endogenous ligands bind with these receptors, they regulate the transcription of genes involved in energy homeostasis. However, the exact molecular mechanism of PPARs in energy metabolism remains unclear. In this review, we summarize the importance of PPAR signals in multiple organs and focus on the pivotal roles of PPAR signals in cellular and whole-body energy homeostasis.

Keywords: PPARs; energy homeostasis; fatty acid oxidation; glucose-lipid metabolism

---

1. Introduction

Energy is essential for the survival of all living organisms, and energy metabolism describes the process of generating energy from nutrients. In humans, dietary-derived glucose and long-chain fatty acids are used as sources of energy. Energy demand in cells is fulfilled by oxidative metabolism in mitochondria. Demand and supply within cells of differing physiological states are controlled by a transcriptional regulatory network in both normal and induced cells, for example, when exercising or fasting. Peroxisome proliferator-activated receptors (PPARs) are members of a nuclear receptor superfamily within this network that regulate nutrient-dependent transcription. These receptors were first identified in the 1990s in rodents and named after their property of peroxisome proliferation [1–3]. In more recent work, it has become clear that PPARs also regulate gene transcription of eicosanoids and fatty acids (FAs) [4]. Moreover, PPARs have been established as a group of structurally diverse chemicals associated with transcriptional activation of the peroxisome FA β-oxidation system [5].

Similar to the other nuclear receptor family members, PPARs have a canonical domain structure. They possess an amino terminal region, which comprises a DNA binding domain and a ligand-independent transactivation domain, AF-1. At the carboxyl terminal region is a dimerization and ligand-binding domain with a ligand-dependent transactivation domain, AF-2 [6,7]. Different from other nuclear receptors, the ligand binding pocket of PPARs is unusually large and can accommodate a variety of endogenous lipids, including FAs, eicosanoids, oxidized and nitrated FAs, and derivatives of linoleic acids [8].

Three isoforms of PPAR, α, β/δ, and γ, have been identified and are each expressed in various tissues. PPARγ may be further classified as PPARγ-1, γ-2, and γ-3 [2]. PPARγ-2 is generated by alternative splicing and contains 28 additional amino acids at the N-terminal region compared to PPARγ-1. PPARγ-3 is a splicing variant of PPARγ-1 that gives rise to the same protein [9]. Three PPAR isoforms exhibit 80% homology and are more divergent in the ligand-binding domain, explaining their different responses to various ligands [10]. PPARs act as FA sensors to control many metabolic activities and they are involved in various biological processes, including adipogenesis, lipid metabolism, insulin sensitivity, inflammation, reproduction, and cell growth and differentiation [8,11,12]. They regulate this function upon activation of target genes by endogenous ligands. Binding of endogenous ligands to the ligand binding domain of the receptor causes a conformational change that facilitates PPARs to heterodimerize with the retinoid X receptor. This conformational change helps with binding and the release of small accessory molecules that are essential for transcription. The heterodimerized complex now assembled at PPAR response elements (PPREs) causes the transactivation of target genes of mitochondria and peroxisomes. This series of events regulates a network of proteins that are involved in systemic energy homeostasis [3,11,12].

PPARα is highly expressed in hepatocytes, enterocytes, as well as vascular and immune cell types, such as monocytes/macrophages, endothelial cells, smooth muscle cells, lymphocytes, and non-neuronal cells, such as microglia and astroglia. PPARα activates genes encoding enzymes involved in fatty acid oxidation (FAO), which include carnitine palmitoyltransferase 1 (CPT1), medium-chain acyl CoA dehydrogenase, acyl-CoA oxidase, fatty acyl-CoA synthase, FA transport proteins, and their derivatives to enter into the β-oxidation pathway [13]. In the liver, it plays a crucial role in FAO, thereby providing energy for peripheral tissues and elevating mitochondrial and peroxisomal fatty acid β-oxidation rates. PPARα is also involved in ketogenesis, by lowering plasma triglyceride levels and increasing plasma high-density lipoprotein (HDL) levels. PPARα is activated by several molecules such as long-chain unsaturated fatty acids, eicosanoids, and hypolipidemic drugs [9]. PPARγ is expressed in skeletal muscle, liver, heart, and intestine. Among the three types of PPARγ, PPARγ1 is expressed in a broad range of tissues, whereas PPARγ2 is limited to the adipose tissue. PPARγ3 is abundantly found in macrophages, large intestine, and white adipose tissue (WAT). In adipose tissue, PPARγ controls FA uptake, adipogenesis, adipokine production, lipid partitioning to fat, in addition to increasing insulin sensitivity. PPARβ/δ is expressed in skeletal muscle, adipocytes, macrophages, lungs, brain, and skin. It promotes FA metabolism and obesity resistance, improves insulin sensitivity, helps to form oxidative muscle fibers through exercise physiology, and suppresses macrophage-derived inflammation [3,6,8,12]. PPARβ/δ activators have been proposed for treating metabolic disease and are currently under clinical trials [9].

All three PPAR isotypes play essential roles in lipid and FA metabolism by directly binding to, and modulating, genes involved in fat metabolism [1]. Although they share similarities in function and mechanism of action, PPAR isotypes display important physiological and pharmacological differences. The metabolic effects of PPARβ/δ and PPARα are similar in promoting energy dissipation; in contrast, PPARγ promotes energy storage. PPARβ/δ enhances FAO in several tissues and normalizes plasma lipid levels. PPARγ and PPARβ/δ enhance insulin sensitivity, whereas PPARα is not involved in this process. PPARβ/δ-mediated glucose handling is not similar to that of PPARγ, but PPARγ and PPARβ/δ both are involved in skeletal muscle fiber type distribution, hepatic glucose metabolism, and pancreatic islet function [12]. PPARα promotes FAO under lipid catabolism, in events such as fasting, and PPARγ promotes lipogenesis during anabolism by acting on adipose tissue [4]. This review will discuss the role of PPARs in energy metabolism within various parts of the body.

2. PPAR Signals in Liver

Liver is the primary organ involved in whole-body energy metabolism because it can metabolize FAs and glucose. Among the three isoforms, PPARα is predominantly expressed in the liver where it regulates energy metabolism by FAO [5]. During fasting, it regulates FA uptake, ketogenesis, and β-oxidation [14].

In a previous study, it was demonstrated that FA uptake and FAO became suppressed in PPARα knockout mice. In addition to this, ketogenesis and gluconeogenesis were impaired in PPARα knockout mice. A different isotype, PPARβ/δ, has been shown to possess a different role in energy metabolism regulation in the liver. Overexpression of PPARβ/δ upregulates genes involved in energy metabolism, and deletion of PPARβ/δ reduces the expression of genes that are responsible for lipogenesis and utilization of glucose [1]. There was a significant decrease in the blood glucose level of PPARα-deficient mice after 24 h of fasting. Upregulation of TRB3 (an inhibitor of Akt/protein kinase B and a positive regulator of the cellular response to insulin) by the direct transcriptional control of PPARα has a negative effect on insulin signaling. It suggests that PPARα is important for glucose homeostasis in the liver [15]. FAO by PPARα in the liver also has an important role in ketosis, which fulfils the energy requirement in fasting [14].

PPARα enhances the expression of mitochondrial acyl-CoA dehydrogenase and, thus, it increases FA oxidation and acetyl-CoA enzyme production [12]. In the case of fasting, uptake and mitochondrial transport of FAs from adipose tissue is increased by PPARα by enhancing levels of mitochondrial HMG-CoA synthase, which converts acetyl-CoA to ketone bodies. PPARα modulates levels of glycoprotein CD36, which is responsible for FA uptake. PPARα regulates the enzymes involved in the degradation of straight chain FAs in the peroxisome. Hepatic enzymes, such as glycerol-3-phosphate dehydrogenase (GPDH) and glycerol kinase, which converts glycerol to glucose, are regulated by PPARα [15]. In the case of feeding, PPARα directs de novo lipogenesis to supply FAs that are stored as triglycerides and can be utilized in starvation [12].

Expression of PPARγ in the liver of mice causes liver steatosis. PPARs are considered as the target molecules of non-alcoholic fatty liver disease (NAFLD) and non-alcoholic steatohepatitis (NASH) that might cause liver cirrhosis. NASH is involved in the misregulation of PPAR signaling accompanied by PPARγ and SREBP-1c-mediated metabolic disorders. Administration of PPARγ ligand aggravates concanavalin A-induced liver injury. Abnormal stimulation of PPARα generates hepatocellular carcinoma through fatty liver [16].

3. PPAR Signals in Adipose Tissue

Adipose tissue is essential for energy homeostasis in the body. There are two functional types: WAT and brown adipose tissue (BAT). WAT acts as a caloric reservoir for other organs. In conditions of excess nutrition, it stores nutrients as lipids. During starvation, it releases energy through lipolysis. BAT is specialized for storage of lipids and increases energy expenditure by production of heat. Adipose tissues perform endocrinal functions, and secrete various hormones, cytokines, and metabolites called adipokines that signal for systemic energy metabolism. They regulate energy balance by obtaining signals from the central nervous system and metabolic activity in peripheral tissues [17–19]. PPARγ is extensively expressed in both types of adipose tissue. It is involved in the induction of genes that are essential for FA uptake and storage, as well as adipose tissue differentiation [20]. Ectopic expression of PPARγ in non-adipogenic cells converts them into adipocytes effectively [21]. Knockout of PPARγ in embryonic fibroblasts abolishes their differentiation into adipocytes [22]. A previous in vivo model has shown that PPARγ is essential for adipocyte generation and survival in animals. Heterozygous, dominant negative PPARγ mutations cause lipodystrophy in humans [8,23]. PPARα is highly expressed in BAT, but not in WAT, and it functions to regulate the expression of mitochondrial uncoupling proteins, UCP1 (Uncoupling protein 1) and PGC1α. Knockout of PPARα reduces the expression of these mitochondrial proteins under normal and cold exposure conditions. However, FA metabolism in BAT remains unaffected. When PPARα is activated in human and mouse adipocytes, it induces FAO gene expression and increases energy expenditure. PPARβ/δ is also expressed in both BAT and WAT. It plays an important role in the regulation of FAO and thermogenesis in BAT. When PPARβ/δ is ectopically expressed in adipose tissue, it dramatically induces the expression of genes involved in FAO, oxidative phosphorylation (OXPHOS), and thermogenesis. Furthermore, deletion of PPARβ/δ in BAT reduces the expression of FAO and thermogenic genes. The role of PPARβ/δ in WAT remains to be explored [1]. In rodents, BAT plays an important role in protection against obesity and obesity-associated metabolic problems. Activation of

PPARγ in adipose tissue induces the expression of genes for fatty acid transport and storage as well as promotes de novo adipogenesis so that PPARγ activator thiazolidinediones (TZD) has been widely used in treatment of type II diabetes [8].

4. PPAR Signals in Skeletal Muscle

Skeletal muscle covers approximately 40% of the total body mass and is an important site for glycogen storage, insulin mediated glucose use, lipid metabolism, FAO, and glucose metabolism. In addition, it is also involved in the regulation of cholesterol and HDL levels. As a result, it has a significant role in insulin sensitivity and lipid metabolism. PPARβ/δ expression is dominant in skeletal muscle and it regulates gene expression involved in energy metabolism by relying on FAs as an energy source [14,24–26]. It regulates genes for triglyceride hydrolysis, lipid uptake, and FA oxidation, as well as activating uncoupling proteins to provide energy for OXPHOS. It also encodes mitochondrial protein CPT1 to regulate long chain FAO. PPARβ/δ activates FOXO1, a transcription factor for metabolic adaptation, and pyruvate dehydrogenase kinase 4 (PDK4), which inactivates the pyruvate dehydrogenase complex and is, therefore, a rate-limiting step in muscle carbohydrate oxidation. PDK4 acts on several genes that code for lipid efflux and energy expenditure [25]; it also upregulates fatty acid β-oxidation. Furthermore, glucose metabolism was shown to be increased in PPARβ/δ transgenic mice [24]. To control muscle FA metabolism, PPARβ activates gene transcription of lactate dehydrogenase B (LDHB), which is important for glucose oxidation, by converting glucose and lactate into pyruvate for mitochondrial oxidation [27].

Energy metabolism in skeletal muscle is regulated by PPARγ coactivator-1α (PGC-1α), a regulator of mitochondrial biogenesis [28], involved in the catabolic process to synthesize aerobic adenosine triphosphate (ATP). PGC-1α expression is directly activated by PPARβ/δ to regulate skeletal muscle metabolism by increasing the expression of mitochondrial proteins [29,30]. PGC-1α stimulates the expression of genes responsible for glucose and lipid metabolism, energy transfer, and muscle contractile function. Furthermore, PGC-1α knockout mice have shown defects in skeletal muscle energetics, and have decreased mitochondrial biogenesis and oxidative function [31]. In skeletal muscle, the increment in lipid oxidation and reduction of glucose utilization is conducted by the activation of PPARδ. In the nucleus, transcription factor EB (TFEB) induces the expression of genes involved in lysosomal biogenesis and lipid metabolism through PGC-1α during fasting [32].

5. PPAR Signals in Kidney

All three isoforms of PPARs (PPARα, PPARβ/δ, and PPARγ) are found in the kidney. PPARα is highly expressed in the renal proximal tubules and the medullary thick ascending limbs of Henle [33]. PPARγ is mainly found in the medullary collecting duct with low expression in glomeruli and proximal tubules [34]. The nuclear receptors, PPARα and PPARγ, are concerned with the control of FAs and glucose metabolism. FAs are the main source of fuel for energy production in kidney cortex tissue [33]. PPARγ alters large numbers of target genes involved in peripheral glucose and FA metabolism leading to improved insulin sensitivity and glycemic control [34]. PPARα is the master regulator of lipid metabolism by controlling the transcription of its target genes such as acyl-CoA oxidase, acyl-CoA, CPT1a, PGC1α, UCP2, and UCP3 [35]. It regulates renal FA β-oxidation [33,36], which provides the source of ATP in proximal renal tubular cells. PPARs regulate FAO and control energy homeostasis, as well as lipid and glucose metabolism by gluconeogenesis, stimulating ketone body synthesis and adipogenesis [33]. In renal proximal tubule cells, FA metabolites derived from arachidonic or linoleic acids via cyclooxygenase or lipoxygenase pathways activate PPARα. Mouse kidney cortex cells use polyunsaturated FAs as the primary source of energy production. Mitochondrial biogenesis is controlled by PPARα through OXPHOS, FA metabolism, and the tricarboxylic acid (TCA) cycle [37].

Moreover, the kidney has a role in energy balance because of its vast gluconeogenic enzyme activities including that of PDK4 and its contribution to glucose during fasting. Furthermore, fasting induces high levels of PGC-1α along with its regulating partners, estrogen-related receptors (ERRs) in the kidney, which are involved in the TCA cycle and mitochondrial OXPHOS [38]. PPAR agonists and antagonists may approach to modulate renal diseases like glomerulonephritis, glomerulosclerosis and diabetic nephropathy [39].

6. PPAR Signals in Heart

The heart consumes ATP to maintain its contractile function [40] and FAs are the main source of energy [41]. Around 70% of ATP used by the heart is obtained from FAO. Cardiac FAO is regulated at different stages such as FA uptake, triglyceride formation and storage, triglyceride lipolysis to release unesterified FAs, transfer of FA into mitochondria for FAO, and ATP production. Most of the proteins are transcriptionally regulated by PPARα [42]. PPARβ/δ in the myocardium controls glucose and lipid utilization, and promotes insulin sensitivity. The activity of PPARs in the heart is regulated by PGC-1α, which is responsible for mitochondrial biogenesis and metabolism [43].

PGC-1α is a highly expressed gene in the heart. PGC-1α interacts with PPARα, PPARγ, ERR, the retinoid X receptor, and nuclear respiratory factors to co-activate the transcription factors. Overexpression of PGC-1α significantly increases nuclear- and mitochondrial-related gene expression that changes the metabolic energy substrate from glucose to FA. The G-protein-coupled receptor kinase interacting protein-1 (GIT1) is a regulator of cardiac mitochondrial biogenesis that helps PGC-1α-regulated gene expression [44]. Under mild stress conditions like exercise, the level of FAO is increased due to oxidation of palmitate. However, the ATP produced by FAO is more than that produced by glucose oxidation because glucose oxidation needs oxygen, thus making it an efficient mode of cardiac energy production [45].

PPARα regulates cardiac FAO by activating genes in FA metabolism pathways such as FA uptake and β-oxidation, but not in the TCA cycle. Mitochondrial OXPHOS genes regulated by PGC-1α and ERRs in the heart are suppressed by the activation of PPARα, and thus, PPARα reduces glucose import and glycolysis by inducing cellular FA uptake and β-oxidation. Moreover, the importance of PPARα in regulating FAO was confirmed when PPARα knockout showed reduced FA uptake and β-oxidation. In addition, overexpression of PPARβ/δ induces FAO by upregulating genes in mitochondrial FA transport and β-oxidation. However, PPARβ/δ overexpression does not cause lipid accumulation and cardiac dysfunction. This may be due to high glucose utilization. The deletion of PPARβ/δ downregulates FAO genes and causes cardiac hypertrophy by lipid accumulation [1].

Patients with metabolic syndrome and aortic stenosis express high level of PPARγ in heart which is strongly correlated cardiac lipid accumulation and poor cardiac function. When the level of PPARγ is high under certain pathological conditions, it may cause cardiomyopathy [8].

7. PPAR Signals in Brain

All three PPARs (α, β/δ and γ) are expressed in the central nervous system (CNS) [46,47]. Among them, PPARγ is a key neuronal isoform used to regulate energy homeostasis [47–52]. It regulates genes involved in FA metabolism like acetyl-coenzyme A carboxylase (ACC), fatty acid synthase (FAS), and CPT1. It is expressed within the ventromedial nucleus (VMN) and the arcuate nucleus of the hypothalamus (ARC) of the brain. Overexpression of central PPARγ increases food intake, abdominal fat, activity of neuropeptide Y (NPY), and the expression of pro-opiomelanocortin (POMC) in the ARC. Conversely, the roles of PPARα and PPARβ/δ in energy metabolism are less understood. Knockdown of PPARβ/δ showed a decrease in leptin sensitivity with no change in food intake, but an increase in the expression of genes that are responsible for lipid uptake, lipid synthesis, and FA oxidation in the hypothalamus [51].

Recent studies suggest that activation of PPARα and/or PPARγ contribute to weight gain and obesity. Knockout of PPARγ in neurons and the hypothalamus prevents the development of diet-induced obesity (DIO). PPARα activation in the hypothalamus corrected the hypophagic phenotype in a model of increased CNS fatty acid sensing. Studies using rodent models suggest that the hypothalamic lipid accumulation is associated with obesity, and this may be due to the role of PPARβ/δ in the regulation of genes coded for lipid oxidation in the CNS [47]. The identification of PPARγ expression in dopaminergic neurons of the ventral tegmental area of the brain has helped to investigate a surprising role between food and other stimuli. ARC neurons, such as NPY/AgRP and POMC neurons with nuclear PPARγ, play important roles in the sensing of signals related to nutritional state, such as leptin, insulin, ghrelin, glucose, and FAs and transduce these signals to affect food intake, energy expenditure, and insulin sensitivity [53]. Thus, the maintenance of glucose homeostasis and food intake is controlled by central signaling of glucose, regulated by PPARs [51].

PPARγ agonist have shown their effect in Parkinson disease, Alzheimer disease, brain injury and amyotrophic lateral sclerosis. They are effective in suppressing the development of animal models of CNS inflammatory and neurodegenerative disorders [6].

8. PPAR Signals in Pancreatic β-Cells

PPARβ/δ is abundantly expressed in the pancreatic tissue of rats and human. PPARβ/δ is highly expressed in β-cells, but PPAR-α and -γ are relatively lowly expressed here [54–57]. PPARγ reduction leads to abnormal glucose metabolism in islets, meaning that it is required to maintain glucose metabolism [56]. PPARα and PPARγ play important roles in FA metabolism by regulating genes in FAO and energy uncoupling in mitochondria, such as CPT1 and UCP2. PPARβ/δ regulates mitochondrial energy metabolism and insulin secretion in β-cells [54,55,58], and increases the activation of FA β-oxidation enzyme genes, long chain acyl-CoA dehydrogenase (LCAD), PDK4, and UCP2. PPARβ/δ upregulates the mRNA level of PDK4 and increases the utilization of FAs, thus reducing insulin secretion. UCP2 is the bridge between mitochondrial energy metabolism and insulin secretion function [54]. The treatment of db/db mice with a PPARβ/δ agonist decreased blood glucose levels and improved insulin sensitivity and pancreatic islet function. It suggests that PPARβ/δ contributes as a FA sensor and to improve insulin secretion in β-cells [59]. Recent studies have shown that PPARα is ectopically expressed in INS-1 cells that could induce lipid accumulation alone with an increase in β-oxidation. PPARγ promotes FA disposal in pancreatic β-cells [56].

9. PPAR Signals in Intestine

PPARα and PPARβ/δ are highly expressed in the intestine [60,61]. In the lumen of the colon, short chain FAs (SCFAs) such as acetic acid, propionic acid, and butyric acid are produced. A recent study showed that propionate lowers FA content in the plasma and reduces food intake. Dietary triglyceride (TG) is hydrolyzed into free FAs in the lumen of the intestine. These free FAs are taken up by intestinal epithelial cells to the endoplasmic reticulum where they are resynthesized into TG. This intestinal TG metabolism process is very important for systemic energy homeostasis [61].

Animal studies have demonstrated relationships between intestinal colonization, energy utilization, and weight gain. The mechanism of this process involves regulation of angiopoietin-like protein 4 (ANGPTL4) expression in the intestinal epithelium. ANGPTL4 is a secreted protein that regulates lipid and glucose homeostasis. The amino terminal domain of ANGPTL4 inhibits lipoprotein lipase activity and decreases triglyceride uptake and storage. In addition, it induces lipolysis and results in the elevation of circulating triglyceride levels. Deletion of ANGPTL4 results in changes in metabolism, decreased intestinal absorption of oils, and thickening of the intestinal mucosa. PPARγ is involved in regulating FA metabolism through β-oxidation. PPARγ regulates ANGPTL4 expression and PPRE within the third intron of the ANGPTL4 gene. SCFAs activate PPARγ and are the products of dietary fibers and main energy sources for colonocytes [62].

PPARα agonist Wy-14643 induces the protein expression of enzymes involved in FAO and ketogenesis such as CPT1A and mitochondrial 3-hydroxy-3-methylglutaryl-CoA synthase in the small intestine [63]. PPARα regulates various transporters and phase I enzymes involved in FA uptake and oxidation. Nutritional-activated PPARα controls FAO and cholesterol and glucose transport [64]. During fasting, PPARα plays an important role in regulating transporter and phage I/II metabolism genes in the small intestine [65].

Similarly, administration of another PPARα modulator, K-877, regulates intestinal FAO and apolipoprotein mRNA expression and reduces plasma TG levels. K-877 administration significantly reduces

Npc1l1 expression and increases Abca1 expression. Npc1l1 is a rate-limiting transporter for cholesterol absorption in the small intestine of mice, whereas Abca1 is an important molecule involved in HDL-C production by transporting intracellular cholesterol from the small intestine. Intestinal Abca1 deficiency leads to deficient HDL biogenesis and therefore reduces cholesterol influx in to the circulation [66].

10. Co-Regulators of PPAR in Energy Homeostasis

Balanced energy homeostasis is the result of high pathway interconnectivity and feedback control. The Nuclear Receptor Signaling Atlas has reported around 320 nuclear receptor co-regulators, and there have been 38 co-regulators identified for PPARs alone. Not only do PPARs contribute to systemic energy homeostasis on their own, but crosstalk of PPARs with various pathways also has an effect [67]. Co-activators and co-repressors collectively regulate mitochondrial energy balance. PPARγ and PGC-1α are the co-regulators for induction of mitochondrial oxidative metabolism. Nuclear co-repressor 1 (NCOR1) antagonizes the effect of PGC1α on mitochondria. Knocking out NCOR1 phenotypically mimics PGC-1α overexpression. PGC-1α participates in the transcriptional response of ERR and PPARs. Nuclear receptor interacting protein 1 (NRIP1) binds to the PPAR nuclear receptors, as well as ERR, and represses the expression of target genes that are involved in energy consumption. NIRP1-deficient mice are lean, and show increased insulin sensitivity and glucose tolerance, and resistance to diet-induced obesity [1,67,68]. Under different nutritional conditions, hepatocyte nuclear factor α (HNFα), Hes6, and the PPARs balance the expression of each other and regulate the transcription cascade in metabolism [23,69]. PPARγ with the transcription factor, CCAAT/enhancer-binding protein α (C/EBPα), is an important driver in the late stage of adipogenesis. Mice with the liver specific knockout of mediator complex subunit 1 (MED1) were shown to have impaired PPARα and PPARγ activities. This suggests that MED1 plays an important role in energy homeostasis via PPARs [67].

PPARs regulate lipid and glucose metabolism and are involved in a variety of diseases, ranging from metabolic disorder to cancer [9,70]. They have a significant, energetic, plastic, and signaling roles in the pathophysiology of cancer cells. Most cancer cells show increased aerobic glycolysis and use PPAR signaling pathways to generate ATP as a main source of energy. Stimulated peroxisomal β-oxidation increases free radical oxygen species that may increase oxidative stress. This significantly contributes to the carcinogenic properties of PPAR ligand in rodents, particularly in the liver. Activation of PPARs (α, β/δ, γ) by natural or synthetic agonists can inhibit growth and induce differentiation or death of tumor cells. Synthetic ligands of PPARs show an important link with cancer. PPARγ and PPARα ligands have been shown to promote the differentiation of various tumor cell lines, including breast, lung, prostate, leukemia, colon, melanoma, and liver cancers [71].

PPARs are involved in controlling the genes responsible for not only energy homeostasis but also cell proliferation, apoptosis, tumorigenesis, and metabolic disease development [72]. A previous study showed that ANGPTL4 and PPARs play potential synergistic roles in the crosstalk between metabolic syndromes and cancer [10]. PPAR transcriptional activity can be modulated through cross-talk with phosphates and kinases, including ERK1/2, P38-MAPK, PKC, AMPK, and GSK3. PPARs activate the transcription of genes involved in anticancer effects in a variety of human tumors. PPARγ appears to be mostly involved in tumorigenesis regulation [73]. The shortage of vitamin D and decreased level of PPARγ may be involved in obesity and cancer development [74]. PPARβ/δ is involved in the initiation and promotion of mammary tumorigenesis by regulating metabolism, inflammation, and immune tolerance [75]. All PPARs, including α, β/δ, and γ, have been shown to be important in lung cancer biology. PPARα activation inhibits tumorigenesis through its antiangiogenic and anti-inflammatory effects. Activated PPARγ is also anti-tumorigenic and anti-metastatic, regulating several function of cancer cells and controlling the tumor microenvironment [76]. Among the synthetic ligands of PPARγ, thiazolidinediones, which are used to treat diabetes mellitus type 2, increase the risk of bladder cancer [77]. FAs from conjugated linoleic acid-enriched egg yolks (EFA-CLA) act as a potential ligand for PPAR receptors in the breast cancer cell line MCF-7. PPAR-responsive genes can be regulated by EFA-CLA, leading to reduced tumor cell proliferation, which has a greater influence than non-enriched FAs or single synthetic CLA isomers [78]. PPAR modulators may have beneficial effects as chemo-preventive agents. However, it remains unclear whether PPARs act as oncogenes or tumor suppressors [9]. The co-regulators of PPARs in carcinogenic process is summarized in Figure A2. Further studies are needed to develop new approaches for treating neoplasia.

PPARs are involved in various pathways for energy homeostasis in different organs. These pathways are affected in disease conditions and cause the metabolic energy imbalance. Thus, PPARs can provide therapeutic targets for different diseases such as dyslipidemia, diabetes, obesity, inflammation, neurodegenerative disorders and cardiomyopathy [6,8].

11. Conclusions

Thus, PPARs are crucial transcriptional factors involved in energy metabolism for the whole body and the three isotypes have complementary and distinct metabolic activities. PPARs also act with other co-regulators in the maintenance of energy homeostasis. The overview of PPARs in cellular and whole body energy homeostasis is illustrated in Figure A1. However, the exact molecular mechanism of PPARs within energy metabolism remains unclear. Future research in this field should be oriented towards the molecular mechanism, to ensure the use of PPAR as a therapeutic targets.

Acknowledgments: This work was supported by a grant from the National Research Foundation (NRF-2015M3A9B4066493 and NRF-2015M3A9B4051053).

Author Contributions: Shreekrishna Lamichane, Babita Dahal Lamichane and Sang-Mo Kwon have equal contribution for this review.

Conflicts of Interest: The authors declare no conflict of interest.

Abbreviations

PPARs	Peroxisome proliferator-activated receptors
FAs	Fatty acids
PPREs	PPAR response elements
FAO	Fatty acid oxidation
CPT1	Carnitine palmitoyltransferase 1
HDL	High-density lipoprotein
WAT	White adipose tissue
BAT	Brown adipose tissue
GPDH	Glycerol-3-phosphate dehydrogenase
OXPHOS	Oxidative phosphorylation
PDK4	Pyruvate dehydrogenase kinase 4
LDHB	Lactate dehydrogenase B
PGC-1α	PPARγ coactivator-1α
UCP1	Uncoupling protein 1
ATP	Adenosine triphosphate
TFEB	Transcription factor EB
TCA	Tricarboxylic acid
GIT1	G-protein-coupled receptor kinase interacting protein-1
ACC	Acetyl-coenzyme A carboxylase
FAS	Fatty acid synthase
VMN	Ventromedial nucleus
ARC	Arcuate nucleus of the hypothalamus
NPY	Neuropeptide Y

POMC	Pro-opiomelanocortin
CNS	Central nervous system
DIO	Diet induced obesity
LCAD	Long chain acyl-CoA dehydrogenase
NCOR1	Nuclear co-repressor 1
HNFα	Hepatocyte nuclear factor α
C/EBPα	CCAAT/enhancer-binding protein α
MED1	Mediator complex subunit 1
ANGPTL4	Angiopoietin-like protein 4
TG	Triglyceride
HMG-CoAS2	Mitochondrial 3-hydroxy-3-methylglutaryl-CoA synthase
SCFAs	Short chain fatty acids
EFA-CLA	Fatty acids from conjugated linoleic acid-enriched egg yolks
CLA	Conjugated linoleic acid
NAFLD	Non-alcoholic fatty liver disease
NASH	Non-alcoholic steatohepatitis
TZD	Thiazolidinediones
ERK1/2	Extracellular signal-regulated kinase type 1 and 2
P38-MAPK	Mitogen-activated protein kinase p38
PKC	Protein kinase C
AMPK	5′Adenosine monophosphate-activated protein kinase
GSK3	Glycogen synthase kinase 3

Appendix A

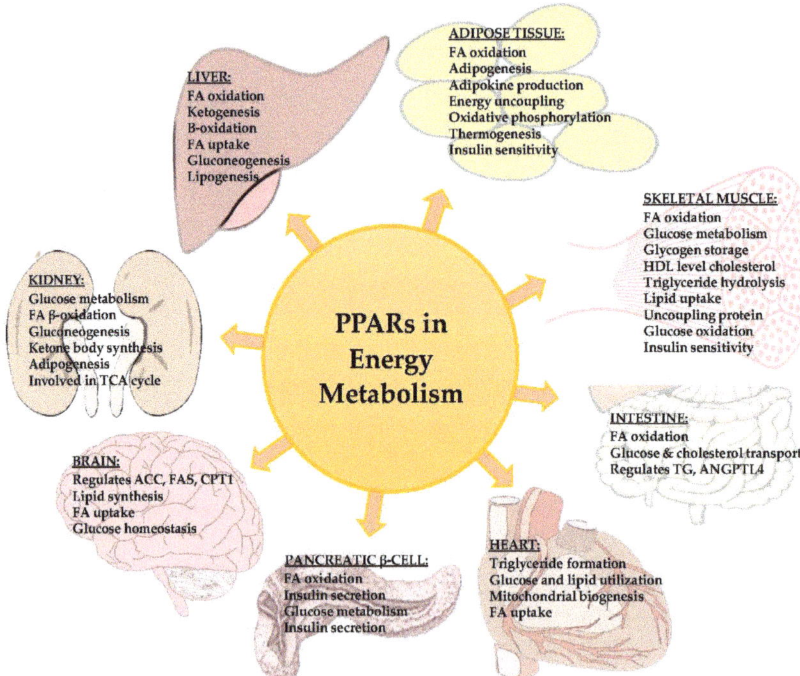

Figure A1. A schematic overview of role of PPARs in energy metabolism in various body organs.

Figure A2. Co-regulators of PPARs in carcinogenic process.

References

1. Fan, W.; Evans, R. PPARs and ERRs: Molecular mediators of mitochondrial metabolism. *Curr. Opin. Cell Biol.* **2015**, *33*, 49–54. [CrossRef] [PubMed]
2. Kota, B.P.; Huang, T.H.-W.; Roufogalis, B.D. An overview on biological mechanisms of PPARs. *Pharmacol. Res.* **2005**, *51*, 85–94. [CrossRef] [PubMed]
3. Ahmed, W.; Ziouzenkova, O.; Brown, J.; Devchand, P.; Francis, S.; Kadakia, M.; Kanda, T.; Orasanu, G.; Sharlach, M.; Zandbergen, F.; et al. PPARs and their metabolic modulation: New mechanisms for transcriptional regulation? *J. Intern. Med.* **2007**, *262*, 184–198. [CrossRef] [PubMed]
4. Kersten, S.; Desvergne, B.; Wahli, W. Roles of PPARs in health and disease. *Nature* **2000**, *405*, 421–424. [CrossRef] [PubMed]
5. Pyper, S.R.; Viswakarma, N.; Yu, S.; Reddy, J.K. PPARα: Energy combustion, hypolipidemia, inflammation and cancer. *Nucl. Recept. Signal.* **2010**, *8*, e002. [CrossRef] [PubMed]
6. Tyagi, S.; Gupta, P.; Saini, A.S.; Kaushal, C.; Sharma, S. The peroxisome proliferator-activated receptor: A family of nuclear receptors role in various diseases. *J. Adv. Pharm. Technol. Res.* **2011**, *2*, 236–240. [CrossRef] [PubMed]
7. Ferré, P. The Biology of Peroxisome Proliferator-Activated Receptors: Relationship with Lipid Metabolism and Insulin Sensitivity. *Diabetes* **2004**, *53*, S43–S50. [CrossRef] [PubMed]
8. Wang, Y.X. PPARs: Diverse regulators in energy metabolism and metabolic diseases. *Cell Res.* **2010**, *20*, 124–137. [CrossRef] [PubMed]
9. Tachibana, K.; Yamasaki, D.; Ishimoto, K.; Doi, T. The Role of PPARs in Cancer. *PPAR Res.* **2008**, *2008*. [CrossRef] [PubMed]
10. La Paglia, L.; Listi, A.; Caruso, S.; Amodeo, V.; Passiglia, F.; Bazan, V.; Fanale, D. Potential Role of ANGPTL4 in the Cross Talk between Metabolism and Cancer through PPAR Signaling Pathway. *PPAR Res.* **2017**, *2017*. [CrossRef] [PubMed]
11. Wu, J.; Chen, L.; Zhang, D.; Huo, M.; Zhang, X.; Pu, D.; Guan, Y. Peroxisome proliferator-activated receptors and renal diseases. *Front. Biosci. (Landmark Ed.)* **2009**, *14*, 995–1009. [CrossRef] [PubMed]
12. Dubois, V.; Eeckhoute, J.; Lefebvre, P.; Staels, B. Distinct but complementary contributions of PPAR isotypes to energy homeostasis. *J. Clin. Investig.* **2017**, *127*, 1202–1214. [CrossRef] [PubMed]
13. Feingold, K.R.; Wang, Y.; Moser, A.; Shigenaga, J.K.; Grunfeld, C. LPS decreases fatty acid oxidation and nuclear hormone receptors in the kidney. *J. Lipid Res.* **2008**, *49*, 2179–2187. [CrossRef] [PubMed]

14. Dressel, U.; Allen, T.L.; Pippal, J.B.; Rohde, P.R.; Lau, P.; Muscat, G.E. The peroxisome proliferator-activated receptor beta/delta agonist, GW501516, regulates the expression of genes involved in lipid catabolism and energy uncoupling in skeletal muscle cells. *Mol. Endocrinol.* **2003**, *17*, 2477–2493. [CrossRef] [PubMed]
15. Lefebvre, P.; Chinetti, G.; Fruchart, J.C.; Staels, B. Sorting out the roles of PPARα in energy metabolism and vascular homeostasis. *J. Clin. Investig.* **2006**, *116*, 571–580. [CrossRef] [PubMed]
16. Kondo, Y.; Uno, K.; Machida, K.; Terajima, M. PPARs and liver disease. *PPAR Res.* **2013**, *2013*. [CrossRef] [PubMed]
17. Choe, S.S.; Huh, J.Y.; Hwang, I.J.; Kim, J.I.; Kim, J.B. Adipose Tissue Remodeling: Its Role in Energy Metabolism and Metabolic Disorders. *Front. Endocrinol. (Lausanne)* **2016**, *7*, 30. [CrossRef] [PubMed]
18. Birsoy, K.; Festuccia, W.T.; Laplante, M. A comparative perspective on lipid storage in animals. *J. Cell Sci.* **2013**, *126*, 1541–1552. [CrossRef] [PubMed]
19. Rosen, E.D.; Spiegelman, B.M. Adipocytes as regulators of energy balance and glucose homeostasis. *Nature* **2006**, *444*, 847–853. [CrossRef] [PubMed]
20. Tontonoz, P.; Hu, E.; Graves, R.A.; Budavari, A.I.; Spiegelman, B.M. mPPAR gamma 2: Tissue-specific regulator of an adipocyte enhancer. *Genes Dev.* **1994**, *8*, 1224–1234. [CrossRef] [PubMed]
21. Tontonoz, P.; Hu, E.; Spiegelman, B.M. Stimulation of adipogenesis in fibroblasts by PPAR gamma 2, a lipid-activated transcription factor. *Cell* **1994**, *79*, 1147–1156. [CrossRef]
22. Rosen, E.D.; Hsu, C.H.; Wang, X.; Sakai, S.; Freeman, M.W.; Gonzalez, F.J.; Spiegelman, B.M. C/EBPα induces adipogenesis through PPARgamma: A unified pathway. *Genes Dev.* **2002**, *16*, 22–26. [CrossRef] [PubMed]
23. Mullican, S.E.; Dispirito, J.R.; Lazar, M.A. The orphan nuclear receptors at their 25-year reunion. *J. Mol. Endocrinol.* **2013**, *51*, T115–T140. [CrossRef] [PubMed]
24. Manickam, R.; Wahli, W. Roles of Peroxisome Proliferator-Activated Receptor beta/delta in skeletal muscle physiology. *Biochimie* **2017**, *136*, 42–48. [CrossRef] [PubMed]
25. Cho, S.Y.; Jeong, H.W.; Sohn, J.H.; Seo, D.B.; Kim, W.G.; Lee, S.J. An ethanol extract of Artemisia iwayomogi activates PPARdelta leading to activation of fatty acid oxidation in skeletal muscle. *PLoS ONE* **2012**, *7*, e33815. [CrossRef] [PubMed]
26. Periasamy, M.; Herrera, J.L.; Reis, F.C.G. Skeletal Muscle Thermogenesis and Its Role in Whole Body Energy Metabolism. *Diabetes Metab. J.* **2017**, *41*, 327–336. [CrossRef] [PubMed]
27. Gan, Z.; Burkart-Hartman, E.M.; Han, D.H.; Finck, B.; Leone, T.C.; Smith, E.Y.; Ayala, J.E.; Holloszy, J.; Kelly, D.P. The nuclear receptor PPARβ/δ programs muscle glucose metabolism in cooperation with AMPK and MEF2. *Genes Dev.* **2011**, *25*, 2619–2630. [CrossRef] [PubMed]
28. Schnuck, J.K.; Sunderland, K.L.; Gannon, N.P.; Kuennen, M.R.; Vaughan, R.A. Leucine stimulates PPARbeta/delta-dependent mitochondrial biogenesis and oxidative metabolism with enhanced GLUT4 content and glucose uptake in myotubes. *Biochimie* **2016**, *128*, 1–7. [CrossRef] [PubMed]
29. Perez-Schindler, J.; Svensson, K.; Vargas-Fernandez, E.; Santos, G.; Wahli, W.; Handschin, C. The coactivator PGC-1α regulates skeletal muscle oxidative metabolism independently of the nuclear receptor PPARbeta/delta in sedentary mice fed a regular chow diet. *Diabetologia* **2014**, *57*, 2405–2412. [CrossRef] [PubMed]
30. Thach, T.T.; Lee, C.K.; Park, H.W.; Lee, S.J.; Lee, S.J. Syringaresinol induces mitochondrial biogenesis through activation of PPARbeta pathway in skeletal muscle cells. *Bioorg. Med. Chem. Lett.* **2016**, *26*, 3978–3983. [CrossRef] [PubMed]
31. Cho, Y.; Hazen, B.C.; Russell, A.P.; Kralli, A. Peroxisome proliferator-activated receptor gamma coactivator 1 (PGC-1)- and estrogen-related receptor (ERR)-induced regulator in muscle 1 (Perm1) is a tissue-specific regulator of oxidative capacity in skeletal muscle cells. *J. Biol. Chem.* **2013**, *288*, 25207–25218. [CrossRef] [PubMed]
32. Kong, X.Y.; Feng, Y.Z.; Eftestol, E.; Kase, E.T.; Haugum, H.; Eskild, W.; Rustan, A.C.; Thoresen, G.H. Increased glucose utilization and decreased fatty acid metabolism in myotubes from Glmp(gt/gt) mice. *Arch. Physiol. Biochem.* **2016**, *122*, 36–45. [CrossRef] [PubMed]
33. Li, S.; Nagothu, K.K.; Desai, V.; Lee, T.; Branham, W.; Moland, C.; Megyesi, J.K.; Crew, M.D.; Portilla, D. Transgenic expression of proximal tubule peroxisome proliferator-activated receptor-α in mice confers protection during acute kidney injury. *Kidney Int.* **2009**, *76*, 1049–1062. [CrossRef] [PubMed]
34. Yang, J.; Zhou, Y.; Guan, Y. PPARγ as a therapeutic target in diabetic nephropathy and other renal diseases. *Curr. Opin. Nephrol. Hypertens.* **2012**, *21*, 97–105. [CrossRef] [PubMed]

35. Suzuki, M.; Nakamura, F.; Taguchi, E.; Nakata, M.; Wada, F.; Takihi, M.; Inoue, T.; Ohta, S.; Kawachi, H. 4′,6-Dimethoxyisoflavone-7-*O*-β-D-glucopyranoside (wistin) is a peroxisome proliferator-activated receptor α (PPARα) agonist in mouse hepatocytes. *Mol. Cell. Biochem.* **2018**, 1–7. [CrossRef] [PubMed]
36. Negishi, K.; Noiri, E.; Maeda, R.; Portilla, D.; Sugaya, T.; Fujita, T. Renal L-type fatty acid-binding protein mediates the bezafibrate reduction of cisplatin-induced acute kidney injury. *Kidney Int.* **2008**, *73*, 1374–1384. [CrossRef] [PubMed]
37. Lopez-Hernandez, F.J.; Lopez-Novoa, J.M. Potential utility of PPARα activation in the prevention of ischemic and drug-induced acute renal damage. *Kidney Int.* **2009**, *76*, 1022–1024. [CrossRef] [PubMed]
38. Teng, C.T.; Li, Y.; Stockton, P.; Foley, J. Fasting induces the expression of PGC-1α and ERR isoforms in the outer stripe of the outer medulla (OSOM) of the mouse kidney. *PLoS ONE* **2011**, *6*, e26961. [CrossRef] [PubMed]
39. Guan, Y.; Breyer, M.D. Peroxisome proliferator-activated receptors (PPARs): Novel therapeutic targets in renal disease. *Kidney Int.* **2001**, *60*, 14–30. [CrossRef] [PubMed]
40. Roe, N.D.; Standage, S.W.; Tian, R. The Relationship Between KLF5 and PPARα in the Heart: It's Complicated. *Circ. Res.* **2016**, *118*, 193–195. [CrossRef] [PubMed]
41. Ravingerova, T.; Adameova, A.; Carnicka, S.; Nemcekova, M.; Kelly, T.; Matejikova, J.; Galatou, E.; Barlaka, E.; Lazou, A. The role of PPAR in myocardial response to ischemia in normal and diseased heart. *Gen. Physiol. Biophys.* **2011**, *30*, 329–341. [CrossRef] [PubMed]
42. Drosatos, K.; Pollak, N.M.; Pol, C.J.; Ntziachristos, P.; Willecke, F.; Valenti, M.C.; Trent, C.M.; Hu, Y.; Guo, S.; Aifantis, I.; et al. Cardiac Myocyte KLF5 Regulates Ppara Expression and Cardiac Function. *Circ. Res.* **2016**, *118*, 241–253. [CrossRef] [PubMed]
43. Mora, C.; Pintado, C.; Rubio, B.; Mazuecos, L.; Lopez, V.; Fernandez, A.; Salamanca, A.; Barcena, B.; Fernandez-Agullo, T.; Arribas, C.; et al. Central leptin regulates heart lipid content by selectively increasing PPAR beta/delta expression. *J. Endocrinol.* **2018**, *236*, 43–56. [CrossRef] [PubMed]
44. Pang, J.; Xu, X.; Getman, M.R.; Shi, X.; Belmonte, S.L.; Michaloski, H.; Mohan, A.; Blaxall, B.C.; Berk, B.C. G protein coupled receptor kinase 2 interacting protein 1 (GIT1) is a novel regulator of mitochondrial biogenesis in heart. *J. Mol. Cell Cardiol.* **2011**, *51*, 769–776. [CrossRef] [PubMed]
45. Arumugam, S.; Sreedhar, R.; Thandavarayan, R.A.; Karuppagounder, V.; Watanabe, K. Targeting fatty acid metabolism in heart failure: Is it a suitable therapeutic approach? *Drug Discov. Today* **2016**, *21*, 1003–1008. [CrossRef] [PubMed]
46. Di Giacomo, E.; Benedetti, E.; Cristiano, L.; Antonosante, A.; d'Angelo, M.; Fidoamore, A.; Barone, D.; Moreno, S.; Ippoliti, R.; Ceru, M.P.; et al. Roles of PPAR transcription factors in the energetic metabolic switch occurring during adult neurogenesis. *Cell Cycle* **2017**, *16*, 59–72. [CrossRef] [PubMed]
47. Kocalis, H.E.; Turney, M.K.; Printz, R.L.; Laryea, G.N.; Muglia, L.J.; Davies, S.S.; Stanwood, G.D.; McGuinness, O.P.; Niswender, K.D. Neuron-specific deletion of peroxisome proliferator-activated receptor delta (PPARdelta) in mice leads to increased susceptibility to diet-induced obesity. *PLoS ONE* **2012**, *7*, e42981. [CrossRef] [PubMed]
48. Kouidhi, S.; Seugnet, I.; Decherf, S.; Guissouma, H.; Elgaaied, A.B.; Demeneix, B.; Clerget-Froidevaux, M.S. Peroxisome proliferator-activated receptor-gamma (PPARγ) modulates hypothalamic Trh regulation in vivo. *Mol. Cell. Endocrinol.* **2010**, *317*, 44–52. [CrossRef] [PubMed]
49. Stump, M.; Guo, D.F.; Lu, K.T.; Mukohda, M.; Cassell, M.D.; Norris, A.W.; Rahmouni, K.; Sigmund, C.D. Nervous System Expression of PPARgamma and Mutant PPARgamma Has Profound Effects on Metabolic Regulation and Brain Development. *Endocrinology* **2016**, *157*, 4266–4275. [CrossRef] [PubMed]
50. Lu, M.; Sarruf, D.A.; Talukdar, S.; Sharma, S.; Li, P.; Bandyopadhyay, G.; Nalbandian, S.; Fan, W.; Gayen, J.R.; Mahata, S.K.; et al. Brain PPAR-γ promotes obesity and is required for the insulin-sensitizing effect of thiazolidinediones. *Nat. Med.* **2011**, *17*, 618–622. [CrossRef] [PubMed]
51. Rijnsburger, M.; Belegri, E.; Eggels, L.; Unmehopa, U.A.; Boelen, A.; Serlie, M.J.; la Fleur, S.E. The effect of diet interventions on hypothalamic nutrient sensing pathways in rodents. *Physiol. Behav.* **2016**, *162*, 61–68. [CrossRef] [PubMed]
52. Ryan, K.K.; Li, B.; Grayson, B.E.; Matter, E.K.; Woods, S.C.; Seeley, R.J. A role for central nervous system PPAR-γ in the regulation of energy balance. *Nat. Med.* **2011**, *17*, 623–626. [CrossRef] [PubMed]

53. Sarruf, D.A.; Yu, F.; Nguyen, H.T.; Williams, D.L.; Printz, R.L.; Niswender, K.D.; Schwartz, M.W. Expression of peroxisome proliferator-activated receptor-γ in key neuronal subsets regulating glucose metabolism and energy homeostasis. *Endocrinology* **2009**, *150*, 707–712. [CrossRef] [PubMed]
54. Wan, J.; Jiang, L.; Lu, Q.; Ke, L.; Li, X.; Tong, N. Activation of PPARδ up-regulates fatty acid oxidation and energy uncoupling genes of mitochondria and reduces palmitate-induced apoptosis in pancreatic β-cells. *Biochem. Biophys. Res. Commun.* **2010**, *391*, 1567–1572. [CrossRef] [PubMed]
55. Jiang, L.; Wan, J.; Ke, L.Q.; Lu, Q.G.; Tong, N.W. Activation of PPARδ promotes mitochondrial energy metabolism and decreases basal insulin secretion in palmitate-treated β-cells. *Mol. Cell. Biochem.* **2010**, *343*, 249–256. [CrossRef] [PubMed]
56. Li, L.; Li, T.; Zhang, Y.; Pan, Z.; Wu, B.; Huang, X.; Zhang, Y.; Mei, Y.; Ge, L.; Shen, G.; et al. Peroxisome proliferator-activated receptorβ/δ activation is essential for modulating p-Foxo1/Foxo1 status in functional insulin-positive cell differentiation. *Cell Death Dis.* **2015**, *6*, e1715. [CrossRef] [PubMed]
57. Bendlova, B.; Vejrazkova, D.; Vcelak, J.; Lukasova, P.; Burkonova, D.; Kunesova, M.; Vrbikova, J.; Dvorakova, K.; Vondra, K.; Vankova, M. PPARγ2 Pro12Ala polymorphism in relation to free fatty acids concentration and composition in lean healthy Czech individuals with and without family history of diabetes type 2. *Physiol. Res.* **2008**, *57* (Suppl S1), S77–S90. [PubMed]
58. Roduit, R.; Morin, J.; Masse, F.; Segall, L.; Roche, E.; Newgard, C.B.; Assimacopoulos-Jeannet, F.; Prentki, M. Glucose down-regulates the expression of the peroxisome proliferator-activated receptor-α gene in the pancreatic β-cell. *J. Biol. Chem.* **2000**, *275*, 35799–35806. [CrossRef] [PubMed]
59. Cohen, G.; Riahi, Y.; Shamni, O.; Guichardant, M.; Chatgilialoglu, C.; Ferreri, C.; Kaiser, N.; Sasson, S. Role of lipid peroxidation and PPAR-δ in amplifying glucose-stimulated insulin secretion. *Diabetes* **2011**, *60*, 2830–2842. [CrossRef] [PubMed]
60. Vrins, C.L.; van der Velde, A.E.; van den Oever, K.; Levels, J.H.; Huet, S.; Oude Elferink, R.P.; Kuipers, F.; Groen, A.K. Peroxisome proliferator-activated receptor delta activation leads to increased transintestinal cholesterol efflux. *J. Lipid Res.* **2009**, *50*, 2046–2054. [CrossRef] [PubMed]
61. Higashimura, Y.; Naito, Y.; Takagi, T.; Uchiyama, K.; Mizushima, K.; Yoshikawa, T. Propionate Promotes Fatty Acid Oxidation through the Up-Regulation of Peroxisome Proliferator-Activated Receptor α in Intestinal Epithelial Cells. *J. Nutr. Sci. Vitaminol. (Tokyo)* **2015**, *61*, 511–515. [CrossRef] [PubMed]
62. Korecka, A.; de Wouters, T.; Cultrone, A.; Lapaque, N.; Pettersson, S.; Dore, J.; Blottiere, H.M.; Arulampalam, V. ANGPTL4 expression induced by butyrate and rosiglitazone in human intestinal epithelial cells utilizes independent pathways. *Am. J. Physiol. Gastrointest. Liver Physiol.* **2013**, *304*, G1025–G1037. [CrossRef] [PubMed]
63. Karimian Azari, E.; Leitner, C.; Jaggi, T.; Langhans, W.; Mansouri, A. Possible role of intestinal fatty acid oxidation in the eating-inhibitory effect of the PPAR-α agonist Wy-14643 in high-fat diet fed rats. *PLoS ONE* **2013**, *8*, e74869. [CrossRef] [PubMed]
64. de Vogel-van den Bosch, H.M.; Bunger, M.; de Groot, P.J.; Bosch-Vermeulen, H.; Hooiveld, G.J.; Muller, M. PPARα-mediated effects of dietary lipids on intestinal barrier gene expression. *BMC Genom.* **2008**, *9*, 231. [CrossRef] [PubMed]
65. van den Bosch, H.M.; Bunger, M.; de Groot, P.J.; van der Meijde, J.; Hooiveld, G.J.; Muller, M. Gene expression of transporters and phase I/II metabolic enzymes in murine small intestine during fasting. *BMC Genom.* **2007**, *8*, 267. [CrossRef] [PubMed]
66. Takei, K.; Nakagawa, Y.; Wang, Y.; Han, S.I.; Satoh, A.; Sekiya, M.; Matsuzaka, T.; Shimano, H. Effects of K-877, a novel selective PPARα modulator, on small intestine contribute to the amelioration of hyperlipidemia in low-density lipoprotein receptor knockout mice. *J. Pharmacol. Sci.* **2017**, *133*, 214–222. [CrossRef] [PubMed]
67. Lempradl, A.; Pospisilik, J.A.; Penninger, J.M. Exploring the emerging complexity in transcriptional regulation of energy homeostasis. *Nat. Rev. Genet.* **2015**, *16*, 665–681. [CrossRef] [PubMed]
68. Yamamoto, H.; Williams, E.G.; Mouchiroud, L.; Canto, C.; Fan, W.; Downes, M.; Heligon, C.; Barish, G.D.; Desvergne, B.; Evans, R.M.; et al. NCoR1 is a conserved physiological modulator of muscle mass and oxidative function. *Cell* **2011**, *147*, 827–839. [CrossRef] [PubMed]
69. Martinez-Jimenez, C.P.; Kyrmizi, I.; Cardot, P.; Gonzalez, F.J.; Talianidis, I. Hepatocyte nuclear factor 4α coordinates a transcription factor network regulating hepatic fatty acid metabolism. *Mol. Cell. Biol.* **2010**, *30*, 565–577. [CrossRef] [PubMed]

70. Mello, T.; Materozzi, M.; Galli, A. PPARs and Mitochondrial Metabolism: From NAFLD to HCC. *PPAR Res.* **2016**, *2016*. [CrossRef] [PubMed]
71. Scatena, R.; Bottoni, P.; Giardina, B. Mitochondria, PPARs, and Cancer: Is Receptor-Independent Action of PPAR Agonists a Key? *PPAR Res.* **2008**, *2008*. [CrossRef] [PubMed]
72. Vitale, S.G.; Lagana, A.S.; Nigro, A.; La Rosa, V.L.; Rossetti, P.; Rapisarda, A.M.; La Vignera, S.; Condorelli, R.A.; Corrado, F.; Buscema, M.; et al. Peroxisome Proliferator-Activated Receptor Modulation during Metabolic Diseases and Cancers: Master and Minions. *PPAR Res.* **2016**, *2016*. [CrossRef] [PubMed]
73. Fanale, D.; Amodeo, V.; Caruso, S. The Interplay between Metabolism, PPAR Signaling Pathway, and Cancer. *PPAR Res.* **2017**, *2017*. [CrossRef] [PubMed]
74. Bandera Merchan, B.; Tinahones, F.J.; Macias-Gonzalez, M. Commonalities in the Association between PPARG and Vitamin D Related with Obesity and Carcinogenesis. *PPAR Res.* **2016**, *2016*. [CrossRef] [PubMed]
75. Glazer, R.I. PPARδ as a Metabolic Initiator of Mammary Neoplasia and Immune Tolerance. *PPAR Res.* **2016**, *2016*. [CrossRef] [PubMed]
76. Lakshmi, S.P.; Reddy, A.T.; Banno, A.; Reddy, R.C. PPAR Agonists for the Prevention and Treatment of Lung Cancer. *PPAR Res.* **2017**, *2017*. [CrossRef] [PubMed]
77. Chiu, M.; McBeth, L.; Sindhwani, P.; Hinds, T.D. Deciphering the Roles of Thiazolidinediones and PPARγ in Bladder Cancer. *PPAR Res.* **2017**, *2017*. [CrossRef] [PubMed]
78. Koronowicz, A.A.; Banks, P.; Master, A.; Domagala, D.; Piasna-Slupecka, E.; Drozdowska, M.; Sikora, E.; Laidler, P. Fatty Acids of CLA-Enriched Egg Yolks Can Induce Transcriptional Activation of Peroxisome Proliferator-Activated Receptors in MCF-7 Breast Cancer Cells. *PPAR Res.* **2017**, *2017*. [CrossRef] [PubMed]

International Journal of
Molecular Sciences

Review

PPAR Agonists and Metabolic Syndrome: An Established Role?

Margherita Botta [1,†], Matteo Audano [1,†], Amirhossein Sahebkar [2,3,4], Cesare R. Sirtori [5], Nico Mitro [1,*] and Massimiliano Ruscica [1]

1 Dipartimento di Scienze Farmacologiche e Biomolecolari, Università degli Studi di Milano, 20133 Milan, Italy; margherita.botta@unimi.it (M.B.); matteo.audano@unimi.it (M.A.); massimiliano.ruscica@unimi.it (M.R.)
2 Biotechnology Research Center, Pharmaceutical Technology Institute, Mashhad University of Medical Sciences, Mashhad 9177948564, Iran; amir_saheb2000@yahoo.com
3 Neurogenic Inflammation Research Center, Mashhad University of Medical Sciences, Mashhad 9177948564, Iran
4 School of Pharmacy, Mashhad University of Medical Sciences, Mashhad 9177948564, Iran
5 Centro Dislipidemie, Azienda Socio Sanitaria Territoriale Grande Ospedale Metropolitano Niguarda, 20162 Milan, Italy; cesare.sirtori@icloud.com
* Correspondence: nico.mitro@unimi.it
† These authors contributed equally to this work.

Received: 27 March 2018; Accepted: 11 April 2018; Published: 14 April 2018

Abstract: Therapeutic approaches to metabolic syndrome (MetS) are numerous and may target lipoproteins, blood pressure or anthropometric indices. Peroxisome proliferator-activated receptors (PPARs) are involved in the metabolic regulation of lipid and lipoprotein levels, i.e., triglycerides (TGs), blood glucose, and abdominal adiposity. PPARs may be classified into the α, β/δ and γ subtypes. The PPAR-α agonists, mainly fibrates (including newer molecules such as pemafibrate) and omega-3 fatty acids, are powerful TG-lowering agents. They mainly affect TG catabolism and, particularly with fibrates, raise the levels of high-density lipoprotein cholesterol (HDL-C). PPAR-γ agonists, mainly glitazones, show a smaller activity on TGs but are powerful glucose-lowering agents. Newer PPAR-α/δ agonists, e.g., elafibranor, have been designed to achieve single drugs with TG-lowering and HDL-C-raising effects, in addition to the insulin-sensitizing and antihyperglycemic effects of glitazones. They also hold promise for the treatment of non-alcoholic fatty liver disease (NAFLD) which is closely associated with the MetS. The PPAR system thus offers an important hope in the management of atherogenic dyslipidemias, although concerns regarding potential adverse events such as the rise of plasma creatinine, gallstone formation, drug–drug interactions (i.e., gemfibrozil) and myopathy should also be acknowledged.

Keywords: metabolic syndrome; PPARs; pemafibrate; elafibrinor

---

1. Introduction

The incidence of metabolic syndrome (MetS), representing a global public health issue, has been estimated to vary from 20 to 27% in developing countries [1,2] to 35% in the USA [3]. MetS is a cluster of cardiometabolic risk factors, from high triglycerides (TGs), to elevated waist circumference (WC), high blood pressure (BP) and insulin resistance [4].

Following the first definition of MetS by the World Health Organization (WHO), several expert panels attempted to introduce stricter diagnostic criteria. In 2001, the National Cholesterol Education Program (NCEP) Adult Treatment Panel III (ATP III) [5] recognized that the clustering of the metabolic risk factors included in the syndrome were indeed cardiovascular (CV) risk factors.

In 2003, the American Association of Clinical Endocrinologists (AACE) modified the ATP III criteria highlighting the central role of insulin resistance in the pathogenesis of the syndrome [6]. In 2005, the International Diabetes Federation (IDF) issued a consensus document aimed at introducing a clinically useful definition of MetS in order to identify individuals at high risk of CV disease (CVD) and type 2 diabetes mellitus (T2D) on a worldwide basis [7]. In the same year, the American Heart Association (AHA)/National Heart, Lung and Blood Institute (NHLBI) suggested more specific criteria for the diagnosis of MetS [7]. Finally, a joint statement of IDF, NHLBI, AHA, World Heart Federation and International Association for the Study of Obesity, best known as the "Harmonization definition", has been introduced and now represents the most commonly recognized criterion for the clinical diagnosis of MetS [4] (Table 1).

Table 1. Risk factors for the clinical diagnosis of metabolic syndrome.

	Value	Alternative Indicator
Waist circumference	* >94 cm in males, >80 cm in females ** >102 cm in males, >88 cm in females	
Raised blood pressure	Systolic ≥130 and/or diastolic ≥85 mm Hg	Treatment of previously diagnosed hypertension
Raised FPG	≥100 mg/dL (5.6 mmol/L)	Previously diagnosed of T2DM
Raised TG	>150 mg/dL (1.7 mmol/L)	Specific pharmacological treatment
Reduced HDL-C	<40 mg/dL (1.0 mmol/L) in males <50 mg/dL (1.3 mmol/L) in females	Specific pharmacological treatment

* Based on the International Diabetes Federation (IDF) threshold for Europid population. ** Based on the AHA/NHLBI (ATP III) threshold for USA population. FPG, Fasting Plasma Glucose; TG, triglycerides; HDL-C, High-Density Lipoprotein-Cholesterol; T2DM, Type 2 Diabetes Mellitus. Conversion factors: (i) mg/dL cholesterol = mmol/L × 38.6; (ii) mg/dL triglycerides = mmol/L × 88.5 and (iii) mg/dL glucose = mmol/L × 18. Reproduced with permission [8].

Carriers of MetS are at higher risk of developing atherosclerotic CVD, a condition worsened by the so called "atherogenic dyslipidemia". This mixed dyslipidemia has emerged as the most clinically relevant "competitor" of elevated low-density lipoprotein cholesterol (LDL-C) among lipid risk factors. It is characterized by hypertriglyceridemia, low high-density lipoprotein (HDL)-cholesterol levels, and the prevalence of small, dense low-density lipoprotein (LDL) particles as well as an accumulation of cholesterol-rich remnant particles [9].

In this metabolic derangement, peroxisome proliferator-activated receptors (PPARs), i.e., nuclear receptors involved in the regulation of metabolic homeostasis, represent a valuable therapeutic target. PPAR activators have provided significant benefit in patients with primary hypertriglyceridemia (i.e., fibrates and omega-3 fatty acids, both PPAR-α agonists), as well as in cases of mixed hyperlipidemias with raised TGs and low HDL-C; conversely, PPAR-γ activators have become choice drugs in T2D.

PPAR agonists are generally recognized as effective pharmacological tools for the management of MetS [10,11]. A growing interest in PPAR activators has been acknowledged in recent years as they have been used in the more and more frequent occurrence of non-alcoholic fatty liver disease [12], the hepatic manifestation of MetS. However, post-marketing adverse effects should be recalled [13], i.e., weight gain, fluid retention, congestive heart failure, liver and gallbladder disease, renal effects, bone fractures, myopathy and rhabdomyolysis; these two last particularly with fibrates with an added risk when co-administered with statins [14,15].

Hence, the present review was aimed at discussing available evidence on new PPAR agonists in the clinical setting as well as at describing molecular mechanisms underlying the effects of these drugs. This review also attempts to reduce the current disagreement on the interpretation of outcomes of clinical trials with fibrates. To this end, we have revised and updated the available English-language studies relevant to the key clinical questions, published up to April 2018.

2. Peroxisome Proliferator-Activated Receptor: Key Players in Energy Homeostasis

PPARs are a subfamily of three ligand-inducible transcription factors, belonging to the superfamily of nuclear hormone receptors. In mammals, three different isoforms of PPARs have been described so far: PPAR-α, PPAR-β/δ and PPAR-γ. PPARs belong to the nuclear hormone receptor superfamily and, by binding to PPAR-responsive regulatory elements (PPRE), heterodimerise with the retinoid X receptor (RXR) and control a group of genes involved in adipogenesis, lipid metabolism, inflammation and maintenance of metabolic homeostasis [16–20]. PPAR-α is the first identified member and is mainly expressed in energy-demanding tissues that show high rates of β-oxidation (i.e., liver, kidney, heart and muscle). On the other hand, PPAR-β/δ is ubiquitously expressed in humans, whereas in mice it is expressed to a higher extent in the gastrointestinal duct, specifically stomach, large and small intestine. PPAR-γ is expressed at high levels in the adipose tissue.

PPARs are activated by fatty acids and eicosanoids [21], as well as by small molecules, such as fibrates for PPAR-α, GW501516, GW0742, bezafibrate and Telmisartan for PPAR-β/δ, and glitazones for PPAR-γ. PPAR-α mediates the hypolipidemic function of fibrates in the treatment of hypertriglyceridemia and hypoalphalipoproteinemia [22], being the main regulator of intra- and extracellular lipid metabolism. Indeed, fibrates downregulate hepatic apolipoprotein C-III (ApoCIII) and stimulate lipoprotein lipase gene expression, thus being key players in TG metabolism [23].

Moreover, PPAR-α activation raises plasma HDL-C via induction of hepatic apolipoprotein A-I and apolipoprotein A-II expression in humans. On the other hand, glitazones exert hypotriglyceridemic activity by activating PPAR-γ, in turn inducing lipoprotein lipase expression in adipose tissue [23]. Finally, PPARs exert their function on intracellular lipid metabolism by regulating key proteins involved in the conversion of fatty acids to acyl-CoA esters, fatty acid import into mitochondria and peroxisomal and mitochondrial fatty acid oxidation [24,25]. Major roles and functions of the PPAR isotypes are depicted in Figure 1.

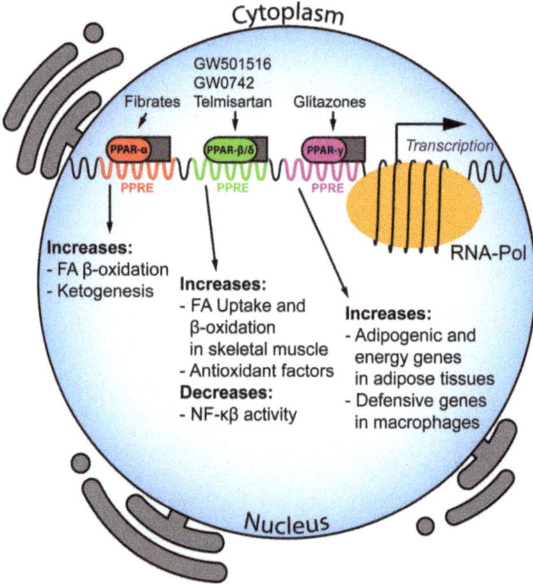

Figure 1. Major roles of different peroxisome proliferator-activated receptors (PPARs) isotypes. PPARs are a class of nuclear transcription factors that heterodimerize with retinoid X receptor (RXR, gray boxes) upon physiological (i.e., fatty acids) and synthetic activation (i.e., fibrates, glitazones etc.) to regulate the specific indicated pathways. FA, Fatty Acids; NFκB, Nuclear Factor-κB.

2.1. PPAR-α

PPAR-α (also called NR1C1) activation occurs mainly under energy deprivation. This leads to the upregulation of intracellular energy metabolism, ultimately inducing ATP production from oxidative phosphorylation. PPAR-α mRNA is upregulated in mouse liver during fasting, whereas PPAR-α knock-out (KO) fasted mice display significant hypoglycemia, hypoketonemia, hypothermia, and increased plasma free fatty acids, thus suggesting an inhibition of fatty acid uptake and oxidation [26]. PPAR-α-mediated fatty acid catabolism is crucial for the synthesis of several metabolites to be used as energy sources by other tissues such as ketone bodies in the brain [27]. Classical genes regulated by this nuclear receptor are the β-oxidative enzymes, e.g., carnitine palmitoyltransferase 1A and 2 (*CPT1A* and *2*), acyl-CoA dehydrogenase very long chain (*ACADVL*), hydroxyacyl-CoA dehydrogenase trifunctional multienzyme complex subunit-α (*HADHA*); similarly, important ketogenic genes like 3-hydroxy-3-methylglutaryl-CoA synthase 2 (*HMGCS2*), 3-hydroxymethyl-3-methylglutaryl-CoA lyase (*HMGCL*) and acetyl-CoA acetyltransferase 1 (*ACAT1*) are stimulated by PPAR-α. This last seems to control liver glucose metabolism as well. Administration of fenofibrate to mice decreases expression levels of glucokinase and flux through this enzyme, suggesting a lower liver glucose uptake upon PPAR-α activation [28]. In another study, PPAR-α was found to induce pyruvate dehydrogenase kinase 4 (*PDK4*) expression, suppressing pyruvate transition to acetyl-CoA. Conversely, PPAR-α KO pups displayed a primary defect in gluconeogenesis, specifically from glycerol, leading to significant hypoglycemia [29].

PPAR-α induction also exerts an anti-inflammatory activity in mouse models, even though contrasting data are reported. The first evidence of PPAR-α involvement in the regulation of inflammation was provided by the group of Wahli more than 20 years ago [30]. The authors demonstrated that leukotriene B4 acts as a ligand for PPAR-α transcription, and the inflammatory response is prolonged in PPAR-α KO mice [31]. More recent studies have confirmed this association, showing that PPAR-α activation in mouse liver downregulates the CCAAT/enhancer binding protein β (C/EBPβ) as well as alpha (C/EBPα) and nuclear factor-κB (NFκB) protein expression, leading to lower levels of C-reactive protein, interleukin-6 and prostaglandins [32]. Conversely, dietary treatment with PPAR-α agonists increased lipopolysaccharide-induced plasma tumor necrosis factor α (TNF-α) levels that is instead reduced in PPAR-α-deficient mice, highlighting a possible role for PPAR-α also as a pro-inflammatory factor [33].

2.2. PPAR-β/δ

PPAR-β/δ (also called NR1C2) is the least well characterized isotype among PPARs. Nevertheless, it plays an important role in the metabolic adaptation of numerous tissues to environmental stimuli [34]. Physiologically, this isotype is activated by long-chain fatty acids, saturated and unsaturated, and by prostacyclin [35,36]. Notably, specific PPAR-β/δ activation leads to increased levels of fatty acid β-oxidation [37]. Moreover, PPAR-β/δ expression is upregulated specifically in skeletal muscle during fasting. These data were confirmed in PPAR-β/δ agonist treated L6 rat myocytes, showing increased fatty acid uptake and β-oxidation compared to controls [37].

Further studies suggest that PPAR-β/δ and physical exercise are tightly related; indeed, endurance training (6 weeks) boosted an up to 2.6-fold PPAR-β/δ protein expression in the tibialis anterior muscle [38]. The indispensable role of PPAR-β/δ in cell energy metabolism has been also shown in rat breast adenocarcinoma cells [39]. Specifically, the authors demonstrated that high PPAR-β/δ protein was associated with increased cancer cell growth in vitro and in vivo, suggesting that PPAR-β/δ favors breast cancer cell survival by regulating specific metabolic pathways.

PPAR-β/δ ligands have also been proposed as potential anti-inflammatory drugs [40,41]. Pharmacological activation of PPAR-β/δ in endothelial cells is associated with a potent anti-inflammatory effect, possibly by involving antioxidative genes and release of nuclear corepressors [42]. Moreover, investigation of PPAR-β/δ role in the modulation of NF-κB-driven inflammatory response confirmed the anti-inflammatory activity of this isotype [43]. The highly selective PPAR-β/δ agonist GW0742 in rats effectively antagonized lethality consequent to cecal ligation and puncture: drug treated animals had reduced release of pro-inflammatory cytokines and neutrophil infiltration in lung, liver, and cecum.

2.3. PPAR-γ

PPAR-γ (also called NR1C3) is mainly expressed in adipocytes (brown and white) and plays a major role in cell differentiation and energy metabolism [44,45]. Upregulation of PPAR-γ activity in vivo leads to bone loss and higher bone marrow adiposity, whereas downregulation leads to elevated bone mass [46]. The PPAR-γ agonists thiazolidinediones (TZDs) have a hypoglycemic action in ob/ob mice and improve insulin action in several models of obesity and diabetes [47,48]. A high correlation between the hypoglycemic activity of TZDs and their affinity for PPAR-γ has been repeatedly shown. While PPAR-γ-KO animals show embryonic lethality dying at 10.5–11.5 days postcoitum due to placental dysfunction, PPAR-γ heterozygote KO are characterized by higher insulin sensitivity and resistance to high-fat diet-induced insulin resistance [49–51].

PPAR-γ Pro12Ala partial loss-of-function mutation in humans leads to decreased body mass index, higher insulin sensitivity and protection from T2D [52]. The most accepted hypothesis is that PPAR-γ facilitates energy storage after high-fat diet challenge, partly by suppressing leptin expression in adipocytes [49]. In addition, PPAR-γ directly binds the promoter of almost all adipogenic genes, such as factors involved in glucose and fatty acid metabolism, indicating that this isotype is fundamental for the activation of metabolic programs during adipogenesis [53]. While PPAR-γ is not needed for macrophage differentiation, it is fundamental for a proper anti-inflammatory activity in adipose tissue macrophages [54]. This specific feature of PPAR-γ is due to cell type specific DNA binding, as shown in macrophages, where PPAR-γ cooperates mainly with the spleen focus forming virus (SFFV) proviral integration into the oncogene transcription factor [55]. On the other hand, it is well established that CCAAT/enhancer binding protein (C/EBP) α and β transcription factors are major partners of PPAR-γ during adipocyte differentiation [53].

Comparative analysis of PPAR-γ binding patterns in adipocytes and macrophages indicates that it binds to immune defense genes in macrophages, whereas the only cluster of genes shared with adipocytes is the one encoding metabolic factors [55]. Notably, myeloid lineage-specific PPAR-γ KO mice display lower alternative macrophage activation and a strong metabolic phenotype, characterized by diet-induced obesity, insulin resistance, and glucose intolerance [55]. In addition, PPAR-γ activation decreases T-lymphocyte-dependent inflammation of adipose tissue and development of insulin resistance in diet-induced obese mice [56]. Recently, it has been demonstrated that the PPAR-γ agonist pioglitazone may prevent or delay aortic aneurysm progression in patients [57]. Treated patients show decreased macrophage infiltration into the retroperitoneal periaortic fat, as well as tumor necrosis factor α (*TNFα*) and matrix metallopeptidase 9 (*MMP9*) gene expression. On the other hand, treatment increased adiponectin expressions in both tissues compared to controls. Neurons are protected by pioglitazone treatment [58]; indeed, neuron and axon susceptibility to both nitric-oxide donor-induced and microglia-derived nitric oxide-induced toxicity are reduced, whereas catalase expression is raised [58].

3. PPAR-α: Fibrates and Omega-3 Fatty Acids in the Metabolic Syndrome

The characteristic features of MetS, i.e., increased triglyceridemia, abdominal obesity, reduced HDL-C levels and increased glycemia, in addition to raised blood pressure, clearly indicate that PPAR-α agonists have an ideal profile to control most of these features [59].

3.1. Fibrates

The role of fibrates in the clinical management of disorders characterized by elevated TGs is now well established [60]. Fibrates, activators of the PPAR system, mainly PPAR-α, have shown significant benefit in clinical trials of CV prevention, i.e., reducing the occurrence of nonfatal myocardial infarction, particularly when restricting evaluation to patients with concomitant TG elevation and HDL-C reduction [61].

Clinical findings, at times disappointing (FIELD—Fenofibrate Intervention and Event Lowering in Diabetes—study) [62], have been hampered by inappropriate patient selection and, possibly, by the use of non-optimal drug formulations [63,64]. On the other hand, long-term re-evaluation of most recent studies, particularly the ACCORD—Action to Control Cardiovascular Risk in Diabetes—trial with fenofibrate, has clearly indicated that the agent has a heterogeneous response, but may have a valid indication in reducing CVD in appropriately selected patients, i.e., those with diabetes, hypertriglyceridemia and low HDL-C [65]. In these subjects, a definite benefit on the CV outcomes may be predicted.

When projected to the growing field of lipid-lowering treatments in clinical practice, these observations have fostered the development of newer pharmacological approaches, partially based on the PPAR target, but providing additional or alternative mechanisms which may lead to lipoprotein changes and altered glycemic parameters, not found with the established agents. Hence, this pharmacological approach may be of potential value in the clinical approach to the more and more frequent occurrence of MetS [66].

A recent meta-analysis from 22 RCTs involving a total of 11,402 subjects showed that fibrate administration decreased fasting plasma glucose (−5 mg/dL), insulin levels (−0.56 μIU/mL), and insulin resistance (HOMA-IR: −1.09), but not HbA1C. This latter evidence may be due to the short-term duration—less than three months—of the included studies. Moreover, the reduction in glucose, although statistically significant, was small and, thus, not clinically relevant [67]. These findings have been well characterized from studies with fenofibrate and bezafibrate. This latter, differently from the most selective PPAR-α modulator fenofibrate, acts as a pan PPAR activator for all three PPAR isoforms (α, γ and δ) [68]. Bezafibrate, similar to the selective PPAR-γ agonists, i.e., thiazolidinedione (TZDs) that are specifically used in T2D treatment, may exert a glucose-lowering activity but apparently without causing water retention, weight gain and peripheral edema that are potential side effects of glitazones [69,70]. Bezafibrate, compared with other fibrates, reduces the incidence of T2D, with a better reduction of blood glucose, HbA1C and insulin resistance [71].

Fibrates: Evidence from the Most Recent Clinical Trials

Pemafibrate (formerly known as K-877) is one of the newest members of the selective PPAR-α modulators, being >2000-fold more selective for PPAR-α vs. either PPAR-γ or -δ (delta) [72]. Pemafibrate has been recently approved in Japan for the treatment of hyperlipidemias, with a recommended dosage of 0.1 mg bid with the possibility of reaching a maximum of 0.2 mg bid [73].

The long-term efficacy of pemafibrate has been recently reported in a phase 3 multicenter, placebo-controlled, randomized, double-blind, parallel-group study (JapicCTI-142412) on T2D patients (HbA1c $\geq$ 6.2%) with fasting TGs $\geq$ 150 mg/dL, not on statins. The primary endpoint was the percentage change in fasting TG levels from baseline and secondary endpoints were changes in fasting and postprandial lipoproteins and glycemic parameters. Among the 167 eligible participants (mean age 60.5 years), 54 and 55 were randomized to pemafibrate 0.1 or 0.2 mg bid, respectively, and 57 to a placebo. Twenty-four weeks of treatment led to a significant 45% decrement of TGs regardless of dose; fasting TGs $\leq$ 150 mg/dL were achieved by 81.5% and 70.9% of patients on 0.2 and 0.4 mg/day, respectively; statistically significant when compared to the placebo group. In addition, non-HDL-C, cholesterol remnants, ApoB100, ApoB48, and ApoCIII levels were reduced with a concomitant rise of HDL-C and ApoA-I (Table 2) [74]. The reduction of ApoCIII levels confirmed the general activity of PPAR activators on this variable, in line with what reported with statin treatments [75].

Notably, pemafibrate led to a more anti-atherogenic profile, i.e., higher levels of medium, small and very small HDL particles vs very large and large particles at baseline. Although no changes were seen in LDL-C levels, a significant increment of large LDL and a reduction of small and very small particles were found in the pemafibrate arms. Modest changes were seen in glycemic parameters: only the 0.2 mg dose significantly reduced the HOMA-insulin resistance score with no significant changes in fasting glucose, insulin, glycated albumin and HbA1c. Both pemafibrate doses significantly raised circulating levels of FGF-21 [74]. All groups displayed comparable rates of adverse events and drug reactions, i.e., serum creatinine and liver enzyme increases.

The efficacy of pemafibrate over that of fenofibrate was reported in a 24-week, randomized, double blind, active-controlled, phase 3 trial. Patients with fasting TG $\geq$ 150 mg/dL as well as HDL-C $\leq$50 mg/dL for men and $\leq$55 mg/dL for women were randomly assigned to pemafibrate 0.1 (n = 73) or 0.2 (n = 74) mg bid or to fenofibrate 106.6 mg qd (n = 76). The primary efficacy analysis, i.e., percent change in fasting TG from baseline, demonstrated that pemafibrate treatments reduced TGs around −46% vs. −39.7% for fenofibrate. These findings can be translated into a further −6.5% and −6.2% difference in TG reduction for patients on 0.2 and 0.4 mg/day pemafibrate vs. fenofibrate. TC, non-HDL-C, ApoB, and VLDL-C were significantly decreased, and HDL-C, ApoAI, and ApoAII increased by both agents with no significant differences among treatment groups. Conversely, FGF-21 levels were raised to a greater extent in the 0.4 mg/day pemafibrate group vs. fenofibrate [76] (Table 2). Adverse drug reactions, such as rises in liver enzymes and serum creatinine, were observed in the fenofibrate group, not in the pemafibrate groups.

The non-inferiority of pemafibrate over fenofibrate was confirmed in a 12-week phase 3 trial enrolling 489 patients with TG $\geq$ 200 mg/dL and HDL-C $\leq$ 50 mg/dL. The TG lowering effects of pemafibrate were dose dependent −46.3% (0.1 mg/day), −46.7% (0.2 mg/day) and −51.8% (0.4 mg/day) and non-inferior to those of fenofibrate, −38.3% (100 mg/day) and −51.5% (200 mg/day) (Table 2). Adverse events were less frequent than with fenofibrate 200 mg/day [77].

The long-term efficacy of pemafibrate to treat residual hypertriglyceridaemia during statin treatment has been recently evaluated in two randomized, double-blind, placebo-controlled phase 2 trials. The primary endpoint was the percentage changes in fasting TGs from baseline [78]. The first trial study enrolled 188 patients with residual dyslipidemia (fasting TGs from 347 to 382 mg/dL) on a pitavastatin background with LDL-C in the range of 116–125 mg/dL. The 12-week pemafibrate administration (0.1, 0.2 or 0.4 mg/day) significantly reduced TG levels by −46.1%, −53.4% and −52%, respectively. Conversely, no TG reduction was observed in the pitavastatin monotherapy group. Combination therapy led to a significant rise of HDL-C (range: +12.7–19.7%), ApoAI (range: +1.5–6.6%) and ApoAII (range: +18.5–27.6%) and a reduction of non-HDL-C (range: −10.7–13.1%) and ApoB (range: −7.9–8.6%) (Table 2). Notably, all of these changes were statistically significant when compared to pitavastatin alone. Pemafibrate as an add-on therapy resulted in a more anti-atherogenic lipoprotein profile, i.e., increment of cholesterol in medium, small, and very small HDL subclasses and in large and medium LDL subclasses [79].

In the second trial, pemafibrate (0.2 mg/day) was given for 24 weeks to 423 patients with residual dyslipidemia (TGs ranging from 325 to 333 mg/dL) on statins (most commonly atorvastatin, rosuvastatin and pitavastatin); LDL-C was around 108 mg/dL. Notably, if TGs were $\geq$ 150 mg/dL after 12 weeks, pemafibrate was up-titrated to 0.4 mg/dL. Regardless of statin background, combination therapy with pemafibrate 0.2 or 0.4 mg/day led to TG reductions of about 50% from baseline [79] (Table 2). Compared to the monotherapy arm, patients receiving pemafibrate showed a significant decrement in non-HDL, ApoB and ApoCIII, as well as an increment in HDL-C, ApoAI and ApoAII [79]. As previously described, the addition of pemafibrate led to an increment in medium and small lipid-poor HDL, more efficient in reverse cholesterol efflux [80]. In both of these last two studies, the incidence of adverse events during treatment was similar across all groups. The proportion of patients experiencing elevated alanine transaminase (ALT), creatine kinase (CK) and serum creatinine were comparable.

The clear definition of efficacy of pemafibrate on the lipid profile, i.e., TG reduction and HDL-C increment, in preclinical studies as well as in phase 1 and phase 2 clinical trials (previously reviewed [73,81]) led to the planning of the PROMINENT (pemafibrate to reduce cardiovascular outcomes by reducing triglycerides in patients with diabetes) trial (registered as NCT03071692).

The primary objective of this phase 3 study is evaluation in T2D patients already on statin (fasting TGs: ≥200 to <500 mg/dL; HDL-C ≤ 40 mg/dL), testing whether pemafibrate (0.2 mg bid) can delay the time of the first occurrence of nonfatal myocardial infarction (MI), nonfatal ischemic stroke, hospitalization for unstable angina requiring unplanned coronary revascularization, and CV death. Changes in lipid end-points including ApoAI, ApoCIII, ApoE and non-fasting remnant cholesterol are listed as secondary outcomes [82].

The efficacy of fibrates on CV prevention has been disputed, mainly on the ground of studies on non-selected patients [63,64]. Long-term re-evaluation of some of the most recent trials has clearly confirmed that fenofibrate in particular may have a valid indication in reducing CVD in patients with diabetes, hypertriglyceridemia and low HDL-C [65]. The recently available low-dosage pemafibrate seems to provide a higher activity on the HDL system and the ongoing studies on vascular prevention will provide further data on the link between biochemical markers of the MetS and CV risk.

3.2. Omega-3

Fatty acids of the n-3 series (i.e., with multiple double bonds, the first one being in the n-3 position from the terminal methyl group) have provided an important addition to the dietary treatment in syndromes characterized by elevated TGs. Omega-3s act as "fraudulent fatty acids" [83] i.e., they, somewhat similar to drugs with a fatty acid-like structure, particularly fibrates, do not enter the liver metabolic handling by the classical fatty acetylCoA oxidative mechanism, with carnitine mediated transport to mitochondria [84]. They are instead handled by a non-mitochondrial regulated pathway, differently expressed both in the liver and other tissues [85]. Peroxisomal associated receptors are stimulated in the presence of those fatty acids catabolized not only by the classical mitochondrial pathway [86]. Both fibrates and omega-3s, thus, will not act as classical substrates of mitochondrial metabolism, but rather stimulate the metabolism of fatty acids coming from diet or end products of, e.g., TG metabolism by the PPAR-α mediated pathway. Peroxisomal stimulation is less extensive than in the case of fibrates, but it can well stimulate fatty acid oxidation. An additional mechanism, more closely related to the plasma glucose elevation in MetS, is the activation of tissue glucose uptake by the GLUT4 transporter in adipocytes; this mechanism appears to be mediated by the GPR120 protein, functioning as an omega-3 fatty acid receptor/sensor [87].

Controlled trials in patients given relatively elevated daily doses of omega-3 in the form of TGs or, more recently, of ethyl esters of eicosapentaenoic (EPA) and docosahexaeneoic (DHA) acids, as well as with novel formulations of separate fatty acids [88], have repeatedly confirmed a TG reduction in hypertriglyceridemic conditions associated or not with diabetes [89,90]. Controlled studies indicate lowering, in general, of 20–30% of fasting triglyceridemia in these conditions [91], with a moderate rise of HDL cholesterolemia as well as an increment in reverse cholesterol transport mainly by influencing HDL remodeling and promoting hepatobiliary sterol excretion [92].

A general review on the mechanism of omega-3, improving abnormalities characteristic of MetS, may in addition to the classical activation of fatty acid metabolism involve increased adipocyte differentiation, reduced lipolysis and lipogenesis, as well as a significant activity on low grade inflammation, including reduced adipokines and specialized pro-resolving lipid mediators [93]. By the activation of fat metabolism and consequent energy expenditure, it being peroxisomal and to some extent also mitochondrial, positive effects on obesity may be observed. Indeed, PPAR-α KO-obese mice show, in fact, a clear worsening of obesity that may be instead improved by omega-3 administration in different diet-induced conditions [94]. Reduced lipogenesis and low-grade inflammation may be of value in the treatment of complex metabolic disorders.

In the case of adipose tissue biology, the most recent evidence points out to the importance of both white and brown adipose tissue (WAT and BAT) function. "Healthy adipocytes" in WAT are relatively small fat cells with a high capacity for mitochondrial oxidative phosphorylation, TG/FA cycling and de novo lipogenesis [94]. These cells, with a flexible phenotype, may provide beneficial local and systemic effects by protecting against inflammatory responses during lipolysis, preventing fat accumulation and dyslipidemia caused by increased liver VLDL-TG synthesis.

Indeed, dietary omega-3s may indeed redirect adipose tissue to a "healthy phenotype" [94]. These polyunsaturated fatty acids appear to stimulate the "G protein coupled receptor" GPR120 [95], promoting BAT activation [96], thus inducing brown as well as beige adipocyte differentiation and thermogenic activation which seems to be linked to an increase in blood FGF-21 levels. Characteristically, in animals devoid of GPR120, adipose tissue thermogenic activation is not achieved [97]. These observations are of ethnological significance, in view of the high consumption of omega-3 from fish in individuals living in cold areas such as the Eskimos [98]. Interestingly, in in vitro systems only BAT and not WAT cells synthesize DHA [99].

Inflammatory changes in the adipose tissue are characteristic of obesity. They are driven by rises in circulating endotoxins and infiltrate immune cell populations. This will lead to an increased secretion of inflammatory adipokines (e.g., IL-6, TNF–α, monocyte chemoattractant protein (MCP) and chemokine (C-C motif) ligand (CCL) from multiple cellular sources [100]. The end result of the increased secretion of inflammatory mediators is the development of insulin resistance. A characteristic reduction in chemokine secretion from LPS stimulated co-cultures of omega-3 fed rodents is followed by a reduced secretion of IL-6 (−42%) and TNF-α (−67%), as well as by a similar reduction of other inflammatory mediators. Concomitantly, omega-3s increase the mRNA expression of negative regulators of inflammatory signaling, such as the monocyte chemoattractant 1-induced protein (MCPIP; +9.3-fold) and the suppressor of cytokine signaling 3 (SOCS3; +1.7-fold) [101]. In patients with MetS a reduction of high sensitivity CRP levels [102] has been found, although data from a number of clinical studies on this topic have been inconsistent [91,103].

It can thus be concluded that, in the adipose tissue, besides the stimulation of fatty acid catabolism, well-characterized anti-inflammatory and anti-chemotactic effects can be exerted by omega-3s. All these effects, recognized at the cellular level, may be followed by biochemical changes in patients with MetS and diabetes mellitus. In these patients, statistically significant TG reductions, compared to placebo, have been observed with, however, somewhat differential effects on LDL-C. Apparently, products containing DHA may increase LDL-C levels, whereas those containing EPA only products do not lead to a similar consequence [104,105].

Omega-3: Evidence from the Most Recent Clinical Trials

Most recently, the effect of omega-3 fatty acids (2 g daily) in reducing TGs and other lipid concentrations in patients with severe hypertriglyceridemia (TG > 500 mg/dL and <2500 mg/dL) was evaluated in the EVOLVEII (Epanova® for lowering very high triglycerides II) trial, a double-blind, randomized, olive oil-controlled study. After an 8-week screening period for patients who required washout or stabilization of lipid-lowering therapy (e.g., statin or cholesterol-absorption inhibitors), they were randomized to receive two 1 g soft gelatin capsules with omega-3 (550 mg EPA + approximately 200 mg DHA in a new formulation) or olive oil once a day for 12 weeks. Notably, stratification was carried out based on TG levels, i.e., $\geq$ 500 $\leq$ 885 mg/dL or > 885 < 2500 mg/dL. Omega-3 capsules reduced TG by −28.1% vs. −10.2% (olive oil) in the group with TGs $\geq$ 500 and by −37.5% vs. −9.3% (olive oil) in the group with TGs > 885 mg/dL. In the whole population, TG differences between the two treatment groups were −14.2%. Non-HDL-C percentage changes were instead −8.8% (omega-3) vs. +0.4% (olive oil) in the group with TGs $\geq$ 500, with more marked differences in those with TG > 885 (−14% vs. +3.1%), and an overall −9% non-HDL-C reduction (Table 2). Omega-3 supplementation led to a significant lowering of VLDL-C, both when compared to baseline or to the olive oil arm. The decrease of VLDL-C concentrations was similar to that of TGs. HDL-C were modestly raised by both treatments, with no extra benefit given by omega-3 [106].

The results of this trial are in line with those found in the previous EVOLVE (Epanova for lowering very high triglycerides) double-blind, randomized, parallel, 4-arm study. In subjects with severe hypertriglyceridemia (TGs $\geq$ 500 mg/dL but <2000 mg/dL), administration of omega-3-FA 2 g/die (plus olive oil 2 g/day), omega-3FA 3 g/die (plus olive oil 1 g/day), or omega-3-FA 4 g/day for 12 weeks in combination with diet and lifestyle changes led to a −31% reduction in fasting TG in the group receiving omega-3-FFA 4 g/die vs. 25% in the other two treatment groups. A minimal TG reduction (−4.3%) was found in patients receiving olive oil 4g/day. A similar trend was found for non-HDL-C, with a maximal −9.6% reduction with omega-3-FA 4 g/day, vs. a +2.5% increment in the olive oil group (Table 2). HDL-C were not significantly changed at any dosage [107].

The effect of omega-3 as an add-on therapy to a statin background was evaluated in the ESPRIT (Epanova combined with a statin in patients with hypertriglyceridemia to reduce non-HDL cholesterol) trial, on persistently hypertriglyceridemic patients already on a maximally tolerated dose of statin or statin + ezetimibe, with TG levels $\geq$ 200 mg/dL and <500 mg/dL). Compared to olive oil (4 g/day), omega-3-FA 2 g/day or omega-3-FA 4 g/day administration led to a significant reduction in non-HDL-C (−3.9% and −6.9%, respectively) and TG (−14.6% and −20.6%, respectively) (Table 2) [108].

A limited number of studies have selectively evaluated the efficacy of omega-3 fatty acids on CV outcomes, both in primary and secondary prevention. Positive outcomes were reported from the GISSI (Gruppo Italiano per lo studio della sopravvivenza nell'infarto miocardico) [109] study which dealt mainly with patients with an acute coronary syndrome, and from the JELIS (Japan eicosapentaenoic acid lipid intervention study) [110] testing the efficacy of omega-3 (1800 mg/day) in primary prevention moderately hypercholesterolemic patients, mainly on statins. In contrast, the large alpha omega trial on coronary patients on a smaller daily intake of EPA + DHA (400 mg/day) failed to reach the targeted reduction of CV events [111]. While these last authors, in a recent meta-analysis on 77,917 patients in 10 different studies, appeared to confirm the lack of a significant impact of omega-3 on CV endpoints [112], at present, the effect of omega-3 supplementation on CV outcomes, i.e., any component of the composite of major adverse cardiac events (MACE), is being evaluated in the STRENGTH (study to assess statin residual risk reduction with Epanova in high cardiovascular risk patients with hypertriglyceridemia) and REDUCE-IT (reduction of cardiovascular events with icosapent ethyl–intervention) trials [113] targeting particularly hypertriglyceridemic patients.

Unquestionably, the efficacy of omega-3 intake on metabolic parameters and on inflammatory changes is now well established, with a convincing series of mechanistic studies, but the efficacy of these nutritional supplements on CV outcomes is at present unsettled. The ongoing studies should shed light on this last issue, particularly as pertains to patients with hypertriglyceridemia associated risk.

4. PPAR-γ Agonists

Agonists of PPAR-γ belong to the thiazolidinedione (TZDs) drug class and are currently in use for T2D [114]. Pioglitazone and rosiglitazone are the only two drugs available. Following epidemiological data indicative of a raised CV risk after rosiglitazone [115,116], the drug was taken off the market in Europe. Pioglitazone appears instead to reduce CV events [117]. Clinical trials with pioglitazone, e.g., PERISCOPE, PROactive and CHICAGO, have, in fact, demonstrated that in addition to beneficial effects in reducing TGs and increasing HDL-C levels, pioglitazone can reduce CV risk in T2D patients [118].

The PERISCOPE trial (pioglitazone effect on regression of intravascular sonographic coronary obstruction prospective evaluation) recruited patients with baseline HbA1c $\geq$ 6.0% to $\leq$9.0% (if on a glucose-lowering medication) or $\geq$6.5% to $\leq$10% (if not on drug therapy) with positive coronary angiogram (at least 1 angiographic stenosis with at least 20% narrowing). Pioglitazone, compared with glimepiride, raised HDL-C by +5.7 mg/dL vs. 0.9 mg/dL, whereas TG levels were decreased by −16.3 mg/dL vs. a rise of +3.3mg/dL. Fasting insulin levels were decreased by pioglitazone and raised by glimepiride. The primary end-point, namely the percent atheroma volume change measured by intravascular ultrasound (IVUS), was reduced by −0.16% in the pioglitazone arm vs. a +0.73% rise in the glimepiride group (Table 2); these between group changes were statistically significant [119]. Interestingly, a post-hoc analysis showed a greater relative increase in HDL-C (+14.2% vs. 7.8%) with a concomitant relative reduction of TGs (−13.3% vs. −1.9%), TG/HDL-C ratio (−22.5% vs. −9.9%) and

HbA1C (−0.6% vs. −0.3%) upon pioglitazone administration, possibly responsible for the atheroma regression [120].

The CHICAGO (carotid intima-media thickness in atherosclerosis using pioglitazone) trial tested the hypothesis that pioglitazone would have a beneficial effect in reducing carotid intima-media thickness (IMT) progression compared with glimepiride [121]. The reported reduced progression of IMT appeared to be associated with a rise of HDL-C (+14%) following pioglitazone (Table 2) [122]. The positive effect of pioglitazone on carotid IMT is thus independent of its glucose lowering effect. Of note, weight and body mass index were higher in the pioglitazone arm.

The IRIS (insulin resistance intervention after stroke) trial showed that pioglitazone reduced the occurrence of fatal and non-fatal stroke or MI in insulin resistant patients without diabetes, also halving the occurrence of diabetes (Table 2). Notably, this latter occurred in 3.8% of patients on pioglitazone vs. 7.7% of those assigned to placebo (hazard ratio: 0.48; 95% CI: 0.33–0.69) [123,124]. Evaluation of safety outcomes indicated that pioglitazone led to (i) a weight gain of 4.5 kg (52.2% of patients) vs. +13.6 kg (11.4% of patients) for placebo [123]; with (ii) an increment in the absolute risk of fractures risk by 1.6% vs. 4.9%, depending on fracture classification [125]; and (iii) a higher incidence of edema (+35.6% vs. +24.9%) [123], this last being an as yet poorly understood frequent side effect of glitazones [126].

The clinical outcome studies on PPAR-γ agonists have been focused mainly on pioglitazone, particularly in view of the better tolerability. These have concluded that this treatment may be associated both with a reduced atheroma progression and a lower incidence of diabetes. Further, the IRIS study provided clear evidence of the efficacy of pioglitazone in preventing CV outcomes in insulin resistant patients [123,124].

5. PPAR Dual Agonists

Dual PPAR agonists or partial agonists, e.g., dual α/γ, α/δ or β/δ [127] were developed with the aim of achieving the TG-lowering and HDL-raising effects of PPAR-α activators as well as the insulin-sensitizing and antihyperglycemic effects of TZDs with a single drug. Such a combination of effects would be ideal for the treatment of T2D, MetS and NAFLD, which all share as common features atherogenic dyslipidemia and insulin resistance [128].

Of particular interest is the case of NAFLD, for which dual PPAR-α/δ agonists offer significant hope. Elafibranor (formerly known as GFT-505), with preferential α (EC50 = 6 nM) and complementary δ (EC50 = 47 nM) receptor agonist activity, is targeted to the liver, where it is converted to the main active metabolite, GFT-1007, in a dose-dependent manner. Elafibranor has been shown to be effective in disease models of NAFLD/NASH and liver fibrosis [129], as well as in T2D patients for whom a reduction in TG and LDL-C levels and improved insulin sensitivity were reported [130]. Recent results from the Phase 2b GOLDEN trial (NCT01694849) indicated that elafibranor treatment leads to a substantial histological improvement of NASH, including resolution of steatohepatitis and reduced CV risk. NASH was resolved without fibrosis worsening in 23% and 21% of patients assigned to receive either 80 mg or 120 mg/day elafibranor vs. 17% in the placebo arm; no significant differences between groups were found. When a more stringent definition of NASH was considered, changes in NASH resolution were 19% after elafibranor administration vs. 12% in the placebo group ($p = 0.045$) (Table 2) [131].

For activators of PPAR $\alpha + \gamma$ receptor, only preliminary data are available for saroglitazar, i.e., positive effects on the lipid profile, blood pressure, atherosclerosis, inflammation, and clotting [132]. Saroglitazar is being tested in an ongoing phase 3 trial in non-cirrhotic biopsy-proven NASH patients, in order to evaluate a possible improvement in NASH histology without worsening of fibrosis [133]. Interestingly, this agent is currently approved in India for the treatment of diabetic dyslipidemia [134]. Other glitazars, i.e., tesaglitazar and rasaglitazar have been discontinued from clinical development due to renal side effects, anaemia and leukopenia (tesaglitazar) and bladder tumor development (ragaglitazar) [135].

At present, dual agonists show an attractive metabolic profile, including an activity on MetS associated liver abnormalities. Clinical outcome studies are awaited with interest.

Table 2. Effect of PPARs on the features of metabolic syndrome—evidence from clinical trials.

PPAR-α Agonist	Clinic Study	Major Findings
Pemafibrate	Phase 3 (JapicCTI-142412; clinicaltrials.jp) follow-up: 24 weeks subjects: 166 [74]	1. Reduction in TGs: −45% 2. Decrement in non-HDL 3. Increase in HDL cholesterol
	Phase 3 (JapicCTI-142620; clinicaltrials.jp) follow-up: 24 weeks subjects: 225 [76]	1. Reduction in TGs: −46.2% 2. A further −6.5% TG reduction compared to fenofibrate
	Phase 3 (JapicCTI-121764; clinicaltrials.jp) follow-up: 12 weeks subjects: 489 [77]	1. TGs: −46.3% (0.1 mg/day), −46.7% (0.2 mg/day) and −51.8% (0.4 mg/day) vs. −38.3% (fenofibrate 100 mg/day) and −51.5% (fenofibrate 200 mg/day)
	Phase 2 follow-up: 12 weeks subjects: 188 [79]	1. Reduction in TGs: range from −46.1% to −53.4%
	Phase 2 follow-up: 24 weeks subjects: 423 [79]	1. Reduction in TGs: range from −46.8% to −50.8%
	On going Phase 3 trial PROMINENT (Pemafibrate to Reduce Cardiovascular OutcoMes by Reducing Triglycerides IN patiENts with diabeTes)—NCT03071692	Outcomes: First occurrence of nonfatal myocardial infarction, nonfatal ischemic stroke, hospitalization for unstable angina requiring unplanned coronary revascularization, and CV death.
Omega-3	EVOLVEII (Epanova® for Lowering Very High Triglycerides II)—NCT02009865 Phase 3 follow-up: 12 weeks subjects: 162 [106]	1. Reduction in TGs: −14.2% 2. Reduction in non-HDL-C: −9%
	EVOLVE (The EpanoVa fOr Lowering Very high triglyceridEs)—NCT01242527 Phase 3 follow-up: 12 weeks subjects: 399 [107]	1. Reduction in TGs: range −25.5%/−30.9% 2. Reduction in non-HDL-C: range from −6.9% to −9.6%
	ESPRIT (EPANOVA Combined with a STATIN in PATIENTS With HYPERTRIGLYCERIDEMIA to Reduce Non-HDL CHOLESTEROL)—NCT01408303. Phase 3 follow-up: 6 weeks subjects: 647 [108]	1. Reduction in TGs: range from −14.6% to −20.6% 2. Reduction in non-HDL-C: range from −3.9% to −6.9%
	On going phase 3 trials: (i) STRENGTH (Study to assess statin residual risk Reduction with Epanova in high cardiovascular risk patients with Hypertriglyceridemia)—NCT02104817 (ii) REDUCE-IT (Reduction of Cardiovascular Events with Icosapent Ethyl–Intervention)—NCT01492361 [113]	Outcomes: First occurrence of cardiovascular death, nonfatal MI, nonfatal stroke, emergent/elective coronary revascularization, or hospitalization for unstable angina

Table 2. *Cont.*

PPAR-α Agonist	Clinic Study	Major Findings
Pioglitazone	The PERISCOPE Trial (Pioglitazone Effect on Regression of Intravascular Sonographic Coronary Obstruction Prospective Evaluation)—NCT00225277 Phase 3 follow-up: 18 months subjects: 543 [119]	1. Percent atheroma volume change: −0.16% 2. Raise in HDL-C: +5.7 mg/dL 3. Decrement in TGs: −16.3 mg/dL
	The CHICAGO (Carotid Intima-Media Thickness in Atherosclerosis Using Pioglitazone) trial—NCT00225264 Phase 3 follow-up: 72 weeks subjects: 462 [121,122]	1. Progression of mean CIMT: −0.013 mm vs. glimepiride 2. Progression of maximum CIMT: −0.024 mm vs. glimepiride 3. HDL-C: +14%
	The IRIS (Insulin Resistance Intervention after Stroke)—NCT00091949 Phase 3 follow-up: 4.8 years subjects: 3876 [123,124]	1. Reduction of stroke or MI in insulin resistant patients 2. Reduction in recurrence of diabetes: −52%
Elafibranor	GOLDEN trial—NCT01694849 Phase 2b follow-up: 52 weeks subjects: 256 [131]	1. NASH resolution in 19% of patients

All percentage changes are vs. baseline otherwise differently indicated. CIMT, Carotid Intima-Media Thickness; CV, cardiovascular; HDL-C, high-density lipoprotein cholesterol; MI, Miocardial Infarction; NASH, nonalcoholic steatohepatitis; TG, triglyceride.

6. Conclusions

The management of MetS represents one of the major targets in atherosclerosis prevention. While treatment of hypercholesterolemia or diabetes can be successfully handled with drugs targeting cholesterol biosynthesis or beta-islet-cell function, MetS is characterized by a number of diverse metabolic abnormalities which are more difficult to pursue. While lifestyle modifications [136], including changes in diet and increased exercise, can provide help to a small number of patients, an improved knowledge of drugs affecting PPAR system has led to more frequent and better focused treatment choices. PPAR agonist "fraudulent fatty acids", i.e., fibrates and omega-3 fatty acids, find a growing role in the handling of hypertriglyceridemia and, in the case of fibrates, also positively affecting HDL-cholesterol and the consequently raised CV risk. PPAR-γ agonists are instead targeted to the glycemic abnormalities of MetS; they may, however, lead to weight increase. The causal role of hypertriglyceridemia as a CV risk factor, confirmed by Mendelian randomization studies, has brought clinicians back to this somewhat forgotten risk marker, now rated by many as an unmet need [137]. The HDL-C raising approach has also become of high interest after the preventive failure of drugs such as the cholesteryl ester transfer protein inhibitors [138], disclosing the as yet not fully clarified CV protective mechanism of HDL [139].

Acknowledgments: This work was supported by Fondazione Cariplo grant No. 2015-0552; Linea 2, Azione (Intramural grant) received by Università degli Studi di Milano and by the Fondazione Carlo Sirtori.

Author Contributions: Margherita Botta and Matteo Audano wrote the manuscript; Amirhossein Sahebkar and Cesare R. Sirtori critically revised the text; Nico Mitro and Massimiliano Ruscica specifically examined the molecular aspects of PPAR activation.

Conflicts of Interest: The authors declare no conflict of interest.

References

1. De Carvalho Vidigal, F.; Bressan, J.; Babio, N.; Salas-Salvado, J. Prevalence of metabolic syndrome in Brazilian adults: A systematic review. *BMC Public Health* **2013**, *13*, 1198. [CrossRef] [PubMed]
2. Li, R.; Li, W.; Lun, Z.; Zhang, H.; Sun, Z.; Kanu, J.S.; Qiu, S.; Cheng, Y.; Liu, Y. Prevalence of metabolic syndrome in Mainland China: A meta-analysis of published studies. *BMC Public Health* **2016**, *16*, 296. [CrossRef] [PubMed]
3. Aguilar, M.; Bhuket, T.; Torres, S.; Liu, B.; Wong, R.J. Prevalence of the metabolic syndrome in the United States, 2003-2012. *JAMA* **2015**, *313*, 1973–1974. [CrossRef] [PubMed]
4. Alberti, K.G.; Eckel, R.H.; Grundy, S.M.; Zimmet, P.Z.; Cleeman, J.I.; Donato, K.A.; Fruchart, J.C.; James, W.P.; Loria, C.M.; Smith, S.C., Jr.; et al. Harmonizing the metabolic syndrome: A joint interim statement of the International Diabetes Federation Task Force on Epidemiology and Prevention; National Heart, Lung, and Blood Institute; American Heart Association; World Heart Federation; International Atherosclerosis Society; and International Association for the Study of Obesity. *Circulation* **2009**, *120*, 1640–1645. [PubMed]
5. National Cholesterol Education Program (NCEP); Expert Panel on Detection, Evaluation, and Treatment of High Blood Cholesterol in Adults. Third Report of the National Cholesterol Education Program (NCEP) Expert Panel on Detection, Evaluation, and Treatment of High Blood Cholesterol in Adults (Adult Treatment Panel III). *JAMA* **2001**, *285*, 2486–2497.
6. Einhorn, D.; Reaven, G.M.; Cobin, R.H.; Ford, E.; Ganda, O.P.; Handelsman, Y.; Hellman, R.; Jellinger, P.S.; Kendall, D.; Krauss, R.M.; et al. American College of Endocrinology position statement on the insulin resistance syndrome. *Endocr. Pract.* **2003**, *9*, 237–252. [PubMed]
7. Alberti, K.G.; Zimmet, P.; Shaw, J.; IDF Epidemiology Task Force Consensus Group. The metabolic syndrome—A new worldwide definition. *Lancet* **2005**, *366*, 1059–1062. [CrossRef]
8. Ferri, N.; Ruscica, M. Proprotein convertase subtilisin/kexin type 9 (PCSK9) and metabolic syndrome: Insights on insulin resistance, inflammation, and atherogenic dyslipidemia. *Endocrine* **2016**, *54*, 588–601. [CrossRef] [PubMed]
9. Tenenbaum, A.; Fisman, E.Z.; Motro, M.; Adler, Y. Atherogenic dyslipidemia in metabolic syndrome and type 2 diabetes: Therapeutic options beyond statins. *Cardiovasc. Diabetol.* **2006**, *5*, 20. [CrossRef] [PubMed]

10. Chapman, M.J.; Redfern, J.S.; McGovern, M.E.; Giral, P. Niacin and fibrates in atherogenic dyslipidemia: Pharmacotherapy to reduce cardiovascular risk. *Pharmacol. Ther.* **2010**, *126*, 314–345. [CrossRef] [PubMed]
11. Mansour, M. The roles of peroxisome proliferator-activated receptors in the metabolic syndrome. *Prog. Mol. Biol. Transl. Sci.* **2014**, *121*, 217–266. [PubMed]
12. Diehl, A.M.; Day, C. Cause, Pathogenesis, and Treatment of Nonalcoholic Steatohepatitis. *N. Engl. J. Med.* **2017**, *377*, 2063–2072. [CrossRef] [PubMed]
13. Bortolini, M.; Wright, M.B.; Bopst, M.; Balas, B. Examining the safety of PPAR agonists—Current trends and future prospects. *Expert Opin. Drug Saf.* **2013**, *12*, 65–79. [CrossRef] [PubMed]
14. Corsini, A.; Bellosta, S.; Davidson, M.H. Pharmacokinetic interactions between statins and fibrates. *Am. J. Cardiol.* **2005**, *96*, 44K–49K; discussion 34K–35K. [CrossRef] [PubMed]
15. Magni, P.; Macchi, C.; Morlotti, B.; Sirtori, C.R.; Ruscica, M. Risk identification and possible countermeasures for muscle adverse effects during statin therapy. *Eur. J. Intern. Med.* **2015**, *26*, 82–88. [CrossRef] [PubMed]
16. Dubois, V.; Eeckhoute, J.; Lefebvre, P.; Staels, B. Distinct but complementary contributions of PPAR isotypes to energy homeostasis. *J. Clin. Investig.* **2017**, *127*, 1202–1214. [CrossRef] [PubMed]
17. Chen, L.; Yang, G. PPARs Integrate the Mammalian Clock and Energy Metabolism. *PPAR Res.* **2014**, *2014*, 653017. [CrossRef] [PubMed]
18. Schoonjans, K.; Staels, B.; Auwerx, J. The peroxisome proliferator activated receptors (PPARS) and their effects on lipid metabolism and adipocyte differentiation. *Biochim. Biophys. Acta* **1996**, *1302*, 93–109. [CrossRef]
19. Moore, K.J.; Rosen, E.D.; Fitzgerald, M.L.; Randow, F.; Andersson, L.P.; Altshuler, D.; Milstone, D.S.; Mortensen, R.M.; Spiegelman, B.M.; Freeman, M.W. The role of PPAR-gamma in macrophage differentiation and cholesterol uptake. *Nat. Med.* **2001**, *7*, 41–47. [CrossRef] [PubMed]
20. Saluja, I.; Granneman, J.G.; Skoff, R.P. PPAR delta agonists stimulate oligodendrocyte differentiation in tissue culture. *Glia* **2001**, *33*, 191–204. [CrossRef]
21. Cermenati, G.; Audano, M.; Giatti, S.; Carozzi, V.; Porretta-Serapiglia, C.; Pettinato, E.; Ferri, C.; D'Antonio, M.; De Fabiani, E.; Crestani, M.; et al. Lack of sterol regulatory element binding factor-1c imposes glial Fatty Acid utilization leading to peripheral neuropathy. *Cell Metab.* **2015**, *21*, 571–583. [CrossRef] [PubMed]
22. Staels, B.; Dallongeville, J.; Auwerx, J.; Schoonjans, K.; Leitersdorf, E.; Fruchart, J.C. Mechanism of action of fibrates on lipid and lipoprotein metabolism. *Circulation* **1998**, *98*, 2088–2093. [CrossRef] [PubMed]
23. Schoonjans, K.; Peinado-Onsurbe, J.; Lefebvre, A.M.; Heyman, R.A.; Briggs, M.; Deeb, S.; Staels, B.; Auwerx, J. PPARalpha and PPARgamma activators direct a distinct tissue-specific transcriptional response via a PPRE in the lipoprotein lipase gene. *EMBO J.* **1996**, *15*, 5336–5348. [PubMed]
24. Riserus, U.; Sprecher, D.; Johnson, T.; Olson, E.; Hirschberg, S.; Liu, A.; Fang, Z.; Hegde, P.; Richards, D.; Sarov-Blat, L.; et al. Activation of peroxisome proliferator-activated receptor (PPAR)delta promotes reversal of multiple metabolic abnormalities, reduces oxidative stress, and increases fatty acid oxidation in moderately obese men. *Diabetes* **2008**, *57*, 332–339. [CrossRef] [PubMed]
25. Lee, W.J.; Kim, M.; Park, H.S.; Kim, H.S.; Jeon, M.J.; Oh, K.S.; Koh, E.H.; Won, J.C.; Kim, M.S.; Oh, G.T.; et al. AMPK activation increases fatty acid oxidation in skeletal muscle by activating PPARalpha and PGC-1. *Biochem. Biophys. Res. Commun.* **2006**, *340*, 291–295. [CrossRef] [PubMed]
26. Kersten, S.; Seydoux, J.; Peters, J.M.; Gonzalez, F.J.; Desvergne, B.; Wahli, W. Peroxisome proliferator-activated receptor alpha mediates the adaptive response to fasting. *J. Clin. Investig.* **1999**, *103*, 1489–1498. [CrossRef] [PubMed]
27. Kersten, S. Integrated physiology and systems biology of PPARalpha. *Mol. Metab.* **2014**, *3*, 354–371. [CrossRef] [PubMed]
28. Oosterveer, M.H.; Grefhorst, A.; van Dijk, T.H.; Havinga, R.; Staels, B.; Kuipers, F.; Groen, A.K.; Reijngoud, D.J. Fenofibrate simultaneously induces hepatic fatty acid oxidation, synthesis, and elongation in mice. *J. Biol. Chem.* **2009**, *284*, 34036–34044. [CrossRef] [PubMed]
29. Wu, P.; Peters, J.M.; Harris, R.A. Adaptive increase in pyruvate dehydrogenase kinase 4 during starvation is mediated by peroxisome proliferator-activated receptor alpha. *Biochem. Biophys. Res. Commun.* **2001**, *287*, 391–396. [CrossRef] [PubMed]
30. Keller, H.; Wahli, W. Peroxisome proliferator-activated receptors A link between endocrinology and nutrition? *Trends Endocrinol. Metab.* **1993**, *4*, 291–296. [CrossRef]

31. Devchand, P.R.; Keller, H.; Peters, J.M.; Vazquez, M.; Gonzalez, F.J.; Wahli, W. The PPARalpha-leukotriene B4 pathway to inflammation control. *Nature* **1996**, *384*, 39–43. [CrossRef] [PubMed]
32. Kleemann, R.; Gervois, P.P.; Verschuren, L.; Staels, B.; Princen, H.M.; Kooistra, T. Fibrates down-regulate IL-1-stimulated C-reactive protein gene expression in hepatocytes by reducing nuclear p50-NFkappa B-C/EBP-beta complex formation. *Blood* **2003**, *101*, 545–551. [CrossRef] [PubMed]
33. Hill, M.R.; Clarke, S.; Rodgers, K.; Thornhill, B.; Peters, J.M.; Gonzalez, F.J.; Gimble, J.M. Effect of peroxisome proliferator-activated receptor alpha activators on tumor necrosis factor expression in mice during endotoxemia. *Infect. Immun.* **1999**, *67*, 3488–3493. [PubMed]
34. Braissant, O.; Foufelle, F.; Scotto, C.; Dauca, M.; Wahli, W. Differential expression of peroxisome proliferator-activated receptors (PPARs): Tissue distribution of PPAR-alpha, -beta, and -gamma in the adult rat. *Endocrinology* **1996**, *137*, 354–366. [CrossRef] [PubMed]
35. Amri, E.Z.; Bonino, F.; Ailhaud, G.; Abumrad, N.A.; Grimaldi, P.A. Cloning of a protein that mediates transcriptional effects of fatty acids in preadipocytes. Homology to peroxisome proliferator-activated receptors. *J. Biol. Chem.* **1995**, *270*, 2367–2371. [CrossRef] [PubMed]
36. Hertz, R.; Berman, I.; Keppler, D.; Bar-Tana, J. Activation of gene transcription by prostacyclin analogues is mediated by the peroxisome-proliferators-activated receptor (PPAR). *Eur. J. Biochem.* **1996**, *235*, 242–247. [CrossRef] [PubMed]
37. Holst, D.; Luquet, S.; Nogueira, V.; Kristiansen, K.; Leverve, X.; Grimaldi, P.A. Nutritional regulation and role of peroxisome proliferator-activated receptor delta in fatty acid catabolism in skeletal muscle. *Biochim. Biophys. Acta* **2003**, *1633*, 43–50. [CrossRef]
38. Luquet, S.; Lopez-Soriano, J.; Holst, D.; Fredenrich, A.; Melki, J.; Rassoulzadegan, M.; Grimaldi, P.A. Peroxisome proliferator-activated receptor delta controls muscle development and oxidative capability. *FASEB J.* **2003**, *17*, 2299–2301. [CrossRef] [PubMed]
39. Wang, X.; Wang, G.; Shi, Y.; Sun, L.; Gorczynski, R.; Li, Y.J.; Xu, Z.; Spaner, D.E. PPAR-delta promotes survival of breast cancer cells in harsh metabolic conditions. *Oncogenesis* **2016**, *5*, e232. [CrossRef] [PubMed]
40. Kilgore, K.S.; Billin, A.N. PPARbeta/delta ligands as modulators of the inflammatory response. *Curr. Opin. Investig. Drugs* **2008**, *9*, 463–469. [PubMed]
41. Wahli, W.; Michalik, L. PPARs at the crossroads of lipid signaling and inflammation. *Trends Endocrinol. Metab.* **2012**, *23*, 351–363. [CrossRef] [PubMed]
42. Fan, Y.; Wang, Y.; Tang, Z.; Zhang, H.; Qin, X.; Zhu, Y.; Guan, Y.; Wang, X.; Staels, B.; Chien, S.; et al. Suppression of pro-inflammatory adhesion molecules by PPAR-delta in human vascular endothelial cells. *Arterioscler. Thromb. Vasc. Biol.* **2008**, *28*, 315–321. [CrossRef] [PubMed]
43. Zingarelli, B.; Piraino, G.; Hake, P.W.; O'Connor, M.; Denenberg, A.; Fan, H.; Cook, J.A. Peroxisome proliferator-activated receptor {delta} regulates inflammation via NF-{kappa}B signaling in polymicrobial sepsis. *Am. J. Pathol.* **2010**, *177*, 1834–1847. [CrossRef] [PubMed]
44. Siersbaek, R.; Nielsen, R.; Mandrup, S. PPARgamma in adipocyte differentiation and metabolism—Novel insights from genome-wide studies. *FEBS Lett.* **2010**, *584*, 3242–3249. [CrossRef] [PubMed]
45. Chawla, A.; Schwarz, E.J.; Dimaculangan, D.D.; Lazar, M.A. Peroxisome proliferator-activated receptor (PPAR) gamma: Adipose-predominant expression and induction early in adipocyte differentiation. *Endocrinology* **1994**, *135*, 798–800. [CrossRef] [PubMed]
46. Zhuang, H.; Zhang, X.; Zhu, C.; Tang, X.; Yu, F.; Shang, G.W.; Cai, X. Molecular Mechanisms of PPAR-gamma Governing MSC Osteogenic and Adipogenic Differentiation. *Curr. Stem Cell Res. Ther.* **2016**, *11*, 255–264. [CrossRef] [PubMed]
47. Fujiwara, T.; Yoshioka, S.; Yoshioka, T.; Ushiyama, I.; Horikoshi, H. Characterization of new oral antidiabetic agent CS-045. Studies in KK and *ob/ob* mice and Zucker fatty rats. *Diabetes* **1988**, *37*, 1549–1558. [CrossRef] [PubMed]
48. Olefsky, J.M. Treatment of insulin resistance with peroxisome proliferator-activated receptor gamma agonists. *J. Clin. Investig.* **2000**, *106*, 467–472. [CrossRef] [PubMed]
49. Kubota, N.; Terauchi, Y.; Miki, H.; Tamemoto, H.; Yamauchi, T.; Komeda, K.; Satoh, S.; Nakano, R.; Ishii, C.; Sugiyama, T.; et al. PPAR gamma mediates high-fat diet-induced adipocyte hypertrophy and insulin resistance. *Mol. Cell* **1999**, *4*, 597–609. [CrossRef]

50. Barak, Y.; Nelson, M.C.; Ong, E.S.; Jones, Y.Z.; Ruiz-Lozano, P.; Chien, K.R.; Koder, A.; Evans, R.M. PPAR gamma is required for placental, cardiac, and adipose tissue development. *Mol. Cell* **1999**, *4*, 585–595. [CrossRef]
51. Miles, P.D.; Barak, Y.; He, W.; Evans, R.M.; Olefsky, J.M. Improved insulin-sensitivity in mice heterozygous for PPAR-gamma deficiency. *J. Clin. Investig.* **2000**, *105*, 287–292. [CrossRef] [PubMed]
52. Deeb, S.S.; Fajas, L.; Nemoto, M.; Pihlajamaki, J.; Mykkanen, L.; Kuusisto, J.; Laakso, M.; Fujimoto, W.; Auwerx, J. A Pro12Ala substitution in PPARgamma2 associated with decreased receptor activity, lower body mass index and improved insulin sensitivity. *Nat. Genet.* **1998**, *20*, 284–287. [CrossRef] [PubMed]
53. Lefterova, M.I.; Zhang, Y.; Steger, D.J.; Schupp, M.; Schug, J.; Cristancho, A.; Feng, D.; Zhuo, D.; Stoeckert, C.J., Jr.; Liu, X.S.; et al. PPARgamma and C/EBP factors orchestrate adipocyte biology via adjacent binding on a genome-wide scale. *Genes Dev.* **2008**, *22*, 2941–2952. [CrossRef] [PubMed]
54. Odegaard, J.I.; Ricardo-Gonzalez, R.R.; Goforth, M.H.; Morel, C.R.; Subramanian, V.; Mukundan, L.; Red Eagle, A.; Vats, D.; Brombacher, F.; Ferrante, A.W.; et al. Macrophage-specific PPARgamma controls alternative activation and improves insulin resistance. *Nature* **2007**, *447*, 1116–1120. [CrossRef] [PubMed]
55. Lefterova, M.I.; Steger, D.J.; Zhuo, D.; Qatanani, M.; Mullican, S.E.; Tuteja, G.; Manduchi, E.; Grant, G.R.; Lazar, M.A. Cell-specific determinants of peroxisome proliferator-activated receptor gamma function in adipocytes and macrophages. *Mol. Cell. Biol.* **2010**, *30*, 2078–2089. [CrossRef] [PubMed]
56. Foryst-Ludwig, A.; Hartge, M.; Clemenz, M.; Sprang, C.; Hess, K.; Marx, N.; Unger, T.; Kintscher, U. PPARgamma activation attenuates T-lymphocyte-dependent inflammation of adipose tissue and development of insulin resistance in obese mice. *Cardiovasc. Diabetol.* **2010**, *9*, 64. [CrossRef] [PubMed]
57. Motoki, T.; Kurobe, H.; Hirata, Y.; Nakayama, T.; Kinoshita, H.; Rocco, K.A.; Sogabe, H.; Hori, T.; Sata, M.; Kitagawa, T. PPAR-gamma agonist attenuates inflammation in aortic aneurysm patients. *Gen. Thorac. Cardiovasc. Surg.* **2015**, *63*, 565–571. [CrossRef] [PubMed]
58. Gray, E.; Ginty, M.; Kemp, K.; Scolding, N.; Wilkins, A. The PPAR-gamma agonist pioglitazone protects cortical neurons from inflammatory mediators via improvement in peroxisomal function. *J. Neuroinflammation* **2012**, *9*, 63. [CrossRef] [PubMed]
59. Nikolic, D.; Castellino, G.; Banach, M.; Toth, P.P.; Ivanova, E.; Orekhov, A.N.; Montalto, G.; Rizzo, M. PPAR Agonists, Atherogenic Dyslipidemia and Cardiovascular Risk. *Curr. Pharm. Des.* **2017**, *23*, 894–902. [CrossRef] [PubMed]
60. Xiao, C.; Dash, S.; Morgantini, C.; Hegele, R.A.; Lewis, G.F. Pharmacological Targeting of the Atherogenic Dyslipidemia Complex: The Next Frontier in CVD Prevention Beyond Lowering LDL Cholesterol. *Diabetes* **2016**, *65*, 1767–1778. [CrossRef] [PubMed]
61. Saha, S.A.; Kizhakepunnur, L.G.; Bahekar, A.; Arora, R.R. The role of fibrates in the prevention of cardiovascular disease—A pooled meta-analysis of long-term randomized placebo-controlled clinical trials. *Am. Heart J.* **2007**, *154*, 943–953. [CrossRef] [PubMed]
62. Keech, A.; Simes, R.J.; Barter, P.; Best, J.; Scott, R.; Taskinen, M.R.; Forder, P.; Pillai, A.; Davis, T.; Glasziou, P.; et al. Effects of long-term fenofibrate therapy on cardiovascular events in 9795 people with type 2 diabetes mellitus (the FIELD study): Randomised controlled trial. *Lancet* **2005**, *366*, 1849–1861. [CrossRef]
63. Sirtori, C.R. Mechanisms of action of absorbable hypolipidemic drugs. *Adv. Exp. Med. Biol.* **1985**, *183*, 241–252. [PubMed]
64. Sirtori, C.R. The FIELD study. *Lancet* **2006**, *367*, 1141–1142; author reply 1142–1143. [CrossRef]
65. Elam, M.B.; Ginsberg, H.N.; Lovato, L.C.; Corson, M.; Largay, J.; Leiter, L.A.; Lopez, C.; O'Connor, P.J.; Sweeney, M.E.; Weiss, D.; et al. Association of Fenofibrate Therapy With Long-term Cardiovascular Risk in Statin-Treated Patients With Type 2 Diabetes. *JAMA Cardiol.* **2017**, *2*, 370–380. [CrossRef] [PubMed]
66. Ferri, N.; Corsini, A.; Sirtori, C.; Ruscica, M. PPAR-alpha agonists are still on the rise: An update on clinical and experimental findings. *Expert Opin. Investig. Drugs* **2017**, *26*, 593–602. [CrossRef] [PubMed]
67. Simental-Mendia, L.E.; Simental-Mendia, M.; Sanchez-Garcia, A.; Banach, M.; Atkin, S.L.; Gotto, A.M., Jr.; Sahebkar, A. Effect of fibrates on glycemic parameters: A systematic review and meta-analysis of randomized placebo-controlled trials. *Pharmacol. Res.* **2017**, in press. [CrossRef] [PubMed]
68. Tenenbaum, A.; Fisman, E.Z. Balanced pan-PPAR activator bezafibrate in combination with statin: Comprehensive lipids control and diabetes prevention? *Cardiovasc. Diabetol.* **2012**, *11*, 140. [CrossRef] [PubMed]

69. Vamecq, J.; Latruffe, N. Medical significance of peroxisome proliferator-activated receptors. *Lancet* **1999**, *354*, 141–148. [CrossRef]
70. Brown, J.D.; Plutzky, J. Peroxisome proliferator-activated receptors as transcriptional nodal points and therapeutic targets. *Circulation* **2007**, *115*, 518–533. [CrossRef] [PubMed]
71. Flory, J.H.; Ellenberg, S.; Szapary, P.O.; Strom, B.L.; Hennessy, S. Antidiabetic action of bezafibrate in a large observational database. *Diabetes Care* **2009**, *32*, 547–551. [CrossRef] [PubMed]
72. Raza-Iqbal, S.; Tanaka, T.; Anai, M.; Inagaki, T.; Matsumura, Y.; Ikeda, K.; Taguchi, A.; Gonzalez, F.J.; Sakai, J.; Kodama, T. Transcriptome Analysis of K-877 (a Novel Selective PPARalpha Modulator (SPPARMalpha))-Regulated Genes in Primary Human Hepatocytes and the Mouse Liver. *J. Atheroscler. Thromb.* **2015**, *22*, 754–772. [CrossRef] [PubMed]
73. Blair, H.A. Pemafibrate: First Global Approval. *Drugs* **2017**, *77*, 1805–1810. [CrossRef] [PubMed]
74. Araki, E.; Yamashita, S.; Arai, H.; Yokote, K.; Satoh, J.; Inoguchi, T.; Nakamura, J.; Maegawa, H.; Yoshioka, N.; Tanizawa, Y.; et al. Effects of Pemafibrate, a Novel Selective PPARalpha Modulator, on Lipid and Glucose Metabolism in Patients With Type 2 Diabetes and Hypertriglyceridemia: A Randomized, Double-Blind, Placebo-Controlled, Phase 3 Trial. *Diabetes Care* **2018**, *41*, 538–546. [CrossRef] [PubMed]
75. Sahebkar, A.; Simental-Mendia, L.E.; Mikhailidis, D.P.; Pirro, M.; Banach, M.; Sirtori, C.R.; Ruscica, M.; Reiner, Z. Effect of statin therapy on plasma apolipoprotein CIII concentrations: A systematic review and meta-analysis of randomized controlled trials. *J. Clin. Lipidol.* **2018**, in press. [CrossRef] [PubMed]
76. Ishibashi, S.; Arai, H.; Yokote, K.; Araki, E.; Suganami, H.; Yamashita, S.; Group, K.S. Efficacy and safety of pemafibrate (K-877), a selective peroxisome proliferator-activated receptor alpha modulator, in patients with dyslipidemia: Results from a 24-week, randomized, double blind, active-controlled, phase 3 trial. *J. Clin. Lipidol.* **2018**, *12*, 173–184. [CrossRef] [PubMed]
77. Arai, H.; Yamashita, S.; Yokote, K.; Araki, E.; Suganami, H.; Ishibashi, S.; Group, K.S. Efficacy and Safety of Pemafibrate Versus Fenofibrate in Patients with High Triglyceride and Low HDL Cholesterol Levels: A Multicenter, Placebo-Controlled, Double-Blind, Randomized Trial. *J. Atheroscler. Thromb.* **2018**. [CrossRef] [PubMed]
78. Camejo, G. Phase 2 clinical trials with K-877 (pemafibrate): A promising selective PPAR-alpha modulator for treatment of combined dyslipidemia. *Atherosclerosis* **2017**, *261*, 163–164. [CrossRef] [PubMed]
79. Arai, H.; Yamashita, S.; Yokote, K.; Araki, E.; Suganami, H.; Ishibashi, S.; Group, K.S. Efficacy and safety of K-877, a novel selective peroxisome proliferator-activated receptor alpha modulator (SPPARMalpha), in combination with statin treatment: Two randomised, double-blind, placebo-controlled clinical trials in patients with dyslipidaemia. *Atherosclerosis* **2017**, *261*, 144–152. [CrossRef] [PubMed]
80. Kontush, A.; Lindahl, M.; Lhomme, M.; Calabresi, L.; Chapman, M.J.; Davidson, W.S. Structure of HDL: Particle subclasses and molecular components. *Handb. Exp. Pharmacol.* **2015**, *224*, 3–51. [PubMed]
81. Fruchart, J.C. Pemafibrate (K-877), a novel selective peroxisome proliferator-activated receptor alpha modulator for management of atherogenic dyslipidaemia. *Cardiovasc. Diabetol.* **2017**, *16*, 124. [CrossRef] [PubMed]
82. NCT03071692. Pemafibrate to Reduce Cardiovascular OutcoMes by Reducing Triglycerides IN patiENts with diabeTes (PROMINENT). Available online: https://www.clinicaltrials.gov/ct2/results?cond=&term=NCT03071692&cntry=&state=&city=&dist= (accessed on 13 April 2018).
83. Sirtori, C.R.; Galli, C.; Franceschini, G. Fraudulent (and non fraudulent) fatty acids for human health. *Eur. J. Clin. Investig.* **1993**, *23*, 686–689. [CrossRef]
84. Shi, L.; Tu, B.P. Acetyl-CoA and the regulation of metabolism: Mechanisms and consequences. *Curr. Opin. Cell Biol.* **2015**, *33*, 125–131. [CrossRef] [PubMed]
85. Calder, P.C. Mechanisms of action of (n-3) fatty acids. *J. Nutr.* **2012**, *142*, 592S–599S. [CrossRef] [PubMed]
86. Huang, J.; Jia, Y.; Fu, T.; Viswakarma, N.; Bai, L.; Rao, M.S.; Zhu, Y.; Borensztajn, J.; Reddy, J.K. Sustained activation of PPARalpha by endogenous ligands increases hepatic fatty acid oxidation and prevents obesity in *ob/ob* mice. *FASEB J.* **2012**, *26*, 628–638. [CrossRef] [PubMed]
87. Oh, D.Y.; Talukdar, S.; Bae, E.J.; Imamura, T.; Morinaga, H.; Fan, W.; Li, P.; Lu, W.J.; Watkins, S.M.; Olefsky, J.M. GPR120 is an omega-3 fatty acid receptor mediating potent anti-inflammatory and insulin-sensitizing effects. *Cell* **2010**, *142*, 687–698. [CrossRef] [PubMed]
88. Pirillo, A.; Catapano, A.L. Update on the management of severe hypertriglyceridemia—Focus on free fatty acid forms of omega-3. *Drug Des. Dev. Ther.* **2015**, *9*, 2129–2137.

89. Pirillo, A.; Catapano, A.L. Omega-3 polyunsaturated fatty acids in the treatment of hypertriglyceridaemia. *Int. J. Cardiol.* **2013**, *170* (Suppl. S1), S16–S20. [CrossRef] [PubMed]
90. Burke, M.F.; Burke, F.M.; Soffer, D.E. Review of Cardiometabolic Effects of Prescription Omega-3 Fatty Acids. *Curr. Atheroscler. Rep.* **2017**, *19*, 60. [CrossRef] [PubMed]
91. Sperling, L.S.; Nelson, J.R. History and future of omega-3 fatty acids in cardiovascular disease. *Curr. Med. Res. Opin.* **2016**, *32*, 301–311. [CrossRef] [PubMed]
92. Pizzini, A.; Lunger, L.; Demetz, E.; Hilbe, R.; Weiss, G.; Ebenbichler, C.; Tancevski, I. The Role of Omega-3 Fatty Acids in Reverse Cholesterol Transport: A Review. *Nutrients* **2017**, *9*, 1099. [CrossRef] [PubMed]
93. Pahlavani, M.; Ramalho, T.; Koboziev, I.; LeMieux, M.J.; Jayarathne, S.; Ramalingam, L.; Filgueiras, L.R.; Moustaid-Moussa, N. Adipose tissue inflammation in insulin resistance: Review of mechanisms mediating anti-inflammatory effects of omega-3 polyunsaturated fatty acids. *J. Investig. Med.* **2017**, *65*, 1021–1027. [CrossRef] [PubMed]
94. Kuda, O.; Rossmeisl, M.; Kopecky, J. Omega-3 fatty acids and adipose tissue biology. *Mol. Aspects Med.* **2018**, in press. [CrossRef] [PubMed]
95. Hasan, A.U.; Ohmori, K.; Hashimoto, T.; Kamitori, K.; Yamaguchi, F.; Noma, T.; Igarashi, J.; Tsuboi, K.; Tokuda, M.; Nishiyama, A.; et al. GPR120 in adipocytes has differential roles in the production of pro-inflammatory adipocytokines. *Biochem. Biophys. Res. Commun.* **2017**, *486*, 76–82. [CrossRef] [PubMed]
96. Bargut, T.C.; Silva-e-Silva, A.C.; Souza-Mello, V.; Mandarim-de-Lacerda, C.A.; Aguila, M.B. Mice fed fish oil diet and upregulation of brown adipose tissue thermogenic markers. *Eur. J. Nutr.* **2016**, *55*, 159–169. [CrossRef] [PubMed]
97. Quesada-Lopez, T.; Cereijo, R.; Turatsinze, J.V.; Planavila, A.; Cairo, M.; Gavalda-Navarro, A.; Peyrou, M.; Moure, R.; Iglesias, R.; Giralt, M.; et al. The lipid sensor GPR120 promotes brown fat activation and FGF21 release from adipocytes. *Nat. Commun.* **2016**, *7*, 13479. [CrossRef] [PubMed]
98. Parkinson, A.J.; Cruz, A.L.; Heyward, W.L.; Bulkow, L.R.; Hall, D.; Barstaed, L.; Connor, W.E. Elevated concentrations of plasma omega-3 polyunsaturated fatty acids among Alaskan Eskimos. *Am. J. Clin. Nutr.* **1994**, *59*, 384–388. [CrossRef] [PubMed]
99. Qin, X.; Park, H.G.; Zhang, J.Y.; Lawrence, P.; Liu, G.; Subramanian, N.; Kothapalli, K.S.; Brenna, J.T. Brown but not white adipose cells synthesize omega-3 docosahexaenoic acid in culture. *Prostaglandins Leukot. Essent. Fatty Acids* **2016**, *104*, 19–24. [CrossRef] [PubMed]
100. Ruscica, M.; Baragetti, A.; Catapano, A.L.; Norata, G.D. Translating the biology of adipokines in atherosclerosis and cardiovascular diseases: Gaps and open questions. *Nutr. Metab. Cardiovasc. Dis.* **2017**, *27*, 379–395. [CrossRef] [PubMed]
101. Monk, J.M.; Liddle, D.M.; De Boer, A.A.; Brown, M.J.; Power, K.A.; Ma, D.W.; Robinson, L.E. Fish-oil-derived n-3 PUFAs reduce inflammatory and chemotactic adipokine-mediated cross-talk between co-cultured murine splenic $CD8^+$ T cells and adipocytes. *J. Nutr.* **2015**, *145*, 829–838. [CrossRef] [PubMed]
102. Bays, H.E.; Ballantyne, C.M.; Braeckman, R.A.; Stirtan, W.G.; Doyle, R.T., Jr.; Philip, S.; Soni, P.N.; Juliano, R.A. Icosapent Ethyl (Eicosapentaenoic Acid Ethyl Ester): Effects Upon High-Sensitivity C-Reactive Protein and Lipid Parameters in Patients With Metabolic Syndrome. *Metab. Syndr. Relat. Disord.* **2015**, *13*, 239–247. [CrossRef] [PubMed]
103. Buoite Stella, A.; Gortan Cappellari, G.; Barazzoni, R.; Zanetti, M. Update on the Impact of Omega 3 Fatty Acids on Inflammation, Insulin Resistance and Sarcopenia: A Review. *Int. J. Mol. Sci.* **2018**, *19*, 218. [CrossRef] [PubMed]
104. Wei, M.Y.; Jacobson, T.A. Effects of eicosapentaenoic acid versus docosahexaenoic acid on serum lipids: A systematic review and meta-analysis. *Curr. Atheroscler. Rep.* **2011**, *13*, 474–483. [CrossRef] [PubMed]
105. Fialkow, J. Omega-3 Fatty Acid Formulations in Cardiovascular Disease: Dietary Supplements are Not Substitutes for Prescription Products. *Am. J. Cardiovasc. Drugs* **2016**, *16*, 229–239. [CrossRef] [PubMed]
106. Stroes, E.S.G.; Susekov, A.V.; de Bruin, T.W.A.; Kvarnstrom, M.; Yang, H.; Davidson, M.H. Omega-3 carboxylic acids in patients with severe hypertriglyceridemia: EVOLVE II, a randomized, placebo-controlled trial. *J. Clin. Lipidol.* **2018**, *12*, 321–330. [CrossRef] [PubMed]
107. Kastelein, J.J.; Maki, K.C.; Susekov, A.; Ezhov, M.; Nordestgaard, B.G.; Machielse, B.N.; Kling, D.; Davidson, M.H. Omega-3 free fatty acids for the treatment of severe hypertriglyceridemia: The EpanoVa fOr Lowering Very high triglyceridEs (EVOLVE) trial. *J. Clin. Lipidol.* **2014**, *8*, 94–106. [CrossRef] [PubMed]

108. Maki, K.C.; Orloff, D.G.; Nicholls, S.J.; Dunbar, R.L.; Roth, E.M.; Curcio, D.; Johnson, J.; Kling, D.; Davidson, M.H. A highly bioavailable omega-3 free fatty acid formulation improves the cardiovascular risk profile in high-risk, statin-treated patients with residual hypertriglyceridemia (the ESPRIT trial). *Clin. Ther.* **2013**, *35*, 1400–1411. [CrossRef] [PubMed]
109. GISSI-Prevenzione Investigators. Dietary supplementation with n-3 polyunsaturated fatty acids and vitamin E after myocardial infarction: Results of the GISSI-Prevenzione trial. Gruppo Italiano per lo Studio della Sopravvivenza nell'Infarto miocardico. *Lancet* **1999**, *354*, 447–455.
110. Yokoyama, M.; Origasa, H.; Matsuzaki, M.; Matsuzawa, Y.; Saito, Y.; Ishikawa, Y.; Oikawa, S.; Sasaki, J.; Hishida, H.; Itakura, H.; et al. Effects of eicosapentaenoic acid on major coronary events in hypercholesterolaemic patients (JELIS): A randomised open-label, blinded endpoint analysis. *Lancet* **2007**, *369*, 1090–1098. [CrossRef]
111. Kromhout, D.; Giltay, E.J.; Geleijnse, J.M.; Alpha Omega Trial Group. n-3 fatty acids and cardiovascular events after myocardial infarction. *N. Engl. J. Med.* **2010**, *363*, 2015–2026. [CrossRef] [PubMed]
112. Aung, T.; Halsey, J.; Kromhout, D.; Gerstein, H.C.; Marchioli, R.; Tavazzi, L.; Geleijnse, J.M.; Rauch, B.; Ness, A.; Galan, P.; et al. Associations of Omega-3 Fatty Acid Supplement Use With Cardiovascular Disease Risks: Meta-analysis of 10 Trials Involving 77917 Individuals. *JAMA Cardiol.* **2018**, *3*, 225–234. [CrossRef] [PubMed]
113. Bhatt, D.L.; Steg, P.G.; Brinton, E.A.; Jacobson, T.A.; Miller, M.; Tardif, J.C.; Ketchum, S.B.; Doyle, R.T., Jr.; Murphy, S.A.; Soni, P.N.; et al. Rationale and design of REDUCE-IT: Reduction of Cardiovascular Events with Icosapent Ethyl-Intervention Trial. *Clin. Cardiol.* **2017**, *40*, 138–148. [CrossRef] [PubMed]
114. Ruscica, M.; Baldessin, L.; Boccia, D.; Racagni, G.; Mitro, N. Non-insulin anti-diabetic drugs: An update on pharmacological interactions. *Pharmacol. Res.* **2017**, *115*, 14–24. [CrossRef] [PubMed]
115. Nissen, S.E.; Wolski, K. Effect of rosiglitazone on the risk of myocardial infarction and death from cardiovascular causes. *N. Engl. J. Med.* **2007**, *356*, 2457–2471. [CrossRef] [PubMed]
116. Nissen, S.E.; Wolski, K. Rosiglitazone revisited: An updated meta-analysis of risk for myocardial infarction and cardiovascular mortality. *Arch. Intern. Med.* **2010**, *170*, 1191–1201. [CrossRef] [PubMed]
117. Lincoff, A.M.; Wolski, K.; Nicholls, S.J.; Nissen, S.E. Pioglitazone and risk of cardiovascular events in patients with type 2 diabetes mellitus: A meta-analysis of randomized trials. *JAMA* **2007**, *298*, 1180–1188. [CrossRef] [PubMed]
118. Betteridge, D.J. CHICAGO, PERISCOPE and PROactive: CV risk modification in diabetes with pioglitazone. *Fundam. Clin. Pharmacol.* **2009**, *23*, 675–679. [CrossRef] [PubMed]
119. Nissen, S.E.; Nicholls, S.J.; Wolski, K.; Nesto, R.; Kupfer, S.; Perez, A.; Jure, H.; De Larochelliere, R.; Staniloae, C.S.; Mavromatis, K.; et al. Comparison of pioglitazone vs glimepiride on progression of coronary atherosclerosis in patients with type 2 diabetes: The PERISCOPE randomized controlled trial. *JAMA* **2008**, *299*, 1561–1573. [CrossRef] [PubMed]
120. Nicholls, S.J.; Tuzcu, E.M.; Wolski, K.; Bayturan, O.; Lavoie, A.; Uno, K.; Kupfer, S.; Perez, A.; Nesto, R.; Nissen, S.E. Lowering the triglyceride/high-density lipoprotein cholesterol ratio is associated with the beneficial impact of pioglitazone on progression of coronary atherosclerosis in diabetic patients: Insights from the PERISCOPE (Pioglitazone Effect on Regression of Intravascular Sonographic Coronary Obstruction Prospective Evaluation) study. *J. Am. Coll. Cardiol.* **2011**, *57*, 153–159. [PubMed]
121. Mazzone, T.; Meyer, P.M.; Feinstein, S.B.; Davidson, M.H.; Kondos, G.T.; D'Agostino, R.B., Sr.; Perez, A.; Provost, J.C.; Haffner, S.M. Effect of pioglitazone compared with glimepiride on carotid intima-media thickness in type 2 diabetes: A randomized trial. *JAMA* **2006**, *296*, 2572–2581. [CrossRef] [PubMed]
122. Davidson, M.; Meyer, P.M.; Haffner, S.; Feinstein, S.; D'Agostino, R., Sr.; Kondos, G.T.; Perez, A.; Chen, Z.; Mazzone, T. Increased high-density lipoprotein cholesterol predicts the pioglitazone-mediated reduction of carotid intima-media thickness progression in patients with type 2 diabetes mellitus. *Circulation* **2008**, *117*, 2123–2130. [CrossRef] [PubMed]
123. Kernan, W.N.; Viscoli, C.M.; Furie, K.L.; Young, L.H.; Inzucchi, S.E.; Gorman, M.; Guarino, P.D.; Lovejoy, A.M.; Peduzzi, P.N.; Conwit, R.; et al. Pioglitazone after Ischemic Stroke or Transient Ischemic Attack. *N. Engl. J. Med.* **2016**, *374*, 1321–1331. [CrossRef] [PubMed]
124. Inzucchi, S.E.; Viscoli, C.M.; Young, L.H.; Furie, K.L.; Gorman, M.; Lovejoy, A.M.; Dagogo-Jack, S.; Ismail-Beigi, F.; Korytkowski, M.T.; Pratley, R.E.; et al. Pioglitazone Prevents Diabetes in Patients With Insulin Resistance and Cerebrovascular Disease. *Diabetes Care* **2016**, *39*, 1684–1692. [CrossRef] [PubMed]

125. Viscoli, C.M.; Inzucchi, S.E.; Young, L.H.; Insogna, K.L.; Conwit, R.; Furie, K.L.; Gorman, M.; Kelly, M.A.; Lovejoy, A.M.; Kernan, W.N.; et al. Pioglitazone and Risk for Bone Fracture: Safety Data from a Randomized Clinical Trial. *J. Clin. Endocrinol. Metab.* **2017**, *102*, 914–922. [CrossRef] [PubMed]
126. Wang, S.; Dougherty, E.J.; Danner, R.L. PPARgamma signaling and emerging opportunities for improved therapeutics. *Pharmacol. Res.* **2016**, *111*, 76–85. [CrossRef] [PubMed]
127. Palomer, X.; Barroso, E.; Pizarro-Delgado, J.; Pena, L.; Botteri, G.; Zarei, M.; Aguilar, D.; Montori-Grau, M.; Vazquez-Carrera, M. PPARbeta/delta: A Key Therapeutic Target in Metabolic Disorders. *Int. J. Mol. Sci.* **2018**, *19*, 913. [CrossRef] [PubMed]
128. Sahebkar, A.; Chew, G.T.; Watts, G.F. New peroxisome proliferator-activated receptor agonists: Potential treatments for atherogenic dyslipidemia and non-alcoholic fatty liver disease. *Expert Opin. Pharmacother.* **2014**, *15*, 493–503. [CrossRef] [PubMed]
129. Staels, B.; Rubenstrunk, A.; Noel, B.; Rigou, G.; Delataille, P.; Millatt, L.J.; Baron, M.; Lucas, A.; Tailleux, A.; Hum, D.W.; et al. Hepatoprotective effects of the dual peroxisome proliferator-activated receptor alpha/delta agonist, GFT505, in rodent models of nonalcoholic fatty liver disease/nonalcoholic steatohepatitis. *Hepatology* **2013**, *58*, 1941–1952. [CrossRef] [PubMed]
130. Cariou, B.; Zair, Y.; Staels, B.; Bruckert, E. Effects of the new dual PPAR alpha/delta agonist GFT505 on lipid and glucose homeostasis in abdominally obese patients with combined dyslipidemia or impaired glucose metabolism. *Diabetes Care* **2011**, *34*, 2008–2014. [CrossRef] [PubMed]
131. Ratziu, V.; Harrison, S.A.; Francque, S.; Bedossa, P.; Lehert, P.; Serfaty, L.; Romero-Gomez, M.; Boursier, J.; Abdelmalek, M.; Caldwell, S.; et al. Elafibranor, an Agonist of the Peroxisome Proliferator-Activated Receptor-alpha and -delta, Induces Resolution of Nonalcoholic Steatohepatitis without Fibrosis Worsening. *Gastroenterology* **2016**, *150*, 1147–1159. [CrossRef] [PubMed]
132. Chatterjee, S.; Majumder, A.; Ray, S. Observational study of effects of Saroglitazar on glycaemic and lipid parameters on Indian patients with type 2 diabetes. *Sci. Rep.* **2015**, *5*, 7706. [CrossRef] [PubMed]
133. Rotman, Y.; Sanyal, A.J. Current and upcoming pharmacotherapy for non-alcoholic fatty liver disease. *Gut* **2017**, *66*, 180–190. [CrossRef] [PubMed]
134. Jani, R.H.; Pai, V.; Jha, P.; Jariwala, G.; Mukhopadhyay, S.; Bhansali, A.; Joshi, S. A multicenter, prospective, randomized, double-blind study to evaluate the safety and efficacy of Saroglitazar 2 and 4 mg compared with placebo in type 2 diabetes mellitus patients having hypertriglyceridemia not controlled with atorvastatin therapy (PRESS VI). *Diabetes Technol. Ther.* **2014**, *16*, 63–71. [PubMed]
135. Balakumar, P.; Rose, M.; Ganti, S.S.; Krishan, P.; Singh, M. PPAR dual agonists: Are they opening Pandora's Box? *Pharmacol. Res.* **2007**, *56*, 91–98. [CrossRef] [PubMed]
136. Sirtori, C.R.; Pavanello, C.; Calabresi, L.; Ruscica, M. Nutraceutical approaches to metabolic syndrome. *Ann. Med.* **2017**, *49*, 678–697. [CrossRef] [PubMed]
137. Madsen, C.M.; Varbo, A.; Nordestgaard, B.G. Unmet need for primary prevention in individuals with hypertriglyceridaemia not eligible for statin therapy according to European Society of Cardiology/European Atherosclerosis Society guidelines: A contemporary population-based study. *Eur. Heart J.* **2018**, *39*, 610–619. [CrossRef] [PubMed]
138. Ferri, N.; Corsini, A.; Sirtori, C.R.; Ruscica, M. Present therapeutic role of cholesteryl ester transfer protein inhibitors. *Pharmacol. Res.* **2018**, *128*, 29–41. [CrossRef] [PubMed]
139. Madsen, C.M.; Nordestgaard, B.G. Is It Time for New Thinking about High-Density Lipoprotein? *Arterioscler. Thromb. Vasc. Biol.* **2018**, *38*, 484–486. [CrossRef] [PubMed]

International Journal of
Molecular Sciences

Review

PPARβ/δ: A Key Therapeutic Target in Metabolic Disorders

Xavier Palomer [1,2,3], Emma Barroso [1,2,3], Javier Pizarro-Delgado [1,2,3], Lucía Peña [1,2,3], Gaia Botteri [1,2,3], Mohammad Zarei [1,2,3], David Aguilar [1,2,3], Marta Montori-Grau [1,2,3] and Manuel Vázquez-Carrera [1,2,3,*]

1 Pharmacology Unit, Department of Pharmacology, Toxicology and Therapeutic Chemistry, Institute of Biomedicine of the University of Barcelona (IBUB), Faculty of Pharmacy and Food Sciences, University of Barcelona, 08028 Barcelona, Spain; xpalomer@ub.edu (X.P.); ebarroso@ub.edu (E.B.); jpizarro@ub.edu (J.P.-D.); lucia.pena.moreno@gmail.com (L.P.); gaia.btt@gmail.com (G.B.); biotech.zarei@gmail.com (M.Z.); d.aguilarrecarte@gmail.com (D.A.); mmontori@ub.edu (M.M.-G.)

2 Pediatric Research Institute-Hospital Sant Joan de Déu, 08950 Esplugues de Llobregat, Spain

3 Spanish Biomedical Research Center in Diabetes and Associated Metabolic Diseases (CIBERDEM), Instituto de Salud Carlos III, C/Monforte de Lemos 3-5, Pabellón 11, Planta 0, 28029 Madrid, Spain

* Correspondence: mvazquezcarrera@ub.edu; Tel.: +34-93-402-4531; Fax: +34-93-403-5982

Received: 29 January 2018; Accepted: 17 March 2018; Published: 20 March 2018

Abstract: Research in recent years on peroxisome proliferator-activated receptor (PPAR)β/δ indicates that it plays a key role in the maintenance of energy homeostasis, both at the cellular level and within the organism as a whole. PPARβ/δ activation might help prevent the development of metabolic disorders, including obesity, dyslipidaemia, type 2 diabetes mellitus and non-alcoholic fatty liver disease. This review highlights research findings on the PPARβ/δ regulation of energy metabolism and the development of diseases related to altered cellular and body metabolism. It also describes the potential of the pharmacological activation of PPARβ/δ as a treatment for human metabolic disorders.

Keywords: PPARβ/δ; obesity; dyslipidaemia; type 2 diabetes mellitus; non-alcoholic fatty liver disease

1. Introduction

Acquired metabolic disorders, particularly obesity and its associated co-morbidities currently pose a risk to human health on a global scale. These metabolic disorders are closely related to adipose tissue dysfunction, one of the primary defects observed in obesity that may link this condition to its co-morbidities such as non-alcoholic fatty liver disease (NAFLD), atherogenic dyslipidaemia, type 2 diabetes mellitus and cardiovascular disease [1–3]. In fact, up to a third of obese subjects are metabolically healthy and they do not develop obesity-related metabolic or cardiovascular disorders [4], probably because of the preservation of normal adipose tissue architecture and function. Adipose tissue dysfunction in obese patients results from the interactions of genetic and environmental factors that lead to the presence of hypertrophic adipocytes, which have a pro-inflammatory, insulin-resistant phenotype compared to small adipocytes [5]. In addition, a critical factor contributing to the difference between metabolically healthy and unhealthy obese individuals is the anatomical distribution of the adipose tissues. Expansion of the visceral adipose tissue, which is considered a dysfunctional adipose tissue unable to store excessive levels of lipids, to a greater extent than that of subcutaneous adipose tissue, is associated with metabolic alterations [6]. Failure to store surplus lipids into visceral adipose tissue causes a chronic elevation of circulating fatty acids (FA), which can reach toxic levels in non-adipose tissues, such as skeletal muscle, the liver and the pancreas [7]. The deleterious effect of lipid accumulation in non-adipose tissues is known as lipotoxicity. This surplus of fatty acids (FAs), especially saturated FA and their derived metabolites, such as diacylglycerol and ceramides, induces chronic low-grade inflammation and has harmful effects on multiple organs and systems.

Given the enormous stress on global health services caused by the increasing incidence of obesity and its co-morbidities on global health services, there is a need to better understand the mechanisms behind the relationship between obesity and the development of metabolic disorders to prevent and to improve the outcomes of these diseases. Peroxisome proliferator-activated receptor (PPAR)β/δ is a nuclear receptor that exerts many metabolic effects. Its activation may prevent and improve the outcome of obesity-related metabolic disorders. In this review, we will summarize the molecular features of PPARβ/δ and the benefits of using its agonists to treat obesity and its related co-morbidities.

2. Basic Overview of the Molecular Features of PPARβ/δ

PPARβ/δ is a member of the nuclear receptor (NR) superfamily of ligand-inducible transcription factors and belongs to the PPAR family, which comprises three isotypes: PPARα (NR1C1, according to the unified nomenclature system for the NR superfamily); PPARβ/δ (NR1C2); and PPARγ (NR1C3) [8,9]. PPARβ/δ was initially called PPARβ when it was first cloned in *Xenopus laevis*. However, when cloned in other species it was not clearly identified as being homologous to the Xenopus PPARβ and it was alternatively called NUC-1 in humans and PPARδ in mice. Currently, it is accepted that Xenopus PPARβ is homologous to murine PPARδ, giving rise to the terminology PPARβ/δ [8]. PPARβ/δ consists of four major functional domains: The N-terminal ligand-independent transactivation domain (A/B domain), often known as activation function 1 (AF-1); the DNA binding domain (DBD or C domain); the hinge region (D domain); and the carboxy-terminal E domain or AF-2, which includes the ligand-binding domain and the ligand-dependent transactivation domain [8,9]. The major physiological functions of PPARβ/δ result from its activity as a transcription factor, modulating the expression of specific target genes. Through this mechanism, PPARβ/δ regulates lipid metabolism and glucose homeostasis [8–11]. In addition, PPARβ/δ can regulate inflammation [12]. The involvement of PPARβ/δ in all these functions depends on its tissue distribution, ligand binding and the recruitment of co-activators or co-repressors.

PPARβ/δ is ubiquitously expressed, although it is most abundant in metabolically active tissues, especially in those organs/cells associated with FA metabolism, such as skeletal and cardiac muscle, hepatocytes and adipocytes. It has also been particularly characterized in macrophages. Compared to other NRs, PPARs present a large ligand-binding pocket ($\approx$1300 $Å^3$) [8], which directly contributes to the ability of PPARs to bind a great variety of endogenous and synthetic ligands. FAs are considered endogenous PPAR ligands but they show little selectivity for the different PPAR isoforms. Although all-trans retinoic acid has been reported to be a PPARβ/δ agonist [13], this has not been confirmed by other groups [14,15] and therefore remains controversial. To elucidate PPARβ/δ functions, synthetic ligands with high affinity and specificity (GW501516, GW0742 and L-165041) that only activate PPARβ/δ at very low concentrations both in vivo and in vitro have been developed [8]. At present, there are no clinically available drugs targeting PPARβ/δ but three PPARβ/δ agonists have reached clinical trials: Seladelpar (MBX-8025) (CymaBay Therapeutics) [16]; KD-3010 (Kalypsys) [17]; and CER-002 (Cerenis).

To activate transcription, PPARβ/δ forms an obligate heterodimer with retinoid X receptor (RXR or NR2B) and binds to peroxisome proliferator response elements (PPREs) located at the promoter regions of target genes, thereby increasing gene transcription in a ligand-dependent manner (transactivation) [8]. In the absence of a ligand, the PPARβ/δ-RXR heterodimer is bound by nuclear co-repressor proteins, which block transcriptional activation by preventing the binding of the heterodimer to the promoter. Ligand binding induces a conformational change within PPARβ/δ, resulting in the dissociation of the co-repressors and the recruitment of co-activators, which subsequently lead to PPARβ/δ-RXR binding to PPREs to initiate transcription [8]. PPARβ/δ also regulates gene expression independently of DNA binding, via cross-talk with other types of transcription factors, thus influencing their function through a mechanism termed receptor-dependent transrepression [8]. Most of the anti-inflammatory effects of PPARs probably occur through this mechanism [12].

3. PPARβ/δ as a Major Regulator of Metabolic Disorders

3.1. Obesity

PPARβ/δ-deficient mice exhibited a marked reduction in adiposity compared to wild-type mice levels [18]. This effect, however, cannot be reproduced in mice harbouring an adipose tissue-specific deletion of PPARβ/δ, indicating that PPARβ/δ elicits peripheral functions in systemic lipid metabolism. In fact, PPARβ/δ activation prevents weight gain in diet- or genetically-induced animal models of obesity by increasing fat burning in different tissues [19,20] or switching muscle fibre type, which, in turn, increases the muscle oxidative capacity [21] (Figure 1). A recent study also suggests that intestinal PPARβ/δ protects against diet-induced obesity, since intestinal epithelial cell-specific deletion of PPARβ/δ in mice results in increased amounts of omental white adipose tissue [22].

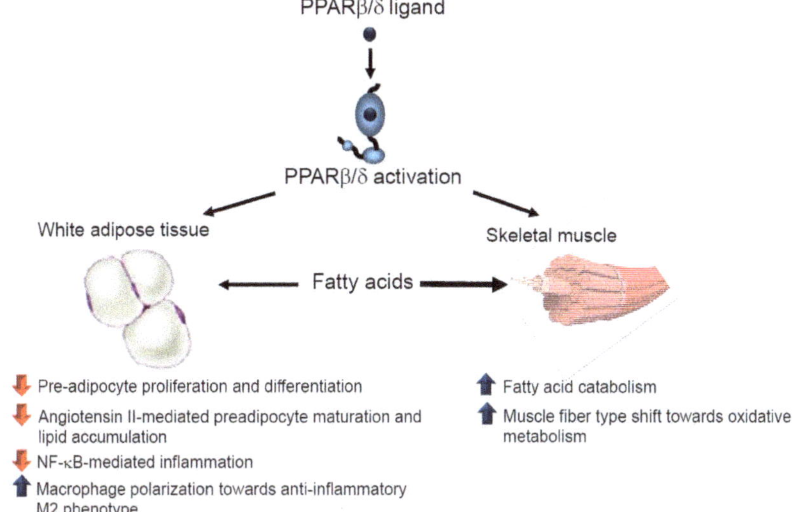

Figure 1. PPARβ/δ activation prevents obesity through several mechanisms. PPARβ/δ activation reduces pre-adipocyte proliferation and differentiation, attenuates angiotensin II-mediated dysfunctional hypertrophic adipogenesis and inhibits inflammation in adipose tissue. PPARβ/δ ligands reduce the availability of fatty acids to be stored in adipose tissue since these drugs induce fat burn in skeletal muscle by either increasing fatty acid oxidation or switching muscle fibre type towards oxidative metabolism. Blue arrow: increases. Red arrow: decreases.

Additional PPARβ/δ-mediated mechanisms can also contribute to the reduction in adiposity, since PPARβ/δ also regulates preadipocyte proliferation and differentiation through different mechanisms [23–25], such as by regulating of the expression of PPARγ, a key regulator of terminal adipocyte differentiation. Moreover, PPARβ/δ ligands prevent angiotensin II-induced adipocyte growth and lipid accumulation [26]. Angiotensin II increases levels of reactive oxygen species (ROS), which attenuate the canonical Wnt signalling pathway, leading to dysfunctional hypertrophic adipogenesis. PPARβ/δ agonists prevent oxidative stress and the reduction in Wnt signalling pathway induced by angiotensin II by increasing the expression of heme oxygenase 1 in adipose tissue. Consequently, PPARβ/δ activation delays preadipocyte maturation and lipid accumulation, leading to increased numbers of smaller adipocytes with an improved adipocytokine profile. Thus, overall,

PPARβ/δ activation prevents oxidative stress and dysfunctional adipogenesis under conditions of overactive renin-angiotensin system [26].

In humans, PPARβ/δ expression is reduced in both the subcutaneous and in visceral adipose tissues of morbidly obese patients compared to non-obese subjects [27]. This might result in adipose tissue dysregulation since PPARβ/δ has anti-inflammatory effects in white adipose tissue. PPARβ/δ activation inhibits lipopolysaccharide (LPS)-induced cytokine expression and secretion by preventing nuclear factor (NF)-κB activation in adipocytes via the activation of mitogen-activated protein kinase (MAPK)–extracellular signal-regulated kinase (ERK)1/2 (MEK1/2) activation [28]. Furthermore, adipose tissue inflammation is characterized by increased infiltration and an altered polarization of the macrophages from the anti-inflammatory M2 phenotype towards the pro-inflammatory M1 phenotype, with PPARβ/δ a crucial signalling molecule that activates polarization towards the anti-inflammatory M2 phenotype [29]. Consistent with the effects of PPARβ/δ in adipose tissue, it has been reported that overweight patients with mixed dyslipidaemia who were administered the PPARβ/δ agonist MBX-8025 for 8 weeks presented favourable trends in their body fat percentage, lean body mass and waist circumference, although the differences did not reach statistical significance [30].

3.2. Dyslipidaemia

Atherogenic dyslipidaemia, often observed in patients with obesity, insulin resistance, metabolic syndrome and type 2 diabetes mellitus, is a significant risk factor for cardiovascular disease. This dyslipidaemia is characterized by the presence of low high-density lipoprotein (HDL) cholesterol levels, elevated triglyceride (TG)-rich very low-density lipoprotein (VLDL) amounts and an increased proportion of small and dense low-density lipoprotein (LDL) particles. It is presently accepted that atherogenic dyslipidaemia is initiated by insulin resistance through the overproduction of TG-rich VLDL [31]. Under conditions of insulin resistance, adipose tissue lipolysis is enhanced, leading to an increase in plasma non-esterified FA (NEFA). The subsequent increase in the flux of NEFA into the liver overcomes the oxidative capacity of hepatocytes and NEFA are then esterified for TG production, causing hepatic steatosis and VLDL over secretion in the plasma [31]. Finally, in the presence of hypertriglyceridemia, the cholesterol-ester content of LDL and HDL decreases, whereas these lipoproteins are enriched in their TG content through the activity of cholesteryl ester transfer protein. These TG-enriched particles are then hydrolysed by hepatic lipase, leading to the formation of small, dense LDL and to the decrease in HDL-cholesterol levels [31].

PPARβ/δ agonists show a strong TG-lowering action in vivo. Given that the main factor affecting hepatic TG secretion is FA availability, the hypotriglyceridaemic effect of PPARβ/δ activators has been attributed, at least in part, to their ability to induce FA β-oxidation in the liver [32] and other tissues [21,33] (Figure 2). In the liver, this role of PPARβ/δ involves the increased expression of the genes involved in FA oxidation via amplification of the lipin 1/PPARγ-coactivator 1α (PGC-1α)/PPARα signalling system and increased levels of the hepatic endogenous ligand for PPARα, 16:0/18:1-phosphatidylcholine [32]. Moreover, the increased FA β-oxidation caused by PPARβ/δ activators might be due to the activation of AMP kinase (AMPK), probably through an increase in the AMP:ATP ratio in hepatocytes [32]. Likewise, the effects of PPARβ/δ on the expression of several genes (*VldlR*, *ApoA5*, *ApoA4* and *ApoC1*) involved in lipoprotein metabolism can contribute to its hypotriglyceridaemic effect [32,34]. In accordance with these effects of PPARβ/δ, mice deficient in this receptor fed a high-fat diet (HFD) show increased plasma TG levels due to hepatic VLDL overproduction. Moreover, these mice also exhibited reduced activity of the enzyme lipoprotein lipase, which catalyses the hydrolysis of the TG component of circulating chylomicrons and VLDL.

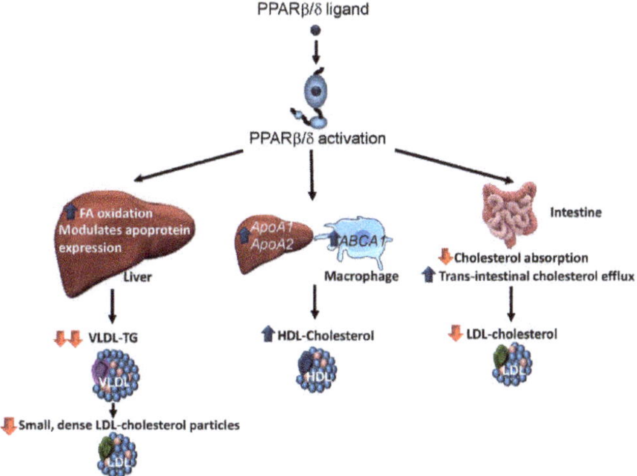

Figure 2. Effects of PPARβ/δ activation in dyslipidaemia. PPARβ/δ activation ameliorates atherogenic dyslipidaemia by reducing the amounts of very low-density lipoprotein (VLDL)-triglyceride (TG) and small dense low-density lipoprotein (LDL) particles and increasing the levels of high-density lipoprotein (HDL)-cholesterol. PPARβ/δ ligands reduce VLDL-TG by increasing hepatic fatty acid (FA) oxidation, which decreases the availability of this lipid for TG synthesis and changing the expression of several apoproteins. PPARβ/δ ligands increase HDL-cholesterol levels by elevating the amounts of the main apopoproteins of these lipoproteins (ApoA1 and ApoA2) in the liver and raising the levels of ATP-binding cassette A1 (ABCA1) in macrophages. Reduced LDL-cholesterol levels results from a decrease in cholesterol absorption and an increase in faecal excretion that are mediated by PPARβ/δ activation. Blue arrow: increases. Red arrow: decreases.

PPARβ/δ agonists also increase plasma HDL-cholesterol levels and reduce LDL-cholesterol and NEFA levels in both animal models and humans [30,33–36]. The increase in plasma HDL-cholesterol levels following PPARβ/δ activation has been linked to an increased expression of the two major apolipoproteins of HDL, *ApoA1* and *ApoA2* [36], in the liver. In addition, PPARβ/δ activation increases the expression in macrophages of the reverse cholesterol transporter, ATP-binding cassette A1 (*ABCA1*) [36,37], which is crucial for the formation of HDL particles through its transport of cholesterol and phospholipid to apolipoprotein acceptors in the bloodstream. Furthermore, PPARβ/δ agonists also regulate the expression of hepatic phospholipid transfer protein (*Pltp*), which regulates the size and the composition of HDL and plays an important role in controlling plasma HDL levels [38]. More recently, it has been reported that the absence of intestinal PPARβ/δ abolishes the ability of its agonists to increase HDL-cholesterol plasma levels [22].

Regarding the reduction in LDL-cholesterol levels, PPARβ/δ agonists have been shown to decrease the efficiency of intestinal cholesterol absorption possibly by reducing the intestinal abundance of the cholesterol absorption protein, Niemann-Pick C1-like 1 (NPC1L1) [39]. Furthermore, PPARβ/δ activation also stimulates faecal excretion of cholesterol in mice, primarily by the two-fold increase in trans-intestinal cholesterol efflux [40], a non-hepatobiliary-related route that transports cholesterol from the blood to the intestinal lumen directly via enterocytes.

The assessment of PPARβ/δ agonists in several small-scale clinical trials mainly for the treatment of atherogenic dyslipidaemia has confirmed that these drugs reduce plasma TG levels, increase the amounts of HDL-cholesterol and decrease the levels of small dense LDL particles in humans, indicating that treatment with these drugs initiates a transition towards a less atherogenic lipoprotein profile [30,37,41–43].

3.3. Type 2 Diabetes Mellitus

More than 90% of patients with type 2 diabetes mellitus are overweight or obese, since obesity is associated with insulin resistance. Insulin resistance, which is defined as a defect in the ability of insulin to drive glucose into its target tissues, predicts and precedes the development of type 2 diabetes mellitus [44]. However, patients with insulin resistance do not develop hyperglycaemia and type 2 diabetes mellitus until the pancreatic β cells fail to secrete sufficient amounts of insulin to meet the increased metabolic demand for this hormone. Adipose tissue expansion in obese individuals releases increased amounts of NEFAs, hormones, pro-inflammatory cytokines and other factors that contribute to the development of insulin resistance. Most of these molecules cause a chronic low-level inflammation, which contributes to insulin resistance and type 2 diabetes mellitus [45].

PPARβ/δ agonists improve glucose tolerance and insulin sensitivity in animal models [20,46]. The antidiabetic effects of these drugs are exerted in different tissues. For instance, macrophage infiltration into adipose tissue and polarization towards the pro-inflammatory M1 phenotype promotes inflammation and correlates with the degree of insulin resistance [47]. As mentioned above, PPARβ/δ activates polarization towards the anti-inflammatory M2 phenotype in macrophages [29] (Figure 3). In accordance with this, myeloid-specific PPARβ/$\delta^{-/-}$ mice show adipocyte dysfunction and insulin resistance [29]. Interestingly, a link exists between metabolism and function in macrophages. Thus, M2 macrophages require oxidative metabolism for their responses, whereas M1 macrophages depend on aerobic glycolysis [48,49]. In fact, blocking oxidative metabolism leads to the polarization of macrophages from the M2 to the M1 phenotype. Similarly, forcing oxidative metabolism in an M1 macrophage potentiates the M2 phenotype [50,51]. Given that PPARβ/δ activation increases β-oxidation in macrophages [52], this effect might also contribute to the polarization to the M2 phenotype caused by the agonists of this receptor. Macrophages also play a key function in a specialized phagocytic process called efferocytosys [53] that contributes to promoting the resolution of inflammation and PPARβ/δ activation enhances this process [53,54]. Since a defective efferocytosys has emerged as a causal factor in the etiopathogenesis of atherosclerosis [55], the increase in this process caused by PPARβ/δ activation might contribute to its beneficial effects in atherosclerosis.

Interleukin (IL)-6 is one of the inflammatory mediators released by adipose tissue that correlates most strongly with obesity and insulin resistance, predicting the development of type 2 diabetes mellitus [56]. PPARβ/δ activation prevents IL-6-induced insulin resistance by inhibiting the signal transducer and activator of transcription 3 (STAT3) pathway in adipocytes, whereas this pathway is over activated in PPARβ/δ-null mice compared to wild-type animals [57].

Skeletal muscle is the primary site of insulin resistance in obesity and type 2 diabetes mellitus since it displays the highest level of insulin-stimulated glucose utilization [5]. Increased plasma levels of saturated NEFAs, caused by the expansion of adipose tissue, promote inflammation and insulin resistance through several mechanisms: the synthesis of FA-derived complex lipids such as diacylglycerol and ceramides; the impairment of the function of cellular organelles (endoplasmic reticulum [ER] stress and mitochondrial dysfunction); and the activation of pro-inflammatory pathways through membrane receptors, such as toll-like receptor 4 (TLR4). PPARβ/δ activation the decrease in insulin sensitivity by suppressing the FA-induced increase in diacylglycerol levels and the subsequent activation of protein kinase C (PKC)θ and NF-κB by enhancing the expression of the genes involved in FA oxidation via PGC-1α and by increasing AMPK phosphorylation [58,59]. Furthermore, PPARβ/δ overexpression in the skeletal muscle of mice has been reported to promote the interaction between PPARβ/δ and AMPK, which enhances glucose uptake, FA oxidation and insulin sensitivity [60]. PPARβ/δ agonists also prevent palmitate-induced ER stress in myotubes through a mechanism involving AMPK activation [61]. Overall, it is believed that PPARβ/δ activation in skeletal muscle produces changes that resemble the effects of exercise training [61], making it a potential candidate for mimicking the effects of exercise to treat metabolic diseases [62].

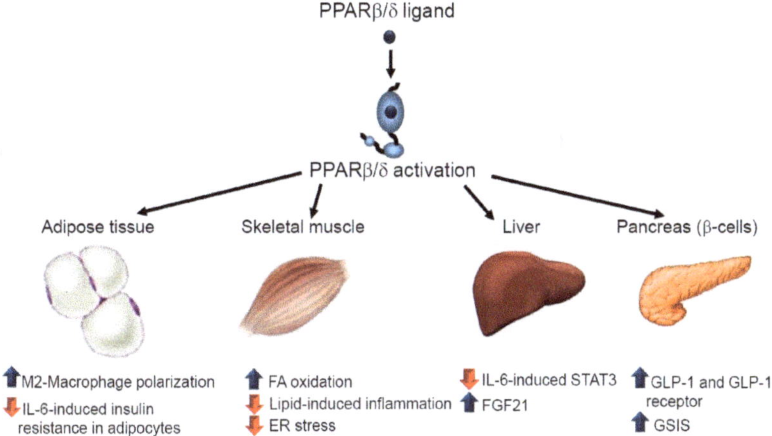

Figure 3. Effects of PPARβ/δ in type 2 diabetes mellitus. This figure depicts the effects of PPARβ/δ ligands in adipose tissue, skeletal muscle, the liver and pancreatic β cells that contribute to the attenuation of type 2 diabetes mellitus. In adipose tissue, PPARβ/δ activation switches macrophage polarization towards the anti-inflammatory M2 phenotype and prevents IL-6-induced insulin resistance by inhibiting STAT3. In skeletal muscle, PPARβ/δ ligands induce FA oxidation, reducing their availability for the synthesis of deleterious complex lipids involved in inflammation and prevent endoplasmic reticulum (ER) stress by activating AMPK. PPARβ/δ activation in hepatocytes blocks the effects of IL-6 by inhibiting the STAT3 pathway through several mechanisms and increasing FGF21 levels. PPARβ/δ activators promote the beneficial effects of GLP-1 in the pancreas and enhance GSIS.ER, endoplasmic reticulum; FA, fatty acid; GLP-1, glucagon-like peptide 1; GSIS, glucose-stimulated insulin secretion; IL-6, interleukin 6; STAT3, signal transducer and activator of transcription 3. Blue arrow: increases. Red arrow: decreases.

In the liver, IL-6 induces insulin resistance by activating STAT3-suppressor of cytokine signalling 3 (SOCS3) pathway [63]. We have previously reported that PPARβ/δ activation prevents the attenuation of the insulin signalling pathway in human liver cells by preventing IL-6-induced STAT3 activation through a mechanism that inhibits ERK1/2 phosphorylation and suppresses the reduction in phospho-AMPK levels [64]. More recently, the inhibitory effect of PPARβ/δ on STAT3 was confirmed, with a new mechanism described involving a T cell protein tyrosine phosphatase 45 (TCPTP45) isoform [65]. According to this recent study, short-term PPARβ/δ activation prevents IL-6-induced insulin resistance as a result of PPARβ/δ forming a complex with nuclear TCPTP45 and retaining it in the nucleus, thereby deactivating the STAT3-SOCS3 signalling [65]. Fibroblast growth factor 21 (FGF21) is a liver-derived circulating hormone that has emerged as an important regulator of glucose and lipid metabolism, making it a promising agent for the treatment of insulin resistance and type 2 diabetes mellitus [66]. Since PPARβ/δ activators increase the plasma levels of FGF21 in humans [67], some of the antidiabetic effects of these drugs might be mediated by the increased levels of this protein.

β-cell failure, a result of the progressive decline in pancreatic β cell function and mass, impairs insulin secretion and contributes to the development of type 2 diabetes mellitus [68]. PPARβ/δ activation in the small intestine potentiates the production of glucagon-like peptide (GLP)-1, which preserves β cell morphology and function, thereby increasing systemic insulin sensitivity [69]. This is consistent with a recent study reporting that intestinal PPARβ/δ protects against diet-induced obesity and insulin resistance [26]. Moreover, PPARβ/δ activation protects pancreatic β cells from palmitate-induced apoptosis by upregulating the expression of the receptor for GLP-1 [70]. PPARβ/δ agonists also increase mitochondrial oxidation in β cells, enhance glucose-stimulated insulin secretion (GSIS) from pancreatic islets and protect GSIS from the adverse effects of prolonged FA exposure [71].

In fact, PPARβ/δ is critical for the expression of the genes involved in mitochondrial function and consequently ATP production, in β cells, which is required for GSIS [72].

3.4. Non-Alcoholic Fatty Liver Disease (NAFLD)

NAFLD encompasses a spectrum of liver disorders ranging from simple steatosis (non-alcoholic fatty liver, NAFL) to non-alcoholic steatohepatitis (NASH) and liver fibrosis. It is closely linked to obesity and metabolic syndrome, predisposing susceptible individuals to cirrhosis, hepatocellular carcinoma and cardiovascular disease [73]. At present, there are no approved pharmacological therapies for NAFLD. In animal models, long-term treatment with PPARβ/δ agonists attenuates hepatic steatosis by enhancing FA oxidation, reducing lipogenesis and improving insulin sensitivity [74–77]. PPARβ/δ activation and overexpression inhibit lipogenesis in hepatocytes by inducing the expression of insulin-induced gene-1 (*INSIG-1*), an ER protein that blocks the activation of sterol regulatory element-binding protein-1 (SREBP-1), a pivotal transcription factor controlling lipogenesis in hepatocytes [77]. However, short treatments with PPARβ/δ agonists might result in a transient increase in hepatic TG levels [78] but without hepatotoxicity, since PPARβ/δ increases the number of monounsaturated FAs but reduces the levels of saturated FAs [79]. In addition, PPARβ/δ might affect hepatic TG levels by regulating the abundance of the VLDL receptor [80]. In humans, PPARβ/δ agonists reduce hepatic fat content and elicit improvements in the plasma markers of liver function [30,33]. These and additional findings point to PPARβ/δ, similarly to PPARα, as a master regulator of hepatic intermediary metabolism. During fasting conditions, hepatic metabolism is programmed to oxidize FA and both, PPARα and PPARβ/δ are thought to promote ketogenesis by inducing FGF21 [67] and the expression of genes involved in fatty acid oxidation [81] in rodents. Interestingly, as mentioned before, PPARβ/δ activation in mice increases the hepatic levels of the hepatic endogenous ligand for PPARα, 16:0/18:1-phosphatidylcholine, leading to amplification of the PGC-1α-PPARα pathway [32], suggesting the presence of a cooperation between both nuclear receptors in the regulation of hepatic metabolism.

The progression from NAFL to NASH involves the development of inflammation and signs of hepatocellular damage [65]. Both the hepatic expression of inflammatory genes [74,76,82] and hepatic ER stress [76], which contribute to the activation of inflammatory pathways, are reduced by PPARβ/δ ligands. This hepatoprotective effect of PPARβ/δ ligands might also involve Kupffer cells, resident liver macrophages that play a critical role in maintaining liver functions. Hematopoietic deficiency of PPARβ/δ selectively impairs the alternative activation of Kupffer cells in obese mice, leading to reduced oxidative metabolism and hepatic dysfunction [83]. Additional studies are required to conclusively determine the effects of PPARβ/δ agonists on liver fibrosis due to the inconsistent results currently reported in the literature [17,84].

4. Safety of PPARβ/δ Agonists

Despite the promising data of PPARβ/δ agonists in metabolic disorders in preclinical studies, the discovery that mice treated with GW501516 developed adenocarcinoma [85] halted further development of this drug and undermined the potential use of these drugs in human therapeutics. However, subsequent attempts to assess the role of PPARβ/δ in cancer have demonstrated that this receptor both inhibits and promotes tumorigenesis, as it has been extensively reviewed previously [10,86,87], becoming one of the most controversial effects of PPARβ/δ. The conflicting results about PPARβ/δ in cancer might indicate that the activity of this receptor in cancer development is influenced by the mutational status of the tumour cell and the tumour environment [10]. Moreover, it has been proposed that the high-level expression of PPARβ/δ in normal cells suggest an antitumour effect for this receptor but the reduction in its expression or the presence of endogenous antagonists or inverse agonists might lead to a protumorigenic role for PPARβ/δ [86].

Although there are only a few clinical trials assessing the safety of PPARβ/δ agonists, the administration of these drugs to humans seems to be safe and generally well-tolerated, at least for the short periods evaluated [30]. Thus, no study subjects were withdrawn because of adverse effects from a study evaluating the administration of seladelpar for 8 weeks to dyslipidaemic overweight patients at doses showing beneficial metabolic effects [30]. Treatment with this PPARβ/δ agonist slightly but significantly decreased red blood cell count, haemoglobin and haematocrit. In a more recent study, the effect of seladelpar was assessed in patients with primary biliary cholangitis [88]. Drug treatment for 12 wk. elicited anti-cholestatic effect but three patients showed rapidly reversible alanine aminotransferase elevations and the study was interrupted before completion. The authors of the study suggest that this effect might be specific for these patients with primary biliary cholangitis due to the biliary excretion of seladelpar ant its metabolites, leading to increased hepatic drug concentrations.

5. Conclusions

Over the past 20 years, a substantial body of preclinical evidence has demonstrated that PPARβ/δ activation is a promising therapeutic strategy for treating obesity-associated co-morbidities. This has led to the assessment of PPARβ/δ agonists in clinical trials, although these studies on the efficacy and safety of PPARβ/δ agonists in humans are scarce. Safety issues have been raised regarding the role of PPARβ/δ ligands in carcinogenesis [87]. However, there are conflicting findings on the role of PPARβ/δ as a tumour suppressor or tumour promoter [87], the latter being mostly observed in animal models. Further studies are required to obtain conclusive data on the role of PPARβ/δ in human cancer given that although mouse models are invaluable tools for investigating basic tumour biology, they show significant limitations when compared to human beings. For instance, PPARs are expressed at lower levels in human than in rodent cells and gene expression is also regulated differently by PPARs in human and rodent cells [9]. These differences might explain why long-term treatment with PPARα ligands have been shown to induce carcinogenesis in rodents but carcinogenesis has not been observed in humans treated with the PPARα agonists fibrates for dyslipidaemia over the decades [89]. However, this needs to be studied for PPARβ/δ too.

In summary, although further studies are required to confirm the safety of PPARβ/δ ligands, these drugs have demonstrated that modulation of PPARβ/δ activity shows efficacy in preclinical studies and in a few clinical trials in the treatment of dyslipidaemia, type 2 diabetes mellitus and NAFLD.

Acknowledgments: The authors were funded by grants from the Ministerio de Economía, Industria y Competitividad of the Spanish Government (SAF2015-64146-R), la Marató de TV3 and from CIBER de Diabetes y Enfermedades Metabólicas Asociadas (CIBERDEM). CIBERDEM is an initiative of the Instituto de Salud Carlos III (ISCIII)—Ministerio de Economía, Industria y Competitividad. We would like to thank the Language Services at the University of Barcelona for revising the manuscript.

Conflicts of Interest: The authors declare no conflict of interest.

References

1. Birsoy, K.; Festuccia, W.T.; Laplante, M. A comparative perspective on lipid storage in animals. *J. Cell Sci.* **2013**, *126*, 1541–1552. [CrossRef] [PubMed]
2. Swinburn, B.A.; Sacks, G.; Hall, K.D.; McPherson, K.; Finegood, D.T.; Moodie, M.L.; Gortmaker, S.L. The global obesity pandemic: Shaped by global drivers and local environments. *Lancet* **2011**, *378*, 804–814. [CrossRef]
3. Wyatt, S.B.; Winters, K.P.; Dubbert, P.M. Overweight and obesity: Prevalence, consequences, and causes of a growing public health problem. *Am. J. Med. Sci.* **2006**, *331*, 166–174. [CrossRef] [PubMed]
4. Stefan, N.; Häring, H.U.; Hu, F.B.; Schulze, M.B. Metabollically healthy obesity: Epidemiology, mechanisms, and clinical implications. *Lancet Diabetes Endocrinol.* **2013**, *1*, 152–162. [CrossRef]
5. Gustafson, B.; Hedjazifar, S.; Gogg, S.; Hammarstedt, A.; Smith, U. Insulin resistance and impaired adipogenesis. *Trends Endocrinol. Metab.* **2015**, *26*, 193–200. [CrossRef] [PubMed]
6. Després, J.P.; Lemieux, I. Abdominal obesity and metabolic syndrome. *Nature* **2006**, *444*, 881–887. [CrossRef] [PubMed]
7. Virtue, S.; Vidal-Puig, A. It's not how fat you are, it's what you do with it that counts. *PLoS Biol.* **2008**, *6*, e237. [CrossRef] [PubMed]

8. Giordiano Attianese, G.M.P.; Desvergne, B. Integrative and systemic approaches for evaluating PPARβ/δ (PPARD) function. *Nucl. Recept. Signal.* **2015**, *13*, e001.
9. Neels, J.G.; Grimaldi, P.A. Physiological functions of peroxisome proliferator-activated receptor β. *Physiol. Rev.* **2014**, *94*, 795–858. [CrossRef] [PubMed]
10. Tan, N.S.; Vázquez-Carrera, M.; Montagner, A.; Sng, M.K.; Guillou, H.; Wahli, W. Transcriptional control of physiological and pathological processes by the nuclear receptor PPARβ/δ. *Prog. Lipid Res.* **2016**, *64*, 98–122. [CrossRef] [PubMed]
11. Vázquez-Carrera, M. Unraveling the Effects of PPARβ/δ on Insulin Resistance and Cardiovascular Disease. *Trends Endocrinol. Metab.* **2016**, *27*, 319–334. [CrossRef] [PubMed]
12. Bishop-Bailey, D.; Bystrom, J. Emerging roles of peroxisome proliferator-activated receptor-β/δ in inflammation. *Pharmacol. Ther.* **2009**, *124*, 141–150. [CrossRef] [PubMed]
13. Schug, T.T.; Berry, D.C.; Shaw, N.S.; Travis, S.N.; Noy, N. Opposing effects of retinoic acid on cell growth result from alternate activation of two different nuclear receptors. *Cell* **2007**, *129*, 723–733. [CrossRef] [PubMed]
14. Borland, M.G.; Foreman, J.E.; Girroir, E.E.; Zolfaghari, R.; Sharma, A.K.; Amin, S.; Gonzalez, F.J.; Ross, A.C.; Peters, J.M. Ligand activation of peroxisome proliferator-activated receptor-β/δ inhibits cell proliferation in human HaCaT keratinocytes. *Mol. Pharmacol.* **2008**, *74*, 1429–1442. [CrossRef] [PubMed]
15. Rieck, M.; Meissner, W.; Ries, S.; Müller-Brüsselbach, S.; Müller, R. Ligand-mediated regulation of peroxisome proliferator-activated receptor (PPAR)β/δ: A comparative analysis of PPAR-selective agonists and all-trans retinoic acid. *Mol. Pharmacol.* **2008**, *74*, 1269–1277. [CrossRef] [PubMed]
16. Billin, A.N. PPAR-β/δ agonists for type 2 diabetes and dyslipidemia: An adopted orphan still looking for a home. *Expert Opin. Investig. Drugs* **2008**, *17*, 1465–1471. [CrossRef] [PubMed]
17. Iwaisako, K.; Haimerl, M.; Paik, Y.H.; Taura, K.; Kodama, Y.; Sirlin, C.; Yu, E.; Yu, R.T.; Downes, M.; Evans, R.M.; et al. Protection from liver fibrosis by a peroxisome proliferator-activated receptor δ agonist. *Proc. Natl. Acad. Sci. USA* **2012**, *109*, 1369–1376. [CrossRef] [PubMed]
18. Barak, Y.; Liao, D.; He, W.; Ong, E.S.; Nelson, M.C.; Olefsky, J.M.; Boland, R.; Evans, R.M. Effects of peroxisome proliferator-activated receptor δ on placentation, adiposity, and colorectal cancer. *Proc. Natl. Acad. Sci. USA* **2002**, *99*, 303–308. [CrossRef] [PubMed]
19. Wang, Y.X.; Lee, C.H.; Tiep, S.; Yu, R.T.; Ham, J.; Kang, H.; Evans, R.M. Peroxisome-proliferator-activated receptor δ activates fat metabolism to prevent obesity. *Cell* **2003**, *113*, 159–170. [CrossRef]
20. Tanaka, T.; Yamamoto, J.; Iwasaki, S.; Asaba, H.; Hamura, H.; Ikeda, Y.; Watanabe, M.; Magoori, K.; Ioka, R.X.; Tachibana, K.; et al. Activation of peroxisome proliferator-activated receptor δ induces fatty acid β-oxidation in skeletal muscle and attenuates metabolic syndrome. *Proc. Natl. Acad. Sci. USA* **2003**, *100*, 15924–15929. [CrossRef] [PubMed]
21. Wang, Y.X.; Zhang, C.L.; Yu, R.T.; Cho, H.K.; Nelson, M.C.; Bayuga-Ocampo, C.R.; Ham, J.; Kang, H.; Evans, R.M. Regulation of muscle fiber type and running endurance by PPARδ. *PLoS Biol.* **2004**, *2*, e294. [CrossRef] [PubMed]
22. Doktorova, M.; Zwarts, I.; Zutphen, T.V.; Dijk, T.H.; Bloks, V.W.; Harkema, L.; Bruin, A.; Downes, M.; Evans, R.M.; Verkade, H.J.; et al. Intestinal PPARδ protects against diet-induced obesity, insulin resistance and dyslipidemia. *Sci. Rep.* **2017**, *12*, 846. [CrossRef] [PubMed]
23. Bastie, C.; Holst, D.; Gaillard, D.; Jehl-Pietri, C.; Grimaldi, P.A. Expression of peroxisome proliferator-activated receptor PPARδ promotes induction of PPARgamma and adipocyte differentiation in 3T3C2 fibroblasts. *J. Biol. Chem.* **1999**, *274*, 21920–21925. [CrossRef] [PubMed]
24. Bastie, C.; Luquet, S.; Holst, D.; Jehl-Pietri, C.; Grimaldi, P.A. Alterations of peroxisome proliferator-activated receptor δ activity affect fatty acid-controlled adipose differentiation. *J. Biol. Chem.* **2000**, *275*, 38768–38773. [CrossRef] [PubMed]
25. Hansen, J.B.; Zhang, H.; Rasmussen, T.H.; Petersen, R.K.; Flindt, E.N.; Kristiansen, K. Peroxisome proliferator-activated receptor δ (PPARδ)-mediated regulation of preadipocyte proliferation and gene expression is dependent on cAMP signaling. *J. Biol. Chem.* **2001**, *276*, 3175–3182. [CrossRef] [PubMed]
26. Sodhi, K.; Puri, N.; Kim, D.H.; Hinds, T.D.; Stechschulte, L.A.; Favero, G.; Rodella, L.; Shapiro, J.I.; Jude, D.; Abraham, N.G. PPARδ binding to heme oxygenase 1 promoter prevents angiotensin II-induced adipocyte dysfunction in Goldblatt hypertensive rats. *Int. J. Obes.* **2014**, *38*, 456–465. [CrossRef] [PubMed]

27. Bortolotto, J.W.; Margis, R.; Ferreira, A.C.; Padoin, A.V.; Mottin, C.C.; Guaragna, R.M. Adipose tissue distribution and quantification of PPARβ/δ and PPARgamma1-3 mRNAs: Discordant gene expression in subcutaneous, retroperitoneal and visceral adipose tissue of morbidly obese patients. *Obes. Surg.* **2007**, *17*, 934–940. [CrossRef] [PubMed]
28. Rodriguez-Calvo, R.; Serrano, L.; Coll, T.; Moullan, N.; Sánchez, R.M.; Merlos, M.; Palomer, X.; Laguna, J.C.; Michalik, L.; Wahli, W.; et al. Activation of peroxisome proliferator- activated receptor β/δ inhibits lipopolysaccharide-induced cytokine production in adipocytes by lowering nuclear factor-κB activity via extracellular signal-related kinase 1/2. *Diabetes* **2008**, *57*, 2149–2157. [CrossRef] [PubMed]
29. Kang, K.; Reilly, S.M.; Karabacak, V.; Gangl, M.R.; Fitzgerald, K.; Hatano, B.; Lee, C.H. Adipocyte-derived Th2 cytokines and myeloid PPARδ regulate macrophage polarization and insulin sensitivity. *Cell Metab.* **2008**, *7*, 485–895. [CrossRef] [PubMed]
30. Bays, H.E.; Schwartz, S.; Littlejohn, T., 3rd; Kerzner, B.; Krauss, R.M.; Karpf, D.B.; Choi, Y.J.; Wang, X.; Naim, S.; Roberts, B.K. MBX-8025, a novel peroxisome proliferator receptor-δ agonist: Lipid and other metabolic effects in dyslipidemic overweight patients treated with and without atorvastatin. *J. Clin. Endocrinol. Metab.* **2011**, *96*, 2889–2897. [CrossRef] [PubMed]
31. Taskinen, M.R.; Borén, J. New insights into the pathophysiology of dyslipidemia in type 2 diabetes. *Atherosclerosis* **2015**, *239*, 483–495. [CrossRef] [PubMed]
32. Barroso, E.; Rodríguez-Calvo, R.; Serrano-Marco, L.; Astudillo, A.M.; Balsinde, J.; Palomer, X.; Vázquez-Carrera, M. The PPARβ/δ activator GW501516 prevents the down-regulation of AMPK caused by a high-fat diet in liver and amplifies the PGC-1/–lipin 1–PPAR/pathway leading to increased fatty acid oxidation. *Endocrinology* **2011**, *152*, 1848–1859. [CrossRef] [PubMed]
33. Risérus, U.; Sprecher, D.; Johnson, T.; Olson, E.; Hirschberg, S.; Liu, A.; Fang, Z.; Hegde, P.; Richards, D.; Sarov-Blat, L.; et al. Activation of peroxisome proliferator-activated receptor (PPAR)δ promotes reversal of multiple metabolic abnormalities, reduces oxidative stress, and increases fatty acid oxidation in moderately obese men. *Diabetes* **2008**, *57*, 332–339. [CrossRef] [PubMed]
34. Sanderson, L.M.; Boekschoten, M.V.; Desvergne, B.; Müller, M.; Kersten, S. Transcriptional profiling reveals divergent roles of PPARalpha and PPARβ/δ in regulation of gene expression in mouse liver. *Physiol. Genomics* **2010**, *41*, 42–52. [CrossRef] [PubMed]
35. Leibowitz, M.D.; Fiévet, C.; Hennuyer, N.; Peinado-Onsurbe, J.; Duez, H.; Bergera, J.; Cullinan, C.A.; Sparrow, C.P.; Baffic, J.; Berger, G.D.; et al. Activation of PPARδ alters lipid metabolism in db/db mice. *FEBS Lett.* **2000**, *473*, 333–336. [CrossRef]
36. Oliver, W.R., Jr.; Shenk, J.L.; Snaith, M.R.; Russell, C.S.; Plunket, K.D.; Bodkin, N.L.; Lewis, M.C.; Winegar, D.A.; Sznaidman, M.L.; Lambert, M.H.; et al. A selective peroxisome proliferator-activated receptor δ agonist promotes reverse cholesterol transport. *Proc. Natl. Acad. Sci. USA* **2001**, *98*, 5306–5311. [CrossRef] [PubMed]
37. Sprecher, D.L.; Massien, C.; Pearce, G.; Billin, A.N.; Perlstein, I.; Willson, T.M.; Hassall, D.G.; Ancellin, N.; Patterson, S.D.; Lobe, D.C.; et al. Triglyceride:high-density lipoprotein cholesterol effects in healthy subjects administered a peroxisome proliferator activated receptor δ agonist. *Arterioscler. Thromb. Vasc. Biol.* **2007**, *27*, 359–365. [CrossRef] [PubMed]
38. Chehaibi, K.; Cedó, L.; Metso, J.; Palomer, X.; Santos, D.; Quesada, H.; Naceur Slimane, M.; Wahli, W.; Julve, J.; Vázquez-Carrera, M.; et al. PPAR-β/δ activation promotes phospholipid transfer protein expression. *Biochem. Pharmacol.* **2015**, *94*, 101–108. [CrossRef] [PubMed]
39. Van der Veen, J.N.; Kruit, J.K.; Havinga, R.; Baller, J.F.; Chimini, G.; Lestavel, S.; Staels, B.; Groot, P.H.; Groen, A.K.; Kuipers, F. Reduced cholesterol absorption upon PPARδ activation coincides with decreased intestinal expression of NPC1L1. *J. Lipid Res.* **2005**, *46*, 526–534. [CrossRef] [PubMed]
40. Vrins, C.L.; van der Velde, A.E.; van den Oever, K.; Levels, J.H.; Huet, S.; Oude Elferink, R.P.; Kuipers, F.; Groen, A.K. Peroxisome proliferator-activated receptor δ activation leads to increased transintestinal cholesterol efflux. *J. Lipid Res.* **2009**, *50*, 2046–2054. [CrossRef] [PubMed]
41. Ehrenborg, E.; Skogsberg, J. Peroxisome proliferator-activated receptor δ and cardiovascular disease. *Atherosclerosis* **2013**, *231*, 95–106. [CrossRef] [PubMed]
42. Olson, E.J.; Pearce, G.L.; Jones, N.P.; Sprecher, D.L. Lipid effects of peroxisome proliferator-activated receptor-δ agonist GW501516 in subjects with low high-density lipoprotein cholesterol: Characteristics of metabolic syndrome. *Arterioscler. Thromb. Vasc. Biol.* **2012**, *32*, 2289–2294. [CrossRef] [PubMed]

43. Choi, Y.J.; Roberts, B.K.; Wang, X.; Geaney, J.C.; Naim, S.; Wojnoonski, K.; Karpf, D.B.; Krauss, R.M. Effects of the PPAR-δ agonist MBX-8025 on atherogenic dyslipidemia. *Atherosclerosis* **2012**, *220*, 470–476. [CrossRef] [PubMed]
44. Samuel, V.T.; Shulman, G.I. The pathogenesis of insulin resistance: Integrating signaling pathways and substrate flux. *J. Clin. Investig.* **2016**, *126*, 12–22. [CrossRef] [PubMed]
45. Schenk, S.; Saberi, M.; Olefsky, J.M. Insulin sensitivity: Modulation by nutrients and inflammation. *J. Clin. Investig.* **2008**, *118*, 2992–3002. [CrossRef] [PubMed]
46. Lee, C.H.; Olson, P.; Hevener, A.; Mehl, I.; Chong, L.W.; Olefsky, J.M.; Gonzalez, F.J.; Ham, J.; Kang, H.; Peters, J.M.; et al. PPARδ regulates glucose metabolism and insulin sensitivity. *Proc. Natl. Acad. Sci. USA* **2006**, *103*, 3444–3449. [CrossRef] [PubMed]
47. Glass, C.K.; Olefsky, J.M. Inflammation and lipid signaling in the etiology of insulin resistance. *Cell Metab.* **2012**, *15*, 635–645. [CrossRef] [PubMed]
48. Jha, A.K.; Huang, S.C.; Sergushichev, A.; Lampropoulou, V.; Ivanova, Y.; Loginicheva, E.; Chmielewski, K.; Stewart, K.M.; Ashall, J.; Everts, B.; et al. Network integration of parallel metabolic and transcriptional data reveals metabolic modules that regulate macrophage polarization. *Immunity* **2015**, *42*, 419–430. [CrossRef] [PubMed]
49. O'Neill, L.A.; Pearce, E.J. Immunometabolism governs dendritic cell and macrophage function. *J. Exp. Med.* **2016**, *213*, 15–23. [CrossRef] [PubMed]
50. Rodríguez-Prados, J.C.; Través, P.G.; Cuenca, J.; Rico, D.; Aragonés, J.; Martín-Sanz, P.; Cascante, M.; Boscá, L. Substrate fate in activated macrophages: A comparison between innate, classic, and alternative activation. *J. Immunol.* **2010**, *185*, 605–614. [CrossRef] [PubMed]
51. Vats, D.; Mukundan, L.; Odegaard, J.I.; Zhang, L.; Smith, K.L.; Morel, C.R.; Wagner, R.A.; Greaves, D.R.; Murray, P.J.; Chawla, A. Oxidative metabolism and PGC-1β attenuate macrophage-mediated inflammation. *Cell Metab.* **2006**, *4*, 13–24. [CrossRef] [PubMed]
52. Bojic, L.A.; Sawyez, C.G.; Telford, D.E.; Edwards, J.Y.; Hegele, R.A.; Huff, M.W. Activation of peroxisome proliferator-activated receptor δ inhibits human macrophage foam cell formation and the inflammatory response induced by very low-density lipoprotein. *Arterioscler. Thromb. Vasc. Biol.* **2012**, *32*, 2919–2928. [CrossRef] [PubMed]
53. Korns, D.; Frasch, S.C.; Fernandez-Boyanapalli, R.; Henson, P.M.; Bratton, D.L. Modulation of Macrophage Efferocytosis in Inflammation. *Front. Immunol.* **2011**, *2*, 57. [CrossRef] [PubMed]
54. Mukundan, L.; Odegaard, J.I.; Morel, C.R.; Heredia, J.E.; Mwangi, J.W.; Ricardo-Gonzalez, R.R.; Goh, Y.P.; Eagle, A.R.; Dunn, S.E.; Awakuni, J.U. PPAR-δ senses and orchestrates clearance of apoptotic cells to promote tolerance. *Nat. Med.* **2009**, *15*, 1266–1272. [CrossRef] [PubMed]
55. Tajbakhsh, A.; Rezaee, M.; Kovanen, P.T.; Sahebkar, A. Efferocytosis in atherosclerotic lesions: Malfunctioning regulatory pathways and control mechanisms. *Pharmacol. Ther.* **2018**. [CrossRef] [PubMed]
56. Qu, D.; Liu, J.; Lau, C.W.; Huang, Y. IL-6 in diabetes and cardiovascular complications. *Br. J. Pharmacol.* **2014**, *171*, 3595–3603. [CrossRef] [PubMed]
57. Serrano-Marco, L.; Rodríguez-Calvo, R.; El Kochairi, I.; Palomer, X.; Michalik, L.; Wahli, W.; Vázquez-Carrera, M. Activation of peroxisome proliferator-activated receptor-β/-δ (PPAR-β/-δ) ameliorates insulin signaling and reduces SOCS3 levels by inhibiting STAT3 in interleukin-6-stimulated adipocytes. *Diabetes* **2011**, *60*, 1990–1999. [CrossRef] [PubMed]
58. Coll, T.; Alvarez-Guardia, D.; Barroso, E.; Gómez-Foix, A.M.; Palomer, X.; Laguna, J.C.; Vázquez-Carrera, M. Activation of peroxisome proliferator-activated receptor-δ by GW501516 prevents fatty acid-induced nuclear factor-kB activation and insulin resistance in skeletal muscle cells. *Endocrinology* **2010**, *151*, 1560–1569. [CrossRef] [PubMed]
59. Kleiner, S.; Nguyen-Tran, V.; Baré, O.; Huang, X.; Spiegelman, B.; Wu, Z. PPARδ agonism activates fatty acid oxidation via PGC-1α but does not increase mitochondrial gene expression and function. *J. Biol. Chem.* **2009**, *284*, 18624–18633. [CrossRef] [PubMed]
60. Gan, Z.; Burkart-Hartman, E.M.; Han, D.H.; Finck, B.; Leone, T.C.; Smith, E.Y.; Ayala, J.E.; Holloszy, J.; Kelly, D.P. The nuclear receptor PPARβ/δ programs muscle glucose metabolism in cooperation with AMPK and MEF2. *Genes Dev.* **2011**, *25*, 2619–2630. [CrossRef] [PubMed]

61. Salvadó, L.; Barroso, E.; Gómez-Foix, A.M.; Palomer, X.; Michalik, L.; Wahli, W.; Vázquez-Carrera, M. PPARβ/δ prevents endoplasmic reticulum stress-associated inflammation and insulin resistance in skeletal muscle cells through an AMPK-dependent mechanism. *Diabetologia* **2014**, *57*, 2126–2135. [CrossRef] [PubMed]
62. Fan, W.; Waizenegger, W.; Lin, C.S.; Sorrentino, V.; He, M.X.; Wall, C.E.; Li, H.; Liddle, C.; Yu, R.T.; Atkins, A.R.; et al. PPARδ Promotes Running Endurance by Preserving Glucose. *Cell Metab.* **2017**, *25*, 1186–1193. [CrossRef] [PubMed]
63. Senn, J.J.; Klover, P.J.; Nowak, I.A.; Zimmers, T.A.; Koniaris, L.G.; Furlanetto, R.W.; Mooney, R.A. Suppressor of cytokine signaling-3 (SOCS-3), a potential mediator of interleukin-6-dependent insulin resistance in hepatocytes. *J. Biol. Chem.* **2003**, *278*, 13740–13746. [CrossRef] [PubMed]
64. Serrano-Marco, L.; Barroso, E.; El Kochairi, I.; Palomer, X.; Michalik, L.; Wahli, W.; Vázquez-Carrera, M. The peroxisome proliferator-activated receptor (PPAR) β/δ agonist GW501516 inhibits IL-6-induced signal transducer and activator of transcription 3 (STAT3) activation and insulin resistance in human liver cells. *Diabetologia* **2012**, *55*, 743–751. [CrossRef] [PubMed]
65. Yoo, T.; Ham, S.A.; Lee, W.J.; Hwang, S.I.; Park, J.A.; Hwang, J.S.; Hur, J.; Shin, H.C.; Han, S.G.; Lee, C.H.; et al. Ligand-dependent Interaction of PPARδ with T Cell Protein Tyrosine Phosphatase 45 Enhances Insulin Signaling. *Diabetes* **2017**, *67*, 360–371. [CrossRef] [PubMed]
66. Markan, K.R.; Naber, M.C.; Ameka, M.K.; Anderegg, M.D.; Mangelsdorf, D.J.; Kliewer, S.A.; Mohammadi, M.; Potthoff, M.J. Circulating FGF21 is liver derived and enhances glucose uptake during refeeding and overfeeding. *Diabetes* **2014**, *63*, 4057–4063. [CrossRef] [PubMed]
67. Christodoulides, C.; Dyson, P.; Sprecher, D.; Tsintzas, K.; Karpe, F. Circulating fibroblast growth factor 21 is induced by peroxisome proliferator-activated receptor agonists but not ketosis in man. *J. Clin. Endocrinol. Metab.* **2009**, *94*, 3594–3601. [CrossRef] [PubMed]
68. Vetere, A.; Choudhary, A.; Burns, S.M.; Wagner, B.K. Targeting the pancreatic β-cell to treat diabetes. *Nat. Rev. Drug Discov.* **2014**, *13*, 278–289. [CrossRef] [PubMed]
69. Daoudi, M.; Hennuyer, N.; Borland, M.G.; Touche, V.; Duhem, C.; Gross, B.; Caiazzo, R.; Kerr-Conte, J.; Pattou, F.; Peters, J.M.; et al. PPARβ/δ activation induces enteroendocrine L cell GLP-1 production. *Gastroenterology* **2011**, *140*, 1564–1574. [CrossRef] [PubMed]
70. Yang, Y.; Tong, Y.; Gong, M.; Lu, Y.; Wang, C.; Zhou, M.; Yang, Q.; Mao, T.; Tong, N. Activation of PPARβ/δ protects pancreatic β cells from palmitate-induced apoptosis by upregulating the expression of GLP-1 receptor. *Cell Signal.* **2014**, *26*, 268–278. [CrossRef] [PubMed]
71. Ravnskjaer, K.; Frigerio, F.; Boergesen, M.; Nielsen, T.; Maechler, P.; Mandrup, S. PPARδ is a fatty acid sensor that enhances mitochondrial oxidation in insulin-secreting cells and protects against fatty acid-induced dysfunction. *J. Lipid Res.* **2010**, *51*, 1370–1379. [CrossRef] [PubMed]
72. Tang, T.; Abbott, M.J.; Ahmadian, M.; Lopes, A.B.; Wang, Y.; Sul, H.S. Desnutrin/ATGL activates PPARδ to promote mitochondrial function for insulin secretion in islet β cells. *Cell Metab.* **2013**, *18*, 883–895. [CrossRef] [PubMed]
73. Haas, J.T.; Francque, S.; Staels, B. Pathophysiology and Mechanisms of Nonalcoholic Fatty Liver Disease. *Annu. Rev. Physiol.* **2016**, *78*, 181–205. [CrossRef] [PubMed]
74. Nagasawa, T.; Inada, Y.; Nakano, S.; Tamura, T.; Takahashi, T.; Maruyama, K.; Yamazaki, Y.; Kuroda, J.; Shibata, N. Effects of bezafibrate, PPAR pan-agonist, and GW501516, PPARδ agonist, on development of steatohepatitis in mice fed a methionine- and choline-deficient diet. *Eur. J. Pharmacol.* **2006**, *536*, 182–191. [CrossRef] [PubMed]
75. Wu, H.T.; Chen, C.T.; Cheng, K.C.; Li, Y.X.; Yeh, C.H.; Cheng, J.T. Pharmacological activation of peroxisome proliferator-activated receptor δ improves insulin resistance and hepatic steatosis in high fat diet-induced diabetic mice. *Horm. Metab. Res.* **2011**, *43*, 631–635. [CrossRef] [PubMed]
76. Bojic, L.A.; Telford, D.E.; Fullerton, M.D.; Ford, R.J.; Sutherland, B.G.; Edwards, J.Y.; Sawyez, C.G.; Gros, R.; Kemp, B.E.; Steinberg, G.R.; et al. PPARδ activation attenuates hepatic steatosis in *Ldlr*$^{-/-}$ mice by enhanced fat oxidation, reduced lipogenesis, and improved insulin sensitivity. *J. Lipid Res.* **2014**, *55*, 1254–1266. [CrossRef] [PubMed]
77. Qin, X.; Xie, X.; Fan, Y.; Tian, J.; Guan, Y.; Wang, X.; Zhu, Y.; Wang, N. Peroxisome proliferator-activated receptor-δ induces insulin-induced gene-1 and suppresses hepatic lipogenesis in obese diabetic mice. *Hepatology* **2008**, *48*, 432–441. [CrossRef] [PubMed]

78. Garbacz, W.G.; Huang, J.T.; Higgins, L.G.; Wahli, W.; Palmer, C.N. PPARα Is Required for PPARδ Action in Regulation of Body Weight and Hepatic Steatosis in Mice. *PPAR Res.* **2015**, *2015*, 927057. [CrossRef] [PubMed]
79. Liu, S.; Hatano, B.; Zhao, M.; Yen, C.C.; Kang, K.; Reilly, S.M.; Gangl, M.R.; Gorgun, C.; Balschi, J.A.; Ntambi, J.M.; et al. Role of peroxisome proliferator-activated receptor δ/β in hepatic metabolic regulation. *J. Biol. Chem.* **2011**, *286*, 1237–1247. [CrossRef] [PubMed]
80. Zarei, M.; Barroso, E.; Palomer, X.; Dai, J.; Rada, P.; Quesada-López, T.; Escolà-Gil, J.C.; Cedó, L.; Zali, M.R.; Molaei, M.; et al. Hepatic regulation of VLDL receptor by PPARβ/δ and FGF21 modulates non-alcoholic fatty liver disease. *Mol. Metab.* **2017**. [CrossRef] [PubMed]
81. Nakamura, M.T.; Yudell, B.E.; Loor, J.J. Regulation of energy metabolism by long-chain fatty acids. *Prog. Lipid Res.* **2014**, *53*, 124–144. [CrossRef] [PubMed]
82. Lee, M.Y.; Choi, R.; Kim, H.M.; Cho, E.J.; Kim, B.H.; Choi, Y.S.; Naowaboot, J.; Lee, E.Y.; Yang, Y.C.; Shin, J.Y.; et al. Peroxisome proliferator-activated receptor δ agonist attenuates hepatic steatosis by anti-inflammatory mechanism. *Exp. Mol. Med.* **2012**, *44*, 578–585. [CrossRef] [PubMed]
83. Odegaard, J.I.; Ricardo-Gonzalez, R.R.; Red Eagle, A.; Vats, D.; Morel, C.R.; Goforth, M.H.; Subramanian, V.; Mukundan, L.; Ferrante, A.W.; Chawla, A. Alternative M2 activation of Kupffer cells by PPARδ ameliorates obesity-induced insulin resistance. *Cell Metab.* **2008**, *7*, 496–507. [CrossRef] [PubMed]
84. Kostadinova, R.; Montagner, A.; Gouranton, E.; Fleury, S.; Guillou, H.; Dombrowicz, D.; Desreumaux, P.; Wahli, W. GW501516-activated PPARβ/δ promotes liver fibrosis via p38-JNK MAPK-induced hepatic stellate cell proliferation. *Cell Biosci.* **2012**, *2*, 34. [CrossRef] [PubMed]
85. Gupta, R.A.; Wang, D.; Katkuri, S.; Wang, H.; Dey, S.K.; DuBois, R.N. Activation of nuclear hormone receptor peroxisome proliferator-activated receptor-δ accelerates intestinal adenoma growth. *Nat. Med.* **2004**, *10*, 245–247. [CrossRef] [PubMed]
86. Peters, J.M.; Gonzalez, F.J.; Müller, R. Establishing the Role of PPARβ/δ in Carcinogenesis. *Trends Endocrinol. Metab.* **2015**, *26*, 595–607. [CrossRef] [PubMed]
87. Müller, R. PPARβ/δ in human cancer. *Biochimie* **2017**, *136*, 90–99. [CrossRef] [PubMed]
88. Jones, D.; Boudes, P.F.; Swain, M.G.; Bowlus, C.L.; Galambos, M.R.; Bacon, B.R.; Doerffel, Y.; Gitlin, N.; Gordon, S.C.; Odin, J.A.; Sheridan, D.; et al. Seladelpar (MBX-8025), a selective PPAR-δ agonist, in patients with primary biliary cholangitis with an inadequate response to ursodeoxycholic acid: A double-blind, randomised, placebo-controlled, phase 2, proof-of-concept study. *Lancet Gastroenterol. Hepatol.* **2017**, *2*, 716–726. [CrossRef]
89. Youssef, J.; Badr, M. Peroxisome proliferator-activated receptors and cancer: Challenges and opportunities. *Br. J. Pharmacol.* **2011**, *164*, 68–82.

Review

The CD36-PPARγ Pathway in Metabolic Disorders

Loïze Maréchal [1,2], Maximilien Laviolette [1,3], Amélie Rodrigue-Way [1,3], Baly Sow [1,3], Michèle Brochu [2], Véronique Caron [1] and André Tremblay [1,3,4,5,*]

1 Research Center, CHU Sainte-Justine, Montréal, QC H3T 1C5, Canada; loize.marechal@umontreal.ca (L.M.); maximilien.laviolette-brassard@umontreal.ca (M.L.); amelierway@aol.com (A.R.-W.); baly.sow@umontreal.ca (B.S.); veronique.caron.4@gmail.com (V.C.)

2 Department of Physiology, Faculty of Medicine, University of Montreal, Montréal, QC H3T 1J4, Canada; michele.brochu@umontreal.ca

3 Department of Biochemistry and Molecular Medicine, Faculty of Medicine, University of Montreal, Montréal, QC H3T 1J4, Canada

4 Centre de Recherche en Reproduction et Fertilité, University of Montreal, Saint Hyacinthe, QC J2S 7C6, Canada

5 Department of Obstetrics & Gynecology, Faculty of Medicine, University of Montreal, Montréal, QC H3T 1C5, Canada

* Correspondence: andre.tremblay.1@umontreal.ca; Tel.: +1-514-345-4931

Received: 8 April 2018; Accepted: 16 May 2018; Published: 21 May 2018

Abstract: Uncovering the biological role of nuclear receptor peroxisome proliferator-activated receptors (PPARs) has greatly advanced our knowledge of the transcriptional control of glucose and energy metabolism. As such, pharmacological activation of PPARγ has emerged as an efficient approach for treating metabolic disorders with the current use of thiazolidinediones to improve insulin resistance in diabetic patients. The recent identification of growth hormone releasing peptides (GHRP) as potent inducers of PPARγ through activation of the scavenger receptor CD36 has defined a novel alternative to regulate essential aspects of lipid and energy metabolism. Recent advances on the emerging role of CD36 and GHRP hexarelin in regulating PPARγ downstream actions with benefits on atherosclerosis, hepatic cholesterol biosynthesis and fat mitochondrial biogenesis are summarized here. The response of PPARγ coactivator PGC-1 is also discussed in these effects. The identification of the GHRP-CD36-PPARγ pathway in controlling various tissue metabolic functions provides an interesting option for metabolic disorders.

Keywords: scavenger receptor; PPAR nuclear receptors; PGC-1; fatty acid oxidation; energy metabolism; GHRP; hexarelin; atherosclerosis; insulin resistance

---

1. Introduction

In years to come, metabolic defects are predicted to remain one of the principal causes of death and disability in industrialized countries, and their occurrence is seen to be increasing in several developing countries. Excess body weight is considered a major risk factor for metabolic disorders, and the epidemic of pre-obese and obese conditions and type 2 diabetes and their increasing prevalence in children indicate that these pathologies will continue to impact human health [1,2]. Hence, the mechanisms underlying excessive fat storage and its clinical complications remain a challenge to understand and treat.

The liver, skeletal muscle and fat tissue are known as the major sites for the central control of adaptive metabolic regulation of fatty acids (FA) in the body, playing a critical role in maintaining normal glucose and lipid homeostasis. In the condition of surpassed lipid storage, the normal fatty acid metabolism is disrupted and consequent build-up of fat accumulation occurs in non-adipose depots such as the liver, pancreatic islets, muscle, and myocardium. Such accumulation contributes to eliciting a number of metabolic defects, such as dyslipidemia, atherosclerosis, hypertension, and type

2 diabetes [3–5]. While numerous therapeutic strategies are being developed and used in clinics in our attempt to correct the various conditions associated to metabolic dysfunctions, targeting the peroxisome proliferator-activated receptors (PPARs) undoubtedly remains an important option of treatment.

2. The Peroxisome Proliferator-Activated Receptors (PPARs): Fatty Acid Sensors Controlling Metabolism

The PPARs consist of three isotypes, PPARα (NR1C1), PPARβ/δ (NR1C2), and PPARγ (NR1C3), which belong to the nuclear receptor family of ligand-activated transcription factors [6]. PPARs variously bind mono- and polyunsaturated fatty acids and derivatives such as eicosanoids to control the transcription of many genes that govern lipid metabolism [7]. Once activated, they heterodimerize with the nuclear receptor RXR (NR2B family) to bind DNA and modulate target gene transcription [8]. PPARα is a target of the hypolipidemic fibrate drugs and a major activator of FA oxidation in the liver, heart and brown adipose tissue [9,10]. PPARβ/δ is ubiquitously expressed and shares similar functions with PPARα in promoting FA oxidation in metabolic tissues such as skeletal muscle, liver and heart [10,11]. PPARγ is most highly expressed in metabolic tissues including white and brown adipose tissue, where it is a master regulator of whole-body lipid metabolism, adipogenesis, and insulin sensitivity [9,12]. Compared to the other PPARs, PPARγ responds poorly to native fatty acids, while oxidized fatty acid derivatives contained in circulating oxidized low-density lipoproteins (oxLDL) elicit a strong PPARγ activation [13]. PPARγ activity is regulated by transcriptional coactivators such as PGC-1, and also by post-translational modifications often independent of ligand binding, such as phosphorylation, ubiquitination, and SUMOylation [14,15]. In addition to its role in lipid and glucose metabolism, PPARγ has been involved in macrophage cholesterol metabolism and inflammatory response and also plays a major role in mitochondrial physiology and energy metabolism [16–18].

Because of its potent insulin-sensitizing activity, PPARγ has been recognized as a major therapeutic target with the identification of thiazolidinediones (TZDs) as high-affinity ligands [19,20]. TZDs are currently used to correct circulating glucose levels in type 2 diabetes patients [21,22]. However, the clinical efficacy of TZDs on insulin sensitivity has become limited [9,23]. This is partly due to their side effect of stimulating adiposity by upregulating PPARγ target genes, such as fatty acid synthase (FAS) and scavenger receptor CD36 involved in FA formation and storage [24,25]. More importantly, serious health issues have restricted the use of TZDs lately. As a result, some TZDs have been withdrawn from clinics due to life-threatening hepatic toxicity, while serious safety warnings were recently issued for others [26–28]. While strategies to develop safer PPAR pan/dual agonists are of continuous interest [29,30], it has become a fundamental priority to identify other treatment strategies in order to avoid the adverse effects of PPAR ligands while keeping the benefits of correcting whole body glucose and cardiovascular dysfunctions. Our recent identification of PPARγ as a new target of CD36 signaling might feed into the development of potential alternatives in the beneficial control of lipid metabolism.

3. The Growth Hormone Releasing Peptide (GHRP) Family

Growth hormone releasing peptides (GHRP; also known as growth hormone secretagogues) are a family of synthetic peptides and peptidomimetic agonists initially designed to promote growth hormone secretion in GH-deficient patients. However, despite tremendous effort at designing efficacious GHRPs that will exhibit elevated oral bioavailability and induce the pulsatile release of GH, low-cost recombinant GH remains the treatment of choice for GH-deficient patients. Yet, the use of GHRPs in human subjects appears relatively safe, highlighting positive effects on children growth velocity, increased lean mass, decreased bone turnover, and improved cardiac function [31–33]. However, studies are still needed to address the long-term impact of GHRPs and their benefits in diverse clinical scenarios.

GHRP-6 was the first GH-releasing efficient hexapeptide designed, which was then modified as GHRP-2, but their poor oral bioavailability and short-lasting effect have limited their use [34,35]. To address this drawback, additional compounds were designed, including MK-0677, a non-peptidic sulfonamide derivative [36], and hexarelin, also referred to as examorelin or EP-23905 [37,38].

Hexarelin (His-D-2MeTrp-Ala-Trp-D-Phe-Lys-NH_2) differs from GHRP-6 by having D-Trp substituted by D-2-methyl-Trp, making hexarelin biologically more stable with greater GH release activity than GHRP-6 and the first orally active GHRP. Studies in humans have shown that hexarelin was efficient and well tolerated, eliciting a substantial and dose-dependent elevation in plasma GH concentrations, while causing minor sleep problems as side effects [33,39]. Because of the highly vascularized nasal cavity, intranasal administration was also implemented for hexarelin that further improved its bioavailability and efficacy as a therapeutic tool for GH deficiency [40]. Mainly due to their potent GH-releasing activities, hexarelin and other GHRPs, such as GHRP-2 and GHRP-6, are used to enhance athletic performance. This resulted in the implementation of routine GHRP screening since the 2014 Olympics in Sochi and the banning of its use by the World Anti-Doping Agency [41,42].

4. Central vs. Peripheral Actions of GHRPs

The receptor that mediates the response to GHRPs was initially identified as the growth hormone secretagogue receptor GHS-R1a, a member of the G protein-coupled receptor family [43]. GHS-R1a exhibits high-affinity binding toward GHRPs and is highly expressed in the anterior pituitary gland and hypothalamus, consistent with its role in regulating central GH release. A second isoform, GHS-R1b, was also identified but represents a truncated GHS-R receptor devoid of signal transduction activity and thought to act as a dominant-negative form of GHS-R1a through heterodimer formation [44]. Interestingly, ghrelin was later discovered as the endogenous ligand of GHS-R1a, which was then renamed the ghrelin receptor [45]. Ghrelin is an acetylated 28 amino acid hormone initially isolated from the stomach, which promotes central release of GH in somatotroph cells and induces orexigenesis [46–48]. Also consistent with a role in fat and energy metabolism [31,49], decreased circulating ghrelin levels were reported in obese children, increasing their prevalence to insulin resistance and metabolic syndrome [50–52].

Peripheral distribution of GHS-R1a has been reported, supporting physiological effects of GHRPs independently from GH release. Tissues such as vascular endothelium, heart, adrenals, monocytes/macrophages, β pancreatic cells, and bone were shown to express GHS-R1a [53–55]. Consistent with such a GH-independent role, peripheral ghrelin actions have been linked to clinical implications of cardiovascular disease, insulin resistance, and obesity [31,56–59]. Likewise, GH-independent effects on fat metabolism, cardioprotection, hemodynamic control, and bone cell differentiation have been reported for GHRPs [60–66]. Hence, such peripheral effects of GHRPs are thought to play important roles in energy homeostasis, adiposity and vascular integrity and identify the GHRPs as highly promising therapeutic targets in metabolic diseases.

5. Scavenger Receptor CD36, a Target of Hexarelin

Besides interacting with GHS-R1a, hexarelin was also identified as a high-affinity ligand for scavenger receptor CD36 based on experiments using rat cardiac membranes [67]. Furthermore, CD36 binding was more specific for hexarelin than other GHRPs, since compounds such as MK-0677 and EP51389 were unable to compete with hexarelin binding. Such findings correlated well with initial observations highlighting the cardioprotective properties of hexarelin in GH-deficient rats [68,69] and the different tissue binding pattern of hexarelin compared to that of MK-0677 or ghrelin [70]. Scavenger receptor CD36 is a surface glycoprotein originally known as fatty acid translocase (FAT). CD36 topology predicts for two transmembrane domains separated by a large extracellular domain with multiple *N*-linked glycosylation sites, and two short cytoplasmic tails required for intracellular signaling (Figure 1). The extracellular loop also contains a proline-rich domain and a hydrophobic stretch thought to loop back into the membrane bilayer. Despite the short length of the two cytoplasmic domains, each contains known sites for modification, such as palmitoylation that guides CD36 to membrane lipid rafts [71] and ubiquitination to sort the receptor for lysosomal degradation [72].

CD36 has been extensively studied for its role in facilitating long-chain fatty acid uptake and oxidation, positioning CD36 as a key player in FA metabolism [73–75]. However, its wide expression pattern and numerous identified ligands identify CD36 as a multi-functional receptor. Indeed, CD36 is expressed in a variety of cell types and tissues, including but not limited to adipose tissue, macrophages, platelets, endothelial cells, heart, skeletal muscle and liver. Besides long chain fatty acids, it is recognized by thrombospondin, collagen, malaria-infected erythrocytes, lipolysacharrides, anionic phospholipids, and oxidized lipoproteins (e.g., oxLDL) [76–79]. Therefore, CD36 can function in a wide range of processes not always related to FA uptake, including apoptosis, angiogenesis, phagocytosis, thrombosis, inflammation, and atherosclerosis [80,81]. Whether all these numerous and seemingly unrelated effects share a common underlying mechanism and/or signaling event associated with CD36 remains unclear. However, adding the GHRPs to the list of high affinity ligands for scavenger receptor CD36 certainly provides an additional layer in CD36 complexity of regulation and most importantly give us new opportunities on the clinical value of GHRPs.

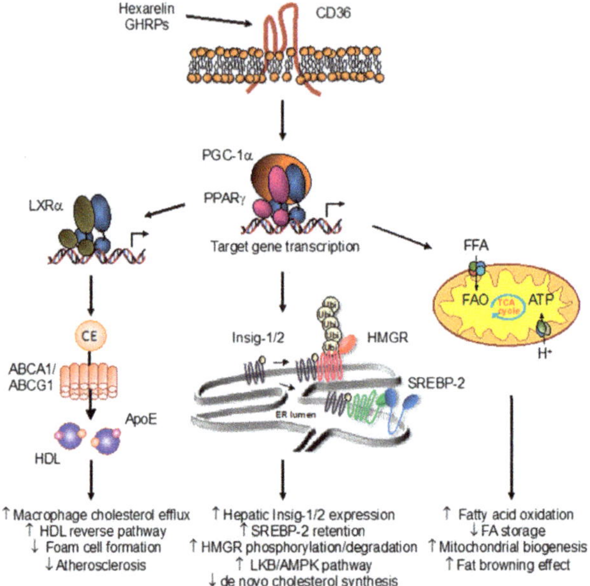

Figure 1. Overview of the growth hormone releasing peptide (GHRP)-peroxisome proliferator-activated receptor-gamma (PPARγ) pathway in lipid and energy metabolism. The interaction of hexarelin with the scavenger receptor CD36 promotes the transcriptional activation of nuclear receptor PPARγ and target gene profiling involved in metabolism. In macrophages, hexarelin and other GHRPs induce a molecular cascade involving nuclear liver X receptor LXRα and expression of apolipoprotein E (apoE) and sterol transporters ABCA1 and ABCG1. Such activation of the PPARγ-LXRα-ABC metabolic pathway increases cholesterol efflux, resulting in enhanced HDL reverse cholesterol transport and regression of atherosclerosis. In hepatocytes, CD36 activation by hexarelin reduces *de novo* cholesterol synthesis. Activation of the LKB-AMPK pathway resulted in the inhibition of the rate-limiting 3-hydroxy-3-methylglutaryl coenzyme A reductase (HMGR) enzyme. Also, induction of Insig1/2 expression through PPARγ/PGC-1α activation led to HMGR degradation and SREBP-2 retention in the endoplasmic reticulum (ER), thereby blunting the homeostatic response to sterol depletion. In adipocytes, CD36 activation with hexarelin promotes mitochondrial activity and biogenesis through enhanced PPARγ and co-activator PGC-1α transcriptional activity. Induction of key genes involved in fatty acid utilization and energy production in mitochondria results in an increased fatty acid β-oxidation and thermogenic-like profile indicative of a browning effect of white fat (adapted from Refs [82,83]).

CD36 and Atherosclerosis

The original observation that cultured macrophages were able to internalize modified low-density lipoproteins at a much higher rate than native LDL particles, resulting in foam cell transformation, has led to the identification of scavenger receptors [84]. Since then, several scavenger receptors have been identified and classified based on their structural features, their capacity to bind modified LDL particles (e.g., acetylated, oxidized) and their contribution to atherogenesis. Each family member possesses distinct properties, although their ligand-recognition specificity often overlaps, which complicates our understanding of their specific role and downstream actions. However, important physiological roles of scavenger receptors have been identified in body protection from infection, clearance of apoptotic cells and removal of modified lipoproteins that might be potentially harmful.

Scavenger receptor CD36 is a member of the B subtype that also includes SR-B1, which functions as a receptor that binds high-density lipoproteins (HDL) particles and is involved in the reverse cholesterol pathway. On the other hand, through its strong ability to capture oxLDL, CD36 has clearly been established as a critical component for macrophage foam cell formation and a major pro-atherogenic factor. Atherosclerosis is a complex disease consisting of the infiltration and accumulation of LDL and cellular debris within the intima of medium and large arteries following vascular injury or inflammation [85]. Oxidation of LDL particles (oxLDL) is considered a priming step for the development of the atherosclerotic plaque, with subsequent and excessive engulfment of oxLDL by macrophages, which then become foam cells loaded with lipids resulting in fatty streaks and plaque formation. Activation of macrophages and constant recruitment of immune cells to the inflammatory site results in increased cytokine secretion and continuous oxidation of LDL. OxLDL are no longer recognized by the LDL receptor and become high-affinity ligands for scavenger receptors, principally CD36 present on macrophages [86]. Normally, this process allows macrophages to clear the neointima from the harmful abundance of oxLDL. However, in conditions where macrophages become overwhelmed by oxLDL, unbalanced uptake vs clearance of lipids is taking place, resulting in lipid-laden macrophages or foam cells. Enhanced inflammation, cellular necrosis, and thinning of the fibrotic plaque eventually ensue, leading to plaque rupture and thrombosis.

At the molecular level, internalized oxLDL provide oxidized fatty acids that serve as ligands to PPARγ thereby inducing genes such as CD36 and LXRα (NR1H3) with a subsequent increase in HDL production and reverse cholesterol transport [87]. Therefore, the role of CD36 is central to the pro-atherogenic effect of modified LDL particles.

Studies using apolipoprotein (apo) E-null mice as a model of fatty streak lesions and atherosclerosis have shown that CD36 was essential in that process. When crossed with apoE-negative mice, CD36 null mice were resistant to developing atherosclerosis [88]. CD36-null murine peritoneal macrophages also exhibited impaired binding and uptake of oxLDL, suggesting that CD36 represents the predominant macrophage receptor for oxLDL [89]. Prior studies in humans had already assessed the critical role of CD36 in the uptake of oxLDL and its abundant and specific expression in atherosclerotic plaques [90,91]. CD36 genetic variants were also identified in humans characterized by high serum triglycerides, low HDL levels, and hyperglycemia with insulin resistance, all considered clinical features of metabolic syndrome [75,92,93]. Patients also demonstrated signs of cardiomyopathy, probably due to impaired uptake of long-chain fatty acids essential to maintaining normal heart function. Population studies have also identified several *CD36* polymorphisms linked to increased risk of metabolic syndrome, acute myocardial infarction and type 2 diabetes [81,94–98], which might support their determination in the context of personalized therapeutic strategies. In particular, polymorphisms found to impair LDL-binding domain of CD36 were correlated with increased cardiovascular risk factors and unstable plaque formation. The potential of CD36 as a therapeutic target for atherosclerosis and other complications of metabolic syndrome is therefore emphasized by our increasing knowledge of its mode of action and certainly warrants the development of novel alternatives aimed to correct for these metabolic defects.

6. The GHRP-PPARγ Pathway in Macrophages

Scavenging oxLDL has been defined as a beneficial role of CD36 to liberate intima from cholesterol depots but is also instrumental in early steps of atherogenesis [88,99]. Using conditions to prevent GH release, we have determined that long-term treatment with GHRPs markedly decreased plaque formation in apoE-null mice fed a high-fat diet, a model known to develop atherosclerosis [100,101]. In particular, GHRP EP80317, a CD36 specific ligand, and hexarelin were both potent in strongly reducing atherosclerotic lesion areas [100,101]. The interaction of the GHRPs with CD36 was suggested to initiate an intracellular signaling resulting in the activation of the PPARγ-LXRα-ABC metabolic cascade involved in reverse cholesterol pathway (Figure 1). Treatment of mouse peritoneal macrophages as well as differentiated human THP-1 macrophages with hexarelin resulted in an increase in cholesterol efflux. Such cholesterol removal from cells correlated with a rise in the expression of LXRα, ApoE, ABCA1 and ABCG1, all critical players promoting the HDL-mediated cholesterol efflux pathway.

Considering that expression of LXRα gene can be upregulated by PPARγ ligands [102] and that oxLDL internalization through CD36 results in PPARγ activation with the entry of oxidized fatty acids [12,103], we thus analyzed the effect of hexarelin on PPARγ transcriptional potential. Using cell-based assays, the interaction of hexarelin with either CD36 or GHS-R1a was shown to induce PPARγ transcriptional potential [101]. In addition, the response to hexarelin was strongly impaired in peritoneal macrophages from PPARγ heterozygote mice, suggesting a critical role of PPARγ. These findings highlight the potential of hexarelin to promote a metabolic cascade involving PPARγ and LXRα as an attempt to efficiently remove oxLDL deleterious actions from the vessel wall and shunt free cholesterol into the HDL reverse cholesterol pathway, thus providing a protective effect in condition of plaque formation in vivo.

The beneficial effect of hexarelin on PPARγ activation appears to be balanced with the coordinated induction of LXRα and downstream target genes achieving optimal lipid efflux. Consistent with this, the activation of PPARγ by hexarelin did not result in an increase in CD36 expression, as opposed to oxLDL-induced PPARγ activity which upregulates CD36, leading to subsequent positive autoregulatory loops being considered pro-atherogenic [87,103]. The exact mechanism for such distinct regulation remains unclear, but we found that the ligand binding domain was not necessary for PPARγ activation by hexarelin, thereby avoiding any effect of exogenous PPARγ ligands (e.g., oxidized fatty acids) that might arise from oxLDL entry. This also supports a role for the N-terminal AF-1 domain that might mediate PPARγ transcriptional activation in response to hexarelin-elicited intracellular transduction pathways. In support of this, PPARγ phosphorylation was strongly induced by hexarelin, providing a molecular basis of PPARγ response to hexarelin signaling [83,101]. GHS-R1a activation by hexarelin also increased PPARγ activity and may therefore suggest a concerted role of GHS-R1a to signal PPARγ [101]. Interestingly, activation of GHS-R1a receptor by hexarelin or its natural ligand ghrelin led to enhanced PPARγ phosphorylation through the coordinated action of Fyn and Akt kinases in macrophages [104]. Whether such concerted response of both GHSR-1a and CD36 receptors is required in the overall beneficial effects of hexarelin on atherosclerosis remains to be further explored.

A more recent study has also described the suppressive effect of hexarelin on plaque formation. Using a vitamin-D3 induced rat model of atherosclerosis, hexarelin was shown to reduce foam cell formation, aortic calcium sedimentation, and vascular smooth muscle cell growth [105]. With its ability to promote ligand-independent PPARγ activation, to interfere with the pro-atherogenic regulatory loop resulting from CD36 upregulation, and to increase overall cholesterol efflux from cells, hexarelin represents a potent regulator to correct for pathological imbalance between sterol uptake and efflux that usually leads to foam cell formation.

7. The CD36-PPARγ Axis in Adipocytes

Primary defects in energy balance that produce visceral adiposity are sufficient to result in the development of insulin resistance and vascular disease. Current knowledge has implied a role for fat-derived adipokines, such as leptin, tumor necrosis factor TNFα, adiponectin, adipsin and resistin, as important regulators of insulin sensitivity, defining fat tissue not just as a passive storage depot but also as an endocrine organ [106,107]. PPARγ is recognized as a major regulator of adipokine synthesis in mature adipocytes and as such, it has become a therapeutic target of TZD actions [19,20]. Because of their potent insulin-sensitizing activity, TZDs are currently used to correct circulating glucose levels in type 2 diabetes patients but under increasing restricted conditions [9,21–23].

Activation of PPARγ in white adipocytes is known to promote FA storage, triglyceride (TG) synthesis and glucose uptake involving upregulation of key target genes related to fatty acid metabolism. In addition, the induction of expression and secretion of insulin-sensitizing adipokines, such as adiponectin, will dictate a decrease in lipid accumulation and an increase of glucose uptake and fatty acid oxidation in other tissues. These actions are part of the mechanism by which the TZDs improve insulin resistance in diabetic patients [23]. PPARγ is also a master regulator of adipogenesis. Studies of fat-specific PPARγ knockout mice revealed that PPARγ is essential for differentiation and survival of fat cells [108,109]. Consistent with the dual effect of PPARγ to ameliorate insulin sensitivity while promoting fat differentiation, genetic studies have revealed that a partial loss-of-function Pro12Ala variant improved insulin sensitivity, while the gain-of-function Pro115Gln mutation was associated to obesity and insulin resistance in humans [110]. Therefore, it becomes essential to consider a PPARγ selective modulator that might exhibit a better insulin sensitizing profile as compared to a full agonist.

Several studies have shown that mature adipocytes do express CD36, whereas expression of GHS-R1a remains unclear despite a functional response to ghrelin [111,112]. However, the mechanism by which CD36 may affect the overall metabolic activity of fat storage and mobilization is not completely defined. Based on evidence that CD36 activation with hexarelin resulted in PPARγ activation in macrophages, it was expected that a similar activation of PPARγ and subsequent downstream effects could take place in adipocytes.

Indeed, we found that hexarelin promoted beneficial effects in white adipose tissue, resulting in a striking thermogenic profile of FA oxidation and mitochondria biogenesis in cultured adipocytes and in epididymal fat of treated mice [113]. These effects were translated through PPARγ and required CD36, establishing a functional CD36-PPARγ pathway in fat [83]. Interestingly, gene profile analysis has revealed that many of the genes upregulated by hexarelin were shared with TZD treatment, indicating a common effect on PPARγ activation. However, not all established PPARγ targets were upregulated by hexarelin, including CD36 itself [113]. This was also observed in macrophages, suggesting a similar mechanism for CD36 gene regulation in response to hexarelin (Figure 1). Gene expression and functional studies have indicated that adipocytes respond to hexarelin with an increased mobilization of fatty acids rather than the expected adipogenic effect of PPARγ activation seen with TZDs, revealing an unexpected effect of hexarelin to promote the β-oxidation of fatty acids [113]. Whether this indicates that hexarelin may serve as an energy deficit signal that prevents fat utilization during deprivation and promotes its use in excess is not certain, but if true, such a scenario has clear implications for obesity-related metabolic defects. Consistent with this, the induction of key markers of fatty acid oxidation and mitochondrial activity, including Cpt1b, Acaa1 and 2, and several subunits of the cytochrome c oxidase (COX) complex, were increased in response to hexarelin. Interestingly, a recent study has implicated the metabolic response of white fat tissue to hexarelin in correcting abnormal lipid metabolic states of insulin-resistant mice through modulation of genes related to fatty acid uptake and oxidation [114]. Given that PPARα also plays a pivotal role in FA metabolism by regulating genes related to mitochondrial and peroxisomal β-oxidation pathways in high oxidative tissues, such as liver, heart and brown fat [115,116], the metabolic response of fat to hexarelin strongly suggests also a role for PPARα activation. Consistent with this, we found that both PPARα and PPARβ/δ were activated in response to hexarelin, supporting a cellular response to CD36 activation that might implicate the various PPAR isotypes [83,101].

The preferred redirection of FA toward mitochondrial oxidation process was accompanied by noticeable changes in mitochondrial morphology in white adipose tissue of hexarelin-treated mice. Increases in the intramitochondrial matrix surface and cristae formation observed were typical of tissues with high oxidative potential, such as brown fat, suggesting a browning effect of hexarelin [113]. This was also consistent with the induction of key thermogenic markers PGC-1α and uncoupling protein (UCP)-1, which rose from low normal levels usually found in white fat cells to those mainly characteristic of brown fat. PGC-1α and UCP-1 are highly expressed in brown fat and play critical roles in thermogenesis and energy expenditure with enhanced oxidative metabolism and mitochondrial biogenesis [117–119]. The ability of hexarelin to upregulate PGC-1α provides a clue by which CD36 signaling might control the fine-tuning of mitochondrial function towards FA oxidation and energy balance. This suggests that such increase in mitochondrial activity and biogenesis by hexarelin might thus provide a benefit to defects associated to mitochondrial diseases. Consistent with our findings, a recent study also reported a protective effect of hexarelin on mitochondria function using a rat model of cachexia [120]. The authors reported an increase of mitochondrial markers such as PGC-1α at the protein levels, supporting the potential of hexarelin to induce a mitochondrial response, but the mechanism involved and the role of CD36 were not addressed in this context. Interestingly, besides PPARγ, PGC-1α upregulation by hexarelin and CD36 activation might also affect other known nuclear receptors coregulated by PGC-1, such as the estrogen-related receptors (ERRs) involved in mitochondrial function and biogenesis [121–123]. Therefore, investigating their contribution is certainly an interesting avenue to pursue.

8. The Hexarelin-PPARγ Axis in Hepatocytes

Although considered highly expressed in insulin-sensitizing tissues, PPARγ is found at low levels in the liver, and therefore its influence on hepatic function is not fully understood. In fact, much negative attention was given to hepatic PPARγ with the hepatotoxicity effect of TZD troglitazone, resulting in its withdrawal from the market [26,27]. Part of the noxious effects of troglitazone in liver was associated with the production of toxic reactive metabolites and signs of mitochondrial DNA damage, mitochondrial defects and cell death [124,125], which emphasizes anti-oxidant strategies [126]. However, some evidence indicates that the toxic effect of troglitazone might be independent of PPARγ activity [127]. Recent studies have reported beneficial hepatic effects of PPARγ agonists in reversing nonalcoholic steatohepatitis (NASH) in patients, reducing liver inflammation, fibrosis and triglyceride content [128,129]. Interestingly, in condition of PPARγ overexpression triggered by insulin or oleic acid treatment in hepatocytes, or induced in mice fed a high-fat diet, there was the expected increase in PPARγ lipogenic genes but also of PPARα target genes involved in FA oxidation [130–132]. Such induction of hepatic PPARγ might therefore represent an adaptive response to promote beneficial lipid utilization.

The role of GHRPs on liver function has not been fully characterized and given their ability to promote macrophage cholesterol reverse transport through CD36 receptor, one would expect that the CD36-PPARγ axis might play a role of regulation on sterol metabolism in hepatic cells. We have recently reported that hexarelin regulates hepatic cholesterol homeostasis by repressing de novo cholesterol synthesis through enhanced 3-hydroxy-3-methylglutaryl coenzyme A reductase (HMGR) degradation and sterol regulatory element-binding protein (SREBP)-2 retention in the endoplasmic reticulum [82]. Elegant work from Brown and Goldstein has detailed the mechanism responsible for maintaining hepatic cholesterol homeostasis [133,134]. The rate-limiting HMGR is under a tight control by available cellular cholesterol content both at the gene level, through regulation of expression by sterol regulatory element-binding protein SREBP-2, and at the protein level, through enzyme phosphorylation and degradation. Our findings have demonstrated that CD36 activity reduced cholesterol levels in liver cells by impeding the compensatory activation of HMGR and decreasing SREBP-2 transactivation normally occurring in cells during sterol depletion (Figure 1). Interestingly, this potential of CD36 to inhibit cholesterol synthesis was associated with activation of the LKB/AMPK energetic pathway, known to play an imperative role in energy homeostasis by regulating a plethora of pathways for the main purpose of saving energy and access readily available fuel for the cell. The AMPK activation by hexarelin resulted in the phosphorylation of HMGR, achieving a rapid inhibition of its activity in hepatocytes, similar to the inhibition triggered by statin compounds. With the role of CD36 in internalizing long chain fatty acids and cholesterol derivatives, the immediate activation of AMPK by hexarelin is believed to promote a need to preserve energy in liver cells. Similarly, fatty acid-induced AMPK activation has also been reported in the heart to promote CD36 regulation and adjust for fatty acid usage and oxidation [135,136]. Although the exact role remains to be determined, our findings suggest a metabolic cascade between CD36 and the LKB/AMPK pathway, providing a role of CD36 to regulate downstream AMPK targets involved in energy metabolism.

The CD36-PPARγ pathway appears to be functional in hepatocytes with the activation of PPARγ by hexarelin, which identified Insig-1 and Insig-2 genes as PPARγ-responsive genes [82]. Insig-1 and Insig-2 were reported to promote HMGR ubiquitination and degradation [137], and also to prevent the transit of SREBP-2 to the Golgi for its processing and activation [133,138]. Therefore, this provides a mechanism by which genes encoding key enzymes involved in cholesterol synthesis and under the control of SREBP-2 remained unresponsive to sterol depletion in the context of CD36 activation by hexarelin [82]. The rapid Insig-mediated degradation of HMGR protein and the retention of SREBP-2 in the endoplasmic reticulum represent two checkpoints of regulation of CD36 signaling to prevent sterol accumulation in liver cells.

Interestingly, the coactivation potential of PGC-1α was enhanced in response to hexarelin, accompanied by an increase in PGC-1α recruitment to PPARγ [82]. This suggests that CD36 can signal PGC-1α to induce PPARγ coactivation in hepatocytes. Consistent with this, the recruitment of PGC-1α to activated AMPKα was enhanced by hexarelin, leading to Sirt1-mediated deacetylation and PGC-1α transcriptional activation. Such metabolic activation of PGC-1α has also been described in adipocytes whereby CD36 promoted increases in PGC-1α and downstream effectors, such as UCP-1 and ATP synthase [113]. Given a similar increase in PGC-1α activity and UCP-1 expression in hepatocytes, and the prominent role of PGC-1 in cellular energy homeostasis, FA oxidation, hepatic gluconeogenesis, and mitochondrial biogenesis [117,139,140], the role of CD36 is likely to be extended to different pathways of regulation involved in liver metabolism and function.

9. Concluding Remarks

While the molecular events by which CD36 and GHRPs exert their actions are not completely understood, increasing evidence supports a prominent role of scavenger receptor CD36 to initiate profound changes in key metabolic pathways, especially pertaining to PPARγ-controlled critical steps. Also, with its potential to promote PGC-1α transcriptional competence and related key functions of fatty acid usage, glucose homeostasis and mitochondrial activity, we might expect that GHRP-CD36 signaling may expand to other metabolic pathways and involve additional nuclear receptors. Given the increasing prevalence of metabolic defects associated with deregulated glucose and lipid metabolism and with mitochondria dysfunction, targeting CD36 with GHRPs appears to be a safe option for the treatment of metabolic disorders.

Acknowledgments: We thank members of the lab for useful comments and suggestions. This work was supported by the Canadian Institutes of Health Research and the Canadian Diabetes Association (to André Tremblay). Loïze Maréchal is supported by a doctoral award form the Fonds de la Recherche du Québec en Santé (FRQS) and by the Faculté des Études Supérieures de l'Université de Montréal (FESP), Maximilien Laviolette and Baly Sow are supported by the FESP. and Amélie Rodrigue-Way was supported by a doctoral award from the Natural Sciences and Engineering Research Council of Canada .

Conflicts of Interest: The authors declare no conflict of interest.

References

1. Suliga, E. Visceral adipose tissue in children and adolescents: A review. *Nutr. Res. Rev.* **2009**, *22*, 137–147. [CrossRef] [PubMed]
2. Wittcopp, C.; Conroy, R. Metabolic Syndrome in Children and Adolescents. *Pediatr. Rev.* **2016**, *37*, 193–202. [CrossRef] [PubMed]

3. Hajer, G.R.; van Haeften, T.W.; Visseren, F.L. Adipose tissue dysfunction in obesity, diabetes, and vascular diseases. *Eur. Heart J.* **2008**, *29*, 2959–2971. [CrossRef] [PubMed]
4. Eckel, R.H.; Alberti, K.G.; Grundy, S.M.; Zimmet, P.Z. The metabolic syndrome. *Lancet* **2010**, *375*, 181–183. [CrossRef]
5. Grundy, S.M. Overnutrition, ectopic lipid and the metabolic syndrome. *J. Investig. Med.* **2016**, *64*, 1082–1086. [CrossRef] [PubMed]
6. Mangelsdorf, D.J.; Thummel, C.; Beato, M.; Herrlich, P.; Schutz, G.; Umesono, K.; Blumberg, B.; Kastner, P.; Mark, M.; Chambon, P.; et al. The nuclear receptor superfamily: The second decade. *Cell* **1995**, *83*, 835–839. [CrossRef]
7. Evans, R.M.; Barish, G.D.; Wang, Y.X. PPARs and the complex journey to obesity. *Nat. Med.* **2004**, *10*, 355–361. [CrossRef] [PubMed]
8. Ricote, M.; Valledor, A.F.; Glass, C.K. Decoding transcriptional programs regulated by PPARs and LXRs in the macrophage: Effects on lipid homeostasis, inflammation, and atherosclerosis. *Arterioscler. Thromb. Vasc. Biol.* **2004**, *24*, 230–239. [CrossRef] [PubMed]
9. Ahmadian, M.; Suh, J.M.; Hah, N.; Liddle, C.; Atkins, A.R.; Downes, M.; Evans, R.M. PPARγ signaling and metabolism: The good, the bad and the future. *Nat. Med.* **2013**, *19*, 557–566. [CrossRef] [PubMed]
10. Poulsen, L.; Siersbaek, M.; Mandrup, S. PPARs: Fatty acid sensors controlling metabolism. *Semin. Cell Dev. Biol.* **2012**, *23*, 631–639. [CrossRef] [PubMed]
11. Barish, G.D.; Narkar, V.A.; Evans, R.M. PPARδ: A dagger in the heart of the metabolic syndrome. *J. Clin. Investig.* **2006**, *116*, 590–597. [CrossRef] [PubMed]
12. Tontonoz, P.; Spiegelman, B.M. Fat and beyond: The diverse biology of PPARγ. *Ann. Rev. Biochem.* **2008**, *77*, 289–312. [CrossRef] [PubMed]
13. Nagy, L.; Tontonoz, P.; Alvarez, J.G.A.; Chen, H.; Evans, R.M. Oxidized LDL regulates macrophage gene expression through ligand activation of PPARγ. *Cell* **1998**, *93*, 229–240. [CrossRef]
14. Hauser, S.; Adelmant, G.; Sarraf, P.; Wright, H.M.; Mueller, E.; Spiegelman, B.M. Degradation of the peroxisome proliferator-activated receptor γ is linked to ligand-dependent activation. *J. Biol. Chem.* **2000**, *275*, 18527–18533. [CrossRef] [PubMed]
15. Pascual, G.; Fong, A.L.; Ogawa, S.; Gamliel, A.; Li, A.C.; Perissi, V.; Rose, D.W.; Willson, T.M.; Rosenfeld, M.G.; Glass, C.K. A SUMOylation-dependent pathway mediates transrepression of inflammatory response genes by PPARγ. *Nature* **2005**, *437*, 759–763. [CrossRef] [PubMed]
16. Kubota, N.; Terauchi, Y.; Miki, H.; Tamemoto, H.; Yamauchi, T.; Komeda, K.; Satoh, S.; Nakano, R.; Ishii, C.; Sugiyama, T.; et al. PPARγ mediates high-fat diet-induced adipocyte hypertrophy and insulin resistance. *Mol. Cell* **1999**, *4*, 597–609. [CrossRef]
17. Tontonoz, P.; Hu, E.; Spiegelman, B.M. Stimulation of adipogenesis in fibroblasts by PPARγ2, a lipid-activated transcription factor. *Cell* **1994**, *79*, 1147–1156. [CrossRef]
18. Corona, J.C.; Duchen, M.R. PPARγ as a therapeutic target to rescue mitochondrial function in neurological disease. *Free. Radic. Biol. Med.* **2016**, *100*, 153–163. [CrossRef] [PubMed]
19. Lehmann, J.M.; Moore, L.B.; Smith-Oliver, T.A.; Wilkinson, W.O.; Wilson, T.M.; Kliewer, S.A. An antidiabetic thiazolidinedione is a high affinity ligand for peroxisome-activated receptor γ. *J. Biol. Chem.* **1995**, *270*, 12953–12956. [CrossRef] [PubMed]
20. Lambe, K.G.; Tugwood, J.D. A human peroxisome-proliferator-activated receptor-γ is activated by inducers of adipogenesis, including thiazolidinedione drugs. *Eur. J. Biochem.* **1996**, *239*, 1–7. [CrossRef] [PubMed]
21. Yki-Jarvinen, H. Thiazolidinediones. *N. Engl. J. Med.* **2004**, *351*, 1106–1118. [CrossRef] [PubMed]
22. Leahy, J.L. Thiazolidinediones in prediabetes and early type 2 diabetes: What can be learned about that disease's pathogenesis. *Curr. Diab. Rep.* **2009**, *9*, 215–220. [CrossRef] [PubMed]
23. Soccio, R.E.; Chen, E.R.; Lazar, M.A. Thiazolidinediones and the promise of insulin sensitization in type 2 diabetes. *Cell Metab.* **2014**, *20*, 573–591. [CrossRef] [PubMed]
24. Yang, X.; Smith, U. Adipose tissue distribution and risk of metabolic disease: Does thiazolidinedione-induced adipose tissue redistribution provide a clue to the answer? *Diabetologia* **2007**, *50*, 1127–1139. [CrossRef] [PubMed]
25. De Souza, C.J.; Eckhardt, M.; Gagen, K.; Dong, M.; Chen, W.; Laurent, D.; Burkey, B.F. Effects of pioglitazone on adipose tissue remodeling within the setting of obesity and insulin resistance. *Diabetes* **2001**, *50*, 1863–1871. [CrossRef] [PubMed]

26. Anonymous. Troglitazone withdrawn from market. *Am. J. Health Syst. Pharm.* **2000**, *57*, 834.
27. Cleland, J.G.; Atkin, S.L. Thiazolidinediones, deadly sins, surrogates, and elephants. *Lancet* **2007**, *370*, 1103–1104. [CrossRef]
28. Starner, C.I.; Fenrick, B.; Coleman, J.; Wickersham, P.; Gleason, P.P. Rosiglitazone prior authorization safety policy: A cohort study. *J. Manage. Care Pharm.* **2012**, *18*, 225–233. [CrossRef] [PubMed]
29. Rosenson, R.S.; Wright, R.S.; Farkouh, M.; Plutzky, J. Modulating peroxisome proliferator-activated receptors for therapeutic benefit? Biology, clinical experience, and future prospects. *Am. Heart J.* **2012**, *164*, 672–680. [CrossRef] [PubMed]
30. Maqbool, F.; Safavi, M.; Bahadar, H.; Rahimifard, M.; Niaz, K.; Abdollahi, M. Discovery Approaches for Novel Dyslipidemia Drugs. *Curr. Drug Discov. Technol.* **2015**, *12*, 90–116. [CrossRef] [PubMed]
31. Mosa, R.M.; Zhang, Z.; Shao, R.; Deng, C.; Chen, J.; Chen, C. Implications of ghrelin and hexarelin in diabetes and diabetes-associated heart diseases. *Endocrine* **2015**, *49*, 307–323. [CrossRef] [PubMed]
32. Cao, J.M.; Ong, H.; Chen, C. Effects of ghrelin and synthetic GH secretagogues on the cardiovascular system. *Trends Endocrinol. Metab.* **2006**, *17*, 13–18. [CrossRef] [PubMed]
33. Sigalos, J.T.; Pastuszak, A.W. The Safety and Efficacy of Growth Hormone Secretagogues. *Sex. Med. Rev.* **2018**, *6*, 45–53. [CrossRef] [PubMed]
34. Pandya, N.; DeMott-Friberg, R.; Bowers, C.Y.; Barkan, A.L.; Jaffe, C.A. Growth hormone (GH)-releasing peptide-6 requires endogenous hypothalamic GH-releasing hormone for maximal GH stimulation. *J. Clin. Endocrinol. Metab.* **1998**, *83*, 1186–1189. [CrossRef] [PubMed]
35. Smith, R.G. Development of growth hormone secretagogues. *Endocr. Rev.* **2005**, *26*, 346–360. [CrossRef] [PubMed]
36. Jacks, T.; Smith, R.; Judith, F.; Schleim, K.; Frazier, E.; Chen, H.; Krupa, D.; Hora, D., Jr.; Nargund, R.; Patchett, A.; et al. MK-0677, a potent, novel, orally active growth hormone (GH) secretagogue: GH, insulin-like growth factor I, and other hormonal responses in beagles. *Endocrinology* **1996**, *137*, 5284–5289. [CrossRef] [PubMed]
37. Deghenghi, R.; Cananzi, M.M.; Torsello, A.; Battisti, C.; Muller, E.E.; Locatelli, V. GH-releasing activity of Hexarelin, a new growth hormone releasing peptide, in infant and adult rats. *Life Sci.* **1994**, *54*, 1321–1328. [CrossRef]
38. Ghigo, E.; Arvat, E.; Gianotti, L.; Imbimbo, B.P.; Lenaerts, V.; Deghenghi, R.; Camanni, F. Growth hormone-releasing activity of hexarelin, a new synthetic hexapeptide, after intravenous, subcutaneous, intranasal, and oral administration in man. *J. Clin. Endocrinol. Metab.* **1994**, *78*, 693–698. [PubMed]
39. Imbimbo, B.P.; Mant, T.; Edwards, M.; Amin, D.; Dalton, N.; Boutignon, F.; Lenaerts, V.; Wuthrich, P.; Deghenghi, R. Growth hormone-releasing activity of hexarelin in humans. A dose-response study. *Eur. J. Clin. Pharmacol.* **1994**, *46*, 421–425. [CrossRef] [PubMed]
40. Laron, Z.; Frenkel, J.; Deghenghi, R.; Anin, S.; Klinger, B.; Silbergeld, A. Intranasal administration of the GHRP hexarelin accelerates growth in short children. *Clin. Endocrinol. (Oxford)* **1995**, *43*, 631–635. [CrossRef]
41. Sobolevsky, T.; Krotov, G.; Dikunets, M.; Nikitina, M.; Mochalova, E.; Rodchenkov, G. Anti-doping analyses at the Sochi Olympic and Paralympic Games 2014. *Drug Test Anal.* **2014**, *6*, 1087–1101. [CrossRef] [PubMed]
42. WADA. The World Anti-Doping Code: Prohibited List. 2018. Available online: https://www.wada-ama.org (accessed on 10 January 2018).
43. Howard, A.D.; Feighner, S.D.; Cully, D.F.; Arena, J.P.; Liberator, P.A.; Rosenblum, C.I.; Hamelin, M.; Hreniuk, D.L.; Palyha, O.C.; Anderson, J.; et al. A receptor in pituitary and hypothalamus that functions in growth hormone release. *Science* **1996**, *273*, 974–977. [CrossRef] [PubMed]
44. Leung, P.K.; Chow, K.B.; Lau, P.N.; Chu, K.M.; Chan, C.B.; Cheng, C.H.; Wise, H. The truncated ghrelin receptor polypeptide (GHS-R1b) acts as a dominant-negative mutant of the ghrelin receptor. *Cell Signal* **2007**, *19*, 1011–1022. [CrossRef] [PubMed]
45. Kojima, M.; Hosoda, H.; Date, Y.; Nakazato, M.; Matsuo, H.; Kangawa, K. Ghrelin is a growth-hormone-releasing acylated peptide from stomach. *Nature* **1999**, *402*, 656–660. [CrossRef] [PubMed]
46. Lazarczyk, M.A.; Lazarczyk, M.; Grzela, T. Ghrelin: A recently discovered gut-brain peptide. *Int. J. Mol. Med.* **2003**, *12*, 279–287. [CrossRef] [PubMed]
47. Van der Lely, A.J.; Tschop, M.; Heiman, M.L.; Ghigo, E. Biological, physiological, pathophysiological, and pharmacological aspects of ghrelin. *Endocr. Rev.* **2004**, *25*, 426–457. [CrossRef] [PubMed]

48. Coll, A.P.; Farooqi, I.S.; O'Rahilly, S. The hormonal control of food intake. *Cell* **2007**, *129*, 251–262. [CrossRef] [PubMed]
49. Wells, T. Ghrelin-Defender of fat. *Prog. Lipid. Res.* **2009**, *48*, 257–274. [CrossRef] [PubMed]
50. Pacifico, L.; Poggiogalle, E.; Costantino, F.; Anania, C.; Ferraro, F.; Chiarelli, F.; Chiesa, C. Acylated and nonacylated ghrelin levels and their associations with insulin resistance in obese and normal weight children with metabolic syndrome. *Eur. J. Endocrinol.* **2009**, *161*, 861–870. [CrossRef] [PubMed]
51. Pedrosa, C.; Oliveira, B.M.; Albuquerque, I.; Simoes-Pereira, C.; Vaz-de-Almeida, M.D.; Correia, F. Metabolic syndrome, adipokines and ghrelin in overweight and obese schoolchildren: Results of a 1-year lifestyle intervention programme. *Eur. J. Pediatr.* **2011**, *170*, 483–492. [CrossRef] [PubMed]
52. Razzaghy-Azar, M.; Nourbakhsh, M.; Pourmoteabed, A.; Nourbakhsh, M.; Ilbeigi, D.; Khosravi, M. An Evaluation of Acylated Ghrelin and Obestatin Levels in Childhood Obesity and Their Association with Insulin Resistance, Metabolic Syndrome, and Oxidative Stress. *J. Clin. Med.* **2016**, *5*, 61. [CrossRef] [PubMed]
53. Katugampola, S.D.; Pallikaros, Z.; Davenport, A.P. [125I-His(9)]-ghrelin, a novel radioligand for localizing GHS orphan receptors in human and rat tissue: Up-regulation of receptors with athersclerosis. *Br. J. Pharmacol.* **2001**, *134*, 143–149. [CrossRef] [PubMed]
54. Gnanapavan, S.; Kola, B.; Bustin, S.A.; Morris, D.G.; McGee, P.; Fairclough, P.; Bhattacharya, S.; Carpenter, R.; Grossman, A.B.; Korbonits, M. The tissue distribution of the mRNA of ghrelin and subtypes of its receptor, GHS-R, in humans. *J. Clin. Endocrinol. Metab.* **2002**, *87*, 2988. [CrossRef] [PubMed]
55. Maccarinelli, G.; Sibilia, V.; Torsello, A.; Raimondo, F.; Pitto, M.; Giustina, A.; Netti, C.; Cocchi, D. Ghrelin regulates proliferation and differentiation of osteoblastic cells. *J. Endocrinol.* **2005**, *184*, 249–256. [CrossRef] [PubMed]
56. St-Pierre, D.H.; Karelis, A.D.; Coderre, L.; Malita, F.; Fontaine, J.; Mignault, D.; Brochu, M.; Bastard, J.P.; Cianflone, K.; Doucet, E.; et al. Association of acylated and nonacylated ghrelin with insulin sensitivity in overweight and obese postmenopausal women. *J. Clin. Endocrinol. Metab.* **2007**, *92*, 264–269. [CrossRef] [PubMed]
57. Garcia, E.A.; Korbonits, M. Ghrelin and cardiovascular health. *Curr. Opin. Pharmacol.* **2006**, *6*, 142–147. [CrossRef] [PubMed]
58. Tesauro, M.; Schinzari, F.; Iantorno, M.; Rizza, S.; Melina, D.; Lauro, D.; Cardillo, C. Ghrelin improves endothelial function in patients with metabolic syndrome. *Circulation* **2005**, *112*, 2986–2992. [CrossRef] [PubMed]
59. Chuang, J.C.; Sakata, I.; Kohno, D.; Perello, M.; Osborne-Lawrence, S.; Repa, J.J.; Zigman, J.M. Ghrelin directly stimulates glucagon secretion from pancreatic α-cells. *Mol. Endocrinol.* **2011**, *25*, 1600–1611. [CrossRef] [PubMed]
60. Marleau, S.; Mulumba, M.; Lamontagne, D.; Ong, H. Cardiac and peripheral actions of growth hormone and its releasing peptides: Relevance for the treatment of cardiomyopathies. *Cardiovasc. Res.* **2006**, *69*, 26–35. [CrossRef] [PubMed]
61. De Gennaro Colonna, V.; Rossoni, G.; Bernareggi, M.; Muller, E.E.; Berti, F. Cardiac ischemia and impairment of vascular endothelium function in hearts from growth hormone-deficient rats: Protection by hexarelin. *Eur. J. Pharmacol.* **1997**, *334*, 201–207. [CrossRef]
62. Locatelli, V.; Rossoni, G.; Schweiger, F.; Torsello, A.; De, G.C.V.; Bernareggi, M.; Deghenghi, R.; Muller, E.E.; Berti, F. Growth hormone-independent cardioprotective effects of hexarelin in the rat. *Endocrinology* **1999**, *140*, 4024–4031. [CrossRef] [PubMed]
63. Huang, J.; Li, Y.; Zhang, J.; Liu, Y.; Lu, Q. The growth hormone secretagogue hexarelin protects rat cardiomyocytes from in vivo ischemia/reperfusion injury through interleukin-1 signaling pathway. *Int. Heart J.* **2017**, *58*, 257–263. [CrossRef] [PubMed]
64. Zhang, X.; Qu, L.; Chen, L.; Chen, C. Improvement of cardiomyocyte function by in vivo hexarelin treatment in streptozotocin-induced diabetic rats. *Physiol. Rep.* **2018**, *6*. [CrossRef] [PubMed]
65. Mao, Y.; Tokudome, T.; Kishimoto, I. The cardiovascular action of hexarelin. *J. Geriatr. Cardiol.* **2014**, *11*, 253–258. [PubMed]
66. Zhao, Y.; Zhang, X.; Chen, J.; Lin, C.; Shao, R.; Yan, C.; Chen, C. Hexarelin protects rodent pancreatic β-cells function from cytotoxic effects of streptozotocin involving mitochondrial signalling pathways in vivo and in vitro. *PLoS ONE* **2016**, *11*, e0149730. [CrossRef] [PubMed]

67. Bodart, V.; Febbraio, M.; Demers, A.; McNicoll, N.; Pohankova, P.; Perreault, A.; Sejlitz, T.; Escher, E.; Silverstein, R.L.; Lamontagne, D.; et al. CD36 mediates the cardiovascular action of growth hormone-releasing peptides in the heart. *Circ. Res.* **2002**, *90*, 844–849. [CrossRef] [PubMed]
68. Berti, F.; Muller, E.; De Gennaro Colonna, V.; Rossoni, G. Hexarelin exhibits protective activity against cardiac ischaemia in hearts from growth hormone-deficient rats. *Growth Horm. IGF Res.* **1998**, *8* (Suppl. B), 149–152. [CrossRef]
69. Rossoni, G.; De Gennaro Colonna, V.; Bernareggi, M.; Polvani, G.L.; Muller, E.E.; Berti, F. Protectant activity of hexarelin or growth hormone against postischemic ventricular dysfunction in hearts from aged rats. *J. Cardiovasc. Pharmacol.* **1998**, *32*, 260–265. [CrossRef] [PubMed]
70. Papotti, M.; Ghe, C.; Cassoni, P.; Catapano, F.; Deghenghi, R.; Ghigo, E.; Muccioli, G. Growth hormone secretagogue binding sites in peripheral human tissues. *J. Clin. Endocrinol. Metab.* **2000**, *85*, 3803–3807. [CrossRef] [PubMed]
71. Thorne, R.F.; Ralston, K.J.; de Bock, C.E.; Mhaidat, N.M.; Zhang, X.D.; Boyd, A.W.; Burns, G.F. Palmitoylation of CD36/FAT regulates the rate of its post-transcriptional processing in the endoplasmic reticulum. *Biochim. Biophys. Acta* **2010**, *1803*, 1298–1307. [CrossRef] [PubMed]
72. Smith, J.; Su, X.; El-Maghrabi, R.; Stahl, P.D.; Abumrad, N.A. Opposite regulation of CD36 ubiquitination by fatty acids and insulin: Effects on fatty acid uptake. *J. Biol. Chem.* **2008**, *283*, 13578–13585. [CrossRef] [PubMed]
73. Ma, X.; Bacci, S.; Mlynarski, W.; Gottardo, L.; Soccio, T.; Menzaghi, C.; Iori, E.; Lager, R.A.; Shroff, A.R.; Gervino, E.V.; et al. A common haplotype at the CD36 locus is associated with high free fatty acid levels and increased cardiovascular risk in Caucasians. *Hum. Mol. Genet.* **2004**, *13*, 2197–2205. [CrossRef] [PubMed]
74. Bell, J.A.; Reed, M.A.; Consitt, L.A.; Martin, O.J.; Haynie, K.R.; Hulver, M.W.; Muoio, D.M.; Dohm, G.L. Lipid partitioning, incomplete fatty acid oxidation, and insulin signal transduction in primary human muscle cells: Effects of severe obesity, fatty acid incubation, and fatty acid translocase/CD36 overexpression. *J. Clin. Endocrinol. Metab.* **2010**, *95*, 3400–3410. [CrossRef] [PubMed]
75. Love-Gregory, L.; Sherva, R.; Sun, L.; Wasson, J.; Schappe, T.; Doria, A.; Rao, D.C.; Hunt, S.C.; Klein, S.; Neuman, R.J.; et al. Variants in the CD36 gene associate with the metabolic syndrome and high-density lipoprotein cholesterol. *Hum. Mol. Genet.* **2008**, *17*, 1695–1704. [CrossRef] [PubMed]
76. Nicholson, A.C.; Han, J.; Febbraio, M.; Silversterin, R.L.; Hajjar, D.P. Role of CD36, the macrophage class B scavenger receptor, in atherosclerosis. *Ann. N. Y. Acad. Sci.* **2001**, *947*, 224–228. [CrossRef] [PubMed]
77. Silverstein, R.L.; Febbraio, M. CD36, a scavenger receptor involved in immunity, metabolism, angiogenesis, and behavior. *Sci Signal.* **2009**, *2*, re3. [CrossRef] [PubMed]
78. Le Foll, C.; Dunn-Meynell, A.A.; Levin, B.E. Role of FAT/CD36 in fatty acid sensing, energy, and glucose homeostasis regulation in DIO and DR rats. *Am. J. Physiol. Regul. Integr. Comp. Physiol.* **2015**, *308*, R188–R198. [CrossRef] [PubMed]
79. Berger, E.; Heraud, S.; Mojallal, A.; Lequeux, C.; Weiss-Gayet, M.; Damour, O.; Geloen, A. Pathways commonly dysregulated in mouse and human obese adipose tissue: FAT/CD36 modulates differentiation and lipogenesis. *Adipocyte* **2015**, *4*, 161–180. [CrossRef] [PubMed]
80. Febbraio, M.; Hajjar, D.P.; Silverstein, R.L. CD36, a class B scavenger receptor involved in angiogenesis, atherosclerosis, inflammation, and lipid metabolism. *J. Clin. Investig.* **2001**, *108*, 785–791. [CrossRef] [PubMed]
81. Gautam, S.; Banerjee, M. The macrophage Ox-LDL receptor, CD36 and its association with type II diabetes mellitus. *Mol. Genet. Metab.* **2011**, *102*, 389–398. [CrossRef] [PubMed]
82. Rodrigue-Way, A.; Caron, V.; Bilodeau, S.; Keil, S.; Hassan, M.; Levy, E.; Mitchell, G.A.; Tremblay, A. Scavenger receptor CD36 mediates inhibition of cholesterol synthesis via activation of the PPARγ/PGC-1α pathway and Insig1/2 expression in hepatocytes. *FASEB J.* **2014**, *28*, 1910–1923. [CrossRef] [PubMed]
83. Demers, A.; Rodrigue-Way, A.; Tremblay, A. Hexarelin Signaling to PPARγ in Metabolic Diseases. *PPAR Res.* **2008**, *2008*, 364784. [CrossRef] [PubMed]
84. Goldstein, J.L.; Ho, Y.K.; Basu, S.K.; Brown, M.S. Binding site on macrophages that mediates uptake and degradation of acetylated low density lipoprotein, producing massive cholesterol deposition. *Proc. Natl. Acad. Sci. USA* **1979**, *76*, 333–337. [CrossRef] [PubMed]
85. Moore, K.J.; Freeman, M.W. Scavenger receptors in atherosclerosis: Beyond lipid uptake. *Arterioscler. Thromb. Vasc. Biol.* **2006**, *26*, 1702–1711. [CrossRef] [PubMed]

86. Nicholson, A.C. Expression of CD36 in macrophages and atherosclerosis: The role of lipid regulation of PPARγ signaling. *Trends Cardiovasc. Med.* **2004**, *14*, 8–12. [CrossRef] [PubMed]
87. Chawla, A.; Boisvert, W.A.; Lee, C.H.; Laffitte, B.A.; Barak, Y.; Joseph, S.B.; Liao, D.; Nagy, L.; Edwards, P.A.; Curtiss, L.K.; et al. A PPARγ-LXR-ABCA1 pathway in macrophages is involved in cholesterol efflux and atherogenesis. *Mol. Cell* **2001**, *7*, 161–171. [CrossRef]
88. Febbraio, M.; Podrez, E.A.; Smith, J.D.; Hajjar, D.P.; Hazen, S.L.; Hoff, H.F.; Sharma, K.; Silverstein, R.L. Targeted disruption of the class B scavenger receptor CD36 protects against atherosclerotic lesion development in mice. *J. Clin. Investig.* **2000**, *105*, 1049–1056. [CrossRef] [PubMed]
89. Febbraio, M.; Abumrad, N.A.; Hajjar, D.P.; Sharma, K.; Cheng, W.; Pearce, S.F.; Silverstein, R.L. A null mutation in murine CD36 reveals an important role in fatty acid and lipoprotein metabolism. *J. Biol. Chem.* **1999**, *274*, 19055–19062. [CrossRef] [PubMed]
90. Nozaki, S.; Kashiwagi, H.; Yamashita, S.; Nakagawa, T.; Kostner, B.; Tomiyama, Y.; Nakata, A.; Ishigami, M.; Miyagawa, J.; Kameda-Takemura, K.; et al. Reduced uptake of oxidized low density lipoproteins in monocyte-derived macrophages from CD36-deficient subjects. *J. Clin. Investig.* **1995**, *96*, 1859–1865. [CrossRef] [PubMed]
91. Nakata, A.; Nakagawa, Y.; Nishida, M.; Nozaki, S.; Miyagawa, J.; Nakagawa, T.; Tamura, R.; Matsumoto, K.; Kameda-Takemura, K.; Yamashita, S.; et al. CD36, a novel receptor for oxidized low-density lipoproteins, is highly expressed on lipid-laden macrophages in human atherosclerotic aorta. *Arterioscler. Thromb. Vasc. Biol.* **1999**, *19*, 1333–1339. [CrossRef] [PubMed]
92. Yamashita, S.; Hirano, K.; Kuwasako, T.; Janabi, M.; Toyama, Y.; Ishigami, M.; Sakai, N. Physiological and pathological roles of a multi-ligand receptor CD36 in atherogenesis; insights from CD36-deficient patients. *Mol. Cell Biochem.* **2007**, *299*, 19–22. [CrossRef] [PubMed]
93. Zhao, L.; Varghese, Z.; Moorhead, J.F.; Chen, Y.; Ruan, X.Z. CD36 and lipid metabolism in the evolution of atherosclerosis. *Br. Med. Bull.* **2018**. [CrossRef] [PubMed]
94. Love-Gregory, L.; Abumrad, N.A. CD36 genetics and the metabolic complications of obesity. *Curr. Opin. Clin. Nutr. Metab. Care* **2011**, *14*, 527–534. [CrossRef] [PubMed]
95. Rac, M.E.; Suchy, J.; Kurzawski, G.; Kurlapska, A.; Safranow, K.; Rac, M.; Sagasz-Tysiewicz, D.; Krzystolik, A.; Poncyljusz, W.; Jakubowska, K.; et al. Polymorphism of the CD36 gene and cardiovascular risk factors in patients with coronary artery disease manifested at a young age. *Biochem. Genet.* **2012**, *50*, 103–111. [CrossRef] [PubMed]
96. Melis, M.; Carta, G.; Pintus, S.; Pintus, P.; Piras, C.A.; Murru, E.; Manca, C.; Di Marzo, V.; Banni, S.; Tomassini Barbarossa, I. Polymorphism rs1761667 in the CD36 gene is associated to changes in fatty acid metabolism and circulating endocannabinoid levels distinctively in normal weight and obese subjects. *Front Physiol.* **2017**, *8*, 1006. [CrossRef] [PubMed]
97. Plesnik, J.; Sery, O.; Khan, A.S.; Bielik, P.; Khan, N.A. The rs1527483, but not rs3212018, CD36 polymorphism associates with linoleic acid detection and obesity in Czech young adults. *Br. J. Nutr.* **2018**, *119*, 472–478. [CrossRef] [PubMed]
98. Rac, M.E.; Safranow, K.; Garanty-Bogacka, B.; Dziedziejko, V.; Kurzawski, G.; Goschorska, M.; Kuligowska, A.; Pauli, N.; Chlubek, D. CD36 gene polymorphism and plasma sCD36 as the risk factor in higher cholesterolemia. *Arch. Pediatr.* **2018**, *25*, 177–181. [CrossRef] [PubMed]
99. Yuasa-Kawase, M.; Masuda, D.; Yamashita, T.; Kawase, R.; Nakaoka, H.; Inagaki, M.; Nakatani, K.; Tsubakio-Yamamoto, K.; Ohama, T.; Matsuyama, A.; et al. Patients with CD36 deficiency are associated with enhanced atherosclerotic cardiovascular diseases. *J. Atheroscler. Thromb.* **2012**, *19*, 263–275. [CrossRef] [PubMed]
100. Marleau, S.; Harb, D.; Bujold, K.; Avallone, R.; Iken, K.; Wang, Y.; Demers, A.; Sirois, M.G.; Febbraio, M.; Silverstein, R.L.; et al. EP 80317, a ligand of the CD36 scavenger receptor, protects apolipoprotein E-deficient mice from developing atherosclerotic lesions. *FASEB J.* **2005**, *19*, 1869–1871. [CrossRef] [PubMed]
101. Avallone, R.; Demers, A.; Rodrigue-Way, A.; Bujold, K.; Harb, D.; Anghel, S.; Wahli, W.; Marleau, S.; Ong, H.; Tremblay, A. A growth hormone-releasing peptide that binds scavenger receptor CD36 and ghrelin receptor up-regulates sterol transporters and cholesterol efflux in macrophages through a PPARγ-dependent pathway. *Mol. Endocrinol.* **2006**, *20*, 3165–3178. [CrossRef] [PubMed]
102. Laffitte, B.A.; Joseph, S.B.; Walczak, R.; Pei, L.; Wilpitz, D.C.; Collins, J.L.; Tontonoz, P. Autoregulation of the human liver X receptor α promoter. *Mol. Cell Biol.* **2001**, *21*, 7558–7568. [CrossRef] [PubMed]

103. Calkin, A.C.; Tontonoz, P. Transcriptional integration of metabolism by the nuclear sterol-activated receptors LXR and FXR. *Nat. Rev. Mol. Cell Biol.* **2012**, *13*, 213–224. [CrossRef] [PubMed]
104. Demers, A.; Caron, V.; Rodrigue-Way, A.; Wahli, W.; Ong, H.; Tremblay, A. A Concerted kinase interplay identifies PPARγ as a molecular target of ghrelin signaling in macrophages. *PLoS ONE* **2009**, *4*, e7728. [CrossRef] [PubMed]
105. Pang, J.; Xu, Q.; Xu, X.; Yin, H.; Xu, R.; Guo, S.; Hao, W.; Wang, L.; Chen, C.; Cao, J.M. Hexarelin suppresses high lipid diet and vitamin D3-induced atherosclerosis in the rat. *Peptides* **2010**, *31*, 630–638. [CrossRef] [PubMed]
106. Kershaw, E.E.; Flier, J.S. Adipose tissue as an endocrine organ. *J. Clin. Endocrinol. Metab.* **2004**, *89*, 2548–2556. [CrossRef] [PubMed]
107. Rosen, E.D.; Spiegelman, B.M. Adipocytes as regulators of energy balance and glucose homeostasis. *Nature* **2006**, *444*, 847–853. [CrossRef] [PubMed]
108. He, W.; Barak, Y.; Hevener, A.; Olson, P.; Liao, D.; Le, J.; Nelson, M.; Ong, E.; Olefsky, J.M.; Evans, R.M. Adipose-specific peroxisome proliferator-activated receptor γ knockout causes insulin resistance in fat and liver but not in muscle. *Proc. Natl. Acad. Sci. USA* **2003**, *100*, 15712–15717. [CrossRef] [PubMed]
109. Imai, T.; Takakuwa, R.; Marchand, S.; Dentz, E.; Bornert, J.M.; Messaddeq, N.; Wendling, O.; Mark, M.; Desvergne, B.; Wahli, W.; et al. Peroxisome proliferator-activated receptor γ is required in mature white and brown adipocytes for their survival in the mouse. *Proc. Natl. Acad. Sci. USA* **2004**, *101*, 4543–4547. [CrossRef] [PubMed]
110. Jeninga, E.H.; Gurnell, M.; Kalkhoven, E. Functional implications of genetic variation in human PPARγ. *Trends Endocrinol. Metab.* **2009**, *20*, 380–387. [CrossRef] [PubMed]
111. Thompson, N.M.; Gill, D.A.; Davies, R.; Loveridge, N.; Houston, P.A.; Robinson, I.C.; Wells, T. Ghrelin and des-octanoyl ghrelin promote adipogenesis directly in vivo by a mechanism independent of the type 1a growth hormone secretagogue receptor. *Endocrinology* **2004**, *145*, 234–242. [CrossRef] [PubMed]
112. Callaghan, B.; Furness, J.B. Novel and conventional receptors for ghrelin, desacyl-ghrelin, and pharmacologically related compounds. *Pharmacol. Rev.* **2014**, *66*, 984–1001. [CrossRef] [PubMed]
113. Rodrigue-Way, A.; Demers, A.; Ong, H.; Tremblay, A. A growth hormone-releasing peptide promotes mitochondrial biogenesis and a fat burning-like phenotype through scavenger receptor CD36 in white adipocytes. *Endocrinology* **2007**, *148*, 1009–1018. [CrossRef] [PubMed]
114. Mosa, R.; Huang, L.; Wu, Y.; Fung, C.; Mallawakankanamalage, O.; LeRoith, D.; Chen, C. Hexarelin, a growth hormone secretagogue, improves lipid metabolic aberrations in nonobese insulin-resistant male MKR mice. *Endocrinology* **2017**, *158*, 3174–3187. [CrossRef] [PubMed]
115. Feige, J.N.; Gelman, L.; Michalik, L.; Desvergne, B.; Wahli, W. From molecular action to physiological outputs: Peroxisome proliferator-activated receptors are nuclear receptors at the crossroads of key cellular functions. *Prog. Lipid. Res.* **2006**, *45*, 120–159. [CrossRef] [PubMed]
116. Bensinger, S.J.; Tontonoz, P. Integration of metabolism and inflammation by lipid-activated nuclear receptors. *Nature* **2008**, *454*, 470–477. [CrossRef] [PubMed]
117. Puigserver, P.; Spiegelman, B.M. Peroxisome proliferator-activated receptor-γ coactivator 1α (PGC-1α): Transcriptional coactivator and metabolic regulator. *Endocr. Rev.* **2003**, *24*, 78–90. [CrossRef] [PubMed]
118. Spiegelman, B.M.; Heinrich, R. Biological Control through Regulated Transcriptional Coactivators. *Cell* **2004**, *119*, 157–167. [CrossRef] [PubMed]
119. Kazak, L.; Chouchani, E.T.; Stavrovskaya, I.G.; Lu, G.Z.; Jedrychowski, M.P.; Egan, D.F.; Kumari, M.; Kong, X.; Erickson, B.K.; Szpyt, J.; et al. UCP1 deficiency causes brown fat respiratory chain depletion and sensitizes mitochondria to calcium overload-induced dysfunction. *Proc. Natl. Acad. Sci. USA* **2017**, *114*, 7981–7986. [CrossRef] [PubMed]
120. Sirago, G.; Conte, E.; Fracasso, F.; Cormio, A.; Fehrentz, J.A.; Martinez, J.; Musicco, C.; Camerino, G.M.; Fonzino, A.; Rizzi, L.; et al. Growth hormone secretagogues hexarelin and JMV2894 protect skeletal muscle from mitochondrial damages in a rat model of cisplatin-induced cachexia. *Sci. Rep.* **2017**, *7*, 13017. [CrossRef] [PubMed]
121. Vega, R.B.; Kelly, D.P. A role for estrogen-related receptor α in the control of mitochondrial fatty acid β-oxidation during brown adipocyte differentiation. *J. Biol. Chem.* **1997**, *272*, 31693–31699. [CrossRef] [PubMed]

122. Schreiber, S.N.; Emter, R.; Hock, M.B.; Knutti, D.; Cardenas, J.; Podvinec, M.; Oakeley, E.J.; Kralli, A. The estrogen-related receptor α (ERRα) functions in PPARγ coactivator 1α (PGC-1α)-induced mitochondrial biogenesis. *Proc. Natl. Acad. Sci. USA* **2004**, *101*, 6472–6477. [CrossRef] [PubMed]
123. Villena, J.A.; Hock, M.B.; Chang, W.Y.; Barcas, J.E.; Giguere, V.; Kralli, A. Orphan nuclear receptor estrogen-related receptor α is essential for adaptive thermogenesis. *Proc. Natl. Acad. Sci. USA* **2007**, *104*, 1418–1423. [CrossRef] [PubMed]
124. Rachek, L.I.; Yuzefovych, L.V.; Ledoux, S.P.; Julie, N.L.; Wilson, G.L. Troglitazone, but not rosiglitazone, damages mitochondrial DNA and induces mitochondrial dysfunction and cell death in human hepatocytes. *Toxicol. Appl. Pharmacol.* **2009**, *240*, 348–354. [CrossRef] [PubMed]
125. Saha, S.; New, L.S.; Ho, H.K.; Chui, W.K.; Chan, E.C. Direct toxicity effects of sulfo-conjugated troglitazone on human hepatocytes. *Toxicol. Lett.* **2010**, *195*, 135–141. [CrossRef] [PubMed]
126. Lavoie, J.C.; Tremblay, A. Sex-specificity of oxidative stress in newborns leading to a personalized antioxidant nutritive strategy. *Antioxidants (Basel)* **2018**, *7*. [CrossRef] [PubMed]
127. Salomone, S. Pleiotropic effects of glitazones: A double edge sword? *Front Pharmacol.* **2011**, *2*, 14. [CrossRef] [PubMed]
128. Sanyal, A.J.; Chalasani, N.; Kowdley, K.V.; McCullough, A.; Diehl, A.M.; Bass, N.M.; Neuschwander-Tetri, B.A.; Lavine, J.E.; Tonascia, J.; Unalp, A.; et al. Pioglitazone, vitamin E, or placebo for nonalcoholic steatohepatitis. *N. Engl. J. Med.* **2010**, *362*, 1675–1685. [CrossRef] [PubMed]
129. Cusi, K.; Orsak, B.; Bril, F.; Lomonaco, R.; Hecht, J.; Ortiz-Lopez, C.; Tio, F.; Hardies, J.; Darland, C.; Musi, N.; et al. Long-term pioglitazone treatment for patients with nonalcoholic steatohepatitis and prediabetes or type 2 diabetes mellitus: A randomized trial. *Ann. Intern. Med.* **2016**, *165*, 305–315. [CrossRef] [PubMed]
130. Vidal-Puig, A.; Jimenez-Linan, M.; Lowell, B.B.; Hamann, A.; Hu, E.; Spiegelman, B.; Flier, J.S.; Moller, D.E. Regulation of PPARγ gene expression by nutrition and obesity in rodents. *J. Clin. Investig.* **1996**, *97*, 2553–2561. [CrossRef] [PubMed]
131. Patsouris, D.; Reddy, J.K.; Muller, M.; Kersten, S. Peroxisome proliferator-activated receptor α mediates the effects of high-fat diet on hepatic gene expression. *Endocrinology* **2006**, *147*, 1508–1516. [CrossRef] [PubMed]
132. Edvardsson, U.; Ljungberg, A.; Oscarsson, J. Insulin and oleic acid increase PPARγ2 expression in cultured mouse hepatocytes. *Biochem. Biophys. Res. Commun.* **2006**, *340*, 111–117. [CrossRef] [PubMed]
133. Goldstein, J.L.; Debose-Boyd, R.A.; Brown, M.S. Protein sensors for membrane sterols. *Cell* **2006**, *124*, 35–46. [CrossRef] [PubMed]
134. Brown, M.S.; Radhakrishnan, A.; Goldstein, J.L. Retrospective on cholesterol homeostasis: The central role of Scap. *Annu. Rev. Biochem.* **2017**, *87*, 1.1–1.25. [CrossRef] [PubMed]
135. Habets, D.D.; Coumans, W.A.; El Hasnaoui, M.; Zarrinpashneh, E.; Bertrand, L.; Viollet, B.; Kiens, B.; Jensen, T.E.; Richter, E.A.; Bonen, A.; et al. Crucial role for LKB1 to AMPKα2 axis in the regulation of CD36-mediated long-chain fatty acid uptake into cardiomyocytes. *Biochim. Biophys. Acta.* **2009**, *1791*, 212–219. [CrossRef] [PubMed]
136. Abumrad, N.A.; Goldberg, I.J. CD36 actions in the heart: Lipids, calcium, inflammation, repair and more? *Biochim. Biophys. Acta.* **2016**, *1861*, 1442–1449. [CrossRef] [PubMed]
137. Sever, N.; Yang, T.; Brown, M.S.; Goldstein, J.L.; Debose-Boyd, R.A. Accelerated degradation of HMG CoA reductase mediated by binding of insig-1 to its sterol-sensing domain. *Mol. Cell* **2003**, *11*, 25–33. [CrossRef]
138. Jeon, T.I.; Osborne, T.F. SREBPs: Metabolic integrators in physiology and metabolism. *Trends Endocrinol. Metab.* **2012**, *23*, 65–72. [CrossRef] [PubMed]
139. Handschin, C.; Spiegelman, B.M. Peroxisome proliferator-activated receptor γ coactivator 1 coactivators, energy homeostasis, and metabolism. *Endocr. Rev.* **2006**, *27*, 728–735. [CrossRef] [PubMed]
140. Finck, B.N.; Kelly, D.P. PGC-1 coactivators: Inducible regulators of energy metabolism in health and disease. *J. Clin. Investig.* **2006**, *116*, 615–622. [CrossRef] [PubMed]

International Journal of
Molecular Sciences

Review

Insights into the Role of PPARβ/δ in NAFLD

Jiapeng Chen [1,2], Alexandra Montagner [3,4], Nguan Soon Tan [1,2,5,6] and Walter Wahli [1,3,7,*]

1 Lee Kong Chian School of Medicine, Nanyang Technological University, 11 Mandalay Road, Singapore 308232, Singapore; jchen025@e.ntu.edu.sg (J.C.); nstan@ntu.edu.sg (N.S.T.)
2 School of Biological Sciences, Nanyang Technological University, 60 Nanyang Drive, Singapore 637551, Singapore
3 ToxAlim, Research Center in Food Toxicology, National Institute for Agricultural Research (INRA), 180 Chemin de Tournefeuille, 31300 Toulouse, France; alexandra.montagner@inserm.fr
4 Institut National de La Santé et de La Recherche Médicale (INSERM), UMR1048, Institute of Metabolic and Cardiovascular Diseases, 31027 Toulouse, France
5 KK Research Centre, KK Women's and Children Hospital, 100 Bukit Timah Road, Singapore 229899, Singapore
6 Institute of Molecular and Cell Biology, Agency for Science Technology & Research, 61 Biopolis Drive, Proteos, Singapore 138673, Singapore
7 Center for Integrative Genomics, University of Lausanne, Génopode, CH-1015 Lausanne, Switzerland
* Correspondence: walter.wahli@ntu.edu.sg; Tel.: +65-69047012

Received: 15 May 2018; Accepted: 23 June 2018; Published: 27 June 2018

Abstract: Non-alcoholic fatty liver disease (NAFLD) is a major health issue in developed countries. Although usually associated with obesity, NAFLD is also diagnosed in individuals with low body mass index (BMI) values, especially in Asia. NAFLD can progress from steatosis to non-alcoholic steatohepatitis (NASH), which is characterized by liver damage and inflammation, leading to cirrhosis and hepatocellular carcinoma (HCC). NAFLD development can be induced by lipid metabolism alterations; imbalances of pro- and anti-inflammatory molecules; and changes in various other factors, such as gut nutrient-derived signals and adipokines. Obesity-related metabolic disorders may be improved by activation of the nuclear receptor peroxisome proliferator-activated receptor (PPAR)β/δ, which is involved in metabolic processes and other functions. This review is focused on research findings related to PPARβ/δ-mediated regulation of hepatic lipid and glucose metabolism and NAFLD development. It also discusses the potential use of pharmacological PPARβ/δ activation for NAFLD treatment.

Keywords: PPARβ/δ; NAFLD; NASH; steatosis; liver; lipid metabolism

---

1. Introduction

Non-alcoholic fatty liver disease (NAFLD) is an inclusive term describing a broad range of chronic liver pathologies [1]. During the development of this chronic condition, several potentially pathogenic mediators are crucially involved [2]. Risk factors for NAFLD include obesity, insulin resistance, and other features of metabolic syndrome. Steatosis is the initial benign stage, characterized by lipid accumulation in hepatocytes due to impaired triglyceride synthesis and export, and/or reduced fatty acid beta-oxidation. Patients with steatosis may progress to non-alcoholic steatohepatitis (NASH), a more severe form of NAFLD that involves hepatocellular injury and liver inflammation—both drivers of hepatic fibrosis [3]. NASH can lead to more deleterious conditions, such as cirrhosis and hepatocellular carcinoma (HCC) [4]. NASH is rapidly becoming a leading cause of end-stage liver disease and hepatocellular carcinoma, both of which are indications for liver transplantation [5].

As obesity rates have risen, NAFLD has become the most common chronic liver disease in humans and is considered an epidemic disease that constitutes a major global health issue. NAFLD affects

70% of type 2 diabetes patients, and even a greater proportion of obese diabetic individuals [6,7]. Astonishingly, NAFLD affects nearly 30% of the general population worldwide [8–10] and has potentially serious sequelae [11]. Although steatosis is considered a relatively benign condition, about 30% of patients with steatosis will develop NASH, and 30–40% of patients with NASH will progress to fibrosis and cirrhosis. Among patients with cirrhosis, 4% will develop hepatocellular carcinoma with a 10-year mortality rate of 25% [12–14].

Although the majority of affected individuals are asymptomatic, NAFLD can be detected by ultrasound scanning or routine blood testing for elevated plasma levels of the liver enzymes alanine aminotransferase and aspartate aminotransferase, reflecting hepatocyte injury. On the other hand, NASH diagnosis requires a liver biopsy and histological scoring. Individuals who are diabetic or obese, or who suffer from metabolic syndrome, should be suspected as having NAFLD and should be examined accordingly [15–17].

Body weight reduction through increased physical activity and dietary improvement can help with NAFLD management and delay disease progression. However, long-term lifestyle changes may be insufficient in many cases [18–20]. Notably, there is currently no effective FDA-approved therapy for the prevention and/or treatment of NAFLD development and progression, although several drugs are currently being tested in clinical trials [21]. Pharmacological treatments that target insulin resistance, including metformin and thiazolidinediones (TZDs), have been tested in NAFLD patients and those diagnosed with NASH. These studies have not demonstrated that metformin is effective for NAFLD treatment [21,22]. TZDs reportedly lead to decreased hepatic fat and reduced liver injury; however, TZD discontinuation allows NASH recurrence, and long-term TZD treatment can result in medical complications, such as congestive heart failure, osteoporosis, and weight gain in susceptible patients [23,24]. Thus, other than weight loss, there are currently no effective interventions and therapies for NAFLD treatment [18–21].

Peroxisome proliferator-activated receptor (PPAR)β/δ is a nuclear receptor that is closely related to PPARγ, which is activated by TZDs, as well as to PPARα, which is targeted by hypolipidemic agents of the fibrate class. PPARβ/δ exerts a variety of metabolic effects and physiological actions [25–29], and PPARβ/δ activation may inhibit and improve obesity-related metabolic disorders. In the present review, we discuss the involvement of PPARβ/δ in NAFLD, and the effects of PPARβ/δ agonists on this pathology.

2. Hallmark of NAFLD

2.1. Two-Hit Hypothesis

It has been proposed that NAFLD pathogenesis is a "two-hit" process (Figure 1) [30,31]. In this hypothesis, the first hit results from triglyceride accumulation in the hepatocyte cytoplasm due to an imbalance in lipid input and output, which is the hallmark of NAFLD [30]. Four mechanisms can contribute to triglyceride accumulation in hepatocytes: (1) upregulated free fatty acid uptake from blood plasma in the context of increased lipolysis from adipose tissue and/or chylomicrons after high-fat diet consumption [32]; (2) high carbohydrate uptake that increases circulating glucose and insulin levels, thus promoting de novo lipogenesis and contributing to triglyceride accumulation in hepatocytes [33,34]; (3) decreased fatty acid mitochondrial oxidation; and (4) reduced hepatic triglyceride secretion via packaging of apolipoprotein B (ApoB) into very low-density lipoprotein (VLDL) particles, promoting triglyceride accumulation in hepatocytes [33–35]. Overall, aberrations in any lipid metabolism processes, which may involve a large number of genes, can result in NAFLD development [36].

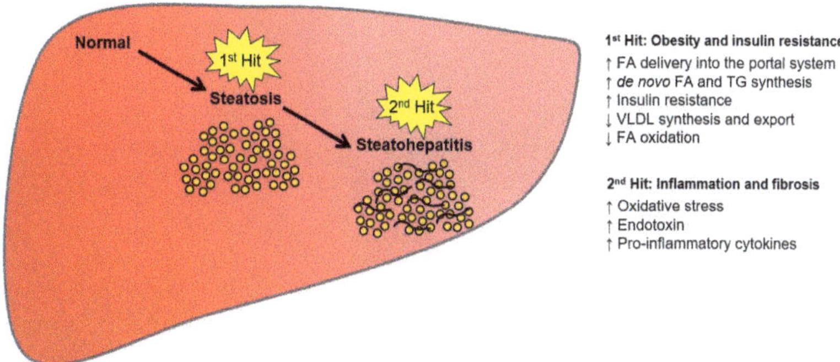

Figure 1. Schematic diagram of the two-hit hypothesis of non-alcoholic fatty liver disease (NAFLD) progression. In the first hit, an imbalance of lipid synthesis, catabolism, and export results in lipid accumulation in liver (steatosis). Obesity and insulin resistance are strongly correlated with liver steatosis. In the second hit, further inflammation processes lead to non-alcoholic steatohepatitis (NASH) and fibrosis, which can evolve into more severe stages, such as cirrhosis and ultimately hepatocellular carcinoma.

The second hit in this NAFLD progression model is an imbalance of pro- and anti-inflammatory factors, resulting in increased inflammation, as seen in NASH [30]. Hence, the most critical and challenging step in NAFLD progression is the transition from relatively benign steatosis to the damaged and inflamed liver in NASH. Any strong chronic inflammation will cause fibrosis, thereby contributing to the development of cirrhosis and eventually hepatocellular carcinoma [37].

2.2. Multiple Parallel Hit Hypothesis

The multiple parallel hit hypothesis considers alterations in the regulation of several factors, including gut nutrient-derived signals, adipokines, and certain pro-inflammatory cytokines (Figure 2) [38]. Insulin resistance leads to alterations of nutrient metabolism and is thus commonly associated with NAFLD development [39]. Elevated levels of inflammatory cytokines, such as interleukin 6 (IL6) and tumor necrosis factor α (TNFα), result in hepatic inflammation [40]. The administration of TNFα antibody into *ob/ob* mice induces steatosis improvement, supporting a role of TNFα in NAFLD progression. Moreover, hepatic steatosis can be induced through primary inflammation in *ob/ob* mice [41]. In humans, inflammation is occasionally observed before steatosis, as seen in patients who have NASH but exhibit lower levels of steatosis [42].

Genome-wide association studies (GWAS) have identified genes that are involved in diseases and that can be targeted for disease treatments. A GWAS of various races found that NAFLD was linked to a polymorphism in the patatin-like phospholipase domain containing 3 (*PNPLA3*) gene [43]. PNPLA3 is a multifunctional enzyme involved in triacylglycerol hydrolysis and acyl-CoA-independent transacylation of acylglycerols [44]. The nonsynonymous rs738409 C/G variant in *PNPLA3* encodes I148M. It is proposed to be the main genetic component of NAFLD and NASH [45]. It reportedly shows the strongest risk effect on NAFLD development, accounting for 5.3% of total variance, and is associated with histological disease severity and NAFLD progression [45,46]. In patients with the single *PNPLA3* nucleotide polymorphism rs738409 G/G, fatty liver progresses directly to NASH [47,48]. Notably, mice with *Pnpla3* deficiency do not develop fatty liver or liver injury [49], and *Pnpla3* knockdown decreases intracellular triglyceride levels in primary hepatocyte cultures [50]. Thus, the function of PNPLA3 in NAFLD warrants further investigation. Interestingly, *Pnpla3* is a downstream target gene of sterol-regulated binding protein 1c (SREBP1c) and can mediate its effect in promoting lipid accumulation. Therefore, PNPLA3 has been suggested as a possible "first hit", preceding other hits that may affect disease progression [51].

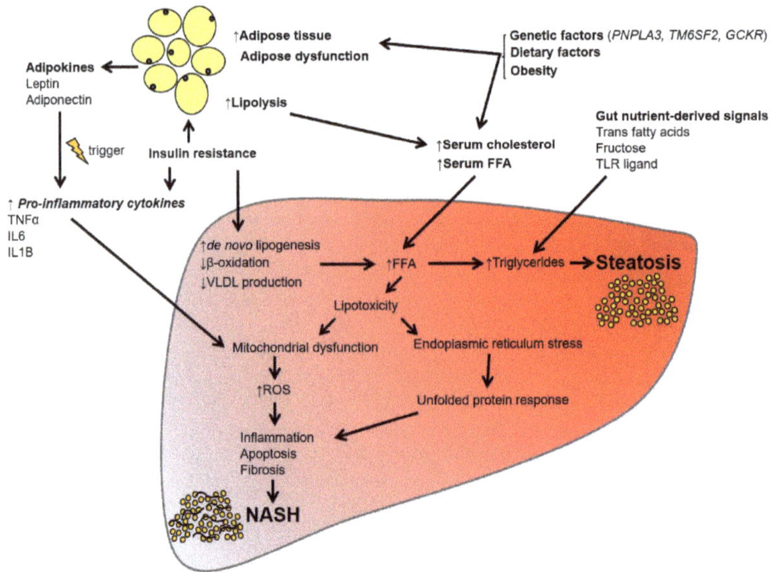

Figure 2. Schematic illustration of the multiple parallel hits hypothesis of NAFLD development. NAFLD develops due to the impaired regulation of several factors, such as gut nutrient-derived signals, adipokines, and certain pro-inflammatory cytokines.

Two other widely studied genetic modifiers of NAFLD are the transmembrane 6 superfamily member 2 (*TM6SF2*) and glucokinase regulator (*GCKR*) genes. TM6SF2 regulates liver fat metabolism, influencing triglyceride secretion and hepatic lipid droplet content [52]. The nonsynonymous rs58542926 variant in *TM6SF2* encodes E167K and is associated with increased liver fat levels [53]. Patients with NAFLD show significantly lower TM6SF2 expression in the liver [54]. With regards to NAFLD risk alleles of *TM6SF2*, the C (Glu167) allele is correlated with higher cardiovascular risk via elevated circulating low-density lipoprotein (LDL)-cholesterol levels [55], and the T (Lys167) allele is associated with NAFLD and NASH [54,56,57]. *GCKR* encodes the glucokinase regulatory protein, which controls the activity and intracellular location of glucokinase, a key enzyme in glucose metabolism [58]. The *GCKR* missense variant rs780094 is significantly associated with histological NAFLD [59,60]. Moreover, *GCKR* mutations reportedly cause maturity-onset diabetes in young individuals with NAFLD risk factors, such as glucose intolerance and insulin resistance [61]. Histological NAFLD is also significantly associated with variants in or near the neurocan (*NCAN)* and lysophospholipase like 1 (*LYPLAL1*), but not protein phosphatase 1 regulatory subunit 3B (*PPP1R3B*) genes [59].

Obesity is another increasingly common global condition that is associated with diseases, including NAFLD, hypertension, type 2 diabetes mellitus, and hyperlipidemia. In fact, hypertension, hypertriglyceridemia, and obesity are predictive risk factors for NAFLD [62]. Over the past decade, visceral obesity has become more common among adults and children worldwide in association with increased consumption of Western-style diets with high fat and fructose contents [63]. Visceral fat accumulation is positively correlated with various organ pathologies, including NAFLD, as well as with insulin resistance in both obese and non-obese individuals. These findings suggest that visceral fat accumulation influences hepatic steatosis, regardless of the degree of obesity [64].

3. Peroxisome Proliferator-Activated Receptor β/δ Expression in Liver

Peroxisome proliferator-activated receptors (PPARs) belong to the nuclear hormone receptor superfamily, which comprises ligand-activated transcription factors. PPARs play important roles in regulating genes involved in fatty acid uptake and oxidation, lipid and carbohydrate metabolism, vascular biology, inflammation, cell proliferation, and senescence [65–67]. To be transcriptionally active, PPARs must heterodimerize with the 9-cis retinoic acid receptor (RXR) (Figure 3) [68]. If an agonist is absent or in the presence of an antagonist, the PPAR-RXR heterodimer associates with co-repressor proteins. This complex occupies the promoter region within a subset of PPAR target genes, and consequently blocks their transcription. Such co-repressor proteins include the well-known silencing mediator of retinoid and thyroid receptors (SMRT), and the nuclear receptor corepressor (NCoR) [68–70].

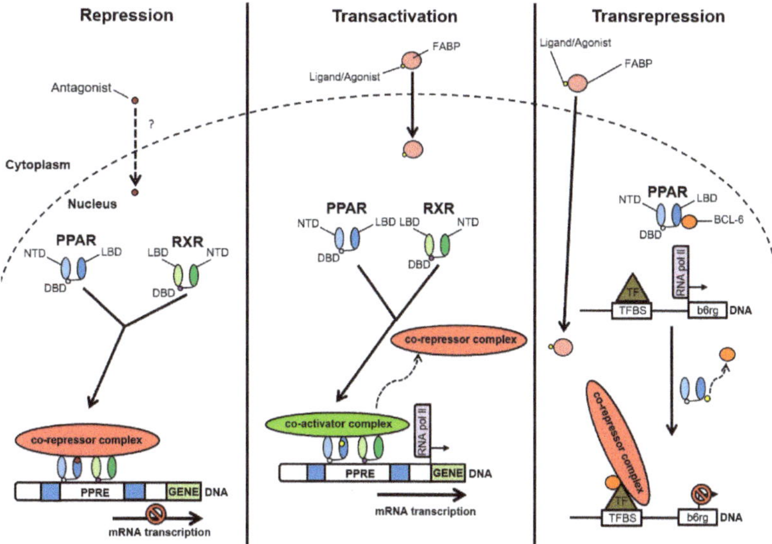

Figure 3. Regulatory mechanisms of gene transcription by peroxisome proliferator-activated receptors (PPARs). Each PPAR structurally comprises an N-terminal domain (NTD), a DNA-binding domain (DBD), and a ligand-binding domain (LBD). In the absence of a ligand or in the presence of an antagonist, the PPAR-RXR heterodimer associates with nuclear receptor co-repressor proteins, leading to repression of PPAR target genes (Repression). Fatty acid-binding protein (FABP) associates with the ligand/agonist to transport it into the cell. Upon ligand binding, a conformational change in PPAR leads to co-repressor dissociation, and co-activators are recruited. The activated PPAR-RXR heterodimer binds the peroxisome proliferator response element (PPRE) and stimulates target gene transcription (Transactivation). In macrophages, endothelial cells, and vascular smooth muscles, in the absence of a PPARβ/δ agonist or ligand, the receptor will scavenge BCL-6 (a PPARβ/δ-associated transcriptional repressor). Once PPARβ/δ neutralizes BCL-6, transcription factors (TFs) bind to TF-binding sites (TFBSs), allows transcription of the genes repressed by BCL-6. However, the binding of a PPARβ/δ ligand to PPARβ/δ will result in BCL-6 dissociation, leading to co-repressor-dependent transrepression of BCL-6 targeted genes, such as *b6rg*, which encodes a sequence-specific transcription repressor (Transrepression). The dashed arrow with a question mark indicates that it is not known how the antagonist is translocated to the cell nucleus. The curvy arrow indicates the dissociation of the co-repressor from the transcription factor.

On the other hand, in the presence of an agonist, PPAR activation results in an exchange within the co-regulator complex. This involves co-activator recruitment upon co-repressor dissociation. Activated PPAR-RXR heterodimers bind to peroxisome proliferator response elements (PPREs) located in the regulatory regions (5′-end region and introns) of PPAR target genes [68,71,72]. This results in altered expression levels of PPAR target genes. PPAR and RXR bind to the 5′ and 3′ half-sites of the PPRE, respectively [73]. The 5′ flanking region of the PPRE contributes to the selectivity of binding of the different PPAR isotypes [74], but the selection of the PPAR target genes to be activated by a given PPAR isotype in vivo is not yet well understood. It is thought that it results from a complex interplay between expression levels of the three isotypes in the cell, ligand and cofactor availability, affinity for a given PPRE, and probably factors binding in the vicinity of the PPRE [72]. Comprehensive studies integrating expression profiling and genome-wide promoter binding by the PPARs are required to better understand the promoter-specific mechanisms of PPAR action. Interestingly, PPAR/RXR heterodimers can induce transcription in response to PPAR or RXR ligand-dependent activation and the relative levels of cofactor expression are important determinants of the specificity of the physiological responses to PPAR or RXR agonists [72]. Studies of PPARs' roles in reducing the expression of a subset of inflammatory response genes have highlighted a repressive molecular mode of action, termed transrepression, through which PPARs impact key transcription factor activity. Transrepression occurs through tethering, in which direct protein–protein interactions inhibit the binding of transcription factors to DNA. The regulation of gene transcription by PPAR can also take place through the sequestration of coactivators or the release of corepressors, which stimulates and represses promoter activity, respectively (Figure 3) [72].

The PPAR family includes three isotypes—PPARα, PPARβ/δ, and PPARγ—which have the canonical nuclear receptor domain organization [68,75]. The N-terminal A/B domain possesses a weak ligand-independent transactivation function known as activation function (AF)-1. The C domain binds DNA via two zinc-finger motifs, and the D domain is a hinge region. The E domain contains the ligand-binding domain (LBD), possesses the ligand-dependent transactivation function termed AF-2, and includes the region for dimerization and interaction with regulatory proteins [76,77]. PPARβ/δ also functions in the regulation of gene expression independently of DNA binding, through cross-talk with other transcription factors, which consequently influences their transrepressor function. For example, PPARβ/δ associates with the transcriptional repressor B-cell lymphoma-6 (BCL-6) (Figure 3) in macrophages, endothelial cells, and vascular smooth muscle cells [78,79]. In the presence of a PPARβ/δ agonist, BCL-6 dissociates from PPARβ/δ and subsequently binds to promoter regions of pro-inflammatory genes, such as vascular cell adhesion molecule-1 (*VCAM-1*) and E-selectin. With the aid of a co-repressor complex, such binding will repress the transcription of these genes [29,80,81].

4. Hepatic Functions of PPARβ/δ Compared to PPARα and PPARγ

As mentioned above, *Pparα*, *Pparβ/δ*, and *Pparγ* encode proteins with a highly conserved structure and molecular mode of action. However, the receptors differ in their tissue distribution patterns and target genes and, therefore, in the biological functions that they regulate. Below, we briefly review the roles of PPARα and PPARγ, and then discuss those of PPARβ/δ in greater detail.

4.1. PPARα

PPARα is predominantly expressed in tissues with high levels of fatty acid catabolism, including the liver, as well as brown adipose tissue, heart, kidney, and skeletal muscle [82–84]. In the liver, PPARα is involved in fatty acid metabolism through transcriptional upregulation of numerous genes that play roles in mitochondrial and peroxisomal fatty acid oxidation, and in phospholipid remodeling [85–87]. PPARα also participates in downregulating hepatic inflammatory processes by reducing the effects of acute exposure to cytokines [88–91].

Preclinical and clinical studies have demonstrated that PPARα can influence NAFLD and NASH development [92–97]. Fasting is sufficient to trigger steatosis in PPARα-null mice, indicating that PPARα activity is required for metabolizing free fatty acids released from adipocytes [98,99]. Since PPARα is expressed and active in many organs, it is possible that the absence of PPARα in these organs might contribute to the development of fasting-induced steatosis. Therefore, we generated a hepatocyte-specific *Pparα*-null mouse and found that hepatocyte-restricted *Pparα* deletion is sufficient to promote steatosis [97]. This mouse shows impaired whole-body fatty acid homeostasis not only during fasting, but also when fed a methionine- and choline-deficient diet or a high-fat diet. Collectively, these data establish PPARα as a relevant drug target in NAFLD [97].

4.2. *PPARγ*

The PPARγ protein has two isoforms: PPARγ1 and PPARγ2. Differential promoter usage and alternate splicing of the PPARγ gene products actually generate three messenger RNAs (mRNAs)—PPARγ1, PPARγ2, and PPARγ3—with the PPARγ1 and PPARγ3 mRNAs both encoding the PPARγ1 protein [100]. PPARγ isoforms γ1 and γ2 are highly expressed in white and brown adipose tissues, where the receptor governs adipocyte differentiation and lipid storage. PPARγ1 is also expressed in the brain, vascular cells, colon, and immune cells [82,83].

PPARγ is weakly expressed in healthy liver, and steatosis is associated with increased hepatic expression of the PPARγ2 isoform, as observed in various mouse models of obesity [101,102]. Accordingly, hepatocyte-specific PPARγ deletion reduces hepatic fat content in mice fed a high-fat diet [103]. Increased PPARγ2 gene expression is also positively correlated with liver steatosis in obese patients [104,105]. Findings in the hepatocyte-specific PPARγ-knockout model indicated that PPARγ directly promotes hepatic fat accumulation by increasing lipid uptake, and by promoting *de novo* lipogenesis [106–110]. More recently, observations in an original mouse model of inducible hepatocyte-specific PPARγ deletion have suggested that PPARγ plays a specific role in fatty acid uptake and diacylglycerol (DAG) synthesis via upregulation of *Cd36* and monoacylglycerol O-acyltransferase 1 (*Mogat1*) [111]. Moreover, PPARγ plays important roles in glucose metabolism by regulating the expression of hexokinase 2 (HK2) and the M2 isoform of pyruvate kinase (PKM2), resulting in massive liver steatosis in phosphatase and tensin homologs deleted on chromosome 10 (PTEN)-null mice [112].

4.3. *PPARβ/δ*

PPARβ/δ is ubiquitously expressed, with the expression level varying among organs, cells, and species. Hepatic expression is low to moderate in adult humans and rats [82,113–116] and moderate to high in mice [117]. Pparβ/δ is highly expressed in hepatocytes, liver sinusoidal endothelial cells (LSECs), and liver-resident macrophages (Kupffer cells) [118]. Pparβ/δ expression is also constitutively high in hepatic stellate cells (HSCs).

In liver tissue of *Pparβ/δ*-null mice, transcriptional profiling revealed downregulation of genes associated with lipoprotein metabolism and glucose utilization pathways, indicating that these genes are positively regulated by PPARβ/δ. On the other hand, genes involved in innate immunity and inflammation were upregulated, suggesting their repression by PPARβ/δ. These transcriptional changes in *Pparβ/δ*-null mice correlated with increased plasma glucose and triglyceride levels, and reduced plasma cholesterol levels [119]. These results suggested important roles of PPARβ/δ in energy metabolism and inflammation, which we discuss below.

4.3.1. PPARβ/δ Roles in Energy Metabolism

In a very informative piece of work, Liu et al. demonstrated that adenovirus-mediated liver-restricted PPARβ/δ overexpression reduced fasting glucose levels in both chow- and high fat-fed mice. In parallel an increased hepatic glycogen and lipid deposition was observed accompanied by an up-regulation of glucose utilization and de novo lipogenesis [28]. PPARβ/δ increased the production of monounsaturated fatty acids (MUFAs), which activate PPARs, while reducing saturated fatty acid levels. Lipid accumulation in the adeno-PPARβ/δ-infected livers reduced cell damage and c-Jun N-terminal kinase (JNK) stress signaling. The authors proposed that the PPARβ/δ-regulated lipogenic program may protect against lipotoxicity, and that altered substrate utilization by PPARβ/δ resulted in AMP-activated protein kinase activation, which may contribute to the glucose-lowering activity of PPARβ/δ. Taken together, this data suggested that PPARβ/δ impacts hepatic energy substrate homeostasis by a coordinated control of fatty acid and glucose metabolism [28].

In line with these findings, PPARβ/δ regulates lipogenic genes during the dark/feeding cycle. Specifically, PPARβ/δ drives MUFA production via stearoyl-CoA desaturase 1 (*Scd1*) upregulation, a process that avoids lipotoxicity by increasing fatty acid oxidation or sequestration of saturated fatty acids. As such, the process inhibits saturated fatty acid-induced cytotoxicity in hepatocytes. Furthermore, long chain acyl-CoA from MUFA production allows esterification into triglycerides [120]. Interestingly, liver-specific PPARβ/δ activation increases fatty acid uptake in muscle, whereas its deletion has an opposite effect. Phosphatidylcholine 18:0/18:1 (PC (18:0/18:1)) was identified as a serum lipid produced in the liver under the control of PPARβ/δ activity, which upon circulating to muscles stimulates fatty acid catabolism through PPARα activation [121].

For a direct comparison of the roles of *Pparα* and *Pparβ/δ* in liver, microarray analysis was being used to compare the liver transcriptome between *Pparα* and *Pparβ/δ*-null mice, revealing a small overlap in the regulation of genes that are both PPARα- and PPARβ/δ-dependent. In the fed state, similar numbers of genes exhibited altered expression in *Pparα* and *Pparβ/δ* deletion. However, during fasting, more genes showed altered expression in *Pparα*-deleted mice compared to *Pparβ/δ*-null mice. Analysis of plasma metabolites, including free fatty acids and β-hydroxybutyrate, supported the notion that PPARα is particularly important during fasting, while PPARβ/δ appears to be important in both the fed and fasted states [119]. Based on functional similarities to PPARα, PPARβ/δ may be a master regulator of hepatic intermediary metabolism. In rodents, both receptors play non-redundant roles in the liver to enhance ketogenesis through induction of *Fgf21* and expression of fatty acid oxidation genes under fasting conditions [122,123]. In fact, PPARα is an important activator of hepatic fatty acid oxidation [97,99,124]. Interestingly, PPARβ/δ cannot compensate for PPARα in *Pparα*-null mice [98].

The differences between PPARα and PPARβ/δ in molecular and biological functions also corresponded with their antiphasic circadian expression profiles. Indeed, PPARα peaks at the end the light/resting period, while PPARβ/δ is highly expressed in the liver during the night/feeding period, according to [86,121], and Montagner et al., unpublished results. Notably, during fasting (usually light period), PPARβ/δ expression decreases while PPARα is highly expressed [125]. In spite of their biphasic expression profile, intra- and inter-organ dialogs between PPARβ/δ and PPARα activities have been described. As mentioned above, increased hepatic PPARβ/δ activity can lead to PPARα activation in muscle tissue via production of the specific PPARα ligand 16:0/18:1-phosphatidylcholine [121]. This mechanism could also occur in the liver [121,126]. Overall, while both PPARα and PPARβ/δ are associated with the regulation of hepatic lipid metabolism [127,128], hepatic PPARβ/δ mainly acts on anabolic metabolic processes and primarily contributes to glucose utilization, MUFA formation, and anti-inflammatory responses [119,129].

Compared with PPARα and PPARγ, less is known about PPARβ/δ in relation to obesity and NAFLD [130]. However, the lipogenic activity of PPARβ/δ raises the question of whether PPARβ/δ activation is associated with steatosis and steatohepatitis. It was recently shown that both PPARβ/δ and PPAR*α* receptors were necessary for adipose tissue reduction driven by the PPARβ/δ agonist GW501516 and subsequent development of hepatic steatosis, with PPARβ/δ working upstream of PPARα [131]. PPARβ/δ is also involved in transforming potentially toxic lipids into less toxic molecules by regulating MUFA synthesis, a process that increases PPARα activity and could protect against NAFLD and promote detoxification. In mice with adenovirus-mediated liver-restricted PPARβ/δ overexpression, examination revealed elevated liver expression of the adiponectin receptor 2 (AdipoR2), leading to enhanced 5′ adenosine monophosphate-activated protein kinase (AMPK) activity [132].

This PPARβ/δ-dependent increase in AMPK activity reportedly suppressed lipogenesis and glycogen synthesis, reduced gluconeogenesis, and increased fatty acid oxidation [25–27]. The AMPK pathway may act as a negative feedback loop for PPARβ/δ, possibly explaining why long-term PPARβ/δ agonist treatment does not lead to liver lipid accumulation [133]. Similarly, PPARβ/δ suppresses lipogenesis by lowering SREBP1c levels, reducing the severity of hepatic steatosis in obese diabetic *db/db* mice via stimulation of the insulin-induced gene-1 (*Insig-1*), the product of which inhibits SREBP1c [134].

Fibroblast growth factor 21 (FGF21) is a circulating hormone derived from the liver, which plays important roles in regulating glucose and lipid metabolism [135,136]. Recent evidence shows that PPARβ/δ and FGF21 exert hepatic regulation of the VLDL receptor, which modulates NAFLD. Liver tissue of *Pparβ/δ*-null mice and *Pparβ/δ−/−* hepatocytes exhibit increased VLDL receptor expression. Moreover, FGF21 neutralizing antibody treatment resulted in triglyceride accumulation in *Pparβ/δ*-null mice [137]. In support of these pre-clinical results, liver biopsies from patients with moderate and severe hepatic steatosis showed increased VLDL receptor levels and reduced PPARβ/δ mRNA levels and DNA-binding activity compared to in control subjects. These findings revealed a novel mechanism in which VLDL receptor levels are controlled by PPARβ/δ and FGF21, impacting hepatic steatosis development [137].

4.3.2. PPARβ/δ Roles in Inflammation

On a high-fat diet, the PPARβ/δ-dependent increase in hepatocyte MUFA production impacts liver-resident macrophages and Kupffer cells—resulting in increased PPARβ/δ activation, and reduced expression of TNFα or interferon gamma (IFNγ) inflammatory markers from these cells—and altering the immune response [28]. Thus, this finding suggests that PPARβ/δ plays an anti-inflammatory role in liver. PPARβ/δ and its ligands are also reportedly associated with anti-inflammatory activities through interference with nuclear factor kappa-light-chain-enhancer of activated B cells (NFκB) signaling [67,138,139] and through interactions with signal transducer and activator of transcription 3 (STAT3) and extracellular-signal-regulated kinase 5 (ERK5) [140,141].

Kupffer cells are also involved in insulin resistance and fatty liver disease [142], and PPARβ/δ plays a role in regulating the alternative activation of these cells [143]. In the presence of IL4 and IL13 stimulation, PPARβ/δ is required for the activation of Kupffer cells to the M2 subtype that has anti-inflammatory activity. Hematopoietic *Pparβ/δ*-deficient obese mice exhibited lower insulin sensitivity and oxidative metabolism, as well as impaired alternative activation of Kupffer cells. This phenotype was validated by three independent lines of experiments. First, *Pparβ/δ* deletion in lean mice resulted in lower expression of genes involved in alternatively activated Kupffer cells, such as arginase 1 (*Arg1*), c-type lectin domain containing 7A (*Clec7a*), jagged 1 (*Jag1*), programmed cell death 1 ligand 2 (*Pdcd1lg2*) and chitinase (*Chia*). However, treatment with PPARβ/δ agonist GW0742 led to increased expression of these genes in liver. Second, replacing the bone marrow of wild-type mice with *Pparβ/δ*-null bone marrow led to insulin resistance and mitochondrial dysfunction in hepatocytes, eliminating the alternative activation of Kupffer cells. Third, direct co-culturing of *Pparβ/δ*-null macrophages with primary hepatocytes induced a significant reduction of oxidative phosphorylation in the parenchymal cells. The study demonstrated the association between *Pparβ/δ*-null Kupffer cells and dysregulation of hepatic metabolism, resulting in increased liver triglycerides [143].

PPARβ/δ is also involved in hepatic stellate cell (HSC) activation; its expression is upregulated in cultures of activated HSCs and in in vivo fibrogenesis [144,145]. Administration of the PPARβ/δ agonist L165041 enhances HSC proliferation, and L165041 administration combined with chronic carbon tetrachloride (CCl_4) treatment leads to higher fibrotic marker expression in rats [146]. These data suggested that PPARβ/δ plays an important role as a signal-transducing factor, leading to HSC proliferation in the event of acute and chronic liver inflammation [146]. In activated HSCs, PPARβ/δ enhances the expression of *Cd36*, which codes for a membrane receptor that facilitates fatty acid uptake. Moreover, upregulated PPARβ/δ expression is associated with elevated expression of proteins involved in retinoid binding and esterification, such as cellular retinol-binding protein 1 (CRBP-1) and lecithin retinol acyltransferase (LRAT). Overall, PPARβ/δ regulates the expression of genes related to vitamin A metabolism in HSCs undergoing activation [144].

Interestingly, CCl_4-induced hepatic fibrotic response requires PPARβ/δ which enhances expression of profibrotic and pro-inflammatory genes in mice. This process results in increased macrophage recruitment and extracellular matrix deposition in the liver [145]. However, this phenotype was not observed in *Pparβ/δ*-null mice treated with CCl_4 alone or with CCl_4 plus GW501516. The same study further demonstrated that GW501516 administration increased HSC proliferation in CCl_4-injured wild-type mice livers, but not in *Pparβ/δ*-null mice with the same treatment. In another study, GW501516-treated *db/db* mice exhibited higher expression of the lipogenic enzyme acetyl-CoA carboxylase β and elevated triglyceride levels in the liver [147]. Moreover, investigations of GW501516 treatment in control and *Pparβ/δ*-knockdown LX-2 human hepatic stellate cells revealed that GW501516-stimulated HSC proliferation occurs via p38 and JNK mitogen-activated protein kinase (MAPK) pathways [145]. However, in the same model of CCl_4-induced liver damage, administration of the PPARβ/δ agonist KD3010 (chemical abstracts service, CAS ID 934760-90-4) ameliorated the CCl_4-induced liver injury with lower deposition of extracellular matrix proteins. KD3010 treatment of primary hepatocytes provided protection from CCl_4-induced cell death or starvation, suggesting that KD3010 administration could have hepatoprotective and antifibrotic effects in animal models of liver fibrosis [148]. Further studies are needed to determine the reasons for the different effects of GW501516 and KD3010 in injured livers [149].

In mice treated with the agonist GW0742, NFκB signaling was attenuated in a PPARβ/δ-dependent manner. Compared to wild-type mice, *Pparβ/δ*-null mice exhibited higher TNFα and αSMA expression in hepatocytes and HSCs, but similar inflammatory signaling in hepatocytes and activation of HSCs [150]. A recent study using the same PPARβ/δ agonist demonstrated that PPARβ/δ upregulates serum high-density lipoprotein (HDL) and HDL phospholipids in NAFLD mice, while this effect is not seen in *Pparβ/δ*-deficient mice [151].

5. Pharmacological Strategies Targeting PPARβ/δ for NAFLD Treatment

5.1. *PPARβ/δ Agonists: GW0742, GW501516*

Preclinical studies have investigated long-term treatment with PPARβ/δ agonists such as GW0742 (CAS ID 317318-84-6) and GW501516 (CAS ID 317318-70-0) in animal models, revealing that PPARβ/δ activation attenuates hepatic steatosis by promoting fatty acid oxidation, reducing lipogenesis, and enhancing insulin sensitivity [134,152–154]. On the contrary, short-term treatment with PPARβ/δ agonists reportedly yields a transient increase in hepatic triglyceride levels [131]. Elevated levels of monounsaturated fatty acids, are accompanied by lower saturated fatty acid levels and no observed hepatotoxicity [28]. Studies involving PPARβ/δ agonist treatment in humans have demonstrated reduced hepatic fat content and improved plasma markers of liver function, including carnitine palmitoyltransferase 1b [155,156]. One study conducted in middle-overweight patients revealed that GW501516 treatment decreased liver lipid content and insulinemia, with no signs of oxidative stress [156]. However, LDL cholesterol plasma level was also reduced. This suggests that the protective effects of PPARβ/δ pharmacological activation are reliant on increased lipid oxidation in muscles.

5.2. *PPAR Dual Agonists: Elafibranor, Saroglitazar*

The PPARα and PPARβ/δ dual agonist elafibranor (also known as GTF-505, CAS ID 923978-27-2) has recently emerged as one of the most promising chemical entities for treatment of NAFLD, especially NASH. Prior studies have demonstrated its efficiency, and it is currently undergoing phase III testing in NASH patients. It has reportedly improved steatosis, inflammation, and fibrosis in mouse models of NAFLD [95], and thus appears to be a good candidate for the treatment of hepatic fibrosis, NAFLD, primary biliary cirrhosis, and NASH. Elafibranor was investigated in a randomized, double-blind, placebo-controlled trial including 274 patients in Europe and the USA (GOLDEN-505 trial; NCT01694849). Post-hoc analysis of those trial results revealed that ALT was significantly reduced after four to 12 weeks of elafibranor treatment among patients who were in the top two quartiles at baseline. Non-cirrhotic patients with NASH did not exhibit any worsening of hepatic fibrosis after 52 weeks of taking elafibranor at 120 mg/day [157]. Liver biopsy analysis in this patient group further revealed disappearance of hepatocellular ballooning, with no or mild lobular inflammation. Elafibranor-treated patients also exhibited improvement in liver enzymes, lipid parameters (triglycerides, low-density lipoprotein, high-density lipoprotein, and cholesterol), serum inflammation biomarkers, steatosis, and fibrosis. Other studies have reported that elafibranor treatment improves glucose homeostasis and insulin resistance in diabetic patients [157,158]. Overall, elafibranor appears to be safe and well-tolerated, with no deaths or cardiovascular incidents reported during treatment. There is currently an ongoing phase III randomized, double-blind, placebo-controlled trial of elafibranor use in 2000 liver biopsy-proven NASH patients, to investigate the efficacy against NASH and the safety regarding fibrosis during longer use (72 weeks) (NCT02704403) [159].

Interestingly, the PPARα/γ dual agonist saroglitazar (CAS ID 495399-09-2) has also exhibited overall beneficial effects in experimental models of NASH [160]. Moreover, saroglitazar treatment induces a significant decrease of ALT levels in subjects with biopsy-proven NASH [21]. Since saroglitazar improves all of the components responsible for NAFLD/NASH in preclinical models, it is also a promising candidate for the management of these conditions. Further studies are needed to examine the possible common and different pathways of action of elafibranor and saroglitazar.

5.3. PPAR Pan-Agonists: Bezafibrate, MHY2013, Lanifibranor

The anti-fibrotic and anti-inflammatory effects of PPARs have inspired growing use of PPAR pan-agonists to treat NAFLD. It is postulated that PPAR pan-agonist may show improved efficacy compared to targeting a single PPAR isotype [161]. The PPAR pan-agonist bezafibrate (CAS ID 41859-67-0), which activates PPARα, PPARβ/δ, and PPARγ, has shown beneficial effects in NASH treatment. In mice fed a methionine- and choline-deficient diet, bezafibrate and GW501516 (selective PPARβ/δ agonist) treatments have resulted in upregulation of β-oxidation and lipid transport genes in hepatocytes. They have inhibited NASH development. These treatments also both resulted in reduced inflammatory gene expression [152]. MHY2013 is another PPAR pan-agonist that also activates all three PPAR isotypes. In aged Sprague-Dawley (SD) rats, MHY2013 treatment improved age-related hepatic lipid accumulation, and resulted in upregulated β-oxidation signaling and lower inflammation in the liver [162]. The PPAR pan-agonist Lanifibranor (CAS ID 927961-18-0) is reportedly effective in experimental skin and lung fibrosis [163,164]. It has been proposed for use as an anti-fibrotic treatment. Lanifibranor is currently being tested in a phase 2b randomized, double-blind, placebo-controlled trial for safety and efficacy in up to 225 patients in 12 European countries (NCT03008070) [165].

6. Conclusions

NAFLD is an alarming health issue that is occurring with rising frequency in developed countries. It is now well documented that PPARβ/δ is involved in regulating glucose and lipid metabolism in the liver. An improved understanding of the physiological roles of PPARs, particularly PPARβ/δ, will likely contribute to the design and development of safe agonists with enhanced therapeutic potential compared to first-generation agonists. Although much remains unknown about the physiological impact of PPARβ/δ, prior research has elucidated highly interesting NAFLD-related functions, as reviewed in this article.

Some results on PPARβ/δ roles seem contradictory, and the reasons for these discrepancies is unclear. It is conceivable that PPARβ/δ exert different functions in a context- and agonist-specific manner. For example, one study reported that PPARβ/δ stimulates the *de novo* lipogenesis pathway, which is accompanied by lipid deposition. Interestingly, this PPARβ/δ-regulated lipogenic program is paralleled by reduced JNK stress signaling, suggesting that it may protect against lipotoxicity [28]. However, it has also been suggested that PPARβ/δ suppresses hepatic lipogenesis. PPARβ/δ overexpression enhanced *Insig-1* expression, which suppressed SREBP-1 activation and thus ameliorated hepatic steatosis in obese *db/db* mice [134]. Similarly, PPARβ/δ agonists GW501516 and KD3010 exerted pro-fibrotic and anti-fibrotic effects, respectively, in CCl_4-injured livers [145,146]. Uncovering the causes for these apparent discrepancies will likely elucidate differentiated responses of PPARβ/δ in specific situations, which will be important for PPARβ/δ as a pharmacological target. We are in the opinion that detail transcriptomic profiling in combination with a better understanding of the pharmacological characteristics of candidate drugs, such as half-life, affinity constant, and bioavailability, may provide insights into their true target and reveal potential off-target effects.

PPARβ/δ also plays an interesting role in the alternative activation of Kupffer cells to the anti-inflammatory macrophage M2 subtype [143], revealing the direct PPARβ/δ-dependent involvement of Kupffer cells in liver lipid metabolism. Based on this beneficial role for alternatively activated Kupffer cells in metabolic syndrome conditions, controlling PPARβ/δ activity in these cells may contribute to delaying NAFLD progression.

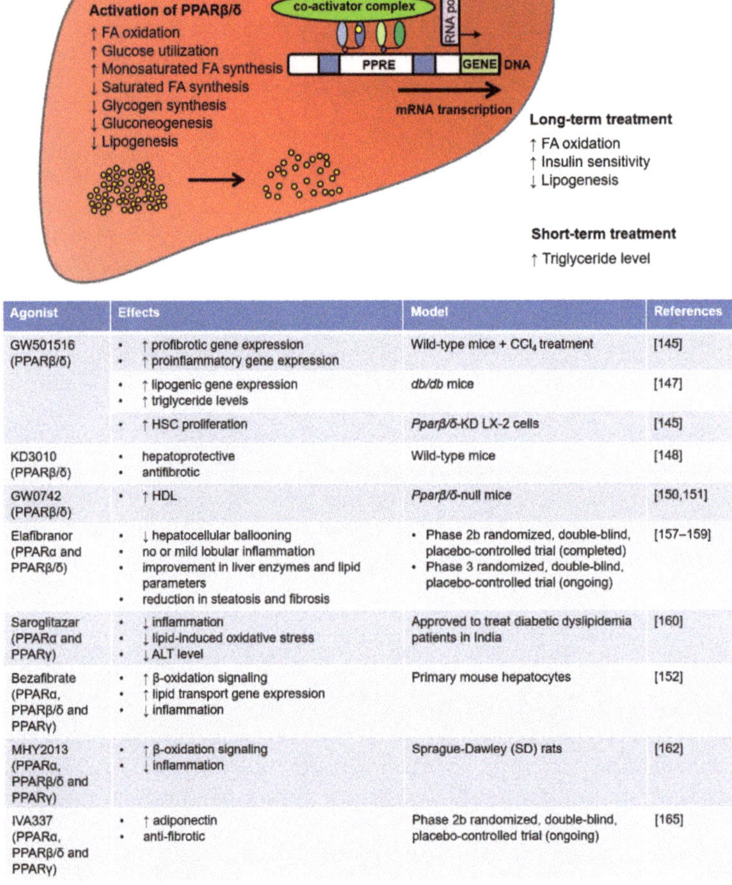

Agonist	Effects	Model	References
GW501516 (PPARβ/δ)	• ↑ profibrotic gene expression • ↑ proinflammatory gene expression	Wild-type mice + CCl_4 treatment	[145]
	• ↑ lipogenic gene expression • ↑ triglyceride levels	*db/db* mice	[147]
	• ↑ HSC proliferation	*Pparβ/δ*-KD LX-2 cells	[145]
KD3010 (PPARβ/δ)	• hepatoprotective • antifibrotic	Wild-type mice	[148]
GW0742 (PPARβ/δ)	• ↑ HDL	*Pparβ/δ*-null mice	[150,151]
Elafibranor (PPARα and PPARβ/δ)	• ↓ hepatocellular ballooning • no or mild lobular inflammation • improvement in liver enzymes and lipid parameters • reduction in steatosis and fibrosis	• Phase 2b randomized, double-blind, placebo-controlled trial (completed) • Phase 3 randomized, double-blind, placebo-controlled trial (ongoing)	[157–159]
Saroglitazar (PPARα and PPARγ)	• ↓ inflammation • ↓ lipid-induced oxidative stress • ↓ ALT level	Approved to treat diabetic dyslipidemia patients in India	[160]
Bezafibrate (PPARα, PPARβ/δ and PPARγ)	• ↑ β-oxidation signaling • ↑ lipid transport gene expression • ↓ inflammation	Primary mouse hepatocytes	[152]
MHY2013 (PPARα, PPARβ/δ and PPARγ)	• ↑ β-oxidation signaling • ↓ inflammation	Sprague-Dawley (SD) rats	[162]
IVA337 (PPARα, PPARβ/δ and PPARγ)	• ↑ adiponectin • anti-fibrotic	Phase 2b randomized, double-blind, placebo-controlled trial (ongoing)	[165]

Figure 4. Overview of research findings regarding PPARβ/δ in hepatic metabolism, and the contrasting effects of various PPARβ/δ agonist treatments in pre-clinical models.

The fine tuning of PPAR-regulated physiological functions in the liver and other organs is influenced by the functional interaction between PPARβ/δ and PPARα [121,131]. PPARβ/δ apparently works upstream of PPARα, controlling the production of MUFAs, as well as PC (18:0/18:1), which activates muscle PPARα to increase muscle energy use [121]. MUFAs also activate PPARα in the liver itself. This regulatory circuit couples ligand production and the activities of two receptors that play key roles in liver energy metabolism.

These complex interactions are certainly of interest for the development of novel PPAR drugs. PPARα/PPARβ/δ dual agonists may have additional beneficial effects due to the integrated roles of these two receptors through the abovementioned regulatory circuit they form together. GFT505 (elafibranor) is the most advanced PPARα/PPARβ/δ dual agonist [158]. It has been tested in several clinical trials and is currently being evaluated in a clinical phase III study [166]. Several other PPAR agonists, dual agonists, and pan-agonists of interest have been investigated, and some are now in clinical studies of safety and efficacy (Figure 4). As PPARs play important roles in regulating genes involved in fatty acid uptake and oxidation [65–67], we propose that targeting PPARs will be one of the best possibilities to treat fatty liver diseases.

Acknowledgments: The work performed in W.W.'s laboratory was supported by Singapore Ministry of Education under its Singapore Ministry of Education Academic Research Fund Tier 1 (2015-T1-001-034) and start-up grant from the Lee Kong Chian School of Medicine, Nanyang Technological University, Singapore. J.C. is a recipient of the Research Scholarship from NTU, Singapore.

Conflicts of Interest: The authors declare no conflict of interest.

Abbreviations

ADIPOR2	adiponectin receptor 2
AF	activation function
APOB	apolipoprotein B
Arg1	arginase 1
CAS	chemical abstracts service
CCL_4	chronic carbon tetrachloride
CHIA	chitinase
CLEC7a	c-type lectin comain containing 7A
DAG	diacylglycerol
DBD	DNA-binding domain
ERK5	extracellular-signal-regulated kinase 5
FABP	fatty acid-binding protein
FGF21	fibroblast growth factor 21
GCKR	glucokinase regulator
GWAS	genome-wide association studies
HCC	hepatocellular carcinoma
HK2	hexokinase 2
HSC	hepatic stellate cell
IFNγ	interferon gamma
IL	interleukin
INSIG-1	insulin-induced gene-1
JAG1	jagged 1
JNK	c-Jun N-terminal kinase
LBD	ligand-binding domain
LDL	low-density lipoprotein
LSEC	liver sinusoidal endothelial cell
LYPLAL1	lysophospholipase like 1
MOGAT1	monoacylglycerol O-acyltransferase 1
MUFA	monounsaturated fatty acids

NAFLD	non-alcoholic fatty liver disease
NASH	non-alcoholic steatohepatitis
NCAN	neurocan
NCoR	nuclear receptor corepressor
NFκB	nuclear factor kappa-light-chain-enhancer of activated B cells
NTD	N-terminal domain
PC	Phosphatidylcholine
Pdcd1lg2	programmed cell death 1 ligand 2
PKM2	M2 isoform of pyruvate kinase
PNAPL3	patatin-like phospholipase domain containing 3
PPAR	peroxisome proliferator-activated receptor
PPP1R3B	protein phosphatase 1 regulatory subunit 3B
PPRE	peroxisome proliferator response element
PTEN	phosphatase and tensin homolog deleted on chromosome 10
RXR	retinoic acid receptor
SCD1	stearoyl-CoA desaturase 1
SMRT	silencing mediator of retinoid and thyroid receptors
SREBP1c	sterol-regulated binding protein 1c
STAT3	signal transducer and activator of transcription 3
TF	transcription factor
TFBS	TF-binding site
TM6SF2	transmembrane 6 superfamily member 2
TNFα	tumor necrosis factor α
TZD	thiazolidinedione
VCAM-1	vascular cell adhesion molecule-1
VLDL	very low-density lipoprotein

References

1. Loomba, R.; Sanyal, A.J. The global NAFLD epidemic. *Nat. Rev. Gastroenterol. Hepatol.* **2013**, *10*, 686–690. [CrossRef] [PubMed]
2. Byrne, C.D.; Targher, G. NAFLD: A multisystem disease. *J. Hepatol.* **2015**, *62*, S47–S64. [CrossRef] [PubMed]
3. Wieckowska, A.; Papouchado, B.G.; Li, Z.; Lopez, R.; Zein, N.N.; Feldstein, A.E. Increased hepatic and circulating interleukin-6 levels in human nonalcoholic steatohepatitis. *Am. J. Gastroenterol.* **2008**, *103*, 1372–1379. [CrossRef] [PubMed]
4. Cohen, J.C.; Horton, J.D.; Hobbs, H.H. Human fatty liver disease: Old questions and new insights. *Science* **2011**, *332*, 1519–1523. [CrossRef] [PubMed]
5. Wong, R.J.; Aguilar, M.; Cheung, R.; Perumpail, R.B.; Harrison, S.A.; Younossi., Z.M.; Ahmed, A. Nonalcoholic Steatohepatitis Is the Second Leading Etiology of Liver Disease Among Adults Awaiting Liver Transplantation in the United States. *Gastroenterology* **2015**, *148*, 547–555. [CrossRef] [PubMed]
6. Targher, G.; Bertolini, L.; Rodella, S.; Tessari, R.; Zenari, L.; Lippi, G.; Arcaro, G. Prevalence of nonalcoholic fatty liver disease and its association with cardiovascular disease among Type 2 diabetic patients. *Diabetes Care* **2007**, *30*, 1212–1218. [CrossRef] [PubMed]
7. Machado, M.; Marques-Vidal, P.; Cortez-Pinto, H. Hepatic histology in obese patients undergoing bariatric surgery. *J. Hepatol.* **2006**, *45*, 600–606. [CrossRef] [PubMed]
8. Angulo, P. Obesity and nonalcoholic fatty liver disease. *Nutr. Rev.* **2007**, *65 Pt 2*, 57–63. [CrossRef]
9. Lazo, M.; Clark, J.M. The epidemiology of nonalcoholic fatty liver disease: A global perspective. *Semin. Liver Dis.* **2008**, *28*, 339–350. [CrossRef] [PubMed]
10. Williams, C.D.; Stengel, J.; Asike, M.I.; Torres, D.M.; Shaw, J.; Contreras, M.; Landt, C.L.; Harrison, S.A. Prevalence of nonalcoholic fatty liver disease and nonalcoholic steatohepatitis among a largely middle-aged population utilizing ultrasound and liver biopsy: A prospective study. *Gastroenterology* **2011**, *140*, 124–131. [CrossRef] [PubMed]

11. Nakamuta, M.; Kohjima, M.; Morizono, S.; Kotoh, K.; Yoshimoto, T.; Miyagi, I.; Enjoji, M. Evaluation of fatty acid metabolism-related gene expression in nonalcoholic fatty liver disease. *Int. J. Mol. Med.* **2005**, *16*, 631–635. [PubMed]
12. Ekstedt, M.; Franzén, L.E.; Mathiesen, U.L.; Thorelius, L.; Holmqvist, M.; Bodemar, G.; Kechagias, S. Long-term follow-up of patients with NAFLD and elevated liver enzymes. *Hepatology* **2006**, *44*, 865–873. [CrossRef] [PubMed]
13. Farrell, G.C.; Larter, C.Z. Nonalcoholic fatty liver disease: From steatosis to cirrhosis. *Hepatology* **2006**, *43* (Suppl. 1), S99–S112. [CrossRef] [PubMed]
14. Bodzin, A.S.; Busuttil, R.W. Hepatocellular carcinoma: Advances in diagnosis, management, and long term outcome. *World J. Hepatol.* **2015**, *7*, 1157–1167. [CrossRef] [PubMed]
15. Marchesini, G.; Bugianesi, E.; Forlani, G.; Cerrelli, F.; Lenzi, M.; Manini, R.; Natale, S.; Vanni, E.; Villanova, N.; Melchionda, N.; et al. Nonalcoholic fatty liver, steatohepatitis, and the metabolic syndrome. *Hepatology* **2003**, *37*, 917–923. [CrossRef] [PubMed]
16. Adams, L.A.; Waters, O.R.; Knuiman, M.W.; Elliott, R.R.; Olynyk, J.K. NAFLD as a risk factor for the development of diabetes and the metabolic syndrome: An eleven-year follow-up study. *Am. J. Gastroenterol.* **2009**, *104*, 861–867. [CrossRef] [PubMed]
17. Dowman, J.K.; Tomlinson, J.W.; Newsome, P.N. Systematic review: The diagnosis and staging of non-alcoholic fatty liver disease and non-alcoholic steatohepatitis. *Aliment. Pharmacol. Ther.* **2010**, *33*, 525–540. [CrossRef] [PubMed]
18. McCarthy, E.M.; Rinella, M.E. The role of diet and nutrient composition in nonalcoholic Fatty liver disease. *J. Acad. Nutr. Diet.* **2012**, *112*, 401–409. [CrossRef] [PubMed]
19. Promrat, K.; Kleiner, D.E.; Niemeier, H.M.; Jackvony, E.; Kearns, M.; Wands, J.R.; Fava, J.L.; Wing, R.R. Randomized controlled trial testing the effects of weight loss on nonalcoholic steatohepatitis. *Hepatology* **2010**, *51*, 121–129. [CrossRef] [PubMed]
20. St George, A.; Bauman, A.; Johnston, A.; Farrell, G.; Chey, T.; George, J. Effect of a lifestyle intervention in patients with abnormal liver enzymes and metabolic risk factors. *J. Gastroenterol. Hepatol.* **2008**, *24*, 399–407. [CrossRef] [PubMed]
21. Oseini, A.M.; Sanyal, A.J. Therapies in non-alcoholic steatohepatitis (NASH). *Liver Int.* **2017**, *37* (Suppl. 1), 97–103. [CrossRef] [PubMed]
22. Haukeland, J.W.; Konopski, Z.; Eggesbø, H.B.; von Volkmann, H.L.; Raschpichler, G.; Bjøro, K.; Haaland, T.; Løberg, E.M.; Birkeland, K. Metformin in patients with non-alcoholic fatty liver disease: A randomized, controlled trial. *Scand. J. Gastroenterol.* **2009**, *44*, 853–860. [CrossRef] [PubMed]
23. Lutchman, G.; Modi, A.; Kleiner, D.E.; Promrat, K.; Heller, T.; Ghany, M.; Borg, B.; Loomba, R.; Liang, T.J.; Premkumar, A.; et al. The effects of discontinuing pioglitazone in patients with nonalcoholic steatohepatitis. *Hepatology* **2007**, *46*, 424–429. [CrossRef] [PubMed]
24. Aithal, G.P.; Thomas, J.A.; Kaye, P.V.; Lawson, A.; Ryder, S.D.; Spendlove, I.; Austin, A.S.; Freeman, J.G.; Morgan, L.; Webber, J. Randomized, placebo-controlled trial of pioglitazone in nondiabetic subjects with nonalcoholic steatohepatitis. *Gastroenterology* **2008**, *135*, 1176–1184. [CrossRef] [PubMed]
25. Berlanga, A.; Guiu-Jurado, E.; Porras, J.A.; Auguet, T. Molecular pathways in non-alcoholic fatty liver disease. *Clin. Exp. Gastroenterol.* **2014**, *7*, 221–239. [CrossRef] [PubMed]
26. Yamauchi, T.; Nio, Y.; Maki, T.; Kobayashi, M.; Takazawa, T.; Iwabu, M.; Okada-Iwabu, M.; Kawamoto, S.; Kubota, N.; Kubota, T.; et al. Targeted disruption of AdipoR1 and AdipoR2 causes abrogation of adiponectin binding and metabolic actions. *Nat. Med.* **2007**, *13*, 332–339. [CrossRef] [PubMed]
27. Narkar, V.A.; Downes, M.; Yu, R.T.; Embler, E.; Wang, Y.X.; Banayo, E.; Mihaylova, M.M.; Nelson, M.C.; Zou, Y.; Juguilon, H.; et al. AMPK and PPARdelta agonists are exercise mimetics. *Cell* **2008**, *134*, 405–415. [CrossRef] [PubMed]
28. Liu, S.; Hatano, B.; Zhao, M.; Yen, C.C.; Kang, K.; Reilly, S.M.; Gangl, M.R.; Gorgun, C.; Balschi, J.A.; Ntambi, J.M.; et al. Role of peroxisome proliferator-activated receptor δ/β in hepatic metabolic regulation. *J. Biol. Chem.* **2011**, *286*, 1237–1247. [CrossRef] [PubMed]
29. Tan, N.S.; Vázquez-Carrera, M.; Montagner, A.; Sng, M.K.; Guillou, H.; Wahli, W. Transcriptional control of physiological and pathological processes by the nuclear receptor PPARβ/δ. *Prog. Lipid Res.* **2016**, *64*, 98–122. [CrossRef] [PubMed]
30. Day, C.P.; James, O.F. Steatohepatitis: A tale of two "hits"? *Gastroenterology* **1998**, *114*, 842–845. [CrossRef]

31. Imajo, K.; Yoneda, M.; Kessoku, T.; Ogawa, Y.; Maeda, S.; Sumida, Y.; Hyogo, H.; Eguchi, Y.; Wada, K.; Nakajima, A. Rodent Models of Nonalcoholic Fatty Liver Disease/Nonalcoholic Steatohepatitis. *Int. J. Mol. Sci.* **2013**, *14*, 21833–21857. [CrossRef] [PubMed]
32. Delarue, J.; Magnan, C. Free fatty acids and insulin resistance. *Curr. Opin. Clin. Nutr. Metab. Care* **2007**, *10*, 142–148. [CrossRef] [PubMed]
33. Kawano, Y.; Cohen, D.E. Mechanisms of hepatic triglyceride accumulation in non-alcoholic fatty liver disease. *J. Gastroenterol.* **2013**, *48*, 434–441. [CrossRef] [PubMed]
34. Fuchs, M. Non-alcoholic Fatty liver disease: The bile Acid-activated farnesoid x receptor as an emerging treatment target. *J. Lipids* **2011**, *2012*, 934396. [CrossRef] [PubMed]
35. Bechmann, L.P.; Hannivoort, R.A.; Gerken, G.; Hotamisligil, G.S.; Trauner, M.; Canbay, A. The interaction of hepatic lipid and glucose metabolism in liver diseases. *J. Hepatol.* **2012**, *56*, 952–964. [CrossRef] [PubMed]
36. Musso, G.; Gambino, R.; Cassader, M. Recent insights into hepatic lipid metabolism in non-alcoholic fatty liver disease (NAFLD). *Prog. Lipid Res.* **2008**, *48*, 1–26. [CrossRef] [PubMed]
37. Argo, C.K.; Northup, P.G.; Al-Osaimi, A.M.; Caldwell, S.H. Systematic review of risk factors for fibrosis progression in non-alcoholic steatohepatitis. *J. Hepatol.* **2009**, *51*, 371–379. [CrossRef] [PubMed]
38. Buzzetti, E.; Pinzani, M.; Tsochatzis, E.A. The multiple-hit pathogenesis of non-alcoholic fatty liver disease (NAFLD). *Metabolism* **2016**, *65*, 1038–1048. [CrossRef] [PubMed]
39. Tilg, H.; Moschen, A.R. Evolution of inflammation in nonalcoholic fatty liver disease: The multiple parallel hits hypothesis. *Hepatology* **2010**, *52*, 1836–1846. [CrossRef] [PubMed]
40. Tomeno, W.; Yoneda, M.; Imajo, K.; Ogawa, Y.; Kessoku, T.; Saito, S.; Eguchi, Y.; Nakajima, A. Emerging drugs for non-alcoholic steatohepatitis. *Expert Opin. Emerg. Drugs* **2013**, *18*, 279–290. [CrossRef] [PubMed]
41. Li, Z.; Yang, S.; Lin, H.; Huang, J.; Watkins, P.A.; Moser, A.B.; Desimone, C.; Song, X.Y.; Diehl, A.M. Probiotics and antibodies to TNF inhibit inflammatory activity and improve nonalcoholic fatty liver disease. *Hepatology* **2003**, *37*, 343–350. [CrossRef] [PubMed]
42. Tiniakos, D.G.; Vos, M.B.; Brunt, E.M. Nonalcoholic fatty liver disease: Pathology and pathogenesis. *Annu. Rev. Pathol.* **2010**, *5*, 145–171. [CrossRef] [PubMed]
43. Romeo, S.; Kozlitina, J.; Xing, C.; Pertsemlidis, A.; Cox, D.; Pennacchio, L.A.; Boerwinkle, E.; Cohen, J.C.; Hobbs, H.H. Genetic variation in PNPLA3 confers susceptibility to nonalcoholic fatty liver disease. *Nat. Genet.* **2008**, *40*, 1461–1465. [CrossRef] [PubMed]
44. Sookoian, S.; Pirola, C.J. PNPLA3, the triacylglycerol synthesis/hydrolysis/storage dilemma, and nonalcoholic fatty liver disease. *World J. Gastroenterol.* **2012**, *18*, 6018–6026. [CrossRef] [PubMed]
45. Sookoian, S.; Pirola, C.J. Meta-analysis of the influence of I148M variant of patatin-like phospholipase domain containing 3 gene (PNPLA3) on the susceptibility and histological severity of nonalcoholic fatty liver disease. *Hepatology* **2011**, *53*, 1883–1894. [CrossRef] [PubMed]
46. Sookoian, S.; Castaño, G.O.; Burgueño, A.L.; Gianotti, T.F.; Rosselli, M.S.; Pirola, C.J. A nonsynonymous gene variant in the adiponutrin gene is associated with nonalcoholic fatty liver disease severity. *J. Lipid Res.* **2009**, *50*, 2111–2116. [CrossRef] [PubMed]
47. Valenti, L.; Alisi, A.; Galmozzi, E.; Bartuli, A.; Del Menico, B.; Alterio, A.; Dongiovanni, P.; Fargion, S.; Nobili, V. I148M patatin-like phospholipase domain-containing 3 gene variant and severity of pediatric nonalcoholic fatty liver disease. *Hepatology* **2010**, *52*, 1274–1280. [CrossRef] [PubMed]
48. Kawaguchi, T.; Sumida, Y.; Umemura, A.; Matsuo, K.; Takahashi, M.; Takamura, T.; Yasui, K.; Saibara, T.; Hashimoto, E.; Kawanaka, M.; et al. Genetic polymorphisms of the human PNPLA3 gene are strongly associated with severity of non-alcoholic fatty liver disease in Japanese. *PLoS ONE* **2012**, *7*, e38322. [CrossRef] [PubMed]
49. Chen, W.; Chang, B.; Li, L.; Chan, L. Patatin-like phospholipase domain-containing 3/adiponutrin deficiency in mice is not associated with fatty liver disease. *Hepatology* **2010**, *52*, 1134–1142. [CrossRef] [PubMed]
50. Hao, L.; Ito, K.; Huang, K.H.; Sae-tan, S.; Lambert, J.D.; Ross, A.C. Shifts in dietary carbohydrate-lipid exposure regulate expression of the non-alcoholic fatty liver disease-associated gene PNPLA3/adiponutrin in mouse liver and HepG2 human liver cells. *Metabolism* **2014**, *63*, 1352–1362. [CrossRef] [PubMed]
51. Qiao, A.; Liang, J.; Ke, Y.; Li, C.; Cui, Y.; Shen, L.; Zhang, H.; Cui, A.; Liu, X.; Liu, C.; et al. Mouse patatin-like phospholipase domain-containing 3 influences systemic lipid and glucose homeostasis. *Hepatology* **2011**, *54*, 509–521. [CrossRef] [PubMed]

52. Mahdessian, H.; Taxiarchis, A.; Popov, S.; Silveira, A.; Franco-Cereceda, A.; Hamsten, A.; Eriksson, P.; van't Hooft, F. TM6SF2 is a regulator of liver fat metabolism influencing triglyceride secretion and hepatic lipid droplet content. *Proc. Natl. Acad. Sci. USA* **2014**, *111*, 8913–8918. [CrossRef] [PubMed]
53. Kozlitina, J.; Smagris, E.; Stender, S.; Nordestgaard, B.G.; Zhou, H.H.; Tybjærg-Hansen, A.; Vogt, T.F.; Hobbs, H.H.; Cohen, J.C. Exome-wide association study identifies a TM6SF2 variant that confers susceptibility to nonalcoholic fatty liver disease. *Nat. Genet.* **2014**, *46*, 352–356. [CrossRef] [PubMed]
54. Sookoian, S.; Castaño, G.O.; Scian, R.; Mallardi, P.; Fernández Gianotti, T.; Burgueño, A.L.; San Martino, J.; Pirola, C.J. Genetic variation in transmembrane 6 superfamily member 2 and the risk of nonalcoholic fatty liver disease and histological disease severity. *Hepatology* **2015**, *61*, 515–525. [CrossRef] [PubMed]
55. Holmen, O.L.; Zhang, H.; Fan, Y.; Hovelson, D.H.; Schmidt, E.M.; Zhou, W.; Guo, Y.; Zhang, J.; Langhammer, A.; Løchen, M.L.; et al. Systematic evaluation of coding variation identifies a candidate causal variant in TM6SF2 influencing total cholesterol and myocardial infarction risk. *Nat. Genet.* **2014**, *46*, 345–351. [CrossRef] [PubMed]
56. Pirola, C.J.; Sookoian, S. The dual and opposite role of the TM6SF2-rs58542926 variant in protecting against cardiovascular disease and conferring risk for nonalcoholic fatty liver: A meta-analysis. *Hepatology* **2015**, *62*, 1742–1756. [CrossRef] [PubMed]
57. Wang, X.; Liu, Z.; Peng, Z.; Liu, W. The TM6SF2 rs58542926 T allele is significantly associated with non-alcoholic fatty liver disease in Chinese. *J. Hepatol.* **2015**, *62*, 1438–1439. [CrossRef] [PubMed]
58. Iynedjian, P.B. Molecular physiology of mammalian glucokinase. *Cell. Mol. Life Sci.* **2009**, *66*, 27–42. [CrossRef] [PubMed]
59. Speliotes, E.K.; Yerges-Armstrong, L.M.; Wu, J.; Hernaez, R.; Kim, L.J.; Palmer, C.D.; Gudnason, V.; Eiriksdottir, G.; Garcia, M.E.; Launer, L.J.; et al. Genome-wide association analysis identifies variants associated with nonalcoholic fatty liver disease that have distinct effects on metabolic traits. *PLoS Genet.* **2011**, *7*, e1001324. [CrossRef] [PubMed]
60. Zain, S.M.; Mohamed, Z.; Mohamed, R. Common variant in the glucokinase regulatory gene rs780094 and risk of nonalcoholic fatty liver disease: A meta-analysis. *J. Gastroenterol. Hepatol.* **2015**, *30*, 21–27. [CrossRef] [PubMed]
61. Dimas, A.S.; Lagou, V.; Barker, A.; Knowles, J.W.; Mägi, R.; Hivert, M.F.; Benazzo, A.; Rybin, D.; Jackson, A.U.; Stringham, H.M.; et al. Impact of type 2 diabetes susceptibility variants on quantitative glycemic traits reveals mechanistic heterogeneity. *Diabetes* **2014**, *63*, 2158–2171. [CrossRef] [PubMed]
62. Tsuneto, A.; Hida, A.; Sera, N.; Imaizumi, M.; Ichimaru, S.; Nakashima, E.; Seto, S.; Maemura, K.; Akahoshi, M. Fatty liver incidence and predictive variables. *Hypertens. Res.* **2010**, *33*, 638–643. [CrossRef] [PubMed]
63. Nobili, V.; Svegliati-Baroni, G.; Alisi, A.; Miele, L.; Valenti, L.; Vajro, P. A 360-degree overview of paediatric NAFLD: Recent insights. *J. Hepatol.* **2013**, *58*, 1218–1229. [CrossRef] [PubMed]
64. Eguchi, Y.; Eguchi, T.; Mizuta, T.; Ide, Y.; Yasutake, T.; Iwakiri, R.; Hisatomi, A.; Ozaki, I.; Yamamoto, K.; Kitajima, Y.; et al. Visceral fat accumulation and insulin resistance are important factors in nonalcoholic fatty liver disease. *J. Gastroenterol.* **2006**, *41*, 462–469. [CrossRef] [PubMed]
65. Kersten, S.; Desvergne, B.; Wahli, W. Roles of ppars in health & disease. *Nature* **2000**, *405*, 421–424. [CrossRef] [PubMed]
66. Burdick, A.D.; Kim, D.J.; Peraza, M.A.; Gonzalez, F.J.; Peters, J.M. The role of peroxisome proliferator-activated receptor-beta/delta in epithelial cell growth and differentiation. *Cell. Signal.* **2006**, *18*, 9–20. [CrossRef] [PubMed]
67. Wahli, W.; Michalik, L. PPARs at the crossroads of lipid signaling and inflammation. *Trends Endocrinol. Metab.* **2012**, *23*, 351–363. [CrossRef] [PubMed]
68. Keller, H.; Dreyer, C.; Medin, J.; Mahfoudi, A.; Ozato, K.; Wahli, W. Fatty acids and retinoids control lipid metabolism through activation of peroxisome proliferator-activated receptor-retinoid X receptor heterodimers. *Proc. Natl. Acad. Sci. USA* **1993**, *90*, 2160–2164. [CrossRef] [PubMed]
69. Hörlein, A.J.; Näär, A.M.; Heinzel, T.; Torchia, J.; Gloss, B.; Kurokawa, R.; Ryan, A.; Kamei, Y.; Söderström, M.; Glass, C.K.; Rosenfeld, M.G. Ligand-independent repression by the thyroid hormone receptor mediated by a nuclear receptor co-repressor. *Nature* **1995**, *377*, 397–404. [CrossRef] [PubMed]
70. Chen, J.D.; Evans, R.M. A transcriptional co-repressor that interacts with nuclear hormone receptors. *Nature* **1995**, *377*, 454–457. [CrossRef] [PubMed]

71. Chinetti, G.; Fruchart, J.C.; Staels, B. Peroxisome proliferator-activated receptors (PPARs): Nuclear receptors at the crossroads between lipid metabolism and inflammation. *Inflamm. Res.* **2000**, *49*, 497–505. [CrossRef] [PubMed]
72. Feige, J.N.; Gelman, L.; Michalik, L.; Desvergne, B.; Wahli, W. From molecular action to physiological outputs: Peroxisome proliferator-activated receptors are nuclear receptors at the crossroads of key cellular functions. *Prog. Lipid Res.* **2006**, *45*, 120–159. [CrossRef] [PubMed]
73. IJpenberg, A.; Jeannin, E.; Wahli, W.; Desvergne, B. Polarity and specific sequence requirements of peroxisome proliferator-activated receptor (PPAR)/retinoid X receptor heterodimer binding to DNA. A functional analysis of the malic enzyme gene PPAR response element. *J. Biol. Chem.* **1997**, *272*, 20108–20117. [CrossRef] [PubMed]
74. Juge-Aubry, C.; Pernin, A.; Favez, T.; Burger, A.G.; Wahli, W.; Meier, C.A.; Desvergne, B. DNA binding properties of peroxisome proliferator-activated receptor subtypes on various natural peroxisome proliferator response elements. Importance of the 5′-flanking region. *J. Biol. Chem.* **1997**, *272*, 25252–25259. [CrossRef] [PubMed]
75. Dreyer, C.; Krey, G.; Keller, H.; Givel, F.; Helftenbein, G.; Wahli, W. Control of the peroxisomal beta-oxidation pathway by a novel family of nuclear hormone receptors. *Cell* **1992**, *68*, 879–887. [CrossRef]
76. Nolte, R.T.; Wisely, G.B.; Westin, S.; Cobb, J.E.; Lambert, M.H.; Kurokawa, R.; Rosenfeld, M.G.; Willson, T.M.; Glass, C.K.; Milburn, M.V. Ligand binding and co-activator assembly of the peroxisome proliferator-activated receptor-gamma. *Nature* **1998**, *395*, 137–143. [CrossRef] [PubMed]
77. Moras, D.; Gronemeyer, H. The nuclear receptor ligand-binding domain: Structure and function. *Curr. Opin. Cell Biol.* **1998**, *10*, 384–391. [CrossRef]
78. Fan, Y.; Wang, Y.; Tang, Z.; Zhang, H.; Qin, X.; Zhu, Y.; Guan, Y.; Wang, X.; Staels, B.; Chien, S.; et al. Suppression of pro-inflammatory adhesion molecules by PPAR-delta in human vascular endothelial cells. *Arterioscler. Thromb. Vasc. Biol.* **2008**, *28*, 315–321. [CrossRef] [PubMed]
79. Zhang, J.; Fu, M.; Zhu, X.; Xiao, Y.; Mou, Y.; Zheng, H.; Akinbami, M.A.; Wang, Q.; Chen, Y.E. Peroxisome proliferator-activated receptor delta is up-regulated during vascular lesion formation and promotes post-confluent cell proliferation in vascular smooth muscle cells. *J. Biol. Chem.* **2002**, *277*, 11505–11512. [CrossRef] [PubMed]
80. Lee, C.H.; Chawla, A.; Urbiztondo, N.; Liao, D.; Boisvert, W.A.; Evans, R.M.; Curtiss, L.K. Transcriptional repression of atherogenic inflammation: Modulation by PPARdelta. *Science* **2003**, *302*, 453–457. [CrossRef] [PubMed]
81. Matsushita, Y.; Ogawa, D.; Wada, J.; Yamamoto, N.; Shikata, K.; Sato, C.; Tachibana, H.; Toyota, N.; Makino, H. Activation of peroxisome proliferator-activated receptor delta inhibits streptozotocin-induced diabetic nephropathy through anti-inflammatory mechanisms in mice. *Diabetes* **2011**, *60*, 960–968. [CrossRef] [PubMed]
82. Braissant, O.; Foufelle, F.; Scotto, C.; Dauça, M.; Wahli, W. Differential expression of peroxisome proliferator-activated receptors (PPARs): Tissue distribution of PPAR-alpha, -beta, and -gamma in the adult rat. *Endocrinology* **1996**, *137*, 354–366. [CrossRef] [PubMed]
83. Michalik, L.; Auwerx, J.; Berger, J.P.; Chatterjee, V.K.; Glass, C.K.; Gonzalez, F.J.; Grimaldi, P.A.; Kadowaki, T.; Lazar, M.A.; O'Rahilly, S.; et al. International Union of Pharmacology. LXI. Peroxisome proliferator-activated receptors. *Pharmacol. Rev.* **2006**, *58*, 726–741. [CrossRef] [PubMed]
84. Mandard, S.; Patsouris, D. Nuclear control of the inflammatory response in mammals by peroxisome proliferator-activated receptors. *PPAR Res.* **2013**, *2013*, 613864. [CrossRef] [PubMed]
85. Mandard, S.; Müller, M.; Kersten, S. Peroxisome proliferator-activated receptor alpha target genes. *Cell. Mol. Life Sci.* **2004**, *61*, 393–416. [CrossRef] [PubMed]
86. Montagner, A.; Korecka, A.; Polizzi, A.; Lippi, Y.; Blum, Y.; Canlet, C.; Tremblay-Franco, M.; Gautier-Stein, A.; Burcelin, R.; Yen, Y.C.; et al. Hepatic circadian clock oscillators and nuclear receptors integrate microbiome-derived signals. *Sci. Rep.* **2016**, *6*, 20127. [CrossRef] [PubMed]
87. Régnier, M.; Polizzi, A.; Lippi, Y.; Fouché, E.; Michel, G.; Lukowicz, C.; Smati, S.; Marrot, A.; Lasserre, F.; Naylies, C.; et al. Insights into the role of hepatocyte PPARα activity in response to fasting. *Mol. Cell. Endocrinol.* **2017**. [CrossRef]
88. Devchand, P.R.; Keller, H.; Peters, J.M.; Vazquez, M.; Gonzalez, F.J.; Wahli, W. The PPARalpha-leukotriene B4 pathway to inflammation control. *Nature* **1996**, *384*, 39–43. [CrossRef] [PubMed]

89. Tailleux, A.; Wouters, K.; Staels, B. Roles of PPARs in NAFLD: Potential therapeutic targets. *Biochim. Biophys. Acta* **2012**, *1821*, 809–818. [CrossRef] [PubMed]
90. Vanden Berghe, W.; Vermeulen, L.; Delerive, P.; De Bosscher, K.; Staels, B.; Haegeman, G. A paradigm for gene regulation: Inflammation, NF-kappaB and PPAR. *Adv. Exp. Med. Biol.* **2003**, *544*, 181–196. [PubMed]
91. Gervois, P.; Kleemann, R.; Pilon, A.; Percevault, F.; Koenig, W.; Staels, B.; Kooistra, T. Global suppression of IL-6-induced acute phase response gene expression after chronic in vivo treatment with the peroxisome proliferator-activated receptor-alpha activator fenofibrate. *J. Biol. Chem.* **2004**, *279*, 16154–16160. [CrossRef] [PubMed]
92. Abdelmegeed, M.A.; Yoo, S.H.; Henderson, L.E.; Gonzalez, F.J.; Woodcroft, K.J.; Song, B.J. PPARalpha expression protects male mice from high fat-induced nonalcoholic fatty liver. *J. Nutr.* **2011**, *141*, 603–610. [CrossRef] [PubMed]
93. Costet, P.; Legendre, C.; Moré, J.; Edgar, A.; Galtier, P.; Pineau, T. Peroxisome proliferator-activated receptor α-isoform deficiency leads to progressive dyslipidemia with sexually dimorphic obesity and steatosis. *J. Biol. Chem.* **1998**, *273*, 29577–29585. [CrossRef] [PubMed]
94. Ip, E.; Farrell, G.C.; Robertson, G.; Hall, P.; Kirsch, R.; Leclercq, I. Central role of PPARalpha-dependent hepatic lipid turnover in dietary steatohepatitis in mice. *Hepatology* **2003**, *38*, 123–132. [CrossRef] [PubMed]
95. Staels, B.; Rubenstrunk, A.; Noel, B.; Rigou, G.; Delataille, P.; Millatt, L.J.; Baron, M.; Lucas, A.; Tailleux, A.; Hum, D.W.; et al. Hepatoprotective effects of the dual peroxisome proliferator-activated receptor alpha/delta agonist, GFT505, in rodent models of nonalcoholic fatty liver disease/nonalcoholic steatohepatitis. *Hepatology* **2013**, *58*, 1941–1952. [CrossRef] [PubMed]
96. Francque, S.; Verrijken, A.; Caron, S.; Prawitt, J.; Paumelle, R.; Derudas, B.; Lefebvre, P.; Taskinen, M.R.; Van Hul, W.; Mertens, I.; et al. PPARα gene expression correlates with severity and histological treatment response in patients with non-alcoholic steatohepatitis. *J. Hepatol.* **2015**, *63*, 164–173. [CrossRef] [PubMed]
97. Montagner, A.; Polizzi, A.; Fouché, E.; Ducheix, S.; Lippi, Y.; Lasserre, F.; Barquissau, V.; Régnier, M.; Lukowicz, C.; Benhamed, F.; Iroz, A.; et al. Liver PPARα is crucial for whole-body fatty acid homeostasis and is protective against NAFLD. *Gut* **2016**, *65*, 1202–1214. [CrossRef] [PubMed]
98. Kersten, S.; Seydoux, J.; Peters, J.M.; Gonzalez, F.J.; Desvergne, B.; Wahli, W. Peroxisome proliferator-activated receptor alpha mediates the adaptive response to fasting. *J. Clin. Investig.* **1999**, *103*, 1489–1498. [CrossRef] [PubMed]
99. Leone, T.C.; Weinheimer, C.J.; Kelly, D.P. A critical role for the peroxisome proliferator-activated receptor alpha (PPARα) in the cellular fasting response: The PPARα-null mouse as a model of fatty acid oxidation disorders. *Proc. Natl. Acad. Sci. USA* **1999**, *96*, 7473–7478. [CrossRef] [PubMed]
100. Fajas, L.; Fruchart, J.C.; Auwerx, J. PPARgamma3 mRNA: A distinct PPARgamma mRNA subtype transcribed from an independent promoter. *FEBS Lett.* **1998**, *438*, 55–60. [CrossRef]
101. Rahimian, R.; Masih-Khan, E.; Lo, M.; van Breemen, C.; McManus, BM.; Dubé, G.P. Hepatic over-expression of peroxisome proliferator activated receptor gamma2 in the ob/ob mouse model of non-insulin dependent diabetes mellitus. *Mol. Cell. Biochem.* **2001**, *224*, 29–37. [CrossRef] [PubMed]
102. Memon, R.A.; Tecott, L.H.; Nonogaki, K.; Beigneux, A.; Moser, A.H.; Grunfeld, C.; Feingold, K.R. Up-regulation of peroxisome proliferator-activated receptors (PPAR-α) and PPAR-γ messenger ribonucleic acid expression in the liver in murine obesity: Troglitazone induces expression of PPAR-γ-responsive adipose tissue-specific genes in the liver of obese diabetic mice. *Endocrinology* **2001**, *141*, 4021–4031. [CrossRef]
103. Morán-Salvador, E.; López-Parra, M.; García-Alonso, V.; Titos, E.; Martínez-Clemente, M.; González-Périz, A.; López-Vicario, C.; Barak, Y.; Arroyo, V.; Clària, J. Role for PPARγ in obesity-induced hepatic steatosis as determined by hepatocyte- and macrophage-specific conditional knockouts. *FASEB J.* **2011**, *25*, 2538–2550. [CrossRef] [PubMed]
104. Westerbacka, J.; Kolak, M.; Kiviluoto, T.; Arkkila, P.; Sirén, J.; Hamsten, A.; Fisher, R.M.; Yki-Järvinen, H. Genes involved in fatty acid partitioning and binding, lipolysis, monocyte/macrophage recruitment, and inflammation are overexpressed in the human fatty liver of insulin-resistant subjects. *Diabetes* **2007**, *56*, 2759–2765. [CrossRef] [PubMed]
105. Pettinelli, P.; Videla, L.A. Up-regulation of PPARγ mRNA expression in the liver of obese patients in parallel with a reinforcement of the lipogenic pathway by SREBP-1c induction. *J. Clin. Endocrinol. Metab.* **2011**, *96*, 1424–1430. [CrossRef] [PubMed]

106. Gavrilova, O.; Haluzik, M.; Matsusue, K.; Cutson, J.J.; Johnson, L.; Dietz, K.R.; Nicol, C.J.; Vinson, C.; Gonzalez, F.J.; Reitman, M.L. Liver peroxisome proliferator-activated receptor gamma contributes to hepatic steatosis, triglyceride clearance, and regulation of body fat mass. *J. Biol. Chem.* **2003**, *278*, 34268–34276. [CrossRef] [PubMed]
107. Matsusue, K.; Haluzik, M.; Lambert, G.; Yim, S.H.; Gavrilova, O.; Ward, J.M.; Brewer, B., Jr.; Reitman, M.L.; Gonzalez, F.J. Liver-specific disruption of PPARgamma in leptin-deficient mice improves fatty liver but aggravates diabetic phenotypes. *J. Clin. Investig.* **2003**, *111*, 737–747. [CrossRef] [PubMed]
108. Matsusue, K.; Aibara, D.; Hayafuchi, R.; Matsuo, K.; Takiguchi, S.; Gonzalez, F.J.; Yamano, S. Hepatic PPARγ and LXRα independently regulate lipid accumulation in the livers of genetically obese mice. *FEBS Lett.* **2014**, *588*, 2277–2281. [CrossRef] [PubMed]
109. Schadinger, S.E.; Bucher, N.L.; Schreiber, B.M.; Farmer, S.R. PPARgamma2 regulates lipogenesis and lipid accumulation in steatotic hepatocytes. *Am. J. Physiol. Endocrinol. Metab.* **2005**, *288*, E1195–E1205. [CrossRef] [PubMed]
110. Zhang, Y.L.; Hernandez-Ono, A.; Siri, P.; Weisberg, S.; Conlon, D.; Graham, M.J.; Crooke, R.M.; Huang, L.S.; Ginsberg, H.N. Aberrant hepatic expression of PPARgamma2 stimulates hepatic lipogenesis in a mouse model of obesity, insulin resistance, dyslipidemia, and hepatic steatosis. *J. Biol. Chem.* **2006**, *281*, 37603–37615. [CrossRef] [PubMed]
111. Wolf Greenstein, A.; Majumdar, N.; Yang, P.; Subbaiah, P.V.; Kineman, R.D.; Cordoba-Chacon, J. Hepatocyte-specific, PPARγ-regulated mechanisms to promote steatosis in adult mice. *J. Endocrinol.* **2017**, *232*, 107–121. [CrossRef] [PubMed]
112. Panasyuk, G.; Espeillac, C.; Chauvin, C.; Pradelli, L.A.; Horie, Y.; Suzuki, A.; Annicotte, J.S.; Fajas, L.; Foretz, M.; Verdeguer, F.; et al. PPARγ contributes to PKM2 and HK2 expression in fatty liver. *Nat. Commun.* **2012**, *3*, 672. [CrossRef] [PubMed]
113. Kliewer, S.A.; Forman, B.M.; Blumberg, B.; Ong, E.S.; Borgmeyer, U.; Mangelsdorf, D.J.; Umesono, K.; Evans, R.M. Differential expression and activation of a family of murine peroxisome proliferator-activated receptors. *Proc. Natl. Acad. Sci. USA* **1994**, *91*, 7355–7359. [CrossRef] [PubMed]
114. Auboeuf, D.; Rieusset, J.; Fajas, L.; Vallier, P.; Frering, V.; Riou, J.P.; Staels, B.; Auwerx, J.; Laville, M.; Vidal, H. Tissue distribution and quantification of the expression of mRNAs of peroxisome proliferator-activated receptors and liver X receptor-alpha in humans: No alteration in adipose tissue of obese and NIDDM patients. *Diabetes* **1997**, *46*, 1319–1327. [CrossRef] [PubMed]
115. Tugwood, J.D.; Aldridge, T.C.; Lambe, K.G.; Macdonald, N.; Woodyatt, N.J. Peroxisome proliferator-activated receptors: Structures and function. *Ann. N. Y. Acad. Sci.* **1996**, *804*, 252–265. [CrossRef] [PubMed]
116. Mukherjee, R.; Jow, L.; Croston, G.E.; Paterniti, J.R., Jr. Identification, characterization, and tissue distribution of human peroxisome proliferator-activated receptor (PPAR) isoforms PPARgamma2 versus PPARgamma1 and activation with retinoid X receptor agonists and antagonists. *J. Biol. Chem.* **1997**, *272*, 8071–8076. [CrossRef] [PubMed]
117. Girroir, E.E.; Hollingshead, H.E.; He, P.; Zhu, B.; Perdew, G.H.; Peters, J.M. Quantitative expression patterns of peroxisome proliferator-activated receptor-beta/delta (PPARbeta/delta) protein in mice. *Biochem. Biophys. Res. Commun.* **2008**, *371*, 456–461. [CrossRef] [PubMed]
118. Hoekstra, M.; Kruijt, J.K.; Van Eck, M.; Van Berkel, T.J. Specific gene expression of ATP-binding cassette transporters and nuclear hormone receptors in rat liver parenchymal, endothelial, and Kupffer cells. *J. Biol. Chem.* **2003**, *278*, 25448–25453. [CrossRef] [PubMed]
119. Sanderson, L.M.; Boekschoten, M.V.; Desvergne, B.; Müller, M.; Kersten, S. Transcriptional profiling reveals divergent roles of PPARalpha and PPARbeta/delta in regulation of gene expression in mouse liver. *Physiol. Genom.* **2010**, *41*, 42–52. [CrossRef] [PubMed]
120. Ricchi, M.; Odoardi, M.R.; Carulli, L.; Anzivino, C.; Ballestri, S.; Pinetti, A.; Fantoni, L.I.; Marra, F.; Bertolotti, M.; Banni, S.; et al. Differential effect of oleic and palmitic acid on lipid accumulation and apoptosis in cultured hepatocytes. *J. Gastroenterol. Hepatol.* **2009**, *24*, 830–840. [CrossRef] [PubMed]
121. Liu, S.; Brown, J.D.; Stanya, K.J.; Homan, E.; Leidl, M.; Inouye, K.; Bhargava, P.; Gangl, M.R.; Dai, L.; Hatano, B.; et al. A diurnal serum lipid integrates hepatic lipogenesis and peripheral fatty acid use. *Nature* **2013**, *502*, 550–554. [CrossRef] [PubMed]

122. Christodoulides, C.; Dyson, P.; Sprecher, D.; Tsintzas, K.; Karpe, F. Circulating fibroblast growth factor 21 is induced by peroxisome proliferator-activated receptor agonists but not ketosis in man. *J. Clin. Endocrinol. Metab.* **2009**, *94*, 3594–3601. [CrossRef] [PubMed]
123. Nakamura, M.T.; Yudell, B.E.; Loor, J.J. Regulation of energy metabolism by long-chain fatty acids. *Prog. Lipid Res.* **2014**, *53*, 124–144. [CrossRef] [PubMed]
124. Rando, G.; Tan, C.K.; Khaled, N.; Montagner, A.; Leuenberger, N.; Bertrand-Michel, J.; Paramalingam, E.; Guillou, H.; Wahli, W. Glucocorticoid receptor-PPARα axis in fetal mouse liver prepares neonates for milk lipid catabolism. *Elife* **2016**, *5*, e11853. [CrossRef] [PubMed]
125. Escher, P.; Braissant, O.; Basu-Modak, S.; Michalik, L.; Wahli, W.; Desvergne, B. Rat PPARs: Quantitative analysis in adult rat tissues and regulation in fasting and refeeding. *Endocrinology* **2001**, *142*, 4195–4202. [CrossRef] [PubMed]
126. Barroso, E.; Rodríguez-Calvo, R.; Serrano-Marco, L.; Astudillo, A.M.; Balsinde, J.; Palomer, X.; Vázquez-Carrera, M. The PPARβ/δ activator GW501516 prevents the down-regulation of AMPK caused by a high-fat diet in liver and amplifies the PGC-1α-Lipin 1-PPARα pathway leading to increased fatty acid oxidation. *Endocrinology* **2011**, *152*, 1848–1859. [CrossRef] [PubMed]
127. Palomer, X.; Barroso, E.; Pizarro-Delgado, J.; Peña, L.; Botteri, G.; Zarei, M.; Aguilar, D.; Montori-Grau, M.; Vázquez-Carrera, M. PPARβ/δ: A Key Therapeutic Target in Metabolic Disorders. *Int. J. Mol. Sci.* **2018**, *19*, 913. [CrossRef] [PubMed]
128. Chakravarthy, M.V.; Pan, Z.; Zhu, Y.; Tordjman, K.; Schneider, J.G.; Coleman, T.; Turk, J.; Semenkovich, C.F. "New" hepatic fat activates PPARalpha to maintain glucose, lipid, and cholesterol homeostasis. *Cell Metab.* **2005**, *1*, 309–322. [CrossRef] [PubMed]
129. Varga, T.; Czimmerer, Z.; Nagy, L. PPARs are a unique set of fatty acid regulated transcription factors controlling both lipid metabolism and inflammation. *Biochim. Biophys. Acta* **2011**, *1812*, 1007–1022. [CrossRef] [PubMed]
130. Videla, L.A.; Pettinelli, P. Misregulation of PPAR Functioning and Its Pathogenic Consequences Associated with Nonalcoholic Fatty Liver Disease in Human Obesity. *PPAR Res.* **2012**, *2012*, 107434. [CrossRef] [PubMed]
131. Garbacz, W.G.; Huang, J.T.; Higgins, L.G.; Wahli, W.; Palmer, C.N. PPARα Is Required for PPARδ Action in Regulation of Body Weight and Hepatic Steatosis in Mice. *PPAR Res.* **2015**, *2015*, 927057. [CrossRef] [PubMed]
132. Horike, N.; Sakoda, H.; Kushiyama, A.; Ono, H.; Fujishiro, M.; Kamata, H.; Nishiyama, K.; Uchijima, Y.; Kurihara, Y.; Kurihara, H.; et al. AMP-activated protein kinase activation increases phosphorylation of glycogen synthase kinase 3beta and thereby reduces cAMP-responsive element transcriptional activity and phosphoenolpyruvate carboxykinase C gene expression in the liver. *J. Biol. Chem.* **2008**, *283*, 33902–33910. [CrossRef] [PubMed]
133. Tanaka, T.; Yamamoto, J.; Iwasaki, S.; Asaba, H.; Hamura, H.; Ikeda, Y.; Watanabe, M.; Magoori, K.; Ioka, R.X.; Tachibana, K.; et al. Activation of peroxisome proliferator-activated receptor delta induces fatty acid beta-oxidation in skeletal muscle and attenuates metabolic syndrome. *Proc. Natl. Acad. Sci. USA* **2003**, *100*, 15924–15929. [CrossRef] [PubMed]
134. Qin, X.; Xie, X.; Fan, Y.; Tian, J.; Guan, Y.; Wang, X.; Zhu, Y.; Wang, N. Peroxisome proliferator-activated receptor-delta induces insulin-induced gene-1 and suppresses hepatic lipogenesis in obese diabetic mice. *Hepatology* **2008**, *48*, 432–441. [CrossRef] [PubMed]
135. Markan, K.R.; Naber, M.C.; Ameka, M.K.; Anderegg, M.D.; Mangelsdorf, D.J.; Kliewer, S.A.; Mohammadi, M.; Potthoff, M.J. Circulating FGF21 is liver derived and enhances glucose uptake during refeeding and overfeeding. *Diabetes* **2014**, *63*, 4057–4063. [CrossRef] [PubMed]
136. Iroz, A.; Montagner, A.; Benhamed, F.; Levavasseur, F.; Polizzi, A.; Anthony, E.; Régnier, M.; Fouché, E.; Lukowicz, C.; Cauzac, M.; et al. A Specific ChREBP and PPARα Cross-Talk Is Required for the Glucose-Mediated FGF21 Response. *Cell Rep.* **2017**, *21*, 403–416. [CrossRef] [PubMed]
137. Zarei, M.; Barroso, E.; Palomer, X.; Dai, J.; Rada, P.; Quesada-López, T.; Escolà-Gil, J.C.; Cedó, L.; Zali, M.R.; Molaei, M.; et al. Hepatic regulation of VLDL receptor by PPARβ/δ and FGF21 modulates non-alcoholic fatty liver disease. *Mol. Metab.* **2018**, *8*, 117–131. [CrossRef] [PubMed]

138. Ding, G.; Cheng, L.; Qin, Q.; Frontin, S.; Yang, Q.J. PPARdelta modulates lipopolysaccharide-induced TNFalpha inflammation signalling in cultured cardiomyocytes. *Mol. Cell. Cardiol.* **2006**, *40*, 821–828. [CrossRef] [PubMed]
139. Rival, Y.; Benéteau, N.; Taillandier, T.; Pezet, M.; Dupont-Passelaigue, E.; Patoiseau, J.F.; Junquéro, D.; Colpaert, F.C.; Delhon, A. PPARalpha and PPARdelta activators inhibit cytokine-induced nuclear translocation of NF-kappaB and expression of VCAM-1 in EAhy926 endothelial cells. *Eur. J. Pharmacol.* **2002**, *435*, 143–151. [CrossRef]
140. Kino, T.; Rice, K.C.; Chrousos, G.P. The PPARdelta agonist GW501516 suppresses interleukin-6-mediated hepatocyte acute phase reaction via STAT3 inhibition. *Eur. J. Clin. Investig.* **2007**, *37*, 425–433. [CrossRef] [PubMed]
141. Woo, C.H.; Massett, M.P.; Shishido, T.; Itoh, S.; Ding, B.; McClain, C.; Che, W.; Vulapalli, S.R.; Yan, C.; Abe, J. ERK5 activation inhibits inflammatory responses via peroxisome proliferator-activated receptor delta (PPARdelta) stimulation. *J. Biol. Chem.* **2006**, *281*, 32164–32174. [CrossRef] [PubMed]
142. Lanthier, N.; Molendi-Coste, O.; Horsmans, Y.; van Rooijen, N.; Cani, P.D.; Leclercq, I.A. Kupffer cell activation is a causal factor for hepatic insulin resistance. *Am. J. Physiol. Gastrointest. Liver Physiol.* **2010**, *298*, G107–G116. [CrossRef] [PubMed]
143. Odegaard, J.I.; Ricardo-Gonzalez, R.R.; Red Eagle, A.; Vats, D.; Morel, C.R.; Goforth, M.H.; Subramanian, V.; Mukundan, L.; Ferrante, A.W.; Chawla, A. Alternative M2 activation of Kupffer cells by PPARdelta ameliorates obesity-induced insulin resistance. *Cell Metab.* **2008**, *7*, 496–507. [CrossRef] [PubMed]
144. Hellemans, K.; Rombouts, K.; Quartier, E.; Dittié, A.S.; Knorr, A.; Michalik, L.; Rogiers, V.; Schuit, F.; Wahli, W.; Geerts, A. PPARbeta regulates vitamin A metabolism-related gene expression in hepatic stellate cells undergoing activation. *J. Lipid Res.* **2003**, *44*, 280–295. [CrossRef] [PubMed]
145. Kostadinova, R.; Montagner, A.; Gouranton, E.; Fleury, S.; Guillou, H.; Dombrowicz, D.; Desreumaux, P.; Wahli, W. GW501516-activated PPARβ/δ promotes liver fibrosis via p38-JNK MAPK-induced hepatic stellate cell proliferation. *Cell Biosci.* **2012**, *2*, 34. [CrossRef] [PubMed]
146. Hellemans, K.; Michalik, L.; Dittie, A.; Knorr, A.; Rombouts, K.; De Jong, J.; Heirman, C.; Quartier, E.; Schuit, F.; Wahli, W.; et al. Peroxisome proliferator-activated receptor-beta signaling contributes to enhanced proliferation of hepatic stellate cells. *Gastroenterology* **2003**, *124*, 184–201. [CrossRef] [PubMed]
147. Lee, C.H.; Olson, P.; Hevener, A.; Mehl, I.; Chong, L.W.; Olefsky, J.M.; Gonzalez, F.J.; Ham, J.; Kang, H.; Peters, J.M.; et al. PPARdelta regulates glucose metabolism and insulin sensitivity. *Proc. Natl. Acad. Sci. USA* **2006**, *103*, 3444–3449. [CrossRef] [PubMed]
148. Iwaisako, K.; Haimerl, M.; Paik, Y.H.; Taura, K.; Kodama, Y.; Sirlin, C.; Yu, E.; Yu, R.T.; Downes, M.; Evans, R.M.; et al. Protection from liver fibrosis by a peroxisome proliferator-activated receptor δ agonist. *Proc. Natl. Acad. Sci. USA* **2012**, *109*, E1369–E1376. [CrossRef] [PubMed]
149. Tan, C.K.; Zhuang, Y.; Wahli, W. Synthetic and natural Peroxisome Proliferator-Activated Receptor (PPAR) agonists as candidates for the therapy of the metabolic syndrome. *Expert Opin. Ther. Targets* **2017**, *21*, 333–348. [CrossRef] [PubMed]
150. Shan, W.; Palkar, P.S.; Murray, I.A.; McDevitt, E.I.; Kennett, M.J.; Kang, B.H.; Isom, H.C.; Perdew, G.H.; Gonzalez, F.J.; Peters, J.M. Ligand activation of peroxisome proliferator-activated receptor beta/delta (PPARbeta/delta) attenuates carbon tetrachloride hepatotoxicity by downregulating proinflammatory gene expression. *Toxicol. Sci.* **2008**, *105*, 418–428. [CrossRef] [PubMed]
151. Chehaibi, K.; Cedó, L.; Metso, J.; Palomer, X.; Santos, D.; Quesada, H.; Naceur Slimane, M.; Wahli, W.; Julve, J.; Vázquez-Carrera, M.; et al. PPAR-β/δ activation promotes phospholipid transfer protein expression. *Biochem. Pharmacol.* **2015**, *94*, 101–108. [CrossRef] [PubMed]
152. Nagasawa, T.; Inada, Y.; Nakano, S.; Tamura, T.; Takahashi, T.; Maruyama, K.; Yamazaki, Y.; Kuroda, J.; Shibata, N. Effects of bezafibrate, PPAR pan-agonist, and GW501516, PPARdelta agonist, on development of steatohepatitis in mice fed a methionine- and choline-deficient diet. *Eur. J. Pharmacol.* **2006**, *536*, 182–191. [CrossRef] [PubMed]
153. Wu, H.T.; Chen, C.T.; Cheng, K.C.; Li, Y.X.; Yeh, C.H.; Cheng, J.T. Pharmacological activation of peroxisome proliferator-activated receptor δ improves insulin resistance and hepatic steatosis in high fat diet-induced diabetic mice. *Horm. Metab. Res.* **2011**, *43*, 631–635. [CrossRef] [PubMed]

154. Bojic, L.A.; Telford, D.E.; Fullerton, M.D.; Ford, R.J.; Sutherland, B.G.; Edwards, J.Y.; Sawyez, C.G.; Gros, R.; Kemp, B.E.; Steinberg, G.R.; et al. PPARδ activation attenuates hepatic steatosis in Ldlr−/− mice by enhanced fat oxidation, reduced lipogenesis, and improved insulin sensitivity. *J. Lipid Res.* **2014**, *55*, 1254–1266. [CrossRef] [PubMed]
155. Bays, H.E.; Schwartz, S.; Littlejohn, T.; Kerzner, B.; Krauss, R.M.; Karpf, D.B.; Choi, Y.J.; Wang, X.; Naim, S.; Roberts, B.K. MBX-8025, a novel peroxisome proliferator receptor-delta agonist: Lipid and other metabolic effects in dyslipidemic overweight patients treated with and without atorvastatin. *J. Clin. Endocrinol. Metab.* **2011**, *96*, 2889–2897. [CrossRef] [PubMed]
156. Risérus, U.; Sprecher, D.; Johnson, T.; Olson, E.; Hirschberg, S.; Liu, A.; Fang, Z.; Hegde, P.; Richards, D.; Sarov-Blat, L.; et al. Activation of peroxisome proliferator-activated receptor (PPAR)delta promotes reversal of multiple metabolic abnormalities, reduces oxidative stress, and increases fatty acid oxidation in moderately obese men. *Diabetes* **2008**, *57*, 332–339. [CrossRef] [PubMed]
157. Ratziu, V.; Harrison, S.A.; Francque, S.; Bedossa, P.; Lehert, P.; Serfaty, L.; Romero-Gomez, M.; Boursier, J.; Abdelmalek, M.; Caldwell, S.; et al. Elafibranor, an Agonist of the Peroxisome Proliferator-Activated Receptor-α and -δ, Induces Resolution of Nonalcoholic Steatohepatitis without Fibrosis Worsening. *Gastroenterology* **2016**, *150*, 1147–1159. [CrossRef] [PubMed]
158. Cariou, B.; Hanf, R.; Lambert-Porcheron, S.; Zaïr, Y.; Sauvinet, V.; Noël, B.; Flet, L.; Vidal, H.; Staels, B.; Laville, M. Dual peroxisome proliferator-activated receptor α/δ agonist GFT505 improves hepatic and peripheral insulin sensitivity in abdominally obese subjects. *Diabetes Care* **2013**, *36*, 2923–2930. [CrossRef] [PubMed]
159. Perazzo, H.; Dufour, J.F. The therapeutic landscape of non-alcoholic steatohepatitis. *Liver Int.* **2017**, *37*, 634–647. [CrossRef] [PubMed]
160. Jain, M.R.; Giri, S.R.; Bhoi, B.; Trivedi, C.; Rath, A.; Rathod, R.; Ranvir, R.; Kadam, S.; Patel, H.; Swain, P.; et al. Dual PPARα/γ agonist saroglitazar improves liver histopathology and biochemistry in experimental NASH models. *Liver Int.* **2017**. [CrossRef] [PubMed]
161. Wettstein, G.; Luccarini, J.M.; Poekes, L.; Faye, P.; Kupkowski, F.; Adarbes, V.; Defrêne, E.; Estivalet, C.; Gawronski, X.; Jantzen, I.; et al. The new-generation pan-peroxisome proliferator-activated receptor agonist IVA337 protects the liver from metabolic disorders and fibrosis. *Hepatol. Commun.* **2017**, *1*, 524–537. [CrossRef] [PubMed]
162. An, H.J.; Lee, B.; Kim, S.M.; Kim, D.H.; Chung, K.W.; Ha, S.G.; Park, K.C.; Park, Y.J.; Kim, S.J.; Yun, H.Y.; et al. A PPAR Pan Agonist, MHY2013 Alleviates Age-Related Hepatic Lipid Accumulation by Promoting Fatty Acid Oxidation and Suppressing Inflammation. *Biol. Pharm. Bull.* **2018**, *41*, 29–35. [CrossRef] [PubMed]
163. Ruzehaji, N.; Frantz, C.; Ponsoye, M.; Avouac, J.; Pezet, S.; Guilbert, T.; Luccarini, J.M.; Broqua, P.; Junien, J.L.; Allanore, Y. Pan PPAR agonist IVA337 is effective in prevention and treatment of experimental skin fibrosis. *Ann. Rheum. Dis.* **2016**, *75*, 2175–2183. [CrossRef] [PubMed]
164. Avouac, J.; Konstantinova, I.; Guignabert, C.; Pezet, S.; Sadoine, J.; Guilbert, T.; Cauvet, A.; Tu, L.; Luccarini, J.M.; Junien, J.L.; et al. Pan-PPAR agonist IVA337 is effective in experimental lung fibrosis and pulmonary hypertension. *Ann. Rheum. Dis.* **2017**, *76*, 1931–1940. [CrossRef] [PubMed]
165. Sumida, Y.; Yoneda, M. Current and future pharmacological therapies for NAFLD/NASH. *J. Gastroenterol.* **2018**, *53*, 362–376. [CrossRef] [PubMed]
166. 10 Studies Found for: GFT505. Available online: https://www.clinicaltrials.gov/ct2/results?term=GFT505&Search=Search (accessed on 15 May 2018).

International Journal of
Molecular Sciences

Review

The Role of PPAR and Its Cross-Talk with CAR and LXR in Obesity and Atherosclerosis

Pengfei Xu [1,2], Yonggong Zhai [1,2,*] and Jing Wang [3,*]

1 Beijing Key Laboratory of Gene Resource and Molecular Development, College of Life Sciences, Beijing Normal University, Beijing 100875, China; pex9@pitt.edu
2 Key Laboratory for Cell Proliferation and Regulation Biology of State Education Ministry, College of Life Sciences, Beijing Normal University, Beijing 100875, China
3 Department of Biology Science and Technology, Baotou Teacher's College, Baotou 014030, China
* Correspondence: ygzhai@bnu.edu.cn (Y.Z.); vicwj@163.com (J.W.); Tel.: +86-10-5880-6656 (Y.Z.); +86-186-482-6849 (J.W.)

Received: 26 March 2018; Accepted: 19 April 2018; Published: 23 April 2018

Abstract: The prevalence of obesity and atherosclerosis has substantially increased worldwide over the past several decades. Peroxisome proliferator-activated receptors (PPARs), as fatty acids sensors, have been therapeutic targets in several human lipid metabolic diseases, such as obesity, atherosclerosis, diabetes, hyperlipidaemia, and non-alcoholic fatty liver disease. Constitutive androstane receptor (CAR) and liver X receptors (LXRs) were also reported as potential therapeutic targets for the treatment of obesity and atherosclerosis, respectively. Further clarification of the internal relationships between these three lipid metabolic nuclear receptors is necessary to enable drug discovery. In this review, we mainly summarized the cross-talk of PPARs-CAR in obesity and PPARs-LXRs in atherosclerosis.

Keywords: PPAR; CAR; LXR; obesity; atherosclerosis

---

1. Introduction

Obesity is a lipid metabolic disturbance that has been growing across the world for nearly half a century. It is a global human health concern. In 2016, more than 1.9 billion adults ($\geq$18 years old) were overweight and, of these, over 650 million were obese. Furthermore, 340 million children and adolescents (5–18 years old) and 41 million children ($\leq$5 years old) were overweight or obese [1,2]. The body mass index (BMI), defined as a person's weight in kilograms divided by the square of their height in meters, is a simple index used to classify overweight and obesity in adults. Obesity is associated with various metabolic disorders and cardiovascular diseases. A high BMI is considered to be an indicator of high body fatness that may lead to a high risk of cardiometabolic syndrome and atherosclerotic vascular disease [3–5]. Atherosclerosis, also known as arteriosclerosis, hardening of the arteries, is a disease in which fatty plaque deposits build up inside the arteries, narrowing them, leading to some serious problems, including coronary artery disease, stroke, or even death [6]. Obesity and atherosclerosis are common chronic lipid metabolic disorder diseases. The treatment and prevention of obesity and atherosclerosis are both major challenges, and studying this problem can help us live longer, healthier lives.

Nuclear receptors (NRs), a class of ligand-activated transcriptional factors, play significant roles in metabolic homeostasis. It is well known that there are 48 and 49 NR genes in humans (*Homo sapiens*) and mice (*Mus musculus*), respectively [7,8]. Most of the NRs contain six functional domains, such as the variable N-terminal regulatory domain (A–B), the conserved DNA-binding domain (DBD) (C), the variable hinge region (D), the conserved ligand binding domain (LBD) (E), and the variable C-terminal domain (F) (Figure 1a) [7,9]. The classical function of NRs is to transcriptionally regulate the expression of cognate target genes through the recruitment of coactivators or corepressors when ligands bind to the receptors [10,11] (Figure 1b). To perform the transcriptional activity, NRs either (1) act as monomers; (2) need to form dimeric complexes (homodimers); or (3) form complexes with the retinoid X receptor (RXR) (heterodimers) and bind to the DNA in the cell nucleus [9]. Recently, many studies have indicated the role of some NRs in the regulation of lipid metabolism. It has been recognized that peroxisome proliferator-activated receptors (PPARs) act as fatty acid sensors, regulating the multiple pathways involved in lipid and glucose metabolism and overall energy metabolism [12,13]. Furthermore, the constitutive androstane receptor (CAR), which was initially characterized as a xenosensor that controls xenobiotic responses, has been recently identified as a therapeutic target for obesity and its related metabolic disorders [14,15], whereas liver X receptors (LXRs) are sterol sensors that mainly regulate cholesterol, fatty acid and glucose homeostasis, they can inhibit atherosclerosis development, but promote lipogenesis in liver [16]. In this review, we briefly summarize the roles of PPARs, CAR and LXRs and their ligands in the treatment of metabolic diseases, obesity and atherosclerosis, and discuss the cross-talk of PPARs-CAR and PPARs-LXRs in lipid metabolism regulation.

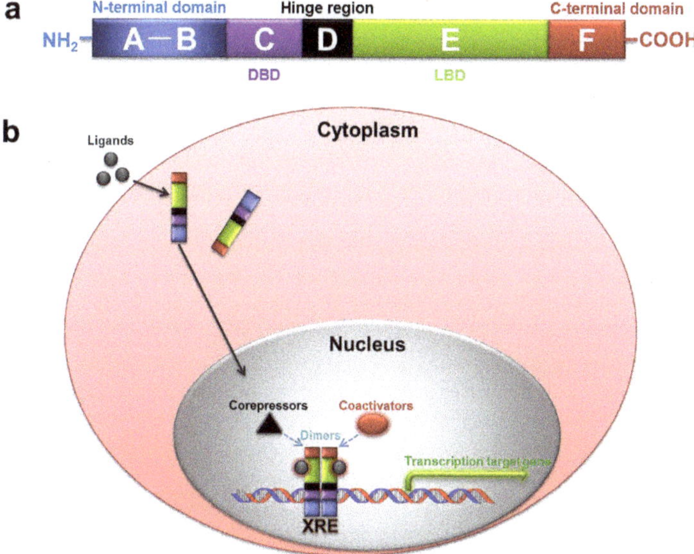

Figure 1. Schematic structure of NRs (nuclear receptors) and model of NR signalling. (**a**) General domain structure of NRs; and (**b**) the mechanism of general NR action. The ligands bind to the LBD (ligand-binding domain) of NRs in the cytoplasm, and translocate to the nucleus. Then the DBD (DNA-binding domain) of NRs bind to the XRE (xenobiotic responsive elements) forming dimeric complexes with RXR and the recruitment of co-activators or co-repressors. Finally, this leads to the transcription of the target genes. This model is applied to type II NRs, including PPARs, CAR, LXRs, and others. The colorful words just match the corresponding colorful shape. The dotted arrows mean different ligands can recruit coactivators or corepressors to form dimers, respectively.

2. The Initial Characterization of PPAR, CAR, and LXR

2.1. Fatty Acids Sensor PPARs

PPARs are molecular sensors of fatty acids and fatty acid derivatives and control energy homeostasis (carbohydrate, lipid, and protein) [17]. There are three types of PPARs which have been identified: PPARα (NR1C1, encoded by *PPARA*), PPARβ/δ (NR1C2, encoded by *PPARD*), and PPARγ (NR1C3, encoded by *PPARG*). They are all lipid sensors that transcriptionally regulate diverse aspects in response to nutritional inputs, and serving as effective therapeutic targets for some types of lipid metabolic syndrome, including obesity, atherosclerosis, dyslipidaemia, type 2 diabetes mellitus (T2DM), and nonalcoholic fatty liver disease (NAFLD) [12,18]. PPARα is highly active in liver, brown adipose tissue (BAT), kidney, heart, and muscle tissue [19], where it regulates the adaptive response to prolonged fasting by controlling the process of ketogenesis, fatty acid transport, fatty acid binding, fatty acid activation and mitochondrial fatty acid β-oxidation [20,21]. Genomic studies have indicated that PPARα, as a master regulator of lipid metabolism, has various target genes; the classical genes include acyl-CoA oxidase, thiolase, fatty acid transport protein (*FATP*), carnitine palmitoyltransferase I (*CPT1*), and peroxisome proliferator-activated receptor gamma coactivator 1-alpha (*PGC-1α*) [20,22]. The expression of PPARβ/δ is highest in adipose tissue, skeletal muscle, macrophages, brain, and skin, but is at low levels in the liver, where it mainly regulates fatty acid catabolism and the glycolytic-to-oxidative muscle fibre-type switching used in improving lipid homeostasis [23–26]. PPARα and PPARβ/δ have been shown to block lipid absorption by upregulating L-type fatty acid binding protein (L-FABP) and cluster of differentiation 36 (CD36) in the small intestine [27]. PPARγ function has mainly been characterized in adipose tissue, macrophages and the colon, and it has three forms: PPARγ1, PPARγ2, and PPARγ3 through alternative splicing [28–30]. PPARγ1 and PPARγ3 encode the same protein, and PPARγ3 is a splicing variant of PPARγ1. PPARγ2 has 28 additional amino acids at the variable N-terminal regulatory domain compared with PPARγ1 [31]. Furthermore, PPARγ1 has been found in nearly all tissues, except muscle, whereas PPARγ2 is mostly found in the adipose tissue and intestine, and PPARγ3 is mainly expressed white adipose tissue, colon, and macrophages [32]. PPARγ was initially known as an inducer during adipocyte differentiation [33,34], and its most famous role is in regulating lipogenic pathways. Genomic studies have revealed that PPARγ controls the expression of the early adipogenic differentiation factors CCAAT-enhancer-binding proteins (C/EBPs) and fatty acid binding protein 4 (FABP4), glucose homeostasis factors glucose transporter type 4 (GLUT4), and catabolite activator protein (*CAP*) genes. Moreover, PPARγ regulates some insulin sensitive adipokines, such as leptin, adiponectin, and tumour necrosis factor α (TNF-α) [35–37]. PPARγ is also involved in the metabolism of long-chain unsaturated fatty acid in the intestinal epithelium [38]. Although there are many similarities in lipid and glucose homeostasis, each of the PPAR isoforms has unique functions in vivo, probably due to their differential tissue distributions, the distinct ligands, and the inherent differences in biochemical characteristics [39].

Many endogenous agonists of PPARs have been identified, including polyunsaturated fatty acids, branched chain fatty acids, nitro/oxidized-fatty acids, phospholipids, eicosanoids, prostaglandin, oleoylethanolamide, carbaprostacyclin, 5HT metabolites, and so on [40–43]. In addition, many natural and synthetic PPAR ligands have been applied to treat lipid and glucose metabolic syndrome in pharmaceutical companies, as shown in Table 1. Fibrate drugs (including bezafibrate, clofibrate, fenofibrate, gemfibrozil, ronifibrate, etc.) are a class of classical PPARα agonists used to treat hyperlipidaemia and increase high-density lipoprotein cholesterol (HDL-c) in clinical settings. Moreover, pemafibrate [44] (approved in Japan in July 2017) and LY518674 [45] (phase II) are selective PPARα modulators used as anti-atherosclerosis agents in clinical trials. PPARβ/δ agonists are currently not used in clinical applications, but seladelpar (MBX-8025) is currently a promising activator for improving mixed dyslipidaemia and normalizing alkaline phosphatase levels, and is in phase 2 clinical development [46]. Additionally, KD-3010 is also a promising PPARβ/δ agonist for the potential treatment of diabetes and obesity in the phase I clinical trial. It shows the protective and anti-fibrotic effects in liver injury induced by carbon tetrachloride (CCl_4) injection or bile duct ligation (BDL) [47]. Thiazolidinediones (generically marked as pioglitazone, rosiglitazone, and lobeglitazone) are potent agonists of PPARγ with powerful insulin sensitizing activity which can be used in the treatment of T2DM. However, they have some undesirable side effects, such as weight gain, osteoporosis, and congestive heart failure [39,48]. Some failed and non-marked thiazolidinediones include troglitazone (marked as Rezulin, which was withdrawn due to adverse liver effects), balaglitazone, ciglitazone, darglitazone, netoglitazone, and rivoglitazone, etc. Recently, several partial agonists of PPARγ have been reported to keep beneficial antidiabetic characteristics with few side effects. Honokiol is a natural compound purified from the bark of *Magnolia officinalis* in traditional Chinese medicine, which has been identified as a novel non-adipogenic partial PPAPγ ligand. It has an anti-hyperglycemic property but does not trigger adipogenesis in vitro and in vivo [48]. Amorfrutins, as selective PPARγ modulators, are also natural products derived from two legumes, *Glycyrrhiza foetida* and *Amorpha fruticose*. They were reported to improve insulin sensitivity and dyslipidemia and protect liver steatosis without a concomitant increase of body weight gain in diet-induced obese and db/db mice [49,50]. In our recent study, Danshensu Bingpian Zhi (DBZ) is a synthetic derivative of the natural compounds *Danshensu* (*tanshinol*) and *Bingpian* (*borneol*), which are used as "sovereign" and "courier" in the traditional Chinese medicine formula Fufang Danshen (FFDS). We found that DBZ is a putative PPARγ partial activator capable of preventing insulin resistance, obesity, and atherosclerosis in mice without significant unwanted effects [51,52]. Along with improving our understanding of the biological roles of PPARs, we suggest that further study of the selectively pleiotropic PPAR agonist is a promising approach for developing further therapies.

Table 1. Different PPAR ligands and their development status regarding the treatment of lipid and glucose metabolic syndrome.

Ligands	Classification	Structure	Indication	Current Stage
Bezafibrate	PPARα agonist	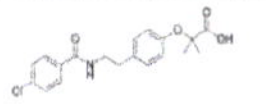	Hyperlipidemia	On the market
Clofibrate	PPARα agonist	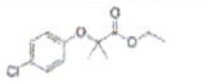	Hyperlipidemia	Discontinued
Fenofibrate	PPARα agonist	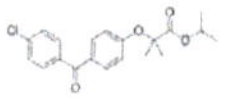	Hypercholesterolemia, mixed dyslipidemia	On the market
Gemfibrozil	PPARα agonist	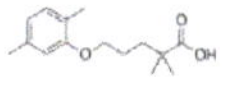	Hyperlipidemia, ischaemic disorder	On the market
Pemafibrate	PPARα agonist	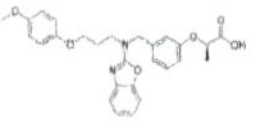	Lipid modifying agent	On the market in Japan
LY518674	PPARα agonist	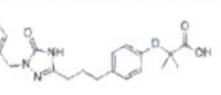	Atherosclerosis	Phase II
Seladelpar (MBX-8025)	PPARβ/δ agonist	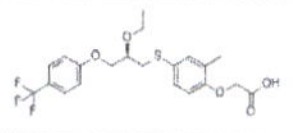	Dyslipidaemia, T2D, NASH	Phase II
KD-3010	PPARβ/δ agonist	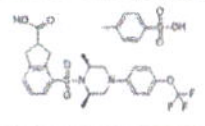	Diabetes, obesity, dyslipidemia	Phase I
Troglitazone	PPARγ agonist	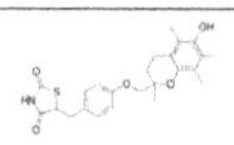	T2D	Withdrawn due to hepatotoxicity
Rosiglitazone	PPARγ agonist	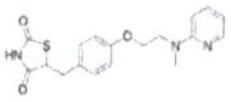	T2D	Withdrawn due to risk of CV events
Pioglitazone	PPARγ agonist	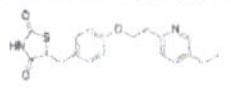	T2D	On the market
Lobeglitazone	PPARα/PPARγ agonist	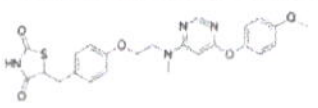	T2D	On the market in Korea

Table 1. *Cont.*

Ligands	Classification	Structure	Indication	Current Stage
Balaglitazone (DRF-2593)	PPARγ agonist	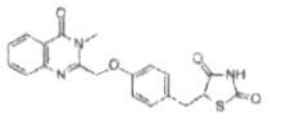	T2D	Phase III Discontinued
Ciglitazone	PPARγ agonist	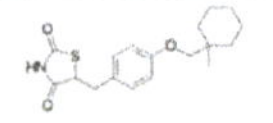	T2D	Phase II Discontinued
Darglitazone	PPARγ inhibitor	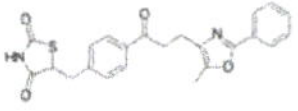	T2D	Phase I Discontinued
Netoglitazone (MCC-555)	PPARα/PPARγ agonist	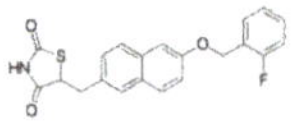	T2D	Phase II Discontinued
Rivoglitazone	PPARγ agonist	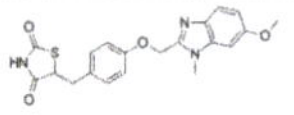	T2D	Phase III Discontinued
Honokiol	PPARγ agonist	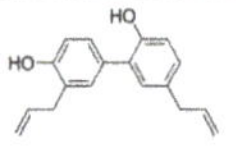	Gingival diseases, anti-hyperglycemic property	Phase III

2.2. Xenobiotic Receptor CAR

CAR is a member of the NR1I3 family of nuclear receptors, initially serves as a xenobiotic nuclear receptor, responding to xenobiotics and drug stress [53,54]. Androstenol, and some isomers of androstanol, androstanes, have been found to be endogenous antagonists of CAR, and dehydroepiandrosterone (DHEA), also an androstane, is an endogenous agonist of CAR. Androstanes, despite acting as ligands, are the basis for the naming of this receptor. The name "constitutive androstane receptor" refers to the unusual, constitutively-active status of this receptor when not occupied by a ligand. CAR is primarily expressed in the liver and small intestine, but is also found in the kidney, heart, and brain [55], and we also detected it in the mammary gland, ovary, and uterus (our unpublished data). It, often along with the pregnane X receptor (PXR) and vitamin D receptor (VDR), regulates the phase I and II xenobiotic metabolizing enzymes (including cytochrome P450s, sulfotransferases, glutathione-*S*-transferases) and other multidrug-resistance associated proteins used to both modulate drug metabolism and bilirubin clearance and prevent hepatotoxicity [56–58]. More recently, CAR has been reported to regulate both lipid and glucose metabolism and has been a potential therapeutic target for several metabolic diseases, such as obesity [15,59], atherosclerosis [60,61], NAFLD [62,63], and T2DM [64,65], due to its ability to balance the endogenous homeostasis of components, including glucose, steroids, bile acids, bilirubin, and thyroid hormone.

Since CAR has a large hydrophobic LBD pocket, a variety of chemical xenobiotics can activate it, such as clinical drugs, insecticides, flavonoids, terpenoids, polyphenols, environmental chemicals, and others [66,67]. Interestingly, CAR exhibits arresting species specificity in the ligand binding recognition between human and rodent, though both species use the same DNA response element sequences to recruit CAR. For example, TCPOBOP (1,4-*bis*[2-(3,5-dichloropyridyloxy)]benzene), is a potent mouse CAR (mCAR) agonist which only activates mouse, but not human, CAR, whereas CITCO (6-(4-chlorophenyl) imidazo [2,1-β] [1,3] thiazole-5-carbaldehyde-*O*-(3,4-dichlorobenzyl) oxime) is only a human CAR (hCAR) agonist, having no effect on mouse CAR [68,69]. Thus, this specificity should be considered when choosing the animal model for studying pharmacologic effects or drug screens targeting CAR. Phenobarbital, also known as phenobarb or phenobarbitone, is the preferred antiepileptic and sedation medicine used clinically, and it can activate both human and mouse CAR. Some early studies have shown that phenobarbital can regulate energy mentalism and improve insulin sensitivity and hepatic lipid homeostasis in ob/ob mice and human patients [70–72]. Activation of CAR reduced sterol regulatory element-binding protein 1 (SREBP-1) levels by inducing the expression of insulin induced gene 1 protein (*INSIG-1*), a protein blocking the proteolytic activation of SREBPs [73]. In a previous study, we reported that activation of CAR inhibited lipogenesis by suppressing LXR ligand-responded recruitment of LXR to the LXR response element (LXRE) and the expression of LXR target genes, whereas activation of LXR inhibited the CAR ligand-induced recruitment of CAR to Cyp2b10 [74]. Although CAR is a potential therapeutic target for lipid metabolic disease, some barriers exist for the clinical use of its agonists: there are concerns around hepatic enlargement and carcinogenesis. CAR also interacts with PPAR and LXR in regulating lipid and glucose homeostasis. Better understanding of these mechanistic properties might help us overcome these barriers in the future.

2.3. Oxysterol Sensor LXRs

LXRs are well-known nuclear oxysterol receptors that have two isotypes: LXRα (NR1H3) and LXRβ (NR1H2). LXRα is highly active in the liver, intestines, kidneys, adipose tissue, lungs, macrophages, and adrenal glands. LXRβ, also named as a ubiquitous receptor, is expressed in almost all tissues and organs [75–77]. Both of them may control cholesterol, fatty acid, and glucose metabolism to protect against atherosclerosis, lipid disorders, diabetes, chronic inflammation, Alzheimer's disease, and even cancer [78–81].

In cholesterol and lipid homeostasis, activation of LXR can stimulate reverse cholesterol transport and reduce the body's cholesterol overload by inducing the sterol metabolism and transporter network, including cytochrome P450 family 7 subfamily A member 1 (CYP7A1), ATP-binding cassette sub-family A member 1 (ABCA1), ABCG1, ABCG5, ABCG8, and apolipoprotein E (ApoE) [82–84]. Furthermore, LXR activation also results in an increase in lipid synthesis in the liver through inducing the expression of SREBP-1c, fatty acid synthase (FAS), acetyl-CoA carboxylase 1 (ACC1), and stearoyl-CoA desaturase 1 (SCD-1) [85–87]. LXRs, as sterol sensors, have a variety of endogenous activators, most of which are oxidation products of cholesterol, such as 27-hydroxycholesterol, 22(*R*)-hydroxycholesterol, 20(*S*)-hydroxycholesterol, 24(*S*)-hydroxycholesterol and 24(*S*), and 25-epoxycholesterol [16,76,88]. Interestingly, these endogenous agonists, unlike natural synthetic LXR activators, do not activate the SREBP signal pathway [89–91]. Several studies have reported that mice treated with synthetic LXR activators, including GW3965 and TO901317, show enhanced hepatic and serous triglyceride levels, and have promoted very low-density lipoprotein (VLDL) secretion [86,92,93]. These shortcomings limit the use of LXR activators in clinical settings. LXRα is the major sensor of dietary cholesterol. Mice lacking LXRα cannot induce transcription of the gene encoding cholesterol 7α-hydroxylase (CYP7A), which is a rate-limiting enzyme in bile acid synthesis. LXR$\alpha^{-/-}$ mice are healthy when fed with a normal chow (low cholesterol) diet. However, they develop enlarged fatty livers with high cholesterol levels, and lead to impaired hepatic function when fed a high-cholesterol diet [94]. LXR-623 (WAY-252623) is the first LXRα-partial/LXRβ-full agonist used for the treatment of atherosclerosis in animal models and has been tested in a phase I clinical trial. However, the trial was terminated due to adverse effects on the central nervous system [95,96]. Similar synthetic agonists, including CS8080, BMS-852927 (also named XL-041) have been terminated for undisclosed reasons, and only BMS-779788 (also named XL-652) has proved safe enough to continue with clinical trials [97,98], the detailed information as shown in Table 2. LXR activators can reduce cholesterol level in blood and liver. They also improve glucose tolerance in mice by decreasing insulin resistance. Human functional and genetic analysis showed that the common LXR promoter SNPs rs35463555 and rs17373080 may regulate sensibility to T2D [99]. We recently reported that DBZ inhibits foam cell formation and protects against atherosclerosis in ApoE$^{-/-}$ mice through activating LXRs [52,100]. DBZ also activates PPARγ and prevents high fat diet-induced obesity, insulin resistance and gut dysbiosis in mice [51]. By clarifying the cross-talk between PPARs and LXRs we may gain a better understanding of their synactic function in cholesterol and lipid homeostasis.

Table 2. Different LXR ligands and their development status regarding anti-atherosclerosis.

Ligands	Classification	Structure	Indication	Current Stage
LXR-623 (WAY-252623)	LXRα-partial LXRβ-full agonist		Atherosclerosis	Phase I Discontinued
BMS-852927 (XL-041)	LXR modulator		Atherosclerosis, hypercholesterolemia	Phase I Discontinued
BMS-779788 (XL-652)	LXR agonist		Atherosclerosis	Phase I

3. Cross-Talk of PPARs and CAR Links to Obesity

PPARs and CAR are both essential lipid metabolic nuclear receptors active in controlling obesity and its related metabolic disorders. PPARs are quite interesting. PPARα and PPARβ/δ are potential targets to prevent obesity [101–103], by the mechanism as mentioned above in Section 2.1. Contrarily, PPARγ is a master regulator of adipocyte differentiation both in vivo and in vitro [104]. A lack of PPARγ results in the inability to develop adipose tissue, as seen in PPARγ knockout mice [105,106]. Thiazolidinediones, as famous PPARγ activators, are a group of anti-diabetic drugs to treat T2MD, but can lead to serious side effects. Weight gain is an unwanted side effect: activation of PPARγ in adipose tissue stimulates the expression of genes leading to lipogenesis, including *AP2*, *CD36*, *SCD-1*, *SREBP-1*, and others, which promote lipid storage [18]. PPARα, as a key nutritional sensor, regulates the metabolism of lipids, carbohydrates, and amino acids [107]. It is a potential therapeutic target for the treatment of obesity, hypertriglyceridemia, NAFLD, and atherogenic dyslipidaemia [108–110]. Oestrogen inhibits the actions of PPARα on obesity and lipid metabolism through its effects on the PPARα-dependent regulation of target genes [111,112]. CAR, as a therapeutic target for obesity, was reported about ten years ago. Activation of CAR also increased faecal bile acid excretion and attenuated atherosclerosis in low-density lipoprotein receptor-deficient ($LDLR^{-/-}$) and $ApoE^{-/-}$ mice by increasing reverse cholesterol transport [60,61]. Recently, we reported that activation of CAR with TCPOBOP inhibited lipogenesis and promoted fibrosis in the mammary gland of adolescent female mice [113]. The classical CAR agonist TCPOBOP has a robust anti-obesity phenotype in high-fat diet-induced obese mouse models. Mechanically, activation of CAR improves insulin sensitivity, inhibits lipogenesis and gluconeogenesis, and increases brown adipose tissue energy expenditure.

The cross-talk between PPARs and CAR in obesity can be achieved through their target gene PGC-1α. PGC-1α, as a transcriptional coactivator, interacts with nuclear receptor PPAR and controls energy metabolism through the regulation of mitochondrial biogenesis [114,115]. CAR regulates the degradation of PGC-1α by recruiting E3 ligase targeting PGC1α and promoting ubiquitination in the liver [116]. During fasting, the PPARα activator WY14643 induces both CAR and its target gene CYP2B expression in a PPARα-dependent manner in rat hepatocytes [117,118]. Meanwhile, Guo et al. reported that synthetic PPARα ligands ciprofibrate, clofibrate, and others drove adenoviral-enhanced green fluorescent protein-CAR into the hepatocyte nucleus in a PPARα- and PPARβ-independent manner in mouse liver in vivo. More interestingly, molecular docking assay showed that PPARα activators, Wy-14643 and ciprofibrate, could fit into the ligand binding pocket of CAR and their binding modes were similar with that of androstanol, an endogenic CAR inverse agonist. PPARα activators interfered with coactivator recruitment to the LBD of CAR and suppressed the constitutive transactivation of

CAR. Mechanistically, the transcription coactivator PPAR-binding protein (PBP) plays a pivotal role in nuclear translocation of CAR in mouse liver, but not the PPAR-interacting protein (PRIP) [119,120]. These results indicated that activation of PPARα by some ligands induced nuclear translocation of CAR. β-oxidation is also controlled by both PPARs and CAR. PPARα regulates mitochondrial fatty acid β-oxidation by inducing the gene expression of *CPT1*, as previously mentioned. Conversely, the CAR ligand pentobarbital inhibits mitochondrial CPT1 expression and β-oxidation, resulting in increasing ketone production in serum [8,121]. However, in BAT, activation of CAR by TCPOBOP significantly increased expression of *PGC-1α* and β-oxidation [15]. Hence, the cross-talk between PPAR and CAR should be separately considered for different tissue types. Above all, the dual functions of PPAR activators have possible cross-talk with CAR through target gene *PGC1α*, coactivator recruitment, and mitochondrial fatty acid β-oxidation in different conditions in energy metabolism.

4. Cross-Talk of PPARS and LXRS in Atherosclerosis

There is a potential cross-talk or interaction between PPARs and LXRs in the prevention and treatment of atherosclerosis. Most nuclear receptors form heterodimers with RXR, including PPAR/RXR, LXR/RXR, CAR/RXR, and others. Ide et al. has elegantly reported that LXR-RXR-PPAR forms a network that regulates fatty acid metabolism and lipid degradation [122]. These compounds enhance binding to their respective target gene promoters. Unsaturated fatty acids increase the expression of LXRα, but not the LXRβ in rat liver cells, both in vivo and in vitro. This upregulated effect of LXRα is associated with the transcriptional rate and binding of PPARα to PPAR response element (PPRE). Meanwhile, a PPRE is found in the human LXRα flanking region [123]. *SREBP-1c*, as a direct target gene regulated by LXR, is crucial in both lipid and sterol biosynthesis. Luciferase assays have proven that the activation of PPARα and PPARγ reduces LXR-induced *SREBP-1c* promoter activity and gel shift assays have demonstrated that PPARs inhibit the binding of LXR/RXR to LXRE [124]. Thus, PPARs and LXRs play opposite roles in regulating triglyceride synthesis in the liver and serum. LXRα also inhibits peroxisome proliferator signalling through cross-talk with PPARα [125]. Moreover, Liduo Yue et al. reported that LXRs could bind to PPARs with different binding affinities in vitro using surface plasmon resonance technology and molecular dynamics simulation [126].

Despite the opposite roles in triglyceride homeostasis, PPARs and LXRs have some common ground in their anti-atherosclerotic effects. In foam cell macrophages, both PPARα and PPARγ (through the LXR-dependent ABC pathway) control cholesterol efflux [127,128], and activation of PPARα and PPARγ both prevent foam cell formation and atherosclerosis development in $ApoE^{-/-}$ and $LDLR^{-/-}$ mice [129,130]. Activation of LXRα also raises the expression of ABCA1 and ABCG1, which accelerate the reverse transport of cholesterol and then deposit in the liver [131]. PPAR-LXR-ABCA1 is an important pathway involved in cholesterol efflux and atherogenesis. In intestine tissue, the activation of LXR also increases the expression of ABCG5 and ABCG8 which regulate absorption of cholesterol and protect against atherosclerosis [79,132]. PPARs activation has performed similar acts inhibiting intestinal cholesterol absorption in rats and mice [133,134]. Taken together, both LXR and PPAR promote the movement of cholesterol from peripheral cells to the feces, which is referred to as reverse cholesterol transport (RCT).

Atherosclerosis is a chronic inflammatory disease; inflammation plays an important role in the pathogenesis and progression of atherosclerosis [135,136]. Recent studies have revealed the mechanism by which PPARs and LXRs regulate the inflammation process through some inflammatory target genes. Activation of PPARs and LXR can inhibit lipopolysaccharide- and cytokine-induced pro-inflammatory gene expression by repressing the toll-like receptor (TLR)-nuclear factor kappa B (NF-κB) signal pathway [137–139]. PPARα increases the expression of inhibitor of kappa B (IκB) to antagonize the NF-κB signalling pathway [140]. PPARβ/δ induces transforming growth factor beta (TGF-β) and inhibits the activation of NF-κB, thus regulating inflammatory processes [141]. Thiazolidinediones (TZDs) induced PPARγ activation also reduced the expression of inflammatory factors, including TNF-α and gelatinase B, in the aortic root, thus inhibiting the development of atherosclerosis [142].

All three PPAR isoforms regulate the immune response through different cell-signalling systems. LXRs repress inflammatory pathways through their transcriptional mechanisms [143,144]. LXRs and PPARγ control immunity by mediating proinflammatory gene transrepression through parallel small ubiquitin-like modifier (SUMO) ylation-dependent pathways [145]. PPARs and LXRs have been a critical interface for inflammation and cholesterol homeostasis. Concurrent activation of LXR and PPAR may have some beneficial effects. Activation of LXR by TO901317 and PPARα by fenofibrate in combination improves glucose tolerance, alleviates insulin resistance, and blocks TO901317-induced hyperlipidaemia, but aggravates hepatic steatosis in high fat diet-induced obese mice [146]. TO901317 and fenofibrate are both potent agonists. Concurrent partial agonists of LXR and PPAR may keep beneficial characteristics with few side effects. In our recently study, DBZ, as a promising therapeutic agent for atherogenesis and obesity in the mouse models, inhibits inflammation, macrophage migration, and foam cell formation, possibly through the partial activation of both PPARγ and LXRs.

5. Conclusions

PPARs, CAR, and LXRs are a part of nuclear hormone receptors that form heterodimers with RXR to regulate lipid metabolism. Ligand binding results in DNA binding and then triggers target gene expression. Obesity and atherosclerosis are both chronic lipid metabolic disorders, which were traditionally regarded as lipid deposition diseases, principally involving triglycerides in adipose tissue and cholesterol ester in arteries. Although they are distinct conditions, obesity is often associated with atherosclerosis. Recent findings have revealed the biological roles and mechanisms of these three NRs in obesity and atherosclerosis. These receptors have been potential therapeutic targets for drug discovery; further clarification and consideration of the internal relationship between them is necessary. In this study, we summarized the interaction of PPARs and CAR in lipid metabolism and obesity-related metabolic syndrome, and the cross-talk between PPARs and LXRs in cholesterol homeostasis and atherosclerosis (Figure 2). Concurrent activation of these NRs may have some beneficial effects in lipid metabolic disease. In recently study, we reported that DBZ prevented high fat diet-induced obesity and related metabolic disorders and attenuated atherosclerosis through concurrent partial activation of both PPARγ and LXRs. Moreover, it had no apparent side effects.

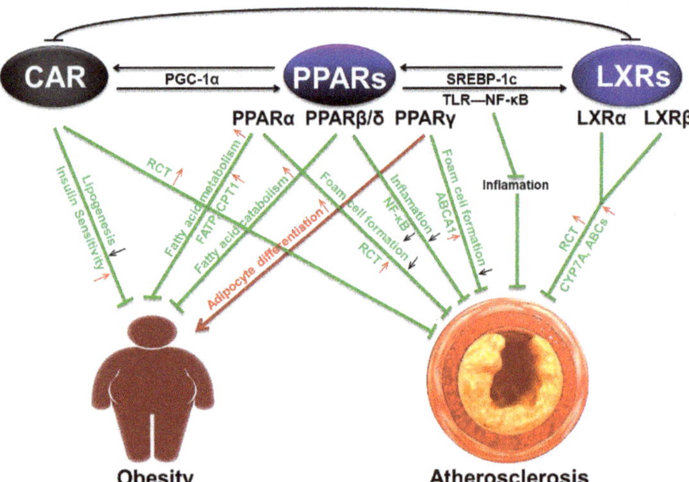

Figure 2. Proposed model of the cross-talks between PPARs and CAR in obesity and PPARs and LXRs in atherosclerosis. Red arrows: promotion; green T-bar: inhibition; red up-arrows: up-regulation; black down-arrows: down-regulation.

Beyond these cross-talks, more NRs, such as PXR, farnesoid X receptor (FXR), aryl hydrocarbon receptor (AhR), and retinoid-related orphan receptors (RORs), are being investigated. Future studies should focus on the complex network between these NRs and how that network affects their functions. We hope that by establishing a better understanding of nuclear receptor cross-talk between metabolic disorder diseases, we can reveal promising therapeutic targets for future research.

Acknowledgments: This study was supported by grants from the National Natural Science Foundation of China (Nos. 31571164 and 31271207 to Yonggong Zhai); the Natural Science Foundation of Inner Mongolia (No. 2017BS0801 to Jing Wang); the Program for Young Talents of Science and Technology in Universities of Inner Mongolia (No. NJYT-18-B29 to Jing Wang); and the Scientific Research Projection of Higher Schools of Inner Mongolia (No. NJZY17294 to Jing Wang).

Conflicts of Interest: The authors declare no conflict of interest.

References

1. WHO. Available online: http://www.who.int/mediacentre/factsheets/fs311/en/ (accessed on 28 February 2018).
2. Xu, P.; Wang, J.; Hong, F.; Wang, S.; Jin, X.; Xue, T.; Jia, L.; Zhai, Y. Melatonin prevents obesity through modulation of gut microbiota in mice. *J. Pineal Res.* **2017**. [CrossRef] [PubMed]
3. Lovren, F.; Teoh, H.; Verma, S. Obesity and atherosclerosis: Mechanistic insights. *Can. J. Cardiol.* **2015**, *31*, 177–183. [CrossRef] [PubMed]
4. Xu, P.; Hong, F.; Wang, J.; Cong, Y.; Dai, S.; Wang, S.; Wang, J.; Jin, X.; Wang, F.; Liu, J.; et al. Microbiome remodeling via the montmorillonite adsorption-excretion axis prevents obesity-related metabolic disorders. *EBioMedicine* **2017**, *16*, 251–261. [CrossRef] [PubMed]
5. Xu, P.F.; Dai, S.; Wang, J.; Zhang, J.; Liu, J.; Wang, F.; Zhai, Y.G. Preventive obesity agent montmorillonite adsorbs dietary lipids and enhances lipid excretion from the digestive tract. *Sci. Rep.* **2016**, *6*. [CrossRef] [PubMed]
6. Bloomgarden, Z.T. International diabetes federation meeting, 1997, and other recent meetings. atherosclerosis and related topics. *Diabetes Care* **1998**, *21*, 1356–1363. [CrossRef] [PubMed]
7. Zhao, Y.; Zhang, K.; Giesy, J.P.; Hu, J. Families of nuclear receptors in vertebrate models: Characteristic and comparative toxicological perspective. *Sci. Rep.* **2015**, *5*. [CrossRef] [PubMed]
8. Xiao, L.; Wang, J.; Jiang, M.; Xie, W.; Zhai, Y. The emerging role of constitutive androstane receptor and its cross talk with liver X receptors and peroxisome proliferator-activated receptor A in lipid metabolism. *Vitam. Horm.* **2013**, *91*, 243–258. [PubMed]
9. Germain, P.; Staels, B.; Dacquet, C.; Spedding, M.; Laudet, V. Overview of nomenclature of nuclear receptors. *Pharmacol. Rev.* **2006**, *58*, 685–704. [CrossRef] [PubMed]
10. Horwitz, K.B.; Jackson, T.A.; Bain, D.L.; Richer, J.K.; Takimoto, G.S.; Tung, L. Nuclear receptor coactivators and corepressors. *Mol. Endocrinol.* **1996**, *10*, 1167–1177. [PubMed]
11. Chai, X.; Zeng, S.; Xie, W. Nuclear receptors PXR and CAR: Implications for drug metabolism regulation, pharmacogenomics and beyond. *Expert Opin. Drug. Metab. Toxicol.* **2013**, *9*, 253–266. [CrossRef] [PubMed]
12. Gross, B.; Pawlak, M.; Lefebvre, P.; Staels, B. PPARs in obesity-induced T2DM, dyslipidaemia and NAFLD. *Nat. Rev. Endocrinol.* **2017**, *13*, 36–49. [CrossRef] [PubMed]
13. Wahli, W.; Michalik, L. PPARs at the crossroads of lipid signaling and inflammation. *Trends Endocrinol. Metab.* **2012**, *23*, 351–363. [CrossRef] [PubMed]
14. Yan, J.; Chen, B.; Lu, J.; Xie, W. Deciphering the roles of the constitutive androstane receptor in energy metabolism. *Acta Pharmacol. Sin.* **2015**, *36*, 62–70. [CrossRef] [PubMed]
15. Gao, J.; He, J.H.; Zhai, Y.G.; Wada, T.R.; Xie, W. The constitutive androstane receptor is an anti-obesity nuclear receptor that improves insulin sensitivity. *J. Biol. Chem.* **2009**, *284*, 25984–25992. [CrossRef] [PubMed]
16. Xiao, L.; Xie, X.; Zhai, Y. Functional crosstalk of CAR-LXR and ROR-LXR in drug metabolism and lipid metabolism. *Adv. Drug Deliv. Rev.* **2010**, *62*, 1316–1321. [CrossRef] [PubMed]
17. Grimaldi, P.A. Peroxisome proliferator-activated receptors as sensors of fatty acids and derivatives. *Cell. Mol. Life Sci.* **2007**, *64*, 2459–2564. [CrossRef] [PubMed]
18. Evans, R.M.; Barish, G.D.; Wang, Y.X. PPARs and the complex journey to obesity. *Nat. Med.* **2004**, *10*, 355–361. [CrossRef] [PubMed]

19. Tyagi, S.; Gupta, P.; Saini, A.S.; Kaushal, C.; Sharma, S. The peroxisome proliferator-activated receptor: A family of nuclear receptors role in various diseases. *J. Adv. Pharm. Technol. Res.* **2011**, *2*, 236–240. [CrossRef] [PubMed]
20. Kersten, S. Integrated physiology and systems biology of PPARalpha. *Mol. Metab.* **2014**, *3*, 354–371. [CrossRef] [PubMed]
21. Kersten, S.; Seydoux, J.; Peters, J.M.; Gonzalez, F.J.; Desvergne, B.; Wahli, W. Peroxisome proliferator-activated receptor alpha mediates the adaptive response to fasting. *J. Clin. Investig.* **1999**, *103*, 1489–1498. [CrossRef] [PubMed]
22. Grygiel-Gorniak, B. Peroxisome proliferator-activated receptors and their ligands: Nutritional and clinical implications—A review. *Nutr. J.* **2014**, *13*. [CrossRef] [PubMed]
23. Luquet, S.; Lopez-Soriano, J.; Holst, D.; Fredenrich, A.; Melki, J.; Rassoulzadegan, M.; Grimaldi, P.A. Peroxisome proliferator-activated receptor delta controls muscle development and oxidative capability. *FASEB J.* **2003**, *17*, 2299–2301. [CrossRef] [PubMed]
24. Wang, Y.X.; Zhang, C.L.; Yu, R.T.; Cho, H.K.; Nelson, M.C.; Bayuga-Ocampo, C.R.; Ham, J.; Kang, H.; Evans, R.M. Regulation of muscle fiber type and running endurance by PPARdelta. *PLoS Biol.* **2004**, *2*. [CrossRef] [PubMed]
25. Schuler, M.; Ali, F.; Chambon, C.; Duteil, D.; Bornert, J.M.; Tardivel, A.; Desvergne, B.; Wahli, W.; Chambon, P.; Metzger, D. PGC1alpha expression is controlled in skeletal muscles by PPARbeta, whose ablation results in fiber-type switching, obesity, and type 2 diabetes. *Cell Metab.* **2006**, *4*, 407–414. [CrossRef] [PubMed]
26. Palomer, X.; Barroso, E.; Pizarro-Delgado, J.; Pena, L.; Botteri, G.; Zarei, M.; Aguilar, D.; Montori-Grau, M.; Vazquez-Carrera, M. PPARbeta/delta: A key therapeutic target in metabolic disorders. *Int. J. Mol. Sci.* **2018**, *19*, 913. [CrossRef] [PubMed]
27. Poirier, H.; Niot, I.; Monnot, M.C.; Braissant, O.; Meunier-Durmort, C.; Costet, P.; Pineau, T.; Wahli, W.; Willson, T.M.; Besnard, P. Differential involvement of peroxisome-proliferator-activated receptors alpha and delta in fibrate and fatty-acid-mediated inductions of the gene encoding liver fatty-acid-binding protein in the liver and the small intestine. *Biochem. J.* **2001**, *355*, 481–488. [CrossRef] [PubMed]
28. Park, Y.K.; Wang, L.; Giampietro, A.; Lai, B.; Lee, J.E.; Ge, K. Distinct roles of transcription factors KLF4, Krox20, and peroxisome proliferator-activated receptor gamma in adipogenesis. *Mol. Cell. Biol.* **2017**, *37*. [CrossRef] [PubMed]
29. Meirhaeghe, A.; Fajas, L.; Gouilleux, F.; Cottel, D.; Helbecque, N.; Auwerx, J.; Amouyel, P. A functional polymorphism in a STAT5B site of the human PPAR gamma 3 gene promoter affects height and lipid metabolism in a French population. *Arterioscler. Thromb. Vasc. Biol.* **2003**, *23*, 289–294. [CrossRef] [PubMed]
30. Fajas, L.; Auboeuf, D.; Raspe, E.; Schoonjans, K.; Lefebvre, A.M.; Saladin, R.; Najib, J.; Laville, M.; Fruchart, J.C.; Deeb, S.; et al. The organization, promoter analysis, and expression of the human PPARgamma gene. *J. Biol. Chem.* **1997**, *272*, 18779–18789. [CrossRef] [PubMed]
31. Tachibana, K.; Yamasaki, D.; Ishimoto, K.; Doi, T. The role of PPARs in cancer. *PPAR Res.* **2008**, *2008*. [CrossRef] [PubMed]
32. Lamichane, S.; Dahal Lamichane, B.; Kwon, S.M. Pivotal roles of peroxisome proliferator-activated receptors (PPARs) and their signal cascade for cellular and whole-body energy homeostasis. *Int. J. Mol. Sci.* **2018**, *19*, 949. [CrossRef] [PubMed]
33. Tontonoz, P.; Spiegelman, B.M. Fat and beyond: The diverse biology of PPARgamma. *Annu. Rev. Biochem.* **2008**, *77*, 289–312. [CrossRef] [PubMed]
34. Brun, R.P.; Tontonoz, P.; Forman, B.M.; Ellis, R.; Chen, J.; Evans, R.M.; Spiegelman, B.M. Differential activation of adipogenesis by multiple PPAR isoforms. *Genes Dev.* **1996**, *10*, 974–984. [CrossRef] [PubMed]
35. Hollenberg, A.N.; Susulic, V.S.; Madura, J.P.; Zhang, B.; Moller, D.E.; Tontonoz, P.; Sarraf, P.; Spiegelman, B.M.; Lowell, B.B. Functional antagonism between CCAAT/Enhancer binding protein-alpha and peroxisome proliferator-activated receptor-gamma on the leptin promoter. *J. Biol. Chem.* **1997**, *272*, 5283–5290. [CrossRef] [PubMed]
36. Hofmann, C.; Lorenz, K.; Braithwaite, S.S.; Colca, J.R.; Palazuk, B.J.; Hotamisligil, G.S.; Spiegelman, B.M. Altered gene expression for tumor necrosis factor-alpha and its receptors during drug and dietary modulation of insulin resistance. *Endocrinology* **1994**, *134*, 264–270. [CrossRef] [PubMed]

37. Iwaki, M.; Matsuda, M.; Maeda, N.; Funahashi, T.; Matsuzawa, Y.; Makishima, M.; Shimomura, I. Induction of adiponectin, a fat-derived antidiabetic and antiatherogenic factor, by nuclear receptors. *Diabetes* **2003**, *52*, 1655–1663. [CrossRef] [PubMed]
38. Duszka, K.; Oresic, M.; Le May, C.; Konig, J.; Wahli, W. PPARgamma modulates long chain fatty acid processing in the intestinal epithelium. *Int. J. Mol. Sci.* **2017**, *18*, 2559. [CrossRef] [PubMed]
39. Ahmadian, M.; Suh, J.M.; Hah, N.; Liddle, C.; Atkins, A.R.; Downes, M.; Evans, R.M. PPARgamma signaling and metabolism: the good, the bad and the future. *Nat. Med.* **2013**, *19*, 557–566. [CrossRef] [PubMed]
40. Xu, H.E.; Lambert, M.H.; Montana, V.G.; Parks, D.J.; Blanchard, S.G.; Brown, P.J.; Sternbach, D.D.; Lehmann, J.M.; Wisely, G.B.; Willson, T.M.; et al. Molecular recognition of fatty acids by peroxisome proliferator-activated receptors. *Mol. Cell* **1999**, *3*, 397–403. [CrossRef]
41. Schupp, M.; Lazar, M.A. Endogenous ligands for nuclear receptors: digging deeper. *J. Biol. Chem.* **2010**, *285*, 40409–40415. [CrossRef] [PubMed]
42. Poulsen, L.; Siersbaek, M.; Mandrup, S. PPARs: Fatty acid sensors controlling metabolism. *Semin. Cell Dev. Biol.* **2012**, *23*, 631–639. [CrossRef] [PubMed]
43. Bensinger, S.J.; Tontonoz, P. Integration of metabolism and inflammation by lipid-activated nuclear receptors. *Nature* **2008**, *454*, 470–477. [CrossRef] [PubMed]
44. Hennuyer, N.; Duplan, I.; Paquet, C.; Vanhoutte, J.; Woitrain, E.; Touche, V.; Colin, S.; Vallez, E.; Lestavel, S.; Lefebvre, P.; et al. The novel selective PPARalpha modulator (SPPARMalpha) pemafibrate improves dyslipidemia, enhances reverse cholesterol transport and decreases inflammation and atherosclerosis. *Atherosclerosis* **2016**, *249*, 200–208. [CrossRef] [PubMed]
45. Khera, A.V.; Millar, J.S.; Ruotolo, G.; Wang, M.D.; Rader, D.J. Potent peroxisome proliferator-activated receptor-alpha agonist treatment increases cholesterol efflux capacity in humans with the metabolic syndrome. *Eur. Heart J.* **2015**, *36*, 3020–3022. [CrossRef] [PubMed]
46. Jones, D.; Boudes, P.F.; Swain, M.G.; Bowlus, C.L.; Galambos, M.R.; Bacon, B.R.; Doerffel, Y.; Gitlin, N.; Gordon, S.C.; Odin, J.A.; et al. Seladelpar (MBX-8025), a selective PPAR-delta agonist, in patients with primary biliary cholangitis with an inadequate response to ursodeoxycholic acid: A double-blind, randomised, placebo-controlled, phase 2, proof-of-concept study. *Lancet Gastroenterol. Hepatol.* **2017**, *2*, 716–726. [CrossRef]
47. Iwaisako, K.; Haimerl, M.; Paik, Y.H.; Taura, K.; Kodama, Y.; Sirlin, C.; Yu, E.; Yu, R.T.; Downes, M.; Evans, R.M.; et al. Protection from liver fibrosis by a peroxisome proliferator-activated receptor delta agonist. *Proc. Natl. Acad. Sci. USA* **2012**, *109*, E1369–E1376. [CrossRef] [PubMed]
48. Atanasov, A.G.; Wang, J.N.; Gu, S.P.; Bu, J.; Kramer, M.P.; Baumgartner, L.; Fakhrudin, N.; Ladurner, A.; Malainer, C.; Vuorinen, A.; et al. Honokiol: A non-adipogenic PPARgamma agonist from nature. *Biochim. Biophys. Acta* **2013**, *1830*, 4813–4819. [CrossRef] [PubMed]
49. Weidner, C.; de Groot, J.C.; Prasad, A.; Freiwald, A.; Quedenau, C.; Kliem, M.; Witzke, A.; Kodelja, V.; Han, C.T.; Giegold, S.; et al. Amorfrutins are potent antidiabetic dietary natural products. *Proc. Natl. Acad. Sci. USA* **2012**, *109*, 7257–7262. [CrossRef] [PubMed]
50. Weidner, C.; Wowro, S.J.; Freiwald, A.; Kawamoto, K.; Witzke, A.; Kliem, M.; Siems, K.; Muller-Kuhrt, L.; Schroeder, F.C.; Sauer, S. Amorfrutin B is an efficient natural peroxisome proliferator-activated receptor gamma (PPARgamma) agonist with potent glucose-lowering properties. *Diabetologia* **2013**, *56*, 1802–1812. [CrossRef] [PubMed]
51. Xu, P.; Hong, F.; Wang, J.; Wang, J.; Zhao, X.; Wang, S.; Xue, T.; Xu, J.; Zheng, X.; Zhai, Y. DBZ is a putative PPARgamma agonist that prevents high fat diet-induced obesity, insulin resistance and gut dysbiosis. *Biochim. Biophys. Acta* **2017**, *1861*, 2690–2701. [CrossRef] [PubMed]
52. Wang, J.; Xu, P.; Xie, X.; Li, J.; Zhang, J.; Wang, J.; Hong, F.; Li, J.; Zhang, Y.; Song, Y.; et al. DBZ (Danshensu Bingpian Zhi), a novel natural compound derivative, attenuates atherosclerosis in apolipoprotein e-deficient mice. *J. Am. Heart Assoc.* **2017**, *6*. [CrossRef] [PubMed]
53. Banerjee, M.; Robbins, D.; Chen, T. Targeting xenobiotic receptors PXR and CAR in human diseases. *Drug. Discov. Today* **2015**, *20*, 618–628. [CrossRef] [PubMed]
54. Yan, J.; Xie, W. A brief history of the discovery of PXR and CAR as xenobiotic receptors. *Acta Pharm. Sin. B* **2016**, *6*, 450–452. [CrossRef] [PubMed]

55. Choi, H.S.; Chung, M.; Tzameli, I.; Simha, D.; Lee, Y.K.; Seol, W.; Moore, D.D. Differential transactivation by two isoforms of the orphan nuclear hormone receptor CAR. *J. Biol. Chem.* **1997**, *272*, 23565–23571. [CrossRef] [PubMed]
56. Tzameli, I.; Moore, D.D. Role reversal: New insights from new ligands for the xenobiotic receptor CAR. *Trends Endocrinol. Metab.* **2001**, *12*, 7–10. [CrossRef]
57. Cherrington, N.J.; Hartley, D.P.; Li, N.; Johnson, D.R.; Klaassen, C.D. Organ distribution of multidrug resistance proteins 1, 2, and 3 (Mrp1, 2, and 3) mRNA and hepatic induction of Mrp3 by constitutive androstane receptor activators in rats. *J. Pharmacol. Exp. Ther.* **2002**, *300*, 97–104. [CrossRef] [PubMed]
58. Ding, X.; Lichti, K.; Kim, I.; Gonzalez, F.J.; Staudinger, J.L. Regulation of constitutive androstane receptor and its target genes by fasting, cAMP, hepatocyte nuclear factor alpha, and the coactivator peroxisome proliferator-activated receptor gamma coactivator-1alpha. *J. Biol. Chem.* **2006**, *281*, 26540–26551. [CrossRef] [PubMed]
59. Gao, J.; Xie, W. Pregnane X receptor and constitutive androstane receptor at the crossroads of drug metabolism and energy metabolism. *Drug Metab. Dispos.* **2010**, *38*, 2091–2095. [CrossRef] [PubMed]
60. Sberna, A.L.; Assem, M.; Xiao, R.; Ayers, S.; Gautier, T.; Guiu, B.; Deckert, V.; Chevriaux, A.; Grober, J.; Le Guern, N.; et al. Constitutive androstane receptor activation decreases plasma apolipoprotein B-containing lipoproteins and atherosclerosis in low-density lipoprotein receptor-deficient mice. *Arterioscler. Thromb. Vasc. Biol.* **2011**, *31*, 2232–2239. [CrossRef] [PubMed]
61. Sberna, A.L.; Assem, M.; Gautier, T.; Grober, J.; Guiu, B.; Jeannin, A.; Pais de Barros, J.P.; Athias, A.; Lagrost, L.; Masson, D. Constitutive androstane receptor activation stimulates faecal bile acid excretion and reverse cholesterol transport in mice. *J. Hepatol.* **2011**, *55*, 154–161. [CrossRef] [PubMed]
62. Dong, B.; Saha, P.K.; Huang, W.; Chen, W.; Abu-Elheiga, L.A.; Wakil, S.J.; Stevens, R.D.; Ilkayeva, O.; Newgard, C.B.; Chan, L.; et al. Activation of nuclear receptor CAR ameliorates diabetes and fatty liver disease. *Proc. Natl. Acad. Sci. USA* **2009**, *106*, 18831–18836. [CrossRef] [PubMed]
63. Lynch, C.; Pan, Y.; Li, L.; Heyward, S.; Moeller, T.; Swaan, P.W.; Wang, H. Activation of the constitutive androstane receptor inhibits gluconeogenesis without affecting lipogenesis or fatty acid synthesis in human hepatocytes. *Toxicol. Appl. Pharmacol.* **2014**, *279*, 33–42. [CrossRef] [PubMed]
64. Dong, B.; Qatanani, M.; Moore, D.D. Constitutive androstane receptor mediates the induction of drug metabolism in mouse models of type 1 diabetes. *Hepatology* **2009**, *50*, 622–629. [CrossRef] [PubMed]
65. Jiang, M.; Xie, W. Role of the constitutive androstane receptor in obesity and type 2 diabetes: A case study of the endobiotic function of a xenobiotic receptor. *Drug Metab. Rev.* **2013**, *45*, 156–163. [CrossRef] [PubMed]
66. Cherian, M.T.; Chai, S.C.; Chen, T. Small-molecule modulators of the constitutive androstane receptor. *Expert Opin. Drug Metab. Toxicol.* **2015**, *11*, 1099–1114. [CrossRef] [PubMed]
67. Chai, S.C.; Cherian, M.T.; Wang, Y.M.; Chen, T. Small-molecule modulators of PXR and CAR. *Biochim. Biophys. Acta* **2016**, *1859*, 1141–1154. [CrossRef] [PubMed]
68. Tzameli, I.; Pissios, P.; Schuetz, E.G.; Moore, D.D. The xenobiotic compound 1, 4-*bis*[2-(3,5-dichloropyridyloxy)]benzene is an agonist ligand for the nuclear receptor CAR. *Mol. Cell. Biol.* **2000**, *20*, 2951–2958. [CrossRef] [PubMed]
69. Maglich, J.M.; Parks, D.J.; Moore, L.B.; Collins, J.L.; Goodwin, B.; Billin, A.N.; Stoltz, C.A.; Kliewer, S.A.; Lambert, M.H.; Willson, T.M.; et al. Identification of a novel human constitutive androstane receptor (CAR) agonist and its use in the identification of CAR target genes. *J. Biol. Chem.* **2003**, *278*, 17277–17283. [CrossRef] [PubMed]
70. Lahtela, J.T.; Arranto, A.J.; Sotaniemi, E.A. Enzyme inducers improve insulin sensitivity in non-insulin-dependent diabetic subjects. *Diabetes* **1985**, *34*, 911–916. [CrossRef] [PubMed]
71. Karvonen, I.; Stengard, J.H.; Huupponen, R.; Stenback, F.G.; Sotaniemi, E.A. Effects of enzyme induction therapy on glucose and drug metabolism in obese mice model of non-insulin dependent diabetes mellitus. *Diabetes Res.* **1989**, *10*, 85–92. [PubMed]
72. Sotaniemi, E.A.; Karvonen, I. Glucose tolerance and insulin response to glucose load before and after enzyme inducing therapy in subjects with glucose intolerance and patients with NIDDM having hyperinsulinemia or relative insulin deficiency. *Diabetes Res.* **1989**, *11*, 131–139. [PubMed]
73. Roth, A.; Looser, R.; Kaufmann, M.; Blattler, S.M.; Rencurel, F.; Huang, W.; Moore, D.D.; Meyer, U.A. Regulatory cross-talk between drug metabolism and lipid homeostasis: constitutive androstane receptor and pregnane X receptor increase Insig-1 expression. *Mol. Pharmacol.* **2008**, *73*, 1282–1289. [CrossRef] [PubMed]

74. Zhai, Y.; Wada, T.; Zhang, B.; Khadem, S.; Ren, S.; Kuruba, R.; Li, S.; Xie, W. A functional cross-talk between liver X receptor-alpha and constitutive androstane receptor links lipogenesis and xenobiotic responses. *Mol. Pharmacol.* **2010**, *78*, 666–674. [CrossRef] [PubMed]
75. Whitney, K.D.; Watson, M.A.; Goodwin, B.; Galardi, C.M.; Maglich, J.M.; Wilson, J.G.; Willson, T.M.; Collins, J.L.; Kliewer, S.A. Liver X receptor (LXR) regulation of the LXRalpha gene in human macrophages. *J. Biol. Chem.* **2001**, *276*, 43509–43515. [CrossRef] [PubMed]
76. Janowski, B.A.; Grogan, M.J.; Jones, S.A.; Wisely, G.B.; Kliewer, S.A.; Corey, E.J.; Mangelsdorf, D.J. Structural requirements of ligands for the oxysterol liver X receptors LXRalpha and LXRbeta. *Proc. Natl. Acad. Sci. USA* **1999**, *96*, 266–271. [CrossRef] [PubMed]
77. Chuu, C.P.; Kokontis, J.M.; Hiipakka, R.A.; Liao, S. Modulation of liver X receptor signaling as novel therapy for prostate cancer. *J. Biomed. Sci.* **2007**, *14*, 543–553. [CrossRef] [PubMed]
78. Jakobsson, T.; Treuter, E.; Gustafsson, J.A.; Steffensen, K.R. Liver X receptor biology and pharmacology: New pathways, challenges and opportunities. *Trends Pharmacol. Sci.* **2012**, *33*, 394–404. [CrossRef] [PubMed]
79. Bonamassa, B.; Moschetta, A. Atherosclerosis: Lessons from LXR and the intestine. *Trends Endocrinol. Metab.* **2013**, *24*, 120–128. [CrossRef] [PubMed]
80. Lee, S.D.; Tontonoz, P. Liver X receptors at the intersection of lipid metabolism and atherogenesis. *Atherosclerosis* **2015**, *242*, 29–36. [CrossRef] [PubMed]
81. Ma, Z.; Deng, C.; Hu, W.; Zhou, J.; Fan, C.; Di, S.; Liu, D.; Yang, Y.; Wang, D. Liver X receptors and their agonists: Targeting for cholesterol homeostasis and cardiovascular diseases. *Curr. Issues Mol. Biol.* **2017**, *22*, 41–64. [CrossRef] [PubMed]
82. Repa, J.J.; Turley, S.D.; Lobaccaro, J.A.; Medina, J.; Li, L.; Lustig, K.; Shan, B.; Heyman, R.A.; Dietschy, J.M.; Mangelsdorf, D.J. Regulation of absorption and ABC1-mediated efflux of cholesterol by RXR heterodimers. *Science* **2000**, *289*, 1524–1529. [CrossRef] [PubMed]
83. Repa, J.J.; Berge, K.E.; Pomajzl, C.; Richardson, J.A.; Hobbs, H.; Mangelsdorf, D.J. Regulation of ATP-binding cassette sterol transporters ABCG5 and ABCG8 by the liver X receptors alpha and beta. *J. Biol. Chem.* **2002**, *277*, 18793–18800. [CrossRef] [PubMed]
84. Laffitte, B.A.; Repa, J.J.; Joseph, S.B.; Wilpitz, D.C.; Kast, H.R.; Mangelsdorf, D.J.; Tontonoz, P. LXRs control lipid-inducible expression of the apolipoprotein E gene in macrophages and adipocytes. *Proc. Natl. Acad. Sci. USA* **2001**, *98*, 507–512. [CrossRef] [PubMed]
85. Repa, J.J.; Liang, G.; Ou, J.; Bashmakov, Y.; Lobaccaro, J.M.; Shimomura, I.; Shan, B.; Brown, M.S.; Goldstein, J.L.; Mangelsdorf, D.J. Regulation of mouse sterol regulatory element-binding protein-1c gene (SREBP-1c) by oxysterol receptors, LXRalpha and LXRbeta. *Genes Dev.* **2000**, *14*, 2819–2830. [CrossRef] [PubMed]
86. Joseph, S.B.; Laffitte, B.A.; Patel, P.H.; Watson, M.A.; Matsukuma, K.E.; Walczak, R.; Collins, J.L.; Osborne, T.F.; Tontonoz, P. Direct and indirect mechanisms for regulation of fatty acid synthase gene expression by liver X receptors. *J. Biol. Chem.* **2002**, *277*, 11019–11025. [CrossRef] [PubMed]
87. Talukdar, S.; Hillgartner, F.B. The mechanism mediating the activation of acetyl-coenzyme a carboxylase-alpha gene transcription by the liver X receptor agonist T0–901317. *J. Lipid Res.* **2006**, *47*, 2451–2461. [CrossRef] [PubMed]
88. Janowski, B.A.; Willy, P.J.; Devi, T.R.; Falck, J.R.; Mangelsdorf, D.J. An oxysterol signalling pathway mediated by the nuclear receptor LXR alpha. *Nature* **1996**, *383*, 728–731. [CrossRef] [PubMed]
89. Spencer, T.A.; Gayen, A.K.; Phirwa, S.; Nelson, J.A.; Taylor, F.R.; Kandutsch, A.A.; Erickson, S.K. 24(S), 25-Epoxycholesterol. Evidence consistent with a role in the regulation of hepatic cholesterogenesis. *J. Biol. Chem.* **1985**, *260*, 13391–13394. [PubMed]
90. Fu, X.; Menke, J.G.; Chen, Y.; Zhou, G.; MacNaul, K.L.; Wright, S.D.; Sparrow, C.P.; Lund, E.G. 27-hydroxycholesterol is an endogenous ligand for liver X receptor in cholesterol-loaded cells. *J. Biol. Chem.* **2001**, *276*, 38378–38387. [CrossRef] [PubMed]
91. Hong, C.; Tontonoz, P. Liver X receptors in lipid metabolism: Opportunities for drug discovery. *Nat. Rev. Drug Discov.* **2014**, *13*, 433–444. [CrossRef] [PubMed]
92. Bradley, M.N.; Hong, C.; Chen, M.; Joseph, S.B.; Wilpitz, D.C.; Wang, X.; Lusis, A.J.; Collins, A.; Hseuh, W.A.; Collins, J.L.; et al. Ligand activation of LXR beta reverses atherosclerosis and cellular cholesterol overload in mice lacking LXR alpha and apoE. *J. Clin. Investig.* **2007**, *117*, 2337–2346. [CrossRef] [PubMed]

93. Kiss, E.; Popovic, Z.; Bedke, J.; Wang, S.; Bonrouhi, M.; Gretz, N.; Stettner, P.; Teupser, D.; Thiery, J.; Porubsky, S.; et al. Suppression of chronic damage in renal allografts by Liver X receptor (LXR) activation relevant contribution of macrophage LXRalpha. *Am. J. Pathol.* **2011**, *179*, 92–103. [CrossRef] [PubMed]
94. Peet, D.J.; Turley, S.D.; Ma, W.; Janowski, B.A.; Lobaccaro, J.M.; Hammer, R.E.; Mangelsdorf, D.J. Cholesterol and bile acid metabolism are impaired in mice lacking the nuclear oxysterol receptor LXR alpha. *Cell* **1998**, *93*, 693–704. [CrossRef]
95. Katz, A.; Udata, C.; Ott, E.; Hickey, L.; Burczynski, M.E.; Burghart, P.; Vesterqvist, O.; Meng, X. Safety, pharmacokinetics, and pharmacodynamics of single doses of LXR-623, a novel liver X-receptor agonist, in healthy participants. *J. Clin. Pharmacol.* **2009**, *49*, 643–649. [CrossRef] [PubMed]
96. Quinet, E.M.; Basso, M.D.; Halpern, A.R.; Yates, D.W.; Steffan, R.J.; Clerin, V.; Resmini, C.; Keith, J.C.; Berrodin, T.J.; Feingold, I.; et al. LXR ligand lowers LDL cholesterol in primates, is lipid neutral in hamster, and reduces atherosclerosis in mouse. *J. Lipid Res.* **2009**, *50*, 2358–2370. [CrossRef] [PubMed]
97. Li, X.; Yeh, V.; Molteni, V. Liver X receptor modulators: A review of recently patented compounds (2007–2009). *Expert Opin. Ther. Pat.* **2010**, *20*, 535–562. [CrossRef] [PubMed]
98. Loren, J.; Huang, Z.; Laffitte, B.A.; Molteni, V. Liver X receptor modulators: A review of recently patented compounds (2009–2012). *Expert Opin. Ther. Pat.* **2013**, *23*, 1317–1335. [CrossRef] [PubMed]
99. Dahlman, I.; Nilsson, M.; Gu, H.F.; Lecoeur, C.; Efendic, S.; Ostenson, C.G.; Brismar, K.; Gustafsson, J.A.; Froguel, P.; Vaxillaire, M.; et al. Functional and genetic analysis in type 2 diabetes of liver X receptor alleles—A cohort study. *BMC Med. Genet.* **2009**, *10*. [CrossRef] [PubMed]
100. Xie, X.; Wang, S.; Xiao, L.; Zhang, J.; Wang, J.; Liu, J.; Shen, X.; He, D.; Zheng, X.; Zhai, Y. DBZ blocks LPS-induced monocyte activation and foam cell formation via inhibiting nuclear factor-kB. *Cell. Physiol. Biochem.* **2011**, *28*, 649–662. [CrossRef] [PubMed]
101. Wang, Y.X.; Lee, C.H.; Tiep, S.; Yu, R.T.; Ham, J.; Kang, H.; Evans, R.M. Peroxisome-proliferator-activated receptor delta activates fat metabolism to prevent obesity. *Cell* **2003**, *113*, 159–170. [CrossRef]
102. Grimaldi, P.A. Regulatory functions of PPARbeta in metabolism: Implications for the treatment of metabolic syndrome. *Biochim. Biophys. Acta* **2007**, *1771*, 983–990. [CrossRef] [PubMed]
103. Oliver, W.R., Jr.; Shenk, J.L.; Snaith, M.R.; Russell, C.S.; Plunket, K.D.; Bodkin, N.L.; Lewis, M.C.; Winegar, D.A.; Sznaidman, M.L.; Lambert, M.H.; et al. A selective peroxisome proliferator-activated receptor delta agonist promotes reverse cholesterol transport. *Proc. Natl. Acad. Sci. USA* **2001**, *98*, 5306–5311. [CrossRef] [PubMed]
104. Rosen, E.D.; Sarraf, P.; Troy, A.E.; Bradwin, G.; Moore, K.; Milstone, D.S.; Spiegelman, B.M.; Mortensen, R.M. PPAR gamma is required for the differentiation of adipose tissue in vivo and in vitro. *Mol. Cell* **1999**, *4*, 611–617. [CrossRef]
105. Barak, Y.; Nelson, M.C.; Ong, E.S.; Jones, Y.Z.; Ruiz-Lozano, P.; Chien, K.R.; Koder, A.; Evans, R.M. PPAR gamma is required for placental, cardiac, and adipose tissue development. *Mol. Cell* **1999**, *4*, 585–595. [CrossRef]
106. Kubota, N.; Terauchi, Y.; Miki, H.; Tamemoto, H.; Yamauchi, T.; Komeda, K.; Satoh, S.; Nakano, R.; Ishii, C.; Sugiyama, T.; et al. PPAR gamma mediates high-fat diet-induced adipocyte hypertrophy and insulin resistance. *Mol. Cell* **1999**, *4*, 597–609. [CrossRef]
107. Contreras, A.V.; Torres, N.; Tovar, A.R. PPAR-alpha as a key nutritional and environmental sensor for metabolic adaptation. *Adv. Nutr.* **2013**, *4*, 439–452. [CrossRef] [PubMed]
108. Feng, X.; Gao, X.; Jia, Y.; Zhang, H.; Xu, Y.; Wang, G. PPAR-alpha agonist fenofibrate decreased RANTES levels in type 2 diabetes patients with hypertriglyceridemia. *Med. Sci. Monit.* **2016**, *22*, 743–751. [CrossRef] [PubMed]
109. Lamers, C.; Schubert-Zsilavecz, M.; Merk, D. Therapeutic modulators of peroxisome proliferator-activated receptors (PPAR): A patent review (2008-present). *Expert Opin. Ther. Pat.* **2012**, *22*, 803–841. [CrossRef] [PubMed]
110. Pawlak, M.; Lefebvre, P.; Staels, B. Molecular mechanism of PPARalpha action and its impact on lipid metabolism, inflammation and fibrosis in non-alcoholic fatty liver disease. *J. Hepatol.* **2015**, *62*, 720–733. [CrossRef] [PubMed]
111. Yoon, M. The role of PPARalpha in lipid metabolism and obesity: Focusing on the effects of estrogen on PPARalpha actions. *Pharmacol. Res.* **2009**, *60*, 151–159. [CrossRef] [PubMed]

112. Yoon, M. PPARalpha in obesity: Sex difference and estrogen involvement. *PPAR Res.* **2010**, *2010*. [CrossRef] [PubMed]
113. Xu, P.; Hong, F.; Wang, J.; Dai, S.; Wang, J.; Zhai, Y. The CAR agonist TCPOBOP inhibits lipogenesis and promotes fibrosis in the mammary gland of adolescent female mice. *Toxicol. Lett.* **2018**, *290*, 29–35. [CrossRef] [PubMed]
114. Valero, T. Mitochondrial biogenesis: Pharmacological approaches. *Curr. Pharm. Des.* **2014**, *20*, 5507–5509. [CrossRef] [PubMed]
115. Sanchis-Gomar, F.; Garcia-Gimenez, J.L.; Gomez-Cabrera, M.C.; Pallardo, F.V. Mitochondrial biogenesis in health and disease. Molecular and therapeutic approaches. *Curr. Pharm. Des.* **2014**, *20*, 5619–5633. [CrossRef] [PubMed]
116. Gao, J.; Yan, J.; Xu, M.; Ren, S.; Xie, W. CAR suppresses hepatic gluconeogenesis by facilitating the ubiquitination and degradation of PGC1alpha. *Mol. Endocrinol.* **2015**, *29*, 1558–1570. [CrossRef] [PubMed]
117. Wieneke, N.; Hirsch-Ernst, K.I.; Kuna, M.; Kersten, S.; Puschel, G.P. PPARalpha-dependent induction of the energy homeostasis-regulating nuclear receptor NR1i3 (CAR) in rat hepatocytes: Potential role in starvation adaptation. *FEBS Lett.* **2007**, *581*, 5617–5626. [CrossRef] [PubMed]
118. Wada, T.; Gao, J.; Xie, W. PXR and CAR in energy metabolism. *Trends Endocrinol. Metab.* **2009**, *20*, 273–279. [CrossRef] [PubMed]
119. Guo, D.; Sarkar, J.; Suino-Powell, K.; Xu, Y.; Matsumoto, K.; Jia, Y.; Yu, S.; Khare, S.; Haldar, K.; Rao, M.S.; et al. Induction of nuclear translocation of constitutive androstane receptor by peroxisome proliferator-activated receptor alpha synthetic ligands in mouse liver. *J. Biol. Chem.* **2007**, *282*, 36766–36776. [CrossRef] [PubMed]
120. Guo, D.; Sarkar, J.; Ahmed, M.R.; Viswakarma, N.; Jia, Y.; Yu, S.; Sambasiva Rao, M.; Reddy, J.K. Peroxisome proliferator-activated receptor (PPAR)-binding protein (PBP) but not PPAR-interacting protein (PRIP) is required for nuclear translocation of constitutive androstane receptor in mouse liver. *Biochem. Biophys. Res. Commun.* **2006**, *347*, 485–495. [CrossRef] [PubMed]
121. Kiyosawa, N.; Tanaka, K.; Hirao, J.; Ito, K.; Niino, N.; Sakuma, K.; Kanbori, M.; Yamoto, T.; Manabe, S.; Matsunuma, N. Molecular mechanism investigation of phenobarbital-induced serum cholesterol elevation in rat livers by microarray analysis. *Arch. Toxicol.* **2004**, *78*, 435–442. [CrossRef] [PubMed]
122. Ide, T.; Shimano, H.; Yoshikawa, T.; Yahagi, N.; Amemiya-Kudo, M.; Matsuzaka, T.; Nakakuki, M.; Yatoh, S.; Iizuka, Y.; Tomita, S.; et al. Cross-talk between peroxisome proliferator-activated receptor (PPAR) alpha and liver X receptor (LXR) in nutritional regulation of fatty acid metabolism. II. LXRs suppress lipid degradation gene promoters through inhibition of PPAR signaling. *Mol. Endocrinol.* **2003**, *17*, 1255–1267. [CrossRef] [PubMed]
123. Parikh, M.; Patel, K.; Soni, S.; Gandhi, T. Liver X receptor: A cardinal target for atherosclerosis and beyond. *J. Atheroscler. Thromb.* **2014**, *21*, 519–531. [CrossRef] [PubMed]
124. Yoshikawa, T.; Ide, T.; Shimano, H.; Yahagi, N.; Amemiya-Kudo, M.; Matsuzaka, T.; Yatoh, S.; Kitamine, T.; Okazaki, H.; Tamura, Y.; et al. Cross-talk between peroxisome proliferator-activated receptor (PPAR) alpha and liver X receptor (LXR) in nutritional regulation of fatty acid metabolism. I. PPARs suppress sterol regulatory element binding protein-1c promoter through inhibition of LXR signaling. *Mol. Endocrinol.* **2003**, *17*, 1240–1254. [PubMed]
125. Miyata, K.S.; McCaw, S.E.; Patel, H.V.; Rachubinski, R.A.; Capone, J.P. The orphan nuclear hormone receptor LXR alpha interacts with the peroxisome proliferator-activated receptor and inhibits peroxisome proliferator signaling. *J. Biol. Chem.* **1996**, *271*, 9189–9192. [CrossRef] [PubMed]
126. Yue, L.; Ye, F.; Gui, C.; Luo, H.; Cai, J.; Shen, J.; Chen, K.; Shen, X.; Jiang, H. Ligand-binding regulation of LXR/RXR and LXR/PPAR heterodimerizations: SPR technology-based kinetic analysis correlated with molecular dynamics simulation. *Protein Sci.* **2005**, *14*, 812–822. [CrossRef] [PubMed]
127. Chinetti, G.; Lestavel, S.; Bocher, V.; Remaley, A.T.; Neve, B.; Torra, I.P.; Teissier, E.; Minnich, A.; Jaye, M.; Duverger, N.; et al. PPAR-alpha and PPAR-gamma activators induce cholesterol removal from human macrophage foam cells through stimulation of the ABCA1 pathway. *Nat. Med.* **2001**, *7*, 53–58. [CrossRef] [PubMed]
128. Akiyama, T.E.; Sakai, S.; Lambert, G.; Nicol, C.J.; Matsusue, K.; Pimprale, S.; Lee, Y.H.; Ricote, M.; Glass, C.K.; Brewer, H.B., Jr.; et al. Conditional disruption of the peroxisome proliferator-activated receptor gamma gene in mice results in lowered expression of ABCA1, ABCG1, and apoE in macrophages and reduced cholesterol efflux. *Mol. Cell. Biol.* **2002**, *22*, 2607–2619. [CrossRef] [PubMed]

129. Claudel, T.; Leibowitz, M.D.; Fievet, C.; Tailleux, A.; Wagner, B.; Repa, J.J.; Torpier, G.; Lobaccaro, J.M.; Paterniti, J.R.; Mangelsdorf, D.J.; et al. Reduction of atherosclerosis in apolipoprotein E knockout mice by activation of the retinoid X receptor. *Proc. Natl. Acad. Sci. USA* **2001**, *98*, 2610–2615. [CrossRef] [PubMed]
130. Li, A.C.; Binder, C.J.; Gutierrez, A.; Brown, K.K.; Plotkin, C.R.; Pattison, J.W.; Valledor, A.F.; Davis, R.A.; Willson, T.M.; Witztum, J.L.; et al. Differential inhibition of macrophage foam-cell formation and atherosclerosis in mice by PPARalpha, beta/delta, and gamma. *J. Clin. Investig.* **2004**, *114*, 1564–1576. [CrossRef] [PubMed]
131. Chawla, A.; Boisvert, W.A.; Lee, C.H.; Laffitte, B.A.; Barak, Y.; Joseph, S.B.; Liao, D.; Nagy, L.; Edwards, P.A.; Curtiss, L.K.; et al. A PPAR gamma-LXR-ABCA1 pathway in macrophages is involved in cholesterol efflux and atherogenesis. *Mol. Cell* **2001**, *7*, 161–171. [CrossRef]
132. Lo Sasso, G.; Murzilli, S.; Salvatore, L.; D'Errico, I.; Petruzzelli, M.; Conca, P.; Jiang, Z.Y.; Calabresi, L.; Parini, P.; Moschetta, A. Intestinal specific LXR activation stimulates reverse cholesterol transport and protects from atherosclerosis. *Cell Metab.* **2010**, *12*, 187–193. [CrossRef] [PubMed]
133. Umeda, Y.; Kako, Y.; Mizutani, K.; Iikura, Y.; Kawamura, M.; Seishima, M.; Hayashi, H. Inhibitory action of gemfibrozil on cholesterol absorption in rat intestine. *J. Lipid Res.* **2001**, *42*, 1214–1219. [PubMed]
134. Valasek, M.A.; Clarke, S.L.; Repa, J.J. Fenofibrate reduces intestinal cholesterol absorption via PPARalpha-dependent modulation of NPC1L1 expression in mouse. *J. Lipid Res.* **2007**, *48*, 2725–2735. [CrossRef] [PubMed]
135. Ross, R. Atherosclerosis—An inflammatory disease. *N. Engl. J. Med.* **1999**, *340*, 115–126. [CrossRef] [PubMed]
136. Tuttolomondo, A.; Di Raimondo, D.; Pecoraro, R.; Arnao, V.; Pinto, A.; Licata, G. Atherosclerosis as an inflammatory disease. *Curr. Pharm. Des.* **2012**, *18*, 4266–4288. [CrossRef] [PubMed]
137. Welch, J.S.; Ricote, M.; Akiyama, T.E.; Gonzalez, F.J.; Glass, C.K. PPARgamma and PPARdelta negatively regulate specific subsets of lipopolysaccharide and IFN-gamma target genes in macrophages. *Proc. Natl. Acad. Sci. USA* **2003**, *100*, 6712–6717. [CrossRef] [PubMed]
138. Ogawa, S.; Lozach, J.; Benner, C.; Pascual, G.; Tangirala, R.K.; Westin, S.; Hoffmann, A.; Subramaniam, S.; David, M.; Rosenfeld, M.G.; et al. Molecular determinants of crosstalk between nuclear receptors and toll-like receptors. *Cell* **2005**, *122*, 707–721. [CrossRef] [PubMed]
139. Fessler, M.B. The challenges and promise of targeting the Liver X Receptors for treatment of inflammatory disease. *Pharmacol. Ther.* **2018**, *181*, 1–12. [CrossRef] [PubMed]
140. Delerive, P.; De Bosscher, K.; Vanden Berghe, W.; Fruchart, J.C.; Haegeman, G.; Staels, B. DNA binding-independent induction of IkappaBalpha gene transcription by PPARalpha. *Mol. Endocrinol.* **2002**, *16*, 1029–1039. [PubMed]
141. Bishop-Bailey, D.; Bystrom, J. Emerging roles of peroxisome proliferator-activated receptor-beta/delta in inflammation. *Pharmacol. Ther.* **2009**, *124*, 141–150. [CrossRef] [PubMed]
142. Li, A.C.; Brown, K.K.; Silvestre, M.J.; Willson, T.M.; Palinski, W.; Glass, C.K. Peroxisome proliferator-activated receptor gamma ligands inhibit development of atherosclerosis in LDL receptor-deficient mice. *J. Clin. Investig.* **2000**, *106*, 523–531. [CrossRef] [PubMed]
143. Im, S.S.; Osborne, T.F. Liver x receptors in atherosclerosis and inflammation. *Circ. Res.* **2011**, *108*, 996–1001. [CrossRef] [PubMed]
144. Steffensen, K.R.; Jakobsson, T.; Gustafsson, J.A. Targeting liver X receptors in inflammation. *Expert Opin. Ther. Targets* **2013**, *17*, 977–990. [CrossRef] [PubMed]
145. Ghisletti, S.; Huang, W.; Ogawa, S.; Pascual, G.; Lin, M.E.; Willson, T.M.; Rosenfeld, M.G.; Glass, C.K. Parallel SUMOylation-dependent pathways mediate gene- and signal-specific transrepression by LXRs and PPARgamma. *Mol. Cell* **2007**, *25*, 57–70. [CrossRef] [PubMed]
146. Gao, M.; Bu, L.; Ma, Y.; Liu, D. Concurrent activation of liver X receptor and peroxisome proliferator-activated receptor alpha exacerbates hepatic steatosis in high fat diet-induced obese mice. *PLoS ONE* **2013**, *8*. [CrossRef] [PubMed]

International Journal of
Molecular Sciences

Review

The Role of PPAR-δ in Metabolism, Inflammation, and Cancer: Many Characters of a Critical Transcription Factor

Yi Liu [1], Jennifer K. Colby [1], Xiangsheng Zuo [1], Jonathan Jaoude [1], Daoyan Wei [2] and Imad Shureiqi [1,*]

1 Department of Gastrointestinal Medical Oncology, The University of Texas MD Anderson Cancer Center, 1515 Holcombe Boulevard, Unit 426, Houston, TX 77030-4009, USA; yliu25@mdanderson.org (Y.L.); jkcolby@mdanderson.org (J.K.C.); xzuo@mdanderson.org (X.Z.); Jjaoude@mdanderson.org (J.J.)

2 Department of Gastroenterology, Hepatology, and Nutrition, The University of Texas MD Anderson Cancer Center, Houston, TX 77030, USA; dwei@mdanderson.org

* Correspondence: ishureiqi@mdanderson.org; Tel.: +1-713-792-2828

Received: 2 October 2018; Accepted: 23 October 2018; Published: 26 October 2018

Abstract: Peroxisome proliferator-activated receptor-delta (PPAR-δ), one of three members of the PPAR group in the nuclear receptor superfamily, is a ligand-activated transcription factor. PPAR-δ regulates important cellular metabolic functions that contribute to maintaining energy balance. PPAR-δ is especially important in regulating fatty acid uptake, transport, and β-oxidation as well as insulin secretion and sensitivity. These salutary PPAR-δ functions in normal cells are thought to protect against metabolic-syndrome-related diseases, such as obesity, dyslipidemia, insulin resistance/type 2 diabetes, hepatosteatosis, and atherosclerosis. Given the high clinical burden these diseases pose, highly selective synthetic activating ligands of PPAR-δ were developed as potential preventive/therapeutic agents. Some of these compounds showed some efficacy in clinical trials focused on metabolic-syndrome-related conditions. However, the clinical development of PPAR-δ agonists was halted because various lines of evidence demonstrated that cancer cells upregulated PPAR-δ expression/activity as a defense mechanism against nutritional deprivation and energy stresses, improving their survival and promoting cancer progression. This review discusses the complex relationship between PPAR-δ in health and disease and highlights our current knowledge regarding the different roles that PPAR-δ plays in metabolism, inflammation, and cancer.

Keywords: PPAR-δ; β-oxidation metabolism; inflammation; cancer

---

1. Introduction

Peroxisome proliferator-activated receptor-delta (PPAR-δ, also known as PPAR-β) is a member of the PPAR subgroup in the nuclear receptor superfamily. PPARs act as ligand-activated transcription factors that regulate important cellular metabolic functions [1]. Although PPAR-δ is ubiquitously expressed, its expression level in different tissues varies depending on cell type and disease status [2–5]. Homozygous knockout of murine *Ppard* through constructs targeting exon 4, which codes for the DNA binding domain, leads to embryonic lethality or impaired growth, which indicates that PPAR-δ plays a fundamental role in embryo development [6,7].

Details of PPAR structure and signaling mechanisms have been reviewed in detail in Reference [8] and will only be discussed briefly here. The characteristics of PPAR ligand-binding domains (LBD) allow for interaction of a broad range of potential ligands, including many lipid and lipid-like molecules [8]. Natural ligands for PPAR-δ include polyunsaturated fatty acids (PUFA, e.g., arachidonic and linoleic acid)) and their metabolites (e.g., prostacyclin/PGI_2, 13S-hydroxyoctadecadienoic acid

(13S-HODE), and 15S-hydroxyeicosatetraenoic acid (15S-HETE)) [9–12]. Although PPAR-δ has a narrower LBD relative to PPARs-α and-γ, binding pocket characteristics allow potential interaction with a variety of ligands, albeit many appear to bind at relatively low affinities [13]. While many potential endogenous ligands have been suggested in the literature, there is still some uncertainty about the physiological significance [14]. Selective ligands targeting PPAR-δ have also been developed, although none have been approved for clinical use to date [14,15].

PPAR signaling can be regulated in multiple ways, with outcomes depending upon whether PPAR and its binding partners are bound by ligand or not, ligand type (agonist, antagonist, partial agonist, etc.), and concentration as well as the availability of various coactivators or repressors [8]. The delivery of natural PPAR-δ ligands is facilitated by fatty acid transport proteins (FATPs) and fatty acid translocase (FAT, also known as CD36), which aid in import of extracellular lipids into the cell [16,17] and fatty-acid-binding proteins (FABPs), which transport cytoplasmic lipids within the cell [18,19]. Although most FABPs can bind a number of different lipids, it is unknown whether there is any selectivity in terms of the ligands FABP shuttles to PPARs [18,20,21]. In relation to PPAR-δ, FABP5 (also known as K-FABP or E-FABP) appears to be important for transport of lipid ligands to the nucleus [22]. Interestingly, FABP5 expression largely parallels that of PPAR-δ, and interaction between the two appears to be important in both normal and disease states, including many cancers [19]. Although a more detailed discussion is beyond the scope of this review, the interrelationship between the PPARs, their endogenous ligands, and various lipid transport proteins is complex, and several of these transport proteins are known transcriptional targets of PPARs (reviewed in References [16,19]).

Activation of PPARs by their ligands has been discussed in detail elsewhere and will be described only briefly here [8,23,24]. PPAR-δ activation requires interaction with various partners in the nucleus to transcriptionally regulate gene expression. Like other PPARs, PPAR-δ heterodimerizes with the retinoid X receptor (RXR) to activate or repress expression of downstream target genes by binding to PPAR response elements (PPREs) in their promoters [25,26]. In the absence of ligand binding, PPAR-RXR complexes are associated with corepressive factors and histone deacetylases that prevent transcriptional activation. Binding of an activating ligand to PPAR-δ leads to conformational changes that release corepressors and allow binding of coactivators [8,27]. In addition, PPARs can also engage in transrepression of other transcription factors. For example, in its unliganded state, PPAR-δ has been shown to form a complex with the transcription factor BCL-6, which prevents BCL-6 from repressing proinflammatory cytokine genes; therefore, this interaction promotes inflammation [28,29]. Conversely, binding of PPAR-δ agonist leads to disruption of the complex, and BCL-6 is freed to repress gene expression [28]. PPAR-δ has also been reported to interact with other transcription factors, such as β-catenin or NF-κB, to regulate gene expression [30,31].

Accumulating evidence has demonstrated that PPAR-δ can have distinct roles depending on the context (e.g., healthy vs. diseased, specific type of disease). While PPAR-δ allows normal cells (e.g., muscle cells and pancreatic cells) to better cope with adverse nutrient and energy pressures, PPAR-δ overexpression or hyperactivation can lead to promotion of inflammation and tumorigenesis. We will address some of the known discrepancies concerning PPAR-δ's putative roles in metabolism, inflammation, and cancer in this review.

2. Metabolic Regulation by PPAR-δ

Modulation of cellular energy consumption is a major function of PPAR-δ. In muscle cells, ligand activation of PPAR-δ switches energy production from glycolysis to fatty acid oxidation as an alternative energy source, which can enhance muscle endurance [32]. In skeletal muscle cells, PPAR-δ activation by fatty acids increases fatty acid uptake and catabolism via β-oxidation [33]. Genetically targeting PPAR-δ overexpression in skeletal muscle cells increases succinate-dehydrogenase-positive muscle fibers with enhanced fatty acid oxidative capabilities and leads to an overall decrease in body fat [34]. Treatment of insulin-resistant obese monkeys with the PPAR-δ ligand GW501516 increased serum high-density lipoprotein cholesterol while decreasing low density lipoprotein, fasting triglycerides, and insulin [35]. Similarly, in vivo and in vitro transcriptome analyses of rodent muscle showed that PPAR-δ regulated downstream target genes required for fatty acid transport, β-oxidation of fatty acid, and mitochondrial respiration. Transgenic activation of PPAR-δ in mouse adipose tissues upregulated the expression of genes involved in fatty acid β-oxidation and energy dissipation via uncoupling of fatty acid oxidation and ATP production; these effects were observed in both *ob/ob* (mice homozygous for the spontaneous obese mutation (*ob*) in the leptin (*Lep*) gene) and wild-type (WT) mice on high-fat diet [36]. In contrast, PPAR-δ knockout mice were more prone to high-fat-diet-induced obesity [36]. A recent report showed induction of fatty acid oxidation in intestinal stem cells after a 24h fast; this effect was observed in both young and old mice and was mediated by the PPAR-δ target gene carnitine palmitoyltransferase 1 (*Cpt1a*) [37].

In addition to its effects on fatty acid oxidation, PPAR-δ leads to improved blood glucose homeostasis through a number of mechanisms. PPAR-δ is strongly expressed in pancreatic islet beta cells, promoting insulin secretion [38,39]. It is also protective against insulin resistance through effects on hepatic and peripheral energy substrate utilization [40–42]. Treatment of *ob/ob* mice, which have a genetic predisposition to obesity and diabetes, with GW50516 attenuated the ability of high-fat diet to induce obesity and insulin resistance and improved diabetes [43].

These salutary PPAR-δ functions in normal cells are thought to protect against metabolic-syndrome-related diseases, such as obesity, dyslipidemia, insulin resistance, hepatosteatosis, and atherosclerosis [44,45]. Therefore, highly selective synthetic PPAR-δ agonists (e.g., GW0742 [46], GW501516 [35]) were developed and tested clinically. However, improving cellular tolerance to an inhospitable metabolic microenvironment could also promote the survival of cancer cells (Figure 1). For example, overexpression of PPAR-δ was shown to improve breast cancer cell survival during low-glucose or hypoxic cell culture conditions through multiple mechanisms (e.g., enhanced antioxidant signaling, AKT/protein kinase B activation), and increased cell survival was inhibited with PPAR-δ antagonists [47]. Other studies have demonstrated that PPAR-δ promotion of fatty acid oxidation can lead to increased ATP production, contributing not only to the survival of breast cancer cells [48] but also other cancer cells, such as chronic lymphocytic leukemia cells [49]. Concerns regarding the potential protumorigenic effects of PPAR-δ have led to halting of the clinical development of PPAR-δ agonists [50,51].

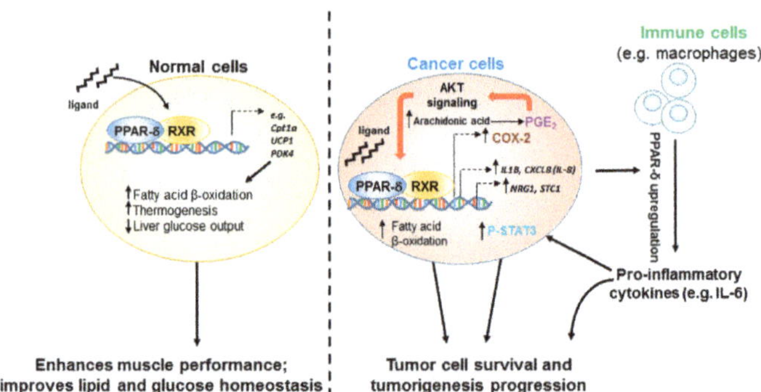

Figure 1. Ligand-dependent actions of PPAR-δ in normal versus cancer cells. Binding of PPAR-δ agonists in normal cells (**left**) leads to the upregulation of genes associated with a switch to using fatty acids as an energy source (increased β-oxidation). It is also associated with systemic improvements in serum glucose regulation through effects on multiple tissues, including pancreas, adipose, liver, and muscle. In cancer cells (**right**), this capacity for PPAR-δ to promote use of fatty acid substrates as an energy source can enhance cell survival and proliferation under harsh metabolic conditions frequently found in tumors. In addition, both COX-2 and PI3K/AKT signaling pathways are often upregulated in tumor cells. Interaction of activated PPAR-δ with these key signaling hubs leads to establishment of a feed-forward circuit promoting cancer development and progression through upregulation of additional factors that enhance neoplastic processes in cancer cells themselves as well as noncancer cells (e.g., tumor-associated macrophages) that make up the tumor microenvironment. See text for additional details.

3. PPAR-δ in Inflammation-Related Diseases

Many studies have revealed that PPARs are involved in regulation of inflammation. Initially, PPARs were generally believed to have anti-inflammatory functions, and current research has more clearly defined such roles for PPAR-α and PPAR-γ [52,53]. PPAR-δ's relationship with inflammation seems to be much different and still needs to be fully elucidated. In some contexts, PPAR-δ has been reported to have anti-inflammatory functions. For example, it was reported that the selective PPAR-δ agonist GW0742 alleviated inflammation in experimental autoimmune encephalomyelitis (EAE), while knockout of PPAR-δ aggravated EAE severity [54,55]. PPAR-δ's antidiabetic functions also appear to be associated with reduced inflammatory signaling. In a rat model of type 2 diabetes, GW0742 was shown to reduce the proinflammatory cytokines tumor necrosis factor-α (TNF-α) and monocyte chemoattractant protein-1 (MCP-1) in liver tissues, in conjunction with reduced hepatic fat accumulation [56]. GW0742 was also shown to inhibit streptozotocin-induced diabetic nephropathy in mice through a reduction of inflammatory mediators, including MCP-1 and osteopontin [57]. In addition, a more recent study using both the *db/db* (homozygous for the spontaneous *db* mutation in the leptin receptor gene (*Lepr*)) and high-fat-diet-induced obese diabetic mouse models showed that PPAR-δ is a key mediator in exercise-induced reduction of vascular inflammation; PPAR-δ knockout mice did not exhibit this improvement with exercise [58].

In contrast to the reports mentioned above, PPAR-δ signaling appears to promote inflammation in other contexts. PPAR-δ expression is increased in patients with psoriasis, a common immune-mediated disease primarily affecting the skin [59]. In a transgenic mouse model, induction of PPAR-δ activation in the epidermis led to development of a psoriasis-like skin condition, which was correlated with increased IL-1 signaling and phosphorylation of STAT3 [59]. PPAR-δ signaling may also promote inflammation in some forms of arthritis. Mesenchymal stem cells (MSCs) have immunomodulatory properties that can limit inflammation [60]. In a collagen-induced mouse model of arthritis, mice receiving MSCs with reduced PPAR-δ activity (MSCs harvested from PPAR-δ knockout mice or WT PPAR-δ MSCs pretreated with the PPAR-δ antagonist GSK3787) had better suppression of inflammatory immune responses, leading to improvements in arthritis scores [61]. In the same study, inhibition of PPAR-δ with GSK3787 in human MSCs enhanced their ability to limit proliferation of peripheral blood mononuclear cells in coculture experiments [61].

4. Modulation of Inflammatory Actions in Immune Cells

Mechanistically, fatty acid oxidation is a central metabolic factor impacting immune cell differentiation and activation; for example, the behavior of inflammatory and immunosuppressive T cell and macrophage subsets can be affected by the balance of fatty acid oxidation and synthesis [62–64]. PPAR-δ's roles as a fatty acid sensor and regulator of immune responses suggest a novel role for this receptor in metabolic modulation of immune cell differentiation and activity.

PPAR-δ appears to regulate inflammation through its effects on immune cells, particularly macrophages. Deletion of *Ppard* in liver-specific macrophages (Kupffer cells) led to reduced sensitivity to interleukin-4 and, therefore, impaired polarization to an alternative (M2-like) state with reduced inflammatory potential relative to M1-like macrophages [65]. This failure of differentiation to the M2 state ultimately led to hepatic dysfunction and insulin resistance, which was associated with altered fatty acid metabolism [65]. In another study using macrophage-specific PPAR-δ knockout mice, the deletion of *Ppard* was shown to decrease the phagocytosis of apoptotic cells and inhibit anti-inflammatory cytokine production by macrophages, leading to increased susceptibility to autoimmune kidney disease [66]. However, other evidence has revealed PPAR-δ's modulation of inflammation may be even more complex. In primary human monocyte-derived macrophages (MDMs), PPAR-δ ligands were reported to repress inflammation-associated NF-κB and signal transducer and activator of transcription 1 (STAT1)-targeted genes (e.g., *CXCL8* (IL-8) and *CXCL1*), yielding an M2-like macrophage phenotype [31]. Interestingly, they also observed reduced expression of factors associated with suppression of immune responses, such as indoleamine 2,3-dioxygenase 1 (*IDO1*), which plays an important role in limiting T-cell activation through its metabolism of tryptophan to kynurenine. In coculture experiments with MDMs and autologous T cells, MDMs exposed to PPAR-δ ligands increased $CD8^+/IFN\gamma^+$ T cell differentiation and limited IDO-1 mRNA and protein expression [31]. In addition, both kynurenine and the checkpoint inhibitor PD-1 ligand were also reduced in PPAR-δ ligand-treated MDMs [31]. Pathway analysis of RNA-Seq data from MDMs revealed upregulation of canonical PPAR-δ target genes involved in lipid metabolism, but exactly how they are engaged in immune regulation remains unclear [31]. Although the reports cited shed some light on the functions of PPAR-δ in immune cells, many questions remain concerning the mechanisms involved. The data reported above suggest that regulation of pro- and anti-inflammatory factor production and/or modulation of sensitivity to inflammatory stimuli may both be important in determining the overall outcome of PPAR-δ activation. The complexity of PPAR-δ's effects on inflammation is likely to be related to its different modes of function as a transcription factor, which in turn are dependent on specific interactions with target gene sequences (PPREs) as well as the presence or absence of relevant PPAR-δ ligands and coactivators/repressors [67].

5. The Role of PPAR-δ in Cancer

5.1. PPAR-δ Crosstalk between Inflammation and Cancer

The role of chronic inflammation in promoting tumorigenesis is well recognized and considered to be a hallmark of cancer development [68,69]. Colitis-associated colon cancer (CAC) is one of best established examples of chronic inflammation's role in increasing risk of cancer development, and its effects have been clearly demonstrated in preclinical and clinical studies [69,70]. PPAR-δ's impact on inflammation-promoted tumorigenesis has been studied by various groups, especially in relation to lipid signaling, as in the case of prostaglandin E_2 (PGE_2) [71]. PGE_2 is an eicosanoid lipid mediator generated through the actions of cyclooxygenases (COX-1 and -2). COX-2/ PGE_2 signaling is frequently upregulated in tumors and is especially important in the context of inflammation-driven tumorigenesis; it has been particularly well-studied in the case of colonic tumorigenesis [72]. PGE_2 enhanced PPAR-δ transcriptional activity via PI3 kinase/AKT activation to promote colon cancer cell survival in vitro and intestinal tumorigenesis in APC^{Min} mice [73]. In addition, targeted overexpression of PPAR-δ in intestinal epithelial cells promoted development of azoxymethane (AOM)/dextran sodium sulfate (DSS)-induced CAC in mice via upregulation of IL-6/STAT3 [74]. Activation of PPAR-δ by the synthetic ligand GW501516 in colon cancer cell lines or primary mouse intestinal epithelial cells upregulated COX-2 expression and PGE_2 production, subsequently increasing macrophage production of proinflammatory cytokines (e.g., CXCL1,CXCL2,CXCL4, and IL-1β) [75]. Both chemically and genetically induced colitis and CAC were markedly suppressed in a PPAR-δ knockout mouse model targeting PPAR-δ's DNA binding domain (deleting exons 4 and 5), therefore blocking its function as a transcription factor [75]. Considered together, these findings suggest a positive feedback loop between PPAR-δ and COX-2 that orchestrates a proinflammatory microenvironment to enhance tumorigenesis. However, in studies using a different PPAR-δ knockout mouse model targeting the C-terminal portion of the protein (*Ppard* exon 8), female KO mice treated with DSS showed significant increases in some clinical colitis scores (weight loss and colon length) as well as levels of IFN-γ, TNF-α, IL-6, and worsened histopathological scores compared to WT mice [76,77]. These discordant data have been suggested to be secondary to the *Ppard* exon 8 genetic deletion strategy producing a hypomorphic PPAR-δ protein with some retained/altered function as opposed to a complete loss of function [75]. Interestingly, prostatic-epithelial-targeted genetic deletion of PPAR-δ's DNA binding domain [7] has been recently reported to increase cellularity; in this setting, PPAR-δ was proposed to suppress prostate cancer via DNA-binding-dependent but ligand-binding-independent mechanisms [78].

PPAR-δ's promotion of inflammation and tumorigenesis is not limited to the colon. Tumorigenesis repurposes various components of the inflammatory machinery to create a microenvironment conducive to tumor growth. For example, PPAR-δ is upregulated in tumor-associated macrophages (TAMs) in ovarian-cancer-associated ascites [79], and additional work has shown that macrophages associated with ovarian cancer tend to exhibit protumorigenic properties (e.g., immunosuppression, growth promotion) [80]. In a mouse model of breast cancer, mammary-epithelium-targeted PPAR-δ overexpression promotes tumorigenesis, which is further augmented by treatment with the PPAR-δ agonist GW501516 [81]. Tumor development in this model is associated with upregulation of proinflammatory genes, including COX-2, and activation of AKT signaling [81], reminiscent of the positive association between PPAR-δ, COX-2, and AKT signaling in colorectal tumorigenesis as discussed above. Likewise, a similar positive feedback loop between PPAR-δ and COX-2 signaling has also been demonstrated to enhance the survival of hepatocellular carcinoma cell lines [82].

Overall, while the genetic PPAR-δ deletion models generated via targeting exons encoding the PPAR-δ C-terminal region versus the DNA binding domain have yielded discordant data, the preponderance of evidence supports a model in which PPAR-δ strongly enforces inflammation-driven promotion of tumorigenesis through the enhancement of proinflammatory and protumorigenic mechanisms, especially as illustrated in the case of the positive feedback between PPAR-δ and COX-2 (Figure 1).

5.2. *PPAR-δ Promotion of Cancer*

While PPAR-δ's relationship to cancer remains controversial [5,83], as more data continue to be published, we should see better clarification of PPAR-δ's specific contributions to the tumorigenic process. PPAR-δ is overexpressed in various human cancers, including colorectal cancer (CRC) [84–86], where it can be upregulated even in early stages (e.g., in adenomas) [84]. Similarly, PPAR-δ is upregulated in other human malignancies, including pancreatic cancer, where its upregulation is correlated with higher pathological grade and increased risk of metastasis [87]. PPAR-δ is also known to be expressed in human lung cancer [88]. Of interest, while PPAR-δ protein expression as assessed by immunohistochemistry has only been observed in the nuclei of normal cells, it becomes nuclear and cytoplasmic in cancer cells [85,86]. The significance of this shift in PPAR-δ distribution in relation to its function is unknown. Generally, high PPAR-δ expression in human cancers is associated with negative survival outcomes [67,86].

The controversy regarding PPAR-δ's role in tumorigenesis primarily stems from preclinical studies as illustrated by the following examples. First, a study using a PPAR-δ knockout mouse model (c-terminal KO/exon 8) showed that *Ppard* germline deletion increased the formation of colon tumors when the mice were bred with APC^{Min} mice or treated with AOM [89]. Later, a study by another group showed opposite results; in this case, *Ppard* germline deletion (DNA binding domain KO/exons 4–5) reduced the formation of colon tumors when the mice were bred with APC^{Min} mice [90]. Studies by our group showed that *Ppard* deletion genetically targeted to the intestinal epithelium profoundly inhibited AOM-induced colonic tumorigenesis [91]. While informative, the clinical relevance of using PPAR-δ knockout models of colorectal tumorigenesis in which PPAR-δ expression is reduced to levels below constitutive levels in normal cells is limited because PPAR-δ is typically upregulated in human colorectal cancer [11,84,86,92]. Modeling PPAR-δ's influence on CRC by targeting its overexpression to the intestinal epithelial cells better simulated its upregulation in human colon cancer tissues and allowed clear demonstration that PPAR-δ overexpression strongly promoted AOM-induced colorectal tumorigenesis [93].

Interestingly, when PPAR-δ knockout mice with the c-terminal (exon 8) deletion, which led to increased colonic tumorigenesis in the initial study [89], were backcrossed to MMTV-COX-2 transgenic mice on the FVB/N background, COX-2-induced mammary tumorigenesis was markedly suppressed [94]. Furthermore, subsequent studies in which syngeneic PPAR-δ-expressing B16-F10 melanoma cells or Lewis lung cancer cells were implanted into the same PPAR-δ knockout mouse model also showed inhibition of tumorigenesis [95]. These contradictory findings using the same PPAR-δ knockout construct have been interpreted as suggesting that PPAR-δ has different roles depending on where it is expressed—specifically, that PPAR-δ expressed in noncancerous cells in the tumor microenvironment promotes tumorigenesis, whereas PPAR-δ expressed in cancer cells suppresses tumorigenesis [95]. However, in experiments where B16-F10 mouse melanoma cells in which PPAR-δ was downregulated using shRNA were subcutaneously injected into syngeneic (C57BL/6) WT or PPAR-δ germline knockout mice [74], PPAR-δ downregulation in either cancer or noncancer cells inhibited metastasis, although this effect was stronger when PPARδ expression was suppressed in cancer cells [86]. In addition, downregulation of PPAR-δ expression in cancer cells strongly suppressed metastases in orthotopic injection mouse models of multiple human cancers (e.g., colon, lung, melanoma, breast, and pancreas) via downregulation of important prometastatic genes (e.g., *NRG1*, *CXCL8* (encoding IL-8), and *STC1*) in cancer cells and suppression of critical metastatic events including angiogenesis, epithelial-mesenchymal transition (EMT), and cancer cell invasion and migration [86] (Figure 1). PPAR-δ upregulation in human colon, lung, and breast cancers is also correlated with reduced metastasis-free survival [86]. PPAR-δ has also been reported to promote progression of melanoma via Snail expression [96] and prostate cancer via tumorigenic redirection of transforming growth factor β1 (TGF-β1) signaling [97]. A recently published study using unbiased global transcriptome analysis identified PPAR-δ activation as a driver of intestinal stem cell transformation and tumor promotion in APC^{Min} mice maintained on a high-fat diet, suggesting it may play a mechanistic role in obesity-driven cancers [98]. Overall, a scenario is emerging in which PPAR-δ significantly contributes to both the initiation and progression of tumorigenesis in multiple tissues/tumor types.

6. Conclusions

PPAR-δ is a transcription factor that profoundly influences important cellular functions regulating metabolism and inflammation. PPAR-δ's ability to act as a metabolic switch, shifting cellular energy utilization from glycolysis to fatty acid β-oxidation and thereby improving systemic glycemic control and lipid metabolism, make it an attractive target for prevention or treatment of metabolic-syndrome-related diseases (e.g., obesity, dyslipidemia, diabetes). Nevertheless, emerging data suggest that these same PPAR-δ-triggered mechanisms that help normal cells endure environmental metabolic challenges can also be exploited by cancer cells to promote their survival and, ultimately, cancer progression (Figure 1). Therefore, future therapeutic agents targeting PPAR-δ activation to treat metabolic diseases must be carefully designed and evaluated for any potential risks and off-target effects, including unintended promotion of preneoplastic or neoplastic lesions. Furthermore, our current understanding of the roles of PPAR-δ's interactions with its endogenous ligands, lipid transporters, and other nuclear receptors, coactivators, and repressors remains incomplete. While some exciting discoveries have been made, more studies are still needed to better define the roles and mechanistic actions of PPAR-δ in different physiological and pathophysiological conditions.

Author Contributions: Conception and design: Y.L., J.K.C., and I.S. Conceptual feedback and writing of the manuscript: Y.L., J.K.C., X.Z., J.J., D.W., and I.S.

Funding: This work was supported by grants R01CA142969, R01CA195686, and R01CA206539 awarded to I.S. from the National Cancer Institute at the National Institutes of Health, and grant RP140224 from the Cancer Prevention and Research Institute of Texas (to I.S).

Conflicts of Interest: The authors declare no competing interests.

Abbreviations

AKT	Cellular homolog of viral *akt* gene, also known as protein kinase B
AOM	azoxymethane
APC	Adenomatous polyposis coli
BCL-6	B cell lymphoma 6
CAC	colitis-associated cancer
CRC	Colorectal cancer
COX-2	cyclooxygenase-2
CXCL1	Chemokine (C-X-C motif) ligand 1
CXCL2	Chemokine (C-X-C motif) ligand 2
CXCL4	Chemokine (C-X-C motif) ligand 4
DSS	dextran sodium sulfate
EAE	experimental autoimmune encepaholmyelitis
EMT	Epithelial mesenchymal transition
FABP	Fatty acid binding protein
FAT	Fatty acid translocase
FATP	Fatty acid transport protein
IDO-1	indoleamine 2,3-dioxygenase 1
IFN-γ	Interferon-γ
IL-1β	Interleukin-1β
IL-6	Interleukin-6
IL-8	Interleukin-8
MCP	monocyte chemoattractive protein
MDM	monocyte-derived macrophage
MSC	mesenchymal stem cell
MMTV	Mouse mammary tumor virus
NRG1	Neuregulin 1
NF-κB	Nuclear factor kappa-light-chain-enhancer of activated B cells
PGE_2	prostaglandin E_2
PI3K	phosphatidyl inositol 3 kinase
PPRE	PPAR response element
PUFA	polyunsaturated fatty acid
RXR	retinoid X receptor
STAT1/3	Signal Transducer and Activator of Transcription 1/3

STC1	Stanniocalcin 1
TAM	tumor-associated macrophage
TGF-β1	transforming growth factor-β 1
TNF-α	tumor necrosis factor α

References

1. Wagner, K.D.; Wagner, N. Peroxisome proliferator-activated receptor beta/delta (PPARbeta/delta) acts as regulator of metabolism linked to multiple cellular functions. *Pharmacol. Ther.* **2010**, *125*, 423–435. [CrossRef] [PubMed]
2. Braissant, O.; Foufelle, F.; Scotto, C.; Dauca, M.; Wahli, W. Differential expression of peroxisome proliferator-activated receptors (PPARs): Tissue distribution of PPAR-alpha, -beta, and -gamma in the adult rat. *Endocrinology* **1996**, *137*, 354–366. [CrossRef] [PubMed]
3. Kliewer, S.A.; Forman, B.M.; Blumberg, B.; Ong, E.S.; Borgmeyer, U.; Mangelsdorf, D.J.; Umesono, K.; Evans, R.M. Differential expression and activation of a family of murine peroxisome proliferator-activated receptors. *Proc. Natl. Acad. Sci. USA* **1994**, *91*, 7355–7359. [CrossRef] [PubMed]
4. Auboeuf, D.; Rieusset, J.; Fajas, L.; Vallier, P.; Frering, V.; Riou, J.P.; Staels, B.; Auwerx, J.; Laville, M.; Vidal, H. Tissue distribution and quantification of the expression of mRNAs of peroxisome proliferator-activated receptors and liver X receptor-alpha in humans: No alteration in adipose tissue of obese and NIDDM patients. *Diabetes* **1997**, *46*, 1319–1327. [CrossRef] [PubMed]
5. Xu, M.; Zuo, X.; Shureiqi, I. Targeting peroxisome proliferator-activated receptor-beta/delta in colon cancer: How to aim? *Biochem. Pharmacol.* **2013**, *85*, 607–611. [CrossRef] [PubMed]
6. Nadra, K.; Anghel, S.I.; Joye, E.; Tan, N.S.; Basu-Modak, S.; Trono, D.; Wahli, W.; Desvergne, B. Differentiation of trophoblast giant cells and their metabolic functions are dependent on peroxisome proliferator-activated receptor beta/delta. *Mol. Cell. Boil.* **2006**, *26*, 3266–3281. [CrossRef] [PubMed]
7. Barak, Y.; Liao, D.; He, W.; Ong, E.S.; Nelson, M.C.; Olefsky, J.M.; Boland, R.; Evans, R.M. Effects of peroxisome proliferator-activated receptor delta on placentation, adiposity, and colorectal cancer. *Proc. Natl. Acad. Sci. USA* **2002**, *99*, 303–308. [CrossRef] [PubMed]
8. Harmon, G.S.; Lam, M.T.; Glass, C.K. PPARs and lipid ligands in inflammation and metabolism. *Chem. Rev.* **2011**, *111*, 6321–6340. [CrossRef] [PubMed]
9. Xu, H.E.; Lambert, M.H.; Montana, V.G.; Parks, D.J.; Blanchard, S.G.; Brown, P.J.; Sternbach, D.D.; Lehmann, J.M.; Wisely, G.B.; Willson, T.M.; et al. Molecular Recognition of Fatty Acids by Peroxisome Proliferator–Activated Receptors. *Mol. Cell* **1999**, *3*, 397–403. [CrossRef]
10. Shureiqi, I.; Jiang, W.; Zuo, X.; Wu, Y.; Stimmel, J.B.; Leesnitzer, L.M.; Morris, J.S.; Fan, H.Z.; Fischer, S.M.; Lippman, S.M. The 15-lipoxygenase-1 product 13-S-hydroxyoctadecadienoic acid down-regulates PPAR-delta to induce apoptosis in colorectal cancer cells. *Proc. Natl. Acad. Sci. USA* **2003**, *100*, 9968–9973. [CrossRef] [PubMed]
11. Gupta, R.A.; Tan, J.; Krause, W.F.; Geraci, M.W.; Willson, T.M.; Dey, S.K.; DuBois, R.N. Prostacyclin-mediated activation of peroxisome proliferator-activated receptor delta in colorectal cancer. *Proc. Natl. Acad. Sci. USA* **2000**, *97*, 13275–13280. [CrossRef] [PubMed]
12. Naruhn, S.; Meissner, W.; Adhikary, T.; Kaddatz, K.; Klein, T.; Watzer, B.; Muller-Brusselbach, S.; Muller, R. 15-hydroxyeicosatetraenoic acid is a preferential peroxisome proliferator-activated receptor beta/delta agonist. *Mol. Pharmacol.* **2010**, *77*, 171–184. [CrossRef] [PubMed]
13. Xu, H.E.; Lambert, M.H.; Montana, V.G.; Plunket, K.D.; Moore, L.B.; Collins, J.L.; Oplinger, J.A.; Kliewer, S.A.; Gampe, R.T., Jr.; McKee, D.D.; et al. Structural determinants of ligand binding selectivity between the peroxisome proliferator-activated receptors. *Proc. Natl. Acad. Sci. USA* **2001**, *98*, 13919–13924. [CrossRef] [PubMed]
14. Wu, C.C.; Baiga, T.J.; Downes, M.; La Clair, J.J.; Atkins, A.R.; Richard, S.B.; Fan, W.; Stockley-Noel, T.A.; Bowman, M.E.; Noel, J.P.; et al. Structural basis for specific ligation of the peroxisome proliferator-activated receptor delta. *Proc. Natl. Acad. Sci. USA* **2017**, *114*, E2563–E2570. [CrossRef] [PubMed]
15. Tan, C.K.; Zhuang, Y.; Wahli, W. Synthetic and natural Peroxisome Proliferator-Activated Receptor (PPAR) agonists as candidates for the therapy of the metabolic syndrome. *Expert Opin. Ther. Targets* **2017**, *21*, 333–348. [CrossRef] [PubMed]

16. Glatz, J.F.; Luiken, J.J.; Bonen, A. Membrane fatty acid transporters as regulators of lipid metabolism: Implications for metabolic disease. *Physiol. Rev.* **2010**, *90*, 367–417. [CrossRef] [PubMed]
17. Nakamura, M.T.; Yudell, B.E.; Loor, J.J. Regulation of energy metabolism by long-chain fatty acids. *Prog. Lipid Res.* **2014**, *53*, 124–144. [CrossRef] [PubMed]
18. Makowski, L.; Hotamisligil, G.S. Fatty acid binding proteins—The evolutionary crossroads of inflammatory and metabolic responses. *J. Nutr.* **2004**, *134*, 2464S–2468S. [CrossRef] [PubMed]
19. Amiri, M.; Yousefnia, S.; Seyed Forootan, F.; Peymani, M.; Ghaedi, K.; Nasr Esfahani, M.H. Diverse roles of fatty acid binding proteins (FABPs) in development and pathogenesis of cancers. *Gene* **2018**, *676*, 171–183. [CrossRef] [PubMed]
20. Tan, N.S.; Shaw, N.S.; Vinckenbosch, N.; Liu, P.; Yasmin, R.; Desvergne, B.; Wahli, W.; Noy, N. Selective cooperation between fatty acid binding proteins and peroxisome proliferator-activated receptors in regulating transcription. *Mol. Cell. Boil.* **2002**, *22*, 5114–5127. [CrossRef]
21. Tan, N.S.; Vazquez-Carrera, M.; Montagner, A.; Sng, M.K.; Guillou, H.; Wahli, W. Transcriptional control of physiological and pathological processes by the nuclear receptor PPARbeta/delta. *Prog. Lipid Res.* **2016**, *64*, 98–122. [CrossRef] [PubMed]
22. Armstrong, E.H.; Goswami, D.; Griffin, P.R.; Noy, N.; Ortlund, E.A. Structural basis for ligand regulation of the fatty acid-binding protein 5, peroxisome proliferator-activated receptor beta/delta (FABP5-PPARbeta/delta) signaling pathway. *J. Biol. Chem.* **2014**, *289*, 14941–14954. [CrossRef] [PubMed]
23. Zoete, V.; Grosdidier, A.; Michielin, O. Peroxisome proliferator-activated receptor structures: Ligand specificity, molecular switch and interactions with regulators. *Biochim. Biophys. Acta* **2007**, *1771*, 915–925. [CrossRef] [PubMed]
24. Batista, F.A.; Trivella, D.B.; Bernardes, A.; Gratieri, J.; Oliveira, P.S.; Figueira, A.C.; Webb, P.; Polikarpov, I. Structural insights into human peroxisome proliferator activated receptor delta (PPAR-delta) selective ligand binding. *PLoS ONE* **2012**, *7*, e33643. [CrossRef] [PubMed]
25. Feige, J.N.; Gelman, L.; Tudor, C.; Engelborghs, Y.; Wahli, W.; Desvergne, B. Fluorescence imaging reveals the nuclear behavior of peroxisome proliferator-activated receptor/retinoid X receptor heterodimers in the absence and presence of ligand. *J. Biol. Chem.* **2005**, *280*, 17880–17890. [CrossRef] [PubMed]
26. Matsusue, K.; Miyoshi, A.; Yamano, S.; Gonzalez, F.J. Ligand-activated PPARbeta efficiently represses the induction of LXR-dependent promoter activity through competition with RXR. *Mol. Cell. Endocrinol.* **2006**, *256*, 23–33. [CrossRef] [PubMed]
27. Viswakarma, N.; Jia, Y.; Bai, L.; Vluggens, A.; Borensztajn, J.; Xu, J.; Reddy, J.K. Coactivators in PPAR-Regulated Gene Expression. *PPAR Res.* **2010**, *2010*. [CrossRef] [PubMed]
28. Lee, C.H.; Chawla, A.; Urbiztondo, N.; Liao, D.; Boisvert, W.A.; Evans, R.M.; Curtiss, L.K. Transcriptional repression of atherogenic inflammation: Modulation by PPARdelta. *Science* **2003**, *302*, 453–457. [CrossRef] [PubMed]
29. Chinetti-Gbaguidi, G.; Staels, B. PPARbeta in macrophages and atherosclerosis. *Biochimie* **2017**, *136*, 59–64. [CrossRef] [PubMed]
30. Scholtysek, C.; Katzenbeisser, J.; Fu, H.; Uderhardt, S.; Ipseiz, N.; Stoll, C.; Zaiss, M.M.; Stock, M.; Donhauser, L.; Bohm, C.; et al. PPARbeta/delta governs Wnt signaling and bone turnover. *Nat. Med.* **2013**, *19*, 608–613. [CrossRef] [PubMed]
31. Adhikary, T.; Wortmann, A.; Schumann, T.; Finkernagel, F.; Lieber, S.; Roth, K.; Toth, P.M.; Diederich, W.E.; Nist, A.; Stiewe, T.; et al. The transcriptional PPARbeta/delta network in human macrophages defines a unique agonist-induced activation state. *Nucleic Acids Res.* **2015**, *43*, 5033–5051. [CrossRef] [PubMed]
32. Fan, W.; Waizenegger, W.; Lin, C.S.; Sorrentino, V.; He, M.X.; Wall, C.E.; Li, H.; Liddle, C.; Yu, R.T.; Atkins, A.R.; et al. PPARdelta Promotes Running Endurance by Preserving Glucose. *Cell Metab.* **2017**, *25*, 1186–1193. [CrossRef] [PubMed]
33. Holst, D.; Luquet, S.; Nogueira, V.; Kristiansen, K.; Leverve, X.; Grimaldi, P.A. Nutritional regulation and role of peroxisome proliferator-activated receptor δ in fatty acid catabolism in skeletal muscle. *Biochim. Biophys. Acta Mol. Cell Boil. Lipids* **2003**, *1633*, 43–50. [CrossRef]
34. Luquet, S.; Lopez-Soriano, J.; Holst, D.; Fredenrich, A.; Melki, J.; Rassoulzadegan, M.; Grimaldi, P.A. Peroxisome proliferator-activated receptor δ controls muscle development and oxidative capability. *FASEB J.* **2003**, *17*, 2299–2301. [CrossRef] [PubMed]

35. Oliver, W.R., Jr.; Shenk, J.L.; Snaith, M.R.; Russell, C.S.; Plunket, K.D.; Bodkin, N.L.; Lewis, M.C.; Winegar, D.A.; Sznaidman, M.L.; Lambert, M.H.; et al. A selective peroxisome proliferator-activated receptor delta agonist promotes reverse cholesterol transport. *Proc. Natl. Acad. Sci. USA* **2001**, *98*, 5306–5311. [CrossRef] [PubMed]
36. Wang, Y.-X.; Lee, C.-H.; Tiep, S.; Yu, R.T.; Ham, J.; Kang, H.; Evans, R.M. Peroxisome-Proliferator-Activated Receptor δ Activates Fat Metabolism to Prevent Obesity. *Cell* **2003**, *113*, 159–170. [CrossRef]
37. Mihaylova, M.M.; Cheng, C.W.; Cao, A.Q.; Tripathi, S.; Mana, M.D.; Bauer-Rowe, K.E.; Abu-Remaileh, M.; Clavain, L.; Erdemir, A.; Lewis, C.A.; et al. Fasting Activates Fatty Acid Oxidation to Enhance Intestinal Stem Cell Function during Homeostasis and Aging. *Cell Stem Cell* **2018**, *22*, 769–778. [CrossRef] [PubMed]
38. Ravnskjaer, K.; Frigerio, F.; Boergesen, M.; Nielsen, T.; Maechler, P.; Mandrup, S. PPARdelta is a fatty acid sensor that enhances mitochondrial oxidation in insulin-secreting cells and protects against fatty acid-induced dysfunction. *J. Lipid Res.* **2010**, *51*, 1370–1379. [CrossRef] [PubMed]
39. Iglesias, J.; Barg, S.; Vallois, D.; Lahiri, S.; Roger, C.; Yessoufou, A.; Pradevand, S.; McDonald, A.; Bonal, C.; Reimann, F.; et al. PPARbeta/delta affects pancreatic beta cell mass and insulin secretion in mice. *J. Clin. Investig.* **2012**, *122*, 4105–4117. [CrossRef] [PubMed]
40. Lee, C.H.; Olson, P.; Hevener, A.; Mehl, I.; Chong, L.W.; Olefsky, J.M.; Gonzalez, F.J.; Ham, J.; Kang, H.; Peters, J.M.; et al. PPARdelta regulates glucose metabolism and insulin sensitivity. *Proc. Natl. Acad. Sci. USA* **2006**, *103*, 3444–3449. [CrossRef] [PubMed]
41. Tang, T.; Abbott, M.J.; Ahmadian, M.; Lopes, A.B.; Wang, Y.; Sul, H.S. Desnutrin/ATGL activates PPARdelta to promote mitochondrial function for insulin secretion in islet beta cells. *Cell Metab.* **2013**, *18*, 883–895. [CrossRef] [PubMed]
42. Doktorova, M.; Zwarts, I.; Zutphen, T.V.; Dijk, T.H.V.; Bloks, V.W.; Harkema, L.; Bruin, A.; Downes, M.; Evans, R.M.; Verkade, H.J.; et al. Intestinal PPARdelta protects against diet-induced obesity, insulin resistance and dyslipidemia. *Sci. Rep.* **2017**, *7*, 846. [CrossRef] [PubMed]
43. Tanaka, T.; Yamamoto, J.; Iwasaki, S.; Asaba, H.; Hamura, H.; Ikeda, Y.; Watanabe, M.; Magoori, K.; Ioka, R.X.; Tachibana, K.; et al. Activation of peroxisome proliferator-activated receptor δ induces fatty acid β-oxidation in skeletal muscle and attenuates metabolic syndrome. *Proc. Natl. Acad. Sci. USA* **2003**, *100*, 15924–15929. [CrossRef] [PubMed]
44. Chen, J.; Montagner, A.; Tan, N.; Wahli, W. Insights into the Role of PPARβ/δ in NAFLD. *Int. J. Mol. Sci.* **2018**, *19*, 1893. [CrossRef] [PubMed]
45. Reilly, S.M.; Lee, C.H. PPAR delta as a therapeutic target in metabolic disease. *FEBS Lett.* **2008**, *582*, 26–31. [CrossRef] [PubMed]
46. Sznaidman, M.L.; Haffner, C.D.; Maloney, P.R.; Fivush, A.; Chao, E.; Goreham, D.; Sierra, M.L.; LeGrumelec, C.; Xu, H.E.; Montana, V.G.; et al. Novel selective small molecule agonists for peroxisome proliferator-activated receptor delta (PPARdelta)—Synthesis and biological activity. *Bioorg. Med. Chem. Lett.* **2003**, *13*, 1517–1521. [CrossRef]
47. Wang, X.; Wang, G.; Shi, Y.; Sun, L.; Gorczynski, R.; Li, Y.J.; Xu, Z.; Spaner, D.E. PPAR-delta promotes survival of breast cancer cells in harsh metabolic conditions. *Oncogenesis* **2016**, *5*, e232. [CrossRef] [PubMed]
48. Carracedo, A.; Weiss, D.; Leliaert, A.K.; Bhasin, M.; de Boer, V.C.J.; Laurent, G.; Adams, A.C.; Sundvall, M.; Song, S.J.; Ito, K.; et al. A metabolic prosurvival role for PML in breast cancer. *J. Clin. Investig.* **2012**, *122*, 3088–3100. [CrossRef] [PubMed]
49. Li, Y.J.; Sun, L.; Shi, Y.; Wang, G.; Wang, X.; Dunn, S.E.; Iorio, C.; Screaton, R.A.; Spaner, D.E. PPAR-delta promotes survival of chronic lymphocytic leukemia cells in energetically unfavorable conditions. *Leukemia* **2017**, *31*, 1905–1914. [CrossRef] [PubMed]
50. Sahebkar, A.; Chew, G.T.; Watts, G.F. New peroxisome proliferator-activated receptor agonists: Potential treatments for atherogenic dyslipidemia and non-alcoholic fatty liver disease. *Expert Opin. Pharmacother.* **2014**, *15*, 493–503. [CrossRef] [PubMed]
51. Cox, R.L. Rationally designed PPARdelta-specific agonists and their therapeutic potential for metabolic syndrome. *Proc. Natl. Acad. Sci. USA* **2017**, *114*, 3284–3285. [CrossRef] [PubMed]
52. Daynes, R.A.; Jones, D.C. Emerging roles of PPARs in inflammation and immunity. *Nat. Rev. Immunol.* **2002**, *2*, 748–759. [CrossRef] [PubMed]
53. Choi, J.M.; Bothwell, A.L. The nuclear receptor PPARs as important regulators of T-cell functions and autoimmune diseases. *Mol. Cells* **2012**, *33*, 217–222. [CrossRef] [PubMed]

54. Kanakasabai, S.; Chearwae, W.; Walline, C.C.; Iams, W.; Adams, S.M.; Bright, J.J. Peroxisome proliferator-activated receptor delta agonists inhibit T helper type 1 (Th1) and Th17 responses in experimental allergic encephalomyelitis. *Immunology* **2010**, *130*, 572–588. [CrossRef] [PubMed]
55. Dunn, S.E.; Bhat, R.; Straus, D.S.; Sobel, R.A.; Axtell, R.; Johnson, A.; Nguyen, K.; Mukundan, L.; Moshkova, M.; Dugas, J.C.; et al. Peroxisome proliferator-activated receptor delta limits the expansion of pathogenic Th cells during central nervous system autoimmunity. *J. Exp. Med.* **2010**, *207*, 1599–1608. [CrossRef] [PubMed]
56. Lee, M.Y.; Choi, R.; Kim, H.M.; Cho, E.J.; Kim, B.H.; Choi, Y.S.; Naowaboot, J.; Lee, E.Y.; Yang, Y.C.; Shin, J.Y.; et al. Peroxisome proliferator-activated receptor delta agonist attenuates hepatic steatosis by anti-inflammatory mechanism. *Exp. Mol. Med.* **2012**, *44*, 578–585. [CrossRef] [PubMed]
57. Matsushita, Y.; Ogawa, D.; Wada, J.; Yamamoto, N.; Shikata, K.; Sato, C.; Tachibana, H.; Toyota, N.; Makino, H. Activation of peroxisome proliferator-activated receptor delta inhibits streptozotocin-induced diabetic nephropathy through anti-inflammatory mechanisms in mice. *Diabetes* **2011**, *60*, 960–968. [CrossRef] [PubMed]
58. Cheang, W.S.; Wong, W.T.; Zhao, L.; Xu, J.; Wang, L.; Lau, C.W.; Chen, Z.Y.; Ma, R.C.; Xu, A.; Wang, N.; et al. PPARdelta Is Required for Exercise to Attenuate Endoplasmic Reticulum Stress and Endothelial Dysfunction in Diabetic Mice. *Diabetes* **2017**, *66*, 519–528. [CrossRef] [PubMed]
59. Romanowska, M.; Reilly, L.; Palmer, C.N.; Gustafsson, M.C.; Foerster, J. Activation of PPARbeta/delta causes a psoriasis-like skin disease in vivo. *PLoS ONE* **2010**, *5*, e9701. [CrossRef] [PubMed]
60. Bernardo, M.E.; Fibbe, W.E. Mesenchymal stromal cells: Sensors and switchers of inflammation. *Cell Stem Cell* **2013**, *13*, 392–402. [CrossRef] [PubMed]
61. Luz-Crawford, P.; Ipseiz, N.; Espinosa-Carrasco, G.; Caicedo, A.; Tejedor, G.; Toupet, K.; Loriau, J.; Scholtysek, C.; Stoll, C.; Khoury, M.; et al. PPARbeta/delta directs the therapeutic potential of mesenchymal stem cells in arthritis. *Ann. Rheum. Dis.* **2016**, *75*, 2166–2174. [CrossRef] [PubMed]
62. O'Neill, L.A.J.; Kishton, R.J.; Rathmell, J. A guide to immunometabolism for immunologists. *Nat. Rev. Immunol.* **2016**, *16*, 553–565. [CrossRef] [PubMed]
63. Andrejeva, G.; Rathmell, J.C. Similarities and Distinctions of Cancer and Immune Metabolism in Inflammation and Tumors. *Cell Metab.* **2017**, *26*, 49–70. [CrossRef] [PubMed]
64. Gerriets, V.A.; Kishton, R.J.; Nichols, A.G.; Macintyre, A.N.; Inoue, M.; Ilkayeva, O.; Winter, P.S.; Liu, X.; Priyadharshini, B.; Slawinska, M.E.; et al. Metabolic programming and PDHK1 control CD4+ T cell subsets and inflammation. *J. Clin. Investig.* **2015**, *125*, 194–207. [CrossRef] [PubMed]
65. Odegaard, J.I.; Ricardo-Gonzalez, R.R.; Red Eagle, A.; Vats, D.; Morel, C.R.; Goforth, M.H.; Subramanian, V.; Mukundan, L.; Ferrante, A.W.; Chawla, A. Alternative M2 activation of Kupffer cells by PPARdelta ameliorates obesity-induced insulin resistance. *Cell Metab.* **2008**, *7*, 496–507. [CrossRef] [PubMed]
66. Mukundan, L.; Odegaard, J.I.; Morel, C.R.; Heredia, J.E.; Mwangi, J.W.; Ricardo-Gonzalez, R.R.; Goh, Y.P.; Eagle, A.R.; Dunn, S.E.; Awakuni, J.U.; et al. PPAR-delta senses and orchestrates clearance of apoptotic cells to promote tolerance. *Nat. Med.* **2009**, *15*, 1266–1272. [CrossRef] [PubMed]
67. Muller, R. PPARbeta/delta in human cancer. *Biochimie* **2017**, *136*, 90–99. [CrossRef] [PubMed]
68. Hanahan, D.; Weinberg, R.A. Hallmarks of Cancer: The Next Generation. *Cell* **2011**, *144*, 646–674. [CrossRef] [PubMed]
69. Jess, T.; Rungoe, C.; Peyrin-Biroulet, L. Risk of colorectal cancer in patients with ulcerative colitis: A meta-analysis of population-based cohort studies. *Clin. Gastroenterol. Hepatol.* **2012**, *10*, 639–645. [CrossRef] [PubMed]
70. Terzić, J.; Grivennikov, S.; Karin, E.; Karin, M. Inflammation and Colon Cancer. *Gastroenterology* **2010**, *138*, 2101–2114. [CrossRef] [PubMed]
71. Wang, D.; DuBois, R.N. PPARdelta and PGE2 signaling pathways communicate and connect inflammation to colorectal cancer. *Inflamm. Cell Signal.* **2014**, *1*. [CrossRef]
72. Backlund, M.G.; Mann, J.R.; Dubois, R.N. Mechanisms for the prevention of gastrointestinal cancer: The role of prostaglandin E2. *Oncology* **2005**, *69*, 28–32. [CrossRef] [PubMed]
73. Wang, D.; Wang, H.; Shi, Q.; Katkuri, S.; Walhi, W.; Desvergne, B.; Das, S.K.; Dey, S.K.; DuBois, R.N. Prostaglandin E2 promotes colorectal adenoma growth via transactivation of the nuclear peroxisome proliferator-activated receptor δ. *Cancer Cell* **2004**, *6*, 285–295. [CrossRef] [PubMed]

74. Mao, F.; Xu, M.; Zuo, X.; Yu, J.; Xu, W.; Moussalli, M.J.; Elias, E.; Li, H.S.; Watowich, S.S.; Shureiqi, I. 15-Lipoxygenase-1 suppression of colitis-associated colon cancer through inhibition of the IL-6/STAT3 signaling pathway. *FASEB J.* **2015**, *29*, 2359–2370. [CrossRef] [PubMed]
75. Wang, D.; Fu, L.; Ning, W.; Guo, L.; Sun, X.; Dey, S.K.; Chaturvedi, R.; Wilson, K.T.; DuBois, R.N. Peroxisome proliferator-activated receptor delta promotes colonic inflammation and tumor growth. *Proc. Natl. Acad. Sci. USA* **2014**, *111*, 7084–7089. [CrossRef] [PubMed]
76. Peters, J.M.; Lee, S.S.; Li, W.; Ward, J.M.; Gavrilova, O.; Everett, C.; Reitman, M.L.; Hudson, L.D.; Gonzalez, F.J. Growth, adipose, brain, and skin alterations resulting from targeted disruption of the mouse peroxisome proliferator-activated receptor beta(delta). *Mol. Cell. Boil.* **2000**, *20*, 5119–5128. [CrossRef]
77. Hollingshead, H.E.; Morimura, K.; Adachi, M.; Kennett, M.J.; Billin, A.N.; Willson, T.M.; Gonzalez, F.J.; Peters, J.M. PPARbeta/delta protects against experimental colitis through a ligand-independent mechanism. *Dig. Dis. Sci.* **2007**, *52*, 2912–2919. [CrossRef] [PubMed]
78. Martín-Martín, N.; Zabala-Letona, A.; Fernández-Ruiz, S.; Arreal, L.; Camacho, L.; Castillo-Martin, M.; Cortazar, A.R.; Torrano, V.; Astobiza, I.; Zúñiga-García, P.; et al. PPARδ Elicits Ligand-Independent Repression of Trefoil Factor Family to Limit Prostate Cancer Growth. *Cancer Res.* **2018**, *78*, 399–409. [CrossRef] [PubMed]
79. Reinartz, S.; Finkernagel, F.; Adhikary, T.; Rohnalter, V.; Schumann, T.; Schober, Y.; Nockher, W.A.; Nist, A.; Stiewe, T.; Jansen, J.M.; et al. A transcriptome-based global map of signaling pathways in the ovarian cancer microenvironment associated with clinical outcome. *Genome Biol.* **2016**, *17*, 108. [CrossRef] [PubMed]
80. Colvin, E.K. Tumor-Associated Macrophages Contribute to Tumor Progression in Ovarian Cancer. *Front. Oncol.* **2014**, *4*, 137. [CrossRef] [PubMed]
81. Yuan, H.; Lu, J.; Xiao, J.; Upadhyay, G.; Umans, R.; Kallakury, B.; Yin, Y.; Fant, M.E.; Kopelovich, L.; Glazer, R.I. PPARdelta induces estrogen receptor-positive mammary neoplasia through an inflammatory and metabolic phenotype linked to mTOR activation. *Cancer Res.* **2013**, *73*, 4349–4361. [CrossRef] [PubMed]
82. Xu, L.; Han, C.; Lim, K.; Wu, T. Cross-talk between Peroxisome Proliferator-Activated Receptor δ and Cytosolic Phospholipase A2α/Cyclooxygenase-2/Prostaglandin E2 Signaling Pathways in Human Hepatocellular Carcinoma Cells. *Cancer Res.* **2006**, *66*, 11859–11868. [CrossRef] [PubMed]
83. Peters, J.M.; Gonzalez, F.J.; Muller, R. Establishing the Role of PPARbeta/delta in Carcinogenesis. *Trends Endocrinol. Metab.* **2015**, *26*, 595–607. [CrossRef] [PubMed]
84. Takayama, O.; Yamamoto, H.; Damdinsuren, B.; Sugita, Y.; Ngan, C.Y.; Xu, X.; Tsujino, T.; Takemasa, I.; Ikeda, M.; Sekimoto, M.; et al. Expression of PPAR[delta] in multistage carcinogenesis of the colorectum: Implications of malignant cancer morphology. *Br. J. Cancer* **2006**, *95*, 889–895. [CrossRef] [PubMed]
85. Yoshinaga, M.; Taki, K.; Somada, S.; Sakiyama, Y.; Kubo, N.; Kaku, T.; Tsuruta, S.; Kusumoto, T.; Sakai, H.; Nakamura, K.; et al. The expression of both peroxisome proliferator-activated receptor delta and cyclooxygenase-2 in tissues is associated with poor prognosis in colorectal cancer patients. *Dig. Dis. Sci.* **2011**, *56*, 1194–1200. [CrossRef] [PubMed]
86. Zuo, X.; Xu, W.; Xu, M.; Tian, R.; Moussalli, M.J.; Mao, F.; Zheng, X.; Wang, J.; Morris, J.S.; Gagea, M.; et al. Metastasis regulation by PPARD expression in cancer cells. *JCI Insight* **2017**, *2*, e91419. [CrossRef] [PubMed]
87. Abdollahi, A.; Schwager, C.; Kleeff, J.; Esposito, I.; Domhan, S.; Peschke, P.; Hauser, K.; Hahnfeldt, P.; Hlatky, L.; Debus, J.; et al. Transcriptional network governing the angiogenic switch in human pancreatic cancer. *Proc. Natl. Acad. Sci. USA* **2007**, *104*, 12890–12895. [CrossRef] [PubMed]
88. Pedchenko, T.V.; Gonzalez, A.L.; Wang, D.; DuBois, R.N.; Massion, P.P. Peroxisome proliferator-activated receptor beta/delta expression and activation in lung cancer. *Am. J. Respir. Cell Mol. Boil.* **2008**, *39*, 689–696. [CrossRef] [PubMed]
89. Harman, F.S.; Nicol, C.J.; Marin, H.E.; Ward, J.M.; Gonzalez, F.J.; Peters, J.M. Peroxisome proliferator-activated receptor-delta attenuates colon carcinogenesis. *Nat. Med.* **2004**, *10*, 481–483. [CrossRef] [PubMed]
90. Wang, D.; Wang, H.; Guo, Y.; Ning, W.; Katkuri, S.; Wahli, W.; Desvergne, B.; Dey, S.K.; DuBois, R.N. Crosstalk between peroxisome proliferator-activated receptor delta and VEGF stimulates cancer progression. *Proc. Natl. Acad. Sci. USA* **2006**, *103*, 19069–19074. [CrossRef] [PubMed]
91. Zuo, X.; Peng, Z.; Moussalli, M.J.; Morris, J.S.; Broaddus, R.R.; Fischer, S.M.; Shureiqi, I. Targeted Genetic Disruption of Peroxisome Proliferator-Activated Receptor-{delta} and Colonic Tumorigenesis. *J. Natl. Cancer Inst.* **2009**, *101*, 762–767. [CrossRef] [PubMed]
92. He, T.C.; Chan, T.A.; Vogelstein, B.; Kinzler, K.W. PPARdelta is an APC-regulated target of nonsteroidal anti-inflammatory drugs. *Cell* **1999**, *99*, 335–345. [CrossRef]

93. Zuo, X.; Xu, M.; Yu, J.; Wu, Y.; Moussalli, M.J.; Manyam, G.C.; Lee, S.I.; Liang, S.; Gagea, M.; Morris, J.S.; et al. Potentiation of colon cancer susceptibility in mice by colonic epithelial PPAR-delta/beta overexpression. *J. Natl. Cancer Inst.* **2014**, *106*, dju052. [CrossRef] [PubMed]
94. Ghosh, M.; Ai, Y.; Narko, K.; Wang, Z.; Peters, J.M.; Hla, T. PPARδ is pro-tumorigenic in a mouse model of COX-2-induced mammary cancer. *Prostaglandins Other Lipid Mediat.* **2009**, *88*, 97–100. [CrossRef] [PubMed]
95. Muller-Brusselbach, S.; Komhoff, M.; Rieck, M.; Meissner, W.; Kaddatz, K.; Adamkiewicz, J.; Keil, B.; Klose, K.J.; Moll, R.; Burdick, A.D.; et al. Deregulation of tumor angiogenesis and blockade of tumor growth in PPARbeta-deficient mice. *EMBO J.* **2007**, *26*, 3686–3698. [CrossRef] [PubMed]
96. Ham, S.A.; Yoo, T.; Hwang, J.S.; Kang, E.S.; Lee, W.J.; Paek, K.S.; Park, C.; Kim, J.H.; Do, J.T.; Lim, D.S.; et al. Ligand-activated PPARdelta modulates the migration and invasion of melanoma cells by regulating Snail expression. *Am. J. Cancer Res.* **2014**, *4*, 674–682. [PubMed]
97. Her, N.G.; Jeong, S.I.; Cho, K.; Ha, T.K.; Han, J.; Ko, K.P.; Park, S.K.; Lee, J.H.; Lee, M.G.; Ryu, B.K.; et al. PPARdelta promotes oncogenic redirection of TGF-beta1 signaling through the activation of the ABCA1-Cav1 pathway. *Cell Cycle* **2013**, *12*, 1521–1535. [CrossRef] [PubMed]
98. Beyaz, S.; Mana, M.D.; Roper, J.; Kedrin, D.; Saadatpour, A.; Hong, S.J.; Bauer-Rowe, K.E.; Xifaras, M.E.; Akkad, A.; Arias, E.; et al. High-fat diet enhances stemness and tumorigenicity of intestinal progenitors. *Nature* **2016**, *531*, 53–58. [CrossRef] [PubMed]

International Journal of
Molecular Sciences

Review

Peroxisome Proliferator-Activated Receptors (PPAR)γ Agonists as Master Modulators of Tumor Tissue

Daniel Heudobler [1], Michael Rechenmacher [1], Florian Lüke [1], Martin Vogelhuber [1], Tobias Pukrop [1], Wolfgang Herr [1], Lina Ghibelli [2], Christopher Gerner [3] and Albrecht Reichle [1,*]

1 Department of Internal Medicine III, University Hospital Regensburg, Hematology and Oncology, 93042 Regensburg, Germany; daniel.heudobler@ukr.de (D.H.); michael.rechenmacher@kr.de (M.R.); florian.lueke@ukr.de (F.L.); martin.vogelhuber@ukr.de (M.V.); tobias.pukrop@ukr.de (T.P.); wolfgang.herr@ukr.de (W.H.)

2 Department Biology, Universita' di Roma Tor Vergata, 00173 Rome, Italy; ghibelli@uniroma2.it

3 Institut for Analytical Chemistry, Faculty Chemistry, University Vienna, A-1090 Vienna, Austria; christopher.gerner@univie.ac.at

* Correspondence: albrecht.reichle@ukr.de

Received: 28 September 2018; Accepted: 6 November 2018; Published: 9 November 2018

Abstract: In most clinical trials, thiazolidinediones do not show any relevant anti-cancer activity when used as mono-therapy. Clinical inefficacy contrasts ambiguous pre-clinical data either favoring anti-tumor activity or tumor promotion. However, if thiazolidinediones are combined with additional regulatory active drugs, so-called 'master modulators' of tumors, i.e., transcriptional modulators, metronomic low-dose chemotherapy, epigenetically modifying agents, protein binding pro-anakoinotic drugs, such as COX-2 inhibitors, IMiDs, etc., the results indicate clinically relevant communicative reprogramming of tumor tissues, i.e., anakoinosis, meaning 'communication' in ancient Greek. The concerted activity of master modulators may multifaceted diversify palliative care or even induce continuous complete remission in refractory metastatic tumor disease and hematologic neoplasia by establishing novel communicative behavior of tumor tissue, the hosting organ, and organism. Re-modulation of gene expression, for example, the up-regulation of tumor suppressor genes, may recover differentiation, apoptosis competence, and leads to cancer control—in contrast to an immediate, 'poisoning' with maximal tolerable doses of targeted/cytotoxic therapies. The key for uncovering the therapeutic potential of Peroxisome proliferator-activated receptor γ (PPARγ) agonists is selecting the appropriate combination of master modulators for inducing anakoinosis: Now, anakoinosis is trend setting by establishing a novel therapeutic pillar while overcoming classic obstacles of targeted therapies, such as therapy resistance and (molecular-)genetic tumor heterogeneity.

Keywords: anakoinosis; communicative reprogramming; nuclear transcription factors; metronomic low-dose chemotherapy; glitazones; all-trans retinoic acid; COX-2 inhibitor; master modulators; undruggable targets; therapy pillar; peroxisome proliferator-activated receptors (PPARs); energy homeostasis; metabolic regulations; organ cross-talk; cancer and reprogramming of energy metabolism; systems biology

---

1. Introduction

Peroxisome-proliferator-activated receptors (PPARs) line up in the group of nuclear receptors and encompass three receptors PPARα, PPARγ, and PPARδ, which concertedly and multifaceted have impact on regulating tumor growth [1]. From metabolic disease, the resolution of insulin resistance by PPARγ and combined PPARα/γ agonists, as well as long-term outcome in patients with type II diabetes, we learned a lot on simultaneous PPARα/γ stimulation. A specific PPARγ agonist has been withdrawn from the market, as rosiglitazone was associated with a significant increase in the risk of death from cardiovascular causes, from myocardial infarction [2]. The beneficial effects of the dual PPARα/γ agonist, pioglitazone, namely the reduction of mortality, including non-fatal myocardial infarction and stroke in patients with type 2 diabetes who are at high risk concerning macro-vascular events, shed light on the multi-level concerted activity profile of PPARα and PPARγ in diabetes [3]. Particularly, these clinical trials in patients with diabetes type II highlight the striking anti-inflammatory component of PPARα. The initial hypothesis that efficacious anti-inflammatory therapy may also control advanced cancer could be confirmed by introducing pioglitazone in treatment of refractory metastatic cancer [4,5]. From pre-clinical data, the appropriate PPARα agonist for cancer treatment has to be defined, yet [1].

Nuclear receptors (NRs) encompass a huge heterogeneous group of ligand-controlled transcription factors, endocrine, orphan and adopted receptors [6]. Schedules for cancer treatment using ligand-mediated modulation of NRs are well-established, particularly the blockade of endocrine NRs in prostate and breast cancer or the stimulation with high-dose glucocorticoids in lymphoma or multiple myeloma [7–9]. In contrast, adopted NR agonists only hesitantly found their way into cancer treatment, e.g., retinoid X receptor (RXR) and retinoic acid receptor (RAR) receptor agonists for treatment of T-cell lymphoma and promyelocytic leukemia, respectively [10,11].

A further NR agonist, pioglitazone, a dual peroxisome-proliferator-activated receptor (PPAR) α/γ agonist, is now starting to blaze the trail for therapy of metastatic tumor diseases and it will be discussed in more detail [12].

Therapeutically intended stimulation of adopted NRs for tumor control is in striking methodologic contrast to blocking endocrine NRs with antagonists or inducing direct cytotoxicity with high-dose glucocorticoids, opens a novel view on tumor pathophysiology, and finally, implies a change of treatment paradigms [13].

Following oncogenic events, dysregulated homeostatic pathways and transcription factors in tumor tissues are communication-technically accessible via endocrine, orphan, and adopted NR agonists or more general, via master modulators, a term summarizing regulatory active, less toxic drugs administered at regulatory active dose levels [12,14]. 'Master modulators' of tumors, i.e., transcriptional modulators, metronomic low-dose chemotherapy, epigenetically modifying agents, protein binding pro-anakoinotic drugs, such as COX-2 inhibitors, IMiDs, etc., are aiming at attenuation of cancer-associated hallmarks or at establishing novel biologic hallmarks linked to tumor control. 'Master modulators' deploy therapeutic activity via regulatory accessible structures, functions, and hubs in tumor tissue, thereby e.g., reestablishing differentiation and apoptosis competence (Table 1) [12].

NRs are involved in regulating multifold biologic processes in normal and tumor tissue [15–17]. Clinical trials have shown that 'normalization' of dysregulated transcription factors with NR agonists belongs to a pivotal, clinically relevant concept, and it finally constitutes a novel therapeutic pillar for treatment of (refractory) metastatic cancer [12]. However, in relation to the multitude of orphan and adopted nuclear receptors, the clinical impact of corresponding nuclear receptor agonists has not been nearly exploited in the clinical setting.

Multi-level activity profiles on single cell compartments, tissues, or the whole organism are characteristic for NR agonists [12]. Multifaceted clinically beneficial changes in tumor behavior are based on the ubiquitous availability of NRs in tissues. The distribution of single NRs, however, is tissue-specifically varying, implicating tissue-, and as shown, cancer-specific activity profiles [6,18]. Moreover, the kind of ligand, i.e., a synthetic or natural hormone and/or lipophilic drug, additionally, has major impact on multi-level outcome [19,20].

Table 1. Explanation of communication-associated terms.

Communication-Associated Terms	Explanation
Anakoinosis	Anakoinosis is a novel paradigm for cancer treatment based on a key role for communicative reprogramming of tumor systems. Building on a systems biology approach to cancer, anakoinosis utilizes a range of non-cancer and cancer drugs in combination to treat advanced tumor disease, such as pioglitazone. In contrast to standard therapies, anakoinosis protocols are characterized by low toxicity and a good safety profile, with encouraging responses in a number of clinical trials to date. The use of drug repurposing, that is the use of non-cancer drugs as cancer treatments, is especially a notable feature of this approach.
Pro-anakoinotic therapeutic tools (examples)	Transcriptional modulators, nuclear receptor agonists and antagonists, metronomic low-dose chemotherapy, cyclooxygenase-2 inhibitors, IMiDs, arsenic trioxide, liposomal encapsulated small oligonucleotide encoding small activating RNAs, etc.
Metronomic tumor therapy	Metronomic tumor therapy may be defined as the frequent administration of (repurposed) drugs at doses significantly below the maximum tolerated dose with no prolonged drug-free breaks, or as the minimum biologically effective dose of an agent given as a continuous dosing regimen with no prolonged drug-free breaks that still leads to anti-tumor activity.
Rationalizations	Describe the physical organization of tumor-associated normative notions (e.g., hallmarks of cancer); are to some degree histology- and genotype-independent; may be re-directed and reorganized by anakoinosis.
Metabolism of evolution	The sum of extrinsically, i.e., therapeutically, and intrinsically inducible evolutionary processes within the tumor environment (tumor stroma, hosting organ, distant organ sites).
Modularity	Modularity describes the degree and specificity to which systems' objects, i.e., cells, pathways, molecules, therapeutic targets etc. may be communicatively rededicated by anakoinosis.
Validity and denotation	Validity of systems objects, functions and hubs: Availability on demand at distinct systems stages; denotation: Current functional impact at a distinct systems stage, e.g., of potentially tumor-promoting pathways. In the bio-world, presence and functioning of an object (e.g., an enzyme), respectively.

The ligand induced physical activity profile of NRs is dependent on multifold system-specific co-variables [21]. Their receptor and non-receptor mediated activity profile may explain and provide an insight in the multiplicity of biologic effects based on the predominantly regulatory and coordinating cell and tissue activities of NRs [22–24]. Therefore, interpretation and prediction of biologic outcome on the different observation levels within an organism is difficult and only accessible by application of novel technologies for monitoring ligand mediated biologic activities while treating metastatic tumors with nuclear receptor agonists.

Ligand induced structural changes of NRs facilitate binding at nuclear receptor response elements (NRREs) across the genome, but also the recruitment of co-regulators and interaction with other transcription factors, which may be again context-dependently activated or inhibited [25–27]. The obvious communication guided activity profile of NRs explicates why biologic read-outs may be contradictory depending on respective boundary conditions or systems stages, particularly in diseased organs [12].

When considering the context-dependent regulatory activity profile of NRs and the fact that cancer is constituted by complex dysregulation of transcription factors and homeostatic pathways, the following question arises: what kind of paradigms must be assumed for introducing NR agonists as attractive clinical targets, here, in particular, the α and γ variant of the peroxisome-proliferator-activated receptors (PPARs)? Secondly, may be the novel treatment approach, including agonists of nuclear receptors, universally applicable for treatment of metastatic, and refractory cancer and hematologic neoplasia?

2. Peroxisome Proliferator-Activated Receptor γ (PPARγ)/Cyclooxygenase-2 (COX-2) Expression in Tumors

Modulating COX-2 activity influences the local availability of PPAR ligands. Therefore, COX-2 indirectly modulates PPAR activity. Though acting on different signaling pathways, COX-2 and PPARγ modulate common molecular targets. Thus, COX-2 and PPARγ may concertedly inhibit cancer development [28]. Thereby, COX-2 inhibitors may act as partial PPARγ agonists [29], or PPARγ agonists as partial COX-2 inhibitors and suppressors of PGE2 synthesis [30,31].

Because of the close interaction of COX-2 and PPARγ, the differential expression in many human tumors, and the emerging possibilities to use them as targets for tumor therapy, we studied the correlation of PPARγ/COX-2 immunoreactivity with tissue microarrays (TMA) in a broad spectrum of histologic tumor types in comparison to normal tissue. In malignant melanoma, we focused on the correlation between clinic-pathologic features and outcome of patients with malignant melanoma (MM) [18].

TMA consisted of normal and tumor tissues ($n = 3448$) from 47 organs and tissue entities, including skin neoplasms ($n = 323$) of melanocytic (MM, benign nevi) and non-melanocytic origin (squamous cell carcinomas, basal cell carcinomas, Kaposi sarcomas, histiocytomas, capillary hemangiomas, sebaceous adenomas) [18].

COX-2 and PPARγ expression assays showed differential expression in almost every tissue type as well as in normal vs. neoplastic tissue: i.e., a continuous increase in COX2 expression from prostatic hyperplasia to prostatic intraepithelial neoplasia (PIN), to organ-confined prostate cancer, to castration-resistant prostate cancer, and to metastatic disease. In contrast, PPARγ expression decreases from the organ confined to the metastatic stage and increases again to the castration-resistant stage.

It could not be confirmed that COX-2 and PPARγ are inversely expressed in the human breast cancers, as breast cancer histologies are quite heterogeneous and differentially express COX-2 and PPARγ [32]. Activation of PPARγ may cause COX-2 inhibition or the down-regulation of COX-2 expression [33], whereas the inhibition of COX-2 resulted in PPARγ activation [34] or up-regulation of PPARγ expression [35].

Additional series of TMAs consisted of 88 MM with follow-up data, 101 MM metastases, and 161 benign nevi. A further TMA ($n = 194$) consisted of MM metastases from 36 patients with metastatic stage IV melanoma who had participated in a randomized phase II trial using a stroma-directed biomodulatory approach combining COX-2/ PPARγ-targeting with metronomic low-dose chemotherapy [18].

COX-2 and PPARγ immunoreactivity were paralleled and significantly increased from benign nevi (51%/0%) to primary MM (86%/22%) and MM metastases (91%/33%; $p < 0.001$, respectively). In the case of primary MM, positive COX-2 staining was associated with advanced Clark levels ($p = 0.004$) and shorter recurrence free survival ($p = 0.03$). However, PPARγ expression in primary MM was not associated with any of the clinic-pathologic characteristics or tumor progression and overall survival [18].

On the other hand, patients ($n = 36$) with PPARγ positive MM metastases who had been treated either with pro-anakoinotic metronomic low-dose chemotherapy (trofosfamide) alone or combined with COX-2/ PPARγ -targeting drugs, i.e., rofecoxib and pioglitazone, showed a significant advantage concerning progression-free survival ($p = 0.044$), but not overall survival ($p = 0.179$). Expression of COX-2 (score 2+–3+) in the metastases, however, was not associated with overall and progression-free survival, respectively [36].

We conclude that the expression of COX-2 and PPARγ is a frequent finding in the progression of MM. Regarding primary MM, the expression of COX-2 indicates an increased risk of tumor recurrence, i.e., melanoma progression.

In metastatic MM, the expression of PPARγ may serve as positive predictive marker of potential responsiveness to anakoinosis-inducing stroma-targeted therapy [36].

3. PPARγ Expression in Tumor Stroma

Apart from specifically stroma cells targeting drugs, some well-established pro-anakoinotic drugs, among them NR agonists, have revealed antitumor activity by unfolding pleiotropic biological effects. In this context thiazolidinedione derivatives such as pioglitazone are of special interest as they exert both a direct anti-tumor and a broad spectrum of stromal activities, including modulation of immune response, angiogenesis, and inflammation [37].

Stroma cell-specific NR signatures have to be suggested to collectively influencing tumor proliferation and metastasis [38]. Compartment specific NR expression and their context-dependent interaction with coregulators of NRs facilitate a complex dysregulated communicative network of transcription factors supporting multifold biologic hallmarks and tumor growth. On this presumably stage- and tumor-dependent background of NR expression, the profiling of NRs in stroma cells is urgently warranted for providing further rationales for combined transcriptional modulation in a therapeutic setting.

4. Induction of Anakoinosis with Master Modulators

Expression patterns of PPARγ in histologic different tumor tissues, both in tumor cells and adjacent stroma cells indicate histology and even tumor stage specific characteristic patterns, even, as shown, with predictive impact. Tumor-specific patterns of PPARγ expression support that PPARγ is strongly involved in maintaining homeostatic processes by adapting lipid and carbohydrate metabolism to respective tumor specific conditions, and by controlling tumor suppressor gene expression for keeping homeostatic pathways under tumor growth-promoting conditions, such as Wnt, Hippo-YAP pathway, etc. [27,39–41].

Consecutively, many experimental data indicate that PPARγ agonists may modulate multifold biologic hallmarks in cancer: Cell cycle, differentiation, proliferation, apoptosis, and oxidative stress, innate immunity, angiogenesis, and inflammation [42–44].

However, in most trials, thiazolidinediones (TZD) do not show any clinically relevant anti-cancer activity when used in mono-therapy (Table 2). Therefore, clinical inefficacy contrasts ambiguous pre-clinical data mostly favoring anti-tumor activity, but also tumor promotion. Thus, most review papers come to no consistent conclusion about the clinical use of PPARγ agonists for cancer treatment.

In contrast, if thiazolidinediones are combined with additional regulatory active drugs, so-called 'master modulators' of tumors, i.e., transcriptional modulators, metronomic low-dose chemotherapy, epigenetically modifying agents, protein binding pro-anakoinotic drugs, such as COX-2 inhibitors, IMiDs etc., clinical results indicate the relevant communicative reprogramming of tumor tissues, i.e., anakoinosis, meaning 'communication' in ancient Greek (Table 1).

Table 2. Glitazones including treatment schedules in metastatic cancer or hematologic neoplasia.

Glitazones in Refractory Tumors or Hematologic Neoplasia						
Neoplasia	**No pts**	**Chemotherapy (* = Metronomic)**	**Transcriptional Modulators**	**Small Molecule**	**Best Response**	**Reference**
Sarcomas						
Liposarcomas, intermediate to high-grade (case reports)	-	-	Troglitazone	-	Histological and biochemical differentiation	[45]
Liposarcoma	3	Trofosfamide *	Troglitazone	-	Lineage-appropriate differentiation can be induced pharmacologically in a human solid tumor.	[46]
Liposarcoma (Phase II study)	12	-	Rosiglitazone	-	Rosiglitazone is not effective as an antitumoral drug in the treatment of liposarcomas	[47]
Kaposi sarcoma, refractory	1	Trofosfamide *	Pioglitazone	COX-2 inhibitor	Partial remission	[48]
(Hem-)angiosarcomas	12	Trofosfamide *	Pioglitazone	COX-2 inhibitor	Continuos complete remission	[49]
Breast cancer						
Refractory breast cancer (Phase II study)	22	-	Troglitazone	-	No significant effect	[50]
Melanoma						
Melanoma III (versus DTIC), phase II ClinicalTrials.gov:NCT01614301	6	Trofosfamide *	Pioglitazone	Temsirolimus COX-2 inhibitor	Partial remission, Resolution of cachexia	[51]
Melanoma (randomized)						
Melanoma II Arm M	35	Trofosfamide *	Pioglitazone	-	Stable disease	[52]
Arm A/M	32	Trofosfamide *	Pioglitazone	COX-2 inhibitor	Partial remission	
Hepatocellular carcinoma						
Hepatocellular carcinoma	38	Capecitabine *	Pioglitazone	COX-2 inhibitor	Partial remission	[4]
Cholangiocellular carcinoma						
Cholangiocellular carcinoma	21	Trofosfamide *	Pioglitazone	COX-2 inhibitor	Partial remission	[18]
Colorectal cancer						
Chemotherapy-resistant metastatic colorectal cancer (phase II study)	25	-	Troglitazone	-	Not active for the treatment of metastatic colorectal cancer	[53]
Renal clear cell carcinoma (historic comparison)						
Renal clear cell carcinoma, relapsed	18	Capecitabine *	Pioglitazone	COX-2 inhibitor	Partial remission	[54]
Renal clear cell carinoma, relapsed	33	Capecitabine *	Pioglitazone Interferon-alpha	COX-2 inhibitor	Continuous complete remission	[5]
Prostate cancer						
Prostate cancer	41	-	Troglitazone	-	Lengthened stabilisation of prostate-specific antigen	[55]
Castration-resistant prostate cancer	61	Treosulfan *	Pioglitazone, Dexamethasone	COX-2 inhibitor Imatinib	Long-term tumor control at minimal disease	[56]
Castration-resistant prostate cancer	36	Capecitabine *	Pioglitazone, Dexamethasone	COX-2 inhibitor	Long-term tumor control	[57,58]

Table 2. *Cont.*

Glitazones in Refractory Tumors or Hematologic Neoplasia						
Neoplasia	**No pts**	**Chemotherapy (* = Metronomic)**	**Transcriptional Modulators**	**Small Molecule**	**Best Response**	**Reference**
Prostate carcinoma (randomized)						
Rising serum prostate-specific antigen level after radical prostatectomy and/or radiation therapy	106	-	Rosiglitazone *Versus* Placebo		Rosiglitazone did not increase PSA doubling time or prolong the time to disease progression	[59]
Gastric cancer (randomized) Gastric cancer Arm A/M Arm M	21 21	Capecitabine * Capecitabine *	Pioglitazone	COX-2 inhibitor	Partial remission Pioglitazone no impact	[60]
Glioblastoma Glioblastoma, refractory	14	Capecitabine *	Pioglitazone	COX-2 inhibitor	Disease stabilization	[61]
Multiple myeloma Multiple myeloma, third-line Clinicaltrials.gov, NCT001010243	6	Treosulfan *	Pioglitazone, Dexamethasone	Lenalidomide	Complete remission	[62]
Langerhans cell histiocytosis Langerhans cell histiocytosis, refractory	2 + 7	Trofosfamide *	Pioglitazone Dexamethasone	COX-2 inhibitor	Continuous complete remission	[13,63,64]
Hodgkin's lymphoma Hodgkin lymphoma, refractory	3	Treosulfan *	Pioglitazone, Dexamethasone	COX-2 inhibitor Everolimus	Continuous complete remission	[65]
Chronic myelocytic leukemia Chronic myelocytic leukemia without moleclar CR	24	-	Pioglitazone	Imatinib	Molecular complete remission (54%)	[66]
Acute myelocytic leukemia Acute myelocytic leukemia Refractory (on-going trial)	5 + 7	Azacitidine	Pioglitazone All-trans retinoic acid		Molecular complete remission Myelodysplastic synrome with phagocytically active blasts	[67,68]

5. Keys for Uncovering the Therapeutic Potential of PPARγ Agonists: Selecting the Appropriate, Histology-Independent Combination of Master Modulators

Clinical data reveals that most regulatory active drugs, i.e., master modulators of tumor tissues, exert only a modest or no monoactivity in cancer treatment (Table 2). Also, metronomic low-dose chemotherapy has just modest activity in randomized comparisons [69–75].

However, combining master modulators in 17 different histologic tumor entities leads to impressive, and, interestingly, highly diversified tumor responses up to continuous complete remission (Table 2). Moreover, single combinatory schedules of master modulators, including pioglitazone, are cross-responsive among quite different tumor histologies [12,51]. Cross-responsiveness now clearly indicates that different tumor histologies share identical patterns of hallmarks of cancer and constitute similar physical organizations of hallmarks, so called rationalizations of hallmarks, despite underlying (molecular-) genetic tumor heterogeneity (Table 1).

Thus, the clinically used top-down approaches reveal that tumor phenotypes are not dominantly minted and are associated with multifold recessively developing tumor features, which may be accessible for the concerted activity of regulatory active drugs. The specific therapeutic and clinically relevant access to tumor systems prompted us to choose for the procedure the term 'anakoinosis', communicative reprogramming [12]. The term anakoinosis reflects how regulatory active drugs may concertedly induce major tumor response, obviously by altering validity, i.e., availability on demand at distinct time points, and denotation, i.e., current functional impact at a distinct systems stage of tumor-promoting pathways (Table 1, Figure 1).

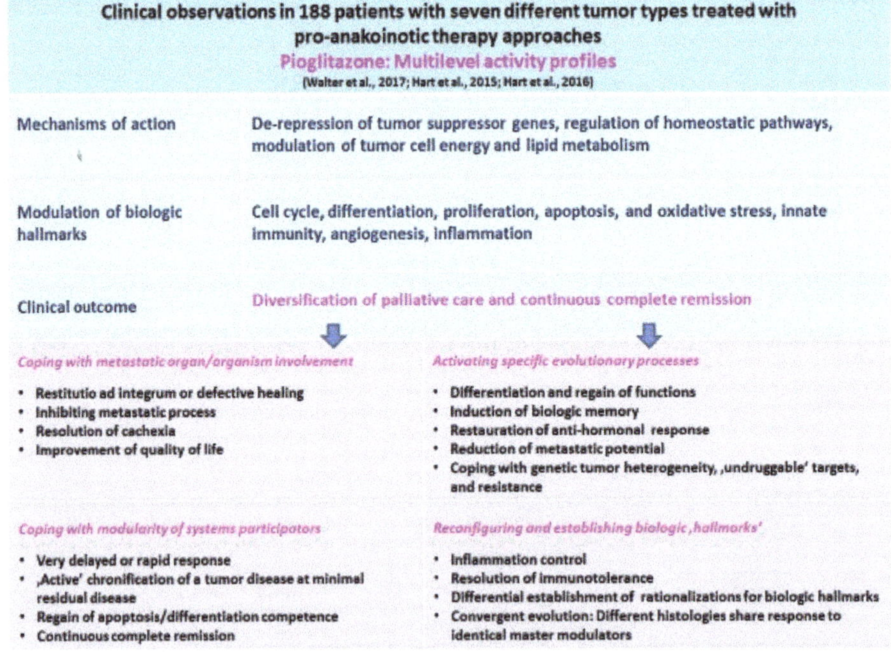

Clinical observations in 188 patients with seven different tumor types treated with pro-anakoinotic therapy approaches
Pioglitazone: Multilevel activity profiles
(Walter et al., 2017; Hart et al., 2015; Hart et al., 2016)
Mechanisms of action
De-repression of tumor suppressor genes, regulation of homeostatic pathways, modulation of tumor cell energy and lipid metabolism
Modulation of biologic hallmarks
Cell cycle, differentiation, proliferation, apoptosis, and oxidative stress, innate immunity, angiogenesis, inflammation
Clinical outcome
Diversification of palliative care and continuous complete remission
Coping with metastatic organ/organism involvement
• Restitutio ad integrum or defective healing
• Inhibiting metastatic process
• Resolution of cachexia
• Improvement of quality of life
Activating specific evolutionary processes
• Differentiation and regain of functions
• Induction of biologic memory
• Restauration of anti-hormonal response
• Reduction of metastatic potential
• Coping with genetic tumor heterogeneity, ‚undruggable' targets, and resistance
Coping with modularity of systems participators
• Very delayed or rapid response
• ‚Active' chronification of a tumor disease at minimal residual disease
• Regain of apoptosis/differentiation competence
• Continuous complete remission
Reconfiguring and establishing biologic ‚hallmarks'
• Inflammation control
• Resolution of immunotolerance
• Differential establishment of rationalizations for biologic hallmarks
• Convergent evolution: Different histologies share response to identical master modulators

Figure 1. Pioglitazone in tumor therapy regulates the communicative interface of transcriptional modulation, lipid and carbohydrate metabolism, particularly in combination with additional master modulators. Thus, tumor-promoting pathways can be functionally attenuated without direct blocking tumor-promoting pathways or by shutting off tumor-associated cellular compartments. Clinical equivalents are diversification of palliative care, even continuous complete remission.

The fact that communication rules may change validity and denotation of systems objects may be generally attributed to communication.

The successful concerted administration of pro-anakoinotic drug combinations, including PPARγ agonists in the clinical setting, may now explain multiple, from the clinical point of view cumulatively vague, as always context-dependent and often opposing results on the function of PPARγ agonists [1,6,24,76–84]. The missing conception for integrating pre-clinical results in clinical practice underlines missing communication-based therapeutic paradigms provided by an evolution-adjusted tumor pathophysiology and implies an unjustified hesitant introduction of master modulators, including PPARα/γ agonists, like pioglitazone in tumor therapy (Table 1).

Pre-clinically synergistic activities of PPARγ have been reviewed, particularly combinations with chemotherapy, besides RXR ligands and statins [82].

5.1. Poor Monoactivity of PPARγ Agonists Across Different Tumor Histologies

Monoactivity of glitazones in cancer patients is very modest, whereas strong activity is well established in single tumor histologies for dexamethasone, all-trans retinoic acid, and bexarotene [11,85].

Metabolically active drugs, such as metformin or PPARγ/α agonists, are considered as chemopreventive agents [86,87]. Metformin may prolong survival in cancer patients following surgery, but only in distinct histologic tumor types, as retrospective studies are indicating [88]. Nevertheless, very recent data shows a mechanistic link between glucose metabolism and cancer being mediated by TET2-function [89].

Agonists of 'adopted' orphan receptors commonly have poor monoactivity in interventional cancer trials [59,90], in contrast to hormones and cytokines [91,92]. Particularly, dexamethasone plays a decisive role in the induction treatment for acute lymphocytic leukemia or multiple myeloma [90].

5.2. PPARγ Agonists in Pro-Anakoinotic Combination Therapy with Master Modulators

Stromal cells or normal epithelial cells are not equipped for directly sensing tumor promoting genetic or molecular-genetic aberrations in neighboring malignant transformed cells. Adjacent non-tumor cells, however, sense dysregulations in homeostatic pathways. Thus, it is not surprising that hair follicle epithelia may spontaneously eliminate malignant transformed counterparts, irrespective of the underlying oncogenic events by sensoring dysregulated homeostasis [93].

Therefore, our commonly used therapeutic procedure, based on 'sensing' and consecutively blocking oncogenic pathways, is completely different from the pathophysiological based in vivo recognition of equivalences of malignancy by non-tumor cells, i.e., dysregulated homeostatic processes. All tumor-associated (molecular-) genetic aberrations are profoundly involved in dysregulations of homeostatic pathways [14]. Thus, tumors can be considered as a big dysregulated network of transcription factors. Just the communicatively evolving transcriptional system irregularities may be recognized as therapeutic target for master modulators. Master modulators are equipped with the capacity for 'normalizing' homeostatic networks on quite different topographic levels: the tumor's different cell compartments, the tumor and the tumor-harboring organ, and finally, the tumor and the whole organism (Table 1) [12].

Dysregulated homeostatic pathways represent, even if complex for pre-clinical evaluation, a pivotal therapeutic tool for 'normalizing' dysregulated homeostatic processes via master modulators, including agonists of nuclear transcription factors. NR antagonists are well integrated in clinical use and are here excluded from consideration. The review, particularly, concentrates on pioglitazone, a dual receptor agonist for PPARα/γ.

As shown to some extent, tumor-associated transcriptional dysregulation provides access for specific pro-anakoinotic effects via master modulators, including NR agonists. Moreover, histologically different tumor types share distinct communication-derived dysregulations, independent of the oncogenic background and show cross-reactivity to distinct systems adapted combinations of master modulators [12].

Cancer-specific impressive transcriptional dysregulation in comparison to the homeostatically well-balanced repertoire of transcription factors in normal organ tissue might be responsible for the modest toxicity profile of therapies, including combinations of master modulators. Therapeutic effects of combinations of master modulators should be to some degree neglectable in homeostatically well balanced, normal tissues, as they do not lay themselves open to therapeutic attack with master modulators selected for special evolution-related operative conditions in tumor tissue (Table 1) [12].

The top-down approach only has established how agonists of nuclear transcription factors, or generally master modulators, might communicatively interact for diversifying palliative care or even for inducing continuous complete remission. Additionally, maximal tolerable doses can be yield up, as pro-anakoinotic acting, lower doses are sufficient for achieving a therapeutically relevant response (Figure 2).

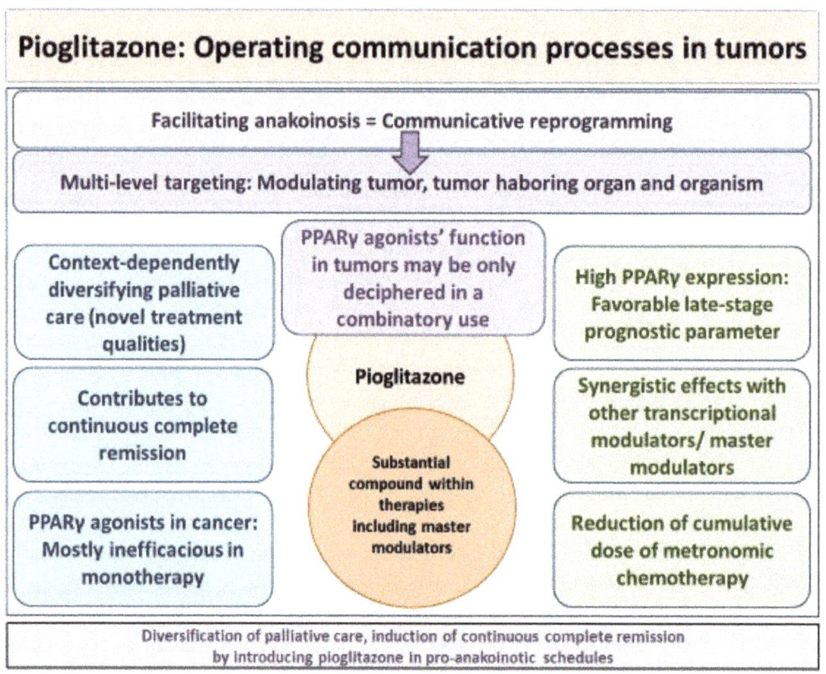

Figure 2. Pioglitazone, operating communication processes in tumors: Clinical relevance.

5.2.1. PPARγ Agonists Combined with Metronomic Low-Dose Chemotherapy/Demethylating Agents

Metronomic low-dose chemotherapy is still not established in routine therapy of neoplasia, as randomized comparisons often show no advantage for metronomically scheduled chemotherapy [69–75]. However, pre-clinical and clinical data give hints that the addition of classic targeted therapies or master modulators may improve outcome, even may diversify palliative care, and may contribute to continuous complete remission [94].

There are several reasons to include metronomic low-dose chemotherapy in the group of master modulators of tumor tissues. By adding pioglitazone and a COX-2 inhibitor, or an additional transcriptional regulator, such as a glucocorticoid, all-trans retinoic acid or interferon-α, outcome in refractory metastatic tumor disease could be improved up to continuous complete remission (Table 2). Additionally, chemotherapy doses could be reduced up to a quarter or third of the respective cumulative dose, which would be administered as pulsed therapy every three to four weeks, without loss of efficacy. Therefore, currently the question remains unanswered, which is the lowest, still regulatory active dose of metronomic chemotherapy when combining several master modulators [12].

In metastatic melanoma, the addition of pioglitazone to metronomic low-dose chemotherapy and COX-2 inhibitor has important therapeutic impact on outcome, as indicated in the paragraph 'PPARγ agonist plus COX-2 inhibitor'.

An important link between pioglitazone and metronomic chemotherapy may be physically explained. Pioglitazone sensitizes metronomic low-dose chemotherapy response by up-regulation of both, the receptor for the angiogenesis inhibitor thrombospondin 1, CD 36, and the phosphatase and tensin homolog PTEN [95–98].

5.2.2. PPARγ Agonists Plus Dexamethasone

Interacting with transcription factors as well as other cell-signaling systems nuclear receptors are important regulators in innate and adaptive immunity. PPARs, LXRs, and the glucocorticoid receptor (GR) may act together and thereby integrate local and systemic responses to inflammation by p65/IRF3-independent mechanisms [99]. Cooperating with the GR PPARs und LXRs synergistically transrepress distinct subsets of toll-like receptor-responsive genes. Thus, the combinatorial control of homeostasis and immune responses by nuclear receptors may specify the response and suggest novel approaches for treatment of pro-inflammatory tumor diseases [99].

In a series of quite different tumor histologies, the cross-responsiveness to dual transcriptional modulation with pioglitazone and glucocorticoid could be nicely shown, when added to metronomic low-dose chemotherapy. The concept has been tested in multiple myeloma, Hodgkin disease, and Langerhans cell histiocytosis, all inflammation-triggered diseases. C-reactive protein control in peripheral blood was indicative for response [12,63].

Preclinical data show that thiazolidinediones induce growth arrest and apoptosis of Waldenström's macroglobulinemia cells, at concentrations that are relevant to those achieved in previous clinical uses of these drugs [100].

From pre-clinical data on prostate cancer, PPARγ agonists may be acting, in part, by inhibiting transactivation of androgen-responsive genes [101]: Peroxisome proliferator-activated receptor γ agonists may down-regulate prostate-specific antigen expression in human prostate cancer [102].

Positive correlation between PPARγ and fatty acid synthase (FASN) protein in prostate cancer cell lines and synergism between TZDs and FASN blockers could be shown in prostate cancer cell viability reduction and apoptosis induction. [103].

Androgen receptor and Wnt/β-catenin/Tcf are cross-regulated. RAR/RXR, GR, thyroid receptor (TR), vitamin D receptor (VDR), estrogen receptor (ER), and PPAR modulate canonical Wnt signaling in dynamic manner with striking cell line- and tissue-specific differences indicating selective therapeutic access and requiring deciphering for combined transcriptional modulation in a therapeutic setting [40]. This fact may give hints for the combinatorial use of receptor agonists and antagonists.

Dual transcriptional modulation with glucocorticoids and pioglitazone in combination with metronomic low-dose chemotherapy and COX-2 inhibitor improved in a historic comparison overall survival in high-risk patients with castration-resistant prostate cancer from 19 months to more than three years. The addition of imatinib had no impact in this trial [56].

Thus, rapidly progressive castration-resistant prostate cancer responded to the same therapy principle as refractory Hodgkin disease, multiple myeloma, and Langerhans cell histiocytosis, but the communication-technically provided dysregulated systems targets seem to be different. Castration-resistant prostate cancer is only in rare cases that are associated with pro-inflammatory systems reaction, and C-reactive response in serum was no indicator for response as in refractory Hodgkin disease, multiple myeloma, and Langerhans cell histiocytosis [54,63–65]

5.2.3. PPARγ Agonists Plus All-Trans Retinoic Acid

The combination of azacitidine plus all-trans retinoic acid and pioglitazone may induce ex vivo granulocytic differentiation in more of 50% of blasts from acute myelocytic leukemia [67]. Moreover, these granulocytes regain phagocytic activity, when exposed to *E. coli* (unpublished data). Clinically, it is possible to induce continuous complete remission in acute myelocytic leukemia with the triple combination, while using only about 50% of the recommended dose of azacitidine [67,104].

A randomized trial in refractory acute myeloid leukemia (AML) is on-going, comparing the approved dose of azacitidine in comparison to the dose per square meter plus all-trans retinoic acid and pioglitazone.

Synergistic activity of dual transcriptional modulation has been well established in pre-clinical studies, for example, for pioglitazone and all-trans retinoic acid in tumor cell lines of different histology [68,105–108], but also for glitazones in combination with chemotherapy [76]. Clinical trial designs translated these pre-clinical results comparatively hesitantly.

5.2.4. PPARγ Agonists Plus Interferon-α

In renal clear cell carcinoma (RCC), IL-6 is a prognostic factor for survival [109]. Vice versa, in anakoinosis-inducing trials, including pioglitazone, C-reactive protein response to anakoinosis-inducing therapy is indicating tumor response [110].

Interferon-α is an approved drug in RCC and acts strongly anti-inflammatory by inducing circulating tumor necrosis factor receptor p55 and mediates a rapid and strong C-reactive protein (CRP) decrease by inhibiting TNFα. RCC is a tumor, producing directly CRP, not only mediated via liver [111].

In a first trial, pioglitazone combined with metronomic chemotherapy and COX-2 inhibitor relatively poor response, mainly stable disease could be observed in > third line situation. The addition of low-dose interferon-α opened the possibility to induce histologically proven remission in resistant metastatic RCC, which translated in continuous complete remission, now lasting > 10 years in single patients [5].

Interestingly, interferon-α is active in renal cell carcinoma, both in combination with retinoids or pioglitazone [5,112,113].

5.2.5. PPARγ Agonists Plus COX-2 Inhibitor

COX-2 inhibition is tightly regulating cellular levels of fatty acids and their derivatives, which are mainly derived from the lipoxygenase and cyclooxygenase pathways. Modulating COX-2 activity influences the local availability of PPAR ligands, therefore indirectly PPAR activity [114].

Inhibiting the canonical Wnt signaling pathway, nonsteroidal anti-inflammatory drugs as well as PPARγ agonists are candidate agents for chemoprevention. Celecoxib suppresses cancer stemness and the progression of hepatocellular carcinoma via activation of PPARγ and up-regulation of PTEN [115]. COX-2 and peroxisome proliferator-activated receptor delta are involved in important growth promoting signaling pathways in human hepatocellular carcinoma [116]. The non-steroidal anti-inflammatory drugs (NSAID)-dependent inhibiton of COX-2 and activation of PPARγ has been shown to suppress cancer stem cells in colon cancer [97]. Celecoxib, for instance, induces up-regulation of PTEN in N1-S1 cells. This process can be enhanced by rosiglitazone. Moreover, it has also been shown that celecoxib increases PPARγ expression and PTEN activity in wild-type and COX-2-deleted Huh7 cells [117]. Concerning the mechanism, within the PTEN promoter, two putative PPARγ binding sites have been identified [96].

Anti-tumor-effects of a cyclooxygenase-2 inhibitor and a peroxisome proliferator-activated receptor γ agonist have been also demonstrated in an in vivo mouse model of spontaneous breast cancer [43,118].

In a series of clinical trials, we used pioglitazone combined with rofecoxib or etoricoxib. From one randomized trial in metastatic melanoma, at least the impact of pioglitazone in addition to COX-2 inhibitor and metronomic low-dose chemotherapy may be delineated. High PPARγ expression in melanoma cells is a favorable prognostic factor for progression-free survival. PPARγ is a late stage predictive marker in metastatic melanoma, and PFS is significantly improved by adding pioglitazone to a pro-anakoinotic schedule, including metronomic low-dose chemotherapy and COX-2 inhibitor (Table 2).

It is not possible to directly estimate the clinical impact of the COX-2 inhibitor from single arm pioglitazone and COX-2 inhibitor, including trials in addition to metronomic low-dose chemotherapy.

5.2.6. PPARγ Agonists and IMiDs

In an animal model, pomalidomide enhances the expression of PPARγ and CCAAT/enhancer binding protein α (C/EBPα), as well as the activity of lipoprotein lipase (LPL) and fatty acid synthetase (FAS). The pro-inflammatory activity of TNFα has the opposite effect on the biochemical indexes and genes that are related to lipid deposition in the liver [119].

Additional experimental data on tumor growth inhibition implicate thalidomide as being involved in the PPARγ pathway. Thalidomide and pomalidomide increase PPARγ protein dose-dependently, also activity of peroxisome proliferator response element [120].

In a clinical trial on multiple myeloma, we successfully used pioglitazone and lenalidomide plus low-dose metronomic chemotherapy and glucocorticoid for rescuing patients following failure of lenalidomide containing regimens in > third line therapy. All of these modulating activities justify for including IMiDs to master modulators of tumor tissue [63].

5.3. PPARγ Agonists in Pro-Anakoinotic Combination Therapy Combined with Targeted Therapy

5.3.1. Pioglitazone and Imatinib

Pioglitazone with imatinib in CML may reduce minimal residual disease. PPARγ agonists target chronic myeloid leukemia (CML) quiescent stem cells in vitro by decreasing transcription of STAT5. A fact that was also shown in multiple myeloma for STAT3. A phase III trial is on-going in France when comparing imatinib versus imatinib plus pioglitazone, as front-line therapy for CML [121].

The addition of imatinib in prostate cancer had no impact on outcome, although there are strong pre-clinical results indicating an impact of imatinib on potentially clinical relevant PDGFR inhibition in prostate cancer [57].

5.3.2. PPARγ and Mechanistic Target of Rapamycin (mTOR) Inhibitor

An additive or synergistic activity of thiazolidinediones and mTOR inhibitors can be suggested from pre-clinical data. Activation of PPARγ by thiazolidinediones leads to inhibition of cell growth and proliferation via key pathways of the Insulin/IGF axis, such as PI3K/mTOR, mitogen-activated protein kinase (MAPK), and GSK3-β/Wnt/β-catenin cascades. This signal pathways regulate cancer cell survival, cell reprogramming, and differentiation [84]. The inhibitory effect of rosiglitazone on non-small cell lung cancer (NSCLC) cell growth was enhanced by the mTOR inhibitor rapamycin. Rosiglitazone, via up-regulation of the PTEN/AMPK and down-regulation of the Akt/mTOR/p70S6K signal cascades, inhibits NSCLC cell proliferation through PPARγ-dependent and PPARγ-independent signals [122].

In refractory Hodgkin disease and MM, an mTOR inhibitor was introduced in addition to pioglitazone, metronomic low-dose chemotherapy, and COX-2 inhibitor; in Hodgkin lymphoma, a glucocorticoid was used, additionally. Metastatic uveal melanomas responded with long-term disease stabilization, improvement of Eastern Cooperative Oncology Group (ECOG) status and resolution of cachexia. In fourth-line PET negative complete remissions were achieved in Hodgkin lymphomas. All patients received allogeneic blood stem cell transplantation in first complete remission [51,66].

Both neoplasia are poorly responding to mTOR inhibitors, only. Thus, also the use of classic targeted therapy, here, the mTOR inhibitor may be repurposed [123].

6. Specific Methodological Aspects of Anakoinosis Inducing Therapies

6.1. Communication Tools

The successful use of pro-anakoinotic therapy approaches gives hints that generally available evolutionary strategies of single cells and tissues may be therapeutically recalled and accessed, particularly in the diseased stage, characterized by transcriptional dysfunctions.

As clinically shown, operating communication tools, including master modulators, evolves therapeutic capacity for biologically 'neutralizing' tumor promoting systems features without blocking tumor-relevant pathways or without targeted elimination of cell compartments of the tumor. The suggestion of communication tools seems to oppose molecular-biologic thinking in networking pathways supposing structures, functions, and hubs with simplistically presumed invariant validity and denotation. However, those classic pathway paradigms disregard that each systems object, also that in a tumor, whatever it will represent physically, a structure, such as a molecule or cell, a function or hub, may be intrinsically or extrinsically, namely therapeutically, nudged by communication derived impulses for context-dependently changing its validity and denotation. Secondly, the identity of structures, functions, or hubs is always communicatively mediated, and necessarily includes and depends on the environmental conditions, functioning as boundary conditions, and integrates the scientific point of view, which is invariably subjected, even if it can be commonly objectively backtraced.

Communication within biologic systems works with the implicit understanding that (1) validity and denotation of systems objects, molecules, cells etc., is always context-dependent, (2) and may be therapeutically redeemed by master modulators via systems-immanent communication tools, which are determined by descriptively accessible communicative systems textures, including inter-systemic exchange processes. The difference between theory, the activity profile of systems participators under invariant 'standard conditions', and practice, the situative evolution-adjusted activity profile, may be bridged by operating communication tools inducing evolutionarily conserved and therapeutically retrievable evolutionary processes (Table 1, Figure 1) [124].

With the introduced paradigmatic changes, the circle can be closed, between multifaceted and contradictory pre-clinical results on the action of PPARγ agonists and unambiguous, reproducible clinical observations resulting from the combined use of master modulators, including NR agonists.

The clinical observations on therapies with master modulators also support experimental data that tumor development and progression is not only a matter of oncogenic events, but of the disease stage, an observation that is also supported by PPARγ expression and predictivity for progression-free survival in metastatic melanoma (Figure 2) [18].

NR agonists develop context and ligand dependent activity profiles. Therapeutic top-down approaches for treating refractory metastatic tumors and hematologic neoplasia indicate that the PPARγ agonists' clinical function may be only deciphered in a combinatory use. Only by introducing several master modulators in therapeutic schedules, including, for example, PPARγ agonists, master modulators develop the capacity for mutually specifying and enhancing response, now up to a clinically relevant level, which can be hardly achieved with mono-therapy, as shown by the missing monoactivity of PPARγ agonists in cancer treatment [12].

Clinical read-outs following combined administration of master modulators are also multifaceted, but reproducible, and they are resulting in diversified, clinically meaningful, palliative care, or response may even disembogue in continuous complete remission. Situative and stage-dependently varying communication features on the respective topographic levels, tumor tissue, tumor-harboring organ, and organism represent the therapeutic counterpart to the diversified context-dependent pre-clinical observations (Figure 1). In case of cachexia, cachexia may be resolved in metastatic melanoma with PPARγ agonist, including schedules with master modulators [51,125].

Importantly, the combined activity profile of PPARγ agonists plus further master modulators is highly specific. However, tumors may either share the communicative systems contexts and therefore, also the therapeutic accessibility towards distinct combinations of master modulators, or may be in the worst case unresponsive, due to the presence of alternative communicative systems contexts, or alternative constitutions of identical hallmarks of cancer.

At this step, individualization of pro-anakoinotic therapy could take place by describing the evolution-adjusted tumor pathophysiology, for example, via serum proteomics and metabolomics [125,126].

6.2. What Is the Appropriate Model System: From Histology to 'Evolution-Adjusted' Tumor Pathophysiology?

The key for uncovering the therapeutic potential of PPARγ agonists is selecting the appropriate combination(s) of master modulators for inducing anakoinosis: Now, anakoinosis is trend setting by establishing a novel therapeutic pillar while overcoming classic obstacles of targeted therapies, such as therapy resistance and (molecular-)genetic tumor heterogeneity.

The clue is that different histologic tumor types share response to distinct combinations of master modulators. That means histologic systematics, in any case reaching its operational limitations in clinics, even by including molecular-pathology and molecular genetics, may be newly unlocked and re-systematized. For this purpose, the systematic specification of tumor-specific communication networks may be adducted, based on the ubiquitously available communication tools, and the evaluation of diversified rationalizations, i.e., physical constitutions of biologic hallmarks, including the hallmarks of cancer. Diversified rationalizations may constitute identical normative notions, for example, rapidly displacing growth of acute leukemias in bone marrow (Table 1). Thus, profound systematics of tumor-specific communication routes, not to be mixed up with the context-independently discussed tumor-promoting pathways, and knowledge about the situative physical constitution of rationalizations results in an 'evolution-adjusted' tumor pathophysiology, which may be prerequisite for the targeted selection of combinations with master modulators (Table 1). By operating anakoinosis in tumors, therapy may cope with the situative relativity of biologic systems, i.e., situative validity and denotation of systems objects in biologic systems. Master modulators may be successfully therapeutically applied for exploiting the possibilities of palliative care and for inducing continuous complete remission.

All communication guiding, validity, and denotation modulating structural, functional tools, including tuning of hubs, are principally therapeutically accessible with pro-anakoinotic drug cocktails, as shown for multiple histologic quite different tumor entities.

6.3. What Is the Appropriate Dosage of Pro-Anakoinotic Therapy?

Single dosages of master modulators, so the postulate, must sufficiently equip the tumor system with reprogramming capacity for attenuating tumor growth. The appropriate regulatory active dosage cannot be pharmacokinetically defined, yet. As clinically indicated, the combination of PPARγ agonists with metronomic chemotherapy facilitates dose-reduction of the cumulative chemotherapy dosage to a quarter or third of the pulsed dose given every three to four weeks without the loss of clinical efficacy.

The two different dose levels of pioglitazone 60 or 45 mg daily seem not to have any impact on response. In addition, patients with both, reduced doses of metronomic chemotherapy, and with dose-reduction of pioglitazone achieved significant clinical response [12].

Within the combined pro-anakoinotic therapy schedules, typical, but modest side effects, can be attributed to the administration of pioglitazone [57,58]. Peripheral edema Grade I to II occurred in 52.4 to 58.5%, including few Grade III toxicities in hepatocellular carcinoma due to pre-existing liver disease. Renal failure Grade I to II was observed in 13.2% of patients with hepatocellular carcinoma. Adverse events leading to dose adjustment or temporary interruption of therapy in the prostate cancer trial occurred in 13.8%, permanent discontinuation in 1.2%. In hepatocellular carcinoma, dose adjustments of pioglitazone (starting dose 60 mg daily) were performed in 33% of patients, no permanent discontinuation.

6.4. Pro-Anakoinotic Therapy Schedules: Indications and Diagnostics

Up-to-now, anakoinosis inducing therapies, including pioglitazone, have been administered in metastatic and refractory cancer and hematologic malignancies. From the results of these trials (Table 2) a proof of principle can be delineated, namely activity of anakoinosis inducing therapies in poor-risk patient populations. First steps in the direction of combining classic targeted therapy with anakoinosis-inducing schedules were successful (Table 2). When considering tumor response as a timely multi-step biologic approach, during those reprogramming and classic targeted steps are repetitively and/or simultaneously necessary for inducing long-term tumor response, differential valuable clinical endpoints may be biologically accessible, such as induction of biologic memory, 'active' long-term chronification of tumor disease, or, in the best case, continuous complete remission. Anakoinosis inducing therapies could be perspectival integrated in ideal manner in classic targeted approaches. Classic targeted therapies may be even repurposed with many possible implications for additional clinical approvals [123].

Monitoring of such anakoinosis-inducing therapies must be completely reorganized in comparison to targeted therapies, where the availability of the target on the tumor or stroma cell is used as indicator for possible tumor response. Now, multiple parameter analysis derived from proteome and metabolome analytics from serum or plasma might be helpful, before and during therapy.

7. Conclusions

A long way of failures accompanied the introduction of PPARγ agonists in tumor therapy. In contrast to NRs that are activated by hormones and the prompt incipient activity of hormones, adopted NRs have an intrinsic tissue- and stage-dependent pro-anakoinotic activity profile, which is pointed in tumor tissues with their severe dysregulation of transcription factors. However, PPARγ agonists are clinically irrelevant, as far as, for example, PPARγ agonists are used in mono-therapy. With respect to pioglitazone, the tumor systems related activity profile may be exclusively focused and up graded to a clinically meaningful range by introducing additional NR agonists, as shown for glucocorticoids, all-trans retinoic acid, or the transcriptional modulator interferon-α, or more generally, by adding master modulators. Under conditions of concerted activity of master modulators, it should be generally possible to elaborate and adopt combinations of master modulators inducing response in metastatic, refractory neoplasia, irrespective of the histologic origin.

Thus, with induction of anakoinosis, a novel therapy pillar may be introduced providing several advantages compared to classic targeted therapies:

Anakoinotic processes may cope with fundamental obstacles of classic targeted therapies, with tumor heterogeneity and poor risk parameters, with context-dependent validity and denotation of tumor-promoting aberrations and targets, with drug resistance or undruggable targets by targeting dynamic evolutionary processes, for example, multifaceted biologic steps that are necessary for establishing 'active' long-term tumor control or continuous complete remission [12]. Pro-anakoinotic therapies may inhibit further metastatic progression in case of metastatic disease (Table 1, Figure 1) [127].

Auspicious 'personalized' tumor therapy is now supplemented by a novel treatment methodology, which is at its beginnings but multifaceted adaptable to tumor systems stages. Importantly, metastatic tumors of quite different histologic origin may share communication features and may be reprogrammed with identical combinations of master modulators operating communication tools. Thus, in future, an evolution-adjusted tumor pathophysiology could be the driving force for specifying combinations of NR agonists and antagonists. Studies on proteomics and metabolomics in serum and plasma will provide new information on on-going systems changes induced by pro-anakoinotic therapy approaches.

A randomized trial introducing in the experimental arm metronomic low-dose chemotherapy, a selective inhibitor of the enzyme steroid-17α-hydroxylase (CYP17A1), which catalysis steps in the testosterone and estrogen biosynthesis, and dual transcriptional modulation with glucocorticoid and pioglitazone is ongoing in castration-resistant prostate cancer. In a second on-going trial, pioglitazone and all-trans retinoic acid are combined with azacitidine in the experimental arm for treating refractory acute myelocytic leukemia.

Author Contributions: D.H. and A.R. wrote the manuscript. M.R., F.L., M.V., T.P., W.H., L.G., and C.G. gave critical comments, read and approved the final version of the manuscript.

Funding: This research received no external funding.

Acknowledgments: The authors thank the patients for participating in the trials and the investigators for their contributions.

Conflicts of Interest: The authors declare no conflict of interest.

References

1. Gou, Q.; Gong, X.; Jin, J.; Shi, J.; Hou, Y. Peroxisome proliferator-activated receptors (PPARs) are potential drug targets for cancer therapy. *Oncotarget* **2017**, *8*, 60704–60709. [CrossRef] [PubMed]
2. Dormandy, J.A.; Charbonnel, B.; Eckland, D.J.A.; Erdmann, E.; Massi-Benedetti, M.; Moules, I.K.; Skene, A.M.; Tan, M.H.; Lefèbvre, P.J.; Murray, G.D.; et al. Secondary prevention of macrovascular events in patients with type 2 diabetes in the PROactive Study (PROspective pioglitAzone Clinical Trial In macroVascular Events): A randomised controlled trial. *Lancet* **2005**, *366*, 1279–1289. [CrossRef]
3. Nissen, S.E.; Wolski, K. Effect of rosiglitazone on the risk of myocardial infarction and death from cardiovascular causes. *N. Engl. J. Med.* **2007**, *356*, 2457–2471. [CrossRef] [PubMed]
4. Walter, I.; Schulz, U.; Vogelhuber, M.; Wiedmann, K.; Endlicher, E.; Klebl, F.; Andreesen, R.; Herr, W.; Ghibelli, L.; Hackl, C.; et al. Communicative reprogramming non-curative hepatocellular carcinoma with low-dose metronomic chemotherapy, COX-2 inhibitor and PPAR-γ agonist: A phase II trial. *Med. Oncol.* **2017**, *34*, 192. [CrossRef] [PubMed]
5. Walter, B.; Schrettenbrunner, I.; Vogelhuber, M.; Grassinger, J.; Bross, K.; Wilke, J.; Suedhoff, T.; Berand, A.; Wieland, W.F.; Rogenhofer, S.; et al. Pioglitazone, etoricoxib, interferon-α, and metronomic capecitabine for metastatic renal cell carcinoma: Final results of a prospective phase II trial. *Med. Oncol.* **2012**, *29*, 799–805. [CrossRef] [PubMed]
6. Dhiman, V.K.; Bolt, M.J.; White, K.P. Nuclear receptors in cancer—Uncovering new and evolving roles through genomic analysis. *Nat. Rev. Genet.* **2018**, *19*, 160–174. [CrossRef] [PubMed]
7. Capper, C.P.; Rae, J.M.; Auchus, R.J. The metabolism, analysis, and targeting of steroid hormones in breast and prostate cancer. *Horm. Cancer* **2016**, *7*, 149–164. [CrossRef] [PubMed]
8. Kfir-Erenfeld, S.; Yefenof, E. Non-genomic events determining the sensitivity of hemopoietic malignancies to glucocorticoid-induced apoptosis. *Cancer Immunol. Immunother.* **2014**, *63*, 37–43. [CrossRef] [PubMed]
9. Sionov, R.V.; Spokoini, R.; Kfir, R.S.; Cohen, O.; Yefenof, E. Mechanisms regulating the susceptibility of hematopoietic malignancies to glucocorticoid-induced apoptosis. *Adv. Cancer Res.* **2008**, *101*, 127–248. [PubMed]
10. Photiou, L.; van der Weyden, C.; McCormack, C.; Miles Prince, H. Systemic treatment options for advanced-stage mycosis fungoides and sézary syndrome. *Curr. Oncol. Rep.* **2018**, *20*, 32. [CrossRef] [PubMed]
11. Platzbecker, U.; Avvisati, G.; Cicconi, L.; Thiede, C.; Paoloni, F.; Vignetti, M.; Ferrara, F.; Divona, M.; Albano, F.; Efficace, F.; et al. Improved Outcomes With Retinoic Acid and Arsenic Trioxide Compared With Retinoic Acid and Chemotherapy in Non-High-Risk Acute Promyelocytic Leukemia: Final Results of the Randomized Italian-German APL0406 Trial. *J. Clin. Oncol.* **2017**, *35*, 605–612. [CrossRef] [PubMed]
12. Hart, C.; Vogelhuber, M.; Wolff, D.; Klobuch, S.; Ghibelli, L.; Foell, J.; Corbacioglu, S.; Rehe, K.; Haegeman, G.; Thomas, S.; et al. Anakoinosis: Communicative Reprogramming of Tumor Systems—for Rescuing from Chemorefractory Neoplasia. *Cancer Microenviron.* **2015**, *8*, 75–92. [CrossRef] [PubMed]
13. Reichle, A.; Hildebrandt, G.C. Principles of modular tumor therapy. *Cancer Microenviron.* **2009**, *2*, 227–237. [CrossRef] [PubMed]

14. Bradner, J.E.; Hnisz, D.; Young, R.A. Transcriptional Addiction in Cancer. *Cell* **2017**, *168*, 629–643. [CrossRef] [PubMed]
15. Vallée, A.; Lecarpentier, Y. Crosstalk Between Peroxisome Proliferator-Activated Receptor γ and the Canonical WNT/β-Catenin Pathway in Chronic Inflammation and Oxidative Stress During Carcinogenesis. *Front. Immunol.* **2018**, *9*, 745. [CrossRef] [PubMed]
16. Michalik, L.; Desvergne, B.; Wahli, W. Peroxisome-proliferator-activated receptors and cancers: Complex stories. *Nat. Rev. Cancer* **2004**, *4*, 61–70. [CrossRef] [PubMed]
17. Kersten, S.; Desvergne, B.; Wahli, W. Roles of PPARs in health and disease. *Nature* **2000**, *405*, 421–424. [CrossRef] [PubMed]
18. Reichle, A. (Ed.) *From Molecular to Modular Tumor Therapy*; Springer: Dordrecht, The Netherlands, 2010.
19. Tan, C.K.; Zhuang, Y.; Wahli, W. Synthetic and natural Peroxisome Proliferator-Activated Receptor (PPAR) agonists as candidates for the therapy of the metabolic syndrome. *Exp. Opin. Ther. Targets* **2017**, *21*, 333–348. [CrossRef] [PubMed]
20. Bosscher, K. Selective Glucocorticoid Receptor modulators. *J. Steroid Biochem. Mol. Biol.* **2010**, *120*, 96–104. [CrossRef] [PubMed]
21. Dasgupta, S.; Lonard, D.M.; O'Malley, B.W. Nuclear receptor coactivators: Master regulators of human health and disease. *Ann. Rev. Med.* **2014**, *65*, 279–292. [CrossRef] [PubMed]
22. Akbiyik, F.; Ray, D.M.; Gettings, K.F.; Blumberg, N.; Francis, C.W.; Phipps, R.P. Human bone marrow megakaryocytes and platelets express PPARγ, and PPARγ agonists blunt platelet release of CD40 ligand and thromboxanes. *Blood* **2004**, *104*, 1361–1368. [CrossRef] [PubMed]
23. Lin, C.; Yang, L.; Tanasa, B.; Hutt, K.; Ju, B.-g.; Ohgi, K.; Zhang, J.; Rose, D.W.; Fu, X.-D.; Glass, C.K.; et al. Nuclear receptor-induced chromosomal proximity and DNA breaks underlie specific translocations in cancer. *Cell* **2009**, *139*, 1069–1083. [CrossRef] [PubMed]
24. Koeffler, H.P. Peroxisome proliferator-activated receptor γ and cancers. *Clin. Cancer Res.* **2003**, *9*, 1–9. [PubMed]
25. Danielian, P.S.; White, R.; Lees, J.A.; Parker, M.G. Identification of a conserved region required for hormone dependent transcriptional activation by steroid hormone receptors. *EMBO J.* **1992**, *11*, 1025–1033. [CrossRef] [PubMed]
26. Sorrentino, G.; Ruggeri, N.; Zannini, A.; Ingallina, E.; Bertolio, R.; Marotta, C.; Neri, C.; Cappuzzello, E.; Forcato, M.; Rosato, A.; et al. Glucocorticoid receptor signalling activates YAP in breast cancer. *Nat. Commun.* **2017**, *8*, 14073. [CrossRef] [PubMed]
27. Katoh, M.; Katoh, M. WNT signaling pathway and stem cell signaling network. *Clin. Cancer Res.* **2007**, *13*, 4042–4045. [CrossRef] [PubMed]
28. Michael, M.S.; Badr, M.Z.; Badawi, A.F. Inhibition of cyclooxygenase-2 and activation of peroxisome proliferator-activated receptor-γ synergistically induces apoptosis and inhibits growth of human breast cancer cells. *Int. J. Mol. Med.* **2003**, *11*, 733–736. [CrossRef] [PubMed]
29. Lehmann, J.M.; Lenhard, J.M.; Oliver, B.B.; Ringold, G.M.; Kliewer, S.A. Peroxisome proliferator-activated receptors α and γ are activated by indomethacin and other non-steroidal anti-inflammatory drugs. *J. Biol. Chem.* **1997**, *272*, 3406–3410. [CrossRef] [PubMed]
30. Gelman, L.; Fruchart, J.C.; Auwerx, J. An update on the mechanisms of action of the peroxisome proliferator-activated receptors (PPARs) and their roles in inflammation and cancer. *Cell. Mol. life Sci.* **1999**, *55*, 932–943. [CrossRef] [PubMed]
31. Subbaramaiah, K.; Lin, D.T.; Hart, J.C.; Dannenberg, A.J. Peroxisome proliferator-activated receptor γ ligands suppress the transcriptional activation of cyclooxygenase-2. Evidence for involvement of activator protein-1 and CREB-binding protein/p300. *J. Biol. Chem.* **2001**, *276*, 12440–12448. [CrossRef] [PubMed]
32. Badawi, A.F.; Badr, M.Z. Expression of cyclooxygenase-2 and peroxisome proliferator-activated receptor-γ and levels of prostaglandin E2 and 15-deoxy-delta12,14-prostaglandin J2 in human breast cancer and metastasis. *Int. J. Cancer* **2003**, *103*, 84–90. [CrossRef] [PubMed]
33. Lee, C.J.; Han, J.S.; Seo, C.Y.; Park, T.H.; Kwon, H.C.; Jeong, J.S.; Kim, I.H.; Yun, J.; Bae, Y.S.; Kwak, J.Y.; et al. Pioglitazone, a synthetic ligand for PPARγ, induces apoptosis in RB-deficient human colorectal cancer cells. *Apoptosis* **2006**, *11*, 401–411. [CrossRef] [PubMed]

34. Clay, C.E.; Namen, A.M.; Atsumi, G.; Willingham, M.C.; High, K.P.; Kute, T.E.; Trimboli, A.J.; Fonteh, A.N.; Dawson, P.A.; Chilton, F.H. Influence of J series prostaglandins on apoptosis and tumorigenesis of breast cancer cells. *Carcinogenesis* **1999**, *20*, 1905–1911. [CrossRef] [PubMed]
35. Nagahara, T.; Okano, J.; Murawaki, Y. Mechanisms of anti-proliferative effect of JTE-522, a selective cyclooxygenase-2 inhibitor, on human liver cancer cells. *Oncol. Rep.* **2007**, *18*, 1281–1290. [CrossRef] [PubMed]
36. Meyer, S.; Vogt, T.; Landthaler, M.; Berand, A.; Reichle, A.; Bataille, F.; Marx, A.H.; Menz, A.; Hartmann, A.; Kunz-Schughart, L.A.; et al. Cyclooxygenase 2 (COX2) and peroxisome proliferator-activated receptor γ (PPARG) are stage-dependent prognostic markers of malignant melanoma. *PPAR Res.* **2009**, *2009*, 848645. [PubMed]
37. Bundscherer, A.; Reichle, A.; Hafner, C.; Meyer, S.; Vogt, T. Targeting the tumor stroma with peroxisome proliferator activated receptor (PPAR) agonists. *ACAMC* **2009**, *9*, 816–821. [CrossRef]
38. Knower, K.C.; Chand, A.L.; Eriksson, N.; Takagi, K.; Miki, Y.; Sasano, H.; Visvader, J.E.; Lindeman, G.J.; Funder, J.W.; Fuller, P.J.; et al. Distinct nuclear receptor expression in stroma adjacent to breast tumors. *Breast Cancer Res. Treat.* **2013**, *142*, 211–223. [CrossRef] [PubMed]
39. Basu-Roy, U.; Han, E.; Rattanakorn, K.; Gadi, A.; Verma, N.; Maurizi, G.; Gunaratne, P.H.; Coarfa, C.; Kennedy, O.D.; Garabedian, M.J.; et al. PPARγ agonists promote differentiation of cancer stem cells by restraining YAP transcriptional activity. *Oncotarget* **2016**, *7*, 60954–60970. [CrossRef] [PubMed]
40. Mulholland, D.J.; Dedhar, S.; Coetzee, G.A.; Nelson, C.C. Interaction of nuclear receptors with the Wnt/β-catenin/Tcf signaling axis: Wnt you like to know? *Endocr. Rev.* **2005**, *26*, 898–915. [CrossRef] [PubMed]
41. Vallée, A.; Lecarpentier, Y.; Guillevin, R.; Vallée, J.-N. Opposite Interplay Between the Canonical WNT/β-Catenin Pathway and PPARγ: A Potential Therapeutic Target in Gliomas. *Neurosci. Bull.* **2018**, *34*, 573–588. [CrossRef] [PubMed]
42. Maniati, E.; Bossard, M.; Cook, N.; Candido, J.B.; Emami-Shahri, N.; Nedospasov, S.A.; Balkwill, F.R.; Tuveson, D.A.; Hagemann, T. Crosstalk between the canonical NF-κB and Notch signaling pathways inhibits Pparγ expression and promotes pancreatic cancer progression in mice. *J. Clin. Investig.* **2011**, *121*, 4685–4699. [CrossRef] [PubMed]
43. Hong, O.-Y.; Youn, H.J.; Jang, H.-Y.; Jung, S.H.; Noh, E.-M.; Chae, H.S.; Jeong, Y.-J.; Kim, W.; Kim, C.-H.; Kim, J.-S. Troglitazone inhibits matrix metalloproteinase-9 expression and invasion of breast cancer cell through a peroxisome proliferator-activated receptor γ-dependent mechanism. *J. Breast Cancer* **2018**, *21*, 28–36. [CrossRef] [PubMed]
44. Ge, Y.; Domschke, C.; Stoiber, N.; Schott, S.; Heil, J.; Rom, J.; Blumenstein, M.; Thum, J.; Sohn, C.; Schneeweiss, A.; et al. Metronomic cyclophosphamide treatment in metastasized breast cancer patients: Immunological effects and clinical outcome. *Cancer Immunol. Immunother.* **2012**, *61*, 353–362. [CrossRef] [PubMed]
45. Tontonoz, P.; Singer, S.; Forman, B.M.; Sarraf, P.; Fletcher, J.A.; Fletcher, C.D.; Brun, R.P.; Mueller, E.; Altiok, S.; Oppenheim, H.; et al. Terminal differentiation of human liposarcoma cells induced by ligands for peroxisome proliferator-activated receptor γ and the retinoid X receptor. *Proc. Natl. Acad. Sci. USA* **1997**, *94*, 237–241. [CrossRef] [PubMed]
46. Demetri, G.D.; Fletcher, C.D.M.; Mueller, E.; Sarraf, P.; Naujoks, R.; Campbell, N.; Spiegelman, B.M.; Singer, S. Induction of solid tumor differentiation by the peroxisome proliferator-activated receptor-γ ligand troglitazone in patients with liposarcoma. *Proc. Natl. Acad. Sci. USA* **1999**, *96*, 3951–3956. [CrossRef] [PubMed]
47. Debrock, G.; Vanhentenrijk, V.; Sciot, R.; Debiec-Rychter, M.; Oyen, R.; van Oosterom, A. A phase II trial with rosiglitazone in liposarcoma patients. *Br. J. Cancer* **2003**, *89*, 1409–1412. [CrossRef] [PubMed]
48. Coras, B.; Hafner, C.; Reichle, A.; Hohenleutner, U.; Szeimies, R.-M.; Landthaler, M.; Vogt, T. Antiangiogenic therapy with pioglitazone, rofecoxib, and trofosfamide in a patient with endemic kaposi sarcoma. *Arch. Dermatol.* **2004**, *140*, 1504–1507. [CrossRef] [PubMed]
49. Vogt, T.; Hafner, C.; Bross, K.; Bataille, F.; Jauch, K.-W.; Berand, A.; Landthaler, M.; Andreesen, R.; Reichle, A. Antiangiogenetic therapy with pioglitazone, rofecoxib, and metronomic trofosfamide in patients with advanced malignant vascular tumors. *Cancer* **2003**, *98*, 2251–2256. [CrossRef] [PubMed]

50. Burstein, H.J.; Demetri, G.D.; Mueller, E.; Sarraf, P.; Spiegelman, B.M.; Winer, E.P. Use of the peroxisome proliferator-activated receptor (PPAR) γ ligand troglitazone as treatment for refractory breast cancer: A phase II study. *Breast Cancer Res. Treat.* **2003**, *79*, 391–397. [CrossRef] [PubMed]
51. Hart, C.; Vogelhuber, M.; Hafner, C.; Landthaler, M.; Berneburg, M.; Haferkamp, S.; Herr, W.; Reichle, A. Biomodulatory metronomic therapy in stage IV melanoma is well-tolerated and may induce prolonged progression-free survival, a phase I trial. *J. Eur. Acad. Dermatol. Venereol.* **2016**, *30*, e119–e121. [CrossRef] [PubMed]
52. Reichle, A.; Vogt, T.; Coras, B.; Terheyden, P.; Neuber, K.; Trefzer, U.; Schultz, E.; Berand, A.; Bröcker, E.B.; Landthaler, M.; et al. Targeted combined anti-inflammatory and angiostatic therapy in advanced melanoma: A randomized phase II trial. *Melanoma Res.* **2007**, *17*, 360–364. [CrossRef] [PubMed]
53. Kulke, M.H.; Demetri, G.D.; Sharpless, N.E.; Ryan, D.P.; Shivdasani, R.; Clark, J.S.; Spiegelman, B.M.; Kim, H.; Mayer, R.J.; Fuchs, C.S. A phase II study of troglitazone, an activator of the PPARγ receptor, in patients with chemotherapy-resistant metastatic colorectal cancer. *Cancer J.* **2002**, *8*, 395–399. [CrossRef] [PubMed]
54. Reichle, A.; Grassinger, J.; Bross, K.; Wilke, J.; Suedhoff, T.; Walter, B.; Wieland, W.-F.; Berand, A.; Andreesen, R. C-reactive protein in patients with metastatic clear cell renal carcinoma: An important biomarker for tumor-associated inflammation. *Biomark. Insights* **2007**, *1*, 87–98. [CrossRef] [PubMed]
55. Mueller, E.; Smith, M.; Sarraf, P.; Kroll, T.; Aiyer, A.; Kaufman, D.S.; Oh, W.; Demetri, G.; Figg, W.D.; Zhou, X.P.; et al. Effects of ligand activation of peroxisome proliferator-activated receptor γ in human prostate cancer. *Proc. Natl. Acad. Sci. USA* **2000**, *97*, 10990–10995. [CrossRef] [PubMed]
56. Vogelhuber, M.; Feyerabend, S.; Stenzl, A.; Suedhoff, T.; Schulze, M.; Huebner, J.; Oberneder, R.; Wieland, W.; Mueller, S.; Eichhorn, F.; et al. Biomodulatory treatment of patients with castration-resistant prostate cancer: A phase II study of imatinib with pioglitazone, etoricoxib, dexamethasone and low-dose treosulfan. *Cancer Microenviron.* **2015**, *8*, 33–41. [CrossRef] [PubMed]
57. Walter, B.; Rogenhofer, S.; Vogelhuber, M.; Berand, A.; Wieland, W.F.; Andreesen, R.; Reichle, A. Modular therapy approach in metastatic castration-refractory prostate cancer. *World J. Urol.* **2010**, *28*, 745–750. [CrossRef] [PubMed]
58. Vogt, T.; Coras, B.; Hafner, C.; Landthaler, M.; Reichle, A. Antiangiogenic therapy in metastatic prostate carcinoma complicated by cutaneous lupus erythematodes. *Lancet Oncol.* **2006**, *7*, 695–697. [CrossRef]
59. Smith, M.R.; Manola, J.; Kaufman, D.S.; George, D.; Oh, W.K.; Mueller, E.; Slovin, S.; Spiegelman, B.; Small, E.; Kantoff, P.W. Rosiglitazone versus placebo for men with prostate carcinoma and a rising serum prostate-specific antigen level after radical prostatectomy and/or radiation therapy. *Cancer* **2004**, *101*, 1569–1574. [CrossRef] [PubMed]
60. Reichle, A.; Lugner, A.; Ott, C.; Klebl, F.; Vogelhuber, M.; Berand, A.; Andreesen, R. Control of cancer-associated inflammation and survival: Results from a prospective randomized phase II trial in gastric cancer. *J. Clin. Oncol.* **2009**, *27*, e15584.
61. Hau, P.; Kunz-Schughart, L.; Bogdahn, U.; Baumgart, U.; Hirschmann, B.; Weimann, E.; Muhleisen, H.; Ruemmele, P.; Steinbrecher, A.; Reichle, A. Low-dose chemotherapy in combination with COX-2 inhibitors and PPAR-γ agonists in recurrent high-grade gliomas—A phase II study. *Oncology* **2007**, *73*, 21–25. [CrossRef] [PubMed]
62. Reichle, A.; Hart, C.; Grube, M.; Andreesen, R. Anti-inflammatory, immuno-modulatory and angiostatic treatment as third-line therapy for multiple myeloma (MM)—A combined treatment setting of lenalidomide with pioglitazone, dexamethasone and low-dose treosulfan (phase I/II). *Blood* **2012**, *120*, 5029.
63. Heudobler, D.; Rehe, K.; Foell, J.; Corbacioglu, S.; Hildebrandt, G.; Herr, W.; Reichle, A.; Vogelhuber, M. Biomodulatory metronomic therapy shows remarkable activity in chemorefractory multi-system langerhans cell histiocytosis. *Blood* **2016**, *128*, 4254.
64. Reichle, A.; Vogt, T.; Kunz-Schughart, L.; Bretschneider, T.; Bachthaler, M.; Bross, K.; Freund, S.; Andreesen, R. Anti-inflammatory and angiostatic therapy in chemorefractory multisystem Langerhans' cell histiocytosis of adults. *Br. J. Haematol.* **2005**, *128*, 730–732. [CrossRef] [PubMed]
65. Ugocsai, P.; Wolff, D.; Menhart, K.; Hellwig, D.; Holler, E.; Herr, W.; Reichle, A. Biomodulatory metronomic therapy induces PET-negative remission in chemo- and brentuximab-refractory Hodgkin lymphoma. *Br. J. Haematol.* **2016**, *172*, 290–293. [CrossRef] [PubMed]

66. Prost, S.; Relouzat, F.; Spentchian, M.; Ouzegdouh, Y.; Saliba, J.; Massonnet, G.; Beressi, J.-P.; Verhoeyen, E.; Raggueneau, V.; Maneglier, B.; et al. Erosion of the chronic myeloid leukaemia stem cell pool by PPARγ agonists. *Nature* **2015**, *525*, 380–383. [CrossRef] [PubMed]
67. Thomas, S.; Schelker, R.; Klobuch, S.; Zaiss, S.; Troppmann, M.; Rehli, M.; Haferlach, T.; Herr, W.; Reichle, A. Biomodulatory therapy induces complete molecular remission in chemorefractory acute myeloid leukemia. *Haematologica* **2015**, *100*, 4–6. [CrossRef] [PubMed]
68. Heudobler, D.; Klobuch, S.; Thomas, S.; Hahn, J.; Herr, W.; Reichle, A. Cutaneous leukemic infiltrates successfully treated with biomodulatory therapy in a rare case of therapy-related high risk MDS/AML. *Front. Pharmacol.* **2018**. [CrossRef]
69. Simkens, L.H.J.; van Tinteren, H.; May, A.; Tije, A.J.; Creemers, G.-J.M.; Loosveld, O.J.L.; Jongh, F.E.; Erdkamp, F.L.G.; van der Torren, A.M.; Tol, J. Maintenance treatment with capecitabine and bevacizumab in metastatic colorectal cancer (CAIRO3): A phase 3 randomised controlled trial of the Dutch Colorectal Cancer Group. *Lancet* **2015**, *385*, 1843–1852. [CrossRef]
70. Pramanik, R.; Agarwala, S.; Gupta, Y.K.; Thulkar, S.; Vishnubhatla, S.; Batra, A.; Dhawan, D.; Bakhshi, S. Metronomic Chemotherapy vs Best Supportive Care in Progressive Pediatric Solid Malignant Tumors: A Randomized Clinical Trial. *JAMA Oncol.* **2017**, *3*, 1222–1227. [CrossRef] [PubMed]
71. Rochlitz, C.; Bigler, M.; Moos, R.; Bernhard, J.; Matter-Walstra, K.; Wicki, A.; Zaman, K.; Anchisi, S.; Küng, M.; Na, K.-J.; et al. SAKK 24/09: Safety and tolerability of bevacizumab plus paclitaxel vs. bevacizumab plus metronomic cyclophosphamide and capecitabine as first-line therapy in patients with HER2-negative advanced stage breast cancer—A multicenter, randomized phase III trial. *BMC Cancer* **2016**, *16*, 780. [CrossRef] [PubMed]
72. Kummar, S.; Wade, J.L.; Oza, A.M.; Sullivan, D.; Chen, A.P.; Gandara, D.R.; Ji, J.; Kinders, R.J.; Wang, L.; Allen, D.; et al. Randomized phase II trial of cyclophosphamide and the oral poly (ADP-ribose) polymerase inhibitor veliparib in patients with recurrent, advanced triple-negative breast cancer. *Investig. New Drugs* **2016**, *34*, 355–363. [CrossRef] [PubMed]
73. Bottini, A.; Generali, D.; Brizzi, M.P.; Fox, S.B.; Bersiga, A.; Bonardi, S.; Allevi, G.; Aguggini, S.; Bodini, G.; Milani, M.; et al. Randomized phase II trial of letrozole and letrozole plus low-dose metronomic oral cyclophosphamide as primary systemic treatment in elderly breast cancer patients. *J. Clin. Oncol.* **2006**, *24*, 3623–3628. [CrossRef] [PubMed]
74. Clarke, J.L.; Iwamoto, F.M.; Sul, J.; Panageas, K.; Lassman, A.B.; DeAngelis, L.M.; Hormigo, A.; Nolan, C.P.; Gavrilovic, I.; Karimi, S.; et al. Randomized phase II trial of chemoradiotherapy followed by either dose-dense or metronomic temozolomide for newly diagnosed glioblastoma. *J. Clin. Oncol.* **2009**, *27*, 3861–3867. [CrossRef] [PubMed]
75. Senerchia, A.A.; Macedo, C.R.; Ferman, S.; Scopinaro, M.; Cacciavillano, W.; Boldrini, E.; Lins de Moraes, V.L.; Rey, G.; Oliveira, C.T.; Castillo, L.; et al. Results of a randomized, prospective clinical trial evaluating metronomic chemotherapy in nonmetastatic patients with high-grade, operable osteosarcomas of the extremities: A report from the Latin American Group of Osteosarcoma Treatment. *Cancer* **2017**, *123*, 1003–1010. [CrossRef] [PubMed]
76. Konopleva, M.; Andreeff, M. Role of peroxisome proliferator-activated receptor-γ in hematologic malignancies. *Curr. Opin. Hematol.* **2002**, *9*, 294–302. [CrossRef] [PubMed]
77. Elrod, H.A.; Sun, S.-Y. PPARγ and Apoptosis in Cancer. *PPAR Res.* **2008**, *2008*, 704165. [CrossRef] [PubMed]
78. Nemenoff, R.A.; Winn, R.A. Role of nuclear receptors in lung tumourigenesis. *Eur. J. Cancer* **2005**, *41*, 2561–2568. [CrossRef] [PubMed]
79. Rumi, M.A.K.; Ishihara, S.; Kazumori, H.; Kadowaki, Y.; Kinoshita, Y. Can PPARγ ligands be used in cancer therapy? *Curr. Med. Chem.* **2004**, *4*, 465–477.
80. Schmidt, M.V.; Brüne, B.; Knethen, A. The nuclear hormone receptor PPARγ as a therapeutic target in major diseases. *Sci. World J.* **2010**, *10*, 2181–2197. [CrossRef] [PubMed]
81. Youssef, J.; Badr, M. Peroxisome proliferator-activated receptors and cancer: Challenges and opportunities. *Br. J. Haematol.* **2011**, *164*, 68–82. [CrossRef] [PubMed]
82. Skelhorne-Gross, G.; Nicol, C.J.B. The Key to Unlocking the Chemotherapeutic Potential of PPARγ Ligands: Having the Right Combination. *PPAR Res.* **2012**, *2012*, 946943. [CrossRef] [PubMed]

83. Polvani, S.; Tarocchi, M.; Tempesti, S.; Bencini, L.; Galli, A. Peroxisome proliferator activated receptors at the crossroad of obesity, diabetes, and pancreatic cancer. *World J. Gastroenterol.* **2016**, *22*, 2441–2459. [CrossRef] [PubMed]
84. Vella, V.; Nicolosi, M.L.; Giuliano, S.; Bellomo, M.; Belfiore, A.; Malaguarnera, R. PPAR-γ Agonists As Antineoplastic Agents in Cancers with Dysregulated IGF Axis. *Front. Endocrinol.* **2017**, *8*, 31. [CrossRef] [PubMed]
85. Querfeld, C.; Nagelli, L.V.; Rosen, S.T.; Kuzel, T.M.; Guitart, J. Bexarotene in the treatment of cutaneous T-cell lymphoma. *Exp. Opin. Pharmacother.* **2006**, *7*, 907–915. [CrossRef] [PubMed]
86. Fröhlich, E.; Wahl, R. Chemotherapy and chemoprevention by thiazolidinediones. *BioMed Res. Int.* **2015**, *2015*, 845340. [CrossRef] [PubMed]
87. Higurashi, T.; Hosono, K.; Takahashi, H.; Komiya, Y.; Umezawa, S.; Sakai, E.; Uchiyama, T.; Taniguchi, L.; Hata, Y.; Uchiyama, S.; et al. Metformin for chemoprevention of metachronous colorectal adenoma or polyps in post-polypectomy patients without diabetes: A multicentre double-blind, placebo-controlled, randomised phase 3 trial. *Lancet Oncol.* **2016**, *17*, 475–483. [CrossRef]
88. Coyle, C.; Cafferty, F.H.; Vale, C.; Langley, R.E. Metformin as an adjuvant treatment for cancer: A systematic review and meta-analysis. *Ann. Oncol.* **2016**, *27*, 2184–2195. [CrossRef] [PubMed]
89. Di, W.; Di, H.; Chen, H.; Shi, G.; Fetahu, I.S.; Wu, F.; Rabidou, K.; Fang, R.; Tan, L.; Xu, S.; et al. Glucose-regulated phosphorylation of TET2 by AMPK reveals a pathway linking diabetes to cancer. *Nature* **2018**, *559*, 637–641. [CrossRef] [PubMed]
90. Di Masi, A.; Leboffe, L.; Marinis, E.; Pagano, F.; Cicconi, L.; Rochette-Egly, C.; Lo-Coco, F.; Ascenzi, P.; Nervi, C. Retinoic acid receptors: From molecular mechanisms to cancer therapy. *Mol. Aspects Med.* **2015**, *41*, 1–115. [CrossRef] [PubMed]
91. Mitchell, C.D.; Richards, S.M.; Kinsey, S.E.; Lilleyman, J.; Vora, A.; Eden, T.O.B. Benefit of dexamethasone compared with prednisolone for childhood acute lymphoblastic leukaemia: Results of the UK Medical Research Council ALL97 randomized trial. *Br. J. Haematol.* **2005**, *129*, 734–745. [CrossRef] [PubMed]
92. McDermott, D.F.; Regan, M.M.; Clark, J.I.; Flaherty, L.E.; Weiss, G.R.; Logan, T.F.; Kirkwood, J.M.; Gordon, M.S.; Sosman, J.A.; Ernstoff, M.S.; et al. Randomized phase III trial of high-dose interleukin-2 versus subcutaneous interleukin-2 and interferon in patients with metastatic renal cell carcinoma. *J. Clin. Oncol.* **2005**, *23*, 133–141. [CrossRef] [PubMed]
93. Brown, S.; Pineda, C.M.; Xin, T.; Boucher, J.; Suozzi, K.C.; Park, S.; Matte-Martone, C.; Gonzalez, D.G.; Rytlewski, J.; Beronja, S.; et al. Correction of aberrant growth preserves tissue homeostasis. *Nature* **2017**, *548*, 334–337. [CrossRef] [PubMed]
94. Tang, T.C.; Man, S.; Xu, P.; Francia, G.; Hashimoto, K.; Emmenegger, U.; Kerbel, R.S. Development of a resistance-like phenotype to sorafenib by human hepatocellular carcinoma cells is reversible and can be delayed by metronomic UFT chemotherapy. *Neoplasia* **2010**, *12*, 928–940. [CrossRef] [PubMed]
95. Cao, L.; Wang, X.; Wang, Q.; Xue, P.; Jiao, X.; Peng, H.; Lu, H.; Zheng, Q.; Chen, X.; Huang, X.; et al. Rosiglitazone sensitizes hepatocellular carcinoma cell lines to 5-fluorouracil antitumor activity through activation of the PPARγ signaling pathway. *Acta Pharmacol. Sin.* **2009**, *30*, 1316–1322. [CrossRef] [PubMed]
96. Cao, L.-q.; Chen, X.-l.; Wang, Q.; Huang, X.-h.; Zhen, M.-C.; Zhang, L.-J.; Li, W.; Bi, J. Upregulation of PTEN involved in rosiglitazone-induced apoptosis in human hepatocellular carcinoma cells. *Acta Pharmacol. Sin.* **2007**, *28*, 879–887. [CrossRef] [PubMed]
97. Yap, R.; Veliceasa, D.; Emmenegger, U.; Kerbel, R.S.; McKay, L.M.; Henkin, J.; Volpert, O.V. Metronomic low-dose chemotherapy boosts CD95-dependent antiangiogenic effect of the thrombospondin peptide ABT-510: A complementation antiangiogenic strategy. *Clin. Cancer Res.* **2005**, *11*, 6678–6685. [CrossRef] [PubMed]
98. Biziota, E.; Briasoulis, E.; Mavroeidis, L.; Marselos, M.; Harris, A.L.; Pappas, P. Cellular and molecular effects of metronomic vinorelbine and 4-O-deacetylvinorelbine on human umbilical vein endothelial cells. *Anti-Cancer Drugs* **2016**, *27*, 216–224. [CrossRef] [PubMed]
99. Ogawa, S.; Lozach, J.; Benner, C.; Pascual, G.; Tangirala, R.K.; Westin, S.; Hoffmann, A.; Subramaniam, S.; David, M.; Rosenfeld, M.G.; et al. Molecular determinants of crosstalk between nuclear receptors and toll-like receptors. *Cell* **2005**, *122*, 707–721. [CrossRef] [PubMed]
100. Mitsiades, C.S.; Mitsiades, N.; Richardson, P.G.; Treon, S.P.; Anderson, K.C. Novel biologically based therapies for Waldenstrom's macroglobulinemia. *Semin. Oncol.* **2003**, *30*, 309–312. [CrossRef] [PubMed]

101. Hisatake, J.I.; Ikezoe, T.; Carey, M.; Holden, S.; Tomoyasu, S.; Koeffler, H.P. Down-Regulation of prostate-specific antigen expression by ligands for peroxisome proliferator-activated receptor γ in human prostate cancer. *Cancer Res.* **2000**, *60*, 5494–5498. [PubMed]
102. Narayanan, S.; Srinivas, S.; Feldman, D. Androgen-glucocorticoid interactions in the era of novel prostate cancer therapy. *Nat. Rev. Urol.* **2016**, *13*, 47–60. [CrossRef] [PubMed]
103. Mansour, M.; Schwartz, D.; Judd, R.; Akingbemi, B.; Braden, T.; Morrison, E.; Dennis, J.; Bartol, F.; Hazi, A.; Napier, I.; et al. Thiazolidinediones/PPARγ agonists and fatty acid synthase inhibitors as an experimental combination therapy for prostate cancer. *Int. J. Oncol.* **2011**, *38*, 537–546. [CrossRef] [PubMed]
104. Papi, A.; Guarnieri, T.; Storci, G.; Santini, D.; Ceccarelli, C.; Taffurelli, M.; Carolis, S.; Avenia, N.; Sanguinetti, A.; Sidoni, A.; et al. Nuclear receptors agonists exert opposing effects on the inflammation dependent survival of breast cancer stem cells. *Cell Death Differ.* **2012**, *19*, 1208–1219. [CrossRef] [PubMed]
105. Papi, A.; Rocchi, P.; Ferreri, A.M.; Orlandi, M. RXRγ and PPARγ ligands in combination to inhibit proliferation and invasiveness in colon cancer cells. *Cancer Lett.* **2010**, *297*, 65–74. [CrossRef] [PubMed]
106. Papi, A.; Storci, G.; Guarnieri, T.; Carolis, S.; Bertoni, S.; Avenia, N.; Sanguinetti, A.; Sidoni, A.; Santini, D.; Ceccarelli, C.; et al. Peroxisome proliferator activated receptor-α/hypoxia inducible factor-1α interplay sustains carbonic anhydrase IX and apoliprotein E expression in breast cancer stem cells. *PLoS ONE* **2013**, *8*, e54968. [CrossRef] [PubMed]
107. Papi, A.; Tatenhorst, L.; Terwel, D.; Hermes, M.; Kummer, M.P.; Orlandi, M.; Heneka, M.T. PPARγ and RXRγ ligands act synergistically as potent antineoplastic agents in vitro and in vivo glioma models. *J. Neurochem.* **2009**, *109*, 1779–1790. [CrossRef] [PubMed]
108. Konopleva, M.; Elstner, E.; McQueen, T.J.; Tsao, T.; Sudarikov, A.; Hu, W.; Schober, W.D.; Wang, R.-Y.; Chism, D.; Kornblau, S.M.; et al. Peroxisome proliferator-activated receptor γ and retinoid X receptor ligands are potent inducers of differentiation and apoptosis in leukemias. *Mol. Cancer Ther.* **2004**, *3*, 1249–1262. [PubMed]
109. Thiounn, N.; Pages, F.; Flam, T.; Tartour, E.; Mosseri, V.; Zerbib, M.; Beuzeboc, P.; Deneux, L.; Fridman, W.H.; Debré, B. IL-6 is a survival prognostic factor in renal cell carcinoma. *Immunol. Lett.* **1997**, *58*, 121–124. [CrossRef]
110. Tilg, H.; Vogel, W.; Dinarello, C.A. Interferon-α induces circulating tumor necrosis factor receptor p55 in humans. *Blood* **1995**, *85*, 433–435. [CrossRef]
111. Jabs, W.J.; Busse, M.; Krüger, S.; Jocham, D.; Steinhoff, J.; Doehn, C. Expression of C-reactive protein by renal cell carcinomas and unaffected surrounding renal tissue. *Kidney Int.* **2005**, *68*, 2103–2110. [CrossRef] [PubMed]
112. Buer, J.; Probst, M.; Ganser, A.; Atzpodien, J. Response to 13-cis-retinoic acid plus interferon alfa-2a in two patients with therapy-refractory advanced renal cell carcinoma. *J. Clin. Oncol.* **1995**, *13*, 2679–2680. [CrossRef] [PubMed]
113. Aviles, A.; Neri, N.; Fernandez-Diez, J.; Silva, L.; Nambo, M.-J. Interferon and low doses of methotrexate versus interferon and retinoids in the treatment of refractory/relapsed cutaneous T-cell lymphoma. *Hematology* **2015**, *20*, 538–542. [CrossRef] [PubMed]
114. Finch, E.R.; Tukaramrao, D.B.; Goodfield, L.L.; Quickel, M.D.; Paulson, R.F.; Prabhu, K.S. Activation of PPARγ by endogenous prostaglandin J2 mediates the antileukemic effect of selenium in murine leukemia. *Blood* **2017**, *129*, 1802–1810. [CrossRef] [PubMed]
115. Chu, T.-H.; Chan, H.-H.; Kuo, H.-M.; Liu, L.-F.; Hu, T.-H.; Sun, C.-K.; Kung, M.-L.; Lin, S.-W.; Wang, E.-M.; Ma, Y.-L.; et al. Celecoxib suppresses hepatoma stemness and progression by up-regulating PTEN. *Oncotarget* **2014**, *5*, 1475–1490. [CrossRef] [PubMed]
116. Xu, L.; Han, C.; Lim, K.; Wu, T. Cross-talk between peroxisome proliferator-activated receptor delta and cytosolic phospholipase A(2)α/cyclooxygenase-2/prostaglandin E(2) signaling pathways in human hepatocellular carcinoma cells. *Cancer Res.* **2006**, *66*, 11859–11868. [CrossRef] [PubMed]
117. Zheng, Z.; Zhou, L.; Gao, S.; Yang, Z.; Yao, J.; Zheng, S. Prognostic role of C-reactive protein in hepatocellular carcinoma: A systematic review and meta-analysis. *Int. J. Med. Sci.* **2013**, *10*, 653–664. [CrossRef] [PubMed]
118. Mustafa, A.; Kruger, W.D. Suppression of tumor formation by a cyclooxygenase-2 inhibitor and a peroxisome proliferator-activated receptor γ agonist in an in vivo mouse model of spontaneous breast cancer. *Clin. Cancer Res.* **2008**, *14*, 4935–4942. [CrossRef] [PubMed]

119. Liu, D.; Mai, K.; Zhang, Y.; Xu, W.; Ai, Q. Tumour necrosis factor-α inhibits hepatic lipid deposition through GSK-3β/β-catenin signaling in juvenile turbot (*Scophthalmus maximus* L.). *Gen. Comp. Endocrinol.* **2015**, *228*, 1–8. [CrossRef] [PubMed]
120. DeCicco, K.L.; Tanaka, T.; Andreola, F.; Luca, L.M. The effect of thalidomide on non-small cell lung cancer (NSCLC) cell lines: Possible involvement in the PPARγ pathway. *Carcinogenesis* **2004**, *25*, 1805–1812. [CrossRef] [PubMed]
121. Rousselot, P.; Prost, S.; Guilhot, J.; Roy, L.; Etienne, G.; Legros, L.; Charbonnier, A.; Coiteux, V.; Cony-Makhoul, P.; Huguet, F.; et al. Pioglitazone together with imatinib in chronic myeloid leukemia: A proof of concept study. *Cancer* **2017**, *123*, 1791–1799. [CrossRef] [PubMed]
122. Han, S.; Roman, J. Rosiglitazone suppresses human lung carcinoma cell growth through PPARγ-dependent and PPARγ-independent signal pathways. *Mol. Cancer Ther.* **2006**, *5*, 430–437. [CrossRef] [PubMed]
123. Hafner, C.; Reichle, A.; Vogt, T. New indications for established drugs: Combined tumor-stroma-targeted cancer therapy with PPARγ agonists, COX-2 inhibitors, mTOR antagonists and metronomic chemotherapy. *CCDT* **2005**, *5*, 393–419. [CrossRef]
124. Reichle, A. *Evolution-adjusted Tumor Pathophysiology*; Springer: Dordrecht, The Netherlands, 2013.
125. Muqaku, B.; Eisinger, M.; Meier, S.M.; Tahir, A.; Pukrop, T.; Haferkamp, S.; Slany, A.; Reichle, A.; Gerner, C. Multi-omics Analysis of Serum Samples Demonstrates Reprogramming of Organ Functions Via Systemic Calcium Mobilization and Platelet Activation in Metastatic Melanoma. *Mol. Cell. Proteom.* **2017**, *16*, 86–99. [CrossRef] [PubMed]
126. Hanash, S.M.; Pitteri, S.J.; Faca, V.M. Mining the plasma proteome for cancer biomarkers. *Nature* **2008**, *452*, 571–579. [CrossRef] [PubMed]
127. Reichle, A. Tumor Systems Need to be Rendered Usable for a New Action-Theoretical Abstraction: The Starting Point for Novel Therapeutic Options. In *From Molecular to Modular Tumor Therapy*; Reichle, A., Ed.; Springer Netherlands: Dordrecht, The Netherlands, 2010; pp. 9–28.

International Journal of
Molecular Sciences

Review

The Involvement of PPARs in the Peculiar Energetic Metabolism of Tumor Cells

Andrea Antonosante [1], Michele d'Angelo [1], Vanessa Castelli [1], Mariano Catanesi [1], Dalila Iannotta [1], Antonio Giordano [2,3], Rodolfo Ippoliti [1], Elisabetta Benedetti [1] and Annamaria Cimini [1,2,4,*]

1 Department of Life, Health and Environmental Sciences, University of L'Aquila, 67100 L'Aquila, Italy; andrea.antonosante@gmail.com (A.A.); dangelo-michele@hotmail.com (M.d.); castelli.vane@gmail.com (V.C.); Mariano.catanesi86@gmail.com (M.C.); iannottadalila@gmail.com (D.I.); rodolfo.ippoliti@univaq.it (R.I.); elisabetta.benedetti@univaq.it (E.B.)
2 Sbarro Institute for Cancer Research and Molecular Medicine, Department of Biology, Temple University, Philadelphia, PA 19122, USA; giordano12@unisi.it
3 Department of Medicine, Surgery and Neuroscience, University of Siena, 53100 Siena, Italy
4 National Institute for Nuclear Physics (INFN), Gran Sasso National Laboratory (LNGS), 67100 Assergi, Italy
* Correspondence: annamaria.cimini@univaq.it; Tel.: +39-0862-433289

Received: 30 April 2018; Accepted: 24 June 2018; Published: 29 June 2018

Abstract: Energy homeostasis is crucial for cell fate, since all cellular activities are strongly dependent on the balance between catabolic and anabolic pathways. In particular, the modulation of metabolic and energetic pathways in cancer cells has been discussed in some reports, but subsequently has been neglected for a long time. Meanwhile, over the past 20 years, a recovery of the study regarding cancer metabolism has led to an increasing consideration of metabolic alterations in tumors. Cancer cells must adapt their metabolism to meet their energetic and biosynthetic demands, which are associated with the rapid growth of the primary tumor and colonization of distinct metastatic sites. Cancer cells are largely dependent on aerobic glycolysis for their energy production, but are also associated with increased fatty acid synthesis and increased rates of glutamine consumption. In fact, emerging evidence has shown that therapeutic resistance to cancer treatment may arise from the deregulation of glucose metabolism, fatty acid synthesis, and glutamine consumption. Cancer cells exhibit a series of metabolic alterations induced by mutations that lead to a gain-of-function of oncogenes, and a loss-of-function of tumor suppressor genes, including increased glucose consumption, reduced mitochondrial respiration, an increase of reactive oxygen species, and cell death resistance; all of these are responsible for cancer progression. Cholesterol metabolism is also altered in cancer cells and supports uncontrolled cell growth. In this context, we discuss the roles of peroxisome proliferator-activated receptors (PPARs), which are master regulators of cellular energetic metabolism in the deregulation of the energetic homeostasis, which is observed in cancer. We highlight the different roles of PPAR isotypes and the differential control of their transcription in various cancer cells.

Keywords: nuclear receptors; energy metabolism; cancer metabolism

---

1. Introduction

Mammalian cellular activities require a significant energy source, which is produced by specific mechanisms involved in the regulation of cellular energy homeostasis. The correct balance between catabolic and anabolic pathways strongly influence cellular fate, since they are involved in biochemical reactions that drive ATP (adenosine triphosphate) production/consumption. Oxidative glucose metabolism by OXPHOS (oxidative phosphorylation) produces up to 36 ATP per mole of glucose, whereas non-oxidative glucose metabolism by glycolysis results in two ATP per mole of glucose [1]. Hence, oxygen availability provides an optimal cellular condition to produce high levels of energy, while hypoxia determines a less efficient cellular condition in which the cells prefer to use glycolysis to produce energy. Another way to meet cellular energy demands is lipid metabolism by the peroxisomal and mitochondrial β-oxidation of fatty acids (FAs), which provides energy in the form of redox potential [2,3]. Regarding lipid metabolism, many cell types present cytosolic lipid deposits, also called lipid droplets (LDs). These are dynamic organelles that contain triacylglycerols (TAGs) and cholesteryl esters, and present several functions such as reducing lipotoxicity, lipid storage, and lipid metabolism, and they are directly involved in cellular physiology [4–7]. Unlike normal cells, cancer cells exhibit uncontrolled proliferation that needs energy metabolism adjustments in order to ensure their cell growth and division. The high proliferation rate in tumor cells leads to significant metabolic changes that are closely related to the environmental conditions and genetic/epigenetic characteristics of the tissue from which tumor arises. To safeguard their survival, cancer cells metabolically switch from less efficient energy pathways to higher performing energy pathways in order to cope with the considerable energy demands of tumor bulk. Meanwhile, neoplastic cells show altered glucose and lipid metabolism in association with unstable OXPHOS and glutamine metabolism; accordingly, PPARs play a key role in regulating these metabolic switch events. Therefore, our purpose in this review is to describe recent observations concerning the pivotal role of PPARs in promoting or preventing the characteristic metabolic switch that provides the energy for tumor survival. The main metabolic mechanisms adopted by tumor cells that are under the control of PPARs will be briefly described below.

1.1. Glucose Metabolism and OXPHOS in Cancer Cells

Although in normoxia, healthy cells use the degradation of glucose to pyruvate and later the TCA (tricarboxylic acid) cycle to produce ATP, neoplastic cells prefer to use glycolysis to produce energy rather than oxidative phosphorylation. The first observation of this phenomenon is about 88 years old, when Otto Warburg noticed that tumor cells switch toward a glycolytic metabolism with high lactate production, even in aerobic conditions, and mitochondrial metabolism suppression. This metabolic adaptation is called "aerobic glycolysis" or the "Warburg effect" [8]. Despite aerobic glycolysis not being influenced by oxygen levels, in hypoxic conditions, tumor cells present an overexpression of the genes involved in the glycolytic pathway. Usually, in solid tumors, near the core, there is a hypoxic area, and this hypoxic environment supports glycolytic metabolism and provides chemotherapy resistance as well as an optimal niche for the maintenance of CSCs (cancer stem cells) [9–11]. It was also observed that many types of cancer cells (glioma, hepatoma, and breast) are able to obtain ATP from OXPHOS, and they can pass from a fermentative to an oxidative metabolism and vice versa, and glucose is directly involved in this switch [12–15]. On the other hand, tumor cells can perform a glucose-dependent suppression of mitochondrial respiration, which is called the "Crabtree effect" [16]. This effect is reversible and collaborates with the "Warburg effect" to ensure cancer cell survival, independently from the presence of oxygen [17]. In a recent study, using a mathematical computational model, Epstein and collaborators [18] explored the coexistence between glycolytic and oxidative pathways in cancer cells, starting from the assumption that cancer cells quickly need ATP, but at the same time need to maintain baseline levels of ATP, mainly during moments of apparent standby. Consequently, in relation to fluctuating energy demands and assuming that tumor cells exist in a heterogeneous environment, they can use a glycolytic pathway to produce ATP quickly in short-term energy requests; conversely, baseline levels of energy are obtained through OXPHOS [18]. In addition, although lactate is a waste product of aerobic glycolysis, it is recycled by subpopulations of cancer cells and directed toward the TCA cycle [19]. These evidences lead to the belief that there is cooperation between different types of cells within the tumor, which could be a key mechanism for tumor progression. Several genes involved in the glycolytic pathway regulate the adjustment of cancer cells to the metabolic switch; some of them are oncogenes. Among them, PI3K/Akt signaling induces the expression of proteins related to glucose transport (GLUTs) in association with high hexokinase II (HKII) activity. HKII is able to bind to the voltage-dependent anion channel (VDAC) on the outer mitochondrial membrane to protect cells from apoptosis [20,21]. Moreover, altered c-Myc (cancer-myelocytomatosis) regulation affects the expression of the genes that are related to glutamine metabolism and aerobic glycolysis (HKII, lactate dehydrogenase (LDH), pyruvate kinase isoenzyme M2 (PKM2), phosphofructokinase 1 (PFK1), and GLUT1) [21,22], and PKM2 plays a central role in the shift of cellular metabolism to aerobic glycolysis in cancer cells. PKM2 is the specific isoform that is mainly expressed in tumor cells [23]. Whereas PKM1 is a constitutively active tetrameric enzyme, the 22 amino acid differences in PKM2 create a fructose 1,6-bisphosphate (FBP) binding pocket that renders it dependent on the allosteric binding of FBP for formation of an active tetramer. PKM2 activity is more flexible than PKM1 activity, which is why PKM2 is more suitable to guarantee the metabolic switch in cancer cells. In addition, PKM2 presents a low activity index, probably allowing the storage of glycolytic metabolites to ensure macromolecule biosynthesis [24]. In this context, the hypoxic environment provides an additional incentive to trigger the transcription of genes linked to the Warburg effect, and they are directly under the transcriptional control of hypoxia inducible factor-1α (HIF-1α) [25]. However, aerobic glycolysis is also essential for the macromolecule biosynthesis, in order to provide the structural components for cell proliferation. An increased flux of pyruvate provides the carbon source for the anabolic process, such as the de novo synthesis of nucleotides, lipids, and proteins. At the same time, the synthesis of macromolecules in cancer cells is necessary to produce reducing equivalents, such as NADH (nicotinamide adenine dinucleotide H) and NADPH (nicotinamide adenine dinucleotide phosphate H); in turn, they are essential for ensuring glucose metabolism, biosynthesis, and the degradation of macromolecules [26].

1.2. Lipid, Cholesterol, and Glutamine Metabolism in Cancer Cells

Fatty acids synthesis is typically reactivated in cancer cells by the upregulation of lipogenic enzymes to provide monomeric components for membrane building, lipid signaling, and post-translational protein modification [27]. Breast and prostate cancer show an increased expression of fatty acids synthase (FAS) and enzymes involved in the elongation of very long-chain fatty acids such as ELOVL1-7 (elongation of very long chain fatty acids protein 1-7) [28,29]. The stability and fluidity of cellular membranes are cholesterol-dependent, and lipid rafts (which are involved in the regulation of intracellular transduction signals) are mainly composed of cholesterol [30]. Furthermore, the mevalonate pathway (MVA), which is responsible for cholesterol synthesis, is linked to the production of intermediates that are crucial for post-translational modifications of Rho, Ras, and other small GTPase (isoprenylation, farnesylation, and genarylation) [31]. Interestingly, statins, which are drugs that are used to decrease plasma cholesterol levels in hypercholesterolemic conditions, inhibit HMG-CoA (3-hydroxy-3-methyl-glutaryl-coenzime A) reductase (HMGCR); this is the rate-limiting step of MVA. In support of the lipid importance in tumors, it was demonstrated that statins are able to decrease the proliferative index in breast cancer and acute myeloid leukemia cells, and make colorectal cancer cells more sensitive to chemotherapy [32–34]. Moreover, prostate cancer cells showed high levels of cholesterol [35]. The excess quantity of LDs in cancer cells are further evidence that FAs and cholesterol accumulate in many types of cancer. Label-free Raman spectroscopy imaging of high-grade prostate cancer and metastasis revealed the accumulation of abnormal LDs associated with PTEN (phosphatase and tensin homolog) loss and PI3K/Akt activation [36]; similar evidences were observed in the breast cancer cell line [37] and colon cancer stem cells [38]. Meanwhile, in gliomas, a higher amount of LDs was directly proportional to the degree of tumor aggressiveness [39]. As previously mentioned, FAs derived from free triacylglycerides or intracellular deposits can be metabolized to produce energy in the form of redox fuel. The early phases of this process occur in cytoplasm (triglyceride and monoacylglycerol lipases), and the late phases occur in mitochondria and are called fatty acid β-oxidation (FAO), but can occur also in the lumen of peroxisomes. The end products of lipid decomposition, such as NADH, $FADH_2$, and acetyl-CoA, are directed toward the TCA cycle; this is the reason why some non-glycolytic cancers, such as prostate cancer and large B-cell lymphoma, need FAO to meet their energetic demands [40–43]. In spite of this, even some glycolytic tumors, under certain conditions require FAO to produce energy [43], while in glioblastoma, FAO contributes to protect the cells from oxidative stress by upregulation of detoxification enzymes, such as glutathione (GSH) [44]. Unlike aerobic glycolysis, where a hypoxic condition increases glucose utilization, lipid biosynthesis is not encouraged by oxygen lack, resulting in lipid accumulation into LDs [45,46]. In this scenario, the carbon source to synthesize lipid compounds is supplied by glutamine; isocitrate dehydrogenase-1 (IDH1) activity releases citrate in the cytosol after carboxylation of glutamine-derived α-ketoglutarate [47–49]. Moreover, Ras oncogene together with hypoxia induces the pyruvate dehydrogenase kinase 1 (PDK1), which in turn inhibits pyruvate dehydrogenase (PDH) and forces cells to implement glutamine-dependent anaplerotic behavior [45,48,50]. This phenomenon restores the TCA cycle under specific conditions and highlights the key role of glutamine metabolism in cancer cell growth. Beyond anaplerotic involvement, glutamine catabolism provides nitrogen to synthetize the nucleotide glutathione, resulting in the major energy source in some transformed cells [51]. Cancer cells rely on glutamine uptake to ensure a further pathway to support their accelerated metabolism. Gao et al. [22] showed that c-Myc stimulates glutaminase (GLS) expression through the suppression of miR-23a/b, while the inhibition of Rho-GTPase by a small compound determines the reduction of glutaminase activity, which is dependent of NFκB (nuclear factor kappa-light-chain-enhancer of activated B cells) in breast cancer and B lymphoma cells [52]. In addition, DeBerardinis and collaborators [53] observed that glioblastoma cells performed aerobic glycolysis associated with elevated glutamine catabolism to obtain redox energy and TCA cycle intermediates in order to support biosynthetic activity, mainly FAs. Interestingly, highly invasive ovarian cancer cells showed more remarkable glutamine dependence than low-invasive ovarian cancer cells; this feature is related to glutamine-mediated STAT3 (signal transducer and activator of transcription 3) modulation [54].

2. PPARs

Peroxisome proliferator-activated receptors (PPARs) are ligand-activated transcription factors belonging to the nuclear hormone receptor superfamily. PPARα (NR1C1) was the first described as the receptor mediating peroxisome proliferation in rodent hepatocytes in 1990; later, two related isotypes, PPARβ/δ (NR1C2) and PPARγ (NR1C3) were found and characterized. PPARα is mainly expressed in tissues presenting high fatty acid catabolism activity, such as the liver, the heart, the brown adipose tissue, the kidney, and the intestine; it is also involved in regulating lipoprotein synthesis. Regarding PPARγ, there are two isoforms: γ1 and γ2, which are obtained by alternative splicing. Both isoforms act in the white and brown adipose tissue to promote adipocyte differentiation and lipid storage, while only PPARγ1 is expressed in other tissues, such as the gut or immune cells. PPARγ transcriptional targets are also involved in regulating inflammatory processes, the cell cycle, and glucose metabolism by improving insulin sensitivity; in fact, it is a useful target for type 2 diabetes therapy. PPARβ/δ is ubiquitously expressed, and it has important functions in the skeletal muscle, adipose tissue, skin, gut, and the brain, including fatty acid oxidation regulation, keratinocyte differentiation, and wound healing [55–58].

Ordinarily, PPARs are active at transcriptional levels only in presence of their specific ligands, and each ligand is able to trigger a specific PPARs response; conversely, some findings demonstrated the basal activity of PPARs in the absence of ligands [59]. Unlike the steroid hormone receptors (nuclear receptors class 1) that function as homodimers, PPARs (nuclear receptors class 2) are active when they heterodimerize with retinoid x receptors (RXR); then, each monomer binds a specific DNA sequence, called PPREs (peroxisomes proliferator response elements). PPREs are direct repetitions that are located in the promoter region of the target gene as single or multiple copies [58,60,61]. As mentioned above, specific PPARs transcriptional activities are strictly related to lipid ligand type, and consequently, a wide range of natural or synthetic lipids can bind to the LBD (ligand-binding domain) of PPARs. These ligands can be obtained from diet or intracellular signaling pathways, among which FAs from prostaglandins and leukotrienes, as well as synthetic ligands, are described. Peculiar fatty acid-binding proteins (FABPs) allow the ligand delivery toward the nucleus, where

PPARs reside [62]. Long-chain unsaturated FAs, eicosanoids, and hypolipidemic drugs (fibrates) can activate the PPARα, while thiazolidinediones (TZDs) are able to active PPARγ and increase insulin sensitivity. In this regard, PPARs are considered important therapeutic targets, mainly for metabolic diseases [63,64]. Given their role as master regulators of cellular energy pathways and considering the metabolic alterations in tumor cells, PPARs' modulation can be involved in the specific metabolic plan undertaken by neoplastic cells. The central debate is whether the transcriptional activity of PPARs promotes or hinders tumorigenesis and tumor progression. To date, research activity has yielded conflicting evidence in this regard. There are several evidences about the tumor suppression role of PPARα and PPARγ [65–71], but there are also several evidences about their cancer promotion activity [72–74]; instead, regarding PPARβ/δ, the majority of study conducted shows its oncogenic role [75–77]. Although PPARs can have a dual role that is oncogenic as well as oncosuppressive, their behavior is severely influenced by the tissue type from which the tumor arises and by the tumor microenvironment.

2.1. *PPARα and Cancer Metabolism*

The process of tumorigenesis can be described by a series of molecular features, among which the alteration of cellular metabolism has recently emerged. This metabolic rewiring fulfills the energy and biosynthetic demands of fast proliferating cancer cells and amplifies their metabolic reserves to survive and proliferate in the poorly oxygenated and nutrient-deprived tumor microenvironment. In these harsh environmental conditions, the deregulation of glucose and glutamine metabolism, alterations of lipid synthesis and FAO, and a complex rewiring of mitochondrial and peroxisomal function are required. However, mitochondria and peroxisomes display close relationships; in fact, it was recently reported that in glioblastoma, an increase of peroxisomes leads to the increase of mitochondria [78].

PPARα mainly regulates the gene expression of specific proteins that are involved in mitochondrial and peroxisomal functions, such as fatty acids' β-oxidation, glucose metabolism, and fatty acid transport [56,58]. The relationship between gene transcription regulated by PPARα and tumor metabolism can determine oncogenic or oncosuppressive effects. PPARα activation and tumor suppression was reported in melanoma [79] and glioblastoma [80]; on the other hand, PPARα activation demonstrated a positive role in stimulating the proliferation of breast and renal carcinoma cell lines [81,82], while *PPARα*-null mice were insensitive to hepatic carcinogenesis induced by PPARα agonist [83].

Several evidences support the paradigm that tumors originate from cancer stem cells and/or cancer stem progenitor cells, namely, tumor initiating cells or cancer stem-like cells (CSCs). CSCs represent a small population of cancer cells that exhibit self-renewal and differentiation features similar to normal stem cells, although they differ in the regulation of their self-renewal pathways. Based on the CSC presence, they are responsible for tumor formation, progression, metastasis, and relapse, as well as drug resistance. Even if it is generally known that tumor cells, particularly CSCs, show glucose and lipid metabolism alterations, the specific metabolic pathways and their regulation are still poorly understood [11]. Due to their crucial roles in energetic metabolism, the PPARs have been investigated by many authors regarding their involvement in tumorigenesis, showing an upregulation of the α isotype in several tumors and CSCs.

Recently, we demonstrated decreased tumor proliferation with an alteration of glucose and lipid tumor metabolism, antagonizing PPARα by synthetic ligand (GW6471) in glioblastoma stem cells (GSCs) [84]. GSCs are responsible for drug resistance and relapse; they reside in intratumoral perivascular and necrotic/hypoxic niches, which provide the GSCs with the optimal environment to keep their stemness features. Hypoxia is associated with glioblastoma progression and plays a crucial role in stem cells' biology. HIF proteins regulate the cellular response to hypoxia or variable oxygen concentrations by upregulating the genes related to tumor progression, angiogenesis, drug resistance, and the phenotype maintenance of GSCs. Between HIF proteins, HIF-1α triggers the expression of genes related to tumor metabolic switch, which in turn induces glucose uptake, glycolytic enzyme activity, lactate production, and the anaerobic production of ATP. However, it is also able to control synthetic pathways (fatty acids and glycogen synthesis), stimulating the expression of anabolic enzymes, which are related to glucose–glycogen conversion [11,85,86]. In addition, we demonstrated that glioblastoma and GSCs in hypoxic condition show higher levels of PPARα compared with the normoxic condition [87], while PPARγ levels are downregulated under hypoxia [84]. In GSCs, glycogen storage appeared more abundant in hypoxia than in normoxia, since hypoxic cells need glucose to quickly produce ATP through glycolysis, and the glycogen storages are essential to maintain this fast energetic process. Moreover, HIF-1α stimulates the expression of genes involved in glycogen synthesis, as glycogen synthase kinase 3β (GSK3β). When GSK3β is phosphorylated at Ser 9, it is then inactive and unable to phosphorylate glycogen synthase, thus allowing the start of the anabolic process [88].

GW6471 treatment decreased the viability of GSCs, the number and size of neurospheres, and induced apoptosis, which was associated with low glycogen supplies. Increasing glycogen degradation was due to the upregulation of glycogen phosphorylase (GPBB) and downregulation of phosphorylated GSK3β at Ser 9. Furthermore, a decreased amount of GLUT3 and glucose uptake in hypoxic-treated GSCs have been reported. Moreover, it was also demonstrated higher amounts of LDs in cancer cells, mainly in the hypoxic environment, which is in line with previous evidence [21,39,87,89]. FABP7 (fatty acid binding protein 7) transports the fatty acids toward the nucleus; in the same way, it supplies the LDs to promote tumor growth [90,91], and it appears increased by hypoxia. GW6471 treatment induced the loss of LDs amounts, cholesterol supply, and the transcriptional activity of genes encoding for mevalonate pathway enzymes. However, FABP7 levels appeared decreased only in antagonist-treated hypoxic GSCs, since the inhibition of PPARα transcriptional activity in hypoxic GSCs adversely affects fatty acid and cholesterol amounts. The MVA pathway plays a central role in glioblastoma survival; besides, its inhibition by PPARα antagonist is linked to cell death and tumor suppression. This effect is similar to the downregulation of the MVA pathway together with the upregulation of PPARγ induced by statins. These results seem to emphasize the key role of PPARα in the metabolic switch that occurs in cancer hypoxic cells, such as GSGs. In harsh environmental conditions PPARα was upregulated and could result in metabolic directives to ensure energy for tumor cells. In this regard, the antagonist GW6471 was able to reduce the synthetic processes, such as glycogen synthesis and LDs biogenesis, which normally ensure fast-acting energy for cancer cells (as summarized in Figure 1A).

In another study, Abu Aboud et al. [82] used the same PPARα antagonist (GW6471) to treat two cell lines (Caki-1 and 786-O cell line) of renal cell carcinoma (RCC). They observed that PPARα levels were higher in high-grade RCC tissue compared with low-grade tissue, linking PPARα protein levels to RCC aggressiveness. High-grade RCC presents more energy demands than low-grade RCC, and therefore requires active fatty acid oxidation (FAO), which is regulated upstream by PPARα [92]. Both the antagonist and siRNA directed against the PPARα showed the capability of reducing c-Myc protein levels, which is likely by PPARα-mediated alteration of oncoprotein stabilization. This event was associated with the downregulation of cyclin D1/CDK4 and the G1/S transition block with cell cycle arrest in G0/G1 phase [93,94]. The authors hypothesized that the transcriptional activity of PPARα was inhibited, which was the reason why the renal carcinoma cells were unable to use FAO by converging on glycolysis to obtain energy. In fact, GW6471 effects were more pronounced in media with low glucose concentrations than media with normal glucose concentrations. Furthermore, 2-Deoxy-D-glucose (2-DG), which is an inhibitor of glycolysis, acted in synergy with GW6471 to induce tumor death. Regarding that, by blocking PPARα in the RCC cell line, the researchers demonstrated the reduction of cell viability with a marked reduction of c-Myc, cyclin D1, and CDK4 protein levels in synergy with glycolysis inhibition (as shown in Figure 1A).

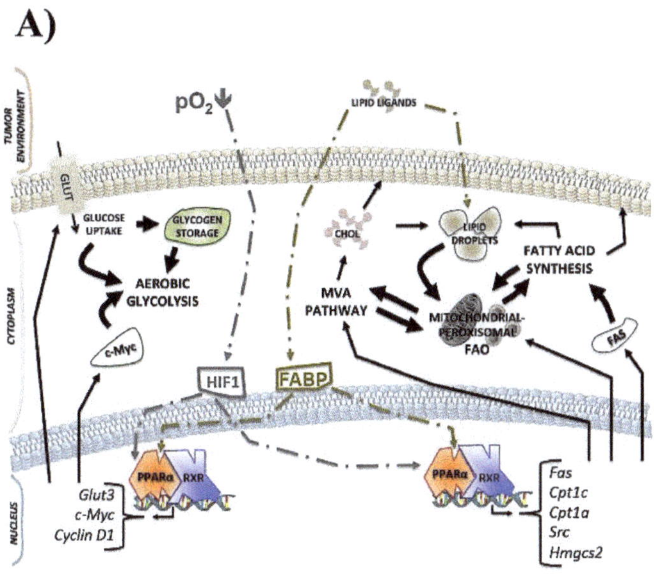

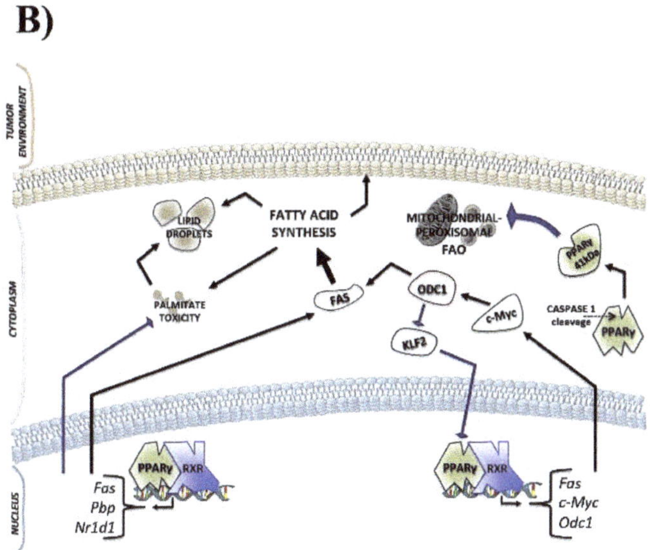

Figure 1. *Cont.*

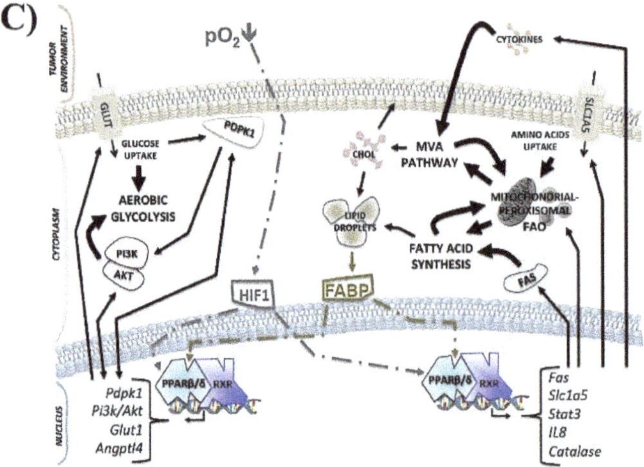

Figure 1. Schematic representation of PPARs-dependent oncogenic metabolic pathways highlighted in this review. The representation concerns the metabolic mechanisms that are activated/inhibited in tumor cells under the transcriptional control of PPARs. These hypotheses of molecular mechanisms are based on evidence obtained by different cancer types. For each PPAR isotype, the specific activated/inhibited metabolic pathways are reported together with some of the PPARs' target genes. (**A**) Hypoxia-inducible factor-1 (HIF-1) can active PPARα, which in turn activates the transcription of specific genes resulting in high glycolysis, high glycogen storage, and high proliferation rate (glucose transporter 3 (GLUT3), c-Myc, and cyclin D1). However, PPARα activation is also related to the induction of fatty acid oxidation (FAO) by upregulation of carnitine palmitoyltransferase 1 (CPT1). In addition, PPARα induces fatty acid synthesis by upregulation of fatty acid synthase (FAS) enzymes. It is noteworthy that mitochondrial 3-hydroxy-3-methylglutaryl-CoA synthase (HMGCS2) is upregulated by PPARα; besides, HMGCS2 can form a heterodimeric complex with PPARα to induce Src expression. The phosphorylation of Src triggers the mevalonate (MVA) pathway, resulting in high levels of cholesterol (CHOL). Lipid components and cholesterol are useful for membrane synthesis, and their large amounts are confined in lipid droplets. Extracellular lipids and some intracellular lipids (from lipid droplets) can be PPARα ligands; they are delivered to the nucleus by fatty acid binding protein (FABP). (**B**) PPARγ transcriptional activity activates some proteins related to fatty acid synthesis, such as FAS, c-Myc, PBP (PPARγ-binding protein), NR1D1 (nuclear receptor subfamily 1, group D, number 1), and ODC1 (ornithine decarboxylase 1). ODC1 is able to inhibit krüppel-like factor 2 (KLF2), which in turn is unable to inhibit PPARγ. Other PPARγ-dependent mechanisms are able to reduce palmitate toxicity by confining it into lipid droplets. Moreover, PPARγ 41 kDa fragment, which is derived from caspase 1 cleavage, is able to inhibit FAO. (**C**) PPARβ/δ stimulates glycolysis by the overexpression of GLUT1, angiopoietin-like 4 (ANGPTL4), phosphoinositide-dependent protein kinase 1 (PDPK1), and PI3K/Akt; likewise, PDPK1 and PI3K/Akt can activate PPARβ/δ expression. Fatty acid synthesis and FAO are activated by PPARβ/δ transcriptional activity on FAS and SLC1A5 (solute carrier 1 A5) genes. SLC1A5 is linked to the uptake of amino acids; thus, anaplerosis is also positively affected by PPARβ/δ. Anaplerosis also supports FAO. Interesting, PPARβ/δ upregulates cytokines expression; for example, interleukin 8 (IL8) and cytokines in concert with PPARβ/δ induce STAT3 overexpression. The MVA pathway is a downstream process triggered by STAT3. The thin black continuous lines with arrows indicate upregulation events. The thick black continuous lines with arrows indicate a stimulation of the metabolic pathway. The thin blue continuous lines with bars indicate inhibition events. The thick blue continuous lines with bars indicate the inhibition of a metabolic pathway. The HIF-1-mediated upregulation of PPARs is represented by a grey dash dot and arrow at the end, while FABP-mediated ligand-dependent activation of PPARs is represented by a gold dash dot and arrow at the end.

It has been ascertained that the most of oncogenes are involved in the metabolic reprogramming of tumor cells [95,96]. Among them, cyclin D1, contrary to what has just been mentioned, was demonstrated to inactivate the PPARα-mediated gene expression of enzymes related to FAO in hepatocytes as well as hepatocellular and breast cancer-derived cell lines [97]. Previous evidences have demonstrated the role of cyclin D1 in the regulation of androgen receptors, estrogen receptors, thyroid hormone receptor, and PPARγ [98,99] in different cell types. Several pieces of evidence about the cyclin D1 regulation of cell metabolism, via the inhibition of PPARα transcription factor has been provided, while the overexpression of cyclin D1 induced the low expression of genes related to FAO. On the other hand, knockdown of cyclin D1 promoted FAO enzymes expression, but PPARα gene silencing weakened this effect. These results highlight the role of cyclin D1 in affecting FAO in a PPARα-dependent manner; for instance, a mitogen-stimulated cancer cell line showed low PPARα and FAO activity, indicating that the transition from a quiescent state to a proliferation state requires less energy from fatty acid [99]. Data reported in the paper of Kamarajugadda et al. [97] suggest that cyclin D1 blocks the binding of PPARα on the PPRE of specific FAO enzymes in a not clear way. At the same time, cyclin D1 could disturb the association of specific co-activators with PPARα, and then determine some changes in chromatin conformation; besides, cyclin D1 controls the expression of CBP/p300 [100,101].

Fatty acid synthase (FAS) is upregulated in a tumor of the urinary tract, such as RCC, and the downstream intermediates of fatty acid synthesis are endogenous ligands of PPARα, while the inhibition of FAS in the liver of mice provides rodents with PPARα dysfunction [102–104]. As mentioned above [82], a histological grade of RCC is directly linked to PPARα levels, and its inhibition leads to cell cytotoxicity, cell cycle arrest with glycolysis, and FAO deregulation. Recently, in RCC cell lines (Caki-1 and 786-O) and normal human kidney cells (NHK), it was reported that the inhibition of glycolysis triggered FAO and OXPHOS, even though PPARα inactivation reversed this metabolic pattern. Moreover, in normal cells, PPARα antagonist did not inhibit the glycolysis; conversely, in RCC, cell line glycolysis was attenuated, which was likely due to a difference of c-Myc protein levels between cancer cells and normal cells [105]. FAO can be considered an alternative metabolic pathway to produce energy when the glycolysis is obstructed. In fact, the RCC cell line showed increased levels of palmitate 24 h following 2-DG administration. Instead, co-administrations with GW6471 involve the decay of palmitate levels. Usually, fatty acid β-oxidation provided the acetyl-CoA groups to supply TCA cycle and OXPHOS, which in turn are more active with glycolysis inhibition. When RCC cell lines were treated with a combination of 2-DG and PPARα antagonist (GW6471), the OXPHOS activity levels showed no significant differences compared to the control cells, while GW6471 alone was able to impair oxidative phosphorylation, but not FAO. Therefore, PPARα antagonist adversely affected the levels of oncogene c-Myc in the RCC cell line, which is involved in the overactivation of protein related to glucose uptake and glycolysis. Most likely, PPARα controls glycolysis via c-Myc at least in RCC cell lines, and the simultaneous administration of 2-DG also induces FAO inhibition. This double effect is detrimental to the main metabolic pathways that are normally used by the RCC cell line [82,105].

Human hepatocellular carcinoma (HCC) tissue showed increased mRNA levels of the gene involved in FAO and glucose metabolism, among which PPARα, carnitine palmitoyl transferase 1A (CPT1A is the rate-limiting enzyme of FAO), glyceraldehyde 3-phosphate dehydrogenase (G3PDH), and the upregulation of cyclin D1 mRNA. Although increased levels of PPARα were associated with the deregulation of metabolic pathways that trigger carcinogenesis, there has not been evidence of HCC incidence in human patients who were exposed to peroxisome proliferators [106].

Regarding carnitine palmitoyl transferase enzyme, the possible regulatory role of PPARα-CPT1C axis in tumor proliferation and senescence was recently demonstrated [107]. As mentioned above, as CPT1A, CPT1C is also a rate-limiting enzyme in FAO, and the enzymatic reaction allows the acylation of a long fatty acid chain with subsequent entry into the mitochondria. In cancer cells, the CPT1 enzyme's family is upregulated [108–110]. Moreover, it was identified as PPRE in the first exon of the CPT1B gene [111,112]. In order to investigate the possible relationships between CPT1 genes and PPARα, were performed some analyses on two different cancer cell lines, MDA-MB-231 (breast cancer cell line) and PANC-1 (pancreas cancer cell line) with knockdown or overexpression of the PPARα gene. Dual-luciferase reporter gene assays showed *CPT1C* active transcription by PPARα in association with cell proliferation and senescence interruption. The effects were completely different when the PPARα gene was depleted; an increase in senescence with low proliferation rate was observed, indicating that the CPT1C gene is regulated by PPARα. This is further evidence of the ability of PPARα to modulate cancer cell metabolism (see also Figure 1A) [107].

During carbohydrate deprivation, the cells can adopt ketogenesis to ensure lipid-derived energy; this process is essential for tumor initiation and metastasis [113]. Mitochondrial 3-hydroxy-3-methylglutaryl-CoA synthase (HMGCS2) belongs to the HMG-CoA family, and catalyzes the first enzymatic reaction in ketogenesis. Several proteins related to the ketogenesis pathway were overexpressed in prostate cancer cells [114], among which HMGCS2 was included; on this basis, some researchers demonstrated the direct interaction between PPARα and HMGCS2 [115], resulting in Src activation and the promotion of malignancy and invasion. This study demonstrated the correlation between the increased mRNA levels of HMGCS2 and poor clinical outcomes as well as grade malignancy in colorectal cancer (CRC) and oral squamous cell carcinoma (OSCC) tumor biopsy from affected patients. The demonstration of a direct interaction at the nuclear level between HMGCS2 and PPARα is interesting; besides, other analyses confirmed that the heterodimeric complex binds the *Src* promoter region and induced genes linked to tumor invasion (Figure 1A) [115].

Chronic lymphocytic leukemia (CLL) patients present poor clinical outcomes, and the most effective therapy is based on high dose of glucocorticoids (GCs) with or without monoclonal antibodies. Nevertheless, this therapeutic protocol is not curative, and is characterized by progressive tumor resistance to GCs [116]. Glucocorticoids have immunosuppressive effects, inhibiting glucose metabolism and increasing FAO in tissue under starvation condition. Tung et al. [117] found in CLL that primary culture from patient's blood increased PPARα expression mediated by GCs with pronounced tumor dependence on FAO. Lipid oxidation ensures tumor survival, providing an alternative mechanism to the metabolic limitations dictated by GCs. PPARα antagonist impaired the tumor chemoresistance mechanism of GCs. Pyruvate kinase M2 (PKM2) activity was downregulated at the transcriptional and protein level by dexamethasone (DEX); despite this, acetate levels were kept constant, suggesting an increase in FAO activity linked to DEX. PPARα and PPARβ/δ mRNA levels were increased after DEX administration, while the downregulation of PKM2 occurred before the PPARα upregulation; it is likely that the nuclear receptor did not affect pyruvate kinase gene transcription. Nevertheless, the pyruvate dehydrogenase kinase 4 (PDK4) gene is under the transcriptional control of PPARα and PPARβ/δ; then, PDK4 phosphorylates and inhibits pyruvate dehydrogenase. Thus, pyruvate is useful for FAO rather than for OXPHOS [118]. Moreover, in order to understand the role of DEX in FAO and related chemoresistance triggering, the effects of DEX administration in association with FAO substrates were investigated. About that, CLL cells were co-cultured with OP-9-derived adipocytes in order to obtain an in vitro model in which lipids were derived from cells with an adipocyte phenotype. This model was used to mimic an in vivo tumor environment, where CLL cells are close to the adipocyte, and the high amount of lipids in the surrounding environment could improve tumor resistance to drugs by feeding FAO [119,120]. CLL showed greater resistance to DEX when cultured with adipocytes compared with CLL cells in serum-free media, and the effects were the same with conditioned media from an OP-9-derived adipocyte. These results highlight that lipids secreted from OP-9-derived adipocytes conferred chemoresistance. This experimental evidence demonstrated the direct involvement of PPARα in GCs tumor resistance, since it is upregulated by DEX and is a well-known FAO regulator; in addition, PPARα antagonists revoked these effects and sensitized CLL cells to DEX [117].

Contrary to what is stated, PPARα activity could be useful to counteract tumor progression in some tissue, as evidenced in melanoma [79]. In addition, PPARα is able to decrease the transcription of fatty acid synthesis genes and positively affect the transcription of FAO enzymes. In this regard, Chandran et al. reported the protective roles of the clofibrate, which is a PPARα agonist, in counteracting breast cancer inflammation and invasion [121]. The researchers used two triple negative breast cancer cell lines, SUM149PT and SUM1315MO2; the first from an invasive ductal carcinoma of a patient with inflammatory breast cancer, and the second from a highly invasive breast cancer specimen of a patient with skin metastasis. These two cell lines showed an increased expression of PPARα with respect to primary human mammary epithelial cells (HMEC). Clofibrate was able to reduce inflammation by decreasing the levels of COX-2 (cyclooxygenase-2) and 5LO (5-lipoxygenase) in association with the inhibition of growth tumor. Early events of cancer development require the upregulation of fatty acid synthesis, which is dramatically exacerbated during the late events of tumor progression [122]. FAS activity was attenuated by clofibrate, which in turn downregulated the expression of HMG-CoA synthase 2, acyl-CoA oxidase, and the sterol regulatory element binding protein 1c (SREBP-1c) gene. HMG-CoA synthase 2 and acyl-CoA oxidase are involved in the mevalonate pathway, while SREBP-1c is a transcription factor acting on sterol regulatory element DNA sequences. SREBP-1c (sterol regulatory element binding protein 1c) plays a key role in regulating de novo fatty acid synthesis, while its cognate SREBP-2 regulates the genes of the cholesterol metabolism [123]; SREBP's pathway has a significant role in the de novo fatty acid synthesis of prostate cancer cells [124]. As reported by Chandran et al. [121], the activation of PPARα by clofibrate was able to impair the gene expression of SREBPs and reduce the NFκB and Erk1/2 (extracellular signal-regulated kinase 1/2) protein levels in breast cancer cells derived from high metastatic inflammatory tumor specimens. Conversely, in the same cancer cells, clofibrate was linked to the CPT-1a (first enzyme in FAO) upregulation (as reviewed in Figure 2A) [121].

Some evidence has indicated PPARα activation as a possible trigger of ineffective tumor metabolism. It was reported that the administration of fenofibrate (a PPARα agonist), on cell lines and a mouse model of oral cancer, supported hexokinase II and VDAC (voltage-dependent anion channel) dissociation. This event destabilizes the Warburg effect and provides a metabolic switch to OXPHOS. Furthermore, in these in vivo and in vitro oral cancer models, the activity of fenofibrate affected hexokinase II, PDH, and VDAC protein levels, as indicated in Figure 2A [125–127]. Recently, Huang and Chang [128] studied, through proteomic analysis, the differences between normal and cancer oral tissue from mice, relating it to enzymes involved in the Warburg effect. At the same time, they investigated the role of PPARα in the fibrate-dependent metabolic changes of the oral cancer cell line. Proteomic analyses were performed in a basic isoelectric point (pI) range, because the enzymes of glycolysis, the TCA cycle, and OXPHOS show mainly alkaline pI [128,129]. Seven proteins showed decreased levels in tumor tissue compared with normal tissue; they were triosephosphate isomerase and pyruvate dehydrogenase E1 component subunit beta for glycolysis, IDH3 and aconitate hydratase for the TCA cycle, NADH dehydrogenase [ubiquinone] 1 alpha subcomplex subunit 10 and cytochrome c1 for the respiratory chain. Considering oral cancer cells' dependence on the Warburg effect, the researchers evaluated the effect induced by fibrate. PPARα activation induced the reduction of hexokinase II protein levels, ATP levels, and enhanced PDH activity, alongside reducing cell viability. Interestingly, they observed a significant increase in TCA cycle metabolites after fenofibrate administrations in primary cell culture from mouse tongue tumor tissue. Probably, PPARα agonist increased PDH activity; accordingly, pyruvate was decarboxylated to acetyl-CoA, and TCA cycle was encouraged. Otherwise, fenofibrate could increase FAO, resulting in high acyl group levels that are useful for TCA cycle reactions (Figure 2A) [127,128].

Regarding the Warburg effect and related aerobic glycolysis, the repression activity of PPARα on the GLUT1 gene with reduced glucose uptake was reported; these evidences were obtained in different cancer cell lines (HCT-116, SW480, MCF-7, and HeLa) (as indicated in Figure 2A) [71].

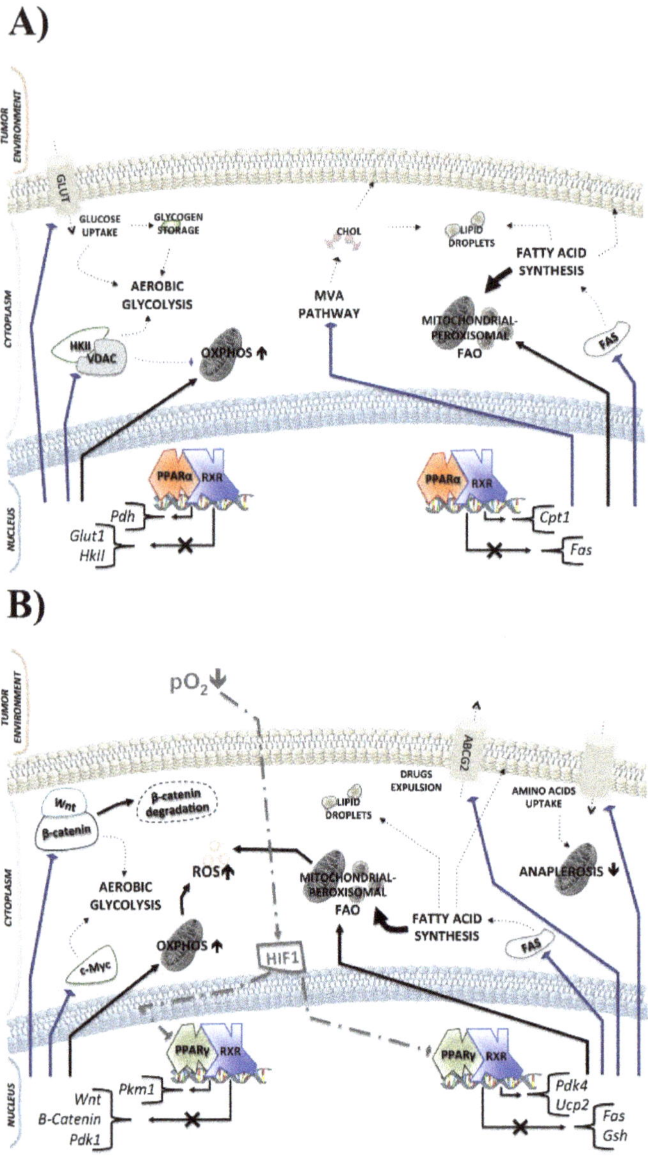

Figure 2. *Cont.*

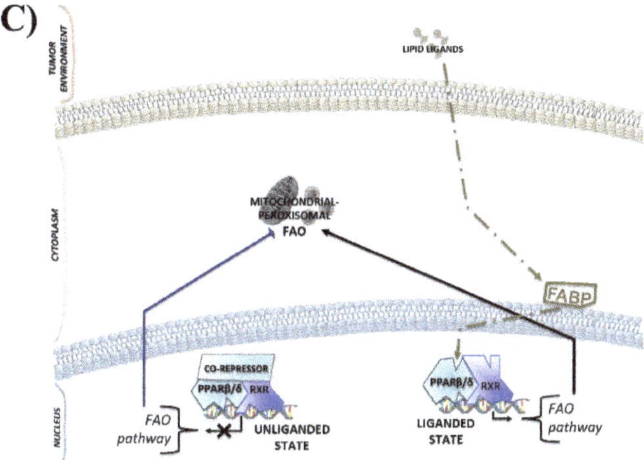

Figure 2. Schematic representation of PPARs-dependent oncosuppressive metabolic pathways highlighted in this review. The representation concerns the metabolic mechanisms that are activated/inhibited in tumor cells under the transcriptional control of PPARs. These hypotheses of molecular mechanisms are based on evidences obtained by different cancer types. For each PPAR isotype, the specific activated/inhibited metabolic pathways are reported together with some PPAR target genes. (**A**) Aerobic glycolysis is inhibited by the PPARα's transcriptional repression of glucose transporter 1 (GLUT1) and hexokinase II (HKII) genes. Meanwhile, the complex between the voltage-dependent anion channel (VDAC) complex and HKII is destroyed by PPARα activity, thus adversely affecting glycolysis and increasing oxidative phosphorylation (OXPHOS). In addition, pyruvate dehydrogenase (PDH) is upregulated by PPARα to promote OXPHOS. Impairment in fatty acid synthesis by the downregulation of fatty acid synthase (FAS) and impairment of the mevalonate (MVA) pathway are due to effects adversely exerted by PPARα on specific target genes. Conversely, carnitine palmitoyl transferase 1 (CPT1) is upregulated by PPARα; this condition promotes fatty acid oxidation (FAO). Despite the reduced activity of fatty acid synthesis, FAO depletes insufficient lipid reserves and impairs cancer cells for life. (**B**) PPARγ downregulates the c-Myc/Wnt/β-catenin axis and stimulates β-catenin proteasome degradation. Further downregulation of pyruvate dehydrogenase kinase 1 (PDK1) and upregulation of pyruvate kinase isoenzyme M1 by PPARγ promotes OXPHOS and impairs aerobic glycolysis. Fatty acid synthesis, amino acid uptake, and anaplerosis are adversely affected by PPARγ activity in concert with increased levels of FAO. High FAO levels are related to the upregulation of PDK4 and mitochondrial uncoupling protein 2 (UCP2). Moreover, PPARγ activity negatively affects ATP binding cassette G2 (ABCG2) and prevents chemoresistance; this is associated with the high sensitivity of tumor cells to ROS, whose levels are increased through FAO and OXPHOS metabolic pathways. In addition, there is glutathione (GSH) downregulation, while hypoxia inducible factor-1 (HIF-1) is able to inhibit PPARγ activity. (**C**) In the absence of ligands, PPARβ/δ acts as a repressor, which is probably due to the strong interaction between PPARβ/δ/RXR heterodimer and a co-repressor. However, the repressor complex is able to downregulate the genes involved in FAO, this condition is abolished in the presence of exogenous or endogenous PPARβ/δ ligands. The thin black continuous lines with arrows indicate upregulation events. The thick black continuous lines with arrows indicate the stimulation of a metabolic pathway. The thin black dashed lines with arrows indicate a reduction activity of metabolic pathways. The thin blue continuous lines with bars indicate inhibition events. The thin blue dashed lines with bars indicate a reduction of the inhibition of the metabolic pathway. The HIF-1-mediated downregulation of PPARs is represented by a grey dash dotted line with a bar at the end, while the FABP-mediated ligand-dependent activation of PPARs is represented by a gold dash dotted line with an arrow at the end.

2.2. PPARγ and Cancer Metabolism

Several cell types express PPARγ, which is involved in different mechanisms that are essential to sustain normal cell life. Adipose tissue, liver tissue, muscle, brain, and immune cells (mainly macrophages) require PPARγ activation to meet energy demands and regulate glucose and lipid metabolism, insulin sensitivity, and cell fate. PPARγ plays a key role in adipocytes and the differentiation of macrophages [130–132]. As previously mentioned for PPARα, as well as for PPARγ, there have been several demonstrations about its role in tumorigenesis, some of them related to the antiproliferative effects of PPARγ activation, such as in breast [133], hepatic [134], lung [135], and colorectal cancer [136]. Moreover, PPARγ activation negatively affects the epithelial mesenchymal transition (EMT) [137]. However, there is other proof of the tumorigenic potential of PPARγ activation, such as in colorectal cancer [138–140], breast cancer [141,142], and urological cancer [143]. Both roles of PPARγ are strictly tumor tissue-dependent and tumor microenvironment-dependent.

Several types of epithelial cancers show a common feature: deregulation of the Wnt/β-catenin pathway, resulting in the upregulation of enzymes related to aerobic glycolysis. The availability of Wnt ligands triggers the nuclear translocation of the β-catenin, where it is able to bind specific target genes, including pyruvate dehydrogenase kinase (PDK), monocarboxylate lactate transporter-1 (MCT-1), c-Myc, cyclin D1, and COX-2. Without Wnt ligands, β-catenin is phosphorylated and then demolished by proteasome. In this view, PPARγ downregulation is associated with Wnt/β-catenin upregulation; on the other hand the inhibition of Wnt/β-catenin is mediated by PPARγ activation (see also Figure 2B). Accordingly, it is not inconceivable to think about a mechanism of interconnection between Wnt/β-catenin and PPARγ, in which each one is able to prevent the pathway of the other, as already demonstrated [144]. PDK1 acts as a phosphorylating pyruvate dehydrogenase, and then pyruvate is transformed in lactate by the activation of lactate dehydrogenase. Meanwhile, MCT-1 is involved in lactate secretion outside the cytoplasm. These two events enable improving the angiogenesis and biosynthesis of macromolecules, thus providing a unique and favorable tumor microenvironment [21]. In this context, PPARγ suppresses *PDK1* gene transcription, resulting in an ineffective Wnt/β-catenin pathway (Figure 2B) [145].

Studies conducted on PPARγ agonist or with PPARγ overexpressing cells, support the idea that PPARγ activation is useful to counteract tumor progression; in fact, thiazolidinediones (TZDs) show the ability to contain tumor growth in vitro and in vivo models of lung cancer. In addition, it was reported that the overexpression of PPARγ in a group of non-small lung cancer cells and its activation affect some genetic pattern underlying the tumor metabolic demands [146]. Srivastava and collaborators [135] reported in two lung adenocarcinoma cell lines (NCI-H2347 and NCI-H1993) that PPARγ activation compromised glucose, fatty acid, and glutamine metabolism, which are associated with increased ROS (reactive oxygen species) and hypophosphorylated RB (retinoblastoma protein). Dephosphorylated RB is opposed to the cell cycle progression. Unlike what was previously mentioned, in this work, the researchers found an upregulation of PDK4 expression by pioglitazone, and the central role of PDK4 in inducing the metabolic switch from glucose oxidation to fatty acid oxidation was suggested. PDK4 knockdown abolished the effect induced by pioglitazone related to RB hypophosphorylation and ROS levels; simultaneously, the same results were achieved in cell lines and xenograft mice models by inhibiting FAO with chemical compounds. These results suggested that PDK4 upregulation, by pioglitazone, compromised glucose utilization and triggered FAO with a subsequent increase of ROS levels, which in turn induced RB hypophosphorylation. Moreover, the researchers reported alterations in glutamine metabolism, an impairment of glutaminolysis, and downregulation of reduced glutathione (GSH) levels; therefore, tumor cells were unable to carry out ROS detoxification processes (as reported in Figure 2B) [135].

One common feature of several tumors such as non-small cell lung cancer (NSCLC) is resistance to radiation and chemotherapy, but the specific mechanisms are not entirely understood. However, it is well-known that hypoxia supports the malignancy and expression of ATP-binding cassette (ABC) transporters, which drive chemotherapeutic agents outside the cells [147,148]. The hypoxic condition is also combined with the downregulation of mitochondrial uncoupling protein 2 (UCP2) in NSCLC cells, as highlighted in a recent work [149]. UCP2 is a mitochondrial protein that is involved in the detoxification process by reducing ROS levels, because the cells are more sensible to superoxide anion released after proton force development by the electron transport chain. Moreover, a double role was suggested for UCP2: the reduction of ROS levels, and the metabolic regulation of glycolysis, fatty acid, and glutamine oxidation [150]. Downregulation of UCP2 by hypoxia was associated with PPARγ repression, the upregulation of the ABC transporter and ATP binding cassette G2 (ABCG2), and an increase of aerobic glycolysis and chemoresistance. HIF-1 was directly involved in PPARγ and FAO downregulation; this condition negatively affected the *UCP2* transcription. Conversely, glucose consumption was stimulated and established a progressive increase of ROS in concert with ABCG2 upregulation, as indicated in Figure 2B [149,151].

Several studies show the ability of ATRA (all-trans retinoic acid) to induce the differentiation of some myelocytic cell lines (HL-60, U937, and NB4) into mature phagocytic cells. ATRA administration is useful for the therapy of acute promyelocytic leukemia (APL), but the permanent administration of ATRA causes high resistance at differentiation, because there is overexpression of cytosolic retinoic acid binding proteins [152,153]. In this regard, the association between ATRA and PPARγ ligands was demonstrated to be synergistic in the differentiation effect on myelocytic leukemia cell lines [130]. The synergistic effect also concerned the enhancement of lipogenesis, as evidenced in the NB4 cell line by an accumulation of lipid droplets. Therefore, an induction of differentiation by ATRA and pioglitazone results in a high activity of triacylglycerol synthesis in human myelocytic leukemia cell lines [154].

An induction of PPARγ activity and concomitant autophagic cell death in human chronic myeloid leukemia (CML) cell lines (K562 and KCL-22) was reported by Shinohara et al. [155]. By docking analysis, they observed that anti-cancer fatty-acid derivative, called AIC-47, was able to bind PPARγ, making it transcriptionally active, and indirectly reducing c-Myc protein levels, since PPARγ activation is related to the proteasome degradation of β-catenin, as already mentioned [144]. Other interesting results also demonstrated the involvement of AIC-47/PPARγ in the deregulation of the glycolytic pathway. In fact, the upregulation of c-Myc is a cause and a consequence of aerobic glycolysis in tumor cells. As previously demonstrated, c-Myc can induce the overexpression of three heterogeneous nuclear ribonucleoproteins (hnRNPs), and, in turn, they can suppress the alternative splicing of pyruvate kinase isoenzyme M1 (PKM1), which is the less present isoform in cancer cells. Unlike other isoforms of PK that need allosteric regulation to be active, PKM1 is a tetrameric stable and active enzyme; for this reason, cancer cells prefer to use PKM2 for their metabolic purposes. PKM2 shows slow activity in cancer cells, because it also allows the biosynthetic pathways; consequently, in cancer cells, the PKM1/PKM2 ratio is low and c-Myc-dependent [23,24,156]. PPARγ activation AIC-47-dependent induced c-Myc downregulation, resulting in β-catenin inactivation with an increase of the PKM1/PKM2 ratio and the metabolic switch from glycolysis to the TCA cycle; simultaneously, ROS levels increase, which results in autophagy induction (Figure 2B) [155].

Survival in hepatocellular carcinoma (HCC) patients is related to the expression patterns of some genes, including ODC1 (ornithine decarboxylase 1). Its overexpression is associated with reduced patients survival [157]. The OCD1 enzyme catalyzes the first reaction in the biosynthesis pathway of polyamine; its mRNA and protein levels are increased together with c-Myc activity in HCC tissue compared with normal tissue [158]. An impairment of OCD1 expression by gene silencing was related to cell cycle interruption and apoptotic cell death; besides, phenotypic alterations occurred through a characteristic deregulation of 119 genes. Among them, it was interesting that the downregulation of PPARγ gene and lipogenesis were both linked to the upregulation of KLF2 (krüppel-like factor 2) oncogene. It was reported that the siRNA of ODC1 gene induced the upregulation of the KLF2 gene, which, in turn, negatively affected PPARγ expression, thus causing a downregulation of lipogenic enzymes, such as FAS and ACC2 (acetyl-CoA carboxylase 2), as already highlighted; see also Figure 1B [159,160].

Regarding de novo fatty acid synthesis, in ERBB2 (erythroblastic oncogene B)-positive breast cancer cells, a remarkable amount of lipid droplets was observed. ERBB2 cells assumed this metabolic behavior under the transcriptional control of PPARγ, and the inhibition of PPARγ decreased tumor cell viability. By RNA interference screening, some genes that are required for fatty acid metabolism and tumor cell survival were identified [161]. Within this group of genes, two were associated with PPARγ activity: PBP (PPARγ-binding protein) and NR1D1 (nuclear receptor subfamily 1, group D, number 1). Both were identified as activators of PPARγ expression; it is likely that PBP was a co-activator, and NR1D1 was the target gene [162]. The gene sequence of *PBP* and *NR1D1* are located in the ERBB2 amplicon, and in breast cancer, mutations in this gene locus are linked to high lipid synthesis and PBD, NR1D1 overexpression. PBD and NR1D1 activity is aimed at the regulation of FAS (fatty acid synthase), ACLY (ATP citrate lyase), and ACACA (acetyl-coenzyme A carboxylase α) gene expression [162,163]. In this regard, palmitate, the last metabolic product of the fatty acid synthesis pathway, was described as lipotoxic agent, likely by ROS induction [164]. Some researchers identified a protective role of PPARγ against palmitate-induced lipotoxicity in ERBB2-positive breast cancer cell lines (BT474 and MDA-MB-361), but not in other types of breast cancer (MCF-7) and normal cells. The PPARγ activity allowed the induction of triacylglycerol synthesis, in order to remove the excess of fatty acid and enclose them in specific stores (LDs) to relieve lipotoxicity. Moreover, PPARγ played a central role in keeping the FAS active by confinement of palmitate in specific stores. The inhibition of PPARγ by antagonist abolished the protective mechanism, and ERBB2 cells were more sensible to palmitate-dependent toxicity (Figure 1B) [165]. A pertinent work showed the suppressive effects of PPARγ antagonism in populations of cancer stem cells (CSCs) derived from ERBB2-positive breast cancer cell lines (BT474 and SKRB3). These cell lines expressed high levels of ALDH (aldehyde dehydrogenase) activity with greater lipid storage than ERBB2-negative cells. Also in this case, the tumor suppressive effects were related to increased ROS levels and a damaged lipogenesis pathway. The researchers' assumption was that the epigenetic pattern of ACLY (ATP citrate lyase) could be altered by PPARγ inactivation, considering the ACLY gene a PPARγ target gene. In fact, acetylation levels of H3 and H4 histone were found to be different between ERBB2-positive cells and control cells [166].

Recently, an interesting approach, called sleeping beauty (SB), was used to mainly find genes leading to prostate cancer metastatic events. Briefly, this approach is based on transposons, which can induce somatic mutations, and the expression of transposase enzymes could be tissue-specific or ubiquitous [167]. Most of the analyses were conducted on PTEN-null mice, because patients with poor prognosis presented low PTEN levels and conversely high PPARγ and FAS levels. Noteworthy, the insertion of mutations within the PPARγ gene established greater tumor aggressiveness in PTEN-deleted mice than in mice without insertion. Also in this study, the PPARγ overexpression determined the upregulation of enzymes involved in de novo fatty acid synthesis, and conversely, this effect was abolished by PPARγ knockout and downregulation [168].

Tumor-associated macrophages (TAMs) have a close relationship with the tumor microenvironment and encourage tumor progression. Several evidences support the idea that stromal cells play a key role in tumor maintenance, since tumor cells exploit them by using their energy resource, in the form of metabolic intermediates or end products (lactate, ketones, glutamine, and fatty acids). Concerning this scenario, the ability of caspase-1 to cut PPARγ in a 41-kDa fragment was reported. Afterwards, this fragment translocates into mitochondria to dampen MCDA activity. Medium-chain acyl-CoA dehydrogenase (MCDA) contributes to fatty acid β-oxidation [169]; its inactivation was demonstrated to be linked to lipid synthesis, and LDs increase with concomitant TAMs differentiation. Considering the caspase-1/PPARγ/MCDA axis as an important mechanism to improve TAMs differentiation and tumor aggressiveness, when this axis was damaged with a caspase-1 inhibitor, TAMs cells suffered a specific commitment that negatively affected tumor progression (Figure 1B) [170].

2.3. PPARβ/δ and Cancer Metabolism

PPARβ/δ, similar to other PPAR isotypes, regulates the transcription of genes that are required for the main metabolic processes, such as glucose and fatty acid catabolism, although its regulatory role is also implicated in cell proliferation, cell differentiation, wound healing, and inflammation [55,171,172]. Several scientific evidences reported the pro-tumorigenic role of PPARβ/δ, but to date, there has been conflicting information on the exact role of PPARβ/δ in carcinogenesis [75,173]. This aspect was especially investigated in breast cancer with conflicting results, showing that the estrogen receptor was involved in the effects induced by PPARβ/δ activity modulation. In fact, proliferation in the MCF-7 cell line (estrogen receptor positive, ER^+) was increased by PPARβ/δ overexpression; conversely, the MDA-MB-231 cell line (estrogen receptor negative, ER^-) showed no effect on the cell proliferation rate. Unfortunately, these results are not consistent with other evidences, showing that in MCF-7 cells, the overexpression of PPARβ/δ induced differentiation and cell cycle interruption [174–176]. On the other hand, the negative effect of PPARβ/δ activation on tumor survival in MCF-7 and MDA-MB-231 cell lines has also been reported [177].

Tumor progression in non-small cell lung cancer (NSCLC) was associated with PPARβ/δ upregulation, an increase in VEGF (vascular endothelial growth factor) levels and activation of the PI3K/Akt pathway [178]. PPARβ/δ could be considered an upstream regulator of PI3K/Akt activity. PI3K/Akt signaling is able to reduce PTEN levels and increase PDPK1 (3-phosphoinositide-dependent protein kinase-1) expression [179]. Since the PDPK1 gene presents PPRE specific for PPARβ/δ, as already demonstrated [180], an interesting analysis was conducted on mammary tumorigenesis in an in vivo model. In this regard, transgenic mice carrying the PDPK1 gene under the transcriptional control of mouse mammary tumor virus (MMTV-mice) were used. Nevertheless, the expression was limited to the mammary gland [181]. Transgenic mice showed higher PPARβ/δ expression levels than control mice; the expression was further increased in MMTV-mice fed a diet containing PPARβ/δ agonist. Mammary carcinogenesis was promoted in both wild-type and transgenic mice under feeding treatment, especially in transgenic mice. The researchers emphasized the differences between wild-type and MMTV mice regarding the treatment response, because mice bearing the PDPK1 transgene and treated with PPARβ/δ agonist were more prone to tumor initiation, which might have been due to differences in the involved metabolic pathway. Regarding that, the PI3K/Akt pathway is able to phosphorylate and activate ATP citrate lyase; simultaneously, PDPK1 slows down the pyruvate flow into oxidative phosphorylation and the Acss2 (Acyl-coenzyme A synthetase short-chain family member 2) supports the conversion of lactate to pyruvate. These three proteins work in concert to raise the acetyl-CoA amount in order to promote glycolysis and fatty acid synthesis, and the PPARβ/δ agonist increases their efficiency. Although PDPK1 expression alone was not able to induce carcinogenesis, its association with the active PPARβ/δ triggered a malignancy molecular pathway that was more aggressive in transgenic mice than in wild-type mice treated with PPARβ/δ agonist. Therefore, two different metabolic mechanisms can be activated, whereby PDPK1 induces the expression of PPARβ/δ and vice versa; this loop in turn supports the transcription and the activity of genes related to glycolysis and lipid synthesis. Fatty acid synthesis could be useful for supplying PPARβ/δ endogenous ligands and continuing to feed PDK1-PPARβ/δ loop activity (see also Figure 1C) [181].

Despite the maintenance of hematopoietic stem cells (HSCs) and endurance of muscle cells establishing an unfavorable metabolic condition, they are safeguarded through PPARβ/δ activity. It is likely that PPARβ/δ triggers specific molecular mechanisms related to the metabolic switch to allow the cell life cycle [182,183]. As already demonstrated by Tung and collaborators [117], PPARβ/δ transcription was promoted when leukemic cells were stressed by glycolysis inhibitors. The same results were obtained in a recent paper, but in breast cancer cell lines. When the cells grow in standard culture conditions for 10 days without medium replacement, the overexpressing-PPARβ/δ cells continued to proliferate much better than control cells. Conversely, cells with the PPARβ/δ knockdown, through CRISP/Cas9, showed a proliferation rate comparable to the control levels. However, the low glucose culture conditions induced a more pronounced PPARβ/δ upregulation in transfected cells compared to standard culture conditions, confirming the central role of PPARβ/δ in tumor metabolic modulation. Furthermore, these events were associated with increased levels of catalase and Akt protein, as well as an upregulation of the antioxidant defenses (Figure 1C) [184].

As mentioned above, PPARβ/δ plays, in concert with FAO, a key role in the preservation of HSCs, also in the presence of harsh environmental conditions. Regarding that, PPARβ/δ-FAO pathway undergoes an upstream regulation by the PML (promyelocytic leukemia) protein; which is codified by a tumor-suppressor gene. For example, Ito and colleagues [183] demonstrated that HSCs with *Pml* gene deletion were less inclined to asymmetric division with significant variation of the asymmetric/symmetric division ratio, and there are similar results also in breast cancer cells that sustain this observation [185]. These evidences provide further support regarding PPARβ/δ-FAO pathway regulation by PML upstream control. Therefore, abolishing the oxidative metabolism of fatty acids could damage cancer stem cells and more differentiated scaffold cells [183]. In this regard, a similar effect was observed in chronic lymphocytic leukemia (CLL) cells (Daudi cell line and primary culture), where the stressful environmental conditions stimulated PPARβ/δ expression by triggering a protective mechanism in cancer cells. Various kinds of harsh conditions were tested: low glucose, hypoxia, exposure to glucocorticoids, and cytotoxic agents. In any case, the response of tumor cells was to improve antioxidant activity and make better use of energy supplies through a proper metabolic pathway [186]. More recently, the involvement of PPARβ/δ signaling in the survival of CLL cell lines was reported. This event was associated with increased cholesterol and plasma membrane biosynthesis. Exposure to PPARβ/δ agonists was found to induce high cholesterol levels and interferon-dependent STAT phosphorylation. Cytokines stimulated the specific pathway related to cholesterol synthesis, while the inability of cytokines to upregulate PPARβ/δ was also demonstrated. On the other hand, PPARβ/δ could stimulate the cytokines expression in order to maintain the tumor microenvironment (as reported in Figure 1C) [187].

Consistent with these results, the direct role of PPARβ/δ in the IL-8 gene transcription was also observed in colon cancer cells, mainly in a hypoxic environment [188]. Unlike PPARα and γ, which present both pro-tumor and anti-tumor effects in colorectal cancer, different experimental evidences showed the pro-tumorigenic role of PPARβ/δ, mainly through its involvement in the APC/β-catenin/K-Ras oncogenic pathway [189,190]. The upregulation of PPARβ/δ was observed in human HCT116 colon cancer cells in a hypoxic environment. Whereas p300/PPARβ/δ interaction was triggered by HIF-1. p300 is an all-purpose co-activator for the nuclear receptor that contributes to the formation of transcriptional complex. The authors reported high levels of tumor angiogenesis by IL-8 and VEGF overexpression linked to hypoxic conditions that in turn induce the p300/PPARβ/δ complex. This complex strongly affected the expression of inflammatory cytokines. At the same time, PPARβ/δ is upstream regulated by PI3K/Akt, but PPARβ/δ itself is able to regulate PI3K and Akt expression; thus, a permanently active closed loop could be generated, as indicated in Figure 1C [188].

As mentioned above, PPARγ is directly involved in the differentiation of TAMs [170]; it is also worth mentioning that macrophages can assume two specific phenotypes: M1 (inflammatory) and M2 (anti-inflammatory). However, TAMs present a mix of both phenotypes [191]. PPARγ and PPARβ/δ are able to regulate the final fate of macrophages in a tumor environment. The transcriptional control of PPARβ/δ on genes linked to TAMs was observed in ovarian cancer cells. $CD14^+$ monocytes cells from ovarian carcinoma ascites were used as TAMs in vitro model [192]. This study evaluated which genes related to TAMs were under PPARβ/δ transcriptional control. The overall results confirmed the regulation of metabolic pathway genes and inflammatory/migration pathway genes. The upregulation of these genes was also found in the presence of PUFA (poly-unsaturated fatty acid) ligands. Therefore, transcriptional regulation by PPARβ/δ could be associated not only with the maintenance of TAMs, but also with tumor progression. The upregulation of genes encoding for soluble mediators of cancer progression, such as ANGPTL4 (angiopoietin-like 4), could be under the transcriptional control of the PPARβ/δ. ANGPTL4 is a lipoprotein lipase regulator; it is essential for tumor-metastatic progression. In fact, angiopoietin-like 4 prevents the cell death by anoikis [193,194]. However, the preservation of

TAMs was dependent of PPARβ/δ activation, which in turn induced the transcription of downstream elements, such as ANGPTL4 and PDK4, in order to allow a metabolic switch to aerobic glycolysis (Figure 1C). In fact, high lactate levels were detected, and cells from ascites resulted in having high fatty acid ligands for PPARβ/δ; thus, the nuclear receptor activity in TAMs maintenance was greatly facilitated by the tumor microenvironment [192].

Since the metabolic fate undertaken by cancer cells is a response depending on the cell phenotypic/genotypic characteristics, and on the specific microenvironment around the neoplastic bulk. PPARβ/δ behavior also undergoes this specific tumor conditioning. However, the microenvironment affects the ability of cancer cells to acquire nutrients from extracellular compartments to cytoplasm by transmembrane transporter proteins. In this regard, Zhang et al. [77] reported the direct binding of PPARβ/δ on the PPRE of *Glut1* and *Slc1-a5* genes, and highlighted their upregulation by PPARβ/δ activation, in order to ensure glucose and amino acids for tumor growth. Transfected SW480 cells (cell line from colon adenocarcinoma) with PPARβ/δ transgenes showed high mRNA and protein levels of GLUT1 and SLC1-A5 (solute carrier family 1 member 5), and as a consequence, lactate increases, and there is also high glucose and glutamine consumption. All of these results were abolished by PPARβ/δ knockdown or through the use of an antagonist. Moreover, the overexpression of GLUT1 and SLC1-A5, with contemporary PPARβ/δ silencing, caused an increase in the proliferation rate, which was abolished in cells with a specific deletion of the transporter protein genes and overexpression of PPARβ/δ. Considering these results, it is conceivable to hypothesize that there is a PPARβ/δ-dependent molecular pathway leading to GLUT1 and SLC1-A5 upregulation, resulting in the modulation of metabolic patterns suitable for tumor growth (Figure 1C) [77].

Unlike the evidence reported so far, the oncosuppressive activity of PPARβ/δ in prostate cancer was recently demonstrated [195]. In a tumor tissue biopsy of prostate cancer, low mRNA levels of PPARβ/δ were observed compared with benign tissue. The same results were obtained in prostate cancer cell lines (DU145, PC3, LNCAP, VCAP, C4-2, and 22RV); thus, the downregulation of PPARβ/δ was associated with high aggressiveness. In the absence of ligands, the PPARβ/δ could exist as a transcriptional repressor [172]; in fact, the inhibition of FAO was demonstrated when PPARβ/δ was overexpressed, but only in PC3 and LNCaP cell lines. The repressive effect on FAO was abolished by PPARβ/δ agonist, without affecting PPARβ/δ transcription levels. The results obtained confirm the suppressive role of PPARβ/δ in its unliganded form; see also Figure 2C [195].

3. Conclusions

Although there is no clear view on the exact role of PPARs in carcinogenesis, and considering that most of the experimental proofs are mutually conflicting, the key role of PPARs in the metabolic modulation faced by cancer cells to ensure their own survival has been accepted. Each cancer cell exhibits a specific metabolic signature that is related to its tissue-specific genotypic and phenotypic features. Nevertheless, the specific cell phenotype has a close relationship with the microenvironment around the tumor bulk; thus, tumor phenotypic manifestations are the result of the effects induced by the tumor microenvironment on cellular transcription events. Regarding that, the transcriptional activity of PPARs on specific target genes is deeply correlated to the tissue type from which the tumor arises and to the tumor microenvironment. For this reason, each PPAR isotype establishes different effects in various tumor cell types. Overall, all of these factors determine whether PPARs promote tumorigenesis and tumor progression or counteract cancer survival. Moreover, the tumor microenvironment provides PPAR ligands; consequently, the extracellular environment can directly modulate the activities of PPARs.

The most recent evidences reported in this review demonstrate the involvement of PPARs in a metabolic switch that occurs in different cancer types. The oncogenic metabolic pathway of PPARα is characterized by high glycolysis in concert with c-Myc and cyclin D1 upregulation, as well as high levels of lipid and glycogen synthesis. In addition, an increase of LDs is observed that is associated with the upregulation of the MVA pathway, while, less frequently, PPARα oncogenic activity can be connected to the induction of OXPHOS and FAO. Moreover, an increase in fatty acid oxidation was reported to confer chemoresistance, i.e., against glucocorticoids [116]. Hypoxia exerts its oncogenic role through a stimulation of PPARα transcriptional activity (Figure 1A). Oncogenic metabolic behavior related to PPARγ activity mainly triggers an increase in lipid synthesis and reduces FAO. Meanwhile, lipotoxicity related to a high amount of palmitate is arrested by PPARγ, which drives palmitate confinement into LDs. The positive role of PPARγ in the differentiation of TAMs is intriguing; the behavior of tumor stromal cells is affected by the PPARγ-mediated inhibition of FAO and induction of lipid synthesis (Figure 1B). Unlike other two PPAR isotypes, most of the evidence regarding PPARβ/δ activity highlights its oncogenic role. Environmental stress, such as hypoxia and low glucose, triggers the tumor metabolic pathway under PPARβ/δ transcriptional control; thus, aerobic glycolysis, lipid synthesis, anaplerosis, and FAO are stimulated. It is noteworthy in leukemia cells that the upstream regulation of cytokines by PPARβ/δ is related to high cholesterol levels and malignancy (Figure 1C).

Under certain circumstances, the transcriptional activity of PPARs is aimed at suppressing specific tumor metabolic pathways. PPARα can inhibit lipid and cholesterol synthesis in concert with FAO induction. Glycolysis is obstructed by the PPARα-dependent destruction of the hexokinase II/VDAC complex, leading to metabolic switch and high OXPHOS levels, as demonstrated in oral cancer cells (Figure 2A) [127,128]. Unlike PPARα, in some tissues, the hypoxia inducible factor downregulates PPARγ, leading to the loss of its tumor suppression activity. In normoxic conditions, PPARγ represses the expression of the gene related to glycolysis (Wnt/β-catenin, c-Myc), glutamine anaplerosis, chemoresistance, and antioxidant defenses. Conversely, PPARγ transcriptional activity encourages the expression of genes involved in tumor differentiation, TCA cycle, and FAO, which are all in agreement with the PKM1/PKM2 ratio increase (Figure 2B). Among the scant evidence supporting the oncosuppressive role of PPARβ/δ, its ability to decrease FAO and disrupt tumor proliferation in prostate cancer cells is accepted, but only in absence of its ligands (Figure 2C).

Overall, this review highlights the central role of PPARs in tumor metabolic decisions, which are in turn affected by the genetic signature of tumor cells and the specific tumor microenvironment. In this regard, epigenetic events could play a key role in the regulation of PPAR activities in tumor metabolic response, while the possible relationship between the three PPARs isotypes in tumor metabolism should be taken in consideration, as already described in the pathogenesis of neurodegenerative diseases [196]. However, in order to fully understand the exact role of PPARs in cancer metabolism, studying the epigenetic effects related to PPARs and the relationship between the three isotypes could be interesting, in order to efficiently target the complex machinery that achieves the energy demands of cancer cells.

Funding: This research was supported by RIA (University Relevant Interest) funds.

Conflicts of Interest: The authors declare no conflict of interest.

References

1. Nelson, D.; Lehninger, A.; Cox, M. *Lehninger Principles of Biochemistry*; W.H. Freeman: New York, NY, USA, 2008.
2. Poirier, Y.; Antonenkov, V.D.; Glumoff, T.; Hiltunen, J.K. Peroxisomal β-oxidation—A metabolic pathway with multiple functions. *Biochim. Biophys. Acta* **2006**, *1763*, 1413–1426. [CrossRef] [PubMed]
3. Santos, C.R.; Schulze, A. Lipid metabolism in cancer. *FEBS J.* **2012**, *279*, 2610–2623. [CrossRef] [PubMed]
4. Young, S.G.; Zechner, R. Biochemistry and pathophysiology of intravascular and intracellular lipolysis. *Genes Dev.* **2013**, *27*, 459–484. [CrossRef] [PubMed]
5. Cermelli, S.; Guo, Y.; Gross, S.P.; Welte, M.A. The lipid-droplet proteome reveals that droplets are a protein-storage depot. *Curr Biol.* **2006**, *16*, 1783–1795. [CrossRef] [PubMed]
6. Walther, T.C.; Farese, R.V., Jr. Lipid droplets and cellular lipid metabolism. *Annu. Rev. Biochem.* **2012**, *81*, 687–714. [CrossRef] [PubMed]
7. Khor, V.K.; Shen, W.J.; Kraemer, F.B. Lipid droplet metabolism. *Curr. Opin. Clin. Nutr. Metab. Care* **2013**, *16*, 632–637. [CrossRef] [PubMed]

8. Warburg, O. On the origin of cancer cells. *Science* **1956**, *123*, 309–314. [CrossRef] [PubMed]
9. Trédan, O.; Galmarini, C.M.; Patel, K.; Tannock, I.F. Drug resistance and the solid tumor microenvironment. *J. Natl. Cancer Inst.* **2007**, *99*, 1441–1454. [CrossRef] [PubMed]
10. Persano, L.; Rampazzo, E.; Della Puppa, A.; Pistollato, F.; Basso, G. The three-layer concentric model of glioblastoma: Cancer stem cells, microenvironmental regulation, and therapeutic implications. *Sci. World J.* **2011**, *11*, 1829–1841. [CrossRef] [PubMed]
11. Fidoamore, A.; Cristiano, L.; Antonosante, A.; d'Angelo, M.; Di Giacomo, E.; Astarita, C.; Giordano, A.; Ippoliti, R.; Benedetti, E.; Cimini, A. Glioblastoma Stem Cells Microenvironment: The Paracrine Roles of the Niche in Drug and Radioresistance. *Stem Cells Int.* **2016**, *2016*, 6809105. [CrossRef] [PubMed]
12. Martin, M.; Beauvoit, B.; Voisin, P.J.; Canioni, P.; Guérin, B.; Rigoulet, M. Energetic and morphological plasticity of C6 glioma cells grown on 3-D support; effect of transient glutamine deprivation. *J. Bioenergy Biomembr.* **1998**, *30*, 565–578. [CrossRef]
13. Guppy, M.; Leedman, P.; Zu, X.; Russell, V. Contribution by different fuels and metabolic pathways to the total ATP turnover of proliferating MCF-7 breast cancer cells. *Biochem. J.* **2002**, *364*, 309–315. [CrossRef] [PubMed]
14. Pasdois, P.; Deveaud, C.; Voisin, P.; Bouchaud, V.; Rigoulet, M.; Beauvoit, B. Contribution of the phosphorylable complex I in the growth phase-dependent respiration of C6 glioma cells in vitro. *J. Bioenergy Biomembr.* **2003**, *35*, 439–450. [CrossRef]
15. Rossignol, R.; Gilkerson, R.; Aggeler, R.; Yamagata, K.; Remington, S.J.; Capaldi, R.A. Energy substrate modulates mitochondrial structure and oxidative capacity in cancer cells. *Cancer Res.* **2004**, *64*, 985–993. [CrossRef] [PubMed]
16. Crabtree, H.G. Observations on the carbohydrate metabolism of tumours. *Biochem. J.* **1929**, *23*, 536–545. [CrossRef] [PubMed]
17. Smolková, K.; Plecitá-Hlavatá, L.; Bellance, N.; Benard, G.; Rossignol, R.; Ježek, P. Waves of gene regulation suppress and then restore oxidative phosphorylation in cancer cells. *Int. J. Biochem. Cell Biol.* **2011**, *43*, 950–968. [CrossRef] [PubMed]
18. Epstein, T.; Gatenby, R.A.; Brown, J.S. The Warburg effect as an adaptation of cancer cells to rapid fluctuations in energy demand. *PLoS ONE* **2017**, *12*, e0185085. [CrossRef] [PubMed]
19. Feron, O. Pyruvate into lactate and back: From the Warburg effect to symbiotic energy fuel exchange in cancer cells. *Radiother. Oncol.* **2009**, *92*, 329–333. [CrossRef] [PubMed]
20. Majewski, N.; Nogueira, V.; Bhaskar, P.; Coy, P.E.; Skeen, J.E.; Gottlob, K.; Chandel, N.S.; Thompson, C.B.; Robey, R.B.; Hay, N. Hexokinase-mitochondria interaction mediated by Akt is required to inhibit apoptosis in the presence or absence of Bax and Bak. *Mol. Cell* **2004**, *16*, 819–830. [CrossRef] [PubMed]
21. DeBerardinis, R.J.; Lum, J.J.; Hatzivassiliou, G.; Thompson, C.B. The biology of cancer: Metabolic reprogramming fuels cell growth and proliferation. *Cell Metab.* **2008**, *7*, 11–20. [CrossRef] [PubMed]
22. Gao, P.; Tchernyshyov, I.; Chang, T.C.; Lee, Y.S.; Kita, K.; Ochi, T.; Zeller, K.I.; De Marzo, A.M.; Van Eyk, J.E.; Mendell, J.T.; et al. c-Myc suppression of miR-23a/b enhances mitochondrial glutaminase expression and glutamine metabolism. *Nature* **2009**, *458*, 762–765. [CrossRef] [PubMed]
23. Christofk, H.R.; Vander Heiden, M.G.; Harris, M.H.; Ramanathan, A.; Gerszten, R.E.; Wei, R.; Fleming, M.D.; Schreiber, S.L.; Cantley, L.C. The M2 splice isoform of pyruvate kinase is important for cancer metabolism and tumour growth. *Nature* **2008**, *452*, 230–233. [CrossRef] [PubMed]
24. Dayton, T.L.; Jacks, T.; Vander Heiden, M.G. PKM2, cancer metabolism, and the road ahead. *EMBO Rep.* **2016**, *17*, 1721–1730. [CrossRef] [PubMed]
25. Semenza, G.L. Defining the role of hypoxia-inducible factor 1 in cancer biology and therapeutics. *Oncogene* **2010**, *29*, 625–634. [CrossRef] [PubMed]
26. Liberti, M.V.; Locasale, J.W. The Warburg Effect: How Does it Benefit Cancer Cells? *Trends Biochem. Sci.* **2016**, *41*, 211–218. [CrossRef] [PubMed]
27. Beloribi-Djefaflia, S.; Vasseur, S.; Guillaumond, F. Lipid metabolic reprogramming in cancer cells. *Oncogenesis* **2016**, *5*, e189. [CrossRef] [PubMed]
28. Menendez, J.A.; Lupu, R. Fatty acid synthase and the lipogenic phenotype in cancer pathogenesis. *Nat. Rev. Cancer* **2007**, *7*, 763–777. [CrossRef] [PubMed]

29. Tamura, K.; Makino, A.; Hullin-Matsuda, F.; Kobayashi, T.; Furihata, M.; Chung, S.; Ashida, S.; Miki, T.; Fujioka, T.; Shuin, T.; et al. Novel lipogenic enzyme ELOVL7 is involved in prostate cancer growth through saturated long-chain fatty acid metabolism. *Cancer Res.* **2009**, *69*, 8133–8140. [CrossRef] [PubMed]
30. Lingwood, D.; Simons, K. Lipid rafts as a membrane-organizing principle. *Science* **2010**, *327*, 46–50. [CrossRef] [PubMed]
31. Konstantinopoulos, P.A.; Karamouzis, M.V.; Papavassiliou, A.G. Post-translational modifications and regulation of the RAS superfamily of GTPases as anticancer targets. *Nat. Rev. Drug Discov.* **2007**, *6*, 541–555. [CrossRef] [PubMed]
32. Gray-Bablin, J.; Rao, S.; Keyomarsi, K. Lovastatin induction of cyclin-dependent kinase inhibitors in human breast cells occurs in a cell cycle-independent fashion. *Cancer Res.* **1997**, *57*, 604–609. [PubMed]
33. Newman, A.; Clutterbuck, R.D.; Powles, R.L.; Catovsky, D.; Millar, J.L. A comparison of the effect of the 3-hydroxy-3-methylglutaryl coenzyme A (HMG-CoA) reductase inhibitors simvastatin, lovastatin and pravastatin on leukaemic and normal bone marrow progenitors. *Leuk Lymphoma.* **1997**, *24*, 533–537. [CrossRef] [PubMed]
34. Kodach, L.L.; Jacobs, R.J.; Voorneveld, P.W.; Wildenberg, M.E.; Verspaget, H.W.; van Wezel, T.; Morreau, H.; Hommes, D.W.; Peppelenbosch, M.P.; van den Brink, G.R.; et al. Statins augment the chemosensitivity of colorectal cancer cells inducing epigenetic reprogramming and reducing colorectal cancer cell 'stemness' via the bone morphogenetic protein pathway. *Gut* **2011**, *60*, 1544–1553. [CrossRef] [PubMed]
35. Hager, M.H.; Solomon, K.R.; Freeman, M.R. The role of cholesterol in prostate cancer. *Curr. Opin. Clin. Nutr. Metab. Care* **2006**, *9*, 379–385. [CrossRef] [PubMed]
36. Yue, S.; Li, J.; Lee, S.Y.; Lee, H.J.; Shao, T.; Song, B.; Cheng, L.; Masterson, T.A.; Liu, X.; Ratliff, T.L.; et al. Cholesteryl ester accumulation induced by PTEN loss and PI3K/AKT activation underlies human prostate cancer aggressiveness. *Cell Metab.* **2014**, *19*, 393–406. [CrossRef] [PubMed]
37. Abramczyk, H.; Surmacki, J.; Kopeć, M.; Olejnik, A.K.; Lubecka-Pietruszewska, K.; Fabianowska-Majewska, K. The role of lipid droplets and adipocytes in cancer. Raman imaging of cell cultures: MCF10A, MCF7, and MDA-MB-231 compared to adipocytes in cancerous human breast tissue. *Analyst* **2015**, *140*, 2224–2235. [CrossRef] [PubMed]
38. Tirinato, L.; Liberale, C.; Di Franco, S.; Candeloro, P.; Benfante, A.; La Rocca, R.; Potze, L.; Marotta, R.; Ruffilli, R.; Rajamanickam, V.P.; et al. Lipid droplets: A new player in colorectal cancer stem cells unveiled by spectroscopic imaging. *Stem Cells* **2015**, *33*, 35–44. [CrossRef] [PubMed]
39. Benedetti, E.; Galzio, R.; Laurenti, G.; D'Angelo, B.; Melchiorre, E.; Cifone, M.G.; Fanelli, F.; Muzi, P.; Coletti, G.; Alecci, M.; et al. Lipid metabolism impairment in human gliomas: Expression of peroxisomal proteins in human gliomas at different grades of malignancy. *Int. J. Immunopathol. Pharmacol.* **2010**, *23*, 235–246. [CrossRef] [PubMed]
40. Liu, Y. Fatty acid oxidation is a dominant bioenergetic pathway in prostate cancer. *Prostate Cancer Prostatic Dis.* **2006**, *9*, 230–234. [CrossRef] [PubMed]
41. Liu, Y.; Zuckier, L.S.; Ghesani, N.V. Dominant uptake of fatty acid over glucose by prostate cells: A potential new diagnostic and therapeutic approach. *Anticancer Res.* **2010**, *30*, 369–374. [PubMed]
42. Caro, P.; Kishan, A.U.; Norberg, E.; Stanley, I.A.; Chapuy, B.; Ficarro, S.B.; Polak, K.; Tondera, D.; Gounarides, J.; Yin, H.; et al. Metabolic signatures uncover distinct targets in molecular subsets of diffuse large B cell lymphoma. *Cancer Cell* **2012**, *22*, 547–560. [CrossRef] [PubMed]
43. Khasawneh, J.; Schulz, M.D.; Walch, A.; Rozman, J.; Hrabe de Angelis, M.; Klingenspor, M.; Buck, A.; Schwaiger, M.; Saur, D.; Schmid, R.M.; et al. Inflammation and mitochondrial fatty acid β-oxidation link obesity to early tumor promotion. *Proc. Natl. Acad. Sci. USA* **2009**, *106*, 3354–3359. [CrossRef] [PubMed]
44. Pike, L.S.; Smift, A.L.; Croteau, N.J.; Ferrick, D.A.; Wu, M. Inhibition of fatty acid oxidation by etomoxir impairs NADPH production and increases reactive oxygen species resulting in ATP depletion and cell death in human glioblastoma cells. *Biochim. Biophys. Acta* **2011**, *1807*, 726–734. [CrossRef] [PubMed]
45. Kim, J.W.; Tchernyshyov, I.; Semenza, G.L.; Dang, C.V. HIF-1-mediated expression of pyruvate dehydrogenase kinase: A metabolic switch required for cellular adaptation to hypoxia. *Cell Metab.* **2006**, *3*, 177–185. [CrossRef] [PubMed]

46. Gimm, T.; Wiese, M.; Teschemacher, B.; Deggerich, A.; Schödel, J.; Knaup, K.X.; Hackenbeck, T.; Hellerbrand, C.; Amann, K.; Wiesener, M.S.; et al. Hypoxia-inducible protein 2 is a novel lipid droplet protein and a specific target gene of hypoxia-inducible factor-1. *FASEB J.* **2010**, *24*, 4443–4458. [CrossRef] [PubMed]
47. Mullen, A.R.; Wheaton, W.W.; Jin, E.S.; Chen, P.H.; Sullivan, L.B.; Cheng, T.; Yang, Y.; Linehan, W.M.; Chandel, N.S.; DeBerardinis, R.J. Reductive carboxylation supports growth in tumour cells with defective mitochondria. *Nature* **2011**, *481*, 385–388. [CrossRef] [PubMed]
48. Metallo, C.M.; Gameiro, P.A.; Bell, E.L.; Mattaini, K.R.; Yang, J.; Hiller, K.; Jewell, C.M.; Johnson, Z.R.; Irvine, D.J.; Guarente, L.; et al. Reductive glutamine metabolism by IDH1 mediates lipogenesis under hypoxia. *Nature* **2011**, *481*, 380–384. [CrossRef] [PubMed]
49. Wise, D.R.; Ward, P.S.; Shay, J.E.; Cross, J.R.; Gruber, J.J.; Sachdeva, U.M.; Platt, J.M.; DeMatteo, R.G.; Simon, M.C.; Thompson, C.B. Hypoxia promotes isocitrate dehydrogenase-dependent carboxylation of α-ketoglutarate to citrate to support cell growth and viability. *Proc. Natl. Acad. Sci. USA* **2011**, *108*, 19611–19616. [CrossRef] [PubMed]
50. Le, A.; Lane, A.N.; Hamaker, M.; Bose, S.; Gouw, A.; Barbi, J.; Tsukamoto, T.; Rojas, C.J.; Slusher, B.S.; Zhang, H.; et al. Glucose-independent glutamine metabolism via TCA cycling for proliferation and survival in B cells. *Cell Metab.* **2012**, *15*, 110–121. [CrossRef] [PubMed]
51. Fan, J.; Kamphorst, J.J.; Mathew, R.; Chung, M.K.; White, E.; Shlomi, T.; Rabinowitz, J.D. Glutamine-driven oxidative phosphorylation is a major ATP source in transformed mammalian cells in both normoxia and hypoxia. *Mol. Syst. Biol.* **2013**, *9*, 712. [CrossRef] [PubMed]
52. Wang, J.B.; Erickson, J.W.; Fuji, R.; Ramachandran, S.; Gao, P.; Dinavahi, R.; Wilson, K.F.; Ambrosio, A.L.; Dias, S.M.; Dang, C.V.; et al. Targeting mitochondrial glutaminase activity inhibits oncogenic transformation. *Cancer Cell* **2010**, *18*, 207–219. [CrossRef] [PubMed]
53. DeBerardinis, R.J.; Mancuso, A.; Daikhin, E.; Nissim, I.; Yudkoff, M.; Wehrli, S.; Thompson, C.B. Beyond aerobic glycolysis: Transformed cells can engage in glutamine metabolism that exceeds the requirement for protein and nucleotide synthesis. *Proc. Natl. Acad. Sci. USA* **2007**, *104*, 19345–19350. [CrossRef] [PubMed]
54. Yang, L.; Moss, T.; Mangala, L.S.; Marini, J.; Zhao, H.; Wahlig, S.; Armaiz-Pena, G.; Jiang, D.; Achreja, A.; Win, J.; et al. Metabolic shifts toward glutamine regulate tumor growth, invasion and bioenergetics in ovarian cancer. *Mol. Syst. Biol.* **2014**, *10*, 728. [CrossRef] [PubMed]
55. Tan, N.S.; Michalik, L.; Noy, N.; Yasmin, R.; Pacot, C.; Heim, M.; Flühmann, B.; Desvergne, B.; Wahli, W. Critical roles of PPARβ/δ in keratinocyte response to inflammation. *Genes Dev.* **2001**, *15*, 3263–3277. [CrossRef] [PubMed]
56. Mandard, S.; Müller, M.; Kersten, S. Peroxisome proliferator-activated receptor α target genes. *Cell. Mol. Life Sci.* **2004**, *61*, 393–416. [CrossRef] [PubMed]
57. Lehrke, M.; Lazar, M.A. The many faces of PPARγ. *Cell* **2005**, *123*, 993–999. [CrossRef] [PubMed]
58. Feige, J.N.; Gelman, L.; Michalik, L.; Desvergne, B.; Wahli, W. From molecular action to physiological outputs: Peroxisome proliferator-activated receptors are nuclear receptors at the crossroads of key cellular functions. *Prog. Lipid Res.* **2006**, *45*, 120–159. [CrossRef] [PubMed]
59. Feige, J.N.; Gelman, L.; Tudor, C.; Engelborghs, Y.; Wahli, W.; Desvergne, B. Fluorescence imaging reveals the nuclear behavior of peroxisome proliferator-activated receptor/retinoid X receptor heterodimers in the absence and presence of ligand. *J. Biol. Chem.* **2005**, *280*, 17880–17890. [CrossRef] [PubMed]
60. Juge-Aubry, C.; Pernin, A.; Favez, T.; Burger, A.G.; Wahli, W.; Meier, C.A.; Desvergne, B. DNA binding properties of peroxisome proliferator-activated receptor subtypes on various natural peroxisome proliferator response elements. Importance of the 5′-flanking region. *J. Biol. Chem.* **1997**, *272*, 25252–25259. [CrossRef] [PubMed]
61. Olefsky, J.M. Nuclear receptor minireview series. *J. Biol. Chem.* **2001**, *276*, 36863–36864. [CrossRef] [PubMed]
62. Tan, N.S.; Shaw, N.S.; Vinckenbosch, N.; Liu, P.; Yasmin, R.; Desvergne, B.; Wahli, W.; Noy, N. Selective cooperation between fatty acid binding proteins and peroxisome proliferator-activated receptors in regulating transcription. *Mol. Cell. Biol.* **2002**, *22*, 5114–5127. [CrossRef] [PubMed]
63. Lehmann, J.M.; Moore, L.B.; Smith-Oliver, T.A.; Wilkison, W.O.; Willson, T.M.; Kliewer, S.A. An antidiabetic thiazolidinedione is a high affinity ligand for peroxisome proliferator-activated receptor γ (PPARγ). *J. Biol. Chem.* **1995**, *270*, 12953–12956. [CrossRef] [PubMed]

64. Krey, G.; Braissant, O.; L'Horset, F.; Kalkhoven, E.; Perroud, M.; Parker, M.G.; Wahli, W. Fatty acids, eicosanoids, and hypolipidemic agents identified as ligands of peroxisome proliferator-activated receptors by coactivator-dependent receptor ligand assay. *Mol. Endocrinol.* **1997**, *11*, 779–791. [CrossRef] [PubMed]
65. Pozzi, A.; Ibanez, M.R.; Gatica, A.E.; Yang, S.; Wei, S.; Mei, S.; Falck, J.R.; Capdevila, J.H. Peroxisomal proliferator-activated receptor-α-dependent inhibition of endothelial cell proliferation and tumorigenesis. *J. Biol. Chem.* **2007**, *282*, 17685–17695. [CrossRef] [PubMed]
66. Hou, Y.; Moreau, F.; Chadee, K. PPARγ is an E3 ligase that induces the degradation of NFκB/p65. *Nat. Commun.* **2012**, *3*, 1300. [CrossRef] [PubMed]
67. Hou, Y.; Gao, J.; Xu, H.; Xu, Y.; Zhang, Z.; Xu, Q.; Zhang, C. PPARγ E3 ubiquitin ligase regulates MUC1-C oncoprotein stability. *Oncogene* **2014**, *33*, 5619–5625. [CrossRef] [PubMed]
68. Skrypnyk, N.; Chen, X.; Hu, W.; Su, Y.; Mont, S.; Yang, S.; Gangadhariah, M.; Wei, S.; Falck, J.R.; Jat, J.L.; et al. PPARα activation can help prevent and treat non-small cell lung cancer. *Cancer Res.* **2014**, *74*, 621–631. [CrossRef] [PubMed]
69. Gao, J.; Liu, Q.; Xu, Y.; Gong, X.; Zhang, R.; Zhou, C.; Su, Z.; Jin, J.; Shi, H.; Shi, J.; et al. PPARα induces cell apoptosis by destructing Bcl2. *Oncotarget* **2015**, *6*, 44635–44642. [CrossRef] [PubMed]
70. Xu, Y.; Jin, J.; Zhang, W.; Zhang, Z.; Gao, J.; Liu, Q.; Zhou, C.; Xu, Q.; Shi, H.; Hou, Y.; et al. EGFR/MDM2 signaling promotes NF-κB activation via PPARγ degradation. *Carcinogenesis* **2016**, *37*, 215–222. [CrossRef] [PubMed]
71. You, M.; Jin, J.; Liu, Q.; Xu, Q.; Shi, J.; Hou, Y. PPARα Promotes Cancer Cell Glut1 Transcription Repression. *J. Cell. Biochem.* **2017**, *118*, 1556–1562. [CrossRef] [PubMed]
72. Kaipainen, A.; Kieran, M.W.; Huang, S.; Butterfield, C.; Bielenberg, D.; Mostoslavsky, G.; Mulligan, R.; Folkman, J.; Panigrahy, D. PPARα deficiency in inflammatory cells suppresses tumor growth. *PLoS ONE* **2007**, *2*, e260. [CrossRef] [PubMed]
73. Spaner, D.E.; Lee, E.; Shi, Y.; Wen, F.; Li, Y.; Tung, S.; McCaw, L.; Wong, K.; Gary-Gouy, H.; Dalloul, A.; et al. PPAR-alpha is a therapeutic target for chronic lymphocytic leukemia. *Leukemia* **2013**, *27*, 1090–1099. [CrossRef] [PubMed]
74. Messmer, D.; Lorrain, K.; Stebbins, K.; Bravo, Y.; Stock, N.; Cabrera, G.; Correa, L.; Chen, A.; Jacintho, J.; Chiorazzi, N.; et al. A Selective Novel Peroxisome Proliferator-Activated Receptor (PPAR)-α Antagonist Induces Apoptosis and Inhibits Proliferation of CLL Cells In Vitro and In Vivo. *Mol. Med.* **2015**, *21*, 410–419. [CrossRef] [PubMed]
75. Michalik, L.; Desvergne, B.; Wahli, W. Peroxisome-proliferator-activated receptors and cancers: Complex stories. *Nat. Rev. Cancer* **2004**, *4*, 61–70. [CrossRef] [PubMed]
76. You, M.; Yuan, S.; Shi, J.; Hou, Y. PPARδ signaling regulates colorectal cancer. *Curr. Pharm. Des.* **2015**, *21*, 2956–2959. [CrossRef] [PubMed]
77. Zhang, W.; Xu, Y.; Xu, Q.; Shi, H.; Shi, J.; Hou, Y. PPARδ promotes tumor progression via activation of Glut1 and SLC1-A5 transcription. *Carcinogenesis* **2017**, *38*, 748–755. [CrossRef] [PubMed]
78. Frattini, V.; Pagnotta, S.M.; Tala; Fan, J.J.; Russo, M.V.; Lee, S.B.; Garofano, L.; Zhang, J.; Shi, P.; Lewis, G.; et al. A metabolic function of FGFR3-TACC3 gene fusions in cancer. *Nature* **2018**, *553*, 222–227. [CrossRef] [PubMed]
79. Grabacka, M.; Plonka, P.M.; Urbanska, K.; Reiss, K. Peroxisome proliferator-activated receptor α activation decreases metastatic potential of melanoma cells in vitro via down-regulation of Akt. *Clin. Cancer Res.* **2006**, *12*, 3028–3036. [CrossRef] [PubMed]
80. Liu, D.C.; Zang, C.B.; Liu, H.Y.; Possinger, K.; Fan, S.G.; Elstner, E. A novel PPAR alpha/gamma dual agonist inhibits cell growth and induces apoptosis in human glioblastoma T98G cells. *Acta Pharmacol. Sin.* **2004**, *25*, 1312–1319. [PubMed]
81. Suchanek, K.M.; May, F.J.; Robinson, J.A.; Lee, W.J.; Holman, N.A.; Monteith, G.R.; Roberts-Thomson, S.J. Peroxisome proliferator-activated receptor α in the human breast cancer cell lines MCF-7 and MDA-MB-231. *Mol. Carcinog.* **2002**, *34*, 165–171. [CrossRef] [PubMed]
82. Abu Aboud, O.; Wettersten, H.I.; Weiss, R.H. Inhibition of PPARα induces cell cycle arrest and apoptosis, and synergizes with glycolysis inhibition in kidney cancer cells. *PLoS ONE* **2013**, *8*, e71115. [CrossRef] [PubMed]
83. Peters, J.M.; Cattley, R.C.; Gonzalez, F.J. Role of PPAR alpha in the mechanism of action of the nongenotoxic carcinogen and peroxisome proliferator Wy-14,643. *Carcinogenesis* **1997**, *18*, 2029–2033. [CrossRef] [PubMed]

84. Fidoamore, A.; Cristiano, L.; Laezza, C.; Galzio, R.; Benedetti, E.; Cinque, B.; Antonosante, A.; d'Angelo, M.; Castelli, V.; Cifone, M.G.; et al. Energy metabolism in glioblastoma stem cells: PPARα a metabolic adaptor to intratumoral microenvironment. *Oncotarget* **2017**, *8*, 108430–108450. [CrossRef] [PubMed]
85. Yang, L.; Lin, C.; Wang, L.; Guo, H.; Wang, X. Hypoxia and hypoxia-inducible factors in glioblastoma multiforme progression and therapeutic implications. *Exp. Cell Res.* **2012**, *318*, 2417–2426. [CrossRef] [PubMed]
86. Persano, L.; Rampazzo, E.; Basso, G.; Viola, G. Glioblastoma cancer stem cells: Role of the microenvironment and therapeutic targeting. *Biochem. Pharmacol.* **2013**, *85*, 612–622. [CrossRef] [PubMed]
87. Laurenti, G.; Benedetti, E.; D'Angelo, B.; Cristiano, L.; Cinque, B.; Raysi, S.; Alecci, M.; Cerù, M.P.; Cifone, M.G.; Galzio, R.; et al. Hypoxia induces peroxisome proliferator-activated receptor α (PPARα) and lipid metabolism peroxisomal enzymes in human glioblastoma cells. *J. Cell. Biochem.* **2011**, *112*, 3891–3901. [CrossRef] [PubMed]
88. Pelletier, J.; Bellot, G.; Gounon, P.; Lacas-Gervais, S.; Pouysségur, J.; Mazure, N.M. Glycogen Synthesis is induced in Hypoxia by the Hypoxia-Inducible Factor and Promotes Cancer Cell Survival. *Front. Oncol.* **2012**, *2*, 18. [CrossRef] [PubMed]
89. Guo, D.; Bell, E.H.; Chakravarti, A. Lipid metabolism emerges as a promising target for malignant glioma therapy. *CNS Oncol.* **2013**, *2*, 289–299. [CrossRef] [PubMed]
90. Morihiro, Y.; Yasumoto, Y.; Vaidyan, L.K.; Sadahiro, H.; Uchida, T.; Inamura, A.; Sharifi, K.; Ideguchi, M.; Nomura, S.; Tokuda, N.; Kashiwabara, S.; et al. Fatty acid binding protein 7 as a marker of glioma stem cells. *Pathol. Int.* **2013**, *63*, 546–553. [CrossRef] [PubMed]
91. Ta, M.T.; Kapterian, T.S.; Fei, W.; Du, X.; Brown, A.J.; Dawes, I.W.; Yang, H. Accumulation of squalene is associated with the clustering of lipid droplets. *FEBS J.* **2012**, *279*, 4231–4244. [CrossRef] [PubMed]
92. Perroud, B.; Ishimaru, T.; Borowsky, A.D.; Weiss, R.H. Grade-dependent proteomics characterization of kidney cancer. *Mol. Cell. Proteom.* **2009**, *8*, 971–985. [CrossRef] [PubMed]
93. Shah, Y.M.; Morimura, K.; Yang, Q.; Tanabe, T.; Takagi, M.; Gonzalez, F.J. Peroxisome proliferator-activated receptor α regulates a microRNA-mediated signaling cascade responsible for hepatocellular proliferation. *Mol. Cell. Biol.* **2007**, *27*, 4238–4247. [CrossRef] [PubMed]
94. Wang, C.; Lisanti, M.P.; Liao, D.J. Reviewing once more the c-myc and Ras collaboration: Converging at the cyclin D1-CDK4 complex and challenging basic concepts of cancer biology. *Cell Cycle* **2011**, *10*, 57–67. [CrossRef] [PubMed]
95. Cairns, R.A.; Harris, I.S.; Mak, T.W. Regulation of cancer cell metabolism. *Nat. Rev. Cancer* **2011**, *11*, 85–95. [CrossRef] [PubMed]
96. Diaz-Moralli, S.; Tarrado-Castellarnau, M.; Miranda, A.; Cascante, M. Targeting cell cycle regulation in cancer therapy. *Pharmacol. Ther.* **2013**, *138*, 255–271. [CrossRef] [PubMed]
97. Kamarajugadda, S.; Becker, J.R.; Hanse, E.A.; Mashek, D.G.; Mashek, M.T.; Hendrickson, A.M.; Mullany, L.K.; Albrecht, J.H. Cyclin D1 represses peroxisome proliferator-activated receptor alpha and inhibits fatty acid oxidation. *Oncotarget* **2016**, *7*, 47674–47686. [CrossRef] [PubMed]
98. Fu, M.; Rao, M.; Bouras, T.; Wang, C.; Wu, K.; Zhang, X.; Li, Z.; Yao, T.P.; Pestell, R.G. Cyclin D1 inhibits peroxisome proliferator-activated receptor γ-mediated adipogenesis through histone deacetylase recruitment. *J. Biol. Chem.* **2005**, *280*, 16934–16941. [CrossRef] [PubMed]
99. Hanse, E.A.; Mashek, D.G.; Becker, J.R.; Solmonson, A.D.; Mullany, L.K.; Mashek, M.T.; Towle, H.C.; Chau, A.T.; Albrecht, J.H. Cyclin D1 inhibits hepatic lipogenesis via repression of carbohydrate response element binding protein and hepatocyte nuclear factor 4α. *Cell Cycle* **2012**, *11*, 2681–2690. [CrossRef] [PubMed]
100. Fu, M.; Wang, C.; Rao, M.; Wu, X.; Bouras, T.; Zhang, X.; Li, Z.; Jiao, X.; Yang, J.; Li, A.; et al. Cyclin D1 represses p300 transactivation through a cyclin-dependent kinase-independent mechanism. *J. Biol. Chem.* **2005**, *280*, 29728–29742. [CrossRef] [PubMed]
101. Knudsen, K.E. Cyclin D1 goes metabolic: Dual functions of cyclin D1 in regulating lipogenesis. *Cell Cycle* **2012**, *11*, 3533–3534. [CrossRef] [PubMed]
102. Chakravarthy, M.V.; Pan, Z.; Zhu, Y.; Tordjman, K.; Schneider, J.G.; Coleman, T.; Turk, J.; Semenkovich, C.F. "New" hepatic fat activates PPARα to maintain glucose, lipid, and cholesterol homeostasis. *Cell Metab.* **2005**, *1*, 309–322. [CrossRef] [PubMed]

103. Chakravarthy, M.V.; Zhu, Y.; López, M.; Yin, L.; Wozniak, D.F.; Coleman, T.; Hu, Z.; Wolfgang, M.; Vidal-Puig, A.; Lane, M.D.; et al. Brain fatty acid synthase activates PPARα to maintain energy homeostasis. *J. Clin. Investig.* **2007**, *117*, 2539–2552. [CrossRef] [PubMed]
104. Horiguchi, A.; Asano, T.; Asano, T.; Ito, K.; Sumitomo, M.; Hayakawa, M. Fatty acid synthase over expression is an indicator of tumor aggressiveness and poor prognosis in renal cell carcinoma. *J. Urol.* **2008**, *180*, 1137–1140. [CrossRef] [PubMed]
105. Abu Aboud, O.; Donohoe, D.; Bultman, S.; Fitch, M.; Riiff, T.; Hellerstein, M.; Weiss, R.H. PPARα inhibition modulates multiple reprogrammed metabolic pathways in kidney cancer and attenuates tumor growth. *Am. J. Physiol. Cell Physiol.* **2015**, *308*, C890–C898. [CrossRef] [PubMed]
106. Kurokawa, T.; Shimomura, Y.; Bajotto, G.; Kotake, K.; Arikawa, T.; Ito, N.; Yasuda, A.; Nagata, H.; Nonami, T.; Masuko, K. Peroxisome proliferator-activated receptor α (PPARα) mRNA expression in human hepatocellular carcinoma tissue and non-cancerous liver tissue. *World J. Surg. Oncol.* **2011**, *9*, 167. [CrossRef] [PubMed]
107. Chen, Y.; Wang, Y.; Huang, Y.; Zeng, H.; Hu, B.; Guan, L.; Zhang, H.; Yu, A.M.; Johnson, C.H.; Gonzalez, F.J.; et al. PPARα regulates tumor cell proliferation and senescence via a novel target gene carnitine palmitoyltransferase 1C. *Carcinogenesis* **2017**, *38*, 474–483. [CrossRef] [PubMed]
108. Wolfgang, M.J.; Kurama, T.; Dai, Y.; Suwa, A.; Asaumi, M.; Matsumoto, S.; Cha, S.H.; Shimokawa, T.; Lane, M.D. The brain-specific carnitine palmitoyltransferase-1c regulates energy homeostasis. *Proc. Natl. Acad. Sci. USA* **2006**, *103*, 7282–7287. [CrossRef] [PubMed]
109. Reilly, P.T.; Mak, T.W. Molecular pathways: Tumor cells Co-opt the brain-specific metabolism gene CPT1C to promote survival. *Clin. Cancer Res.* **2012**, *18*, 5850–5855. [CrossRef] [PubMed]
110. Nath, A.; Chan, C. Genetic alterations in fatty acid transport and metabolism genes are associated with metastatic progression and poor prognosis of human cancers. *Sci. Rep.* **2016**, *6*, 18669. [CrossRef] [PubMed]
111. Mascaró, C.; Acosta, E.; Ortiz, J.A.; Marrero, P.F.; Hegardt, F.G.; Haro, D. Control of human muscle-type carnitine palmitoyltransferase I gene transcription by peroxisome proliferator-activated receptor. *J. Biol. Chem.* **1998**, *273*, 8560–8563. [CrossRef] [PubMed]
112. Brandt, J.M.; Djouadi, F.; Kelly, D.P. Fatty acids activate transcription of the muscle carnitine palmitoyltransferase I gene in cardiac myocytes via the peroxisome proliferator-activated receptor α. *J. Biol. Chem.* **1998**, *273*, 23786–23792. [CrossRef] [PubMed]
113. Martinez-Outschoorn, U.E.; Lin, Z.; Whitaker-Menezes, D.; Howell, A.; Lisanti, M.P.; Sotgia, F. Ketone bodies and two-compartment tumor metabolism: Stromal ketone production fuels mitochondrial biogenesis in epithelial cancer cells. *Cell Cycle* **2012**, *11*, 3956–3963. [CrossRef] [PubMed]
114. Saraon, P.; Cretu, D.; Musrap, N.; Karagiannis, G.S.; Batruch, I.; Drabovich, A.P.; van der Kwast, T.; Mizokami, A.; Morrissey, C.; Jarvi, K.; et al. Quantitative proteomics reveals that enzymes of the ketogenic pathway are associated with prostate cancer progression. *Mol. Cell. Proteom.* **2013**, *12*, 1589–1601. [CrossRef] [PubMed]
115. Chen, S.W.; Chou, C.T.; Chang, C.C.; Li, Y.J.; Chen, S.T.; Lin, I.C.; Kok, S.H.; Cheng, S.J.; Lee, J.J.; Wu, T.S.; et al. HMGCS2 enhances invasion and metastasis via direct interaction with PPARα to activate Src signaling in colorectal cancer and oral cancer. *Oncotarget* **2017**, *8*, 22460–22476. [PubMed]
116. Spaner, D.E. Oral high-dose glucocorticoids and ofatumumab in fludarabine-resistant chronic lymphocytic leukemia. *Leukemia* **2012**, *26*, 1144–1145. [CrossRef] [PubMed]
117. Tung, S.; Shi, Y.; Wong, K.; Zhu, F.; Gorczynski, R.; Laister, R.C.; Minden, M.; Blechert, A.K.; Genzel, Y.; Reichl, U.; et al. PPARα and fatty acid oxidation mediate glucocorticoid resistance in chronic lymphocytic leukemia. *Blood* **2013**, *122*, 969–980. [CrossRef] [PubMed]
118. Schulze, A.; Downward, J. Flicking the Warburg switch-tyrosine phosphorylation of pyruvate dehydrogenase kinase regulates mitochondrial activity in cancer cells. *Mol. Cell* **2011**, *44*, 846–848. [CrossRef] [PubMed]
119. Pond, C.M. Adipose tissue and the immune system. *Prostaglandins Leukot. Essent. Fatty Acids* **2005**, *73*, 17–30. [CrossRef] [PubMed]
120. Wolins, N.E.; Quaynor, B.K.; Skinner, J.R.; Tzekov, A.; Park, C.; Choi, K.; Bickel, P.E. OP9 mouse stromal cells rapidly differentiate into adipocytes: Characterization of a useful new model of adipogenesis. *J. Lipid Res.* **2006**, *47*, 450–460. [CrossRef] [PubMed]
121. Chandran, K.; Goswami, S.; Sharma-Walia, N. Implications of a peroxisome proliferator-activated receptor alpha (PPARα) ligand clofibrate in breast cancer. *Oncotarget* **2016**, *7*, 15577–15599. [CrossRef] [PubMed]

122. Lupu, R.; Menendez, J.A. Targeting fatty acid synthase in breast and endometrial cancer: An alternative to selective estrogen receptor modulators? *Endocrinology* **2006**, *147*, 4056–4066. [CrossRef] [PubMed]
123. Rakhshandehroo, M.; Knoch, B.; Müller, M.; Kersten, S. Peroxisome proliferator-activated receptor α target genes. *PPAR Res.* **2010**, *2010*, 612089. [CrossRef] [PubMed]
124. Lacasa, D.; Le Liepvre, X.; Ferre, P.; Dugail, I. Progesterone stimulates adipocyte determination and differentiation 1/sterol regulatory element-binding protein 1c gene expression. potential mechanism for the lipogenic effect of progesterone in adipose tissue. *J. Biol. Chem.* **2001**, *276*, 11512–11516. [CrossRef] [PubMed]
125. Vander Heiden, M.G.; Cantley, L.C.; Thompson, C.B. Understanding the Warburg effect: The metabolic requirements of cell proliferation. *Science* **2009**, *324*, 1029–1033. [CrossRef] [PubMed]
126. Grabacka, M.; Pierzchalska, M.; Reiss, K. Peroxisome proliferator activated receptor α ligands as anticancer drugs targeting mitochondrial metabolism. *Curr. Pharm. Biotechnol.* **2013**, *14*, 342–356. [CrossRef] [PubMed]
127. Jan, C.I.; Tsai, M.H.; Chiu, C.F.; Huang, Y.P.; Liu, C.J.; Chang, N.W. Fenofibrate Suppresses Oral Tumorigenesis via Reprogramming Metabolic Processes: Potential Drug Repurposing for Oral Cancer. *Int. J. Biol. Sci.* **2016**, *12*, 786–798. [CrossRef] [PubMed]
128. Huang, Y.P.; Chang, N.W. Proteomic analysis of oral cancer reveals new potential therapeutic targets involved in the Warburg effect. *Clin. Exp. Pharmacol. Physiol.* **2017**, *44*, 880–887. [CrossRef] [PubMed]
129. Li, X.M.; Patel, B.B.; Blagoi, E.L.; Patterson, M.D.; Seeholzer, S.H.; Zhang, T.; Damle, S.; Gao, Z.; Boman, B.; Yeung, A.T. Analyzing alkaline proteins in human colon crypt proteome. *J. Proteome Res.* **2004**, *3*, 821–833. [CrossRef] [PubMed]
130. Tontonoz, P.; Nagy, L.; Alvarez, J.G.; Thomazy, V.A.; Evans, R.M. PPARγ promotes monocyte/macrophage differentiation and uptake of oxidized LDL. *Cell* **1998**, *93*, 241–252. [CrossRef]
131. Rosen, E.D.; Sarraf, P.; Troy, A.E.; Bradwin, G.; Moore, K.; Milstone, D.S.; Spiegelman, B.M.; Mortensen, R.M. PPARγ is required for the differentiation of adipose tissue in vivo and in vitro. *Mol. Cell* **1999**, *4*, 611–617. [CrossRef]
132. Desvergne, B.; Wahli, W. Peroxisome proliferator-activated receptors: Nuclear control of metabolism. *Endocr. Rev.* **1999**, *20*, 649–688. [CrossRef] [PubMed]
133. Kotta-Loizou, I.; Giaginis, C.; Theocharis, S. The role of peroxisome proliferator-activated receptor-γ in breast cancer. *Anticancer Agents Med. Chem.* **2012**, *12*, 1025–1044. [CrossRef] [PubMed]
134. Pang, X.; Wei, Y.; Zhang, Y.; Zhang, M.; Lu, Y.; Shen, P. Peroxisome proliferator-activated receptor-γ activation inhibits hepatocellular carcinoma cell invasion by upregulating plasminogen activator inhibitor-1. *Cancer Sci.* **2013**, *104*, 672–680. [CrossRef] [PubMed]
135. Srivastava, N.; Kollipara, R.K.; Singh, D.K.; Sudderth, J.; Hu, Z.; Nguyen, H.; Wang, S.; Humphries, C.G.; Carstens, R.; Huffman, K.E.; et al. Inhibition of cancer cell proliferation by PPARγ is mediated by a metabolic switch that increases reactive oxygen species levels. *Cell Metab.* **2014**, *20*, 650–661. [CrossRef] [PubMed]
136. Girnun, G.D.; Smith, W.M.; Drori, S.; Sarraf, P.; Mueller, E.; Eng, C.; Nambiar, P.; Rosenberg, D.W.; Bronson, R.T.; Edelmann, W.; et al. APC-dependent suppression of colon carcinogenesis by PPARγ. *Proc. Natl. Acad. Sci. USA* **2002**, *99*, 13771–13776. [CrossRef] [PubMed]
137. Reka, A.K.; Kurapati, H.; Narala, V.R.; Bommer, G.; Chen, J.; Standiford, T.J.; Keshamouni, V.G. Peroxisome proliferator-activated receptor-γ activation inhibits tumor metastasis by antagonizing Smad3-mediated epithelial-mesenchymal transition. *Mol. Cancer Ther.* **2010**, *9*, 3221–3232. [CrossRef] [PubMed]
138. Saez, E.; Tontonoz, P.; Nelson, M.C.; Alvarez, J.G.; Ming, U.T.; Baird, S.M.; Thomazy, V.A.; Evans, R.M. Activators of the nuclear receptor PPARγ enhance colon polyp formation. *Nat. Med.* **1998**, *4*, 1058–1061. [CrossRef] [PubMed]
139. Jansson, E.A.; Are, A.; Greicius, G.; Kuo, I.C.; Kelly, D.; Arulampalam, V.; Pettersson, S. The Wnt/β-catenin signaling pathway targets PPARγ activity in colon cancer cells. *Proc. Natl. Acad. Sci. USA* **2005**, *102*, 1460–1465. [CrossRef] [PubMed]
140. Yang, K.; Fan, K.H.; Lamprecht, S.A.; Edelmann, W.; Kopelovich, L.; Kucherlapati, R.; Lipkin, M. Peroxisome proliferator-activated receptor γ agonist troglitazone induces colon tumors in normal C57BL/6J mice and enhances colonic carcinogenesis in $Apc^{1638\,N/+}$ $Mlh1^{+/-}$ double mutant mice. *Int. J. Cancer* **2005**, *116*, 495–499. [CrossRef] [PubMed]
141. Burstein, H.J.; Demetri, G.D.; Mueller, E.; Sarraf, P.; Spiegelman, B.M.; Winer, E.P. Use of the peroxisome proliferator-activated receptor (PPAR) γ ligand troglitazone as treatment for refractory breast cancer: A phase II study. *Breast Cancer Res. Treat.* **2003**, *79*, 391–397. [CrossRef] [PubMed]

142. Saez, E.; Rosenfeld, J.; Livolsi, A.; Olson, P.; Lombardo, E.; Nelson, M.; Banayo, E.; Cardiff, R.D.; Izpisua-Belmonte, J.C.; Evans, R.M. PPARγ signaling exacerbates mammary gland tumor development. *Genes Dev.* **2004**, *18*, 528–540. [CrossRef] [PubMed]
143. Egerod, F.L.; Nielsen, H.S.; Iversen, L.; Thorup, I.; Storgaard, T.; Oleksiewicz, M.B. Biomarkers for early effects of carcinogenic dual-acting PPAR agonists in rat urinary bladder urothelium in vivo. *Biomarkers* **2005**, *10*, 295–309. [CrossRef] [PubMed]
144. Lecarpentier, Y.; Claes, V.; Duthoit, G.; Hébert, J.L. Circadian rhythms, Wnt/beta-catenin pathway and PPAR alpha/gamma profiles in diseases with primary or secondary cardiac dysfunction. *Front. Physiol.* **2014**, *5*, 429. [CrossRef] [PubMed]
145. Abbot, E.L.; McCormack, J.G.; Reynet, C.; Hassall, D.G.; Buchan, K.W.; Yeaman, S.J. Diverging regulation of pyruvate dehydrogenase kinase isoform gene expression in cultured human muscle cells. *FEBS J.* **2005**, *272*, 3004–3014. [CrossRef] [PubMed]
146. Jeong, Y.; Xie, Y.; Lee, W.; Bookout, A.L.; Girard, L.; Raso, G.; Behrens, C.; Wistuba, I.I.; Gadzar, A.F.; Minna, J.D.; et al. Research resource: Diagnostic and therapeutic potential of nuclear receptor expression in lung cancer. *Mol. Endocrinol.* **2012**, *26*, 1443–1454. [CrossRef] [PubMed]
147. Reck, M.; Heigener, D.F.; Mok, T.; Soria, J.C.; Rabe, K.F. Management of non-small-cell lung cancer: Recent developments. *Lancet* **2013**, *382*, 709–719. [CrossRef]
148. Vadlapatla, R.K.; Vadlapudi, A.D.; Pal, D.; Mitra, A.K. Mechanisms of drug resistance in cancer chemotherapy: Coordinated role and regulation of efflux transporters and metabolizing enzymes. *Curr. Pharm. Des.* **2013**, *19*, 7126–7140. [CrossRef] [PubMed]
149. Wang, M.; Li, G.; Yang, Z.; Wang, L.; Zhang, L.; Wang, T.; Zhang, Y.; Zhang, S.; Han, Y.; Jia, L. Uncoupling protein 2 downregulation by hypoxia through repression of peroxisome proliferator-activated receptor γ promotes chemoresistance of non-small cell lung cancer. *Oncotarget* **2017**, *8*, 8083–8094. [CrossRef] [PubMed]
150. Bouillaud, F. UCP2, not a physiologically relevant uncoupler but a glucose sparing switch impacting ROS production and glucose sensing. *Biochim. Biophys. Acta* **2009**, *1787*, 377–383. [CrossRef] [PubMed]
151. Moon, E.J.; Giaccia, A. Dual roles of NRF2 in tumor prevention and progression: Possible implications in cancer treatment. *Free Radic. Biol. Med.* **2015**, *79*, 292–299. [CrossRef] [PubMed]
152. Cornic, M.; Delva, L.; Guidez, F.; Balitrand, N.; Degos, L.; Chomienne, C. Induction of retinoic acid-binding protein in normal and malignant human myeloid cells by retinoic acid in acute promyelocytic leukemia patients. *Cancer Res.* **1992**, *52*, 3329–3334. [PubMed]
153. Hu, Z.B.; Ma, W.; Uphoff, C.C.; Lanotte, M.; Drexler, H.G. Modulation of gene expression in the acute promyelocytic leukemia cell line NB4. *Leukemia* **1993**, *7*, 1817–1823. [PubMed]
154. Yasugi, E.; Horiuchi, A.; Uemura, I.; Okuma, E.; Nakatsu, M.; Saeki, K.; Kamisaka, Y.; Kagechika, H.; Yasuda, K.; You, A. Peroxisome proliferator-activated receptor γ ligands stimulate myeloid differentiation and lipogenensis in human leukemia NB4 cells. *Dev. Growth Differ.* **2006**, *48*, 177–188. [CrossRef] [PubMed]
155. Shinohara, H.; Taniguchi, K.; Kumazaki, M.; Yamada, N.; Ito, Y.; Otsuki, Y.; Uno, B.; Hayakawa, F.; Minami, Y.; Naoe, T.; et al. Anti-cancer fatty-acid derivative induces autophagic cell death through modulation of PKM isoform expression profile mediated by bcr-abl in chronic myeloid leukemia. *Cancer Lett.* **2015**, *360*, 28–38. [CrossRef] [PubMed]
156. David, C.J.; Chen, M.; Assanah, M.; Canoll, P.; Manley, J.L. HnRNP proteins controlled by c-Myc deregulate pyruvate kinase mRNA splicing in cancer. *Nature* **2010**, *463*, 364–368. [CrossRef] [PubMed]
157. Lee, J.S.; Chu, I.S.; Heo, J.; Calvisi, D.F.; Sun, Z.; Roskams, T.; Durnez, A.; Demetris, A.J.; Thorgeirsson, S.S. Classification and prediction of survival in hepatocellular carcinoma by gene expression profiling. *Hepatology* **2004**, *40*, 667–676. [CrossRef] [PubMed]
158. Gan, F.Y.; Gesell, M.S.; Alousi, M.; Luk, G.D. Analysis of ODC and c-myc gene expression in hepatocellular carcinoma by in situ hybridization and immunohistochemistry. *J. Histochem. Cytochem.* **1993**, *41*, 1185–1196. [CrossRef] [PubMed]
159. Banerjee, S.S.; Feinberg, M.W.; Watanabe, M.; Gray, S.; Haspel, R.L.; Denkinger, D.J.; Kawahara, R.; Hauner, H.; Jain, M.K. The Krüppel-like factor KLF2 inhibits peroxisome proliferator-activated receptor-γ expression and adipogenesis. *J. Biol. Chem.* **2003**, *278*, 2581–2584. [CrossRef] [PubMed]
160. Choi, Y.; Oh, S.T.; Won, M.A.; Choi, K.M.; Ko, M.J.; Seo, D.; Jeon, T.W.; Baik, I.H.; Ye, S.K.; Park, K.U.; et al. Targeting ODC1 inhibits tumor growth through reduction of lipid metabolism in human hepatocellular carcinoma. *Biochem. Biophys. Res. Commun.* **2016**, *478*, 1674–1681. [CrossRef] [PubMed]

161. Kourtidis, A.; Jain, R.; Carkner, R.D.; Eifert, C.; Brosnan, M.J.; Conklin, D.S. An RNA interference screen identifies metabolic regulators NR1D1 and PBP as novel survival factors for breast cancer cells with the ERBB2 signature. *Cancer Res.* **2010**, *70*, 1783–1792. [CrossRef] [PubMed]
162. Zhu, Y.; Qi, C.; Jain, S.; Rao, M.S.; Reddy, J.K. Isolation and characterization of PBP, a protein that interacts with peroxisome proliferator-activated receptor. *J. Biol. Chem.* **1997**, *272*, 25500–25506. [CrossRef] [PubMed]
163. Bertucci, F.; Borie, N.; Ginestier, C.; Groulet, A.; Charafe-Jauffret, E.; Adélaïde, J.; Geneix, J.; Bachelart, L.; Finetti, P.; Koki, A.; et al. Identification and validation of an ERBB2 gene expression signature in breast cancers. *Oncogene* **2004**, *23*, 2564–2575. [CrossRef] [PubMed]
164. Schaffer, J.E. Lipotoxicity: When tissues overeat. *Curr. Opin. Lipidol.* **2003**, *14*, 281–287. [CrossRef] [PubMed]
165. Kourtidis, A.; Srinivasaiah, R.; Carkner, R.D.; Brosnan, M.J.; Conklin, D.S. Peroxisome proliferator-activated receptor-γ protects ERBB2-positive breast cancer cells from palmitate toxicity. *Breast Cancer Res.* **2009**, *11*, R16. [CrossRef] [PubMed]
166. Wang, X.; Sun, Y.; Wong, J.; Conklin, D.S. PPARγ maintains ERBB2-positive breast cancer stem cells. *Oncogene* **2013**, *32*, 5512–5521. [CrossRef] [PubMed]
167. March, H.N.; Rust, A.G.; Wright, N.A.; ten Hoeve, J.; de Ridder, J.; Eldridge, M.; van der Weyden, L.; Berns, A.; Gadiot, J.; Uren, A.; et al. Insertional mutagenesis identifies multiple networks of cooperating genes driving intestinal tumorigenesis. *Nat. Genet.* **2011**, *43*, 1202–1209. [CrossRef] [PubMed]
168. Ahmad, I.; Mui, E.; Galbraith, L.; Patel, R.; Tan, E.H.; Salji, M.; Rust, A.G.; Repiscak, P.; Hedley, A.; Markert, E.; et al. Sleeping Beauty screen reveals Pparg activation in metastatic prostate cancer. *Proc. Natl. Acad. Sci. USA* **2016**, *113*, 8290–8295. [CrossRef] [PubMed]
169. Kompare, M.; Rizzo, W.B. Mitochondrial fatty-acid oxidation disorders. *Semin. Pediatr. Neurol.* **2008**, *15*, 140–149. [CrossRef] [PubMed]
170. Niu, Z.; Shi, Q.; Zhang, W.; Shu, Y.; Yang, N.; Chen, B.; Wang, Q.; Zhao, X.; Chen, J.; Cheng, N.; et al. Caspase-1 cleaves PPARγ for potentiating the pro-tumor action of TAMs. *Nat. Commun.* **2017**, *8*, 766. [CrossRef] [PubMed]
171. Peters, J.M.; Lee, S.S.; Li, W.; Ward, J.M.; Gavrilova, O.; Everett, C.; Reitman, M.L.; Hudson, L.D.; Gonzalez, F.J. Growth, adipose, brain, and skin alterations resulting from targeted disruption of the mouse peroxisome proliferator-activated receptor β (δ). *Mol. Cell. Biol.* **2000**, *20*, 5119–5128. [CrossRef] [PubMed]
172. Chong, H.C.; Tan, M.J.; Philippe, V.; Tan, S.H.; Tan, C.K.; Ku, C.W.; Goh, Y.Y.; Wahli, W.; Michalik, L.; Tan, N.S. Regulation of epithelial-mesenchymal IL-1 signaling by PPARβ/δ is essential for skin homeostasis and wound healing. *J. Cell Biol.* **2009**, *184*, 817–831. [CrossRef] [PubMed]
173. Peters, J.M.; Shah, Y.M.; Gonzalez, F.J. The role of peroxisome proliferator-activated receptors in carcinogenesis and chemoprevention. *Nat. Rev. Cancer* **2012**, *12*, 181–195. [CrossRef] [PubMed]
174. Stephen, R.L.; Gustafsson, M.C.; Jarvis, M.; Tatoud, R.; Marshall, B.R.; Knight, D.; Ehrenborg, E.; Harris, A.L.; Wolf, C.R.; Palmer, C.N. Activation of peroxisome proliferator-activated receptor δ stimulates the proliferation of human breast and prostate cancer cell lines. *Cancer Res.* **2004**, *64*, 3162–3170. [CrossRef] [PubMed]
175. Aung, C.S.; Faddy, H.M.; Lister, E.J.; Monteith, G.R.; Roberts-Thomson, S.J. Isoform specific changes in PPARα and β in colon and breast cancer with differentiation. *Biochem. Biophys. Res. Commun.* **2006**, *340*, 656–660. [CrossRef] [PubMed]
176. Buttitta, L.A.; Edgar, B.A. Mechanisms controlling cell cycle exit upon terminal differentiation. *Curr. Opin. Cell Biol.* **2007**, *19*, 697–704. [CrossRef] [PubMed]
177. Yao, P.L.; Morales, J.L.; Zhu, B.; Kang, B.H.; Gonzalez, F.J.; Peters, J.M. Activation of peroxisome proliferator-activated receptor-β/δ (PPAR-β/δ) inhibits human breast cancer cell line tumorigenicity. *Mol. Cancer Ther.* **2014**, *13*, 1008–1017. [CrossRef] [PubMed]
178. Genini, D.; Garcia-Escudero, R.; Carbone, G.M.; Catapano, C.V. Transcriptional and Non-Transcriptional Functions of PPARβ/δ in Non-Small Cell Lung Cancer. *PLoS ONE* **2012**, *7*, e46009. [CrossRef] [PubMed]
179. Pedchenko, T.V.; Gonzalez, A.L.; Wang, D.; DuBois, R.N.; Massion, P.P. Peroxisome proliferator-activated receptor β/δ expression and activation in lung cancer. *Am. J. Respir. Cell Mol. Biol.* **2008**, *39*, 689–696. [CrossRef] [PubMed]
180. Di-Poï, N.; Michalik, L.; Tan, N.S.; Desvergne, B.; Wahli, W. The anti-apoptotic role of PPARβ contributes to efficient skin wound healing. *J. Steroid Biochem. Mol. Biol.* **2003**, *85*, 257–265. [CrossRef]

181. Pollock, C.B.; Yin, Y.; Yuan, H.; Zeng, X.; King, S.; Li, X.; Kopelovich, L.; Albanese, C.; Glazer, R.I. PPARδ activation acts cooperatively with 3-phosphoinositide-dependent protein kinase-1 to enhance mammary tumorigenesis. *PLoS ONE* **2011**, *6*, e16215. [CrossRef] [PubMed]
182. Narkar, V.A.; Downes, M.; Yu, R.T.; Embler, E.; Wang, Y.X.; Banayo, E.; Mihaylova, M.M.; Nelson, M.C.; Zou, Y.; Juguilon, H.; et al. AMPK and PPARdelta agonists are exercise mimetics. *Cell* **2008**, *134*, 405–415. [CrossRef] [PubMed]
183. Ito, K.; Carracedo, A.; Weiss, D.; Arai, F.; Ala, U.; Avigan, D.E.; Schafer, Z.T.; Evans, R.M.; Suda, T.; Lee, C.H.; et al. A PML–PPAR-δ pathway for fatty acid oxidation regulates hematopoietic stem cell maintenance. *Nat. Med.* **2012**, *18*, 1350–1358. [CrossRef] [PubMed]
184. Wang, X.; Wang, G.; Shi, Y.; Sun, L.; Gorczynski, R.; Li, Y.J.; Xu, Z.; Spaner, D.E. PPAR-delta promotes survival of breast cancer cells in harsh metabolic conditions. *Oncogenesis* **2016**, *5*, e232. [CrossRef] [PubMed]
185. Carracedo, A.; Weiss, D.; Leliaert, A.K.; Bhasin, M.; de Boer, V.C.; Laurent, G.; Adams, A.C.; Sundvall, M.; Song, S.J.; Ito, K.; et al. A metabolic prosurvival role for PML in breast cancer. *J. Clin. Investig.* **2012**, *122*, 3088–3100. [CrossRef] [PubMed]
186. Li, Y.J.; Sun, L.; Shi, Y.; Wang, G.; Wang, X.; Dunn, S.E.; Iorio, C.; Screaton, R.A.; Spaner, D.E. PPAR-delta promotes survival of chronic lymphocytic leukemia cells in energetically unfavorable conditions. *Leukemia* **2017**, *31*, 1905–1914. [CrossRef] [PubMed]
187. Sun, L.; Shi, Y.; Wang, G.; Wang, X.; Zeng, S.; Dunn, S.E.; Fairn, G.D.; Li, Y.J.; Spaner, D.E. PPAR-delta modulates membrane cholesterol and cytokine signaling in malignant B cells. *Leukemia* **2018**, *32*, 184–193. [CrossRef] [PubMed]
188. Jeong, E.; Koo, J.E.; Yeon, S.H.; Kwak, M.K.; Hwang, D.H.; Lee, J.Y. PPARδ deficiency disrupts hypoxia-mediated tumorigenic potential of colon cancer cells. *Mol. Carcinog.* **2014**, *53*, 926–937. [CrossRef] [PubMed]
189. He, T.C.; Chan, T.A.; Vogelstein, B.; Kinzler, K.W. PPARδ is an APC-regulated target of nonsteroidal anti-inflammatory drugs. *Cell* **1999**, *99*, 335–345. [CrossRef]
190. Zuo, X.; Peng, Z.; Moussalli, M.J.; Morris, J.S.; Broaddus, R.R.; Fischer, S.M.; Shureiqi, I. Targeted genetic disruption of peroxisome proliferator-activated receptor-δ and colonic tumorigenesis. *J. Natl. Cancer Inst.* **2009**, *101*, 762–767. [CrossRef] [PubMed]
191. Qian, B.Z.; Pollard, J.W. Macrophage diversity enhances tumor progression and metastasis. *Cell* **2010**, *141*, 39–51. [CrossRef] [PubMed]
192. Schumann, T.; Adhikary, T.; Wortmann, A.; Finkernagel, F.; Lieber, S.; Schnitzer, E.; Legrand, N.; Schober, Y.; Nockher, W.A.; Toth, P.M.; et al. Deregulation of PPARβ/δ target genes in tumor-associated macrophages by fatty acid ligands in the ovarian cancer microenvironment. *Oncotarget* **2015**, *6*, 13416–13433. [CrossRef] [PubMed]
193. Zhu, P.; Tan, M.J.; Huang, R.L.; Tan, C.K.; Chong, H.C.; Pal, M.; Lam, C.R.; Boukamp, P.; Pan, J.Y.; Tan, S.H.; et al. Angiopoietin-like 4 protein elevates the prosurvival intracellular O_2^-:H_2O_2 ratio and confers anoikis resistance to tumors. *Cancer Cell* **2011**, *19*, 401–415. [CrossRef] [PubMed]
194. Zhu, P.; Goh, Y.Y.; Chin, H.F.; Kersten, S.; Tan, N.S. Angiopoietin-like 4: A decade of research. *Biosci. Rep.* **2012**, *32*, 211–219. [CrossRef] [PubMed]
195. Martín-Martín, N.; Zabala-Letona, A.; Fernández-Ruiz, S.; Arreal, L.; Camacho, L.; Castillo-Martin, M.; Cortazar, A.R.; Torrano, V.; Astobiza, I.; Zúñiga-García, P.; et al. PPARδ Elicits Ligand-Independent Repression of Trefoil Factor Family to Limit Prostate Cancer Growth. *Cancer Res.* **2018**, *78*, 399–409. [CrossRef] [PubMed]
196. Aleshin, S.; Strokin, M.; Sergeeva, M.; Reiser, G. Peroxisome proliferator-activated receptor (PPAR)β/δ, a possible nexus of PPARα- and PPARγ-dependent molecular pathways in neurodegenerative diseases: Review and novel hypotheses. *Neurochem. Int.* **2013**, *63*, 322–330. [CrossRef] [PubMed]

International Journal of
Molecular Sciences

Article

The Role of PPARβ/δ in Melanoma Metastasis

Jonathan Chee Woei Lim [1,2,†], Yuet Ping Kwan [1,†], Michelle Siying Tan [1], Melissa Hui Yen Teo [1], Shunsuke Chiba [3], Walter Wahli [1,4,*] and Xiaomeng Wang [1,5,6,7,*]

1. Lee Kong Chian School of Medicine, Nanyang Technological University Singapore, 59 Nanyang Drive, Singapore 636921, Singapore; cheewoei@upm.edu.my (J.C.W.L.); kwanyuetping@ntu.edu.sg (Y.P.K.); michelle.siying@ntu.edu.sg (M.S.T.); teohy@imcb.a-star.edu.sg (M.H.Y.T.)
2. Department of Medicine, Faculty of Medicine and Health Sciences, Universiti Putra Malaysia, Serdang 43400, Selangor, Malaysia
3. Division of Chemistry and Biological Chemistry, School of Physical and Mathematical Sciences, Nanyang Technological University, 21 Nanyang Link, Singapore 637371, Singapore; Shunsuke@ntu.edu.sg
4. Center for Integrative Genomics, University of Lausanne, Le Génopode, 1015 Lausanne, Switzerland
5. Institute of Molecular and Cell Biology, Agency for Science, Technology and Research (A*STAR), Proteos, 61 Biopolis Dr, Singapore 138673, Singapore
6. Singapore Eye Research Institute, The Academia, 20 College Road, Discovery Tower Level 6, Singapore 169856, Singapore
7. Institute of Ophthalmology, University College London, 11-43 Bath Street, London EC1V 9EL, UK

* Correspondence: walter.wahli@ntu.edu.sg (W.W.); wangxiaomeng@ntu.edu.sg (X.W.); Tel.: +65-6592-3927 (W.W.); +65-6592-3840 (X.W.)

† These authors contributed equally to this work.

Received: 31 August 2018; Accepted: 18 September 2018; Published: 20 September 2018

Abstract: Background: Peroxisome proliferator–activated receptor (PPAR) β/δ, a ligand-activated transcription factor, is involved in diverse biological processes including cell proliferation, cell differentiation, inflammation and energy homeostasis. Besides its well-established roles in metabolic disorders, PPARβ/δ has been linked to carcinogenesis and was reported to inhibit melanoma cell proliferation, anchorage-dependent clonogenicity and ectopic xenograft tumorigenicity. However, PPARβ/δ's role in tumour progression and metastasis remains controversial. Methods: In the present studies, the consequence of PPARβ/δ inhibition either by global genetic deletion or by a specific PPARβ/δ antagonist, 10h, on malignant transformation of melanoma cells and melanoma metastasis was examined using both in vitro and in vivo models. Results: Our study showed that 10h promotes epithelial-mesenchymal transition (EMT), migration, adhesion, invasion and trans-endothelial migration of mouse melanoma B16/F10 cells. We further demonstrated an increased tumour cell extravasation in the lungs of wild-type mice subjected to 10h treatment and in $Ppar\beta/\delta^{-/-}$ mice in an experimental mouse model of blood-borne pulmonary metastasis by tail vein injection. This observation was further supported by an increased tumour burden in the lungs of $Ppar\beta/\delta^{-/-}$ mice as demonstrated in the same animal model. Conclusion: These results indicated a protective role of PPARβ/δ in melanoma progression and metastasis.

Keywords: melanoma; peroxisome proliferator–activated receptor β/δ; migration; EMT; invasion; metastasis

1. Introduction

Melanoma is the deadliest and the most aggressive form of skin cancer that originates from the pigment making melanocytes in epidermis. In contrast to the declining incidence for many types of cancers, the number of annually diagnosed melanoma cases has doubled in the past decade [1]. Furthermore, melanoma is more likely to spread than any other types of skin cancers and the lung is the most common site of distant metastases [2]. Although surgical treatment of early-stage melanoma shows a 90% cure rate, therapeutic options for advanced melanoma are very limited. To make things worse, melanoma cells are particularly adept at rewiring themselves and will inevitably evolve resistance to treatments [3]. The median survival for patients with metastatic malignant melanoma ranges from 6–15 months [4] and the 5-year survival rate is only around 5% [5]. To date, effective therapies for metastatic melanoma remain a significant challenge.

Peroxisome proliferator–activated receptors (PPARs) are ligand-activated transcription factors that belong to the nuclear receptor superfamily. They regulate many biological processes, including cell proliferation, migration and differentiation, immune response, and energy homeostasis [6]. Upon ligand binding, PPARs undergo conformational change which allows the release of corepressors, recruitment of coactivators and subsequent activation of downstream target genes. There are three distinct members of the PPAR family: PPARα, PPARβ/δ, and PPARγ [7]. The biological roles of PPARα and PPARγ have been extensively studied, whereas, PPARβ/δ's function is less clear or even controversial. Several studies linked PPARβ/δ to tumour growth and progression with both promoting and inhibitory effects reported [8–11]. PPARβ/δ was previously shown to be expressed in melanocytes [12], suggesting a potential physiological role in melanocyte activity and function. Although the activation of PPARβ/δ has no impact on melanocyte proliferation [12], it significantly inhibits the proliferation of melanoma cells, prevents anchorage-dependent clonogenicity and attenuates ectopic xenograft tumorigenicity [13]. On the contrary, shRNA-mediated knockdown of PPARβ/δ in a highly malignant mouse melanoma cell line B16-F10 demonstrated reduced lung metastasis and tumour burden [9]. In this study, we showed that PPAβ/δ inhibition by either the PPAβ/δ antagonist 10h, or gene deletion promotes the transformation of melanoma cells towards a more malignant phenotype, leading to increased melanoma cell extravasation and tumour burden in the lungs.

2. Results

2.1. 10h Induces Phenotypic Changes of B16/F10 Mouse Melanoma Cells

Although melanoma cells do not show classic epithelial or mesenchymal characteristics, they undergo a coordinated phenotypic switch which is necessary for tumour cells to disperse from the primary site [14]. To evaluate the role of PPARβ/δ in melanoma cell function, we used a novel PPARβ/δ antagonist 10h [15,16]. 10h has been shown to efficiently antagonized agonist-mediated transcriptional activation of PPARβ/δ. On the contrary, it had no significant effect on ligand activated PPARα or PPARγ [15]. To test the efficacy of 10h-mediated PPARβ/δ inhibition, we first evaluated the expression of known PPARβ/δ target genes in 10h-treated B16/F10 cells and demonstrated a significant reduction in Angiopoietin-like 4 (*Angptl4*) and *Cpt1* expression (Figure 1A). ANGPTL4 was previously demonstrated to prevent tumour metastasis by inhibiting tumour cell motility and invasiveness [17]. Consistent with this observation, 10h-treated B16/F10 cells underwent a drastic change in morphology and were converted from a typical "cuboidal" shape into an elongated mesenchymal like structure (Figure 1B). This phenotypic change was associated with an apparent depigmentation in both the 10 h-treated B16/F10 cells (Figure 1C) and conditioned medium of these cells (Figure 1D), which are characteristic features of transformed invasive melanoma cells [18]. Microphthalmia-associated transcription factor (Mitf) drives the expression of a number of genes involved in melanocyte pigmentation [19]. The expression of this factor is stimulated by the α-melanocyte-stimulating hormone (α-MSH), an endogenous peptide hormone that plays a critical role in melanogenesis. Our study showed that 10h significantly attenuated both basal and α-MSH-induced Mitf expression in B16/F10 cells (Figure 1E). Consistently, there was a significant reduction in the α-MSH-induced melanin secretion after 10h treatment (Figure 1F). Transforming growth factor (TGF) β1 is a potent stimulator of epithelial to mesenchymal transition (EMT) during tumour invasion and metastasis [20]. Similarly to TGFβ1, 10h significantly induced the expression of the specific mesenchymal markers Fibronectin and N-cadherin in B16/F10 cells (Figure 1G). Together, our study showed that 10h induces the switch of melanoma cells towards a more transformed phenotype.

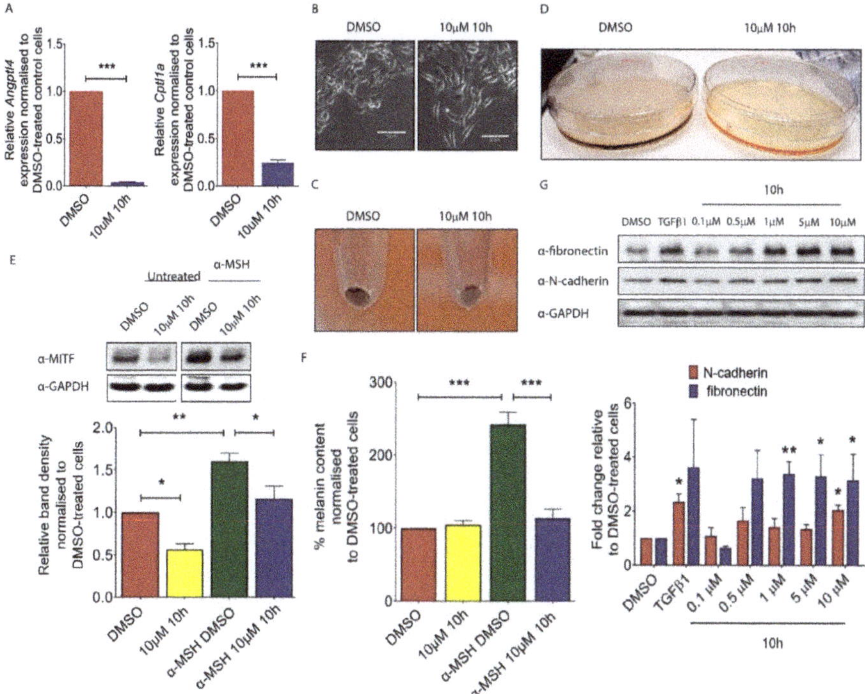

Figure 1. Effect of 10h on B16/F10 mouse melanoma cells. (**A**) *Angptl4* and *Cpt1* gene expression measured using real-time quantitative PCR analysis. (**B**) Morphology of B16/F10 cells after treatment with 10 μM of 10h in 5% serum supplemented DMEM compared to 0.05% DMSO-treated control cells. Scale bar: 50 μm. Representative picture of trypsinized B16/F10 cell pellets (**C**) and conditioned medium (**D**) after 72 h treatment with 10 μM of 10h. (**E**) Representative images and quantitative analysis of western blot for MITF in α-MSH and/or 10h-treated B16/F10 melanoma cells. (**F**) Percentage of melanin content in α-MSH and/or 10h-treated B16/F10 melanoma cells. (**G**) Representative images and quantitative analysis of western blot for fibronectin, N-cadherin, and GAPDH in 10h-treated B16/F10 cells. Data are presented as mean ± s.e.m of three independent experiments. Statistical analysis was performed using one-way ANOVA followed by Turkey's post hoc analysis or two-tailed, unpaired student's *t*-test; * $p < 0.05$, ** $p < 0.01$, *** $p < 0.001$.

2.2. 10h Promotes Melanoma Cell Migration and Invasion

To understand the functional consequences of the 10h-induced morphological transformation of melanoma cells, we carried out the Transwell migration assay and demonstrated an increased motility of 10 μM of 10h-treated B16/F10 cells as compared to vehicle-treated control cells (Figure 2A). Next, to mimic the invasion process, 10h-treated B16/F10 cells were seeded on top of a Matrigel coated Transwell membrane. Consistent with the increased motility, 10h significantly increased the invasiveness of B16/F10 cells (Figure 2B). During invasion, epithelial-derived tumour cells move from the lamina-enriched basal membrane to the collagen and fibronectin-enrich connective tissue region [21,22]. The ability of tumour cells to adapt to this abrupt change in microenvironment contributes to their metastatic and invasive behaviour. Consistently, our study showed a promoting effect of 10h on the capability of B16/F10 cells to adhere to fibronectin-coated cell culture plates

(Figure 2C). A critical prerequisite for metastatic tumour cells to invade the surrounding tissue is their capacity to degrade extracellular matrix (ECM) by the action of matrix metalloproteinases (MMPs) [23–25]. Among all MMPs, MMP9 is particularly important for melanoma progression [26], and increased expression and activity of these MMPs were observed in invasive melanoma cell lines [27,28]. Our study showed that both transcript (Figure 2D) and protein (Figure 2E) levels of MMP9 were induced in 10h-treated B16/F10 cells. Together, our data showed a promoting effect of 10h on B16/F10 melanoma cell motility, invasion, and MMP9 expression, all critical characteristics for melanoma progression and metastasis.

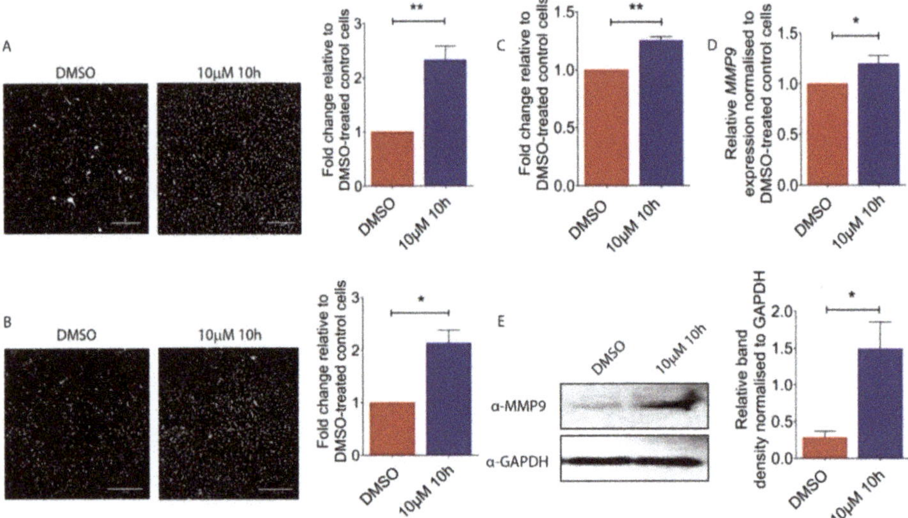

Figure 2. Effect of 10h on B16/F10 cell migration and invasion. (**A**) Representative images and quantitative analysis of migrated B16/F10 cells after 10h and DMSO treatments. (**B**) Representative images and quantitative analysis of invading B16/F10 cells after respective treatments. Scale bar: 200 μm. (**C**) Quantitative analysis of B16/F10 cells attached to fibronectin coated plate normalised to 0.05% DMSO-treated controls. (**D**) MMP9 gene expression in 10 μM 10h or DMSO-treated B16/F10 cells as determined by real-time quantitative PCR analysis. (**E**) Representative Western blot and densitometry analysis of MMP9 in 10 μM 10h or 0.05% DMSO-treated control B16/F10 cells. All images are representative. Data are presented as the mean ± s.e.m of three independent experiments. Statistical analysis was performed using two-tailed, unpaired student's *t*-test; * $p < 0.05$, ** $p < 0.01$.

2.3. 10h Increases B16/F10 Melanoma Cell Adhesion to the Endothelium and Trans-Endothelial Migration

To reach distant sites, cancer cells must be able to adhere to endothelial cells (ECs) and migrate across the endothelium [29,30]. To understand PPARβ/δ's role in this process, we applied 10h treated B16/F10 cells to pulmonary microvascular endothelial cells (HPMECs) and observed an increased number of 10h-treated B16/F10 cells adhered to a HPMEC monolayer (Figure 3A). In vitro trans-endothelial migration assay is commonly used to mimic the process of tumour cells to cross the endothelium, a critical step in metastatic spread. Similarly, 10h significantly increased the number of 10h-treated B16/F10 cells that migrated across the HPMEC monolayer as compared to those treated with DMSO (Figure 3B). We then investigated the effects of PPARβ/δ inhibition in HPMECs on melanoma cell behaviour. B16/F10 cells were layered on top of the HPMECs pre-treated with 10h. Similarly to the above observation, there was an increased number of melanoma cells that adhered to (Figure 3C) and migrated across the 10h-treated HPMECs (Figure 3D). Together, these data demonstrated that PPARβ/δ inactivation by 10h either in tumour cells or in ECs facilitates tumour cell dissemination.

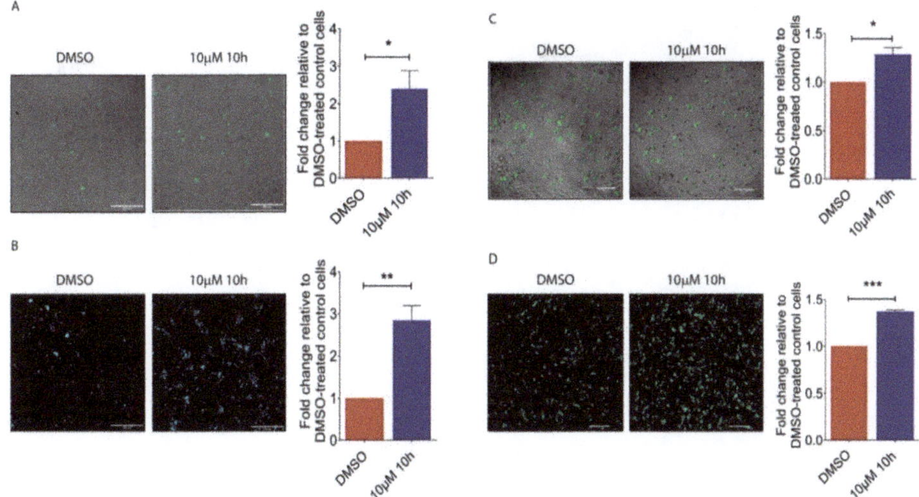

Figure 3. Impact of 10h on B16/F10 cell adhesion to endothelial cells and transendothelial migration. (**A**) Representative images and quantitative analysis of the number of CMFDA-positive B16/F10 cells adhered on the HPMEC monolayer. Scale bar: 200 μm. (**B**) Representative images and quantitative analysis of the number of CMFDA-positive cells that migrated across the HPMEC monolayer. Scale bar: 200 μm. (**C**) Representative images and quantitative analysis of the number of CMFDA-positive B16/F10 cells adhered on the 0.05% DMSO or 10 μM 10h-treated HPMEC monolayer. (**D**) Representative images and quantitative analysis of the number of CMFDA-positive cells that migrated across 0.05% DMSO- or 10 μM 10h-treated HPMEC monolayers. Scale bar: 200 μm. All images are representative. Data are presented as mean ± s.e.m of three independent experiments. Statistical analyses were performed using two tailed, unpaired student's *t*-test. * $p < 0.05$; ** $p < 0.01$; *** $p < 0.001$.

2.4. *PPARβ/δ Inhibition Promotes Lung Metastasis of Melanoma Cells In Vivo*

Having established an important role of PPARβ/δ in attenuating the morphological and functional transition of B16/F10 cells towards a more aggressive phenotype, we next investigated its involvement in lung metastasis in vivo. We observed an increased pulmonary extravasation of 10h-treated B16/F10 cells in an experimental C57BL/6 mouse model of blood-borne pulmonary metastasis, as compared to mice injected with DMSO-treated control cells (Figure 4A). Similarly, there was an increased extravasation of B16/F10 cells in *Pparβ/δ*$^{-/-}$ mice as compared to that in wild-type controls (Figure 4B). Furthermore, the lungs of *Pparβ/δ*$^{-/-}$ mice were leakier than WT counterparts as demonstrated by the Mile's assay (Figure 4C). We further demonstrated an increased tumour burden in the lungs of *Pparβ/δ*$^{-/-}$ mice (Figure 4D,E). Together, these data showed that PPARβ/δ inhibition in tumour cells or *Pparβ/δ* deletion in the host promoted melanoma metastasis.

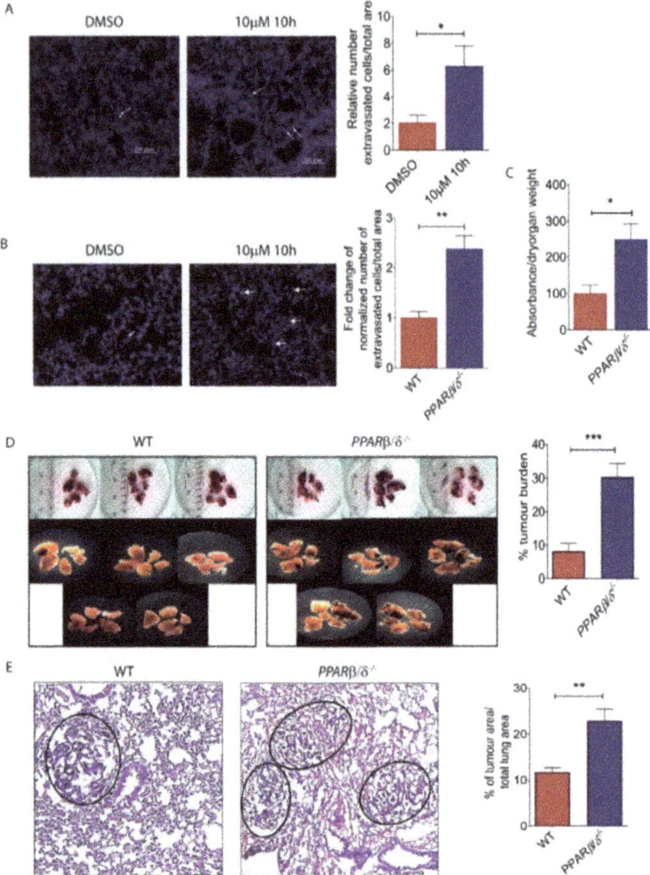

Figure 4. PPARβ/δ inhibition promotes pulmonary lung metastasis of B16/F10 melanoma cells. (**A**) Representative images and quantitative analysis of extravasated B16/F10 cells in C57BL/6 mice subjected to intravenous delivery of 10h- or vehicle-treated B16/F10 cells (n = 3). (**B**) Representative images and quantification of extravasated B16/F10 cells in the lungs of wild type (n = 6) or $Ppar\beta/\delta^{-/-}$ mice (n = 7). Scale bar: 60 μm. Arrows indicate extravasated tumour cells. (**C**) Quantification of Evans blue dye extravasation in the lungs of wild type (n = 4) and $Ppar\beta/\delta^{-/-}$ (n = 4) 30 min after Evans blue injection. (**D**) Representative images and quantification of the metastatic burden in the lungs of wild type (n = 8) and $Ppar\beta/\delta^{-/-}$ mice (n = 8) subjected to intravenous delivery of B16/F10 cells. (**E**) Representative images and quantification of Hematoxylin & Eosin staining showing metastatic tumour nodules in the lungs of wild type (n = 5) and $Ppar\beta/\delta^{-/-}$ mice (n = 5). All images are representative. Data are presented as the mean ± s.e.m. Significance was determined by two tailed, unpaired, Student's t-test; * $p < 0.05$; ** $p < 0.01$; *** $p < 0.001$.

3. Discussion

Melanoma is the deadliest form of skin cancer. It is well known for its aggressive, metastatic and invasive properties [31], and its capability of rewiring itself to develop resistance to treatments [3]. Currently, treatments for metastatic melanoma are very limited and great efforts have been made towards developing novel targeted therapies.

PPARs are ligand-activated transcription factors that are members of the nuclear receptor superfamily. There are three related isotypes of PPARs, PPARα, PPARβ/δ, PPARγ, which share high levels of sequence homology but have distinct ligand specificity and tissue distribution patterns [32,33]. PPARs play numerous physiological functions and their dys-regulation is implicated in various pathological conditions, including cancers [34]. The roles of PPARβ/δ in cancer development and progression are very complicated with both promoting and protective roles reported [35,36]. For example, PPARβ/δ knockdown by shRNA in B16/F10 cells has been shown to significantly attenuate lung metastasis [9] and the ligand-mediated PPARβ/δ activation by GW501516 is able to increase the migration and invasion of the highly metastatic human melanoma cell line A375SM [31]. However, GW501516 was also reported to reduce metastasis and migration of pancreatic cancer cells, whereas short hairpin RNA-mediated inhibition of PPARβ/δ has been shown to promote the invasiveness of pancreatic cancer cells [37]. Furthermore, GW501516 is able to inhibit the expression of MMP-9, an important matrix remodelling proteinase that is involved in degrading pericellular and stromal compartments to facilitate metastasis [38]. Our study suggests a protective instead of promoting role of PPARβ/δ in melanoma progression and metastasis. Inhibition of PPARβ/δ signalling by 10h leads to phenotypic transformation of melanoma cell B16/F10 to elongated mesenchymal like shaped mimicking the initiation of EMT, an important process for tumour cells to achieve further differentiation and progress to an advanced stage [39]. This observation is supported by a reduced expression of melanocyte lineage-specific transcription factor, Mitf, a key regulator of melanin synthesis [40,41]. In line with this observation, 10h-treated B16/F10 cells and conditioned medium from these cells showed a significant reduction in pigmentation. This is associated with a concomitant increase of mesenchymal markers, fibronectin and N-cadherin. Besides its role in phenotypic expression of the melanocytic lineage, Mitf is reported to act as a suppressor of melanoma invasion and metastasis [42–45]. Consistent with the reduced Mitf expression, 10h promotes B16/F10 cell migration, expression of MMP9, invasion into the Matrigel, and adhesion to fibronectin. To metastasize to distant organs, cancerous cells need to extravasate from the vascular system into the surrounding tissue [46]. Our data demonstrate that PPARβ/δ inactivation in both B16/F10 cells and PMVECs promotes melanoma cell adhesion to and transmigration through the endothelium. We further demonstrated increased extravasation of 10h-treated B16/F10 cells in C57/BL6 mice in an experimental mouse model of blood-borne pulmonary metastasis. Similar observation was made in *Pparβ/δ*$^{-/-}$ injected with B16/F10 cells. Finally, *Pparβ/δ* deletion in host tissue caused increased pulmonary vessel leakage and tumour burden in the lungs. In summary, our data support a critical role of PPARβ/δ in attenuating melanoma progression and metastasis in various in vitro and in vivo models.

However, each experimental model has its own limitations and could not faithfully recapitulate all characteristic features of tumour development and progression, particularly the role played by the different cell types, such as the predominant cancer-associated fibroblasts (CAFs), tumour associated macrophages (TAMs), tumour-infiltrating lymphocytes (TILs) and pericytes in tumour progression [47]. Furthermore, levels of PPARβ/δ inhibition or activation by small molecule drugs, combination of ligand target and off-target effects, subtle differences in experimental design, may also contribute to the controversial observations on PPARβ/δ's role in tumourigenesis. Much more research will be necessary to clarify the multifaceted and intriguing pro- and anti-tumorigenesis roles of PPARβ/δ in different cancers.

4. Material and Methodology

4.1. Animals

Pparβ/δ-null mice (mixed genetic background of Sv129/C56BL/6) were kind gifts from Prof. Walter Wahli (University of Lausanne, Lausanne, Switzerland). Wild type mice (C57BL/6) were purchased from the InVivos Pte Ltd., Singapore. All animal procedures were reviewed and approved by the Nanyang Technological University Institutional Animal Care and Use Committee (IACUC, Project number: A0269), Singapore.

4.2. Cell Culture

Mouse melanoma cell line B16/F10 was cultured in Dulbecco's modified Eagle's medium (DMEM; Gibco, Carlsbad, CA, USA) supplemented with 10% fetal bovine serum (FBS; Gibco), 2 mM of L-glutamine (Gibco), 100 U/mL of penicillin and 100 μg/mL of streptomycin (Nacalai Tesque, Kyoto, Japan). Human Pulmonary Microvascular Endothelial Cells (HPMEC) cells were cultured in Endothelial Cell Medium-2 supplemented with endothelial cell growth medium bulletkits (Lonza, Cologne, Germany). All cell lines were maintained in 5% CO_2 atmosphere at 37 °C.

4.3. Synthesis of 10h

Compound 10h (IUPAC: methyl 3-(*N*-(4-(isopentylamino)-2-methoxyphenyl)sulfamoyl)-thiophene-2-carboxylate; PPARβ/δ antagonist) was synthesised in-house (Shunsuke Chiba) according to as shown in Toth et al. 2012 [15]. The synthesis scheme can be found in Sng et al. 2018 [16].

4.4. RNA Extraction and Quantitative Real-time PCR

Total RNA was extracted using RNAzol RT (Molecular Research Centre, Cincinnati, OH, USA) and cDNA was synthesised using qScript cDNA Supermix (Quantabio, Beverly, MA, USA) according to manufacturer protocols. Quantitative real-time PCR was performed in a total of 20 μL volume containing SYBER green (PrimerDesign Precision, Camberley, UK) on a QuantiStudio 6 Flex Real-time PCR system. The primers used were as follows (Table 1):

Table 1. Sequences of the forward and reverse primers utilized for gene expression analysis.

Target (Mouse)	Forward Sequence (5′–3′)	Reverse Sequence (5′–3′)
Angptl4 (NM_020581.2)	GGGACCTTAACTGTGCCAAGAG	GGAAGTATTGTCCATTGAGATTGGA
Cpt1a (NM_013495.2)	ACACCATCCACGCCATACTG	TCCCAGAAGACGAATAGGTTTGAG
Mmp9 (NM_013599.4)	GCCCTGGAACTCACACGACA	TTGGAAACTCACACGCCAGAAG
Gapdh (NM_001289726.1)	ACTGAGGACCAGGTTGTCTCC	CTGTAGCCGTATTCATTGTCATACC

4.5. Epithelial Mesenchymal Transition (EMT)

The B16/F10 cells (0.5×10^5 cells/well) was seeded in a well of a 6-well plate and serum starved in serum free DMEM for 8 h before treatment with DMSO (0.05% *v*/*v*) as control or 10h (10 μM). After 72 h, the cells were harvested for western blot analysis to detect EMT marker proteins, MITF and MMP-9 expression or seeded for migration and invasion assays.

4.6. Measurement of Extracellular Melanin Content

For melanin content analysis, B16/F10 cells were cultured in phenol red-free DMEM (Gibco) containing 10% FBS. The cells were cultured with α-MSH at 0.1 μM (Sigma, St. Louis, MO, USA) treatment in the absence or presence of 10h for 3 days. The cultured cells or media were harvested, and pellets were dissolved in 1N NaOH (Merck, Kenilworth, NJ, USA) containing 10% DMSO (Sigma) at 80 °C for 1 h. The melanin content was measured at 475 nm wavelength using an ELISA plate reader (BioTek, Winooski, VT, USA).

4.7. Cell Adhesion and Tumour-Endothelial Assay

For cell adhesion assay, B16/F10 cells (6×10^4 cells/well) were seeded in a fibronectin coated 96-well tissue culture plate and incubated for 30 min at 37 °C in a 5% CO_2 incubator. Non-adherent cells were aspirated, and adherent cells were washed thrice with PBS. The remaining cells were then fixed with ethanol and stained with crystal violet. Excess dye was removed by washing with PBS and the dye absorbed by adherent cell nuclei was extracted with 0.5% Triton X-100. Quantification was by measuring the absorbance at 570 nm using an ELISA plate reader (BioTek).

For tumour-endothelial assay, B16/F10 cells were pre-treated with 0.05% DMSO or 10 μM 10h and pulsed with 25 μM Cell Tracker Green CMFDA (Invitrogen, Carlsbad, CA, USA) dye for 30 min and harvested to obtain a cell suspension. B16/F10 cells (2×10^4) were seeded onto a monolayer of HPMEC and incubated for 15 min at 37 °C in a 5% CO_2 incubator. Non-adherent cells were aspirated, and adherent cells were washed 3 times with PBS. Remaining adherent cells were fixed using 1% paraformaldehyde (PFA) for 15 min and the adherent cancer cells tagged with green signal were counted under an inverted fluorescence microscope (Nikon, Tokyo, Japan).

4.8. Trans-Well Migration and Invasion Assay

The cell migration assay was carried out using a transwell system with pore size of 8 μm (Corning Costar, Corning, NY, USA). The cells (0.8×10^5 cells per well) were seeded into the upper chamber in serum free DMEM, while conditioned medium from NIH-3T3 was applied to the lower chamber. After 4 h incubation, migrated cells on the bottom surface of the membrane were fixed with 1% PFA, non-migrated cells on the top of the membrane were gently removed by cotton bud while the migrated cells at the bottom of the membrane were stained with DAPI. Migrated cells were counted in five random microscopic fields (original magnification, 10×). The cell invasion assay was carried out using a transwell system with pore size of 8 μm (Corning Costar, USA), coated with Matrigel (BD Biosciences, San Jose, CA, USA). Similarly, the cells (1×10^5 cells per well) were seeded into the upper chamber in DMEM supplemented with 1% FBS, while medium with 10% FBS was applied to the lower chamber. After 24 h, invading cells were quantified as described above.

4.9. Trans-Endothelial Migration

1×10^5 HPMEC cells were seeded in the upper chamber of a collagen-coated transwell insert and allowed to grow in complete endothelial medium to confluence (up to 4 days with P2 (HPMEC); fresh medium was added at day 2). Pre-treated B16/F10 cells with 0.05% DMSO or 10 μM 10h were pulsed with 25 μM CellTracker Green CMFDA dye (Invitrogen) for 30 min before being detached by 0.25% trypsin. 0.8×10^5 of pre-treated B16/F10 cells were seeded onto the endothelial monolayer and allowed to migrate for 18 h at 37 °C in 5% CO_2. Conditioned NIH-3T3 medium was used as a chemoattractant. The transwells were then fixed in 4% PFA and cells on the apical side of each insert were scraped off, and the transwell membrane mounted onto glass slides. Migration to the basolateral side of the membrane was visualized with a immunofluorescent microscope at 10× magnification. An average of five random fields were quantified using the NIH ImageJ analysis software (http://rsbweb.nih.gov/ij/). For trans-endothelial migration for HPMEC treated with 10h, the experimental design is similar to the above with the exception that the endothelial monolayer was treated with 0.05% DMSO or with 10h up to 4 days to confluence and green signal tagged B16/F10 cells were seeded onto the endothelial monolayer.

4.10. In Vivo Metastasis Assay

For experimental metastasis assays, pre-treated 5×10^5 B16/F10 cells were injected into the tail vein of 6- to 8-week-old WT and *Pparβ/δ*$^{-/-}$ mice as described [48]. After 2 weeks, the animals were sacrificed to harvest the lungs. The area or the number of metastases in the lungs was measured on photographs using the ImageJ software (http://rsbweb.nih.gov/ij/).

4.11. Extravasation Assay

1×10^6 B16/F10 cells labelled with CellTracker Green CMFDA (Invitrogen) were injected into the tail vein of 6- to 8-week-old C57BL/6 female mice. Twenty-four hours later, mice were sacrificed, and lungs were harvested. Lung tissues were subjected to fixation using 4% PFA for 3 h and were cryoprotected in 15% sucrose and then 30% sucrose in PBS overnight. The lungs were then frozen using Shandon™ Cryomatrix™ embedding resin (Thermo Scientific, Waltham, MA, USA) and cryostat-cut in 6-μm thick sections. Immunofluorescent staining for blood vessels was conducted with anti-CD31 RFP conjugated primary antibody (1:100 dilution) (BD Biosciences, Bedford, MA, USA) and anti-DAPI (1:1000 dilution). The specimens were then examined using an upright widefield fluorescence microscope (Leica, Wetzlar, Germany).

4.12. Evans Blue Assay

For evans blue extravasation assay, 200 μL of 0.5% Evans Blue Solution was injected into the tail vein of 6- to 8- week-old WT and *Pparβ/δ*$^{-/-}$ mice. After 30 min, the mice were euthanized before perfused transcardially with 5 mL of PBS. The lungs were then excised, washed with PBS and allowed to dry at 37 °C for 48 h. The dry weights were recorded and 500 μL of formamide (Sigma, St. Louis, MO, USA) was added. The samples were then incubated for 24 h at 55 °C. The amount of extravasated evans blue was measured by the absorbance of 620 nm using an ELISA plate reader (BioTek).

4.13. Statistics

One-way ANOVA followed by Turkey's post hoc analysis or two-tailed, unpaired Student's *t*-test was used to determine whether the experimental samples were significantly different from the control samples. Differences were considered statistically significant at * $p < 0.05$; ** $p < 0.01$; *** $p < 0.001$.

Author Contributions: Conceptualization, X.W. and W.W.; Data curation, X.W., Y.P.K. and J.C.W.L.; Formal analysis, X.W., Y.P.K. and J.C.W.L.; Funding acquisition, X.W. and W.W.; Investigation, Y.P.K., J.C.W.L., M.S.T. and M.H.Y.T.; Methodology, X.W, Y.P.K., J.C.W.L., M.S.T. and M.H.Y.T.; Project administration, X.W.; Resources, S.C.; Supervision, X.W.; Validation, Y.P.K. and J.C.W.L.; Writing—original draft, X.W., Y.P.K. and J.C.W.L.; Writing—review & editing, W.W.

Funding: This project was supported by MOE Academic Research Fund Tier 2 (MOE2014-T2-1-036) to X.W. and W.W. and by the Lee Kong Chian School of Medicine, Nanyang Technological University Singapore Start-Up Grant to X.W. and W.W.

Conflicts of Interest: The authors declare no conflict of interest.

References

1. Apalla, Z.; Lallas, A.; Sotiriou, E.; Lazaridou, E.; Ioannides, D. Epidemiological trends in skin cancer. *Dermatol. Pract. Concept.* **2017**, *7*, 1–6. [CrossRef] [PubMed]
2. Tas, F. Metastatic behavior in melanoma: Timing, pattern, survival, and influencing factors. *J. Oncol.* **2012**, *2012*, 647684. [CrossRef] [PubMed]
3. Pawlik, T.M.; Sondak, V.K. Malignant melanoma: Current state of primary and adjuvant treatment. *Crit. Rev. Oncol. Hematol.* **2003**, *45*, 245–264. [CrossRef]
4. Balch, C.M.; Gershenwald, J.E.; Soong, S.J.; Thompson, J.F.; Atkins, M.B.; Byrd, D.R.; Buzaid, A.C.; Cochran, A.J.; Coit, D.G.; Ding, S.; et al. Final version of 2009 AJCC melanoma staging and classification. *J. Clin. Oncol.* **2009**, *27*, 6199–6206. [CrossRef] [PubMed]
5. Sandru, A.; Voinea, S.; Panaitescu, E.; Blidaru, A. Survival rates of patients with metastatic malignant melanoma. *J. Med. Life* **2014**, *7*, 572–576. [PubMed]
6. Magadum, A.; Engel, F. PPARβ/δ: Linking metabolism to regeneration. *Int. J. Mol. Sci.* **2018**, *19*, 2013. [CrossRef] [PubMed]
7. Dreyer, C.; Krey, G.; Keller, H.; Givel, F.; Helftenbein, G.; Wahli, W. Control of the peroxisomal beta-oxidation pathway by a novel family of nuclear hormone receptors. *Cell* **1992**, *68*, 879–887. [CrossRef]
8. Wang, X.; Wang, G.; Shi, Y.; Sun, L.; Gorczynski, R.; Li, Y.J.; Xu, Z.; Spaner, D.E. Ppar-delta promotes survival of breast cancer cells in harsh metabolic conditions. *Oncogenesis* **2016**, *5*, e232. [CrossRef] [PubMed]
9. Zuo, X.; Xu, W.; Xu, M.; Tian, R.; Moussalli, M.J.; Mao, F.; Zheng, X.; Wang, J.; Morris, J.S.; Gagea, M.; et al. Metastasis regulation by PPARD expression in cancer cells. *JCI Insight* **2017**, *2*, e91419. [CrossRef] [PubMed]
10. Muller-Brusselbach, S.; Komhoff, M.; Rieck, M.; Meissner, W.; Kaddatz, K.; Adamkiewicz, J.; Keil, B.; Klose, K.J.; Moll, R.; Burdick, A.D.; et al. Deregulation of tumor angiogenesis and blockade of tumor growth in PPARbeta-deficient mice. *EMBO J.* **2007**, *26*, 3686–3698. [CrossRef] [PubMed]

11. Montagner, A.; Delgado, M.B.; Tallichet-Blanc, C.; Chan, J.S.K.; Sng, M.K.; Mottaz, H.; Degueurce, G.; Lippi, Y.; Moret, C.; Baruchet, M.; et al. Src is activated by the nuclear receptor peroxisome proliferator-activated receptor beta/delta in ultraviolet radiation-induced skin cancer. *EMBO Mol. Med.* **2014**, *6*, 80–98. [CrossRef] [PubMed]
12. Kang, H.Y.; Chung, E.; Lee, M.; Cho, Y.; Kang, W.H. Expression and function of peroxisome proliferator-activated receptors in human melanocytes. *Br. J. Dermatol.* **2004**, *150*, 462–468. [CrossRef] [PubMed]
13. Borland, M.G.; Yao, P.L.; Kehres, E.M.; Lee, C.; Pritzlaff, A.M.; Ola, E.; Wagner, A.L.; Shannon, B.E.; Albrecht, P.P.; Zhu, B.K.; et al. Editor's highlight: PPAR beta/delta and PPAR gamma inhibit melanoma tumorigenicity by modulating inflammation and apoptosis. *Toxicol. Sci.* **2017**, *159*, 436–448. [CrossRef] [PubMed]
14. Kim, J.E.; Leung, E.; Baguley, B.C.; Finlay, G.J. Heterogeneity of expression of epithelial-mesenchymal transition markers in melanocytes and melanoma cell lines. *Front. Genet.* **2013**, *4*, 97. [CrossRef] [PubMed]
15. Toth, P.M.; Naruhn, S.; Pape, V.F.S.; Dorr, S.M.A.; Klebe, G.; Muller, R.; Diederich, W.E. Development of improved PPAR ss/d inhibitors. *Chemmedchem* **2012**, *7*, 159–170. [CrossRef] [PubMed]
16. Sng, M.K.; Chan, J.S.K.; Teo, Z.; Phua, T.; Tan, E.H.P.; Wee, J.W.K.; Koh, N.J.N.; Tan, C.K.; Chen, J.P.; Pal, M.; et al. Selective deletion of PPARbeta/delta in fibroblasts causes dermal fibrosis by attenuated LRG1 expression. *Cell Discov.* **2018**, *4*, 15. [CrossRef] [PubMed]
17. Galaup, A.; Cazes, A.; Le Jan, S.; Philippe, J.; Connault, E.; Le Coz, E.; Mekid, H.; Mir, L.M.; Opolon, P.; Corvol, P.; et al. Angiopoietin-like 4 prevents metastasis through inhibition of vascular permeability and tumor cell motility and invasiveness. *Proc. Natl. Acad. Sci. USA* **2006**, *103*, 18721–18726. [CrossRef] [PubMed]
18. Carreira, S.; Goodall, J.; Denat, L.; Rodriguez, M.; Nuciforo, P.; Hoek, K.S.; Testori, A.; Larue, L.; Goding, C.R. Mitf regulation of Dia1 controls melanoma proliferation and invasiveness. *Genes Dev.* **2006**, *20*, 3426–3439. [CrossRef] [PubMed]
19. D'Mello, S.A.; Finlay, G.J.; Baguley, B.C.; Askarian-Amiri, M.E. Signaling pathways in melanogenesis. *Int. J. Mol. Sci.* **2016**, *17*, 1144. [CrossRef] [PubMed]
20. Wendt, M.K.; Allington, T.M.; Schiemann, W.P. Mechanisms of the epithelial-mesenchymal transition by TGF-beta. *Future Oncol.* **2009**, *5*, 1145–1168. [CrossRef] [PubMed]
21. Nelson, C.M.; Bissell, M.J. Of extracellular matrix, scaffolds, and signaling: Tissue architecture regulates development, homeostasis, and cancer. *Annu. Rev. Cell Dev. Biol.* **2006**, *22*, 287–309. [CrossRef] [PubMed]
22. Gaggioli, C.; Hooper, S.; Hidalgo-Carcedo, C.; Grosse, R.; Marshall, J.F.; Harrington, K.; Sahai, E. Fibroblast-led collective invasion of carcinoma cells with differing roles for RHOGTPases in leading and following cells. *Nat. Cell Biol.* **2007**, *9*, 1392–1400. [CrossRef] [PubMed]
23. Kessenbrock, K.; Plaks, V.; Werb, Z. Matrix metalloproteinases: Regulators of the tumor microenvironment. *Cell* **2010**, *141*, 52–67. [CrossRef] [PubMed]
24. Gialeli, C.; Theocharis, A.D.; Karamanos, N.K. Roles of matrix metalloproteinases in cancer progression and their pharmacological targeting. *FEBS J.* **2011**, *278*, 16–27. [CrossRef] [PubMed]
25. Radisky, E.S.; Radisky, D.C. Matrix metalloproteinase-induced epithelial-mesenchymal transition in breast cancer. *J. Mammary Gland. Biol. Neoplasia* **2010**, *15*, 201–212. [CrossRef] [PubMed]
26. Orgaz, J.L.; Pandya, P.; Dalmeida, R.; Karagiannis, P.; Sanchez-Laorden, B.; Viros, A.; Albrengues, J.; Nestle, F.O.; Ridley, A.J.; Gaggioli, C.; et al. Diverse matrix metalloproteinase functions regulate cancer amoeboid migration. *Nat. Commun.* **2014**, *5*, 4255. [CrossRef] [PubMed]
27. Zhao, W.H.; Liu, H.; Xu, S.; Entschladen, F.; Niggemann, B.; Zänker, K.S.; Han, R. Migration and metalloproteinases determine the invasive potential of mouse melanoma cells, but not melanin and telomerase. *Cancer Lett.* **2001**, *162*, S49–S55. [CrossRef]
28. Shellman, Y.G.; Makela, M.; Norris, D.A. Induction of secreted matrix metalloproteinase-9 activity in human melanoma cells by extracellular matrix proteins and cytokines. *Melanoma Res.* **2006**, *16*, 207–211. [CrossRef] [PubMed]
29. Cavallaro, U.; Christofori, G. Cell adhesion in tumor invasion and metastasis: Loss of the glue is not enough. *Biochim. Biophys. Acta* **2001**, *1552*, 39–45. [CrossRef]
30. Onken, M.D.; Mooren, O.L.; Mukherjee, S.; Shahan, S.T.; Li, J.; Cooper, J.A. Endothelial monolayers and transendothelial migration depend on mechanical properties of the substrate. *Cytoskeleton* **2014**, *71*, 695–706. [CrossRef] [PubMed]

31. Ham, S.A.; Yoo, T.; Hwang, J.S.; Kang, E.S.; Lee, W.J.; Paek, K.S.; Park, C.; Kim, J.H.; Do, J.T.; Lim, D.S.; et al. Ligand-activated PPARδ modulates the migration and invasion of melanoma cells by regulating Snail expression. *Am. J. Cancer Res.* **2014**, *4*, 674. [PubMed]
32. Berger, J.; Moller, D.E. The mechanisms of action of PPARs. *Annu. Rev. Med.* **2002**, *53*, 409–435. [CrossRef] [PubMed]
33. Feige, J.N.; Gelman, L.; Michalik, L.; Desvergne, B.; Wahli, W. From molecular action to physiological outputs: Peroxisome proliferator-activated receptors are nuclear receptors at the crossroads of key cellular functions. *Prog. Lipid Res.* **2006**, *45*, 120–159. [CrossRef] [PubMed]
34. Youssef, J.; Badr, M. Peroxisome proliferator-activated receptors and cancer: Challenges and opportunities. *Br. J. Pharmacol.* **2011**, *164*, 68–82. [CrossRef] [PubMed]
35. Muller, R. PPARbeta/delta in human cancer. *Biochimie* **2017**, *136*, 90–99. [CrossRef] [PubMed]
36. Michalik, L.; Wahli, W. PPARs mediate lipid signaling in inflammation and cancer. *PPAR Res.* **2008**, *2008*, 134059. [CrossRef] [PubMed]
37. Coleman, J.D.; Thompson, J.T.; Smith, R.W., 3rd.; Prokopczyk, B.; Vanden Heuvel, J.P. Role of peroxisome proliferator-activated receptor beta/delta and B-cell lymphoma-6 in regulation of genes involved in metastasis and migration in pancreatic cancer cells. *PPAR Res.* **2013**, *2013*, 121956. [CrossRef] [PubMed]
38. Falzone, L.R.; Travali, S.; Scalisi, A.; McCubrey, J.A.; Candido, S. Libra M1. MMP-9 overexpression is associated with intragenic hypermethylation of MMP9 gene in melanoma. *Aging* **2016**, *8*, 933. [CrossRef] [PubMed]
39. Heerboth, S.; Housman, G.; Leary, M.; Longacre, M.; Byler, S.; Lapinska, K.; Willbanks, A.; Sarkar, S. EMT and tumor metastasis. *Clin. Transl. Med.* **2015**, *4*, 6. [CrossRef] [PubMed]
40. Steingrimsson, E.; Copeland, N.G.; Jenkins, N.A. Melanocytes and the microphthalmia transcription factor network. *Annu. Rev Genet.* **2004**, *38*, 365–411. [CrossRef] [PubMed]
41. Vance, K.W.; Goding, C.R. The transcription network regulating melanocyte development and melanoma. *Pigment Cell Res.* **2004**, *17*, 318–325. [CrossRef] [PubMed]
42. Pinner, S.; Jordan, P.; Sharrock, K.; Bazley, L.; Collinson, L.; Marais, R.; Bonvin, E.; Goding, C.; Sahai, E. Intravital imaging reveals transient changes in pigment production and Brn2 expression during metastatic melanoma dissemination. *Cancer Res.* **2009**, *69*, 7969–7977. [CrossRef] [PubMed]
43. Thurber, A.E.; Douglas, G.; Sturm, E.C.; Zabierowski, S.E.; Smit, D.J.; Ramakrishnan, S.N.; Hacker, E.; Leonard, J.H.; Herlyn, M.; Sturm, R.A. Inverse expression states of the BRN2 and MITF transcription factors in melanoma spheres and tumour xenografts regulate the NOTCH pathway. *Oncogene* **2011**, *30*, 3036–3048. [CrossRef] [PubMed]
44. Bell, R.E.; Khaled, M.; Netanely, D.; Schubert, S.; Golan, T.; Buxbaum, A.; Janas, M.M.; Postolsky, B.; Goldberg, M.S.; Shamir, R.; et al. Transcription factor/microRNA axis blocks melanoma invasion program by miR-211 targeting NUAK1. *J. Investig. Dermatol.* **2014**, *134*, 441–451. [CrossRef] [PubMed]
45. Cheli, Y.; Giuliano, S.; Fenouille, N.; Allegra, M.; Hofman, V.; Hofman, P.; Bahadoran, P.; Lacour, J.P.; Tartare-Deckert, S.; Bertolotto, C.; et al. Hypoxia and MITF control metastatic behaviour in mouse and human melanoma cells. *Oncogene* **2012**, *31*, 2461–2470. [CrossRef] [PubMed]
46. Ombrato, L.; Malanchi, I. The EMT universe: Space between cancer cell dissemination and metastasis initiation. *Crit. Rev. Oncog.* **2014**, *19*, 349–361. [PubMed]
47. Tan, N.S.; Vazquez-Carrera, M.; Montagner, A.; Sng, M.K.; Guillou, H.; Wahli, W. Transcriptional control of physiological and pathological processes by the nuclear receptor PPAR beta/delta. *Prog. Lipid Res.* **2016**, *64*, 98–122. [CrossRef] [PubMed]
48. Nadra, K.; Anghel, S.I.; Joye, E.; Tan, N.S.; Basu-Modak, S.; Trono, D.; Wahli, W.; Desvergne, B. Differentiation of trophoblast giant cells and their metabolic functions are dependent on peroxisome proliferator-activated receptor beta/delta. *Mol. Cell. Boil.* **2006**, *26*, 3266–3281.

MDPI
St. Alban-Anlage 66
4052 Basel
Switzerland
Tel. +41 61 683 77 34
Fax +41 61 302 89 18
www.mdpi.com

International Journal of Molecular Sciences Editorial Office
E-mail: ijms@mdpi.com
www.mdpi.com/journal/ijms

www.ingramcontent.com/pod-product-compliance
Lightning Source LLC
Chambersburg PA
CBHW051700210326
41597CB00032B/5318

* 9 7 8 3 0 3 8 9 7 4 6 1 1 *